Handbook of GC-MS

Handbook of GC-MS

Fundamentals and Applications

Fourth Edition

Hans-Joachim Hübschmann

WILEY-VCH

Author

Dr. Hans-Joachim Hübschmann
HANS Analytical Solutions
Mainz
Germany

Osaka
Japan

Cover Image: Courtesy of Hans-Joachim Hübschmann

All books published by **WILEY-VCH** are carefully produced. Nevertheless, authors, editors, and publisher do not warrant the information contained in these books, including this book, to be free of errors. Readers are advised to keep in mind that statements, data, illustrations, procedural details or other items may inadvertently be inaccurate.

Library of Congress Card No.: applied for

British Library Cataloguing-in-Publication Data
A catalogue record for this book is available from the British Library.

Bibliographic information published by the Deutsche Nationalbibliothek
The Deutsche Nationalbibliothek lists this publication in the Deutsche Nationalbibliografie; detailed bibliographic data are available on the Internet at <http://dnb.d-nb.de>.

© 2025 WILEY-VCH GmbH, Boschstraße 12, 69469 Weinheim, Germany

The manufacturer's authorized representative according to the EU General Product Safety Regulation is Wiley-VCH GmbH, Boschstr. 12, 69469 Weinheim, Germany, e-mail: Product_Safety@wiley.com.

All rights reserved (including those of translation into other languages, text and data mining and training of artificial technologies or similar technologies). No part of this book may be reproduced in any form – by photoprinting, microfilm, or any other means – nor transmitted or translated into a machine language without written permission from the publishers. Registered names, trademarks, etc. used in this book, even when not specifically marked as such, are not to be considered unprotected by law.

Print ISBN: 978-3-527-35403-0
ePDF ISBN: 978-3-527-84763-1
ePub ISBN: 978-3-527-84762-4
oBook ISBN: 978-3-527-84764-8

Typesetting Straive, Chennai, India
Druck und Bindung: CPI books GmbH Leck, Germany

For Mika

Contents

Foreword *xvii*
Preface to the Fourth Edition *xix*

1 **Introduction** *1*
1.1 The Historical Development of the GC-MS Technique *4*
 References *7*

2 **Fundamentals** *9*
2.1 Sample Preparation *9*
2.1.1 QuEChERS Sample Preparation *11*
2.1.2 Dispersive Liquid/Liquid Microextraction *17*
2.1.3 Solid Phase Extraction *19*
2.1.3.1 Online Solid Phase Extraction *24*
2.1.3.2 Micro Solid Phase Extraction *25*
2.1.4 Solid Phase Microextraction *27*
2.1.4.1 Solid Phase Microextraction Devices *28*
2.1.4.2 Solid Phase Microextraction Operation for GC-MS *36*
2.1.4.3 Solid Phase Microextraction Sorbent Materials *42*
2.1.5 Static Headspace Technique *49*
2.1.5.1 Measures for Improved Headspace Response *53*
2.1.5.2 Quantitation by Multiple Headspace Extraction *56*
2.1.5.3 Headspace Analysis Operation *59*
2.1.6 Dynamic Headspace – Purge & Trap Technique *62*
2.1.6.1 Coupling of Purge and Trap with GC-MS Systems *63*
2.1.6.2 Modes of Operation of Purge and Trap Systems *64*
2.1.6.3 Static Headspace vs. Purge and Trap *70*
2.1.7 Dynamic Headspace – In-Tube Extraction *76*
2.1.8 Adsorptive Enrichment and Thermal Desorption *77*
2.1.8.1 Sample Collection *81*
2.1.8.2 Calibration *82*
2.1.8.3 Desorption *84*
2.1.9 Stir Bar Sorptive Extraction *87*
2.1.10 Pyrolysis *88*

2.1.10.1	Foil Pyrolysis	90
2.1.10.2	Curie Point Pyrolysis	92
2.1.10.3	Micro-Furnace Pyrolysis	94
2.1.11	Thermal Extraction (Outgassing)	95
2.1.12	Liquid Chromatography Clean-up	98
2.1.13	Pressurized Liquid Extraction	100
2.1.13.1	In-Cell Clean-up	103
2.1.13.2	In-Cell Hydrocarbon Oxidation	105
	References	105
2.2	Gas Chromatography	128
2.2.1	Sample Inlet Systems	128
2.2.2	Carrier Gas Regulation	129
2.2.2.1	Forward Pressure Regulation	130
2.2.2.2	Back Pressure Regulation	131
2.2.2.3	Carrier Gas Saving	132
2.2.3	Injection Port Septa	134
2.2.3.1	Septum Purge	136
2.2.3.2	The MicroSeal Septum	136
2.2.4	Injection Port Liner	138
2.2.4.1	Split Injection	138
2.2.4.2	Splitless Injection	139
2.2.4.3	Liner Activity and Deactivation	140
2.2.4.4	Liner Geometry	142
2.2.5	Hot Split/Splitless Sample Injection Techniques	143
2.2.5.1	Hot Needle Thermospray Injection Technique	144
2.2.5.2	Cold Needle Liquid Band Injection Technique	147
2.2.5.3	Filled Needle Injections	147
2.2.5.4	Split Injection	148
2.2.5.5	Splitless Injection (Total Sample Transfer)	148
2.2.5.6	Concurrent Solvent Recondensation	150
2.2.5.7	Concurrent Backflush	151
2.2.6	Temperature Programmable Injectors	155
2.2.6.1	PTV Injection Modes	159
2.2.6.2	Cryofocusing	166
2.2.7	Non-Vaporizing Injection Techniques	168
2.2.7.1	On-Column Injection	168
2.2.7.2	PTV On-Column Injection	171
2.2.7.3	LC-GC Coupling	171
2.2.8	Capillary Column Choice and Separation Optimization	173
2.2.8.1	Choice of Carrier Gas	174
2.2.8.2	Optimization of the Carrier Gas Flow	197
2.2.8.3	Sample Capacity	200
2.2.8.4	Internal Diameter	200
2.2.8.5	Film Thickness	202
2.2.8.6	Column Length	204

2.2.8.7	Properties of Column Phases	205
2.2.8.8	Ionic Liquid Phases	209
2.2.9	Chromatography Parameters	212
2.2.9.1	The Chromatogram and its Meaning	214
2.2.9.2	Capacity Factor k′	216
2.2.9.3	Chromatographic Resolution	216
2.2.9.4	Factors Affecting the Resolution	218
2.2.9.5	Maximum Sample Capacity	222
2.2.9.6	Peak Symmetry	222
2.2.9.7	Effect of Oven Temperature Ramp Rate	225
2.2.10	Fast Gas Chromatography Solutions	227
2.2.10.1	Fast Chromatography	227
2.2.10.2	Vacuum Outlet (Low Pressure) Chromatography	233
2.2.10.3	Ultra-Fast Chromatography	235
2.2.10.4	Flow-Field Thermal Gradient Gas Chromatography	237
2.2.11	Multi-Dimensional Gas Chromatography	239
2.2.11.1	Heart Cutting	241
2.2.11.2	Comprehensive GC – GC × GC	241
2.2.11.3	Modulation	245
2.2.11.4	Detection	246
2.2.11.5	Data Handling	248
2.2.11.6	Moving Capillary Stream Switching	249
2.2.12	Classical Detectors for GC-MS Systems	251
2.2.12.1	Atomic Emission Detector (AED)	253
2.2.12.2	Electron Capture Detector (ECD)	254
2.2.12.3	Electrolytical Conductivity Detector (ELCD)	256
2.2.12.4	Flame-Ionization Detector (FID)	257
2.2.12.5	Flamephotometric Detector (FPD)	258
2.2.12.6	Helium Ionization Detector (HID)	258
2.2.12.7	Nitrogen-Phosphorous Detector (NPD)	259
2.2.12.8	Pulsed Discharge Detector (PDD)	261
2.2.12.9	Photo Ionization Detector (PID)	262
2.2.12.10	Sulfur Chemiluminescence Detector (SCD)	265
2.2.12.11	Thermal Conductivity Detector (TCD)	265
2.2.12.12	Vacuum Ultra Violet Detector (VUV)	266
2.2.12.13	Olfactometry	268
2.2.12.14	Classical Detectors Parallel to the Mass Spectrometer	268
2.2.12.15	Microchannel Devices	271
	References	273
2.3	Mass Spectrometry	291
2.3.1	Ionization	292
2.3.1.1	Electron Ionization	292
2.3.1.2	Chemical Ionization	295
2.3.2	Mass Analysis	313
2.3.2.1	Resolving Power and Resolution in Mass Spectrometry	314

2.3.2.2	Quadrupole and Quadrupole Ion Trap Mass Spectrometer	325
2.3.2.3	Sector Field Mass Spectrometer	327
2.3.2.4	Orbitrap Mass Spectrometer	329
2.3.2.5	Time-of-Flight Analyzer	333
2.3.2.6	Ion Mobility Analyzer	336
2.3.2.7	High and Low Mass Resolution in the Case of Dioxin Analysis	339
2.3.3	Isotope Ratio Monitoring GC-MS	344
2.3.3.1	The Principles of Isotope Ratio Monitoring	345
2.3.3.2	Notations in irm-GC-MS	346
2.3.3.3	Isotopic Fractionation	347
2.3.3.4	irm-GC-MS Technology	349
2.3.3.5	The Open Split Interface	353
2.3.3.6	Compound Specific Isotope Analysis	353
2.3.3.7	Online Combustion for $\delta^{13}C$ and $\delta^{15}N$ Determination	354
2.3.3.8	The Oxidation Reactor	355
2.3.3.9	The Reduction Reactor	355
2.3.3.10	Water Removal	355
2.3.3.11	The Liquid Nitrogen Trap	357
2.3.3.12	Online High Temperature Conversion for $\delta^{2}H$ and $\delta^{18}O$ Determination	357
2.3.3.13	Mass Spectrometer for Isotope Ratio Analysis	359
2.3.3.14	Injection of Reference Gases	361
2.3.3.15	Isotope Reference Materials	361
2.3.4	Acquisition Techniques in GC-MS	363
2.3.4.1	Detection of the Complete Mass Spectrum (Full Scan)	363
2.3.4.2	Recording Individual Masses (SIM)	364
2.3.4.3	High Resolution Accurate Mass SIM Data Acquisition	374
2.3.4.4	MS/MS – Tandem Mass Spectrometry	380
2.3.5	Mass Calibration	388
2.3.6	Vacuum Systems	396
	References	400
3	**Evaluation of GC-MS Analyses**	**413**
3.1	Display of Chromatograms	413
3.1.1	Total Ion Current Chromatograms	413
3.1.2	Mass Chromatograms	415
3.2	Substance Identification	418
3.2.1	Reading Mass Spectra	418
3.2.2	Extraction of Mass Spectra	420
3.2.2.1	Manual Spectrum Subtraction	420
3.2.2.2	Deconvolution of Mass Spectra	425
3.2.3	The Retention Index	430
3.2.4	Libraries of Mass Spectra	434
3.2.4.1	Universal Mass Spectral Libraries	436
3.2.4.2	Application Libraries of Mass Spectra	438

3.2.5	Library Search Programs	443
3.2.5.1	The NIST Search Procedure	445
3.2.6	Interpretation of Mass Spectra	451
3.2.6.1	Isotope Patterns	461
3.2.6.2	Fragmentation and Rearrangement Reactions	468
3.2.6.3	DMOX Derivatives for Location of Double Bond Positions	471
3.2.7	Mass Spectroscopic Features of Selected Substance Classes	472
3.2.7.1	Volatile Halogenated Hydrocarbons	472
3.2.7.2	Benzene/Toluene/Ethylbenzene/Xylenes (BTEX, Alkylaromatics)	475
3.2.7.3	Polyaromatic Hydrocarbons	478
3.2.7.4	Phenols	480
3.2.7.5	Pesticides	482
3.2.7.6	Polychlorinated Biphenyls	492
3.2.7.7	Polychlorinated Dioxins/Furans (PCDDs/PCDFs)	495
3.2.7.8	Drugs	496
3.2.7.9	Explosives	498
3.2.7.10	Chemical Warfare Agents	498
3.2.7.11	Brominated Flame Retardants (BFRs)	504
3.3	Quantitation	505
3.3.1	Acquisition Rate	506
3.3.2	Decision Limit	507
3.3.3	Detection Limit	509
3.3.4	Limit of Quantitation	512
3.3.5	Sensitivity	514
3.3.6	The Calibration Function	514
3.3.7	Quantitation and Standardization	516
3.3.7.1	External Standardization	516
3.3.7.2	Internal Standardization	517
3.3.7.3	Standard Addition	522
3.4	Frequently Occurring Impurities	523
	References	531
4	**Applications**	**541**
4.1	Air Analysis According to U.S. EPA Method TO-14	541
4.1.1	Introduction	541
4.1.2	Analysis Conditions	543
4.1.3	Limit of Detection	544
4.1.4	Results	544
4.2	BTEX in Surface Water as of U.S. EPA Method 8260	549
4.2.1	Introduction	549
4.2.2	Sample Preparation	549
4.2.3	Experimental Conditions	550
4.2.4	Analysis Conditions	550
4.2.5	Results	551
4.2.6	Conclusions	553

4.3	Volatile Priority Pollutants	554
4.3.1	Introduction	554
4.3.2	Analysis Conditions	556
4.3.3	Results	557
4.4	MAGIC 60 – VOC Analysis	560
4.4.1	Introduction	560
4.4.2	Analysis Conditions	561
4.4.3	Results	569
4.5	irm-GC-MS of Volatile Organic Compounds	569
4.5.1	Introduction	569
4.5.2	Analysis Conditions	570
4.5.3	Results	571
4.6	Organotin Compounds in Water	572
4.6.1	Introduction	572
4.6.2	Analysis Conditions	573
4.6.3	Experimental Conditions	574
4.6.4	Results	577
4.7	Analysis of Dithiocarbamate Pesticides	577
4.7.1	Introduction	577
4.7.2	Analysis Conditions	578
4.7.3	Sample Preparation	580
4.7.4	Preparation of Standard Solutions and Reaction Mixture	580
4.7.4.1	Carbon Disulfide Standard Solution	580
4.7.4.2	Standard Solution of Thiram	580
4.7.4.3	Preparation of Reaction Mixture	580
4.7.4.4	Calibration Standards	580
4.7.5	Experimental Conditions	581
4.7.6	Sample Measurements	581
4.7.7	Results	583
4.7.7.1	Sensitivity	583
4.7.7.2	Recovery	583
4.7.7.3	Accuracy	583
4.7.8	General Guidelines for DTC Analysis	583
4.7.9	Conclusions	584
4.8	Pesticides Multi-Method by Single Quadrupole MS	584
4.8.1	Introduction	584
4.8.2	Analysis Conditions	592
4.8.3	Results	593
4.9	QuEChERSER Analysis of Pesticides	594
4.9.1	Introduction	594
4.9.2	Analysis Conditions	595
4.9.3	Analysis	597
4.9.4	Results	598
4.10	Pesticide Analysis with Ethyl Acetate Extraction and Automated Micro-SPE Clean-up	599

4.10.1	Introduction *599*
4.10.2	Analysis Conditions *600*
4.10.3	Ethyl Acetate Extraction *601*
4.10.3.1	Micro-SPE Clean-up *601*
4.10.4	Results *601*
4.11	Multi-Residue Pesticides Analysis in Ayurvedic Churna *603*
4.11.1	Introduction *603*
4.11.2	Analysis Conditions *604*
4.11.3	Sample Preparation *605*
4.11.4	Experimental Conditions *605*
4.11.5	Results *612*
4.11.6	Conclusions *612*
4.12	Polar Aromatic Amines by SPME *614*
4.12.1	Introduction *614*
4.12.2	Analysis Conditions *614*
4.12.3	Results *616*
4.13	Phthalates in Liquors *619*
4.13.1	Introduction *619*
4.13.2	Analysis Conditions *619*
4.13.3	Sample Preparation *620*
4.13.4	Experimental Conditions *621*
4.13.5	Sample Measurements *621*
4.13.6	Results *622*
4.13.7	Quantitation *623*
4.13.8	Sensitivity *624*
4.13.9	Method Precision and Recovery *625*
4.13.10	Conclusions *625*
4.14	Natural Spice Ingredients Capsaicin, Piperine, Thymol, and Cinnamaldehyde *626*
4.14.1	Introduction *627*
4.14.2	Analysis Conditions *627*
4.14.3	Experimental Conditions *628*
4.14.4	Sample Measurements *629*
4.14.5	Results *631*
4.14.6	Conclusions *632*
4.15	Geosmin and Methylisoborneol in Drinking Water *632*
4.15.1	Introduction *632*
4.15.2	Analysis Conditions *633*
4.15.3	SPME Method *634*
4.15.4	Results *634*
4.16	Flavor and Fragrance Profiling by Dual-Column GC-MS *637*
4.16.1	Introduction *637*
4.16.2	Analysis Conditions *638*
4.16.3	Sample Preparation *638*
4.16.4	Experimental Setup *639*

4.16.5	Results *641*
4.16.6	Conclusions *644*
4.17	Aroma Profiling of Cheese *644*
4.17.1	Introduction *644*
4.17.2	Analysis Conditions *645*
4.17.3	Sample Preparation *646*
4.17.4	Experimental Conditions *647*
4.17.4.1	Dynamic Headspace Sampling *647*
4.17.4.2	Thermal Desorption *647*
4.17.5	Sample Measurements *647*
4.17.6	Results *648*
4.17.7	Conclusions *649*
4.18	Allergens *649*
4.18.1	Introduction *649*
4.18.2	Analysis Conditions *649*
4.18.3	Sample Preparation *651*
4.18.3.1	Extraction *651*
4.18.4	Experimental Conditions *651*
4.18.5	Results *651*
4.18.6	Conclusions *656*
4.19	Metabolite Profiling *656*
4.19.1	Introduction *656*
4.19.1.1	Workflow Phase I: Discovery *656*
4.19.1.2	Workflow Phase II: Targeted Quantitation *658*
4.19.2	Analysis Conditions *658*
4.19.3	Sample Preparation *659*
4.19.4	Results *660*
4.19.5	Conclusion *664*
4.20	Extractables and Leachables *665*
4.20.1	Introduction *665*
4.20.2	Analysis Conditions *666*
4.20.3	Sample Preparation *668*
4.20.4	Experimental Conditions *668*
4.20.5	Data Processing and Results *668*
4.20.6	AMDIS Chromatogram Deconvolution *668*
4.20.7	Mass Frontier Spectrum Interpretation *673*
4.20.8	Conclusions *675*
4.21	Volatiles in Car Interiors *677*
4.21.1	Introduction *677*
4.21.2	Analysis Conditions *679*
4.21.3	Sample Measurements *680*
4.21.3.1	Sample Preparation *680*
4.21.3.2	VOC testing *680*
4.21.3.3	SVOC Testing *681*
4.21.4	Conclusion *683*

4.22	Azo Dyes in Leather and Textiles	683
4.22.1	Introduction	683
4.22.2	Analysis Conditions	685
4.22.3	Results	688
4.23	Fast GC of 16 Priority PAHs	691
4.23.1	Introduction	691
4.23.2	Analysis Conditions	694
4.23.3	Results	695
4.24	Environmental Contaminants in Fish	697
4.24.1	Introduction	698
4.24.2	Analysis conditions	698
4.24.3	Sample Preparation	700
4.24.4	Sample Measurements	700
4.24.5	Results	705
4.24.6	Method Limitations	711
4.24.7	Conclusions	712
4.25	Fast GC of PCBs	712
4.25.1	Introduction	712
4.25.2	Analysis Conditions	714
4.25.3	Results	716
4.26	Dioxin Screening in Food and Feed	718
4.26.1	Introduction	718
4.26.2	Analysis Conditions	720
4.26.3	Sample Preparation	721
4.26.4	Experimental Conditions	721
4.26.5	Results	724
4.26.6	Conclusions	729
4.27	Confirmation Analysis of Dioxins and Dioxin-like PCBs	730
4.27.1	Introduction	730
4.27.2	Analysis Conditions	732
4.27.2.1	Chromatographic Analysis	733
4.27.3	Results	736
4.28	U.S. EPA 1614 Brominated Flame Retardants PBDEs	739
4.28.1	Introduction	739
4.28.2	Analysis Conditions	739
4.28.3	Results	740
4.29	PBB Analysis by SPME	747
4.29.1	Introduction	747
4.29.2	Analysis Conditions	748
4.29.3	Results	749
4.30	THC-A in Urine by NCI	751
4.30.1	Introduction	751
4.30.2	Analysis Conditions	752
4.30.3	Sample Preparation	753
4.30.3.1	Hydrolysis	753

4.30.3.2	Extraction	754
4.30.3.3	Derivatization	754
4.30.4	Experimental Conditions	754
4.30.5	Sample Measurements	754
4.30.5.1	Reproducibility of Retention Times	754
4.30.5.2	Limit of Detection and Limit of Quantification	754
4.30.5.3	Recovery and Calibration	754
4.30.6	Results	755
4.30.6.1	Mass Spectra and GC Separation	755
4.30.7	Quality Control Samples	757
4.30.8	Conclusions	757
4.31	Comprehensive Drug Screening and Quantitation	757
4.31.1	Introduction	757
4.31.2	Analysis Conditions	758
4.31.3	Sample Preparation	759
4.31.4	Experimental Conditions	759
4.31.5	Sample Measurements	759
4.31.6	Results	761
4.31.7	Conclusions	761
4.32	Drugs of Abuse	762
4.32.1	Introduction	762
4.32.2	Analysis Conditions	763
4.32.3	Results	765
4.33	Structure Elucidation by CI and MS/MS	766
4.33.1	Introduction	766
4.33.2	Analysis Conditions	767
4.33.3	Experimental Conditions	768
4.33.4	Sample Measurements	768
4.33.5	Results	769
	References	770

Glossary 787
Further Reading 858

Author Index 859
Subject Index 861
Compound Index 881

Foreword

"The goal is to turn data into information, and information into insight."
Carly Fiorina
Chief Executive Officer of Hewlett-Packard from 1999 to 2005

Our lives can truly be said to be surrounded by chemistry. Globally, there are an estimated 200 000–350 000 chemicals that could be regulated in some way, depending on their use, production levels, and associated risks. While these chemicals bring significant benefits to our daily lives, there's also a considerable concern about ensuring they are properly managed. Among them, the number of tightly regulated substances – such as high risk chemicals – is estimated to be around 40 000–60 000. In this context, it's evident that only efficient analytical control can deliver the necessary insights to minimize their drawbacks.

A full-coverage Gas Chromatography-Mass Spectrometry (GC-MS) system is estimated to analyze about 30–40% of these chemicals, which comes out to roughly 12 000–24 000 substances. This underscores the importance of GC-MS in identifying and measuring harmful, misused, or unusual levels, as well as assessing the product quality of chemical compounds across multiple domains.

As a consequence of the awareness in controlling these compounds and following the well-known precautionary principle of "better safe than sorry", the number of samples analyzed using this technique keeps growing, reaching hundreds of thousands every year worldwide.

This situation has greatly contributed to the impressive improvement in the performance of this analytical technique. The constant progress is pretty remarkable, looking back, especially, to 20 years ago, the changes and advancements are truly surprising. That aligns with a steady rise in scientific publications, increasing from just 5 000 to an impressive 10 000 annually over the past decade.

Such improvements are related to all the main parts of the instrument: injector, chromatography, ion source, analyzer, detector, data treatment, and user-friendly software. With such enhancements achieving sensitivities at the ppt level, and enabling the evaluation of hundreds of compounds in a short time with remarkable precision and robustness. A major breakthrough has been the introduction of high resolution GC-MS, which enables non-target analysis and metabolomics studies while providing access to accurate mass libraries.

Last but not least, sample treatment procedures have also shown significant evolution. This progress addresses the challenges of selectively extracting target compounds present at extremely low levels compared to other substances in the sample. Additionally, it considers the need to transfer compounds from a hydrophilic phase to the one compatible with GC-MS analysis. These complementary analytical devices and procedures to GC-MS have undoubtedly facilitated the growth in applications, significantly expanding its use.

This book expertly presents the latest technical and scientific advancements in a clear and accessible manner, with chapters authored by one of the most renowned experts in the field.

Finally, to say we are far from witnessing the end of GC-MS's evolution. Exciting advancements lie ahead, such as leveraging artificial intelligence and machine learning for enhanced data interpretation and adopting green chromatography to reduce solvent use and energy consumption.

Here's to the enduring legacy of GC-MS!

Amadeo R. Fernandez-Alba
Full Professor in Analytical Chemistry
University of Almeria, Spain

Preface to the Fourth Edition

"We must trust to nothing but facts: These are presented to us by nature, and cannot deceive. We ought, in every instance, to submit our reasoning to the test of experiment, and never to search for truth but by the natural road of experiment and observation."
Antoine-Laurent de Lavoisier, 1743–1794, French chemist

Lavoisier forever changed the practice and concepts of chemistry by forging laboratory analyses that would bring order to the chaotic centuries of Greek philosophy and medieval alchemy. Lavoisier's work in framing the principles of modern chemistry led future generations to regard him as a founder of analytical science.
American Chemical Society, 1999

"Mass spectrometry has developed fantastically. And chromatography has progressed from a very hands-on process to an extremely advanced tool with very sophisticated measurements: It does more, and it works better," Peter Schoenmakers stated in the recent interview about the most significant developments across analytical science over the past decade (Hulme and van Geel, 2023).

A decade in the analytical industry is a long time indeed, changing perspectives, challenges, and requirements dramatically. Developments happened in many analytical areas, most importantly in sample preparation, chromatography, and mass spectrometry, the essential topics of this *Handbook of GC-MS*. Green analytical chemistry, miniaturization, automation, sample throughput, and multi-residue methods are main subjects, not to underestimate the outstanding achievements in detection selectivity and sensitivity. Seeking green solutions by reducing sample and solvent volumes, replacing hazardous solvents and reagents, avoiding human bias and errors, and reducing energy consumption, critical waste, and consumables became high level development goals in analytical workflows and instrumentation. Notable advancements were introduced and found their application in routine analysis. The ongoing method refinements and a dedication to special MS analyzer types, exploiting their particular strengths for different gas chromatography–mass spectrometry (GC-MS) applications, expanded the scope of fit-to-purpose GC-MS applications significantly. With the recent advancements in hyphenated sample preparation, using miniaturization and automation, faster and comprehensive GC-MS analyses were accomplished.

The first edition of this *Handbook of GC-MS* was published in 1996: "At last there is a comprehensive manual for the practical use of GC/MS." From the very beginning, the *Handbook of GC-MS* not only cover the necessary fundamentals of GC and MS but also gave many practical examples for key applications suggesting optimum measurement conditions. It became a valuable resource with much convenient information for successful GC-MS operation. After more than ten years since its third edition with encouraging feedback by many readers, it is the right time to present this fourth edition of the *Handbook of GC-MS* – a major revamp and update, keeping the solid fundamentals but getting into detail with newly introduced technologies and opening the floor for the new potential of *green analytical chemistry*. Sample preparation became more and more integrated into GC-MS workflows facilitated by a significant reduction of sample volumes. The application of microextraction techniques in GC-MS delivers the basis for online and also the regular offline automation. Using the normal phase liquid chromatography (LC) for online separation and clean-up provides a high potential for integrated routine methods. After many years in an analytical shadowy existence, LC-GC hyphenation became indispensable replacing many formerly laborious manual procedures with high productivity and high sample throughput solutions for complex sample matrices. It is the miniaturization in sample preparation that invites and facilitates automated workflows, not vice versa.

On the GC side, faster analysis, the resurrection of low pressure GC, and the many successful examples of replacing helium as the only carrier gas while still staying compatible with the various MS types prove this most important synergy. New low bleed polar phases further benefit MS detection. The comprehensive GC × GC analysis with newly designed modulation devices arrived in routine analysis for many complex GC-MS applications.

On the MS side, the most influential developments are the move to MS/MS techniques, high resolution and accurate mass, and ion mobility MS applications. The targeted multi-residue analyses benefit from the high selectivity provided by MS/MS and accurate mass capabilities provided by new and faster triple quadrupole, Orbitrap or QToF instrumentation. In addition, the single quadrupole application remains important as the workhorse for many standard applications, having replaced many conventional detectors. But still, the selective "classical" detectors in parallel to the mass selectivity of the MS, including the olfactory port, deliver valuable complementary information.

Less visible but of paramount importance for processing the growing flood of data, the sophisticated control of complex MS detection in real time and powerful data processing and comprehensive reporting are strongly supporting such innovative analytical developments.

I am grateful to many contributors for their support on many GC-MS topics over the years. My special thanks for providing excellent background material for the current update, valuable comments, and discussions go to Peter Boeker of HyperChrom in Germany, Elefteria Psillakis of ExtraTECH Analytical Solutions in Greece, and Stefan Cretnik and Günter Böhm of CTC Analytics AG in Switzerland. My particular thanks go to my long-standing colleagues and companions Christian Soulier and Frank Theobald for in-depth discussions and collaboration.

Science in general and analytical science in particular are constantly facing challenges. In particular, decisions about exceeding maximum concentrations often entail legal consequences (the founder of analytical science Lavoisier was guillotined for his research!). Solid, reproducible data from suitable methods is the goal of all analytics. In this sense, this *Handbook of GC-MS* is intended to provide a practical contribution to analytical science. It is hoped that this vastly updated *Handbook* will be useful for novice and experienced routine users as well, both in offering the background and fundamental information for an in-depth understanding of newly applied technologies and in giving the reader guidance with many current examples for high performing, green, and reproducible GC-MS analyses.

August 2024 *Hans-Joachim Hübschmann*
Mainz/Osaka

References

American Chemical Society (1999) *The Chemical Revolution*, National Historic Chemical Landmarks program. https://www.acs.org/education/whatischemistry/landmarks/lavoisier.html (accessed 28 June 2024).

Hulme, G. and van Geel, F. (2023) Ten Year Views: With Peter Schoenmakers. *The Analytical Scientist*, 25 August 2023. https://theanalyticalscientist.com/business-education/ten-year-views-with-peter-schoenmakers (accessed 28 June 2024).

1

Introduction

"Food safety is a fundamental need for life, and ideally, humans would be trusted to follow the moral imperative set into laws designed to protect our ecosystem and produce safe food for consumption. However, human nature and past transgressions have demonstrated that testing is needed to verify good agricultural and food safety practices."

Steven J. Lehotay, 2024
USDA Agricultural Research Service, Wyndmoor, PA, USA

Detailed knowledge of the chemical processes in plants, animals, and in our environment with air, water, and soil, about the safety of food and products, has been made possible only through the power of modern instrumental analysis. In an increasingly short time span, more and more data are being collected. The detection limits for organic substances are down in the attomole region, and counting individual molecules per unit time has already become a reality. In food safety and environmental analysis, we achieve measurements at the level of background contamination. However, samples subjected to chemical trace analysis carry high matrix. With the demand for decreasing detection limits by legal regulations, in the future effective sample preparation and separation procedures in association with highly selective detection techniques will be of critical importance for analysis. In addition, the number of substances requiring detection is increasing, and with the broadening possibilities for analysis, so is the number of samples. Even there is the concern of "the inadequacy of current regulations in effectively controlling food contact materials (FCM)" for food safety (Diaz-Galiano et al., 2024). The increase in analytical sensitivity is exemplified with the persistent organic pollutants (POPs) in the case of the "dioxins" with 2,3,7,8-tetrachlorodibenzodioxin (TCDD), the most potent cancer-promoting and teratogenic congener of the polychlorinated dibenzodioxins (PCDDs), still continuously analyzed as contamination in food and feed (Table 1.1).

Capillary gas chromatography with mass spectrometry detection (GC-MS) is today the most important analytical method in organic chemical analysis for the determination of individual low molecular substances in complex mixtures. Mass spectrometry (MS) as the detection method gives the most meaningful data, arising from the direct determination of the substance molecule or of fragments. The

Handbook of GC-MS: Fundamentals and Applications, Fourth Edition. Hans-Joachim Hübschmann.
© 2025 WILEY-VCH GmbH. Published 2025 by WILEY-VCH GmbH.

Table 1.1 Sensitivity progress in mass spectrometry.

Year	Instrumental technique	Limit of detection (pg)
1967	GC-FID (packed column)	500
1973	GC-MS (quadrupole, packed column)	300
1976	GC-MS-SIM (magnetic instrument, capillary column)	200
1977	GC-MS (magnetic sector instrument)	5
1983	GC-HRMS (double focusing magnetic sector MS)	0.15
1984	GC-MSD/SIM (quadrupole mass selective detector)	2
1986	GC-HRMS (double focusing magnetic sector MS)	0.025
1989	GC-HRMS (double focusing magnetic sector MS)	0.010
1990	GC-HRMS required for PCDD/Fs by the US EPA	Method 1613 Rev.A
1992	GC-HRMS (double focusing magnetic sector MS)	0.005
2006	GC×GC-HRMS (using comprehensive GC)	0.0003
2010	Cryogenic zone compression (t-CZC) GC-HRMS	0.0002
2011	t-CZC GC-HRMS reports low attogram levels in serum (Patterson, 2011; Patterson et al., 2011)	<0.0001
2014	GC-MS/MS EU approval for PCDD confirmation in food	EU No. 589/2014
2017	GC-Orbitrap reports U.S. EPA 1613 compliance	0.0001
2018	GC-MS/MS reports low femtogram sensitivity	0.0006
2018	APGC-MSMS reports attogram LODs in fly ash	0.0001

APGC, atmospheric pressure gas chromatography; t-CZC, time controlled cryogenic zone compression; FID, flame ionization detector; GC, gas chromatography; HRMS, high resolution mass spectrometry; LOD, limit of detection; MS, mass spectrometry; MS/MS triple quadrupole analyzer; MSD, mass selective detector; SIM, selected ion monitoring; and US EPA, United States Environmental Protection Agency.

results of MS are therefore used as a reference for other indirect detection processes and finally for confirmation of the facts. The complete integration of MS and gas chromatography (GC) into a single GC-MS system has shown itself to be synergistic in every respect.

It was Fred W. McLaffert who pioneered the technique of coupling a gas chromatograph with a mass spectrometer with Roland Gohlke at Dow Chemical Co., developing a GC-Time-of-Flight (TOF)-MS instrument "capable of rapidly characterizing organic chemical mixtures boiling below 350 °C" (Gohlke 1959). Still at the beginning of the 1980s, MS was considered to be expensive, complicated, and time-consuming or personnel-intensive. At the beginning of the 1990s, MS became more widely recognized and furthermore an indispensable detection method for GC. There is now hardly a GC laboratory which is not equipped with a GC-MS system. The simple construction, clear function, and an operating procedure, which has become easy because of modern computer systems, have resulted in the fact that GC-MS is widely used alongside traditional spectroscopic methods. The universal detection technique, together with high selectivity and very high sensitivity, has made GC-MS indispensable for a broad spectrum of applications. With recent developments, even higher selectivity is provided by the structure selective MS/MS and the elemental formula providing accurate mass technologies for modern multi-residue methods with short sample preparation and clean-up steps. Benchtop GC-MS systems have completely replaced in many applications the stand-alone GC with selective detectors. GC-MS/MS has found its way to routine replacing many single quadrupole systems today, and accurate mass detection follows on its heels.

The control of the chromatographic separation process still contributes significantly to the exploitation of the analytical performance of the GC-MS system (or according to Koni Grob: "Chromatography takes place in the column!"). The analytical prediction capabilities of a GC-MS system are, however, dependent upon mastering the spectrometry. The evaluation and assessment of the data are leading to increasingly greater challenges with decreasing detection limits and the increasing number of compounds sought or found. As quantification is the main application in trace analysis today, the appropriate data processing requires additional measures for confirmation of results provided by mass spectrometric methods.

The high performance of GC lies in the separation of substance mixtures and providing the transient signals for data deconvolution. With the introduction of fused silica columns, GC has become the most important and powerful separation method of analyzing complex mixtures of products. GC-MS accommodates the current trend toward multi-methods or multi-component analyses (e.g. of pesticides, environmental contaminations, fragrances, drugs, and beyond) in an ideal way. Even isomeric compounds, which are present, for example, in essential oils, metabolic profiling, polychlorinated biphenyls (PCBs), or dioxins, are separated by GC, while in many cases their mass spectra are indistinguishable. The high efficiency as a routine process is achieved through the high speed of analysis and the short turnaround time and thus guarantees high productivity with a high sample throughput. Adaptation and optimization for different tasks only require a quick change of column. In many cases, however, and here, the analyst relies on the explanatory power of the mass spectrometer, one type of medium polar column can be used for different applications by adapting the sample injection technique and modifying the method parameters.

The area of application of GC and GC-MS is limited to substances that are volatile enough to be analyzed by GC. The further development of column technology in recent years has been very important for application to the analysis of high boiling compounds. Temperature-stable phases now allow elution temperatures of up to 500 °C for stable compounds. A pyrolizer in the form of a stand-alone sample injection system extends the area of application to involatile substances by separation and detection of thermal decomposition products. A typical example of current interest for GC-MS analysis of high boiling compounds is the determination of polyaromatic hydrocarbons, which has become a routine process using the most modern column material.

The coupling of GC with MS using fused silica capillary columns has played an important role in achieving the current high performance level in chemical analysis. In particular in the areas of environmental analysis, analysis of residues, and forensic science, the high information content of GC-MS analyses has brought chemical analysis into focus through, sometimes, sensational results. For example, it has been used for the determination of process contaminants in food and feed or the accumulation of persistent organic pollutants in the food chain. With the current state of knowledge, GC-MS is an important method for monitoring the introduction, the location and fate of man-made forever-chemicals in the environment, foodstuffs, chemical processes, and biochemical processes in the human body. GC-MS has also made its contribution in areas such as the atmospheric ozone depletion, the safeguarding of quality standards in foodstuffs production, in the study of the metabolism of pharmaceuticals or plant protection agents, or in the investigation of polychlorinated dioxins and furans produced in certain chemical even natural processes, to name but a few.

The technical realization of GC-MS coupling occupies a very special position in instrumental analysis. Fused silica columns are easy to handle, can be changed rapidly, and are available in many high quality forms. New microfluidic switching technologies extend the application without compromising performance for flow switching or parallel detection solutions. The optimized carrier gas streams show good compatibility with mass spectrometers, which is true today for both carrier gases, helium and hydrogen. Coupling can therefore take place easily by directly connecting the GC column to the ion source of the mass spectrometer.

The obvious challenges of GC and GC-MS lie where actual samples contain involatile components (matrix). In this case, the sample must be processed before the analysis appropriately, or suitable column-switching devices need to be considered for backflushing of high boiling matrix components. The clean-up is generally associated with the enrichment of trace components and the separation from incompatible matrix. In many methods, there is a trend toward integrating sample preparation and enrichment in a single instrument. Headspace and purge and trap techniques, thermodesorption, or the micro-SPE clean-up and solid phase microextraction (SPME) are coupled online with GC-MS and integrated into the data systems for seamless control.

Future developments will continue with green analytical chemistry in mind, miniaturized, highly productive with automated sample preparation for multi-compound trace analysis and quantitation of regulated target analytes. In addition, to comply with the aspects of food safety and product safety requirements, non-targeted analytical techniques for the identification of potentially hazardous contaminants will evolve applying combined full scan and accurate mass capabilities.

1.1 The Historical Development of the GC-MS Technique

The foundation work in both GC and MS, which led to the current realization, was published at the end of the 1950s. At the end of the 1970s and the beginning of the 1980s, a rapid increase in the use of GC-MS in all areas of organic analysis began. The instrumental technique has now achieved a mature level for the once much-specialized operation to become an indispensable routine analysis method.

1910: The physicist J.J. Thompson developed the first mass spectrometer and proved for the first time the existence of isotopes (^{20}Ne and ^{22}Ne). He wrote in his book Rays of Positive Electricity and their Application to Chemical Analysis: "I have described at some length the application of positive rays to chemical analysis: one of the main reasons for writing this book was the hope that it might induce others, and especially chemists, to try this method of analysis. I feel sure that there are many problems in chemistry which could be solved with far greater ease by this than any other method." Cambridge 1913. In fact, Thompson developed the first isotope ratio mass spectrometer (IRMS).

1910: In the same year, Mikhail S. Tsvet, a Russian-Italian botanist, published his book in Warsaw on "Chromophores in the Plant and Animal World." With this, he may be considered to be the discoverer of chromatography.

1918: Arthur J. Dempster used electron impact ionization for the first time.

1920: Francis William Aston continued the work of Thompson with his own mass spectrometer equipped with a photoplate as detector. The results verified the existence of isotopes of stable elements (e.g. ^{35}Cl and ^{37}Cl) and confirmed the results of Thompson.

1929: Walter Bartky and Arthur J. Dempster developed the theory for a double-focusing mass spectrometer with electrostat and magnetic sector.

1934: Josef Mattauch and Richard F. K. Herzog published the calculations for an ion optics system with perfect focusing over the whole length of a photoplate.

1935: Arthur J. Dempster published the latest elements to be measured by MS, platinum (Pt), and iridium (Ir). Aston thus regarded MS to have come to the end of its development.

1936: Kenneth T. Bainbridge and Edward B. Jordan determined the mass of nuclides to six significant figures, the first accurate mass application.

1937: Lincoln G. Smith determined the ionization potential of methane (as the first organic molecule).

1938: A. Hustrulid published the first spectrum of benzene.

1941: Archer J.P. Martin and Richard L.M. Synge published a paper on the principle of gas-liquid chromatography (GLC).

1946: W.E. Stephens proposed a TOF mass spectrometer: *Velocitron*.

1947: The US National Bureau of Standards (NBS) began the collection of mass spectra as a result of the use of MS in the petroleum industry.

1948: John A. Hipple described the ion cyclotron principle, known as the Omegatron that now forms the basis of the current ion cyclotron resonance (ICR) instruments.

1950: Roland S. Gohlke published for the first time the coupling of a gas chromatograph (packed column) with a mass spectrometer (Bendix TOF).

1950: The Nobel Prize for chemistry was awarded to Martin and Synge for their work on GLC (1941).

1950: Fred W. McLafferty, Klaus Biemann, and John H. Beynon applied MS to organic substances (natural products) and transferred the principles of organic chemical reactions to the formation of mass spectra.

1952: Erika Cremer and co-workers presented an experimental gas chromatograph to the ACHEMA in Frankfurt; parallel work was carried out by Jaroslav Janák in Czechoslovakia.

1952: Archer J.P. Martin and A.T. James published the first applications of GLC.

1953: Walter H. Johnson and Alfred O.C. Nier published an ion optic with a 90° electric and 60° magnetic sector, which, because of the outstanding focusing properties, was to become the basis for many high resolution, organic mass spectrometers (Nier/Johnson analyzer).

1954: Wolfgang Paul published his fundamental work on the quadrupole analyzer.

1955: W.C. Wiley and I.H. McLaren developed a prototype of the present TOF mass spectrometer.

1955: Denis H. Desty presented the first GC of the present construction type with a syringe injector and thermal conductivity detector. The first commercial instruments were supplied by Burrell Corp., Perkin Elmer, and Podbielniak Corp.

1956: A German patent was granted for the QUISTOR (quadrupole ion storage device) together with the quadrupole mass spectrometer.

1958: Wolfgang Paul published about his research on the quadrupole mass filter as
– a filter for individual ions
– a scanning device for the production of mass spectra
– a filter for the exclusion of individual ions.

1958: Ken Shoulders manufactured the first 12 quadrupole mass spectrometers at Stanford Research Institute, California.

1958: Marcel J.E. Golay reported for the first time on the use of open tubular columns for GC.

1958: James Lovelock developed the argon ionization detector as a forerunner of the electron capture detector (ECD, J. Lovelock and S.R. Lipsky, 1960).

1962: Ulf von Zahn designed the first hyperbolic quadrupole mass filter.

1964: The first commercial quadrupole mass spectrometers were developed as residual gas analyzers (Quad 200 RGAs) by Robert Finnigan and P.M. Uthe at EAI (Electronic Associates Inc., Paolo Alto, California).

1966: Milan S.B. Munson and Frank H. Field published the principle of chemical ionization (CI).

1968: The first commercial quadrupole GC-MS system for organic analysis was supplied by Finnigan Instruments Corporation to the Stanford Medical School Genetics Department.

1978: Raymond D. Dandenau and E.H. Zerenner introduced the technique of fused silica capillary columns.

1978: Richard A. Yost and Chris G. Enke introduced the triple-quadrupole technique.

1982: Robert Finnigan obtained the first patents on ion trap technology for the mode of selective mass instability and presented the ion trap detector as the first universal MS detector with a PC data system (IBM XT).

1989: Prof. Wolfgang Paul, Bonn University, Germany, received the Nobel Prize for physics for work on ion traps, together with Prof. Hans G. Dehmelt, University of Washington, Seattle, and Prof. Norman F. Ramsay, Harvard University, USA.

2000: Alexander Makarov published a completely new mass analyzer concept called *Orbitrap* suitable for accurate mass measurements of low ion beams.

2005: Introduction of a new type of hybrid Orbitrap mass spectrometer by Thermo Electron Corporation, Bremen, Germany, for MS/MS; very high mass resolution and accurate mass measurement on the chromatographic time scale.

2009: Amelia Peterson *et al.*, University Wisconsin, Prof. Josh Coon group, first published the results on the implementation of an EI/CI interface on a hybrid Orbitrap system for ultra-high resolution GC-MS using a GC-Quadrupole-Orbitrap configuration for full scan, selected ion monitoring (SIM), MS/MS, and selected reaction monitoring (SRM) at the American Society of Mass Spectrometry (ASMS) conference.

2011: Agilent Technologies Inc., Santa Clara, CA, USA, introduces Gas Chromatography - Quadrupole Time-of-Flight (GC-QTOF) systems for the high sensitivity detection and analysis of unknown molecules in complex mixtures.

2015: Introduction of the first high resolution accurate mass GC-MS system using Orbitrap technology by Thermo Fisher Scientific, Austin, TX, USA, covering routine GC-MS applications.

2020: Newly introduced Orbitrap GC-MS technology provides 240 000 mass resolution power at m/z 200 and sub ppm mass accuracy.

References

Díaz-Galiano, F.J., Murcia-Morales, M., Gómez-Ramos, M.J., del Mar Gómez-Ramos, M., and Fernández-Alba, A.R. (2024) Economic poisons: a review of food contact materials and their analysis using mass spectrometry. *Trends Anal. Chem.*, **117550**. doi: 10.1016/j.jsamd.2023.100613.

Eliuk, S. and Makarov, A. (2015) Evolution of orbitrap mass spectrometry instrumentation. *Ann. Rev. Anal. Chem.*, **8**, 61–80. doi: 10.1146/annurev-anchem-071114-040325.

Gohlke, R.S. (1959) Time-of-flight mass spectrometry and gas-liquid partition chromatography. *Anal. Chem.*, **31** (4), 535–541. doi: 10.1021/ac50164a024.

Lehotay, S.J. (2024) Food safety analysis 2.0. *Anal. Bioanal. Chem.*, **416** (3), 609–610. doi: 10.1007/s00216-023-05036-4.

Patterson, D.G., Jr. (2011) Human Biomonitoring: Attogram Level Sensitivity and Consequences for Analytical Standards Purity. Application Note 35, Cambridge Isotope Laboratories, Inc., Tewksbury, MA, USA.

Patterson, D.G., Jr., Welch, S.M., Turner, W.E., Sjödin, A., and Focant, J.F. (2011) Cryogenic zone compression for the measurement of dioxins in human serum by isotope dilution at the attogram level using modulated gas chromatography coupled to high resolution magnetic sector mass spectrometry. *J. Chrom. A*, **1218** (21), 3274–3281. doi: 10.1016/j.chroma.2010.10.084.

2

Fundamentals

2.1 Sample Preparation

In chemical analysis, the sample treatment is the most critical step to achieve reliable and reproducible data. Sample preparation is considered the main bottleneck of analytical procedures, consuming most of the available resources with more than 60% of the laboratory time. With a 30% share, the sample preparation procedures are a main source of errors besides the human bias of 19% and other contributors of lower impact. The influence of sample preparation on data quality is often falsely neglected (Majors, 2015). The process of sampling, homogenization, preparation of the sample with extraction, clean-up, and concentration, and then the gas chromatography–mass spectrometry (GC-MS) analysis with injection, separation, detection, and data processing define the total analysis workflow. Low level analytes of different chemical nature are embedded in complex matrices and need to become isolated and concentrated. The traditional manual extraction methods, such as liquid/liquid extraction (LLE), solid phase extraction (SPE), or Soxhlet extraction, are often time-consuming and use large amounts of organic solvents. The renaissance of solid/liquid and liquid/liquid extractions comes with the miniaturization to low sample and solvent volumes with the trend toward green analytical chemistry (GAC). For instance, the well-known U.S. Environmental Protection Agency (EPA) method 8270E for semi-volatile compound (SVOC) analysis by LLE is available fully automated for only 15 mL of sample and online gas chromatography – tandem mass spectrometry (GC-MS/MS) analysis (Lim Sin Yee and Zou, 2024). This development of automated and often hyphenated extraction techniques improves the accuracy and precision of analytical results and the speed of results out and the sample throughput (Qiu and Raynie, 2017). The mass spectrometer in current GC-MS systems became the standard detector replacing classical detectors in multicomponent analysis. A current trend is the seamless integration of previously manual sample preparation steps into automated instrumental workflows with GC-MS.

We see such integrated instrumental sample preparation solutions realized with static and dynamic headspace, solid phase microextraction (SPME), micro-SPE (μSPE), or thermal desorption (TD), only to name a few of the available options. The current trend is even more directed to automated instrumental techniques supporting the green analytical chemistry principles by limiting expensive often

Handbook of GC-MS: Fundamentals and Applications, Fourth Edition. Hans-Joachim Hübschmann.
© 2025 WILEY-VCH GmbH. Published 2025 by WILEY-VCH GmbH.

hazardous solvent, reagents and standards use and reducing the time-consuming error-prone manual workload to the essential (Valcárcel et al., 2014; Ramos et al., 2023; Rodríguez-Delgado et al., 2024).

Avoiding analyte dilution and the concentration steps in sample preparation are of particular importance for coupling with capillary GC-MS. In trace analysis, the limited sample capacity of capillary columns must be compensated for. It is, therefore, necessary that both overloading of the stationary phase by matrix is avoided and the detection limits of mass spectrometric detection are taken into consideration. To optimize separation on a capillary column, incompatible matrix and strongly interfering components must be removed before applying the extract. The primarily universal character of the mass spectrometer sets requirements on the preparation of a sample which are to some extent more demanding than those of an element-specific detector, such as electron capture detector (ECD) or nitrogen/phosphorus detector (NPD) unless highly selective techniques, such as MS/MS or high resolution accurate measurements, are applied. Also, ion source matrix effects resulting in signal quenching must be considered. The clean-up and analyte concentration, which forms part of sample preparation, must therefore always be regarded as a necessary preparative step for GC-MS analysis. The differences in the concentration range between various samples, differences between the volatility of the analytes and that of the matrix, and the varying chemical nature of the substances are important for the choice of a suitable sample preparation procedure (da Silva Oliveira et al., 2024).

Off-line techniques (as opposed to online coupling or hyphenated techniques) have the logistic advantage that samples can be processed in parallel and extracts can be subjected to different analytical processes besides GC-MS. Online techniques have the special advantage of sequential processing of the samples without intermediate manual steps using x,y,z-robotic sample preparation platforms. For instance, the online clean-up of an extract or sample derivatization offers the advantage of overlapping the necessary sample preparation steps with an ongoing GC-MS analysis run. This gives each sample an identical preparation time, improving reproducibility as no sample needs to wait for the samples in sequence to be completed. This overlapping automated operation permits maximum use of the instrument and increases sample throughput.

Online processes generally offer the potential for higher analytical quality through lower contamination from glassware or consumables, smaller sample sizes, less solvent use, and lower detection limits. Frequently, total sample transfer is possible without taking aliquots or diluting also by injection techniques using larger extract volumes. Volatility differences between the sample and the matrix allow, for example, the use of extraction techniques such as the static or dynamic headspace, or SPME techniques as typical GC-MS coupling techniques. These are already in use as online techniques in many laboratories. In case of analytes with insufficient volatility, other extraction procedures, for example, liquid or thermal extraction, pyrolysis, or SPE are increasingly used online. SPE in the form of microextraction, gel permeation chromatography (GPC), liquid chromatography-GC coupling

(LC-GC), or extraction with pressurized or supercritical fluids show high analytical potential here.

Future developments in automated online sample preparation are expected from further miniaturization of manual standard methods, the replacement of hazardous solvents, and the application of alternative deep eutectic solvents or nanomaterials for the establishment of GAC (Scida et al., 2011; Dheyab et al., 2021; Bintanel-Cenis et al., 2023).

2.1.1 QuEChERS Sample Preparation

> *The QuEChERS method is a streamlined approach that makes it easier and less expensive for analytical chemists to examine pesticide residues in food.*
> Steven Lehotay, 2013

The sample preparation and clean-up processes for pesticide analysis have not been standardized in the past, but a clear trend over the recent years indicates a strong preference for laboratories using the QuEChERS extraction method. QuEChERS describes a buffered LLE with acetonitrile with dispersive SPE clean-up for extracting multi-residue pesticides, initially developed for fruits and vegetables, but in general food use and beyond today. The name is formed as an acronym for "Quick, Easy, Cheap, Effective, Rugged and Safe" (Anastassiades, 2024). The high (buffer) salt addition during the LLE leads to spontaneous phase separation of the otherwise water-miscible acetonitrile from the aqueous sample used for further clean-up. This new sample preparation method was first presented during the fourth European Pesticide Residue Workshop in Rome followed by the publication in the *Journal of AOAC International* (Anastassiades et al., 2002, 2003). As the sample preparation methods for pesticides in the past have been optimized for different matrices and mirrored the individual experience and knowledge of many trace chemical laboratories, QuEChERS today is taking the lead for an increasingly standardized methodology, creating an international pool of experience for the growing number of pesticides to analyze on residue level (Cunha et al., 2007; Krol et al., 2014). This trend is impressively pictured with the feedback from EU proficiency tests indicating the used sample preparation method. Figure 2.1 shows the statistics starting from 2006 over the following 6 years. While traditional methods such as Luke, Specht, and solvent extractions steadily decline, the number of labs using QuEChERS is strongly increasing and is dominating the types of sample preparation methods applied in pesticide residue labs. The trend continues to this day.

The steadily growing international use of the QuEChERS method is based on what the method wants to convey: speed, ease of use, minimized use of solvents, and low cost per sample. It combines several sample preparation steps and replaces older, tedious extraction methods with a comprehensive approach of good performance for an increasing number of analytes. Reported recoveries are typically high in the range of 90–110%, as shown in Figure 2.2 (Lehotay, 2007, 2013). Another driving aspect for further widespread use is the compatibility with the requirements

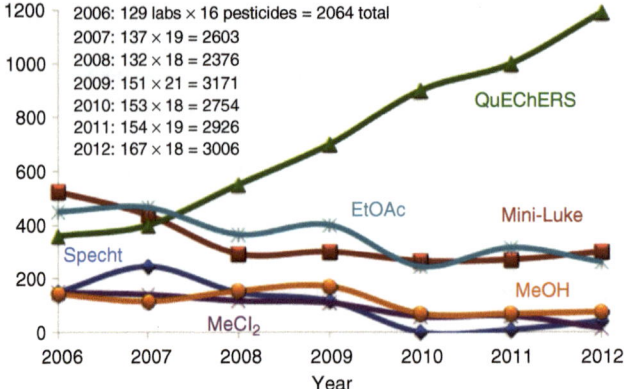

Figure 2.1 Number of results reported in EU PT samples by method (Lehotay, 2013, adapted from Paula Medina Pastor, EURL-FV, Rejczak and Tuzimski, 2015).

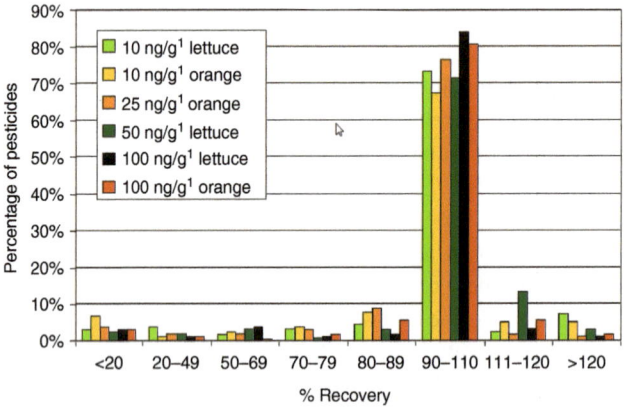

Figure 2.2 Recoveries in the QuEChERS method, 229 pesticides analyzed by GC-MS and LC-MS/MS (Lehotay, 2013/Separation Science).

of high throughput laboratories. Although performed mostly manually, there are already automated instrumental solutions available (Settle *et al.*, 2010; Kaewsuya *et al.*, 2013; Teledyne Tekmar, 2013). For homogeneous samples, the complete procedure from LLE, clean-up to GC-MS injection can be fully automatically performed in 2 mL vials on x,y,z-robotic sampling systems installed on a GC-MS unit (Chong and Huebschmann, 2021).

QuEChERS further expands outside of pesticide analysis. More applications including persistent organic pollutants (POPs) and environmental and veterinary drug residue methods are reported to widen significantly the scope of applications (Luetjohann *et al.*, 2009; Brondi *et al.*, 2011; Usui *et al.*, 2012; Rejczak and Tuzimski, 2015; Sun *et al.*, 2017). A wide variety of matrices that QuEChERS was applied to are reported, including animal products such as meat, fish, kidney, chicken, or milk, cereals, honey, wines, juices, and more.

QuEChERS stands for

- Quick — Less time is required to process samples, compared to previous techniques allowing a higher sample throughput, typically 15 samples can be manually prepared within 1 h.
- Easy — Requires less handling of lower solvent volumes than other techniques; fewer steps are required.
- Cheap — Less sorbent material is needed for smaller sample sizes.
- Effective — The technique gives high and reproducible recovery levels for a wide range of different compound types. The pH can be buffered as needed.
- Rugged — The method can extract a large number of pesticides and other contaminants from different also fat-containing commodities even beyond food, including charged and polar pesticides.
- Safe — Unlike other techniques, it does not require the use of chlorinated solvents. The extraction is typically carried out using acetonitrile or ethyl acetate, which are both GC and acetonitrile LC compatible.

Are there limitations to using QuEChERS? Steve Lehotay summarized the status in his presentation "Revisiting the Advantages of the QuEChERS Approach to Sample Preparation" (Lehotay, 2013). Although there is this one name of the methodology used in common, there are two modified versions in use, partially customized for specific purposes. There are also distinct differences in the original and the currently approved U.S. based AOAC method (AOAC, 2023) and the European method (European Standard EN15662, 2008) that need to be recognized for their effects on recoveries (see Table 2.1). In general, spices, tea, cereals, and fatty matrices require a pre-treatment with the addition of water to the dry sample. Figure 2.3 shows the drop in recovery with increasing fat content of the sample. While polar compounds appear less affected, the non-polar lipophilic compounds drop in recovery significantly above 10–15% fat content. Individual pesticide problems are reported with captan, folpet, and captafol, which can be addressed by using ethyl acetate instead (Schürmann et al., 2023). When acetonitrile is poured into the extraction tube containing the homogenized sample, an exothermic reaction occurs between magnesium sulfate and water. This step may lead to reduced recoveries of the volatile pesticides. To overcome this effect, the sample can be weighed directly into an empty centrifuge tube followed by the addition of acetonitrile (MeCN). The tube can then be immersed in an ice bath with slow addition of salts. Miniaturization of the sample and solvent sizes minimizes such observed effects.

The high matrix content of QuEChERS extracts requires a clean-up before GC-MS analysis. Many different clean-up methods are published for the resulting extracts depending on the different sample matrices. The common goal of laboratories for increased sample throughput and extended GC, LC, and MS maintenance cycles is counteracted by the recommendations of matrix-specific clean-up procedures. The potential for high sample throughput using a quick and easy QuEChERS extraction is impeded by laborious maintenance work (Kowalski and Cochran, 2012; Hildmann et al., 2015). Several matrix-dependent clean-up strategies are

2 Fundamentals

Table 2.1 QuEChERS methods comparison.

Original	AOAC 2007.01	CEN 15662
Ref.: (Anastassiades et al., 2003)	Ref.: (Lehotay et al., 2005; AOAC, 2023)	Ref.: (Anastassiades et al., 2007)
10–15 g sub sample	10–15 g sub sample	10–15 g sub sample
10–15 mL MeCN	10–15 mL 1% HOAc in MeCN	10–15 mL MeCN
Shake	Shake	Shake
0.4 g/mL anh. $MgSO_4$	0.4 g/mL anh. $MgSO_4$	0.4 g/mL anh. $MgSO_4$
0.1 g/mL NaCl	0.1 g/mL NaOAc	0.1 g/mL NaCl
		0.1 g/mL $Na_3Cit·2 H_2O$
		0.05 g/mL $Na_2Cit·1.5 H_2O$
Shake	Shake	Shake
Centrifuge	Centrifuge	Centrifuge
150 mg/mL anh. $MgSO_4$	150 mg/mL anh. $MgSO_4$	150 mg/mL anh. $MgSO_4$
25 mg/mL PSA	50 mg/mL PSA	
Shake and centrifuge	Shake and centrifuge	Shake and centrifuge
Options:	Options:	
+ 50 mg C18	+ 50 mg C18	
+ 7.5 mg GCB	+ 2.5–7.5 mg GCB	
Comments:	*Comments:*	*Comments:*
Sodium chloride is used to reduce polar interferences.	Employs 1% acetic acid in acetonitrile and acetate buffer to protect base sensitive analytes from degradation.	The European method includes sodium chloride to limit polar interferences and several buffering reagents to preserve base sensitive analytes.
Provides the cleanest extraction because it uses fewer reagents.		
Does not use acetic acid which may be problematic in GC-MS analysis.	A USDA study has demonstrated that this method provides superior recovery for pH sensitive compounds when compared to the other two QuEChERS methods.	Sodium hydroxide used in the citrus step should be avoided as it can add impurities to the extract as well as damage the sorbent used in the clean-up step.
Uses dispersive clean-up procedures.		
GCB is helpful removing chlorophyll but can be critical with the retention and low recovery of planar pesticides and PCBs.	The approach uses acetic acid in the extraction step. The acetic acid can overload the PSA sorbent used in the clean-up step making it ineffective and possibly causing GC resolution issues.	Sample preparation and extraction: Freeze samples to −20 °C. Homogenize with dry ice until a free-flowing powder is formed. The sample is then extracted into solvent. Dispersive or cartridge SPE is used for clean-up.
	GCB is helpful removing chlorophyll but can be critical with the retention and low recovery of planar pesticides and PCBs.	

(Lehotay, 2013/Separation Science).

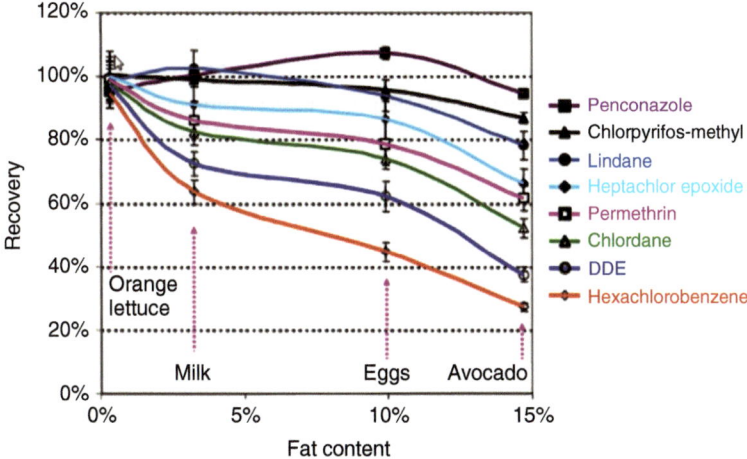

Figure 2.3 Pesticide recoveries versus fat content (Lehotay, 2013/Separation Science).

found in the literature to overcome matrix effects (Anastassiades, 2013; UCT, 2013). "As for clean-up by dSPE, type and amount of sorbent and MgSO₄, and their selectivity were the main problematic issues" (Rejczak and Tuzimski, 2015). With the versatility of the QuEChERS extraction step an equally matrix independent and robust clean-up solution for extracts is highly desirable. The development of the μSPE clean-up, as described in more detail in Section 2.1.3.2, provides the required matrix independent and efficient extract clean-up for pesticide analysis by GC-MS and LC-MS analysis using the QuEChERS extraction (Morris and Schriner, 2015; Michlig and Lehotay, 2022). While the automated μSPE clean-up is usually described for LLEs with acetonitrile (QuEChERS), it is also reported to be equally efficient for the clean-up of the extracts of the classical ethyl acetate extraction (*aka* Swedish ethyl acetate extraction "SweEt", Philström *et al.*, 2007; Schürmann *et al.*, 2023). The μSPE clean-up is typically automated on x,y,z-robotic platforms with the option of online injection (also refer to Section 4.9 for the QuEChERSER analysis of pesticides).

Acetonitrile as solvent for GC is often discussed as having limitations with the popular use of less polar column phases in GC-MS such as the very common apolar 5% phenyl phase. The initial QuEChERS publication used a mid-polar 35% phenyl-65% dimethyl arylene column phase with higher affinity to acetonitrile as solvent (Anastassiades *et al.*, 2003). But, by applying appropriate GC oven programming using 5% phenyl phase columns with an isothermal phase at a temperature above the boiling point (BP) at and after injection excellent peak shapes are achieved. The BP of acetonitrile is 82 °C at ambient temperature and c. 92 °C at 100 kPa GC head pressure. 95 °C initial isothermal oven temperature is often reported. The large vapor volume of evaporated acetonitrile needs to be recognized, in particular with splitless injections. It also needs to be considered that acetonitrile dissolves and carries up to 10% water. Time to let pass the acetonitrile solvent peak in the isothermal phase is recommended. Also, a programmed temperature vaporizer (PTV) injection in solvent

vent mode, the use of a pre-column, or simply a small addition of a more compatible GC solvent such as ethyl acetate or toluene are proven solutions (Hetmanski et al., 2010; Mastovska, 2013). Large-volume GC injectors (SSL-Concurrent Solvent Recondensation (CSR), PTV) allow the application of higher diluted extract volumes compensating for a potential dilution effect (Hoh et al., 2007). On the LC-MS side, the high matrix load gives increased rise to matrix effects resulting in potential analyte quenching during ionization, which can be addressed by dilution of the extract aka "dilute-and-shoot" (Stahnke et al., 2012).

The QuEChERS Reagents and Their Function (UTC, 2011)

Extraction

Magnesium sulfate, anhydrous	Facilitates solvent partitioning, increases ionic strength, improves recovery of polar analytes, and supports the acetonitrile phase separation.
Acetic acid	Used to adjust pH, see buffers.
Acetonitrile	The organic solvent providing the most suitable characteristics for extracting the broadest range of pesticides with the least number of co-extractables. Amenable for both LC and GC analysis.
Buffers	Prevents degradation of pH-sensitive analytes by maintaining optimal pH. Adjusts the sample pH value.
Sodium chloride	Increases the ionic strength, reduces the amount of polar interferences, and leads to the acetonitrile phase separation.

Clean-up

Aminopropyl	Removes sugars and fatty acids. Serves the same function as PSA (primary secondary amine), but is less likely to catalyze degradation of base-sensitive analytes. Aminopropyl has a lower capacity for clean-up than PSA.
ChloroFiltr™	A polymeric sorbent for selective removal of chlorophyll from acetonitrile extracts without loss of polar aromatic pesticides. It is designed to replace GCB for the efficient removal of chlorophyll without loss of planar analytes.
C18	Removes long-chain fatty compounds, sterols, and other non-polar interferences.
GCB	Is a strong sorbent, graphitized carbon black, for removing pigments, polyphenols, and other polar compounds. Planar (polar aromatic) pesticides may be removed: such as chlorothalonil, coumaphos, hexachlorobenzene, thiabendazole, terbufos, and quintozene.
Magnesium sulfate, anhydrous	Removes water from the organic phase.
PSA	Is used in the removal of sugars and fatty acids, organic acids, lipids, and some pigments. When used in combination with C18, additional lipids and sterols can be removed.

2.1.2 Dispersive Liquid/Liquid Microextraction

This fundamentally new liquid microextraction technique was introduced in the early 2000s with a constantly increasing number of applications today. The dispersive liquid/liquid micro/extraction (DLLME) was first published in 2006 by Mohammad Rezaee and Yaghoub Assadi *et al.* as "a very simple and rapid method for extraction and pre-concentration" of organic compounds such as the demonstrated polyaromatic hydrocarbons (PAHs) extraction from water (Rezaee *et al.*, 2006). In the same year, the publication by Sana Berijani and Yaghoub Assadi *et al.* extended the application range to the extraction of organochlorine pesticides (OCPs) from water (Berijani *et al.*, 2006). DLLME brings great potential for the GAC with high concentration rates from very low solvent use. It works most efficiently with high injection speeds of a solvent mix to get dispersed into an aqueous sample. The concept allows a complete automation of the extraction process for high sample throughput (Hutchinson and Carrier, 2017). DLLME is used for extraction, clean-up, and concentration for trace analysis allowing concentration factors of more than 100 times, reported to be up to 272-fold for PAHs and phthalate esters from water for direct analysis by GC-MS (Shi and Lee, 2010; Guo and Lee, 2014).

DLLME is based on a three-component solvent/sample system. For applications in aqueous samples, a few microliters of a water-immiscible solvent (extractant) are dissolved in a disperser solvent and rapidly injected into the aqueous sample. The extractant solvent should have a high affinity for the target analytes (partition coefficient). The disperser must be miscible with both, the extractant solvent and the sample, which was water in the first reported studies. The injected extractant/disperser solvent mixture forms a cloud of microdroplets generating a high surface for analyte extraction. The dispersive solvent ensures the miscibility of the organic phase and the aqueous sample phase and allows the immiscible solvent to form an emulsion with a high surface area for the hydrophobic extractant. Due to this very high surface area, the time to reach a maximum recovery is short. Low polar analytes get efficiently extracted from water into the hydrophobic extractant phase. The extraction process is instantaneous but can be supported by vortexing or sonication of the vial. A pH adjustment and the salting-out effect are additionally applied to further increase analyte recoveries. Following this extraction step, the sample vial is centrifuged. The centrifugation step separates the emulsion.

Tapered vials of c. 10 mL volume with a conical bottom are used, also named "center drain" vials. Special dual-use DLLME vials are commercially available with a conical bottom and a narrow top neck, most suitable for the aspiration of a low density extract in the narrow neck. Typically, high density, often chlorinated solvents are used as extractants that sediment after centrifugation at the vial bottom. Typical disperser solvents include isopropanol, acetone, methanol, or acetonitrile. The enriched high density extractant phase can be taken up from the bottom layer with a robotic autosampler using the bottom-sensing function with a regular GC injection syringe (Figure 2.4). Light hydrocarbon extractants are taken from the top layer (Figure 2.5). The extract can then be directly analyzed by GC-MS without further clean-up or pre-concentration. Non-chlorinated solvents such as

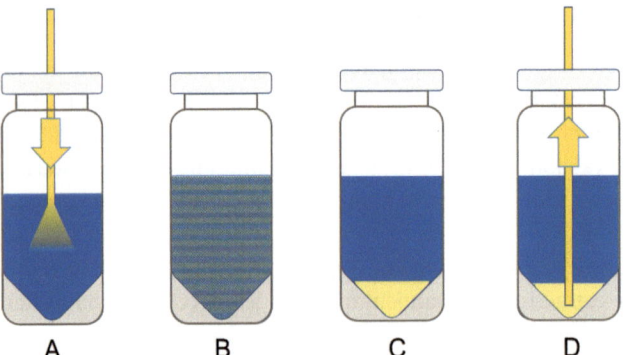

Figure 2.4 DLLME steps with an aqueous sample and high density extraction solvent.
A Injection of high density extractant/disperser solvent mixture
B Emulsion formation
C After phase separation/centrifugation
D Extract take-up with bottom sensing

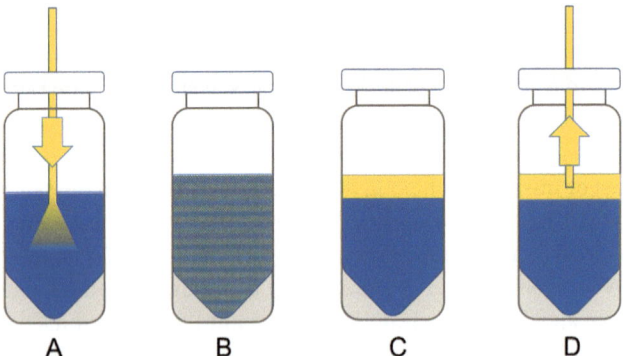

Figure 2.5 DLLME steps with an aqueous sample and low density extraction solvent.
A Injection of low density extractant/disperser solvent mixture
B Emulsion formation
C After phase separation/centrifugation
D Extract take-up from the upper layer

the light hydrocarbons, toluene, octanol, or hexane form an upper layer above the aqueous sample after centrifugation. Such hydrocarbon extracts are well suited for applications with GC-ECD detection.

The numerous DLLME method advantages including its simplicity, rapidity of operation, low consumption of organic solvents, low cost, and high enrichment factors from low volumes of samples made DLLME an attractive green method for analytical laboratories (Mudiam *et al.*, 2012; Rai *et al.*, 2016; Schettino *et al.*, 2023). Routine labs benefit from the increased productivity and ease of automation with x,y,z-robotic samplers. DLLME can be carried out online with GC-MS or LC-MS on the chromatographic timescale in the prep-ahead mode during the analysis run of a previous sample. The flexibility of the three solvents phase system allows the use of the full range of aqueous immiscible organic solvents including ionic liquids as extracting solutions customizing DLLME for a wide range of analytes/matrix combinations (Rykowska et al., 2018).

2.1.3 Solid Phase Extraction

Modern SPE originated in 1974 by Reginald Adams, Thomas Good, and Michael Telepchak (Telepchak et al., 2004). From the mid of the 1980s, this new extraction technique called solid phase extraction (SPE) began to revolutionize the enrichment, extraction, and clean-up of analytical samples. Following the motto "The separating funnel is a museum piece", the time-consuming and arduous separating funnel LLE has increasingly been displaced from the analytical laboratory (Bundt et al., 1991). The main advantage of SPE over the classical liquid/liquid partition is the low consumption of expensive and sometimes harmful solvents, today even further extended by miniaturization to the micro-SPE scale (µSPE). The amount of apparatus and space required is low for SPE. Besides an efficient clean-up, a concentration of the analyte required for GC-MS is achieved. The SPE extracts can be used directly or after concentration by evaporation for GC and GC-MS

In SPE, a stationary phase, typically a solid sorbent material, is packed into a column made of polymer or glass material, shown with its general components in Figure 2.6. The maximum sample volume that can be applied is limited by the breakthrough volume (BTV) of the analyte which needs to be determined separately. Elution volumes should be as small as possible to reduce or prevent subsequent solvent evaporation. In analytical practice, two modes of SPE operation have become established for the isolation of the analytes of interest: the "classical" enrichment and a "scavenging" mode. The basic SPE steps involved in the enrichment mode illustrated in Figure 2.7 (aka load-wash-elute, or catch & release) are as follows:

1) Conditioning: The solid phase sorbent is first conditioned with a solvent to remove any impurities and activate the sorbent.
2) Sample Loading: The sample is then introduced onto the solid phase sorbent, which selectively retains the target analyte(s).

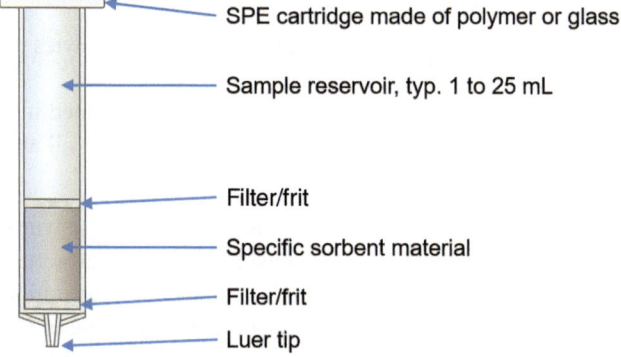

Figure 2.6 Components of an SPE column.

2 Fundamentals

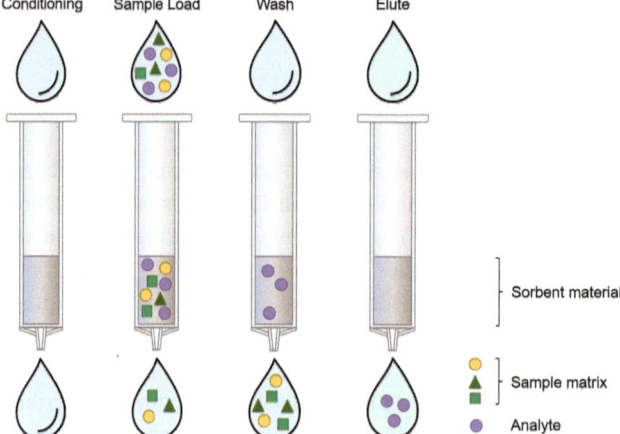

Figure 2.7 SPE in the classical "Enrichment" mode (*aka* Load-Wash-Elute mode). The analytes are retained on the cartridge, matrix washed away, and the isolated analytes are eluted.

3) Washing: Any unwanted compounds, e.g. the matrix, are removed by washing the sorbent with an appropriate solvent while keeping analytes trapped in the sorbent material.
4) Elution: The elution of the target analyte(s) is achieved by changing the solvent properties. For this, there must be a stronger interaction between the elution solvent and the analyte than between the latter and the solid phase.
5) Evaporation: The eluate is often evaporated close to dryness, and the residue is reconstituted in a suitable solvent prior to analysis.

In the scavenging mode, the analyte(s) are not retained on the sorbent material but the matrix is kept behind as shown in Figure 2.8. The basic steps involved in the scavenging mode SPE are as follows:

1) Conditioning of the sorbent and scavenger material. Often, the conditioning step is skipped as the dead volume of the cartridges dilutes the exact.
2) Loading of the sample onto the cartridge.
3) Collection of the passing analyte fraction.
4) Additional solvent can be carefully applied for further elution of the analytes up to the breakthrough of the matrix.
5) Also here, the eluate can be evaporated close to dryness and reconstituted in a suitable solvent prior to analysis. Often, the volume of the eluate is small and does not need evaporation.

If the GC-MS analysis reveals high contents of plasticizers or other polymer additives, the plastic material of the packed columns must first be considered, and in some cases, a change to glass columns must be made. For sample preparation using slurries or turbid water, which rapidly leads to deposits and clogging of the packed cartridges, SPE disks should be used for these kind of samples as displayed in Figure 2.9. Their use is similar to that of cartridges.

2.1 Sample Preparation | 21

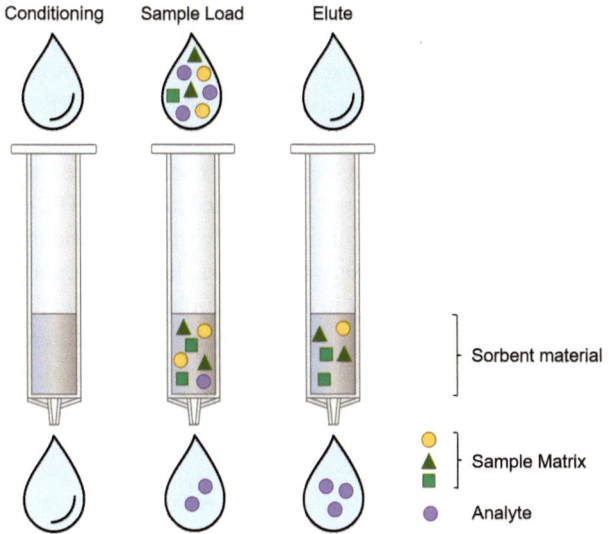

Figure 2.8 SPE in "scavenging" mode. Matrix is retained on the cartridge, the analytes elute.

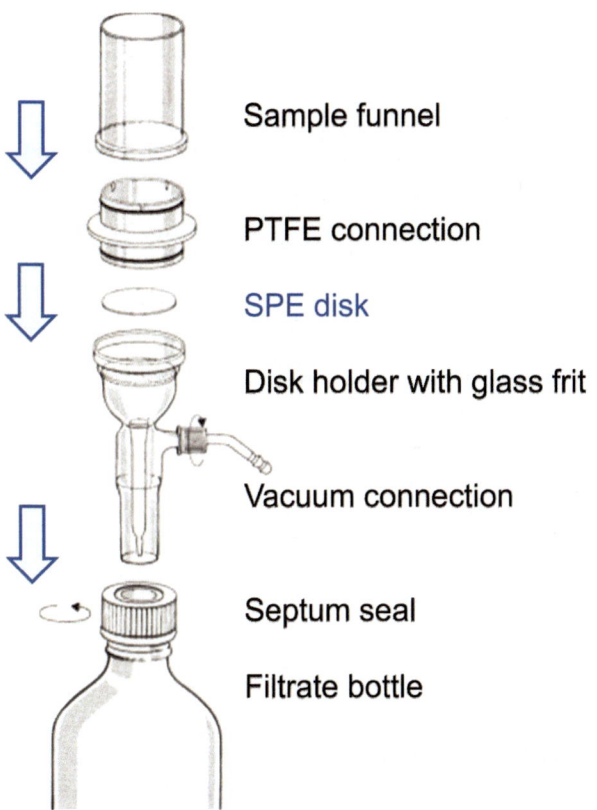

Figure 2.9 Apparatus for solid phase extraction using disks (J.T. Baker).

The choice of sorbent materials permits the exploitation of different separating mechanisms of adsorption chromatography, normal-phase and reversed-phase chromatography, and also ion exchange and size exclusion chromatography. A large variety of different analyte/matrix/sorbent interactions are exploited for SPE. An overview for organic samples and analytes amenable to GC-MS analysis is given in Figure 2.10. Selective extractions can be achieved by a suitable choice of adsorption materials. In case of the analysis of dioxins and polychlorinated biphenyls (PCBs), for example, a silica gel column charged with sulfuric acid for chemical oxidation of the non-halogenated organic matter is necessary as the first clean-up step. Due to their distinctive properties and the flexibility in tailored surface modifications also carbon-based nanomaterials (CNMs) have found increasing use in different sample preparation procedures. They can be used as selective adsorbents by direct interaction with the analyte as reported in Table 2.2 (Scida et al., 2011; Zhang et al., 2013).

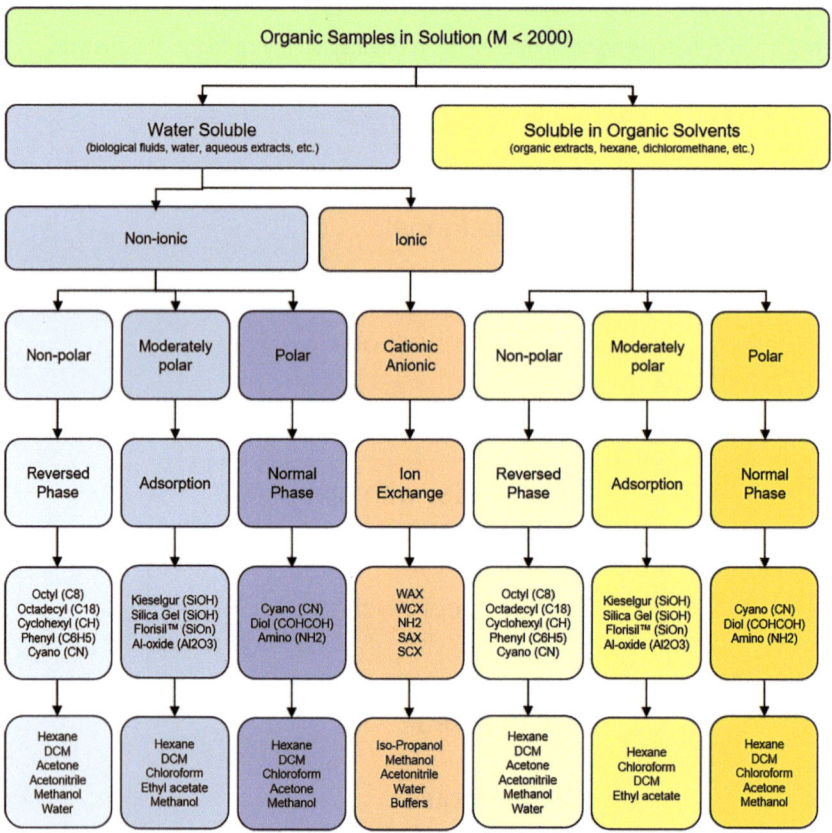

Figure 2.10 SPE selection guide to sorbent materials and eluents. The choice of the SPE phase depends on the molecular solubility of the sample in a particular medium and on its polarity.

Table 2.2 Comparison of carbon-based nanomaterials in sample preparation.

Materials	Special material characteristics	Derivatization methods	Sample preparation potential	Current status[*]	Future perspective[*]
Graphene	1. Both available sides for adsorption 2. Easily synthesized in lab 3. Easily modified with functional groups	Modified via graphene oxide	SPE, MSPE, SPME, LDI substrate, μSPE, PMME, MSDP, SRSE	*****	*****
Carbon nanotubes	1. Mostly used carbon nanomaterials 2. Be easily covalently or non-covalently functionalized	Oxidized CNTs can be grafted via creation of amide bonds, N2-plasma and radical addition	SPE, MSPE, SPME, LDI substrate, μSPE, SRSE	*****	*****
Carbon nanofibers	1. Easily available on a large scale 2. Larger dimensions without coating or functionalization	Polar groups can be introduced by treating with the concentrated nitric acid	SPE, SPME, on-line μSPE	***	****
Fullerenes	First used carbon nanomaterials	Covalently bonded to other reagents	SPE, on-line SPE, MSPE, SPME, LLE, SFE, ion-pair precipitation	****	
Nanodiamonds	1. Chemical inertness and hardness 2. More expensive than others	Functionalized with H/D-terminated, halogenated, aminated, hydroxylated, and carboxylated surfaces	SPE, LDI substrate	***	***
Carbon nanocones, disks and nanohorns	1. Multiple structures 2. Lower aggregation tendency	Functionalized by microwave (for SWNHs)	SPE, SPME	**	**

[*]The star number represents the research status and perspective which is mainly based on properties of materials, the number of published articles and publication year.
LDI, laser desorption ionization; MSPE, magnetic solid phase extraction; PMME, poly (methyl methacrylate); matrix solid phase dispersion (MSDP); stir rod sorptive extraction (SRSE); liquid/liquid extraction (LLE); singlewall carbon nanohorn (SWNH)
(Zhang et al., 2013/with permission of ELSEVIER).

2.1.3.1 Online Solid Phase Extraction

The hyphenation of SPE with GC-MS analysis allows the miniaturization of the sample preparation to the microliter scale making the online approach a true green analysis concept. Important advantages are the reduced processing time, considerably reduced solvent consumption, and significantly higher sensitivity for lower limits of quantitation (LOQs) or less sample volume. Manual handling steps are avoided. The cartridge conditioning, sample load, and washing are automatically performed (Mejía-Carmona et al., 2019). The valve configuration of on-line SPE-GC-MS is illustrated in Figure 2.11a and b. A sample loop defines the sample volume provided either from direct flow inlets or a suitable autosampler. A loading pump transfers the sample plug from the loop to the SPE cartridge mounted to a second valve. The injection pump uses a GC-compatible solvent to elute the focused sample plug in the reverse direction to the GC injector. Either large volume injection (LVI) or the CSR techniques are used (refer to Section 2.2.5.6). The drying of the cartridge, eluate collection in vials, solvent evaporation for concentration, and further reconstitution for chromatographic analysis are not required. On-line SPE does not only save

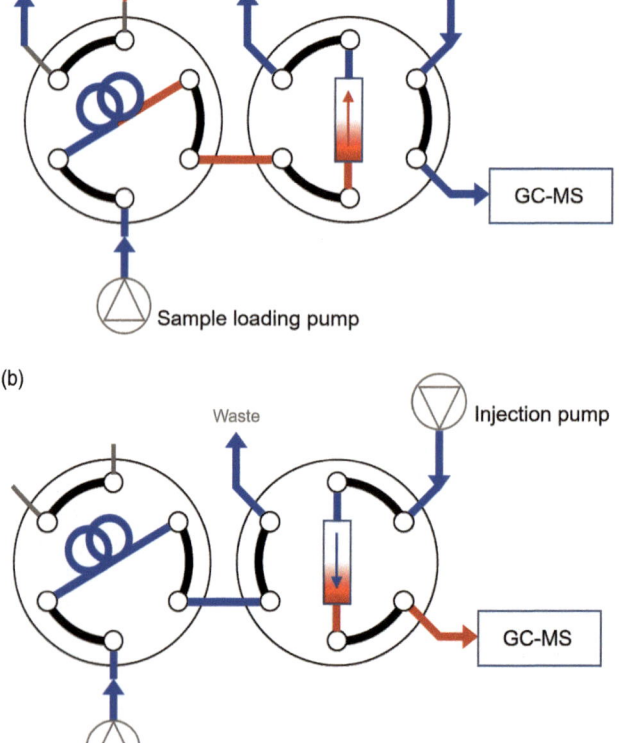

Figure 2.11 Valve configuration for on-line SPE GC-MS. (a) Sample load onto SPE cartridge from sample loop. (b) On-line GC injection from SPE cartridge.

substantial laboratory workload but also reduces sources of potential human error and variability. Critical analytical steps like the use of additional glassware, potential evaporation to dryness, the known sources of losses and contamination, are avoided. On-line SPE injects without splitting the complete undiluted analyte fraction. On the practical side, less sample is required for smaller cartridges, and higher sensitivity can be achieved (Noij and van der Kooi, 1995; de Koning et al., 1998; Wittsiepe et al., 2014; Gerstel, 2015), even the automated analysis of continuous water streams. Aqueous samples are enriched typically on a 10 mm × 2.0 mm inner diameter (ID) cartridge containing C18, hydrophilic-lipophilic balance (HLB), or similar sorbent materials (Tintrop et al., 2023).

A fully automated online-SPE-LVI-GC-MS setup is described for the analysis of raw and drinking water for SVOCs using a robotic x,y,z-sampler (Li et al., 2000). The sample plug is eluted from a C18 SPE cartridge online via LVI into GC-MS. The achieved method detection limits (MDLs) were lower than 0.1 µg/L with linear calibrations ranging from 0.01 to 2.5 µg/L. Recoveries were achieved with 70 to 130%. Comparing the cost of analysis with the manual U.S. EPA Method 525.2 savings of up to 80% could be achieved. In addition, due to the high degree of automation, accuracy, precision, and sensitivity could be improved. The similar U.S. EPA method 543, although directed for online SPE-LC-MS/MS, is of importance for the setup and operation of online SPE applications in water analysis for a wider range of analytes (U.S. EPA, 2015).

2.1.3.2 Micro Solid Phase Extraction

Microextraction techniques also in SPE have emerged as alternatives to traditional extraction techniques. The classical standard protocols have been successfully scaled down to true GAC methods. A format of SPE cartridges with sorbent volumes in the range of up to 100 µL was introduced in 2005 by Kim Gamble, Instrument Top Sample Preparation, Suwanee, GA, USA (Gamble, 2005). A liquid syringe providing a positive pressure operation is used for conditioning, sample load, washing, and analyte elution. Both the classical wash-and-elute and the matrix scavenging modes are in use. Basically, the matrix/analyte separation is achieved comparable to a micro LC column with sharp elution profiles when using low microliter per second load and elution flows. Robotic x,y,z-autosamplers are in use for workflow automation and improved the reproducibility of the clean-up process. The new micro-SPE sample preparation soon found applications in automated food safety, water, and forensic analysis (Morris et al., 2014; Morris and Schriner, 2015; Korenková et al., 2017; Lehmann et al., 2019).

A further improved µSPE cartridge design optimized for automated processing was introduced in 2022 by CTC Analytics AG, Zwingen, Switzerland, and evaluated for routine pesticide analysis by GC-MS (Michlig and Lehotay, 2022; Manzano Sanchez et al., 2023). The novel µSPE cartridge is septumless and consists of two parts enclosing the sorbent material between filters, as shown in Figure 2.12. The polymer material of the cartridge is chemically inert and compatible with acetonitrile (MeCN), ethyl acetate (EtOAc), methanol (MeOH), dichloromethane (DCM), hexane, and aqueous pH of 1 to 12. The dimensions are 35 ± 2 mm in height and

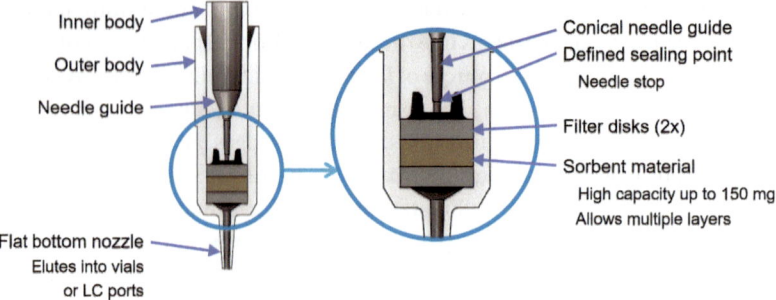

Figure 2.12 Cross section of the novel µSPE cartridge for automated sample preparation (Courtesy of CTC Analytics AG).

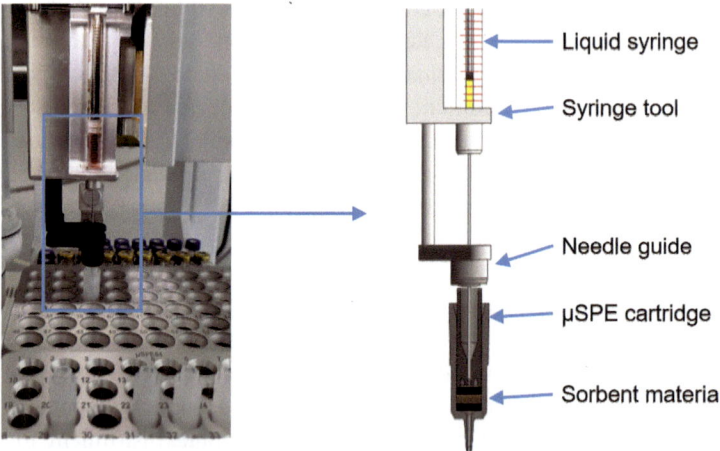

Figure 2.13 Principle of the automated µSPE operation on x,y,z-robotic systems (Courtesy of CTC Analytics AG).

8.5 mm in outer diameter. The cartridge body provides critical functionality with a predefined and reproducible compression of the sorbent/filter layers and a precise needle guide for safe and always upright positioning. The movement of cartridges on automated x,y,z-robotic systems can be achieved by syringe needle or pipette tip transport (Figure 2.13). Of key importance are the resulting high pressure resistance and a leak-free seal using syringe needles of gauge 22 with flat tip for increased and highly reproducible load speeds of up to 500 µL/min.

The automated µSPE clean-up workflows show their versatility for large-scale food safety, environmental or forensic analyses using QuEChERS extractions with acetonitrile (Michlig and Lehotay, 2022; Carrera et al., 2023; Poulsen et al., 2023) or ethyl acetate (Schürmann et al., 2023). Alternative sorbent materials have been evaluated for an improved clean-up in pesticide analysis (Ninga et al., 2024). Beyond clean-up, recent developments also showed the feasibility of automation for the complete extraction and clean-up workflow for homogeneous samples. The application of the complete QuEChERS extraction, clean-up, and GC-MS injection was demonstrated for orange juice (Chong and Huebschmann, 2023).

2.1.4 Solid Phase Microextraction

"The simplification of sample preparation and its integration with both sampling and the convenient introduction of extracted components to analytical instruments are a significant challenge and an opportunity for the contemporary analytical chemist" writes the inventor Prof. Janusz Pawliszyn in his preface to his "Handbook of SPME" (Pawliszyn, 2012).

Since its inception in the 1990s by Catherine L. Arthur and Janusz Pawliszyn at the University of Waterloo, Canada, solid phase microextraction (SPME) paved the way for GAC in many application areas (Arthur and Pawliszyn, 1990). Because of its simplicity in sample preparation, ease of automated operation, and its high sensitivity SPME became popular for a large number of GC-MS applications. Many publications since then cover the analysis of volatile organic compounds (VOCs), flavors and fragrances, and the trace analysis of pesticides, drugs, and other non-purgeable polar compounds (SVOCs) (Potter and Pawliszyn, 1992; Zhang *et al.*, 1994; Boyd-Boland and Pawliszyn, 1995; Boyd-Boland *et al.*, 1996; Strano-Rossi *et al.*, 2005; Pragst, 2007; IOFI, 2010; Reyes-Garcés *et al.*, 2017), in metabolomics (Fiori *et al.*, 2018), and soon became International Standards Organization (ISO) standardized for water analysis (ISO17943, 2014).

The solvent-free microextraction technique SPME is a true GAC technique and a fundamental step ahead toward the miniaturization, instrumentation, and automation of the SPE technique for sample preparation and online introduction to GC-MS (Zhang *et al.*, 1994; Eisert and Pawliszyn, 1997; Lord and Pawliszyn, 1998). Initially, it involved exposing a fused silica fiber coated with a sorbent phase to a sample containing the compounds to be analyzed. Also, for less GC-amenable polar analytes derivatization steps can be coupled to the extraction process improving efficiencies for non-purgeable analytes (Pan *et al.*, 1997). The sample contained in a capped vial is incubated at a constant temperature before and during the sampling process and vigorously stirred (if liquid) to achieve maximum recovery and precision for quantitative assays (Gorecki and Pawliszyn, 1997). Also, field sampling applications are feasible for subsequent analysis in the lab.

The basic principle of SPME extraction is the partitioning of analytes between the sample matrix and the sorbent phase of the SPME device (Pawliszyn, 2009). Both headspace and direct immersion sampling, as shown in Figure 2.14, are applicable depending on the analytes of interest. Headspace SPME sampling (HS-SPME) is suitable for solid, liquid, or gaseous samples. Volatile analytes diffuse to the sorbent surface and partition into the coating material. It needs to be noted at this point that HS-SPME involves a triple-phase distribution between sample, headspace, and the sorbent material. The partition coefficient of analytes into the sorbent material is independent from the analyte partition from the sample to the headspace. Lower extraction temperatures favor the sampling efficiency on the sorbent material. The sample incubation temperature needs to be balanced for the less volatile components

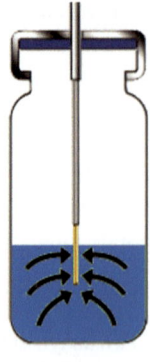 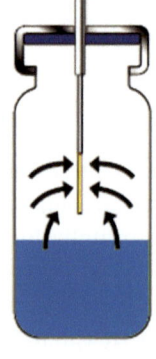

Figure 2.14 SPME principle for immersion and headspace sampling.

Direkt Immersion
DI-SPME

Headspace
HS-SPME

considering for instance the water/gas phase partition of the analytes of interest for an optimum sensitivity of the assay.

Direct immersion SPME sampling (DI-SPME) is used for less volatile and polar analytes from aqueous media. Strong shaking of the liquid sample during immersion extraction is mandatory. A depletion of analytes from the boundary layer of water and matrix around the SPME phase impairs extraction efficiency. Molecules in liquid media, e.g. in water, travel several orders of magnitude slower than in the gas phase. It is recommended to use high stirring rates even using cycloidal stirring patterns to optimize the analytical response (Hübschmann, 2021). Significantly longer extraction times are typical in DI-SPME compared to gas-phase extractions. Adverse effects on the immersion extraction efficiency in larger sample series can be caused by non-volatile matrices like humic acids or proteins present in the sample and need to be considered in the workflow. PDMS overcoated devices are recommended for sampling from biological materials (see Section 2.1.4.3). Wash steps before GC injection, or before advancing to the next sample keep up the extraction performance.

Both HS-SPME and DI-SPME offer distinct advantages for automated sample preparation including a reduced analysis time per sample and improved reproducibility by identical treatment of all samples. Less manual sample handling results in an increased sample throughput. Most importantly for GAC and a cost reduction in the routine laboratory, SPME is a complete solventless technology (Berlardi and Pawliszyn, 1989; Arthur et al., 1992; Pawliszyn, 1997; Crucello et al., 2023).

Another extension of the idea of a miniaturized sorbent extraction is realized in the stir bar sorptive extraction device (SBSE) using a glass-covered, magnetic stir bar for headspace and immersion extractions (refer to Section 2.1.9).

2.1.4.1 Solid Phase Microextraction Devices

Commercial SPME devices are available as SPME fibers, SPME arrows, or high capacity probes with an outside sorptive coating and, as an in-tube device, with a coating inside of a capillary. Although of different design, they share the sorbent

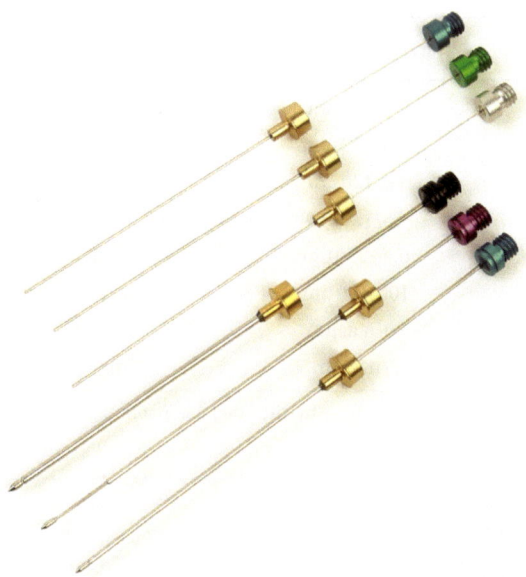

Figure 2.15 SPME assemblies for installation into holders, top: fibers, bottom: arrows. Courtesy CTC Analytics AG.

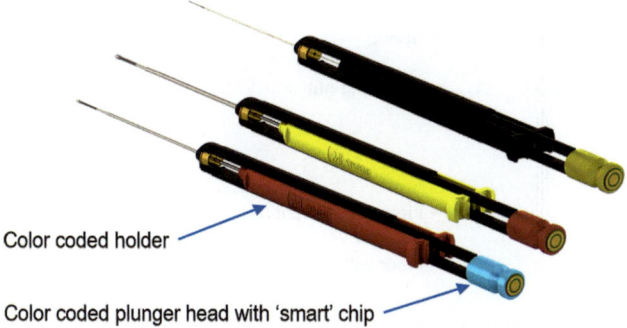

Figure 2.16 New generation SPME devices (Courtesy CTC Analytics AG). SPME Arrow 1.1 mm, SPME Arrow 1.5 mm, SPME Fiber 23 G.

characteristics and the potential for online GC-MS analysis and automation (Figures 2.15, 2.16, and 2.21). Each SPME fiber/arrow is color-coded indicating the type of coating as shown in Table 2.3, and carries a notched hub marking the rod diameter.

Dedicated SPME holders are in use for manual and automated operation. The holder gets the SPME assemblies installed and protects the sorbent-coated section pulled into a sheath during handling (Figure 2.17). For sampling and desorption, the sorbent tip gets exposed to the sample and for desorption in the GC inlet. Commercial SPME devices are compatible with industry standard 10 mL and 20 mL sample vials.

Table 2.3 SPME color codes.

SPME Phase	Thickness Fiber/Arrow	Color Code
Acrylate (Polyacrylate)	85/100 µm	Silver
Carbon WR/PDMS	95/120 µm	Blue
DVB/Carbon WR/PDMS	80/120 µm	Dark gray
DVB/PDMS	65/120 µm	Violet
PDMS	n.a./250 µm	Black
PDMS	100/100 µm	Red
PDMS	30/n.a. µm	Yellow
PDMS	7/n.a. µm	Green

WR, wide range; DVB, divinylbenzene; PDMS, polydimethylsiloxane; n.a. not available.

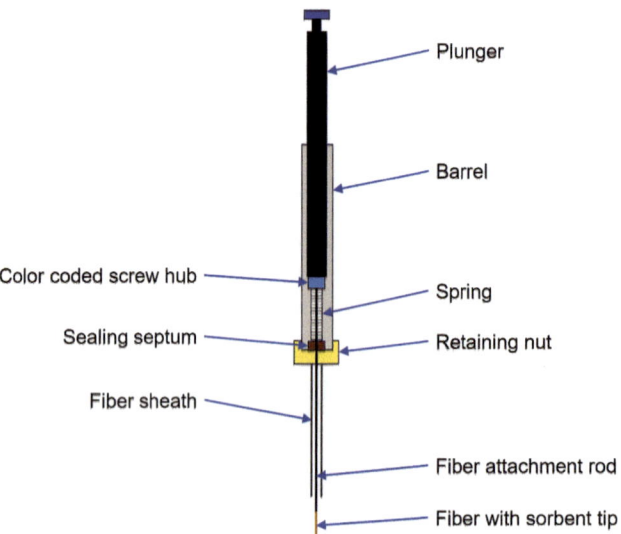

Figure 2.17 SPME fiber holder for autosampler use.

A new generation of SPME devices for automated operation comes pre-installed in a syringe-like holder for ease of handling. An identification chip is embedded in the plunger top. SPME fibers and arrows are clearly distinguished by the color of the shaft: black the SPME fiber with outer diameter (OD) 0.64 mm (23 G), red the SPME arrow with OD 1.1 mm, and yellow the SPME arrow with OD 1.5 mm, illustrated in Figure 2.16. The color of the plunger head identifies the sorbent material as in Table 2.3.

The so-called "smart technology" with an embedded data chip in the plunger head delivers the device identification to the equivalently equipped autosampler. Coded are the important operational parameters like sorbent type, thickness, arrow or fiber type, together with the maximum permitted operation temperatures. The chip also enables critical parameter tracking for each unique SPME device for quality control and traceability. This concept allows the automatic setting of the correct operation parameters for the individual SPME device, avoids detrimental handling errors, and delivers process safety for automated operation.

Solid Phase Microextraction Fiber An SPME fiber device consists of a short length of fused silica fiber coated with a phase similar to those immobilized inside chromatography columns. Also SPME fiber assemblies using a coated less breakable nitinol core, a metal alloy of nickel and titanium providing a shape memory effect and superelasticity, are available. The silicon phase can be mixed with solid adsorbents, for example, divinylbenzene (DVB) polymers, templated resins, or porous carbon materials. The fiber is attached to a stainless-steel plunger in a protective syringe-like holder, used manually or in a specially prepared autosampler. The plunger on the syringe retracts the fiber for storage and piercing septa and exposes the fiber for extraction and desorption of the sample. A spring in the assembly keeps the fiber retracted, reducing the risk of getting damaged (Figure 2.17). The fiber sheath diameter is usually given in the familiar injection syringe gauge values. Standard diameters are 23 G (0.642 mm), also 24 G (0.566 mm) sizes are available. Typical for the SPME fiber devices is the open sheath at its end. The open rod can cause septum coring with repeated injections by autosamplers. Septum particles that fall into the GC injector liner can cause ghost peaks of organic silicon in the chromatogram (also refer to Section 3.4 Frequently Occurring Impurities).

A portable SPME holder design is available with a sealing mechanism providing flexibility and ease of use for on-site sampling. The sample can be stored by placing the tip of the fiber sheath in a septum. SPME is also used as a passive indoor air-sampling device for GC and GC-MS analysis. For this technique to be successful, the sample must be stable to storage.

Solid Phase Microextraction Arrow The SPME Arrow is a most recent development with significant mechanical and analytical improvements. The name derives from the arrow tip of the device. Instead of using fused silica the sorbent material is applied into a recess behind an arrow tip on a steel rod. This makes the SPME arrows rugged and unbreakable for automated routine operations. The larger diameter of the steel rods allows the use of larger phase volumes and surfaces of the sorbent material compared to fused silica fibers making SPME arrow applications significantly more sensitive than SPME fibers (Figure 2.18) (Kremser et al., 2016; Ferracane et al., 2022; Agilent Technologies Inc., 2023).

SPME arrows are available in diameters of 1.1 and 1.5 mm, though the 1.1 mm version is the most suitable and serves the most applications. However, the wider

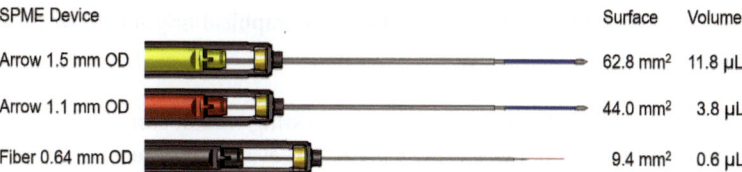

Figure 2.18 SPME device comparison by sorbent phase surface and volume.

sheath diameter of 1.5 mm is advantageous for applications with "on-fiber" derivatization (the same expression "on-fiber" is used in the community also for the SPME arrow derivatization). Derivatization reactions performed for accumulated analytes on the sorbent material often lead to a slight swelling of the sorbent. The wider sheath dimension takes care of such effects without harming the sorbent layer upon closing the sheath after the derivatization reaction is completed.

The larger diameter of the arrow sheaths requires a small modification of the GC inlets. All commercial GC septum caps and the insert assembly drill holes are prepared to accept standard injection syringes up to approx. Gauge 22 G (0.718 mm OD). This fits the SPME fibers well but needs to be modified for use with the wider sheath of the SPME arrows. Suitable kits for SPME arrow use are offered by all major GC manufacturers. Despite the larger arrow diameter, the penetration of septa with the sharp arrow tip is significantly smoother. Septum coring is not observed, even with thick inlet septa, avoiding the typical coring by the open SPME fiber rods. Measuring the force to penetrate a regular headspace vial septum on a balance revealed that the open fiber rod required a force of 1190 g while the sharp arrow tips went through the septum with 800 g for the 1.1 mm (25% less) and even only 900 g for the 1.5 mm diameter arrow sheath.

Vacuum-Assisted Solid Phase Microextraction The application of vacuum to headspace vials before SPME extraction (Vac-SPME) expands the application range of SPME significantly toward higher molecular, less volatile, and thermolabile compounds. Extractions under vacuum conditions allow milder heating, avoiding the risk of sample degradation along with higher extraction efficiencies.

Vacuum tight closures are used for regular crimp top headspace vials (Figure 2.19). The closures keep a gas-tight seal over a long period of time allowing the preparation of samples series even with waiting times on an autosampler tray. The Vac-SPME closures are designed for use with both, SPME fibers, and another type for use with SPME arrows. The closures use a thick GC-type silicon septum for evacuation and SPME operation. The solution is compatible with and allows automated headspace and SPME extractions on x,y,z-robotic samplers.

The effectiveness of Vac-HS-SPME was evaluated compared to conventional HS-SPME with the analysis of virgin olive oil. A 20 mL vial was evacuated to 7 mbar vacuum before sample introduction; 1.5 g of olive oil was added and after 5 min extracted at 30 and 43 °C while agitated. The obtained GC-MS total ion chromatogram (TIC) chromatograms are shown in Figure 2.20. The blue Vac-HS-SPME is much richer, particularly in semi-volatile compounds. Gentle heating further improved the overall extraction yield for semi-volatiles (Purcaro, 2024). Vac-HS-SPME was used also for the study of volatile and semi-volatile biomarkers of honey for the determination of the geographical origin (Mamedova and Alimzhanova, 2023).

Total Vaporization Solid Phase Microextraction An interesting variant compared to the total evaporation headspace analysis (also called "full evaporation technique") is the "total vaporization SPME" analysis (TV-SPME) for sensitive low level detection.

Figure 2.19 Headspace vial closures for Vac-HS-SPME (Courtesy of ExtraTech).

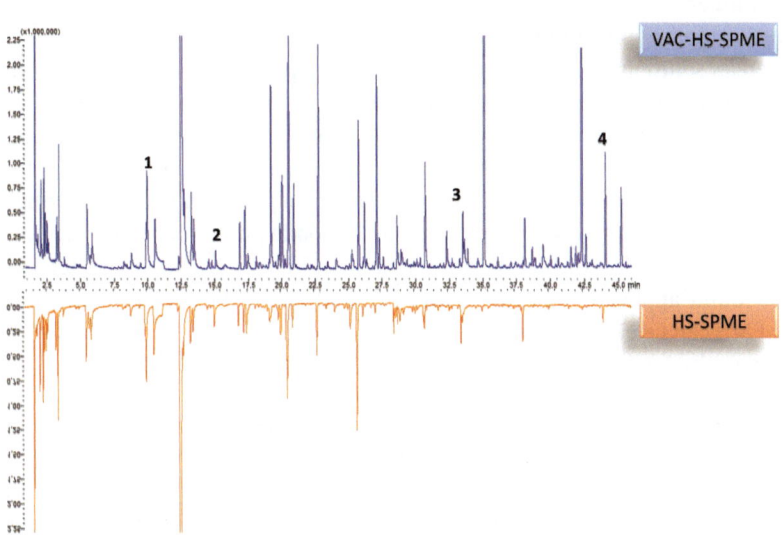

Figure 2.20 Comparison between regular SPME and Vac-HS-SPME (GC-MS TIC, SPME fiber DVB/CAR/PDMS, 50/30 µm, incubation 5 min, 30 or 43 °C with agitation). Marked compounds: (1) hexenal, M 98; (2) 2-heptenal, M 112; (3) decenal, M 154; (4) farnesen, M 204 (Purcaro, 2024/ ExtraTECH Analytical Solutions).

The total sample vaporization skips the partition between a sample matrix and the headspace (Markelov and Guzowski, 1993). The complete analyte content of a sample is considered to be evaporated into the headspace of the sample vial.

TV-SPME sampling takes place directly from the gas phase as shown for illicit drugs from beverages or urine samples (Davis, 2019). Due to this two-phase system sample matrix effects are eliminated. Only small sample volumes of a few microliters are required and placed into a regular 20 mL headspace vial, then heated for total evaporation of the analytes. In case of the reported analysis of illicit amphetamine samples, the used polydimethylsiloxane (PDMS)/DVB fiber was first exposed to a reagent vial containing N,O-Bis(trimethylsilyl)trifluoroacetamide (BSFTA) + 1% trimethylchlorosilane (TMCS) before sampling from the incubated vial for 10 min, followed by GC-MS analysis. Limits of detection (LOD) of 50 ng/mL in human urine were reported.

A similar application describes the vacuum-assisted TV-SPME analysis of PAHs (VA-TF-SPME) from polluted water samples; 100 μL of sample volume was evaporated and extracted at 55 °C using a sampling time of 25 min. LODs were obtained in the range of 0.3 to 5 ng/mL with a good quantitative precision of 6.1 to 7.5% RSD (Beiranvand and Ghiasvand, 2020).

In-tube Solid Phase Microextraction The term "in-tube" SPME, also known as solid phase dynamic extraction (SPDE), describes the coating of the extraction phase inside of a section of fused-silica or steel tubing rather than coated on the outside of a rod as described for the conventional SPME devices (Gou *et al.*, 2000; Ridgway *et al.*, 2006). SPDE probes are used in the same way in headspace (HS-SPDE) or direct immersion (DI-SPDE) extraction procedures (Bicchi *et al.*, 2008), also extending its applications into LC-MS (Grecco *et al.*, 2022).

Commercial solutions are available in which a gas-tight syringe is equipped with a stainless-steel cannula lined on the inside with an approx. 50 μm thick stationary phase. The extraction process allows the sample to be repeatedly pulled through this needle for enrichment. For desorption, the cannula is inserted into the injector of a GC system. The analyte injection is supported by a gas flush from the syringe (Lipinski, 2001). As the syringe can pull up liquid mobile phase such SDPE devices are handy for desorption to high performance liquid chromatography (HPLC) as well, and found application in the online LC analysis of polar compounds too. In-tube SPME can be easily coupled to HPLC by using regular injection valves with appropriately sized loops (Eisert and Pawliszyn, 1997; Gou *et al.*, 2000).

Solid Phase Microextraction High Capacity Probe Further increased sorbent volume of approx. 65 μL and a corresponding surface expansion are provided with the development of the HiSorb™ probe for headspace and immersion extraction, shown in Figure 2.21 (Hearn *et al.*, 2022). The sorbent phase is coated onto a robust steel rod allowing larger sorption capacity than SPME fibers or SPME arrows. The available HiSorb phases and phase combinations are PDMS, DVB/PDMS, carbon wide range (CWR)/PDMS and DVB/CWR/PDMS.

HiSorb probes can be used manually and automated in a dedicated unit for all steps of sample extraction, including heating and agitation, washing and drying of

Figure 2.21 HiSorb sorptive extraction probes with the enlarged view of the sorbent phase section (Courtesy of Markes).

probes, and finally the probe desorption. For desorption, the HiSorb probe is inserted into a standard thermodesorption tube with analytes transferred into a focusing trap (refer to Section 2.1.8). A pre-purge step is used to remove oxygen and prevent reactions with analytes and potential damage to the sorbent material before ramping to the set desorption temperature.

A unique possibility for the detection of very low analyte concentrations in samples was introduced with a multiple-step enrichment (MSE) method, for instance with the quantification of low 1,4-dioxane levels in drinking water. The multiple extraction from one or several sample vials and concentration of the analytes of interest using an intermediate cryofocusing trap offers the possibility to reach the lowest detection limits not accessible without pre-concentration (Markes, 2022).

Thin-Film Solid Phase Microextraction A further increase in sorbent surface area and volume was achieved by the most recent development of the thin film SPME (TF-SPME), also known as thin film microextraction (TFME). Carbon fiber mesh membranes are applied to carry the sorbent materials in a high density PDMS (Grandy et al., 2007). The thin-film device can be used for active sampling from vials as illustrated in Figure 2.22, but is also ideally suited for passive and on-site sampling e.g. from water and air (Risticevic et al., 2009; Emmons et al., 2019). The higher extraction rate of the TF-SPME technique allows a higher extraction efficiency and sensitivity without sacrificing analysis time (Bruheim et al., 2003).

Figure 2.22 SPME thin film membrane used for immersion extraction.

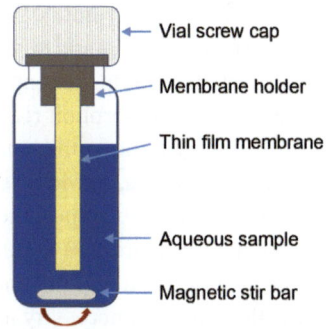

In contrast to the thicker sorbent layers of the SPME fibers and arrows, analytes can be extracted in a shorter time. Typical applications are extractions of polar compounds from aqueous matrices (Pawlizsyn Research Group, 2024). TF-SPME today is semi-automated. After in-vial extraction, or on-site sampling, the thin film is transferred to the racks of an x,y,z-robotic system for automated TD into a GC injector (Stuff et al., 2018). A hydrophilic-lipophilic balanced (HLB)/PDMS phase on TF-SPME film delivered the highest intensities for late eluting fatty acids in almond milk, while the important polar flavor compounds dominate with a DVB/PDMS film (Kfoury et al., 2021).

2.1.4.2 Solid Phase Microextraction Operation for GC-MS

Conditioning At first use, a pre-conditioning of the SPME device is recommended to be performed at about 20 °C above the planned desorption temperature for 15 to 30 min within the applicable temperature range. The conditioning of the SPME probe can take place in a GC injector at the recommended temperature. If a separate GC injector is not available, this way of operation is practical only for manual operation. Automated SPME analyses require an inert gas-purged and heated module to provide the conditioning as part of a robotic autosampler configuration. In case of immersion extraction, also rinsing with solvents of the sorbent phase may become necessary before high temperature conditioning to remove inorganic or high molecular layers before. A recommendation for SPME condition, rinsing, and operation temperatures of different sorbent materials is given with Table 2.4.

During automated SPME workflows, a short conditioning of the SPME device before and after application to a sample is advisable. Typically, the conditioning during the automated measurement of sample series using a separate conditioning module for an automated system is strongly recommended to achieve the conditioning at temperatures above the set GC injector desorption temperature.

Extraction In general, SPME can be seen as an equilibrium rather than as an exhaustive extraction process (Pawliszyn, 1999). An ideal approach would be to wait for the SPME extraction to reach the equilibrium between the sample and the sorbent. Long extraction times over hours are only suitable for field sampling. In the laboratory practice for qualitative and quantitative analytical SPME, the practical goal is to keep the extraction times in the range of GC separation runtimes. Right after the exposure to the sample, the sorption rises quickly in a first nearly linear phase, then it takes time to approach the equilibrium condition asymptotically as shown in Figure 2.23. For routine analysis, providing sufficient sensitivity for quantitative assays and a reasonable sample throughput, the extraction is stopped in the pre-equilibrium phase upon reaching approx. 60 to 80% of the equilibrium condition.

The precise time control and optimum reproducibility for quantitative SPME extraction are then necessary and best achieved with automated systems (O'Reilly

Table 2.4 SPME operation temperature and solvent rinsing guide.

Sorbent phase	SPME type	Phase thickness	Maximum temperature [°C]	Recommended operating temp. [°C]	Conditioning time *min | max | recom. [min]	Recommended rinsing solvents	Rinsing time min | max | recom. [min]
PDMS	Fiber	7 μm	340	200–320	1 | 60 | 5	Water | Acetontrile | Alcohols	0.5 | 10 | 2
PDMS	Fiber	30 μm	280	200–280	1 | 60 | 5	Water | Acetontrile | Alcohols	0.5 | 10 | 2
PDMS	Fiber	100 μm	280	200–280	1 | 60 | 5	Water | Acetontrile | Alcohols	0.5 | 10 | 2
PDMS	Arrow	100 μm	300	200–280	1 | 30 | 5	Water | Acetontrile | Alcohols	0.5 | 10 | 2
PDMS	Arrow	250 μm	300	200–280	1 | 30 | 10	Water | Acetontrile | Alcohols	0.5 | 10 | 2
Acrylate/Polyacrylate	Fiber	85 μm	280	200–280	1 | 30 | 5	Water | MeOH | aliphatic HC	0.5 | 10 | 2
Acrylate/Polyacrylate	Arrow	100 μm	280	200–250	1 | 30 | 5	Water | MeOH | aliphatic HC	0.5 | 10 | 2
Polyethylene glycol (PEG)	Fiber	60 μm	250	200–240	1 | 30 | 5	MeOH | Water w 15% NaCl	0.5 | 10 | 2
Carbon WR/PDMS	Fiber	95 μm	300	220–300	1 | 60 | 10	MeOH | EtOH | iProp	0.5 | 10 | 2
Carbon WR/PDMS	Arrow	120 μm	300	200–300	1 | 30 | 5	MeOH | EtOH | iProp	0.5 | 10 | 2
DVB/PDMS	Fiber	65 μm	300	220–300	1 | 60 | 10	MeOH | EtOH | iProp	0.5 | 10 | 2
DVB/PDMS	Arrow	120 μm	300	220–270	1 | 30 | 10	MeOH | EtOH | iProp	0.5 | 10 | 2
DVB/Carbon WR/PDMS	Fiber	80 μm	300	220–300	1 | 60 | 10	MeOH | EtOH | iProp	0.5 | 10 | 2
DVB/Carbon WR/PDMS	Arrow	120 μm	300	220–270	1 | 30 | 10	MeOH | EtOH | iProp	0.5 | 10 | 2

a) Conditioning temperature within analytical runs, for initial conditioning 15–30 min are recommended. Alcohols: MeOH = methanol, EtOH = ethanol, IPA = isopropanol [2-propanol]. Aliphatic HC = aliphatic hydrocarbons, e.g. n-hexane.

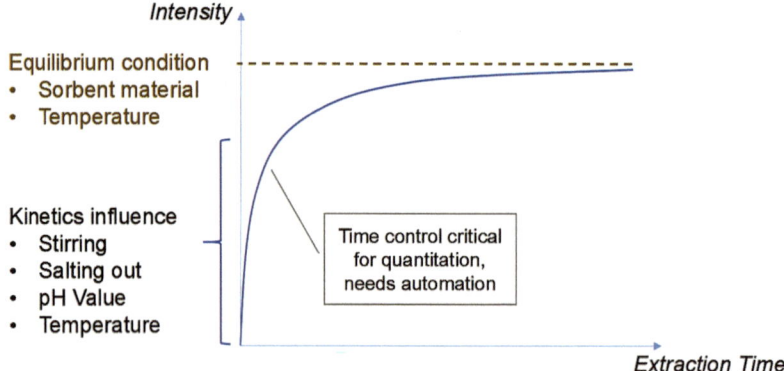

Figure 2.23 SPME extraction – Physical factors affecting sample recovery.

et al., 2005). Also, the penetration depth into the vial should be considered and kept constant for measurements during the non-equilibrium phase. The sorbent-to-sample distance is reported to be of influence for quantitative HS-SPME analyses as well. Short extraction times command for a close distance to the sample for optimum recovery (Nzekoue et al., 2020).

The amount of analyte extracted from the headspace of a sample is dependent upon the partition coefficient (Zhang and Pawliszyn, 1993). Analytes with low partition coefficients K out of the sample into the gas phase (high analyte concentration in the gas phase) are best extracted using HS-SPME, those with high coefficients (higher analyte concentration in the sample) benefit from the DI-SPME extraction. For the partition coefficient K of a volatile compound, also refer to Section 2.1.5 Headspace Techniques, the following eq. (2.1) should be considered.

$$K = \frac{c_S}{c_G} = \frac{\text{concentration in the sample}}{\text{concentration in the gas phase}} \qquad (2.1)$$

Analytes with a lower concentration in the gas phase c_G (higher partition coefficient K) determine the optimum extraction time for the sampling method due to their flatter extraction profile. As a representative example, Figure 2.24 shows the extraction time profiles of compounds with low (bromoform, bromodichloromethane) and higher partition coefficient (1,2-dichlorobenzene, 1,3-dichloropropene). The optimum extraction time for these analytes can be determined with 30 min (Cervera et al., 2011).

The partition coefficient K is temperature dependent. The incubation temperature has a strong effect on the analyte partition and equilibrium level. Also, the partition of the analytes from the gas phase into the sorbent material needs to be considered when optimizing the extraction temperature. A constantly heated agitating device is mandatory to support a reproducible and fast extraction.

Analyte recovery can further be improved by increasing the ionic strength in an aqueous sample using the "salting-out" effect, and often by adjusting a suitable pH value (Myung et al., 1998; Zuba et al., 2002; Agilent Technologies Inc., 2022). As described for static headspace analysis in Section 2.1.5 the addition of salt to the

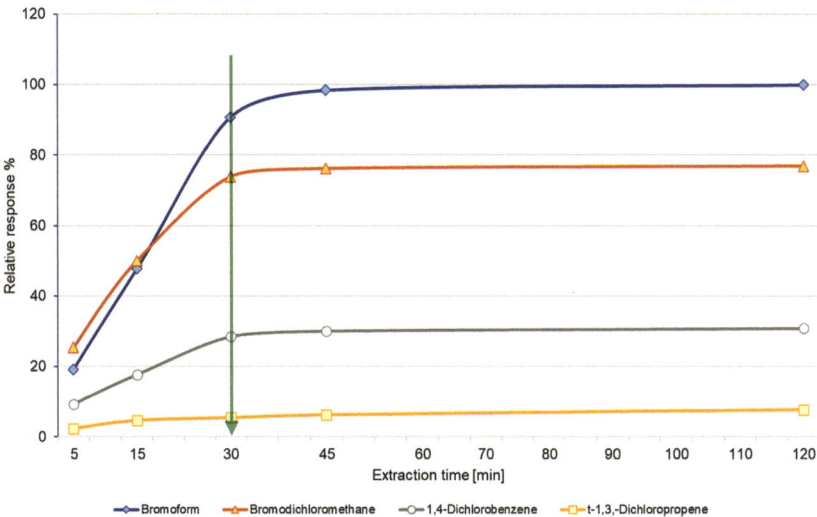

Figure 2.24 SPME extraction time profiles for determination of the optimum extraction time.

sample matrix enhances the amount of analyte extracted by the SPME fiber coating (Steffen and Pawliszyn, 1996).

The extraction situation for DI-SPME is very different as the movement of molecules in liquids is orders of magnitude lower than in the gas phase. And, polar liquids like water create a strong boundary layer around the sorbent section. Intense shaking to overcome the limiting boundary layer and replenish with the analyte is required. Also, suitable extraction times can be longer. Stirring rates up to 1500 rpm using stir bars or the use of cycloid stirring patterns are suggested (Ziegler and Schmarr, 2019). High salt concentrations in water samples are reported to interfere with adsorption binding sites on polar sorbent materials. For the DI-SPME of PAHs, the addition of sodium thiosulfate led to the required low LOQs in drinking water (Coelho et al., 2008; Lim Yee Sin et al., 2019).

Derivatization The analysis of polar compounds from a DI-SPME extraction by GC-MS often requires derivatization. The SPME sorbents do not limit the choice of derivatization agents. All typical reagents for GC-MS analyses are in use and described in many publications (Jurado et al., 2000; Lord and Pawliszyn, 2000; Kim and Shin, 2011; Crucello et al., 2023).

A straightforward SPME derivatization procedure and most suitable for automated workflows is the so-called "on-fiber" derivatization as illustrated in Figure 2.25 for an online flow-cell sampling from a water stream. Depending on the application, three scenarios are in use:

- Perform the sample extraction first, then expose the collected analytes to the derivatization agent (post-derivatization).
- Loading the sorbent phase with derivatization agent first, then proceed with the sample extraction.

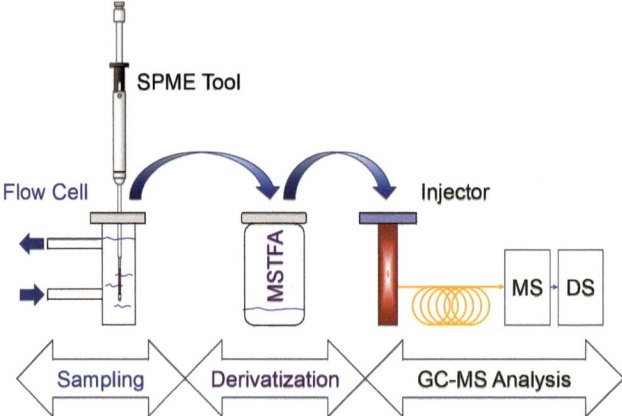

Figure 2.25 SPME on-line flow cell sampling and derivatization for GC-MS analysis.

- Combining both – load the derivatization agent first, extract the analytes, and then expose the probe to the derivatization agent again.

Alternatively, in-situ derivatizations reduce the processing steps and are in many applications in use to get polar analytes easier accessible to HS-SPME or DI-SPME extractions for GC-MS, for instance for short-chain fatty acids using pentafluorobenzyl bromide (PFB-Br) (Lenzi *et al.*, 2023), acetylation of chlorophenols (Regueiro *et al.*, 2009), or by ethyl chloroformate for drug residues in wastewater (Clairmont Feitosa, 2011).

Quantification The application of internal standards or the use of the standard addition method for quantitation is strongly recommended for achieving the required quantitative precision with low relative standard deviations (Poerschmann *et al.*, 1997; Zhu *et al.*, 2015; Lee *et al.*, 2017). The standard addition method is the preferred SPME calibration method as it uses the identical analytes from reference standard mixes in the identical sample matrix (see also Section 3.3.7). Potential matrix effects can be excluded. Automated SPME analyses using state-of-the-art x,y,z robotic samplers facilitate such routine quantification procedures significantly. The use of just one single internal standard may not be sufficient to compensate for matrix effects of analytes of different chemical nature (Nolvachai *et al.*, 2023).

Also, analog to the quantitation by multiple headspace extraction (MHE) a multiple SPME extraction is described (refer to Section 2.1.5.2). SPME can be performed in consecutive runs with the same sample in both immersion and headspace modes (Tena and Carrillo, 2007). Multiple headspace SPME (MHS-SPME) results from the combination of MHE and SPME. The MHS-SPME peak areas are used to calculate the amount of analyte of an exhaustive extraction (Martínez-Uruñuela *et al.*, 2005; Hakkarainen, 2007).

Desorption After headspace or immersion extraction, the SPME sorbent phase is thermally desorbed in the GC injector for analysis. The injector temperature may not exceed the specified maximum permitted temperature to avoid irreversible damage to the sorbent material, as recommended in Table 2.4. In contrast to extraction, the desorption times are fast, typically from less than 1 min to up to 2 min for SPME fibers and arrows, depending on the sorbent phase thickness (Ilias *et al.*, 2006).

SPME applications benefit during desorption in the GC inlet from narrow ID inlet liners. A high gas speed and reduced dilution in carrier gas improve peak shapes and sensitivity. Dedicated SPME inlet liners are available for the major GC types and should replace the regular split/spitless liners used for liquid injection. The choice of internal diameter should best fit the diameter of the SPME device in use, as exemplified in Table 2.5. In practice, the 1.7 mm ID liner shows good injection performance and peak shape and is a general recommendation. The desorption can

Table 2.5 Recommended GC inlet liner ID for SPME Fiber and SPME Arrow desorption.

	ID [mm]	OD [mm]	L [mm]	SPME Fiber OD 0.64 mm	SPME Arrow OD 1.1 mm	SPME Arrow OD 1.5 mm
Injector Inlet Liner	0.8	as of GC type below		**best fit** ①		①
	0.8	as of GC type below		**best fit** ①		①
	1.3	as of GC type below		②	**best fit**	①
	1.3	as of GC type below		②	**best fit**	①
	1.7	as of GC type below		②	**recommended**	**best fit**
	1.7	as of GC type below		②	**recommended**	**best fit**
	2.0	as of GC type below		②	**good fit**	**recommended**
	2.0	as of GC type below		②	**good fit**	**recommended**
Conditioning Station	1.7	6.3	78.5	③	④	④
GC Type Inlet Liner Dimensions						
Agilent 6890/7890/8890	s.o.	6.3	78.5			
Shimadzu 2010/2030	s.o.	5.0	95.0			
Thermo TRACE 1300/1310	s.o.	6.3	78.5			
Thermo TRACE Ultra	s.o.	8.0	105.0			

recommended Takes into account a potential swelling of the sorbent, or bending for automated operation.
① **WARNING:** Use of this liner will lead to serious damage of SPME Arrow and injector!
② This liner type can be used as well with slight loss of GC peak performance.
③ Bended fibers can contact the liner surface, replace upon visible contamination or fiber damage.
④ Check monthly. Replace upon baseline problems, at least once a year with continuous operation.

take place using small split ratios with the split valve open in the desired split ratio until completion of the desorption process.

The penetration depth into the GC inlet liner has a significant impact on the desorption process and is subject of individual optimization (Arthur and Pawliszyn, 1990). As different GC brands use different liner length and vary much in the injector heating profiles, the general rule is to position the tip of the SPME probe to the evenly heated zone which usually is in the lower liner section. SPME arrows allow a maximum injector penetration depth of 70 mm. It should be noted that SPME inlet liners cannot be used for split or splitless liquid injections. The maximum available liner volume is 216 µL only with a typical 1.7 mm ID liner!

Caution has to be taken with the use of bended SPME fibers or arrows during penetration into the injector liner as the sorbent phase may touch and adhere to the hot liner. Retraction of the SPME device may tear off the sorbent layer.

2.1.4.3 Solid Phase Microextraction Sorbent Materials

Polydimethylsiloxane (PDMS) is the common basis of most of the SPME coatings. PDMS is used as is but also as the "binder" for different solid sorbent materials. The standard sorbents are equally available for all commercial SPME devices, also in different phase thicknesses. There has been a general color coding established identifying the available sorbent materials as the phases itself, except carbon, are hardly visually distinguishable.

As usual, the principle "like goes with like" is a good guidance in SPME as well when choosing a suitable sorbent phase. Moreover, sorbent materials used in SPME exhibit adsorption and absorption characteristics which need to be considered when choosing SPME probes for certain applications. Absorption dissolves the analyte into the sorbent material. The analytical capacity involves the entire volume of the coating. In contrast, adsorption keeps the analytes adhering to the surface of the adsorbent. As it is a surface-based process, space for analyte extraction is limited and competition with components or matrix of similar chemical nature can occur (Wang and Pawliszyn, 2023). Sorbent volumes and surface values of SPME devices are compared in Figure 2.18. SPME coatings can be classified beyond the coating thickness by the prevailing absorption or adsorption characteristics and polarity (see Table 2.6). Coating type and thickness also plays an important role in the recommended choice of SPME probes as illustrated in Figure 2.26.

The SPME probes of different polarity provide extraction selectivity and increase the recovery of such analytes that match the sorbent polarity, as it is indicated in Figure 2.27. Also, potentially interfering matrix can be reduced, for instance by extracting polar compounds from organic matrices. The combination of sorbent materials with PDMS, or DVB/Carbon, combines the analytical properties. In contrast to the individual selective sorbent materials, the so-called "triple phase" of DVB/Carbon/PDMS is considered to be of general non-targeted use.

Often, during method development or the use of different selectivities for samples, the change of the SPME sorbent is desired. The automated exchange of SPME fibers within an analytical sequence can be achieved by a tool change or a multi fiber exchange (MFX) accessory for a PAL x,y,z-robotic system. MFX allows both

Table 2.6 Type of coating extraction mechanism and polarity.

SPME sorbent	Mechanism	Polarity
Polydimethylsiloxane (PDMS)	*Absorbent*	Nonpolar
Polyacrylate (PA)	*Absorbent*	Polar
Carbowax (PEG)	*Absorbent*	Polar
DVB/Carbowax	*Absorbent*	Polar
DVB/PDMS	**Adsorbent**	Bipolar
Carbopack/PDMS	**Adsorbent**	Bipolar
Carboxen/PDMS	**Adsorbent**	Bipolar
DVB/Carboxen/PDMS	**Adsorbent**	Bipolar

(Pawliszyn, 2012/with permission of ELSEVIER).

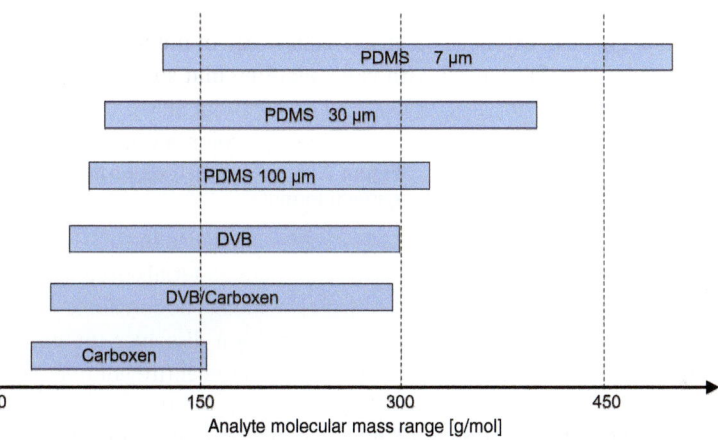

Figure 2.26 Molecular mass extraction range for selected SBME fiber coatings (after Pawliszyn).

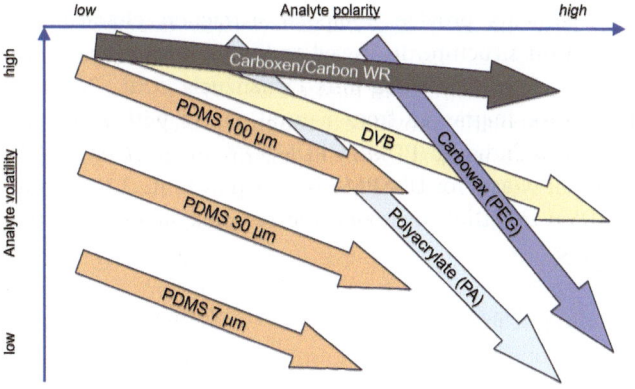

Figure 2.27 SPME sorbent material properties related to analyte volatility and polarity (after Shirey, 2009).

analysis conditions and SPME fibers to be automatically changed at any time within a method or a sequence. MFX reduces the risk of fiber breakage during manual handling, simplifies method development, and enhances the productivity of routine SPME analysis (Barbu and Teaca, 2018; SRA Instruments, 2022).

Polydimethylsiloxane (PDMS) PDMS is a widely used non-polar siloxane polymer of the general formula shown in Figure 2.28. The typical application range is mainly the extraction of non-polar analytes from samples, for instance, PAHs or terpenes.

The PDMS polymerization degree used in SPME devices is a rubber-like viscous liquid, insoluble in water or other polar solvents like alcohol. Alcohols and polar solvents such as methanol, glycerol, and water do not swell the material, but acetone, 1-propanol, and pyridine expand the PDMS layer to a small extent. However, non-polar organic solvents like ethyl acetate, methylethylketone, or toluene can be absorbed and also swell the material. Solvents such as chloroform, ethyl ether, and tetrahydrofuran (THF) largely swell PDMS. Diisopropylamine is known to swell PDMS to the greatest extent (Lee et al., 2003). All non-polar solvents or derivatization reagents should be used with care, avoiding long exposure times at higher temperatures.

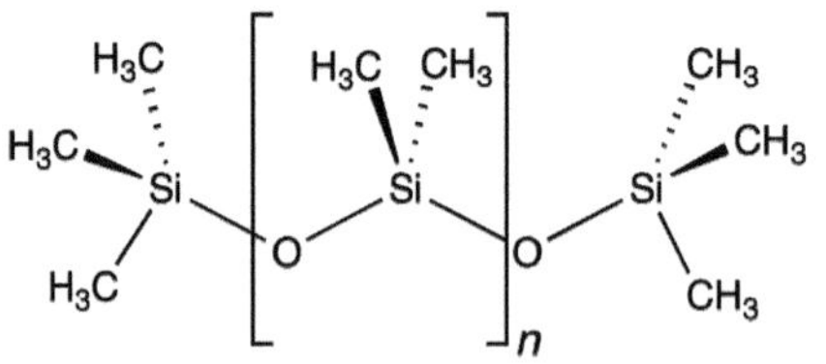

Figure 2.28 Polydimethylsiloxane (PDMS) general formula.

Divinylbenzene (DVB)/Polydimethylsiloxane (PDMS) Divinylbenzene (DVB) is a reactive monomer that polymerizes autocatalytically (Figure 2.29). Porous DVB particles are blended into PDMS. The pores of the DVB material are ideal for trapping C_6 to C_{15} analytes and retain smaller analytes better than PDMS alone.

SPME DVB-containing sorbents provide a bipolar extraction characteristic based on the aromatic electron structure. It proved to be the most efficient, for instance, for the polar VOCs from fermented milk (Valenzuela et al., 2024), or for the determination of musk fragrances from natural waters with recoveries ranking as DVB/PDMS > PA > Carboxen/PDMS > PDMS (Winkler et al., 2000). The selectivity of such sorbent phases for DI-SPME is shown in Table 2.7. Acrylate and the combination of DVB in PDMS increased the response of most of such non-purgeable polar analytes.

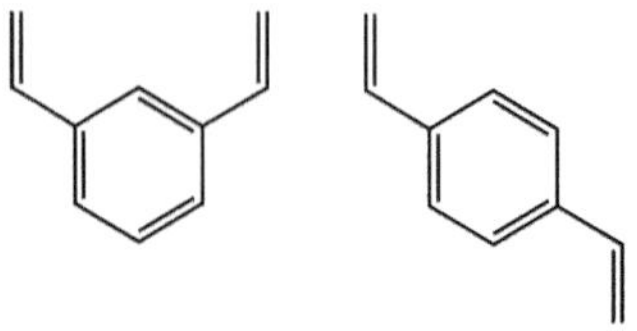

Figure 2.29 Divinylbenzene (DVB) monomers.

Table 2.7 Comparison of SPME fibers for extraction of non-purgeable polar analytes. Data are factors relative to ethanol on PDMS.

Component	PDMS	Acrylate	DVB/PDMS	DVB/Carbowax
Methylamine	0.30	0.28	0.88	0.30
Methanol	0.58	0.68	0.53	0.70
Dimethylamine	0.48	0.58	1.80	0.43
Ethanol	1.00	0.85	0.68	0.98
Acetonitrile	1.08	2.28	1.00	1.63
Isopropanol	1.55	1.73	1.23	1.65
Propanol	0.80	4.28	0.65	4.25
Diethylamine	0.43	0.90	2.53	0.28
2-Methoxyethanol	0.08	0.03	0.05	0.01
Ethylene glycol	0	0.01	0.10	0
Triethylamine	10.68	0.68	8.18	1.35
Dimethyl sulfoxide (DMSO)	0.05	0.05	0.05	0.03

Conditions: Analytes in water, 25 ppm, pH 7, 27% NaCl, immersion, 15 min, rapid stirring, desorption 270 °C, 5 min.
(Pawliszyn, 2012/Royal Society of Chemistry).

Table 2.8 Comparison of SPME coatings for extracting C_2 to C_6 hydrocarbons (in area cts).

Analyte	PDMS	DVB/PDMS	Carboxen/PDMS
Ethane (C_2)	0	0	750
Propane (C_3)	0	0	20 000
Butane (C_4)	0	340	72 000
Pentane (C_5)	230	2150	108 000
Hexane (C_6)	460	9300	106 000

(Pawliszyn, 1999/Royal Society of Chemistry).

Carboxen/Polydimethylsiloxane Carboxen™ or carbon wide range (CWR) of different pore characteristics distribution with micro, meso, and macro pores are blended into PDMS. Molecules in the size of C_2 to C_{12} are the preferred application range (Table 2.8). Higher desorption temperatures are required to release in particular large molecules from the carbon pores. Volatile organic sulfur compounds (VOSs) are most efficiently extracted using a Carboxen-PDMS fiber, for instance as markers of insect infestation in rice (Mendivelso-Pérez and Shimelis, 2023).

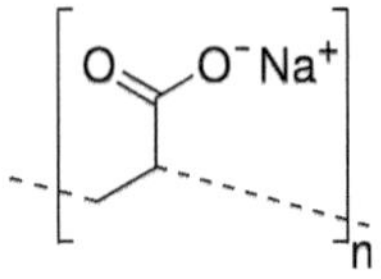

Figure 2.30 Polyacrylate (PA), general formula.

Polyacrylate Polyacrylate (PA) is a highly polar polymer consisting of acrylate monomer chains (Figure 2.30). The mechanical properties depend on the degree of polymerization as viscous liquid or even as solid material. Used in SPME probes it is a highly viscous rubber that turns liquid at desorption temperatures. PA extracts like PDMS analytes via absorption.

Besides headspace applications, polar analytes are extracted using PA by DI-SPME. A typical application is the GC-MS analysis of phthalate plasticizers like dibutylphthalate (DBP) and benzylbutylphthalate (BBP), migrating from plastic food containers into food (Moreira et al., 2014).

Polyethylene glycol Carbowax™ is the trademark of polyethylene glycols (PEGs) with different polymerization degrees (Figure 2.31). PEGs tend to swell in water and are even soluble. For sampling and rinsing the addition of NaCl is required to reduce swelling. For rinsing PEG fibers can be immersed in hydrocarbon solvents and will not swell. Carbowax is sensitive to oxygen at elevated temperatures. The sorbent layer can be oxidized in the presence of air and oxygen. Injection port desorption temperatures are recommended in the range of 180 to 240 °C. A PEG fiber is recommended compared to DVB/PDMS for pesticide analysis from water with superior results (Pereira et al., 2014).

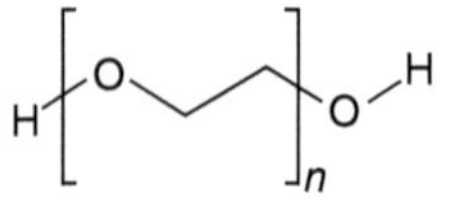

Figure 2.31 General formula of polyethylene glycols.

Overcoated fiber materials A major shortcoming when sampling biological materials was observed with a fouling process of the sorbent material during direct immersion in complex matrices, for instance with high protein or sugar content. The overcoating of the regular polar sorbents with thin layers of PDMS improved the multiple use of SPME fibers significantly without observing losses in recoveries. The PDMS outer layer thickness for the optimized coating was determined to be approximately 25 to 30 μm (Souza Silva and Pawliszyn, 2012).

The robustness of a PDMS/DVB/PDMS fiber in DI-SPME mode was studied e.g. for triazole pesticides in grape pulp. DI-SPME using a single overcoated fiber performed well in 130 consecutive extractions. Also, extraction capabilities toward pesticides extractions from water samples were proven to be similar to those exhibited by a non-coated DVB/PDMS probe. The rugged performance of overcoated sorbent materials showed its advantages in particular for automated DI-SPME analyses in complex matrices.

General Recommendations for SPME Analysis

- Use a "triple phase" sorbent for non-targeted analysis (DVB/Carbon/PDMS).
- Consider "*like-goes-with-like*" sorbents for selective analyses.
- With PDMS a thicker film is required to retain small molecules and a thinner film is recommended for larger low volatility molecules.
- Derivatize polar and non-volatile compounds in situ or after extraction for GC-MS analysis.
- Condition the SPME probe before sampling and after analysis according to the manufacturer's guidance.
- Analyze the blank SPME probe before the extraction of a sample series.
- Analyze blanks from an empty vial randomly within a series depending on a potential matrix interference, and to prove the absence of carryover.
- Consider the salting-out effect and add salt (manual, or in an automated setup saturated salt solution) to aqueous samples to enhance the extraction yield.
- The recommended temperature for aqueous samples is 60 °C. High extraction temperatures reduce the absorption to PDMS phases.
- After DI-SPME in aqueous samples, the drying of the sorbent phase is recommended to prevent excess water injected to GC.
- Replace the GC split/splitless inlet liner with the recommended SPME ID liner.
- Do not inject liquids into an SPME liner.

For Quantitation

- Use a standard mix of the analytes with known concentrations.
- The required extraction time for HS-SPME is determined by the less volatile compounds to reach sufficient response.
- For quantitation, an individual calibration curve must be established for each compound.
- Use the standard addition method in case of results close to regulated limits, and to compensate for matrix effects on the results.
- Fill sample and calibration vials to the same volumes for quantitative analyses.
- Incubate the sample at a constant temperature for all samples.
- The incubation temperature and extraction time must be identical for all samples and calibration solutions for quantitative analysis.
- Sample agitation/stirring must be identical for all samples and calibration solutions.
- Regular standard mix runs within series confirm the extraction performance.

About Maintenance

- SPME phases deteriorate with use due to high temperature and matrix exposure, check performance with standards.
- Probes used for DI-SPME should be rinsed before conditioning.

(Continued)

(Continued)
- Replace even slightly bent SPME probes.
- GC injector septa can be cored by the rod of SPME fibers and must be replaced frequently to avoid silicon ghost peaks and in worst case septum leakage.
- At the same time, the inlet liner must be replaced to remove potential septa particles.
- The septum must be replaced more frequently with SPME fiber use compared to liquid injections.

New Coating Developments A shortcoming of SPME in the past was the lack of fibers that are polar enough to extract very polar or ionic species from aqueous solutions. New coatings have been developed to close the gap of applications for low volatile polar compounds for instance in food, environmental, and metabolomics analyses using HS-SPME and DI-SPME techniques (Risticevic *et al.*, 2009; Bojko *et al.*, 2012; Souza Silva and Pawliszyn, 2012).

Instead of selective properties, many applications require non-selective extractions for a wide range of analytes. The evaluation of hydrophilic-lipophilic balanced sorbent material (HLB, a divinylbenzene N-vinylpyrrolidone copolymer) in SPME opens new potential for water analysis by the headspace extraction of analytes with large differences in polarity (Grandy *et al.*, 2018). The highest extraction efficiency of a set of analytes ranging from phenol to alkanes with air-water distribution ratios (K_{aw}) from 5.0 to 2.5 and *n*-octanol/water partition coefficients (K_{ow}) of 1.5 to 7.5 was achieved by a 5 µm HLB arrow compared to regular DVB-CWR-PDMS arrow (Tintrop *et al.*, 2023). Also, the mix of different sorbent materials of HLB with weak cationic exchange (HLB-WCX), or weak anionic exchange (HLB-WAX) using a polyacrylonitrile (PAN) binder confirmed the good extraction performance for polar and nonpolar analytes, as well as positively and negatively charged compounds (Zhou *et al.*, 2024).

Ionic liquids (IL) and in particular polymer ionic liquids (PILs) offer high thermal and chemical stability and became of interest for use in different techniques in sample pretreatment. SPME with PIL coatings of high polarity and low volatility were developed for instance for headspace extraction of low molecular mass alcohols from aqueous samples, pesticides for food safety or food metabolomics applications with improved recoveries (Carda-Broch *et al.*, 2010; Yu *et al.*, 2013; Gionfriddo *et al.*, 2018).

New capabilities in SPME are also expected from the use of nanomaterials. Graphene was immobilized on a SPME fiber. The nanomaterial showed improved extraction efficiency, higher mechanical and thermal stability, and with more than 250 stable extractions a significantly longer life span than commercial PDMS or PDMS/DVB-coated fibers (Arthur and Pawliszyn, 1990; Huang *et al.*, 2012). An ionic-liquid-mediated multi-walled carbon nanotube (CNT)-poly(dimethylsiloxane) hybrid coating was used as a solid phase microextraction adsorbent for methyl-*t*-butyl ether (MTBE) in water. This innovative fiber as reported has a high thermal stability of >320 °C and a long lifespan with over 210 analyses (Vatani and Yazdi, 2014).

Octanol/water partition coefficient

The partition coefficient between *n*-octanol and water (K_{ow}) is an often-stated measure of the lipophilicity (fat solubility) or hydrophilicity (water solubility) of an analyte. It is calculated from the ratio of the equilibrium concentrations in an *n*-octanol/water system: $K_{ow} = c_{n\text{-octanol}}/c_{H_2O}$

The coefficient is typically given as logarithmic value, where positive log K_{ow} values stand for lipophilic substances and negative values for hydrophilic substances. Amphiphilic substances, chemical compounds possessing both hydrophilic (water-soluble, polar) and lipophilic (fat-soluble, non-polar) properties have log K_{ow} values close to zero (Sangster, 1997).

2.1.5 Static Headspace Technique

One of the most elegant possibilities for automated instrumental sample preparation and sample transfer to GC-MS systems is the use of the headspace technique (Figure 2.32). Here all the frequently expensive steps, such as extraction of the

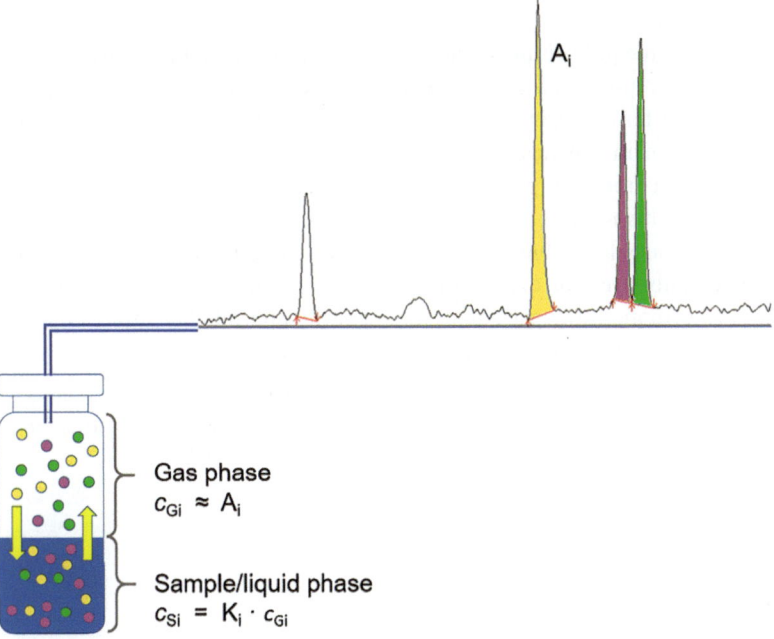

Gas phase
$c_{Gi} \approx A_i$

Sample/liquid phase
$c_{Si} = K_i \cdot c_{Gi}$

Figure 2.32 Principle of headspace analysis: A_i area of the GC signal of the *i*th component. c_{Gi}, c_{Si} concentrations in the gas and sample phases of component i. K_i partition coefficient of component i (Kolb, 1980/John Wiley & Sons).

sample, clean-up, and concentration are dispensed with (Hachenberg and Schmidt, 1977; Ettre and Kolb, 1991). The term headspace analysis was coined in the early 1960s when the analysis of substances with odors in the headspace of canned food was developed (Buttery and Roy Teranishi 1961). The equilibrium of volatile substances between a sample matrix and the gas phase above it in a closed static system is the basis for static headspace gas chromatography (HS-GC). The term headspace is often used without the word static, for example, "headspace analysis", "headspace sampler".

The volatile compounds to be analyzed are distributed between the sample and the gas phase in a closed sample vial. At the set incubation temperature of the sample, a distribution equilibrium is reached. The concentration of the substances in the gas phase then remains constant. Using the headspace technique, the volatile substances in the sample are separated from the matrix. The latter is not volatile under the conditions of the analysis. The tightly closed sample vessels, which are used for the static headspace procedure, can frequently even be filled at the sampling location. The risk of false results (loss of analyte) as a result of transportation and further processing is thus reduced.

The extraction of the analytes is based on the partition of the very and moderately volatile substances between the matrix and the gas phase above the sample. The concentration of the analyte in the gas phase above the sample is proportional to the concentration in a liquid sample, as described by Henry's law. A very comprehensive compilation of water/air partition coefficients is provided by Rolf Sander which may be helpful in method development, available at http://www.henrys-law.org (Sander, 2015).

After the partition equilibrium is reached, the gas phase contains a qualitatively and quantitatively representative cross-section of the analytes in the sample, and hence can be used for analyzing the components to be determined. All involatile components remain in the headspace vial and are not analyzed. At this point, an aliquot is taken from the gas phase and transferred into the GC-MS system (Figure 2.33). For this reason, the coupling of headspace instruments and GC-MS systems is particularly favorable. Since the interfering organic matrix is not transfered, a longer duty cycle of the instrument and high detection sensitivities are achieved. Furthermore, headspace analyses are easily and reliably automated and achieve a high sample throughput in a 24 h operating period.

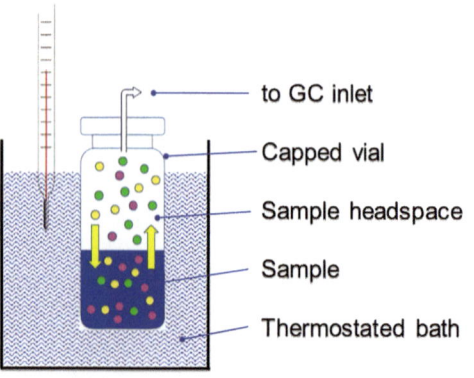

Figure 2.33 Sample handling in the static headspace method (Hachenberg, 1988/Springer Nature).

2.1 Sample Preparation

> **Static Headspace Analysis**
>
> In static headspace analysis, the samples are taken from a closed static system (closed headspace vial) after the thermodynamic equilibrium (partition) between the liquid/solid matrix and the headspace above it has been established.

In static headspace analysis, the partition coefficient of the analytes is used to assess and plan the method (Kolb and Ettre, 2006). The partition coefficient K of a volatile compound is defined by the following eq. (2.1):

$$K = \frac{c_S}{c_G} = \frac{\text{concentration in the sample}}{\text{concentration in the gas phase}} \tag{2.1}$$

The partition coefficient at equilibrium depends on the sample temperature. The incubation temperature must therefore be kept constant for all measurements.

Rearranging the equation for the analyte concentration in the sample gives:

$$c_S = K \cdot c_G \tag{2.2}$$

As the peak area A determined in the GC-MS analysis is proportional to the concentration of the substance in the gas phase c_G it follows:

$$A \approx c_G = \frac{1}{K} \cdot c_S \tag{2.3}$$

The quantity M_0 of a volatile substance in the original sample is divided at equilibrium into a portion in the gas phase M_G and the part still in the sample matrix M_S. To be able to calculate the concentration of a substance c_0 in the original sample, the mass M_0 at equilibrium must be referred to:

$$M_0 = M_S + M_G \tag{2.4}$$

Replacing M with the product of the concentration c and volume V gives:

$$c_0 \cdot V_0 = c_S \cdot V_S + c_G \cdot V_G \tag{2.5}$$

By using the definition of the partition coefficient in Eq. (2.2), the unknown parameter c_S can be replaced by $K \cdot c_G$:

$$c_0 \cdot V_0 = K \cdot c_G \cdot V_S + c_G \cdot V_G \tag{2.6a}$$

$$c_0 = c_G \cdot \frac{V_S}{V_0} \cdot \left(K + \frac{V_G}{V_S} \right) \tag{2.6b}$$

The starting concentration c_0 of the sample, assuming $V_0 = V_S$, is given by:

$$c_0 = c_G \cdot \left(K + \frac{V_G}{V_S} \right) \tag{2.7}$$

As the determined peak area is proportional to the concentration of the volatile substance in the gas phase c_G, the following equation is valid, corresponding to the proportionality between the gas phase and the peak area:

$$c_0 \approx A \cdot (K + \beta) \tag{2.8}$$

where by β is the phase ratio V_G/V_S (see Eq. (2.7)) and therefore describes the degree of filling the headspace vial (Ettre and Kolb, 1991).

The effects on the sensitivity of a static headspace analysis can then be easily derived from the ratio given in Eq. (2.8):

$$A \approx c_0 \cdot \frac{1}{K + \beta} \quad \text{with} \quad \beta = \frac{V_G}{V_S} \qquad (2.9)$$

The peak area determined from a given concentration c_0 of a component depends on the partition coefficient K and the sample volume V_S (through the phase ratio β).

For polar substances in aqueous media with high partition coefficients K (e.g. ethanol, isopropyl alcohol, dioxane) the contribution of the sample volume V_S in eq. (2.9) can be neglected. However, for substances with small partition coefficients K (e.g. cyclohexane, trichloroethylene, and xylene in water) the phase ratio β determines the headspace analysis sensitivity. Doubling the quantity of the sample leads to a twofold increase in the peak area. Figure 2.34 illustrates the change of response of cyclohexane (K = 0.05) and 1,4-dioxane (K = 642) with sample volumes of 1 and 5 mL. As a consequence, for all quantitative determinations of compounds with small partition coefficients, an exact filling volume must be maintained.

The static headspace method is an indirect analysis procedure, requiring special care in performing quantitative determinations. The position of the equilibrium depends on the analysis parameters (e.g. temperature, sample volume) and on the sample matrix itself. The matrix dependence of the response can be counteracted

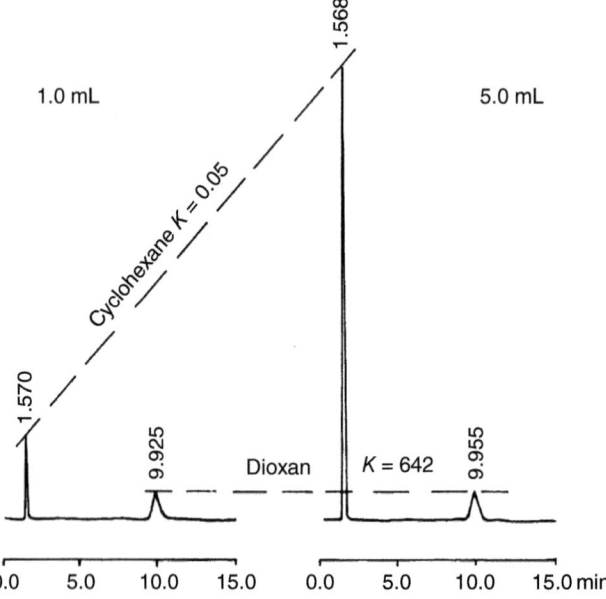

Figure 2.34 Sample volume and sensitivity in the static headspace for cyclohexane (K = 0.05) and 1,4-dioxane (K = 642) with sample volumes of 1 and 5 mL, respectively (Adapted from Kolb, 1992).

in various ways. The matrix can be standardized, for example, by the addition of Na_2SO_4 or Na_2CO_3. Other possibilities used for quantification include the standard addition method (see Section 3.3.7.3), internal standardization, or the MHE procedure (see Section 2.1.5.2) (Kolb and Ettre, 1991; Zhu et al., 2005). For the coupling of HS-GC with MS, the internal standard procedure is recommended for quantitative analyses. Besides the headspace-specific effects, possible variations in the MS detection are also compensated for. The best possible precision is thus achieved for the whole analysis procedure.

There are limitations to the coupling of headspace analysis with GC-MS systems if a lot of water vapor is extracted from the sample vial as well. In certain cases, water can impair the focusing of volatile components at the beginning of the GC column. Impairment of the GC resolution can be counteracted by choosing crosslinked polar and thickfilm GC columns which elute water as a solvent peak before ramping the column oven for analyte separation. The typical incubation temperature when using aqueous samples for headspace analysis is 60 °C.

It is also known that water affects the stability of the MS ion source, which nowadays is becoming ever smaller. In the case of repeated analyses, the effects are manifested by a marked loss in response and poor reproducibility. In such cases, special precautions must be taken with the choice of ion source parameters, in particular the source temperature.

The headspace technique is very flexible and can be applied to the most widely differing sample qualities. Liquid or solid sample matrices are generally used, but gaseous samples can also be analyzed precisely using this method. Even the water content in food or pharmaceutical products can be determined by headspace GC (Kolb, 1993). Both qualitative and quantitative determinations are carried out coupled to GC-MS. Static HS analysis is typical also in the characterization of food volatile flavor profiles in ion mobility MS (Zhu et al., 2024).

There are different methods possible for analysis of the headspace volatiles of a sample which lead to very different requirements concerning the instrumentation: the static and dynamic headspace techniques. The areas of use overlap partially but the strengths of the two methods are demonstrated in the different types of applications further discussed in Sections 2.1.6 and 2.1.7.

2.1.5.1 Measures for Improved Headspace Response

The sensitivity of detection in static and dynamic headspace analysis can be improved by different measures. For compounds with a low partition coefficient K, the fill volume of the headspace vial is important. Large sample volumes deliver higher sensitivity for instance for BTEX (benzene/toluene/ethylbenzene/xylene) analysis.

One obvious option is the lowering of the partition coefficient K by the incubation at a higher temperature. Substances with high partition coefficients benefit particularly from higher equilibration temperatures (Table 2.9, e.g. alcohols in water). Raising the temperature is, however, limited by increased matrix evaporation. Temperatures above the recommended 60 to max. 80 °C for aqueous samples, closer to the BP of water risk bursting the sample vial or the seal.

Table 2.9 Air/water partition coefficients K of selected compounds.

Substance	40 °C	60 °C	80 °C
Cyclohexane	0.07	0.05	0.02
n-Hexane	0.14	0.04	<0.01
Tetrachloroethane	1.5	1.3	0.9
Tetrachloroethene	1.5	1.3	0.9
1,1,1-Trichloroethane	1.6	1.5	1.2
1,1,1-Trichloromethane	1.7	1.5	1.3
o-Xylene	2.4	1.3	1.0
Toluene	2.8	1.6	1.3
Benzene	2.9	2.3	1.7
Chlorobenzene	4.3	2.6	1.9
Dichloromethane	5.7	3.3	2.1
n-Butyl acetate	31.4	13.6	7.6
Methyl isobutyl ketone	54.3	22.8	11.8
Methyl isobutyl ketone	54.3	22.8	11.8
Ethyl acetate	62.4	29.3	17.5
Methyl ethyl ketone	140.0	69.0	35.0
2-Butanone	145.0	69.0	37.0
n-Butanol	647.0	238.0	99.0
Isopropyl alcohol	825.0	286.0	117.0
Ethanol	1355.0	511.0	216.0
1,4-Dioxane	1618.0	642.0	288.0

(Kolb, 1992 and Kolb and Ettre, 2006).

Changes in the sample matrix also affect the partition coefficient. For samples with high water contents, an electrolyte can be added for the so-called "salting out effect" (Figure 2.35). The addition of inorganic salt to samples standardizes the matrix effects. The improved response increases with the analyte polarity. For example, for the determination of ethanol in water, the addition of NH_4Cl gives a twofold increase in the sensitivity of detection, and the addition of K_2CO_3 even gives an eightfold increase. The strength of the salting-out effect with chlorides in water follows the order $NH_4Cl > NaCl > CaCl_2$ and increases with increasing salt concentration (Banat et al., 1999; Majors, 2009). A systematic investigation of the salting-out effect on alcohols from blood also showed K_2CO_3 as the most efficient, followed in the distance by $(NH_4)_2SO_4$, Na_2CO_3 and $MgSO_4$, Na_2SO_4, and NaCl, with the lowest effect $MgCl_2$ (Zuba et al., 2002). The addition of NaCl as a solid or saturated solution works well as a general recommendation.

The headspace sensitivity can also be raised by the addition of nonelectrolytes. In the determination of residual monomers in polystyrene dissolved in DMF, adding

Figure 2.35 Matrix effects in static headspace. The effect of salting out on polar substances (Adapted from Kolb, 1992).

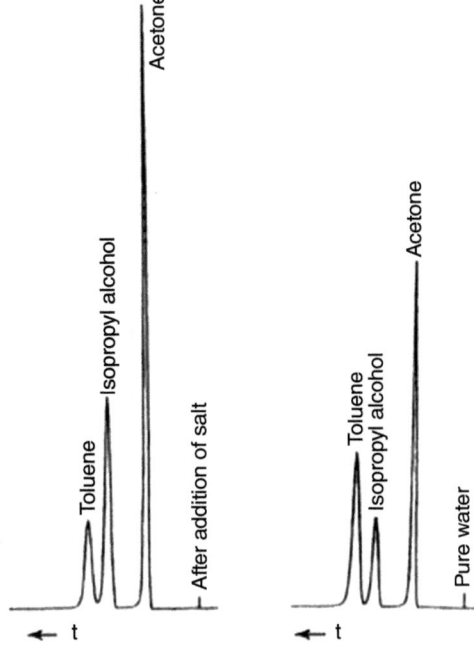

Figure 2.36 Increasing water concentrations in the determination of styrene as a residual monomer lead to a sharp increase in response (Hachenberg, 1988/Springer Nature).

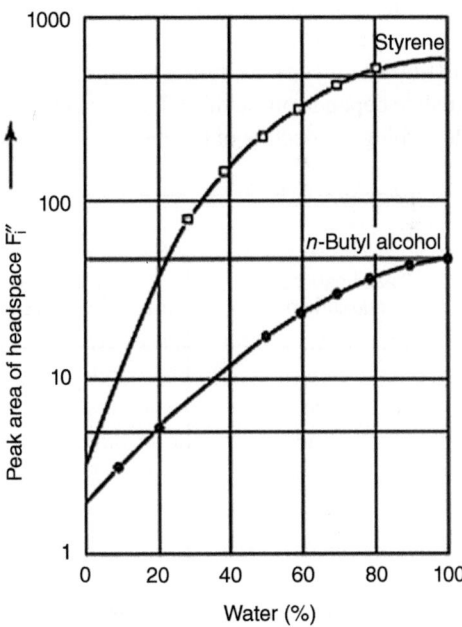

increasing quantities of water, for example, in the determination of styrene and butanol, leads to an increase in peak area for styrene by a factor of 160 and for n-butyl alcohol by a factor of 25 (Figure 2.36).

The full evaporation technique (FET), first published by Markelov and Guzowski, addresses the known problems of sensitivity with changing and unpredictable sample matrices (Markelov and Guzowski, 1993). FET also extends the accessible BP range of analytes for HS-GC (Kialengila et al., 2013). Only a small sample aliquot is prepared into standard capped HS vials, using a common workflow with an agitator/incubator for equilibration. As only small volumes are dispensed, the necessary evaporation times are short. A rapid determination of ethanol in fermentation products uses only 50 µL of an aqueous sample dispensed into an empty capped HS vial (Li, 2009). The small sample is fully evaporated at an incubation temperature of 105 °C, above the BP of water. After a short incubation time of only 3 min, an aliquot of the headspace is taken for GC analysis. Although the sample size is small the relative GC response is significantly increased. Due to the fast processing, independency from matrix and enhanced sensitivity, the FET is the ideal approach for automated quality control of large sample series.

Additional possibilities arise from in-situ derivatization reactions, as described as well for SPME in Section 2.1.4, which make HS-GC introduction very robust and suitable for routine analysis of less volatile analytes with high sample throughput (Alzaga et al., 2007).

2.1.5.2 Quantitation by Multiple Headspace Extraction

The concentration of an analyte in a sample can be determined by MHE from only one vial, as compared to the static and dynamic procedures in Figure 2.37 (Maggio et al., 1991; Kolb, 1992). This is possible because the decline in peak area follows a first-order kinetics according to Eq. (2.10). At any time t, the concentration of the analyte depends on the initial concentration c_0 and the constant q in the exponent, describing the decline of the measured peak areas (Kolb and Ettre, 1991).

$$c = c_0 \times e^{-qt} \tag{2.10}$$

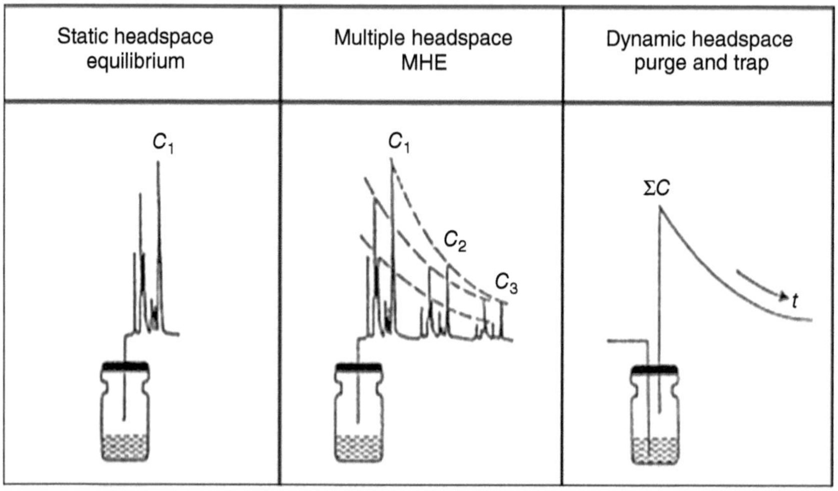

Figure 2.37 Gas extraction techniques (Adapted from Kolb, 1992).

The MHE procedure is carried out in a series of consecutive runs from one and the same sample vial. Immediately after the first headspace analysis, the vial is depressurized to the atmosphere and incubated with the same method again for the next analysis. This procedure can also be accomplished by many types of headspace autosamplers. In case of a headspace syringe autosampler, an optional "MHE device" is required, which is used to depressurize the vial (Vulpius and Baltensperger, 2009). For quantification, the use of internal standards can be applied to MHE as well.

The stepwise MHE procedure Eq. (2.10) can be rewritten in Eq. (2.11) by replacing the concentration with peak areas (with i the number of injections, A_i the peak area after i injections and A_1 the peak area of the first injection), and for the time t with the number of analyses from the same vial. The measurements of consecutive extractions show an exponential decrease of the peak areas.

$$A_i = A_1 \times e^{-q(i-1)} \tag{2.11}$$

Plotting the peak areas logarithmic against the number of extractions, a linear relationship can be expressed in Eq. (2.12).

$$\ln A_i = \ln A_1 - q(i-1) \tag{2.12}$$

In practice, we do not need a large number of injections. Due to the linear relationship in Eq. (2.12), the total area can be calculated from the area of the first run. The slope q, which is the decline in peak areas, is taken from the slope expressed in the regression calculation of the logarithmic graph. The total peak area, which is equivalent to the total concentration of the analyte in the sample, is given by the summation of all potential extractions A_i in Eq. (2.13).

$$\sum_{i=1}^{\infty} A_i = \frac{A_1}{1 - e^q} \tag{2.13}$$

As an example the analysis of ethylene oxide (EO) in foodstuffs and food contact materials serves as a very typical application for MHE-HS-GC-MS (Arya and Dhyani, 2023). The quantitation of EO in polyvinylchloride (PVC) by MHE demonstrates the

Table 2.10 MHE peak area results of six consecutive headspace runs.

Run#	Area (cts)	ln (Area)
1	151 909	11.93
2	63 127	11.05
3	26 802	10.20
4	10 963	9.30
5	5768	8.66
6	2240	7.71

above-described calculation steps (Petersen, 2008). A series of six measurements is performed providing the peak area data given in Table 2.10, plotted in Figure 2.38. Plotting the areas logarithmic against the number of extractions provides Figure 2.39 with the slope of the linear relationship as of Eq. (2.12).

The calculation for the total analyte peak area in the sample as of Eq. (2.13) with $A_1 = 151909$ from Run 1 from Table 2.10 and the slope $q = -0.833$ from the regression calculation in Figure 2.39, and $e^{-0.833} = 0.4347$, gives:

$$\sum_{i=1}^{6} A_i = \frac{A_1}{1-e^q} = \frac{151909}{1-0.4347} = 268743 \text{ cts} \tag{2.14}$$

This result is calculated using all six data points. With only two measurements, the slope can be calculated with a good and acceptable precision compared to the detailed experiment as shown in Table 2.11. The difference in the calculated result

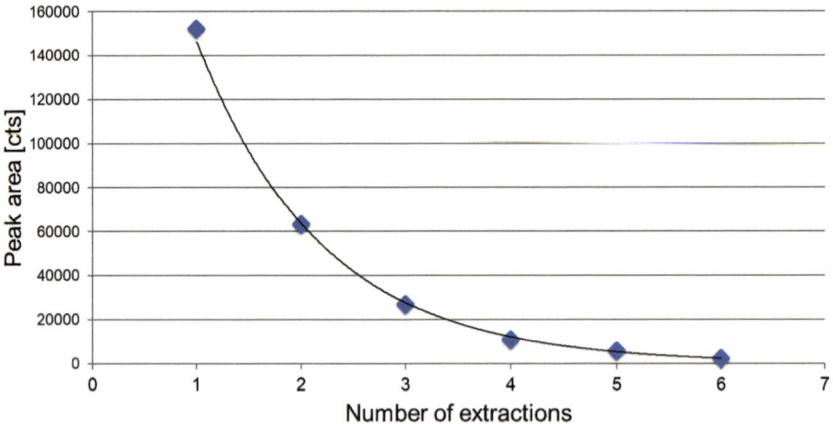

Figure 2.38 Graphical representation of the MHE measurements of Table 2.10, peak area vs. run number.

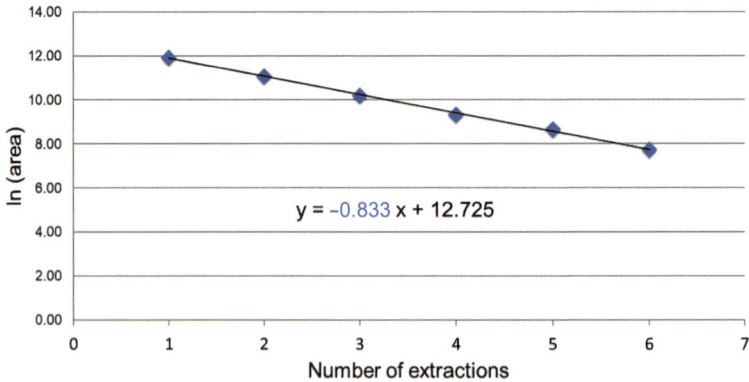

Figure 2.39 Logarithmic plot of the MHE area results of Table 2.10, ln (peak area) versus run number, with regression formula.

Table 2.11 Comparison of the MHE experiment using six to the first two measurements.

	6 Runs	2 Runs	Delta
Area A_1 (cts)	151 909	151 909	0
Slope q	−0.8330	−0.8781	−0.0451
$\exp(q)$	0.4347	0.4156	0.0192
Total area A_i (cts)	268 743	259 928	3.3%

is low with only 3.3% in this illustrated experiment. For the final determination of the analyte concentration in the sample, a regular response calibration with external or internal standards is required.

2.1.5.3 Headspace Analysis Operation

The mixing of samples during equilibration time is a standard feature of up-to-date automated headspace samplers. For liquid samples, vigorous mixing constantly replenishes the depleting phase boundary and guarantees rapid and reproducible delivery of analyte-rich sample material. Teflon-coated stirring rods have not proved successful because of losses through adsorption and potential cross-contamination. However, shaking devices that mix the sample in the headspace vial through vertical or rotational movement have proved effective (Table 2.12). More advanced shakers use adjustable excitation frequencies to achieve optimal mixing of the contents at each degree of filling and viscosity, as shown in Figure 2.40. The resulting equilibration times are in the region of 10 to 30 min using shaking devices compared to 45 to 60 min without mixing. In addition, quantitative measurements become more reproducible. The use of shaking devices in static headspace techniques can reduce relative standard deviations to less than 2%.

Table 2.12 Comparison of the analyses with/without shaking of the headspace sample (volatile halogenated hydrocarbons), average values in parts per billion and precision in standard deviations.

Substance	Without shaking			With shaking		
Ethylbenzene	353	18	(5.2%)	472	8	(1.7%)
Toluene	336	20	(5.9%)	411	4	(1.0%)
o-Xylene	324	13	(4.1%)	400	7	(1.8%)
Benzene	326	18	(5.4%)	372	5	(1.3%)
1,3-Dichlorobenzene	225	13	(5.6%)	255	5	(2.1%)
1,2,4-Trichlorobenzene	207	9	(4.2%)	225	6	(2.5%)
Bromobenzene	213	11	(5.2%)	220	5	(2.1%)

(after Tekmar).

60 2 Fundamentals

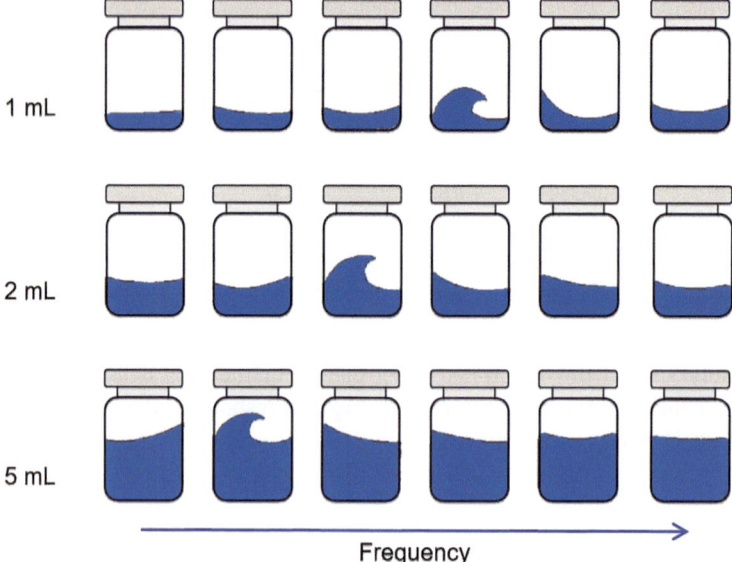

Figure 2.40 Effect of variable shaking frequencies (2 to 10 Hz) on different filling volumes in headspace vessels (Perkin-Elmer).

Static Headspace Injection Techniques

- Pressure Balanced Injection (Figure 2.41):
  Variable quantity injected: can be controlled (programmed) by the length of the injection process, injection volume = injection time × flow rate of the GC column, vial pressurization with carrier gas, no drop in pressure to atmospheric pressure, depressurizing to initial pressure of column.
  Change in pressure in the headspace vial: reproducible pressure build-up with carrier gas pressure, mixing, initial pressure in column maintained during sample injection. Sample losses: none known.

- Syringe Injection (Figure 2.42):
  Variable injection volumes through syringe pump action: easily controlled, syringes of different volumes available, vial pressurization by injection of inert (carrier) gas necessary, pressure drops to the atmosphere on transferring the syringe to the injector.
  Change in pressure in the headspace vial: reproducible pressurization by injection of a carrier gas volume equal or exceeding the sample volume to be extracted, mixing.
  Sample losses: potential minor losses due to pressure drop from the syringe to atmosphere after it has been removed from the sample vial, glass surface contact of the sample plug.

2.1 Sample Preparation

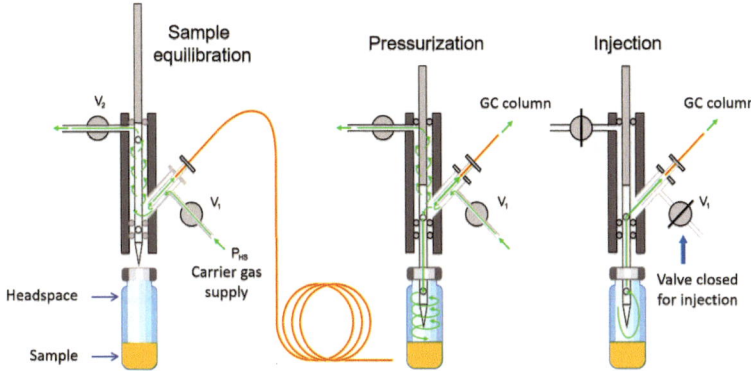

Figure 2.41 Injection techniques for static headspace: the principle of pressure balanced injection. (a) Equilibration, (b) pressurization, and (c) injection phase (Perkin-Elmer).

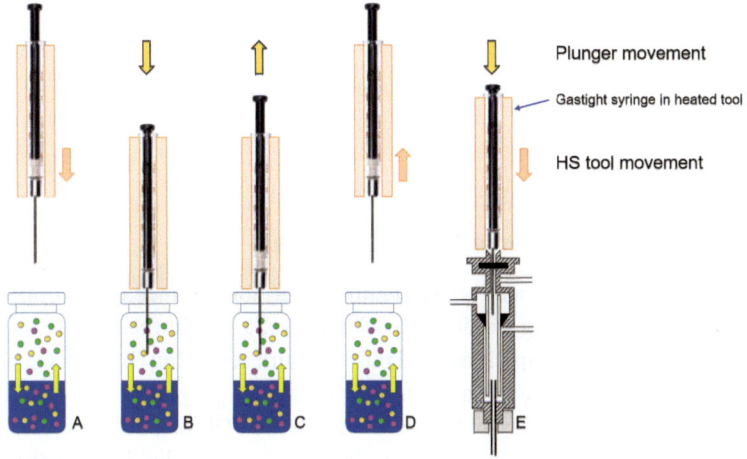

Figure 2.42 Headspace analysis using a heated gastight syringe: A incubation, B pressurization/mixing, C sampling, D GC injection (CTC Analytics).

- Sample Loop (Figure 2.43):
  Injection volume fixed in the instrument by a sample loop, changes in the volume injected require a change of the sample loop.
  Vial pressurization with carrier gas, drop in pressure to atmospheric or variable backpressure on filling the sample loop. The expansion volume must be in excess of the sample loop to be able to fill it reproducibly.
  Sample losses: for larger sample loops attention must be paid to dilution with carrier gas during the pressure build-up phase, analytes in contact with hot steel surfaces.

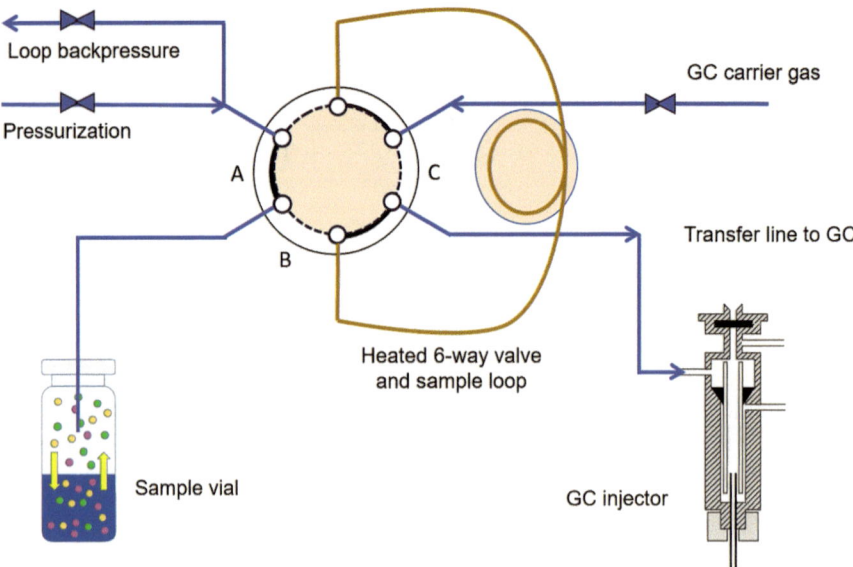

Figure 2.43 Headspace injection using a heated sample loop: A pressurization, B loop filling, C GC injection.

2.1.6 Dynamic Headspace – Purge & Trap Technique

The trace analysis of VOCs is a very typical requirement, for instance in environmental monitoring, in outgassing studies on packaging material, in the analysis of flavors and fragrances, or in occupational and public health screening (Bicchi et al., 2008). The dynamic headspace technique, initially known as *purge and trap (PAT, P&T)*, is the process in which highly and moderately volatile VOCs are continuously extracted from a matrix and concentrated in an adsorption trap (Wylie, 1987, 1988). The substances in a second step get evaporated from the internal trap by thermal desorption and reach the GC as a concentrated sample plug where they are finally separated and detected (Tekmar, 1991). Early P&T applications in the 1960s already involved the analysis of body fluids (Seibel, 2015). In the 1970s, P&T became known because of the increasing requirement for the testing of drinking water for volatile halogenated hydrocarbons. The analysis of drinking water for the determination of a large number of these compounds in ppq quantities (concentrations of less than 1 µg/L) became possible only through the dynamic headspace extraction and concentration process at that time. In particular, the coupling with GC-MS systems allows its use as a multi-component process for the automatic analysis of a large series of samples. For quantitation, internal standards are in use. Also, a method for quantification of the total amount using multiple runs was discussed in the past (Westendorf, 1985; Lin et al., 1993).

An analytically interesting variant of the procedure is the fine dispersion of liquid samples into a carrier gas stream. This so-called *spray-and-trap* process uses the high surface area of the sample droplets for an effective gas extraction. The procedure also works very well with foaming samples. A detergent content of up to 0.1% does not affect the extraction result. The spray-and-trap process is particularly suitable for use with mobile GC-MS analyses (Matz and Kesners, 1993).

2.1.6.1 Coupling of Purge and Trap with GC-MS Systems

There are two fundamentally different possibilities for the installation of a P&T instrument coupled to a GC-MS system, which depend on the areas of use and the number of P&T samples to be processed per day.

In many laboratories, the most flexible arrangement involves connecting a P&T system to the gas supply of the GC injector. The P&T concentrator is connected in such a way that either manual or automatic syringe injection can still be carried out. The carrier gas is passed from the GC carrier gas regulation (EPC) to the central 6-port valve of the P&T system (see Figure 2.44). In standby, the carrier gas flows back through the continuously heated transfer line to the injector of the gas chromatograph. A particular advantage of this type of installation is that the injector can still be used for liquid samples. This allows the manual injection of control samples or the operation of liquid autosamplers. In addition, it guarantees that the connecting tubing remains free from contamination even when the P&T instrument is on standby.

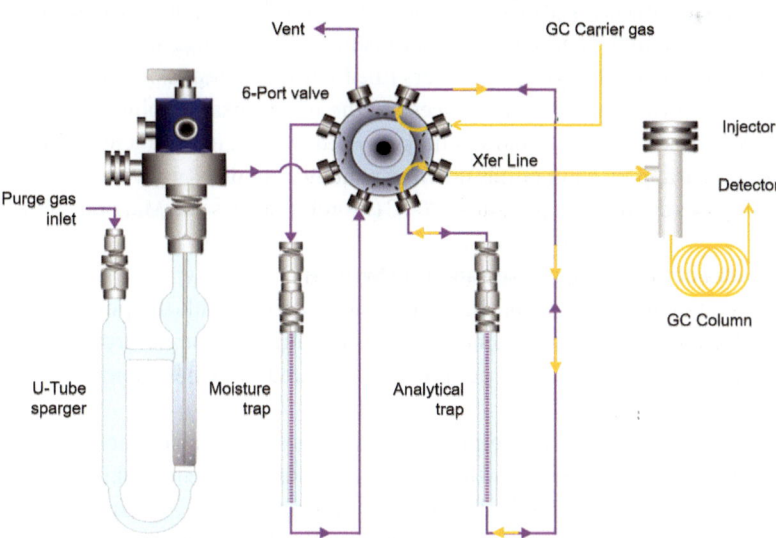

Figure 2.44 Gas flow schematics of the P&T-GC coupling. Purple: purge and trap phase. Yellow: desorption phase to GC ans standy. The 6-port valve switches the phases.

The purge gas is connected directly to the P&T unit, it can be He carrier gas as well, or N_2. Hydrogen cannot be used due to safety concerns. The desorption gas flow is provided by the GC by switching the 6-port valve into the desorption mode.

When the sample is transferred to the GC column, it should be ensured that a focus of the analytes when entering the GC column is achieved. For this purpose, GC columns are in use with film thicknesses of about 1.8 μm or more, which are used generally for the analysis of volatile halogenated hydrocarbons with GC-MS systems. The injector is operated using a small split ratio, which can be selected depending on the detection limit required.

A complete splitless injection of the transferred plug of analytes onto the GC column can only be achieved with injector or on-column cryofocusing (see also Section 2.2.6.2). For the on-column cryofocusing, the GC column is connected to the transfer line or the central 6-port valve of the P&T unit in such a way that the beginning of the column in the GC oven passes through a zone about 10 cm long which can be cooled by an external cooling agent to −30 to −150 °C such as liquid CO_2 or N_2. All components that reach the GC column after desorption from the trap can be frozen out in a narrow band at the beginning of the analytical column. This enables the highest sensitivities to be achieved, in particular with mass spectrometers (Brettell and Groh, 1986; Westendorf, 2000). As large quantities of water are also focused together with the analytes in the case of moist or aqueous samples, removal of moisture in the desorption phase is particularly important with cryofocusing techniques. Insufficient removal of water leads to the deposition of ice and blockage of the capillary. The consequences are poor focus or complete failure of the analysis.

More elegant and free from risks of blocking the column flow is the use of temperature programmable injectors (PTV) for cryofocusing providing the necessary reliability for automated routine analyses (Poy and Cobelli, 1985; Kolahgar and Pfannkoch, 2002). For cold trapping mostly insert liners filled with glass beads or Tenax are in use. As coolant closed loop cooling devices are the most versatile solution. Liquid CO_2 from cylinders with an internal standpipe is used to reach injector temperatures to about −70 °C trapping compounds down to pentane. Liquid N_2 is in use for trapping organic gases down to propane at −180 °C (Moreira Novaes and Marriott, 2021).

2.1.6.2 Modes of Operation of Purge and Trap Systems

The analysis procedure consists of three main steps starting with the purge phase with the simultaneous moisture removal and concentration, the desorption with transfer to the GC, and a final bake-out phase to prepare for the next sample.

The Purge Phase During the purge phase, the volatiles (purgeables) are stripped from the solid or liquid sample. In the case of aqueous samples (drinking water, wastewater) the purge gas is finely dispersed by passing through a glass frit in the base of a U-tube (frit sparger, Figure 2.45a). Usually, 5 mL U-tubes are used for wastewater and 25 mL U-tubes for drinking water samples. The extraction surface of the liquid can be greatly increased by the generation of very small gas bubbles so that the contact between the liquid sample and the purge gas is maximized. Solid

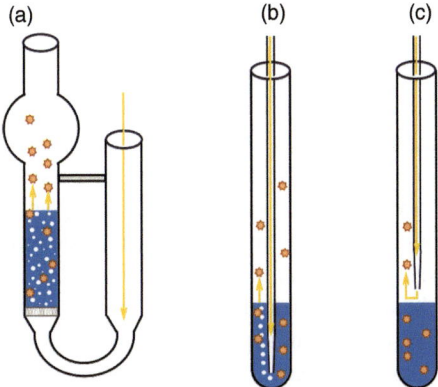

Figure 2.45 Purge and trap glassware. (a) U-tube with/without frits (fritless/frit sparger) for water samples. (b) Sample vessel (needle sparger) for water and soil samples (solids). (c) Sample vessel (needle sparger) for foaming samples (yellow = carrier gas flow).

samples are analyzed in a special needle sparger, as shown in Figure 2.45b into which a needle with side hole is immersed. For foaming samples, the headspace sweep technique as shown in Figure 2.45c can be used.

The purge gas strips the analytes from the sample and transports them via moisture removal to the internal analyte trap. The analytes are retained in this trap and concentrated while the purge gas passes out through the vent. The purge efficiency is defined as the quantity of the analytes, which is purged from a sample with a defined quantity of gas. It depends upon various factors. Among them are: purge volume, sample temperature and nature of the sparger (frit or needle), the nature of the substances to be analyzed, and the matrix.

The total quantity of VOCs extracted from the sample depends on the purge volume. The purge volume is the product of the purge flow rate and the purge time. Many environmental samples are analyzed at a purge volume of 440 mL. This value is achieved using a flow rate of 40 mL/min and a purge time of 11 min. A purge flow rate of 40 mL/min provides optimum purge efficiency, recommended for aqueous samples using the commercial U-tube sparger. Changes in the purge volume should consequently be made by adjusting the purge time. Although a purge volume of 440 mL is optimal in most cases, some samples may require larger purge volumes to achieve adequate sensitivity.

Control of Purge Gas Pressure during the Purge Phase The adsorption and chromatographic separation of volatile halogenated hydrocarbons is improved significantly by regulating the pressure of the purge gas during the purge phase. Additional back pressure control during this phase forms a sharp adsorption band in the trap. The very volatile components (VVOCs) benefit most as a broader distribution band is avoided. During the desorption phase, the narrow adsorption band determines the quality of the sample transfer to the capillary column. The result is an improved peak symmetry, and thus better GC resolution and an improvement in sensitivity. The risk of the analytes passing through the trap is excluded under the given conditions.

Water Removal Water is evaporated and trapped from aqueous or moist samples as well. As the peak shape and resolution of VOCs on capillary columns would be impaired, additional devices are used to remove excess water from the purge gas stream. In particular, where the P&T technique is used with ECD or MS as detectors, reliable water removal is mandatory. Different technical solutions working in-line before trapping or after the desorption tube are in use with the P&T instruments of different manufacturers (Figure 2.44). Also, the choice of sorbent material for use with aqueous samples is important. Tenax, Carbopack, and Carboxen materials (Vocarb™) are hydrophobic and reduce the trapping of water vapors. These sorbents can be dry purged. Traps with silica gel show a high affinity to water and cannot be dry purged.

Water removal devices use a temperature-controlled steel capillary loop working as a moisture trap. If the dew point for water is reached a stationary water phase is formed. BTEX and volatile halogenated hydrocarbon analytes pass unaffected. Polar components such as alcohols can also pass the water trap at moderate temperatures but with a retardation by the condensed water phase. Standards added to the sample should not be dissolved in methanol as high concentrations affect the chromatography in the early elution range. The methanol peak can be avoided by using polyethylene glycol as the solvent for the preparation of standard solutions (see Section 4.4). After the desorption phase, the tubing of the moisture trap is dried by baking out in a carrier gas backflush. An example of VOC analysis showing good efficiency and retention of the residual moisture during the desorption process is shown in Figure 2.46 (Johnson and Madden, 1990).

Another approach for water removal was patented by O.I. Analytical (Xylem Inc., College Station, TX, USA). A Cyclone Water Management™ system takes advantage of a cyclonic effect exploiting the propensity of heavy molecules to travel nearer the center of a gas tube and lighter ones to travel farther out (a similar effect was used in former glass jet separators coupling packed columns to mass spectrometer ion sources). "Cooling" the H_2O molecules by interaction with carrier gas allows them to "settle" in the bottom of the fan-cooled device by gravity, while the analyte gas stream continues to the GC for measurement with about 96% less water. The principle of operation is demonstrated in Figure 2.47. The response of polar molecules is improved as there is no passage of a stationary water phase required.

The Dry Purge Phase To remove the remaining water from a hydrophobic adsorption trap (e.g. with a Tenax™ filling), a dry purge phase on the yet cold trap can be introduced optionally. During this step, most of the water condensed in the trap is blown out by dry carrier gas. Purge times of about 6 min are typical.

The Desorption Phase During the desorption phase, the trap is heated and subjected to a backflush (BKF) with carrier gas. The reversal of the direction of the gas flow is

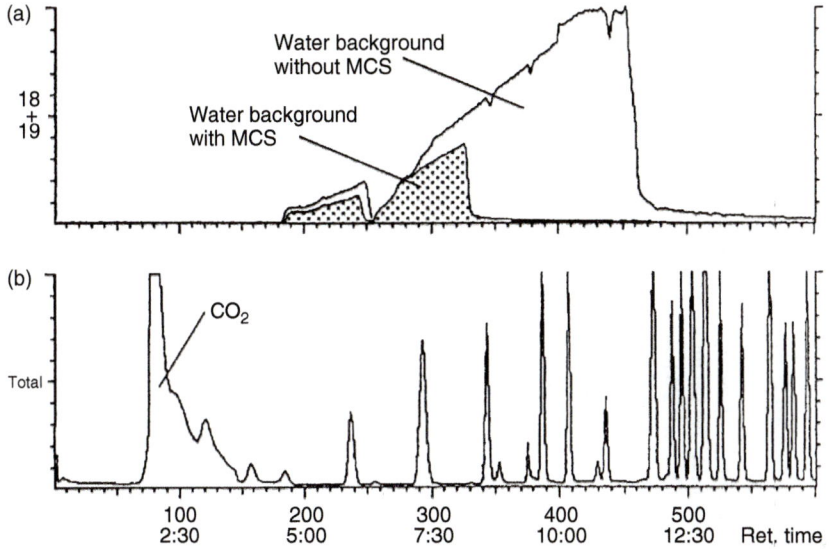

Figure 2.46 GC-MS analysis using purge and trap. (a) Mass chromatogram for water (m/z 18 + 19) with and without moisture removal. (b) Total ion chromatogram with volatile substances (volatile halogenated hydrocarbons) after removal of the water.

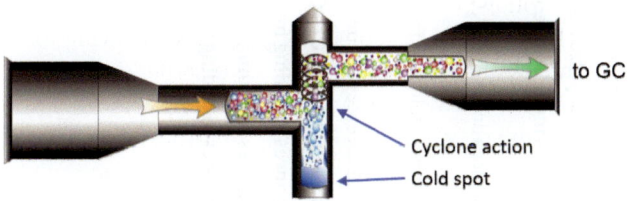

Figure 2.47 Cyclonic water removal during the purge and trap desorption phase (Courtesy O.I. Analytical).

important in order to desorb the analytes in the opposite direction to the concentration step. In this way, narrow peak bands are obtained.

The duration and temperature of the desorption phase affect the chromatography of the substances to be analyzed. The desorption time should be as short as possible but sufficient to transfer the components quantitatively onto the GC column. Most of the analytes are transferred to the GC column during the first minute of the desorption step. The desorption time is generally 4 min for Tenax traps.

The trap temperature set during the desorption step depends on the type of adsorbent in the trap (Table 2.13). The most widely used adsorbent, Tenax TA (*trapping agent*) desorbs very efficiently at 180 °C without forming decomposition products. The peak shape of compounds eluting early can be improved by inserting

Table 2.13 List of trap materials used in purge and trap analysis with details of applications and recommended analysis parameters.

Trap number	Adsorbent	Application	Drying possible?	Drying time [min]	Desorb pre-heat [°C]	Desorb temperature [°C]	Bake-out temperature [°C]	Bake-out time [min]	Remarks
1	Tenax TA/GR	All substances down to CH_2Cl_2	Yes	2-6	220	320	330	7-10	1
2	Tenax, silica gel (Type B)	All substances except freons	No	–	220	225	230	10-12	1, 2
3	Tenax, silica gel, activated charcoal	All substances including freons	No	–	220	225	230	10-12	1, 2, 3
4	Tenax, activated charcoal	All substances down to CH_2Cl_2 and gases	No	–	220	225	230	7-10	1, 3
5	OV-1,Tenax, silica gel, activated charcoal	All substances including freons	No	–	220	225	230	10-12	1, 2, 3
6	OV-1, Tenax, silica gel (Type F)	All substances except freons	No	–	220	225	230	10-12	1, 2
7	OV-1, Tenax	All substances down to CH_2Cl_2	Yes	2-6	220	300	320	7-10	1, 4
8	Carbosieve SIII	All substances including freons	Yes	11	245	330	350	4-10	5
9	Carbopack B, Carbosieve SIII	All substances including freons	Yes	11	245	250	260	4-10	5, 6
10	VOCARB 3000 (Trap K) (Carbopack B, Carboxen 1000, Carboxen 1001)	All substances including freons	Yes	1-3	245	250	260	4	6, 7

| 11 | VOCARB 4000 (Trap I) (Carbopack C, Carbopak B, Carboxen 1000, Carboxen 1001) | All substances including freons | Yes | 1–3 | 245 | 250 | 260 | 4 | 6, 7, 8, 9 |
| 12 | BTEXTRAP (Trap J) (Carbopack B, Carbopack C) | All substances including freons | Yes | 1–3 | 245 | 250 | 260 | 4 | 6, 10 |

Remarks:
1 Tenax is ideal for non-polar compounds, hydrophobic, does not trap water. Polar compounds are less strongly retained, low response of brominated compounds. GR type with graphite and higher BTV. Breakdown background with benzene, toluene and ethylbenzene.
2 Silica gel is excellent for polar and highly volatile compounds, hydrophilic, traps water, dry purge of no effect.
3 Charcoal is hydrophobic, ideal for very volatile components (VVOCs), can be a source of CO_2.
4 OV-1 is a polydimethylsiloxane, hydrophobic, background of methyl silicone breakdown.
5 Carbosieve is ideal for VVOCs, hydrophobic, traps Carbopack breakthrough, excellent thermal stability
6 Carbopack, or graphitized carbon black (GCB), is hydrophobic, does not retained well VVOCs, excellent thermal stability.
7 Carboxen is ideal for VVOCs, hydrophobic, excellent thermal stability.
8 Low response with chlorinated compounds, high back pressure, quantitative losses of chloroethyl vinyl ethers
9 Ideal for less volatile and higher molecular compounds.
10 Ideal for gasoline range organics (GRO), hydrophobic, low water and polar compounds trapping, excellent thermal stability. (Brown and Shirey, 2001; Restek, 2003; Markes, 2012).

a desorb-preheat step. Here the trap is preheated to a temperature near the desorption temperature before the valve is switched for desorption and before the gas flows through the trap. Gas is not passed through the trap during the preheating step, but the analytes are released from the carrier material. When the gas stream is passed through the trap after switching the 6-way valve, it purges the substances from the trap in a concentrated carrier gas cloud. Highly volatile compounds that are not focused at the beginning of the column thus give rise to a narrower peak shape. A preheating temperature of 5 °C below the desorption temperature has been found to be favorable. Tenax GR contains 30% graphite and allows denser packaging for an improved retention capacity.

Vocarb materials, the combination of Carbopack and Carboxen sorbents, are used in applications with low and higher molecular mass compounds including the volatile halogenated hydrocarbons. Desorption temperatures can be higher than Tenax with up to 290 °C (Table 2.13). At higher desorption temperatures, however, the possibility of catalytic decomposition of some substances must be taken into account (Supelco, 1999).

Bake Out Phase The analytical trap can be subjected to a baking out phase after the desorption of the analytes as a clean-up for the next sample. During bake-out at elevated temperatures, VOCs of lower volatility are released and the analytical trap is prepared for the following analysis, a recommended step with automated systems. The maximum bake-out temperatures of the sorbent materials must be observed.

2.1.6.3 Static Headspace vs. Purge and Trap

Both instrumental extraction techniques have specific advantages and limitations when coupled to GC and GC-MS. In particular, the nature of the sample material, the concentration range for the measurement and the effort required to automate the analyses for larger sample series play a significant role. The partition coefficient, the recovery, and thus the sensitivity to be achieved, are relevant to the analytical assessment of the procedure. For both procedures, it must be possible to vaporize the analytes of interest below 150 °C to partition them in the gas phase. The vapor pressure and solubility of the analytes in the sample matrix, as well as the extraction temperature, affect both procedures (Figure 2.48).

How then do the techniques differ? For this, the terms *recovery* and *sensitivity* must be considered. For both methods, the recovery depends on the vapor pressure, the solubility, and the sample temperature. The effects of temperature can be dealt with because it is easy to increase the vapor pressure of a compound by raising the temperature during the vaporization step. The P&T technique is an exhaustive evaporation technique, until in the ideal case the analyte concentration in the purge gas c_G goes down to zero, here the term *percentage recovery* is used (Figure 2.48). This is the amount of an analyte that reaches the GC for analysis relative to the amount that was originally present in the sample. If a sample contains 100 pg benzene and 90 pg reaches the GC column, the percentage recovery is 90%. The static headspace analysis in contrast is an equilibrium technique. A simple recovery expression cannot be used. The distribution constant K between the sample matrix and gas phase controls

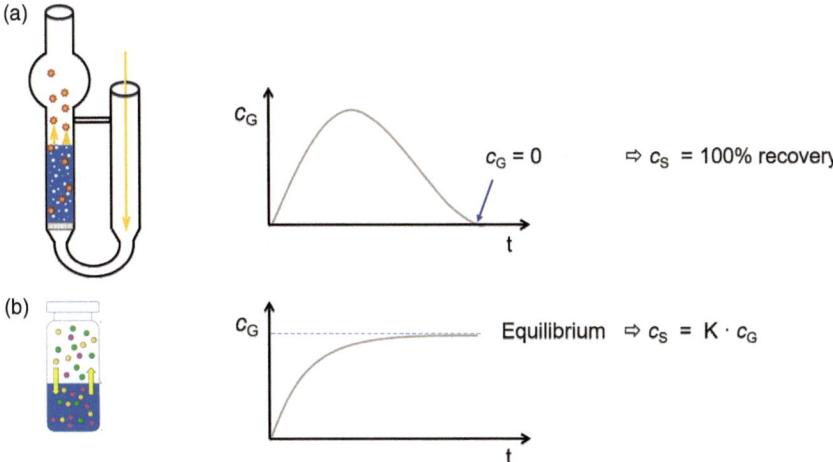

Figure 2.48 Comparison of the purge and trap and static headspace techniques: (a) purge & trap and (b) static headspace.

the analyte concentration in the gas phase c_G. After reaching equilibrium an aliquot of the headspace is taken for analysis.

The commonly used term connected with the static headspace method is the partition coefficient K, as mentioned earlier. K is defined as the quantity of a compound in the sample c_S divided by the quantity in the vapor phase c_G. Therefore, the smaller the partition coefficient is (c_G is large), the higher the sensitivity. It should be noted that a partition coefficient is valid only for the analysis parameters at which it has been determined. These include the temperature, the size of the sample vial, the quantity of sample (weight and volume), the nature of the matrix, and the volume of the headspace. The quantity injected is the next available parameter affecting headspace sensitivity. The quantity injected is limited by a range of factors. For example, only a limited gas volume can be removed from the headspace of a closed vessel even using pressurization. Attempts to remove a larger quantity of sample vapor would lead to a partial vacuum in the sample vial. As a rule of thumb, the same amount of carrier gas is injected into the headspace vial (pressurization step), e.g. by a syringe sampler, that is programmed to be withdrawn (Figure 2.42). Also, loop type sampler pressurize the sample vials before loop filling by releasing the vial pressure against a set backpressure (Figure 2.43). Furthermore, only a limited volume can be injected onto a GC column without causing peak broadening. Thick film capillary columns require cold trapping at injection quantities of more than about 200 µL, hence split injections are often used. For larger quantities of sample and for splitless injection, injector cryofocusing is necessary.

An alternative injection system involves the "pressure-balanced injection" (Figure 2.42). This technique pressurizes the vial by the carrier gas pressure. The injection releases the vial pressure directly into the GC carrier gas stream. The quantity injected is controlled by the injection time from the vial. Because this technique is controlled by "injection time" to inject the sample, the absolute volume of the sample is unknown. It is impossible to measure the exact volume injected; however, the reproducibility of this method is very high. The headspace sample is introduced onto the column without using a gas syringe, thus avoiding fractionation due to pressure changes. Since the needle is sealed, there are no losses of headspace gas during transfer. Limitations typically arise from the necessary sample transfer line. This includes a potential for sample carryover and the fact that the injection port is always occupied by the transfer line (or the column is connected as a transfer line directly to the HS unit bypassing the injector), and therefore not available for manual or liquid autosampler use.

First Example: Volatile Halogenated Hydrocarbons

A comparison with actual concentration values clearly illustrates the differences between the static headspace and P&T techniques. The percentage recovery for the P&T technique is, for example, for an environmental sample of volatile halogenated hydrocarbons: 95% for chloroform, 92% for bromodichloromethane, 87% for chlorodibromomethane and 71% for bromoform. For a sample of 5 mL, which contains 1 ppb of each substance (i.e. a total quantity of 5 ng of each compound), 4.75, 4.6, 4.35, and 3.55 ng are recovered. In a typical static headspace system with a sample vessel of 20 mL containing 15 mL of sample at 70 °C, the partition coefficients for the corresponding volatile halogenated hydrocarbons are 0.3, 0.9, 1.5, and 3.0. This means that the quantities in the 5 mL of headspace are: 11.5, 7.9, 6.0, and 3.8 ng. On injection of 20 µL of the headspace gas mixture onto a standard capillary column, the quantities injected are 0.05, 0.03, 0.02, and 0.015 ng. For a larger injection of 500 µL using cryofocusing, the quantities injected are 1.2, 0.8, 0.6, and 0.4 ng. The P&T technique is consequently more sensitive than the static headspace procedure for these volatile halogenated hydrocarbons by factors of 4.1, 5.8, 7.2, and 9.3.

Second Example: Edible Oils

For compounds with lower recoveries, for example, in the analysis of free aldehydes in edible oils, the difference is even more pronounced. At 150 °C, the P&T analysis gives recoveries of 47%, 59% and 55% for butanal, 2-hexenal, and nonane. For a sample of 500 µL containing 100 ppb of the compounds, 24, 30, and 28 ng are recovered. In the static headspace analysis, the partition coefficients for these compounds at 200 °C are all higher than 200. Assuming the value of 200, the quantity of each of these compounds in 5 mL of headspace is 0.7 ng. For an injection of 0.5 mL, the quantity injected is therefore only 0.07 ng. The sensitivity differences favoring P&T analysis are 343, 428, and 400!

Third Example: Residual Solvents in Food Packaging Materials

Comparable ratios are obtained in the analysis of a solid sample, for example, the analysis of residual solvents in packaging materials. A run using the P&T technique and 10 g of sample at 150 °C gave a recovery of 63% for toluene. The sample contained 1.6 ppm, which corresponds to a quantity of 101 ng. The partition coefficient in the static headspace technique at 150 °C is 95. The quantity of residual solvent in 19 mL of headspace is therefore 17 ng. For an injection of 500 µL of headspace, 0.4 ng is injected. The quantity injected is therefore smaller by a factor of 250 compared to the P&T analysis. Furthermore, the reproducibility of this analysis was 7% for the P&T technique and 32% RSD for the static headspace analysis.

The theoretically achievable or effectively necessary sensitivity is not the only factor deciding the choice of method. The specific interactions between the analytes and the matrix, the nature of the sample with for instance the potential of foaming, the performance of the detectors available, and the legally required detection limits play a more important role. U.S. EPA methods are mainly based on P&T use while European methods usually apply static headspace methods for similar applications.

Besides the sensitivity, there are other aspects which must be taken into account when comparing the P&T and static headspace techniques: The static headspace technique is the most simple and quick sample preparation method and is easily automated. The procedure is well documented in the literature, and for many applications, the sensitivity is more than adequate, so its use is favored over the P&T technique which requires significantly higher instrumental and maintenance efforts. There are many applications where good results are obtained with the static headspace technique which cannot be improved upon by the P&T method. These include the forensic determination of alcohol in blood, free fatty acids in cell cultures, ethanol in fermentation units or drinks, and residual solvents in pharmaceuticals (USP 467).

However, for many other samples, specific limitations besides sensitivity arise on the use of the static headspace technique, which can be overcome using P&T and other dynamic HS techniques. In every static headspace system, all the compounds present in the headspace are injected, not only the organic analytes. This means that an air peak is also obtained. All air contaminants are visible and oxygen can impair the service life of the capillary column at high temperatures. In addition, from aqueous samples a larger quantity of water is injected by static headspace than in the P&T method. For drinks containing carbonic acid, CO_2 can lead to the build-up of excess pressure. Even at room temperature, an undesirably high pressure can build up in the sample vials, which can lead to flooding of the GC column during the analysis. In addition, a very large quantity of CO_2 is injected. Heating the sample enhances these effects. For safety reasons, sample vials with safety caps to guard against excess pressure have to be used. Dust particles lead to special precautions in the case of powder samples. To achieve the necessary sensitivity for an analysis, the sample often has to be heated to a temperature that is higher than that necessary in the P&T technique. This can lead to thermal decomposition, which is frequently observed with foodstuffs. Low temperature HS is required e.g. for dairy samples. In addition, oxygen cannot be eliminated from the sample before heating in a static headspace system. It requires sampling under and into N_2 or Ar atmosphere. During the thermostatting phase, contaminations can arise as a result of the septum used. Substances leaching from the septum (e.g. CS_2 from butyl rubber septa) can falsify the chromatogram.

Advantages of the Static Headspace Technique
- The static headspace can be easily automated. All commercial headspace samplers operate automatically for only a few or large numbers of samples.

- For manual preliminary and qualitative samples, a gas-tight syringe heated up in the GC oven is satisfactory.
- Samples, which tend to foam or contain unexpectedly high concentrations of analyte, do not generally lead to faults or cross-contamination.
- All sample matrices, solid, liquid, or gaseous, can be used directly with minimal sample preparation.
- Headspace samplers are often readily portable and, when required, can be rapidly connected to different types of GC instruments.
- The headspace vials with caps are intended only for single use. There is no additional workload of cleaning glass equipment and hence cross-contamination does not occur.
- Headspace vials can be filled and sealed at the sampling point outside the laboratory. This dispenses with the transfer of the sample material and eliminates the possibility of loss of the sample. Moreover, the risk of inclusion of contamination from the laboratory environment is reduced.
- Because of the high degree of automation, the cost of analyzing an individual sample is kept low.

Limitations of the Static Headspace Technique
- On filling headspace vials, a corresponding quantity of air is enclosed in the vial unless filling is carried out under an inert gas atmosphere. During thermostatting, undesirable side reactions can occur as a result of atmospheric oxygen in the sample.
- During injection from headspace vials, air regularly gets into the GC system, passes the column at low temperature, but can affect sensitive column phases.
- In the case of moist or aqueous samples, a considerable quantity of water vapor gets injected into the column. This requires special measures to ensure the integrity of the early eluting peaks. Headspace samplers do not provide devices for moisture removal as is standard on P&T systems.
- When mass spectrometers are used, special attention must be paid to the stability of ion sources. On insufficient heating, the surface of the source and the lens system become increasingly coated with moisture in the course of the work caused by the injection of water vapor. As a result, the focus of the mass spectrometer can change and this can impair quantitative work.
- Quantitation is matrix-dependent. Standardization measures are necessary, such as salt addition, internal standards, and MHE procedures. Standard addition methods are recommended for matrix independent quantitation.
- The ability of substances to be analyzed is limited by the maximum possible filling of the headspace vials and by the partition coefficients. For compounds with high partition coefficients, larger quantities of sample do not lead to an increase in sensitivity (see Section 2.1.5). If the maximum possible equilibration temperature is being used, it is almost impossible to increase the sensitivity further. Trapping several headspace extractions, as some commercial HS samplers offer, or the use of dynamic HS devices like the in-tube extraction (ITEX DHS), as some automated systems offer, are practical solutions to overcome sensitivity limitations.

- The injection volume is limited. If the sensitivity is insufficient injector cryofocusing is necessary.
- Undesired background contamination can be obtained through contaminated laboratory air which gets into the headspace vial or from bleeding of septum caps.
- Excess vial pressure by high incubation temperatures above solvent boiling points can cause headspace vials to burst e.g. with drinks containing carbonic acid or high equilibration temperatures. This puts instruments out of operation for long periods and results in considerable clean-up costs. Only the use of special vial caps with spring rings can release excess pressure to prevent bursting.

Advantages of the Purge and Trap Technique
- The sample volume can easily be adapted by appropriate glassware of 5 to 25 mL to achieve the required sensitivity.
- A pre-purge step can remove atmospheric oxygen from the sample at room temperature, if required.
- No septum or other permeable material is placed in the sample path.
- Substances with high partition coefficients in the static headspace can be determined with good yields.
- There are no excess pressure problems. The entire gas stream is passed pressure-free through the analytical trap to the vent.
- Water vapor can be kept largely out of the GC-MS system by means of a dry purge step and a moisture removal system.
- In automated operations adding an internal standard without manual measures is optionally possible.

Limitations of the Purge and Trap Technique
- Foaming samples require special treatment, the headspace sweep technique (see Figure 2.45).
- The purge vessels (made of glass) are reused and must be cleaned carefully. Carryover in automated systems requires special care of cleaning after highly contaminated samples. Economical single use vessels are currently available in polymer materials.
- Larger quantities of sample require longer purge times.
- For highly contaminated samples, the breakthrough volume (BTV) of the trap must be taken into consideration.
- Contamination of the system gas lines requires expensive maintenance procedures.
- If the bake-out step is inadequate, the danger of a carryover exists.
- Coupling with capillary GC necessitates the use of small split ratios or cryofocusing.
- The high instrumental effort for P&T analyses is justified only for U.S. EPA methods with the mandatory use of U-tubes with sparger vessels.
- Automated dynamic HS devices as e.g. the in-tube-extraction (ITEX DHS, see Section 2.1.7) can analytically replace all P&T methods except the U.S. EPA methods with the mandatory use of U-tubes with sparger vessels.

2.1.7 Dynamic Headspace – In-Tube Extraction

As an alternative to the purge & trap technique using U-tube sparger vessels the in-tube dynamic headspace extraction (ITEX DHS) combines the versatility and ease of use of the static headspace technique with a dynamic extraction and analyte trapping (Laaks et al., 2015). The dynamic headspace analysis with in-tube extraction facilitates the purge and trap analysis with a tool for industry standard x,y,z-robotic autosampler (Figure 2.49). The re-routing of the carrier gas lines during installation, the need for additional bench space, handling fragile glassware is not required. Most importantly for routine operation, the risk of system contamination is entirely avoided. ITEX DHS uses a dedicated headspace syringe of approx. 1300 µL volume with a sideport in the upper part of the glass barrel. The analytical trap of about 160 µL of sorbent material is the upper part of the syringe needle, as shown in Figure 2.50, and easily exchangeable. The syringe trap is enclosed by a fast direct heater, which is cooled in a few minutes back to room temperature.

Applications cover the same wide range and analytical performance as known from the purge & trap and outgassing techniques, and beyond, but with significantly less instrumentation and installation requirements (Jochmann et al., 2008; Michiu

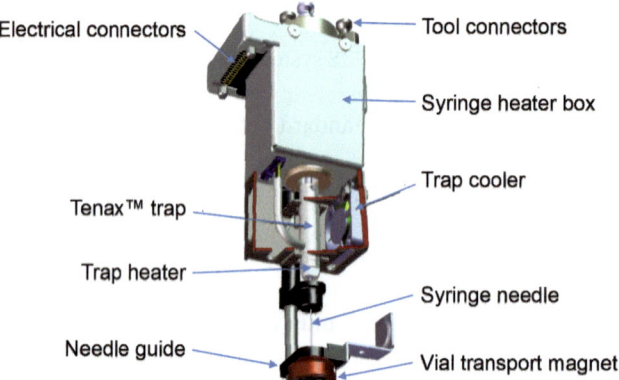

Figure 2.49 The in-tube dynamic headspace extraction (ITEX DHS) tool cutaway (Courtesy CTC Analytics AG).

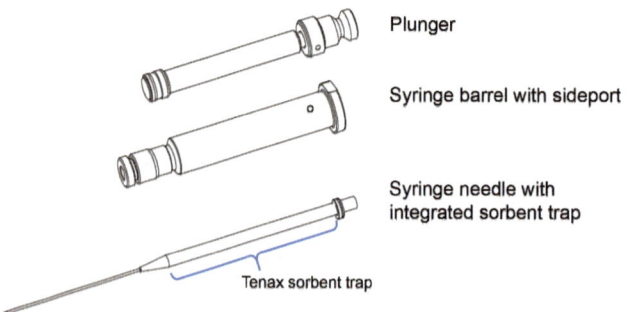

Figure 2.50 Components of the ITEX DHS syringe trap (Courtesy CTC Analytics AG).

et al., 2012; Zapata et al., 2012; Hu et al., 2017; Kaziur-Cegla et al., 2021). The outgassing of additives from plastic samples including phthalates, bisphenols, adipates, citrates, benzophenones, organophosphorus compounds, among others is reported with recoveries between 70 and 128% and quantitation limits below 0.1 µg/g for most of such compounds (Concha-Graña et al., 2024).

A recent development with the application of vacuum to the sample vial extends the extraction capabilities uniquely to low volatile substances not assessable by other techniques, the vacuum ITEX technology (vITEX) (Fuchsmann et al., 2019; Cretnik and Fuchsmann, 2023). The vITEX technique finds its use in nutrivolatilomics and metabolomics for lower volatile polar compounds (Fuchsmann et al., 2020; Meng et al., 2021; Bütikofer et al., 2022).

The well-known and most used sorbent materials Tenax TA and Tenax GR, as well as Carboxen, Carbopack, Carbosieve, or combinations of those, are available as trapping materials. An auxiliary purge gas flow is used via the side port of the syringe barrel for moisture removal or flushing the syringe. The GC injection by a quick thermal desorption of the analytical trap is a regular syringe injection and does not require any modifications or a transfer line on the GC side. The ITEX DHS syringe is a closed system, an invaluable benefit with a low risk of contamination. In the worst case, only a syringe needs to be cleaned and can be reused.

2.1.8 Adsorptive Enrichment and Thermal Desorption

In the analysis of air or outgas samples, an extremely large number of volatile and the so-called very volatile organic compounds (VOCs, VVOCs) of widely differing chemical classes must be considered over a wide concentration range. GC-MS is the method of choice for the determination of such highly volatile compounds. Usually, the concentration of the substances of interest is too low for the direct measurement of an air sample, and therefore enrichment on suitable sorbent materials is necessary. The concentration on solid adsorption material allows the accumulation of organic components from large volumes of gas. Typical areas of use include flavors of food, beverages and packaging, workplace air monitoring, gases from landfill sites, urban air pollution, or indoor air analysis. Compounds to be analyzed cover the wide range of GC amenable compounds with a boiling range up to typically C40. Most organic and permanent gases like O_2, O_3, CO, CO_2, SO_2, NO_2, and also methane cannot be accumulated. Exceptions hereto are H_2S, CS_2, N_2O, and SF_6 which are well absorbed.

Besides providing sufficient storage capacity with a sampling device, the risk of analyte breakthrough (BTV) needs to be considered. The BTV for compounds must be greater than the sampling volumes. The adsorption material is expected to have a low affinity for water (air moisture). This is not only important for GC-MS analysis. Neither water nor CO_2 should have a negative effect on the BTV of the organic components (Knobloch and Engewald, 1995; Cosnier et al., 2006). Ambient air generally has a high moisture content. Particular precautions in this respect must be taken in the case of combustion gases. Finally for analysis, the desorption of the enriched components from the carrier materials should be complete and without thermal decomposition.

78 | *2 Fundamentals*

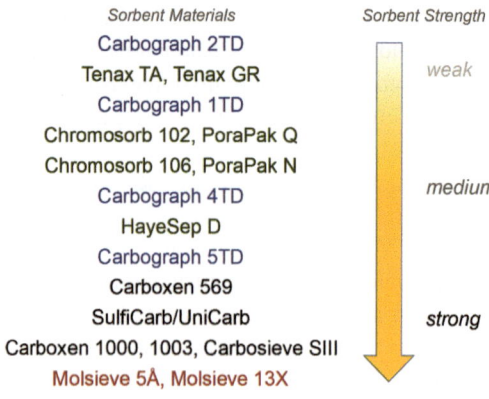

Figure 2.51 Sorbent material strength (Markes International Ltd.). Blue = Graphitized carbon blacks Green = Porous polymers Black = Carbonized molecular sieves Red = Zeolite molecular sieves

The selection of the most suitable sorbent material or a combination of adsorbents for sampling and release of the analytes is of the highest importance when developing a TD method (Figure 2.51). The choice of sorbent principally depends on the vapor pressure of the analyte. The more volatile the analyte to be trapped, the stronger the sorbent must be. As a rule of thumb, the BP of the analytes can be used as a guideline. The sorbent must quantitatively retain the compounds of interest from the volume of air sampled. For the desorption process, the sorbent material must release those compounds efficiently and must provide high thermal stability not to contribute to the background of the analysis. The expected adsorption and stability properties should not change for a large number of analyses, even with repeated use of the sampling tube (Brown and Shirey, 2001; Markes, 2012a). In air analysis, adsorption materials, such as Tenax, graphitized carbon blacks, molecular sieves, and XAD resins, are of general use (see Figure 2.52 and Table 2.14) (Sigma-Aldrich,

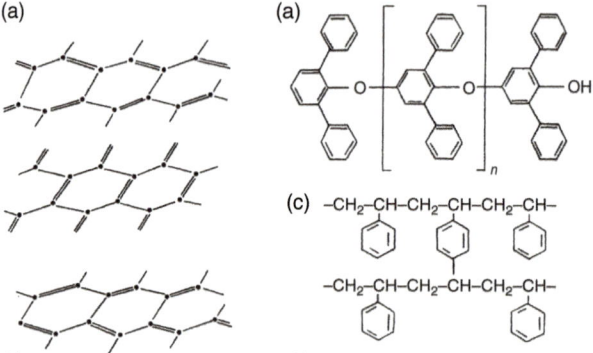

Figure 2.52 Surface model for common sorbent materials (Supelco): (a) Carbotrap, surface area about 100 m^2/g, uniform charge distribution over all carbon atom centers, (b) Tenax (2,6-diphenyl-p-phenylene oxide), surface area about 24 m^2/g, non-uniform charge distribution, the charge is essentially localized on the oxygen atoms. Tenax TA has replaced former Tenax GC, a new material of higher purity; Tenax GR is a graphitized modification of higher density. (c) Amberlite XAD-2, surface area about 300 m^2/g, non-uniform charge distribution, less polar than Tenax (XAD-4, about 800 m^2/g).

2014). Activated charcoal is an outstanding adsorbent for many organic compounds; however, charcoal can only be sufficiently desorbed using liquid solvents (Krebs et al., 1991). Complete TD of organic analytes from charcoal requires extremely high temperatures (> 600 °C). This will lead to pyrolytic decomposition of the organic compounds and is not used in TD practice.

Thermally stable materials offering a wide range of sorbent strengths have been developed over the last 50 years for use in TD applications (Brown, 2013). These include standard carbon blacks (Carbographs™ and Carbopacks™/Carbotraps™), carbonized molecular sieves such as the Carboxen™ series and modified carbon blacks (Carbograph 5 TD and Carbopack X) which offer much of the strength of carbonized molecular sieves but with minimal water retention. All of them are in widespread use as sorbents in TD tubes (Supelco, 1986, 1988; Betz et al., 1989).

Graphitized carbon black materials (GCBs), Carboxen and Carbosieve™ materials show a low affinity for water, like Tenax. These adsorption materials can be dry purged with a pure gas stream, e.g. the GC carrier gas, in the direction of adsorption, without significant loss of analytes (Gawlowski and Gierczak, 2000).

VOCARB™ traps are combinations of the adsorption materials GCB and carbonized molecular sieve. These combinations have been optimized in the form of VOCARB 3000 and VOCARB 4000 for volatile and less volatile compounds respectively, corresponding to the U.S. EPA methods 624/1624 and 542.2 (Supelco, 1992). VOCARB 4000 provides higher absorptivity for less volatile components, such as naphthalenes and trichlorobenzenes. However, it shows catalytic activity toward 2-chloroethyl vinyl ether (complete degradation!), 2,2-dichloropropane, bromoform, and methyl bromide.

Volatile polar components are enriched on highly polar absorbent materials. The combination of Tenax TA and silica gel has proved particularly successful for the enrichment of polar compounds. Water absorption by silica gel needs to be taken care of in the application for air analysis.

To cover a wider range of molecular sizes, sorbent materials are combined in multibed arrangements of increasing sorbent strength. For example, the Carbotrap multibed adsorption tube (Figure 2.53) consists of three materials: Carbopack C is used to enrich high molecular mass components, such as alkylbenzenes, PAHs or PCBs, as the first trap in the sampling direction.

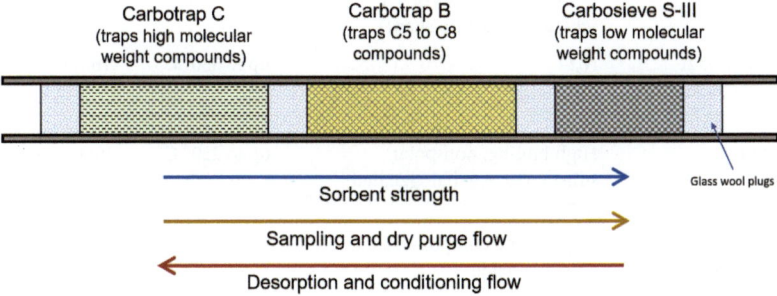

Figure 2.53 Multibed adsorption/desorption tube Carbotrap 300 (after Supelco).

Table 2.14 Adsorption materials and application areas of frequent use.

Sorbent material (type)	Most frequent use	Conditioning	Desorption
Carbosieve SIII (carbonized molecular sieve)	Light hydrocarbons to C5 Ethane Ethylene Fumigants PH_3, SO_2F_2 and CH_3Br Light hydrocarbons, BP – 60 °C to 80 °C	up to 380 °C	up to 360 °C
Carbograph, various (graphitized carbon black)	Light hydrocarbons Ketones, alcohols, aldehydes Non-polar components Perfluorocarbon tracer Alkylbenzenes Hydrocarbons to C20 SVOCs	up to 380 °C	up to 360 °C
Carbotrap, various (graphitized carbon black)	For boiling points 15 to 120 °C C_1/C_2-chlorinated hydrocarbons Volatile halogenated hydrocarbons BTEX Styrene	up to 380 °C	up to 360 °C
Carboxen, various (carbonized molecular sieve)	VOCs, e.g. vinyl chloride Permanent gases Ultralight hydrocarbons, BP – 60 °C to 80 °C	up to 380 °C	up to 360 °C
Chromosorb (porous polymer)	Benzene Volatile hydrocarbons, BP 50 °C to 120 °C Oxygenated compounds	up to 220 °C	up to 200 °C
HayeSep D (porous polymer)	Volatile hydrocarbons, C5 to C12, BP 50 °C to 120 °C GB/GE derivatives of VX (chemical warfare agent)	up to 280 °C	up to 260 °C
Molecular sieve 5 Å (zeolite molecular sieve)	N_2O	up to 350 °C	165 °C
Porapak (porous polymer)	High boiling, non-polar substances, e.g. halogenated narcotics	up to 220 °C	up to 190 °C
SulfiCarb/UniCarb (carbonized molecular sieve)	Oxygenated compounds, e.g. ethylene oxide	up to 380 °C	up to 360 °C

Table 2.14 (Continued)

Sorbent material (type)	Most frequent use	Conditioning	Desorption
Tenax TA, Tenax GR	For analytes BP 80–200 °C	up to 330 °C	up to 320 °C
	C_1/C_2-chlorinated hydrocarbons		
	Volatile halogenated hydrocarbons		
	BTEX		
	Aromatics		
	Non-polar compounds, BP > 100 °C		
	Polar compounds, BP >150 °C		
	Phenols		
	SVOCs incl. chemical warfare agents		
XAD-4 (crosslinked DVB)	C_1/C_2-chlorinated hydrocarbons	up to 200 °C	up to 150 °C
	Freon R11		
	Halogenated narcotics		
	Vinyl chloride		
	Ethylene oxide		
	Styrene		

(Brown and Shirey, 2001; Camsco, 2009; Markes, 2012a).

The more volatile substances pass through to the subsequent layers. VOCs are excellently adsorbed on the next Carbotrap layer. Smaller C_2-hydrocarbons are trapped efficiently on Carbosieve S-III material at the end of the adsorption tube. Carbon molecular sieves have lower water retention than regular molecular sieves allowing them to be used in atmospheres with high moisture contents. Many regulated methods recommend the application of certain approved adsorbent materials. A selection of internationally used methods is listed in Table 2.15.

2.1.8.1 Sample Collection

Sample collection can be either passive or active. For passive collection, diffusion tubes with special dimensions are used. The concentration of substances in the surrounding air is integrated, taking the collection time into account. Active collection devices require a calibrated pump with which a predetermined volume is drawn through the adsorption tube. Having estimated the expected concentrations, for example, for indoor air 100 mL/min and for outdoor air 1000 mL/min, the air is drawn through the prepared adsorption tube over a period of 4 h. After sample collection, the adsorption tube must be closed tightly to exclude additional uncontrolled contamination. The storage performance of thermodesorption sampling tubes was tested to be stable over several weeks using Tenax TA and other sorbent materials under different storage conditions (Brown et al., 2014).

Sampling on thermodesorption tubes needs to consider the BTV as a function of temperature, humidity, and flow (Manura, 1995; Brown and Purnell, 1979; Brown

Table 2.15 Thermal desorption tube selection by regulated method.

Method	Thermal desorption tube in use
ASTM D6196	Carbotrap™ 100, Carbopack™ B, Chromosorb® 106
U.S. EPA TO-1	Tenax® TA
U.S. EPA TO-2	Carbosieve™ SIII
U.S. EPA TO-14	Air toxics
U.S. EPA TO-17	Carbotrap 217, Carbotrap 300, Carbotrap 317
U.S. EPA 1P-1B	Tenax TA, Carbotrap 349
U.S. EPA 0030, U.S. EPA 0031, U.S. EPA SW-846	VOST stack sampling tubes: Tenax TA (35/60 mesh), Tenax TA: Petroleum Charcoal (2:1)
MDHS 72	Chromosorb 106
NIOSH 2549	Carbotrap 349

(Sigma-Aldrich, 2024/Merck KGaA).

et al., 2014). Brown and Purnell carried out thorough investigations on the determination of BTVs. The latter generally vary widely with the collection rate. On use of Tenax as the sorbent the ideal collection rate is 50 mL/min, in any case. However, performance remains optimum provided flow rates remain in the range of 20 to 200 mL/min (i.e. BTVs remain constant). Flow rates up to 500 mL/min can also be used for short-duration sampling (up to 15 min). Note that pump flow rates below 5 or 10 mL/min are subject to unacceptable error due to diffusive ingress and should not be used. Moisture does not affect the BTVs with Tenax unlike other porous materials. Furthermore, sample collection is greatly affected by temperature. An increase in temperature increases the BTV (about every 10 K doubles the volume) (Brown and Purnell, 1979; Figge et al., 1987; Manura, 1995).

2.1.8.2 Calibration

For calibration of TD tubes, the same conditions should predominate as in sample collection. Methods such as the liquid application of a calibration solution to the adsorption materials or the comparison with direct injections have been shown to be unsatisfactory. Alternatively, certified reference standards (CRSs) are available (e.g. Markes Int., UK). CRS tubes are recommended in many key standard methods (e.g. U.S. EPA Method TO-17) for auditing purposes and as a means of establishing analytical quality control. CRS tubes are often certified traceable to primary standards and have a minimum shelf life of typically six months. They are available ready for use with concentrations varying from 10 ng to 100 µg per component. Chromatograms from a shipping blank, and an example analysis of a CRS tube should be supplied along with the CRS certificate. CRS tubes are available loaded with benzene, toluene and xylene (BTX) at levels of 100 ng or 1 µg per component; TO-17 standards at 25 ng per component of benzene, toluene, xylene, dichloromethane, 1,1,1-trichloroethane,

1,2,4-trimethylbenzene, MTBE, butanol, ethyl acetate and methylethyl ketone. Custom CRS tubes are also available on a variety of sorbents for a wide range of compounds from different vendors.

Customized calibration tubes can be prepared by using standard atmospheres. The process for the preparation of standard atmospheres by continuous injection into a regulated air stream has been described in standardized guidelines (Figure 2.54) (Kommission Reinhaltung der Luft im VDI, 1981; Health and Safety Laboratory, 1990). In this process, an individual compound, or a mixture of the substances to be determined is continuously charged to an injector through which air is passed (complementary gas), using a thermostatted syringe burette. The air quantity (up to 500 mL/min) is adjusted using a mass flow meter. The complementary gas can be diluted by mixing with a second air stream (dilution gas up to 10 L/min). Moistening the gas can be carried out inside or outside the apparatus. In this way, concentrations in the ppm range can be generated. For further dilution, for example, for calibration of pollution measurements, a separate dilution stage is necessary. The gas samples to be tested are drawn out of the calibration station into a glass tube with several outlets. In this case, active or passive sample collection is possible. Continuous injection has the advantage that the preparation of mixtures is very flexible (Tables 2.16 and 2.17). However, accurate standard atmospheres are difficult to generate and maintain. For this reason, many modern TD standards specify calibration of TD methods by the introduction of small volumes of gas or liquid phase standards onto the sampling end of TD tubes in a stream of inert carrier gas. For example, see EN ISO 16017 or ASTM D6196.

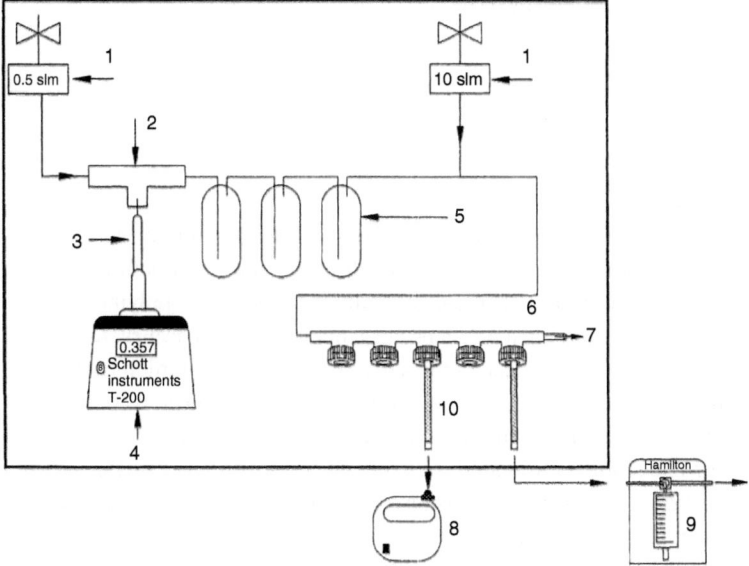

Figure 2.54 Principle of introducing measured volumes of standard atmosphere onto thermal desorption tubes (Tschickard, 1993).

Table 2.16 Evaluation of test tubes which were prepared using the calibration unit by continuous injection (n = 10).

Component	Mean value	Standard deviation	Relative standard deviation [%]
1,1,1-Trichlorethane	562.6	4.95	0.88
Dichloromethane	538.6	7.93	1.47
Benzene	753.3	9.48	1.25
Trichloroethylene	627.3	16.7	2.67
Chloroform	626.6	12.1	1.94
Tetrachloroethylene	698.2	6.11	0.88
Toluene	1074.0	6.88	0.64
Ethylbenzene	358.1	2.91	0.81
p-Xylene	736.3	4.85	0.66
m-Xylene	731.6	4.77	0.65
Styrene	755.8	4.60	0.61
o-Xylene	389.5	3.49	0.89

Table 2.17 Analysis results of BTEX determination of certified samples after calibration with the calibration unit.

Mass [µg]	Benzene	Toluene	m-Xylene
Measured value	1.071	1.136	1.042
Required value (certified)	1.053	1.125	1.043
Standard deviation (certified)	0.014	0.015	0.015

2.1.8.3 Desorption

The elution of the organic compounds collected involves extraction by a solvent (displacement) or TD. Pentane, CS_2, and benzyl alcohol are used as extraction solvents. CS_2 is very suitable for activated charcoal, but cannot be used with polymeric materials, such as Tenax or Amberlite XAD-2, because decomposition occurs. As a result of displacement with solvents, the sample is extensively diluted, which can lead to problems with the detection limits on mass spectrometric detection. With solvents, additional contamination can occur. The extracts are usually applied as solutions for liquid injection. The readily automated static headspace technique can also be used for sample injection. This procedure has also proved to be effective for desorption using polar solvents, such as benzyl alcohol or ethylene glycol monophenyl ether using a 1% solution in water (Krebs *et al.*, 1991).

The thermal desorption has become a routine procedure because of program-controlled samplers. The concentrated volatile components are released by rapid heating of the adsorption tube and after preliminary focusing, are injected into the GC-MS system for analysis. Automated TD gives better sensitivity, precision, and accuracy in the analysis. For the sequential processing of a large number of samples, autosamplers with capacities of up to 100 adsorption tubes are commercially available. Through frequent reuse of the adsorption tube and complete elimination of solvents from the analysis procedure, a significant lowering of cost per sample is achieved.

Before and after sample collection, the adsorption tubes need to be hermetically sealed. This can be achieved with long-term storage caps comprising 2-piece, 1/4-inch metal compression fittings and combined polytetrafluoroethylene (PTFE) fittings. These have been evaluated for long-term storage by independent reference laboratories and found to ensure sample stability for many months (even years) at room temperature. Sorbent tubes also must be kept sealed throughout automated TD analysis – both to preserve the integrity of sampled tubes and to prevent the ingress of contaminants onto desorbed/analyzed tubes. During automated TD, tubes are fitted with temporary Teflon caps. Sampling tubes also can be sealed by use of the DiffLok™ end-cap technique at either end of the tube (Figure 2.55). Diffusion-locking is a patented technology that keeps the sample tubes sealed at ambient pressures but allows gas to flow through the tubes whenever pressure is applied (Woolfenden and Cole, 2003). Diffusion-locking does not involve any kind of valve or other moving parts and is thus inherently simple and robust. With the inlet and outlet tube of a sampler being sufficiently narrow and long, the process of diffusion of vapors into or out of an attached sorbent tube can be reduced close to zero. The sealing of the tubes protects the sample from ingress or loss of volatiles at all stages of the monitoring process (Markes, 2012b).

Tubes using check valves are not suitable for diffusive monitoring. In case Teflon caps are used, they are removed from the instrument before measurement and the adsorption tube is inserted into the desorption oven. Before desorption, the tightness of the seal to the instrument is tested as the initial step of the automated desorption process by monitoring an appropriate carrier gas pressure for a short time. In the desorption oven carrier gas is passed through the adsorption tubes in the reverse direction to the adsorption flow at temperatures of up to 400 °C. The components

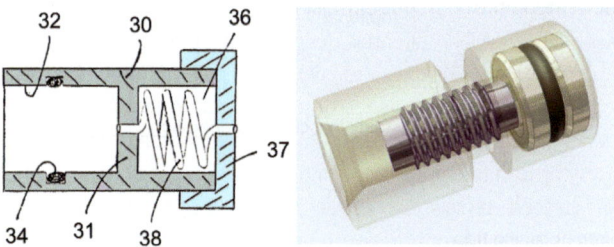

Figure 2.55 Schematic of a DiffLok analytical end-cap for TD sample tubes (Markes, 2003). (30) Cap body, (31) wall, (32) socket, (34) O-ring seal to fit over a sample tube, (36) hollow compartment, (37) closure, and (38) wound capillary tube.

released are collected in a cold trap inside the apparatus. The intermediately trapped sample then is flushed to the GC column by rapidly heating the cold trap. This two-stage desorption and the use of the multiple split technique enable the measuring range to be adapted to a wide range of concentrations. The sample quantity can be adjusted to the capacity of the capillary column used through suitable split ratios both before and after the internal cold trap (Figure 2.56).

The TD process is usually carried out automatically. Until recently, TD was invariably a one-shot process with no sample remaining after desorption to repeat the process if anything went wrong. However, with the development by Markes International SecureTD-Q™ technology it has become possible to re-collect the split portion of the sample quantitatively and routinely for repeat analysis. This overcomes the historical one-shot limitation of TD methods and considerably simplifies method and data validation.

For high throughput applications, an electronic tube tagging using radio frequency identification (RFID) tube tags is available for industry standard sorbent tubes and 4.5″ DAAMS (Depot Area Air Monitoring System) tubes.

Possible and, for certain carrier materials, known decomposition reactions have been reported. For this reason, another sampling method is favored by the U.S. EPA for air analysis. Electropolished and with a nickel chromium oxide (NiCrOx) layer passivated canisters, commonly called SUMMA canister, of 2 L or

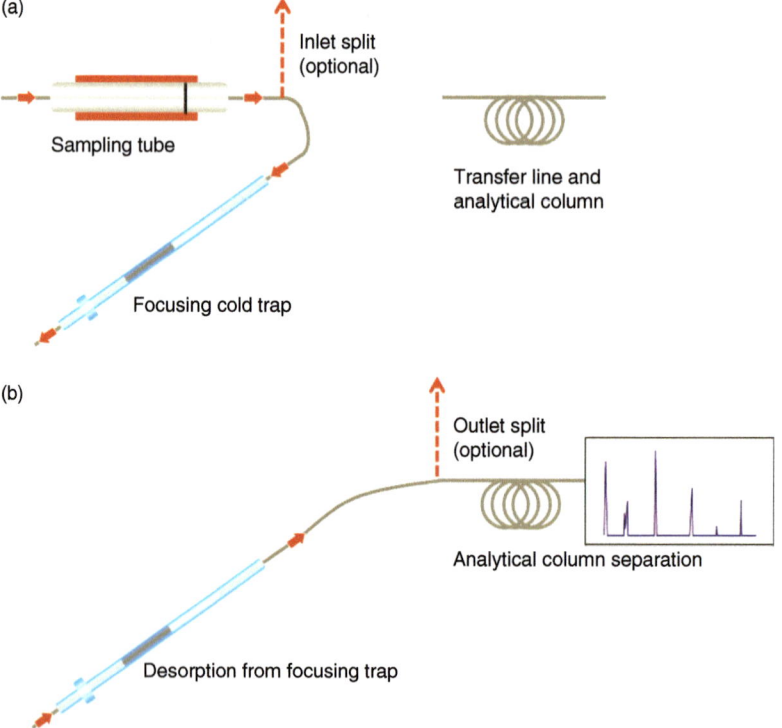

Figure 2.56 Multiple split technique for thermal desorption (Markes).

6 L capacity, maximum of up to 15 L are evacuated for sample collection. Due to risk of degradation of the NiCrOx layer with the release of reactive iron surfaces Silonite™ ceramic coated canisters are recommended instead (Merit, 2018).

The whole air samples collected onsite on opening the canister can be measured several times in the laboratory following U.S. EPA methods such as TO-14 or TO-15, see Section 4.1. Suitable samplers are used, which are connected online with GC-MS. Cryofocusing is used to concentrate the analytes from the volumes collected. If required, the sample can be dried with a semipermeable membrane (Nafion™ drier) or by condensation of the water. Adsorption materials are not used in these processes.

2.1.9 Stir Bar Sorptive Extraction

The principle of SPME found its extension in the SBSE technique developed by Pat Sandra and team for aqueous samples (Baltussen *et al.*, 1999; David and Sandra, 2007) and became soon commercialized as "Twister™" by the German company GERSTEL GmbH (Figure 2.57). In 2000, one year after SBSE was developed, sorptive extraction was applied to headspace analysis coining the name of headspace sorptive extraction (HSSE) (Bicchi *et al.*, 2000; Sandra *et al.*, 2000; Tienpont *et al.*, 2000).

The SBSE/HSSE extraction is performed solventless with a glass-covered, approx. 10 to 40 mm long and 0.5 mm OD magnetic stir bar, which is coated with a layer of approx. 0.5 mm PDMS or other extraction phases (Figure 2.58). After removing the stir bar with tweezers from the sample it is rinsed with water and dried. A thermal or liquid desorption follows for GC-MS or LC-MS analysis (Hoffmann *et al.*, 2000). A fully automated SBSE workflow, called "AutoTwister™", is available with seamless automation for online analysis of large sample series. The x,y,z-robot inserts the PDMS covered AutoTwister rod into a septum-capped sample vial, rinses the rod after the extraction phase in a dedicated wash and dry station and then transfers the loaded SBSE rod to the TD inlet for GC-MS analysis (Gerstel, 2024).

The main difference between SPME and SBSE is the much higher volume and surface of the PDMS sorbent phase resulting in high recoveries (Bicchi *et al.*, 2009). While with the SPME fiber an extraction phase volume of about 15 µL is used, with SBSE a significantly enlarged volume of up to 125 µL is available. The larger volume of sorption phase provides a better phase ratio with increased recovery resulting in up to 250-fold lowered detection limits (Majedi and Lee, 2017). Many samples can be extracted at the same time using a separate multi-position magnetic stir plate.

Figure 2.57 Twister stir bar (Courtesy of GERSTEL GmbH).

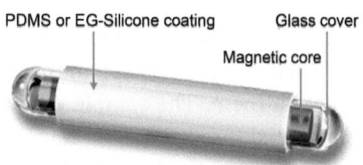

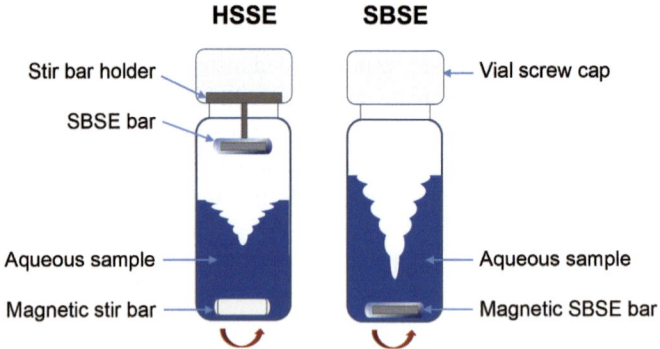

Figure 2.58 Liquid (SBSE) and headspace (HSSE) application of the magnetic stir bar.

Extraction times are significantly higher than SPME with typically up to 60 min and more to exploit the full capacity. The analysis by GC is achieved by transfer into a TD unit being part of an automated x,y,z-robotic injection and a temperature programmable injection unit for cold trapping the analytes and sharp injection upon flash heating. Thermolabile compounds can be alternatively dissolved in small amounts of GC or LC amenable solvents (SBSE/LD) (Serodio and Nogueira, 2004). The excellent sensitivity and reproducibility have been demonstrated with low level multi-residue detection of, for instance, endocrine-disrupting chemicals in drinking water (Canli et al., 2020).

2.1.10 Pyrolysis

The use of pyrolysis with GC-MS instrumentation (Py-GC-MS) extends the application to non-volatile samples that cannot be analyzed directly by GC because they cannot be desorbed from a matrix or evaporated without decomposition (Wampler, 1995; Brodda et al., 1993). In analytical pyrolysis, a large quantity of energy is passed in the shortest time into a sample so that typical degradation products are formed which can be gas chromatographed reproducibly. The pyrolysis reaction initially involves thermal cleavage of C–C bonds, for example in the case of polymers. Thermally induced chemical reactions within the pyrolysis product are undesired side reactions and can be prevented by reaching the pyrolysis temperature as rapidly as possible (Ericsson, 1980). The reactions initiated by the pyrolysis are temperature-dependent. To produce a reproducible and quantifiable mixture of pyrolysis products, the heating rates, and the pyrolysis temperatures in particular, should be kept constant (Ericsson, 1985). The sample and its composition can be characterized using the resulting chromatogram (pyrogram) and by the mass spectroscopic identification of individual pyrolysis products (Irwin, 1982).

The use of pyrolysis apparatus with GC-MS systems imposes particular requirements on them. The sample quantity applied must correspond to the capacity of commercial fused silica capillary columns. It is usually in the microgram range or less (Hancox et al., 1991). By selecting a suitable split ratio, peak equivalents for the

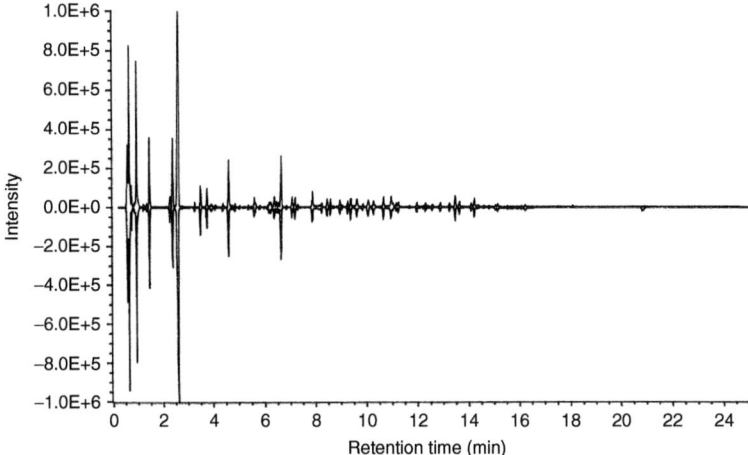

Figure 2.59 Reproducibility of the pyrolysis of an automotive paint on two consecutive days using a foil pyrolizer coupled to GC-MS, mirrored representation (Steger, Audi AG).

pyrolysis products can be achieved in the mid-nanogram range. These lie well within the range of modern GC-MS systems. The small sample quantities are also favorable for analytical pyrolysis in another aspect. The reproducibility of the procedure increases as the sample quantity is lowered as a more rapid heat transport through the sample is possible (Figure 2.59). Side reactions in the sample itself and as a result of reactive pyrolysis products are increasingly eliminated.

Because of the small sample quantities and the high reproducibility of the results, analytical pyrolysis has experienced a renaissance in recent years. Both in the analysis of polymers with regard to quality, composition, and stability (Ericsson, 1978, 1990), and in the areas of environmental analysis, micro- and nanoplastics, foodstuffs analysis and forensic science, pyrolysis has become an important analytical tool, the significance of which has been increased immensely by coupling with GC-MS (Klusmeier et al., 1988; Galletti and Reeves, 1991; Zaikin et al., 1992; Matney et al., 1994; Picó and Barceló, 2020).

Analytical pyrolysis is available using different processes: high frequency pyrolysis (Curie point pyrolysis), furnace, and foil pyrolysis. The processes differ principally through the different means of energy input and the different temperature rise times (TRTs). Pyrolysis instruments can easily be connected to GC and GC-MS systems. The reactors currently used are constructed so they can be placed on top of GC injectors (split operation) and can be installed for short-term use only if required. Also, automated pyrolysis units are available providing high sample throughput (Fischer and Kusch, 1993).

There are various possibilities for the evaluation of pyrograms obtained with a GC-MS system. With classical FID (flame-ionization detector) detection, the pyrogram pattern is compared only with known standards. However, with GC-MS systems, the mass spectra of the individual pyrolysis products can be evaluated. By using mass spectra libraries substances and substance groups can be identified.

GC-MS pyrograms can be selectively investigated for trace components even in complex separating situations by using the characteristic mass fragments of minor components. GC-MS programs, with its full pattern information, can be stored and compared for identity or similarity. Suitable software systems are commercially available providing a numerical measure on similarity instead of only a visual inspection, using, for instance, ChromSync by CI Informatics Ltd. Didcot, UK; MSChromSearch, Axel Semrau GmbH, Sprockhövel, Germany, or GC-LC Concordance, Spectrochrom, Bouc-Bel-Air, France.

In the library search for the spectra of pyrolysis products, special care must be taken. Commercial spectral libraries consist of spectra of particular substances that have been fully characterized, which is not the case for the majority of pyrolysis products. Classical MS fragmentation rules apply for the interpretation of search results (see Section 3.2). Depending on the sample material, however, an extremely large number of reaction products are formed on pyrolysis, which cannot be completely separated even under the best GC conditions. If the detection is sensitive enough, this situation is shown clearly by the mass chromatogram. Also, for polymers, there is a typical appearance of homologous fragments, which must also be taken into account. These series can also be shown easily using mass chromatograms through the choice of suitable fragment ions.

Quantitative determinations using pyrolysis benefit particularly from the selectivity of GC-MS detection. The precision is comparable to that of liquid injections.

Analytical Pyrolysis Procedures

Procedure	Foil	High frequency	Micro-furnace
Carrier	Pt foil	Fe/Ni alloys	Cup
Pyrolysis temperature	Variable, resistively heated up to 1400 °C	Fixed, alloy Curie temperatures up to 1040 °C	Variable, resistively heated furnace up to 1000 °C
TRT	<8 ms	up to 200 ms	up to 3 °C/s

2.1.10.1 Foil Pyrolysis

In foil pyrolysis, the sample is applied to a thin platinum foil (Figure 2.60). The thermal mass of this device is very low. After the direct application of a heating current, any desired temperature up to about 1400 °C can be achieved within milliseconds. The extremely high heating rate results in high reproducibility. The temperature of the Pt foil can be controlled by its resistance. However, the temperature can be measured and controlled more precisely and rapidly from the radiation emitted by the Pt foil. An exact calibration of the pyrolysis temperature can be carried out and the course of the pyrolysis is recorded by this feedback. Besides endothermic pyrolysis, exothermic processes can also be detected and recorded.

Figure 2.60 Scheme of a Pt foil pyrolizer with temperature control by means of fiber optic cable (Pyrola). (a) View of the Pt foil with carrier gas and current inlets. (b) Side view of the pyrolysis cell with the glass cell (about 2 mL volume, Pyrex) and photodiode.

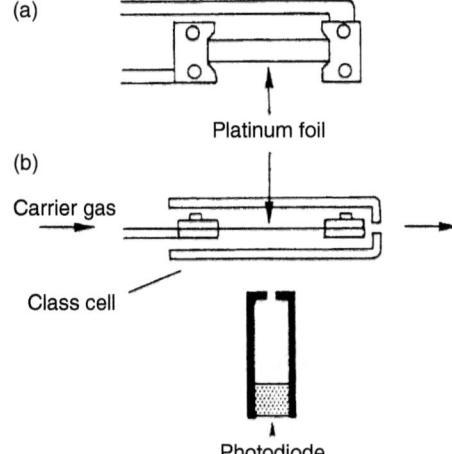

Pyrolysis Nomenclature

Uden (1983) and Ericsson and Lattimer (1989)

- *Pyrolysis*: A chemical degradation reaction initiated by thermal energy alone.
- *Oxidative pyrolysis*: A pyrolysis which is carried out in an oxidative atmosphere (e.g. O_2).
- *Pyrolysate*: The total products of a pyrolysis.
- *Analytical pyrolysis*: The characterization of materials or of a chemical process by instrumental analysis of the pyrolysate.
- *Applied pyrolysis*: The production of commercially usable materials by pyrolysis.
- *Temperature/time profile* (TTP): The graph of temperature against time for an individual pyrolysis experiment.
- *Temperature rise time* (TRT): The time required by a pyrolizer to reach the pre-set pyrolysis temperature from the start time.
- *Flash pyrolysis*: A pyrolysis which is carried out with a short TRT to achieve a constant final temperature.
- *Continuous pyrolizer*: A pyrolizer where the sample is placed in a preheated reactor.
- *Pulse pyrolizer*: A pyrolizer where the sample is placed in a cold reactor and then rapidly heated.
- *Foil pyrolizer*: A pyrolizer where the sample is applied to metal foil or band which is directly heated as a result of its resistance.
- *Curie point pyrolizer*: A pyrolizer with a ferromagnetic sample carrier which is heated inductively to its Curie point.
- *Temperature-programmed pyrolysis*: A pyrolysis where the sample is heated at a controlled rate over a range of temperatures at which pyrolysis occurs.

(Continued)

(Continued)

- *Sequential pyrolysis*: Pyrolysis where the sample is repeatedly pyrolyzed for a short time under identical conditions (kinetic studies).
- *Fractionated pyrolysis*: A pyrolysis where the sample is pyrolyzed sequentially under different conditions in order to investigate different sample fractions.
- *Pyrogram*: The chromatogram (GC, GC-MS) or spectrum (MS) of a pyrolysate.

2.1.10.2 Curie Point Pyrolysis

High frequency pyrolysis uses the known property of ferromagnetic alloys of losing their magnetism spontaneously above the Curie temperature (Curie point). At this temperature, many properties change, such as the electrical resistance or the specific heat. Above the Curie temperature, ferromagnetic substances exhibit paramagnetic properties. The possibility of reaching a defined and constant temperature using the Curie point was first introduced by Simon and Giacobbo (1965). In a high frequency field, a ferromagnetic alloy does not absorb any more energy above its Curie point and remains at this temperature. As the Curie temperature is alloy-dependent, another Curie temperature can be used by changing the alloy composition of the sample carrier. If a sample is applied to a ferromagnetic material and is heated in an energy-rich, high frequency field (Figure 2.61), the pyrolysis takes place at the temperature determined by the choice of alloy and thus its Curie point (Schulten

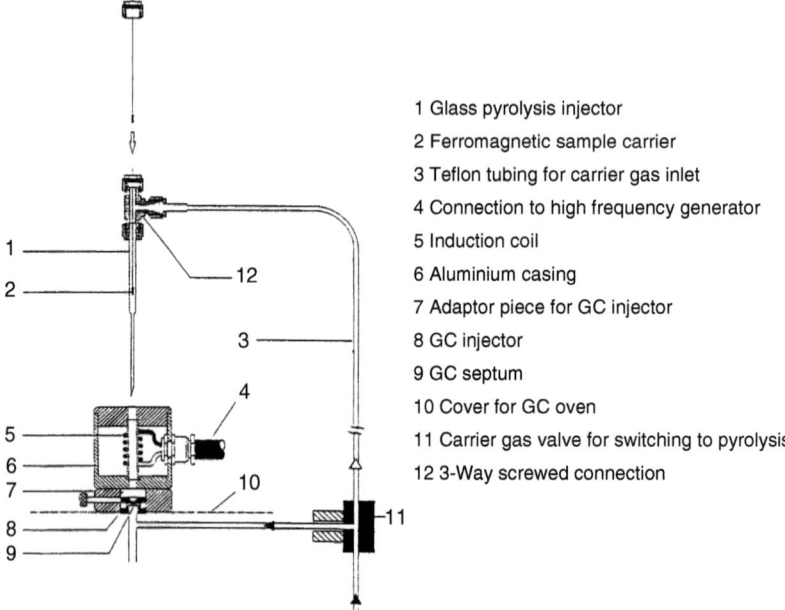

1 Glass pyrolysis injector
2 Ferromagnetic sample carrier
3 Teflon tubing for carrier gas inlet
4 Connection to high frequency generator
5 Induction coil
6 Aluminium casing
7 Adaptor piece for GC injector
8 GC injector
9 GC septum
10 Cover for GC oven
11 Carrier gas valve for switching to pyrolysis
12 3-Way screwed connection

Figure 2.61 GC injector for Curie point pyrolysis (Fischer and Kusch, 1993).

Figure 2.62 Curie temperatures and temperature/time profiles of various ferromagnetic materials (Simon and Giacobbo, 1965/John Wiley & Sons).

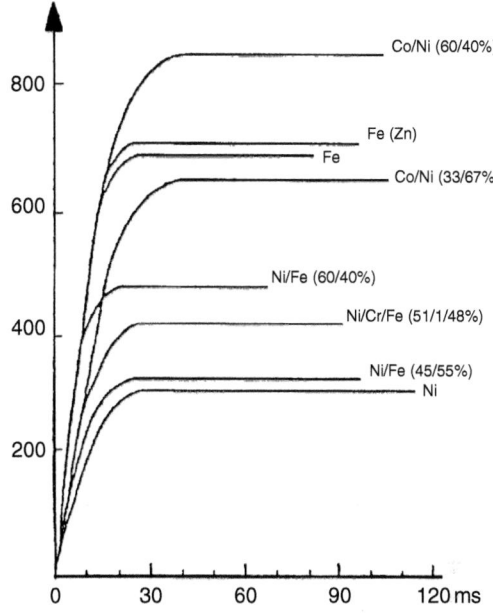

et al., 1987; Dworzanski, 1999; Oguri, 2005). The temperature rise profiles for various metals and alloys are shown in Figure 2.62.

In practice, sample carriers in the form of loops, coils or simple wires made of different alloys at fixed temperature intervals are used. Ferromagnetic metal foils (Pyrofoil™) are available for 21 different pyrolysis temperatures ranging from 160 °C to 1040 °C (JAI Ltd., Tokyo, Japan). In this case, the sample gets crimped into the metal foil using a dedicated tool before analysis (Oguri, 2005).

A potential disadvantage of Curie point pyrolysis is the longer TRT compared with the direct resistively heated foil pyrolysis. The temperature rise time of up to 200 ms to reach the Curie point is significantly slower and depends on the device used. There are also effects due to the not completely inert surface of the sample carrier. They can manifest themselves in the inadequate reproducibility of the pyrolysis process for analytical purposes. Copolymers with thermally reactive functional groups (e.g. free OH, NH_2 or COOH groups) cannot be analyzed by Curie point pyrolysis.

For the analytical assessment of the coupling of pyrolyzer with GC and GC-MS systems, a high boiling mixture of cholesterol with n-alkanes has been proposed (Gassiot-Matas and Julia-Danés, 1976). Figure 2.63 shows the evaporation of cholesterol (500 ng) with the C_{34} and C_{36}-n-alkanes (50 ng of each). Before the intact cholesterol is detected, its dehydration product appears. The intensity of this peak increases with small sample quantities and increasing temperatures in the region of substance transfer in the GC injector. The high peak symmetries of the signals of the alkanes, the cholesterol and its dehydration product indicate that the GC coupling is working well.

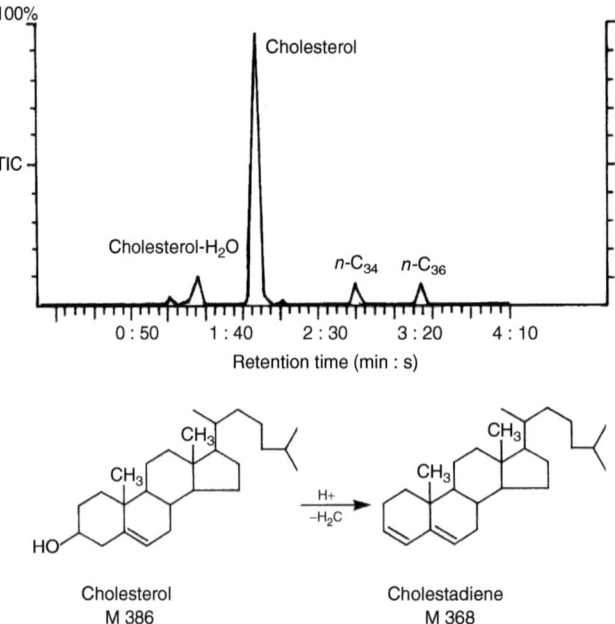

Figure 2.63 Analysis of cholesterol and n-alkanes (C34, C36) for testing the pyrolysis coupling to the GC injector (Richards, 1988). Conditions: GC column 4 m × 0.22 mm ID × 0.25 µm CP-Sil5, 200 to 320 °C, 30 °C/min, GC-ITD.

2.1.10.3 Micro-Furnace Pyrolysis

Another very flexible solution for numerous applications in pyrolysis, outgassing or thermal extraction experiments is the use of a micro-furnace, which can be resistively heated to temperatures above 1000 °C (Roussis and Fedora, 1996). The sample is placed in a small, low, thermal mass cup of 50 to 80 µL volume, which is dropped into the heated pyrolysis furnace consisting of a vertical quartz tube. The furnace tube size is comparable to a GC inlet liner with 4.5 mm inner diameter and 12 mm length and allows programmable temperature profiles with heating rates of up to 200 °C/min (Frontier Lab, 2023).

Typical sample sizes are in the 0.1 mg range, allowing a fast heat transfer to the set desorption or pyrolysis temperature. The transfer of the pyrolysis products is achieved by a carrier gas flow through the heated zone and a needle tip on the bottom of the furnace reaching into the injector of the GC. Desorption of pesticides from samples of mulch films used in agriculture is a recent very descriptive example (Sahai et al., 2024).

Besides the flash pyrolysis of solid samples, the furnace solution allows different sampling devices for a wider bandwidth of analytical experiments (see Figure 2.64). This solution can provide access to additional analytical applications with the introduction of liquid samples using a regular micro syringe, the online pyrolysis for the analysis of high pressure reactions in glass capsules, the TD of small amounts of solid materials, application in microplastic analysis, or the combination with a subsequent second reactor for the reaction with catalyst materials. Unique is the online

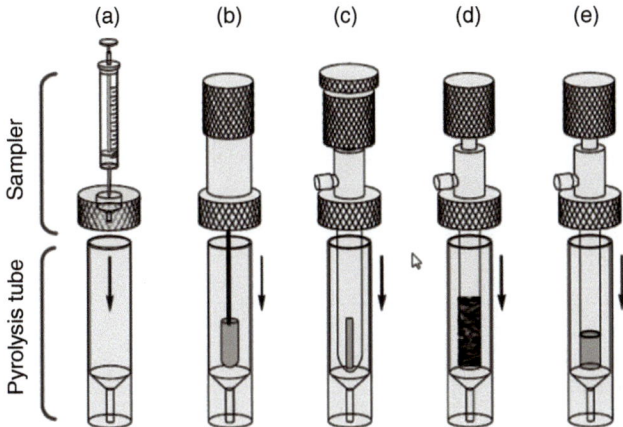

Figure 2.64 Five micro-furnace pyrolysis tubes for different pyrolysis and thermal desorption applications (Frontier Laboratories Ltd): (a) liquid sampler for direct liquid sample injection using a micro syringe, (b) single shot sampler for flash pyrolysis.

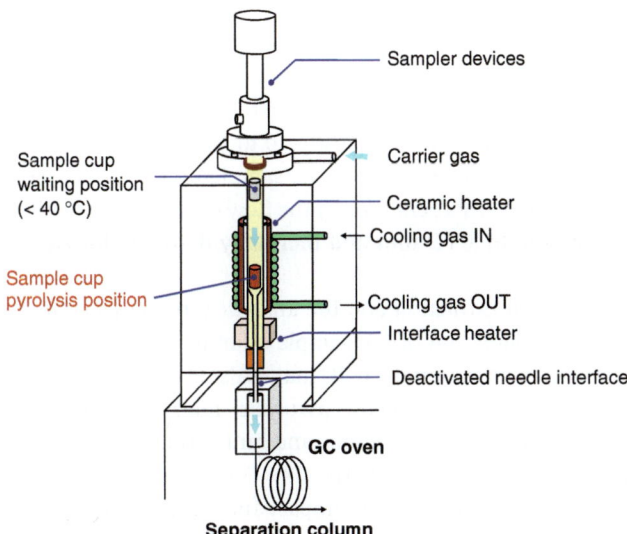

Figure 2.65 Schematic diagram of the micro furnace pyrolizer EGA/PY-3030D (Frontier Laboratories Ltd.).

UV radiation of samples for the analysis of photo, thermal and oxidative degradation products (Figure 2.65).

2.1.11 Thermal Extraction (Outgassing)

Thermal extraction of a sample is performed in a device with a temperature-controlled chamber, which can be operated at a constant temperature for material emissions testing Figure 2.66. Extraction chambers for outgassing are commercially

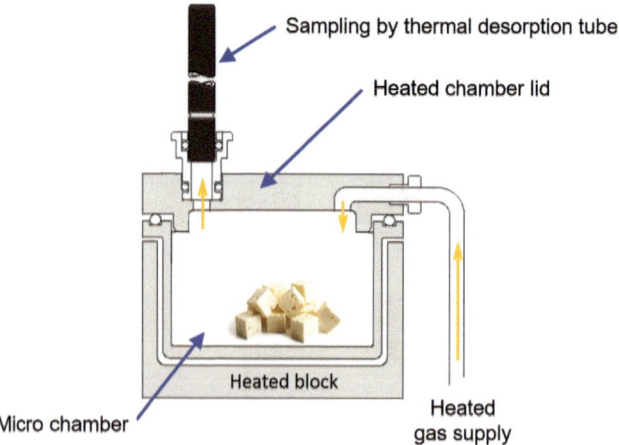

Figure 2.66 Micro-chamber/thermal extractor schematic (Moving Moment/Adobe Stock Photos).

available in different dimensions to accommodate even large sample sizes (Markes, 2015). This allows both the thermal extraction of VOCs and SVOCs from liquid and solid samples and even technical components (U.S. EPA, 1996; Markes, 2012). Preconditioned Tenax or multi-sorbent tubes are attached to the constantly heated desorption chamber. A controlled air flow of typically 10 to 500 mL/min, or inert gas (He, N_2), is passed through. Vapors emitted by the sample under the selected conditions are swept onto an attached sorbent tube by the flow of gas. After volatiles collection, the tube is transferred to a TD unit and thermally desorbed for GC or GC-MS analysis.

The difference between thermal extraction and the analytical pyrolysis systems described earlier lies in the consideration of the sample quantity and the applied temperatures. Even inhomogeneous materials, bulk materials or complete assemblies can be investigated at temperatures below pyrolysis, releasing embedded volatile components. The applications are numerous and range from food produce (Figure 2.67) to semiconductor components (Figure 2.68). They cover, for instance, the emissions from materials that can adversely impact indoor or in-vehicle air quality. National and international reference methods for regulations such as the European Construction Products Regulation, the German protocol for fire-resistant floorings (AgBB, Ausschuss zur gesundheitlichen Bewertung von Bauprodukten) and the Californian CHPS protocol for public school building programs (CHPS) specify the determination of emissions from construction materials using conventional test chambers operated at ambient temperature (Methods ISO/EN 16000-6/-9/-10/-11; ASTM D5116-97; ASTM D7143-05; etc.) in order to certify product performance. However, in this case, smaller micro-chamber/thermal extractor devices provide a useful and complementary quick screening method for factory product control, inhouse comparative checks and tests on raw materials, etc. (ASTM D7706 and ISO 12219-3). Direct thermal extraction is used to measure SVOCs in a wide range of solid, resinous, and liquid materials, and eliminates

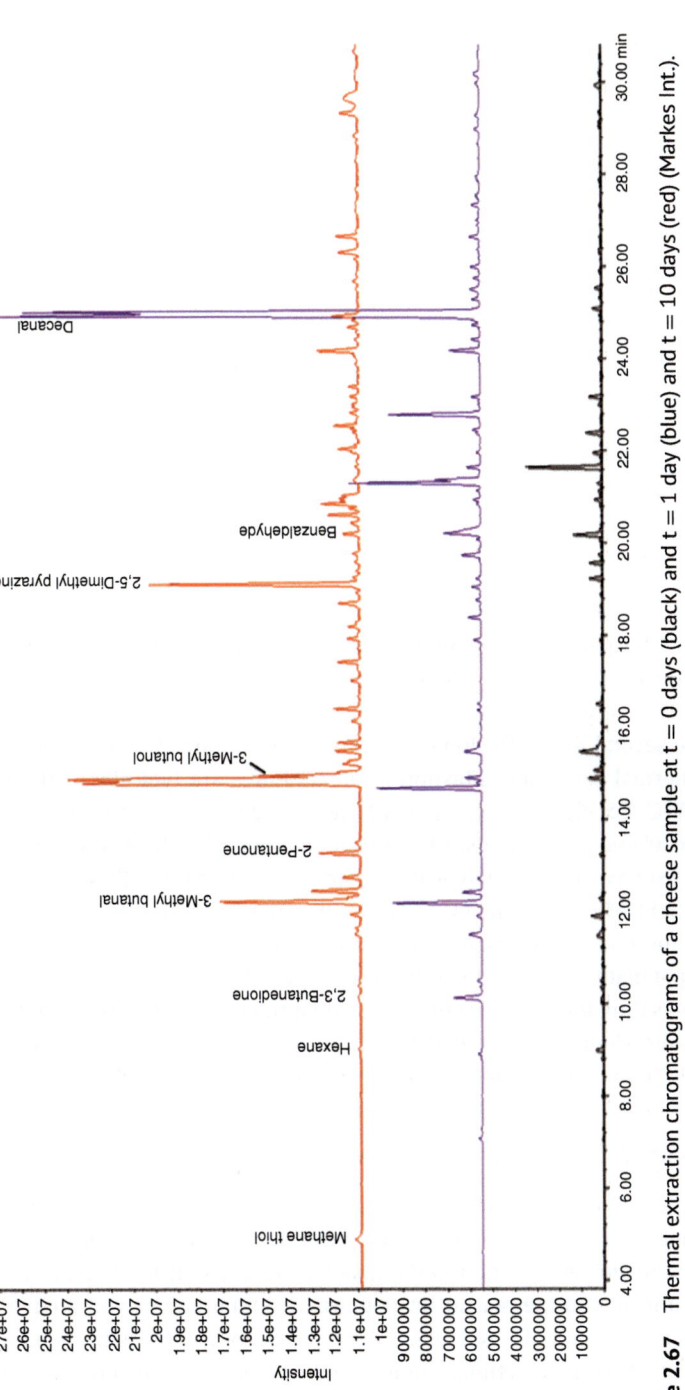

Figure 2.67 Thermal extraction chromatograms of a cheese sample at t = 0 days (black) and t = 1 day (blue) and t = 10 days (red) (Markes Int.).

98 | 2 Fundamentals

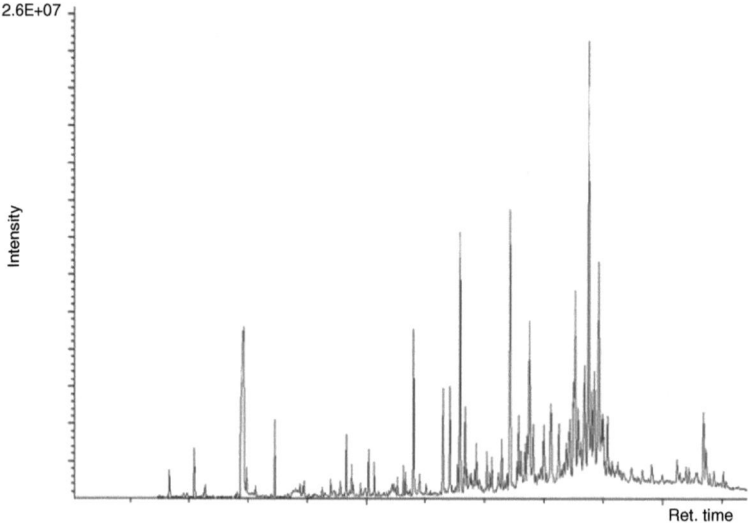

Figure 2.68 Bulk emissions from a sample of printed circuit board (including electronic components) tested at 120 °C. (Markes Int.).

complex liquid extraction steps. It is also used by the paint industry (U.S. EPA Method 311 for paints) for evaluating "low VOC" products and the German automotive industry for testing car interior components (Method VDA 278 2001), see Section 4.21). Outgassing devices are commercially available in different chamber sizes up to several liters of volume and capacity from one to multiple channels for parallel thermal extraction. Maximum temperatures are typically in the range of 200 to 400 °C. For electronic components, as such devices can range in geometry to receive electronic components, silicon wafers or complete hard discs, provided by companies such as Japan Analytical Industry Ltd., Japan (JAI, 2024).

In the field of environmental analysis, thermal extraction is proposed by the U.S. EPA method for the quantitative analysis of semi-volatile compounds from solid sample materials. The U.S. EPA method 8275 is a thermal extraction capillary GC-MS procedure for the rapid and quantitative determination of targeted PCBs and PAHs in soils, sludges, and solid wastes. This method requires extraction temperatures of 340 °C for 3 min for the quantitative desorption of the PCBs (U.S. EPA, 1996).

2.1.12 Liquid Chromatography Clean-up

The necessary effort of sample preparation in trace analysis is driving cost and is limiting productivity not only in the GC-MS laboratory. Especially, a series of manual steps for extraction, evaporation, and transfer are responsible for long analysis times, high cost and is also the reason for potential contaminations and analyte losses. Compared to a manual LLE and SPE, the sample clean-up of analytes by liquid chromatography (HPLC) offers thousand-fold higher separation potential. It is a logical step to make use of this high potential for modern clean-up concepts for GAC with automated sample preparation in mind.

The hyphenation of LC with GC combines the potential for high sample load and selectivity of LC with the sensitivity and versatility of GC-MS. For long a high sensitivity detection in LC was missing. Also, mass spectral libraries were built up from electron impact (EI) mass spectra via solids probe inlets and later mainly via GC-MS. GC was and still is often recognized to be the first choice for small molecule analysis because of its speed of analysis, separation efficiency, and due to the wide range of selective and sensitive detectors. Coupling normal phase LC as a clean-up online with the GC separation offers a hitherto unmatched potential for a significantly improved matrix clean-up, reducing manual sample handling by a full automation, and the reduction of the cost per sample by less solvent and consumables requirements in routine analyses (Serrano et al., 1998; Hogendoorn et al., 1989).

The potential of online coupling of GPC and HPLC as the sample preparation clean-up procedure for GC was first described by Majors with the example of atrazine in sorghum samples (Majors, 1980). The instrumental setup is built by an additional LC system, normal phase micro HPLC separation column and a switching valve connected to the GC inlet as outlined in Figure 2.69 (Munari et al., 1985; Trestianu et al., 1996). An online LC detector is necessary during method development with spiked samples for the determination of the fraction switching times for GC transfer. A first automated system was described for the analysis of pesticides in biological matrices (Ramsteiner, 1987). In 1989, Carlo Erba Strumentatione (Italy) introduced the first commercially available LC-GC instrument "Dualchrom 3000" (Munari and Grob, 1990). The GC is usually configured with a short megabore pre-column (*aka*

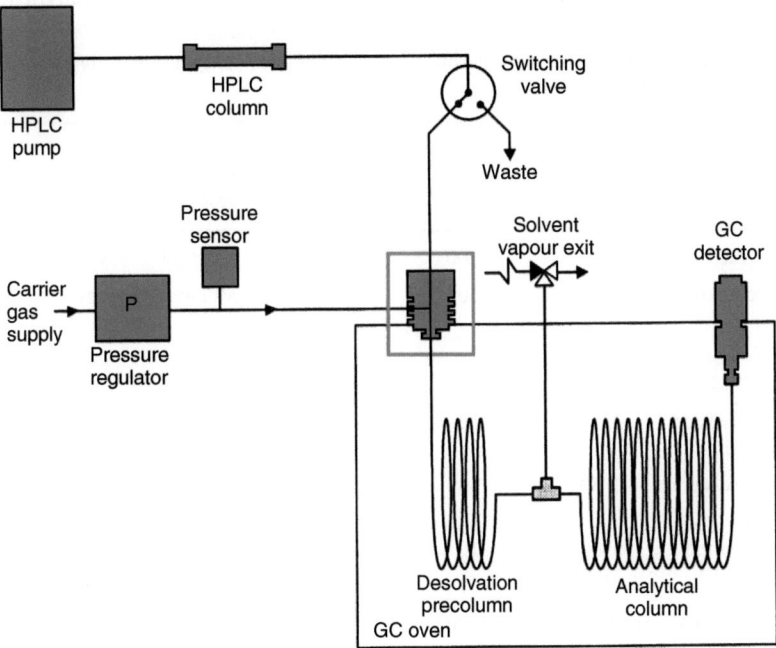

Figure 2.69 Schematic of an on-line LC clean-up GC system (Trestianu et al., 1996).

retention gap, or guard column), connected via a T-piece to the analytical column. A short widebore column leads from the T-piece to a heated solvent vapor exit valve (SVE) shown in Figure 2.105. Several interface solutions to transfer LC fractions into a GC capillary column have been developed since. The PTV, a loop-type interface, or on-column injectors were used (Dugo et al., 2003). A typical six-way transfer valve connection with LC monitor detector, GC carrier gas supply, and the connected pre-column is illustrated in Figure 2.106 in Section 2.2.7.

Commercial LC-GC solutions are established for routine analysis in many laboratories covering a wide range of applications for example for pesticides, PCBs, dioxins, or plasticizers (Munari and Grob, 1990; Hyvönen et al., 1992; Jongenotter et al., 1999; Che et al., 2014). A specific advantage for trace analyses is that the concentrated eluates allow for lower detection limits and improved reproducibility. High extract dilutions causing losses or additional contaminations with the subsequent need for evaporation as known from Soxhlet or SPE extractions are avoided.

2.1.13 Pressurized Liquid Extraction

The trend to fully automated instrumental extraction methods has also reached the time- and solvent-consuming Soxhlet extraction process. The comparison of modern extraction methods such as microwave assisted extraction (MAE), supercritical fluid extraction (SFE), SPME and pressurized liquid extraction (PLE) with the traditional Soxhlet method, reports that PLE has found widespread application due to many of its advantageous features. The most practical advantages are speed, versatility, and reproducibility (Table 2.18). There are both strong analytical

Table 2.18 Comparison of extraction methods for solid samples soil, sediment, fly ash, sludge, and solid wastes.

Extraction method	U.S. EPA method	Avg. volume of solvent [mL]	Avg. extraction time	Instrument cost	Operating cost/sample
Soxhlet extraction	3540C	300	16–24 h	Very low	Very high
Automated Soxhlet extraction	3541	50	2 h	Moderate	Low to moderate
Pressurized liquid extraction (PLE)	3545A	10–30	10–15 min	High	Low
Microwave assisted extraction (MAE)	3546	25–40	10–20 min	Moderate	Low
UltraSonic extraction (USE)	3550C	300	~30 min	Low	High
Supercritical fluid extraction (SFE)	3560/3561/3562	10	20–50 min	Moderate to high	Moderate to high

(Oleszek-Kudlak et al., 2007/John Wiley & Sons).

and economic advantages with PLE: The high extraction efficiency, savings in solvent consumption (and waste generation) and the short extraction times led to a steadily growing number of applications (Richter et al., 1996; Oleszek-Kudlak et al., 2007; Kettle, 2013; Barp et al., 2023). PLE also limits glassware handling, facilitates a safe laboratory working environment, and has become a popular green extraction technology for different classes of organic contaminants, extracted from numerous matrices in environmental, food, feed, biological, and technical samples.

PLE is the general term for the automated technique of extracting solid and semisolid samples with liquid, organic or aqueous solvents used in the official U.S. EPA method 3445A (U.S. EPA, 2007). The PLE method is also marketed under the term *Accelerated Solvent Extraction (ASE)* by former Dionex Corp. (today part of Thermo Fisher Scientific, USA), covered by a number of international patents (Richter et al., 1996; ASE, 2012). Automated PLE instrumentation allows the independent control of temperature and pressure conditions for each sample cell. This control is critical for a high analyte recovery and reproducibility.

PLE, also known as *pressurized fluid extraction* (PFE), or *pressurized solvent extraction* (PSE), is used for extracting the analytes of interest from solid, semisolid or liquid samples using an organic solvent at elevated temperatures up to 200 °C (390 F) and pressures up to 100 bar (1500 psi) in a combination of static and dynamic extraction steps. The high pressure increases the boiling temperature of the solvent and decreases its viscosity, thereby allowing deeper penetration into the matrix and faster extractions to be conducted at higher than ambient temperatures. The benefit of higher temperature extraction is primarily speed. Traditional techniques, such as Soxhlet, can take 24 to 48 h. With PLE, analyte recoveries equivalent to those obtained using traditional extraction methods can be achieved in only 15 to 30 min. Although PLE uses the same aqueous and organic solvents as traditional extraction methods, it uses them more efficiently with less volume. A typical PLE extraction covers a range from 1 to 100 g of sample. Typically, a 10 g sample is used for trace analysis methods and is using 50 to 150 mL of solvent.

Approximately 75% of all PLE extractions can be completed in less than 20 min using the standard extraction conditions (100 °C, 1500 psi). But, PLE extractions are matrix dependent and require further optimization when applied to different matrices. Method development starts with standard conditions and evaluates both the recovery in the first step and the result of a second extraction of the given new sample. Also, the initial application of a standard matrix, for example, sea sand is a helpful step. If these initial parameters do not provide the recoveries desired, the temperature is increased to improve the efficiency of the extraction. Adding a primary static cycle, increasing static time and selecting a different solvent are additional variables that can be used to optimize a method.

At the typical PLE conditions, the solvent characteristics change compared to the traditional Soxhlet technique and allow an improved separation of target analytes from the matrix (Hubert et al., 2001). For instance, the extraction of

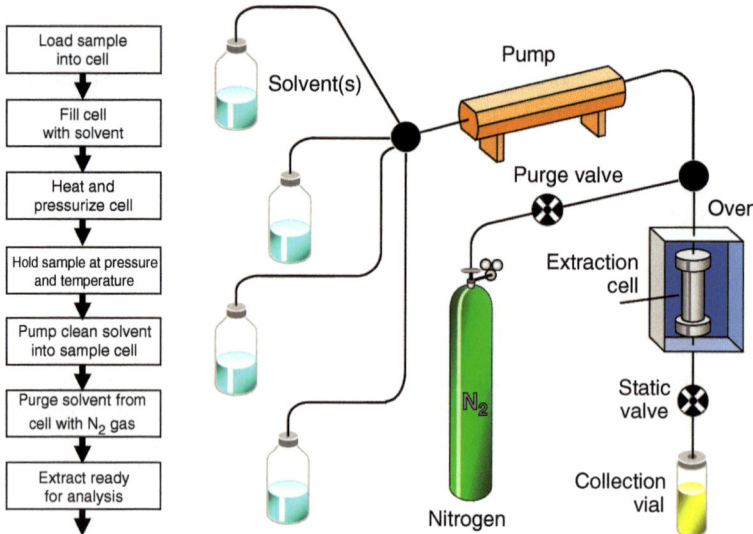

Figure 2.70 Schematics of a pressurized extraction unit (Courtesy of Dionex Corporation).

relatively polar phenols from soil samples is possible using the non-polar solvent n-hexane. PLE extractions typically use the known organic solvents n-hexane, cyclohexane, toluene, dichloromethane, methanol, acetone as applied in Soxhlet methods. Using PLE, the sample is completely and constantly surrounded by solvent; thus, oxidative losses are minimized if the extraction solvent is degassed and oxygen is excluded. Thermal degradation is not reported during PLE extractions. Since the analytes are in the heated zone for only short periods of time, thermal losses do not occur if appropriate temperatures are used for the extractions.

For biological samples and for the extraction of water-soluble components, water-based media in mixtures with organic solvents are applied (Curren and King, 2002). This variation gives access to polar compounds for subsequent LC-MS analysis. Due to a mostly sharp matrix/analyte separation by PLE, the collected extracts require less time-consuming clean-up typically employing SPE or GPC (Figure 2.70).

PLE extraction is internationally accepted by the regulatory bodies, among others by the U.S. EPA SW-846 Method 3545A, which can be used in place of Methods 3540, 3541, 3550 and 8151, by the U.S. Contract Laboratory Program (CLP) statement of work (SOW) OLM04.2, the Chinese method GB/T 19649-2005 and German Method L00.00-34 both for pesticides, also the ASTM Standard Practice D-7210 for additives in polymers and D-7567 for gel content of polyolefins. Method 3545A can be applied to the extraction of base/neutrals and acids (BNAs), chlorinated pesticides and herbicides, PCBs, organophosphorus pesticides, dioxins and furans and total petroleum hydrocarbons (TPHs).

> **Scope and Application of U.S. EPA Method 3545A (U.S. EPA, 2007)**
>
> Method 3545A is a procedure for extracting water-insoluble or slightly water-soluble organic compounds from soils, clays, sediments, sludges, and waste solids. The method uses elevated temperature (100–180 °C) and pressure (1500–2000 psi) to achieve analyte recoveries equivalent to those from Soxhlet extraction, using less solvent and taking significantly less time than the Soxhlet procedure. This procedure was developed and validated on a commercially available, automated extraction system.
>
> This method is applicable to the extraction of semi-volatile organic compounds, organophosphorus pesticides, organochlorine pesticides, chlorinated herbicides, PCBs and PCDDs/PCDFs, which may then be analyzed by a variety of chromatographic procedures.

> **Recommended PLE Extraction Conditions (U.S. EPA, 2007)**
>
> For semi-volatiles, organophosphorus pesticides, organochlorine pesticides, herbicides and PCBs:
>
> | | |
> |---|---|
> | Oven temperature | 100 °C |
> | Pressure | 1500–2000 psi |
> | Static time | 5 min (after 5 min pre-heat equilibration) |
> | Flush volume | 60% of the cell volume |
> | Nitrogen purge | 60 s at 150 psi (purge time may be extended for larger cells) |
> | Static cycles | 1 |
>
> For polychlorinated dioxins (PCDDs) and polychlorinated furans (PCDFs):
>
> | | |
> |---|---|
> | Oven temperature | 150–175 °C |
> | Pressure | 1500–2000 psi |
> | Static time | 5–10 min (after 5 min pre-heat equilibration) |
> | Flush volume | 60–75% of the cell volume |
> | Nitrogen purge | 60 s at 150 psi (purge time may be extended for larger cells) |
> | Static cycles | 2 or 3 |

2.1.13.1 In-Cell Clean-up

Selective PLE can be achieved with specific analyte or matrix-related procedures even derivatization steps (Sanchez-Prado et al., 2010) taking place during the extraction step, saving additional manipulation of the extract for increased productivity (Figure 2.71). Numerous PLE in-cell procedures are known for various applications

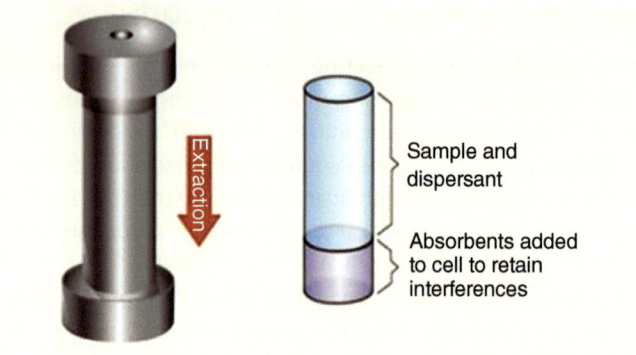

Figure 2.71 PLE in-cell sample preparation procedure with additional adsorbent layer (Thermo Fisher Scientific Inc.).

(Haglund and Spinnel, 2010) also covering trace analysis by GC-MS. The most typical ones with water removal and hydrocarbon matrix oxidation are highlighted in the following in more detail.

Sample drying can be accomplished in several ways such as air drying and oven drying before extraction. However, these approaches are not suited when analyzing volatile or semi-volatile components as they would be removed from the sample before extraction or analysis. The in-cell method can remove moisture when the wet sample is mixed with a water-absorbing polymer (Ullah et al., 2012).

Another common method for moisture removal is by using dehydrated salts such as sodium sulfate, calcium chloride, magnesium sulfate, calcium sulfate and the like. These salts tend to associate with water molecules to form hydrated salts. Of these sodium sulfate cannot be used. It tends to clump together when water is present and is not suitable for in-cell moisture removal and extraction in PLE. Sodium sulfate can dissolve in hot solvent to a certain extent and can precipitate downstream in some instances clogging the outlet frit, tubes, and valves. Moreover, sodium sulfate becomes an aggregate hard lump upon water absorption.

The application of a water-absorbent polymer is the recommended solution for water absorption from a moist sample under high ionic strength conditions. Results also demonstrate that when the polymer is mixed with diatomaceous earth (DE) absorbent, the water-removal efficiency increases significantly due to improved flow characteristics.

The amount of water-absorbing polymer needed is c. 0.20 g for absorbing 1 g of water at room temperature. In PLE, the water-absorbing ability of the polymer decreases with increasing temperature. For example, at 100 °C a 4 g polymer and 4 g DE can remove roughly 10 g of water completely. Hexane can then be used as the extraction solvent. Recoveries for nitro-, alkyl-, or chloro-substituted phenols range from 81% to 98%. New developments of improved water-absorbing polymers can remove up to 5 g of water per gram of the polymer at room temperature.

2.1.13.2 In-Cell Hydrocarbon Oxidation

The modular PLE approach was further developed with an integrated clean-up for dioxin analysis. A matrix retainer module filled with sulfuric acid impregnated silica gel was introduced. In this case, salmon muscle was mixed with sodium sulfate and the carbon trap. This procedure provided extracts clean enough to be analyzed by GC-HRMS (high resolution mass spectrometry) analysis after solvent evaporation to an appropriate final volume, see also Section 4.27 (Haglund and Spinnel, 2010).

The results obtained were in good agreement with those of a Soxhlet reference method. A slightly higher total dioxin equivalency concentration was obtained for the modular PLE (89.3 ± 8.3 pg/g fresh weight) than for Soxhlet extraction (81.3 ± 7.8 pg/g fresh weight), which indicate an improved extraction efficiency of the PLE method (Spinner, 2008).

A miniaturized PLE device has been designed and used for selective fast determination of the endogenous environmentally relevant PCB congeners in fatty foodstuffs (Ramos et al., 2007). Compared with conventional PLE procedures, the developed analytical method reduces sample amounts to about 100 mg, solvent consumption to 3.5 mL and allows complete sample preparation in a single step (reported 17 min per sample), which results in a significant reduction of the analysis cost. The use of SiO_2/H_2SO_4 has been found to be a distinct advantage as compared to previously reported selective PLE protocols, allowing an efficient preliminary fat removal. The miniaturized PLE device allows the quantitative extraction and purification of extracts for instance of PCBs in a single step, requiring lower solvent, sorbent and sample consumption and a reduced analysis cycle time. The achieved RSD was found in the range of 4 to 23% which is like those for reference methods.

References

Section 2.1 Sample Preparation

Bintanel-Cenis, J., Herrero, L., Gómara, B., and Ramos, L. (2023) Green fat removal in the analysis of polychlorinated biphenyls in biotic samples. *Adv. Sample Prep.*, **7** (May), 100081. doi: 10.1016/j.sampre.2023.100081.

da Silva Oliveira, W., Shepelev, I., Dias, F.F.G., and Reineccius, G.A. (2024) Advances in sample preparation for volatile profiling of plant proteins: fundamentals and future perspectives. *Adv. Sample Prep.*, **10** (March), 100111. doi: 10.1016/j.sampre.2024.100111.

Dheyab, A.S., Bakar, M.F.A., Alomar, M., Sabran, S.F., Hanafi, A.F.M., and Mohamad, A. (2021) Deep eutectic solvents (DESs) as green extraction media of beneficial bioactive phytochemicals. *Separations*, **8** (10), 1–24. doi: 10.3390/SEPARATIONS8100176.

Lim Sin Yee, G., and Zou, A. (2024) Automated Sample Preparation Using the PAL3 RTC System for EPA 8270E Semivolatile Organic Analysis by GC/TQ. Application Note 5994-7138EN, Agilent Technologies, Inc.

Majors, R.E. (2015) Overview of sample preparation. *LCGC*, **9** (1), 16–20.

Qiu, C. and Raynie, D.E. (2017) The use of extraction technologies in food safety studies. *LCGC N. Am.*, **35** (3), 662–669.

Ramos, L., Herreros-Martínez, J.M., Gómara, B., Pichon, V., and Psillakis, E. (2023) The potential of new solvents and materials in the context of green analytical chemistry. *Adv. Sample Prep.*, **8** (September), 1–2. doi: 10.1016/j.sampre.2023.100096.

Scida, K., Stege, P.W., Haby, G., Messina, G.A., and García, C.D. (2011) Recent applications of carbon-based nanomaterials in analytical chemistry: critical review. *Anal. Chim. Acta*, **691** (1–2), 6–17. doi: 10.1016/j.aca.2011.02.025.

Rodríguez-Delgado, M.Á., Socas-Rodríguez, B., and Herrera-Herrera, A.V. (eds) (2024) *Microextraction Techniques – Fundamentals, Applications and Recent Developments*, Springer, Cham. doi: 10.1007/978-3-031-50527-0.

Valcárcel, M., Cárdenas, S., and Lucena, R. (2014) Microextraction techniques. *Anal. Bioanal. Chem.*, **406**, 1999–2000. doi: 10.1007/s00216-013-7602-x.

Section 2.1.1 QuEChERS Sample Preparation

Anastassiades, M. Lehotay, S.J., and Stajnbaher, D. (2002) Quick, easy, cheap, effective, rugged and safe (QuEChERS) approach for the determination of pesticide residues Workshop in Rome, Italy, May 28–31, 2002.

Anastassiades, M., Lehotay, S.J., Stajnbaher, D., and Schenck, F.J. (2003) Fast and easy multiresidue method employing acetonitrile extraction/partitioning and "dispersive solid phase extraction" for the determination of pesticide residues in produce. *J. AOAC Int.*, **86**, 412.

Anastassiades, M., Scherbaum, E., Tasdelen, B., and Stajnbaher, D. (2007) Recent developments in QuEChERS methodology for pesticide multiresidue analysis, in *Pesticide Chemistry, Crop Protection, Public Health, Environmental Safety* (eds H. Ohkawa, H. Miyagawa, and P.W. Lee), Wiley-VCH Verlag GmbH & Co. KGaA.

Anastassiades, M. (2024) Home of the QuEChERS Method, *http://quechers.cvua-stuttgart.de* (accessed 8 March 2024).

AOAC (2023) AOAC Official Method 2007.01 – Pesticide residues in foods by acetonitrile extraction and partitioning with magnesium sulfate, in *Official Methods of Analysis of AOAC International*, 22nd edn (ed. G.W. Latimer Jr.), AOAC Publications, New York, 4 January 2023, https://doi.org/10.1093/9780197610145.001.0001 (accessed 2 October 2024).

Brondi, S.H.G., de Macedo, A.N., Vicente, G.H.L., and Nogueira, A.D.A. (2011) Evaluation of the QuEChERS method and gas chromatography – mass spectrometry for the analysis pesticide residues in water and sediment. *Bull. Environ. Contam. Toxicol.*, **86**, 18–22.

Chong, C.M. and Huebschmann, H.-J. (2021) Fully automated QuEChERS extraction and cleanup of organophosphate pesticides in Orange juice. *LCGC N. Am.*, **39** (April), 12–16.

Cunha, S.C., Lehotay, S.J., Mastovska, K., Fernandes, J.O., Beatriz, M., and Oliveira, P.P. (2007) Evaluation of the QuEChERS sample preparation approach for the analysis of pesticide residues in olives. *J. Sep. Sci.*, **30**, 620–632.

European Standard EN15662 (2008) *Foods of Plant Origin – Determination of Pesticide Residues Using GC-MS and/or LC-MS/MS Following Acetonitrile Extraction/Partitioning and Clean-up by Dispersive SPE -QuEChERS Method*, European Standard.

Hetmanski, M.T., Fussel, R., Godula, M., and Hübschmann, H.-J. (2010) Rapid Analysis of Pesticides in Difficult Matrices Using GC-MS/MS. Application Note 51880, Thermo Fisher Scientific.

Hildmann, F., Gottert, C., Frenzel, T., Kempe, G., and Speer, K. (2015) Pesticide residues in chicken eggs – a sample preparation methodology for analysis by gas and liquid chromatography/tandem mass spectrometry. *J. Chrom. A*, **1403**, 1–20. doi: 10.1016/j.chroma.2015.05.024.

Hoh, E. and Mastovska, K. (2008) Large volume injection techniques in capillary gas chromatography. *J. Chrom. A*, **1186**, 2–15. doi: 10.1016/j.chroma.2007.12.001.

Kaewsuya, P., Brewer, W.E., Wong, J., and Morgan, S.L. (2013) Automated QuEChERS tips for analysis of pesticide residues in fruits and vegetables by GC-MS. *J. Agri. Chem.*, **61**, 2299–2314.

Kowalski, J. and Cochran, J. (2012) QuEChERS: a primer. *Sep. Sci.*, **4** (12).

Krol, W., Eitzer, B., Arsenault, T., Incorvia Mattina, M.J., and White, J. (2014) Significant improvements in pesticide residue analysis in food using the QuEChERS method, *Chrom. Online*, 1–8.

Lehotay, S. (2007) Determination of pesticide residues in foods by acetonitrile extraction and partitioning with magnesium sulfate: collaborative study. *J. AOAC Intern.*, **90** (2), 485–520.

Lehotay, S.J. (2013) Revisiting the advantages of the QuEChERS approach to sample preparation, online presentation, webinar Separation Science, *www.sepscience.com* (accessed 14 May 2014).

Lehotay, S.J. (2022) The QuEChERSER mega-method. *LC-GC N. Am.*, **40** (1), 13–19.

Lehotay, S.J., de Kok, A., Hiemstra, M., and van Bodegraven, P. (2005) Validation of a fast and easy method for the determination of residues from 229 pesticides in fruits and vegetables using gas and liquid chromatography and mass spectrometric detection. *J. AOAC Int.*, **88** (2), 595–596.

Luetjohann, J., Zhang, L., Bammann, S., Kuballa, J., and Jantzen, E. (2009) Dioxins Go QuEChERS – a new approach in the analysis of well-known contaminants. Poster RAFA Prague, Czech Republic.

Mastovska, K. (2013) Secrets to successful GC-MS/MS pesticide method development. NACRW, Agilent Users Meeting, July 25, *https://www.agilent.com/cs/library/slidepresentation/public/NACRW_2013_Mastovska_Users_Meeting_Presentation.pdf*.

Michlig, N. and Lehotay, S.J. (2022) Evaluation of a septumless mini-cartridge for automated solid phase extraction cleanup in gas chromatographic analysis of more than 250 pesticides and environmental contaminants in fatty and nonfatty foods. *J. Chrom. A*, **1685**. doi: 10.1016/j.chroma.2022.463596.

Mol, H.G.J., Rooseboom, A., van Dam, R., Roding, M., Arondeus, K., and Sunarto, S. (2007) Modification and revalidation of the ethyl acetate-based multiresidue method for pesticides in produce. *Anal. Bioanal. Chem.*, **89** (6), 1715–1754.

Morris, B.D. and Schriner, R.B. (2015) Development of an automated column solid phase extraction cleanup of QuEChERS extracts, using a zirconia-based sorbent, for pesticide residue analyses by LC-MS/MS. *J. Agric. Food Chem.*, **63**, 5107–5119. doi: 10.1021/jf505539e.

Pihlström, T., Blomkvist, G., Friman, P., Pagard, U., and Österdahl, B.-G. (2007) Analysis of pesticide residues in fruit and vegetables with ethyl acetate extraction using gas and liquid chromatography with tandem mass spectrometric detection. *Anal. Bioanal. Chem.*, **389**, 1773–1789. doi: 10.100I-007-1425-6.

Rejczak, T. and Tuzimski, T. (2015) A review of recent developments and trends in the QuEChERS sample preparation approach. *Open Chem.*, **13**, 980–1010. doi: 10.1515/chem-2015-0109.

Schürmann, A., Crüzer, C., Duss, V., Kämpf, R., Preiswerk, T., and Huebschmann, H.-J. (2024) Automated micro solid phase extraction clean-up and gas chromatography-tandem mass spectrometry analysis of pesticides in foods extracted with ethyl acetate. *Anal. Bioanal. Chem.*, **416**, 689–700. doi: 10.1007/s00216-023-05027-5.

Settle, V., Foster, F., Roberts, P., Stone, P., Stevens, J., Wong, J., and Zhang, K. (2010) Automated QuEChERS Extraction for the Confirmation of Pesticide Residues in Foods Using LC-MS/MS. Application Note 4/2010, Gerstel GmbH & Co. KG, Mülheim.

Stahnke, H., Kittlaus, S., Kempe, G., and Alder, L. (2012) Reduction of matrix effects in liquid chromatography electrospray ionization-mass spectrometry by dilution of the sample extracts: how much dilution is needed? *Anal. Chem.*, **84**, 1474–1482.

Sun, J.-L., Liu, C., and Song, Y. (2017) QuEChERS extraction and analysis of veterinary drugs in products of animal origin. *The Column*, 18–21.

Teledyne Tekmar (2013) Pesticide Analysis Using the AutoMate-Q40: An Automated Solution to QuEChERS Extractions. Application Note April 2013, Teledyne Tekmar, Mason.

Usui, K., Hayashizaki, Y., Hashiyada, M., and Funayama, M. (2012) Rapid drug extraction from human whole blood using a modified QuEChERS extraction method. *Leg. Med.*, **14** (6), 286–296.

UTC (2011) *QuEChERS Pesticide Residue Analysis Manual*, UCT LLC, Bristol, PA, USA.

UCT (2013) *QuEChERS Informational Booklet Pesticide Residues Analysis*, UCT LLC, Bristol, PA, USA.

Section 2.1.2 Dispersive Liquid/Liquid Microextraction

Berijani, S., Assadi, Y., Anbia, M., Hosseini, M.-R.M., and Aghaee, E. (2006) Dispersive liquid–liquid microextraction combined with gas chromatography-flame photometric detection: very simple, rapid and sensitive method for the determination of organophosphorus pesticides in water. *J. Chrom. A*, **1123** (1), 1–9. doi: 10.1016/J.CHROMA.2006.05.010.

Guo, L. and Lee, H.K. (2014) Automated dispersive liquid-liquid microextraction-gas chromatography-mass spectrometry. *Anal. Chem.*, **86** (8), 3743–3749. doi: 10.1021/ac404088c.

Hutchinson, A. and Carrier, D. (2017) Fully Automated Method Using Dispersive Liquid Liquid Micro Extraction (DLLME) for Extractable and Leachable Studies. Technical Note AS182, Anatune Ltd., Cambridge, UK.

Mudiam, M.K.R., Ch, R., Chauhan, A., Manickam, N., Jain, R., and Murthy, R.C. (2012) Optimization of UA-DLLME by experimental design methodologies for the simultaneous determination of endosulfan and its metabolites in soil and urine samples by GC-MS. *Anal. Methods*, **4** (11), 3855–3863. doi: 10.1039/c2ay25432h.

Rai, S., Singh, A.K., Srivastava, A., Yadav, S., Siddiqui, M.H., and Mudiam, M.K.R. (2016) Comparative evaluation of QuEChERS method coupled to DLLME extraction for the analysis of multiresidue pesticides in vegetables and fruits by gas chromatography-mass spectrometry. *Food Anal. Methods*, **9** (4). doi: 10.1007/s12161-016-0445-2.

Rykowska, I., Ziemblińska, J., and Nowak, I. (2018) Modern approaches in dispersive liquid-liquid microextraction (DLLME) based on ionic liquids: a review. *J. Mol. Liq.*, **259** (June), 319–339. doi: 10.1016/j.molliq.2018.03.043.

Schettino, L., Bened, J.L., and Chisvert, A. (2023) Determination of nine prohibited N-nitrosamines in cosmetic products by vortex-assisted dispersive liquid-liquid microextraction prior to gas chromatography-mass spectrometry. *RSC Adv.*, **13**, 2963–2971. doi: 10.1039/d2ra06553c.

Shi, Z.-G. and Lee, H.K. (2010) Dispersive liquid-liquid microextraction coupled with dispersive µ-solid phase extraction for the fast determination of polycyclic aromatic hydrocarbons in environmental water samples. *Anal. Chem.*, **82** (4), 1540–1545. doi: 10.1021/ac9023632.

Rezaee, M., Assadi, Y., Hosseini, M.-R.M., Aghaee, E., Ahmadi, F., and Berijani, S. (2006) Determination of organic compounds in water using dispersive liquid-liquid microextraction. *J. Chrom. A*, **1116** (1–2), 1–9. doi: 10.1016/J.CHROMA.2006.03.007.

Section 2.1.3 Solid Phase Extraction

Bundt, J., Herbel, W., Steinhart, H., Franke, S., and Francke, W. (1991) Structure type separation of diesel fuels by solid phase extraction and identification of the two- and three-ring aromatics by capillary GC-mass spectrometry. *J. High Resolut. Chromatogr.*, **14**, 91–98.

Carrera, M.A., Manzano Sánchez, L., Murcia Morales, M., Fernández-Alba, A.R., and Hernando, M.D. (2023) Method optimisation for large scope pesticide multiresidue analysis in bee pollen: a pilot monitoring study. *Food Chem.*, **137652**. doi: 10.1016/j.foodchem.2023.137652.

Chong, C.M. and Huebschmann, H.-J. (2023) Fully automated QuEChERS extraction and cleanup of organophosphate pesticides in orange juice. *J. Nutr. Food Sci. Techn.*, **4** (2), 1–8, https://unisciencepub.com/journal-of-nutrition-food-science-and-technology-articles-inpress.

de Koning, S., Van Lieshout, M., Janssen, H.-G., and Van Egmond, W. (1998) A VISION to on-line SPE-PTV-GC-MS determination of organic micro pollutants in surface water, in *Proceedings of the 20th International Symposium on Capillary Chromatography, May 26–29, 1998, Riva Del Garda, Italy*, pp. 1–10.

Gamble, K.R. (2005) Sample Collection and Processing Device, U.S. Patent 6,959,615 B2, issued November 1, 2005.

Gerstel (2015) Online SPE with Replaceable Cartridges. Product Information, Gerstel GmbH, Mülheim an der Ruhr, Germany.

Korenková, E., Cepeda-Leucea, M., and MacPherson, K. (2017) The Determination of 1,4-Dioxane in Water by Automated SPE Gas Chromatography-High Resolution Mass Spectrometry (GC-HRMS). LaSB Approved Method, Ministry of the Environment and Climate Change - E3534.

Lehmann, S. et al. (2019) Organ distribution of diclazepam, pyrazolam and 3-fluorophenmetrazine. *Forensic Sci. Int.*, **303** (October), 109959. doi: 10.1016/j.forsciint.2019.109959.

Lehotay, S.J., Han, L., and Sapozhnikova, Y. (2016) Automated mini-column solid phase extraction cleanup for high throughput analysis of chemical contaminants in foods by low pressure gas chromatography-tandem mass spectrometry. *Chromatographia*, **79**, 1113–1130. doi: 10.1007/s10337-016-3116-y.

Li, Y., George, E.J., Hansen, E.M., Thoma, J.J., Werner, M., and Smith, P.D. (2000) An Automated Online SPE-GC-MS System for Water Analysis, in *Proceedings of the NWQMC Monitoring Conference, 10. Austin, TX*, USA: National Water Quality Monitoring Council.

Manzano Sanchez, L., Jesus, F., Ferrer, C., Mar Gomez-Ramos, M., and Fernandez-Alba, A. (2023) Evaluation of automated clean-up for large scope pesticide multiresidue analysis by liquid chromatography coupled to mass spectrometry. *J. Chrom. A*, **1694**, 463906. doi: 10.1016/j.chroma.2023.463906.

Mejía-Carmona, K., Jordan-Sinisterra, M., and Lanças, F.M. (2019) Current trends in fully automated on-line analytical techniques for beverage analysis. *Beverages*, **5** (1). doi: 10.3390/beverages5010013.

Michlig, N. and Lehotay, S.J. (2022) Evaluation of a septumless mini-cartridge for automated solid phase extraction cleanup in gas chromatographic analysis of more than 250 pesticides and environmental contaminants in fatty and nonfatty foods. *J. Chrom. A*, **1685**. doi: 10.1016/j.chroma.2022.463596.

Morris, B.D., Schriner, R.B., Youngblood, R., and Gamble, K. (2014) Use of a robotic solid phase extraction clean-up of QuEChERS extracts to give improved matrix removal for pesticide residue analyses by GC-MS/MS and LC-MS/MS. Presentation at the North American Chemical Residue Workshop.

Morris, B.D. and Schriner, R.B. (2015) Development of an automated column solid phase extraction cleanup of QuEChERS extracts, using a zirconia-based sorbent, for pesticide residue analyses by LC-MS/MS. *J. Agric. Food Chem.*, **63**, 5107–5119. doi: 10.1021/jf505539e.

Ninga, E., Hakme, E., and Poulsen, M.E. (2024) Assessing the performance of various sorbents in micro-solid phase extraction cartridges for pesticide residue analysis in feed. *Anal. Methods*. doi: 10.1039/d4ay00226a.

Noij, T.H.M. and van der Kooi, M.M.E. (1995) Automated analysis of polar pesticides in water by on-line solid phase extraction and gas chromatography using the co-solvent effect. *J. High Res. Chrom.*, **18** (9), 535–539.

Poulsen, M.E., Hakme, E., and Ninga, E. (2023) Evaluation of different μ-SPE clean-up cartridges for pesticides residues in cereals. LAPRW 2023 Presentation, Panama, https://orbit.dtu.dk/en/publications/improvement-in-analytical-performance-from-participation-in-eu-pr.

Schürmann, A., Crüzer, C., Duss, V., Kämpf, R., Preiswerk, T., and Hübschmann, H.-J. (2023) Automated micro solid phase extraction clean-up and gas chromatography-tandem mass spectrometry analysis of pesticides in foods extracted with ethyl acetate. *Anal. Bioanal. Chem.*, **416** (3), 689–700. doi: 10.1007/s00216-023-05027-5.

Scida, K., Stege, P.W., Haby, G., Messina, G.A., and García, C.D. (2011) Recent applications of carbon-based nanomaterials in analytical chemistry: critical review. *Anal. Chim. Acta*, **691** (1–2), 6–17. doi: 10.1016/j.aca.2011.02.025.

Telepchak, M.J., August, T.F., and Chaney, G. (2004) *Forensic and Clinical Applications of Solid Phase Extraction* (ed. S.B. Karch), Humana Press Inc, Totowa, New Jersey, USA.

Tintrop, L.K., Solazzo, L., Salemi, A., Jochmann, M.A., and Schmidt, T.C. (2023) Characterization of a hydrophilic-lipophilic balanced SPME material for enrichment of analytes with different polarities from aqueous samples. *Adv. Sample Prep.*, **8**, 100099. doi: 10.1016/j.sampre.2023.100099.

Wittsiepe, J., Nestola, M., Kohne, M., Zinn, P., and Wilhelm, M. (2014) Determination of polychlorinated biphenyls and organochlorine pesticides in small volumes of human blood by high throughput. *J. Chrom. B*, **945–946**, 217–224. doi: 10.1016/j.jchromb.2013.11.059.

Zhang, B., Zheng, X., Li, H., and Lin, J. (2013) Application of carbon-based nanomaterials in sample preparation: a review. *Anal. Chim. Acta*, **784** (815), 1–17. doi: 10.1016/j.aca.2013.03.054.

Section 2.1.4 Solid Phase Microextraction

Agilent Technologies Inc. (2022) Food Testing & Environmental Analysis. Application Compendium 5994-4783EN.

Agilent Technologies Inc. (2023) Solid Phase Microextraction - Fundamentals. Technical Overview 5994-5775EN.

Arthur, C.L. and Pawliszyn, J. (1990) Solid phase microextraction with thermal desorption using fused silica optical fibers. *Anal. Chem.*, **62**, 2145–2148.

Arthur, C.L., Potter, D.W., Buchholz, K.D., Motlagh, S., and Pawliszyn, J. (1992) Solid phase microextraction for the direct analysis of water: theory and practice. *LC-GC*, **10**, 656–661.

Barbu, L. and Teaca, E. (2018) Sample Preparation Using GERSTEL SPME Multifiber Exchange (MFX), in *International Symposium "The Environment and the Industry"*, pp. 98–99. http://doi.org/10.21698/simi.2018.ab40.

Beiranvand, M. and Ghiasvand, A. (2020) Design and optimization of the VA-TV-SPME method for ultrasensitive determination of the PAHs in polluted water. *Talanta*, **212** (May). doi: 10.1016/j.talanta.2020.120809.

Berlardi, R. and Pawliszyn, J. (1989) The application of chemically modified fused silica fibers in the extraction of organics from water matrix samples and their rapid transfer to capillary columns. *Water Pollut. Res. J. Can.*, **24**, 179.

Bicchi, C., Cordero, C., Liberto, E., Sgorbini, B., and Rubiolo, P. (2008) Headspace sampling of the volatile fraction of vegetable matrices. *J. Chrom. A*, **1184**, 220–233. doi: 10.1016/j.chroma.2007.06.019.

Bojko, B. et al. (2012) SPME – Quo vadis? *Anal. Chim. Acta*, **750**, 132–151. doi: 10.1016/j.aca.2012.06.052.

Boyd-Boland, A. and Pawliszyn, J. (1995) Solid phase microextraction of nitrogen containing herbicides. *J. Chromatogr.*, **704**, 163–172.

Boyd-Boland, A., Magdic, S., and Pawliszyn, J. (1996) Simultaneous determination of 60 pesticides in water using solid phase microextraction and gas chromatography-mass spectrometry. *Analyst*, **121**, 929–937.

Bruheim, I., Liu, X., and Pawliszyn, J. (2003) Thin-film microextraction. *Anal. Chem.*, **75** (4), 1002–1010. doi: 10.1021/ac026162q.

Carda-Broch, S., Ruiz-Angel, M.J., Armstrong, D.W., and Berthod, A. (2010) Ionic liquid based headspace solid phase microextraction-gas chromatography for the determination of volatile polar organic compounds. *Sep. Sci. Technol.*, **45** (16), 2322–2328. doi: 10.1080/01496395.2010.497526.

Cervera, M.I., Beltran, J., Lopez, F.J., and Hernandez, F. (2011) Determination of volatile organic compounds in water by headspace-SPME GC-MS/MS with triple quadrupole analyzer. *Anal. Chim. Acta*, **704** (1–2), 87–97. doi: 10.1016/j.aca.2011.08.012.

De Lima Gomes, P.C.F., Yamashita Barletta, J., Domingues Nazario, C.E., Santos–Neto, A.J., Von Wolff, M.A., Reganhan Coneglian, C.M., Umbuzeiro, G.A., and Lancas, F.M. (2011) Optimization of in situ derivatization SPME by experimental design for GC-MS multi-residue analysis of pharmaceutical drugs in wastewater. *J. Sep. Sci.*, **34** (4), 436–445. doi: 10.1002/jssc.201000708.

Coelho, E., Ferreira, C., and Almeida, C.M.M. (2008) Analysis of polynuclear aromatic hydrocarbons by SPME-GC-FID in environmental and tap waters. *J. Braz. Chem. Soc.*, **19** (6), 1084–1097. doi: 10.1590/S0103-50532008000600006.

Crucello, J. et al. (2023) Automated method using direct immersion solid phase microextraction and on-fiber derivatization coupled with comprehensive two-dimensional gas chromatography high resolution mass spectrometry for profiling naphthenic acids in produced water. *J. Chrom. A*, **1692**, 463844. doi: 10.1016/j.chroma.2023.463844.

Davis, K.E. (2019) Detection of illicit drugs in various matrices via total vaporization solid phase microextraction. Master thesis. Purdue University, Department of Chemistry, Indianapolis, IN, USA.

Eisert, R. and Pawliszyn, J. (1997) Automated in-tube solid phase microextraction coupled to high performance liquid chromatography. *Anal. Chem.*, **69** (16), 3140–3147. doi: 10.1021/ac970319a.

Eisert, R. and Pawliszyn, J. (1997) New trends in solid phase microextraction. *Crit. Rev. Anal. Chem.*, **27**, 103–135.

Emmons, R.V., Tajali, R., and Gionfriddo, E. (2019) Development, optimization and applications of thin film solid phase microextraction (TF-SPME) devices for thermal desorption: a comprehensive review. *Separations*, **6** (3), 39. doi: 10.3390/separations6030039.

Ferracane, A., Manousi, N., Tranchida, P.Q., Zachariadis, G.A., Mondello, L., and Rosenberg, E. (2022) Exploring the volatile profile of whiskey samples using solid phase microextraction arrow and comprehensive two-dimensional gas chromatography-mass spectrometry. *J. Chrom. A*, **1676**. doi: 10.1016/j.chroma.2022.463241.

Fiori, J., Turroni, S., Candela, M., Brigidi, P., and Gotti, R. (2018) Simultaneous HS-SPME GC-MS determination of short chain fatty acids, trimethylamine and trimethylamine N-oxide for gut microbiota metabolic profile. *Talanta*, **189** (November), 573–578. doi: 10.1016/J.TALANTA.2018.07.051.

Gionfriddo, E., Souza-Silva, É.A., Ho, T.D., Anderson, J.L., and Pawliszyn, J. (2018) Exploiting the tunable selectivity features of polymeric ionic liquid-based SPME sorbents in food analysis. *Talanta*, **188**, 522–530. doi: 10.1016/j.talanta.2018.06.011.

Gou, Y., Eisert, R., and Pawliszyn, J. (2000) Automated in-tube solid phase microextraction-high performance liquid chromatography for carbamate pesticide analysis. *J. Chrom. A*, **873** (1), 137–147.

Grandy, J., Boyacı, E., and Pawliszyn, J. (2016) Development of a carbon mesh supported thin film microextraction membrane as a means to lower the detection limits of benchtop and portable GC/MS instrumentation. *Anal. Chem.*, **88** (3), 1760–1767. doi: 10.1021/acs.analchem.5b04008.

Grandy, J.J., Singh, V., Lashgari, M., Gauthier, M., and Pawliszyn, J. (2018) Development of a hydrophilic lipophilic balanced thin film solid phase microextraction device for balanced determination of volatile organic compounds. *Anal. Chem.*, **90** (23), 14072–14080. doi: 10.1021/acs.analchem.8b04544.

Grecco, C.F., Donizeti de Souza, I., Carvalho Oliveira, I.G., and Costa Queiroz, M.E. (2022) In-tube solid phase microextraction directly coupled to mass spectrometric systems: a review. *Separations*, **9** (12). doi: 10.3390/separations9120394.

Gorecki, T. and Pawliszyn, J. (1997) The effect of sample volume on quantitative analysis by SPME. Part I: Theoretical considerations. *Analyst*, **122**, 1079–1086.

Hakkarainen, M. (2007) Developments in multiple headspace extraction. *J. Biochem. Biophys. Methods*, **70** (2), 229–233. doi: 10.1016/j.jbbm.2006.08.012.

Hearn, L., Cole, R., Damiana, N., and Szafnauer, R. (2022) Advances in sample preparation volatile and semi-volatile compounds in flavoured hard seltzer beverages: comparison of high capacity sorptive extraction (HiSorb) methods. *Adv. Sample Prep.*, **3** (June), 100032. doi: 10.10–6/j.sampre.2022.100032.

Huang, G., Li, H.-F., Zhang, B.-T., Ma, Y., and Lin, J.-M. (2012) A vortex solvent bar microextraction combined with gas chromatography – mass spectrometry for the determination of phthalate esters in various sample matrices. *Talanta*, **100**, 64–70.

Hübschmann, H.-J. (2021) *Automated Sample Preparation – Solutions for GC-MS and LC-MS*, Wiley-VCH Verlag GmbH & Co. KGaA, Weinheim, Germany, ISBN 978-3-527-34507-6.

Ilias, Y., Bieri, S., Christen, P., and Veuthey, J.-L. (2006) Evaluation of solid phase microextraction desorption parameters for fast GC analysis of cocaine in coca leaves. *J. Chrom. Sci.*, **44** (7), 394–398. doi: 10.1093/chromsci/44.7.394.

IOFI Working Group on Methods of Analysis (2010) Guidelines for solid phase microextraction (SPME) of volatile flavour compounds for gas-chromatographic analysis, from the working group on methods of analysis of the International Organization of the Flavor Industry (IOFI). *Flavour Fragr. J.*, **25** (6), 404–406.

ISO/TC 147 (2014) ISO/DIS Water Quality - Determination of Volatile Organic Compounds in Water - Method Using Headspace Solid Phase Microextraction (HS-SPME) Followed by Gas Chromatography-Mass Spectrometry (GC-MS). Technical Committee, ISO/DIS 17943:2014(E).

Jurado, C., Gim, M.P., Soriano, T., Menendez, M., and Repetto, M. (2000) Rapid analysis of amphetamine, methamphetamine, MDA, and MDMA in urine using solid phase microextraction, direct on-fiber derivatization, and analysis by GC-MS. *J. Anal. Toxicol.*, **24** (February), 11–16.

Kfoury, N.C., Whitecavage, J.A., and Stuff, J.R. (2021) Comparison of Three Types of Thin Film-Solid Phase Microextraction Phases for Beverage Extractions. Application Note 219, Gerstel Inc., Linthicum, MD, USA.

Kim, H.-J. and Shin, H.-S. (2011) Simple and automatic determination of aldehydes and acetone in water by headspace solid phase microextraction and gas chromatography-mass spectrometry. *J. Sep. Sci.*, **34**, 693–699. doi: 10.1002/jssc.201000679.

Kremser, A., Jochmann, M.A., and Schmidt, T.C. (2016) PAL SPME arrow - evaluation of a novel solid phase microextraction device for freely dissolved PAHs in water. *Anal. Bioanal. Chem.*, **408** (3), 943–952. doi: 10.1007/s00216-015-9187-z.

Lee, J.N., Park, C., and Whitesides, G.M. (2003) Solvent compatibility of poly(dimethylsiloxane)-based microfluidic devices. *Anal. Chem.*, **75** (23), 6544–6554. doi: 10.1007/s10965-018-1550-2.

Lee, I.J., Ahn, J.C., Kim, B., and Chung, D. (2017) Comparative study of extracting fragrance allergens by GC-MS/MS. *Mass Spectrom. Lett.*, **8** (1), 18–22. doi: 10.5478/MSL.2017.8.1.18.

Lenzi, A. *et al.* (2023) HiSorb sorptive extraction for determining salivary short chain fatty acids and hydroxy acids in heart failure patients. *J. Chrom. B: Analyt. Technol. Biomed. Life Sci.*, **1228** (July), 123826. doi: 10.1016/j.jchromb.2023.123826.

Lim Yee Sin, G., Preiswerk, T., Huebschmann, H.-J., Chong, C.M., and Böhm, G. (2019) Robustness of SPME arrow immersion sampling: polycyclic aromatic hydrocarbons in drinking water. Poster at the RAFA Conference Prague, Czech Republic, *https://www.rafa2019.eu*.

Lipinski, J. (2001) Automated solid phase dynamic extraction - extraction of organics using a wall coated syringe needle. *Fres. J. Anal. Chem.*, **369** (1), 57–62. doi: 10.1007/s002160000618.

Lord, H.L. and Pawliszyn, J. (1998) Recent advances in solid phase microextraction. *LCGC Int.*, **12**, 776–785.

Lord, H.L. and Pawliszyn, J. (2000) Microextraction of drugs. *J. Chrom. A*, **902** (1), 17–63, *http://www.ncbi.nlm.nih.gov/pubmed/21637206*.

Mamedova, M. and Alimzhanova, M.B. (2023) Determination of biomarkers in multifloral honey by vacuum-assisted headspace solid phase microextraction. *Food Anal. Methods*, **16**, 1180–1190. doi: 10.1007/s12161-023-02499-0.

Markelov, M. and Guzowski, J.P. (1993) Matrix independent headspace gas chromatographic analysis. This full evaporation technique. *Anal. Chim. Acta*, **276** (2), 235–245. doi: 10.1016/0003-2670(93)80390-7.

Markes (2022) High Throughput Analysis of 1,4-Dioxane in Drinking Water Using SPME Arrow-Trap with Multi-Step Enrichment (MSE). Application Note 282, Markes International Ltd, Llantrisant, UK.

Martínez-Uruñuela, A., González-Sáiz, J.M., and Pizarro, C. (2005) Multiple solid phase microextraction in a non-equilibrium situation: application in quantitative analysis of chlorophenols and chloroanisoles related to cork taint in wine. *J. Chrom. A*, **1089** (1–2), 31–38. doi: 10.1016/j.chroma.2005.06.063.

Mendivelso-Pérez, D. and Shimelis, O. (2023) Headspace-SPME as a versatile monitoring method for early detection of insect infestation in rice. *Analytix Reporter*, **15,** Sigma-Aldrich, *https://www.sigmaaldrich.com/DE/de/technical-documents/technical-article/food-and-beverage-testing-and-manufacturing/chemical-analysis-for-food-and-beverage/early-detection-of-insect-infestation-in-rice* (accessed 27 May 2024).

Moreira, M.A., André, L.C., and Cardeal, Z.L. (2014) Analysis of phthalate migration to food simulants in plastic containers during microwave operations. *Int. J. Environ. Res. Public Health*, **11**, 507–526. doi: 10.3390/ijerph110100507.

Myung, S.W., Min, H.K., Kim, S., Kim, M., Cho, J.B., and Kim, T.-J. (1998) Determination of amphetamine, methamphetamine and dimethamphetamine in human urine by solid phase microextraction (SPME)-gas chromatography/mass spectrometry. *J. Chrom. B: Biomed. Appl.*, **716** (1–2), 359–365. doi: 10.1016/S0378-4347(98)00304-1.

Nzekoue, F.K. *et al.* (2020) Fiber-sample distance, an important parameter to be considered in headspace solid phase microextraction applications. *Anal. Chem.*, **92**, 7478–7484.

Nolvachai, Y., Amaral, M.S.S., Herron, R., and Marriott, P.J. (2023) Solid phase microextraction for quantitative analysis - expectations beyond design? *Green Anal. Chem.*, **4**, 100048. doi: 10.1016/j.greeac.2022.100048.

O'Reilly, J. *et al.* (2005) Automation of solid phase microextraction. *J. Sep. Sci.*, **28** (15), 2010–2022. doi: 10.1002/jssc.200500244.

Pan, L., Chong, J.M., and Pawliszyn, J. (1997) Determination of amines in air and water using derivatization combined with SPME. *J. Chromatogr.*, **773**, 249–260.

Pawliszyn, J.B. (1999) in *Applications of Solid Phase Microextraction (RSC Chromatography Monographs, Volume 5)* (eds J.B. Pawliszyn and R.M. Smith), The Royal Society of Chemistry, Cambridge, UK, ISBN 0-85404-525-2.

Pawliszyn, J. (2009) Theory of solid phase microextraction, in *Handbook of Solid Phase Microextraction* (ed. J. Pawliszyn), Chemical Industry Press, Beijing, P. R. China.

Pawliszyn, J.B. (ed.) (2012) *Handbook of Solid Phase Microextraction*, Elsevier, London, Waltham.

Pawliszyn, J. (1997) *Solid Phase Microextraction: Theory and Practice*, John Wiley & Sons, Inc., New York.

Pawliszyn Research Group (2024) Thin Film Microextraction (TFME), *https://uwaterloo.ca/pawliszyn-group/research/thin-film* (accessed 9 February 2024).

Pereira, A., Silva, E., and Cerejeira, M.J. (2014) Applicability of the new 60 μm polyethylene glycol solid phase microextraction fiber assembly for the simultaneous analysis of six pesticides in water. *J. Chrom. Sci.*, **52** (5), 423–428. doi: 10.1093/chromsci/bmt053.

Poerschmann, J., Zhang, Z., Kopinke, F.-D., and Pawliszyn, J.B. (1997) Solid phase microextraction for determining the distribution of chemicals in aqueous matrices. *Anal. Chem.*, **69**, 597–600.

Potter, D.W. and Pawliszyn, J. (1992) Detection of substituted benzenes in water at the pg/ml level using solid phase microextraction and gas chromatography-ion trap mass spectrometry. *J. Chrom. A*, **625** (2), 247–255. doi: 10.1016/0021-9673(92)85209-C.

Pragst, F. (2007) Application of solid phase microextraction in analytical toxicology. *Anal. Bioanal. Chem.*, **388** (7), 1393–1414. doi: 10.1007/s00216-007-1289-9.

Purcaro, G. (2024) Comparison of Vac-HS-SPME and HS-SPME Coupled to GC-MS for Virgin Olive Oil Aroma Analysis. Application Note, ExtraTECH, Chania-Crete, Greece, www.extratech.gr.

Regueiro, J., Becerril, E., Garcia-Jares, C., and Llompart, M. (2009) Trace analysis of parabens, triclosan and related chlorophenols in water by headspace solid phase microextraction with in situ derivatization and gas chromatography-tandem mass spectrometry. *J. Chrom. A*, **1216** (23), 4693–4702. doi: 10.1016/j.chroma.2009.04.025.

Reyes-Garcés, N. et al. (2017) Advances in solid phase microextraction and perspective on future directions. *Anal. Chem.*, **90** (1), 302–360. doi: 10.1021/acs.analchem.7b04502.

Ridgway, K., Lalljie, S.P.D., and Smith, R.M. (2006) Comparison of in-tube sorptive extraction techniques for non-polar volatile organic compounds by gas chromatography with mass spectrometric detection. *J. Chrom. A*, **1124** (1–2), 181–186. doi: 10.1016/j.chroma.2006.06.105.

Risticevic, S., Niri, V.H., Vuckovic, D., and Pawliszyn, J. (2009) Recent developments in solid phase microextraction. *Anal. Bioanal. Chem.*, **393**, 781–795.

Sangster, J.M. (1997) *Octanol-Water Partition Coefficients: Fundamentals and Physical Chemistry*, Wiley & Sons Ltd, Chichester, UK.

Serodio, P. and Nogueira, J.M.F. (2004) Multi-residue screening of endocrine disrupting chemicals in water samples by stir bar sorptive extraction gas chromatography mass spectrometry detection. *Anal. Chim. Acta*, **517**, 21–32.

Shirey, R.E. (2009) SPME commercial devices and fibre coatings, in *Handbook of Solid Phase Microextraction* (ed. J. Pawliszyn), Chemical Industry Press, Beijing, P. R. China.

Souza Silva, E.A. and Pawliszyn, J. (2012) Optimization of fiber coating structure enables direct immersion solid phase microextraction and high throughput determination of complex samples. *Anal. Chem.*, **84**, 6933–6938. doi: 10.1021/ac301305u.

SRA Instruments (2022) *MFX (Multi Fiber EXchange). Product Brochure*, SRA Instruments, Marcy l'étoile, France, https://www.srainstruments.com/wp-content/uploads/2022/02/Brochure-MFX-for-PAL3.pdf.

Steffen, A. and Pawliszyn, J. (1996) The analysis of flavour volatiles using headspace solid phase microextraction. *J. Agric. Food Chem.*, **44**, 2187–2193.

Strano-Rossi, S., Molaioni, F., and Botri, F. (2005) Application of solid phase microextraction to antidoping analysis: determination of stimulants, narcotics, and other classes of substances excreted free in urine. *J. Anal. Tox.*, **29** (June), 217–222.

Stuff, J.R., Whitecavage, J.A., Grandy, J.J., and Pawliszyn, J. (2018) Analysis of Beverage Samples Using Thin Film Solid Phase Microextraction (TF-SPME) and Thermal Desorption GC/MS. Application Note 200, Gerstel Inc., Linthicum, MD, USA.

Tena, M.T. and Carrillo, J.D. (2007) Multiple solid phase microextraction: theory and applications. *Trends Anal. Chem.*, **26** (3). doi: 10.1016/j.trac.2007.01.008.

Tintrop, L.K., Solazzo, L., Salemi, A., Jochmann, M.A., and Schmidt, T.C. (2023) Characterization of a hydrophilic-lipophilic balanced SPME material for enrichment of analytes with different polarities from aqueous samples. *Adv. Sample Prep.*, **8** (October 2023), 100099. doi: 10.1016/j.sampre.2023.100099.

Valenzuela, J.A., Vázquez, L., Rodriguez, J., Belén Flórez, A., Vasek, O.M., and Mayo, B. (2024) Phenotypic, technological, safety, and genomic profiles of gamma-aminobutyric acid-producing *Lactococcus lactis* and *Streptococcus thermophilus* strains isolated from cow's milk. *Int. J. Mol. Sci.*, **25**, 2328, https://www.mdpi.com/1422-0067/25/4/2328#.

Vatani, H. and Yazdi, A.S. (2014) Ionic liquid-mediated poly(dimethylsiloxane)-grafted carbon nanotube fiber prepared by the sol-gel technique for the head space solid phase microextraction of methyl tert-butyl ether using GC. *J. Sep. Sci.*, **37** (1–2), 127–134.

Wang, Y. and Pawliszyn, J. (2023) Automated sequential solid phase microextraction to address displacement effects in the quantitative analysis of polar compounds. *Green Anal. Chem.*, **6** (June), 100070. doi: 10.1016/j.greeac.2023.100070.

Winkler, M., Headley, J.V., and Peru, K.M. (2000) Optimization of solid phase microextraction for the gas chromatographic – mass spectrometric determination of synthetic musk fragrances in water samples. *J. Chrom. A*, **903**, 203–210.

Yu, H., Ho, T.D., and Anderson, J.L. (2013) Ionic liquid and polymeric ionic liquid coatings in solid phase microextraction. *Trends Anal. Chem.*, **45**, 219–232. doi: 10.1016/j.trac.2012.10.016.

Zhang, Z. and Pawliszyn, J. (1993) Headspace solid phase microextraction. *Anal. Chem.*, **65**, 1843–1852.

Zhang, Z., Yang, M.J., and Pawliszyn, J. (1994) Solid phase microextraction - a new solvent-free alternative for sample preparation. *Anal. Chem.*, **66**, 844A–853A.

Zhu, W., Koziel, J., Cai, L., Özsoy, H.D., and van Leeuwen, J. (2015) Quantification of carbonyl compounds generated from ozone-based food colorants decomposition using on-fiber derivatization-SPME-GC-MS. *Chromatography*, **2** (1), 1–18. doi: 10.3390/chromatography2010001.

Zhou, W., Hu, K., Wang, Y., Jiang, R.W., and Pawliszyn, J. (2024) Embedding mixed sorbents in binder: solid phase microextraction coating with wide extraction coverage and its application in environmental water analysis. *Environ. Sci. Technol.*, **58** (1), 771–779. doi: 10.1021/acs.est.3c07244.

Ziegler, M. and Schmarr, H.G. (2019) Comparison of solid phase microextraction using classical fibers versus mini-arrows applying multiple headspace extraction and various agitation techniques. *Chromatographia*, **82** (2), 635–640. doi: 10.1007/s10337-018-3659-1.

Zuba, D., Parczewski, A., and Reichenbächer, M. (2002) Optimization of solid phase microextraction conditions for gas chromatographic determination of ethanol and other volatile compounds in blood. *J. Chrom. B: Anal. Tech. Biomed. Life Sci.*, **773** (1), 75–82. doi: 10.1016/S1570-0232(02)00143-5.

Section 2.1.5 Static Headspace Techniques

Alzaga, R., Ryan, R.W., Taylor-Worth, K., Lipczynski, A.M., Szucs, R., and Sandra, P. (2007) A generic approach for the determination of residues of alkylating agents in active pharmaceutical ingredients by in situ derivatization-headspace-gas chromatography-mass spectrometry. *J. Pharm. Biomed. Anal.*, **45** (3), 472–479. doi: 10.1016/j.jpba.2007.07.017.

Arya, P. and Dhyani, V. (2023) Estimation of Ethylene Oxide and Ethylene Chlorohydrin in Foodstuffs by HS-GC/MS/MS. Application Note Food & Beverage Testing, 5994-5378EN, Agilent Technologies Inc.

Banat, F.A.O., Abu Al-Rub, F.A., and Simandl, J. (1999) Experimental study of the salt effect in vapor/liquid equilibria using headspace gas chromatography. *Chem. Eng. Techn.*, **22** (9), 761–765. doi: 10.1002/(SICI)1521-4125(199909)22:9<761::AID-CEAT761>3.0.CO;2-U.

Brettell, T.A. and Grob, R.L. (1986) Cryogenic techniques in gas chromatography part two: Cryofocusing and cryogenic trapping. *Int. Lab.*, **April**, 30–48.

Buttery, R.G. and Roy, T. (1961) Gas-liquid chromatography of aroma of vegetables and fruit. *Anal. Chem.*, **33** (10), 1439–1441, https://pubs.acs.org/doi/abs/10.1021/ac60178a052.

Ettre, L.S. and Kolb, B. (1991) Headspace gas chromatography: the influence of sample volume on analytical results. *Chromatographia*, **32** (1/2), 5–12.

Hachenberg, H. and Schmidt, A.P. (1977) *Gas Chromatographic Headspace Analysis*, John Wiley & Sons, Ltd, Chichester.

Hachenberg, H. (1988) *Die Headspace Gaschromatographie als Analysen- und Meßmethode*, DANI- Analysentechnik, Mainz.

Kialengila, D.M., Wolfs, K., Bugalama, J., Van Schepdael, A., and Adams, E. (2013) Full evaporation headspace gas chromatography for sensitive determination of high boiling point volatile organic compounds in low boiling matrices. *J. Chrom. A*, **1315** (November), 167–175. doi: 10.1016/j.chroma.2013.09.058.

Kolb, B. (1980) *Applied Headspace Gas Chromatography*, John Wiley & Sons, Ltd, Chichester.

Kolb, B. and Ettre, L.S. (1991) Theory and practice of multiple headspace extraction. *Chromatographia*, **32** (11–12), 505–513.

Kolb, B. (1992) *Headspace Gas Chromatography a Brief Introduction*, Perkin Elmer GC Applications Laboratory.

Kolb, B. (1993) Die Bestimmung des Wassergehaltes in Lebensmitteln und Pharmaka mittels der gaschromatographischen Headspace-Technik. *Lebensmittel Biotechnol.*, **1**, 17–20.

Kolb, B. and Ettre, L.S. (2006) *Static Headspace - Gas Chromatography: Theory and Practice*, 2nd edn, John Wiley & Sons, Inc., New York.

References

Kolahgar, B. and Pfannkoch, E. (2002) The Use of Different PTV Inlet Liner Types for Trapping Alkanes, Aromatics and Oxygenated Compounds during Thermal Desorption. Technical Note 3, Gerstel GmbH, Mülheim an der Ruhr, Germany

Krebs, G., Schneider, E., and Schumann, A. (1991) Head Space GC - Analytik flüchtiger aromatischer und halogenierter Kohlenwasserstoffe aus Bodenluft. *GIT Fachz. Lab.*, **1**, 19–22.

Lin, D.P., Falkenberg, C., Payne, D.A., Thakkar, J., Tang, C., and Elly, C. (1993) Kinetics of purging for the priority volatile organic compounds in water. *Anal. Chem.*, **65**, 999–1002.

Maggio, A. et al. (1991) Multiple headspace extraction capillary gas chromatography (MHE-CGC) for the quantitative determination of volatiles in contaminated soils, in *13th International Symposium on Capillary Chromatography, Riva del Garda, May 1991* (ed. P. Sandra), Huethig Verlag, pp. 394–405.

Majors, R.E. (2009) Salting out liquid-liquid extraction. *LCGC N. Am.*, **27** (7), 526–533, http://www.chromatographyonline.com/saltingoutliquidliquidextractionsalle.

Petersen, M.A. (2008) Quantification of volatiles in cheese using multiple headspace extraction (MHE). Presentation 2008, University of Copenhagen, Department of Food Science, Quality and Technology.

Poy, F. and Cobelli, L. (1985) Automatic headspace and programmed temperature vaporizer (PTV) operated in cryo-enrichment mode in high resolution gas chromatography. *J. Chrom. Sci.*, **23** (3), 114–119. doi: 10.1093/chromsci/23.3.114.

Sander, R. (2015) Compilation of Henry's law constants (version 5.0.0-Rc.0) for water as solvent. *Atmos. Chem. Phys.*, **15**, 4399–4981. doi: 10.5194/acp-15-4399-2015.

Vulpius, T. and Baltensperger, B. (2009) Multiple Headspace Extraction (MHE) with Syringe Sampling, MSC ApS and CTC Scientific Poster.

Westendorf, R.G. (2000) Cryofocusing, Tekmar Technical Documentation, Tekmar Inc.

Wylie, P.L. (1987) Comparison of headspace with purge and trap techniques for the analysis of volatile priority pollutants, in *8th International Symposium on Capillary Chromatography, Riva del Garda, May 19th–21st 1987* (ed. P. Sandra), Huethig, pp. 482–499.

Wylie, P.L. (1988) Comparing headspace with purge and trap for analysis of volatile priority pollutants. *Res. Technol.*, **8**, 65–72.

Zhu, J.Y. and Chai, X.-S. (2005) Some recent developments in headspace gas chromatography. *Curr. Anal. Chem.*, **1**, 79–83.

Zhu, L. et al. (2024) Detection and comparison of volatile organic compounds in four varieties of hawthorn using HS-GC-IMS. *Separations*, **11** (100). doi: 10.3390/separations11040100.

Zuba, D., Parczewski, A., and Reichenbächer, M. (2002) Optimization of solid phase microextraction conditions for gas chromatographic determination of ethanol and other volatile compounds in blood. *J. Chrom. B Analy. Technol. Biomed. Life Sci.*, **773** (1), 75–82. doi: 10.1016/S1570-0232(02)00143-5.

Section 2.1.6 Dynamic Headspace – Purge & Trap Technique

Bicchi, C., Cordero, C., Liberto, E., Sgorbini, B., and Rubiolo, P. (2008) Headspace sampling of the volatile fraction of vegetable matrices. *J. Chrom. A*, **1184**, 220–233. doi: 10.1016/j.chroma.2007.06.019.

Brown, J. and Shirey, B. (2001) A Tool for Selecting an Adsorbent for Thermal Desorption Applications. SUPELCO, Technical Report. Bellefonte, PA, USA, *https://www.sigmaaldrich.com/content/d'm/sigma-aldrich/docs/Supelco/ Bulletin/11342.pdf*.

Johnson, E. and Madden, A. (1990) Efficient Water Removal for GC-MS Analysis of Volatile Organic Compounds with Tekmar's Moisture Control Module. Finnigan MAT Technical Report No. 616.

Markes (2012) Advice on Sorbent Selection, Tube Conditioning, Tube Storage and Air Sampling. Technical Support, TDTS 5, Markes International Ltd., Llantrisant, UK.

Matz, G. and Kesners, P. (1993) Spray and trap method for water analysis by thermal desorption gas chromatography/mass spectrometry in field applications. *Anal. Chem.*, **65**, 2366–2371.

Moreira Novaes, F., Jr. and Marriott, P.J. (2021) Cryogenic trapping as a versatile approach for sample handling, enrichment and multidimensional analysis in gas chromatography. *J. Chrom. A*, **1644** (May), 462135. doi: 10.1016/j.chroma. 2021.462135.

Restek (2003) *Optimizing the Analysis of Volatile Organic Compounds, Technical Guide*, Restek Corp, Bellefonte, PA, USA.

Seibel, B. (2015) Fundamentals of Purge and Trap, Tekmar Talk Blog, *http://blog .teledynetekmar.com/fundamentals-of-purge-and-trap* (accessed 28 February 2024).

Supelco (1999) Purge-and-Trap System Guide. Bulletin 916, Supelco, Bellefonte, PA, USA.

Tekmar (1991) Fundamentals of Purge and Trap, Tekmar Technical Documentation B 121988, Tekmar Inc.

Westendorf, R.G. (1985) A quantitation method for dynamic headspace analysis using multiple runs. *J. Chromatogr. Sci.*, **23**, 521–524.

Section 2.1.7 Dynamic Headspace – In-Tube Extraction

Bütikofer, U. *et al.* (2022) Serum and urine metabolites in healthy men after consumption of acidified milk and yogurt. *Nutrients*, **14** (22). doi: 10.3390/nu14224794.

Concha-Graña, E., Moscoso-Pérez, C.M., Fernández-González, V., López-Mahía, P., and Muniategui-Lorenzo, S. (2024) New approach for the quantitative analysis of additives from plastic using in-tube extraction dynamic headspace sampling technique coupled to GC-MS/MS. *Anal. Chim. Acta*, **1302**, 342487, doi: *10.1016/j.aca .2024.342487*.

Cretnik, S. and Fuchsmann, P. (2023) Application of the V-ITEX method in different matrices. Agroscope, Federal Department of Economic Affairs, Education and Research EAER, and CTC Analytics AG, SelectScience webinar July 26, 2023.

Fuchsmann, P., Stern, M.T., Bischoff, P., Badertscher, R., Breme, K., and Walther, B. (2019) Development and performance evaluation of a novel dynamic headspace vacuum transfer 'in trap' extraction method for volatile compounds and comparison with headspace solid phase microextraction and headspace in-tube extraction. *J. Chrom. A*, **1601**, 60–70. doi: 10.1016/j.chroma.2019.05.016.

Fuchsmann, P. et al. (2020) Nutrivolatilomics of urinary and plasma samples to identify candidate biomarkers after cheese, milk, and soy-based drink intake in healthy humans. *J. Proteome Res.*, **19** (10), 4019–4033. doi: 10.1021/acs.jproteome.0c00324.

Hu, J., Pagliano, E., Hou, X., Zheng, C., Yang, L., and Mester, Z. (2017) Sub-ppt determination of butyltins, methylmercury and inorganic mercury in natural waters by dynamic headspace in-tube extraction and GC-ICPMS detection. *J. Anal. At. Spectrom.*, **12**, 1–8. doi: 10.1039/C7JA00296C.

Jochmann, M.A., Yuan, X., Schilling, B., and Schmidt, T.C. (2008) In-tube extraction for enrichment of volatile organic hydrocarbons from aqueous samples. *J. Chrom. A*, **1179** (2), 96–105. doi: 10.1016/j.chroma.2007.11.100.

Kaziur-Cegla, W., Wykowski, L., Jochmann, M.A., Molt, K., Bruchmann, A., and Schmidt, T.C. (2021) In-tube dynamic extraction a green solventless alternative for a sensitive headspace analysis of volatile organic compounds in olive oils. *Adv. Sample Prep.*, **1** (December), 100002. doi: 10.1016/j.sampre.2021.100002.

Laaks, J., Jochmann, M.A., Schilling, B., and Schmidt, T.C. (2015) Optimization strategies of in-tube extraction (ITEX) methods. *Anal. Bioanal. Chem.*, **407** (22), 6827–6838. doi: 10.1007/s00216-015-8854-4.

Meng, H.Y., Piccand, M., Fuchsmann, P., Dubois, S., Baumeyer, A., Stern, M.T., and von Ah, U. (2021) Formation of 3-methylbutanal and 3-methylbutan-1-ol recognized as malty during fermentation in Swiss raclette-type cheese, reconstituted milk, and de man, rogosa, and sharpe broth. *J. Agric. Food Chem.*, **69** (2), 717–729. doi: 10.1021/acs.jafc.0c06570.

Michiu, D., Tofana, M., Socaci, S.A., Mudura, E., Salanta, L.C., and Farcas, A.C. (2012) Optimization of ITEX/GC-MS method for beer wort volatile compounds characterisation. *J. Agroaliment. Proc. Technol.*, **18** (229–35), 227.

Zapata, J., Lopez, R., and Ferreira, V. (2012) Quantification of aroma compounds in wine based on the automated multiple headspace in-tube extraction: comparison of release behaviour in different wines. *J. Chromatogr. A*, **5** (4), 7, https://www.sepscience.com/Sectors/Food/Articles/1870-/Quantification-of-Aroma-Compounds-in-Wine-Based-on-The-Automated-Multiple-Headspace-InTube-Extraction-Comparison-of-Release-Behaviour-in-Different-Wines.

Section 2.1.8 Adsorptive Enrichment and Thermal Desorption

Betz, W.R., Hazard, S.A., and Yearick, E.M. (1989) Characterization and utilization of carbon-based adsorbents for adsorption and thermal desorption of volatile, semivolatile and non-volatile organic contaminants in air, water and soil sample matrices. *Int. Labmate*, **XV**, 1.

Brown, J. (2013) Thermal desorption tubes. Choosing the Right Adsorbent for your Thermal Desorption Gas Chromatography Applications, Webinar, 22 October 2013, https://www.sigmaaldrich.com/DE/de/technical-documents/technical-article/environmental-testing-and-industrial-hygiene/air-testing/thermal-desorption (accessed 29 February 2024).

Brown, R.H. and Purnell, C.J. (1979) Collection and analysis of trace vapour pollutants in ambient atmospheres. *J. Chromatogr.*, **178**, 79–90.

Brown, J. and Shirey, B. (2001) A Tool for Selecting an Adsorbent for Thermal Desorption Applications. Technical Report, T402025 EQF, Supelco, Bellefonte, PA, USA.

Brown, V.M., Crump, D.R., Plant, N.T., and Pengelly, I. (2014) Evaluation of the stability of a mixture of volatile organic compounds on sorbents for the determination of emissions from indoor materials and products using thermal desorption/gas chromatography/mass spectrometry. *J. Chrom. A*, **1350**, 1–9.

Camsco (2009) *Sorbent Selection Chart*, Houston, TX, USA, https://camsco.com/pages/sorbent-selection-chart (accessed 7 March 2024).

Cosnier, F., Celzard, A., Furdin, G., Bégin, D., and Marêché, J.F. (2006) Influence of water on the dynamic adsorption of chlorinated VOCs on active carbon: relative humidity of the gas phase versus pre-adsorbed. *Adsorp. Sci. Technol.*, **24**, 193–282.

Figge, K., Rubel, W., and Wieck, A. (1987) Adsorptionsmittel zur Anreicherung von organischen Luftinhaltsstoffen. *Fres. Z. Anal. Chem.*, **327**, 261–278.

Gawlowski, J. and Gierczak, T. (2000) Dry purge for the removal of water from the solid sorbents used to sample volatile organic compounds from the atmospheric air. *Analyst*, **125**, 2112–2117. doi: 10.1039/b004678g.

Health and Safety Laboratory (HSL) (1990) Generation of test atmospheres of organic vapours by the syringe injection technique, ISBN 0-11-885647-2.

Knobloch, T. and Engewald, W. (1995) Sampling and gas chromatographic analysis of volatile organic compounds in hot and extremely humid emissions. *J. High Resol. Chromatogr.*, **18** (10), 635–642.

Kommission Reinhaltung der Luft im VDI und DIN - Normenausschuss KRdL (1981) Measurement of Gases; Calibration Gas Mixtures; Preparation by Continuous Injection Method. Directive 13.040.01, *www.vdi.de* (accessed 14 May 2014).

Krebs, G., Schneider, E., and Schuhmann, A. (1991) Analytik flüchtiger aromatischer und halogenierter Kohlenwasserstoffe aus Bodenluft. *GIT Labor Fachz.*, **35**, 19–22.

Manura, J.J. (1995) Adsorbent Resins - Calculation and Use of Breakthrough Volume Data. Scientific Instrument Services, Ringoes, NJ, Application Note, *http://www.sisweb.com/index/referenc* resin10.htm (accessed 9 May 2014).

Markes (2003) Sampling Device, U.S. Patent 6,564,656.

Markes (2012a) Advice on Sorbent Selection, Tube Conditioning, Tube Storage and Air Sampling. Technical Report TDTS 5, Markes International Ltd., Llantrisant, UK.

Markes (2012b) Diffusion-Locking Technology. Technical Report TDTS 61, Markes International, Llantrisant, UK.

Merit (2018) *Summa is for Wrestling not TO-15 Air Sampling*, Merit Laboratories, https://www.meritlabs.com/blog/2018/2/27/silonite-ceramic-coated-canisters-a-better-option-than-the-outdated-summa-canisters-for-to-15-air-sampling (accessed 6 October 2024).

Sigma-Aldrich (2024) Thermal Desorption Tube Selection Guide, *https://www.sigmaaldrich.com/DE/de/technical-documents/technical-article/environmental-testing-and-industrial-hygiene/air-testing/thermal-desorption-tube-selection* (accessed 7 March 2024).

Supelco (1986) Carbotrap - An Excellent Adsorbent for Sampling Many Airborne Contaminants. GC Bulletin 846C, Supelco Company Publication.

Supelco (1988) Efficiently Monitor Toxic Compounds by Thermal Desorption. GC Bulletin 849C, Supelco Company Publication.

Supelco (1992) New Adsorbent Trap for Monitoring Volatile Organic Compounds in Wastewater. Environmental Notes, Supelco Company Publication, Bellafonte.

Tschickard, M. (1993) Bericht über eine Prüfgasapparatur zur Herstellung von Kalibriergasen nach dem Verfahren der kontinuierlichen Injektion. *Gesch.Zchn: 35-820 Tsch, Mainz, Landesamt für Umweltschutz und Gewerbeaufsicht* (accessed 25 May 1993).

Woolfenden, E.A. and Cole, A. (2003) Sampling Device, U.S. Patent 6564656 B1, May 2003.

Section 2.1.9 Stir Bar Sorptive Extraction

Baltussen, E., Sandra, P., David, F., and Cramers, C. (1999) Stir bar sorptive extraction (SBSE), a novel extraction technique for aqueous samples: theory and principles. *J. Microcol. Sep.*, **11** (10), 737–747.
doi: 10.1002/(SICI)1520-667X(1999)11:10<737::AID-MCS7>3.0.CO;2-4.

Bicchi, C., Cordero, C., Sgorbini, B., Liberto, E., and Rubiolo, P. (2009) Stir-bar sorptive extraction and headspace sorptive extraction: an overview. *LCGC N. Am.*, **27** (5), 376–390.

Bicchi, C., Cordero, C., Iori, C., Rubiolo, P., and Sandra, P. (2000) Headspace sorptive extraction (HSSE) in the headspace analysis of aromatic and medicinal plants. *HRC J. High Res. Chrom.*, **23** (9), 539–546.
doi: 10.1002/1521-4168(20000901)23:9<539::AID-JHRC539>3.0.CO;2-3.

Canli, O., Çetintürk, K., and Öktem, O. (2020) Determination of 117 endocrine disruptors (EDCs) in water using SBSE TD-GC-MS/MS under the European water framework directive. *Anal. Bioanal. Chem.*, **412**, 5169–5178.

David, F. and Sandra, P. (2007) Stir bar sorptive extraction for trace analysis. *J. Chrom. A*, **1152** (1–2), 54–69. doi: 10.1016/j.chroma.2007.01.032.

Gerstel (2024) *AutoTwister. Product Information*, GERSTEL GmbH & Co KG, Mülheim, Germany.

Hoffmann, A., Bremer, R., Sandra, P., and David, F. (2000) A novel extraction technique for aqueous samples. *Gerstel Solutions Worldwide*, **49**, 4–7, Gerstel GmbH, Mülheim, Germany.

Majedi, S.M. and Lee, H.K. (2017) Chapter 14: Microextraction and solventless techniques, in *The Application of Green Solvents in Separation Processes* (eds F. Pena-Pereira and M. Tobiszewski), pp. 415–450, *http://www.sciencedirect.com/science/article/pii/B9780128052976000140*.

Tienpont, B., David, F., Bicchi, C., and Sandra, P. (2000) High capacity headspace sorptive extraction. *J. Microcol. Sep.*, **12** (11), 577–584. doi: 10.1002/1520-667X(2000)12:11<577::AID-MCS30>3.0.CO;2-Q.

Sandra, P., Baltussen, E., David, F., and Hoffmann, A. (2000) A Novel Extraction Technique for Aqueous Samples: Stir Bar Sorptive Extraction. Application Note 1/2000, Gerstel GmbH, Mülheim, Germany.

Serodio, P. and Nogueira, J.M.F. (2005) Development of a stir-bar-sorptive extraction-liquid desorption-large-volume injection capillary gas chromatographic-mass spectrometric method for pyrethroid pesticides in water samples. *Anal. Bioanal. Chem.*, **382** (4), 1141–1151. doi: 10.1007/s00216-005-3210-8.

Section 2.1.10 Pyrolysis

Brodda, B.-G., Dix, S., and Fachinger, J. (1993) Investigation of the pyrolytic degradation of ion exchange resins by means of foil pulse pyrolysis coupled with gas chromatography/mass spectrometry. *Sep. Sci. Technol.*, **28**, 653–673.

Dworzanski, J.P. and Meuzelaar, H.L.C. (2017) Pyrolysis mass spectrometry, methods. In: Eds. J.C. Lindon, G.E. Tranter, D.W. Koppenaal, *Encyclopedia of Spectroscopy and Spectrometry*, 3rd edn. (eds J.C. Lindon, G.E. Tranter, and D.W. Koppenaal) Academic Press, pp. 789–801, doi: 10.1016/B978-0-12-409547-2.11686-5.

Ericsson, I. (1978) Sequential pyrolysis gas chromatographic study of the decomposition kinetics of cis-1,4-polybutadiene. *J. Chromatogr. Sci.*, **16**, 340–344.

Ericsson, I. (1980) Determination of the temperature time profile of filament pyrolyzers. *J. Anal. Appl. Pyrolysis*, **2**, 187–194.

Ericsson, I. (1985) Influence of pyrolysis parameters on results in pyrolysis-gas chromatography. *J. Anal. Appl. Pyrolysis*, **8**, 73–86.

Ericsson, I. (1990) Trace determination of high molecular weight polyvinylpyrrolidone by pyrolysis-gas-chromatography. *J. Anal. Appl. Pyrolysis*, **17**, 251–260.

Ericsson, I. and Lattimer, R.P. (1989) Pyrolysis nomenclature. *J. Anal. Appl. Pyrolysis*, **14**, 219–221.

Fischer, W.G. and Kusch, P. (1993) An automated curie-point pyrolysis-high resolution gas chromatography system. *LCGC*, **6**, 760–763.

Frontier Lab (2023) Multishot Pyrolyzer EGA PY-3030D. Product information. Frontier Laboratories Ltd., *https://www.frontier-lab.com/products/multi-functional-pyrolysis-system/17811* (accessed 8 March 2024).

Gassiot-Matas, M. and Julia-Danés, E. (1976) Pyrolysis gas chromatography of some sterols (II). *Chromatographia*, **9**, 151–156. doi: 10.1007/BF02281661.

Galletti, G.C. and Reeves, J.B. (1991) Pyrolysis-gas chromatography/mass spectrometry of lignocellulosis in forages and by-products. *J. Anal. Appl. Pyrolysis*, **19**, 203–212.

Hancox, R.N., Lamb, G.D., and Lehrle, R.S. (1991) Sample size dependence in pyrolysis: an embarrassment or an utility? *J. Anal. Appl. Pyrolysis*, **19**, 333–347.

Irwin, W.J. (1982) *Analytical Pyrolysis: A Comprehensive Guide*, Marcel Dekker, New York.

Klusmeier, W., Vogler, P., Ohrbach, K.H., Weber, H., and Kettrup, A. (1988) Thermal decomposition of pentachlorobenzene, hexachlorobenzene and octachlorostyrene in air. *J. Anal. Appl. Pyrolysis*, **14**, 25–36.

Matney, M.L., Limero, T.F., and James, J.T. (1994) Pyrolysis-gas chromatography/mass spectrometry analyses of biological particulates collected during recent space shuttle missions. *Anal. Chem.*, **66**, 2820–2828.

Oguri, N. and Kirn, P. (2005) Design and Applications of a Curie Point Pyrolyzer. Technical Report, Japan Analytical Industry Co., Ltd., Musashi, Mizuho, Nishitama Tokyo.

Picó, Y. and Barceló, D. (2020) Pyrolysis gas chromatography-mass spectrometry in environmental analysis: focus on organic matter and microplastics. *Trends Anal. Chem.*, **115964**. doi: 10.1016/j.trac.2020.115964.

Roussis, S.G. and Fedora, J.W. (1996) Use of a thermal extraction unit for furnace-type pyrolysis: suitability for the analysis of polymers by pyrolysis/GC/MS. *Rapid Commun. Mass Spectrom.*, **10** (1), 82–90.
doi: 10.1002/(SICI)1097-0231(19960115)10:1<82::AID-RCM449>3.0.CO;2-N.

Sahai, H., Dolores, M., Jesús, M., Bueno, M., Aguilera, A.M., and Fernandes-Alba, A.R. (2024) Chemosphere evaluation of the sorption/desorption processes of pesticides in biodegradable mulch films used in agriculture. *Chemosphere*, **351** (December 2023), 141183. doi: 10.1016/j.chemosphere.2024.141183.

Schulten, H.-R., Fischer, W., and Wallstab, H.-J. (1987) New automatic sampler for Curie Point pyrolysis its combination with gas chromatography. *HRC CC*, **10**, 467.

Simon, W. and Giacobbo, H. (1965) *Chem. Ing. Technol.*, **37**, 709.

Uden, P.C. (1983) Nomenclature and terminology for analytical pyrolysis (IUPAC recommendations 1993). *Pure Appl. Chem.*, **65**, 2405–2409.

Wampler, T.P. (ed.) (1995) *Applied Pyrolysis Handbook*, Marcel Dekker, New York.

Zaikin, V.G., Mardanov, R.G., Yakovlev, V.A., and Platé, N.A. (1992) Composition and microstructure of butadiene-isoprene copolymers from pyrolysis-gas chromatographic/mass spectrometric data. *J. Anal. Appl. Pyrolysis*, **23**, 33–42.

Section 2.1.11 Thermal Extraction (Outgassing)

JAI (2024) List of Outgas Collectors, *www.jai.co.jp/english/products/outgas/index.html* (accessed 8 March 2024).

Markes (2009b) Direct Desorption of Car Trim Materials for Volatile Organic Compounds (VOC) and Semi-volatile Organic Compounds (SVOC) Analysis in Accordance with Method VDA 278. Application Note TDTS 59.

Markes (2011) Introducing the Micro-Chamber/Thermal Extractor™ (μ-CTE™) for Rapid Screening of Chemicals Released (emitted) by Products and Materials. Application Note 67.

Markes (2012) Rapid Aroma Profiling of Cheese Using a Micro-Chamber/Thermal Extractor with TD-GC-MS Analysis. Application Note TDTS 101, Markes International Ltd, Llantrisant, UK.

Markes (2015) Micro Chamber/Thermal Extractor (μ-CTE). Product information, Markes International Ltd., Llantrisant, UK.

Oguri, N. (2005) Outgas Analysis for Hard Disk Industries. Technical Report, Japan Analytical Industry Co., Ltd.

Roussis, S.G. and Fedora, J.W. (1998) Use of a thermal extraction unit for furnacetype pyrolysis: suitability for the analysis of polymers by pyrolysis/GC-MS. *Rapid Commun. Mass Spectrom.*, **10**, 82–90.

U.S. EPA (1996) Method 8275A. Semivolatile Organic Compounds (PAHs and PCBs) in Soil/Sludges and Solid Wastes Using Thermal Extraction/Gas Chromatography Mass Spectrometry (TE/GC-MS), Rev.1. U.S. Environmental Protection Agency, 1 December 1996.

Section 2.1.12 Liquid Chromatography Clean-up

Che, J., Yu, C., Liang, L., and Hübschmann, H.-J. (2014) Automated Online GPC/GC-MS for the Determination of 181 Pesticides in Vegetables. Application Note 10422, Thermo Fisher Scientific, Shanghai.

Dugo, P., Dugo, G., and Mondello, L. (2003) On-line coupled LC-GC: theory and applications, in *Advanstar Communications - Recent Applications in Multidimensional Chromatography. LCGC Eur.*, **12**, 2–10, www.lcgceurope.com.

Hogendoorn, E.A., van der Hoff, G.R., and van Zoonen, P. (1989) Automated sample clean-up and fractionation of organochlorine pesticides and polychlorinated biphenyls in human milk using NP-HPLC with column-switching. *J. High Res. Chrom.*, **12** (12), 784–789. doi: 10.1002/jhrc.1240121204.

Hyvönen, H., Auvinen, T., Riekkola, M.-L.L., and Himberg, K. (1992) Determination of single PCB congeners in fish by on-line LC-GC. *J. Microcol. Sep.*, **4** (2), 123–127. doi: 10.1002/mcs.1220040203.

Jongenotter, G.A., Kerkhoff, M.A.T., Van der Knaap, H.C.M., and Vandeginste, B.G.M. (1999) Automated on-line GPC- GC-FID involving co-solvent trapping and the on-column interface for the determination of organophosphorus pesticides in olive oils. *J. High. Res. Chro.*, **22**, 17–23.

Majors, R.E. (1980) Multidimensional high performance liquid chromatography. *J. Chromatogr. Sci.*, **18**, 571–577.

Munari, F., Trisciani, A., Mapelli, G., Trestianu, S., Grob, K., and Colin, J.M. (1985) Analysis of petroleum fractions by on-line micro HPLC-HRGC coupling, involving increased efficiency in using retention gaps by partially concurrent solvent evaporation. *J. High Res. Chrom.*, **8** (9), 601–606. doi: 10.1002/jhrc.1240080925.

Munari, F., Dugo, G., and Cotroneo, A. (1990) Automated on-line HPLC-HRGC with gradient elution and multiple GC transfer applied to the characterization of citrus essential oils. *J. High Res. Chrom.*, **13** (1), 56–61. doi: 10.1002/jhrc.1240130112.

Ramsteiner, K.A. (1987) On-line liquid chromatography-gas chromatography in residue analysis. *J. Chrom. A*, **393** (1), 123–131. doi: 10.1016/S0021-9673(01)94210-4.

Schmarr, H.-G., Mosandl, A., and Grob, K. (1989) Early solvent vapor exit in GC for coupled LC-GC involving concurrent eluent evaporation. *J. High Res. Chrom.*, **12** (6), 375–382. doi: 10.1002/jhrc.1240120607.

Serrano, R., Hernández, F., Van Der Hoff, G.R., and Van Zoonen, P. (1998) Sample clean-up and fractionation of organophosphorus pesticide residues in mussels using normal-phase LC. *Int. J. Env. Anal. Chem.*, **70** (1–4), 3–18. doi: 10.1080/03067319808032600.

Trestianu, S., Munari, F., and Grob, K. (1996) *Riva 85 - Riva 96, Ten Years Experience on Creating Instrumental Solutions for Large Volume Sample Injections into Capillary GC Columns*, CE Instruments Publication, Rodano, Italy.

Section 2.1.13 Pressurized Liquid Extraction

Accelerated Solvent Extraction (ASE) (2012) Techniques for In-Line Selective Removal of Interferences. Technical Note 210, Thermo Fisher Scientific.

Barp, L., Višnjevec, A.M., and Moret, S. (2023) Pressurized liquid extraction: a powerful tool to implement extraction and purification of food contaminants. *Foods*, **12** (10), 1–25. doi: 10.3390/foods12102017.

Curren, M.S.S. and King, J.W. (2002) Sampling and sample food analysis preparation, in *Comprehensive Analytical Chemistry, Chap*, Vol. **25**, Elsevier B.V, pp. 873–874, doi: 10.1016/S0166-526X(02)80062-9.

Haglund, P. and Spinnel, E. (2010) A modular approach to pressurized liquid extraction with in-cell clean-up. *LCGC Eur.*, **23**, 292–301.

Hubert, A., Wenzel, K.-D., Engewald, W., and Schüürmann, G. (2001) Accelerated solvent extraction - more efficient extraction of POPs and PAHs from real contaminated plant and soil samples. *Rev. Anal. Chem.*, **20**, 101–144.

Kettle, A. (2013) Recent advances in pressurized fluid extraction. *LCGC supplements special. Issues*, **31** (Nov 01), 28–33, *https://www.chromatographyonline.com/view/recent-advances-pressurized-fluid-extraction*.

Oleszek-Kudlak, S., Shibata, E., Nakamura, T., Lia, X.W., Yua, Y.M., and Donga, X.D. (2007) Review of the sampling and pretreatment methods for dioxins determination in solids, liquids and gases. *J. Chin. Chem. Soc.*, **54**, 245–262.

Ramos, J.J., Dietz, C., Gonzalez, M.J., and Ramosa, L. (2007) Miniaturised selective pressurised liquid extraction of polychlorinated biphenyls from foodstuffs. *J. Chromatogr. A*, **1152**, 254–261.

Richter, B.E., Jones, B.A., Ezzell, J.L., Porter, N.L., Avdalovic, N., and Pohl, C. (1996) Accelerated solvent extraction: a technique for sample preparation. *Anal. Chem.*, **68** (6), 1033–1039. doi: 10.1021/ac9508199.

Sanchez-Prado, L., Lamas, J.P., Lores, M., Garcia-Jares, C., and Llompart, M. (2010) Simultaneous in-cell derivatization pressurized liquid extraction for the determination of multiclass preservatives in leave-on cosmetics. *Anal. Chem.*, **82**, 9384.

Spinner, E. (2008) PLE with integrated cleanup followed by alternative detection steps for cost-effective analysis of dioxins and dioxin-like compounds. PhD thesis. Umeå University, Sweden.

Ullah, S.M.R., Srinivasan, K., and Pohl, C. (2012) Advances in Sample Preparation for Accelerated Solvent Extraction. Technical Report PN70071, Thermo Fisher Scientific, Sunnyvale, CA.

U.S. EPA Method 3545A, Rev. 1 (2007) *https://www.epa.gov/esam/method-3545a-sw-846-pressurized-fluid-extraction-pfe* (accessed 16 February 2024).

2.2 Gas Chromatography

Chromatography, in all its forms, undoubtedly represents the most outstanding development to date in physical chemistry for the separation of molecules. During my lifetime it has revolutionized nearly every field of analytical chemistry.

Denis H. Desty, 1991
1923–1994, British scientist and inventor, known primarily for his work in the fields of chromatography and combustion science

2.2.1 Sample Inlet Systems

Much less attention was paid to this area at the time when packed columns were used. On-column injection (OCI) was the state-of-the-art technique and was in no way a limiting factor for the quality of the chromatographic separation. With the introduction of capillary column techniques in the earlier form of glass capillaries and fused silica capillaries today, high resolution gas chromatography (HRGC) for GC-MS made a great technological advance. Many well-known names in chromatography are associated with important contributions to GC sample introduction: Desty, Ettre, Grob, Halasz, Poy, Schomburg, and Vogt, among others. The exploitation of the high separating capacity of capillary columns now requires perfect control of a problem-orientated sample injection technique.

"*If the column may be described as the heart of chromatography, then sample introduction may, with some justification, be referred to as the Achilles heel! Sample indroduction is, arguably, the least understood and most confused aspect of modern gas chromatography.*" (Pretorius and Bertsch, 1983.)

Sample injection should satisfy the following requirements still valid today (Schomburg et al., 1977; Schomburg, 1983, 1987):

- Achieving the optimal efficiency of the column.
- Achieving a high signal/noise (S/N) ratio through peaks which are as steep as possible in order to be certain of the detection and quantitative determination of trace components at sufficient resolution (no band broadening).
- Avoidance of any change in the quantitative composition of the original sample (systematic errors, accuracy).
- Avoidance of statistical errors that are too high for the absolute and relative peak areas (precision).
- Avoidance of thermal and/or catalytic decomposition or chemical reaction of sample components.
- Sample components that cannot be evaporated must not reach the column or must be removed easily (pre-column backflush). Involatile sample components lead to decreases in separating capacity through peak broadening and shortening of the service life of the capillary column.
- In the area of trace analysis, it is necessary to transfer the substances to be analyzed to the separating system with as little loss as possible. Here the injection of larger sample volumes (up to 100 µL) is desirable.

- Simple handling, service, and preventive maintenance of the sample injection system play an important role in routine applications.
- The possibility of automation of the injection is important, not only for large numbers of samples but also as automatic injection is superior to manual injection for achieving a low standard deviation.

The sample injection is of fundamental importance for the quality of the chromatographic separation with all GC and GC-MS systems. Careless injection of a sample extract frequently overlooks the outstanding possibilities of the capillary technique. Poor injection cannot be compensated for even by the choice of the best column material, or the best choice of detectors. The use of a mass spectrometer as a detector can be much more powerful if the chromatography is of the best quality. In fact, the use of GC-MS systems shows very often that as soon as GC is coupled to MS, the chromatography is rapidly degraded to an inlet route. Effort is put too quickly into the optimization of the parameters of the MS detector without exploiting the much wider potential of the GC side. Each GC-MS system is only as good as the chromatography allows!

The starting point for the discussion of sample inlet systems is the target of creating a sample zone at the top of the capillary column for the start of the separation which is as narrow as possible. This narrow sample band principally determines the quality of the chromatography as the peak shape at the end of the separation cannot be better (narrower, more symmetrical, etc.) than at the beginning. As an explanatory model, chromatography can be described as a chain of distillation plates. The number of separation steps (number of plates) of a column is used by many column producers as a measure of the separating capacity of a column. In this sense, sample injection means the application to the first plate of the column. In capillary GC, the volume of such a plate is less than 0.01 µL. The sample extracts used in trace analysis are generally very dilute, making larger quantities of solvent (≥ 1 µL) necessary. This limited sample capacity of capillary columns shows the importance of the split and splitless sample injection techniques for high resolution gas chromatography with fused silica capillary columns (HRGC). In practice, different types of injectors are used. The sample injection systems in use are classified as hot, cold, or on-column according to their function.

2.2.2 Carrier Gas Regulation

The carrier gas pressure regulation found in commercial GC instrumentation follows two different principles which influence injection modes and the adaptation of automated sample preparation devices. The basic principles are represented with the inflow and outflow of a water bucket as illustrated in Figure 2.72. The water level in the bucket can be considered as the pressure in the GC injector. In the forward regulation scheme, the inflow is regulated to keep a constant water level that stands for the head pressure of the outlet. The outlet flow is constant depending on the regulated inflow (pressure) and line (column) dimensions. The back pressure regulated scheme provides a constant inflow (mass flow controller). A variable outflow in the back controls the water level of the bucket.

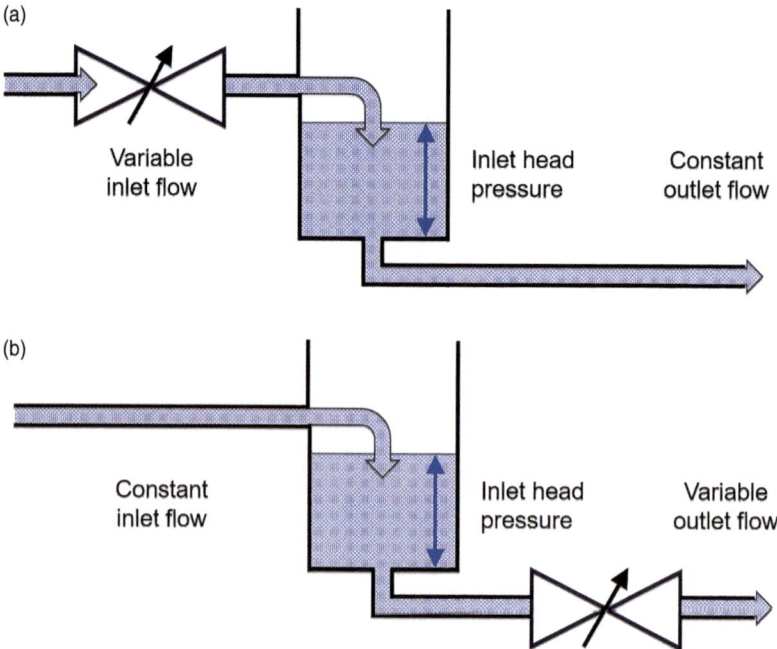

Figure 2.72 Basic operation principles of the forward (a) and back pressure (b) flow regulation.

Most modern GC instruments are equipped with electronic pressure and flow control modules (EPC, electronic pressure control). The pressure sensor control loop must be short and taken special care of for a rapid feedback control upon integrating external devices by cutting and rerouting the carrier gas lines for headspace, purge and trap (P&T), or thermal desorption (TD) units. In these installations, the carrier gas flow is often directed from the GC regulation to the external device and returned again via the capillary transfer line to the GC injector. Internal in-line filter cartridges are typically used to prevent the lines from accumulating trace contaminations in the carrier gas, which needs to be exchanged frequently. Also, when applying large volume injection (LVI) techniques, knowledge of the individual pressure regulation scheme is required for successful operation.

2.2.2.1 Forward Pressure Regulation

The classical control of column flow rates is achieved by a "forward pressure" regulation (Figure 2.72a). The inlet carrier gas flow is regulated by an EPC valve in front of the injector. The pressure sensor is installed before and close to the injector body and hence to the column head for fast feedback on the regulation in the carrier gas supply line to the injector, the septum purge or the split exit line. The split exit usually is regulated further down the line by a separate flow-regulating unit (see Figure 2.73).

The forward injector pressure regulation is the most simple and versatile carrier gas regulation used in all split/splitless injection modes as well as the cold on-column injector. It allows the LVI in the splitless mode using the concurrent

2.2 Gas Chromatography

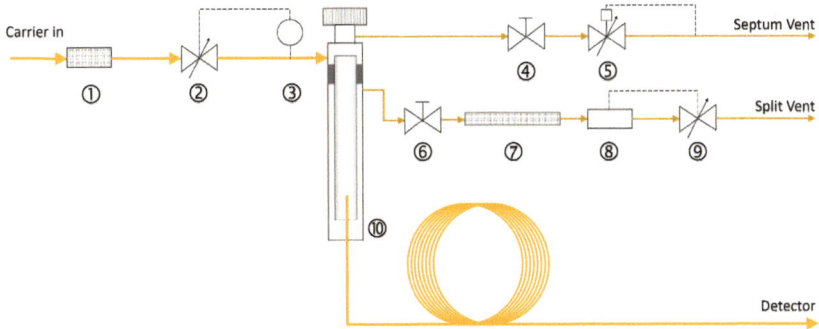

Figure 2.73 Injector forward pressure regulation. 1. Carrier gas inlet filter, 2. proportional control valve, 3. electronic pressure sensor, 4. septum purge on/off valve, 5. septum purge regulator, 6. split on/off valve, 7. split outlet filter cartridge, 8. electronic flow sensor, 9. proportional control valve, and 10. injector body with column installed.

solvent recondensation (CSR) (Magni and Porzano, 2003). Also, split and splitless injections for narrow bore columns, as used in Fast GC, benefit from the forward pressure regulation with a stable and very reproducible regulation at low carrier gas flows that is not usually possible with mechanical or electronic mass flow controllers. Also, the switch of injection techniques in one injector, for example, using a programmed temperature vaporizer (PTV) for regular split/splitless and OCI with a pre-column (retention gap) is possible without any modification of the injector gas flow regulation.

The adaptation of external devices such as headspace, P&T, or TD is straightforward, with the pressure sensor installed close to the injector head which is generally the case with commercial forward-regulated GC instruments. Carrier gas flow and pressure of the external device are controlled by the EPC module of the GC. The position of the pressure sensor in the flow path of an inlet is important for accurate measurement of the inlet pressure to get rapid feedback control.

LVIs are also straightforward using forward pressure-regulated injectors. Typically, the regular split/splitless device can be used for larger volume injections of up to 10 or 50 µL without any hardware modification, exploiting the CSR effect (see Section 2.2.5.6). For injections in the range of up to 30 µL, only the injection volume programming of the autosampler and the length of the first oven isothermal phase have to be adjusted accordingly, letting elute the wider solvent peak before ramping the GC oven.

2.2.2.2 Back Pressure Regulation

The term *back pressure* describes the position of the pressure sensor and EPC valve behind the injector in the split exit line. A mass flow controller is installed in front of the injector (see Figure 2.74). In this widely used carrier gas regulation scheme for split/splitless injectors, the pressure sensor in the septum purge line is installed close to the injector body to ensure a precise pressure measurement at the column head.

The total carrier gas flow to the back pressure-regulated injector is set by the mass flow controller. The total flow into the injector is split into three flow paths of split,

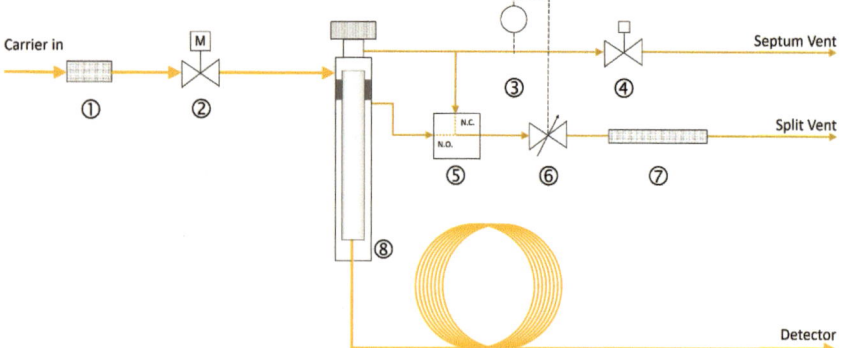

Figure 2.74 Injector back pressure regulation. 1. Carrier gas inlet filter, 2. mass flow regulator, 3. electronic pressure sensor, 4. septum purge regulator, 5. solenoid valve, 6. electronic pressure control valve, 7. split outlet filter cartridge, and 8. injector body with column installed.

septum purge, and column flow. The column flow is adjusted and kept constant by varying the split and septum purge flows according to the required column flow conditions. This setup allows the exact and independent setting of split flows from column flows. However, when the back pressure regulated inlet is used in split mode, a large amount of sample passes through the regulation through the split exit line. The split mode of operation requires special care to prevent the system from flow restrictions by sample deposit, chemical attack and finally clogging.

When adapting an external inlet device, the outlet of the mass flow controller is directed on the shortest route to the external device. A sample transfer line from the external device is returned to the injector either using the regular carrier inlet or by piercing a needle end of the transfer line through the regular septum. The pressure sensor must be relocated from the split line close to the EPC valve to achieve correct flows and sufficient pressure control stability within a short feedback loop due to the additional flow restrictions in the external device. The carrier gas flow through the external device is regulated by the mass flow controller. The GC inlet pressure at the injector and the effective column flow are regulated independently by the EPC module through the EPC valve. Alternatively, an external auxiliary EPC module for feeding the external device is recommended.

2.2.2.3 Carrier Gas Saving

The supply of helium carrier gas for GC-MS became increasingly difficult and expensive in many regions. Saving helium carrier gas is especially important with the expected further increasing cost in the future. Several options to save or replace helium as carrier gas have been discussed for years. While hydrogen is the classical carrier gas in GC and is always used for Flame-Ionization Detector (FID) operation, there are limitations for use with GC-MS due to its reactivity causing analyte hydrogenation and the technical requirements for appropriate vacuum systems to cope with the increased flow rates. The replacement by nitrogen for chromatography can be an option in special cases but extended runtimes and reduced separation power need to be considered (also refer to Section 2.2.8.1). Switching to nitrogen can be a valuable option during the standby operation. Then an acceptable startup time needs to be considered for flushing the GC system before operation with helium can proceed again.

Most of the modern GC systems are equipped with a device for electronic carrier gas control (EPC). The EPC function not only allows the convenient setting of the optimum gas flow parameters during the injection process but also allows a different setting during the separation phase after injection. This applies in particular to the split flow. The split flows are typically high during a split injection or directly after a splitless injection phase in the range of 30 to 100 mL/min. The split flow consumes most of the carrier gas, but without any analytical function after injection other than keeping the injector clean for the next injection. After completing the injection phase, it is not necessary to maintain high split flows, and it is advisable to save carrier gas after the oven program has started. A carrier gas saver function of the EPC can set the split flow to a lower value of 10 to 20 mL/min depending on the injector design. The low split flow is activated about 2 to 3 min after the injection, the sample vapor transfer to the column is completed by then, at the start of the oven temperature program and remains during the complete development of the chromatogram.

Another even more practical solution is available which replaces helium during the injection process completely, serving split ratio, and septum purge, but maintains the low consuming column flow with helium as usual. This helium saver injector requires a different design with two separate gas inlets and electronic gas controls for helium carrier and nitrogen as auxiliary gas (Warner et al., 2022). Figure 2.75

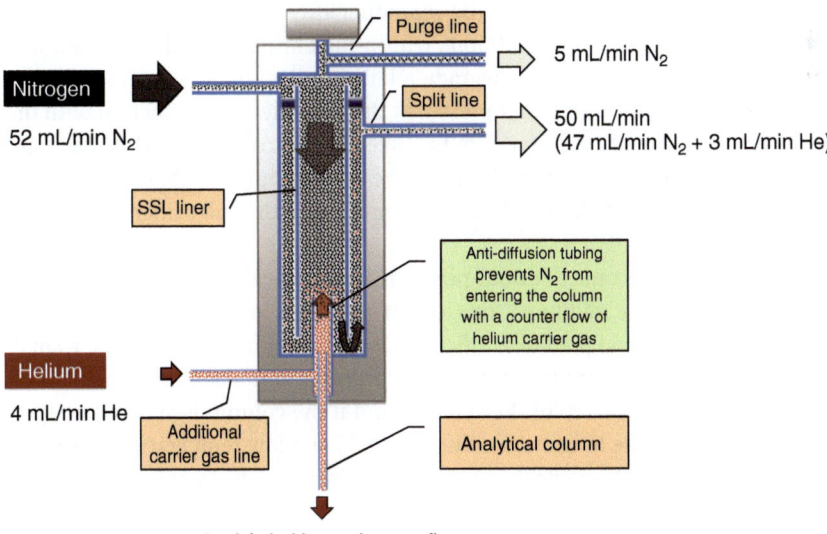

Figure 2.75 Helium saver injector operation principle (Thermo Fisher Scientific Inc.).

describes the basic operation. A diffusion tube at the bottom of the injector body feeds the helium carrier to the analytical column while separating the carrier gas from the auxiliary nitrogen gas. The injection and sample vaporization processes are performed in a nitrogen atmosphere, also serving the septum purge and split lines. All gas flows are EPC controlled to move the sample vapor cloud into the analytical column in either split or splitless operation. The column separation is maintained as usual with the low flow of helium as carrier gas resulting in a significant saving in helium carrier gas. No modifications to the chromatographic methods, mass spectral identification, or a targeted MS/MS detection are required.

2.2.3 Injection Port Septa

The injection port septum is a still unsolved weak point in GC. It is arguably the most important but often overlooked functional part of the injector. The septum is used to seal the injection port, must withstand the inlet pressure, and provides smooth access for the syringe needle to inject the sample. The syringe needle punches the septum repeatedly in a small area causing gradually small leaks, which are detectable with an electronic leak detector already before a noticeable analyte response drop occurs. The risk of leaks increases with use, older septa take longer to re-seal than new ones (Hinshaw, 2008; Lidgett and Grossman, 2015). The septum purge flow is mandatory to flush potential septum leaks, in particular oxygen, and leachables away from the inlet.

The septum material is based on silicon rubber, a comparable polymer that is used for the stationary phases of fused silica columns. The ideal injection port septum should exhibit low bleed, resist leaks, be easy to penetrate for the syringe needle, and re-seal itself after the injection, reliably for a large number of analyses. The septum performance is vital to the reproducibility of results, sensitivity, peak shape, and reproducible flow conditions within the chromatographic system. It is widely accepted that silicon septa degrade relative to the operating temperature of the injector, the operation at higher pressures, and the physical interaction with the syringe needle, especially with modern fast injection autosamplers. This mechanical stress causes shedding of particles from the septum but also abrasion from the outside of the metal needle which can accumulate in the inlet liner (de Zeeuw, 2013).

Septum bleed can become a serious problem during trace analysis, as the leached compounds elute in the middle of the GC run with high intensity. Persistent bleed causes low sensitivity and poor quantification. Once a septum is punctured, a small amount of silicon can be transferred into the injector with each injection. This is quickly transferred to the top of the column, and if the column is at a low temperature, focused here. The longer the column is held at a low temperature, the more intense the presence of bleed peaks in the chromatogram. The major component of septum bleed is a series of siloxanes with increasing molecular mass (see Table 3.16 and Figure 3.169). Other observed bleed components can be of phthalates and hydrocarbons (Warden and Pereira, 2013).

Hot split/splitless injectors of different manufacturers can have very different temperatures at the septum nut. For a given inlet temperature set point, the temperature at the septum is typically lower than the set point but varies among instruments. Inlets with a larger temperature gradient over the injector length and a cooler septum base typically experience fewer problems with septa sticking. Operators of instruments with inlets that are more evenly heated with a smaller temperature gradient and a hotter septum base need to consider using septa that are rated for the highest operation temperature and setting the inlet at the lowest permissible temperature (Grossman, 2013). Septa brands, however, specify a single, maximum operating temperature that corresponds to the inlet set point, not the actual septum nut temperature the septum can withstand and still function properly. All commercial septa today withstand a nominal injector temperature of 350 °C without melting or hardening. Special care is advised when installing a new cold septum to avoid a typical inherent overtightening, as the silicon septa tend to swell significantly with increasing temperature.

Bleed and temperature optimized septa (BTO™, Chromatography Research Supplies, Inc., Louisville, KY, USA) are well suited for trace GC and GC-MS applications which require low bleed and extended sealing capabilities for high injection numbers. As an additional benefit, many septa provide a pre-drilled needle guide for increased septum lifetime, which makes them ideal for use with liquid autosamplers as well as with solid phase microextraction devices (SPME). BTO septa are specified for use with up to 400 °C inlet temperature, are preconditioned, and require very little additional conditioning before use. In routine operation it is recommended to change septa at the end of a working day, so the septa can condition overnight to ensure the lowest bleed for the next-day analyses.

All GC septa may absorb contaminants when handled without gloves or from volatile organics present in the air. Any such contaminants picked up by a septum may elute during analysis and appear as septa bleed or additional ghost peaks. To avoid contamination, septa generally need to be stored in their original glass containers with the lid tightly closed. Forceps are recommended for handling septa, and if the septa need to be handled with fingers, then only the edge should be touched and not the sealing surface.

The reported average lifetime in terms of the number of injections ranges from 150 to 450 injections for silicon rubber septa and up to 2000 injections for the Merlin seal (see Section 2.2.3.2), depending on the septum type and conditions used (Westland et al., 2012). Due to the necessary frequent exchange of the inlet liner, it is recommended to replace both, the septum and insert liner regularly on a preventive maintenance schedule.

A completely septumless injector design of the cooled injection system CIS 4 is provided by the company GERSTEL GmbH, Mülheim an der Ruhr in Germany. Figure 2.76 shows the idle and injection states of the septumless head of the injector. A spring-loaded sealing body located inside the head seals the inlet. A syringe or SPME device easily penetrates pushing the plunger back. Thousands of installations have proven the reliable and silicon background-free injection for automated routine analyses. A septum purge is not required (Gerstel, 2022).

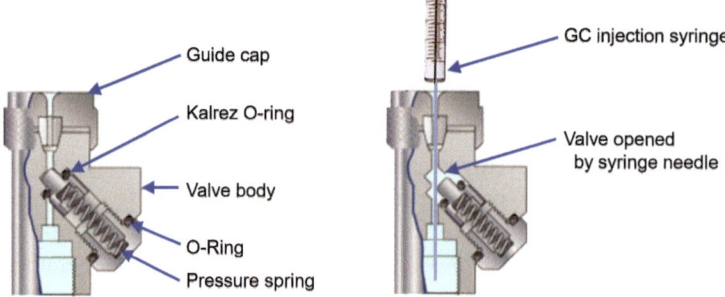

Figure 2.76 Septumless head operation of the GERSTEL CIS 4 injector (Courtesy GERSTEL GmbH & Co KG).

2.2.3.1 Septum Purge

All modern GC instruments offer the injector function called *septum purge*, introduced in 1972 by Kurt Grob. A constant flow in the range of up to 5 mL/min carrier gas to purge any leaking oxygen or leachables from the bottom of the septum has proven to be reliable. This small carrier gas flow prevents leaching contaminations of the septum material from entering into the insert liner (see Figures 2.86 and 2.92).

The septum purge flow is recommended to stay on during the injection phase and analysis. With properly chosen injection conditions keeping the solvent/sample vapor cloud inside of the insert liner, there will be no loss of sample analytes via the septum purge outlet (see Section 2.2.4).

2.2.3.2 The MicroSeal Septum

The Merlin MicroSeal™ Septum is a unique long-life replacement for the conventional silicon septa on split/splitless and PTV injectors. MicroSeal is a trademark of the Merlin Instrument Company, Half Moon Bay, CA, USA. Functionally, the MicroSeal provides a microvalve two-step sealing mechanism (see Figures 2.77 and 2.78). A double O-ring type and a top wiper rib around the syringe needle improve resistance to particulate contamination, and a spring-assisted duckbill seals the

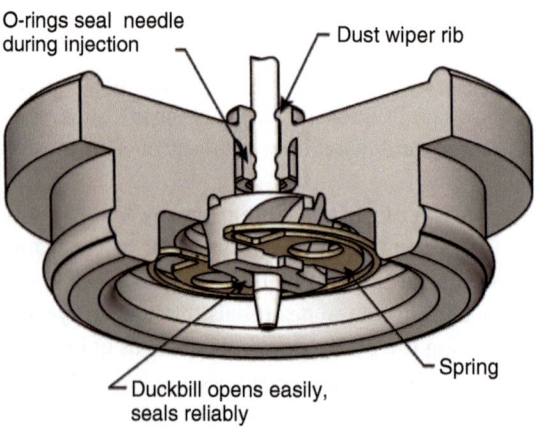

Figure 2.77 MicroSeal cutaway view (Merlin, 2018/Merlin Instrument Company).

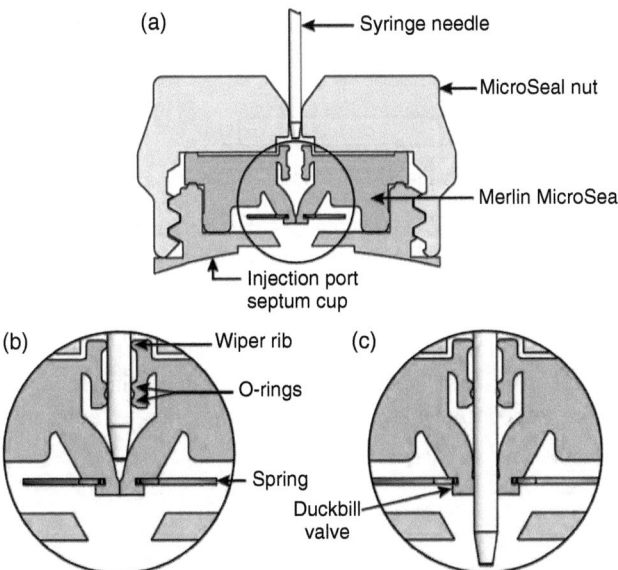

Figure 2.78 MicroSeal injection sequence – a 2-step injection process. (a) MicroSeal installed under regular injector cap, (b) the needle is first penetrating the dust wiper rib, and (c) the needle is penetrating the spring-loaded duck bill seal into the injector liner (Merlin, 2018/Merlin Instrument Company).

Figure 2.79 MicroSeal compatible needle and SPME probe styles.

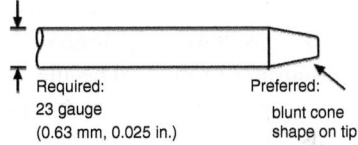

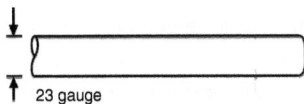

injection port. The MicroSeal can be easily installed on many injector types by removing the existing septa without any modifications. It can be used at injection port temperatures up to 400 °C. The MicroSeal is usable manual or in autosampler applications with 0.63 mm diameter (23 gauge) blunt cone syringes (see Figures 2.79 and 2.80) as well as for SPME fiber (23 gauge) and arrow (26 gauge) probes (Merlin, 2018). The seal is available for operation up to 45 psi (310 kPa) as low pressure MicroSeal or up to 100 psi (690 kPa) with the general purpose series.

The seal is made from Viton™ material, a high temperature-resistant fluorocarbon elastomer providing high resistance to wear. It greatly reduces the shedding of septum particles into the injection port liner. Because the syringe needle does not pierce a septum layer, it eliminates a major source of silicon septum bleed and ghost peaks.

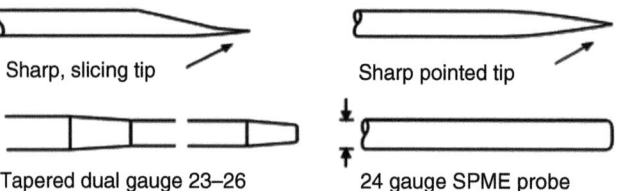

Figure 2.80 Do not use sharp-edged or sharp-pointed needles that can pierce and damage the duckbill seal. Tapered dual gauge and SPME Arrow require a 26 gauge seal.

Due to this sealing mechanism, it is especially suited for trace analysis applications requiring particularly low background conditions. The low syringe insertion force makes it also ideal for manual injections. The longer lifetime for many thousands of injections reduces the chances of septum leaks, especially during extended autosampler runs.

2.2.4 Injection Port Liner

The inlet liners of the injection port are the point of sample evaporation and determine to a high degree the GC performance in terms of sensitivity, reproducibility, and sample integrity. Inlet liners retain nonvolatile matrix components and protect the analytical column from performance degradation. They are consumables necessary to be replaced on a regular basis depending on the sample matrix load on a daily or weekly basis according to a preventive maintenance plan.

The liner choice needs to fit the analytical method setup if split, splitless or SPME injections are planned, or the basic injection techniques hot needle with thermospray or the cold needle with liquid band formation are preferred. Different liner geometries are available to support individual injection and vaporization needs. There is not a single liner that serves all injection modes. Some basic rules help identify the liner of choice for the chosen injection technique.

2.2.4.1 Split Injection

Split injections are used to achieve a "split" of the sample vapor in the injector for a reduction of the amount of sample that enters the analytical column. For concentrated (undiluted) samples, this is necessary due to the limited sample capacity of fused silica columns. The amount of sample vented via the split line relative to the column carrier gas flow is adjusted by the split flow rate. The split ratio is the split flow divided by the column flow at injection, usually normalized to one. Split liners provide high flow speeds inside the liner for a short sample band entering the column. Sufficient space is required between the needle exit and column entry for the evaporation of the solvent with the analytes, and the homogeneous mixing with the carrier gas before the sample cloud "flies" across the column entry. Straight liners with and without glass wool are used for split injections (Restek, 2002). The inside diameter should be low to guarantee high split flows through the injector body out to the split line (see Figure 2.83). The residence times of samples in the liner are short compared to a splitless injection. With low split ratios, labile compounds can also be

2.2.4.2 Splitless Injection

Splitless injections are typical for trace analysis with a total sample transfer into the analytical column. It is mandatory that inlet liners for splitless injection provide sufficient space for the solvent cloud to expand during evaporation (see Table 2.19), hence liners with larger inner diameters are used (see Figure 2.83). The knowledge of liner volume and solvent vapor volume is essential for the optimum parameter setting. Unfortunately, the expansion volume calculated in Table 2.19 does not reflect the diffusion of the solvent vapor against the carrier gas flow, which is considered to be significant. From the practical side the calculated vapor volume of the injected solvent volume should not exceed two third of the liner volume. Using a surge pressure (pressure pulse) with a three to four times higher pressure than the regular column flow head pressure during injection reduces the volume of the vapor cloud.

Special attention must be taken with water as solvent, or water dissolved in solvents. The vapor volume of water is more than 7x higher compared to hexane (Table 2.19). The overfilling of the liner is the major problem when injecting water-containing solutions. Wide liners and small injection volumes must be used. The use of a vapor volume calculator is recommended.

Typically, "goose neck" liners with a tapered bottom end and filled with deactivated quartz wool or a deactivated frit are preferred for splitless injections, directing the sample vapors into the column and preventing analytes from getting in touch with metal surfaces at the bottom of the injector with potential memory and degradation. Overfilling of the liner must be avoided by proper selection of the solvent and the maximum tolerated "at once" injection volume. A comparison of single-tapered

Table 2.19 Solvent expansion volumes in µL.

Injection volume [µL]	0.5	1.0	2.0	3.0	4.0	5.0	M [g/mol]	BP [°C]
H_2O	710	1420	2840	4260	5680	7100	18.0	100
Acetonitrile	245	489	978	1467	1956	2445	41.1	82
Ethanol	219	438	875	1313	1751	2188	46.1	78
CS_2	212	423	846	1270	1690	2120	76.1	46
Dichloromethane	200	401	802	1200	1600	2000	84.9	39
Ethyl acetate	131	262	523	785	1046	1308	88.1	77
Toluene	120	240	480	720	960	1201	92.1	111
Pentane	111	222	443	665	887	1109	72.2	36
Hexane	98	195	390	585	780	975	86.2	69
Iso-Octane	78	155	310	465	620	775	114.2	99

The expansion volumes given here refer to an injection temperature of 250 °C and a column head pressure of 0.7 bar (10 psi). The values for other temperatures and pressures can be calculated according to $V_{Exp.} = 1/P \cdot n \cdot R \cdot T$.

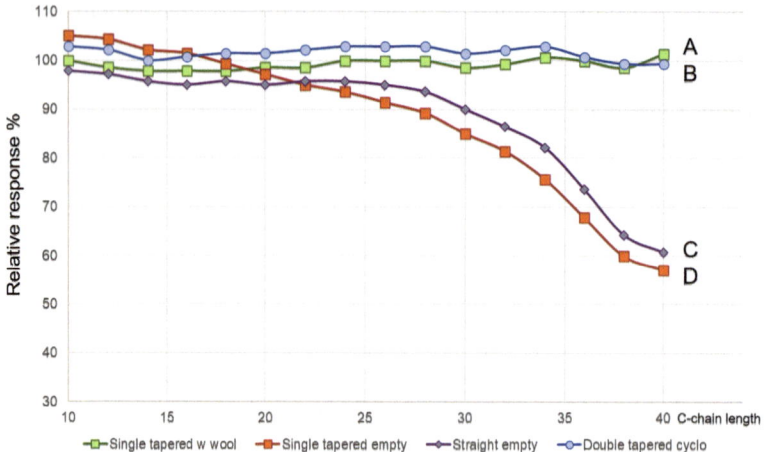

Figure 2.81 Comparison of relative response of alkane hydrocarbons with different splitless liners (Adapted from Waclaski, 2018). Conditions: 5 × 5 ng injection, inlet 280 °C, splitless 1 min, flow 1.5 mL-min, Agilent 7890 GC SSL injector, n=5 liner, FID detector. A single tapered liner with glass wool (green line); B double tapered liner with cyclo insert (blue line); C single tapered liner, empty (red line); D straight liner, empty (purple line).

liners with and without glass wool, an empty straight liner and a double-tapered cyclo liner showed significant differences in the response of high boilers. While the single tapered liners with glass wool and the double tapered cyclo liner provided even response up to C40 alkane hydrocarbons, the empty single tapered and straight liners discriminated high boiling compounds visibly already starting at C20 alkanes down to about 60% response at C40, illustrated in Figure 2.81 (Waclaski, 2018).

The transfer time of analytes from injection and evaporation into the analytical column is the residence time of the solvent/analyte cloud in the inlet liner. Labile compounds require a short residence time in the liner for reduced surface contacts. As a rule of thumb, two liner volumes should be swept into the column for a quantitative transfer. Depending on the column flow the split valve is kept closed until complete transfer (Hinshaw, 1992). Using the surge pressure (pressure pulse) during the transfer phase allows emptying the liner more quickly with shorter splitless times (Grob, 1994). The surge pressure needs to be released through the analytical column after switch off for about 2 min before the split valve can be opened to avoid a backstream and loss of sample from the analytical column.

2.2.4.3 Liner Activity and Deactivation

The activity of inlet liners can be a number one limitation for the analytical performance and is still a constant challenge in GC. Liner activity can be caused by the material itself, an insufficient or degrading deactivation, but is also built up during regular use by the deposit of nonvolatile matrix components, especially when using inlet liners with glass wool or frit. A glass wool plug retains the nonvolatile components of a sample or can become rapidly active and then needs to be replaced. The liner activity changes with the first matrix sample injected. While fulfilling a major task of preventing low or nonvolatile matrix from entering the analytical column, the matrix deposit in the liner is causing adsorption and decomposition effects for subsequent injections. A quality control and preventive maintenance plan is required with regular liner exchange according to the sample matrix burden. The same applies for fritted liners. In comparison with glass wool liners, it should be noted that the deactivation for glass wool is reported to be more efficient than for the frit part (Pack, 2023).

Significant progress is noticeable in recent years in reducing the liner activity by chemically modifying the glass surface or applying high temperature coatings. Commercially deactivated liners are available from different manufacturers using proprietary processes, for instance, with phenylmethyl surface deactivation. These treatments offer a highly efficient deactivation with stability even to high temperature ranges above 400 °C. Specific deactivations are available, for example, for basic compounds such as amines. It is recommended to test different liners for a specific application for inertness, recovery, and robustness, done with a typical matrix and analyte concentrations close to the method detection limits (MDLs).

The liner activity can be measured and compared by using injections of endrin and dichloro-diphenyl-trichloroethane (DDT). The breakdown products are endrin ketone and endrin aldehyde, respectively, dichloro-diphenyl-dichloroethane (DDD). With breakdown percentages less than 15% each, an insert liner is accepted to be inert (e.g. U.S. EPA method 8081b).

Glass wool in the inlet liner supports solvent and analyte evaporation as well as improves injection reproducibility by mixing the resulting vapors with carrier gas. Quartz wool should be preferred as it contains less impurities. Liners with deactivated glass wool are required for the fast cold needle "liquid band" injection (see Section 2.2.5.2). In contrast, and due to its beneficial features the glass wool plug can be a root cause of inlet liner activity by the retained nonvolatile matrix. Approximately 50 mg of deactivated glass wool is typically used in a liner. There are different types of glass wool: regular or base-deactivated borosilicate glass wool, and also deactivated fused silica quartz wool with low alkali content. If the glass wool in the liner becomes the major source of activity, for instance due to matrix deposit, causing decomposition or analyte loss, the only remedy is to change liners more often, or work, if possible, using baffled liners without the wool. Glass wool liners are not recommended for analysis of reactive polar compounds, and thermally labile compounds like phenols, organic acids, pesticides, amines, or drugs of abuse (Agilent, 2021).

Straight liners can be mechanically cleaned with solvents, sonication, or a soft brush (e.g. a pipe cleaner), and reused, but require a new deactivation. Do not scratch or use acids to leach the liner surface; this will increase the liner activity significantly. Several liner deactivation methods are reported in the literature typically offering a gas phase or coated deactivation for instance using dimethyldichlorosilane (DMDCS) (Figure 2.82). For a simple and efficient "in-house" coated deactivation, the treatment with Surfasil™ or Aquasil™ has proven to be very efficient and durable and is used and recommended, for instance, for the multi-residue analysis of pesticides. The immersion procedure for the pre-cleaned liners is easy to perform with only a few steps (Thermo Scientific, 2008). Liners with a glass wool plug can also be

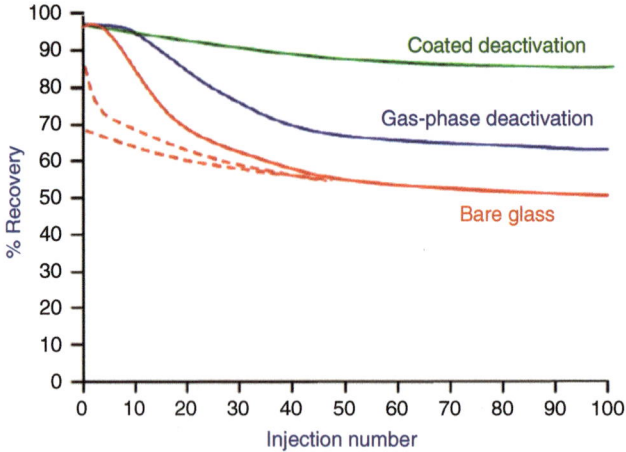

Figure 2.82 Injection port liner activity dependence from the injection number (Klee, 2013).

deactivated using the SurfaSil or AquaSil procedure. Several liners can be treated at once, and stored for later use:

1) Dilute the SurfaSil siliconizing fluid in a non-polar organic solvent such as acetone, toluene, carbon tetrachloride, methylene chloride, chloroform, xylene, or hexane. Typical working concentrations are 1 to 10% mass to volume.
2) Completely immerse or flood the pre-cleaned and dry inlet liner to be coated in the diluted SurfaSil solution for at least 5 to 10 s. Agitate the solution to ensure a uniform coat. A thin film will immediately coat the liner's surface.
3) Rinse the liner with the same solvent in which the reagent was diluted.
4) Rinse the liner with methanol. This rinse is required to prevent interaction of the SurfaSil coating with water and thus, reversing siliconization.
5) Air-dry the liner for 24 h or heat at 100 °C for 20 to 60 min, e.g. in the GC oven.

Other inlet liner deactivation procedures are using gas phase or solution silylation procedures. The gas phase silylation provides a more stable liner deactivation for high temperature use than a liquid process, but is a more laborious procedure (Rood, 2007).

2.2.4.4 Liner Geometry

Split liners are typically narrow and straight tubes, while splitless liners are wider and commonly show a tapered end, see Figure 2.83. The bottom tapers keep the sample cloud in the liner for an optimized transfer into the column head. Tapers prevent possible contact of sample components with the metal (gold) seal at the bottom of the inlet body. With tapered liners, the tip of the column is best positioned just above the tapered section. A thin glass or metal rod pushed upwards in the liner by the column during installation gives valuable assistance in finding the correct position of the column tip. Top tapered liners reduce solvent diffusion into the upper colder part of the injector but do not help when overfilling the liner with solvent vapor.

2.2 Gas Chromatography

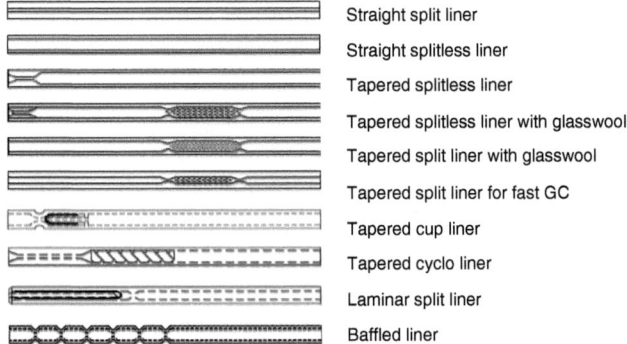

Figure 2.83 Injection port liner types for split and splitless injections.

Other liner types introduce mechanical obstacles for improved vaporization and mixing with carrier gas such as the baffled, cup or cyclo liners. Liners with sintered glass frits can replace glass wool liners by reducing aerosol transfer into the column with improved vaporization. Although those inlet liners offer improved injection conditions, they can hardly be cleaned and reused.

Special liner types are used for headspace and SPME applications. In many cases, the transfer line of a headspace sampler ends in a syringe-type needle. Additional liner dead volume and turbulence must be avoided for the transfer of the sample plug into the analytical column. Special straight "headspace liners" and "SPME liners" with narrow diameters are available for many GC models (see Table 2.5).

Direct OCIs can also be done with a regular syringe using a temperature programmable injector (PTV). This requires a dedicated inlet liner type, which centers a 0.53 mm ID pre-column in the liner, while the top part of the liner serves as a needle guide. Regular needle diameters of c. 0.45 mm OD are used (26G, 26sG). This allows the syringe needle to enter, centered, and deep enough into the wide-bore pre-column for a direct liquid injection. By this way, OCIs can be automated using regular GC autosampler units.

2.2.5 Hot Split/Splitless Sample Injection Techniques

In case liquid samples are applied to GC or GC-MS, the most widely used injection technique evaporates the liquid sample in the inlet liner of a constantly heated injector in order to transfer the analytes into the analytical column. Both, but first the solvent then the dissolved analytes evaporate and the sample vapor is mixed with the carrier gas. Injector temperatures of up to and even above 300 °C are usually applied for evaporation.

The transfer can be only partial for concentrated extracts (split mode), or a total transfer of the sample into the column for trace analysis is performed (splitless mode). Both injection methods require a different parameter setting, choice of inlet liners and oven program start temperature to achieve the optimum performance. Also, the possible injection volumes need to be considered. The operating procedures of split injection and total sample transfer (splitless) differ according

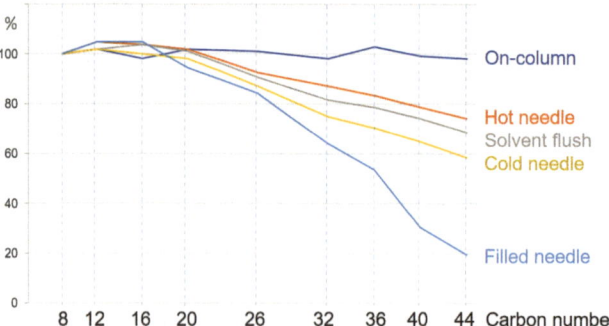

Figure 2.84 Discrimination among alkanes using different SSL injection techniques (Grob, 2001). The peak areas are normalized to C8 = 100%.

to whether there is a partial or complete transfer of the solvent/sample on to the column is planned.

The problem of discrimination of high boiling analytes on injection into hot injectors arises with the question of what is the best injection technique. Figure 2.84 shows the effects of known injection techniques on the discrimination of n-alkanes up to C44. The reference technique is the direct cold OCI avoiding the evaporation step. In the boiling range similar to a hydrocarbon chain length of up to about C20 hardly any differences are observed as illustrated in Figure 2.84. Discrimination for higher boiling compounds can be avoided by a suitable choice of the injection technique. Two vaporizing injection modes can be distinguished: The *hot needle thermospray injection* and the *cold needle liquid band injection*, both with particular analytical advantages and limitations.

2.2.5.1 Hot Needle Thermospray Injection Technique

The hot needle injection is used with hot split/splitless (SSL) injectors both in split and splitless modes. It works well with an empty inlet liner. A glass wool plug is not needed. Common to the hot needle techniques is that the sample plug is pulled up into the barrel of the syringe with a small air gap. A short delay of 3 to 5 s is used after the syringe needle enters the hot injector for heating up the complete needle length. The injection itself is then carried out quickly pushing the plunger rapidly down and removing the needle immediately (see Figure 2.85). Keeping the syringe needle cold with too short waiting time causes measurable discriminations. This injection technique is preferred for all manual injections, and can also be used with programmable autosamplers. The thermospray technique requires a hot injector head for an even heating of the syringe needle (Figure 2.86).

The important benefit of the thermospray injection is the evaporation of analytes in a cloud of solvent vapor. Due to the hot needle, a spray of droplets and vapor leaves the needle. Wall contacts of the analyte to the insert liner are minimized protecting polar and sensitive compounds from adsorption to the glass surface with potential degradation reactions. The thermospray injection is the most gentle injection technique for labile and reactive compounds (Grob, 2001). A practical

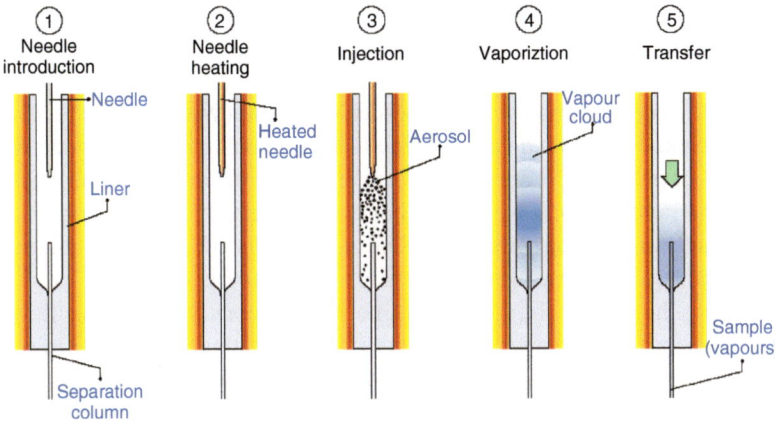

Figure 2.85 Hot needle thermospray injection technique (in splitless mode).

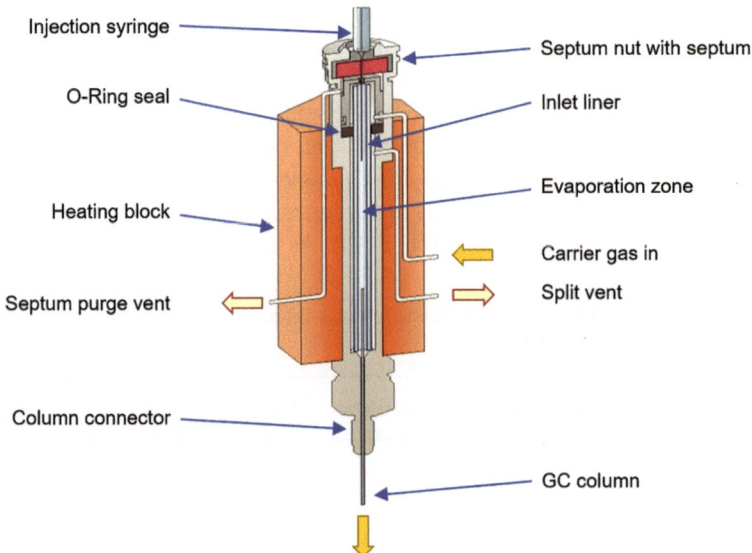

Figure 2.86 Hot split/splitless injector with a straight splitless liner and analytical column installed.

limitation is a potential discrimination toward higher boiling compounds as it is shown in Figure 2.84. The injected liquid volume needs to be adjusted according to the liner dimensions. The generated vapor cloud must stay safely inside the inlet liner. Due to diffusion against the carrier gas flow 60 to 80% of the liner volume limit the applicable solvent volume. Refer to the solvent expansion volumes in Table 2.19 to determine the maximum injection volume of a given solvent for a given inlet liner. Overfilling the liner leads to serious injector contamination.

Solvent Flush Hot Needle Injection This is the injection procedure of choice for hot injectors and provides lower discrimination of high boiling compounds. It is in use for the temperature programmable injectors as well. Approximately 0.5 to 1 µL solvent or a derivatizing agent is first taken up in the syringe, then a small plug of 0.5 µL air, and finally the sample in the desired volume. Using the matrix as the bottom layer provided the best responses for trace pesticide analysis (Westland, 2017). The procedure is also known as *sandwich injection* technique using internal standards as the first solvent plug. 3 to 5 s are used to heat up the needle in the injector, then rapidly inject the sample. The solvent flush hot needle injection can be carried out very well manually or by using a programmable autosampler. The resulting solvent vapor of the total liquid volume injected needs to be recognized.

Understanding the relationships between inlet liner geometry, carrier gas flow, injection volume, and solvent vapor volume is essential for optimizing splitless injections. The expansion volume of the solvent used must be taken into account. The maximum liquid injection volume needs to be adjusted to the volume of the installed inlet liner. Table 2.19 provides the expansion volumes of typical GC solvents for often-used injection volumes at standard inlet conditions. For proper operation, the solvent expansion volume must stay safely within 60 to 80% of the inlet liner capacity. The time required to flush the contents of the liner completely onto the column depends on the column flow. At least twice the liner volume should transfer, corresponding to a yield of 90 to 95%, before opening the split valve.

During the transfer of the sample and solvent vapor cloud into the column, the re-condensation on the column head exploiting the *solvent effect* is required. The oven temperature must be below the *pressure-adjusted* boiling point (BP) of the solvent in use as listed in Table 2.20 (Grob, 1994). As a rule of thumb for every

Table 2.20 Effective solvent boiling points for different inlet pressures in °C, sorted by BP (*most common injector head pressure).

Solvent	Ambient	100 kPa*	200 kPa	300 kPa	400 kPa	500 kPa
iso-Pentane	28	49	65	77	87	98
Diethylether	35	54	72	84	93	101
n-Pentane	35	57	72	84	93	102
Dichloromethane	40	60	75	87	96	105
MTBE	55	72	91	104	118	124
Acetone	56	77	90	100	109	116
Methylacetate	57	72	91	104	115	124
Chloroform	61	81	97	109	121	130
Methanol	65	82	97	107	116	124
n-Hexane	69	95	111	124	136	145
Ethylacetate	77	97	114	126	136	145
Cyclohexane	81	106	122	137	149	159
Acetonitrile	82	105	120	131	140	148
Toluene	110	134	149	161	170	178

10 kPa two degree C can be added to the standard BP. The minimum isothermal time of the oven temperature to stay below the effective solvent boiling is calculated from twice the inlet liner volume by the set carrier gas flow rate, see also Table 2.20.

2.2.5.2 Cold Needle Liquid Band Injection Technique

The cold needle injection technique avoids the heating of the syringe needle in the injector. This prevents potential discrimination caused by the evaporation of the sample in the needle. A band of liquid enters the inlet liner by a rapid injection process. This requires special precaution to prevent the liquid from reaching the bottom of the liner and finally rinsing to the bottom of the injector getting lost for injection. The solution for a liquid band injection is a plug of c. 1 to 2 cm of deactivated fused silica glass wool in the liner. Empty liners cannot be used. Other absorbent materials, even solid materials like Chromosorb™ between small glasswool plugs, can be used as well. The position of the plug depends on the split mode used. A high position of the plug for split injections allows enough mixing with the carrier gas, while a low position of the glass wool plug is used in splitless operation to allow a short distance to the column entry with minimum wall contacts for the evaporating analytes. Using the glass wool plug at the top of the liner allows the syringe needle tip to penetrate the glass wool plug during injection for extracts with a high load of non-volatile matrix material. The injected solvent is held back by the adsorbent plug without rinsing down the liner. The solvent capacity of the glass wool plug can be tested outside of the injector by manually adding increasing volumes of solvent with a syringe, until a solvent rinsing down the liner is observed. Discrimination effects are minimized as the samples reach the glass wool as aspirated from the sample vial. Evaporation starts from the glass wool plug, but cannot be supported by the low thermal mass of the glass wool alone. Evaporating solvent from the adsorbent leads to a local temperature decrease keeping the analytes focused in this region for a concentrated transfer to the column. As the glass wool plug with the sample extract cools down during the evaporation phase, it allows a gentle evaporation process, thereby preventing sudden flash evaporation of the complete sample with the fatal risk of overfilling the liner volume.

The liquid band injection requires fast autosamplers completing the injection phase within at best 100 ms only, preventing the needle from becoming hot. It cannot be performed manually. The needle penetration depth into the injector liner is only short, just penetrating the septum and remaining with the needle tip at the liner entry for injection. Analytically, the liquid band injection is typically used with sample extracts carrying a high matrix load. It provides the highest integrity of the injected sample, preventing matrix deposit and potential discrimination during evaporation from a hot syringe needle. A known limitation is caused by a potentially increasing activity of the glass wool losing its deactivation after a number of injections, and by the accumulating deposit of unevaporated matrix in the glass wool plug. A more frequent liner exchange becomes necessary.

2.2.5.3 Filled Needle Injections

This injection procedure is no longer up-to-date and should be avoided with hot vaporizing injectors. It is associated with certain types of syringes which, on measuring out volumes below 1 µL of the sample extracts, can only allow the liquid plug into the injection needle. A strong discrimination of high boiling components is observed, see Figure 2.84.

2.2.5.4 Split Injection

Using the split injection technique the sample/carrier gas stream is divided after evaporation in the inlet. The larger, easily variable portion leaves the injector via the split exit and the smaller part passes on to the column. The split ratios can be adapted within wide limits to the sample concentration, the sensitivity of the detector, and the capacity of the capillary column used. Typical split ratios are used in the range of up to 1 to 100 or more. The start temperature of a GC oven program is independent of the used type of solvent and allows higher oven temperatures than the solvent BP for shorter runs. In case the oven temperature is kept below the BP of the solvent at the given head pressure, a reconcentration of the solvent at the beginning of the column happens and needs to be considered when calculating the split ratio. In this case, more sample enters the column due to the recondensation effect and consequently, the split ratio is not as calculated by the measured flow ratios.

For concentrated samples, the variation of the split ratio and the volume applied represents the simplest method of matching the quantity of substance to the column load and to the linearity of the detector. This is of great practical benefit in all quality control applications. Even in residue analysis, the split technique is not unimportant. By increasing the split stream, the carrier gas velocity in the injector is increased and this allows an accelerated transport of the sample cloud past the orifice of the column. This permits a very narrow sample zone to be applied to the column. To optimize the process, the possibility of a split injection at a split ratio of less than 1 to 10 should also be considered. In particular, on coupling with static headspace or P&T techniques, better peak profiles and shorter analysis times are achieved. The smallest split ratio that can be used depends on the internal volume of the insert. For liquid injections in hot split mode generally a high carrier gas flow in the liner is required, but also a sufficient evaporation and mixing with carrier gas before the sample cloud passes the column inlet at the split point. Narrow diameter liner increase the flow speed for sharp injection bands, wider liners give a reduced carrier gas speed better suitable for wide bore columns. Too small diameter liners cause insufficient evaporation, especially with the higher flow of widebore columns, and can induce a partial splitting with non-repeatable data. Liners with wool, frit or other flow "obstacles" like baffles or cups improve the mixing process, as in Figure 2.83.

A disadvantage of the split injection technique is an uncontrollable discrimination with regard to the sample composition. This applies particularly to samples with a wide BP range. Quantitation using external standardization is particularly badly affected by this. Because of the deviation of the effective split ratio from that setup, this value should not be used in the calculation. Quantitation with an internal standard or alternatively the standard addition procedure should be used.

2.2.5.5 Splitless Injection (Total Sample Transfer)

The splitless injection technique is applied for the total sample transfer into the analytical column. After injection, the generated solvent vapor must first be kept in the insert liner and may not expand to other parts of the injector. Depending on the insert liner in use and the solvent, there is a maximum injection volume possible. Upon the explosive evaporation of the solvent, liner of a too small volume lead to the expansion of the sample vapor beyond the geometry of the liner into the cold regions of the injector and cause a probable loss by the septum purge as well as continued memory effects. Pressure waves of subsequent injections bring back deposited material as carryover from the split and other gas lines into the next analysis. A compromise made by many manufacturers involves inserting liner volumes in the range of 1 mL for injection volumes of about 1 to 2 µL.

Splitless injections require an optimized needle length as stated by the injector manufacturer that defines the point of evaporation in the hot center of the liner. The septum purge should not be closed even with a splitless injection. With the correct choice of insert, the sample cloud does not reach the septum, so that the low purge flow does not have any effect on the integrity of the ample. With many commercial split/splitless injectors the closure of the septum purge is optional and can be used to enable a pressure surge during injection.

The carrier gas pushes the sample cloud continuously from the inlet liner to the column. This process generally takes about 30 to 90 s for the complete transfer depending on the liner volume and carrier gas flow. There is an exponential decrease in concentration caused by mixing and dilution with the carrier gas. Longer transfer times are generally not advisable because the sample band becomes broadened in the column. Ideal transfer times allow up to three times the volume of the insert to be transferred to the column. This process is favored by high carrier gas flow rates and an increased head pressure during the transfer process (surge pressure). Hydrogen is preferred to helium as carrier gas with respect to the sample cloud transfer into the column. For the same reason, a column diameter of 0.32 mm is preferred for the splitless technique compared to narrower diameters, in order to get an optimal flow regime in the injector.

As the splitless transfer times are longer compared to the split injection and lead to a considerable distribution of the sample cloud on the column, the resulting band broadening must be counteracted by a suitable temperature of the column oven. Splitless injection requires the oven temperature during injection below the actual BP of the solvent for sufficient refocusing of the sample by the so-called *solvent effect* (Grob and Grob, 1978a). As a rule of thumb, the *solvent effect* operates best at an oven temperature of 10 to 15 °C below the pressure-corrected BP of the solvent at the set inlet pressure. A table of the effective BPs at different inlet pressures for solvents typically used in GC is given in Table 2.20. The effective BP at a certain pressure depends on the specific heat of evaporation of the solvent and can be calculated for any solvent using the Clausius-Clapeyron equation.

The solvent condensing on the column walls acts temporarily as an auxiliary phase, accelerates the transfer of the sample cloud to the column by volume contraction due to re-condensation. The solvent phase holds the sample components and focuses them at the beginning of the column while a partial evaporation of solvent at the front of the solvent phase into the carrier gas stream takes place. Here the use of "GC compatible solvents" that are "wettable" on the stationary phase becomes important. "Incompatible" solvents that do not wet the column phase do not generate a suitable solvent effect and need to be avoided. Such solvents can be mixed with a compatible solvent (keeper) or should be injected in split mode. Because of the low oven temperature at injection the splitless technique requires working with oven temperature programs. Due to the largely complete transfer of the sample to the column, the total sample transfer is the method of choice for trace analysis (Grob, 1994, 1995). After the sample transfer into the analytical column is completed, the split valve is opened until the end of the analysis to prevent any further entry of sample material or potential contaminants to the column.

Because of longer residence times of analytes in the injector with the splitless technique, there is an increased risk of thermal or catalytic decomposition of labile components which can usually be addressed by a suitable liner deactivation or a replacement according to a preventive maintenance procedure. There are also losses possible by adsorption on the liner surface or due to non-volatile deposits for instance on the glass wool.

2.2.5.6 Concurrent Solvent Recondensation

The concurrent solvent recondensation technique (CSR) technique permits the injection of larger sample volumes up to 50 µL by using a conventional split/splitless injector. This can be achieved in a very simple and straightforward way, since all processes are self-regulating. Moreover, the technique is robust toward contaminants and therefore it is suitable when complex samples have to be injected in larger amounts (Magni and Porzano, 2003; Biedermann et al., 2004).

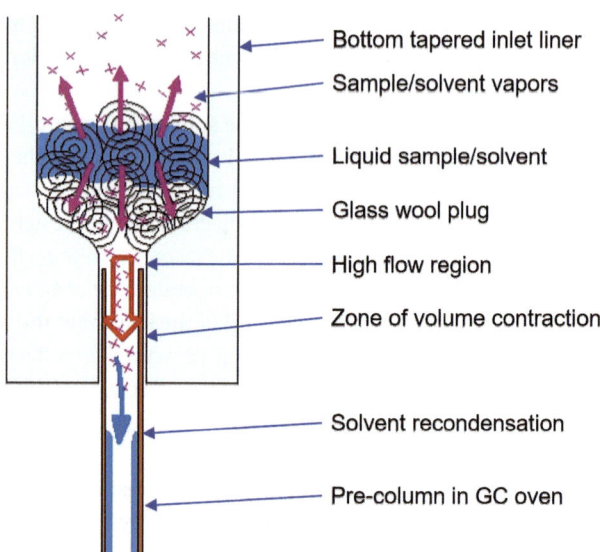

Figure 2.87 Principle of concurrent solvent recondensation in splitless injection (Adapted from Magni and Porzano, 2003).

The recondensation of the solvent vapor inside of the capillary column is causing a volume contraction and hence a pressure drop at the beginning of the column, if the oven temperature at injection is set below the pressure-corrected solvent BP. The pressure difference between column and injector liner significantly speeds up the sample transfer. The solvent vapor with is drawn down from the insert by a strong suction effect due to the lower pressure inside of the column and continuously condensed to form the liquid band at the beginning of the column as illustrated in Figure 2.87. Sample volumes of up to 10 µL can be injected in liquid band mode with silanized glass wool in the insert liner (focus liner) by using the regular 0.25 mm ID capillary columns, while sample volumes of up to 50 µL require the wider diameter of an empty retention gap with 0.32 or 0.53 mm ID to accept the complete amount of liquid injected.

The practical benefit of CSR is the flexibility for the injection of a wider range of diluted sample volumes reducing significantly the extra step and time for extract pre-concentration. The CSR injection is performed with the regular SSL injector hardware without any modifications. Only the autosampler injection volume is adjusted accordingly, when using regular 10 µL syringes. For increased volume injections, only the initial oven isothermal time needs to be extended up to the end of the solvent evaporation from the (pre-)column. The increased solvent vapor cloud needs to leave the condensation region (band formation) before starting the oven program. The remaining oven temperature program remains unchanged.

Concurrent Solvent Recondensation

Key steps are:

- Restricted evaporation rate:
  Injection with liquid band formation into a glass wool plug: The evaporation of the solvent cools the glass wool and leads to a slow evaporation.

- Increased transfer rate:
  The concurrently ongoing recondensation of the solvent in the (pre-)column is generating a strong suction effect.

- Solvent evaporation in the (pre-)column:
  The oven temperature is kept slightly below the pressure-corrected BP until the end of solvent evaporation from the (pre-)column.

2.2.5.7 Concurrent Backflush

The popular multi-component methods with short sample preparation and extract clean-up such as QuEChERS require special care for the chromatographic system in GC and GC-MS. Extraction methods using low polar solvents, for example, ethyl acetate, cyclohexane, acetone or any blend of these, are used for instance in pesticide extractions (Anastassiades et al., 2003; DIN EN 12393-2, 2014; Schürmann et al., 2023). While there is a high recovery reported even for a large number of pesticide components, medium and less polar compounds of high molecular mass get into the extracts. Fruits and vegetables from fresh as well as from processed food give rise to high amounts of a large variety of lipid components as a matrix of high boiling compounds. Although present in the final extracts, the sample extracts often look clear and almost colorless, and no visual quality control is possible.

Once injected into the chromatographic system, high boiling substances persist in the inlet liner of the GC and on the analytical column. While inlet liners typically are exchanged after a certain number of injections, the analytical column gets incrementally contaminated. The elution of high boiling compounds at high oven temperatures used for column baking does not work sufficiently. Matrix compounds accumulate and deliver an increasingly high background level with routine analysis of a large number of samples. One solution could be a reversed column flow after the elution of the last analyte of interest, but this does not prevent the high boiling matrix from entering the column, and adds additional analysis time. Clipping the column and adjustment of retention times is a typical but time-consuming measure.

An optimum solution would be the separation of the analytes during the injection process from the high boiling matrix. The more volatile low molecular mass analytes, for example, pesticides, travel quickly into the analytical column, while high boilers move slower and can be kept in the insert liner and a pre-column. Analytes that reach the analytical column stop motion at the beginning of the column and get focused by the column phase and the solvent effect. At this point, a pre-column can be swept backwards and cleaned-off from higher boiling matrix during the complete analysis run. For maintenance purposes the pre-column, as a cheaper consumable, can be replaced instead of clipping the expensive analytical column. In addition, the backflush setup is highly useful to isolate the injector, while maintaining the carrier flow through the column, for the frequent change of inlet liner or septum as it becomes necessary with the continued measurements of matrix samples (Munari et al., 2000).

Concurrent Backflush Operation The concurrent backflush (BKF) setup consists of four elements: A three-way solenoid valve (BKF valve) in the carrier gas line, a widebore deactivated pre-column, a purged T-connector in the GC oven and a flow restrictor (see Figures 2.88 and 2.89). The analytical column is pushed about 3 to 5 cm into the widebore pre-column to ensure full chromatographic integrity avoiding analyte contact to the surface of the T-piece. The flow restriction connects the carrier gas line of the injector with the BKF line to purge the T-piece. The flow of the restriction is usually factory set to avoid dead volumes.

Injection During injection, the BKF valve is off, as shown in Figure 2.88. The carrier gas is directed in the regular way to the GC injector. A small flow provided by the restrictor grants sufficient purging of the T-connection to avoid the dead volume here. The flow restrictor is designed to provide a purge flow of about 5% of the main flow when the split valve is closed. In the standard configuration, the pre-column consists of a 2 m × 0.53 mm ID uncoated but deactivated fused silica tubing. Coated

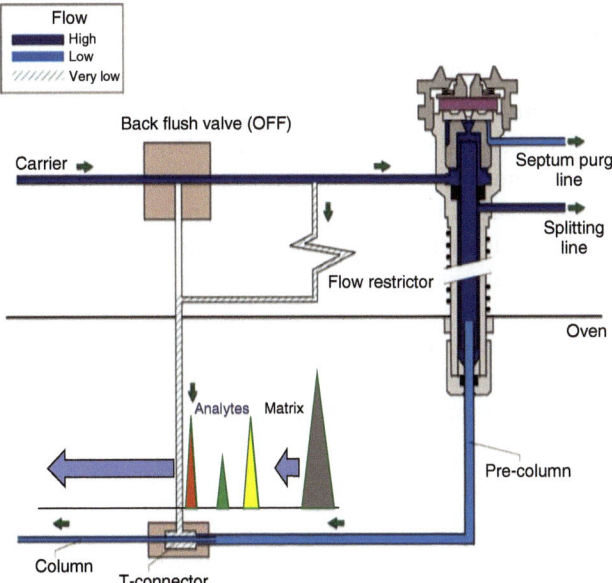

Figure 2.88 Backflush OFF – injection and analyte transfer to pre-column and analytical column.

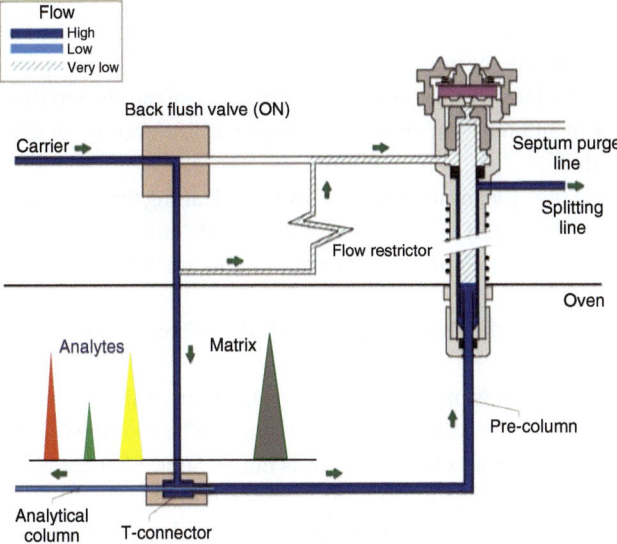

Figure 2.89 Backflush ON – cleaning the pre-column and injector from matrix during the analysis run, also used for LVI-PTV injections.

pre-columns improve inertness but lead to increased transfer times. The injection is done as regular. Smaller molecules (analytes) travel faster in the pre-column than the larger matrix molecules, and reach the analytical column earlier.

Backflush Activated When the BKF valve is activated, the carrier gas flow is diverted directly to the T-connection and the analytical column to proceed with the regular chromatographic separation, as it is illustrated in Figure 2.89. At the same time, the flow through the pre-column is reversed. The carrier gas now enters the inlet from the bottom and is vented through the split line. All high boiling sample matrix residing at this time in the pre-column gets flushed backwards. The pre-column and inlet liner of the injector get cleaned from higher boiling compounds of no analytical interest during the GC run. A small carrier gas stream is provided by the restrictor to the top of the injector, purging the insert liner and septum during the cleaning phase.

The timing of switching the BKF valve is set such that all analytes of interest pass the pre-column to the analytical column. The BKF valve is activated after the last analyte has passed. It needs several runs for optimum adjustment. The switching time is called the *transfer time*.

Concurrent Backflush Setup In a first step, a standard chromatogram using the optimized column temperature program without the use of BKF is acquired as reference. For BKF operation, no changes to the standard oven temperature program are required. In the following step, the BKF operation is optimized with standards starting with a long transfer time.

1. First deactivate the BKF mode.
2. Inject a standard with the optimized oven temperature program required for the complete elution of the analytes of interest.
3. Enable BKF and program a long transfer time. For reference, use the elution time of the last component of interest. For pesticide analysis, a last eluting analyte usually is e.g. deltamethrin.
4. On a PTV injector enable the "Cleaning Phase" and set the cleaning time to match the total analysis time, and the split flow to about 50 to 100 mL/min (Figure 2.90). Using a hot SSL injector, the injector temperature remains unchanged during BKF operation.
5. Inject the sample and compare the chromatogram with the reference obtained at point 2.
6. If the last component of interest is present in the chromatogram, this will be the correct transfer time to set for the BKF. The transfer time can be shortened until the peak of the last component has a lower area or is not present. Then increase the transfer time in 1 min steps (Figure 2.91).
7. The BKF will be activated reproducibly for all subsequent analyses at the same time. If the temperature program and carrier flow conditions are modified, the optimum performance needs to be checked with the injection of a standard solution again.

The analytical benefits provided by the use of a concurrent BKF for sample extracts with matrix content are multiple, supporting the increased productivity of a GC and GC-MS system (Hetmanski, 2009; Hildmann et al., 2013):

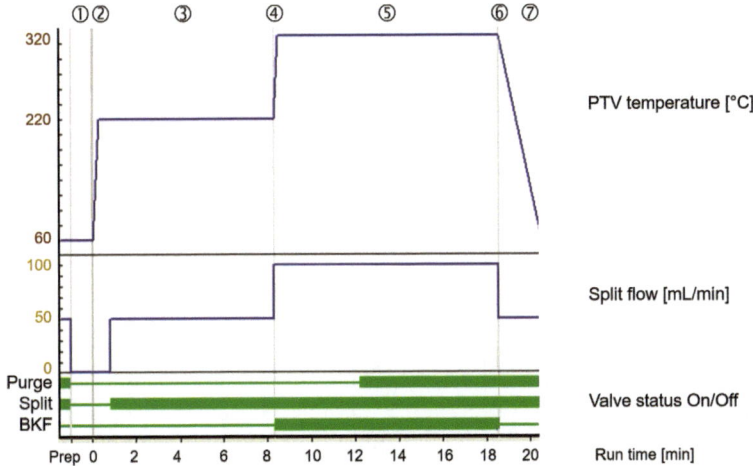

Figure 2.90 PTV operation with backflush and cleaning phase. 1. Sample injection; 2. PTV heating to transfer temperature; 3. transfer time from pre-column into analytical column; 4. switch BKF valve, reverse flow in the pre-column; 5. cleaning phase at elevated temperature on a PTV; 6. switch BKF off for next injection; 7. return PTV temperature for next injection.

- Ease of use: No change to the oven temperature program in use.
- Chromatographic integrity is maintained in particular with difficult matrix samples.
- Time saving by avoiding additional bake-out phases of the column.
- Shorter cycle times – stop the run after the last eluting analyte.
- Lower final oven temperatures improve cool down times to the next analysis.
- The possibility to better select the column phase for the separation of the lighter components.
- Increased column lifetime – no high bake-out temperatures.
- No clipping of the analytical column – stable retention times.
- Thinner column films possible for less matrix and faster separation.
- Compatible with cold PTV and hot SSL injectors.
- Compatible with LVI methods.
- Easy maintenance of the injector by pre-column exchange with BKF on, maintaining the analyte retention times.

2.2.6 Temperature Programmable Injectors

The programmed temperature vaporizer (PTV) for split and splitless operation as it is in use with many GC systems today is based on the systematic works in the early 1980s (Vogt et al., 1979; Poy et al., 1981a, 1982a; Schomburg, 1981). Figure 2.92 illustrates the general design with a low thermal mass body and small inlet liner for a fast direct heating, also using a capable external cooling. The so called "multi-mode inlets" offered from several GC manufacturers are temperature programmable injectors as well, often use larger inlet liners comparable to the design of hot SSL injectors.

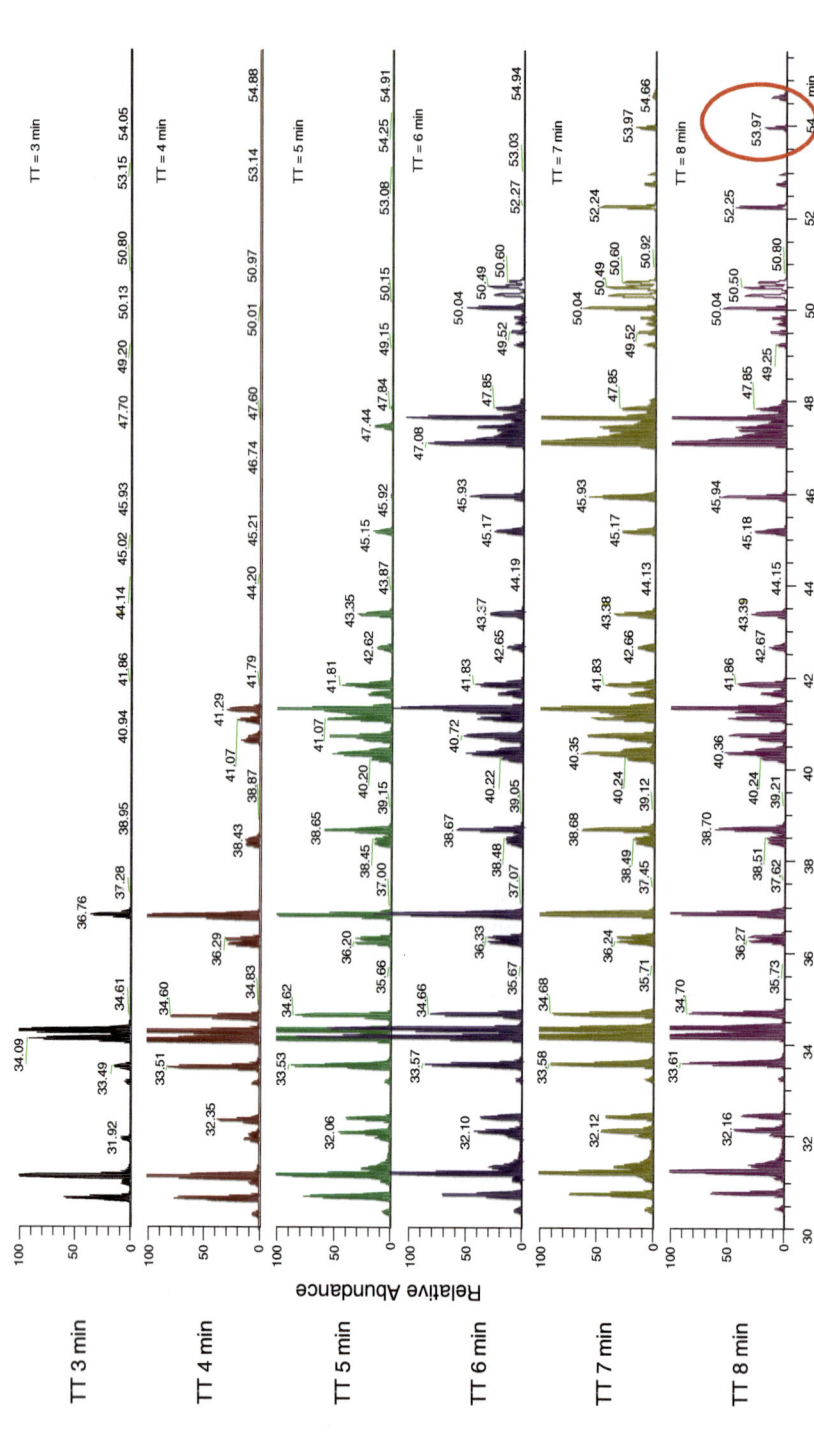

Figure 2.91 Effect of a transfer time (TT) of 3 to 8 min on the late eluting analytes and on retention time stability (Courtesy R. Fussel, FERA York, UK).

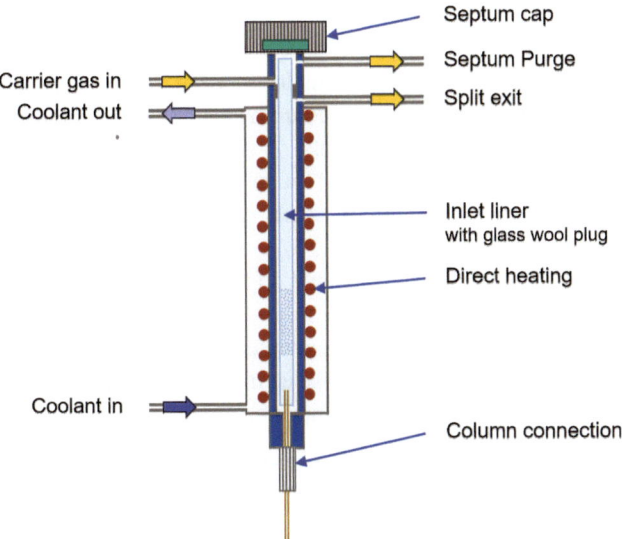

Figure 2.92 PTV split/splitless injector with direct heating and cooling.

The development focused on the precise and accurate execution of quantitative analyses of complex mixtures with a wide BP range (Poy, 1981b, 1982b; Saravalle et al., 1987). Its suitability for low volatile substances avoiding discrimination for substances up to above C_{60}, such as petrochemical and polyaromatic hydrocarbons, could be demonstrated. Examples of the analysis of triglycerides exemplify the injection of samples up to above C_{50} and the analysis of crude oil fractions even up to C_{90}! With its unique features the PTV combines the advantages of the hot SSL and OCI techniques (Grob, 1986; Stan and Müller, 1987; Müller and Stan, 1989).

The intrinsic limitations of the hot SSL injector with the risk of thermal degradation of analytes, high boiler discrimination, and a hot metal syringe needle drove the development of injection alternatives at low temperature like OCI and PTV. In a PTV injector, the sample extract is ejected from the syringe needle in liquid form into the "cold" vaporizer inlet. Heating only begins after the syringe needle has been removed from the injection zone. The injection of a liquid sample into a "cold" inlet liner eliminates the potentially selective evaporation from the hot syringe needle, which leads to discrimination against high boiling components, and adds LVI and solvent split capabilities. There are many advantages of a cold sample injection for split and splitless analyses:

- The extracts enter the inlet liner as a liquid. Discrimination as a result of fractionated evaporation effects from the syringe needle does not occur.
- A defined volume of liquid can be injected reproducibly using regular GC injection syringes.
- The sample evaporates by controlled heating from the glass wool of the liner in the order of their boiling points. The solvent evaporates first and leaves the sample components in the injection area without causing a spatial distribution of the analytes.

- Aerosol and droplet formation is avoided.
- As the inlet liner does not have to take up the complete expansion volume of the injected sample solution at once, inserts with smaller internal volumes can be used. The consequently more rapid transfer to the column lowers band broadening for sharper peaks and thus improves the S/N ratio.
- If the boiling points of the sample components and the solvent differ by more than 100 °C, larger sample volumes (up to more than 100 µL) can be injected using the solvent split mode (see Section 2.2.6.1).
- Impurities and residues which cannot be evaporated do not get on the analytical column and extend the useful lifetime.
- With concentrated samples, the injection can be adapted to the capacity of the column and the dynamic range of the detector by selecting a suitable split ratio.
- Automated OCIs are possible with regular 10 µL injection syringes.

Despite the many advantages of the cold injection system, in practice there are also some limitations to its use (Table 2.21). A typical mishandling of PTV injectors is the injection at high temperatures as it is usual with hot SSL injectors. PTV injectors work with small volume inlet liners which cannot accommodate the resulting solvent vapor cloud, causing poor performance and serious inlet system contamination. For split and splitless injections using PTVs the injector temperature below the pressure-corrected BP of the solvent is mandatory. Only some "multimode" called injectors use larger volume inlet liners.

Other limitations include the analysis of thermally labile substances. Because of the low injection temperature, a cold injection is expected to be particularly suitable for labile substances. However, during the heating phase, the residence times of the substances in the insert are long enough to initiate thermal decomposition.

Table 2.21 Limitations and advantages of common GC injectors.

Characteristics of the sample	Hot split	Hot splitless	PTV split	PTV splitless	PTV solvent split	On column
Concentrated samples	+	−	+	−	−	−
Trace analysis	≈	+	−	+	+	+
Diluted extracts	−	−	−	−	+	+
Large volume injection	−	+	−	+	+	−
Narrow boiling range	+	+	+	+	+	+
Wide boiling range	−	+	+	+	−	+
Volatile substances	+	+	+	+	−	−
Low volatile substances	−	≈	+	+	+	+
With involatile matrix	+	+	+	+	+	−
Thermolabile substances	≈	−	≈	≈	≈	+
Automated injection	+	+	+	+	+	+

+ recommended, ≈ can be used and − not recommended.

For such applications only OCI (on the PTV) can be recommended because it completely avoids external evaporation of the sample for transfer to the column (see Sections 2.2.7.1 and 2.2.7.2). A test for thermal decomposition Donike suggested the injection of a mixture of the same quantities of fatty acid TMS esters (C10 to C22 thermolabile) and n-alkanes (C12 to C32 thermally stable). If no thermal decomposition takes place, all the substances appear with the same peak intensity (Donike, 1973).

The PTV became a standard GC injector today available from all major vendors due to its performance and versatility. The cold injection technique is used in general and in trace analysis for pesticides, pharmaceuticals, polyaromatic hydrocarbons, brominated flame retardants, dioxins or polychlorinated biphenyls (PCB). Beyond of that, the enrichment of volatile halogenated hydrocarbons, the direct analysis of water and the formation of derivatives in the injector also demonstrate the broad usability of PTV type injectors (Bergna et al., 1991; Japp et al., 1991; Mol et al., 1993; Blanch et al., 1994; Gerstel, 1998; Buchgraber et al., 2004; Hoh and Mastovska, 2008; Krumwiede et al., 2008).

2.2.6.1 PTV Injection Modes

PTV Total Sample Transfer (Cold Splitless Injection) The splitless injection for the analysis of a maximum sample equivalent on the column is the standard requirement for residue analysis. The sample is taken up in a suitable solvent and injected at a low PTV temperature. The split valve is closed. The PTV temperature at injection should correspond to the BP of the solvent at ambient pressure. During the injection phase, the oven temperature is kept below the PTV temperature. It must be below the pressure-corrected BP of the solvent (Table 2.20) in order to exploit the necessary solvent effect for focusing the substances at the beginning of the column, as illustrated in Figure 2.93. If the focusing of the substances is insufficient, the peaks of the early eluting components are broad and are detected with a low S/N ratio.

The injector is heated a few seconds after the injection when the solvent has already evaporated and has reached the column (Figure 2.94). Typically, this time interval is between 5 and 30 s. For high boiling substances in particular, longer isothermal times have been found to be favorable. The heating rate should be moderate in order to achieve a smooth evaporation of the sample components required for transfer to the column. Heating rates of about 10 to 14 °C/s have proved to be suitable. The optimal heating rate depends on the dimensions of the insert liner and the flow rate of the carrier gas in the insert.

Using inserts with silanized glass wool has proved effective for the absorption of the sample liquid. However, the glass wool clearly contributes to enlarging the active surface area of the injector and should be used only in the analysis of noncritical compounds. For the injection of polar or basic components special care is required keeping the glass wool deactivated by frequent replacements, or install the empty baffled liner type.

The use of capillary columns of 25 m in length and 0.25 mm internal diameter can be recommended without limitations for cold injection systems. The coupling to a

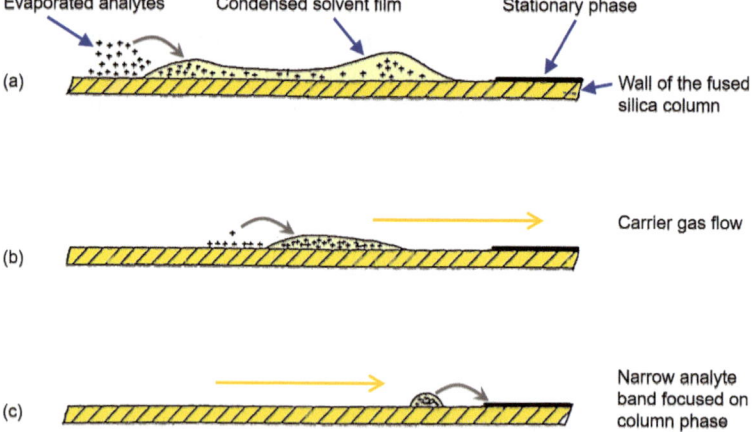

Figure 2.93 Refocusing by the solvent effect in splitless injection (after Grob). (a) Condensation of solvent at the beginning of the capillary column caused by the oven temperature below the actual solvent BP. The condensed solvent acts as a stationary phase and dissolves the evaporated analytes. At the same time the solvent film moves and evaporates from the front side into the carrier gas flow. Solvent saturated carrier gas passes the column as the solvent peak. (b) The continuous evaporation of the solvent film concentrates the analytes on to a narrow ring (band) in the remaining solvent. For this process, no stationary phase is necessary, for instance in an empty pre-column. (c) The concentrated narrow band of analytes meets the stationary phase. The separation starts with a sharp band with the raising oven temperature.

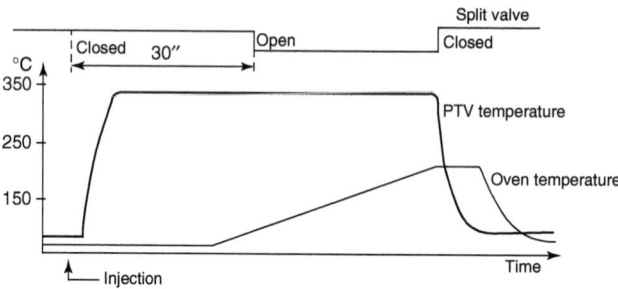

Figure 2.94 PTV total sample transfer (splitless injection).

mass spectrometer is particularly favorable because of the generally low flow rates. Also, the injection of larger sample volumes is straightforward and does not impair the operation of the mass spectrometer. For these reasons the use of cold injection systems for GC-MS is particularly recommended.

The transfer of the analytes to the analytical column for a 1 to 2 µL injection is completed after about 30 to 120 s, depending on the liner dimensions and carrier gas flow. Larger injection volumes require increased transfer times. The temperature program of the GC oven stays isothermal during this transfer phase, and gets started right after completion. At the same time, at the start of the GC oven programming, the injector gets purged of any remaining residues by opening the split valve. If required, a second PTV heating ramp and plateau can follow as cleaning phase to bake out the insert at elevated temperatures. The high injection or cleaning temperature of the PTV is retained until the end of the analysis to keep the injector free from possible adsorptions and the accumulation of impurities from the carrier gas inlet tubing. The start of cooling of the PTV is adjusted so that both the PTV and the oven are ready for the next analysis at the same time".

PTV Total Sample Transfer

- Split valve closed
- PTV at the ambient BP of the solvent
- Oven temperature below the pressure-corrected BP of the solvent
- Start of PTV heating about 5 to 10 s after injection
- Start of the GC temperature program about 30 to 120 s after injection
- PTV remains hot until the end of the analysis

PTV Cold Split Injection In this mode of operation, the split valve is open throughout the analysis (Figure 2.95). In classical applications, this mode of injection is suitable for concentrated solutions. The column loading can be adapted to its capacity and the nature of the detector by regulating the split ratio (Poy, 1982b). Cold injection systems allow a high carrier gas flow rate at the split point. Compared to hot split injectors, smaller split ratios of about 1:5 are possible. It is possible to work in the split mode and thus achieve better S/N ratios and in particular shorter analysis times, compared with total sample transfer, as the injection is performed at high oven temperatures without the need of the focusing solvent effect.

All other PTV adjustments of time and temperature remain the same compared to the total sample transfer! Of particular importance for the split injection using the PTV cold injection system is the fact that the sample is injected into the cold injector (compare to Section 2.2.5.4). The choice of start temperature of the GC oven is now no longer coupled to the BP of the sample solvent. The PTV split mode is particularly useful for the analysis of high boiling compounds (Tipler and Johnson, 1989).

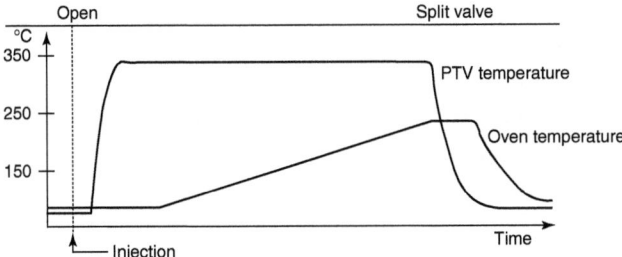

Figure 2.95 PTV split injection.

PTV Split Injection

- Split valve open
- PTV below the ambient BP of the solvent
- Start of PTV heating about 5–10 s after injection
- Oven temperature to be chosen freely, typically above solvent BP.
- PTV remains hot until the end of the analysis.

PTV Large Volume Solvent Split Injection The PTV large volume injection mode (LVI-PTV) allows the automated injection of sample volumes up to 100 µL and even more (Müller et al., 1993; Mol et al., 1994; Hoh and Mastovska, 2008). The large volume injection with solvent split requires that the analytes are significantly less volatile than the used solvent. This technique, also called "solvent elimination injection", is particularly suitable for trace analysis for the analysis of diluted extracts with low concentrations of high boiling analytes (Figure 2.96). The solvent

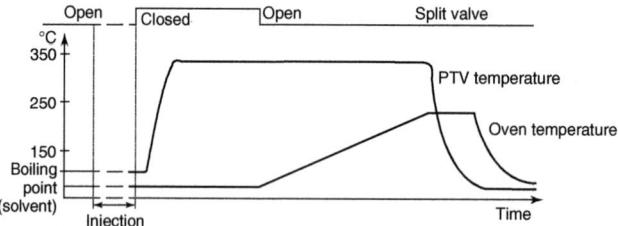

Figure 2.96 PTV solvent split injection.

Table 2.22 Comparison of large volume injection methods.

	LV-on-column with solvent vapor exit	LVI-PTV with solvent split	LV-split/splitless with concurrent solvent recondensation
Typical injection volume	>150 µL[a]	100 µL	50 µL
Robust versus complex matrix	No	Yes	Yes
Volatiles analysis	Yes	Potential losses	Yes
Suitability for thermolabile and actives compounds	High	Medium	Medium
Solvent vented	Yes	Yes	No, passes the column
Requires uncoated pre-column	Yes	No[b]	Yes
Number of parameters	Medium	Large	Small[c]
Software assisted set-up	Yes	No	Yes

a) Several 100 µL are transferred during LC-GC coupling, see 2.2.7.3.
b) It can be necessary for improving volatiles recovery if large amount of solvent is retained.
c) Up to 10 µL with regular 0.25 ID columns.

split mode allows an efficient sample throughput shortening time-consuming pre-concentration steps. With suitable parameter settings larger extract volumes can be applied. The solvent split mode is also suitable for fraction collection from online sample preparation (Staniewski and Rijks, 1991; Mol et al., 1994). An overview of current LVI methods is given in Table 2.22.

For injection of larger sample volumes from c. 10 µL to well over 100 µL, it is advantageous for the insert to be filled at least with silanized glass wool or a sorbent material plug about 0.5 to 1 cm wide of Tenax, Chromosorb, Supelcoport or similar inert material between glass wool plugs. The required solvent capacity of the liner type to be used can easily be tested outside of the injector by adding the intended amount of solvent with the syringe. No rinsing of the solvent may be observed when holding the insert liner upright on a sheet of paper. Depending on the intended extract volume different solutions are feasible.

During LVI the split valve is open, hence the designation as solvent split injection. The PTV is kept at a temperature that corresponds to the ambient BP of the solvent. The oven temperature is below the PTV temperature and thus below the actual BP of the solvent at the carrier gas head pressure. The maximum injection rate depends upon how much solvent per unit time can be evaporated in the insert and eliminated through the split tubing by the carrier gas. The at-once injection of large volumes can best be tested easily by applying the desired amount of solvent with the liner outside of the injector. During injection, a high split flow of >100 mL/min is recommended to focus the analytes on the packing. Only for the ease-of-method development of the injection, the data recording of the GC or GC-MS system is started so that the course of the solvent peak can be followed (Figure 2.97). A broad solvent peak is registered. After the solvent peak is eluted, the split valve is closed and the PTV heated up. A

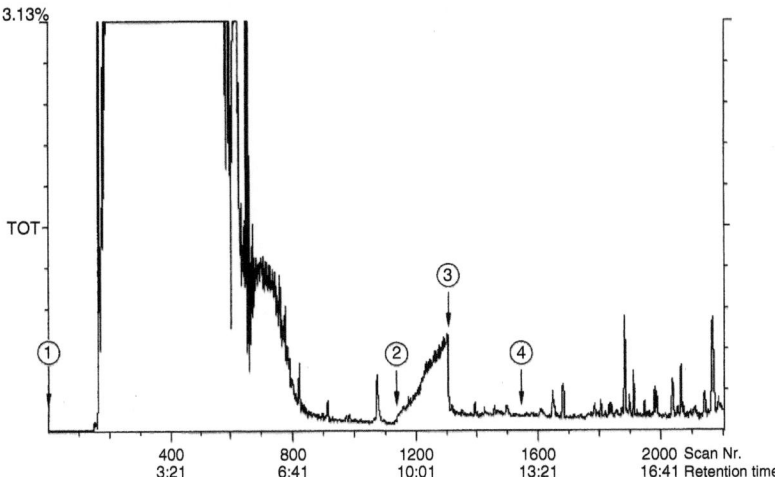

Figure 2.97 The course of an analysis with a large solvent volume of 100 µL hexane at 200 pg/µL PCB. Using the PTV solvent split, insert liner filled with a 1 cm Chromosorb plug. (1) Start of the injection: split open. (2) Start of PTV heating: split closed. (3) Baking out the PTV: split open. (4) Start of the GC oven temperature program: split open, and typically the start of the data acquisition.

smaller roughly triangular solvent peak is produced from remaining solvent in the sorbent material. After that, the split valve is opened again and the oven temperature program starts. The eluting peaks show good resolution and are free from tailing.

When quantities of more than 10 μL are applied, the split valve is closed first when the decline of the solvent peak can be noticed on the display, meaning that the injector is free of most of the solvent. After the split valve has been closed, the PTV injector can be heated up as usual. After about 30 to 120 s the transfer of the enriched sample from the insert to the column is complete and the GC temperature program can be started. The split valve then opens again and the PTV is held at the injection temperature until the end of the analysis.

The PTV solvent split mode also allows the derivatization of analytes in the injector (Knapp, 1979; Pierce, 1982). This procedure simplifies sample preparation considerably and allows automation. The sample extract is injected together with the derivatizing agent e.g. TMSH (trimethylsulfonium hydroxide) for methylation into the PTV. The excess solvent is blown out during the solvent split phase. The derivatization reaction takes place during the heating phase of the PTV and the derivatized substances pass on to the column (Färber et al., 1991; Färber and Schöler, 1992).

A comparison of regular and large volume injection is shown in Figure 2.98 for HxCDF peaks in dioxin analysis. In general, LVI operation is facilitated by the installation of a 5 to 10 m long widebore pre-column of 0.53 mm ID. Rugged metal columns are preferred in handling compared to the brittle widebore fused silica columns.

As outlined in Section 2.2.5.6 the CSR is a very practical way for LVI also on a PTV injector. Here the solvent passes the analytical column. Matrix loaded extracts benefit from a backflush solution (BKF) as illustrated in Figure 2.89.

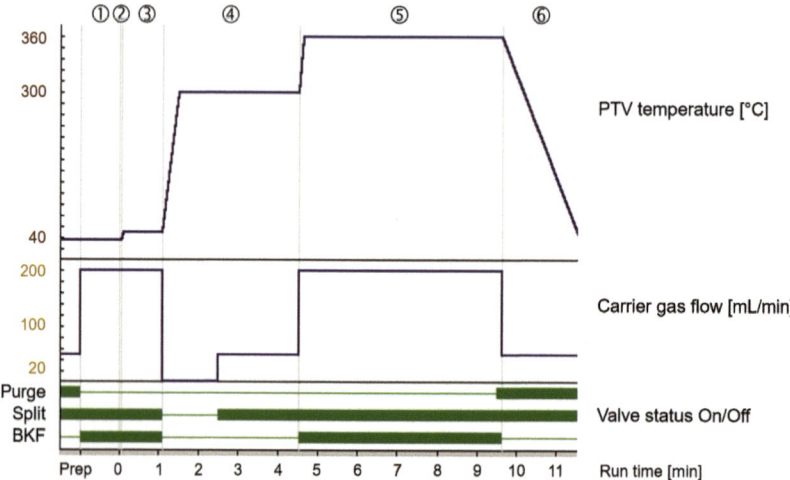

Figure 2.98 LVI-PTV-BKF temperature, flow and valve switching profile using a pre-column backflush. The green bars indicate a ON status of a valve. (1) Equilibration. (2) Injection. (3) Evaporation. (4) Transfer. (5) Cleaning. (6) Analysis.

For the large volume mode, the backflush valve is switched on during injection and evaporation phase so that the solvent vapors do not enter the analytical column. For injection the PTV temperature, carrier gas flow rate and valves are programmed as shown in Figure 2.99 for optimum solvent elimination and analyte transfer.

Also, the direct transfer of large solvent volumes into a pre-column without volatilization of the analytes in a hot injector is a widely used LVI solution. The on-column LVI solution with concurrent solvent evaporation using a solvent vapor exit (SVE) after the pre-column is further described as the LC-GC technique in Section 2.2.7.3.

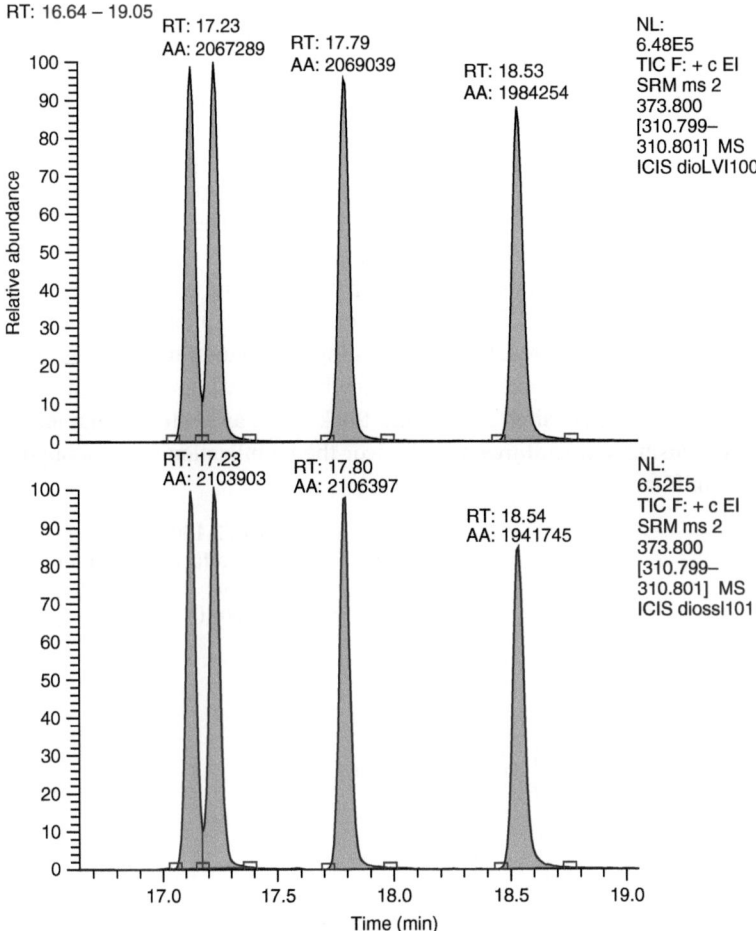

Figure 2.99 Comparison of regular and large volume injection. Identical mass chromatograms of the hexachlorodibenzofuran congener peaks with retention time and peak area of the U.S. EPA 1613 CS3 standard. (a) 80 µL PTV-LVI injection of the 1:80 diluted standard solution. (b) 1 µL PTV splitless injection of the undiluted standard solution. Conditions: Thermo Scientific TSQ Quantum GC, column: TR-5MS, 30 m × 0.25 mm ID × 0.1 µm film, TRACE GC with PTV, packed liner with deactivated glass wool.

2.2.6.2 Cryofocusing

The trace analysis of VVOCs and VOCs often needs a sample enrichment for GC analysis. The cryogenic focusing on the head of the analytical column or in a low volume inlet liner are proven and widely used solutions. The cryofocusing cannot be regarded as an independent injection system, but deserves to be treated individually because the static headspace, P&T, or TD systems can be coupled directly to a cryofocusing unit (Poy and Cobelli, 1985; Kolb et al., 1986; Werkhoff and Bretschneider, 1987; Kolahgar and Pfannkoch, 2002).

On-Column Cryofocusing The analyte concentration in a short section at the head of an empty but deactivated pre-column or the analytical column is achieved by a dedicated cryofocusing unit that allows very fast heating (up to 60 °C/sec) and cooling (GL Sciences, 2015). The capillary column is pushed through a stainless steel tube of the cryofocusing device (Figure 2.100). This tube is firmly welded to a cold finger which is connected to the cooling media supply. The tube is also surrounded by a direct heating coil with temperature control. The reversed flow of the cooling media, liqu. N_2 (to c. −150 °C) or liqu. CO_2 (to c. −50 °C), against the column flow traps the analytes into a narrow band. The analytical column can be connected also to a transfer line of a headspace, P&T, or canister sampler. The chromatographic separation starts with a rapid heating of the trapping region.

Because of on-column focusing such a cryotrap has the same sample capacity as the column for these analytes. A breakthrough of the cryotrap because of a high analyte concentration can be prevented by adjusting the column dimensions. The diagram in Figure 2.101 gives the temperatures at which a breakthrough of analytes must be reckoned in cryofocusing. Larger film thicknesses and internal diameters permit higher focusing temperatures, which favor the mobilization of the analytes from the column film.

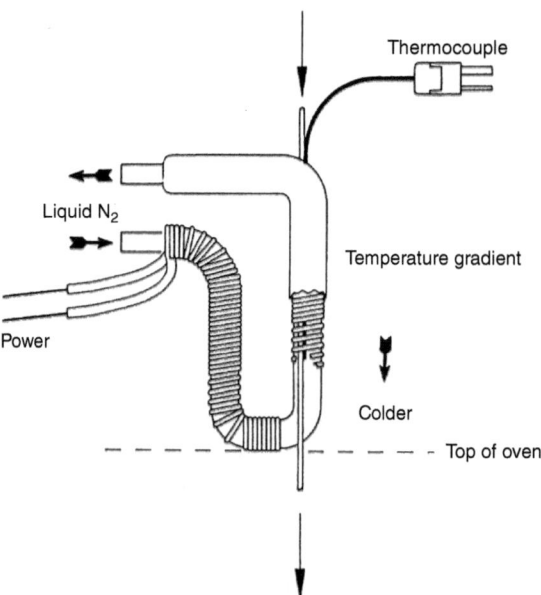

Figure 2.100 Basic principle of a cryofocusing unit for use with fused silica capillary columns (Tekmar).

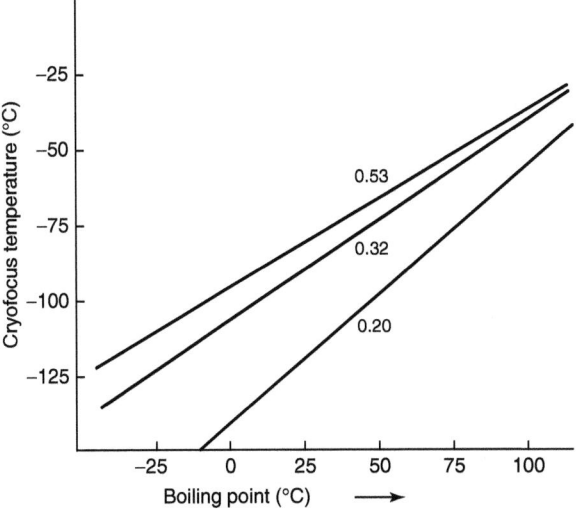

Figure 2.101 Breakthrough temperatures in cryofocusing for various column internal diameters as a function of the boiling point of the analyte (after Teledyne Tekmar). Film thicknesses: 0.25 μm at 0.20 mm internal diameter, 1.0 μm at 0.32 mm internal diameter, and 3.0 μm at 0.53 mm internal diameter.

PTV Cryofocusing A more versatile cryotrapping solution for volatile analysis is using small volume insert or baffled liner in PTV type injectors. This is a central feature with temperature programmable injectors and opens up cryofocusing to much wider areas of application like static and dynamic headspace (DHS), outgassing, pyrolysis, or the TD from sampling tubes or from SBSE extractions. Lower PTV temperatures can be used for improvement of sensitivity by trapping volatiles in the solvent vent mode (Pérez Pavón et al., 2009). The same setup can also be used e.g. for splitless SPME injection allowing multiple injections for improved sensitivity and chromatographic resolution. The PTV cryofocusing approach ensures high inertness, sharp peaks, excellent separation, and best possible recovery and detection limits over a wide boiling range (Pigozzo et al., 1991; Gerstel, 2016, 2022). Peltier elements are slow and of limited capacity, not suitable for the required fast injector cooling. Closed loop cooling devices with adjustable temperature settings are the preferred solution. Liquid CO_2 is used for trapping down to C5 components until approx. −70 °C (BP −78 °C), or liquid N_2 trapping efficiently down to C3 components at approx. −180 °C (BP −196 °C), as common cooling media for PTV cryofocusing (Moreira Novaes and Marriott, 2021).

The insert liner of the PTV is filled with a small quantity (about 1 to 2 mg, up to 3 cm wide) with deactivated glass wool, glass beads, Tenax, Carbotrap, or other low bleed thermally stable trapping material. During the injection, the split valve is open as in the solvent split mode. Like the total sample transfer method, the oven temperature is kept correspondingly low in order to focus the components at the beginning of the analytical column. For transfer of the sample to the GC column, the split valve is closed, the coolant stopped, and the PTV heated rapidly to effect the total

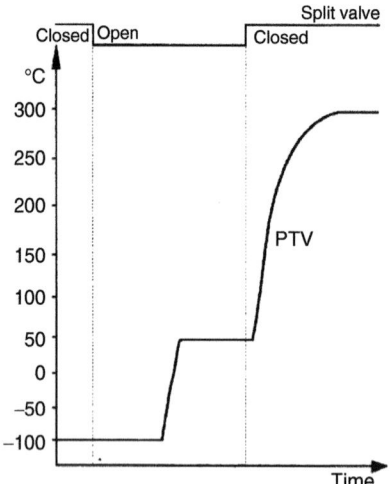

Figure 2.102 PTV cryo-enrichment at −100 °C (with an optional fractionation or drying step at 50 °C) and injection with total sample transfer.

sample transfer of the concentrated analytes. After about 30 to 120 s the temperature program of the oven starts and the split valve is opened again (Figure 2.102). The PTV remains at the same temperature until the end of the analysis. This last step is particularly important in cryo-enrichment because the cooled adsorbent could become enriched with residual impurities from septum or gas lines, which would lead to ghost peaks in the subsequent analysis.

2.2.7 Non-Vaporizing Injection Techniques

GC analysis does not require the initial vaporization of the liquid sample. It has distinct advantages i.e. the dilution by split injection or fast analysis, and has known limitations i.e. the risk of discrimination and degradation. The direct liquid introduction into GC columns circumvents such limitations caused by evaporation in a hot injector and adds additional powerful application areas, but also comes with limitations methods need to take care of.

2.2.7.1 On-Column Injection

The direct transfer of the liquid sample into the analytical column is the simplest technique to introduce a sample "cold" into an analytical column (Grob and Grob, 1978b). In the era of packed columns OCI was the state of the art (although not named as such at the time). The difference between that and the present procedure lies in coping with the small diameters of capillary columns which have caused the term *filigree* to be applied to the on-column technique. Schomburg introduced the first on-column injector for capillary columns in 1977 under the designation of direct injection and described the process of sample injection very precisely (Schomburg et al., 1977). A year later, a variant using syringe injection was developed by Grob. In the upper section of the OC injector is the needle port which allows the introduction of the syringe needle for injection. The injector block is kept at ambient temperature (cold) in all phases of operation by a main air flow. A warm-up of the column head

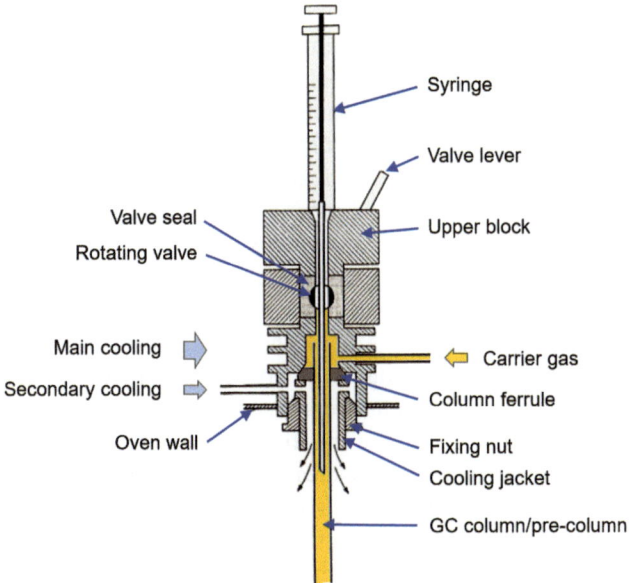

Figure 2.103 On-column injector schematics (Carlo Erba/Thermo Scientific).

by the oven is prevented by a secondary cooling air stream (Figure 2.103). The center section of the on-column injector carries a rotating valve as seal for manual and also automated operation. The use of retention gaps allows autosamplers to be employed for on-column sample injection. Preferred internal column diameter are of 0.32 mm using a 32 gauge needle with 0.24 mm OD. Columns with 0.25 mm inner diameter would require the more sensitive to handle 0.17 mm OD needle (34 gauge). More convenient are syringes with 26 gauge needle for widebore columns or pre-columns with 0.53 mm inner diameter. The needles are 75 mm long for Thermo Fisher/Carlo Erba and 125 mm long for Agilent/HP OC injectors to position the needle tip deep in the (pre-)column at oven temperature level. Removable needles for replacement are available.

On-Column Injection

- Carrier gas flow rate is about 2–3 mL/min for He
- Oven temperature below the pressure corrected BP of the solvent
- Rapid injection of small volumes
- Oven temperature kept isothermal at the evaporation temperature
- Oven ramp started after the elution of the solvent peak.

During the injection the oven temperature must be close to but below the pressure corrected BP of the solvent. This corresponds to a temperature of about 10 to 15 °C above the BP of the solvent under ambient conditions. As a rule of thumb, the BP increases by 1 °C for every 0.1 bar of head pressure (see Table 2.20 for details). Small sample volumes of less than 10 µL can be injected directly on to a capillary column. Larger volumes require a deactivated retention gap of appropriate dimensions. The length of a pre-column per microliter injection volume should be at least 30 cm for 0.32 mm ID or 20 cm for 0.53 mm ID deactivated columns. The length of the flooded zone at about 10 °C below the pressure corrected BP is about 20 cm/µL with 0.32 mm ID and 12 cm/µL with 0.53 mm ID pre-columns. Longer pre-columns do not have a negative impact on the performance of the chromatography.

For injection, the column should be operated at a high flow rate (about 2 to 3 mL/min for He). The injection is carried out by pressing the syringe plunger rapidly down with 15 to 20 µL/s to avoid the liquid being sucked up between the needle and the wall of the column through capillary effects. Otherwise, the subsequent removal of the syringe would result in loss of sample. For larger sample volumes, from about 20 to 30 µL and more, the injection rate must, however, be lowered to about 5 µL/s as the liquid plug causes increasing resistance and reversed movement of the liquid must be avoided (Figure 2.104). The column, or in the same way the retention gap, is wetted over certain length by the sample extract. The solvent evaporates from the beginning of this zone and concentrates the analytes in an increasingly narrower band called "solvent trapping" very similar to the solvent effect shown in Figure 2.93. Upon reaching the stationary phase of the GC column the analytes enter in a narrow band as the starting point for the separation. The oven temperature program with the heating ramp can then begin when the solvent peak is clearly decreasing (Grob, 1983).

The on-column technique has many advantages over a PTV cold injection system:

- The risk of thermal decomposition during the injection is practically eliminated. A substance evaporates at the lowest possible temperature and is heated to its elution temperature at the maximum (Björklund et al., 2004; Binelli et al., 2006).
- Only OCI allows the quantitative and reproducible sample injection on to the column without losses. Low volatile and high boiling substances are transferred totally without discrimination.

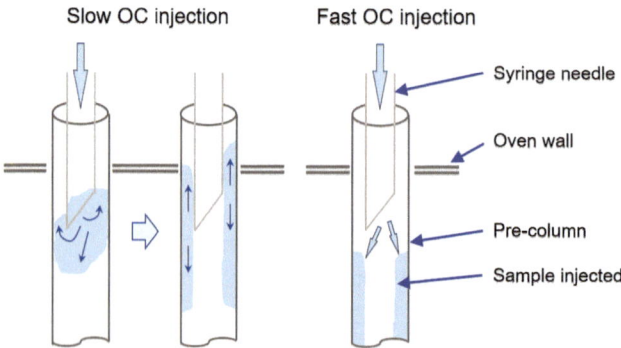

Figure 2.104 Slow and fast on-column injections. With the slow injection, the exterior of the needle is contaminated and part of the sample is lost.

- Defined volumes can be injected with high reproducibility. Standard deviations of about 1% can be achieved.
- The injection volumes can be varied within wide ranges without additional optimization. At the beginning of the heating program, only the duration of elution of the solvent peak needs to be taken into account for the initial isothermal phase.

Systematic limitations to the OCI technique for certain types of samples are:

- Concentrated samples may not be injected on-column. In these cases, preliminary dilution is necessary or the use of the split injector is recommended. A minimum OCI volume is 0.2 µL.
- Samples containing high amounts of matrix rapidly lead to quality depletion. Here a retention gap is absolutely necessary. However, the accumulation of involatile matrix in the retention gap quickly give rise to adsorption and peak broadening because of the small capacity, so that changing it regularly is necessary. For dirty samples, hot or cold PTV sample injection is preferred.
- Samples containing highly volatile components are not easy to control, as focusing by the solvent effect is inadequate under certain circumstances. If changing the solvent does not help, changing to a hot or cold split or splitless system is recommended.

2.2.7.2 PTV On-Column Injection

A regular PTV injector can also be used successfully for OCI. This method allows the straightforward automation of OCIs using regular autosamplers with standard syringes. The special insert liner used needs to have a small ID restrictor at the top as a needle guide into the centered column at the bottom. Syringes with the regular 0.47 mm OD needles (26G, 26sG) can be used allowing the direct injection into widebore columns or pre-columns of equivalent dimensions, for example, 0.53 mm ID. For this purpose, the column inlet is pushed up until it gets positioned and centered at the bottom of the restrictor section of the inlet liner.

In PTV on-column mode, the injector body and the column oven are set for injection to ambient temperature significantly below the solvent BP (see Table 2.20). After a short injection time and solvent peak elution the oven program is started.

2.2.7.3 LC-GC Coupling

The direct transfer of a liquid sample into a GC column, for instance from a LC separation, is the next logical step for large volume OCIs (Munari et al., 1990). The transfer of a solvent plug of up to several hundred microliters into an empty pre-column is the established practice today (Grob et al., 1984; Grob, 1995). Applications use the normal phase LC (NPLC) separation as a higher performing and online sample preparation technique. NPLC solvents are compatible with GC analysis.

The GC configuration for LC-GC operation shown in Figure 2.105 uses a pre-column with the inlet side connected to a Y-piece or 6-way valve for carrier gas supply and the transfer of a LC fraction (Figure 2.106). Inside of the GC oven the pre-column is connected via a T-piece to the analytical column, and a solvent vapor exit line (SVE). The SVE valve is open during sample transfer, and closed during analysis. Pre-column dimensions in use are typical of 0.53 ID megabore

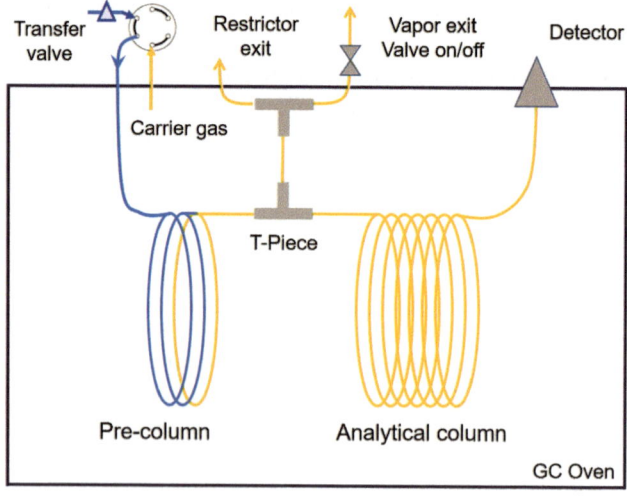

Figure 2.105 GC configuration for LC-GC applications.

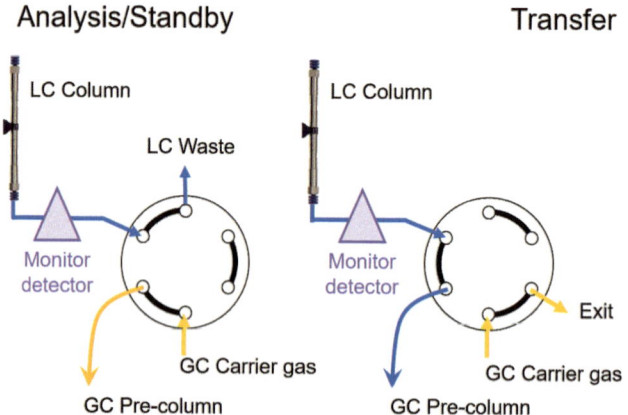

Figure 2.106 LC fraction transfer to GC (= transfer valve).

columns of only 5 to 10 m length. Metal columns are used to avoid the brittleness of megabore fused silica columns providing the required process safety in automated routine operation. Also, and important for a reproducible process, the metal column enables an adequate heat conductance for the solvent evaporation inside of the pre-column. During transfer of the LC sample plug the GC oven temperature is set above the pressure corrected BP of the solvent at the given carrier gas inlet pressure (see Table 2.20).

For online LC-GC coupling the so-called "concurrent solvent evaporation" technique is used (Grob, 1986; Grob and Läubli, 1987; Schmarr et al., 1989). The concurrent solvent evaporation balances the evaporation of the solvent through the open SVE with the transfer flow of liquid. Short pre-columns can be used. In contrast to the OCI technique (Section 2.2.7.1) the volume of the pre-column does not need to match the transferred liquid volume of the solvent fraction. Most of the solvent is evaporated during the transfer phase through the open SVE. The LC solvent transfer rate is adjusted to be in line with the solvent evaporation rate at the set oven temperature. During this sample transfer process into the pre-column the analytical column is idling in a static no-flow mode, with ambient pressure on both ends. Upon closing the SVE valve the T-piece receives the necessary carrier gas pressure for the analytical separation via the pre-column. Residual solvent, together with the analytes, is then swept into the analytical column (*aka* "partial concurrent solvent evaporation") and the analytes get focused at the beginning of the column film in a narrow band (the known "solvent effect"). A regular chromatographic separation follows. Temperatures, switching times, and the LC and GC flows are software controlled parameters of a comprehensive workflow.

The remarkable performance of online LC-GC hyphenation is best described with the recent developments for the authentication and adulteration of expensive edible oils, i.e. the extra virgin olive oil (EVOO), by the analysis of the phytosterol pattern (Aparicio *et al.*, 2013). Online LC-GC is a standard method for the determination of sterols of vegetable oils and fats by NPLC removing triglycerides to separate the analyte classes of alcohols, triterpenes, methyl sterols and sterols (Biedermann *et al.*, 1993), as well as the free sterols/stanols, steryl/stanyl fatty acid esters and other minor lipids (Esche *et al.*, 2013; Steenbergen and Janssen, 2014). Online coupled LC-GC-FID is enabling the detection of 2% of rapeseed, 5% of sunflower, or 10% of soybean or grapeseed oil as potential addition to olive oil (Grob, 1994). Ergosterol is proposed as a marker for olives degraded by yeast or mold, determined by online LC-GC-MS separating it from other higher concentrated sterols present (Boarelli *et al.*, 2020).

A fully automated method using a robotic x,y,z-sampler performing the complete workflow including saponification and extraction steps with injection into an LC-GC system was developed (Nestola and Schmidt, 2016). The automated procedure provides quantitative results with improved precision for campesterol, stigmasterol and Δ-7-stigmastenol in olive oils being well comparable to the manually performed method of the Commission Regulation (EEC) 2568/91 Annex V (The Commission of the European Communities, 1991; The European Commission, 2016). The LC-GC method soon became routine for quality control of edible oils (Bohačenko and Kopicová, 2001; Küchler *et al.*, 2012). Further automated LC-GC applications cover the determination of cholesterol in foods, the adulteration of Arabica coffee with the Robusta variety (Kresse, 2023), or the important food safety analysis of the mineral oils saturated and aromatic hydrocarbons (MOSH/MOAH) or of polyaromatic hydrocarbons (PAHs) (Biedermann and Grob, 2012a,b; Nestola and Schmidt, 2017; Nestola, 2018). The migration of epoxidized soy bean oil (ESBO) from the gaskets of lids into food is another very practical solution by LC-GC analysis (Fankhauser-Noti *et al.*, 2005).

2.2.8 Capillary Column Choice and Separation Optimization

There are no hard-and-fast rules for the choice of columns for GC-MS coupling. The choice of the correct phase is made on the typical criterion: "like dissolves like". If substances exhibit no interaction with the stationary phase, there will be no retention and the substances leave the column at the dead time or get hardly separated from each other. The polarity of the stationary phase should correspond to the polarity of the analytes being separated (see Table 2.23). Less polar substances are better separated on non-polar phases and vice versa.

An exception to this general rule of thumb is given with the newly developed ionic liquid stationary phases. The characteristics and guide to potential applications are discussed in Section 2.2.8.8.

Chose More Polar Stationary Phases

- Weaker retention of non-polar compounds
- Stronger retention of polar compounds
- Shift of compounds with specific interactions
- Exceptions with ionic liquid phases

For coupling with MS the carrier gas flow and the potential column bleed are included in the criteria governing the choice of column. When considering the optimal carrier gas flow, the maximum carrier gas load of the mass spectrometer must be taken into account. The limit for small benchtop mass spectrometers with quadrupole analyzers is in the range of up to 4 mL/min of helium. Larger instruments with high pumping capacity can generally tolerate higher loads, and the use of hydrogen as carrier gas. These conditions limit the column diameter which can be used.

There can be no compromises concerning column bleed in GC-MS. Column bleed generally contributes to chemical noise where MS is used as the mass-dependent detector, and curtails the detection limits. The optimization of a particular S/N ratio can also be improved by selecting principally thermally stable stationary phases with a low tendency to bleed. For use in trace analysis, stationary phases for high temperature applications have proved particularly useful (Figure 2.107 and 2.108). Besides the phase itself, the film thickness also plays an important role. Thinner films and shorter columns exhibit lower column bleed.

2.2.8.1 Choice of Carrier Gas

Helium Helium is a non-renewable resource. It is created by the radioactive decay of thorium and uranium and refined from natural gas. Due to limited supply prices increase. Nevertheless, it is the most popular and of almost general use as carrier gas in GC-MS. The optimum He carrier gas velocity for capillary columns is about 23 cm/s. For speed optimized analysis a higher velocity of 30 cm/s or more is used in a constant pressure setup for up to 30% increase in sample throughput. The increasing cost and availability of He triggers the ongoing discussion of suitable alternatives.

While MS instrument tuning conditions for resolution and sensitivity are standardized for use with He, the change to different carrier gases requires individual optimization.

Table 2.23 Composition of stationary phases for fused silica capillary columns with column designations from different manufacturers.

Columns by Phase Composition

Phase Composition	Agilent (J&W)	Agilent (former Varian/ Chrompack)	Alltech	Macherey-Nagel	Perkin-Elmer	Pheno-menex	Quadrex	Restek	Trajan (former SGE)	Supelco	Thermo Scientific	U.S. Pharma-copeia Designation
Fused Silica wall-coated open tubular (WCOT) columns
100% Dimethyl polysiloxane	DB-1, HP-1	CP-Sil 5 CB	AT-1, EC-1	OPTIMA 1, OPTIMA 1 TG	Elite –1	ZB-1	007–1, 007–1 PHAT	Rtx-1	BP1	SPB-HAP, SPB-1, SPB-1SULFUR		G1, G2, G9, G38
100% Dimethyl polysiloxane (low bleed)	DB-1 ms, DB-1 ms UI, Ultra-1, HP-1 ms, HP-1 ms UI	VF-1 ms, CP-Sil 5, CP Sil 5 CB	AT-1 ms	OPTIMA-1 ms, OPTIMA-1 ms Accent	Elite -1 ms	ZB-1plus, ZB-1 ms	007-1MS	Rtxi-1 ms	SolGel-1 ms	Equity-1	TG-1MS, TG-1MT, TR-1MS	G1, G2, G9, G38
100% Dimethyl polysiloxane (high temp.)	DB-1ht		AT-1ht		Elite -1ht	ZB-1HT, ZB-1HT Inferno	007-1HT	Rxi-1HT	BPX1			
50% n-Octyl 50% methyl siloxane										SPB-Octyl, Petrocol DH Octyl		
5% Phenyl 95% dimethyl polysiloxane	DB-5, HP-5, HP-5 ms, HP-5 ms UI, Ultra-2	VF-5 ms, CP-Sil 8 CB, CP Sil 8 CB MS	AT-5, AT-5 ms, EC-5	OPTIMA 5	Elite-5	ZB-5, ZB-5plus	007-5	Rtx-5/ Rtx-G27	BP5	SAC-5, Equity-5, SPB-5	TR-5, TG-5MS	G27, G36, G41
5% Phenyl 95% dimethyl polysiloxane (base modified)	DB-5 ms EVDX							Rtx-5 Amine		PTA-5	TG-5MS amine	

(Continued)

Table 2.23 (Continued)

Phase Composition	Agilent (J&W)	Agilent (former Varian/Chrompack)	Alltech	Macherey-Nagel	Perkin-Elmer	Pheno-menex	Quadrex	Restek	Trajan (former SGE)	Supelco	Thermo Scientific	U.S. Pharma-copeia Designation
5% Phenyl 95% dimethyl polysilarylene (low bleed)	DB-5 ms, DB-5 ms UI	VF-Xms		OPTIMA-5 ms, OPTIMA-5 ms Accent	Elite-5 ms	ZB -5MS, ZB -5MSplus, ZB-5MSi	007-5MS	Rxi-5 ms		SLB-5 ms	TG-5SILMS, TG-5MT, TR-5MS,	G27, G36
5% Phenyl 95% polysilphenylene-siloxane (low bleed)									BP5MS			
5% Phenyl 95% dimethyl polysilarylene (high temp.)	DB-5ht	VF-5htm VF-5ht Ultimetal		OPTIMA-5 HT	Elite-5ht	ZB-5HT, ZB-5HT Inferno	007-5HT	Rxi-5Sil MS, Rxi-5HT			TG-5HT, TR-5HT, TR-8270	G27, G36
5% Phenyl 95% dimethyl polysilarylene	DB-XLB					ZB-XLB, ZB-XLB-HT Inferno		Rxi-XLB			TG-XLBMS	
5% Phenyl polycarborane-siloxane									HT5	HT-5 (aluminum clad)		
8% Phenyl polycarborane-siloxane									HT8, HT8-PCB		TR-8095	
10% Phenyl 90% dimethyl polysiloxane							007-10, 007-20					G41
14% Phenyl 86% dimethyl polysiloxane		CP-Sil 13 CB										

20% Phenyl 80% dimethyl polysiloxane			AT-20			Rtx-20	SPB-20	G28, G32
35% Phenyl 65% dimethyl polysiloxane	DB-35, HP-35	VF-35 ms	AT-35, AT-35 ms	Elite-35	ZB-35, ZB-35HT, ZB-35HT Inferno	007–35, 007-35MS	Rtx-35, Rtx-35Sil MS	SPB-35	TG-35MS, TR-35MS	G42
35% Phenyl 65% dimethyl polysiloxane (base modified)								TG-35MS amine
35% Phenyl 65% dimethyl arylene siloxane	DB-35 ms, DB-35 ms UI			OPTIMA 35 MS			BPX35, BPX608		G42
50% Phenyl 50% methyl polysiloxane	DB-17 ms		AT-50	OPTIMA 17	Elite-17, Elite-17ht, Elite-17 ms				TG-17MS	G3
50% Phenyl 50% dimethyl polysiloxane		VF-17 ms, CP-Sil 24 CB					Rxi-17Sil MS			G3
5% Diphenyl 95% dimethyl polysiloxane							Rtx-1614, Rtx-DHA, Rtx-5MS		
35% Diphenyl 65% dimethyl polysiloxane							Rtx-25 Amine		
50% Diphenyl 50% dimethyl polysiloxane	DB-17, HP-50+, HP-17				ZB-50		Rxi-17	SPB-50		G3

(Continued)

Table 2.23 (Continued)

Phase Composition	Agilent (J&W)	Agilent (former Varian/ Chrompack)	Alltech	Macherey-Nagel	Perkin-Elmer	Pheno-menex	Quadrex	Restek	Trajan (former SGE)	Supelco	Thermo Scientific	U.S. Pharma-copeia Designation
50% Diphenyl 50% dimethyl polysiloxane (high temp.)	DB-17ht											
50% Diphenyl 50% dimethyl arylene polysiloxane (high temp.)											TG-17SilMS, TR-50MS	
50% Phenyl 50% dimethyl arylene siloxane				OPTIMA 17 MS			007-50HT		BPX50			
65% Phenyl 35% dimethyl polysiloxane		CP-TAP-CB					007-65HT					G17
65% Diphenyl 35% dimethyl polysiloxane								Rtx-65, Rtx-65TG				G17
75% Phenyl 25% dimethyl polysiloxane	HP-50+											G17
100% Methylphenyl polysiloxane								Rtx-50				G3
35% Trifluoropropyl 65% dimethyl polysiloxane	DB-200	VF-200 ms			Elite-200			Rtx-200, Rtx-200MS			TG-200MS	G6

50% Trifluoropropyl 50% dimethyl polysiloxane	DB-210, HP-210	AT-210	OPTIMA 210					G6
6% Cyanopropyl-phenyl 94% dimethyl polysiloxane	DB-1301, DB-624 UI, DB-Select 624 UI 467, HP-Fast GC	CP-1301, VF-1301 ms, VF-624 ms, CP-Select 624 Hexane	AT-624, AT-1301, Method 467 Kits	OPTIMA 1301, OPTIMA 1301 ms, OPTIMA 624, OPTIMA 624 LB	Elite-624, Elite-624 ms, Elite-1301	ZB-624, ZB-624plus	007–1301, 007–624	Rtx-1301/ Rtx-G43, Rtx-624, Rxi-1301Sil MS, Rxi-624Sil MS	BP624	SPB-624, OVI-G43	TG-1301MS, TG-624, TR-V1	G43
6% Cyanopropyl-methyl 94% dimethylarylene polysiloxane								Rxi-1301Sil MS, Rxi-624Sil MS			TG-624SilMS	G43
14% Cyanopropyl-phenyl 86% dimethyl polysiloxane	DB-1701, DB-1701P	CP Sil 19 CB, VF-1701	AT-1701	OPTIMA 1701	Elite-1701	ZB-1701, ZB-1701P	007–1701, 007–17MS	Rtx-1701	BP10 (1701)	Equity-1701	TG-1701MS, TR-1701	G46
14% Cyanopropyl-phenyl 86% dimethyl polysilarylene		VF-1701 ms		OPTIMA 1701 ms								
25% Cyanopropyl 25% dimethyl polysiloxane												G19
33% Cyanopropyl-methyl 67% dimethyl polysiloxane				OPTIMA 240								
50% Cyanopropyl-phenyl 50% dimethyl polysiloxane	DB-225, DB-225 ms, HP-225	CP Sil 43 CB	AT-225		Elite-225		007–225				TG-225MS	G7, G19

(Continued)

Table 2.23 (Continued)

Phase Composition	Agilent (J&W)	Agilent (former Varian/Chrompack)	Alltech	Macherey-Nagel	Perkin-Elmer	Pheno-menex	Quadrex	Restek	Trajan (former SGE)	Supelco	Thermo Scientific	U.S. Pharma-copeia Designation
50% Cyanopropyl 50% dimethyl polysiloxane	DB-23				Elite-23							G5, G8, G48
50% Cyanopropyl-methyl 50% phenylmethyl polysiloxane				OPTIMA 225				Rtx-225		SPB-225		G7, G19
70% Cyanopropyl polysilphenylene-siloxane									BPX70		TR-FAME	
70% biscyanopropyl 30% cyanopropyl-phenyl polysiloxane												
78% Cyanopropyl 22% dimethyl polysiloxane		VF-23 ms	AT-Silar-90									
80% Biscyanopropyl 20% cyanopropyl-phenyl polysiloxane	HP-88									SP-2330		G5, G8, G48
88% Cyanopropyl-aryl polysiloxane		CP-Sil 88										G5
90% Cyanopropyl polysilphenylene-siloxane									BPX90			
90% biscyanopropyl 10% cyanopropyl-phenyl polysiloxane								Rtx-2330		SP-2331, SP-2380	TG-POLAR	G48

Polybiscyanopropyl siloxane			AT-Silar-100			SP-2560, SP-2340	G5
1,2,3-Tris(2-cyanoethoxy) propane	CP-TCEP			Elite-TCEP		TCEP	
PEG - polyethylene glycol	DB-WAXetr, HP-INNOWax	CP-Wax 52 CB, CP-Carbowax 400	AT-Wax, AT-WAXms, AT-AquaWax, EC-WAX	Permabond CW 20 M, OPTIMA WAX, OPTIMA WAXplus	Elite-WAX	ZB-WAX, ZB-WAXplus	007-CW, BTR-CW	Stabilwax, Stabilwax-MS, Rtx-Wax	SolGel-WAX	SPB-PUFA, Omegawax, SUPEL-COWAX 10	TG-WaxMS, TG-WaxMT, TR-Wax, TR-WaxMS	G14, G15, G16, G20, G39, G52
PEG - polyethylene glycol (extended temp. Range)	DB-WAX, DB-WAX UI, DB-FATWAX UI, DB-HeavyWAX, DB-WaxFF	VF-WAXms		Elite-WAX ETR			BP20 (Wax)			G16, G20, G52
PEG - base modified, for amines and basic compounds	CAM	CP-Wax 51	AT-CAM				Stabilwax-DB		Carbowax-Amine	TG-WaxMS B	G50
PEG – nitroterephthalic acid modified, for acidic compounds	DB-FFAP, HP-FFAP	CP-Wax 58 FFAP CB, CP-FFAP CB	AT-1000, EC-1000, AT-AquaWax, AT-AquaWax-DA	Permabond FFAP, OPTIMA FFAP	Elite-FFAP	ZB-FFAP	007-FFAP	Stabilwax-DA	BP21 (FFAP)	Nukol, SP-1000	TG-WaxMS A, TR-FFAP	G25, G35
1,12-di(tripropyl-phosphonium) dodecane bis(tri-fluoromethylsul-fonyl)imide										SLB-IL59, SLB-IL60, SLB-IL82		

(Continued)

Table 2.23 (Continued)

Phase Composition	Agilent (J&W)	Agilent (former Varian/ Chrompack)	Alltech	Macherey-Nagel	Perkin-Elmer	Pheno-menex	Quadrex	Restek	Trajan (former SGE)	Supelco	Thermo Scientific	U.S. Pharma-copeia Designation
1,12-di(tripropyl-phosphonium) dodecane bis(tri-fluoromethyl-sulfonyl)imide trifluoromethyl-sulfonate										SLB-IL61		
Tri(Tripropyl-phosphonium-hexanamido) Triethylamine bis(Trifluoro-methylsulfonyl) imide										SLB-IL76		
1,9-di(3-vinyli-midazolium) nonane bis(tri-fluoromethyl-sulfonyl)imide										SLB-IL100		
1,5-di(2,3-di-methylimida-zolium)pentane bis(trifluoro-methylsulfonyl) imide										SLB-IL111		

Chiral Phases			
Dimethylated β-cyclodextrin	Chiraldex B-DM		
Dimethylated pentylated β-cyclodextrin		HYDRODEX β-3P	
Dimethylated butyldimethylsilylated β-cyclodextrin		HYDRODEX β-6TBDM	
Dimethylated butyldimethylsilylated β-cyclodextrin in 14% cyanopropylphenyl/86% dimethyl polysiloxane			Rt-βDEXsm
Trimethylated β-cyclodextrin		HYDRODEX β-PM	
Diethylated butyldimethylsilylated β-cyclodextrin		HYDRODEX β-6TBDE	
Diethylated butyldimethylsilylated β-cyclodextrin in 14% cyanopropylphenyl/86% dimethyl polysiloxane			Rt-βDEXse

(Continued)

Table 2.23 (Continued)

Phase Composition	Agilent (J&W)	Agilent (former Varian/ Chrompack)	Alltech	Macherey-Nagel	Perkin-Elmer	Pheno-menex	Quadrex	Restek	Trajan (former SGE)	Supelco	Thermo Scientific	U.S. Pharma-copeia Designation
Dipropylated butyldimethyl-silylated β-cyclodextrin in 14% cyanopropylphenyl/86% dimethyl polysiloxane								Rt-βDEXsp				
Diacetylated butyldimethyl-silylated β-cyclodextrin				HYDRODEX β-TBDAc								
Diacetylated butyldimethyl-silylated β-cyclodextrin in 14% cyanopropylphenyl/86% dimethyl polysiloxane								Rt-βDEXsa				
Diacetylated butyldimethyl-silylated γ-cyclodextrin				HYDRODEX γ-TBDAc								

Diacetylated butyldimethyl-silylated γ-cyclodextrin in 14% cyanopropylphenyl/86% dimethyl polysiloxane					Rt-γDEXsa
Dimethoxymethylated butyldimethyl-silylated γ-cyclodextrin			HYDRODEX γ-DI-MOM		
Dialkylated β-cyclodextrin		Chiraldex B-DA			
Permethylated β-cyclodextrin	HP-Chiral β	Chiraldex B-PM		Elite-Betacydex	Cydex-B CHIRALDEX, Supelco DEX
Permethylated β-cyclodextrin in 14% cyanopropylphenyl/86% dimethyl	CP-Cyclodextrin β				Rt-βDEXm
Permethylated hydroxypropyl β-cyclodextrin		Chiraldex B-PH			
Propionylated γ-cyclodextrin		Chiraldex G-PN			
Pentylated α-cyclodextrin			LIPODEX A		
Pentylated β-cyclodextrin			LIPODEX C		Rt-βDEXm Cydex-B CHIRALDEX, Supelco DEX

(Continued)

Table 2.23 (Continued)

Phase Composition	Agilent (J&W)	Agilent (former Varian/ Chrompack)	Alltech	Macherey-Nagel	Perkin-Elmer	Pheno-menex	Quadrex	Restek	Trajan (former SGE)	Supelco	Thermo Scientific	U.S. Pharma-copeia Designation
Pentylated-acetylated α-cyclodextrin				LIPODEX B								
Pentylated-acetylated β-cyclodextrin				LIPODEX D								
Pentylated methylated γ-cyclodextrin				LIPODEX G								
Pentylated butyrylated γ-cyclodextrin				LIPODEX E								
Butyrylated γ-cyclodextrin			Chiraldex G-BP									
Trifluoroacetylated α-cyclodextrin			Chiraldex A-TA									
Trifluoroacetylated β-cyclodextrin			Chiraldex B-TA									
Trifluoroacetylated γ-cyclodextrin			Chiraldex G-TA									
Optical isomers		CP-Cyclo-dextrin-β-2,3,6-M-19										
cyclodextrin-dimethylpoly-siloxane		CP-Chirasil-Dex CB		LIPODEX E								

beta-Cyclodextrin in phenyl-based stationary phase	HP-Chiral β							
30%-heptakis (2,3-di-O-methyl-6-O-t-butyl dimethylsilyl)-B-cyclodextrin in 1701		CycloSil-B		Elite-Betacyclo-dextrin				
1,2,3-tris(cyan-oethoxy)propane			CP-TCEP		Rt-TCEP	TCEP		
D-/L-amino acids in dimethyl-polysiloxane			CP-Chirasil D-Val, Chirasil-L-Val	OPTIMA δ			G49	

Fused Silica porous-layer open tubular columns (PLOT) Columns

Aluminum oxide PLOT	GS-Alumina, HP-PLOT Al₂O₃ M, HP-PLOT Al₂O₃ S,	CP-Al₂O₃		Elite-Alumina PLOT	PLT-AL₂O₃	Rt-Alumina BOND		TR-BOND Alumina (Na₂SO₄/KCl)
Aluminum oxide/KCl PLOT	GS-Alumina KCl, HP-PLOT Al₂O₃ KCl	CP-Al₂O₃/KCl		Elite-Alumina/KCl PLOT			Alumina chloride PLOT	TG-BOND Alumina (KCl)
Aluminum oxide/Na₂SO₄ PLOT		CP-Al₂O₃/Na₂SO₄					Alumina sulfate PLOT	TG-BOND Alumina (Na₂SO₄)
Carboxen							Carboxen-1010 PLOT, Carboxen-1006 PLOT	

(Continued)

Table 2.23 (Continued)

Phase Composition	Agilent (J&W)	Agilent (former Varian/Chrompack)	Alltech	Macherey-Nagel	Perkin-Elmer	Phenomenex	Quadrex	Restek	Trajan (former SGE)	Supelco	Thermo Scientific	U.S. Pharmacopeia Designation
Carbon-based bonded PLOT	CarboBOND, CarboPLOT P7, GS-Carbon PLOT											
DVB 100%					Elite-Cyclosil B PLOT			Rt-Q-BOND, MXT-Q-BOND		Supel-Q-PLOT		
DVB copolymer PLOT	PoraBOND Q						PLT-Q	Rt-S-BOND, MXT-S-BOND				
DVB-ethylene glycol-dimethacrylate PLOT	HP-PLOT U, PoraBOND U, PoraPLOT U	CP-Pora BOND U, CP-Pora PLOT U						Rt-U-BOND				G45
DVB 4-vinylpyridine polymer PLOT	PoraPLOT S	CP-Pora PLOT S										
DVB-polystyrene polymer PLOT	HP-PLOT Q, PoraPLOT Q, PoraPLOT Q-HT	CP-Pora PLOT Q, CP-Pora Bond Q			Elite-Q PLOT		PLT-Q				TR-BOND Q	
Intermediate polarity porous polymer PLOT	GC-Q							Rt-QS-BOND				
Molecular-sieve-coated PLOT	HP-PLOT Molsieve	CP-Molsieve 5 Å			Elite-Molsieve PLOT		PLT-5A	Rt-Msieve 5A, MXT-Msieve 5A		Mol Sieve 5A PLOT	TR-BOND Sieve 5A	
Silica bonded PLOT	GS-GasPro	CP-Silica PLOT										

(Continued)

Table 2.23 (Continued)

Columns by Selected Applications

Application	Agilent (J&W)	Agilent (Varian/ Chrompack)	Alltech	Macherey- Nagel	Perkin- Elmer	Pheno- menex	Quadrex	Restek	Trajan (former SGE)	Supelco	Thermo Scientific	
Volatile amines and ammonia	PoraPLOT Amines	CP- Volamine		OPTIMA 5 Amine	Elite-5 Amine			Rtx-Volatile Amine				
Amine separation				FS-CW 20 M-AM				Rtx-25 Amine		Carbowax Amine	TG-5MS amine, TG-35MS amine	G16
Amines and Basic Compounds								Rtx-5 Amine		PTA-5		G50
Drug analysis		VF-DA	Drug 1, Drug 2, Drug 3			ZB-Drug-1						
Volatile sulfur compounds			AT-Sulfur							SPB- 1SULFUR		
Alcohols in wine and brewing industry		CP-Wax 57 CB										
Residual Solvents in pharmaceuticals								Rtx-G27, Rtx-G43		OV1-G43		G27, G43
Fragrances and flavors					Elite- Volatiles		007-CW	Rt-CW20M F&F				
VOCs, U.S. EPA methods 502.2, 524.2, 601, 602, 624, 8010, 8020, 8240, 8260	DB-VRX, DB-624 UI		AT-502.2		Elite-502.2, Elite-VMS			Rtx-VMS, Rtx-VGC	BPX- VOLA- TILES	VOCOL	TR-524, TG-VMS	G53

(Continued)

Table 2.23 (Continued)

Application	Agilent (J&W)	Agilent (Varian/ Chrompack)	Alltech	Macherey-Nagel	Perkin-Elmer	Pheno-menex	Quadrex	Restek	Trajan (former SGE)	Supelco	Thermo Scientific
VOCs, U.S. EPA methods 502.2, 524.2, 601, 602, 624, 8010, 8020, 8240, 8260					Elite-624, Elite-Volatiles	ZB-624	007–624, 007–502	Rtx-VRX, Rtx-502.2, Rtx-624, Rtx-Volatiles	BP624	SPB-624	TR-525
SVOCs U.S. EPA methods 625, 1625, 8270, CLP	DB-5.625, DB UI 8270					ZB-Semi-volatiles		Rxi-SVOCms			TR-8270
PAHs, 16 EU compounds	DB-EUPAH, Select PAH					ZB-PAH-EU, ZB-PAH-CT		Rxi-PAH			G51
U.S. EPA method 8270D	DB-UI 8270D										
PCBs, U.S. EPA methods 608, 508, and 8080	DB-608								BPX608		
PCB Congeners								Rtx-PCB	HT8-PCB		TR-PCB 8MS
Dioxins and furans (PCDDs, PCDFs) U.S. EPA 1613	DB-Dioxin					ZB-Dioxin		Rtx-Dioxin, Rtx-Dioxin2	BPX-DXN		TR-Dioxin 5MS
EPA pesticides methods by GC-ECD 8081, 8082, 8151 504, 505, 508, and 552	DB-CLP 1, DB-CLP 2				Elite-CL Pesticides, Elite-CL Pesticides2	ZB-CL Pesticides-1, ZB-CL Pesticides-2		Rtx-CL Pesticides2, Stx-CL Pesticides2			

Herbicides EPA method 507							Sub-Herb
OCPs-EPA Methods							
OCPs-EPA Method 8081, 8141A, 608, and CLP Pesticides.		CP Sil 8 CB, CP Sil 19 CB	AT-Pesticide	Elite-608	ZB-5, ZB-35, ZB-1701, ZB-50	007–608, 007–17, 007–1701	Rtx-5, Rtx-35, Rtx-50, Rtx-1701	BP-5, BP-10, BP-608	SPB-608	G3
Pesticides, herbicides, PCBs, PAHs	DB-XLB			Elite-XLB	ZB-Multi Residue-1, ZB-Multi Residue-2		Rtx-440			TG-VRX
Total petroleum hydrocarbons (TPH)	DB-TPH			Elite-TPH			Rtx-Mineral Oil			
ASTM D 2887	DB-2887	CP-SimDist	AT-2887			007–1	Rtx-2887		Petrocol 2887, Petrocol EX2887	
ASTM 3710			AT-3710							
ASTM D 5134, PONA Analysis	DB-Petro, HP-PONA	CP-Sil PONA	AT- Petro	PERMABOND P-100	Elite-PONA	ZB-DHA-PONA	007–1	Rtx-1PONA	BP1 PONA	Petrocol DH, Petrocol DH150, Petrocol DH50.2	G1, G2, G38
ASTM 5501 bioethanol						ZB-Bioethanol				
ASTM D 5623 and 5504	DB-Sulfur SCD									

(Continued)

Table 2.23 (Continued)

Application	Agilent (J&W)	Agilent (Varian/ Chrompack)	Alltech	Macherey-Nagel	Perkin-Elmer	Pheno-menex	Quadrex	Restek	Trajan (former SGE)	Supelco	Thermo Scientific
ASTM D 6584, EN14013, EN14105, EN14106, and EN14110 biodiesel	Biodiesel			OPTIMA BioDiesel M, OPTIMA BioDiesel F, OPTIMA BioDiesel G	Elite Biodiesel TG			Rtx-Biodiesel TG	BPX-BIOD	MET-Biodiesel	
MTBE in reformulated gasolines		CP-Select CB									
ASTM oxygenates methods	GS-OxyPLOT										
Simulated Distillation SIMDIS	DB-HT SimDis				Elite-SimDist	ZB-1XT SimDist		MXT-500 Sim Dist	HT 5	Petrocol 2887, Petrocol EX2887	
Triglycerides				OPTIMA 1-TG, OPTIMA 17-TG							
Fatty Acid Methyl Esters (FAMEs)	DB-FastFAME		AT-FAME			ZB-FAME		FAMEWAX		Omegawax	TR-FAME G1, G2, G3, G38 G52

Blood alcohol analysis	DB-ALC1, DB-ALC2, DB-BAC1 UI, DB-BAC2 UI, HP-Blood Alcohol	Elite-BAC1 Advantage, Elite-BAC2 Advantage	ZB-BAC-1, ZB-BAC-2	Rtx-BAC1, Rtx-BAC2	TG ALC I/II
Amino acids		AT-Amino Acid			
Explosives (8% phenyl poly-carbonate-siloxane)				Rtx-TNT, Rtx-TNT2	TR-8095
Silanes (monomeric, e.g., chlorosilanes)		PERMABOND Silane			

WCOT, wall-coated open tubular, PLOT, porous-layer open tubular columns

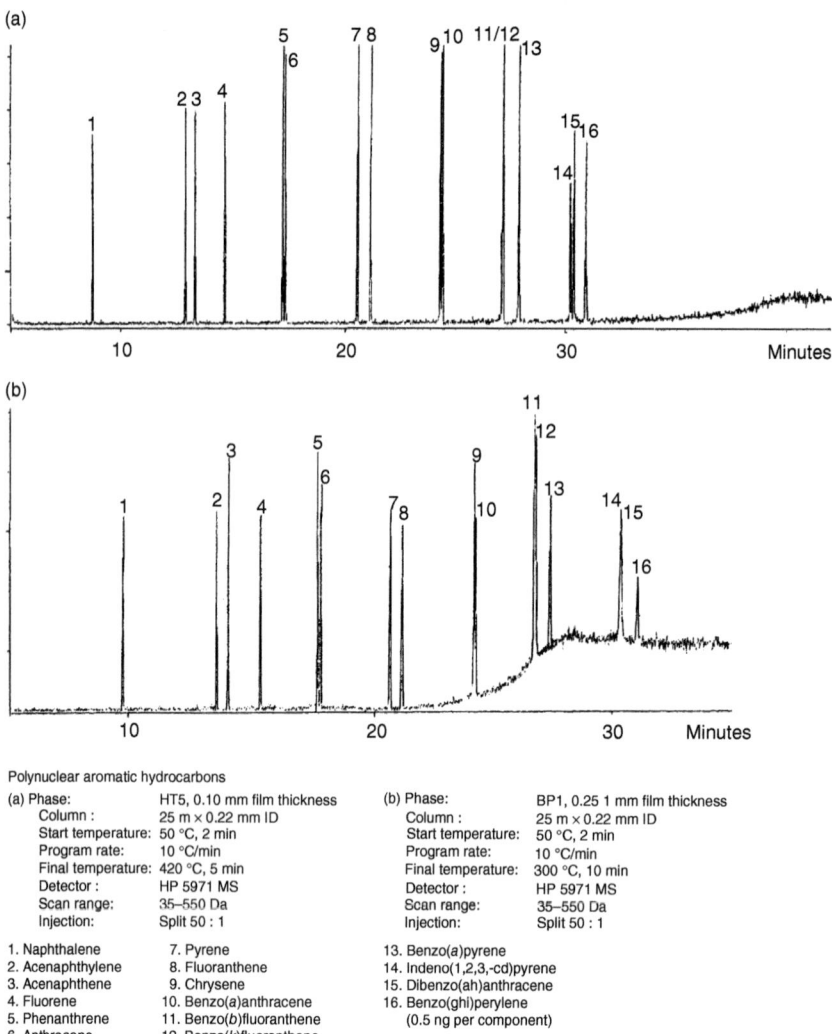

Figure 2.107 Comparison of a high temperature phase: (a) SGE HT5 (siloxane-carborane) with a conventional silicone phase; (b) SGE BP1 (dimethylsiloxane) using a polyaromatic hydrocarbon standard (SGE). The long-temperature program for the high temperature phase up to 420 °C allows the elution of components 14, 15, and 16 during the oven temperature ramp at an elution temperature of about 350 °C. Sharp narrow peaks with low column bleed give better detection conditions for high boiling substances (GC-MS system HP-MSD 5971).

Hydrogen With hydrogen as the carrier gas a much higher velocity with an optimum at 40 cm/s can be used. The viscosity of hydrogen is low and less affected by temperature than helium and nitrogen. The analytical benefit is the significantly shorter analysis time. Furthermore, with hydrogen, the right-hand branch of the van Deemter curve is flatter, so that a further increase in speed is possible without impairing the separation efficiency as illustrated in Figure 2.109 and

2.2 Gas Chromatography | 195

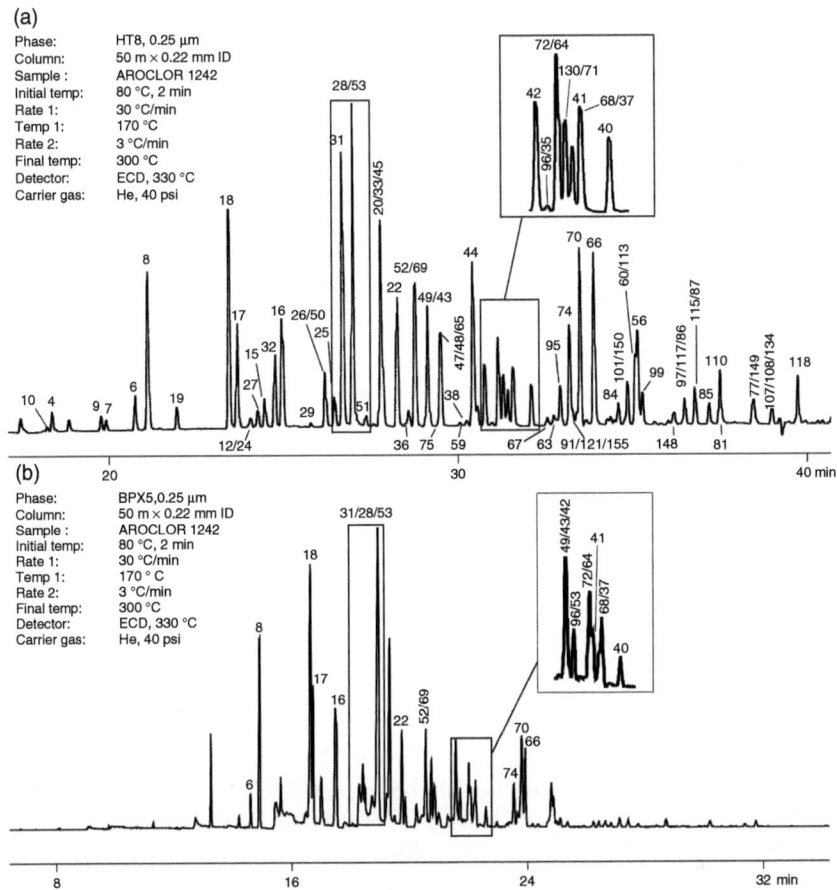

Figure 2.108 An example of the different selectivities of stationary phases for the Aroclor mixture 1242 (see in particular the separation of the critical congeners 31, 28, and 53) (Courtesy SGE). (a) Carborane phase – HT8 50 m × 0.22 mm × 0.25 µm and (b) 5%-phenyl phase – BPX5 50 m × 0.22 × 0.25 µm.

Figure 2.148 (Adam, 2012). Because of the limited pumping capacity in GC-MS systems (see Section 2.3.6) only limited use can be made of these advantages with hydrogen. Smaller ID columns can then be used. The column diameters of 0.20 mm and 30 m length, or less and longer, should be chosen for GC-MS to provide ion source compatible hydrogen carrier gas flows.

Hydrogen is a safe and economical alternative carrier gas also in GC-MS when using hydrogen generators for on-site premium quality supply. Use steel lines and steel fittings, avoid brass and copper materials. It needs to be considered that hydrogen is a reactive gas and mass spectra of the hydrogenated compounds are monitored requiring a dedicated transition setup for target analysis by MS/MS, and a careful interpretation of library search results from with He carrier generated full scan spectra. The analysis of reactive compounds like pesticides delivers comparable performance with hydrogen compared to helium (Cutillas et al., 2024). It is important to

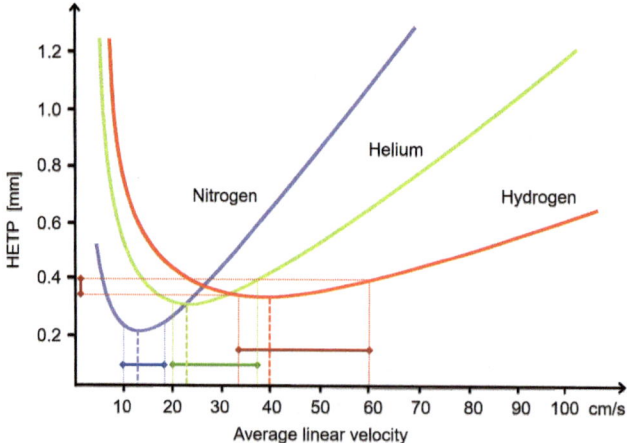

Figure 2.109 Van Deemter curves for nitrogen, helium, and hydrogen carrier gas in isothermal operation. The bars below indicate the recommended optimum linear velocity range for each carrier gas. It is shown for hydrogen that the increase in velocity only has a small effect on the separation power (HETP).

update the selected/multiple reaction monitoring (SRM) transitions for the analyte compounds with the newly determined transitions due to unavoidable hydrogenation reactions in the ion source.

Nitrogen Nitrogen as carrier gas in GC-MS is a viable alternative for many applications with even a slightly lower HETP than He or H_2 (Figure 2.109). The molecular mass of m/z 28 does not interfere with many full scan applications, as the "air region" is typically not scanned, and does not interfere with selected ion monitoring (SIM) and SRM applications. Mass spectra are compatible with He as carrier gas. With the optimum flow below 20 cm/s N_2 could be considered as a "low speed" carrier gas. It is therefore recommended with N_2 to also change to a smaller ID and shorter columns to retain the same separation characteristics as with He (de Zeeuw et al., 2016; Groschke and Becker, 2024). The web-based GC method translators assist well with the move to nitrogen as carrier gas (Fausett, 2023). In addition, nitrogen as carrier gas simplifies the use of the atmospheric pressure chemical ionization (APGC) replacing He completely in this application (Dorman and Stevens, 2022).

Argon With Argon as carrier gas, there is a very slow optimum average linear velocity. It is not common for GC-MS applications. The van Deemter curve for Argon (not shown) is steeper and has a low optimum speed in the range of 7 to 14 cm/s (Agilent, 2024). Analysis times would be much longer than with He even compared to N_2. Ar is a non-reactive carrier gas. It is the best choice when detecting hydrogen using a thermal conductivity detector (TCD) as detector. The thermal conductivity of H_2 (0.1805 W/mK) and He (0.1513 W/mK) are close, and Ar with only 0.01772 W/mK is very much different from H_2. Also, Ar is relatively inexpensive and readily available. Using Ar with GC-MS systems need to consider the higher molecular mass of m/z 40 in scanning conditions.

2.2.8.2 Optimization of the Carrier Gas Flow

The maximum separating capacity of a capillary column can be exploited only with an optimized flow of the chosen carrier gas. The flow of the mobile phase affects the rate of substance transport along the stationary phase. High flow rates allow rapid separation. However, the efficiency is reduced because of the shorter time for exchange of substances, the partition between the stationary and mobile phases (as of Figure 2.138) and leads to peak broadening. But, on the contrary, low flows increase peak broadening by diffusion of the analytes within the mobile phase.

The aim of flow optimization for given column properties at a given temperature program and with a given carrier gas is to find a flow rate which gives either the maximum number of separation steps (smallest HETP) or at adequate efficiency the shortest possible analysis time (see also Section 2.2.9.3).

The number of theoretical plates N depends on the flow rate of the mobile phase v according to the van Deemter graph in Figure 2.109. The linear flow rate v of the mobile phase is calculated in the chromatogram from the dead time t_0 of a non-retarded compound, typically argon or methane is injected:

$$v = \frac{L}{t_0} \tag{2.15}$$

with L = length of the column in cm, t_0 = dead time in s

The following affect the optimum flow rate as graphically illustrated in the van Deemter curve of Figure 2.110:

- The axial diffusion on peak broadening (longitudinal diffusion): This diffusion occurs in and against the direction of flow and decreases with increasing flow rate. The axial diffusion is increased with isothermal oven programming sections, and reduced with temperature programming.

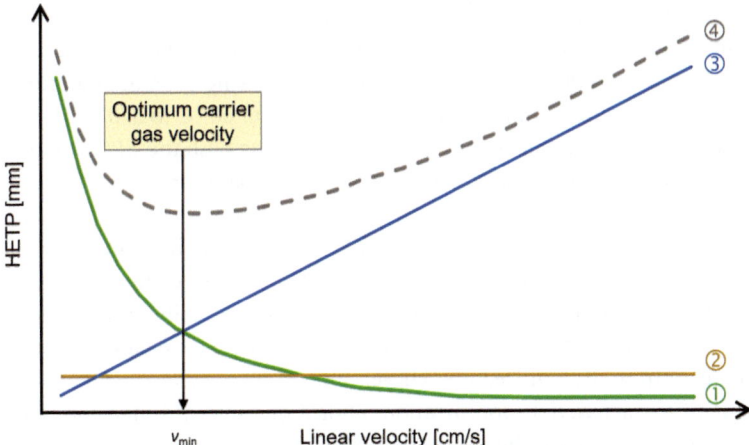

Figure 2.110 Impacts on separation power described by the van Deemter curve. ① Proportion of axial diffusion. ② Proportion of Eddy diffusion. ③ Proportion of partition phenomena. ④ Resulting van Deemter curve HETP(v) with optimum carrier gas velocity.

- The Eddy diffusion on peak broadening: This effect is independent of flow and naturally. The diffusion paths have a strong impact with packed columns shown in Figure 2.137. For open tubular capillary columns the Eddy diffusion is less pronounced, but the flow profile and turbulences need to considered.
- Incomplete partition impact: The transfer of analyte between the stationary and mobile phases only has a finite rate. The contribution to peak broadening increases significantly with an increasing flow rate of the mobile phase.

For the maximum efficiency of the separation, an optimum flow rate must be chosen as a compromise between these opposing effects. The optimum carrier gas velocity is affected by:

- the oven temperature program
- the diameter of the column,
- the quantity of the stationary phase (e.g. film thickness),
- the particle size and density of the packing material (for packed columns),
- and the viscosity and diffusion coefficient of the carrier gas.

A high analytical separating capacity is always effected by a large number of plates (number of separation steps) so the height equivalent of the theoretical plates (HETP) should be small (see also Section 2.2.9.3). In van Deemter curves, the HETP is plotted against the carrier gas velocity, not the volume flow. The minimum of these curves gives the optimal carrier gas flow adjustment for isothermal situations. It needs to be considered that the viscosity of gas increases in course of an oven temperature program.

In practice, with a *constant pressure* setting higher velocities are programmed. In such ramped oven temperature programs the flow will approach the optimum from the right branch of the van Deemter curve. The recommended carrier gas velocities are shown as horizontal ranges in Figure 2.109. Higher velocities at start of a program do not reduce the separation efficiency due to the flatter right branch of the van Deemter curve, in particular with helium, and even more with hydrogen. It is important to take also into consideration that the optimum speed of the carrier gas in temperature programs decreases with the increasing oven temperature, for instance by 30% for ramping up a capillary column from 60 °C to 370 °C (Grob and Tschuor, 1990). Working with a *constant pressure* setting with the EPC of modern GC systems assures the optimum carrier gas flow in programmed temperature analyses. Working with *constant flow* the setting of the optimum flow conditions at start of the program is required. For increased speed and productivity a small loss in column resolution by increased flow rates is more than compensated in many cases by the advantage of a short analysis time which holds true especially for hydrogen. The optimum practical gas flow rate should be set to about 1.5 to maximum twice the optimum taken from the van Deemter curve.

Very helpful in the separation optimization, choice of column dimension, and the change of carrier gases are the online available method translator programs, for instance, the Restek "EZGC Method Translator" (Restek, 2024), or the Agilent "GC Method Translator Software" (Agilent, 2014). In practice, setting up a GC program, carrier gas flows are to be programmed, not velocities. Carrier gas speeds

2.2 Gas Chromatography

Table 2.24 Recommended average carrier gas flows for capillary columns [mL/min].

Column ID	Hydrogen	Helium	Nitrogen
0.10 mm	1.0	0.8	–
0.20 mm	2.0	1.6	0.25
0.25 mm	2.5	2.0	0.5
0.32 mm	3.2	2.6	0.75
0.53 mm	5.3	4.2	1.5

are then calculated, see the "Definition of Chromatographic Parameters" with a non-retained compound. Recommended carrier gas flows for H_2, He and N_2 are listed in Table 2.24. With direct GC-MS coupling, the carrier gas flow in the column is affected by the vacuum on the detector side. In practice, the adjustment is no different from that with classical GC detectors. But, for the MS coupling the maximum tolerated flow need to be taken into account. It should be noted that with increasing oven temperature the carrier gas velocity changes differently with constant flow vs. constant pressure setting (Figure 2.111). Setting the EPC control to constant flow the velocity increases moving further along the right arm of the van Deempter curve. Vice versa constant pressure settings reduce the carrier gas velocity during oven programming, hence a higher pressure setting at program start is favorable not to miss the optimum value of the van Deempter curve.

In comprehensive GC (GC×GC) the flow rate of the second short column is accepted to be significantly above the ideal flow rate using the direct connection to the modulator. A split device would be needed between the modulator and second column for the independent adjustment of the flow rate. GC×GC applications do not use a flow adjustment for the second column because of the cut in substance concentration and hence sensitivity of the method. This is the most important factor, the overwhelming increase in peak separation and S/N ratio by using a high flow compromise.

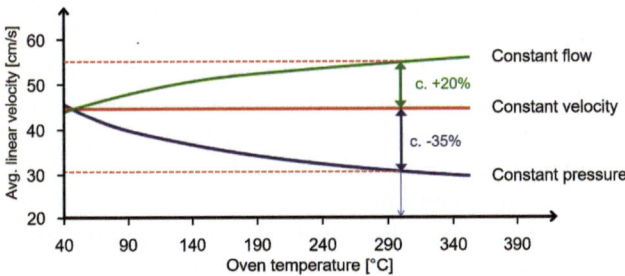

Figure 2.111 Carrier gas velocity changes with increasing oven temperature using different GC carrier gas control settings. Column dimensions 30 m × 0.25 mm ID × 0.25 µm film, He 150 kPa, 45 cm/s at 45 °C.

2.2.8.3 Sample Capacity

The sample capacity is the maximum quantity of an analyte (together with the matrix) which can be "loaded" on the column phase (Table 2.25). An overloaded column exhibits peak fronting, an asymmetrical peak, which has a gradient on the front and a sharp slope on the back side. This effect can increase until a triangular peak shape is obtained, a so-called "shark fin". Overloading occurs rapidly if a column of the wrong polarity is chosen. The capacity of a column depends on the internal diameter, the film thickness, polarity, and the solubility of a substance in the column phase.

Table 2.25 Sample capacities and optimum flow rates for common column diameters.

Internal diameter [mm]	0.18	0.25	0.32	0.53
Film thickness [µm]	0.20	0.25	0.25	1.00
Sample capacity [ng]	<50	50–100	400–500	1000–2000
Theoretical plates/m of column	5300	3300	2700	1600
Optimum flow rate at				
20 cm/s helium [mL/min]	0.3	0.7	1.2	2.6
40 cm/s hydrogen [mL/min]	0.6	1.4	2.4	5.2

Sample Capacity

- Increases with internal diameter
- Increases with film thickness
- Increases with solubility

2.2.8.4 Internal Diameter

The internal diameter of columns used in capillary GC varies from 0.1 mm (microbore capillary), 0.15 and 0.18 mm (narrowbore), the standard columns with 0.25 (normal) and 0.32 mm (widebore), up to 0.53 mm (megabore, halfmil). For the direct coupling with mass spectrometers, in practice only the columns up to 0.32 mm internal diameter are used. An exception is the vacuum outlet GC, aka *low pressure GC* (LPGC), described in Section 2.2.10.2. Megabore columns with a diameter of 0.53 mm are mainly used as pre-columns (aka *retention gap*, deactivated, no stationary phase) or to replace packed 1/8" steel columns in specially engineered GC instruments. Metal columns of types MXT, ProSteel, or SilcoSteel from different manufacturers are preferred in routine use due to the brittleness of fused silica megabore columns, for instance as pre-columns in LC-GC-MS applications.

The internal diameter (ID) affects the resolving power and the analysis time (Figure 2.112). At constant film thicknesses, lower internal diameters are preferred

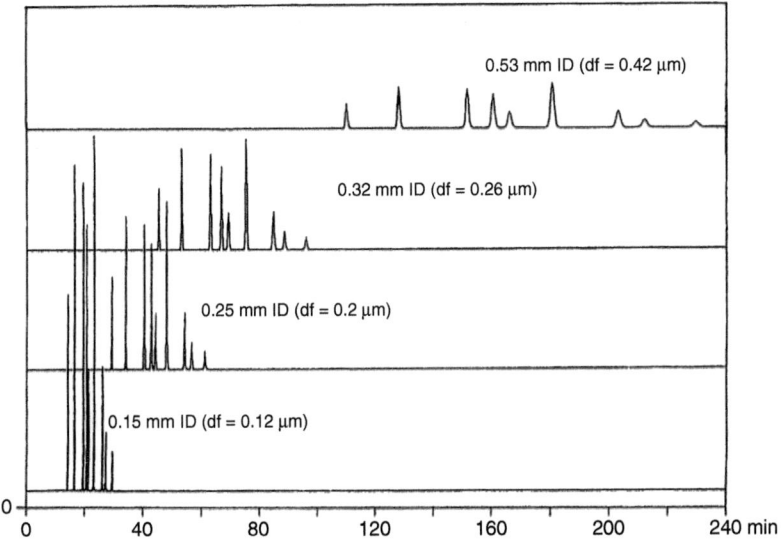

Figure 2.112 Effect of increasing column internal diameter together with increasing film thickness on peak height and retention time. The columns have the same phase ratio! (Chrompack).

to achieve higher chromatographic resolution. As the flow per unit time decreases with higher ID at a particular carrier gas velocity, the analysis time increases. In the case of complex mixtures, changing to a column with a smaller internal diameter improves the separation of critical pairs of compounds (Table 2.26). In practice, it has been shown that even changing from 0.25 mm ID to 0.20 mm allows an improvement in the separation of, for example, PCBs. Another advantage of using narrowbore columns is that the optimal linear velocity of the carrier gas also increases, which allows shorter analysis time. The optimal linear velocity that provides the highest efficiency is for instance 32.7 cm/s for 0.15 mm ID compared to 28.5 cm/s for the conventional 0.25 mm ID GC column (van Ysacker et al., 1993; Khan, 2013) as can be read from the van Deemter plots for capillary GC columns of different diameter in Figure 2.113.

Table 2.26 Effect of column internal diameter and optimum linear carrier gas velocities on the flow rate.

Internal diameter [mm]	Linear carrier gas velocity		Flow rate	
	He [cm/s]	H_2 [cm/s]	He [mL/min]	H_2 [mL/min]
0.18	30–45	45–60	0.5–0.7	0.7–0.9
0.25	30–45	45–60	0.9–1.3	1.3–1.8
0.32	30–45	45–60	1.4–2.2	2.2–2.8
0.54	30–45	45–60	4.0–6.0	6.0–7.9

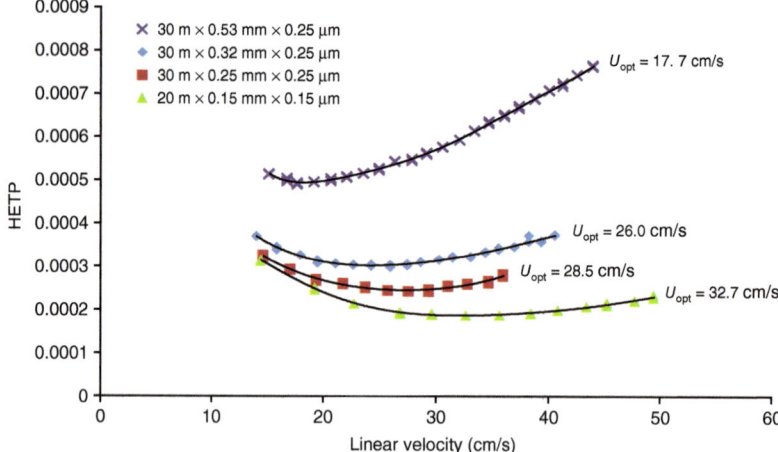

Figure 2.113 Influence of column diameter on optimal gas velocity and separation power (HETP).

The column ID also determines the minimum split ratio that should be taken into account, as with the column ID the column flow rate increases:

Column ID [mm]	Lowest split ratio
0.10	1 : 50 to 1 : 75
0.18	1 : 10 to 1 : 20
0.25	1 : 10 to 1 : 20
0.32	1 : 8 to 1 : 15
0.53	1 : 2 to 1 : 5

Smaller Internal Diameter (at identical film thickness)
- Increases the resolution.
- Decreases the analysis time.
- Reduces total flow to ion source.
- Increases carrier gas velocity.

2.2.8.5 Film Thickness

The variation in the film thickness at a given internal diameter and column length gives the user the possibility for optimization of special separation tasks. As a rule, thick films are used for volatile compounds and thin films for higher boiling analytes.

Thick film columns with coatings of more than 1.0 µm can separate very low boiling compounds well, for example, the VVOCs and VOCs. Through the large increase

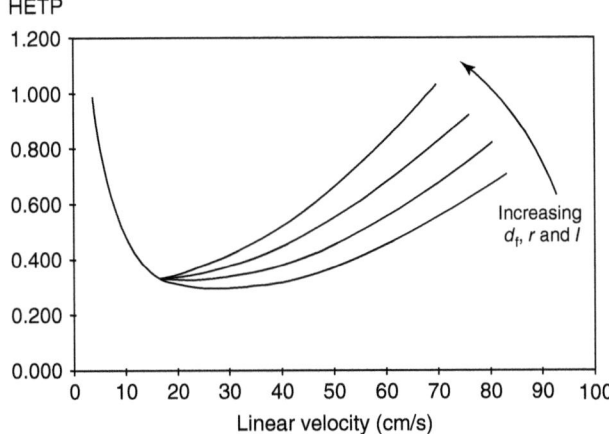

Figure 2.114 Effect of film thickness (df), internal diameter (r), and length (l) on the optimum carrier gas flow (van Deemter curves for helium as the carrier gas).

in capacity with thicker films, it is even possible to apply additional oven cooling or cryofocusing during injection when using headspace, SPME or P&T. However, thick film columns exhibit increased column bleed at elevated temperatures.

For the trace analysis of all other substances, thin film columns with coating thicknesses of about 0.1 μm to 0.25 μm have proved to be very effective in GC-MS. Thin film columns give narrow rapid peaks and can be used in higher temperature ranges without significant column bleed (see Sections 4.23 and 4.25). The elution temperatures of the compounds decrease with thin films and, at the same program duration, the analysis can be extended to compounds with higher molecular masses and thermolabile compounds, for example, the decabromodiphenylether (PBDE 209) (see Section 4.28). The analysis time for a given compound becomes shorter, but the capacity of the column decreases limiting the load for matrix samples (Figure 2.114).

For the analysis of polar compounds, it needs to be considered that thicker films provide higher inertness and less residual column activity by better shielding from any remaining free polar silica silanol groups.

Increasing Film Thickness

- Improves the resolution of volatile compounds
- Increases the analysis time
- Increases the analyte elution temperatures
- Increases column bleed
- Improves inertness for polar compounds

The Relationship between Film Thickness and Internal Diameter The phase ratio ß of a capillary column is determined by the ratio of the volume of the gaseous mobile

Table 2.27 Effect of column diameter and film thickness on the phase ratio ß (separation power).

Internal diameter [mm]	\multicolumn{8}{c}{Film thickness}
	0.10 µm	0.25 µm	0.50 µm	1.0 µm	1.50 µm	2.0 µm	3.0 µm	5.0 µm
0.18	450	180	90	45	30	23	15	9
0.25	625	250	125	63	42	31	21	13
0.32	800	320	160	80	53	40	27	16
0.53	1325	530	265	128	88	66	43	27

phase (V_G, internal column volume) to the volume of the stationary phase (V_L):

$$ß = V_G/V_L = (r_c - d_f)^2 / 2 \cdot r_c \cdot d_f$$

with r_c column radius, d_f film thickness

The phase ratio can be read for each combination of film thickness and column ID (assuming the same film type and the same column length) from Table 2.27. High values of ß point to high separation power. The same ß values show combinations with the same separating performance. For GC-MS, optimal separations can be planned and other conditions, such as carrier gas flow and column bleed can be taken into consideration. To achieve better separation, it is possible to change to a smaller film thickness at the same internal diameter or to keep the film thickness and choose a higher internal diameter. For example, a fast GC column of 0.18 mm ID and 0.10 µm film has almost twice the phase ratio than the commonly used 0.25 mm ID column with 0.25 µm film. The phase ratio is tripled when switching to a 0.25 mm ID column with 0.1 µm film which is typical for trace analysis applications. Using Table 2.27 the separation efficiency can easily be optimized to the required performance.

2.2.8.6 Column Length

The analytical column should be as short as possible and fit to purpose. The most common lengths for standard columns are 30 or 60 m. Fast GC separations go with 10 or 15 m lengths. Greater lengths are usually not necessary in general residue analysis with GC-MS systems, and are reserved for special well-documented separation purposes as known for instance for fatty acid methyl esters (FAMEs), or the complex PCB congener analyses.

Shorter columns would be desirable for simpler separations, but they are with the same diameter at the limit of the maximum flow for the mass spectrometer used. Here the switch to fast GC applications using smaller diameters should be considered. Doubling the column length only results in an improvement in the separation by a factor of 1.4 ($\sqrt{2}$) while the analysis time is doubled (and the cost of the column also!). For isothermal chromatography, the retention time is directly proportional to the column length.

With programmed operation increasing the oven heating rates leads to higher elution temperatures of the compounds as shown in Figure 2.148. The optimum GC oven heating rate for n-alkanes and pesticides in columns with silicone stationary phases is reported to about 10 °C per void time (Blumberg and Klee, 2000). On changing to a longer column, the temperature program should always be optimized again to achieve optimal retention times.

Doubling the Column Length

- Peak resolution increases by a factor of 1.4 only
- Cost is doubled
- Retention times are doubled
- Sample throughput (productivity) is cut by half
- The oven temperature program must be optimized again

2.2.8.7 Properties of Column Phases

GC columns with polysiloxane stationary phases are the most used (Figure 2.115–2.129). The polysiloxane backbone is either 100% methyl substituted, or in large variety modified with other functional groups defining the phase polarity and interaction. Modifications of the backbone as polysilarylene or siloxancarborane improve the thermal stability, and are usually chemically bonded, a criterion of special importance for GC-MS trace analysis applications. The polar polyethylene glycol (PEG) stationary phases are not bonded. As the PEG film is less stable, such polar columns are operated at lower temperature limits to avoid excess column bleed.

Chemically bonded stationary phases are covalently bonded to the fused silica surface, are highly temperature stable, and can be backwards solvent rinsed for cleaning from matrix contaminations. In contrast, *cross-linked* stationary phases are not bonded to the surface, only use chemical links between the polymer chains.

Oxygen degrades the stationary phases. For column storage it is recommended to fill the column with inert gas and seal the ends with a septum (cut in half), and store in a dark place, typically the column box, to prevent stationary phase damage by oxidation, moisture, and UV light, also from scratching the polyimide coating. Not used but installed GC columns should get a low carrier gas flow and low split flow as well (prepare a standby method at low oven temperature). All column manufacturers provide upper and lower temperature limits for use of the individual columns and stationary phases:

- Low temperature limit: Below this temperature only limited phase interaction occurs with loss of peak shape and efficiency.
- Isothermal upper limit: The column film is stable for operation up to this temperature for an extended time with low bleed.
- The isothermal temperature limit needs to be considered when setting the transfer line temperature to the MS. If higher transfer line temperatures are needed, an empty but deactivated guard column is recommended.
- Temperature program upper limit: The column could be used for up to 15 min only with GC oven temperature programs up to this maximum temperature above the isothermal limit. Column bleed increases visibly with a rising baseline, but no

damage to the phase will occur. Extended operation at or above the maximum temperature increases bleed and shortens the column lifetime.

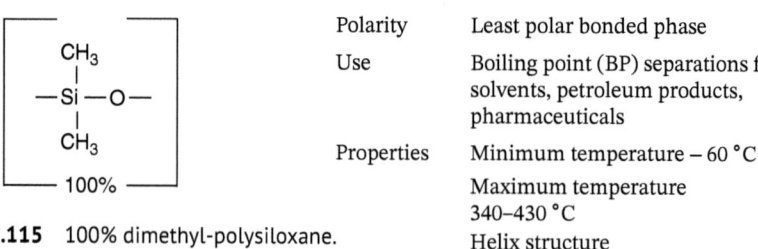

Figure 2.115 100% dimethyl-polysiloxane.

Polarity	Least polar bonded phase
Use	Boiling point (BP) separations for solvents, petroleum products, pharmaceuticals
Properties	Minimum temperature – 60 °C
	Maximum temperature 340–430 °C
	Helix structure

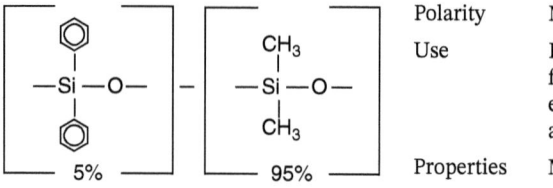

Figure 2.116 5% diphenyl-95% dimethyl-polysiloxane.

Polarity	Nonpolar, bonded phase
Use	Boiling point, point separations for aromatic compounds, environmental samples, flavors, aromatic hydrocarbons
Properties	Minimum temperature – 60 °C
	Maximum temperature 340 °C

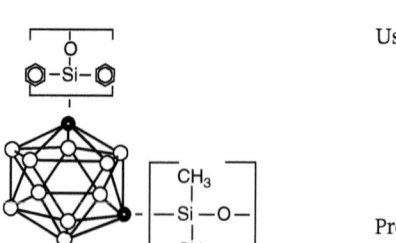

Figure 2.117 Siloxane-carborane, comparable to 5% phenyl.

Use	• Ideal for GC-MS coupling because of very low bleeding
	• All environmental samples
	• All medium and high molecular mass substances
	• Polyaromatic hydrocarbons, PCBs, waxes, triglycerides
Properties	Minimum temperature – 10 °C
	Maximum temperature 480 °C (highest operating temperature of all stationary phases, aluminum coated, 370 °C polyimide coated), high temperature phase

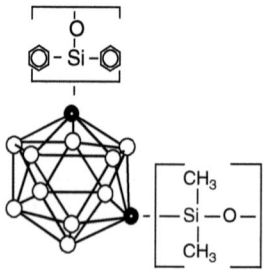

Figure 2.118 Siloxane-carborane, comparable to 8% phenyl.

Polarity	Weakly polar, similar to 8% phenylsiloxane
Use	• Ideal for GC-MS coupling because of very low bleeding
	• All environmental samples, can be used universally
	• Volatile halogenated hydrocarbons, solvents – polyaromatic hydrocarbons, pesticides, only column that separates all PCB congeners
Properties	Minimum temperature – 20 °C
	Maximum temperature 370 °C

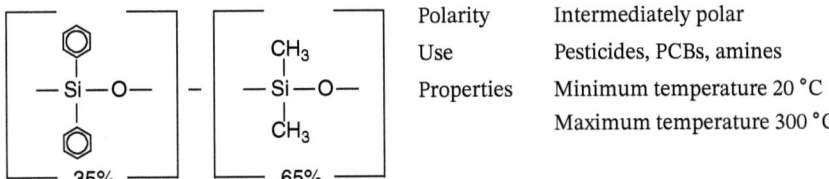

Polarity	Slightly polar
Use	For volatile compounds
Properties	Minimum temperature – 20 °C
	Maximum temperature 310 °C

Figure 2.119 20% Diphenyl-80% dimethyl-polysiloxane.

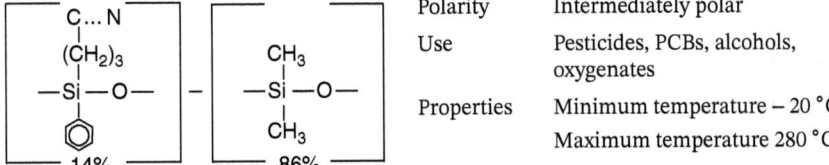

Polarity	Intermediately polar
Use	Pesticides, PCBs, amines
Properties	Minimum temperature 20 °C
	Maximum temperature 300 °C

Figure 2.120 35% Diphenyl-65% dimethyl-polysiloxane.

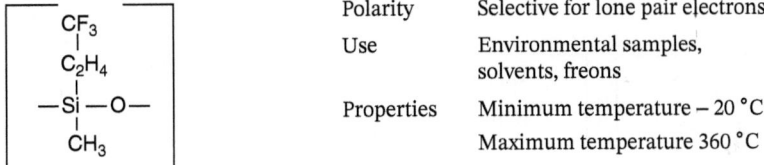

Polarity	Intermediately polar
Use	Pesticides, PCBs, alcohols, oxygenates
Properties	Minimum temperature – 20 °C
	Maximum temperature 280 °C

Figure 2.121 14% Cyanopropylphenyl-86% dimethyl-polysiloxane.

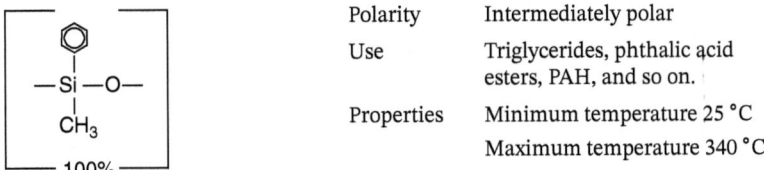

Polarity	Selective for lone pair electrons
Use	Environmental samples, solvents, freons
Properties	Minimum temperature – 20 °C
	Maximum temperature 360 °C

Figure 2.122 100% Trifluoropropylmethyl-polysiloxane.

Polarity	Intermediately polar
Use	Triglycerides, phthalic acid esters, PAH, and so on.
Properties	Minimum temperature 25 °C
	Maximum temperature 340 °C

Figure 2.123 100% Phenyl-methyl-polysiloxane.

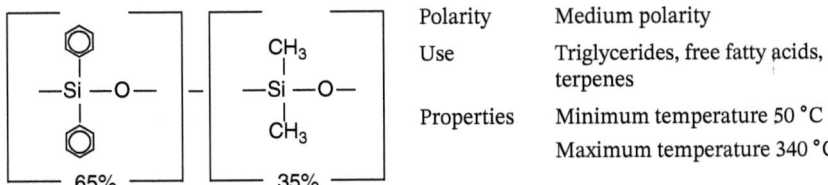

Polarity	Medium polarity
Use	Triglycerides, free fatty acids, terpenes
Properties	Minimum temperature 50 °C
	Maximum temperature 340 °C

Figure 2.124 65% Diphenyl-35% dimethyl-polysiloxane.

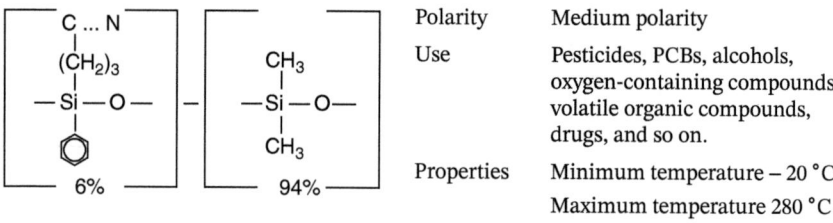

Polarity	Medium polarity
Use	Pesticides, PCBs, alcohols, oxygen-containing compounds, volatile organic compounds, drugs, and so on.
Properties	Minimum temperature − 20 °C Maximum temperature 280 °C

Figure 2.125 6% Cyanoprophylphenyl-94% dimethyl-polysiloxane.

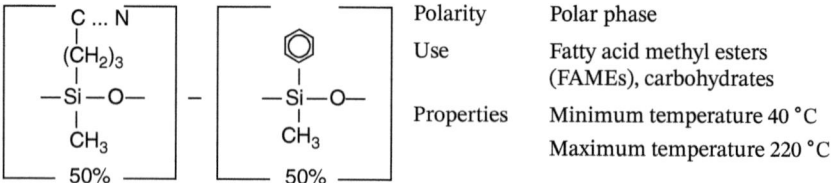

Polarity	Polar phase
Use	Fatty acid methyl esters (FAMEs), carbohydrates
Properties	Minimum temperature 40 °C Maximum temperature 220 °C

Figure 2.126 50% Cyanopropylmethyl-50% phenylmethyl-polysiloxane.

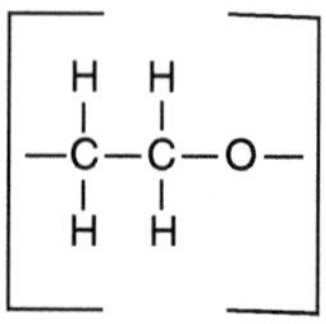

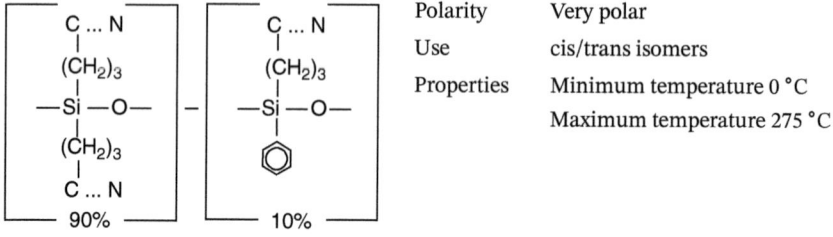

Polarity	Polar phase
Use	FAMEs, terpenes, acids, amines, solvents
Properties	Minimum temperature 40 °C Maximum temperature 280 °C

Figure 2.127 100% Carbowax polyethyleneglycol.

Polarity	Very polar
Use	cis/trans isomers
Properties	Minimum temperature 0 °C Maximum temperature 275 °C

Figure 2.128 90% Biscyanopropyl-10% phenylcyanopropyl-polysiloxane.

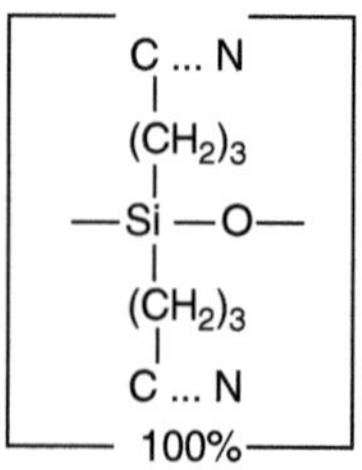

Polarity	Very polar
Use	FAMEs
Properties	Minimum temperature 20 °C Maximum temperature 250 °C

Figure 2.129 100% Biscyanopropyl-polysiloxane.

2.2.8.8 Ionic Liquid Phases

Ionic liquids are a new class of stationary films in capillary GC with remarkable properties and benefits for GC-MS applications as well. Stable room-temperature ionic liquids (RTILs) are organic salts with melting points at or below room temperature (Berthod and Carda-Broch, 2003). They can be applied as fused silica column phases and form a liquid stationary film in which ions are present, but do not contain neutral molecules. The current phases are not (yet) chemically bonded. They contain the equal number of positive and negative ions so that the liquid is electrically neutral. Although not new and already known for various chemical applications, an evaluation for use in GC was performed (Armstrong et al., 1999). The high viscosity, high thermal stability and high polarity make them very suitable as alternative polar phases for GC separations (Vidal et al., 2012). A first commercial ionic liquid GC column was introduced by Supelco in 2008 and further development has yielded additional phases of differing polarity (Talebi et al., 2020). A range of columns is now available encompassing different phase types, from a PEG equivalent polarity with improved thermal stability SLB-ILPAH, SLB-IL59/60 to phases with even higher polarity SLB-IL100/111 (Supelco, 2013).

The main physicochemical properties of RTILs are very compatible with the requirements of capillary GC (Berthod and Carda-Broch, 2004; Berthod et al., 2022). They remain liquid over a temperature range of up to 300 °C, are of high chemical stability, and have practically no vapor pressure, promising a low bleed at high elution temperatures. The properties of RTILs depend on the nature and size of both their cation and anion constituents and offer a high potential for customized phases to special applications. A general structure is shown in Figure 2.130. The combination of high polarity with high temperature stability along with a new selectivity mechanism offers a great potential for future GC, GCxGC and GC-MS applications (Anderson et al., 2006; Armstrong, 2009; Carda-Broch and Ruiz-Angel, 2021).

The retention mechanism is different from the traditional silicon phase systems. It appears from the available application reports that the phase solubility is of low importance (as of "like separates like"). Retention of the analytes is mainly caused by electrostatic and dipole/dipole interactions. With increased phase polarities, for example, the retention times for non-polar hydrocarbons are significantly

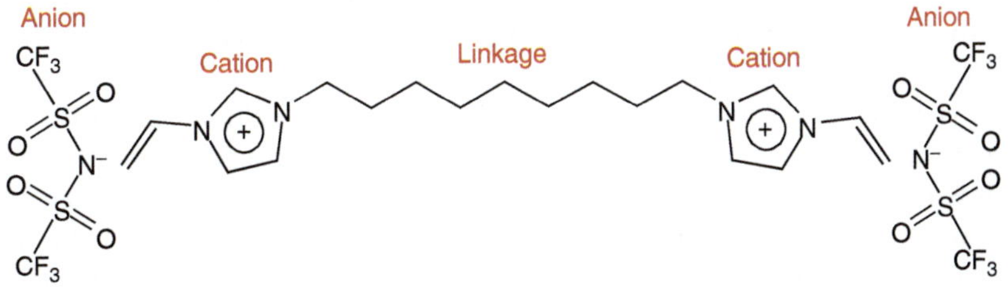

Figure 2.130 General anion/cation structure of ionic liquid phases, 1,9-di(3-vinylimidazolium) nonane bis(trifluoromethyl) sulfonyl imidate phase of the SLB-IL100 column (Supelco).

210 2 Fundamentals

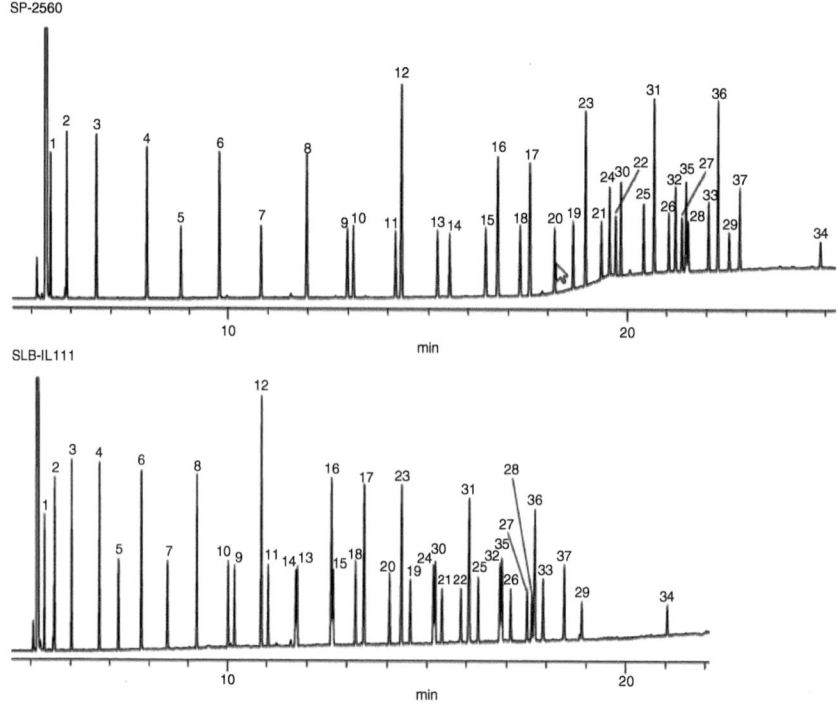

Figure 2.131 Comparison of an ionic liquid phase (SLB-IL111 on bottom) with classical highly polar phase system (poly(biscyanopropyl siloxane phase, top) (Supelco).

reduced, compared to wax or 5% phenyl phases, peak widths increase, due to a poor wetting of the phase surface. Polar analytes are increasingly retained and separated (Whitmarsh, 2012). Ionic liquid stationary phases seem to have a dual nature. They appear to act as a low polarity stationary phase to non-polar compounds. Molecules with strong proton donor groups are strongly retained (Armstrong et al., 1999).

As an example, the high column polarity reduces the retention time for non-polar hydrocarbons, as shown in Figure 2.131 for FAMEs, comparing a traditional polar poly(biscyanopropyl) siloxane phase with the most polar ionic liquid phase in SLB-IL111. Significant is the reduction in analysis time while maintaining the peak separation and the reduced column bleed. With the same temperature program, the compounds are eluting at lower oven temperature. The improvement in productivity occurs to such an extent that a C31 n-alkane elutes at almost half of the retention time on a polar ionic liquid phase (SLB-IL111) compared to a traditional 5% phenyl phase, using identical oven temperature programs (Whitmarsh, 2012).

In food safety FAMEs analysis, the separation of the fatty acid cis/trans compounds is a routine analysis of high importance. For a comparison, a 38-component FAME mix was analyzed on the traditional polar wax and an ionic liquid column under identical conditions. The ion liquid column SLB-IL60 was chosen for the complimentary selectivity to PEG phases, but offers higher maximum temperatures and a significantly lower bleed. The resulting chromatograms are shown in Figure 2.132.

2.2 Gas Chromatography | 211

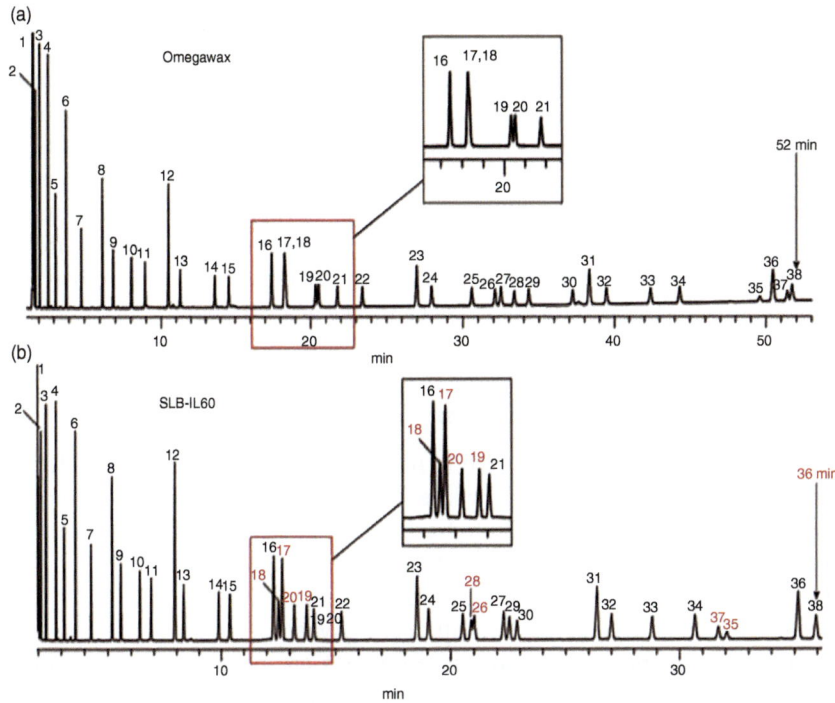

Figure 2.132 FAMEs separation for oleic and linolenic acid cis/trans analysis (#17/18, #19/20) on a traditional wax (a) and ionic liquid phase (b) column. Legend of the FAMEs peak numbers of Figure 2.132 : 1. C4:0, 2. C6:0, 3. C8:0, 4. C10:0, 5. C11:0, 6. C12:0, 7. C13:0, 8. C14:0, 9. C14:1, 10. C15:0, 11. C15:1, 12. C16:0, 13. C16:1, 14. C17:0, 15. C17:1, 16. C18:0, 17. C18:1n9c, 18. C18:1n9t, 19. C18:2n6c, 20. C18:2n6t, 21. C18:3n6, 22. C18:3n3, 23. C20:0, 24. C20:1n9, 25. C20:2, 26. C20:3n6, 27. C21:0, 28. C20:3n3, 29. C20:4n6, 30. C20:5n3, 31. C22:0, 32. C22 : 1n9, 33. C22:2, 34. C23:0, 35. C22:5n3, 36. C24:0, 37. C22:6n3 and 38. C24:1n9 (Supelco).

Analytical conditions for Figure 2.132 FAMEs separation: (a) Omegawax column, bonded phase poly(ethylene glycol), temperature limit 280 °C programmed. (b) SLB-IL60 column (bottom), non-bonded phase 1,12-di(tripropylphosphonium) dodecane bis(trifluoromethylsulfonyl)imide, temperature limit 300 °C programmed. GC conditions: Omegawax, 30 m × 0.25 mm ID, 0.25 µm. SLB-IL60: 30 m × 0.25 mm ID, 0.20 µm. Oven: 170 °C, 1 °C/min to 225 °C, injection temperature: 250 °C, carrier gas: helium, 1.2 mL/min, detector: FID, 260 °C. Injection: 1 µL, 100:1 split, liner: 4 mm ID, split/splitless type, single tapered wool packed FocusLiner™. Sample: Supelco 37-component FAME Mix + C22:5n3, in methylene chloride.

The separation of the FAMEs compounds on the ionic liquid SLB-IL60 column is achieved in a much shorter runtime of 36 min vs. 52 min on a standard Omegawax column, which is a 30% gain in analysis time with improved separation quality (Figure 2.132). The resolution of the *cis/trans* oleic acids C18:1n9c (peak 17) and C18:1n9t (peak 18) is well achieved by the differing selectivity of the ionic liquid column. The *trans* oleic acid C18:2n6t (peak 20) also elutes before the *cis* compound

C18:2n6c (peak 19) and shows a very clear baseline separation (Stenerson et al., 2013, 2014).

Ionic liquids provide a different selectivity from traditional silicon polymer GC phases. They are not susceptible to the same stability issues as siloxane and PEG-based phases, allowing higher maximum temperatures than traditional phases of comparable polarity. The extended temperature operating range is best seen in comparison with the known PEG phases. GC-MS applications benefit from a lower bleed and achieve compound elution at lower oven temperatures. Ionic liquid columns extend the polarity range of GC phases, improve lab productivity by shorter run times, and can be used for a wide variety of applications.

Analytical Benefits of Ionic Liquid Phases

- Greater thermal stability compared to polysiloxane polymers and PEGs. The traditional polar PEG phases show intense bleed in GC-MS applications and are not useful for high temperature trace analysis applications.
- Lower column bleed, longer life, and higher thermal limits.
- Most suitable for the analysis of polar compounds in GC-MS.
- More resistant to damage from moisture and oxygen.
- Numerous combinations of cations and anions are possible, allowing the design of "tailored" selectivities, applications or functions.

2.2.9 Chromatography Parameters

Chromatography in general is a dilution and separation process. All chromatography is based on the multiple repetition of minute separation steps, such as the continuous dynamic partition of the components between two phases (Martin and Synge, 1941; Bock, 1974). Gas chromatography uses a carrier gas as mobile phase and immobilized liquids or porous media as the stationary phase in today's mostly capillary columns.

In a model, chromatography can be regarded as a continuous repetition of equilibrium partition steps (Dettmer-Wilde and Engewald, 2013). The starting point is the partition of a substance between two phases in a separating funnel. Different analytes have different partition coefficients K between the solvents (phases). Suppose a series of classical separation funnels is set up for LLE that all contain the same quantity of phase, i.e. the extraction solvent.

As the extraction solvent remains in the separating funnels, it is called the stationary phase. The sample, dissolved in a second non-miscible auxiliary phase, is placed in the first separation funnel. After establishing equilibrium through shaking, the auxiliary phase is transferred to the next separating funnel. The auxiliary phase thereby becomes the mobile phase. Fresh mobile phase is placed into the first separating funnel, and the process with shaking and transfer to the next separation funnel goes on for a large number of n separation funnels (Figure 2.133).

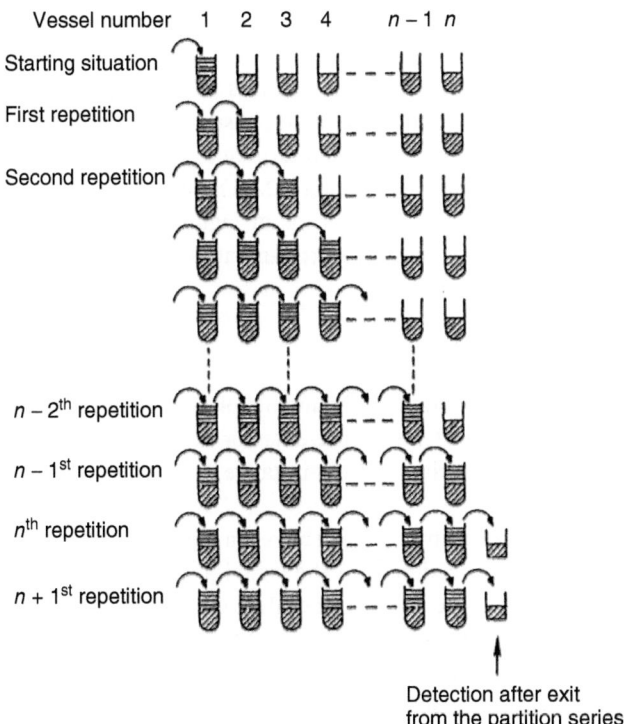

Figure 2.133 Partition series: mode of operation with two phases: stationary and mobile.

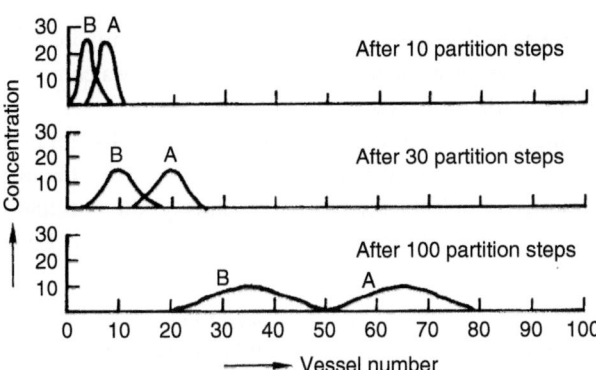

Figure 2.134 Partition of substances A and B after 10, 30, and 100 separating steps. After 10 steps A and B are hardly separated, after 30 steps quite well, and after 100 steps practically completely. The two substances are partitioned among an increasing number of vessels and the concentrations decrease more and more.

The results of this type of partition with 100 vessels and two analytes A and B are shown in Figure 2.134. The prerequisite for this is the validity of the Nernst equation. For detection, the concentrations of A and B in the vessels are determined. With the model described, so many separating steps are carried out that the mobile phase leaves the system of 100 vessels and the individual components A and B are removed, one after the other, from the series of vessels. This process is known as *elution*.

In a chromatographic column of finite length as many as possible "partitions" should happen between the stationary and the mobile phase for optimum separation power. The number of partition steps can be interpreted as the "theoretical plates" of the van Deemter equation (also see Section 2.2.8.1). The smallest HETP should be achieved for optimum separation power (as of Figure 2.109). Analyte peaks at the detector cannot be sharper than the start band at injection (also refer to Figure 2.138).

2.2.9.1 The Chromatogram and its Meaning

The analytes to be separated are transported by the mobile phase and elute to the detector. Their concentrations are registered at the detector ideally as Gaussian curves (peaks). The peaks give qualitative and quantitative information on the mixture investigated.

Qualitative: The retention time is the time elapsing between injection of the sample and the appearance of the peak maximum of the signal at the detector. The retention time of a component is always constant under the same chromatographic conditions. Under defined conditions, the time required for elution of a substance A or B from the start to the end of the separating system, the retention time t_R, is characteristic of the substance. It is measured from sample injection to the peak maximum (Figures 2.135 and 2.136). A peak can therefore be identified by a comparison of the retention time with a standard (pure substance), the typical procedure when using classical detectors.

Quantitative: The height and area of a peak is proportional to the quantity of substance injected. Unknown quantities of substance can be determined by a comparison of the peak areas or heights with known concentrations.

In the ideal case, the eluting peaks are in the shape of a Gaussian distribution (bell-shaped curve, Figure 2.135). A simple explanation of this shape are the different paths taken by molecules through the separating system, e.g. the multipath effect in packed columns, which causes different travel times, is known as "Eddy

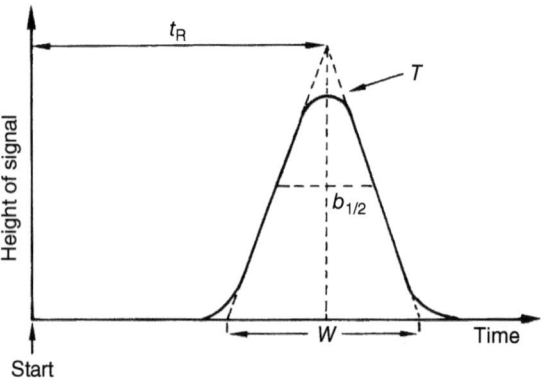

Figure 2.135 Parameters determined for an elution peak with t_R retention time, $t_{1/2}$ half width, and W base width.

2.2 Gas Chromatography | 215

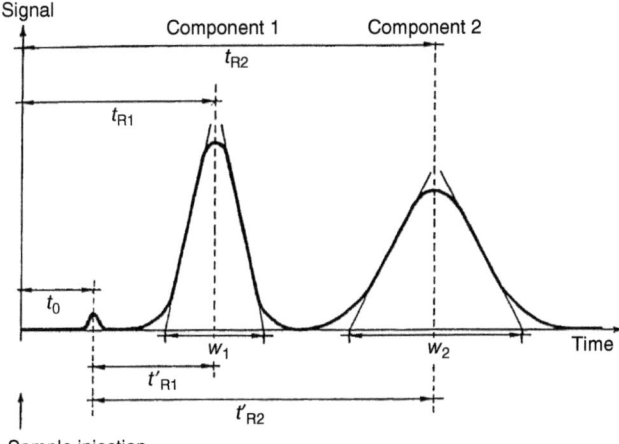

Figure 2.136 The chromatogram and its parameters (after Meyer, 2010/John Wiley & Sons): W Peak width of a peak. $W = 4\sigma$ with σ = standard deviation of the Gaussian peak. t_0 Dead time of the column, i.e. the time that the mobile phase requires to pass through the column. The linear velocity u of the carrier gas in the column is calculated from

$$u = \frac{L}{t_0} \quad \text{with} \quad L = \text{length of the column}$$

A substance that is not retarded, that is, a substance that does not interfere with the stationary phase, appears at t_0 at the detector. t_R Retention time is called the time elapsed between the injection of a sample (start) and the recording of a substance peak maximum. t'_R Net retention time. From the diagram it can be seen that $t_R = t_0 + t'_R$. t_0 is the residence time in the mobile phase. The substances separated differ in their residence times in the stationary phase t'_R. The longer a substance stays dissolved in the stationary phase (partition), the later it is eluted.

diffusion" (Figure 2.137). Other factors like the velocity profile, diffusion and partition phenomena add to this effect, also occurring in open-tubular columns.

At a constant flow rate, t_R is directly proportional to the retention volume V_R.

$$V_R = t_R \cdot F \quad \text{with} \quad F = \text{flow rate in mL/min.} \tag{2.16}$$

The retention volume shows how much mobile phase has passed through the separating system until half of the substance has eluted (peak maximum!).

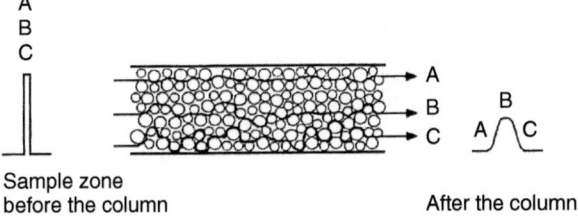

Figure 2.137 Eddy diffusion in a packed column (multipath effect) (Meyer, 2010/John Wiley & Sons).

2.2.9.2 Capacity Factor k'

The retention time t_R depends on the interaction of the analyte with the stationary phase (*like-goes-with-like*), the flow rate of the mobile phase, and the length of the column. If the mobile phase moves slowly or the column is long, t_0 is large and so is t_R. Thus, t_R is not suitable for the comparative characterization of a substance, for example, between two laboratories. It is better to use the capacity factor, also 'known as the *k'* value, which relates the net retention time t'_R to the dead time:

$$k' = \frac{t'_R}{t_0} = \frac{t_R - t_0}{t_0} \tag{2.17}$$

Thus, the *k'* value is independent of the column length and the flow rate of the mobile phase and represents the molar ratio of a particular component in the stationary and mobile phases. Large *k'* values mean long analysis times.

The *k'* value is related to the partition coefficient K as follows:

$$k' = K \cdot \frac{V_1}{V_g} \tag{2.18}$$

where V_1 = volume of the liquid stationary phase and V_g = volume of the mobile phase.

The capacity factor is, therefore, directly proportional to the volume of the stationary phase (or for adsorbents, their specific surface area in m²/g).

The measure of the relative retention α is given by the quotient of the substance partition coefficients K:

$$\alpha = \frac{k'_2}{k'_1} = \frac{K_2}{K_1} \quad (k'_2 > k'_1) \tag{2.19}$$

In the case where $\alpha = 1$, the two components 1 and 2 are not separated because they have the same partition coefficients K. The relative retention α is thus a measure of the selectivity of a column and can be manipulated by choice of a suitable stationary phase. In principle, this is also true for the choice of the mobile phase, but in GC-MS helium or hydrogen are, in fact, used as the standards.

2.2.9.3 Chromatographic Resolution

A second model, the theory of plates, was developed by Martin and Synge in 1941. This is based on the functioning of a fractionating column, then as now a widely used separation technique. It is assumed that the equilibrium between two phases on each plate of the column has been fully established. Using the plate theory, mathematical relationships can be derived from the chromatogram, which are a practical measure of the sharpness of the separation and the resolving power.

The chromatographic column is divided up into theoretical plates, that is, into column sections in the flow direction, the separating capacity of each one corresponding to a theoretical plate. The length of each section of the column is called the *height equivalent to a theoretical plate (HETP)*. The HETP value is calculated from the length of column L divided by the number of theoretical plates N:

$$\text{HETP} = \frac{L}{N} [\text{mm}] \tag{2.20}$$

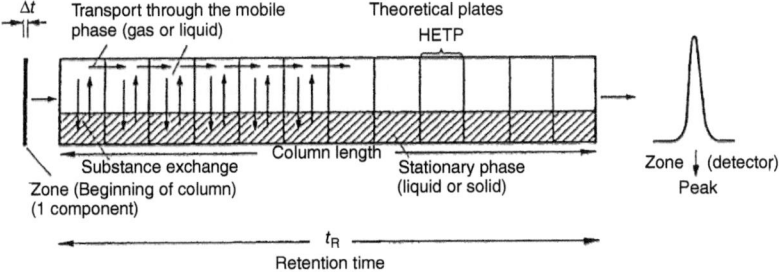

Figure 2.138 Substance exchange and transport in a chromatography column are optimal when there are as many phase transfers as possible with the smallest possible expansion of the given zones (Schomburg, 1981).

The number of theoretical plates N is calculated from the shape of the eluted peak. In the separating funnel model, it is shown that with an increasing number of partition steps the substance partitions itself between a larger number of vessels (as of Figure 2.133). A separation system giving sharp separation concentrates the substance band into a few vessels or plates. The more plates there are per length of a separation system, the sharper the eluted peaks. A column is more effective, the more theoretical plates it has (Figure 2.138) (also refer to Section 2.2.8.1).

The number of theoretical plates N is calculated from the peak profile. The retention time t_R at the peak maximum and the width at the base of the peak measured as the distance between the cutting points of the tangents to the inflection points with the base line are determined from the chromatogram (see Figure 2.135).

$$N = 16 \cdot \left(\frac{t_R}{W}\right)^2 \tag{2.21}$$

where t_R = retention time and W = peak width.

For asymmetric peaks, the half width (the peak width at half height) is used:

$$N = 8 \ln 2 \cdot \left(\frac{t_R}{W_h}\right)^2 \tag{2.22}$$

where t_R = retention time and W_h = peak width at half height.

The width of a peak in the chromatogram determines the resolution of two components at a given distance between the peak maxima (Figure 2.139). The resolution R is used to assess the quality of the separation:

$$R \approx \frac{\text{retention difference}}{\text{peak width}}$$

The resolution R of two neighboring peaks is defined as the quotient of the distance between the two peak maxima, that is, the difference between the two retention times t_R and the arithmetic mean of the two peak widths:

$$R = 2 \cdot \frac{t_{R2} - t_{R1}}{W_1 + W_2} = 1.18 \cdot \frac{t_{R2} - t_{R1}}{W_{h1} + W_{h2}}$$

where W_h = peak width at half height, if the peaks are not baseline resolved.

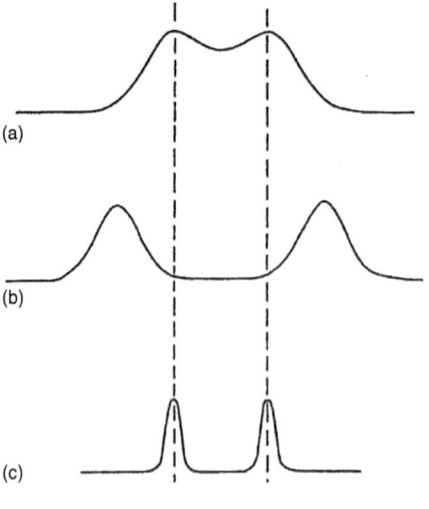

Figure 2.139 Resolution. (a/c) Peaks with the same retention time, (a/b) peaks with the same peak width, and (b/c) separation with the same resolution (after Meyer, 2010/John Wiley & Sons).

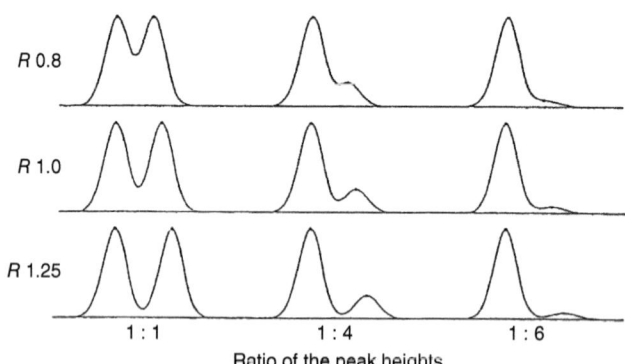

Figure 2.140 Resolution of two neighboring peaks (Snyder and Kirkland, 1979/John Wiley & Sons).

Figure 2.140 shows what can expected optically from a value for R calculated in this way. At a resolution of 1.0, the peaks are not completely separated, but it can be seen that there are two components. The tangents to the inflection points just touch each other and the peak areas only overlap by 2%.

For the precise determination of the peak width, the tangents to the inflection points can be drawn in manually (Figure 2.141). For a critical pair, for example, stearic acid ($C_{18:0}$) and oleic acid ($C_{18:1}$) the construction of the tangents is shown in Figure 2.142.

2.2.9.4 Factors Affecting the Resolution

Rearranging the resolution equation and putting in the capacity factor $k' = (t_R - t_0)/t_0$, the selectivity factor $\alpha = k'_2/k'_1$ and the number of theoretical plates N gives an important basic equation for all chromatographic elution processes. The resolution R is related to the selectivity α (relative retention), the capacity factor k', and by the

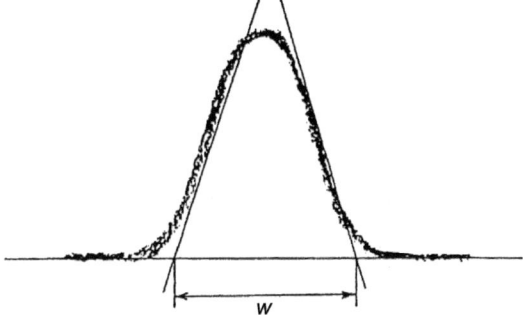

Figure 2.141 Manual determination of the peak width using tangents to the inflection points (after Meyer, 2010/John Wiley & Sons).

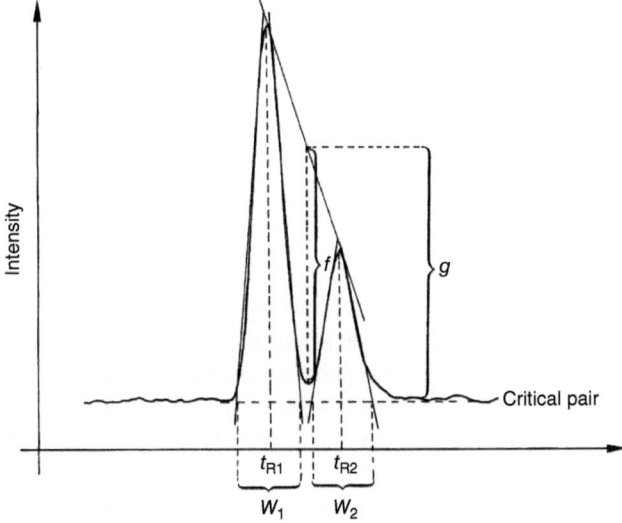

Figure 2.142 Determination of the resolution and peak widths for a critical pair.

number of theoretical plates N:

$$R = \frac{1}{4} \cdot \underbrace{\frac{(\alpha - 1)}{\alpha}}_{A} \cdot \underbrace{\frac{k'}{(1 + k')}}_{B} \cdot \underbrace{\sqrt{N}}_{C} \tag{2.23}$$

Using this fundamental resolution equation (2.23), it is possible to optimize the chromatographic separation by working on the terms A selectivity, B retention, and C dispersion with the number of theoretical plates. Figure 2.143 shows the big impact on the selectivity term α followed in importance for chromatographic resolution by the number of theoretical plates N.

The Selectivity Term A R is directly proportional to the selectivity term $(\alpha-1)/\alpha$. An increase in the ratio of the partition coefficients leads to a sharp improvement

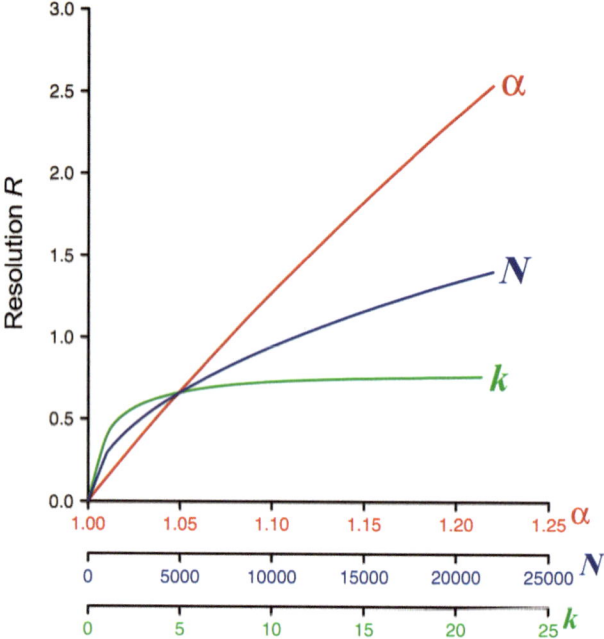

Figure 2.143 Graphical representation of the general resolution equation.

in the resolution, which can be achieved, for example, by changing the polarity of the stationary phase for substances of different polarities. The change in the column selectivity is the most effective of the possible measures for improving the resolution. As shown in Table 2.28, when the relative retention α is small, a significant increase in the plate number N is required to achieve the usually desired resolution of $R = 1.5$.

As the selectivity generally decreases with increasing temperature, difficult separations must be carried out at as low a temperature as possible. Hence, a slow oven temperature ramp is recommended (refer to Figure 2.148).

Table 2.28 Relationship between relative retention α and the chromatographic resolution R.

Relative retention [α]	for $R = 1.0$ [plates]	for $R = 1.5$ [plates]
1.005	650 000	1 450 000
1.01	163 000	367 000
1.05	7100	16 000
1.10	3700	8400
1.25	400	900
1.50	140	320
2.0	65	145

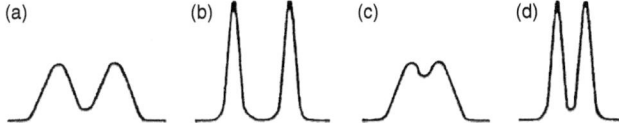

Figure 2.144 Relative retention, number of plates, and resolution (a to d see text).

Figure 2.144 shows the effect of relative retention and number of plates on the separation of two neighboring peaks:

- At high relative retention, the number of theoretical plates in the column does not need to be large to achieve satisfactory resolution (Figure 2.144 a). The column is poor but the system is selective.
- A high relative retention and large number of theoretical plates give a resolution which is higher than the optimum. The analysis is unnecessarily long (Figure 2.144 b).
- At the same (small) number of theoretical plates as in Figure 2.144 a, but at a smaller relative retention, the resolution is strongly reduced (Figure 2.144 c).
- If the relative retention is small, a large number of theoretical plates are required to give a satisfactory resolution (Figure 2.144 d).

The Retention Term B Here the resolution is directly proportional to the residence time of a component in the stationary phase based on the total retention time. If the components stayed only in the mobile phase ($k' = 0!$), there would be no separation.

For very volatile or low molecular mass non-polar substances, there are only weak interactions with the stationary phase. Thus, at a low k' value the denominator $(1 + k')$ of the term is large compared with k' and R is therefore small. This applies to columns with a small quantity of stationary phase, a thin film, and high column temperatures. To improve the resolution, a larger content of stationary phase can be chosen (greater film thickness). Check the phase ratio as outlined in Table 2.27 for a suitable combination of column diameter and film thickness.

The Efficiency Term C The number of plates N characterizes the performance of a column significantly as can be seen in Figure 2.145 with the different measures to improve a separation from an unresolved start situation. However, the resolution R only increases with the square root of N. As N is directly proportional to the column length L, the performance is only proportional to the square root of the column length. Doubling the column length therefore only increases the resolution by a factor of 1.4, but since the retention time t_R is directly proportional to the column length the analysis time is twice as long.

Further measures for improvement of N can be the change to smaller column diameters, optimized linear carrier gas velocity, or the change of carrier gas (Figure 2.109).

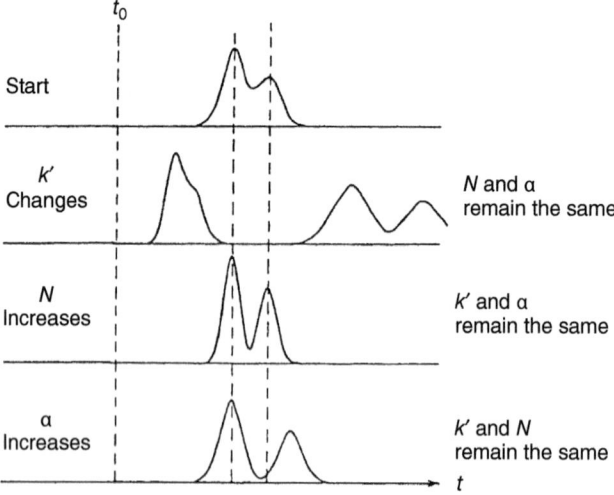

Figure 2.145 Effect of capacity factor, number of plates and relative retention on the chromatogram (Snyder and Kirkland, 1979/John Wiley & Sons).

2.2.9.5 Maximum Sample Capacity

The maximum sample capacity can be derived from the equations concerning the resolution. Under ideal conditions,

$$\frac{f}{g} = 100\% \tag{2.24}$$

where f = the area under the line connecting the peak maxima and g = the height of the connecting line above the base line, measured in the valley between the peaks, (see Figure 2.142).

The maximum sample capacity of a column is reached if f/g falls below 90% for a critical pair. If too much sample material is applied to a column, the k' value and the peak width are no longer independent of the size of the sample, which ultimately affects the identification and the quantitation of the results (Figure 2.146).

2.2.9.6 Peak Symmetry

In exact quantitative work by integration of the peak areas or peak height, a maximum asymmetry must not be exceeded. Symmetrical sharp peaks determine sensitivity and reproducibility. Poor peak symmetries are mainly caused by column overloading leading to fronting (aka shark fins) or active sites in the flow path resulting in peak tailing. Active sites can be successfully counteracted by adding "analyte protectants" to the extract before injection (Anastassiades et al., 2003; Mastovska et al., 2005; Li et al., 2012; Morales and Macherone, 2013).

For practical reasons, the peak symmetry T is determined at a height of 10% of the total peak height (Figure 2.147):

$$T = \frac{b_{0.1}}{a_{0.1}} \tag{2.25}$$

Figure 2.146 Change in the chromatogram with increasing sample size (Snyder and Kirkland, 1979/John Wiley & Sons). (a) Constant k' values, (b–d) increasing changes to the retention behavior through overloading.

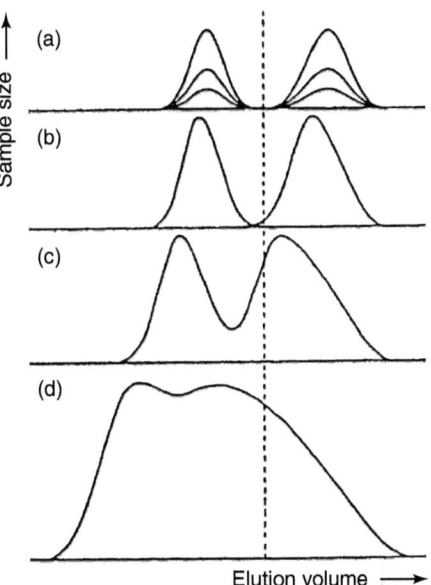

where $a_{0.1}$ = peak width distance from the peak front to the perpendicular at the maximum and $b_{0.1}$ = peak width distance from the maximum to the end of the peak.

T should ideally be 1.0 for a symmetrical peak, but for a practical quality measure, it should not be greater than 2.5. If the tailing exceeds higher values, there will be unacceptable errors in the quantitative area measurement because the point where the peak reaches the base line is difficult and not reproducible to determine which has a large impact on the resulting peak area (Figure 2.110).

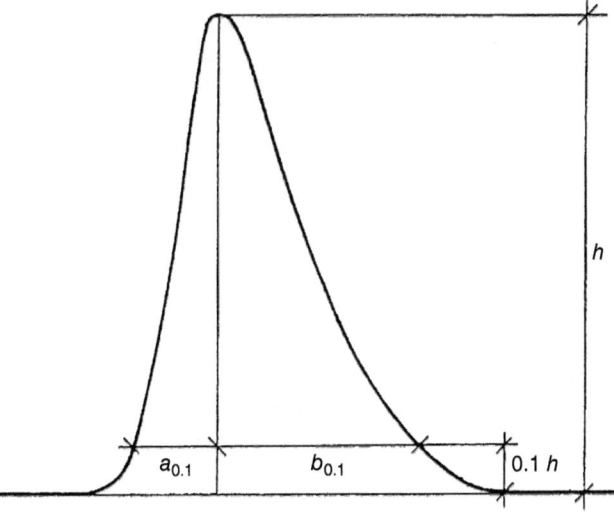

Figure 2.147 Asymmetric peak.

2 Fundamentals

Definition of Chromatographic Parameters

Parameter	Formula	Description
Carrier gas velocity	$v = L/t_0$	The average linear carrier gas velocity has an optimal value for each column with the lowest possible height equivalent to a theoretical plate (see van Deemter); v is independent of temperature, L = length of column and t_0 = dead time.
Partition coefficient	$K = c_l/c_g$	Concentration of the substance in the stationary phase (liquid) is divided by the concentration in the mobile phase (gas). K is constant for a particular substance in a given chromatographic system.
	$K = k' \cdot \beta$	K is also expressed as the product of the capacity ratio (k') and the phase ratio (β).
Capacity ratio (partition ratio)	$k' = K \cdot V_l/V_g$	Determines the retention time of a compound. V_l = volume of the liquid stationary phase and V_g = volume of the gaseous mobile phase,
	$k' = (t_R - t_0)/t_0$	or by the degree of retention of an analyte relative to an unretained peak. t_R = retention time of the analyte and t_0 = dead time.
Phase ratio	$\beta = r/2d_f$	r = internal column radius and d_f = film thickness
Number of theoretical plates	$N = 5.54\,(t_R/W_h)^2$	The number of theoretical plates is a measure of the efficiency of a column. The value depends on the nature of the substance and is valid for isothermal work. N = number of theoretical plates, t_R = retention time of the substance and W_h = peak width at half height.
HETP	$h = L/N$	Is a measure of the efficiency of a column independent of its length. L = length of the column and N = number of theoretical plates.
Resolution	$R = 2\,(t_j - t_i)/(W_j + W_i)$	Gives the resolving power of a column with regard to the separation of components i and j (isothermally). t_i = retention time of substance i; t_j = retention time of substance j and $W_{i,j}$ = peak width at half height of substances i, j.
Separation factor	$\alpha = k'_j/k'_i$	Measure of the separation of the substances i, j.
Trennzahl number	$TZ = \dfrac{t_{R(x+1)} - t_{R(x)}}{W_{h(x+1)} + W_{h(x)}} - 1$	The Trennzahl number is, like the resolution, a means of assessing the efficiency of a column and is also used for temperature-programmed work. TZ gives the number of components which can be resolved between two homologous n-alkanes.

Effective plates	$N_{\text{eff.}} = 5.54$ $((t_{R[i]} - t_0)/W_{h(i)})$	The effective number of theoretical plates takes the dead volume of the column into account.
Retention volume	$V_R = t_R \cdot F$	Gives the carrier gas volume required for elution of a given component. F = carrier gas flow rate.
Kovats index	$KI = 100 \cdot c + 100 \frac{\log(t'_R)_x - \log(t'_R)_c}{\log(t'_R)_{c+1} - \log(t'_R)_c}$	The Kovats index is used for isothermal work. t'_R = corrected retention times for standards and substances $t'_R = t_R - t_0$.
Modified Kovats index	$RI =$ $100 \cdot c + 100 \frac{(t'_R)_x - (t'_R)_c}{(t'_R)_{c+1} - (t'_R)_c}$	The modified Kovats index according to van den Dool and Kratz is used with temperature programming.

2.2.9.7 Effect of Oven Temperature Ramp Rate

The oven temperature ramp rates show a significant impact on the separation performance of a column. Increasing oven ramp rates reduces the column separation power and leads to higher elution temperature of compounds (Figure 2.148). Also refer to the selectivity term A in Section 2.2.9.4. In this case, efficiency cannot be used as a measure of column performance, instead peak width or peak capacity are used. The peak capacity (P_c) decreases by 22% as a temperature ramp rate increases from 10 to 20 °C/min. The practical implication is that the effective elution temperature of late-eluting analytes is increased. The maximum operating temperature of the column may not be high enough to elute compounds. Hence, high oven temperature ramp rates increase the risk that high boiling compounds elute in the final

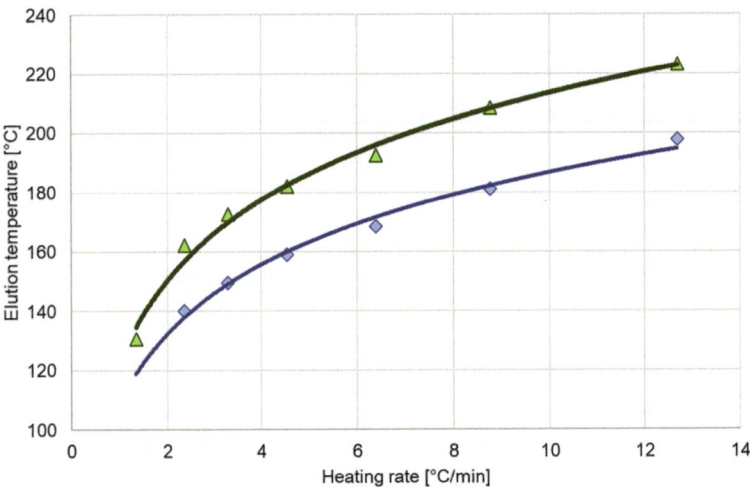

Figure 2.148 The effect of heating rate and carrier gas flow on the elution temperature (retention temperature). Flow conditions: green at 25 mL/min, blue at 75 mL/min (Karasek and Clement, 1988/with permission of ELSEVIER).

isothermal phase with significant peak broadening, increased column bleed, and reduced S/N ratios, even increase the risk of compound decomposition. The oven heating rate needs to be decreased in such cases.

Isothermal phases between oven temperature ramps should be avoided, as this reduces separation power by diffusion. Alternatively reduced heating rates should be considered. Optimized moderate oven ramp rates reduce the effective compound elution temperatures, maintain a high separation performance and the elution within the oven temperature ramp. Thermolabile compounds benefit in response from lower ramp rates with lower elution temperatures.

The example in Figure 2.149 displays the separation of a phenol mix with different GC oven heating rates. The speed of analysis increases with increasing oven temperature ramp rate at the expense of separation power and an increased analyte elution temperature. Each 16 °C/min increase in temperature ramp rate reduces the retention factor by 50%, but at reduced peak resolution. With a ramp rate of 20 °C/min pentachlorophenol (compound 11) elutes at 238 °C, while a ramp of 10 °C/min leads to a 30 °C less elution temperature of 208 °C, but with the trade-off of a 50% increased analysis time. However, if resolution is sufficient then temperature ramp rates should be optimized for increased productivity. In splitless injection for trace analysis, a quick jump from the low oven temperature below the solvent BP to a moderately high oven temperature is preferred, with a following slow heating ramp for compound separation.

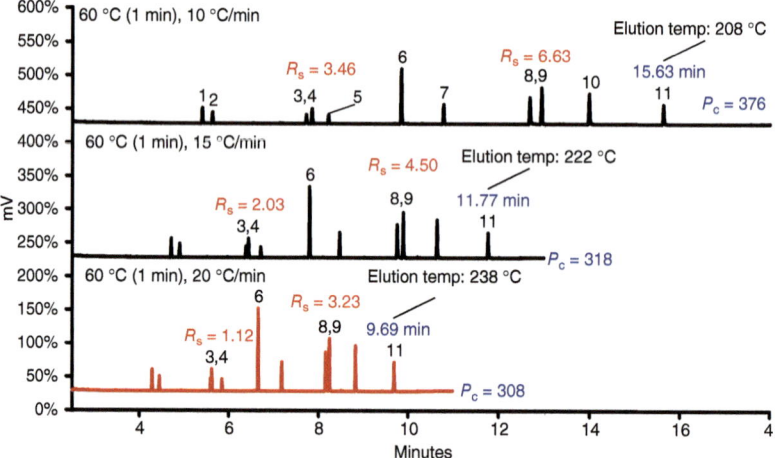

Figure 2.149 Effect of oven temperature ramp rate on analysis time, resolution, and elution temperature (Khan, 2013/Thermo Fisher Scientific Inc.). Experimental conditions: Column type TG-5MS, 30 m × 0.25 mm × 0.25 μm; inlet SSL at 250 °C; carrier gas 1.2 mL/min helium, constant flow; split injection 80:1; injection volume 1.0 μL; detector FID at 280 °C; oven temperature program as indicated with each chromatogram, R_s peak resolution #3,4 and #8,9. P_c peak capacity. Analytes: 1. phenol, 2. 2-chlorophenol, 3. 2-nitrophenol, 4. 2,4-dimethylphenol, 5. 2,4-dichlorophenol, 6. 4-chloro-3-methylphenol, 7. 2,4,6-trichlorophenol, 8. 2,4-dinitrophenol, 9. 4-nitrophenol, 10. 2-methyl-4,6-dinitrophenol and 11. pentachlorophenol.

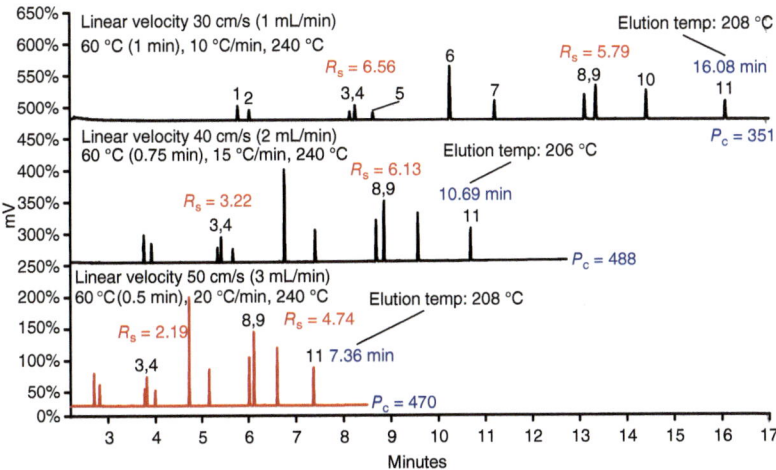

Figure 2.150 Effect of adjusting the linear velocity combined with increased oven temperature ramp rates (Khan, 2013/Thermo Fisher Scientific Inc.). Experimental conditions and analytes as of Figure 2.121, R_s peak resolution #3,4 and #8,9, P_c peak capacity.

Further, optimization of the oven temperature ramp versus the carrier gas flow establishes the desired final separation conditions of short analysis time with still good peak resolution and a reduced elution temperature for the analytes. The example in Figure 2.150 demonstrates the win in analysis time of more than 50% from 16:08 to 7:36 min for pentachlorophenol at a low elution temperature of 208 °C.

2.2.10 Fast Gas Chromatography Solutions

The working pressure in modern analytical laboratories requires for an increased throughput of samples, and hence method development is largely focused on productivity (Figure 2.151). However, speed of chromatographic analysis and enhancement of peak separation generally require analytical method optimization in opposite directions. Basically, two obvious approaches are increasingly being used to combine advancements in speed and peak separation. In the existing conventional GC ovens, the use of narrow bore columns below the standard 0.25 mm ID offers a viable practical "Fast GC" solution for every GC and GC-MS with electronic pressure regulation (EPC). Alternatively, the installation of cartridges for direct column heating are available for "Ultra-Fast" GC analyses.

2.2.10.1 Fast Chromatography

With improved and automated sample preparation methods, GC analysis time is becoming the rate determining step for productivity in many laboratories. The term "*Fast GC*" compares to the conventionally used fused silica columns of

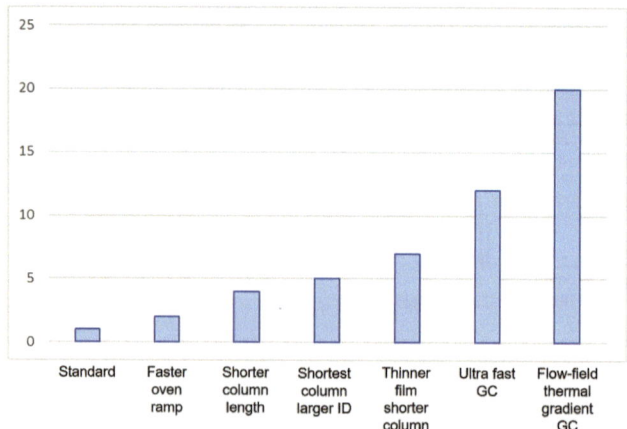

Figure 2.151 Increased sample throughput per hour by progress in separation technology relative to a standard GC analysis.

typical dimensions with lengths of 30 m or longer and inner diameters of 0.25 mm or 0.32 mm. A reduction of GC-MS analysis time is possible using a reduced column ID and length with appropriate film thicknesses and oven temperature programming (Maštovská and Lehotay, 2003; Donato et al., 2007; Khan, 2013). Table 2.29 informs about the equivalence for a 1 : 1 replacement with fast GC capillary columns.

A very good example describing the progress in GC separation technology and advancement in GC productivity is given with an optimization strategy starting from a packed column separation (Facchetti et al., 2002). The sample used for this comparison is the 16-component standard mixture used in U.S. EPA 610 analysis of PAHs (Table 2.30). The U.S. EPA method is antiquated and describes a 45 min analysis using a packed column, 1.8 m long times 2 mm ID glass, packed with 3% OV-17 on Chromosorb (U.S. EPA, 1984). All fused silica capillary columns used in the described example were 5% phenyl-polysilphenylene-siloxane coated TR-5MS

Table 2.29 Capillary column dimensions that can be replaced 1 : 1 in order to achieve fast GC analysis.

Present column	Fast GC column
15 m × 0.25 mm × 0.25 µm	10 m × 0.15 mm × 0.15 µm
30 m × 0.25 mm × 0.25 µm	20 m × 0.15 mm × 0.15 µm
60 m × 0.25 mm × 0.25 µm	40 m × 0.15 mm × 0.15 µm
15 m × 0.32 mm × 0.25 µm	10 m × 0.15 mm × 0.15 µm
30 m × 0.32 mm × 0.25 µm	15 m × 0.15 mm × 0.15 µm
60 m × 0.32 mm × 0.25 µm	30 m × 0.15 mm × 0.15 µm

Table 2.30 Compounds in the U.S. EPA 610 standard on packed column.

Peak	Component	Retention time (min)
1	Naphthalene	4.50
2	Acenaphthylene	10.40
3	Acenaphthene	10.80
4	Fluorene	12.60
5	Phenanthrene	15.90*
6	Anthracene	15.90*
7	Fluoranthene	19.80
8	Pyrene	20.60
9	Benzo(*a*)anthracene	24.70*
10	Chrysene	24.70*
11	Benzo(*b*)fluoranthene	28.00*
12	Benzo(*k*)fluoranthene	28.00*
13	Benzo(*a*)pyrene	29.40
14	Dibenzo(*a,h*)anthracene	36.20*
15	Indeno(1,2,3-*cd*)pyrene	36.20*
16	Benzo(*ghi*)perylene	38.60

Note: Four pairs of peaks remain unresolved using the U.S. EPA 610 packed column technique, marked with *.

columns (Thermo Fisher Scientific, Bellefonte, PA, USA) with dimensions as specified with each chromatogram. Helium is the carrier gas in each case. Flow rates and oven programs were optimized to the column dimensions. The chromatograms were run in either constant flow or constant pressure mode as indicated using a TRACE Ultra GC with a TriPlus autosampler and FID detector (Thermo Fisher Scientific, Milan, Italy).

Initially, capillary GC methods attempted to simulate the packed column method with a run time of 45 min. The example in Figure 2.152 shows a capillary separation where the oven program is slowed down to keep the run time similar to the packed column experience. Helium was substituted for the original nitrogen carrier gas. Several variables within the GC setup can be manipulated to further shorten the run time, bearing in mind the need to maintain resolution, elution order and sensitivity as outlined in Table 2.31. The first strategy for reducing run times should be to modify the oven ramp by increasing the ramp where the peaks are well separated with large retention time spaces and applying a slow ramp where extra separation is required (see Figure 2.153).

Reducing column length is the second strategy for reducing run time. Note the increase the oven ramp rate in order to elute the high boiling materials, while

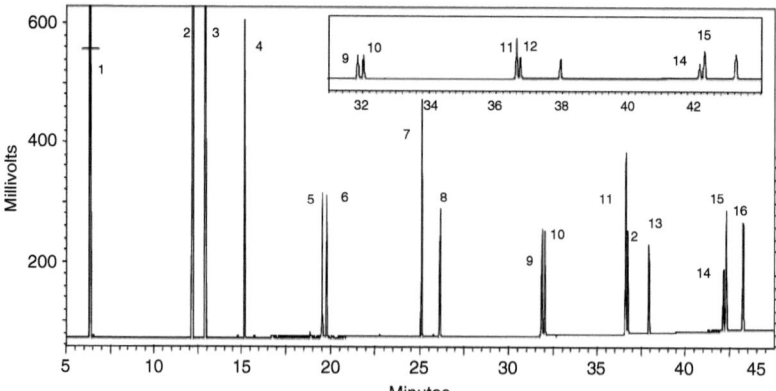

Figure 2.152 Simulation of a packed column separation performed on a 30 m capillary column. Capillary column: 30 m TR-5MS; ID 0.25 mm, film 0.25 μm; initial oven temp: 100 °C; Rate 1: 5 °C/min to 300 °C; carrier gas: He 1.5 mL/min (constant flow); split ratio: 50:1; injection vol.: 1.0 μL; detector: FID. Sample run rate is 1 per hour (includes oven cooling).

Table 2.31 Variables affecting GC run time.

Parameter	Run time effects
Temperature ramp rate	Higher starting temperature and steeper heating rate → peaks elute faster
Column length	Shorter column → shorter run time
Carrier gas flow rate	Each gas type has an optimum linear flow range within which speed can be adjusted. Higher than optimum flow rates are acceptable for productivity increase if critical separations remain compliant.
Film thickness	Thicker film → more interaction of solutes → longer run time
	Thinner film → "Fast Chromatography"
Column internal diameter	Reducing internal diameter increases column efficiency → shorter runtimes using 'Fast Chromatography'
Ultra-fast technology	Combination of short column, small internal diameter, thin film, special ultra-fast heating coil for ultra-fast separations. Requires optional Ultra-Fast (UF) or Low Thermal-Mass (LTM) GC modules.
Vacuum outlet	Short runtimes, higher column load, using a widebore column connected to MS ion source, short restrictor to injector
Thermal flow-field gradient	Very short runtimes, high separation power in short columns by negative temperature gradient

2.2 Gas Chromatography | 231

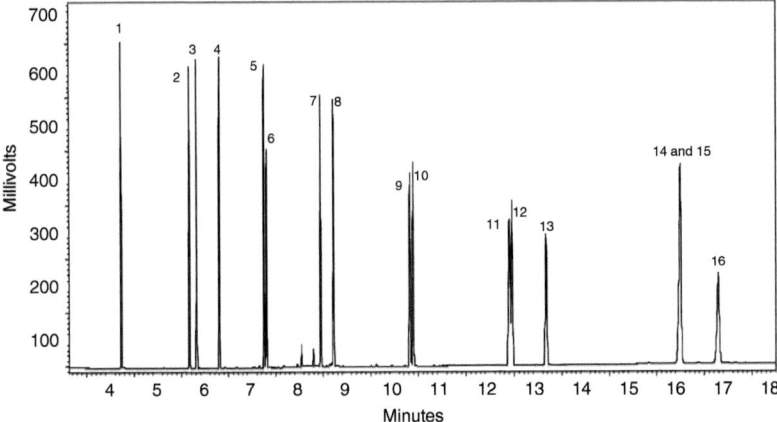

Figure 2.153 Optimization of the oven program has cut the run time by more than half without changing any other parameters. There is some loss of resolution between peaks 14 and 15 compared with the 45 min run; however, the extra specificity of an MS detector would resolve these peaks. Column: 30 m TR-5MS; ID 0.25 mm, film 0.25 µm; carrier gas: He, 1.5 mL/min (constant flow); split ratio: 50:1; injection vol.: 0.5 µL; initial oven temp.: 90 °C for 1 min; rate 1: 25 °C/min to 290 °C; rate 2: 4 °C/min to 320 °C for 5 min; detector: FID. Sample run rate is 2 per h (includes oven cooling).

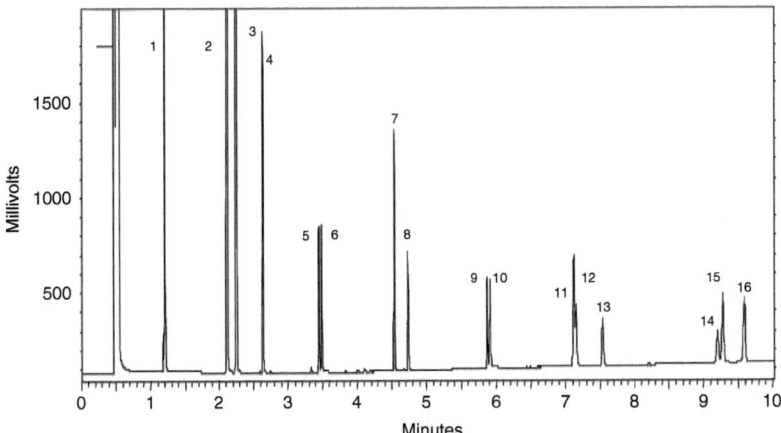

Figure 2.154 Illustration of the effect of simply reducing column length by half reducing the run time to 10 min. Column: 15 m TRACE TR-5MS; ID 0.25 mm, film 0.25 µm; carrier gas: He 1.5 mL/min (constant flow); split ratio: 50:1; injection vol.: 1.0 µL; initial oven temp: 120 °C for 0.2 min; rate 1: 25 °C/min to 260 °C; rate 2: 7 °C/min to 300 °C for 3 min; detector: FID. Sample run rate is 4 per h (includes oven cooling).

retaining peak shape (see Figure 2.154 with the switch to a 15 m, Figure 2.155 to a 7 m column) The finally optimized separation in Figure 2.156 with the longer 10 m × 0.1 µm film column is even faster. In addition, using constant pressure has enhanced the analysis speed to under 5 min. The effect of running at *constant pressure* is to greatly speed up the linear velocity at the beginning of the run where

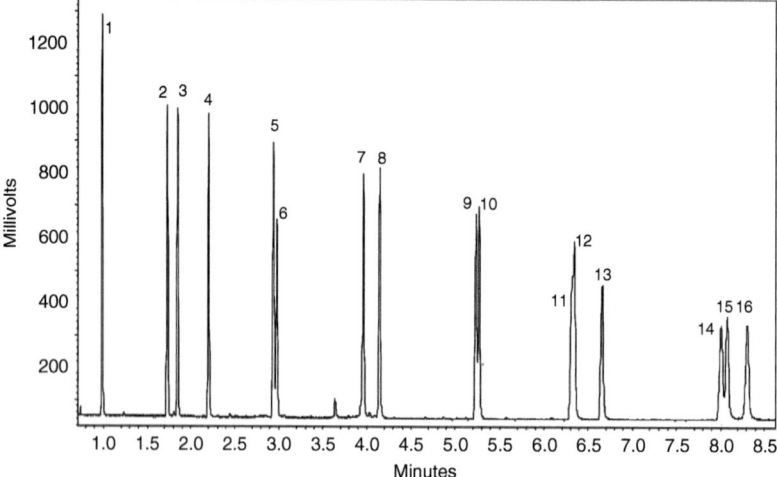

Figure 2.155 Further reduction of the column length by half to 7 m but with larger ID, illustrates the effect of film thickness on run time. The larger ID makes the column less efficient. Therefore, to maintain the resolution, a slower oven program is needed, making the analysis time only slightly shorter than on the 15 m column. Column: 7 m TR-5MS; ID 0.32 mm, film 0.25 μm; carrier gas: He 1.5 mL/min (constant flow); split ratio: 50:1; injection vol.: 0.5 μL; initial oven temp: 120 °C for 1 min; rate 1: 25 °C/min to 250 °C; rate 2: 10 °C/min to 300 °C for 5 min; detector: FID. Sample run rate is 5 per h (includes oven cooling).

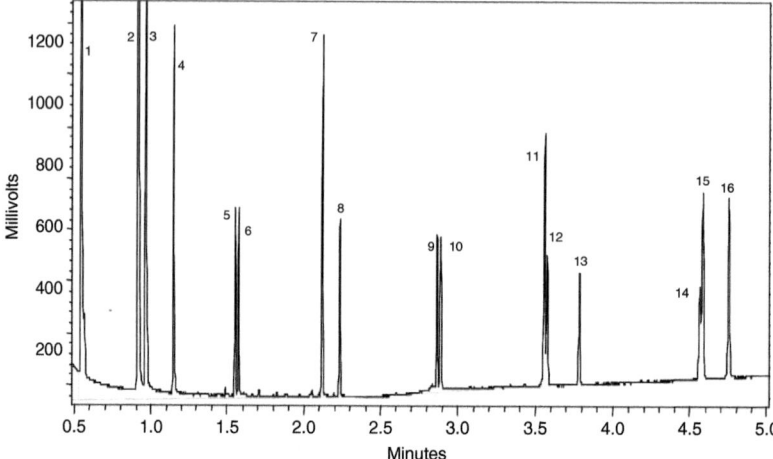

Figure 2.156 The combination of using a narrow 0.1 mm column coated with a thinner film is the strategy for fast analysis.

the peaks are well separated. Later in the run, when the temperature is high, the velocity is lower allowing more dwell time on the column for the later peaks which elute close together. Increasing the linear velocity on a fast GC column of 20 m × 0.15 mm × 0.15 μm film from 30 cm/s to 43 cm/s reduced the total analysis time for the U.S. EPA 625 phenol standard mix by 30% (Khan, 2013a).

2.2.10.2 Vacuum Outlet (Low Pressure) Chromatography

The vacuum outlet GC, aka *low pressure GC* (LPGC) today, takes advantage of the direct coupling of a widebore analytical column to the ion source of a mass spectrometer. For a capillary column the optimum carrier gas flow increases with the low pressure in the column offering fast GC separations. Already in the late 1990s Jaap de Zeeuw coupled a short 0.53 megabore column directly to MS (de Zeeuw et al., 2022). A short narrowbore column between injector and analytical column served as the flow restrictor for an ion source compatible carrier gas volume flow. The concept of LPGC was patented at that time by Varian, today commercially available for any type of GC-MS (de Zeeuw et al., 2000, 2022).

The optimum carrier gas velocity in a 10 m long megabore column in direct MS coupling can reach up to 300 cm/s, a value 10 times higher compared to regular pressurized capillary installations. Separations are significantly faster. Faster eluting peaks result in higher signal intensity and improved S/N values. The analyte elution temperatures are lower benefiting the analysis of thermally labile and high boiling compounds with less column bleed (Peene et al., 2000). Figure 2.157 illustrates the

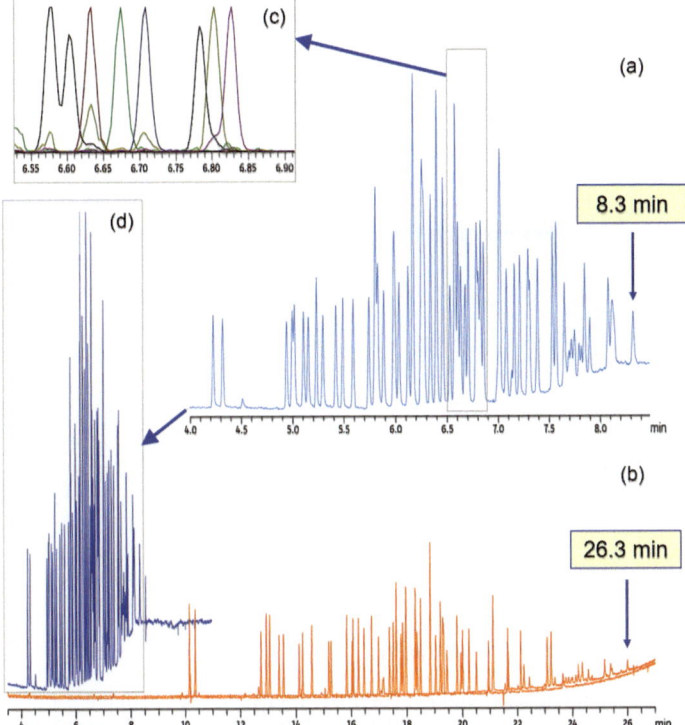

Figure 2.157 Comparison of fast LPGC and regular pesticide analysis chromatography. (a) LPGC TIC, column 15 m × 0.53 mm × 1.0 µm Rtx-5 ms with 5 m × 0.18 mm restrictor on the injector side. (b) Conventional GC-MS TIC, 30 m × 0.25 mm × 0.25 µm Rxi-5 ms. (c) XIC zoom of the separated components into a congested region. (d) APGC overlay over the conventional chromatogram (Restek, 2024/Restek Corporation). Sample run rate is 7 per h (includes column cooling).

practical advantage of shorter analysis times with a threefold shorter runtime for a GC-MS/MS pesticide analysis. Unresolved areas show the differentiation of analytes using the SRM acquisition mode (Restek, 2024). A reduction in carrier gas consumption equivalent to the reduced cycle time is achieved.

With LPGC some separation efficiency is traded for speed of analysis. But, as can be seen in Figure 2.158. Components of complex mixtures that elute at a similar retention time can be separated and quantified independently by MS/MS detection. Applications of LPGC are mainly found in areas of multi-compound target analysis with long retention times with the goal of a higher sample throughput without sacrificing method robustness and sensitivity

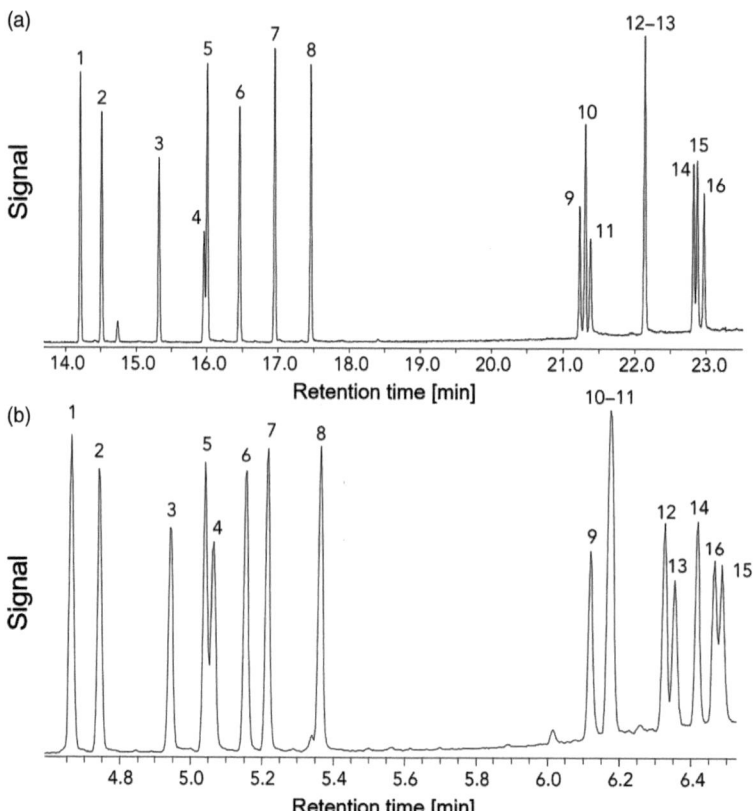

Figure 2.158 Comparison of full scan total ion chromatograms (m/z 50–550) for (a) standard GC-MS with (b) LPGC-MS (Lehotay, 2020/MJH Life Sciences). Conditions: 1 μL splitless injection of a 2 ng/μL pesticide solution. A: 30 m × 0.25 mm × 0.25 μm - 5ms type. Oven: 90 °C for 1 min, 8.5 °C/min to 330 °C for 5 min, 1.4 mL/min He flow rate. B: 5 m × 0.18 mm guard column + 15 m × 0.53 mm × 1 μm - 5 ms type. Oven: 80 °C for 1 min, 45 °C/min to 320 °C for 4 min, 2 mL/min He flow rate. Compounds: 1. diazinon, 2. isazophos, 3. chlorpyrifos-methyl, 4. fenitrothion, 5. pirimiphos-methyl, 6. chlorpyrifos, 7. pirimiphos-ethyl, 8. quinalphos, 9. pyridaphenthion, 10. phosmet, 11. EPN, 12. phosalone, 13. azinphos-methyl, 14. pyrazophos, 15. azinphos-ethyl, and 16. pyraclofos.

(Arrebola et al., 2003; Cajka et al., 2008). The sample capacity and column robustness are increased, also allowing large-volume injections with standard inlets without additional column maintenance. In applications with automated μSPE clean-up (see Sections 2.1.3.2 and 4.9) the cycle times of LPGC ideally meet the also short clean-up procedure for a greatly improved sample throughput, necessary to cope with the increasing demand of results for food safety analysis for pesticides, veterinary drugs, or environmental contaminants (Sapozhnikova and Lehotay, 2015; Sapozhnikova, 2018; Monteiro et al., 2020). "LPGC has been known to be advantageous for nearly 60 years, and for the past 20 years it has been demonstrated that LPGC can be installed in any commercial GC-MS instrument without modification. Except for the analysis of volatiles that are already separated quickly, LPGC-MS is a faster and often better alternative to standard GC-MS" as it is exemplified in Figure 2.158. (Lehotay et al., 2020).

2.2.10.3 Ultra-Fast Chromatography

The term *Ultra-Fast Gas Chromatography* (UFGC) describes gas chromatographic techniques beyond the classical air convection heated column ovens with heating ramps limited to up to about 40 °C/min. UFGC uses different heating concepts with very high temperature programming rates to achieve separation on the seconds timescale instead of minutes.

Direct Column Heating One solution is the direct column-heating technique that is taking advantage of short, narrow bore capillary columns, allowing high temperature programming rates (Bicchi et al., 2005). This technique offers shorter analysis times by a factor up to 30 compared to conventional air heated capillary GC. Such ultra-fast GC conditions typically apply short 0.1 mm ID narrow bore capillary columns of 2.5 to 5 m in length with elevated heating rates of 100 to 1200 °C/min providing peak widths of 100 ms or less, see Figure 2.159. At this point, the limiting factor for sample throughput becomes the ability of the column to cool quickly enough between runs. In addition to a fast heating/cooling regime, a fast conventional or MS detection with an acquisition rate better than 25 Hz for reliable peak identification is required. For example, a restricted full scan, SIM, SRM, or time-of-flight (TOF) modes are used, to achieve the necessary sampling rate across individual peaks. Currently, the fast scanning quadrupole technologies and TOF analyzer are the MS technologies of choice to provide the required speed of detection for an ultra-fast GC-MS at data rates of 50 Hz or better with partial or full mass spectral information.

The special instrumentation required for ultra-fast chromatography comprises a dedicated ultra-fast column module (UFM), see Figure 2.160, or low thermal-mass device (LTM), comprising a specially assembled fused silica column for direct resistive heating, wrapped with a heating element and a temperature sensor, see Figure 2.161 (Facchetti et al., 2002). The assembly is held in a compact metal cage to be installed inside the regular GC oven, which is not active during ultra-fast

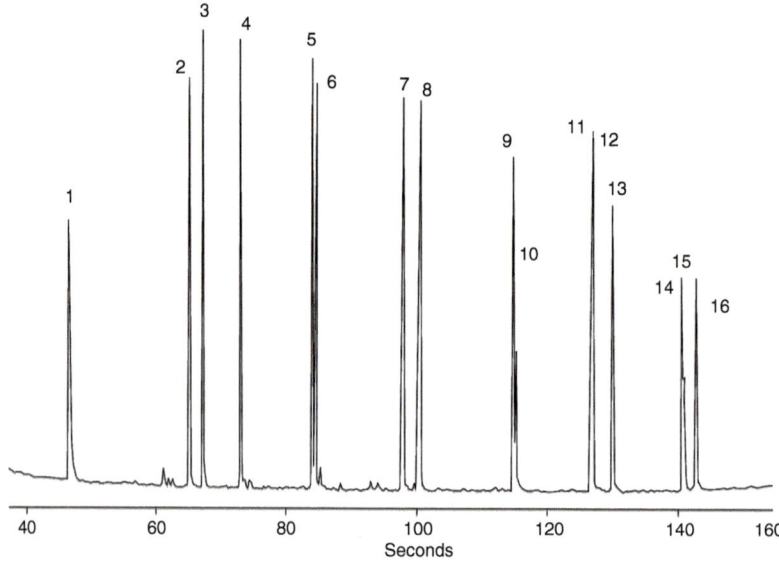

Figure 2.159 Ultra-fast column and detector can run the polyaromatic hydrocarbon (PAH) sample in just 160 s. Column: 5 m Ultra-fast TR5-MS; ID 0.1 mm, film 0.1 μm; carrier gas: H_2 1.0 mL/min; initial temp: 40 °C for 0.3 min; rate 1 : 2 °C/s to 330 °C. Sample run rate is 12 per h (includes column cooling).

Figure 2.160 UFM column module installed in a regular GC oven (Thermo Fisher Scientific, Milan, Italy).

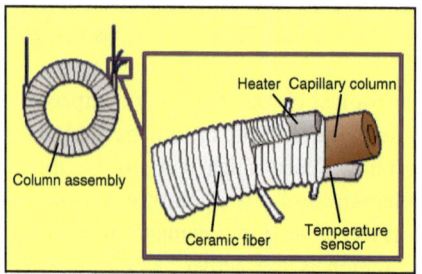

Figure 2.161 Column wrapping detail with heating element and temperature sensor for ultra-fast capillary chromatography.

2.2 Gas Chromatography | 237

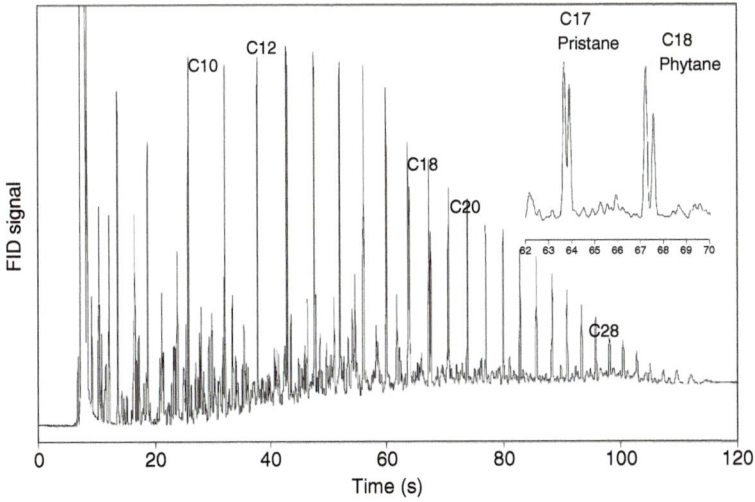

Figure 2.162 Ultra-fast GC-FID chromatogram of a mineral oil sample. The chromatogram was obtained using a 5 m, 0.1 mm ID, 0.1 µm film thickness SE54 column with a temperature program from 40 (12 s) to 350 °C (6 s) at 180 °C/min.

operation. Temperature programming rates can be achieved as high as 1200 °C/min. A notable example in Figure 2.162 shows the ultra-fast analysis of a mineral oil performed in only 2 min still maintaining sufficient resolution to separate pristane and phytane from *n*-C17 and *n*-C18.

2.2.10.4 Flow-Field Thermal Gradient Gas Chromatography

An innovative concept was introduced for ultra-fast GC with the flow-field thermal gradient gas chromatography (FF-TG-GC) by Peter Boeker just in recent years (Boeker and Leppert, 2015). The *"HyperChrom"* called GC has no air heated

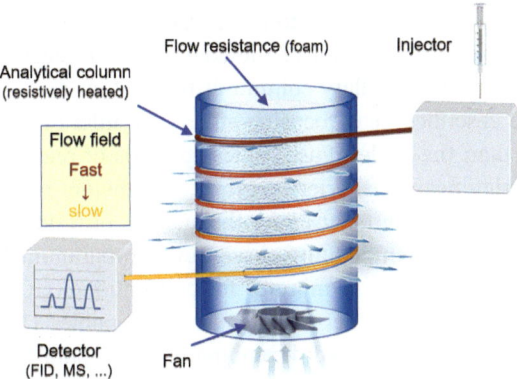

Figure 2.163 Concept of flow field temperature gradient gas chromatography (Adapted from Havelt and Boeker, 2023).

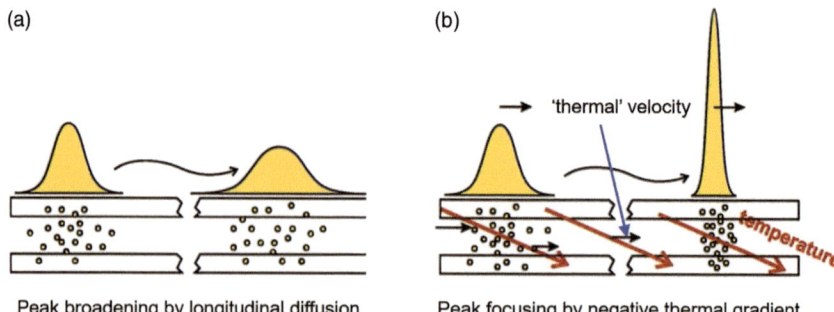

Figure 2.164 Peak broadening in conventional GC (a) compared to peak focusing by negative temperature gradient (b) (Boeker and Leppert, 2015/American Chemical Society).

oven (Figure 2.163). A thermal gradient is formed along the separation column. This design uniquely allows negative temperature gradients along the column. In contrast to classical air heated GC, the peak broadening by longitudinal diffusion is counteracted by this negative gradient peak focusing (Rosenberg et al., 2023). The negative thermal gradient creates a carrier gas velocity gradient within the column where the front is slower than the tail, as it is illustrated in Figure 2.164. The analytical advantage are sharper peaks, significantly faster analysis times, lower elution temperatures and lower LODs. It benefits in particular thermo-labile compounds as it is shown for the analysis of explosives (Leppert et al., 2018).

The separation column is directly heated extremely fast in a thin steel capillary consuming only little energy. The innovative feature of the *HyperChrom* GC is the use of a continuous temperature gradient for peak focusing at reduced elution temperatures. The flow of air in the special helical structure of the tubular central part creates a temperature gradient. A porous foam in the central structure forms a flow resistance for the air blown in from the bottom. Therefore, the cooling effect on the capillary is higher at the bottom end of the column at the dector than on the top injector part. The colder capillary column at the bottom causes a decreasing flow speed with the described peak focusing effect. The inlet of the analytical column is kept hotter than the outlet. Fused silica columns of regular dimensions can be used but in significantly shorter length of few meters only. For instance, an only 2 m long Rxi-624 column is used for the analysis of residual solvents in 90 s (Chopra et al., 2021). The temperature curve can be programmed to a scale of seconds as opposed to the scale of minutes for conventional GCs. Multiple ramps, iso-thermal phases and even negative ramps are possible. The temperature gradient can be increased or decreased depending on the separation requirements.

Applications cover the typical long running GC separations. For instance, the total hydrocarbon analysis (DIN H53/ISO 16703 method) could be made faster with

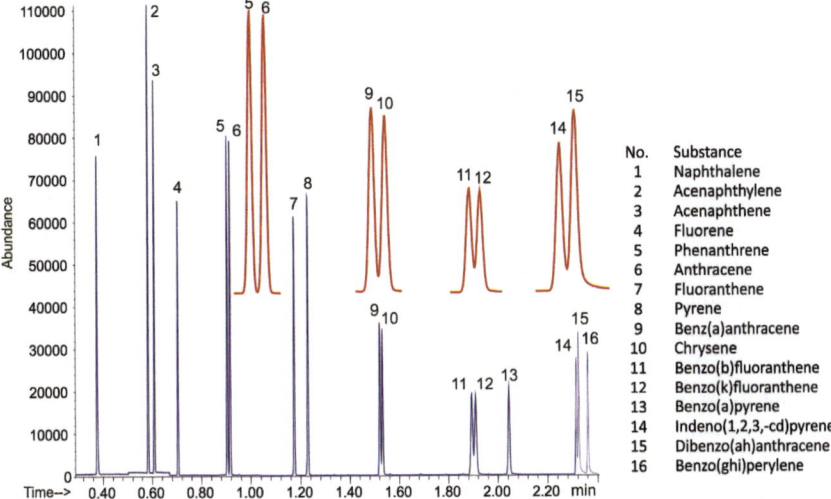

Figure 2.165 Hyperfast separation of the U.S. EPA 16 PAH standard mix within 2.5 min cycle time showing the chromatographic resolution of the critical pairs. Sample run rate is 20 per h (includes oven cooling).

a cycle time of only 70 s with the elution beyond C40 (HyperChrom, 2024). The analysis of PAHs takes 2.5 min cycle time only and resolves well the critical pairs (Figure 2.165).

2.2.11 Multi-Dimensional Gas Chromatography

The search for solutions to improve the at that time limited GC separation performance started already in 1958 by M.C. Simmons and L.R. Snyder with the "Two-Stage Gas/Liquid Cromatography" for the analysis of petroleum products (Simmons and Snyder, 1958). The development of instrumental techniques focused in the beginning on the transfer of specific peak regions of interest, the unresolved "humps", to a second separation column. This technique, widely known today as *heart cutting*, found many successful applications in solving critical research-oriented projects but its use did not become widespread in routine laboratories.

The objective to increase the peak capacity for the resolution of entire complex chromatograms and not only of few discrete congested areas started more than 30 years ago with the introduction of a comprehensive two-dimensional GC (GC × GC) (Liu and Phillips, 1991; Zanella et al., 2021). While the initial ideas of multi-dimensional gas chromatography (MDGC) relied on the use of two regular capillary columns (MDGC, selectivity tuning), current comprehensive two-dimensional chromatography benefits from the alignment with fast GC on the second separation dimension (Bertsch, 1999, 2000; Beens et al., 2000). The result is a significantly improved separation power. With the commercial availability of GC×GC instrumentation this technique found avid utilization for a widespread range of applications within the first few years (Marriott et al., 2003).

Particularly, for the separation of complex mixtures, GC×GC showed far greater resolution and a significant boost in signal to noise, with no increase in analysis times. Further, the hyphenation of GC×GC with MS detection providing three independent analytical dimensions made this technique ideal for the measurement of targeted components within complex samples such as those from food, petrochemical, environmental, and biomedical analysis. The significantly increased chromatographic resolution in GC×GC allowed separation of many previously undetectable components. Fast scanning mass spectrometers are required to maintain the excellent chromatographic resolution, provided today with full scan data by TOF, and fast quadrupole or high resolution MS systems (Hoh and Mastovska, 2008; Vaye et al., 2022). For instance, more than 15 000 peaks could be identified in an ambient air sample from the city of Augsburg, Germany, by TD comprehensive two-dimensional GC with TOF-MS detection (Welthagen et al., 2003).

Multidimensional GC

The great advantage of the combination of multiple chromatographic separation steps is the increase in peak capacity. Peak capacity is the maximum number of peaks that can be resolved in a given retention time frame. The more peaks a combination of techniques is able to resolve, the more complex samples can be analyzed. When a sample is separated using two dissimilar columns, the maximum peak capacity Φ_{max} will be the product of the individual column's peak capacity Φ_n.

$$\Phi_{max} = \Phi_1 \times \Phi_2$$

For example, if each separation mode generates peak capacities of 1100 in the first dimension and 30 in the second, the theoretical peak capacity of the 2D experiment will be 33 000, a huge gain in separation space, which would theoretically compare to the separation power of a 12 000 m column in the normal single dimension analysis. To achieve this gain, however, the two column phases should be totally orthogonal, that is, based upon completely different separation mechanisms.

The comparison of peak capacities in normal, heart cutting, and comprehensive two-dimensional chromatography is illustrated with the following graphics:

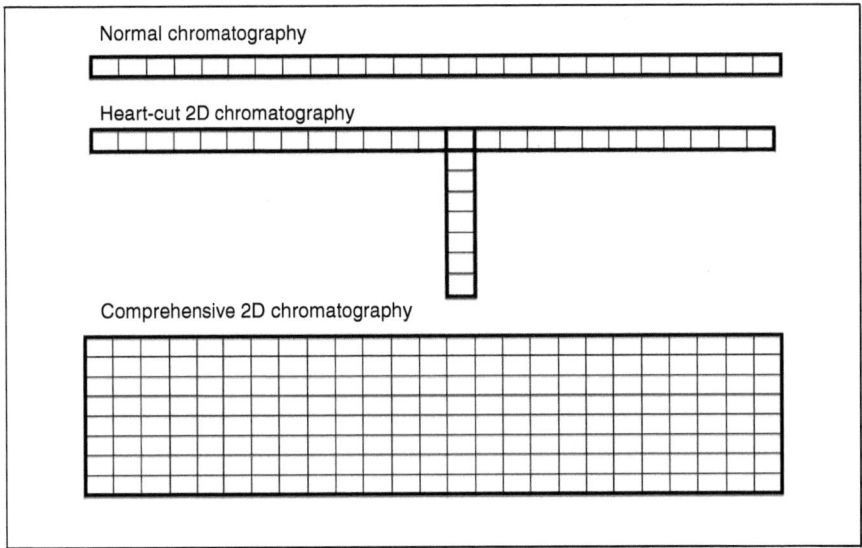

2.2.11.1 Heart Cutting

The goal of heart cutting is the increase of peak capacity for target substances in congested chromatogram regions of unresolved compounds. Two capillary columns are connected in series, typically by means of a valveless flow switch. One or more short retention time slices are sampled onto a second separation column of different separation characteristic. A valveless switching device is ideally suited for high speed and inert flow switching. It is based on the principle of pressure balancing and was introduced by Deans already in 1968. This technique found a wider recognition first with state-of-the-art EPC units, which allowed the integration of column switching as part of regular GC methods for routine applications (Deans, 1968, 1981; McNamara et al., 2003).

Heart cutting requires a sample specific individual setup of the analysis strategy. The sections of interest have to be previously identified and selected by retention time from the first chromatogram to be "cut" into the second column. Several heart cuts can take place within a chromatogram as of analytical need. The choice of column length is application specific and does not interfere with the "cutting" process. The column of the second dimension is required to be of a different polarity to achieve an 'orthogonal' separation. A monitor detector is required, typically a universal FID, which continuously observes the separation on the first dimension. A second detector, often an MS detector, acquires data from the second dimension in now "higher" chromatographic resolution (De Alencastro et al., 2003).

2.2.11.2 Comprehensive GC – GC × GC

True multi-dimensional chromatography requires two independent (orthogonal) separations mechanisms and the conservation of the first separation into the

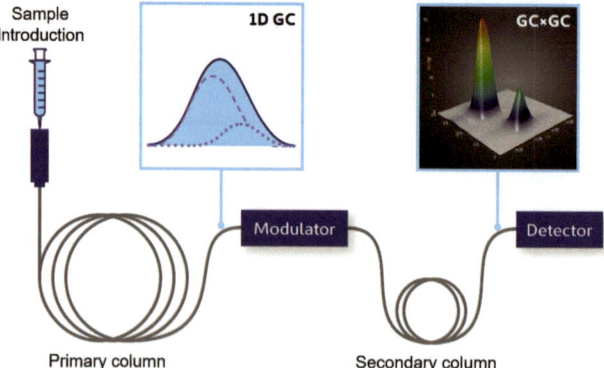

Figure 2.166 GCxGC instrument setup with the modulator connecting the primary and secondary columns (Courtesy of SepSolve).

second dimension aka *comprehensive GC*. The GC × GC technique has been widely accepted and applied to the analysis of complex mixtures. Commercial instrumentation is available at a mature technological standard for routine application in the configuration presented in Figure 2.166.

In contrast to heart cutting, the complete first dimension effluent is continuously separated in slices (which can be interpreted as a continuous sequence of cuts) and forms a three-dimensional data space adding a second separation dimension with retention times in both dimensions. The peak intensities are monitored by classical or MS detection proving the third signal dimension. No monitor detector for the first dimension or prior identification of an individual retention time window is required. The total analysis time with GC × GC analyses does not increase, but the information content significantly. The same sample throughput can be achieved.

The primary and secondary columns are connected by a modulator interface together installed in the same GC oven. The second dimension column should be of a different polarity than the first one to provide different kind of interactions in substance class separations. Typically, non-polar columns are used in the first dimension for a separation along the substance boiling points. Polar film interactions characterize the further enhanced separation on the second dimension, which takes place at a low temperature gradient, almost isothermal due to the speed of elution and of the only small oven temperature program change in this short time. Column lengths in the second dimension are as short of up to 2 m for fast GC conditions.

The operation of the *modulator* interface between the both columns is key to GC × GC (Ong and Marriott, 2002). The modulator operates in an alternating trap and inject mode with a modulation frequency related to the fast second-dimension separation. Short elution sections from the first column are integrated by different ways of trapping and injection to the second-dimension separation. Each injection pulse generates a high speed chromatogram (Bertsch, 1999; Beens et al., 2001).

2.2 Gas Chromatography

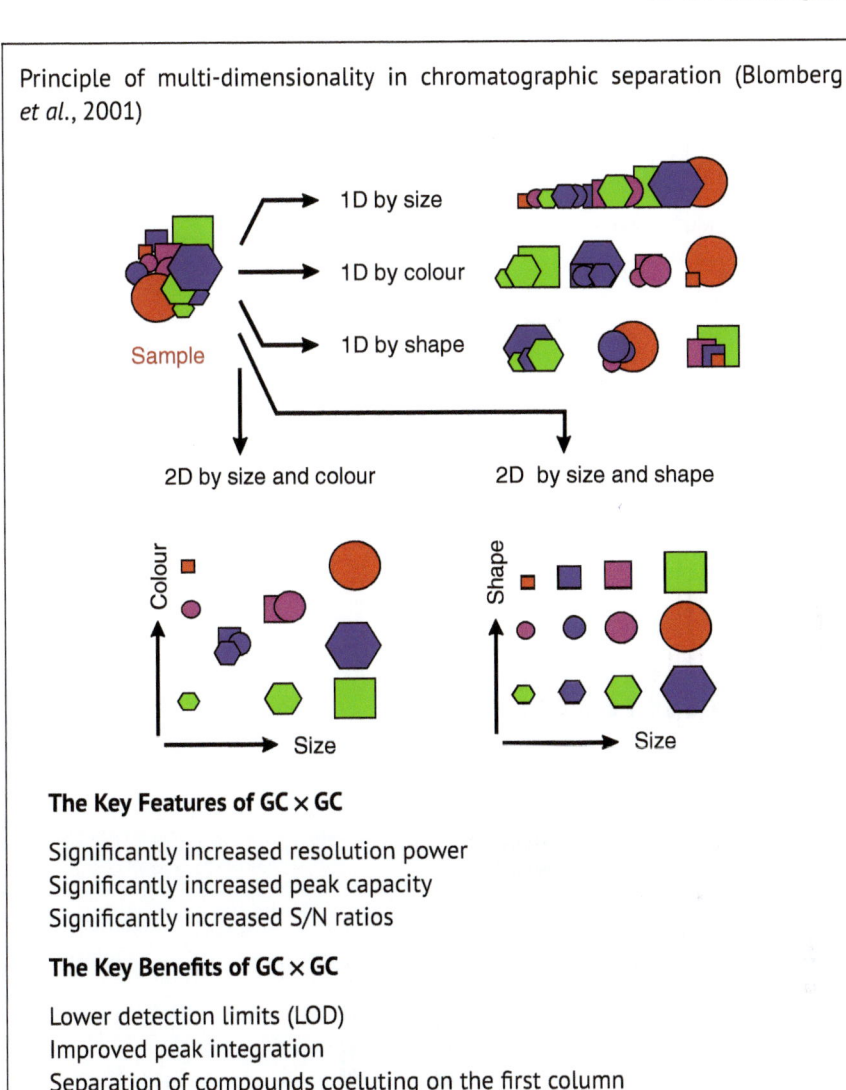

Principle of multi-dimensionality in chromatographic separation (Blomberg et al., 2001)

The Key Features of GC × GC

Significantly increased resolution power
Significantly increased peak capacity
Significantly increased S/N ratios

The Key Benefits of GC × GC

Lower detection limits (LOD)
Improved peak integration
Separation of compounds coeluting on the first column
Separation of substance classes
Productive, no increase of analysis times
Same overall analysis time as conventional separations
Same sample throughput as conventional techniques.

The continued registration of a large series of second-dimension chromatograms results in a large amplification of the column peak capacities as the first-dimension peaks are distributed on a large series of fast second-dimension chromatograms. The effect of increased separation can be visualized best with 3D contour or peak apex plots, illustrating the commonly used term *comprehensive* chromatography, see Figure 2.167. The very high separation power offered by the comprehensive two-dimensional GC allows, even in the case of complex matrices, a good separation

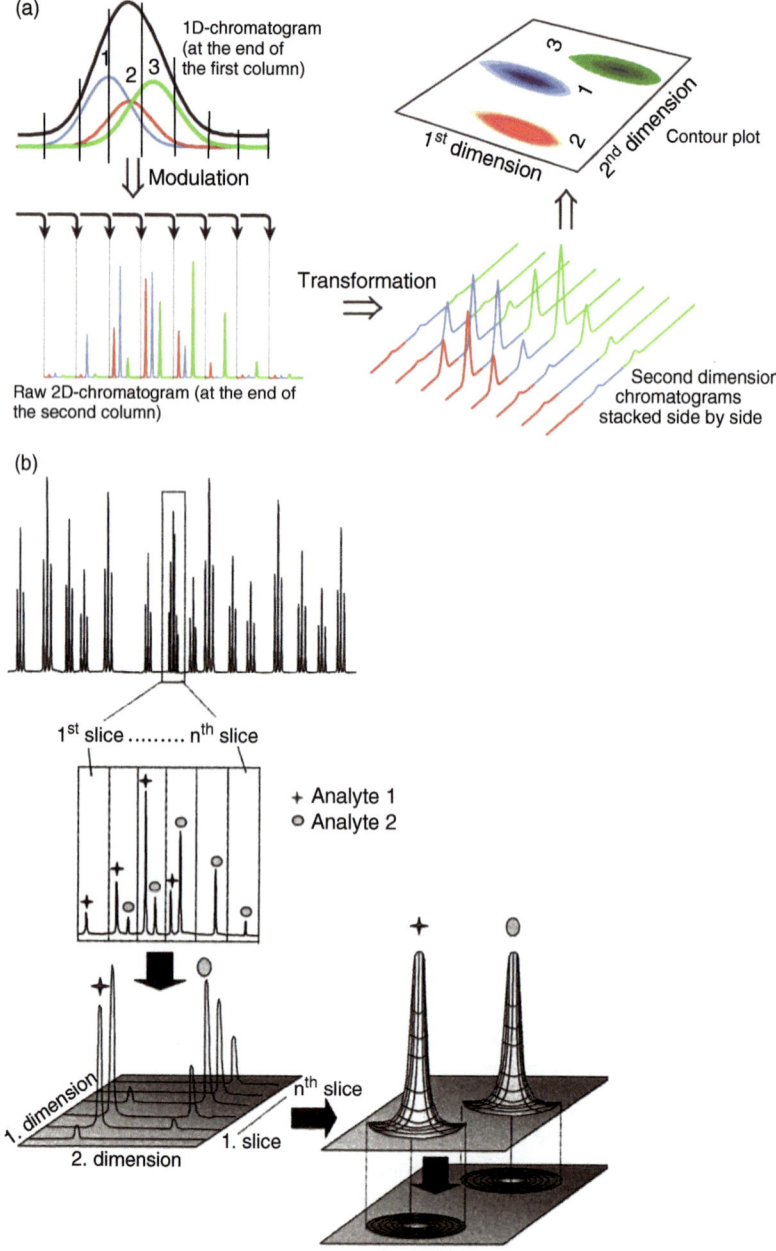

Figure 2.167 (a) Visualization of the principle of multi-dimensional GC separation with the final generation of the contour plot of an unresolved co-elution of three substances in the first dimension (Dallüge, 2003/with permission of ELSEVIER); (b) Construction of a three-dimensional contour plot for analytes 1 and 2 from a modulated chromatogram (Kellner, 2004/American Chemical Society).

of target compounds from the interferences with a significant increase in signal to noise. This emerging technology has substantially enhanced the chromatographic resolution of complex samples and proven its great potential to further expand the capabilities of modern GC-MS systems (Ferracane et al., 2022).

2.2.11.3 Modulation

The key to successful GC×GC is the ability to split, trap and inject fast and efficiently the first column effluent into the second dimension. It is important to modulate faster than the peak width of first dimension peaks, so that multiple second dimension slices are obtained (see Figure 2.167a). The modulator unit has to provide minimized bandwidths in sample transfer to retain the first-dimension resolution and also allow the rapid remobilization. Practically, a high peak compression is achieved by trapping the effluent from the first column on a short band in the modulator, being responsible for the significant increase of S/N values (Patterson et al., 2006). This requires a high frequency handling of the column effluent with a reproducible trapping and controlled substance release. Practically applied modulation frequencies range from 4 to 6 s duration, depending on the second-dimension column parameters.

Different modulator designs are commercially available, most of them as upgrades for any GC model. In general, modulator solutions use thermal- (cryogenic) or flow-based devices (Lee et al., 2000; Ong and Marriott, 2002). Still today there is technical development ongoing:

- Cryogenic zone over a modulator capillary. Substances are released again when the cryofocusing is stopped and resume movement by oven temperature. Also, dual jet cryo modulators have been introduced and commercially available to decouple the processes of collection from the first column and sampling into the second dimension, see Figure 2.168 and 2.169a (Beens et al., 1998, 2000). The operation principle of a dual cryo jet modulator is illustrated in Figure 2.169b (Kellner et al., 2004). Cryo media in use are liquid CO_2 or liquid N_2.
- Loop-based cryogenic-free flow modulators use a differential flow to fill a sample loop and flush the collected slice in the opposite direction onto the second column (SepSolve, 2023). An interchangeable sample loop provides flexibility in

Figure 2.168 Dual jet cryo modulator using a stretched column region (Cavagnino, 2003b/with permission of ELSEVIER).

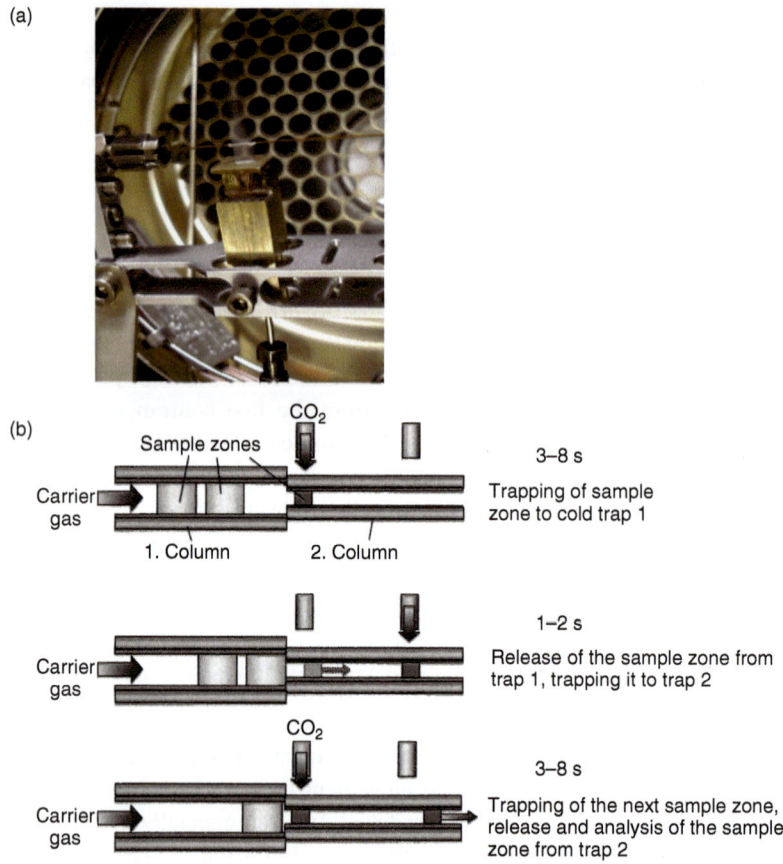

Figure 2.169 (a) Dual jet cryo modulator in trapping operation on the first column (Cavagnino, 2003b/with permission of ELSEVIER). (b) Operation principle the dual jet cryo modulator (Kellner, 2004/American Chemical Society).

method development. Even VVOCs including methane can be modulated without cryofocusing making this solution independent from external cooling media.
- Other cryogenic-free flow modulators use cold and hot air jets to a column section. A two-stage operation is used with a delay loop.

2.2.11.4 Detection

The high speed conditions of the secondary column deliver very sharp peaks of typically 200 ms base width or below, and hence require fast detectors with an appropriate high detection rate. Due to the strong peak band compression obtained during the modulation step, low detection limits are reached. Fast scanning quadrupole MS (scan speed >10 000 Da/s) have been applied by scanning a restricted mass range with an acquisition rate as a cost-effective alternative to TOF-MS for compound identification and quantitation (Mondello et al., 2008). Even magnetic sector mass spectrometers have been used successfully in the fast SIM technique, with excellent sensitivity in environmental trace analysis (Patterson et al., 2006). TOF detection and fast scanning quadrupole instruments in a restricted mass range offer full scan capabilities allowing a detailed peak deconvolution for the extraction of mass spectra from unresolved trace components, see Figure 2.170.

Regarding sensitivity and selectivity, GC × GC/TOF-MS has proven to compare well with the classical HRGC-HRMS methods. Isotope ratio measurements of the most intense ions for both natives and isotopically labeled internal standards ensured the required selectivity (Horii et al., 2004). Potentially interfering matrix compounds are well separated from the compounds to be measured in the two-dimensional chromatographic space (Focant et al., 2004).

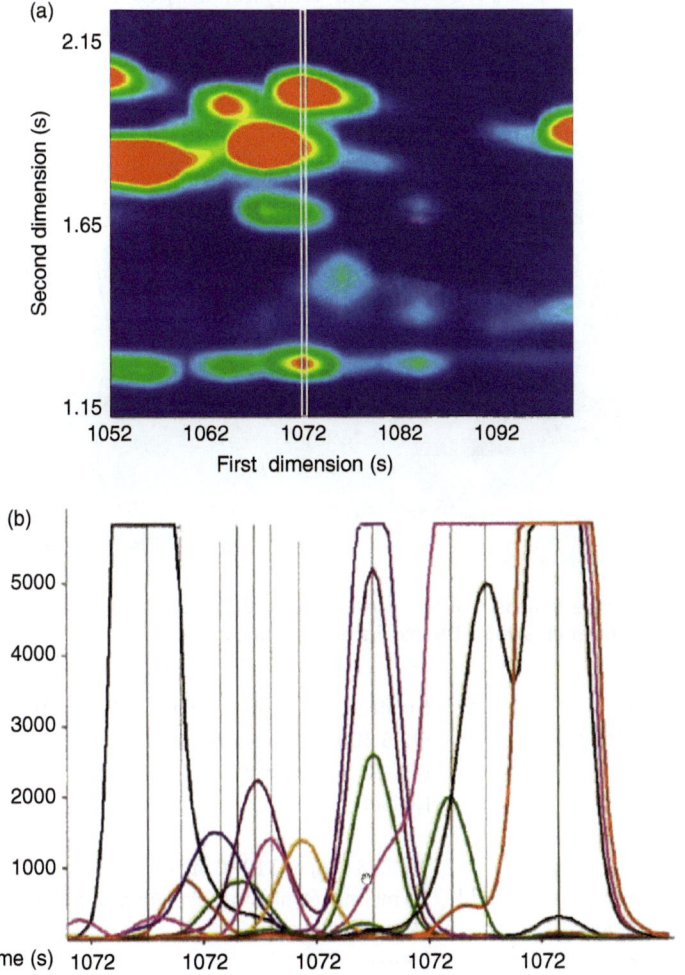

Figure 2.170 Peak deconvolution in full scan GC × GC/TOF-MS. (a) Three-dimensional contour plot, scan 1072 highlighted and (b) scan 1072, each vertical indicates a deconvoluted compound (Reproduced from Dimandja, 2004/with permission of American Chemical Society).

2.2.11.5 Data Handling

Data systems for comprehensive GC require special tools extending the available features of most current GC-MS software suites. Because of the 3D matrix of multiple chromatograms and conventional or MS detection, special tools are required to display and evaluate comprehensive GC separations. The time/response data streams are converted to a matrix format for the three-dimensional contour plot providing a color code of peak intensities (Figure 2.170), or the three-dimensional surface plot generation (Figure 2.171), for visualization.

Integrated software programs are commercially available providing total peak areas also for quantification purposes (Cavagnino et al., 2003a; Reichenbach et al., 2004; GC Image, 2024). Dedicated to the comprehensive qualitative sample characterization with GC×GC-TOF-MS analyses, as well as the quantitative analysis of specific mixture analytes, GC-MS software packages are commercially available. For GC×GC-MS also mass spectral library searching can be used in

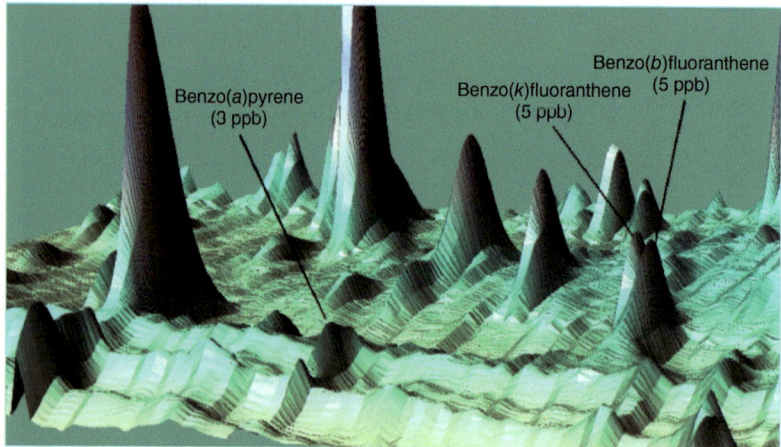

Figure 2.171 Three-dimensional peak view of a GC×GC analysis of a cigarette smoke extract (Cavagnino et al., 2003a/Thermo Fisher Scientific Inc.).

GC Columns:	
Pre-column	5 m deactivated retention gap, 0.32 mm ID
First dimension column	Rtx-5 30 m, 0.32 mm ID, 0.25 μm df
Second dimension column	BPX-50 1 m, 0.1 mm ID, 0.1 μm df
Carrier gas	Helium
Conditions:	
GC temperature program	90 °C (8 min) to 310 °C @ 3 °C/min
Programmed flow	2.5 mL/min (4 min) to 0.8 mL/min @ 5 mL/min/min
Injection volume	30 μL splitless
Splitless time	0.8 min
Moderator	CO_2 dual-jet modulator
Modulation time	6 s
Data system	HyperChrom for acquisition and data processing

conjunction with pattern matching (Hollingsworth et al., 2006). Also, artificial intelligence approaches are getting in use for decision making from comprehensive two-dimensional data (Squara et al., 2023). Those features demonstrate the great potential of GC×GC-MS for unknown compound identification and quantification.

2.2.11.6 Moving Capillary Stream Switching

A valveless flow switching device for heart-cutting purposes or switching the flow direction to different detectors is the moving capillary stream switching system (MCSS) used in similar situations as the Deans switch (see also Section 2.2.11.1). The MCSS is a miniaturized and very effective valveless flow switching system based on the position of capillary column ends in a small transfer tube. The column effluent is not directed by pressure variations but "delivered" to the required column by moving the end of the delivering column close to the column outlet of choice into an auxiliary carrier gas stream as illustrated in Figure 2.172 (Sulzbach, 1991; Sharif et al., 2016; Schlumpberger et al., 2022).

Up to five capillary columns can be installed to a small, a few centimeters long, hollow tip of glass, the "glass dome" which is located inside the GC oven (see Figure 2.172). The column ends are positioned at different locations inside the glass dome. The delivering first column (or pre-column) inlet is movable in position over a length of about 1 cm. One of the installed columns is used for a variable make up flow of carrier gas from a second independent regulation. This make-up gas facilitates the parallel coupling of MS detectors in parallel to classical detectors, olfactometry, or IRMS (isotope ratio mass spectrometry) to prevent the access of the ions source vacuum to the split region. Instead of an additional detector, the

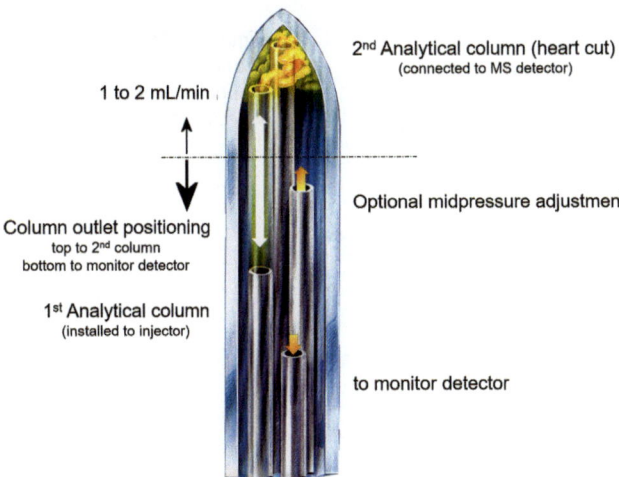

Figure 2.172 Visualization of MCSS glass dome operation. The flow of the movable pre-column is directed to the main column (Courtesy Brechbuehler AG, Schlieren, Switzerland).

additional line can also be used for monitoring the middle pressure between the two columns. Commercial products are available for MDGC covering a variable split range.

The analyte carrier stream fed into the MCSS dome can be split into a variable stream upwards, serving a second analytical column and another stream downwards depending on the position of the delivering column end, for example, to another column outlet, which can lead to a monitor detector. If the delivering column is pushed fully up, the eluate enters the second column and quantitatively transfers the analytes eluted in that position. There is no change of the pressure conditions in the whole switching system during the transfer period. Therefore, the carrier gas flow rates through both columns are always kept stable. Hence, there is no influence of the number or duration of switching processes on the retention times on either column. Analytes can only contact inert glass surfaces, fused silica and the column coating. Furthermore, the system is free of diffusion and maintains the chromatographic resolution of the pre-column. No influences from ambient atmosphere could be observed using ECD, MS or in IRMS detection (Guth, 1996; Horii et al., 2004; Frank and Schieberle, 2022).

The movement of the pre-column is electrically actuated from outside of the GC oven, completely automated and sequence controlled through the time event functions of the GC (see Figure 2.173). The MCSS can be used for several tasks for multi-dimensional heart-cutting GC separations. The range of applications covers the classical MDGC in a single or double oven concept, as well as detector switching, olfactometry, backflushing (BKF), preparative GC and more.

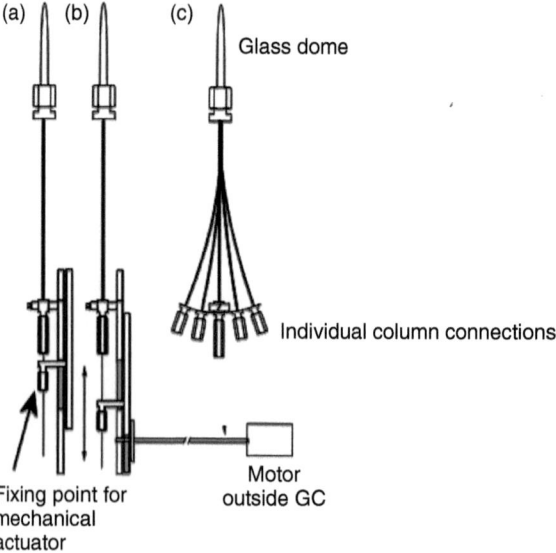

Figure 2.173 MCSS actuation by external stepper motor (a, b side view, c top view).

MCSS Applications

Chiral separations of essential oils (FID/MS)
Classification of fuels (FID)
Matrix exclusion (FID/NPD/ECD/MS)
Sniffing devices in parallel to detectors (MS)
Coplanar PCBs (double ECD/MS/IRMS)
Pesticides (ECD, NPD, FPD)
Preparative sampling (Microprep Trap)
All kinds of GC-IRMS applications
Parallel GC-MS/IRMS detection
and many others.

2.2.12 Classical Detectors for GC-MS Systems

Classical detectors are important for the consideration of GC-MS coupling if an additional selective means of detection is to be introduced parallel to the universal MS detection. The classical detectors can be grouped by the response characteristics into the concentration dependent detectors and the mass flow dependent detectors (Halasz, 1964; Hill and McMinn, 1992). Mass flow detectors respond to the mass of a compound in the detector resulting in a mass/time signal. Concentration detectors respond to the concentration of a compound in the detector volume with a mass/volume signal. For instance, a change of make-up gases does not affect the signal in mass flow dependent detectors (Figure 2.174).

A more application-oriented classification is achieved by detector selectivity. An overview of the most used GC detectors is provided with Table 2.32. Additional and complimentary sample information to MS can be obtained with a parallel detection using element-specific detectors ECD, flamephotometric detectors (FPD), nitrogen-phosphorous detector (NPD) or the atomic emission detector (AED). The inductively coupled plasma MS is coupled to GC as well (GC-ICP-MS), and also in parallel to the here described organic MS, as an element specific 'detector' for the separation and measurement of ultra-trace levels of metals and organo-metallic

Figure 2.174 Detector signal as function of carrier gas flow rate (Adapted from Halasz, 1964).

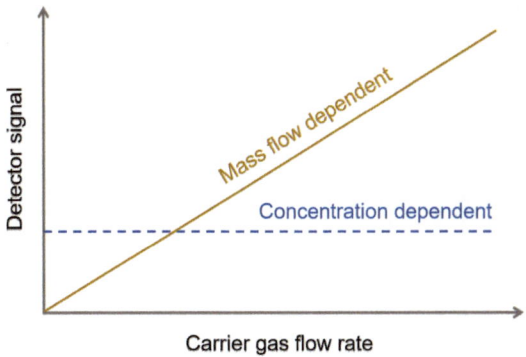

Table 2.32 Important GC detectors – Overview and characteristics.

Detector acronym	Signal depends on	Destructive detection	Response	Characteristics – Limitations	Comments	Gases for operation
AED	Mass flow	Yes	selective	complementary information to MS	parallel to MS with outlet splitter	–
ECD	Concentration	No	selective	can be stacked with FID, NPD	radioactive source*	Ar/CH$_4$, or N$_2$
ELCD	Concentration	No	selective	selectable selectivity		H$_2$O
FID	Mass flow	Yes	universal	but no permanent gases, CCl$_4$, H$_2$O		H$_2$, air (N$_2$ makeup)
FPD	Mass flow	Yes	selective	P and S containing substances		H$_2$, air (N$_2$ makeup)
HID	Concentration	Yes	universal	but not Ne	radioactive source*	He
MS (EI)	Mass flow	Yes	universal	mass and structure selective	online with GC, LC, IC, SFC, direct inlets	CI gases, Ar for MS/MS
NPD	Mass flow	Yes	selective	CN- and P-containing compounds		H$_2$, air (N$_2$ makeup)
O-FID	Mass flow	Yes	selective	oxygen containing compounds, CO, CO$_2$	methanizer in front of FID	H$_2$, air (N$_2$ makeup)
PDD	Concentration	No	universal	three modes of operation		He
PID	Concentration	No	selective	can be stacked with FID, NPD		–
SCD	Mass flow	Yes	selective	complementary information to MS	parallel to MS with outlet splitter	H$_2$, air
TCD	Concentration	No	universal	low sensitivity		Reference is carrier gas
VUV	Mass flow	No	universal	complementary information to MS	parallel to MS with outlet splitter	–

*Shipping, operation, and disposal fall under local governmental regulations.

species (speciation analysis), but not considered as a classical GC detector in the current context (Hu et al., 2017; García-Bellido et al., 2020). Additional conformational analyte information can be obtained using the vacuum ultra violet detector (VUV) detector in parallel. The detection limits which can be achieved with e.g. ECD or a NPD are usually comparable to those attainable using the MS in SIM mode, but different mechanism of selectivity apply. On splitting the carrier gas flow, the ratio of the two parts must be considered when planning such a setup. Normally a larger proportion is passed into the mass spectrometer so that in residue analysis low concentrations of substances do not fall below the detection limit. The use of such flow dividers for quantitative determinations must be checked in an individual case, as a constant split cannot be expected for all BP and flow ranges. A constant split ratio can be attained by using the constant pressure mode. The constant flow mode will change the flow conditions at the split point due to the pressure variation with increasing oven temperature.

Applications can cover the rapid screening, for example, on the intensity of halogenated compounds using an ECD with an intelligent decision for a subsequent MS analysis of the same positively screened sample for a mass selective quantitation. Also, the olfactometric characterization of eluting compounds is a typical application in a flavor and fragrance laboratory parallel to the MS identification.

2.2.12.1 Atomic Emission Detector (AED)

The AED adds multi-element selectivity to the GC detection, not available with any other type of detector (except the isotope fingerprinting with MS, see Section 3.2.6.1).

The AED operation is based on a helium plasma usually microwave induced. The excited elements generate specific emission wavelengths for detection via a holographical diffraction grating and photodiode array. The range of detectable elements is given in Table 2.33. Elements in GC amenable compounds can be determined individual and parallel for a selective quantitative analysis with detection limits in the low pg range (Firor, 1989). The element detection works matrix independent and is typically more sensitive than MS in scan mode. Compared to other GC detectors the AED is more sensitive than the FID for carbon and provides a higher linearity for sulfur than the FPD. For N and P containing compounds a higher selectivity is reported compared to the NPD (JAS, 2009).

Applications cover the environmental analysis, for example, organotin- or organomercury compounds, the food, beverage, or petrochemical analysis for sulfur

Table 2.33 List of AED detectable elements.

Common elements	Carbon, hydrogen, nitrogen, oxygen
Halogens	Bromine, chlorine, fluorine, iodine
Other heteroatoms	Boron, germanium, phosphorus, silicon, sulfur
Metals	Arsenic, iron, lead, manganese, mercury, nickel, selenium, tin, vanadium
Stable isotopes	Carbon-13, nitrogen-15, deuterium

compounds, pesticides, and more (Wylie and Quimby, 1989; van Stee et al., 2002). AED and MS detection provide complementary information.

AED

- Selective, mass-flow dependent detector

Advantages	Highly sensitive element specific detector. Parallel multi-element detection. Complementary information to MS.
Use	For analysis of metal-organic and hetero compounds in food, pharma, environmental and petrochemical applications.
Limitations	Requires high purity He for operation.

2.2.12.2 Electron Capture Detector (ECD)

The operation of the ECD was invented already 1957 by James Lovelock, based on the effect that thermal electrons can be "captured" by electronegative analytes eluting from the analytical column (Lovelock, 1958; Wentworth and Chen, 1981). The ECD consists of an ionization chamber, which contains a nickel plate, on the surface of which a thin layer of the radioactive isotope ^{63}Ni has been applied (100 yrs. half-life, typically 10 to 15 mC, 370 MBq). The makeup gas (N$_2$ or Ar/10% methane) is ionized by the ß-radiation of the ^{63}Ni generating a low energy beta particle spectrum. The free electrons migrate toward the collector electrode and provide the background current of the detector. Substances with electronegative groups reduce the background current by capturing electrons and forming negative molecular ions. The reduction of the background current is proportional to the analyte quantity in the sample. The main reactions in the ECD are dissociative electron capture and electron capture (Table 2.34, see also Section 2.3.1.2 Chemical Ionization). Negative molecular ions can recombine with positive carrier gas ions.

Electron capture is more effective, the slower the electrons move. For this reason, sensitive ECDs are operated using a pulsed DC voltage. By changing the pulse frequency, the current generated by the electrons is kept constant. The pulse frequency thus becomes the actual recorded detector signal. The response rates in the ECD for molecules with different degrees of halogenation are listed in Table 2.35.

Table 2.34 Reactions in the ECD.

Basic reaction	$CG \xrightarrow{\beta} CG^+ + e^-$
Electron capture	$M + e^- \rightarrow M^-$
Dissociative electron capture	$AB + e^- \rightarrow A^0 + B^-$
Recombination	$M^- + CG^+ \rightarrow M^° + CG$

CG, carrier gas. Makeup gas N$_2$ or Ar/10% methane.

Table 2.35 Relative response as of conversion rates in the ECD for molecules with different degrees of halogenation.

Compound	Conversion rate [cm³/mols]	Main product
CH_2Cl_2	1×10^{-11}	Cl^-
$CHCl_3$	4×10^{-9}	Cl^-
CCl_4	4×10^{-7}	Cl^-
CH_3CCl_3	1×10^{-8}	Cl^-
$CH_2ClCHCl_2$	1×10^{-10}	Cl^-
CF_4	7×10^{-13}	M^-
CF_3Cl	4×10^{-10}	Cl^-
CF_3Br	1×10^{-8}	Br^-
CF_2Br_2	2×10^{-7}	Br^-
C_6F_6	9×10^{-8}	M^-
$C_6F_5CF_3$	2×10^{-7}	M^-
$C_6F_{11}CF_3$	2×10^{-7}	M^-
SF_6	4×10^{-7}	M^-
Azulene	3×10^{-8}	M^-
Nitrobenzene	1×10^{-9}	M^-
1,4-Naphthoquinone	7×10^{-9}	M^-

ECD

- Selective, concentration dependent detector
- Non-destructive, can be stacked with FID, NPD
- The ECD reacts with all electronegative elements and functional groups, such as -F, -Cl, -Br, -OCH$_3$ and -NO$_2$ with a high response. All hydrocarbons (generally the matrix) remain transparent, although present.

Advantages	Selectivity for Cl, Br, methoxy and nitro groups.
	Transparency of all hydrocarbons (= matrix).
	High sensitivity.
Disadvantages	Radioactive radiator, therefore local handling authorization necessary.
	Sensitive to misuse.
	Mobile use only under limited conditions.
Use	Typical detector for environmental analysis.
	Ideal for trace analysis.
	Pesticides, OCPs, PCBs, dioxins.
	Volatile halogenated hydrocarbons, freons, nitrous oxides.

(Continued)

(Continued)	
Limitations	Substances with low halogen contents (Cl$_1$, Cl$_2$) only have a low response (Table 2.35).
	Volatile halogenated hydrocarbons with low chlorine levels are better detected with FID or MS.
	Limited dynamic range.
	Multipoint calibration necessary.

2.2.12.3 Electrolytical Conductivity Detector (ELCD)

In the electrolytical conductivity detector (ELCD) the eluate from the GC column passes into a Ni reactor in which all substances are completely oxidized or reduced at an elevated temperature of about 1000 °C. The reaction products dissociate in circulating water which flows through a cell where the conductivity is measured between two Pt spirals. The change in conductivity is the measurement signal. Ionic compounds can be measured without using the reactor, similar the former Hall detector (Piringer and Wolff, 1984; Ewender and Piringer, 1991).

The Hall detector has a comparable function to the ELCD and is a typical detector for packed or halfmil columns of 0.53 mm ID because of the large volume of its measuring cell. Because of the latter significant peak tailing occurs with capillary columns. Special constructions for use with normal bore columns of 0.25 and 0.32 mm ID are available with special instructions.

ELCD

- Selective, concentration dependent detector

Advantages	Can be used for capillary chromatography.
	Selectivity can be chosen for halogens, amines, nitrogen and sulfur.
	High sensitivity at high selectivity.
	Can be used for halogen detection without a source of radioactivity, therefore no authorization necessary.
	Simple calibration, as the response is directly proportional to the number of heteroatoms in the analyte, for example, the proportion of Cl in the molecule.
Use	Environmental analysis, for example, volatile halogenated hydrocarbons, OCPs, PCBs.
	Selective detection of amines, for example, in packaging or foodstuffs.
	Determination of sulfur-containing components.
Limitations	The sensitivity in the halogen mode just reaches that of ECD, so that its use is particularly favorable in association with concentration procedures, such as P&T or thermodesorption.

2.2.12.4 Flame-Ionization Detector (FID)

With the FID, the substances to be detected are burned (oxidized) in a hydrogen flame and are thus partially ionized (Table 2.36). The FID is a mass flow dependent detector. The response of the FID is not affected by the makeup gas flow rate.

For operation, the FID jet is set to a negative potential, hence the produced positive ions are neutralized. The corresponding electrons are captured at a ring-shaped collector electrode to generate the signal current. The electrode is at a potential which is about 200 V more positive than the jet.

Provided that only hydrogen burns in the flame, only radical reactions occur. No ions are formed. If organic substances with C—H and C—C bonds get into the flame, they are first pyrolyzed. The carbon-containing radicals are oxidized by oxygen and OH radicals formed in the flame. The excitation energy leads to the ionization of the oxidation products. Only substances with at least one C—H or C—C bond are detected. The FID response is highest for hydrocarbons and proportional to the number of carbon atoms. Compounds with oxygen, sulfur or a halogen yield smaller responses. Of no response are permanent gases, carbon tetrachloride or water.

Table 2.36 Reactions in the FID.

Pyrolysis	$CH_3^0, CH_2^0, CH^0, C^0$
Exited radicals	O_2^*, OH^*
Ionization	$CH_2^0 + OH^* \rightarrow CH_3O^+ + e^-$
	$CH^0 + OH^* \rightarrow CH_2O^+ + e^-$
	$CH^0 + O_2^* \rightarrow CHO_2^+ + e^-$
	$C^0 + OH^* \rightarrow CHO^+ + e^-$

FID

- Universal, mass flow dependent detector

Advantages	High dynamics up to 10^7, high sensitivity.
	Response is proportional to the number of carbon atoms.
	Robust.
Use	Hydrocarbons, for example, petrochem, fuels, flavor, fragrance, odor substances, BTEX, PAHs, FAMEs, etc.
	Important all-round detector.
Limitations	As it is universal, low selectivity, poor for trace analysis in complex matrices.
	Low or no response for fully oxidized, highly chlorinated or brominated substances, NH_3, N_2.
	Not suitable for permanent gas, except for CO and CO_2 with methanizer (O-FID).

The parallel coupling of an FID does not lead to results which are complementary to those of MS, as both detection processes give practically identical total ion chromatograms. The response factors for most of the organic substances are similar, but not identical.

If a reactor (hydrogenator, methanizer) is connected before the FID, the latter can be converted into an highly sensitive detector for CO and CO_2. The oxygen-specific detector (O-FID) uses two reactors. At first, hydrocarbons are decomposed into carbon, hydrogen and carbon monoxide at above 1300 °C. CO is then converted into methane in the hydrogenation reactor and detected with the FID. With the O-FID, for example, oxygen-containing components in fuels are selectively detected, hydrocarbons give no signals (Schneider et al., 1982; Sironi and Verga, 1995).

2.2.12.5 Flamephotometric Detector (FPD)

Flamephotometric detectors (FPDs) are used as one- or two-flame detectors. In a hydrogen-rich flame, P- and S-containing radicals are in an excited transition state. On passing to the ground state, a characteristic band spectrum is emitted (S : 394 nm, P : 526 nm). The flame emissions initiated the eluting analytes by chemiluminescence are determined using an optical filter and amplified by a photomultiplier.

The analytical capability of the FPD can be expanded by connecting a second photomultiplier tube on the same detector base with a different optical filter, for example, to monitor P and S containing substances in parallel.

FPD

• Selective, mass-flow dependent detector

Advantages	In phosphorus mode, comparable selectivity to FID with high dynamics.
	High selectivity in sulfur mode.
Use	Mostly selective for sulfur and phosphorus compounds e.g., pesticides.
	Detection of sulfur compounds in a complex matrix.
Limitations	Adjustment of the combustion gases important for reproducibility and selectivity.
	In sulfur mode, quenching effect is possible because of too high hydrocarbon matrix (double flame necessary).
	Sensitivity in sulfur mode is not always sufficient for trace analysis.

2.2.12.6 Helium Ionization Detector (HID)

The helium ionization detector (HID) ranks among the most sensitive classical GC detectors. The HID is a universal detector using a radioactive β-emitter to create metastable helium species from the carrier gas. The metastable helium species ionize with an energy of up to 19.8 eV all compounds with lower ionization energy, which covers all organic compounds. The exception is neon with its greater ionization potential of 21.56 eV. It is commonly used to detect gases at the ppb range (Andrawes and Ramsey, 1986).

HID

- Selective, mass flow dependent detector

Advantages	Highly sensitive universal detector for gases and all organic compounds. Requires only He for operation, avoids H_2 with FID for safety reasons.
Use	Mostly used for analysis of gases with no response on FID (NO_x, CO, CO_2, O_2, N_2, and H_2). Complements and can be stacked on TCD for ppb to % range detection.
Limitations	Uses a radioactive source, requires to comply with local regulations. Cannot be used for Ne.

2.2.12.7 Nitrogen-Phosphorous Detector (NPD)

The nitrogen-phosphorous detector basically is a modified FID detector with a rubidium or cesium chloride bead on a Pt wire inside a heater coil, located close to the hydrogen jet and the collector electrode. It is also called the *alkali flame-ionization detector* (AFID). Nitrogen and phosphorus can be selectively detected. The sensitivity for the specific detection of nitrogen or phosphorus can be four orders of magnitude greater than that for carbon.

The alkali beads are heated to red heat both electrically and in the flame and are excited to alkali emission. They are at a negative potential compared with the collector electrode. The heated alkali bead emits electrons by thermionic emissions which are collected at the anode providing the background current of the detector.

In the nitrogen/phosphorous (NP) mode, the hydrogen flow is set just below the minimum required for ignition. The rubidium or cesium bead ignites the hydrogen catalytically, and forms a cold plasma. Excitation of the alkali metal results in the ejection of electrons, which are detected as a background current between anode and cathode. Nitrogen- or phosphorus-containing analytes increase the current. For the detection of phosphorus, the jet is grounded. The electrons emitted by the hydrocarbon parts of the molecule cannot exceed the negative potential of the beads and do not reach the collector electrode. The electrons from the specific alkali reaction reach the collector electrode unhindered (Table 2.37). Since the alkali metal bead is consumed over time, it must be replaced on a regular basis.

Phosphorus-containing substances are first converted in the flame into phosphorus oxides with an uneven number of electrons. Anions formed in the alkali reaction by the addition of an electron are oxidized by OH radicals. The electrons added are released and produce a signal current.

Like phosphorus, nitrogen has an uneven number of electrons. Under the reducing conditions of the flame cyanide and cyanate radicals are formed, which can undergo the alkali reaction (Table 2.38). For this, the input of hydrogen and air is reduced. Instead of the flame the hydrogen burns in the form of a cold plasma around the electrically heated alkali beads.

Table 2.37 Reactions with P compounds in the NPD.

$\overline{O} = \underline{\dot{P}} + A^*$	→	$[\overline{O} = \underline{\overline{P}}]^- + A^+$
$\overline{O} = \dot{P} = \overline{O} + A^*$	→	$[\overline{O} = P = \overline{O}]^- + A^+$
$[\overline{O} = \underline{\overline{P}}]^- + OH°$	→	$HPO_2 + e^-$
$[\underline{\overline{O}} = P = \overline{O}]^- + OH°$	→	$HPO_3 + e^-$
$HPO_3 + H_2O$	→	H_3PO_4

A = alkali.

Table 2.38 Reactions with N compounds in the NPD.

Pyrolysis of CNC compounds	→ $C \equiv N|°$
$CN° + A^*$	→ $CN^- + A^*$
$CN^- + H°$	→ $HCN + e^-$
$CN^- + OH°$	→ $HCNO + e^-$

A = alkali.

In order to form the required cyanide and cyanate radicals, the C—N structure must be present in the molecule. Nitro compounds are detected, but not nitrate esters, ammonia, or nitrogen oxides. By taking part in the alkali reaction the cyanide radical receives an electron. Cyanide ions are formed, which react with other radicals to give neutral species. The electron released provides the detector signal.

NPD

- Specific detector
- Mass flow dependent detector

Advantages	High selectivity and sensitivity.
	Ideal for trace analyses.
Use	Only for CN- and P-containing compounds.
	Plant protection agents.
	Chemical warfare gases, explosives.
	Pharmaceuticals.
Limitations	Additional element specific detector for ECD or MS.
	Quantitative measurements with an internal standard are recommended.
	To some extent, time-consuming optimization of the Rb beads.
	Rb bead needs frequent replacement.

2.2.12.8 Pulsed Discharge Detector (PDD)

The pulsed discharge detector (PDD), aka *pulsed discharge HID* (PDHID), is a universal and highly sensitive nonradioactive and non-destructive detector, also known as *helium photoionization detector*. The PDD can be operated in three different modes: pulsed discharge helium ionization (He-PDPID), pulsed discharge electron capture (PDECD), and helium ionization emission (PDED) (SRI, 2003; Forsyth, 2004). PDDs are based on a pulsed high voltage discharge in helium as its ionization source. The analytes are ionized by photons arising from the transition of diatomic helium to the dissociative ground state, electron capture, or helium ionization.

The response to organic compounds is linear over five orders of magnitude with minimum detectable quantities in the low picogram or femtogram range. The response to permanent gases provides minimum detectable quantities in the low parts per billion range (Li, 2021). The performance of the detector is negatively affected by the presence of impurities in the gas flows (carrier, discharge), therefore, the use of a high quality grade of helium (99.999% pure or better) as carrier and discharge gases is strongly recommended. Because even the highest quality carrier gas may contain some water vapor and other gas impurities, a helium purifier is typically included as part of the detector system.

The PDD detector consists of a quartz cell supplied from the top with ultra-pure helium as discharge gas that reaches the discharge zone consisting of two electrodes connected to a high voltage pulses generator (pulsed discharge module). The eluates from the column, flowing counter to the flow of helium from the discharge zone, are ionized by photons at high energy arising from metastable helium generated into the discharge zone. The resulting electrons are accelerated and measured as electrical signal by the collector electrode. The discharge and carrier gas flows are opposite. For this reason, it is necessary that the discharge gas flow is greater than the carrier gas flow to avoid that the eluates from the column reach the discharge zone with consequent discharge electrode contamination.

The PDD chromatograms show a great similarity to the classical FID detector and offers comparable performance without the use of a flame, radioactive emitter, or combustible gases. The PDD in helium photoionization mode is an excellent replacement for FIDs in petrochemical or refinery environments, where the flame and use of hydrogen can be problematic. In addition, when the helium discharge gas is doped with a suitable noble gas, such as argon, krypton or xenon (depending on the desired cut-off point), the PDD can function as a specific photoionization detector for selective determination of aliphatics, aromatics, amines, as well as other species.

Another very typical application is the area of pure gas analysis. Due to the very high sensitivity, the PDD is able to perform the analysis of impurities in several pure industrial gas mixtures. The availability of a highly sensitive detector (pulse discharge) makes the configured GC system suitable for the determination of compounds in a range of concentrations ranging from hundreds of ppm down to the ppb trace level. The extremely high flexibility and versatility of the system permits, for instance, the characterization of impurities in pure noble gases such as Xe and Kr.

Some PDD detectors also offer an electron capture mode being selective for monitoring high electron affinity compounds such as freons, chlorinated pesticides and other halogen compounds. For such types of compounds, the minimum detectable quantity (MDQ) is at the femtogram or low picogram level. The PDD is similar in sensitivity and response characteristics to a conventional radioactive ECD and can be operated at temperatures up to 400 °C. For operation in this mode, He and CH_4 are introduced just upstream from the column exit.

PDD

- Universal/selective, mass-flow dependent detector

Advantages	Highly sensitive universal detector for gases and all organic compounds. He ionization and electron capture modes. Can replace the radioactive ECD. Requires only He for operation, avoids H_2 with FID for safety reasons.
Use	Replacement for FIDs in petrochemical or refinery environments. Analysis of gases with no response on FID NO_x, CO, CO_2, O_2, N_2, and H_2 Impurities in pure noble gases.
Limitations	Requires high purity He for operation.

2.2.12.9 Photo Ionization Detector (PID)

The first photo ionization detector (PID) Model PI51 was developed by HNU Systems Inc. and introduced in 1976 at the Cleveland, OH, Pittsburgh Conference. The PID operates on the principle of the photon energy absorption emitted by a UV lamp in the detector housing. The absorbed energy leads to the ionization of the molecule by the release of an electron from the excited molecule M* according to Eq. (2.26).

$$M + h \cdot v \rightarrow M^* \rightarrow M^+ + e^- \qquad (2.26)$$

The high selectivity of the PID is based on the different energy levels of the emitted emission lines by the installed UV lamps, see Table 2.39. The different types of lamps are filled either with argon, hydrogen, krypton or with other gases to cover an ionization energy range from 8.4 to 11.8 eV. Ionization occurs only if the ionization potential (IP) of the molecules M is lower than the energy level of the emitted

Table 2.39 Available PID lamp types and applications.

Lamp type [eV]	Typical applications
8.4	Amines, PAH
9.6	Volatile aromatics, BTEX
10.2	General application
11.8	Aldehydes, ketones

UV bands. The number of ions produced is proportional to the absorption coefficient of the molecule and the intensity of the lamp. As the PID is selective and non-destructive it is often found in tandem configuration with the universal FID detector providing comprehensive chromatograms.

The UV lamp is usually removable for exchange to applications requiring different selectivity. While the lamp housing is kept relatively cool (<100 °C), the quartz ionization chamber containing the two measurement electrodes is the heated part of the PID (>300 °C). It is closed by a quartz or alkalifluoride window for entry of UV light.

The PID is mainly used for the analysis of aromatic pollutants, for example, BTEX or PAH, or halogenated compounds in environmental applications. Many U.S. EPA methods, for example, 602, 502 or 503.1 cover priority pollutants in surface or drinking water. Typical for the PID is the detection of impurities in air, also with mobile

Table 2.40 Ionization potentials of selected analytes.

Substance	Ionization potential [eV]
Helium	24.59
Argon	15.76
Nitrogen	15.58
Hydrogen	15.43
Methane	12.98
Ethane	11.65
Ethylene	11.41
2-Chlorobutane	10.65
Acetylene	10.52
n-Hexane	10.18
2-Bromobutane	9.98
n-Butyl acetate	9.97
Iso-Butyraldehyde	9.74
Propene	9.73
Acetone	9.69
Benzene	9.25
Methyl isothiocyanate	9.25
N,N-Dimethylformamide	9.12
2-Iodobutane	9.09
Toluene	8.82
n-Butylamine	8.72
o-Xylene	8.56
Phenol	8.50

detectors. This is due to the fact that the energy of the UV lamp is sufficient to selectively ionize the majority of organic air contaminants, but insufficient to ionize the air components oxygen, nitrogen, water, argon and carbon dioxide.

PID

- Universal and selective, non-destructive, concentration-dependent detector
- The energy-rich radiation of a UV lamp ionizes the substances to be analyzed selectively, depending on the energy content (Table 2.40); measurement of the overall ion flow.

Advantages	The selectivity can be chosen, see Figure 2.175.
	High sensitivity.
	No additional gas supply required.
	Robust, no maintenance required.
	Ideal for mobile and field analyses.
Use	Field and laboratory analysis of VOCs, BTEX, PAHs, and more.
Limitations	Only for substances with low IPs ($\leq$ 10 eV) as of Table 2.40, also consult manufacturers' data.

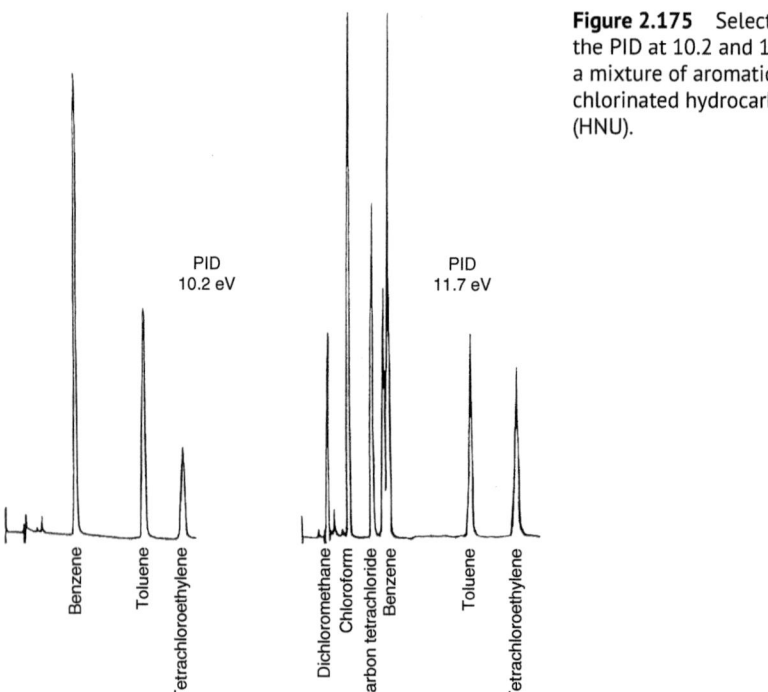

Figure 2.175 Selectivity of the PID at 10.2 and 11.7 eV for a mixture of aromatic and chlorinated hydrocarbons (HNU).

2.2.12.10 Sulfur Chemiluminescence Detector (SCD)

The sulfur chemiluminescence detector (SCD) is sensitive only to sulfur compounds and can selectively detect trace amounts of sulfur compounds in complex sample matrixes.

Compounds eluting from the GC column pass a hydrogen fueled redox cell at about 800 °C. The generated SO radicals transfer into a reaction chamber for collision with ozone from a built-in ozone generator to obtain sulfur dioxide in an excited state. The chemiluminescence is detected when it returns to ground state again. The emitted light in the range of 300 to 400 nm is detected by a photomultiplier. Important with the SCD is its higher sulfur selectivity and the sensitivity with a wider dynamic range compared to the FPD. The response of the SCD is equimolecular and very much independent of the molecular structure of compounds. Hence sulfur compounds can be quantified with just one standard substance. High sensitivities in the low ppb level are reached.

Important applications of the SCD are often reported in parallel to MS detection. Due to the high sensory impact of volatile sulfur compounds (VSCs) especially in food and beverages high detection selectivity and sensitivity is needed (Wiedemer et al., 2023). Characterizing sulfur compounds are often not identified in GC-MS due to their low levels and missing element selectivity (Ochiai et al., 2012; Kawamura et al., 2020). Highly suitable for VSC detection is the combination with automated extraction methods like HS, SPME or DHS for food or environmental applications (Sun et al., 2014; Yu, 2022). Other GC-SCD applications are found in the petrochemical and natural gas industries.

SCD

- Selective, mass-flow dependent detector

Advantages	Highly sensitive sulfur specific detector. Complementary information to MS.
Use	Food and beverage industry for quality control. Environmental applications mainly water and waste water. Petrochemical application.
Limitations	Outlet splitter required for parallel MS detection.

2.2.12.11 Thermal Conductivity Detector (TCD)

The thermal conductivity detector (TCD), aka "*katharometer*" or "*hot wire detector*", usually does not find application in parallel to MS. Reason is the low sensitivity and its main application in gas analysis, which is not the focus of GC-MS.

The TCD consists of filaments in a Wheatstone Bridge with reference channel and the column flow. The difference in conductivity of both channels is monitored. The TCD response is independent on the compound structure. Important note here, the high thermal conductivity of hydrogen compared to helium or nitrogen as possible

Table 2.41 Specific thermal conductivities for gases and solvents.

Compound	Thermal Conductivity [10^5 cal/cm·s·°C]
Hydrogen	49.93
Helium	39.85
Nitrogen	7.18
Ethane	7.67
Water	5.51
Benzene	4.14
Acetone	3.96
Chloroform	2.33

carrier gases makes it a well-suited application along with organic and permanent gases like methane, carbon monoxide, argon, or oxygen (see Table 2.41). Typical applications of the TCD cover gas analysis, but also the determination of water content in solvents and pharmaceuticals (Jayawardhana et al., 2012).

A parallel detection to MS with a thermal conductivity detector is not found often in practice as the mass spectrometric analysis of gases is mostly carried out with specially configured MS systems (RGA, residual gas analyzer, mass range < 100 Da).

TCD

- Universal, non-destructive, concentration dependent detector

Advantages	Most suitable for the analysis of permanent gases Ar, O_2, N_2, CO_2, possible in tandem with FID.
	No additional gas for operation required. Carrier gas can be the reference gas.
Use	Gas mixture and purity analysis, CO_2 contents in beverages, water/moisture contents in solvents and pharmaceutical, petrochemical industry.
Limitations	Moderate sensitivity, not suitable for trace analysis or parallel to MS.

2.2.12.12 Vacuum Ultra Violet Detector (VUV)

The most recent development for GC detection providing complementary data to MS is the VUV. It uses a deuterium lamp with the wavelength range below the classical UV. In the so called "Vacuum UV" (VUV) wavelength range of 100 to 200 nm nearly all organic compounds absorb except He, Ar, and H_2 (Figure 2.176). The high energy radiation induces $\sigma \rightarrow \sigma^*$ and $\pi \rightarrow \pi^*$ electron transitions in most chemical

Figure 2.176 VUV detector wavelength and analyte absorption ranges.

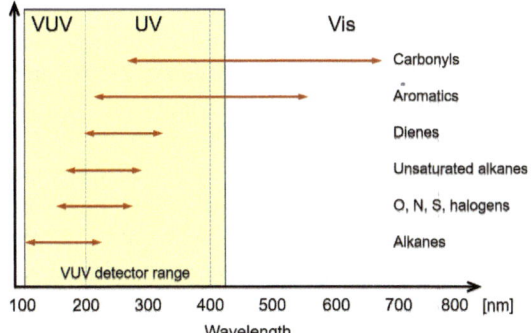

bonds and provide unique spectral fingerprints (Schug et al., 2014). The VUV detector works in the range of 125 nm up to 430 nm for a universal detection delivering three-dimensional data of time, absorbance, and wavelength specific to chemical structures. The strong absorption in the gas phase provides excellent sensitivity in the low to mid pg range, and compound-specific absorption spectra with high selectivity. The acquired absorbance spectra are detected by a CCD device and are specific to the analyte electronic structure and functional group arrangement. The VUV detector works with carrier gases He, H_2, or N_2 in a temperature range up to 430 °C compliant even with high temperature GC applications. Flow cells are available with 10 cm pathlength and 40 or 80 µL cell volume depending on the instrument type. For quantitation a linear dynamic range of more than four orders of magnitude can be reached.

VUV

- Selective, mass-flow dependent detector

Advantages	Highly sensitive analyte structure specific detector.
	Complementary information to MS.
	VUV spectra library and library search.
	Spectra deconvolution.
	No auxiliary gases required for operation.
Use	Universal use.
	Strong for isomer discrimination in food, drug, flavor, and fragrance analysis.
	Petrochemical applications with ASTM approval.
Limitations	No inline installation with MS recommended.
	Outlet splitter for parallel detection required.

The unique compound VUV spectra are stored in spectrum libraries. The spectral fingerprint of unknowns can then be matched with the spectrum library for identification, comparable to MS library searches. A strong feature of the unique VUV spectra is the differentiation of isomeric structures which cannot be distinguished by MS due to the similarity in their mass spectra. The deconvolution of data allows the individual quantitation of isomers complimentary to MS (Schug, 2017). For the parallel installation of the VUV detector to MS an open split and flow division is recommended (see Section 2.2.11.6 and 2.2.12.14). The in-line installation is dissuaded due to the vacuum impact of the MS to the VUV flow cell.

A broad range of industries such as food and beverage (Fan *et al.*, 2016), environmental, forensics (Thomas *et al.*, 2023), fragrances and flavors, or life science use the parallel VUV detection complimentary to MS. It found ASTM approval for the established petrochemical applications of finished gasoline analysis (ASTM D8071), the jet fuel (ASTM D8267), Diesel (ASTM 8368), and verified hydrocarbon (ASTM 8369) analyses.

2.2.12.13 Olfactometry

GC-Olfactometry (GC-O) also called the "sniffing device" describes techniques that allow the human nose to detect and evaluate compounds eluting from a GC separation (Delahunty *et al.*, 2006). Assessors sniff the eluate from a specifically designed odor port parallel to FID or MS detection. GC-O applications have become common not limited to the food and flavor industry to assign specific flavor characteristics to each of the volatile compounds identified. The human nose plays the role of the detector. However, the human nose is often more sensitive than any physical detector, and GC-O exhibits supplementary capabilities that can be applied to any fragrant product.

Olfactometry techniques can be classified into two categories: dilution methods, which are based on successive dilutions of an aroma extract until no odor is perceived at the sniffing port of the GC, and the intensity methods, in which the aroma extract is injected and the assessor records the odor intensity and perception as a function of time. The technical solution is straightforward with a split at the end of a chromatographic column and a heated transfer line to a GC external sniffer port. The eluting compounds are splitted, for example, 1:50 to an FID or MS detector and the sniffing port. The column effluent is combined at the sniffer port with a laminar stream of inert make-up gas, which is heated to a constant temperature and additionally humidified.

For data recording, the peak to its scent correlation is documented by specialized fragrance chemists. Different technical solutions are available to note the odor intensity and characteristics. Usually, an additional channel on an existing data system is connected. Figure 2.177 shows as example the aromagram of a hop essential oil sample with the olfactometry response compared to the FID detection. A strong odor perception can be noted with peaks of only low classical detector response (Plutowska and Wardencki, 2009). Compound identification starts from here using retention index information and full scan MS with library search and classical spectrum interpretation tools.

2.2.12.14 Classical Detectors Parallel to the Mass Spectrometer

Often the parallel detection with a classical detector to MS is desired to obtain comprehensive information on substance identification. In principle, there are two possibilities for operating another detector in parallel. The sample can be already divided

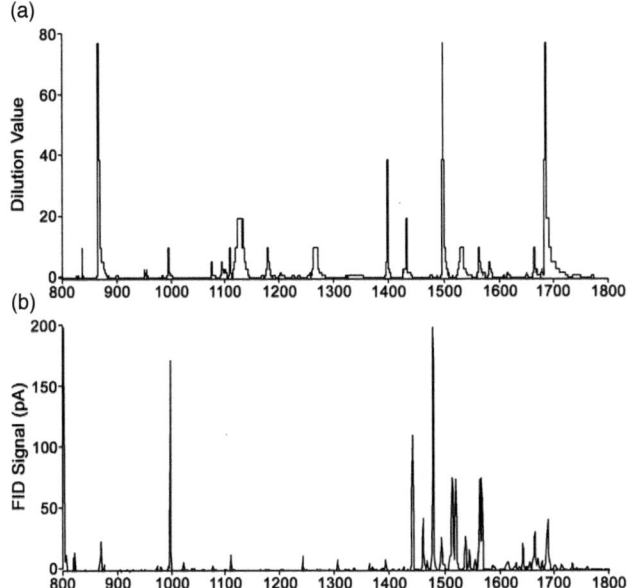

Figure 2.177 Comparison of (a) the GC-O aromagram with (b) the FID chromatogram of a hop essential oil sample (Delahunty, 2006/John Wiley & Sons).

in the injector and pass through two identical columns. The second solution is the split of the eluate at the end of the column. In general, it is advisable due to the constant high transferline temperatures to use a deactivated piece of column (aka *guard column*) as connection from the split point to the ion source. This also allows the use of polar column phases with lower maximum temperatures, reduces the mass spec background of silicon column bleed, and facilitates the exchange of columns without venting the spectrometer vacuum system (ion source off and cool down required!). An easy solution could be the fixation of a mini union with a bracket inside of the GC oven with a permanent guard column of c.1 m length connected to the transfer line.

The division of the sample on to two identical columns can be realized easily and carried out very reliably, but different retention times need to be considered due to the vacuum impact on the MS side if a direct coupling is used for this column. An open split to the transfer line can solve this effect. Since the capacities of the two columns are additive, the quantity of sample injected can be adjusted in order to make use of the sensitivity range of the MS. Two standard columns can be installed for most injectors without further adaptation. In the simplest case, the connection can be made using ferrules with two holes. It must be ensured that there is a good seal and the installation height in the inlet liner is the same. A better connection involves an adaptor piece with a separate screw-in joint for each column. With this construction the independent positioning of the columns in the injector is possible. Suitable adaptors can be obtained for all common injectors.

In GC-MS, the division of the flow at the end of the column requires a considerable higher effort because the direct coupling of the branch to the mass spectrometer causes reduced pressure at the split point. For this reason, the use of a simple Y piece is seldom possible with standard columns. The consequence would be a reversed flow through the detector connected in parallel caused by the wide vacuum impact of the MS into the column. The effect is equivalent to a leakage of air into the MS. A split at the end of the column must therefore be carried out using a makeup gas, and calculate length and diameter of the transfer capillaries as of the required split ratio. Precise split of the eluent is possible, for example, using the glass cap cross divider (Figure 2.178) or the MCSS of similar operation as shown in Figure 2.172. Here, the column, the transfer capillaries to the mass spectrometer, the parallel detector and a makeup gas inlet all meet. By choosing the internal diameter and the position of the end of the column in the glass cap (also known as the *glass dome*), the ratio of the split can easily be adjusted. The advantage of this solution lies in the free choice of column so that small internal diameters with comparatively low flow rates can also be used (Bretschneider and Werkhoff, 1988a,b). The installation of the glass cap cross divider is shown in Figure 2.179 as inlet divider to split the sample on a polar and non-polar column. It is used also as outlet divider in a similar setup replacing the shown pre-column with an analytical column and short transfer columns to the detectors. Important to note here the operation in constant pressure mode.

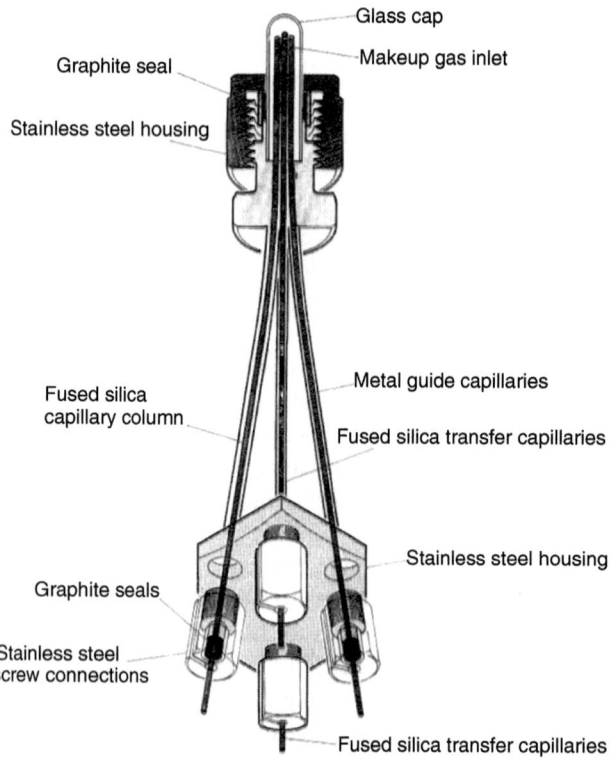

Figure 2.178 Flow divider (Werkhoff Splitter, Glass Cap Cross).

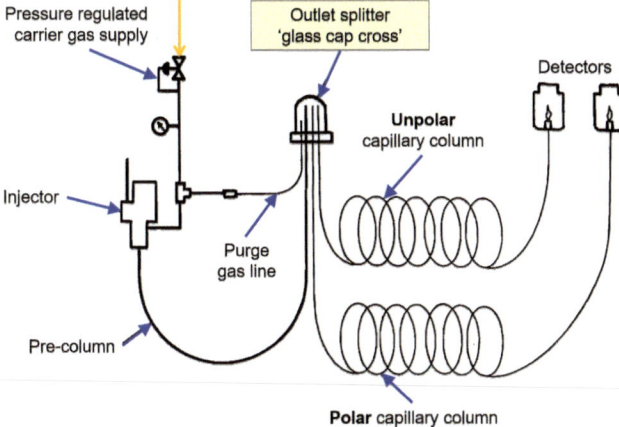

Figure 2.179 Installation and flow diagram of the glass cap cross as inlet splitter (Bretschneider and Werkhoff, 1988b/John Wiley & Sons).

The calculation of gas flow rates through outlet splitters with fixed restrictors is desirable to have a means of estimating the rate of gas flow through a length of fixed restrictor tubing of specified dimensions. Conversely, for other applications it is necessary to estimate the length and ID of restriction tubing required to yield a desired flow rate at a specified head pressure. Within limits, the following calculation known as the Hagen-Poiseulle law for a laminar flow can be used for this purpose.

$$V = \frac{\pi \cdot P \cdot r^4}{8 \cdot L \cdot \eta}$$

with:

V = Volumetric gas flow rate [cm³/min]
P = Pressure differential across the tube [dyne/cm²] (1 psi = 68947.58 dyne/cm²)
R = Tube radius [cm]
L = Tube length [cm]
η = Gas viscosity, poise [P], [dyne-s/cm²]

One of the obvious limitations to this calculation is the critical nature of the radius measurement. Since this is a fourth power term, small errors in the ID measurement can result in relatively large errors in the flow rate. This becomes more critical with smaller diameter tubes (i.e. less than 100 µm). A 100× microscope is a convenient tool for determining the exact ID of the used restrictor capillary. The gas viscosity values can be determined from the graph of gas viscosity against temperature for hydrogen, helium and nitrogen (Figure 2.180), or internet-based gas viscosity calculators.

2.2.12.15 Microchannel Devices

The microchannel or microfluidic devices used for flow switching in GC are recent developments which became available due to progresses made in precise laser machining capabilities of metal films and appropriate metal surface deactivation. The microchannels devices are laser cut into metal sheet (shims) with

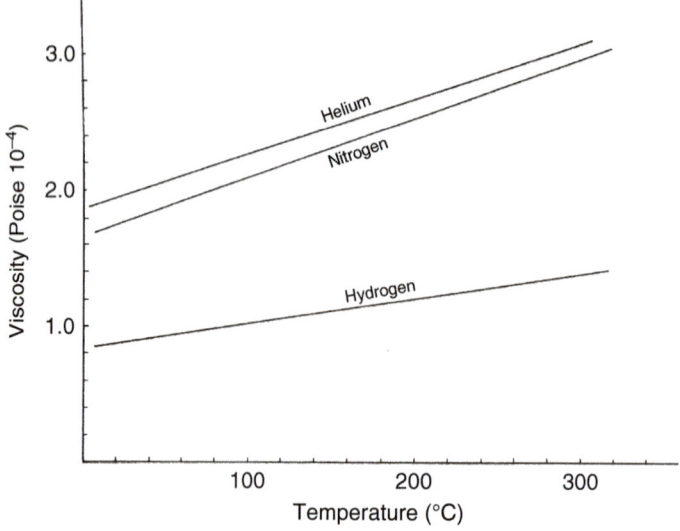

Figure 2.180 Graphs to determine viscosity for hydrogen, helium and nitrogen.

thicknesses from 20 to 500 µm. The resulting channel dimensions are similar to the conventionally fused silica capillary columns, avoiding dead volumes and peak broadening. Shim stacks are bonded together with top and bottom plates for column connections, having low thermal mass for fast GC oven ramp rates. The outer plates are thicker (usually 500 µm) and carry the connection sockets for fused silica analytical or transfer columns. This microchannel concept based on shim stacks allows devices to be designed for instance ideally suitable for flow splitting to different detectors (see Figure 2.181), flow switching such as heart-cutting or the Deans switch applications, also enabling custom designs to be developed for special analytical tasks (Agilent, 2013). The microchannel devices, aka *capillary*

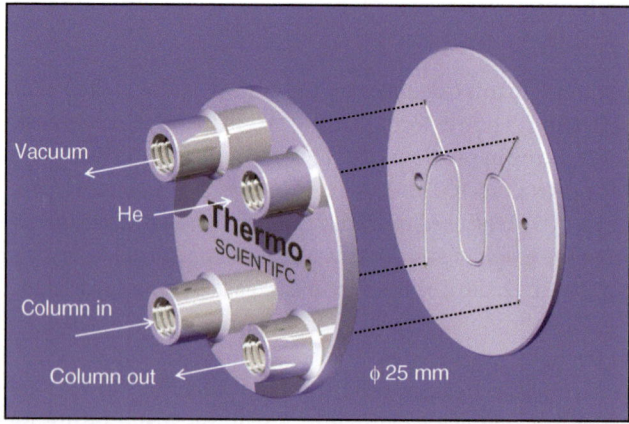

Figure 2.181 Microchannel switching device used for GC-MS detector flow switching (DFS, Dual Data flow switch, Heinz et al., 2016/Thermo Fisher Scientific Inc.).

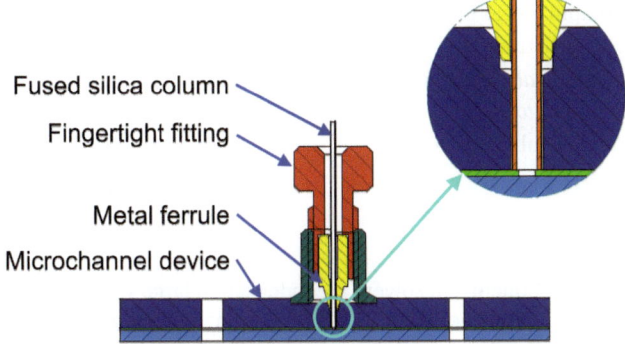

Figure 2.182 Microchannel disk finger tight connection system using SilTite metal ferrules (Courtesy Trajan Scientific and Medical).

flow technology by Agilent Technologies Inc., allows the purged (diluted) split to multiple parallel elemental specific detectors to MS, the backflushing of high boiling compounds from the column for shorter cycle times, or the convenient column change without venting the MS (Quimby et al., 2007; Mezcua et al., 2009).

The metal surface deactivation provides flow channels that are inert even to reactive analytes at trace levels. The inertness achieved is substantially better than can be achieved on conventional inlet liner glass surfaces and is similar to the inertness of deactivated fused silica columns.

The column connection system is special. The microchannels can be blocked if particulate material is allowed to enter the flow path. Graphite ferrules as well as polyimide (Vespel™) ferrules must never be used with microchannel devices. Metal ferrules (e.g. SilTite™) must be used, see Figure 2.182. The metal ferrules have been demonstrated to be extremely reliable with gas-tight seals suitable for MS. Because the thermal coefficient of expansion of the metal ferrules matches the coefficient of thermal expansion of the housing, the seal does not leak with the GC oven temperature cycling unlike polymer ferrules do. The miniature metal ferrules are designed to be used with finger force only on a knurled fitting. Wrenches should never be used. If the finger force only requirement is followed, the ferrules cannot be overtightened avoiding damage to the fitting or tubing.

References

Section 2.2.1 Sample Inlet Systems

Desty, D.H. (1991) A personal account of 40 years of development in chromatography. *LC-GC Int.*, **4** (5), 32.

Pretorius, V. and Bertsch, W. (1983) Sample introduction in capillary gas-liquid chromatography. *J. High Res. Chrom.*, **6** (2), 64–67. doi: 10.1002/jhrc.1240060203.

Schomburg, G., Behlau, H., Dielmann, R., Weeke, F., and Husmann, H. (1977) Sampling techniques in capillary gas chromatography. *J. Chromatogr.*, **142**, 87–102. doi: 10.1016/S0021-9673(01)92028-X.

Schomburg, G. (1983) *Probenaufgabe in der Kapillargaschromatographie. Labo*, **7**, 2–6.

Schomburg, G. (1987) *Gaschromatographie – Grundlagen – Praxis – Kapillartechnik*, 2nd edn, Wiley-VCH Verlag GmbH, Weinheim.

Section 2.2.2 Carrier Gas Regulation

Magni, P. and Porzano, T. (2003) Concurrent solvent recondensation large sample volume splitless injection. *J. Sep. Sci.*, **26**, 1491–1498. doi: 10.1002/jssc.200301553.

Warner, N.A., Ladak, A., Cavagnino, D., and Kutscher, D. (2022) Uncompromised Sensitivity in Polychlorinated Dibenzo-p-Dioxins/Furans Analysis Using Triple Quadrupole GC-MS with Cost-Effective Helium-Saving Technology. Application Note AN001594, Thermo Fisher Scientific, Bremen, Germany.

Warner, N.A., Riccardino, G., Ladak, A., and Cavagnino, D. (2022) Polyaromatic Hydrocarbon Analysis by Gas Chromatography Mass Spectrometry Using HeSaver-H2 Safer Technology. Product Spotlight 001606, Thermo Fisher Scientific, Bremen, Germany.

Section 2.2.3 Injection Port Septa

de Zeeuw, J. (2013) Ghost peaks in gas chromatography part 2: the injection port. *Sep. Sci. Asia Pacific*, **5**, 2–7.

Gerstel (2022) Cooled Injection System CIS. *Product brochure*, GERSTEL GmbH & Co.KG, Mülheim an der Ruhr, Germany.

Grossman, S. (2013) How Hot Is Your Septum? Restek Technical Resource, *www.restek.com* (accessed December 2013).

Hinshaw, J. (2008) Do's and don'ts. *LCGC Eur*, **5**, 332.

Lidgett, D., and Grossman, S. (2015) Preventing Septum Problems. Restek Technical Resource, *https://www.bgb-info.com* (accessed 8 March 2024).

Merlin (2018) *Microseal Users Manual*, Merlin Instrument Company, Half Moon Bay, CA, USA, *https://merlinic.com/wp-content/uploads/2019/06/2018-Microseal-Manual.pdf* (accessed 11 March 2024).

Warden, J. and Pereira, L. (2013) An evaluation of the performance of GC Septa in different commonly used procedures. ASMS Poster, Thermo Fisher Scientific, Runcorn.

Westland, J., Organtini, K., and Dorman, F. (2012) Evaluation of lifetime and analytical performance of gas chromatographic inlet septa for analysis of reactive semivolatile organic compounds. *J. Chromatogr. A.*, **1239**, 72–77.

Section 2.2.4 Injection Port Liner

Agilent (2021) GC Inlet Resource Guide. Product Information, Agilent CrossLab, 5994-2912EN, Agilent Technologies, Inc.

Grob, K. (1994) Injection techniques in capillary GC. *Anal. Chem.*, **66** (20), 1009 A–1019 A. doi: 10.1021/ac00092a716.

Hinshaw, J.V. (1992) Splitless injection: corrections and further information. *LCGC Intl.*, **5** (3), 20–22.

Klee, M. (2013) Liner deactivation. *Sep. Sci. GC. Solutions*, **39**, 1–3.

Pack, E. (2023) Wool vs fritted liners – which is better for controlling dirty matrix? *Blog, Sep*, **28**, 2023, *https://www.restek.com/global/de/chromablography/wool-vs-fritted-liners-which-is-better-for-controlling-dirty-matrix* (accessed 3 June 2024).

Rood, D. (2007) *The Troubleshooting and Maintenance Guide for Gas Chromatographers*, Wiley-VCH Verlag GmbH & Co, KGaA, Weinheim.

Restek (2002) Operating Hints for Using Split/Splitless Injectors. Technical Guide, Restek Corp, Bellefonte, PA, USA.

Thermo Scientific (2008) *AquaSil and SurfaSil Siliconizing Fluids. Instructions*, Thermo Fisher Scientific Inc.

Waclaski, L. (2018) Optimizing splitless GC injections. *The Column*, **14** (8), 11–15, *https://www.chromatographyonline.com/view/optimizing-splitless-gc-injections-0*.

Section 2.2.5 Hot Split/Splitless Sample Injection Techniques

Anastassiades, M., Lehotay, S.J., Stajnbaher, D., and Schenk, F.J. (2003) Fast and easy multiresidue method employing acetonitrile extraction/partitioning and 'dispersive solid phase extraction' for the determination of pesticide residues in produce. *J. AOAC Int.*, **86** (2), 412–431.

Biedermann, M., Fiscalini, A., and Grob, K. (2004) Large volume splitless injection with concurrent solvent recondensation: keeping the sample in place in the hot vaporizing chamber. *J. Sep. Sci.*, **27**, 1157–1165. doi: 10.1002/jssc.200401847.

DIN EN 12393-2 (2014) *Foods of Plant Origin - Multiresidue Methods for the Determination of Pesticide Residues by GC or LC-MS/MS - Part 2: Methods for Extraction and Cleanup*, European Standards, *http://www.en-standard.eu*.

Grob, K. and Grob, K., Jr. (1978a) Splitless injection and the solvent effect. *HRC & CC*, **1**, 57–64.

Grob, K. (1994) Injection techniques in capillary GC. *Anal. Chem.*, **66**, 1009A–1019A.

Grob, K. (1995) *Einspritztechniken in der Kapillar-Gaschromatographie*, Huethig, Heidelberg.

Grob, K. (2001) *Split and Splitless Injection for Quantitative Gas Chromatography, in Concepts, Processes, Practical Guidelines, Sources of Error*, 4th edn, Wiley-VCH Verlag GmbH, Weinheim.

Hetmanski, M.T. (2009) High sensitivity multi-residue pesticide analyses in foods using the TSQ Quantum GC-MS/MS. FERA, York. Presented at the RAFA conference, November 2009, Prague.

Hildmann, F., Kempe, G., and Speer, K. (2013) Application of the precolumn back-flush technology in pesticide residue analysis: a practical view. *J. Sep. Sci.*, **36**, 2128–2135.

Hinshaw, J.W. (1992) Splitless injection: corrections and further information. *LCGC Int.*, **5**, 20–22.

Magni, P. and Porzano, T. (2003) Concurrent solvent recondensation large sample volume splitless injection. *J. Sep. Sci.*, **26**, 1491–1498.

Munari, F., Magni, P., and Facchetti, R. (2000) Improvements in PTV large volume injection and MS detection using a reverse flow injection device, in *Proceedings of the ISCC 2000 Conference, Riva del Garda, Italy*.

Müller, S., Efer, J., Wennrich, L., Engewald, W., and Levsen, K. (1993) Gaschromatographische Spurenanalytik von Methamidophos und Buminafos im Trinkwasser - Einflußgrößen bei der PTV-Dosierung großer Probenvolumina. *Vom Wasser*, **81**, 135–150.

Schürmann, A., Crüzer, C., Duss, V., Kämpf, R., Preiswerk, T., and Huebschmann, H.-J. (2023) Automated micro solid phase extraction clean-up and gas chromatography-tandem mass spectrometry analysis of pesticides in foods extracted with ethyl acetate. *Anal. Bioanal. Chem.*, **416**, 689–700. doi: 10.1007/s00216-023-05027-5.

Westland, J. (2017) Advantages of Reversed Sandwich Injection for Pesticide Residue Analysis. Application Note 5991-7973E, Agilent Technologies, Inc.

Section 2.2.6 Temperature Programmable Injectors

Bergna, M., Banfi, S., and Cobelli, L. (1991) The use of a temperature vaporizer as preconcentrator device in the introduction of large amount of sample, in *12th International Symposium on Capillary Chromatography Riva del Garda* (eds P. Sandra and G. Redant), A. Huethig, pp. 300–309.

Blanch, G.P., Ibanez, E., Herraiz, M., and Reglero, G. (1994) Use of a programmed temperature vaporizer for off-line SFE/GC analysis in food composition studies. *Anal. Chem.*, **66**, 888–892.

Buchgraber, M., Ulberth, F., and Anklam, E. (2004) Interlaboratory evaluation of injection techniques for triglyceride analysis of cocoa butter by capillary gas chromatography. *Chrom. A*, **1036** (2), 197–203. doi: 10.1016/j.chroma.2004.03.011.

Donike, M. (1973) Die temperaturprogrammierte Analyse von Fettsäuremethylsilylestern: Ein kritischer Qualitätstest für gaschromatographische Trennsäulen. *Chromatographia*, **6**, 190.

Färber, H., Peldszus, S., and Schöler, H.F. (1991) Gaschromatographische Bestimmung von aziden Pestiziden in Wasser nach Methylierung mit Trimethylsulfoniumhydroxid. *Vom Wasser*, **76**, 13–20.

Färber, H. and Schöler, H.F. (1991) Gaschromatographische Bestimmung von Harnstoffherbiziden in Wasser nach Methylierung mit Trimethylaniliniumhydroxid oder Trimethylsulfoniumhydroxid. *Vom Wasser*, **77**, 249–262.

Färber, H. and Schöler, H.F. (1992) Gaschromatographische Bestimmung von OH- und NH-aziden Pestiziden nach Methylierung mit Trimethylsulfoniumhydroxid im "Programmed Temperature Vaporizer" (PTV). *Lebensmittelchemie.*, **46**, 93–100.

Gerstel (2016) Thermal Desorption Unit TDU 2. Product information. Gerstel GmbH, Mühlheim an der Ruhr, Germany.

Gerstel (1998) in *Sample Introduction Techniques for Capillary Gas Chromatography. Monograph* (ed. H.-G. Janssen), pp. 1–62.

Gerstel (2022) Cooled Injection System CIS. Product information. Gerstel GmbH, Mühlheim an der Ruhr, Germany.

GL Sciences (2015) CryoFocus-4 - Cryogenic Trapping System for Gas Chromatography. Product information, GL Sciences BV., Eindhoven, The Netherlands.

Grob, K. (1986) *Classical Split and Splitless Injection in Capillary Gas Chromatography: With some Remarks on PTV Injection*, A. Huethig, Heidelberg - New York, ISBN 3778511424.

Hoh, E. and Mastovska, K. (2008) Large volume injection techniques in capillary gas chromatography. *J. Chromatogr. A*, **1186**, 2–15.

Japp, M., Vinall, M., and Osselton, M.D. (1991) Assessment of a programmable temperature vaporizing (PTV) injector. *Europ. Chrom. Anal.*, (February), 5–7.

Knapp, D.R. (1979) *Handbook of Analytical Derivatization Reactions*, John Wiley & Sons, Inc., New York.

Kolahgar, B. and Pfannkoch, E. (2002) The Use of Different PTV Inlet Liner Types for Trapping Alkanes, Aromatics and Oxygenated Compounds During Thermal Desorption. Technical Note 3, Gerstel GmbH, Mülheim an der Ruhr, Germany.

Kolb, B., Liebhardt, B., and Ettre, L.S. (1986) Cryofocusing in the combination of gas chromatography with equilibrium headspace sampling. *Chromatographia*, **21** (6), 305–311. doi: 10.1007/BF02311600.

Krumwiede, D., Munari, F., and Münster, H. (2008) The Application of Large Volume Injection Techniques for Increased Productivity and Sensitivity in Routine POPs Analysis with GC-HRMS and TripleQuad GC-MS. *Organohalogen Compd.*, **70**, 466–468.

Mol, H.G.J., Janssen, H.G.M., and Crainers, C.A. (1993) Use of a temperature programmed injector with a packed liner for direct water analysis and on-line reversed phase LC-GC. *J. High Res. Chrom.*, **16**, 459–463.

Mol, H.G.J., Janssen, H.G.M., Cramers, C.A., and Brinkman, K.A.T. (1994) Large volume sample introduction using temperature programmable injectors - implication of line diameter, in *16th International Symposium on Capillary Chromatography Riva del Garda* (ed. P. Sandra), A. Huethig, pp. 1124–1136.

Moreira Novaes, F.J. and Marriott, P.J. (2021) Cryogenic trapping as a versatile approach for sample handling, enrichment and multidimensional analysis in gas chromatography. *J. Chrom. A*, **1644** (May), 462135.

Müller, H.-M. and Stan, H.-J. (1989) Pesticide residue analysis in food with CGC. Study of long-time stability by the use of different injection techniques, in *12th International Symposium on Capillary Chromatography Riva del Garda* (eds P. Sandra and G. Redant), Huethig, pp. 582–587.

Pierce, A.E. (1982) *Silylation of Organic Compounds*, Pierce Chemical Company, Rockford, Ill.

Pigozzo, F., Munari, F., and Trestianu, S. (1991) Sample transfer from head space into capillary columns, in *Proceedings of the ISCC Conference, Riva Del Garda, Italy*, pp. 409–416.

Saravalle, C.A., Munari, F., and Trestianu, S. (1987) Multi-purpose cold injector for high resolution gas chromatography. *HRC & CC*, **10**, 288–296.

Schomburg, G. (1981) Capillary chromatography, in *4th International Symposium in Hindelang*, Vol. 371, A921 (ed. R. Kaiser), Institut für Chromatographie, Dr. A. Hüthig Publishers, Heidelberg, Germany.

Stan, H.-J. and Müller, H.-M. (1987) Cold Splitless (PTV) and on-column Injektion technique using Capillar gas chromatography for the analysis of organophosphorus pesticides, in *8th International Symposium on Capillary Chromatography Riva del Garda* (ed P. Sandra), Huethig, pp. 406–415.

Stan, H.-J. and Müller, H.-M. (1987) Evaluation of automated and manual hot-splitless, cold-splitless (PTV) and on-column injection technique using capillary gas chromatography for the analysis of organophosphorus pesticides, in *Symposium Proceedings Riva del Garda*, pp. 406–415.

Staniewski, J. and Rijks, J.A. (1991) Potentials and limitations of the liner design for cold temperature programmed large volume injection in capillary GC and for LC-GC interfacing, in *14th International Symposium on Capillary Chromatography Riva del Garda* (ed. P. Sandra), A. Huethig, pp. 1334–1347.

Tipler, A. and Johnson, G.L. (1989) Optimization of conditions for high temperature capillary gas chromatography using a split-mode programmable temperature vaporizing injection system, in *12th International Symposium on Capillary Chromatography Riva del Garda* (eds P. Sandra and G. Redant), A. Huethig, pp. 986–1000.

Pérez Pavón, J.L., Casas Ferreira, A.M., Fernández Laespada, M.E., and Moreno Cordero, B. (2009) Use of a programmed temperature vaporizer and an in situ derivatization reaction to improve sensitivity in headspace-gas chromatography. Application to the analysis of chlorophenols in water. *J. Chrom. A*, **1216** (7), 1192–1199. doi: 10.1016/j.chroma.2008.12.034.

Poy, F., Visani, S., and Terrosi, F. (1981a) A universal sample injection system for capillary column GC using a programmed temperature vaporizer. *J. Chromatogr.*, **217**, 81–90.

Poy, F. (1981b) Practical demonstration, in *4th International Symposium on Capillar Chromatography*, Hindelang, Germany.

Poy, F., Visani, S., and Terrosi, F. (1982a) A universal sample injection system for capillary column GC using a programmed temperature vaporizer (PTV). *HRC & CC*, **5** (7), 355–359.

Poy, F. (1982b) A new temperature programmed injection technique for capillary GC: split-mode with cold introduction and temperature programmed vaporization. *Chromatographia*, **16**, 345.

Poy, F. and Cobelli, L. (1985) Automatic headspace and programmed temperature vaporizer (PTV) operated in cryo enrichment mode in high resolution gas chromatography. *J. Chrom. Sci.*, **23** (3), 114–119.

Vogt, W., Jacob, K., Ohnesorge, A.B., and Obwexer, H.W. (1979) Capillary gas chromatographic injection system for large sample volumes. *J. Chromatogr.*, **186**, 197–205.

Werkhoff, P. and Bretschneider, W. (1987) Dynamic headspace gas chromatography: concentration of volatile components after thermal desorption by intermediate cryofocusing in a cold trap. II. Effect of sampling and desorption parameters on recovery. *J. Chrom. A*, **405** (C), 99–106. doi: 10.1016/S0021-9673(01)81751-9.

Section 2.2.7 Non-Vaporizing injection Techniques

Aparicio, R., Conte, L.S., and Fiebig, H.J. (2013) Olive oil authentication, in *Handbook of Olive Oil: Analysis and Properties*, pp. 589–653. doi: 10.1007/978-1-4614-7777-8_16.

Biedermann, M., Grob, K., and Mariani, C. (1993) Transesterification and on-line LC-GC for determining the sum of free and esterified sterols in edible oils and fats. *Eur. J. Lipid Sci. Technol.*, **95** (4), 127–133. doi: 10.1002/lipi.19930950403.

Biedermann, M. and Grob, K. (2012a) On-line coupled high performance liquid chromatography-gas chromatography for the analysis of contamination by mineral oil. Part 1: method of analysis. *J. Chrom. A*, **1255**, 56–75. doi: 10.1016/j.chroma.2012.05.095.

Biedermann, M. and Grob, K. (2012b) On-line coupled high performance liquid chromatography-gas chromatography for the analysis of contamination by mineral oil. Part 2: migration from paperboard into dry foods: interpretation of chromatograms. *J. Chrom. A*, **1255**, 76–99. doi: 10.1016/J.CHROMA.2012.05.096.

Binelli, A., Roscioli, C., and Guzzella, L. (2006) Improvements in the analysis of decabromodiphenyl ether using on-column injection and electron-capture detection. *J. Chrom. A*, **1136** (2), 243–247. doi: 10.1016/j.chroma.2006.10.047.

Björklund, J., Tollbäck, P., Hiärne, C., Dyremark, E., and Östman, C. (2004) Influence of the injection technique and the column system on gas chromatographic determination of polybrominated diphenyl ethers. *J. Chro. A*, **1041** (1–2), 201–210. doi: 10.1016/j.chroma.2004.04.025.

Boarelli, M.C., Biedermann, M., Peier, M., Fiorini, D., and Grob, K. (2020) Ergosterol as a marker for the use of degraded olives in the production of olive oil. *Food Control*, **112** (January), 107136. doi: 10.1016/j.foodcont.2020.107136.

Bohačenko, I. and Kopicová, Z. (2001) Detection of olive oils authenticity by determination of their sterol content using LC/GC. *Czech J. Food Sci.*, **19** (3), 97–103. https://doi.org/10.17221/6584-cjfs.

Esche, R., Scholz, B., and Engel, K.-H. (2013) Online LC-GC analysis of free sterols/stanols and intact steryl/stanyl esters in cereals. *J. Agric. Food Chem.*, **61** (46), 10932–10939.

Fankhauser-Noti, A., Fiselier, K., Biedermann-Brem, S., and Grob, K. (2005) Epoxidized soy bean oil (ESBO) migrating from the gaskets of lids into food packed in glass jars: analysis by on-line liquid chromatography-gas chromatography (LC-GC). *Eur. Food Res. Technol.*, **221**, 416–422. doi: 10.1007/s00217-005-1194-4.

Grob, K. and Grob, K., Jr. (1978b) On column injection on to glass capillary columns. I. *J. Chromatogr. A*, **151**, 311.

Grob, K. (1983) Guidelines on how to carry out on-column injections. *HRC & CC*, **6**, 581–582.

Grob, K., Fröhlich, D., Schilling, B., Neukom, H.P., and Nägeli, P. (1984) Coupling of high performance liquid chromatography with capillary gas chromatography. *J. Chrom. A*, **295** (C), 55–61. doi: 10.1016/S0021-9673(01)87597-X.

Grob, K., Walder, C., and Schilling, B. (1986) Concurrent solvent evaporation for on-line coupled HPLC-HRGC. *J. High Res. Chrom.*, **9** (2), 95–101. doi: 10.1002/jhrc.1240090208.

Grob, K. and Läubli, T. (1987) Minimum column temperature required for concurrent solvent evaporation in coupled HPLC-GC. *J. High Res. Chrom.*, **10** (8), 435–440. doi: 10.1002/jhrc.1240100803.

Grob, K., Giuffré, A.M., Leuzzi, U., and Mincione, B. (1994) Recognition of adulterated oils by direct analysis of the minor components. *Eur. J. Lipid Sci. Technol.*, **8**, 286–290.

Grob, K. (1995) *Einspritztechniken in der Kapillar-Gaschromatographie*, Huethig, Heidelberg.

Grob, K. (1995) Development of the transfer techniques for on-line high performance liquid chromatography-capillary gas chromatography. *J. Chrom. A*, **703**, 265–276.

Kresse, M. (2023) Sterinanalytik in Lebensmitteln. *Wiley Anal. Sci. Mag.* 20 April 2023, https://analyticalscience.wiley.com/content/article-do/sterinanalytik-lebensmitteln.

Küchler, T., Dümmong, H., Nestola, M., and Tablack, P. (2012) A Fully automated method to determine sterols by coupled LC-GC-FID technique. Poster at the 10th Euro Fed Lipid Congress in Cracow, Poland, Eurofins Analytik GmbH. Hamburg, Germany, https://www.researchgate.net/publication/275648775.

Munari, F. and Grob, K. (1990) Coupling HPLC to GC: why? How? With what instrumentation? *J. Chromatogr. Sci.*, **28**, 61–66.

Nestola, M. and Schmidt, T.C. (2016) Fully automated determination of the sterol composition and total content in edible oils and fats by online liquid chromatography-gas chromatography-flame ionization detection. *J. Chrom. A*, **1463**, 136–143. doi: 10.1016/j.chroma.2016.08.019.

Nestola, M. and Schmidt, T.C. (2017) Determination of mineral oil aromatic hydrocarbons in edible oils and fats by online liquid chromatography–gas chromatography–flame ionization detection–evaluation of automated removal strategies for biogenic olefins. *J. Chrom. A*, **1505**, 69–76. doi: 10.1016/j.chroma.2017.05.035.

Nestola, M. (2018) Reliable determination of PAHs in foodstuff. *eFood-Lab Int.*, **2**, 17–19.

Schmarr, H.-G., Mosandl, A., and Grob, K. (1989) Early solvent vapor exit in GC for coupled LC-GC involving concurrent eluent evaporation. *J. High Res. Chrom.*, **12** (6), 375–382. doi: 10.1002/jhrc.1240120607.

Schomburg, G., Behlau, H., Dielmann, R., Weeke, F., and Husmann, H. (1977) Sampling techniques in capillary gas chromatography. *J. Chromatogr.*, **142**, 87–102. doi: 10.1016/S0021-9673(01)92028-X.

Steenbergen, H. and Janssen, H.G. (2014) A generic method for target (group) analysis in edible oils and fats: combined normal-phase liquid chromatography and capillary gas chromatography. *LC-GC N. Am., Special Issues*, **32** (5), 30–35.

The Commission of the European Communities (1991) Commission Regulation (EEC) No 2568/91 of 11 July 1991 on the Characteristics of Olive Oil and Olive-Residue Oil and on the Relevant Methods of Analysis, Annex V. *O.J.* **L248** (5.9.91), 1–83.

The European Commission (2016) Commission Delegated Regulation (EU) 2016/2095 Amending Regulation (EEC) No 2568/91 on the Characteristics of Olive Oil and Olive-Residue Oil and on the Relevant Methods of Analysis. *O.J.* **L326** (September

2016): 1–6, *https://eur-lex.europa.eu/legal-content/EN/TXT/PDF/?uri=CELEX: 32016R2095&from=PT*.

Section 2.2.8 Capillary Column Choice and Separation

Adam, P. (2012) Hydrogen as an Alternative to Helium for Gas Chromatography. Application Note for Specialty Gases, Linde AG, Linde Gases Division, Pullach, Germany.

Agilent (2024) *GC Calculators and Method Translator Software*, Agilent Technologies, *https://www.agilent.com/en/support/gas-chromatography/gccalculators* (accessed 2 April 2024).

Agilent (2024) KB000306: Can Argon be Utilized as a Carrier Gas? Agilent Technologies, *https://www.agilent.com/en/support/gas-chromatography/kb000306* (accessed 2 April 2024).

Anderson, J.L., Armstrong, D.W., and Wei, G.-T. (2006) Ionic liquids in analytical chemistry. *Anal. Chem.*, **78** (9), 2894–2902.

Armstrong, D.W., He, L., and Liu, Y.S. (1999) Examination of ionic liquids and their interaction with molecules, when used as stationary phases in gas chromatography. *Anal. Chem.*, **71**, 3873–3876.

Armstrong, D.W., Payagala, T., and Sidisky, L.M. (2009) *LCGC N. Am.*, **27**: 596, 598, 600.

Berthod, A. and Carda-Broch, S. (2003) A new class of solvents for CCC: the room temperature ionic liquids. *J. Liqu. Chrom. Rel. Technol.*, **26** (9–10), 1493–1508. doi: 10.1081/JLC-120021262.

Berthod, A. and Carda-Broch, S. (2004) Uses of ionic liquids in analytical chemistry. *Reactions*, **1**, 1–6, *https://www.researchgate.net/publication/28392523*.

Berthod, A., Ruiz-Angel, M.J., and Carda-Broch, S. (2022) Recent advances on ionic liquid uses in separation techniques. *Separations*, **9** (4). doi: 10.3390/separations9040096.

Blumberg, L.M. and Klee, M.S. (2000) Optimal heating rate in gas chromatography. *J. Microcol. Sep.*, **12** (9), 508–514. doi: 10.1002/1520-667X(2000)12:9<508:: AID-MCS5>3.0.CO;2-Y.

Carda-Broch, S. and Ruiz-Angel, M. (2021) Ionic liquids in analytical chemistry. *Anal. Chim. Acta.*, **661** (1), 1–16. doi: 10.1016/j.aca.2009.12.007.

Cutillas, V., Garcia-Gallego, G., Murcia-Morales, M., Ferrer, C., and Fernandez-Alba, A.R. (2024) Beyond helium: hydrogen as a carrier gas in multiresidue pesticide analysis in fruits and vegetables by GC-MS/MS. *Anal. Methods*, **16** (11), 1564–1569. doi: 10.1039/d3ay02119j.

de Zeeuw, J., English, C., Cochran, J., and Nelson, C. (2016) Method translation in gas chromatography: get the same chromatogram when changing flow, pressure drop, column dimensions or carrier gas. *Separation Science*, **1–6**.

Dorman, F. and Stevens, D. (2022) APGC: a better future with nitrogen. *Anal. Scientist*, (Nov 25), *https://theanalyticalscientist.com/techniques-tools/apgc-a-better-future-with-nitrogen*.

Fausett, A. (2023) Method Translation and Evaluation to Implement Nitrogen Carrier Gas in the Dual-Flame Ionization Detector Configuration for Blood Alcohol Analysis. Application Note, 5994-6508EN, Agilent Technologies, Inc.

Grob, K. and Tschuor, R. (1990) Optimal carrier gas velocities at high temperatures in capillary GC. *J. High Res. Chrom.*, **13** (3), 193–194. doi: 10.1002/jhrc.1240130313.

Groschke, M. and Becker, R. (2024) Comparison of carrier gases for the separation and quantification of mineral oil hydrocarbon (MOH) fractions using online coupled high performance liquid chromatography-gas chromatography-flame ionisation detection. *J. Chrom. A*, **1726**, 464946. doi: 10.1016/j.chroma.2024.464946.

Khan, A.I. (2013) Optimizing GC Parameters for Faster Separations with Conventional Instrumentation. Technical Note 20743, Thermo Fisher Scientific, Runcorn, Cheshire, UK.

Restek Corporation (2024) EZGC & EZLC Online Software Suite, *https://ez.restek.com/ezgc-mtfc* (accessed 2 April 2024).

Stenerson, K.K., Halpenny, M.R., Sidisky, L.M., and Buchanan, M.D. (2013) GC analysis of omega-3-fatty acids in fish oil capsules and farm raised salmon. *Supelco Rep.*, **31** (2), 14–15.

Stenerson, K.K., Halpenny, M.R., Sidisky, L.M., and Buchanan, M.D. (2014) Ionic liquid GC column option for the analysis of omega 3 and omega 6 fatty acids. *Supelco Rep.*, **31** (1), 20–22.

Supelco (2013) Supelco Ionic Liquid GC Columns, Introduction to the Technology. Supelco Presentation 2012, updated February 2014, *https://www.sigmaaldrich.com/deepweb/assets/sigmaaldrich/marketing/global/documents/197/160/ionic_liquid_gc_columns.pdf* (accessed 6 January 2025).

Talebi, M., Patil, R.A., and Armstrong, D.W. (2020) Gas chromatography columns using ionic liquids as stationary phase, in *Green Chemistry and Sustainable Technology* (ed. M.B. Shiflett), Springer Nature Switzerland AG, Cham, Switzerland. doi: 10.1201/9780367808310.

Vidal, L., Riekkola, M.-L., and Canals, A. (2012) Ionic liquid-modified materials for solid phase extraction and separation: a review. *Anal. Chim. Acta*, **715**, 19–41.

Whitmarsh, S. (2012) Ionic liquid stationary phases: application in gas chromatographic analysis of polar components of fuels and lubricants. *Chrom. Today*, (Feb/Mar), 12–15.

van Ysacker, P.G., Janssen, H.G., Snijders, H.M.J., van Cruchten, H.J.M., Leclercq, P.A., and Cramers, C.A. (1993) High speed-narrow-bore capillary gas chromatography with ion-trap mass spectrometric detection, in *15th International Symposium on Capillary Chromatography Riva del Garda 1993* (ed. P. Sandra), Huethig.

Section 2.2.9 Chromatography Parameters

Anastassiades, M., Maštovská, K., and Lehotay, S.J. (2003) Evaluation of analyte protectants to improve gas chromatographic analysis of pesticides. *J. Chromatogr. A*, **1015** (May 2002), 163–184. doi: 10.1016/S0021-9673(03)01208-1.

Bock, R. (1974) *Methoden der analytischen Chemie*, Vol. Bd. 1, Trennungsmethoden, Weinheim, Verlag Chemie.

Dettmer-Wilde, K. and Engewald, W. (2013) The chromatographic separation process, in *Practical Gas Chromatography* (eds K. Dettmer-Wilde and W. Engewald), Springer, Heidelberg. doi: 10.1007/978-3-642-54640-2.

Li, Y., Chen, X., Fan, C., and Pang, G. (2012) Compensation for matrix effects in the gas chromatography–mass spectrometry analysis of 186 pesticides in tea matrices using analyte protectants. *J. Chromatogr. A*, **1266**, 131–142.

Karasek, F.W. and Clement, R.E. (1988) *Basic Gas Chromatography Mass Spectrometry*, Elsevier, Amsterdam.

Martin, A.J.P. and Synge, R.L.M. (1941) A new form of chromatogram employing two liquid phases. *J. Biol. Chem.*, **35**, 1358. doi: 10.1042/bj0351358.

Meyer, V. (2010) *Practical High Performance Liquid Chromatography*, 5th edn, Wiley & Sons Inc.

Mastovska, K., Lehotay, S.J., and Anastassiades, M. (2005) Combination of analyte protectants to overcome matrix effects in routine GC analysis of pesticide. *Anal. Chem.*, **77** (24), 8129–8137. doi: 10.1021/AC0515576.

Morales, E. and Macherone, A. (2013) Using Analyte Protectants and Solvent Selection to Maximize the Stability of Organophosphorous Pesticides during GC/MS Analysis. Application Note 5991-1808EN, Agilent Technologies, Inc.

Schomburg, G. (1981) Capillary chromatography, in *4th International Symposium in Hindelang*, Vol. 371, A921 (ed. R. Kaiser), Institut für Chromatographie, Dr. A. Hüthig Publishers, Heidelberg, Germany.

Snyder, L.R. and Kirkland, J.J. (1979) *Introduction to Modern Liquid Chromatography*, 2nd edn, John Wiley & Sons, Inc., New York (3rd Ed. J.J. Dolan, 2010).

Section 2.2.10 Fast Gas Chromatography Solutions

Arrebola, F.J., Martínez Vidal, J.L., González-Rodríguez, M.J., Garrido-Frenich, A., and Sánchez Morito, N. (2003) Reduction of analysis time in gas chromatography: application of low pressure gas chromatography-tandem mass spectrometry to the determination of pesticide residues in vegetables. *J. Chrom. A*, **1005** (1–2), 131–141. doi: 10.1016/S0021-9673(03)00887-2.

Bicchi, C., Brunelli, C., Cordero, C., Rubiolo, P., Galli, M., and Sironi, A. (2005) High speed gas chromatography with direct resistively-heated column (ultra fast module-GC)-separation measure (S) and other chromatographic parameters under different analysis conditions for samples of different complexities and volatilities. *J. Chrom. A*, **1071** (1–2), 3–12. doi: 10.1016/j.chroma.2004.09.051.

Boeker, P. and Leppert, J. (2015) Flow field thermal gradient gas chromatography. *Anal. Chem.*, **87** (17), 9033–9041. doi: 10.1021/acs.analchem.5b02227.

Cajka, T., Hajslova, J., Lacina, O., Mastovska, K., and Lehotay, S.J. (2008) Rapid analysis of multiple pesticide residues in fruit-based baby food using programmed temperature vaporiser injection-low pressure gas chromatography-high resolution time-of-flight mass spectrometry. *J. Chrom. A*, **1186**, 281–294. doi: 10.1016/j.chroma.2007.12.009.

Chopra, M.D., Müller, P.J., Leppert, J., Wüst, M., and Boeker, P. (2021) Residual solvent analysis with hyper-fast gas chromatography-mass spectrometry and a liquid carbon dioxide cryofocusing in less than 90 S. *J. Chrom. A*, **1648** (July), 462179.
doi: 10.1016/j.chroma.2021.462179.

de Zeeuw, J., Peene, J., Jansen, H.-G., and Lou, X. (2000) A simple way to speed up separations by GC-MS using short 0.53 mm columns and vacuum outlet conditions. *J. High Res. Chrom.*, **23** (12), 677–680.
doi: 10.1002/1521-4168(20001201)23:12<677::AID-JHRC677>3.0.CO;2-L.

de Zeeuw, J., Janssen, H.-G., and Lehotay, S.J. (2022) The long and winding road to LPGC-MS. *The Analytical Scientist*, Techniques & Tools, online newsletter, *https://theanalyticalscientist.com/techniques-tools/the-long-and-winding-road-to-lpgc-ms* (accessed 13 April 2024).

Donato, P., Quinto Tranchida, P., Dugo, P., Dugo, G., and Mondello, L. (2007) Review: rapid analysis of food products by means of high speed gas chromatography. *J. Sep. Sci.*, **30**, 508–526.

U.S. EPA (1984) U.S. EPA Method 610 - Polynuclear Aromatic Hydrocarbons, promulgated 1984. Federal Register **49**, 40 CFR Part 136, 43 344, No. 209, *https://www.epa.gov/sites/default/files/2015-10/documents/method_610_1984.pdf* (accessed 5 April 2024).

Facchetti, R., Galli, S., and Magni, P. (2002) Optimization of analytical conditions to maximize separation power in ultra-fast GC, in *Proceedings of 25th International Symposium of Capillary Chromatography, KNL05, Riva del Garda, Italy, May 13–17, 2002* (ed. P. Sandra).

Havelt, T. and Boeker, P. (2023) Hyper fast gas chromatography. *Wiley Analytical Science*, **12**, April, *https://analyticalscience.wiley.com/do/10.1002/was.000600505/full*.

HyperChrom (2024) TPH-Method. Application Note, web site, *https://www.hyperchrom.com/GC/applications* (accessed 8 June 2024).

Khan, A.I. (2013) Optimizing GC Parameters for Faster Separations with Conventional Instrumentation. Technical Note 20743, Thermo Fisher Scientific, Runcorn, Cheshire, UK.

Khan, A.I. (2013a) Fast Analysis of Phenols Using Conventional GC Instrumentation. Application Note 20737, Thermo Fisher Scientific, Runcorn, Cheshire, UK.

Lehotay, S.J., de Zeeuw, J., Sapozhnikova, Y., Michlig, N., Rousova Hepner, J., and Konschnik, J.D. (2020) There is no time to waste: low pressure gas chromatography-mass spectrometry (LPGC-MS) is a proven solution for fast, sensitive, and robust GC-MS analysis. *LCGC N. Am.*, **38** (8), 457–466.

Leppert, J., Härtel, M., Klapötke, T.M., and Boeker, P. (2018) Hyperfast flow-field thermal gradient GC/MS of explosives with reduced elution temperatures. *Anal. Chem.*, **90** (14), 8404–8411. doi: 10.1021/acs.analchem.8b00900.

Maštovská, K. and Lehotay, S.J. (2003) Practical approaches to fast gas chromatography-mass spectrometry. *J. Chrom. A.*, **1000**, 153–180.
doi: 10.1016/S0021-9673(03)00448-5.

Monteiro, S.H., Lehotay, S.J., Sapozhnikova, Y., Ninga, E., and Lightfield, A.R. (2020) High throughput mega-method for the analysis of pesticides, veterinary drugs, and environmental contaminants by UHPLC- MS/MS and robotic mini-SPE cleanup +

LPGC-MS/MS, Part 1: Beef. *J. Agric. Food Chem.*, **69** (4), 1159–1168. doi: 10.1021/acs.jafc.0c00710.

Peene, J., De Zeeuw, J., and De Jong, R. (2000) Low pressure gas chromatography: fast analysis with high sensitivity. *Int. Laboratory*, (September), 41–44.

Restek (2024) *An Introduction to Low Pressure GC-MS (LPGC-MS)*, Restek Corp, Bellefonte, PA USA, *https://www.restek.com/global/ja/articles/an-introduction-to-low-pressure-gc-ms-lpgc-ms* (accessed 12 April 2024).

Rosenberg, E., Klampfl, B., and Müller, R.D. (2023) Negative thermal gradient gas chromatography, in *Novel Aspects of Gas Chromatography and Chemometrics* (eds S.C. Moldoveanu, V. David, and V.D. Hoang), IntechOpen, London, United Kingdom. doi: 10.5772/intechopen.110591.

Sapozhnikova, Y. and Lehotay, S.J. (2015) Review of recent developments and applications in low pressure (vacuum outlet) gas chromatography. *Anal. Chim. Acta*, **899** (November), 13–22. doi: 10.1016/j.aca.2015.10.003.

Sapozhnikova, Y. (2018) High throughput analytical method for 265 pesticides and environmental contaminants in meats and poultry by fast low pressure gas chromatography and ultrahigh performance liquid chromatography tandem mass spectrometry. *J. Chrom. A*, **1572**, 203–211. doi: 10.1016/j.chroma.2018.08.025.

Section 2.2.11 Multi-dimensional Gas Chromatography

Beens, H., Boelwns, R., Tijssen, R., and Blomberg, J. (1998) Simple, non-moving modulation interface for comprehensive two-dimensional gas chromatography. *J. High Res. Chromatogr.*, **21**, 47.

Beens, J., Blomberg, J., and Schoenmakers, P.J. (2000) Proper tuning of comprehensive two-dimensional gas chromatography (GCxGC) to optimize the separation of complex oil fractions. *J. High Resol. Chromatogr.*, **23** (3), 182–188.

Beens, J., Adahchour, M., Vreuls, R.J.J., van Altena, K., and Brinkman, U.A.T. (2001) Simple, non-moving modulation interface for comprehensive two-dimensional gas chromatography. *J. Chromatogr. A*, **919** (1), 127–132.

Bertsch, W.J. (1999) Two-dimensional gas chromatography. Concepts, instrumentation, and applications - Part 1. *J. High Res. Chromatogr.*, **22**, 647–665.

Bertsch, W.J. (2000) Two-dimensional gas chromatography. Concepts, instrumentation, and applications - Part 2: Comprehensive two-dimensional gas chromatography. *J. High Res. Chromatogr.*, **23**, 167–181.

Blomberg, J., Schoenmakers, P., van Zuijlen, M., and Hartog, A. (2001) GCxGC: Structured Information on Complex Mixtures. Presentation, private copy 2008.

Cavagnino, D., Bedini, F., Zilioli, G., and Trestianu, S. (2003a) Improving sensitivity and separation power by using LVSL-GCxGC-FID technique for pollutants' detection at low ppb level. Poster, PittCon Conference, Thermo Fisher Scientific, Rodano, Italy.

Cavagnino, D., Magni, P., Zilioli, G., and Trestianu, S. (2003b) Comprehensive two-dimensional gas chromatography using large sample volume injection for the determination of polynuclear aromatic hydrocarbons in complex matrices. *J. Chrom. A*, **1019** (1–2), 211–220. doi: 10.1016/j.chroma.2003.07.017.

Dallüge, J., Beens, J., and Brinkman, U.A.T. (2003) Comprehensive two-dimensional gas chromatography: a powerful and versatile analytical tool. *J. Chrom. A*, **1000** (1–2), 69–108. doi: 10.1016/s0021-9673(03)00242-5.

De Alencastro, L.F., Grandjean, D., and Tarradellas, J. (2003) Application of multidimensional (heart-cut) gas chromatography to the analysis of complex mixtures of organic pollutants in environmental samples. *Chimia*, **57**, 499–504.

Deans, D.R. (1968) A new technique for heart cutting in gas chromatography. *Chromatographia*, **1**, 18–22.

Deans, D.R. (1981) Use of heart cutting in gas chromatography: a review. *J. Chrom. A*, **203** (C), 19–28. doi: 10.1016/S0021-9673(00)80278-2.

Dimandja, J.M.D. (2004) GCxGC. *Anal. Chem.*, **76** (9), 167A–174A.

Ferracane, A., Manousi, N., Tranchida, P.Q., Zachariadis, G.A., Mondello, L., and Rosenberg, E. (2022) Exploring the volatile profile of whiskey samples using solid phase microextraction arrow and comprehensive two-dimensional gas chromatography-mass spectrometry. *J. Chrom. A*, **1676**. doi: 10.1016/j.chroma.2022.463241.

Focant, J., Sjodin, A., Turner, W., and Patterson, D. (2004) Measurement of selected polybrominated diphenyl ethers, polybrominated and polychlorinated biphenyls, and organochlorine pesticides in human serum and milk using comprehensive two-dimensional gas chromatography isotope dilution time-of-flight mass spectrometry. *Anal. Chem.*, **76** (21), 6313–6320.

Frank, S. and Schieberle, P. (2022) Changes in the major odorants of grape juice during manufacturing of Dornfelder red wine. *J. Agric. Food Chem.*, **70** (43), 13979–13986. doi: 10.1021/acs.jafc.2c06234.

GC Image (2024) *GCxGC Edition Software. Product Information, Web Based, GC Image LLC*, Lincoln, Nebraska, USA, https://www.gcimage.com/gcxgc/index.html.

Guth, H. (1996) Use of the moving capillary switching system (MCSS) in combination with stable isotope dilution analysis (IDA) for the quantification of a trace component in wine. Poster, 18th International Symposium on Capillary Chromatography, Riva del Garda, Italy, May 20–24, 1996.

Hoh, E., Lehotay, S.J., Mastovska, K., and Huwe, J.K. (2007) Evaluation of automated direct sample introduction with comprehensive two-dimensional gas chromatography/time-of-flight mass spectrometry for the screening analysis of dioxins in fish oil. *J. Chromatogr. A*, **1201**, 69–77.

Hollingsworth, B.V., Reichenbach, S.E., Tao, Q., and Visvanathan, A. (2006) Comparative visualization for comprehensive two-dimensional gas chromatography. *J. Chromatogr. A*, **1105**, 51–58.

Horii, Y., Petrick, G., Katase, T., Gamo, T., and Yamashita, N. (2004) Congener specific carbon isotope analysis of technical PCN and PCB preparations using 2DGC IRMS. *Organohalogen Comp.*, **66**, 341–348.

Kellner, R., Mermet, J.-M., Otto, M., Valcarcel, M., and Widmer, H.M. (2004) *Analytical Chemistry*, 2nd edn, Wiley-VCH Verlag GmbH, Weinheim.

Lee, A.L., Lewis, A.C., Bartle, K.D., McQuaid, J.B., and Marriott, P.J. (2000) A comparison of modulating interface technologies in comprehensive two-dimensional gas chromatography (GC×GC). *J. Microcol. Sep.*, **12** (4), 187–193. doi: 10.1002/(SICI)1520-667X(2000)12:4<187::AID-MCS3>3.0.CO;2-P.

Liu, Z. and Phillips, J.B. (1991) Comprehensive two-dimensional gas chromatography using an on-column thermal modulator interface. *J. Chrom. Sci.*, **29** (5), 227–231. doi: 10.1093/chromsci/29.6.227.

Marriott, P.J., Morrison, P.D., Shellie, R.A., Dunn, M.S., Sari, E., and Ryan, D. (2003) Multidimensional and comprehensive two-dimensional gas chromatography. *LCGC Eur.*, **12**, 2–10.

McNamara, K., Leardib, R., and Hoffmann, A. (2003) Developments in 2D GC with heartcutting. *LCGC Eur.*, **12**, 14–22.

Mondello, L., Tranchida, P.Q., Dugo, P., and Dugo, G. (2008) Comprehensive two-dimensional gas chromatography-mass spectrometry: a review. *Mass Spectrom. Rev.*, **27**, 101–124.

Ong, R.C.Y. and Marriott, P.J. (2002) A review of basic concepts in comprehensive two-dimensional gas chromatography. *J. Chrom. Sci.*, **40** (5), 276–291. doi: 10.1093/chromsci/40.5.276.

Patterson, D.G., Welch, S.M., Focant, J.F., and Turner, W.E. (2006) The use of various gas chromatography and mass spectrometry techniques for human biomonitoring studies. *Organohalogen Compounds*, **68**, 932–935. Presented at the Dioxin 2006 Conference, FCC-2602 - 409677, Oslo, Norway, 2006.

Reichenbach, S.E., Ni, M., Kottapalli, V., and Visvanathan, A. (2004) Information technologies for comprehensive two-dimensional gas chromatography. *Chemom. Intell. Lab. Syst.*, **71**, 107–120.

Schlumpberger, P., Stübner, C.A., and Steinhaus, M. (2022) Development and evaluation of an automated solvent-assisted flavour evaporation (aSAFE). *Eur. Food Res. Technol.*, **248** (10), 2591–2602. doi: 10.1007/s00217-022-04072-1.

SepSolve (2023) INSIGHT - Outstanding GC×GC Performance for both Flow and Thermal Modulation. Product information, SepSolve Analytical, Peterborough, UK.

Sharif, K.M., Chin, S.T., Kulsing, C., and Marriott, P.J. (2016) The microfluidic Deans switch: 50 years of progress, innovation and application. *TrAC - Trends Anal. Chem.*, **82** (September), 35–54. doi: 10.1016/j.trac.2016.05.005.

Simmons, M.C. and Snyder, L.R. (1958) Two-stage gas-liquid chromatography. *Anal. Chem.*, **30** (1), 32–35.

Squara, S. *et al.* (2023) Artificial intelligence decision-making tools based on comprehensive two-dimensional gas chromatography data: the challenge of quantitative volatilomics in food quality assessment. *J. Chrom. A*, **1700**, 464041. doi: 10.1016/j.chroma.2023.464041.

Sulzbach, H. (1991) Controlling chromatographic gas streams by adjusting relative positioning off supply columns in connector leading to detector. Patent DE 4017909, Carlo Erba Strumentazione GmbH, Germany.

Vaye, O., Ngumbu, R.S., and Xia, D. (2022) A review of the application of comprehensive two-dimensional gas chromatography MS-based techniques for the analysis of persistent organic pollutants and ultra-trace level of organic pollutants in environmental samples. *Rev. Anal. Chem.*, **41** (1), 63–73. doi: 10.1515/revac-2022-0034.

Welthagen, W., Schnelle-Kreis, J., and Zimmerman, R. (2003) Search criteria and rules for comprehensive two-dimensional gas chromatography-time-of-flight mass spectrometry analysis of airborne particulates. *J. Chromatogr. A*, **1019**, 233–249.

Zanella, D., Focant, J.F., and Franchina, F.A. (2021) 30th anniversary of comprehensive two-dimensional gas chromatography: latest advances. *Anal. Sci. Adv.*, **2** (3–4), 213–224. doi: 10.1002/ansa.202000142.

Section 2.2.12 Classical Detectors for GC-MS Systems

Agilent (2013) Capillary Flow Technology: Deans Switch. Product information, 5989-9384EN, Agilent Technologies Inc.

Andrawes, F. and Ramsey, R. (1986) The helium ionization detector. *J. Chrom. Sci.*, **24** (11), 513–518. doi: 10.1093/chromsci/24.11.513.

Bretschneider, W. and Werkhoff, P. (1988a) Progress in all-glass stream splitting systems in capillary gas chromatography. Part I: Application of a simple "glass-cap-cross" as effluent splitter for splitless and on-column injection. *HRC & CC*, **11**, 543–546.

Bretschneider, W. and Werkhoff, P. (1988b) Progress in all-glass stream splitting systems in capillary gas chromatography, Part II : Application of a simple "glass-cap-cross" as inlet splitter for on-column injection. *HRC & CC*, **11**, 589–592.

Delahunty, C.M., Eyres, G., and Dufour, J.-P. (2006) Gas chromatography-olfactometry. *J. Sep. Sci.*, **29** (14), 2107–2125. doi: 10.1002/jssc.200500509.

Ewender, J. and Piringer, O. (1991) Gaschromatographische Analyse flüchtiger aliphatischer Amine unter Verwendung eines Amin-spezifischen Elektrolytleitfähigkeitsdetektors. *Dt. Lebensm. Rundsch.*, **87**, 5–7.

Fan, H., Smuts, J., Bai, L., Walsh, P., Armstrong, D.W., and Schug, K.A. (2016) Gas chromatography-vacuum ultraviolet spectroscopy for analysis of fatty acid methyl esters. *Food Chem.*, **194** (March), 265–271. doi: 10.1016/j.foodchem.2015.08.004.

Firor, R.L. (1989) Multielement detection using GC-atomic emission spectroscopy. *Int. Lab.*, (September), 44–52.

Forsyth, D.S. (2004) Pulsed discharge detector: theory and applications. *J. Chrom. A*, **1050** (1), 63–68. doi: 10.1016/j.chroma.–004.07.103.

García-Bellido, J., Freije-Carrelo, L., Moldovan, M., and Encinar, J.R. (2020) Recent advances in GC-ICP-MS: focus on the current and future impact of MS/MS technology. *TrAC - Trends Anal. Chem*, **130** (September). doi: 10.1016/j.trac.2020.115963.

Halasz, I. (1964) Concentration and mass flow rate sensitive detectors in gas chromatography. *Anal. Chem.*, **36** (8), 1428–1430. doi: 10.1021/ac60214a009.

Hill, H.H. and McMinn, D.G. (1992) *Detectors for Capillary Gas Chromatography*, John Wiley & Sons Inc., New York.

Hu, J., Pagliano, E., Hou, X., Zheng, C., Yang, L., and Mester, Z. (2017) Sub-ppt determination of butyltins, methylmercury and inorganic mercury in natural waters by dynamic headspace in-tube extraction and GC-ICPMS detection. *J. Anal. At. Spectrom.*, **12**, 1–8. doi: 10.1039/C7JA00296C.

Jayawardhana, D.A., Woods, R.M., Zhang, Y., Wang, C., and Armstrong, D.W. (2012) Rapid, efficient quantification of water in solvents and solvents in water using an ionic liquid-based GC column. *LCGC N. Am.*, **30** (2), 142–158.

JAS (2009) *Atomic Emission Detector.* Product Information, Joint Analytical Systems GmbH, Moers, Germany.

Kawamura, K., Ishii, T., and Takemori, Y. (2020) New Approach to Food Smell Analysis Using Combination of GCMS and GC-SCD (2). Application Note M289, Shimadzu Corporation, Kyoto, Japan.

Li, W. (2021) Analysis of Trace Carbon Dioxide and Permanent Gas Impurities in Fuel Cell Hydrogen and High Purity Hydrogen by GC. Application Note 5994-4415EN, Agilent Technologies, Shanghai, China.

Lovelock, J.E. (1958) A sensitive detector for gas chromatography. *J. Chrom. A*, **1** (C), 35–46. doi: 10.1016/s0021-9673(00)93398-3.

Mezcua, M., Martínez-Uroz, M.A., Wylie, P.L., and Fernández-Alba, A.R. (2009) Simultaneous screening and target analytical approach by gas chromatography-quadrupole-mass spectrometry for pesticide residues in fruits and vegetables. *J. AOAC Int.*, **92** (6), 1790–1806. doi: 10.1093/jaoac/92.6.1790.

Ochiai, N., Sasamoto, K., and MacNamara, K. (2012) Characterization of sulfur compounds in whisky by full evaporation dynamic headspace and selectable one-dimensional/two-dimensional retention time locked gas chromatography-mass spectrometry with simultaneous element-specific detection. *J. Chrom. A*, **1270**, 296–304. doi: 10.1016/j.chroma.2012.11.002.

Plutowska, G. and Wardencki, W. (2009) Headspace solid phase microextraction and gas chromatography-olfactometry analysis of raw spirits of different organoleptic quality. *Flavour Fragr. J.*, **24**, 177–185.

Piringer, O. and Wolff, E. (1984) New electrolytic conductivity detector for capillary gas chromatography - analysis of chlorinated hydrocarbons. *J. Chromatogr.*, **284**, 373–380.

Quimby, B.D., McCurry, J.D., and Norman, W.M. (2007) Capillary flow technology for GC - reinvigorating a mature analytical discipline. *LCGC N. Am.*, **25** (4).

Schneider, W., Frohne, J.C., and Bruddereck, H. (1982) Selektive gaschromatographische Messung sauerstoffhaltiger Verbindungen mittels Flammenionisationsdetektor. *J. Chromatogr.*, **245**, 71. doi: 10.1016/S0021-9673(00)82476-0.

Schug, K.A. *et al.* (2014) Vacuum ultraviolet detector for gas chromatography. *Anal. Chem.*, **86** (16), 8329–8335. doi: 10.1021/ac5018343.

Schug, K.A. *et al.* (2017) The complementarity of vacuum ultraviolet spectroscopy and mass spectrometry for gas chromatography detection. *LCGC - Special Issues*, **35** (8), 511–512, http://www.chromatographyonline.com/complementarity-vacuum-ultraviolet-spectroscopy-and-mass-spectrometry-gas-chromatography-detection.

Sironi, A. and Verga, G.R. (1995) The O-FID and its applications in petroleum product analysis. *J. Chrom. Libr.*, **56**, 143–158. doi: 10.1016/S0301-4770(08)61285-3.

SRI (2003) *Helium Ionization Detector Manual*, SRI Instruments Europe GmbH, Bad Honnef, Germany, *https://srigc.com/public/storage/downloads/HIDman.pdf*.

Sun, J., Hu, S., Sharma, K.R., Keller-Lehmann, B., and Yuan, Z. (2014) An efficient method for measuring dissolved VOSCs in wastewater using GC-SCD with static headspace technique. *Water Res.*, **52** (April), 208–217. doi: 10.1016/j.watres.2013.10.063.

Thomas, S.L., Myers, C., and Schug, K.A. (2023) Comparison of fragrance and flavor components in non-psilocybin and psilocybin mushrooms using vacuum-assisted headspace high capacity solid phase microextraction and gas chromatography-mass spectrometry. *Adv. Sample Prep.*, **8** (September), 100090. doi: 10.1016/j.sampre.2023.100090.

van Stee, L.L.P., Brinkman, U.A.T., and Bagheri, H. (2002) Gas chromatography with atomic emission detection: a powerful technique. *TrAC - Trends Anal. Chem.*, **21** (9–10), 618–626. doi: 10.1016/S0165-9936(02)00810-5.

Wentworth, W.E. and Chen, E.C.M. (1981) in *Electron Capture Theory and Practice in Chromatography* (eds A. Zlatkis and C.F. Poole), Elsevier, New York, p. 27.

Wiedemer, A.M., Mcclure, A.P., and Leitner, E. (2023) Roasting and cacao origin affect the formation of volatile organic sulfur compounds in 100% chocolate. *Molecules*, **28** (3038), 1–13, *https://doi.org/doi:10.3390/molecules28073038*.

Wylie, P.L. and Quimby, B.D. (1989) Applications of gas chromatography with an atomic emission detector. *J. High Res. Chrom.*, **12** (12), 813–818. doi: 10.1002/jhrc.1240121210.

2.3 Mass Spectrometry

> Looking back on my work in MS over the last 40 years, I believe that my major contribution has been to help convince myself, as well as other mass spectrometrists and chemists in general, that the things that happen to a molecule in the mass spectrometer are in fact chemistry, not voodoo; and that mass spectrometrists are, in fact, chemists and not shamans.
>
> Seymour Meyerson, 1994
> 1916–2016, American analytical chemist and
> noted authority on mass spectrometry

Mass spectrometry is certainly the most versatile analytical technique in hyphenation with chromatography due to the vast range of analytes that can be detected, identified, and quantified. The early development of gas chromatography–mass spectrometry (GC-MS) coupling is reviewed by Fred W. McLafferty and John A. Michnowicz (1992). Mass spectrometers ionize the flow of compounds eluting into the ion source and analyze the cloud of ions for mass determination and quantitation. The substances to be analyzed can be introduced to the ion source and ionized by different means. In GC-MS systems, there is continuous transport of substances by the carrier gas into the ion source. Only charged particles can be accelerated into the MS analyzer. Mass spectrometers basically differ in the construction of the analyzer as a continuous ion beam, flight time, or ion storage instrument. Mass separation is achieved on the basis of the ion mass to charge ratio m/z. The SI mass unit used is "u". Also "Da" (Dalton) is in use, for instance to describe large masses of biomolecules, e.g. as kDa. A more correct unit for use on the x axis of a mass spectrum would therefore be u/z, but mass spectrometrists simply choose not to use it. The former atomic mass unit (amu) is depreciated as it is not an SI unit (Busch, 2001). The molecular mass unit of a chemical compound is given with "M". In mass spectrometry with a nominal "M", it is the formula mass using the mass of the most abundant isotopes, not the molar mass as an average of the natural isotope abundance as it is used in chemical reactions. In GC-MS, ions are generally singly charged (few known exceptions for instance with polyaromatic hydrocarbons [PAHs]), so the recorded m/z value directly gives the mass value with $m/1$ of the ion. The performance of a mass spectrometric analyzer is determined by the resolving power for differentiation between masses with small differences, the mass range, and the ion transmission required to achieve high detection sensitivity. Depending on the type of mass analyzer employed, the mass of an ion can be determined as unit mass (low mass resolution, nominal mass resolution) or accurate mass (high mass resolution).

2.3.1 Ionization

2.3.1.1 Electron Ionization

Electron ionization (EI) is the standard process in all GC-MS instruments. An ionization energy of 70 eV is used in all commercial instruments. The energy necessary for ionization of organic molecules is less than 15 eV, much lower than the effective applied energy (Table 2.42). The EI operation of all MS instruments at the high ionization energy of 70 eV was established and is kept at that level as a standard in all commercial instruments with regard to sensitivity (ion yield) and the comparability of the mass spectra obtained for library generation and search purposes.

Some instruments allow the user to reduce the ionization energy for additional identification purposes (Gross, 2004). This allows EI spectra with a high proportion of molecular information to be obtained (low voltage ionization) (Figure 2.183). Today, the low voltage ionization became of increased importance and use for unknown compound characterization in real life sample analysis preventing the known loss in sensitivity with soft ionization techniques. (McGregor and Barden, 2014; Alam et al., 2016).

The process of EI ionization can be explained by a wave or a particle model. The amount of energy transferred in this process is called the *ionization energy*. The current theory is based on the interaction between the energy-rich electron beam with the outer electrons of a molecule. At around 70 eV, the de Broglie wavelength of the electrons matches the length of typical bonds in organic molecules of about 0.14 nm. The energy transfer to an organic analyte molecule is maximized by resonance, leading to ionization and fragmentation reactions. Energy absorption

Table 2.42 First ionization potentials (eV) of selected substances.

Helium	24.6
Nitrogen	15.3
Carbon dioxide	13.8
Oxygen	12.5
Propane	11.07
1-Chloropropane	10.82
Butane	10.63
Pentane	10.34
Nitrobenzene	10.18
Benzene	9.56
Toluene	9.18
Chlorobenzene	9.07
Propylamine	8.78
Aniline	8.32

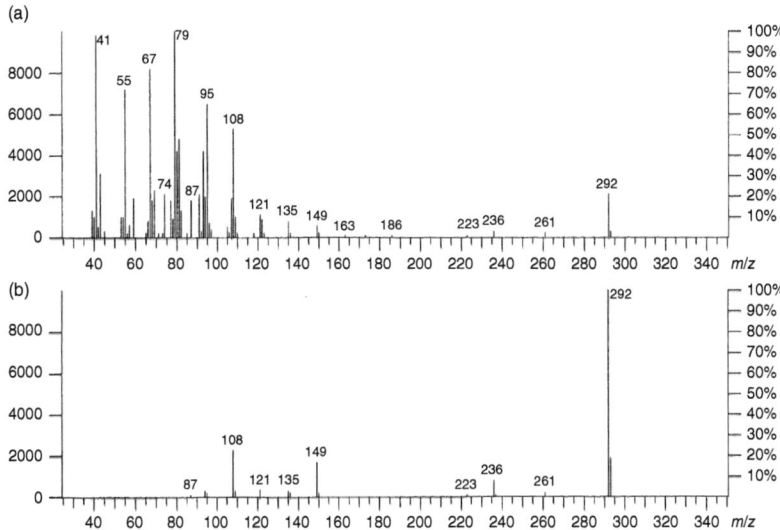

Figure 2.183 EI mass spectra of methyl linolenate, $C_{17}H_{29}COOCH_3$ (Spiteller and Spiteller, 1973). Recording conditions: direct inlet, (a) 70 eV and (b) 17 eV ionization energy.

initially leads to the formation of a molecular ion M+ by loss of an electron. The excess energy causes excitation in the rotational and vibrational energy levels of this radical cation. The subsequent processes of fragmentation depend on the amount of excess energy and the capacity of the molecule for internal stabilization.

The generation of EI mass spectra is well understood with a rich documentation of the fragmentation processes. But in contrast to nuclear magnetic resonance (NMR), mass spectra still cannot be calculated from a theoretical standpoint. A general method to compute EI mass spectra based on a combination of fast quantum chemical methods, molecular dynamics, and stochastic preparation of "hot" ionized species was published for instance by Stefan Grimme. It provides mass spectra that compare well with their experimental counterparts, even in subtle details (Grimme, 2013).

The concept of localized charge empirically describes the fragmentation reactions (Budzikiewicz and Schäfer, 2012). The concept was developed from the observation that molecular ions preferentially fragment bonds near heteroatoms N, O, or S π electron systems. An example of this occurrence is shown in Figure 2.184. This is attributed to the fact that a positive or negative charge is stabilized by an electronegative structure element in the molecule or a favoring mesomerism. Bond breaking can be predicted by subsequent electron migrations or rearrangement. These types of processes include α-cleavage, allyl cleavage, benzyl cleavage, and the McLafferty rearrangement (McLafferty and Turecek, 1993; see also Section 3.2.6).

Assuming a constant substance stream into an ion source, Figure 2.185 shows the intensity of the signal with increasing ionization energy. The steep rise of the signal intensity only begins when the ionization potential (IP) is reached. Low measurable intensities just below the IP are produced as a result of the inhomogeneous

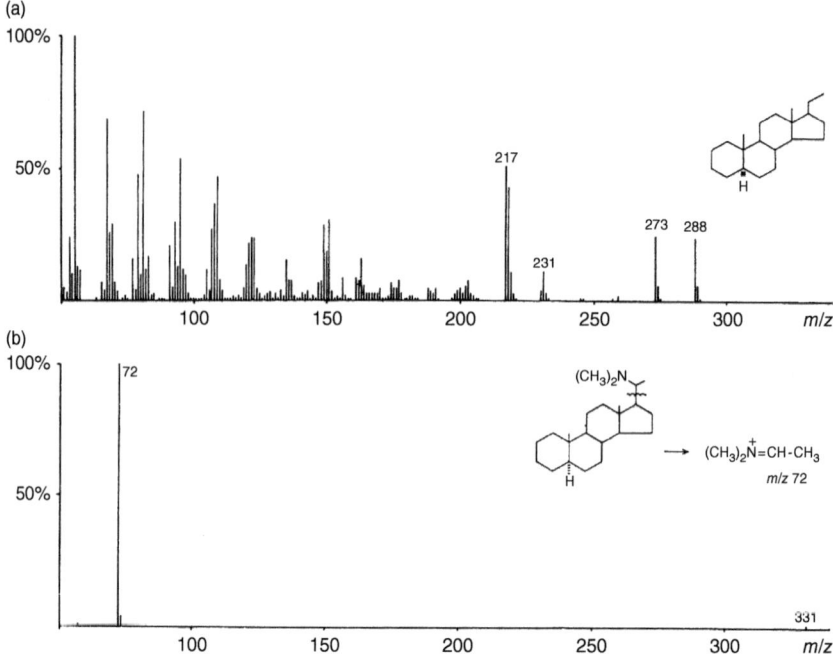

Figure 2.184 Effect of structural features on the appearance of mass spectra – concept of localized charge (Budzikiewicz and Schäfer, 2012): (a) Mass spectrum of 5α-pregnane and (b) mass spectrum of 20-dimethylamino-5α-pregnane. The α-cleavage of the amino group dominates in the spectrum. The information on the structure of the sterane unit is completely absent.

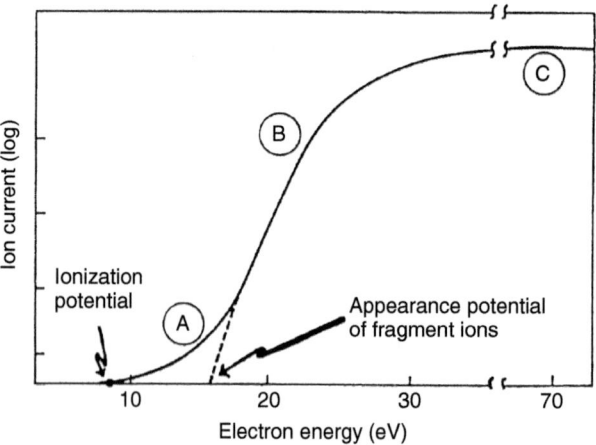

Figure 2.185 Increase in the ion current with increasing electron energy: (A) Threshold region after reaching the appearance potential; molecular ions are mainly produced here. (B) Build-up region with increasing production of fragment ions. (C) Routine operation, stable formation of fragment ions (Frigerio, 1974).

composition of the electron beam. Generally, the increase in signal intensity continues with increasing ionization energy until a plateau is reached. A further increase in the ionization energy is now indicated by a slight decrease in the signal intensity. An electron with an energy of 50 eV has a velocity of 4.2×10^6 m/s and crosses a molecular diameter of a few angstroms in about 10^{-16} s (1 Å = 0.1 nM).

Further increases in the signal intensity are, therefore, not achieved via rising the ionization energy, but by using measures to increase the density and the dispersion of the electron beam. The application of pairs of magnets to the ion source can be used, for example.

The standard ionization energy of 70 eV has been established for many years making mass spectra comparable. At an ionization energy of 70 eV, which is in excess of that required for ionization to M^+, remains as excess energy in the molecule, assuming maximum energy transfer. As a result, bond cleavages with fragmentation reactions occur and lead to an immediate decrease in the concentration of M^+ ions in the ion source. At the same time, stable fragment ions are increasingly formed.

Each mass spectrum is the quantitative analysis of the processes occurring during ionization by the MS analyzer. Recorded ion intensities are presented as a line diagram. The involved fragmentation and rearrangement processes are extensively known today. They serve as fragmentation rules for the manual interpretation of mass spectra and thus for the identification of unknown substances (see Section 3.2.6).

2.3.1.2 Chemical Ionization

In contrast to high energetic EI, soft ionization techniques are used in structure elucidation or for an ionization selective to molecular compositions or structures. In EI, molecular ions M^+ are first produced by bombardment of the molecule M with high energy electrons (70 eV). The high excess energy in M^+ (the IP of organic molecules is below 15 eV) leads to unimolecular fragmentation into fragment ions ($F1^+$, $F2^+$, ...) and uncharged species. The EI mass spectrum shows the fragmentation pattern. But, which line in the EI spectrum is the molecular ion? Only a few stable molecules give dominant M^+ ions, for example, aromatics and their derivatives, such as polychlorinated biphenyls (PCBs) or dioxins. The molecular ion is frequently only present with a low intensity. With the small quantities of analytes applied, as is the case with GC-MS, the molecular information can be identified only with difficulty among the noise (matrix), or it fragments completely and does not appear in the spectrum at all (Howe et al., 1981).

The chemical ionization (CI) is used as a soft ionization technique to reduce the ionization energy and hence prevent excess fragmentation. As an example, Figure 2.186 shows the EI and CI spectra of the phosphoric acid ester tolclofos-methyl. The base peak in the EI spectrum reveals a Cl atom, noticeable by the typical isotope pattern of m/z 265/267. The loss of a methyl group $(M-15)^+$ gives a small signal at m/z 250. Is m/z 265 the nominal molecular mass? The CI spectrum shows m/z 301 for a protonated ion, so the nominal molecular mass could be 300 Da. A chlorine isotope pattern of two Cl atoms is also visible (for the isotope pattern of Cl, refer to Section 3.2.6.1).

296 | 2 Fundamentals

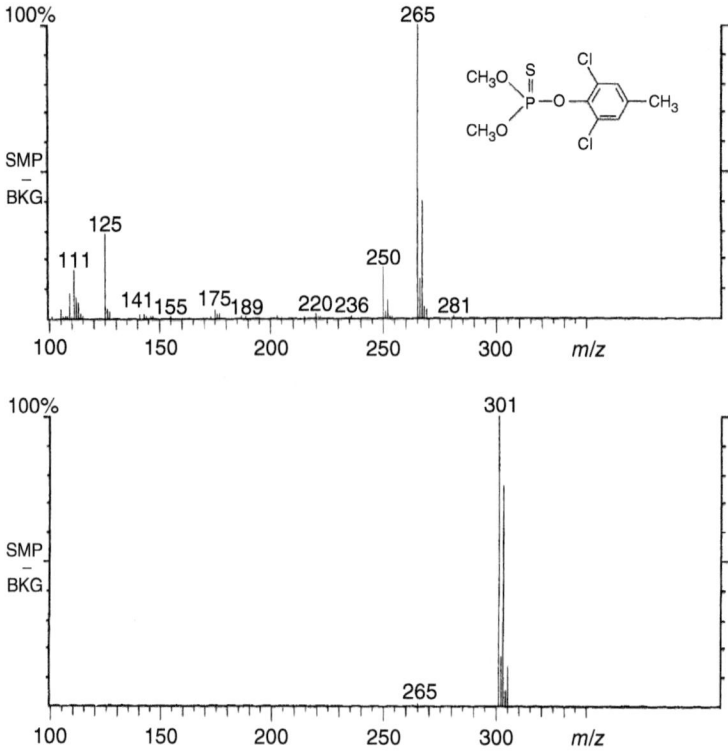

Figure 2.186 EI and CI(NH$_3$) spectra of tolclofos-methyl.

How do the EI and CI spectra complement? Obviously, tolclofos fragments completely in EI by the loss of a Cl atom to m/z 265 as (M−35)$^+$. With CI, this fragmentation does not occur. The attachment of a proton retains the intact molecule with all components of the elemental formula by formation of the cationized molecular ion (M+H)$^+$.

The importance of EI spectra for identification and structure confirmation is related to the fragmentation pattern of the unknown compound. Searches through libraries of spectra are typically based on EI spectra. With the advent of MS/MS instrumentation, also product ion spectra are collected and available for library search. With the introduction of the CI capabilities for internal ionization ion trap systems, a commercial CI library of spectra with more than 300 pesticides was introduced only at that time by Finnigan Corporation.

The term *chemical ionization*, unlike EI, covers all soft ionization techniques which involve an exothermic chemical reaction in the gas phase mediated by a reagent gas and its reagent ions. Stable positive or negative ions are formed as products. Unlike the molecular ions of EI ionization, the cationized molecular ions of CI are not radicals (Harrison, 1992).

The principle of CI was first described by Burnaby Munson and Frank H. Field (1966a, 1966b). CI has developed into a successfully used technique for structural determination and quantitation in GC-MS (Stan and Kellner, 1981). Instead of an open, easily evacuated ion source, a more closed ion volume is necessary for carrying out CI. In the high vacuum environment of the ion source, a reagent gas pressure of about 1 Torr must be maintained to achieve the desired CI reactions. Depending on the construction of the instrument, either changing the ion volume, a change of the whole ion source, or only a software switch is necessary. Through the straightforward technical realization, in the case of combination ion sources, CI has become established in residue and environmental analysis.

The use of CI is helpful in structure determination, confirmation, or determination of molecular masses, and in the determination of significant substructures. Additional selectivity can be introduced into mass spectrometric detection by using the CI reaction of certain reagent gases, for example, the detection of active substances letting a hydrocarbon matrix transparent. Analyses can be quantified selectively, with high sensitivity and unaffected by the low molecular mass matrix by the choice of a less affected quantitation mass in the upper molecular mass range. The spectrum of analytical possibilities with CI is not limited to the basic reactions described briefly here. Furthermore, it opens the whole field of chemical reactions in the gas phase.

The Principle of Chemical Ionization In CI, two reaction steps are always necessary. In the primary reaction, a stable cluster of reagent ions is produced from the reagent gas through electron bombardment. The composition of the reagent gas cluster is typical for the gas used. The cluster formed is displayed for the reagent gas pressure adjustment.

In the secondary reaction, the molecule M of the GC eluate reacts with the ions of the reagent gas cluster. The ionic reaction products are detected and recorded as the CI spectrum. It is the secondary reaction which determines the appearance of the spectrum. Only exothermic reactions give CI spectra. In the case of protonation, this means that the proton affinity PA of M must be higher than that of the reagent gas PA(R) (Figure 2.187 and Table 2.43).

Through the choice of the reagent gas R, the quantity of energy transferred to the molecule M and thus the degree of a potential fragmentation and the question of selectivity can be controlled. If PA(R) is higher than PA(M), no protonation occurs. When a nonspecific hydrocarbon matrix is present, this leads to transparency of the background, while active substances, such as pesticides, appear with high signal-to-noise (S/N) ratios.

For CI, many types of reactions can be used analytically. In gas phase reactions, not only positive, but also negative ions can also be formed (Budzikiewicz, 1981).

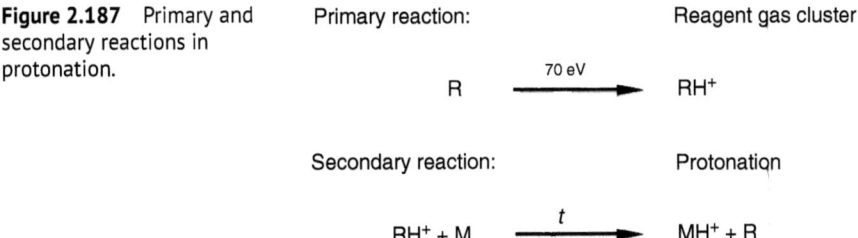

Figure 2.187 Primary and secondary reactions in protonation.

Table 2.43 Proton affinities of selected analyte compounds in [kJ mol^{-1}].

Aliphatic amines			
NH$_3$	857	n-Pr$_2$NH	951
MeNH$_2$	895	i-Pr$_2$NH	957
EtNH$_2$	907	n-Bu$_2$NH	955
n-PrNH$_2$	913	i-Bu$_2$NH	956
i-PrNH$_2$	917	s-Bu$_2$NH	965
n-BuNH$_2$	915	Me$_3$N	938
i-BuNH$_2$	918	Et$_3$N	966
s-BuNH$_2$	922	n-Pr$_3$N	976
t-BuNH$_2$	925	n-Bu$_3$N	981
n-Amyl-NH$_2$	918	Me$_2$EtN	947
Neopentyl-NH$_2$	920	MeEt$_2$N	957
t-Amyl-NH$_2$	929	Et$_2$-n-PrN	970
n-Hexyl-NH$_2$	920	Pyrrolidine	938
Cyclohexyl-NH$_2$	925	Piperidine	942
Me$_2$NH	922	N-methylpyrrolidine	952
MeEtNH	930	N-methylpiperidine	956
Et$_2$NH	941	Me$_3$Si(CH$_2$)$_3$NMe$_2$	966
Oxides and sulfides			
H$_2$O	723	n-Bu$_2$O	852
MeOH	773	i-PrO-t-Bu	873
EtOH	795	n-Pentyl$_2$O	858
n-PrOH	800	Tetrahydrofuran	834
t-BuOH	815	Tetrahydropyran	839
Me$_2$O	807	H$_2$S	738
MeOEt	844	MeSH	788
Et$_2$O	838	Me$_2$S	839
i-PrOEt	850	MeSEt	851
n-Pr$_2$O	848	Et$_2$S	859
i-Pr$_2$O	861	i-Pr$_2$S	875
t-BuOMe	852	H$_2$Se	742
Substituted alkylamines and alcohols			
CF$_3$CH$_2$OH	731	Piperazine	936
CCl$_3$CH$_2$OH	760	1,4-Dioxan	811
CF$_3$NMe$_2$	815	Morpholine	915

Table 2.43 (Continued)

Other N, O, P, S compounds			
Aziridine	902	Ethylene oxide	793
Pyridine	921	Oxetane	823
Piperidine	942	CH_2O	741
HCN	748	MeCHO	790
MeCN	798	$Me_2C=O$	824
EtCN	806	Thiirane	818
Carbonyl compounds, iminoethers, and hydrazines			
EtCHO	800	HCO_2H	764
n-PrCHO	809	$MeCO_2H$	797
Cyclopentanone	835	$EtCO_2H$	808
FCH_2CO_2H	781	n-PrNHCHO	878
$ClCH_2CO_2H$	779	Me_2NCHO	888
CF_3CO_2H	736	$MeNHNH_2$	895
Bases weaker than water			
HF	468	CO_2	530
H_2	422	CH_2	536
O_2	423	N_2O	567
Kr	424	CO	581
N_2	475	C_2H_6	551
Xe	477	—	—

All data in [kJ mol^{-1}] calculated for proton affinities PA(M) at 25 °C corresponding to the reaction $M + H^+ \rightarrow MH^+$ (Aue and Bowers, 1979).

Commercial GC-MS systems usually detect positive ions (positive chemical ionization, PCI) and negative ions (negative chemical ionization, NCI) separately. Specially equipped instruments also allow the simultaneous detection of positive and negative ions produced by CI. The alternating reversal of the ion source and analyzer polarity during scanning (pulsed positive ion negative ion chemical ionization, PPINICI) produces two complementary data files from one analysis (Hunt et al., 1976).

Positive Chemical Ionization Essentially, four types of reactions mainly contribute to the formation of positive ions. As in all CI reactions, reaction partners meet in the gas phase and form a transfer complex M·R⁺. In the following types of reactions, the transfer complex either is retained or reacts further.

Protonation Protonation is the most frequently used reaction in PCI. Protonation leads to the formation of the cationized molecular ion (M+H)⁺, which can then

Table 2.44 Proton affinities of reagent gases for proton transfer.

Gas	Reagent ion	m/z	PA [kJ mol^{-1}]
H_2	H_3^+	3	422
CH_4	CH_5^+	17	527
H_2O	H_3O^+	19	706
CH_3OH	$CH_3OH_2^+$	33	761
$i\text{-}C_4H_{10}$	$t\text{-}C_4H_9^+$	57	807
NH_3	NH_4^+	18	840

undergo fragmentation:

$$M + RH^+ \rightarrow MH^+ + R$$

Methane, water, methanol, isobutane, or ammonia are used as protonating reagent gases (Table 2.44). Methanol occupies a middle position regarding fragmentation and selectivity. Methane is less selective and is designated a "hard" CI gas. Isobutane and ammonia are soft CI gases often used besides other more selective protonating reagent gases (Brodbelt et al., 1991; Landrock et al., 1995).

The CI spectra formed through protonation show the cationized molecular ions $(M+H)^+$. Fragmentations start with this ion. For example, the loss of water shows up as $(M-17)^+$ in the spectrum, formed by $(M+H)^+ - H_2O$. The existence of the cationized molecular ion is often indicated by low signals from addition products of reagent gas ions. In the case of methane, besides $(M+H)^+$, $(M+29)^+$ and $(M+41)^+$ appear (see the section about methane), and for ammonia, besides $(M+H)^+$, $(M+18)^+$ appears with varying intensity (see the section about ammonia).

Hydride Abstraction In this reaction, a hydride ion (H^-) is transferred from the substance molecule to the reagent ion:

$$M + R^+ \rightarrow RH + (M - H)^+$$

This process is observed, for example, in the use of methane when the $C_2H_5^+$ ion (m/z 29) contained in the methane cluster abstracts hydride ions from alkyl chains.

With methane as the reagent gas, both protonation and hydride abstraction can occur, depending on the analyte reaction partner M. The cationized molecular ion obtained is either $(M-H)^+$ or $(M+H)^+$.

Charge Exchange The charge exchange reaction creates a radical molecular ion with an odd number of electrons as in EI. Accordingly, the quality of the fragmentation is comparable to that of an EI spectrum. The extent of fragmentation is determined by the IP of the reagent gas.

$$M + R^+ \rightarrow R + M^{·+}$$

Table 2.45 CI reagent gases for charge exchange reactions by increasing IP.

Gas	Reagent ion	m/z	IP [eV]
C_6H_6	$C_6H_6^+$	78	9.3
Xe	Xe^+	131	12.1
CO_2	CO_2^+	44	13.8
CO	CO^+	28	14.0
N_2	N_2^+	28	15.3
Ar	Ar^+	40	15.8
He	He^+	4	24.6

The IP of organic compounds is below 15 eV. The choice of reagent gas controls whether a dominant molecular ion appears in the spectrum or if and how extensively fragmentation occurs. In the extreme case, spectra similar to those with EI are obtained. Common reagent gases for ionization by charge exchange are benzene (Allgood et al., 1990), nitrogen, carbon monoxide, nitric oxide, or argon, as listed in Table 2.45. When using methane, charge exchange reactions besides protonations can be observed, in particular for molecules with low proton affinities.

Adduct Formation If the transition complex described earlier does not dissociate, the adduct ion is visible in the spectrum:

$$M + R^+ \rightarrow (M + R)^+.$$

This effect is only frequently observed in GC-MS analysis in contrast to the prevailing reaction in ESI-LC-MS (electrospray ionization [ESI]), but must be taken into account when evaluating CI spectra. The enhanced formation of adducts is always observed with intentional protonation reactions where differences in the proton affinity of the participating species are small. High reagent gas pressure in the ion source favors the effect by stabilizing collisions.

Frequently, an $(M+R)^+$ ion is not immediately recognized, but can give information which is as valuable as that from the cationized molecular ion formed by protonation (see the section about methane). Cluster ions of this type, nevertheless sometimes, make the interpretation of spectra more difficult, particularly when the transition complex does not release immediately recognizable neutral species.

Negative Chemical Ionization Negative ions are also formed during ionization in MS even under EI conditions, but their yield is extremely low that it is of no use analytically. The intentional production of negative ions can take place by attachment of thermal electrons (analogous to an electron capture detector [ECD]), by charge exchange, or by extraction of acidic hydrogen atoms (Dougherty, 1981; Horning et al., 1981; DePuy et al., 1982; Stout and Steller, 1984).

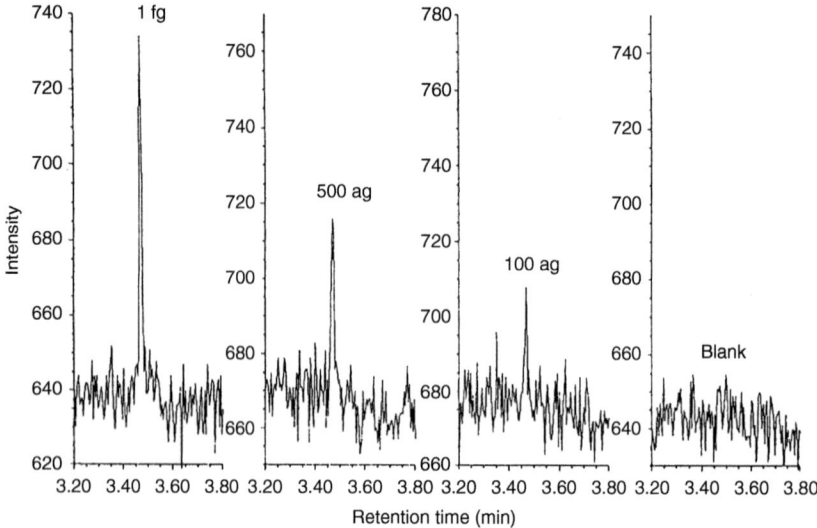

Figure 2.188 GC-MS detection of traces of 10^{-15}–10^{-16} g octafluoronaphthalene by NCI detection of the molecular ion m/z 272 (McLafferty and Turecek, 1993).

Electron Capture Electron capture with the formation of negative ions is the NCI process is most frequently used in the GC-MS analysis. There is a direct analogy with the behavior of substances in the ECD, and the areas of use are similar (Class, 1991; Pizzutti et al., 2012). The commonly used term *ECD-MS* indicates the parallel mechanisms and applications. With NCI, the lowest detection limits in organic MS have been reached (Figure 2.188). The detection of 100 ag octafluoronaphthalene corresponds to the detection of about 200 000 molecules.

At the same energy, electrons in the CI plasma have a much higher velocity (mobility) than those of the heavier positive reagent ions,

$$E = m/2 \cdot v^2 \qquad m(e^-) = 9.12 \cdot 10^{-28} g$$
$$m(CH_5^+) = 2.83 \cdot 10^{-22} g.$$

Electron capture as an ionization method is 100 to 1000 times more sensitive than ion/molecule reactions which are limited by diffusion. For substances with high electron affinities (EAs), higher sensitivities can be achieved than with PCI. Substances which have a high NCI response typically have a high proportion of halogen or nitro groups (Figure 2.189). NCI permits the detection of trace components in complex biological matrices (Figure 2.190). In practice, it has been shown that about five to six halogen atoms in the molecule detection with NCI give a higher specific response than that using EI (Figure 2.191). The decrease in the response in the EI mode with increasing chlorine content is caused by splitting of the overall signal by individual isotopic masses. In the NCI mode, the response increases with increasing chlorine content through the increase in the electronegativity for electron capture (analogous to an ECD, compound dependent). For this reason, in analysis of polychlorinated dioxins and furans, although chlorinated, EI is the predominant method

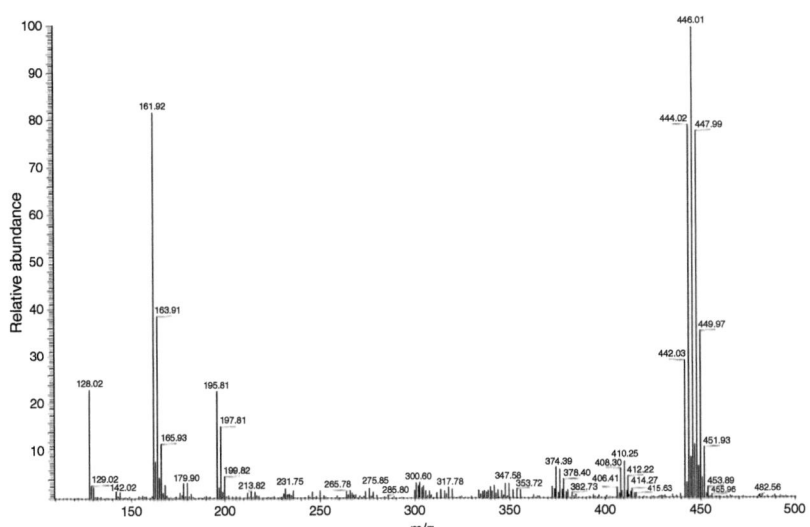

Figure 2.189 NCI spectrum of the toxaphen component Parlar 69, (2,2,5,5,6-exo,8,9,9,10,10-decachlorobornane). The typical fragmentation known from EI ionization is absent, and the total ion current is concentrated on the molecular ion range (Hainzl et al., 1993, 1994; Theobald and Hübschmann, 2007).

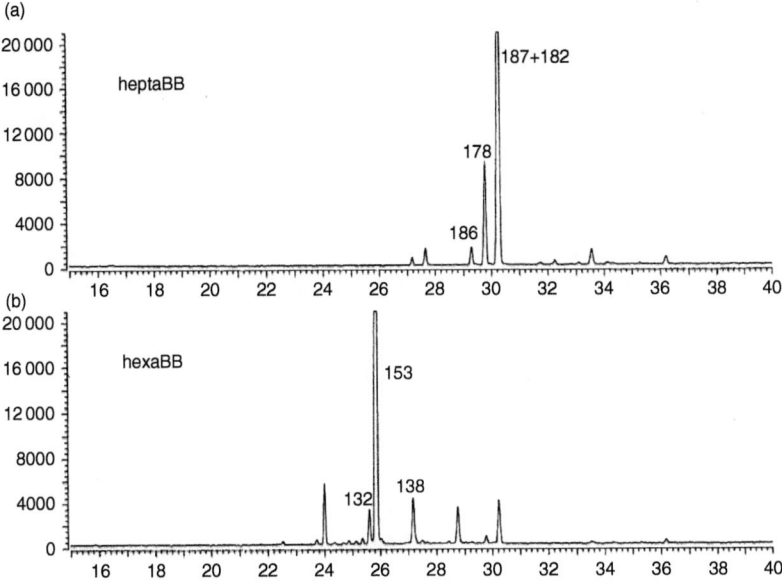

Figure 2.190 Detection of heptabromobiphenylene (a) and hexabromobiphenylene (b) in human milk by NCI (Fürst, 1994).

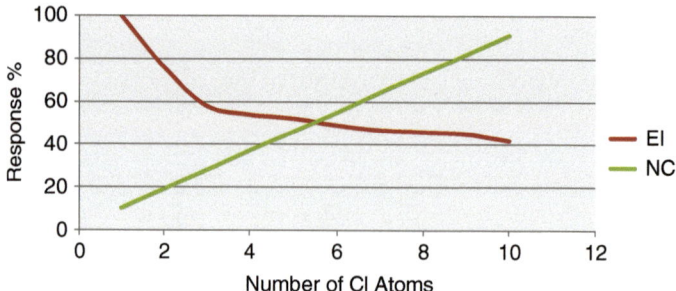

Figure 2.191 Response dependence of the increasing number of Cl atoms in a molecule in the EI and NCI modes.

for the detection of 2,3,7,8-TCDD (tetrachlorodibenzodioxin) in trace analysis (Crow et al., 1981; Buser and Müller, 1994; Chernetsova et al., 2002; Pizzutti et al., 2012).

Another feature of NCI measurements is the fact that, like ECD, the response depends not only on the number of halogen atoms but also on their position in the molecule. Precise quantitative determinations are therefore only possible with defined reference systems via the determination of specific response factors.

The key to sensitive detection of negative ions lies in the production of a sufficiently high population of thermal electrons. The extent of formation of M^- at sufficient electron density depends on the EA of the sample molecule, the energy spectrum of the electron population, and the frequency with which molecular anions collide with neutral particles and become stabilized (collision stabilization of the radical ion).

For residue analysis, derivatization with perfluorinated reagents (e.g. heptafluorobutyric anhydride and perfluorobenzoyl chloride) in association with NCI is important (Knapp, 1979; Yang et al., 2020). Besides being easier to chromatograph, the derivatized analytes show high EAs due to the high number of halogens and allow a very sensitive detection (see Chapter 4.30).

Charge Transfer The ionization of the sample molecule is achieved by the transfer of an electron between the reagent ion and the molecule M,

$$M + R^- \rightarrow M^- + R.$$

The reaction can only take place if the EA of the analyte M is greater than that of the electron donor R,

$$EA(M) > EA(R).$$

In practical analysis, charge transfer to form negative ions is less important unlike the formation of positive ions by charge exchange.

Proton Abstraction Proton transfer in NCI can be understood as proton abstraction from the sample molecule. In this way, all substances with acidic hydrogens, for example, alcohols, phenols, or ketones, can undergo soft ionization,

$$M + R^- \rightarrow (M - H)^- + RH.$$

Proton abstraction only occurs if the proton affinity of the reagent gas ion is higher than that of the conjugate base of the analyte molecule. A strong base, for example, OH⁻, is used as the reagent for ionization. Substances which are more basic than the reagent are not ionized (Smit and Field, 1977).

Reagent gases and organic compounds were arranged in order of gas-phase acidity by Bartmess and McIver (1979). The order corresponds to the reaction enthalpy of the dissociation of their functional groups into a proton and the corresponding base. The gas-phase acidity scale in Figure 2.192 can be used for controlling the selectivity via proton abstraction by choosing suitable reagent gases.

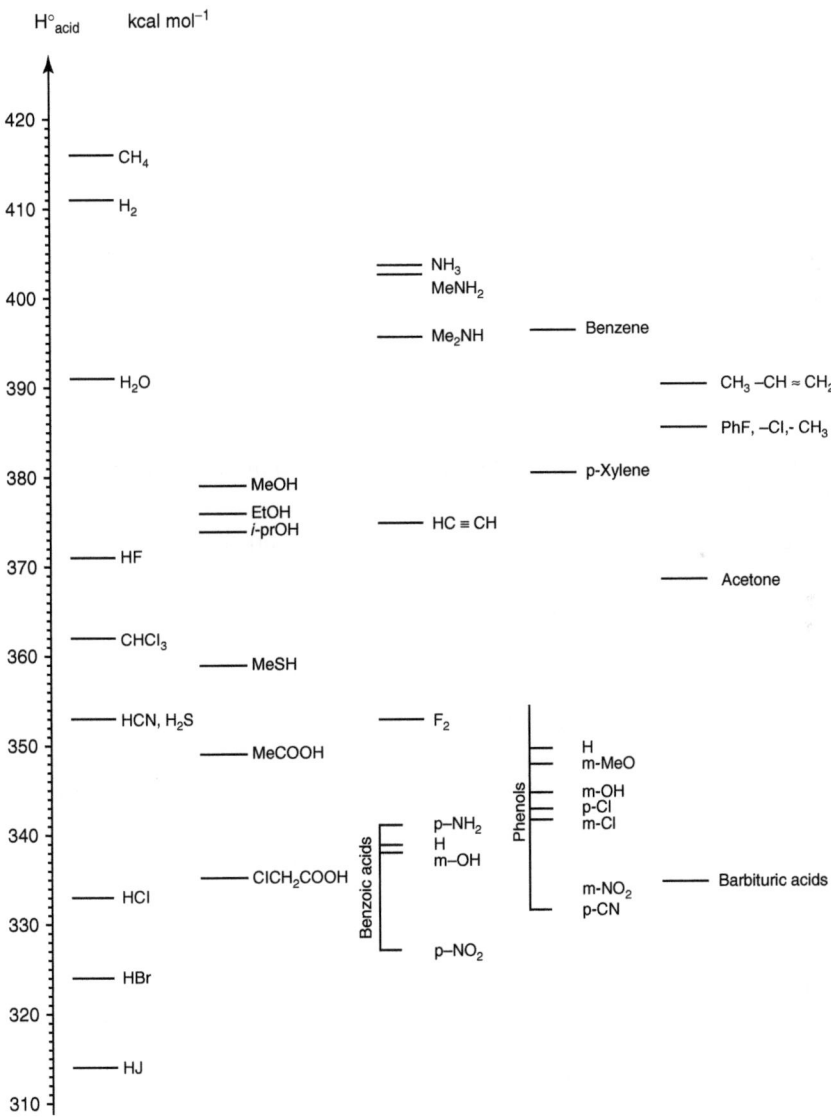

Figure 2.192 Scale of gas phase acidities (after Bartmess and McIver, 1979).

Extensive fragmentation reactions can be excluded by proton abstraction. The energy released in the exothermic reaction is essentially localized in the new compound RH. The anion formed does not contain excess energy for extensive fragmentation.

Reagent Ion Capture The capture of negative reagent gas ions was described in the early 1970s by Manfred von Ardenne and coworkers for the analysis of long-chain aliphatic hydrocarbons with hydroxyl ions (Ardenne et al., 1971),

$$M + R^- \rightarrow M \cdot R^-.$$

Besides associative addition with weak bases, adduct formation can lead to a new covalent bond. The ions formed are more stable than the comparable association products.

Substitution reactions, analogous to an S_N2 substitution in solution, occur more frequently in the gas phase because of poor solvation and low activation energy. Many aromatics give a peak at $(M+15)^-$, which can be attributed to the substitution of H by an O^- radical. Substitution reactions are also known for fluorides and chlorides. Fluoride is the stronger nucleophile and displaces chloride from alkyl halides.

Reagent Gas Systems

Methane Methane is one of the longest known and best-studied reagent gases. It is known and used as a "hard" reagent gas for general CI applications and has been replaced by softer ones in many areas of analysis.

The reagent gas cluster of methane is formed by a multistep reaction, which gives two dominant reagent gas ions with m/z 17 and 29, and in lower intensity an ion with m/z 41,

$$CH_4 \text{ at } 70 \text{ eV} \rightarrow CH_4^+, CH_3^+, CH_2, CH^+ \text{ and others}$$
$$CH_4^+ + CH_4 \rightarrow CH_5^+ + CH_3 \quad m/z\ 17 \quad 50\%$$
$$CH_3^+ + CH_4 \rightarrow C_2H_5^+ + H_2 \quad m/z\ 29 \quad 48\%$$
$$CH_2^+ + CH_4 \rightarrow C_2H_3^+ + H_2 + H$$
$$C_2H_3^+ + CH_4 \rightarrow C_3H_5^+ + H_2 \quad m/z\ 41 \quad 2\%$$

Good CI conditions are achieved if a ratio of m/z 17 to m/z 16 of 10 : 1 is achieved. Experience shows that the correct methane pressure is that at which the ions m/z 17 and 29 dominate in the reagent gas cluster and have approximately the same height with good resolution. The ion m/z 41 should also be recognizable with lower intensity (Figure 2.193).

Methane is used as the reagent gas in protonation reactions, charge exchange processes (PCI) and in pure form or as a mixture with N_2O in the formation of negative ions (NCI). In protonation, methane is causing some fragmentation of the molecular ion. For substances with lower proton affinity, methane frequently provides the final possibility of obtaining CI spectra. The unique adduct ions $(M+C_2H_5)^+ = (M+29)^+$ and $(M+C_3H_5)^+ = (M+41)^+$ formed by the methane cluster confirm the molecular mass interpretation. These adduct ions are easily seen as a mass difference of 28 resp. 40 from the protonated molecular ion $(M+H)^+$.

Figure 2.193 Reagent gas cluster with methane.

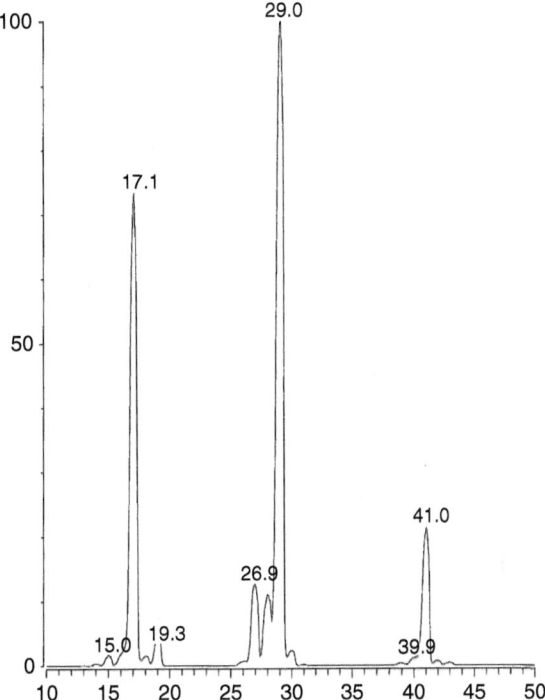

Ammonia Ammonia is a very soft reagent gas for protonation. The selectivity is correspondingly high, which is made use of in the residue analysis of many active substances. Fragmentation reactions only occur to a small extent in ammonia CI.

Adduct formation with NH_4^+ occurs with substances where the proton affinity differs little from that of NH_3 and can be used to confirm the molecular mass interpretation. At higher pressures, adducts of the ammonium ion with ammonia $(NH_3)_n \cdot NH_4^+$ can be formed in the ion source. Interpretation and quantitation can thus be impaired with such compounds. In ion trap instruments, only the ammonium ion NH_4^+ with mass m/z 18 is formed in the reagent gas cluster.

Attention: Very often the ammonium ion is confused with water. Freshly installed reagent gas tubing generally has an intense water background with high intensities at m/z 18 and 19 as H_2O^+/H_3O^+. Clean tubing and correctly adjusted NH_3 CI gas show no intensity at mass m/z 19 (Figure 2.194). To supply the CI system with ammonia, a steel cylinder with a special reducing valve is necessary. Because of the aggressive properties of the gas, the entire tubing system must be made of stainless steel.

Isobutane Like methane, CI with isobutane has been known since long and is well documented. The *t*-butyl cation (m/z 57) is formed in the reagent gas cluster and is responsible for the soft character of the reagent gas (Figure 2.195).

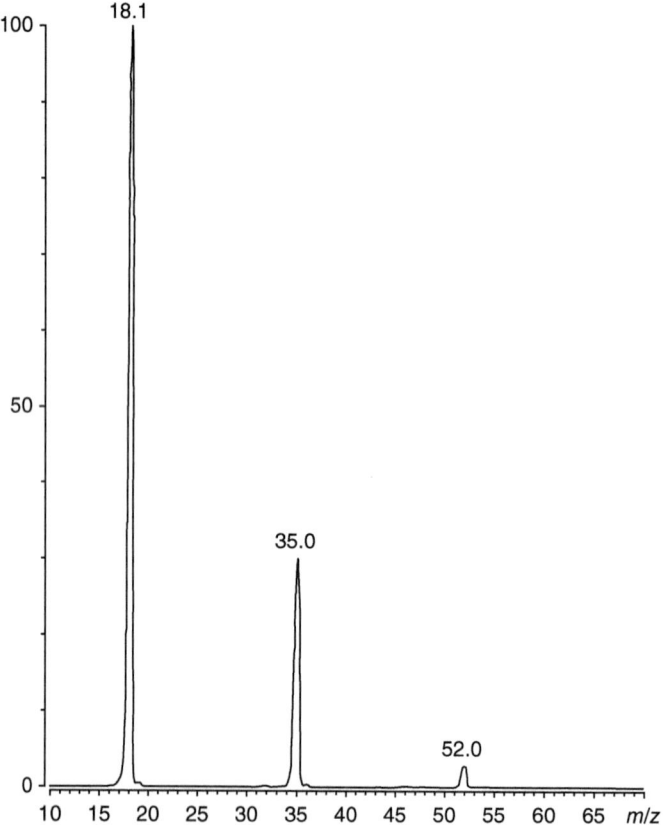

Figure 2.194 Ammonia reagent gas cluster with di- and trimer ions depending on ion source pressure.

Isobutane is used for protonation reactions of multifunctional and polar compounds. Its selectivity is high, and there is very little fragmentation. Even less fragmentation with increased intensities of the molecular ion is observed with isopentane as a reagent gas (McGuire and Munson, 1985).

In practice, with continued use, significant coating of the ion source caused by soot formation has been observed, which can even lead to dousing of the filament. This effect depends on the pressure adjustment and on the instrument. In such cases, ammonia can be used instead.

Liquid CI Gases
Methanol Methanol is a highly suitable CI reagent for soft ionization and used exclusively for protonation. Because of its medium proton affinity, methanol allows a broad spectrum of compound classes to be ionized. The medium proton affinity does not give any pronounced selectivity. However, substances with predominantly alkyl character remain transparent. Fragments have low intensities.

The ion $(CH_3OH \cdot H)^+$ is formed as the protonating ion, which is adjusted to high intensity with good resolution (Figure 2.196). The appearance of a peak at

Figure 2.195 Isobutane reagent gas cluster.

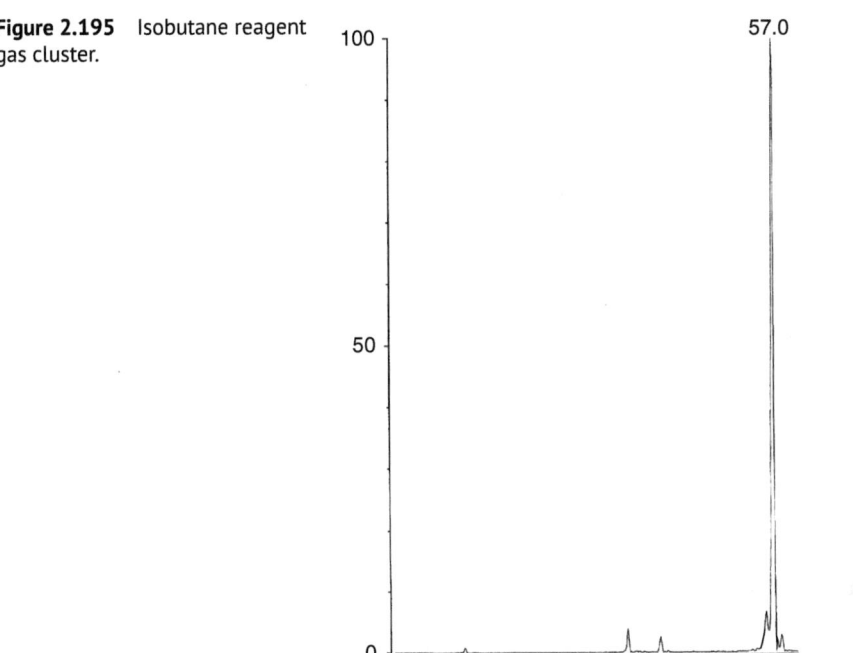

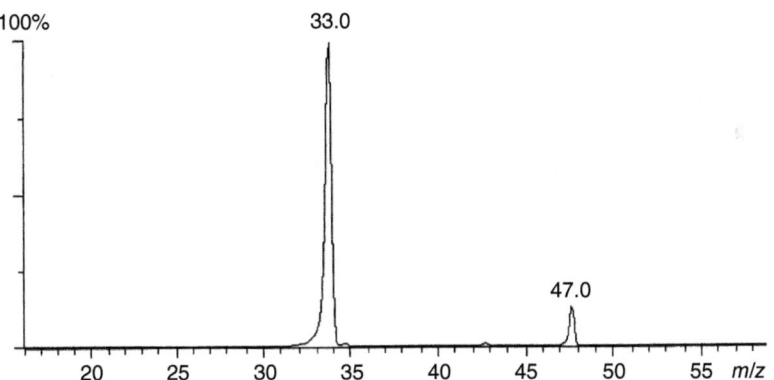

Figure 2.196 Methanol as a reagent gas (ion trap with internal ionization).

m/z 47 shows the dimer formed by the loss of water (dimethyl ether), which is only produced at high methanol concentrations. It does not function as a protonating reagent ion, but its appearance shows that the pressure adjustment is correct.

Because of the low vapor pressure, a gently heated methanol reservoir and heated gas lines are required for a constant "gas" regulation. Methanol is ideal for CI in ion trap instruments with internal ionization. Neither pressure regulators nor cylinders or a long tubing system is required here. The connection of a glass flask or a closed tube containing methanol directly on to the CI inlet is sufficient. Also, for ion traps using external ionization and quadrupole instruments, liquid CI devices are commercially available. In addition, every laboratory has methanol available.

Water The use of water as a reagent gas is universal, but for most mass spectrometrists, water is a problematic substance. However, as a reagent gas, water has extraordinary properties. Because of the high conversion rate into H_3O^+ ions, and due to its low proton affinity, water achieves a high response for many compounds when used as the reagent gas. The spectra obtained usually have few fragments and concentrate the ion beam on a dominant ion.

In the determination, for example, of PAHs, a significant increase in response compared with EI detection is reported. Analytical procedures have also been published for nitroaromatics. Water can also be used successfully as a reagent gas for screening small molecules, for example, volatile halogenated hydrocarbons (industrial solvents), as the reagent ion does not interfere with the low scan range for these substances (Landrock et al., 1995). When water is used as the reagent gas (Figure 2.197), the intensity of the H_3O^+ ion m/z 19 should be as high as possible, still with good mass resolution. A gently heated reservoir and heated gas lines are required for stable adjustments.

Atmospheric Pressure Chemical Ionization A dedicated ion source for the soft ionization under atmospheric pressure for GC-MS was developed already in the early 1970s by Evan C. Horning and team at the Houston, TX, USA, Baylor College of Medicine (Horning et al., 1973). A ^{63}Ni foil element as a β-particle emitter was used for primary ionization of nitrogen. Positive and negative ions can be formed by ion-molecule reactions, e.g. protonation, or charge transfer for positive ions, or, e.g. the capture of thermal electrons, or dissociative electron capture for the generation of negative ions. The characteristic of the atmospheric pressure chemical ionization (APCI) ion source is its high sensitivity, and the concentration of the ion flux on few analyte ions, preferably the molecular or cationized molecular ions, as illustrated in Figure 2.198, with a significant increase in sensitivity. Lower ion source temperatures than EI/CI sources reduce the thermal decomposition of analytes.

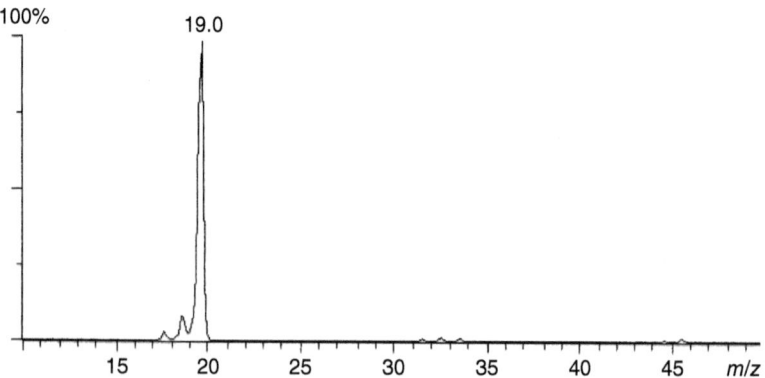

Figure 2.197 Water as a reagent gas (ion trap analyzer with internal ionization).

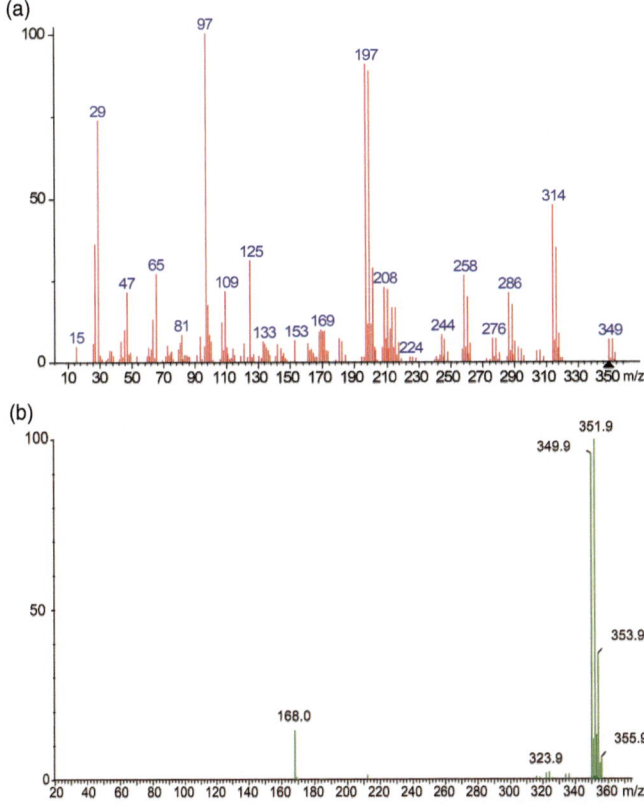

Figure 2.198 Spectra of chlorpyrifos ($C_9H_{11}Cl_3NO_3PS$) M 348.9 Da. (a) EI spectrum, 70 eV (NIST), highly fragmented. (b) APCI spectrum, ion flux concentrated on [M+H]⁺ base peak, no fragmentation.

Today, a corona discharge is used for the primary ionization of nitrogen to form excited species for a further ionization cascade,

$$N_2 + e^- \rightarrow N_2^+ + 2e^-$$
$$N_2^{+*} + 2N_2 \rightarrow N_4^{+*} + N_2.$$

APCI ion sources are mainly in use for the easy hyphenation with LC-MS applications for high resolution time-of-flight (TOF) or quadrupole MS/MS instrumentation. Commercial GC-MS applications were introduced in 2009 when Waters Corporation launched a GC-APCI source under the name of atmospheric pressure gas chromatography (APGC) mainly in use with accurate mass and triple quadruple instrumentation (Ayala-Cabrera et al., 2023; Waters, 2023). Applications are found in low level trace analysis for instance with pesticides (Fang et al., 2019) or dioxins (van Bavel et al., 2015). For the analysis of dioxins and PCBs, it was reported that using APGC-MS/MS can reach lower detection limits than GC-HRMS and meets the requirements of the European Commission (Jones et al., 2018). However, due to generally higher selectivity, GC-HRMS with magnetic sector instrumentation stays the preferred method (ten Dam et al., 2016).

Aspects of Switching Between EI and CI

Quadrupole and Magnetic Sector Instruments An ion source pressure of about 1 Torr in an analyzer environment of 10^{-5}–10^{-7} Torr is necessary for quadrupole and magnetic sector instruments to initiate CI reactions and to guarantee a sufficient conversion rate. For this, the EI ion source is replaced by a special CI source, which must have almost gas-tight connections to the GC column, the electron beam, and the ion exit in order to maintain the pressure in these areas.

Combination sources with mechanical devices for sealing the EI to the CI source have so far only proved successful with magnetic sector instruments. With quadrupole instruments, combination ion sources (or ion volumes) are available as well. The performance in both ionization modes EI and CI is compromised. The response is below optimized sources, and EI/CI mixed spectra may be produced.

Increased effort is required for conversion, pumping, and calibrating the CI source in beam instruments. Because of the high pressure, the reagent gas also leads to rapid contamination of the ion source and thus to additional cleaning measures in order to restore the original sensitivity of the EI system. Readily exchangeable ion volumes have been shown to be ideal for CI applications. This permits a high CI quality to be attained and, after a rapid exchange, unaffected EI conditions to be restored.

Ion Trap Instruments The former ion trap mass spectrometers with internal ionization can use CI without hardware conversion. Because of their mode of operation as storage mass spectrometers, only a very-low reagent gas pressure is necessary for instruments with internal ionization. The pressure is adjusted by means of a special needle valve which is operated at low leak rates and maintains a partial pressure of only about 10^{-5} Torr in the analyzer. The overall pressure of the ion trap analyzer of about 10^{-4}–10^{-3} Torr remains unaffected by it. CI conditions thus set up give rise to the term *low pressure CI*. Compared to the conventional ion source used in high pressure CI, in protonation reactions, for example, a clear dependence of the CI reaction on the proton affinities of the reaction partners is observed. Collision stabilization of the products formed does not occur with low pressure CI. This explains why "high pressure" CI-typical adduct ions are not formed here, which would confirm the identification of the (cationized) molecular ion, e.g. with methane besides $(M+H)^+$, also M+29 and M+41 are expected. The determination of ECD-active substances by electron capture (NCI) is not possible with low pressure CI (Yost, 1988).

The CI reaction is initiated when the reagent ions are generated by changing the operating parameters and a short reaction phase has taken place in the ion trap analyzer. The scan function used in the CI mode with ion trap instruments

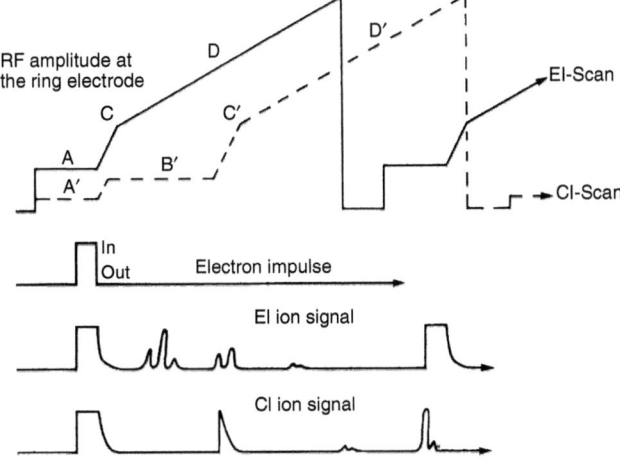

Figure 2.199 Electrical switching between the EI and CI scan functions in the case of an ion trap analyzer with internal ionization (Finnigan). EI scan: (A) Ionization and storage of ions; (C) starting mass; (D) recording of an EI mass spectrum. CI scan: (A') Ionization and storage of reagent gas ions, (B') reaction of reagent gas ions with neutral substance molecules, (C') starting mass, and (D') recording of a CI mass spectrum.

(see Figure 2.199) shows the two plateaus which correspond to the primary and secondary reactions. After the end of the secondary reaction, the product ions, which have been produced and stored, are determined by the mass scan and the CI spectrum registered. Despite the presence of the reagent gas, typical EI spectra can therefore be registered in the EI mode. The desired CI is made possible by simply switching to the CI operating parameters.

2.3.2 Mass Analysis

> "The ideal mass analyzer does not exist!"
>
> *Curt Brunée 1987*
> 1928–2023, German physicist known for his contributions
> to the field of mass spectrometry instrumentation

Ions generated in the ion source get accelerated via an extraction and focusing lens system toward the mass analyzer. Due to their mass and the electrical charge, ions are susceptible to different types of mass analyzers. While the ion source is mainly responsible for sensitivity, the choice of type of mass spectrometer depends on the different requirements for the intended analysis like mass range, mass accuracy, mass resolution, selectivity, the speed of analysis, and for sure the lifetime of ions. In GC-MS, some parameters are already predefined by the chromatography with the compounds amenable and its time axis.

> **The Time Aspect in the Formation and Determination of Ions in MS**
>
> **Ion generation**
> - Flight time of an electron through an organic molecule (70 eV) 10^{-16} s
> - Formation of the molecular ion M$^+$ (EI) 10^{-12} s
> - Fragmentation reactions finished 10^{-9} s
> - Rearrangement reactions finished 10^{-6}–10^{-7} s
> - Lifetime of metastable ions 10^{-3}–10^{-6} s
>
> **Flight times of ions**
> - Magnetic sector analyzer 10^{-5} s
> - Quadrupole analyzer 10^{-4} s
> - Ion trap analyzer (storage times) 10^{-2}–10^{-6} s
> - Orbitrap analyzer 10^{-3}–10^{0} s

> **Mass Resolution**
>
> The smallest mass difference Δm between two equal magnitude peaks so that the valley between them is a specified fraction of the peak height.
>
> **Mass Resolving Power**
>
> In a mass spectrum, the observed mass divided by the difference between two masses that can be separated: $m/\Delta m$.
>
> The procedure by which Δm was obtained and the mass at which the measurement was made shall be reported as the full width at half maximum (FWHM) or 10% valley of two adjacent peaks of similar height (equals 5% peak height of a single isolated mass peak).

2.3.2.1 Resolving Power and Resolution in Mass Spectrometry

The resolving power of a mass spectrometer describes the smallest mass differences which can be separated by the mass analyzer (Webb et al., 2004; Murray, 2022). Resolution and resolving power in MS today are defined differently depending on the analyzer or instrument type and are often stated without the indication of the definition employed. The IUPAC definitions of terms used in MS provide a precise definition.

The mass resolving power R can be calculated by the comparison of two adjacent mass peaks m_1 and m_2 of about equal height. R is defined as

$$\frac{m_1}{\Delta m}$$

with m_1 being the lower value of two adjacent mass peaks. Δm of the two peaks is taken at an overlay (valley) of 10%, or at 50% in the case of the FWHM definition. In the 10% valley definition, the height from the baseline to the junction point of

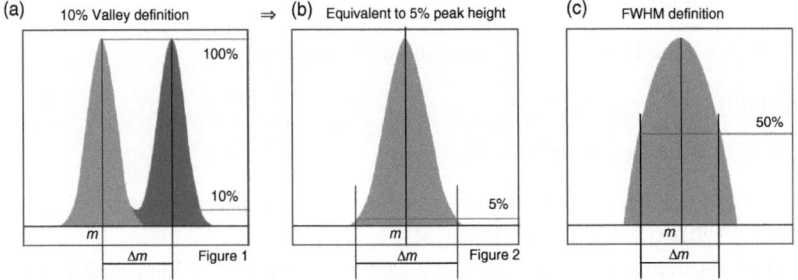

Figure 2.200 Comparison of different mass resolution definitions (Münster and Taylor, 2009). (a) 10% valley definition, (b) single peak 5% height equivalent to 10% valley, and (c) the 50% FWHM definition.

the two peaks is 10% of the full height of the two peaks. Each peak at this point is contributing 5% to the height of the valley.

Because it is difficult to get two mass spectral peaks of equal height adjacent to one another, the practical method of calculating Δm as typically done by the instrument's software is using a single mass peak of a reference compound. The peak width is measured in 5% height for the 10% valley definition and at 50% peak height for the FWHM definition, as shown in Figure 2.200. The resolving power calculated using the FWHM method gives values for R that are about twice of that which is determined by the 10% valley method. This can be checked using the intercept theorem calculation in a triangle in which the ratios of height to width are equal:

$$\frac{h_1}{w_1} = \frac{h_2}{w_2}$$

$$w_2 = \left(\frac{h_2}{h_1}\right) \times w_1,$$

and as the peak height above 5% intensity compares to the half-peak height very close by a factor of 2,

$$h_2 \sim 2 \times h_1,$$

following

$$w_2 \approx \left(2 \times \frac{h_1}{h_1}\right) \times w_1$$

$$\approx 2 \times w_1$$

for

h_1 = half peak height
w_1 = peak width at 50% height
h_2 = peak height above 5% to top
w_2 = peak width at 5%

following for the mass peak width that

$$\Delta m \text{ (5\% peak height)} \approx 2 \times \Delta m \text{ (50\% peak height)}.$$

Consequently,

$$R(10\% \text{ valley}) \times 2 \approx R(\text{FWHM}).$$

In practice, a resolving power R 60 000 at 10% valley compares directly to a specification of R 120 000 at half height (FWHM).

The mass resolving power for magnetic sector instruments is historically given with a 10% valley definition. A mass peak in the high resolution magnetic sector MS is typically triangular with this analyzer type. The peak width in 5% height is of valuable diagnostic use for nonoptimal analyzer conditions. Therefore, this method provides an excellent measure for the quality of the peak shape together with the resolving power which would not be available at the half maximum condition. The typical broad peak base was initially observed in TOF mass spectrometers and caused the 10% valley definition would only calculate poor resolution values for this type of analyzer. The more practical approach to describe the resolving power of TOF analyzers is the measurement at half-peak height, or as it is usually stated, at the FWHM. Both resolving power conditions compare by a factor of 2 as outlined above.

Constant Resolving Power over the Mass Range Double focusing mass spectrometers using both electric and magnetic fields to separate ions operate at constant mass resolving power. At a resolving power of 10 000, these instruments separate ions of m/z 1000 and m/z 1000.1. In this example, Δm is 0.1 and M is 1000; therefore, $R = 1000/0.1$ or 10 000. For practical use, this property of constant resolving power over the entire mass range means that with a resolving power of 1000, values of

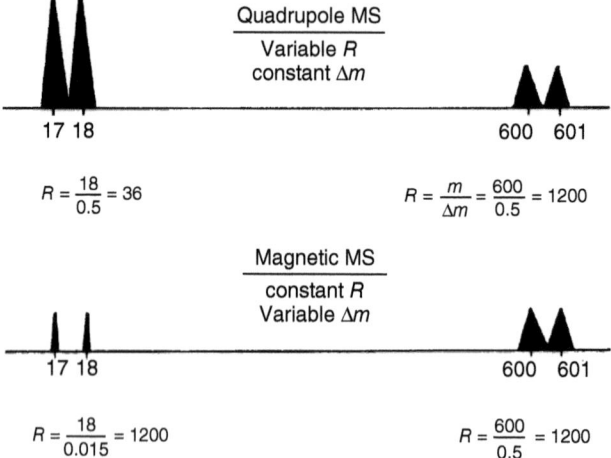

Figure 2.201 Comparison of the resolution power R obtained from quadrupole/ion trap and magnetic sector analyzers. Note the constant peak width of the quadrupole analyzer and the constant resolution power of the magnetic sector instrument over the mass range.

0.01 Da can be separated at m/z 100, that is $R = 10\,000 = m/\Delta m = 100/0.01$. This implies that the visible distance of mass peaks on the m/z scale decreases with increasing mass range, as displayed in Figure 2.201 compared to the quadrupole analyzer. A practical result from dioxin analysis using a high resolution mass analyzer compared to the unit mass resolving quadrupole is shown in Figure 2.202.

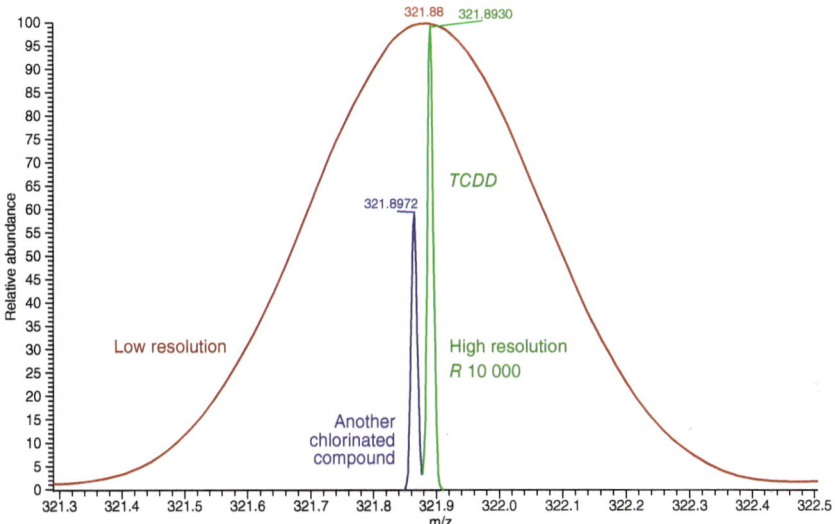

Figure 2.202 Low and high resolved 2,3,7,8-TCDD mass peak, m/z 321.8930 at $R = 10\,000$ (10% valley). The background interference of m/z 321.8972 cannot be resolved at low quadrupole resolution.

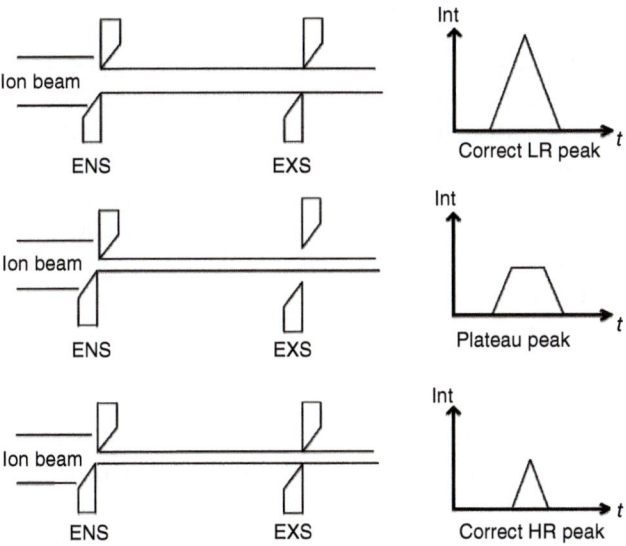

Figure 2.203 Principle of the high resolution adjustment on a magnetic sector MS adjusting the entrance slit (ENS) and the exit slit (EXS).

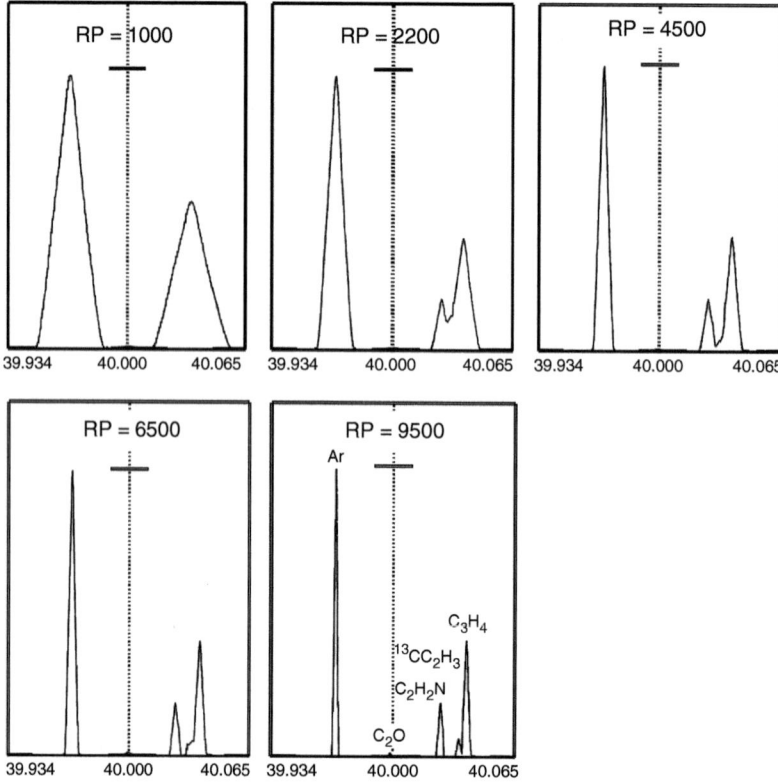

Figure 2.204 Example of the separation and peak widths of the two ions Ar⁺ and $C_3H_4^+$ with increasing resolution power (RP). Ar⁺: 39.9624 Da, C_2O^+: 39.9949 Da, $C_2H_2N^+$: 40.0187 Da, $^{13}C_1C_2H_3^+$: 40.0268 Da, $C_3H_4^+$: 40.0313 Da.

High resolution mass spectrometers are using data system controlled split systems for resolution adjustment at the entrance and exit of the ion beam to and from the analyzer. The adjustment is performed first by closing the entrance slit to get a flat top peak. Next, the exit slit is closed accordingly until the intensity starts to decrease and a triangular peak is formed (see Figure 2.203). No other adjustments in the focusing conditions are necessary to achieve the resolution setting. Figure 2.204 shows the example of the separation and peak widths of the two ions Ar⁺ and $C_3H_4^+$. With increasing resolution power from R 1000 to 9500, the peak width becomes significantly narrower allowing the detection of additional ion species with small mass differences.

Constant Resolution Over the Mass Range Quadrupole and ion trap mass analyzers show constant peak widths and mass differences over the entire mass range. Hence, both analyzer types operate at constant mass resolution with increasing resolving power (see Figure 2.201). This means, that the ability to separate ions at m/z 100 and m/z 1000 is the same. If Δm is 1 mass unit at m/z 100, the resolution at m/z 100 is 1 and the resolving power R is 100/1 or 100. If Δm is 1 at m/z 1000, the resolution at m/z

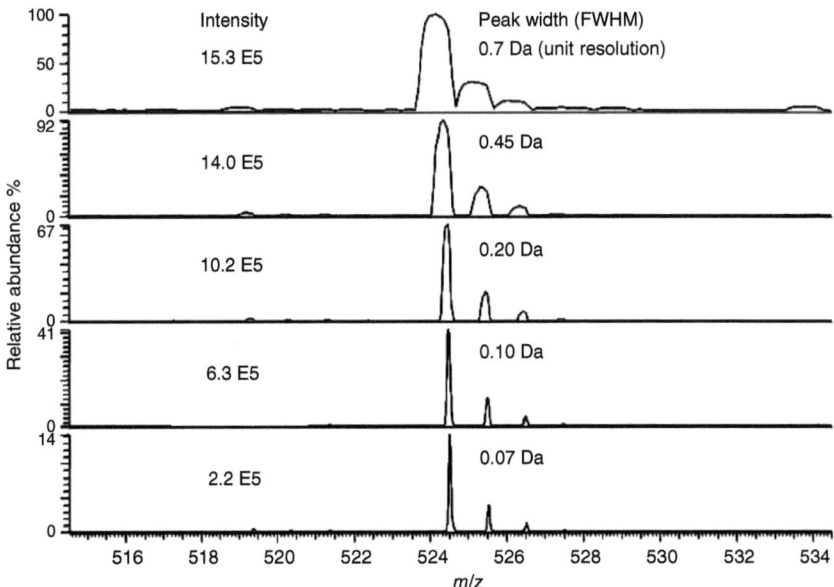

Figure 2.205 Increased peak resolution of hyperbolic quadrupoles as of Figure 2.207. Note the decrease in the peak height of factor 2 at a 0.1 Da peak width compared to unit mass resolution.

1000 is also 1, but the resolving power R at m/z 1000 is 1000/1 or 1000. Consequently, these types of analyzers operate at increasing resolving power with increasing m/z value. Accordingly, the maximum resolving power, which is usually specified with a commercial quadrupole system, is dependent on the maximum specified mass range of the employed quadrupole hardware.

The resolution Δm, which is the visible peak width in Figure 2.212, obtained with commercial quadrupole or ion trap systems is constant throughout the entire mass range. Only hyperbolic-shaped quadrupole rods of a special length and precision machining allow the operation mode for narrower mass peak widths down to 0.1 Da without losing significant ion transmission, as shown in Figure 2.205.

Enhanced Mass Resolution with Quadrupole Analyzers Already in 1968, a publication by Dawson and Whetten dealt with the resolution capabilities of quadrupoles (Dawson and Whetten, 1968). The theoretical investigation covered round rods as well as hyperbolic rods and indicated for the first time the higher resolution potential of quadrupole analyzers using hyperbolic rods (see Figure 2.206). Especially with precisely machined hyperbolic rods, sufficient ion transmission is achieved even at higher resolving power, making this technology especially useful for target compound analysis with increased selectivity. The maximum resolving power of quadrupole analyzers depends on the number of cycles an ion is exposed to the electromagnetic fields inside of the quadrupole assembly, hence on operation frequency and the length of the rods.

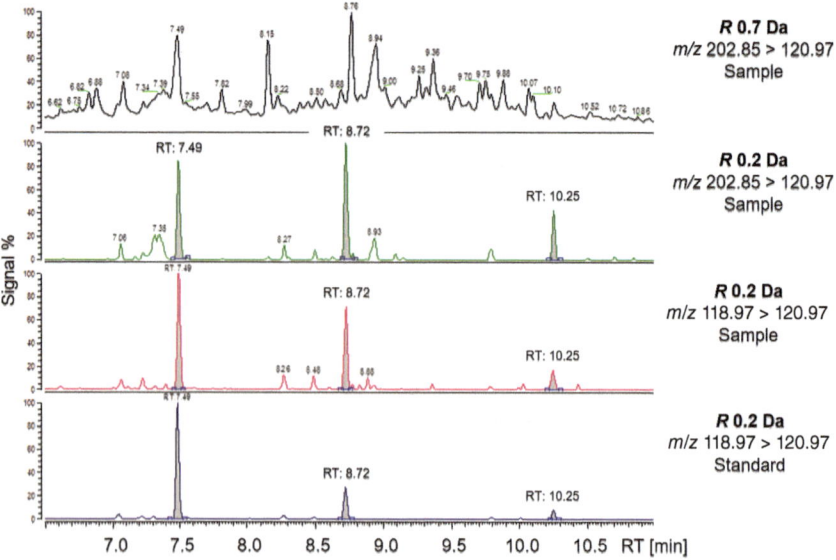

Figure 2.206 Target analysis by the GC-MS/MS MRM mode for tris(dibromopropyl)phosphate flame retardants (Bromkal P, Firemaster) using quadrupole resolution settings 0,7 and 0.2 Da for increased selectivity and quantification. Sample: extract from a children toy (Gummersbach, 2010).

Improved mass resolution of quadrupole analyzers has been published and commercialized for a selected ion monitoring (SIM) technique in MS/MS named selected reaction monitoring "H-SRM" (<0.4 Da) or "U-SRM" (<0.2 Da), by using enhanced mass resolution. Hyperbolic quadrupole rods of a true hyperbolic pole face, high precision four section design with a rod length of 25 cm length are employed in triple quadrupole GC- and LC-MS/MS systems for the selective quantification of target compounds. The increased field radius of 6 mm provides a significantly increased ion transmission for trace analysis (HyperQuad™ technology, Thermo Fisher Scientific). A target analysis for brominated flame retardants using the improved MS/MS selectivity is shown in Figure 2.207.

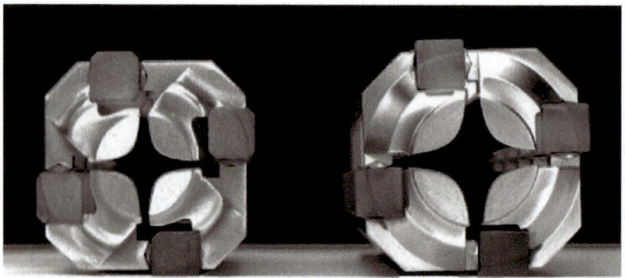

Figure 2.207 Nominal mass resolution (left, TSQ 7000) and enhanced mass resolution (right, TSQ Quantum HyperQuad™) quadrupole analyzers (Thermo Fisher Scientific).

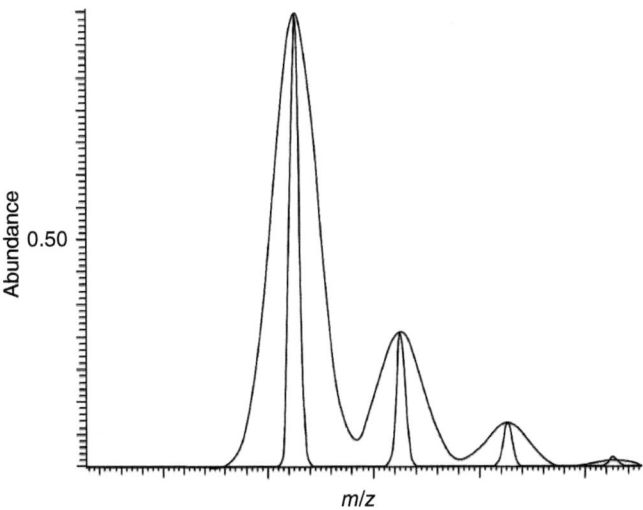

Figure 2.208 Comparison of mass peak widths at 0.7 and 0.1 Da at FWHM.

By operating the first quadrupole in the higher resolution mode, the selectivity for MS/MS analysis of a parent target ion within a complex matrix is significantly increased. Higher S/N values for the product ion peak lead to significantly lowered method detection limits (MDLs). Using the hyperbolic quadrupole rods, the reduction from 0.7 Da peak width to 0.1 Da reduces the overall signal intensity only by half but S/N increases by a factor better than a factor of 10. In many cases, at low levels, it allows to detect the target peak well above the background. Round rod quadrupoles can only be applied for a higher resolved and higher selective detection mode by compromising ion transmission, which leads directly to a low detection sensitivity.

To eliminate the uncertainty associated with possible mass drifts in H-SRM and U-SRM acquisitions, algorithms that correct the calibration table for mass position and peak width have been developed (Jemal and Quyang, 2003; Liu et al., 2006). Data are obtained from U-SRM quantitation at a peak width setting of 0.1 Da at FWHM for the precursor ion. Deviations detected for the mass position or peak width in internal standards is used to internally adjust the mass calibration. The data is then used to determine the deviations if any for all target calibrant ions. The calibration correction method (CCM) is submitted and executed within the sequence of data acquisition. A scan-by-scan CCM has been applied to accommodate a variety of factors that influence precursor ion mass drift and ensure performance eliminating signal roll-off. With real-life samples exhibiting strong matrix backgrounds, a significant increase in S/N is achieved by using the U-SRM technique (Figure 2.208).

Analytical benefits of higher resolving quadrupole systems in GC-MS/MS analysis are the separation of ions with the same nominal m/z value for increased selectivity, in particular the higher resolved precursor ions for the MS/MS mode (Table 2.46).

Table 2.46 Ion transmission for low and enhanced-resolution quadrupoles at 0.2 Da mass peak width FWHM, 100% equals 0.7 Da FWHM (Jemal, 2003).

Quadrupole type	SIM (%)	SRM (%)
Enhanced resolution	70	53
Low resolution	11	0.15

Mass Analyzers for Accurate Mass Measurement

Mass analyzer	Maximum resolving power [FWHM]	Mass accuracy [ppm]	Dynamic range	Inlet methods	MS/MS capability
Time-of-flight, reflectron	5000–50 000	2–5	10^2–10^4	GC-MS, LC-MS	Typical application
Time-of-flight, multipath	1 000 000	0.1	10^5	LC-MS	Typical application
Hyperbolic triple quadrupole	10 000	5	10^4–10^6	GC-MS, LC-MS	Typical application
Magnetic sector	120 000–150 000	2	10^5–10^6	GC-MS, LC-MS, direct inlets	Special applications
FT Orbitrap	>1 000 000	<1	10^4–10^6	GC-MS, LC-MS	Extensive use
FT Ion cyclotron resonance	>1 000 000	<0.2	10^3–10^4	LC-MS	Special applications

Registration of Mass Spectra Ions arriving after mass analysis at the detector generate a signal that is registered by the data system. The detectors are the "eyes" of the instrument (Koppenaal et al., 2005). The MS can record mass spectra in the continuous profile mode and centroid mode.

The *profile* mass spectra recording preserves the shape of the mass peak as ions arrive as of flight time or analyzer voltage parameter with increasing and decreasing intensity during a mass scan or a short scan segment at the detector as time/intensity data. Data systems particularly with high resolution MS allow the recording and display of profile data.

The widely used *centroid* acquisition mode of an MS records the calculated center of the mass peak. The centroid mass is a computed mass describing the position of the mass peak maximum on the mass axis. In case of a close by not resolved interferences, the centroid calculation assigns the position of the mass center of the detected envelope (Urban et al., 2014). The centroid intensity (the later line height in the mass spectrum) represents the normalized mass peak intensity over the data points of the peak profile data. Usually around 15 data points are acquired over a low resolution mass window of 1000 mDa. The centroid acquisition mode is the standard data acquisition mode in low resolving quadrupole and ion trap mass spectrometers, e.g. shown in Figure 3.7 and all library mass spectra. The SIM and selected reaction monitoring (SRM) modes also use centroiding over reduced mass windows. The centroided data files are significantly shorter compared to profile mode data.

The display of line spectra on screen must be considered completely separately from the resolution of the analyzer. By definition, a mass peak with unit mass resolution has a base width of one mass unit or 1000 mDa. On the other hand, the position of the top of the mass peak (centroid) can be computed exactly, independent on any potentially occurring mass interference. Data sometimes given with one or several digits of a mass unit lead to the false impression of a resolution higher than unit mass resolution at the used low resolution MS. Coeluting compounds appearing at the same time at the ion source with signals of the same nominal mass, but slightly differing exact mass, which can naturally occur in GC-MS (because of co-eluates, the matrix, column bleed, etc.) cannot be separated at unit mass resolving power of the quadrupole and ion trap analyzers. The position of a centroid, therefore, cannot be used for any sensible evaluation as compared to accurate mass data. In no case is this the basis for the calculation of a possible empirical formula. Depending on the manufacturer, the labeling of the spectra can be found with pure nominal masses to several decimal places and can usually be altered by the user. It requires special software tools for internal mass calibration and profile type mass acquisition instead of centroiding to exploit the excellent quadrupole mass stability for accurate mass calculations (Gu and Wang, 2008; Wang and Gu, 2010; Abel, 2014). In contrast to that, high mass resolution and accurate mass data of e.g. of the Orbitrap analyzer allow the direct conversion of a low ppm precise mass into sum formula suggestions (see Section 2.3.2.4).

Mass Defect in Mass Spectrometry The "mass defect" in mass spectrometry is defined as the difference between the exact mass and the nominal mass of a compound. There are many uses exploiting the mass defect for the identification of compounds in non-targeted GC-MS and LC-MS analysis (Sleno, 2012; Ubukata et al., 2015). The concept of plotting the mass defect vs. the Kendrick nominal mass by setting the mass of CH_2 to 14.0000, instead as IUPAC defined 14.01565, is used for the identification of homologous compounds in environmental and life sciences applications (Kendrick, 1963).

With targeted analysis by low and high resolution MS, exploiting the mass defect of halogens offers distinct advantages for improved selectivity. The IUPAC masses, for example, for hydrogen, chlorine, and bromine differ significantly in their mass defect, as calculated in Table 2.47. Hydrogen is important despite the small positive mass defect as it occurs in high number in organic molecules. In contrast, halogens show a large negative mass defect and are contained in many target compounds like "dioxins", PCBs, PBDEs, or pesticides. Oxygen and nitrogen only have small mass defects and do not occur in high numbers in molecules, making them negligible in this context (also refer to Table 3.6). Carbon by definition does not have a calculated mass defect.

Table 2.47 Mass defect for hydrogen, chlorine, and bromine, and theoretical example for the detection of 2,3,7,8-TCDD at m/z 320.

Element	Exact mass [g/mol]	Nominal mass [g/mol]	Mass defect [Da]
^1H	1.007825	1	+0.007825
^{35}Cl	34.968855	35	−0.031145
^{37}Cl	36.965896	37	−0.034104
^{79}Br	78.918348	79	−0.081652
^{81}Br	80.916344	81	−0.083656
Example: 2,3,7,8-TCDD: $^{12}C_{12}{}^1H_4{}^{35}Cl_4{}^{16}O_2$	319.89655	320	−0.10345
A hydrocarbon: $^{12}C_{22}{}^1H_{40}{}^{16}O$	320.30792	320	+0.30791

The example in Table 2.47 tells about a potential hydrocarbon background coeluting with 2,3,7,8-TCDD on mass m/z 320. The mass peak maxima are 0.41136 Da apart (−0.10345 to +0.30791). This strong positive mass defect is typical for hydrocarbons in general due to the high content of hydrogens (on a rule of thumb, it is +100 mDa per 100 Da). A similar calculation is true for highly halogenated compounds in general depending on the halogen content.

Low resolving quadrupole instruments benefit from the mass defect in SIM and SRM analysis by setting the SIM or MS/MS precursor mass not to the nominal but the one or two digit rounded exact mass. In the example for TCDD to m/z 319.90, a standard mass resolution window of 0.7 Da then will register masses from m/z 319.55 to 320.25, missing much of the hydrocarbon background. But, setting the quadrupole resolution to a lower peak width of 0.4 Da, it then registers from m/z 319.70 to 320.10 clipping hydrocarbons with a significant increase in selectivity in the presence of unavoidable background. This mass defect situation is exemplified graphically in Figure 2.209. High resolution accurate mass MS (HRAM) register the exact mass to achieve their inherent high selectivity, as also illustrated in Figure 2.202.

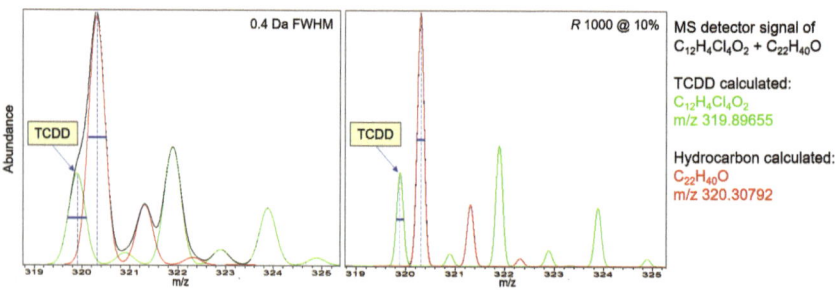

Figure 2.209 TCDD and a hydrocarbon mass peak position on the mass scale at a quadrupole resolution of 0.4 Da and a HRMS resolution at R 1000.

2.3.2.2 Quadrupole and Quadrupole Ion Trap Mass Spectrometer

The quadrupole and quadrupole ion trap analyzers types function on the same mathematical basis (Paul and Steinwedel, 1953, 1956; Paul et al. 1958) and, therefore, show the same resolution properties. A mass spectrum obtained with a quadrupole or ion trap mass spectrometer (Figures 2.210 and 2.211) shows the well-known characteristic with the constant distance of two mass signals and their mass peak widths over the whole mass range (Figure 2.212). How far the instrument can scan to higher masses does not change the peak width; hence, it does not change the available mass resolution (Miller and Denton, 1986). The quadrupole and ion trap peaks of water ($m/z = 17/18$) are of the same width and have the same separation as the masses in the upper mass range (also refer to Figure 2.201).

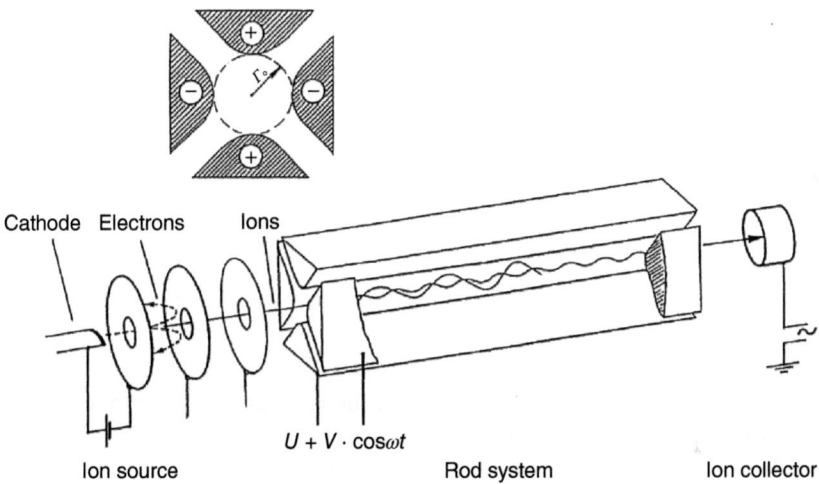

Figure 2.210 Principle of a quadrupole mass spectrometer.

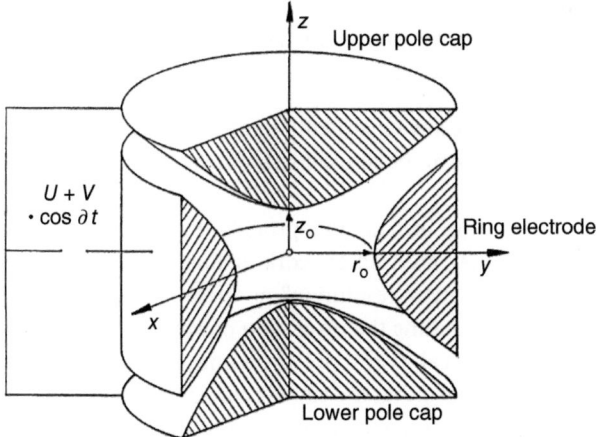

Figure 2.211 Image of the ion trap analyzer (Finnigan).

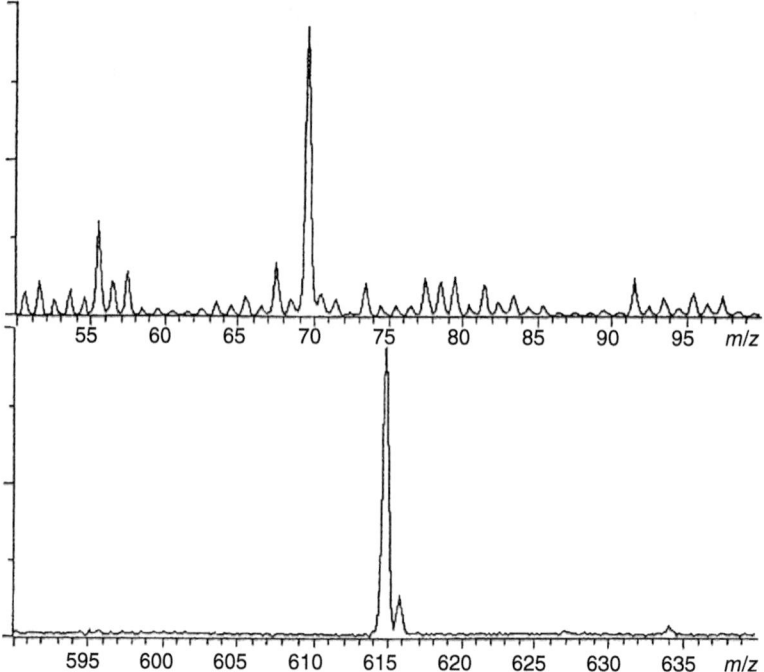

Figure 2.212 Quadrupole mass signals with constant peak widths in the lower and upper mass ranges (calibration gas FC43 with m/z 69 top, and m/z 614 bottom).

Since the distance Δm between two signals, the resolution is constant over the whole mass range for these instrument types, the formula $R = m/\Delta m$ would have the result that the resolving power R would be directly proportional to the highest possible mass m. At a peak separation of one mass unit, the resolving power in the lower mass range would be small using the formula for the resolution (e.g. for water $R \approx 18$) and higher in the upper mass range (e.g. $R \approx 614$ for FC43). But, this calculation cannot be applied for these cases.

The following conclusions may be drawn from these facts concerning the assessment of quadrupole and ion trap instruments:

1) The formula $R = m/\Delta m$ does not give any meaningful figures for quadrupole and ion trap instruments and therefore cannot be used without specifying the mass used for calculating R.
2) Because all quadrupole and ion trap instruments provide the same quality of nominal mass peak resolution, they are said to work at unit mass resolution (see Figure 2.212). The terms "unit mass resolution" or "nominal mass resolution" should be used instead of calculating the resolution power R.
3) The visible optical resolution and peak width are the same in the upper and lower mass ranges. It can be easily seen that the signals corresponding to whole numbers are well separated. This corresponds in practice to the typically used resolution for quadrupole and ion trap instruments.

4) The mass resolution, which is constant over the whole mass range, is set up by the manufacturer in the electronics of the instrument and in common is the same for all types and manufacturers. The peak width is chosen in such a way that the distance between two neighboring nominal mass signals corresponds to one mass unit (1 Da = 1 u = 1000 mu).
5) High mass resolution, as in a magnetic sector or Orbitrap instrument, is not possible for quadrupole and ion trap instruments within the framework of the scan technique used.
6) The mass range of quadrupole and ion trap instruments varies but does not have effect on the mass resolution.

Frequently, the terms *mass range* and *unit mass resolution* are mixed up when giving a quality criterion for a mass range above 1000 Da for a quadrupole instrument. It is also not obvious that a mass range beyond 1000 Da, for example, is always accompanied by unit mass resolution. The effective attainable resolution for a real measurable signal of a reference compound is accurate and meaningful.

2.3.2.3 Sector Field Mass Spectrometer

In a magnetic sector instrument, the flight paths of ions with different m/z values follow a different course caused by the applied magnetic and electric fields (=double focusing) (Figure 2.213). The ion entrance, central, and exit slit systems mask the ion beam and determine with their settings the mass resolution during operation (see also Figure 2.203). Spectra can be recorded by continually changing the operational parameters of the instrument, for example, the acceleration voltage or strength of the magnetic and electrical fields. The width of the ion

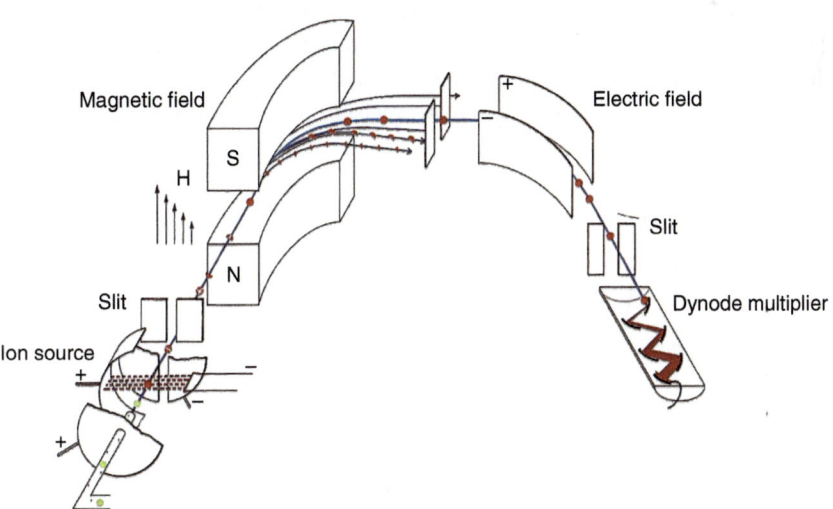

Figure 2.213 Principle of the double focusing magnetic sector mass spectrometer in 'reversed Nier Johnson' geometry BE: in the ion flight path first the magnetic field (B) then the electrical sector (E).

2 Fundamentals

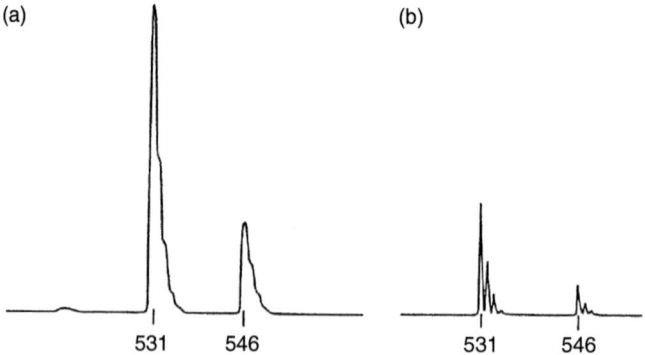

Figure 2.214 Section of a poorly resolved spectrum (a) and of a high resolved one (b) which, however, results in a lower intensity of the signal (Budzikiewicz and Schäfer, 2012).

beam is determined by the source slit. The beams must not overlap, or only to a very small extent so that ions of different masses can be registered consecutively (Figure 2.214).

The resolution power R of neighboring signals (Figure 2.215) for magnetic sector instruments is calculated according to

$$R = \frac{m}{\Delta m} \quad (2.27)$$

with m = mass and Δm = distance to the neighbor mass. According to this formula, the value of the resolution power is dimensionless.

Using high mass resolution, the accurate mass of a molecular ion or of fragment ions can be determined. If the precision of the measurement is high enough, candidates for the empirical formula can be calculated (Figure 2.216). The higher the masses, the more interferences are possible by meaningful elemental compositions and requires even higher mass resolution (see Section 2.3.2.4). An excerpt of the large number of realistic chemical formulae for the small nominal mass m/z 310 ($C_{22}H_{46}$, M 310.3599) is shown in Figure 2.217. A precision in the mass determination of

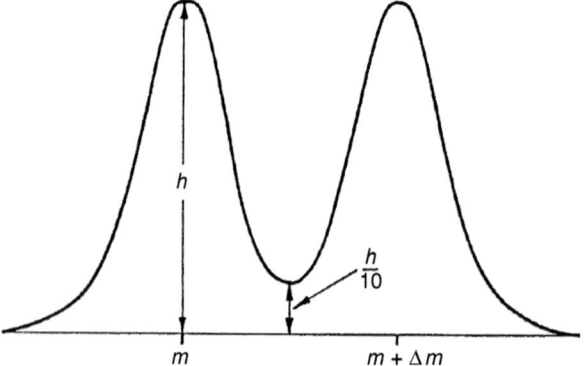

Figure 2.215 General resolution condition for two mass signals as 10% valley definition (Budzikiewicz and Schäfer, 2012).

Figure 2.216 Oscilloscope image of mass signals in high resolution at $R = 61\,000$ on a double focusing magnetic sector instrument (Finnigan MAT 90).

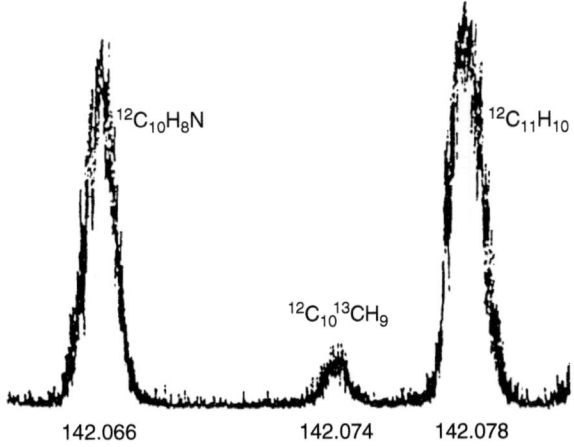

2 ppm or better would be necessary for this example to reduce drastically the number of possible elemental formulas. The maximum resolution and precision that can be achieved characterize the slit system and the quality of the ion optics of a magnetic sector instrument.

Example Calculation of the Necessary Minimum Resolution

What mass resolution is required to obtain the signals of carbon monoxide (CO), nitrogen (N_2), and ethylene (C_2H_4) which are passed as a mix directly into the ion source of a mass spectrometer?

Substance	Nominal mass	Exact mass
CO	m/z 28	m/z 27.994910
C_2H_4	m/z 28	m/z 28.006148
N_2	m/z 28	m/z 28.031296

For MS separation of CO and C_2H_4 in accordance with the formula Eq. (2.27) $R = m/\Delta m$ given above, a resolving power of at least 2500 is necessary (see also Figure 2.204).

All MS systems with low mass resolution need the preliminary GC separation of the mixture. The components CO, C_2H_4, and N_2 would then arrive separated one after the other in the ion source of an MS. This is the application for all quadrupole and ion trap instruments.

2.3.2.4 Orbitrap Mass Spectrometer

The "Orbitrap" is the latest high resolution mass analyzer development invented by Alexander Makarov in continuation of the works of Lidia Gall and Yuri Golikov

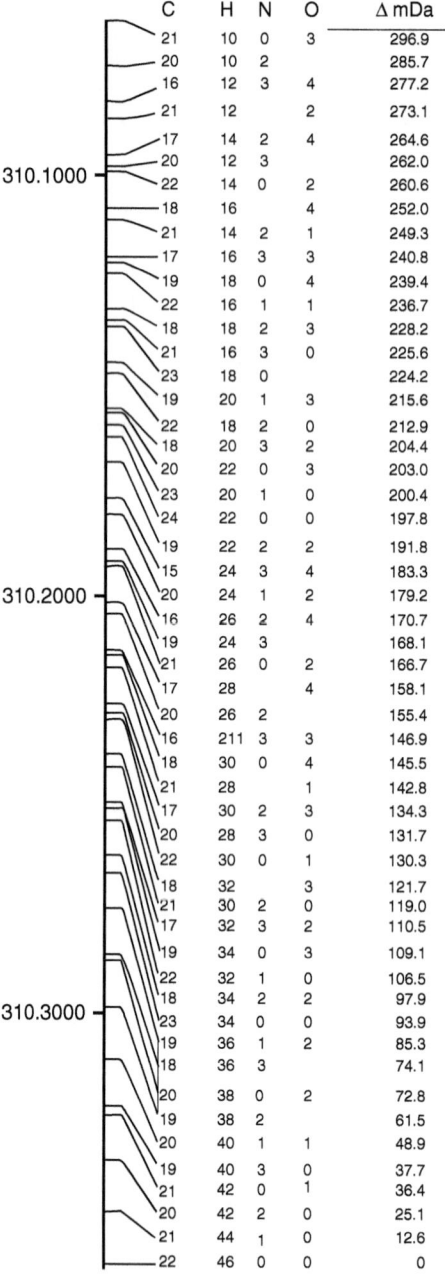

C	H	N	O	Δ mDa
21	10	0	3	296.9
20	10	2		285.7
16	12	3	4	277.2
21	12		2	273.1
17	14	2	4	264.6
20	12	3		262.0
22	14	0	2	260.6
18	16		4	252.0
21	14	2	1	249.3
17	16	3	3	240.8
19	18	0	4	239.4
22	16	1	1	236.7
18	18	2	3	228.2
21	16	3	0	225.6
23	18	0		224.2
19	20	1	3	215.6
22	18	2	0	212.9
18	20	3	2	204.4
20	22	0	3	203.0
23	20	1	0	200.4
24	22	0	0	197.8
19	22	2	2	191.8
15	24	3	4	183.3
20	24	1	2	179.2
16	26	2	4	170.7
19	24	3		168.1
21	26	0	2	166.7
17	28		4	158.1
20	26	2		155.4
16	211	3	3	146.9
18	30	0	4	145.5
21	28		1	142.8
17	30	2	3	134.3
20	28	3	0	131.7
22	30	0	1	130.3
18	32		3	121.7
21	30	2	0	119.0
17	32	3	2	110.5
19	34	0	3	109.1
22	32	1	0	106.5
18	34	2	2	97.9
23	34	0	0	93.9
19	36	1	2	85.3
18	36	3		74.1
20	38	0	2	72.8
19	38	2		61.5
20	40	1	1	48.9
19	40	3	0	37.7
21	42	0	1	36.4
20	42	2	0	25.1
21	44	1	0	12.6
22	46	0	0	0

Figure 2.217 Exact masses of chemically realistic empirical formulae consisting of C, H, N (≤3), and O (≤5) given in deviations (Δ mDa) from the nominal molecular mass of $C_{22}H_{46}$ m/z 310 (McLafferty and Turecek, 1993).

in USSR (Gall et al., 1986). An inner spindle electrode is enclosed in axis by two electrodes in a small device of only a few centimeters in length, as shown in comparison to a Euro coin in Figure 2.218. Ion packets are injected into the Orbitrap by an intermediate C-trap (Figure 2.219). The C-trap collects and limits the number of ions to be injected and analyzed in the Orbitrap analyzer. Ions are generated

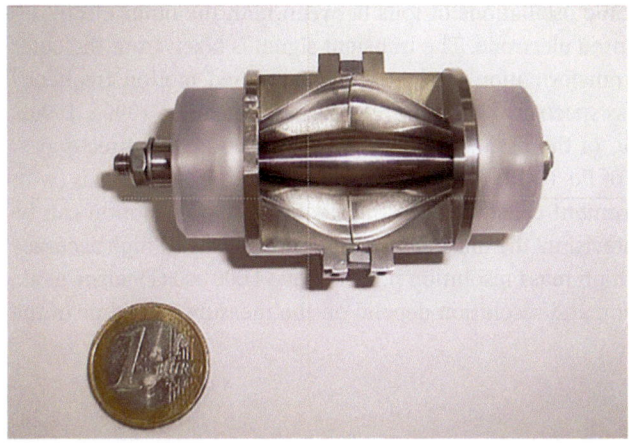

Figure 2.218 How big is the Orbitrap analyzer? Here compared to a one Euro coin (Courtesy Thermo Fisher Scientific).

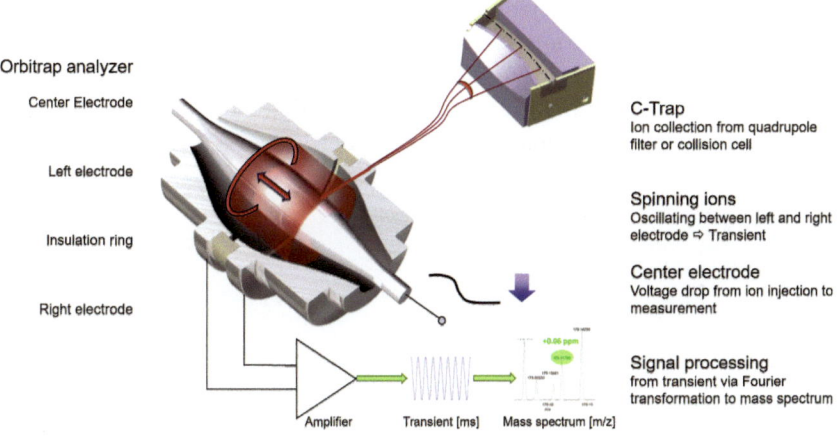

Figure 2.219 Schematics of the Orbitrap analyzer (Courtesy Thermo Fisher Scientific).

in GC-MS by a regular ion source, usually allowing for standard or low energy EI spectra, or CI for soft ionization. A quadrupole filter is used after the ion source in front of the C-trap in order to select a mass range to be analyzed. As typical with ion traps, an overload of the C-trap, resulting in quenching of analyte ions by matrix, must be avoided. The GC EI ion source generated ions are short living and reactive, in contrast to ions generated on LC ESI or APCI sources. Ion/molecule reactions caused by long residence times in the C-trap can occur and are frequently reported. Such situations get counteracted by appropriate analyzer control (Baumeister et al., 2018; De Vooght-Johnson, 2019). A connected linear trap collision cell is used for collision-induced dissociation (CID) processes for MS/MS analysis.

Ions injected form the C-trap into the Orbitrap are forced into a spinning movement around the central spindle. The mass-to-charge ratio of an ion is derived from the frequency of harmonic oscillations of ions between both the outer electrodes along the axis of the central electrode. The transient signal is taken from the outer electrodes. A Fourier transformation of the exactly measured motion frequency of ions delivers the mass spectrum with accurate masses (Makarov, 1999a, 1999b, 2000). The frequency ω_z of the oscillation between the electrodes is directly proportional to mass m as of Eq. (2.28) for the oscillation along the spindle axis (with k and q being the instrument constants). As the frequency of ion motion can be determined with high precision, the mass determination results with high accuracy below 1 ppm and ultra-high mass resolution power above 1 000 000 (Denisov et al., 2012). The mass precision and resolution depend on the measurement time of the trapped ions.

$$\omega_z = \sqrt{\frac{k}{m/q}} \tag{2.28}$$

With its unique capabilities in high mass resolution combined with the uncompromised detection sensitivity and speed, the analytical advantages of the Orbitrap analyzer for GC-MS are developing quickly into a combined targeted and non-targeted trace analysis. Orbitrap MS systems are in GC-MS use for targeted and untargeted analysis, multi-residue trace analyses, and widely applied in LC-MS for proteomics and metabolomics, especially in life sciences (Peterson et al., 2009, 2010; Hecht et al., 2019). A comparison from pesticides analysis illustrating the gain in selectivity by Orbitrap accurate mass detection is shown in Figure 2.220.

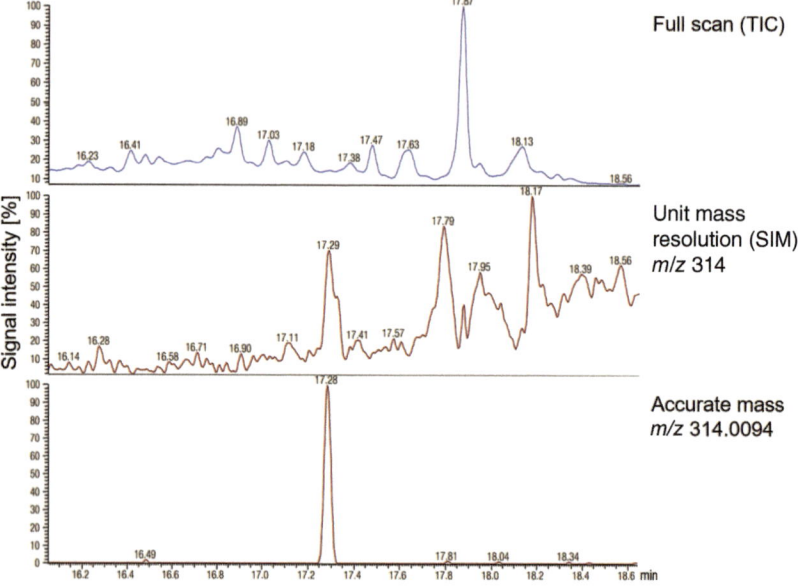

Figure 2.220 Orbitrap pesticides' full scan analysis in a leek sample. Iprodione at 10 ng/g: full scan TIC (top), not detected at nominal mass resolution SIM; accurate mass measurement selects the peak from the matrix with high sensitivity (Cojocariu et al., 2016).

In general, the GC-Orbitrap method development takes place in the full scan mode. This omits the time-consuming preparation of SRM tables for a triple-quadrupole method for hundreds of target compounds (see Section 2.3.4.4). Full scan data also allow the retrospective search for unknowns even years after the measurements. With regard to sensitivity, the GC-Orbitrap compares well with GC-triple-quadrupole analysis due to its accurate mass selectivity in real-life samples (Cojocariu et al., 2016). A negative effect of matrix on sensitivity is reported as of the operation of the C-trap. Matrix ions of a similar mass range can compete with target analyte ions, resulting in a smaller dynamic range or quenching (Belarbi et al., 2021). A mass range restriction on the preceding quadrupole filter, and, as valid for any MS, but even more to exploit the unique features of Orbitraps for identification and quantitation, an effective extract clean-up is mandatory (Mol et al., 2016).

2.3.2.5 Time-of-Flight Analyzer

The concept of TOF MS was proposed already in 1946 by William E. Stephens of the University of Pennsylvania (Borman, 1998). In the TOF analyzer, ions are separated by their velocities as they fly down a field-free drift region toward a collector in the order of their increasing mass-to-charge ratio (see Figure 2.221). With that principle, TOF-MS is probably the simplest method of mass measurement.

TOF instruments were first designed and constructed, starting in the late 1940s. Key advances were made by William C. Wiley and I.H. McLaren of the Bendix Corp., Detroit, MI, USA, the first company to commercialize TOF mass spectrometers (Wiley and MacLaren, 1955; Bendix, 1960). According to pharmacology professor Robert J. Cotter of the Johns Hopkins University School of Medicine, Wiley and McLaren devised a time-lag focusing scheme that improved mass resolution by simultaneously correcting for the initial spatial and kinetic energy distributions of the ions. Mass resolution was also greatly improved by the invention of the reflectron in 1974 by Boris A. Mamyrin of the Physical-Technical Institute, Leningrad, former Soviet Union, which corrects for the effects of the kinetic energy distribution of the ions when leaving the ion source (Cotter, 1992, 1997; Guilhaus, 1995; Mamyrin et al., 1973; Mamyrin, 2001).

Figure 2.221 Time-of-flight MS operating principle: R – reflectron, G – extraction grid, and FT – flight tube entrance grid (McClenathan and Ray, 2004) (Reprinted with permission from the American Chemical Society).

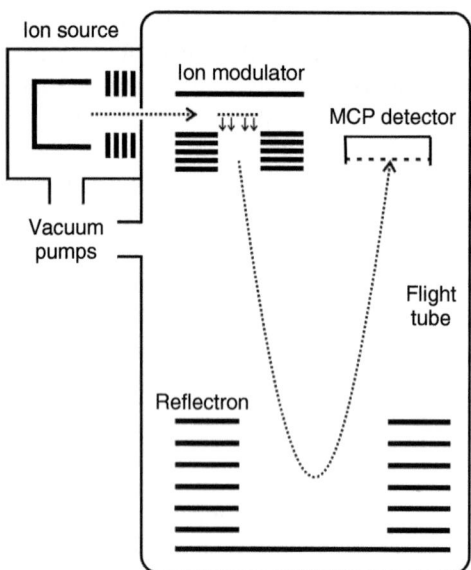

Figure 2.222 TOF mass spectrometer with GC ion source, orthogonal ion modulation into the TOF analyzer, a single reflectron, and MCP detection; the dotted lines indicate the ion flight path.

The schematics of a TOF analyzer are shown in Figure 2.222. Ions are introduced as pulsed packets either directly from the GC or LC ion source of the instrument, or with hyphenated instrumentation, from a previous analyzer stage (e.g. quadrupole-time-of-flight hybrid, Q-TOF). Ions from the ion source are accelerated and focused into a parallel beam that continuously enters the ion modulator region of the TOF analyzer. Initially, the modulator region is field free and ions continue to move in their original direction. A pulsed electric field is applied at a frequency of several kilohertz (kHz) across a modulator gap of several millimeter width, pushing ions in an orthogonal direction (rectangular to their initial movement) into an accelerating section of several kiloelectron volts. The modulator pulses serve as triggers for recording spectra at the detector.

Ideally, all ions in the pulsed ion packet receive the same initial kinetic energy, $E_{kin} = 1/2\ mv^2$. Lighter ions (low m) travel faster and reach the detector earlier. As the ions enter and move down the field-free drift zone, they are separated by their mass-to-charge ratio in time. Commercial GC-MS and LC-MS TOF instruments are usually equipped with at least one reflectron (Mamyrin et al., 1973). This "ion mirror" focuses the kinetic energy distribution, originating from the small initial velocity differences due to different spatial start positions of the ions when getting pulsed from the modulator into the drift tube. The mass resolution in TOF-MS is generally limited by the initial spatial and velocity spreads. The ratio of velocity components in the two orthogonal directions of movement from the source and modulator is selected such that ions are directed to the center of the ion mirror and get focused on a horizontal multichannel plate (MCP) detector plane. The reflectron typically consists of a series of lenses with increasing potential pushing the ions back in a slight angle into the direction of the detector. Ions of higher velocity (energy) penetrate deeper into the mirror. Hence, the ion packet is getting focused in space and time for increased mass resolution. A series of facing reflectrons improve mass resolution. All ions reaching the detector are recorded, explaining the inherent high sensitivity in the full scan mode of TOF analyzers. Duty cycles reached with orthogonal ion acceleration instruments vary in a range of 5–30% depending on the mass range and limited by the slowest ion (highest m recorded) moving across the TOF mass analyzer.

The TOF mass separation is characterized by the following basic equation:

$$m/z = 2e \cdot E \cdot s \, (t/d)^2$$

with

m/z = mass-to-charge ratio,
e = elementary charge,
E = extraction pulse potential,
s = length of ion acceleration, over which E is effective,
t = measured flight time of an ion, triggered by pulse E,
d = length of the field-free drift zone.

Increasing the flight time, for instance by a longer flight tube, improves the resolving power of the TOF analyzer. Newest developments in commercial TOF mass analyzers extend the ion flight path using multiple reflectrons reaching a mass resolving power R of up to 50 000 and with novel "multipath" analyzers up to 1 000 000 FWHM (Verenchikov et al., 2025). Statements on the achievable ion transmission efficiency and a potential loss of detection sensitivity by multiple reflections have not been made available yet (Xian et al., 2012).

The fast data acquisition rates make the TOF analyzer the ideal mass detector for fast GC and comprehensive GC × GC (Dimandja, 2003; Skoczyńska et al., 2008). TOF-MS is fast, offers a high duty cycle in GC-MS with the parallel detection of ions, and has theoretically unlimited mass range. Due to the speed of detection and its inherent sensitivity by the parallel detection of a complete ion packet, it is especially suited for a full scan chromatographic detection. TOF-MS is capable of running fast GC because of its high sampling rates of up to more than hundred spectra per second. It is also widely used in life sciences for the determination of large biomolecules via LC ESI or matrix-assisted laser desorption ionization (MALDI), among many other low and high molecular mass applications. Benchtop GC-TOF instruments provide a low mass resolving power of approx. 1000 at the OFN calibration mass m/z 219, while the hybrid quadrupole-TOF analyzers (GC-Q-TOF-MS), the so-called "high resolution" TOF-MS, provide a mass resolution in the range of 15 000 (FWHM) or better with a mass accuracy of better than 5 ppm for accurate mass determination of chromatographically resolved compounds (Webb et al., 2004; Agilent, 2011a).

Despite the described inherent sensitivity of TOF-MS in full scan analysis, some fundamental trade-offs in terms of response and spectral quality have to be considered when setting up TOF applications. In contrast to the expectation that higher acquisition rates strongly support the deconvolution of coeluting compounds, the average ion abundance is dropping with increased scan rates and limiting the useful dynamic range. With the increase of the scan rate to 50 scans/s,

Table 2.48 Average similarity match values depending on the data acquisition rate of TOF mass spectra versus NIST library for disulfoton and diazinon.

Acquisition rate [spectra/s]	Disulfoton (FIT value)	Diazinon (FIT value)
5	884	852
10	876	851
25	837	827[a]
50	820	623[a]
100	735	583
175	697[b]	626[b]
250	660[b]	586[b]

a) Based on four replicate spectra.
b) Based on three replicate spectra.
The average values of five replicate spectra from independent chromatographic runs are listed, except where low match quality did not provide identification.

the ion abundance significantly drops to about 10% of a 5 scans/s intensity. Minor components may be left unrecognized. Hence, typical reported acquisition rates with current GC-TOF-MS instrumentation are 20–50 Hz, with up to 100 Hz in fast GC or GC×GC applications.

Also, the spectral quality must be taken in account when setting up GC-TOF-MS methods for deconvolution experiments. Spectral skewing, as known from slow scanning quadrupole or sector mass spectrometers due to the increasing or decreasing substance intensity in the transient GC peak, does not occur in the ion package detection of TOF instruments. But, TOF spectral quality is limited by high data acquisition rates. There is still the general valid rule that increased measurement times support the spectrum dynamic range and hence its quality. The effect can be demonstrated when comparing the acquired spectra at increasing scan rate against the reference spectra of the NIST library. The fit values of similarity match drop significantly with scan rates above 25, respectively, 50 scan/s, see Table 2.48 for an experiment on two common pesticides (Figure 2.223).

2.3.2.6 Ion Mobility Analyzer

Ion mobility mass spectrometer (IMS) adds another quality of data to GC-MS analysis. Ions are separated by their mass and structure. Even isomeric isobaric molecules are differentiated, a highly valuable feature in food and life science analysis. Also, IMS serves as a frontend to HRMS (Jiang and Robinson, 2013; Lapthorn et al., 2013; Lanucara et al., 2014; Burnum-Johnson et al., 2019). IMS was pioneered by Earl W. McDaniel of the Georgia Institute of Technology, Atlanta, GA, USA, starting in the 1950s to study gas-phase ion mobilities and reactions (McDaniel and Martin, 1970).

Compared to other MS analyzers, IMS can be performed at ambient pressure and does not require a vacuum pump. The analyzer dimensions are small and allow

2.3 Mass Spectrometry

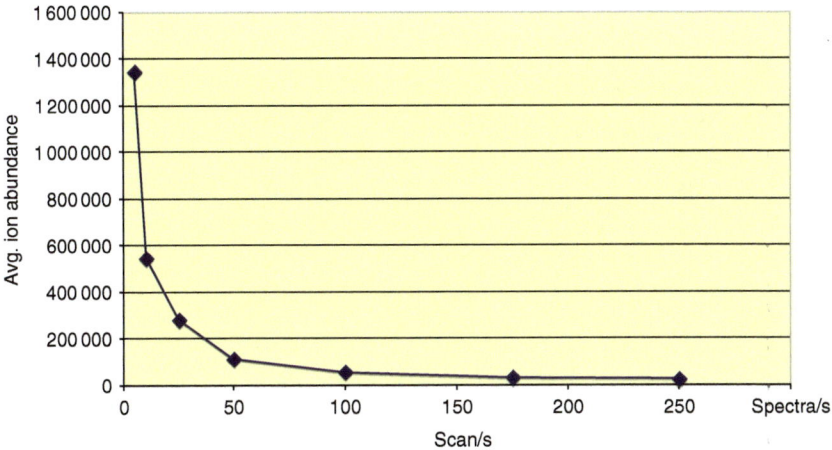

Figure 2.223 Ion abundance dependence from the scan rate in TOF-MS (counts vs. scan rate in spectra/s) (after Meruva et al., 2000).

compact benchtop and mobile instrumentation. Both the direct volatile inlet and the hyphenation with GC are widely used applications. The greatest strength of IMS is the speed of detection in the range of milliseconds and its high sensitivity significant for fast GC and comprehensive GC × GC.

Analytes eluting from the GC column get ionized using a soft CI by a tritium source with a low radiation intensity below the exemption limit of the EU directive 96/29 EURATOM and therefore does not need any licensing (GAS, 2019a). From the ion source, the ions get accelerated into a drift tube against the flow of a drift gas, usually He or just air (drift tube IMS, DT-IMS). A uniform electrical field along the drift tube accelerates the ions toward a Faraday plate for detection (Figure 2.224). During the "drift" time, analyte ions collide with molecules of the drift gas. Lighter ions travel faster than heavier ions. Ions of sterical differences can generate different interactions and hence appear at the detector at different travel times as it is shown in Figure 2.225 (TOFWERK, 2024). The IMS analyzer measures the drift time calibrated against mass. The length of the drift tube and operational parameters define the mass resolution. The analyzer is capable for positive ion but can also be operated with high sensitivity in the negative ion mode.

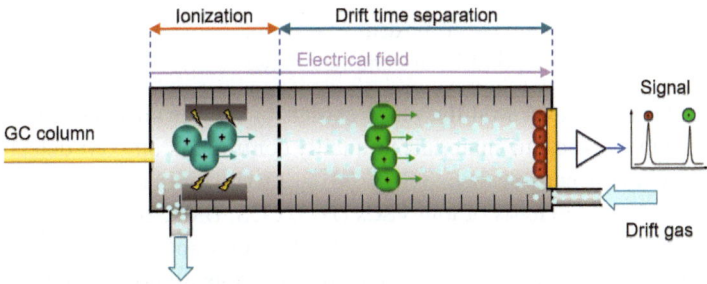

Figure 2.224 Schematics of the IMS analyzer.

338 | 2 Fundamentals

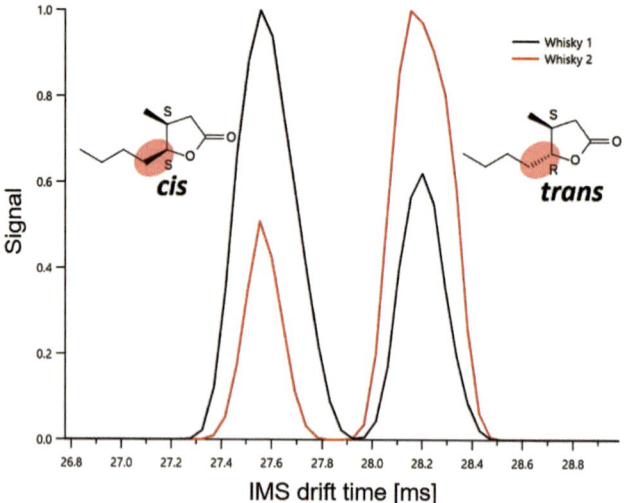

Figure 2.225 IMS stereoisomer separation of an isobaric cis/trans configuration (Courtesy TOFWERK AG).

Resolving power achieved is regular double-digit, in high resolution IMS up to several hundred, but low compared to other MS analyzers (Dodds et al., 2017). The low mass resolution is more than compensated with significant practical and analytical benefits of the IMS technology (GAS, 2019a):

- energy and resource saving potential as a green analytical technology,
- atmospheric pressure ionization,
- detection limits in the low ppbv (µg/m^3) range for volatile organic compounds (VOCs) with heteroatoms like ketones, aldehydes, alcohols, and amines,
- high sensitivity in the negative ion mode for halogenated and sulfur compounds (VSCs),
- selective due to specific analyte ion drift times (2D separation, full orthogonality),
- high sampling frequency for fast GC and GC × GC,
- high reproducibility for spectra comparison,
- automated operation with direct static and dynamic HS,
- standalone operation or GC detector,
- no license for a ^{3}H source required,
- operation with N$_2$, He, or synthetic air drift gases, and
- simplicity, robustness, and point-of-use characteristics.

GC-IMS applications cover well the many aspects of volatile analysis of food or beverages, packaging, and raw materials in quality control and off-odor detection, the authenticity assessment of foods (GAS, 2014, 2019b; Sammarco et al., 2023; Zacometti et al., 2024), analysis of low VSCs (H$_2$S, COS, DMS, and mercaptanes) or halogenated hydrocarbons at a low ppb level. Important examples include adulteration analysis for the discrimination of olive oils from different geographical origins (Gerhardt et al., 2017), in volatilomics (Parastar and Weller, 2024) and environmental the analysis of hazardous chemicals (Moura et al., 2020). Also, high resolution IMS is reported for the separation and identification of isomeric lipids (Groessl et al., 2015) and for drug analysis by distinguishing cannabidiol from Δ^9-tetrahydrocannabinol (THC) (Hädener et al., 2018).

IMS as standalone instrumentation is widely used with thousands of installations for detecting drugs and explosives known from airport security checks.

2.3.2.7 High and Low Mass Resolution in the Case of Dioxin Analysis

Can "dioxin" analysis be carried out with quadrupole, TOF, and Orbitrap instruments? Yes, but there are considerations to discuss regarding resolution power, selectivity, and the regulatory background. As "dioxins" the group of molecules is designated comprising 7 polychlorinated dibenzodioxins (PCDDs), 10 polychlorinated dibenzofurans (PCDFs) and the 14 dioxin-like PCBs (dl-PCBs), as the term is used in the European legislation documents (European Commission, 2014).

The common main challenge in dioxin analysis is the analytical selectivity, precision, and sensitivity in matrices (Focant, 2012). The sensitivity of quadrupole, magnetic sector, and Orbitrap instruments is reported to be adequate. The required selectivity is achieved by either triple quadrupole MS/MS by its structure-specific fragmentation or the high resolution accurate mass (HRAM) methods. GC-TOF instruments are reported to be limited with matrix interferences; hence, the GC × GC separation was applied and showed compliant results (Dindal et al., 2011). In the past also GC-ion trap-MS instruments have been applied successfully for dioxin analysis, also in the MS/MS mode, but such types of MS are discontinued since long (Slayback and Taylor, 1983; Fabrellas et al., 2004; Malavia et al., 2008).

Errors caused by false-positive peak detection and poor real-life detection limits as a result of matrix overlap generally limit the use of single quadrupole mass spectrometers (Eljarrat and Barcelo, 2002). The high resolution and mass precision of the magnetic sector and Orbitrap instruments allow to record the selective accurate masses, e.g. for 2,3,7,8-TCDD at m/z 321.8937 instead of a nominal mass of 322. Table 2.49 shows potential interferences with the nominal mass m/z 322 and the required resolution power for the selective detection. Not taken into account here are the general matrix interferences of a real-life sample. A comparison of the two spectrometric methods of the low and high resolution SIM techniques is shown in Figure 2.226 for the mass traces in the detection of tetrachlorodibenzofuran (TCDF). As a result, using HRAM detection, very high selectivity for very low detection limits of <10 fg is achieved, which gives the necessary assurance for making decisions with a high impact (Figure 2.227). This situation is the fundamental reason why high mass resolution is made mandatory in dioxin analysis by all international regulations to verify positive screening results (see also Section 4.27). The resolution comparison of different analyzer technologies used in dioxin analysis is shown in Figures 2.228–2.230 with simulated ideal mass spectra of the TCDD isotope pattern in the molecular peak region.

Using the Orbitrap for dioxin analysis allows a high resolution full scan mass spectrum acquisition of the mass range of interest compared to the selected ion approach on magnetic sector instruments. Potential interferences on quantitation

2 Fundamentals

Table 2.49 Possible interference with the masses m/z 319.895992 and 321.893042 of 2,3,7,8-TCDD and the minimum analyzer mass resolution required for separation.

Compound	Formula	m/z[a]	Resolving power needed for separation
Tetrachlorobenzyltoluene	$C_{12}H_8Cl_4$	319.9508	5900
Nonachlorobiphenyl	$C_{12}HCl_9$	321.8491	7300
Pentachlorobiphenylene	$C_{12}H_3Cl_5$	321.8677	12 500
Heptachlorobiphenyl	$C_{12}H_3Cl_7$	321.8678	13 000
Hydroxytetrachlorodibenzofuran	$C_{12}H_4O_2Cl_4$	321.8936	Cannot be resolved
DDE	$C_{14}H_9Cl_5$	321.9292	9100
DDT	$C_{14}H_9Cl_5$	321.9292	9100
Tetrachloromethoxybiphenyl	$C_{13}H_8OCl_4$	321.9299	8900

[a] of the interfering ion.
DDE – dichloro-diphenyl-dichloroethylene; DDT – dichlorodiphenyltrichloroethane.

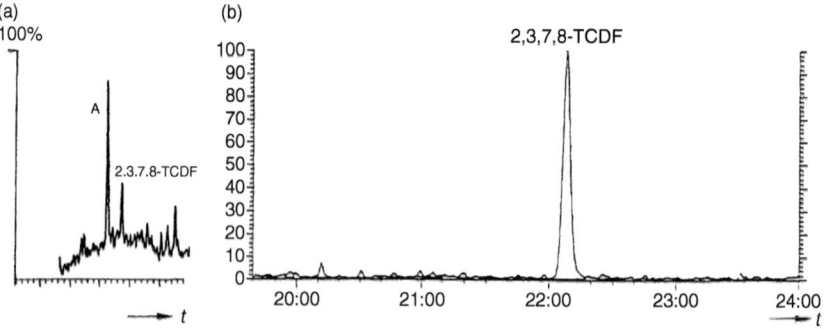

Figure 2.226 Comparison of 2,3,7,8-TCDF mass traces in analyses using a low resolution single quadrupole GC-MS system (SIM trace m/z 319.9) and a high resolution GC-MS system (SIM trace m/z 319.8965). Both chromatograms are from the identical human milk sample. The component marked with A in the quadrupole chromatogram is an interfering component (Fürst and Bernsmann, 2010).

and confirmation ions can be detected. The scan rate of the Orbitrap analyzer is fast and independent on the number of ions acquired. Further information for confirmation can be achieved by the acquisition of full scan and MS/MS data in the same run. With the monitoring of two confirmation ions, full profile MS and MS/MS data, the GC-Orbitrap instrument was able to meet all the method quality assurance/quality control (QA/QC) criteria analyzing a fish sample (Jones and Giesy, 2017). An outstanding Orbitrap selectivity in human blood samples with a mass precision of 4.4 ppm for the accurately measured ion m/z 319.89459 (exact m/z 319.8965) and 4.8 ppm for m/z 321.89151 (exact 321.8936) at a concentration as low as 10 fg 2,3,7,8-TCDD was reported (Peterson et al., 2012).

In practice, for real sample measurements, a significant background contribution must be kept in mind. The characteristics of the quadrupole analyzer for unit mass

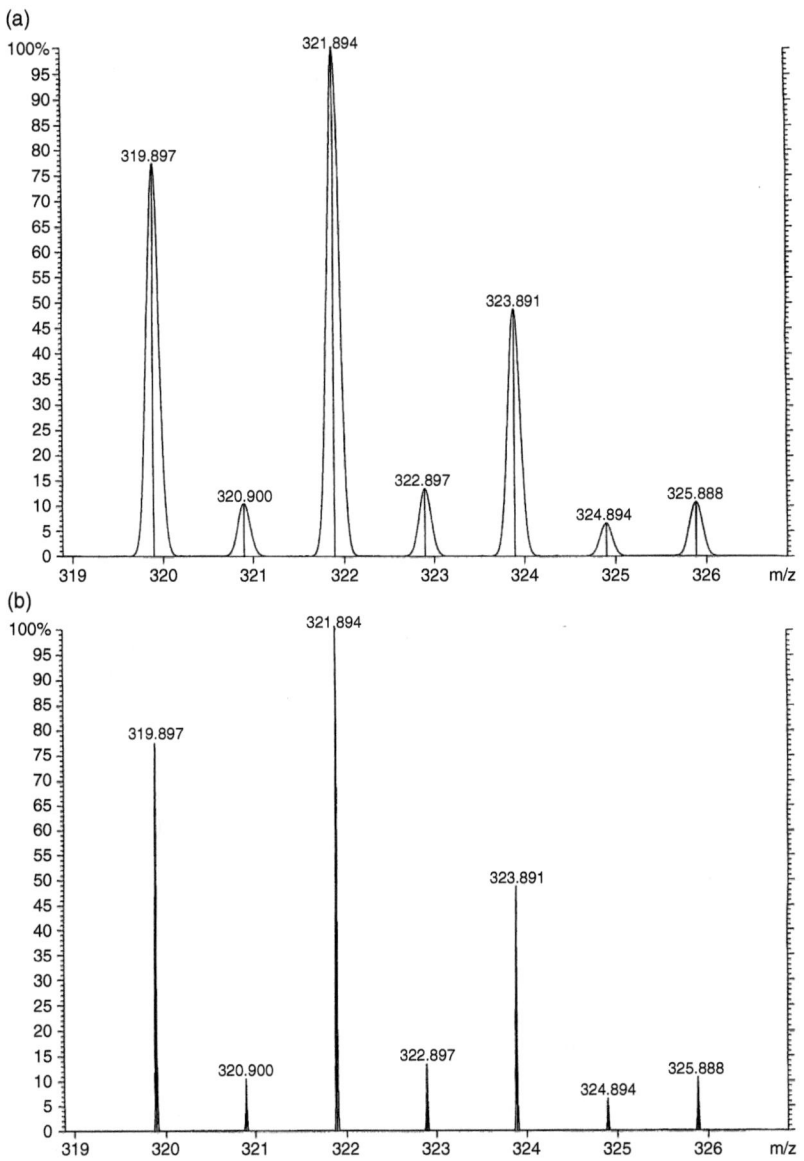

Figure 2.227 The isotope pattern and peak widths of 2,3,7,8-TCDD on a double focusing magnetic sector instrument for mass resolutions R of 1000 (a) and 10 000 (b) (Fürst and Bernsmann, 2010).

resolution in Figure 2.228 show the mass peaks of the isotope pattern with the base peak width of 1 Da (unit mass resolution). Recent enhanced resolution quadrupole technology is able to reduce the peak width down to 0.4 Da with improved mass separation and ion transmission, as illustrated in Figure 2.228. Triple quadrupole mass spectrometers use an enhanced mass resolution of 0.4 Da in the first quadrupole to

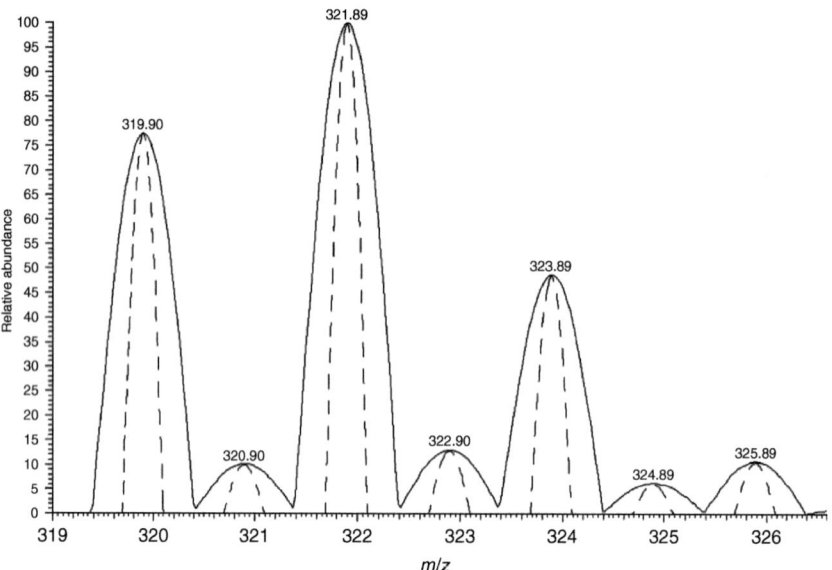

Figure 2.228 Quadrupole mass resolution at 1 and 0.4 Da peak widths (dashed line) in the H-SRM mode (TCDD isotope pattern).

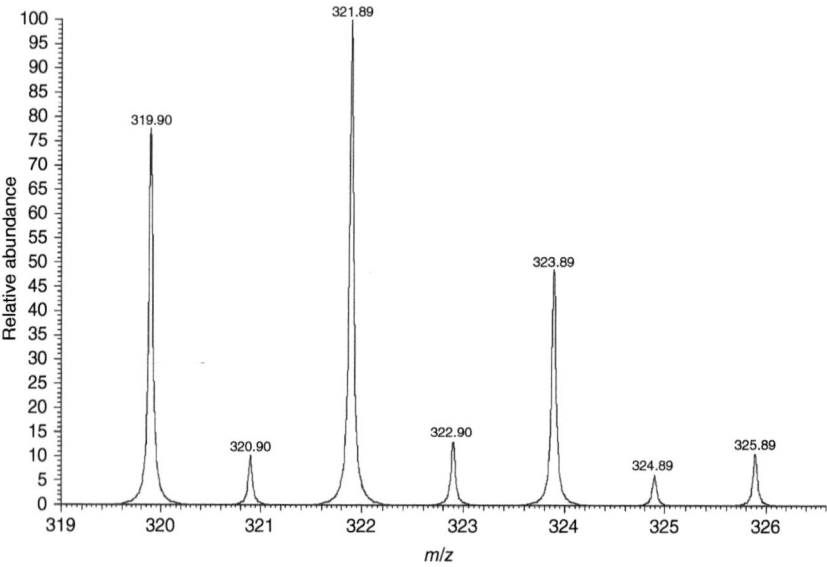

Figure 2.229 TOF mass resolution at R 7000 FWHM (TCDD isotope pattern).

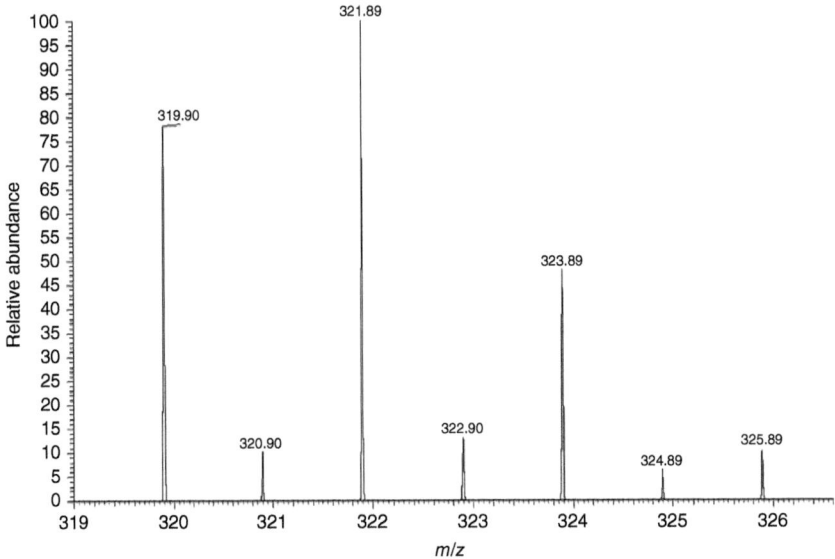

Figure 2.230 High resolution magnetic sector analyzer, mass resolution at R 10 000 at 10% valley (TCDD isotope pattern).

achieve increased selectivity for halogenated compounds against the matrix, exploiting the compounds' negative mass defect compared to the large positive mass defect of hydrocarbons (see in Section 2.3.2.1).

Technical advances in GC-MS/MS triple quadrupole technology have significantly improved the analytical performance with respect to higher sensitivity and selectivity, meeting the requirements for the control of the maximum levels for dioxins (PCDD/Fs and dl-PCBs) in food and feed (Kotz et al., 2009, 2012; Agilent, 2011b, 2017; Cojocariu et al., 2014; Miyagawa, 2016; Takakura et al., 2017, Warner et al., 2023). GC-HRMS and GC-MS/MS are both accepted by the EU Commission Regulation No. 589/2014 as confirmatory methods for dioxin analysis in food and feed. According to this new EU regulation, the following three specific performance criteria must be fulfilled for dioxin analysis with GC-MS/MS, repealing the previous European Commission Regulation No. 1883/2006 and the Commission Regulations No. 252/2012 for food and 278/2012 for feed (Kotz et al., 2012):

- The mass resolution for each quadrupole is required to be set equal to or better than unit mass resolution.
- Two specific precursor ions, each with one specific corresponding product ion for all labeled and unlabeled analytes, must be acquired.
- A maximum tolerance of ±15% is required for the relative ion intensities of each SRM transition of an analyte in comparison to the calculated or measured average values from calibration standards under identical MS/MS conditions, in particular the collision energy and collision gas pressure.

Due to the high structure-related selectivity of triple quadrupole instruments, often no chromatographic baseline is detected (see also Section 4.25). In such situations, the limit of detection (LOD) cannot be determined using the S/N ratio. The calculation of the LOD is suggested using the reproducibility statistics (precision, relative standard deviation [RSD]) of the lowest concentration point of the calibration curve for the determination of the instrument detection limit (IDL, see as well Section 3.3.3) (Wells et al., 2011).

$$IDL = t \times Amount \times \%RSD$$

with

t = student test value at the 99% confidence interval
Amount = amount of analyte (on-column)
%RSD = percentage relative standard deviation for peak areas

TOF analyzers show very tall peaks but typically with a broader base. For that reason, the TOF resolving power is measured at the half peak height. At the very good TOF resolution of 7000 FWHM (frequently named hrTOF), the peak base still is wide over almost one full mass window (Figure 2.229). TOF-MS instrumentation finds application in the comprehensive GC × GC analysis as a complementary method (Focant et al., 2005). The high resolution magnetic sector and Orbitrap instruments provide with distinct difference the resolving power R of 10 000, required by the U.S. EPA method 1613, and better for ultimate selectivity, as represented in Figure 2.230 (Eljarrat and Barcelo, 2002).

2.3.3 Isotope Ratio Monitoring GC-MS

The measurement of isotope ratios was the first mass spectrometry application. In 1907, Thomson for the first time showed the parabola mass spectrum of a neon sample with his newly developed mass spectrometer. Later, in 1919, Aston concluded that the observed lines reveal the isotopes 20 and 22 of neon. He later also discovered the isotope 21 of Ne with only 0.3 at.% abundance. The term *isotope* was coined independently by Frederick Soddy. He observed substances with identical chemical behavior but different atomic masses in the decay of natural radioactive elements. Soddy received the Nobel Prize in Chemistry in 1921 for his investigations into the origin and nature of isotopes. The term *isotope* is derived from the Greek "isos" for equal and "topos" for the place in the table of elements. Consequently, the development in MS in the first half of the last century was dominated by elemental analysis with the determination of the elemental isotope ratios facilitated by further improved mass spectrometer systems with higher resolving power, as introduced by Alfred Nier. He carried out the first measurements on $^{13}C/^{12}C$ abundance ratios in natural samples. Already in 1938, Nier studied bacterial metabolism by using ^{13}C as the tracer.

The following investigations by Alfred Nier formed the foundation for today's high precision isotope ratio mass spectrometry (IRMS). In contrast to mass spectrometric organic structure elucidation and target compound quantitation, IRMS provides a different analytical dimension of precision data. Isotope ratio MS delivers information on the physiochemical history, origin, and authenticity of a sample. IRMS is essential today in multiple areas of research and control such as food, life sciences, forensics, material quality control, geology, or climate research, just to name the most important applications today (Ehleringer and Cerling, 2002; Fry, 2006; Sharp, 2007).

Isotope ratio monitoring techniques using continuous flow sample introduction via GC had been introduced in the 1970s and developed into a mature analytical technology very quickly. Already in 1976, the first approach hyphenating a capillary GC with a magnetic sector MS for the systematic measurement of isotope ratios was published (Sano et al., 1976). The continuous flow determination of individual compound $^{13}C/^{12}C$ ratios was introduced by Matthews and Hayes in 1978, and the term *isotope ratio monitoring GC-MS* was coined, today commonly abbreviated as *irm-GC-MS* (Matthews and Hayes, 1978). The full range of $^{15}N/^{14}N$ and $^{18}O/^{16}O$ and most importantly H/D determinations lasted until its publication in 1998 (Brand et al., 1994; Heuer et al., 1998; Hilkert et al., 1999). Today, irm-GC-MS is the established analytical methodology for delivering the precise ratios of the stable isotopes of the elements H, N, C, and O, which are the major constituents of organic matter. Sulfur is not amenable to irm-GC-MS analysis due to its only low abundance in organic molecules. A compelling example was presented by Ehleringer on the geographical origin of cocaine from South America which was used to determine the distribution of illicit drugs (Ehleringer et al., 2000; Bradley, 2002). An important forensic application reveals the origin of explosives with a reproducible C and N isotope pattern confirming the relationship between reactants and the product (Lock and Meier-Augenstein, 2008). Organic compounds containing these elements are quantitatively converted into simple gases for mass spectrometric analysis, for example, H_2, N_2, CO, CO_2, O_2, and SO_2. The conversion to simple gases integrates and provides the full isotope information of a substance (in contrast to a fragmented organic mass spectrum) also the term *gas isotope ratio mass spectrometry* (GIRMS) is used frequently, also including irm-GC-MS applications.

2.3.3.1 The Principles of Isotope Ratio Monitoring

Isotope ratios, although tabulated with average values for all elements, are not constant. All phase transition processes, transport mechanisms, and enzymatic or chemical reactions are dependent on the physical properties of the reaction partners, that is, most importantly the mass of the molecule involved in the process or chemical reaction, for their kinetic properties.

The high precision quantitative data on isotope ratios obtained are not absolute quantitation data. Natural isotope ratios exhibit only small but meaningful variations that are measured with highest precision relative to a known standard. In most of the applications in stable isotope analysis, the differences in isotopic ratios between samples are of much more interest and significance than the absolute amount in a given sample. The isotope ratio is independent of the amount of material measured (and at the low end only determined in precision by the available ion statistics). The relative ratio measurement can be accomplished in the required and even higher precision, which is at least one order of magnitude higher than the determination of absolute values.

2.3.3.2 Notations in irm-GC-MS

The small abundance differences of stable isotopes are best represented by the delta notation (Eq. 2.29) in which the stable isotope abundance is expressed relative to an isotope standard, measured in the same run for reference,

$$\delta = \left(\frac{R_{sample}}{R_{std}} - 1\right) \times 1000 \, [‰] \tag{2.29}$$

with R being the molar ratio of the heavy to the common (light) isotope of an element, for example, $R = {}^{13}C/{}^{12}C$ or D/H or ${}^{18}O/{}^{16}O$. A graphical representation of the meaning of the δ-value is given in Figure 2.231 for the variation of the ${}^{13}C$ isotope proportion of carbon. δ-values can be positive or negative, indicating a higher or lower abundance of the major isotope in the sample than in the reference.

Other systems in use for expressing isotope ratios are ppm (part per million, relative value), at.% (atom percent, absolute value) and as the most common terminology in biomedical tracer studies (atom percent excess, relative value). The choice is dependent primarily on the specific field of application, for example, with high degrees of enrichments being very much different from natural abundances. Also, typical in this field are historical traditions in the use of different notation systems. Values of at.% and δ-notation can be converted as follows:

$$at.\% = \frac{R_{st} \cdot (\delta/1000 + 1)}{1 + R_{st} \cdot (\delta/1000 + 1)} \cdot 100$$

with R_{st} being the absolute standard ratio.

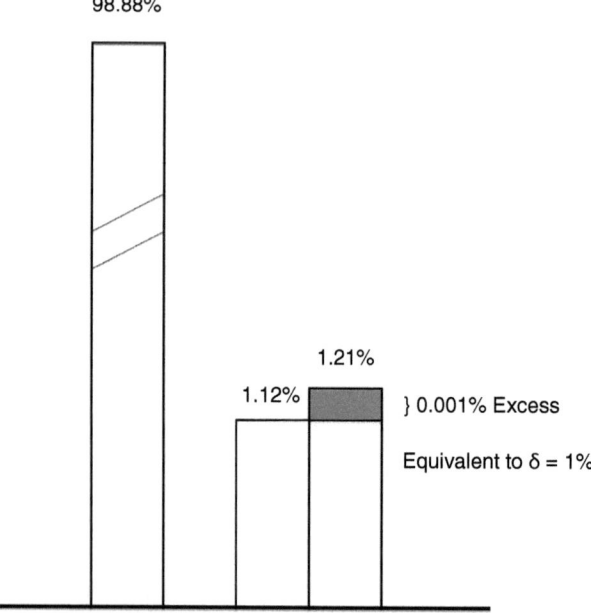

Figure 2.231 Graphical representation of the δ-notation for a ${}^{13}C$ variation of 1‰.

2.3.3.3 Isotopic Fractionation

Isotope effects, as they can be observed in phase transition or dissociation reactions, are usually the result of incomplete processes in diffusive or equilibrium fractionation. This effect is caused by different translational velocities of the lighter and heavier molecule through a medium or across a phase boundary. As the kinetic energy at a given temperature is the same for all gas molecules, the kinetic energy equation

$$E_{kin} = \frac{1}{2} m \cdot v^2$$

applies for both molecules $^{12}CO_2$ and $^{13}CO_2$ with their respective masses 44 and 45 Da:

$$\frac{1}{2}(44 \cdot v_a^2) = \frac{1}{2}(45 \cdot v_b^2),$$

resulting in

$$\frac{v_a}{v_b} = \sqrt{\left(\frac{45}{44}\right)} = 1.0113.$$

The velocity ratio of both CO_2 species explains that the average velocity of $^{12}CO_2$ is 1.13% higher than that of the heavier molecule.

Example Water-Phase Transitions

Boiling water will lose primarily the light $^1H_2^{16}O$ molecules, as can be seen in Figure 2.232. Heavier water molecules will consequently be concentrated in the liquid; the vapor will become depleted. The reversed process is observed during condensation, for example, during the formation of raindrops from humid air. As the extent of this process is temperature dependent, isotopic "thermometers" are formed and ultimately isotopic "signatures" of materials and processes are created.

Equilibrium isotope effects are usually associated with phase transition processes such as evaporation, diffusion, or dissociation reactions. When incomplete phase transition processes occur during sample preparation, they lead to severe alteration of the initial isotope ratio. In fact, and that is of highest importance for all sample preparation steps in IRMS, only complete conversion reactions are acceptable to maintain the integrity of the original isotope ratio of the sample.

Figure 2.232 Alteration of the isotope ratio during the evaporation of water resulting in a different $^{18}O/^{16}O$ ratio at low and high temperatures in vapor as well as in the liquid.

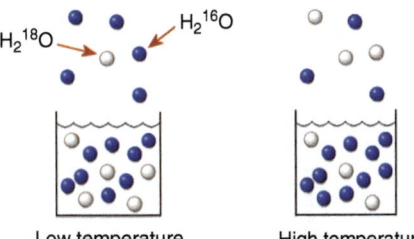

Table 2.50 Natural abundances of light stable isotopes relevant to stable isotope ratio mass spectrometry.

Element	Isotope	Atomic mass	Relative abundance [%]	Elemental relative mass difference	Molecular relative mass difference [%]	Terrestrial range [‰]	Terrestrial range [ppm]	Technical precision [‰]	Technical precision [ppm]
Hydrogen	^{1}H	1.0078	99.9840	D/H	^{1}HD/^{1}H^{1}H	—	—	—	—
Deuterium	^{2}H (D)	2.0141	0.0156	(2/1) 100%	(3/2) 50%	700	109	1	0.16
Boron	^{10}B	10.0129	19.7	^{11}B/^{10}B 10%	—	60	—	—	—
	^{11}B	11.0093	8.3	—	—	—	—	—	—
Carbon	^{12}C	12.0000	98.892	^{13}C/^{12}C	^{13}C^{16}O^{16}O/^{12}C^{16}O^{16}O	—	—	—	—
	^{13}C	13.0034	1.108	(13/12) 8.3%	(45/44) 2.3%	100	1123	0.05	0.56
Nitrogen	^{14}N	14.0031	99.635	^{15}N/^{14}N	^{15}N^{14}N/^{14}N^{14}N	—	—	—	—
	^{15}N	15.0001	0.365	(15/14) 7.1%	(29/28) 3.6%	50	181	0.10	0.72
Oxygen	^{16}O	15.9949	100	^{18}O/^{16}O	^{12}C^{16}O^{18}O/^{12}C^{16}O^{16}O	—	—	—	—
	^{17}O	16.9991	0.037	(18/16) 12.5%	(46/44) 4.6%	—	—	—	—
	^{18}O	17.9992	0204			100	200	0.10	0.20
Silicon	^{28}Si	27.9769	92.21	^{29}Si/^{28}Si	^{29}Si^{19}F$_3$/^{28}Si^{19}F$_3$	6	—	—	—
	^{29}Si	28.9765	4.7	(29/28) 3.6%	(86/85) 1.2%	—	—	—	—
	^{30}Si	29.9738	3.09	—	—	—	—	—	—
Sulfur	^{32}S	31.9721	95.02	^{34}S/^{32}S	^{34}S^{16}O^{16}O/^{32}S^{16}O^{16}O	—	—	—	—
	^{33}S	32.9715	0.76	—	—	—	—	—	—
	^{34}S	33.9769	4.22	(34/32) 6.3%	(66/64) 3.1%	100	4580	0.20	9.16
	^{36}S	35.9671	0.014	—	—	—	—	—	—
Chlorine	^{35}Cl	34.9689	75.77	^{37}Cl/^{35}Cl	^{12}CH$_3$^{37}Cl/^{12}CH$_3$^{35}Cl	10	—	0.10	—
	^{37}Cl	36.9659	24.23	(37/35) 5.7%	(52/50) 4.0%	—	—	—	—

From: Gilles St. Jean, Basic Principles in Stable Geochemistry, IRMS Short Course, 9th Canadian CF-IRMS Workshop 2002 (St. Jean, 2002).

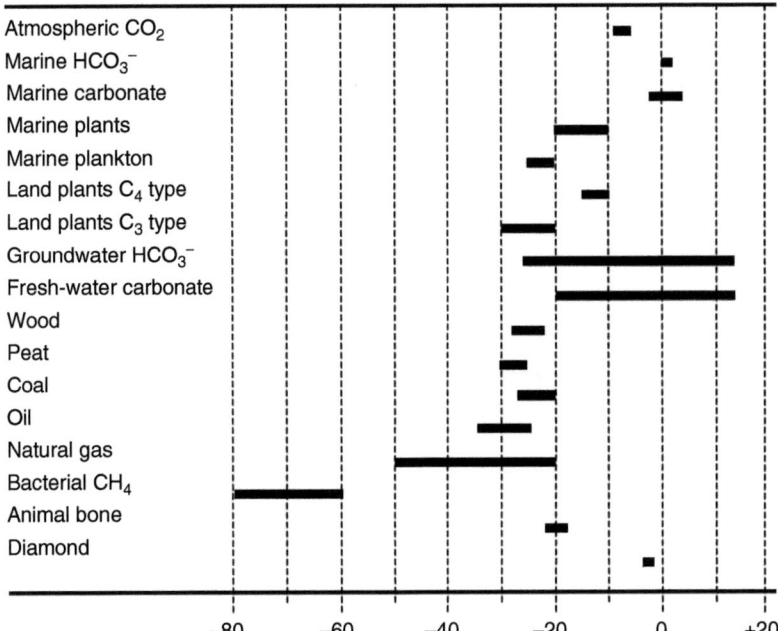

Figure 2.233 δ^{13}C variations in natural compounds (δ^{13}C VPDB, ‰ scale) (de Vries, 2000, courtesy IAEA).

Isotope effects are also observed in chemical and biochemical reactions (kinetic fractionation). Of particular significance is the isotope effect, which occurs during enzymatic reactions with a general depletion in the heavier isotope being key to uncover metabolic processes. Chemical bonds to the heavier isotope are stronger, more stable, and need higher dissociation energies in chemical reactions due to the different vibrational energy levels involved. Hence, the rate of enzymatic reaction is faster with the light isotope, leading to differences in the abundance between substrate and product, unless the substrate is fully consumed, which is not the case in cellular steady state equilibria. Kinetic fractionation effects must also be considered when employing compound derivatization steps (e.g. silylation and methylation) during sample preparation for GC application (Meyer-Augenstein, 1997).

The natural variations in isotopic abundances can be large, depending on the relative elemental mass differences: hydrogen (100%) > oxygen (12.5%) > carbon (8.3%) > nitrogen (7.1%), see Table 2.50 (Rosman and Taylor, 1998; Rossmann, 2001). An overview of the isotopic variations found in natural compounds is given in Figures 2.233–2.236.

2.3.3.4 irm-GC-MS Technology
irm-GC-MS is applied to obtain compound specific data after a GC separation of mixtures in contrast to bulk analytical data from an EA system. The bulk analysis delivers the average isotope ratio within a certain volume of the sample material. The entire sample is converted into simple gases using conventional EAs or high

350 | 2 Fundamentals

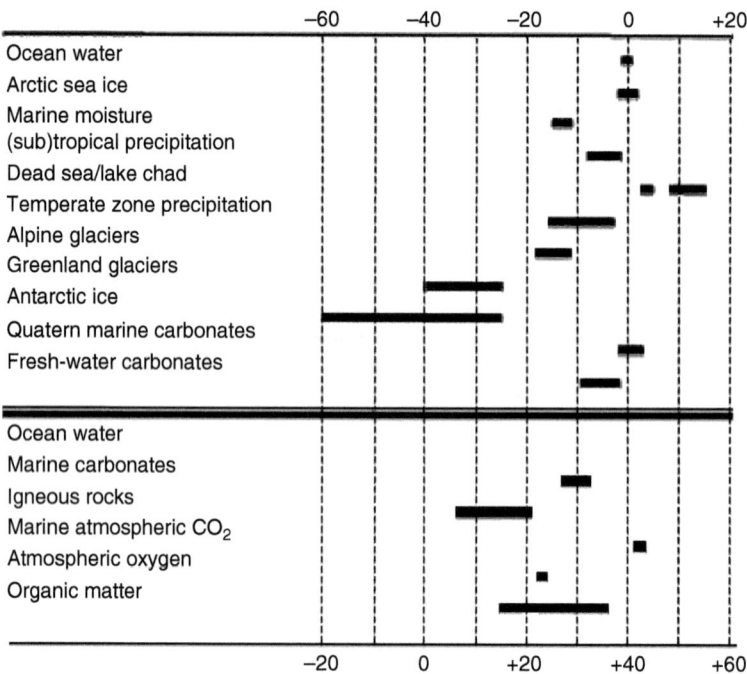

Figure 2.234 $\delta^{18}O$ variations in natural compounds (top: $\delta^{18}O$ VSMOW ‰ scale for waters, $\delta^{18}O$ VPDB ‰ scale for carbonates, bottom: $\delta^{18}O$ VSMOW ‰ scale) (de Vries, 2000, courtesy IAEA).

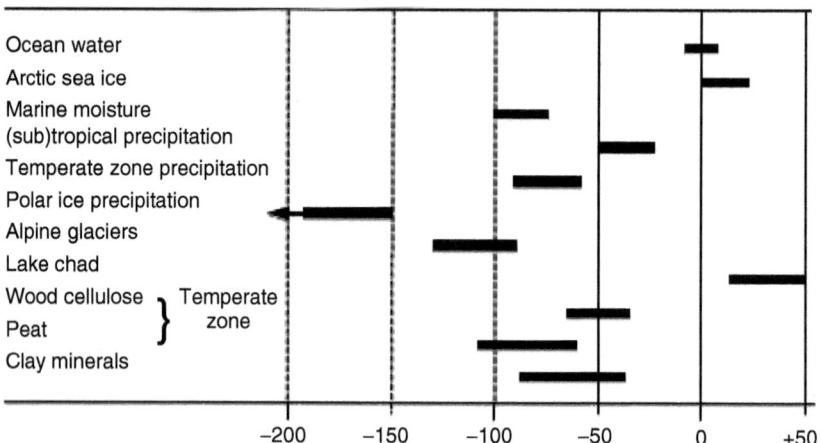

Figure 2.235 $\delta^{2}H$ variations in natural compounds ($\delta^{2}H$ VSMOW ‰ scale) (de Vries, 2000, courtesy IAEA).

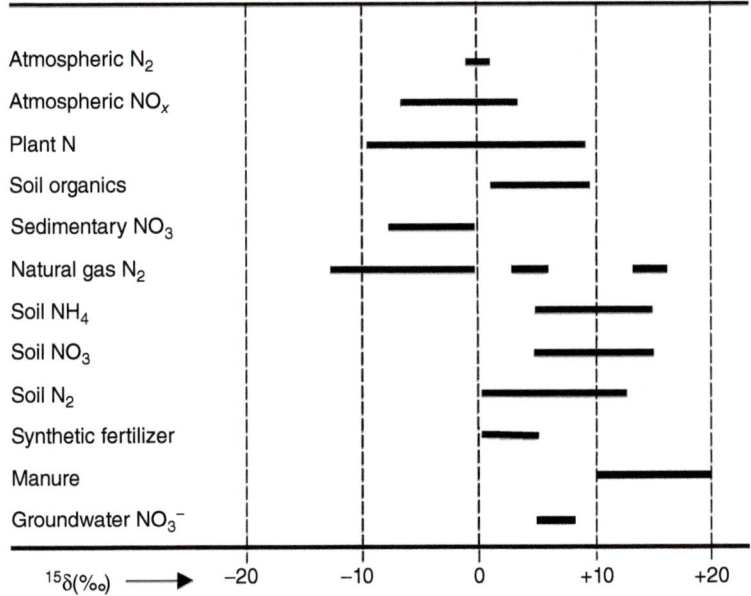

Figure 2.236 $\delta^{15}N$ variations in natural compounds ($\delta^{15}N$ Air ‰ scale) (de Vries, 2000, courtesy IAEA).

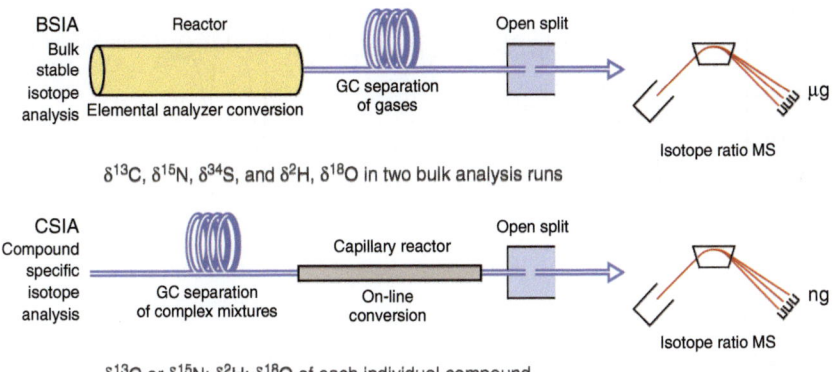

Figure 2.237 Schematics of the bulk sample (BSIA) vs. compound specific (CSIA) isotope analysis (Courtesy Thermo Fisher Scientific).

temperature conversion elemental analyzers (TC/EAs). In contrast to a bulk analysis, irm-GC-MS delivers specific isotope ratio data in the low picomolar range of individual compounds after a conventional capillary GC separation. Figure 2.237 illustrates both the concepts of "bulk stable isotope analysis" (BSIA) and compound specific isotope analysis (CSIA). In BSIA, the IRMS is coupled with an elemental analyzer (EA). The complete sample is first converted to simple gases followed by their chromatographic separation. In CSIA, the IRMS is coupled with a GC. The

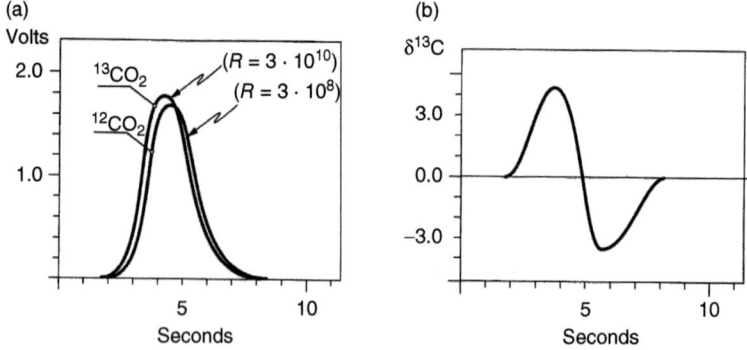

Figure 2.238 Chromatographic elution profiles of CO_2 with natural ^{13}C abundance resulting in the typical S-shaped curvature when displaying the calculated δ-values: (a) the intensity difference of the isotopes with the choice of amplification resistors R in a difference of two orders of magnitude and (b) the typical ratio chromatogram $^{13}C/^{12}C$ (Rautenschlein et al., 1990).

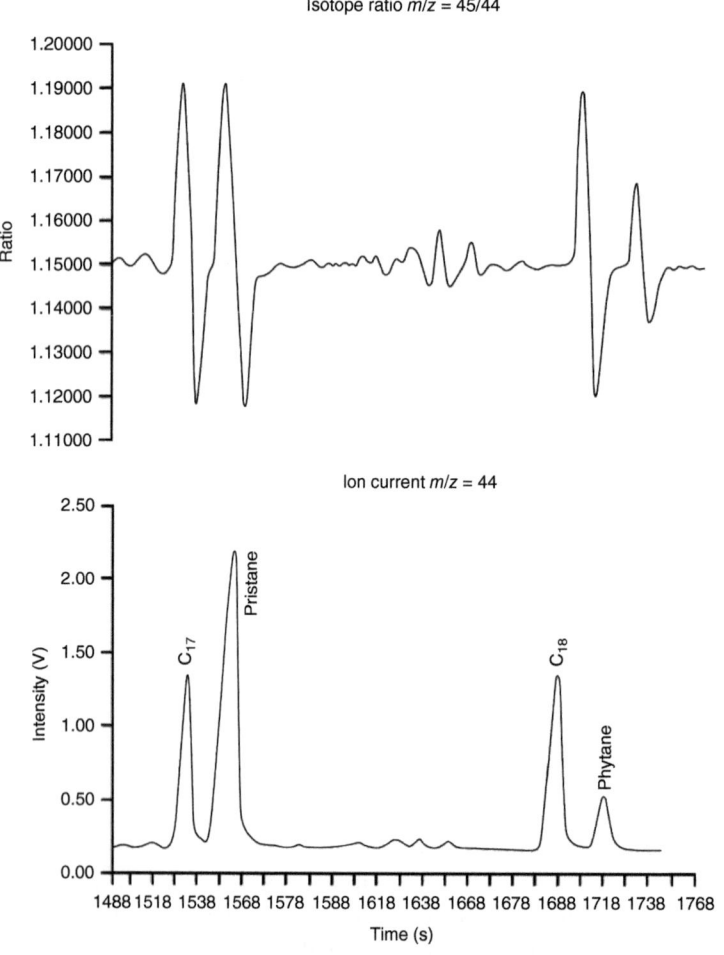

Figure 2.239 irm-GC-MS analysis of the crude oil steroid biomarkers pristane and phytane with the $δ^{13}C$ ratio profile on the top, bottom $^{12}CO_2$ intensity chromatogram.

sample components are first separated by capillary chromatography and then individually converted to the measurement gases.

The most important step in irm-GC-MS is the conversion of the eluting compounds into simple measurement gases such as CO_2, N_2, CO, and H_2. In irm-GC-MS, as a continuous flow application, this conversion is achieved online within the helium carrier gas stream while preserving the chromatographic resolution of the sample. The products of the conversion are then fed by the carrier gas stream into the isotope ratio MS.

It is important to note that during chromatographic separation on regular, fused silica capillary columns, a separation of the different isotopically substituted species of a compound already takes place. Due to the lower molar volume of the heavier components and the resulting differences in mobile/stationary-phase interactions, the heavier components elute slightly earlier (Matucha et al., 1991). This separation effect can also be observed in "organic" GC-MS analyses when having a closer look at the peak top retention times of the ^{13}C carbon isotope mass trace. irm-GC-MS makes special use of the chromatographic isotope separation effect by displaying the typical S-shaped ratio traces during analysis indicating the substances of interest (see Figures 2.238 and 2.239).

2.3.3.5 The Open Split Interface

For high precision isotope ratio determination, the IRMS ion source pressure must be kept absolutely constant. For this reason, it is mandatory in all continuous flow applications coupled to an isotope ratio MS to keep the open split interface at atmospheric pressure. The open split coupling eliminates the isotopic fractionation due to an extended impact of the high vacuum of the mass spectrometer into the chromatography and conversion reactors. The basic principle of an open split is to pick up a part of the helium/sample stream by a transfer capillary from a pressureless environment, that is, an open tube or a wide capillary, and transfer it into the IRMS (see Figure 2.240). The He stream added by a separate capillary ensures protection from ambient air. Retracting the transfer capillary into a zone of pure helium allows cutting off parts of the chromatogram. In all modes, a constant flow of helium into the isotope ratio MS, and consequently constant ion source conditions are maintained. The valve-free open split is inert and does not create any pressure waves during switching.

In addition, the open split interface also offers an automatic peak dilution capability. Due to the transfer of the analyte gases in a helium stream more than one interface can be coupled in parallel for alternate use to an isotope ratio MS via separate needle valves.

2.3.3.6 Compound Specific Isotope Analysis

irm-GC-MS is amenable to all in GC volatile organic compounds down to the low picomole range. A capillary design of oxidation and high temperature conversion reactors is required to guarantee the integrity of GC resolution. Table 2.51 gives an overview of the techniques available today with the achievable precision values by state-of-the-art IRMS technology. CSIA is a new field of isotope analysis, but today

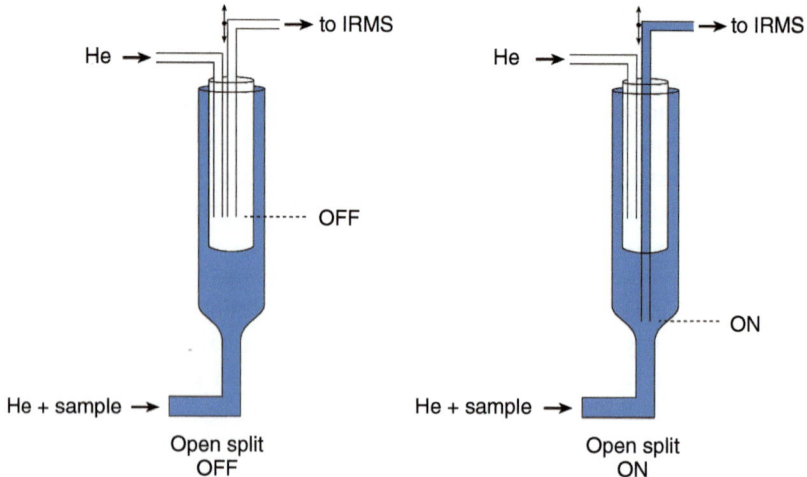

Figure 2.240 Open split interface to IRMS, effected by moving the transfer capillary to IRMS from the column inlet to the He sample flow region (Courtesy Thermo Fisher Scientific).

Table 2.51 Analytical methods used in compound-specific isotope analysis.

Element ratio	Measured species	Method	Temperature	Analytical precision[a]
$\delta^{13}C$	CO_2	Combustion	up to 1000 °C	^{13}C 0.06‰, 0.02‰/nA
$\delta^{15}N$	N_2	Combustion	up to 1000 °C	^{15}N 0.06‰, 0.02‰/nA
$\delta^{18}O$	CO	Pyrolysis	1450 °C	^{18}O 0.15‰, 0.04‰/nA
$\delta^{2}H$	H_2	Pyrolysis	1280 °C	^{2}H 0.50‰, 0.20‰/nA

a) Ten pulses of reference gas (amplitude 3 V, for H_2 5 V) δ notation.

a mature and steadily expanding application. The first GC combustion system for $\delta^{13}C$ determination was commercially introduced in 1988 by Finnigan MAT Bremen, Germany. This technique combines the resolution of capillary GC with the high precision of IRMS. In 1992, the capabilities for analysis of $\delta^{15}N$ and in 1996 for $\delta^{18}O$ were added. Quantitative pyrolysis by high temperature conversion (GC-TC IRMS) for δD analyses was introduced in 1998, introducing an energy discrimination filter in front of the HD collector at m/z 3 for the suppression of $^{4}He^{+}$ ions from the carrier gas interfering with the HD^{+} signal (Hilkert et al., 1999).

2.3.3.7 Online Combustion for $\delta^{13}C$ and $\delta^{15}N$ Determination

All compounds eluting from a GC column are oxidized in a capillary reactor to form CO_2, N_2, and H_2O as a by-product, at 940–1000 °C. NO_x produced in the oxidation reactor is reduced to N_2 in a subsequent capillary reduction reactor. The H_2O formed in the oxidation process is removed by an online Nafion™ dryer, a maintenance-free

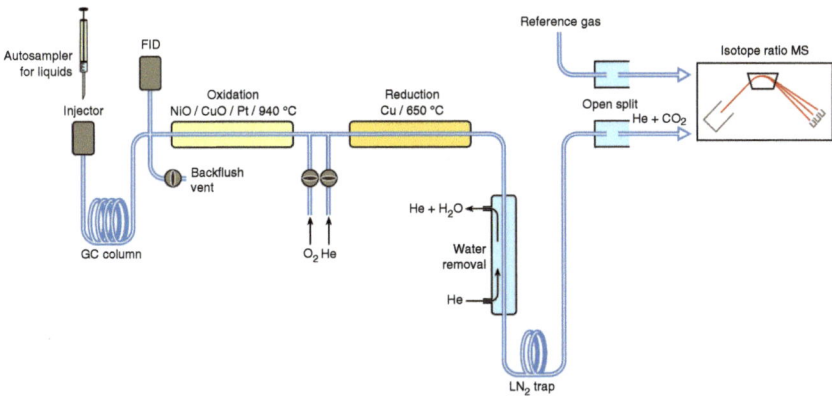

Figure 2.241 Online oxidation of compounds eluting from the GC for the generation of CO_2 and N_2 for IRMS measurement (Courtesy Thermo Fisher Scientific).

water removal system. For the analysis of $\delta^{15}N$, all CO_2 is retained in a liquid nitrogen cryo-trap in order to avoid interferences from CO on the identical masses before transfer into the IRMS through the movable capillary open split. CO is generated in small amounts as a side reaction during the ionization process from CO_2 in the ion source of the MS. A detailed schematic of the online combustion setup is given in Figure 2.241.

2.3.3.8 The Oxidation Reactor

The quantitative oxidation of all organic compounds eluting from the GC column, including the refractory methane, is performed at temperatures up to 1000 °C. The reactor consists of a capillary ceramic tube loaded with twisted Ni, Cu, and Pt wires. The resulting internal volume compares to a capillary column and secures the integrity of the chromatographic separation. The reactor can be charged and recharged automatically with O_2 added to a backflush flow (BKF) every 2–3 days, depending on the operating conditions. The built-in BKF system reverses the flow through the oxidation reactor toward an exit directly after the GC column. The BKF is activated during the analysis to cut off an eluting solvent peak in front of the oxidation furnace by flow switching. All valves must be kept outside of the analytical flow path to maintain optimum GC performance (Figure 2.242).

2.3.3.9 The Reduction Reactor

The reduction reactor typically comprises copper material and is operated at 650 °C to remove any O_2 bleed from the oxidation reactor and to convert any produced NO_x into N_2. It is of the same capillary design as the oxidation reactor.

2.3.3.10 Water Removal

Water produced during the oxidation reaction is removed through a 300 μm inner diameter Nafion™ capillary, which is dried by a counter current of He on the outside (Figure 2.243). The water removal adds no dead volume and is maintenance free.

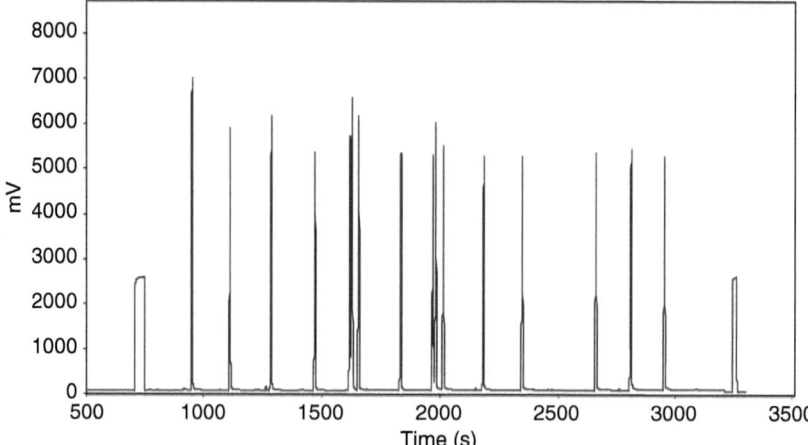

Figure 2.242 irm-GC-MS chromatogram of a fatty acid methyl ester (FAME) sample after online combustion at 940 °C. The rectangular peaks in the beginning and at the end of the chromatogram are the CO_2 reference gas injection peaks (Courtesy Thermo Fisher Scientific).

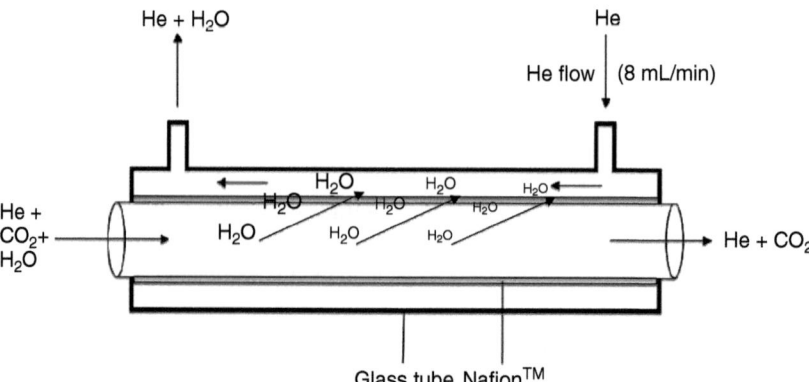

Figure 2.243 Principle of online removal of water from a He stream using a Nafion membrane (Courtesy Thermo Fisher Scientific).

> **Online Water Removal from a He Stream Using Nafion™**
>
> Water is removed from a He sample stream by a gas-tight but hygroscopic Nafion tubing. The sample flow containing He, CO_2, and H_2O passes through the Nafion tubing which is mounted coaxially inside a glass tube. This glass tube, and therefore the outer surface of the Nafion tube, is constantly kept dry by a He flow of 8–10 mL/min. Due to the water gradient through the Nafion membrane wall, any water in the sample flow will move through the membrane. A dry gas comprising of only He and CO_2 resulting from the oxidation is fed to the ion source of the mass spectrometer.

Structure of Nafion:

Properties: Nafion™ is the combination of a stable Teflon™ backbone with acidic sulfonic groups. It is highly conductive to cations, making it ideal for many membrane applications. The Teflon backbone interlaced with ionic sulfonate groups gives Nafion a high operating temperature, for example, up to 190 °C. It is selectively and highly permeable to water. The degree of hydration of the Nafion membrane directly affects its ion conductivity and overall morphology. See also http://de.wikipedia.org/wiki/Nafion.

2.3.3.11 The Liquid Nitrogen Trap

For the analysis of $\delta^{15}N$, all the CO_2 must be removed quantitatively to avoid an interference of CO^+ generated from CO_2 in the ion source with the N_2^+ analyte. This is achieved by immersing the deactivated fused silica capillary between the water removal and the open split into a liquid nitrogen cryo-trap. The trapped CO_2 is easily released after the measurement series with no risk of CO_2 contamination of the ion source by using the movable open split.

2.3.3.12 Online High Temperature Conversion for δ^2H and $\delta^{18}O$ Determination

A quantitative pyrolysis by high temperature conversion of organic matter is applied for the conversion of organic oxygen and hydrogen to form the measurement gases CO and H_2 for the determination of $\delta^{18}O$ or δD (see Figure 2.244). This process

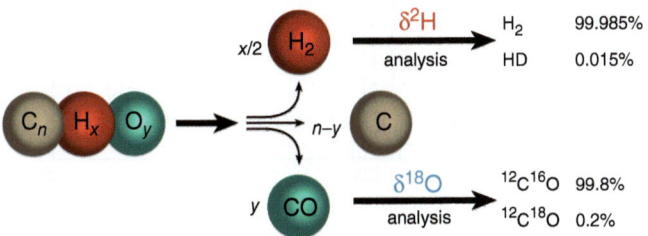

Figure 2.244 Principle of high temperature conversion (Courtesy Thermo Fisher Scientific).

requires an inert and reductive environment to prevent any O- or H-containing material from reacting or exchanging with the analyte.

For the determination of $\delta^{18}O$, the analyte must not contact the ceramic tube that is used to protect against air. For the conversion to CO, the pyrolysis takes place in an inert platinum inlay of the reactor. Due to the catalytic properties of the platinum, the reaction can be performed at 1280 °C. For the determination of δD from organic compounds, the reaction is performed in an empty ceramic tube at 1450 °C. Such high temperatures are required to ensure a quantitative conversion.

Typically, an online high temperature reactor is mounted in parallel to a combustion reactor at the GC oven. The complete setup for online high temperature conversion is given in Figure 2.245. A water removal step has no effect here on the dry analyte gas (Figure 2.246).

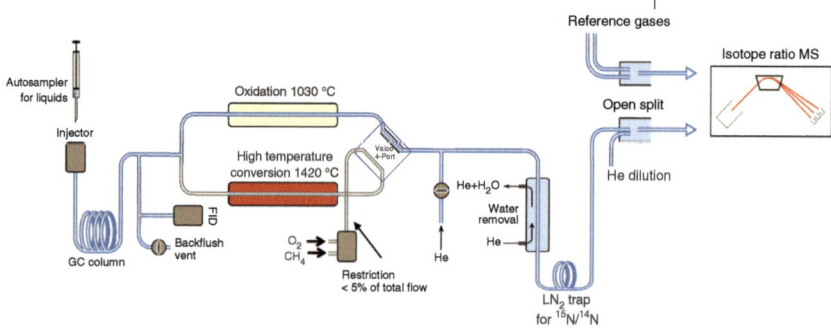

Figure 2.245 Online high temperature conversion of compounds eluting from the GC for the production of CO and H_2 for IRMS measurement (Courtesy Thermo Fisher Scientific).

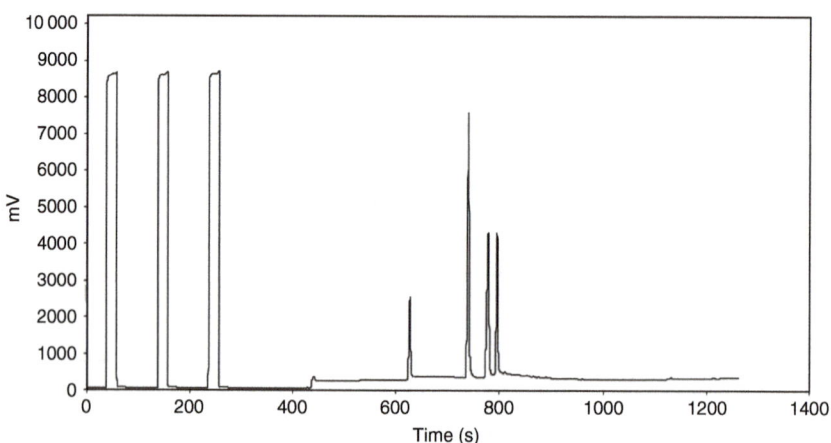

Figure 2.246 irm-GC-TC/MS chromatogram of flavor components using high temperature conversion with three CO reference gas pulses at start (Courtesy Thermo Fisher Scientific).

2.3.3.13 Mass Spectrometer for Isotope Ratio Analysis

Mass spectrometers employed for isotope ratio measurements are dedicated non-scanning, static magnetic sector mass spectrometer systems. The ion source is particularly optimized by a "closed source" design for the ionization of gaseous compounds at very high ion production efficiency of 500–1000 molecules/ion at high response linearity (Brand, 2004). After extraction from the source region, the ions are typically accelerated by 2.5–10 kV to form an ion beam which enters the magnetic sector analyzer through the entrance slit (see Figure 2.247).

The ion currents of the isotopes are measured simultaneously at the individual m/z values by discrete Faraday cups mounted behind a grounded slit (see Figure 2.248). An array of specially designed deep Faraday cups for quantitative measurement is precisely positioned along the optical focus plane representing the typical isotope mass cluster to be determined.

Only the simultaneous measurement of the isotope ion currents with dedicated Faraday cups for each isotope using individual amplifier electronics cancels out ion beam fluctuations due to temperature drifts or electron beam variations and provides the required precision (see Figure 2.249, also refer to Table 2.50). Current instrumentation allows for irm-GC-MS analyses of organic compounds down to the low picomole range.

Isobaric interferences from other isotope species at the target masses require the measurement of more masses than just the targeted isotope ratio for a necessary correction. For the measurement of $\delta^{13}C$, three collectors at m/z 44, 45, and 46 are necessary (see Table 2.51). Algorithms for a fully automated correction of isobaric ion contributions are implemented in modern isotope ratio MS data systems, for example, for the isobaric interferences of ^{17}O and ^{13}C on m/z 46 of CO_2, see Table 2.52 (Craig, 1957). Other possible interferences such as CO on N_2 and N_2O on CO_2 can be taken care of by the interface technology.

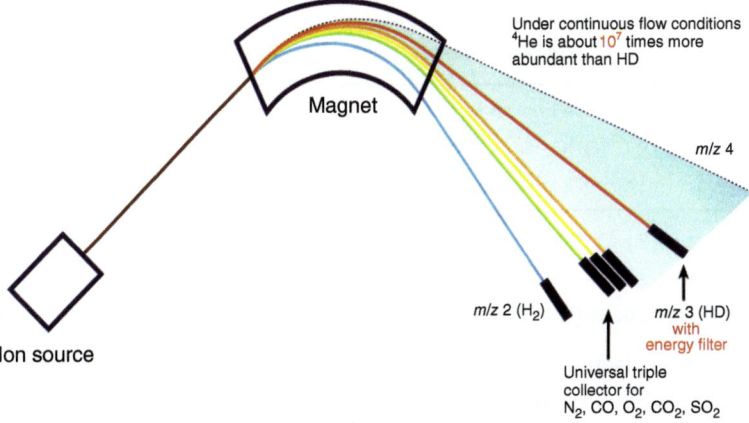

Figure 2.247 Isotope ratio mass spectrometer with ion source for simultaneous isotope detection (Courtesy Thermo Fisher Scientific).

360 | 2 Fundamentals

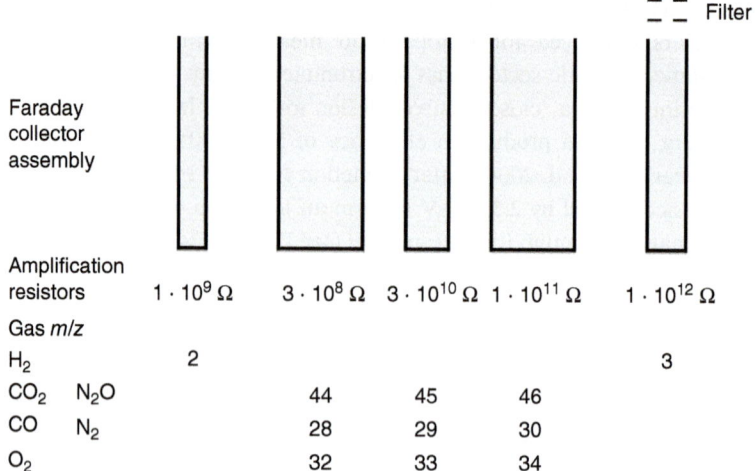

Figure 2.248 Typical Faraday cup arrangement for measurement of the isotope ratios of the most common gas species (below of the cup symbols the values of the amplification resistors are given; the HD cup is shown to be equipped with a kinetic energy filter to prevent from excess $^4He^+$).

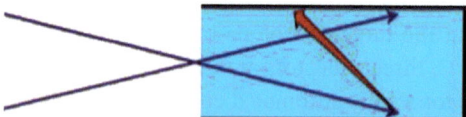

Figure 2.249 Cross section of a Faraday cup for isotope ratio measurement; the large depth of the graphite cup prevents from losses caused by scattering. The arrows represent the ion beam with a focus point at the exit slit in front of the Faraday cup (Courtesy Thermo Fisher Scientific).

Table 2.52 Isobaric interferences when measuring $\delta^{13}C$ from CO_2.

m/z	Ion composition
44	$^{12}C^{16}O^{16}O$
45	$^{13}C^{16}O^{16}O$, $^{12}C^{16}O^{17}O$
46	$^{12}C^{16}O^{18}O$, $^{12}C^{17}O^{17}O$, $^{13}C^{16}O^{17}O$

The Identical Treatment (IT) Principle

Standardization in isotope ratio monitoring measurements should be done exclusively using the principle of "Identical Treatment of reference and sample material", the "IT Principle". Mostly, isotopic referencing is made with a co-injected peak of standard gas.

2.3.3.14 Injection of Reference Gases

The measurement of isotope ratios requires that sample gases be measured relative to a reference gas of a known isotope ratio. For sample-standard referencing in irm-GC-MS, cylinders of calibrated reference gases, the laboratory standard of H_2, CO_2, N_2, or CO is used for an extended period. This referencing procedure turned out to be the most economic and precise compared to the addition of an internal standard to the sample, also providing necessary quality assurance purposes. An inert, fused silica capillary supplies the reference gas in the microliter per minute range into a miniaturized mixing chamber, the reference gas injection port (see Figure 2.250a,b). This capillary is lowered under software control into the mixing chamber, for example, 20 s, creating a He reference gas mixture which is fed into the IRMS source via a second independent gas line. This generates a rectangular, flat-top gas peak without changing any pressures or gas flows in the ion source. Reference gases used are pure nitrogen (N_2), carbon dioxide (CO_2), hydrogen (H_2), and carbon monoxide (CO).

2.3.3.15 Isotope Reference Materials

In IRMS, the measurement of isotope ratios requires the samples to be measured relative to a reference of a known isotope ratio. This is the only means to achieve the required precision level of <1.5 ppm, for example, the $^{13}C/^{12}C$ isotope ratio, which is 0.15‰ in the commonly used δ-notation. The employed isotopic reference scales

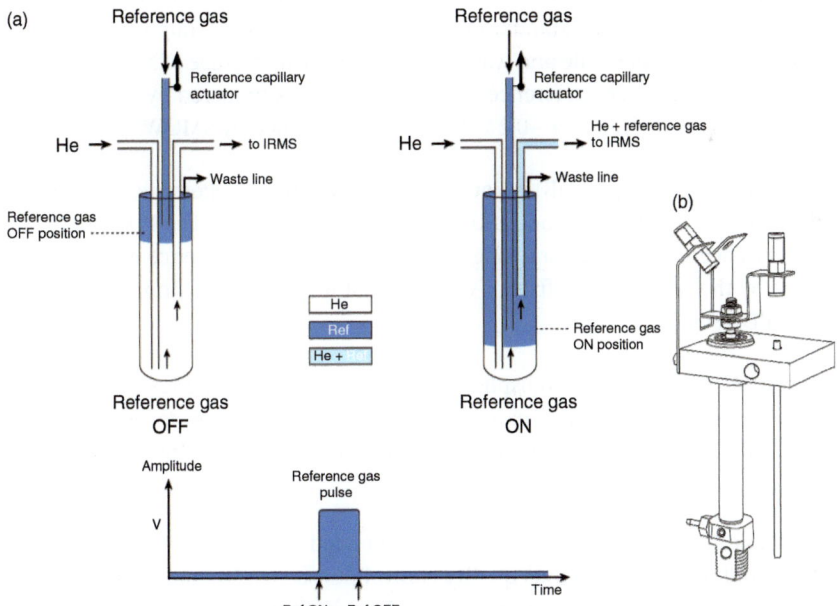

Figure 2.250 (a) Reference gas injection port in irm-GC-MS implemented as an open coupling between GC column and IRMS ion source. (b) Reference gas injection port design: left side pneumatic drive, right side valve tube for fused silica columns, top column connectors for fixed (left) and movable (right) columns (Courtesy Thermo Fisher Scientific).

Table 2.53 IRMS primary reference materials for irm-GC-MS. Absolute isotope ratios R in 1σ standard precision.

H	$R = 0.00015576 \pm 0.000005$	VSMOW	Standard mean ocean water
C	$R = 0.011224 \pm 0.000028$	VPDB	Pee Dee Belemnite[a]
N	$R = 0.0036765$	Air	Atmospheric air
	$R = 0.003663 \pm 0.000005$	NSVEC	N standard by Jung and Svec
O	$R = 0.0020052 \pm 0.000045$	VSMOW	Standard mean ocean water[a]

a) As a general rule, $\delta^{18}O$ data of carbonates and CO_2 gas are reported against VPDB whereas $\delta^{18}O$ data of all other materials should be reported versus VSMOW.

shown in Figures 2.233–2.236 are arbitrarily defined by the community relative to the isotope ratio of a selected primary reference material for a given element.

Primary reference materials are available through the International Atomic Energy Agency (IAEA) in Vienna, Austria. The most used and recognized international reference materials for stable isotope ratio analysis of natural abundances are given in Table 2.53. The abbreviations used with a preceding "V" as in "VSMOW" refer to the reference materials prepared by the IAEA in Vienna compliant with the regular consultants' meetings (for a detailed discussion, see Groening, 2004).

VSMOW Oceans contain almost 97% of the water on Earth and have a uniform isotope distribution. Oceans are the major sink in the hydrological cycle. The Standard Mean Ocean Water (SMOW) standard was initially proposed by Craig 1957 as a concept for the origin of the scale and was calculated from an average of samples taken from different oceans. The reference water VSMOW was prepared by the IAEA in 1968 with the same $\delta^{18}O$ and a −0.2‰ lighter δD as defined by SMOW.

VPDB Pee Dee Belemnite (PDB) is $CaCO_3$ of the rostrum of a bellemnite (Belemnita americana) from the Pee Dee Formation of South Carolina, USA. The PDB reference material has been exhausted for a long time. The new VPDB reference material was anchored by the IAEA with a fixed δ-value so that it corresponds nominally to the previous PDB scale.

Atmospheric Air Nitrogen in atmospheric air has a very homogeneous isotope composition. The atmosphere is the main nitrogen sink and the largest terrestrial nitrogen reservoir. NSVEC is the reference material initially prepared by the Iowa State University, USA (Junk and Svec, 1958).

The available IAEA reference materials are intended to calibrate local laboratory standards and not for continuous quality control purposes (Groening, 2004). Every laboratory should prepare for routine work, long-lasting laboratory specific standards with similar characteristics to the used references and the working standards, which are calibrated against the primary reference materials.

For comparison to results of other laboratories, the raw data of the mass spectrometer are converted into VSMOW or VPDB (Boato, 1960). Due to the relative measurement to a standard ratio, the formula contains a product term besides the sum of the δ-values,

$\delta_3 = \delta_1 + \delta_2 + 10^{-3} \cdot \delta_1 \cdot \delta_2$,
$\delta_1 = \delta$ (sample/reference)   measured in the lab,
$\delta_2 = \delta$ (reference/standard)   known lab calibration,
$\delta_3 = \delta$ (sample/standard)   unknown.

2.3.4 Acquisition Techniques in GC-MS

In data acquisition by the mass spectrometer, there is a significant difference between the detection of the complete spectrum (full scan) and the recording of individual masses by selected/multiple ion monitoring (SIM/MID) or selected/multiple reaction monitoring (SRM/MRM). Particularly with continually operating spectrometers (ion beam instruments: magnetic sector MS, quadrupole MS), there are large differences between these two recording techniques with respect to selectivity, sensitivity, and information content. For spectrometers with storage features (ion storage: ion trap MS, Orbitrap MS), these differences are less strongly pronounced. Besides single stage types of analyzers (GC-MS), multistage mass spectrometers (GC-MS/MS, GC-QTOF) are playing an increasingly important role in residue and target compound analysis as well as structure determination. With the MS/MS technique (tandem MS), which is available in both beam instruments and ion storage mass spectrometers, much more analytical information and the unique structure-related selectivity for target compound quantitation can be obtained.

2.3.4.1 Detection of the Complete Mass Spectrum (Full Scan)

The continuous recording of mass spectra (full scan) and the simultaneous monitoring of the retention time allow the identification of analytes by comparison with libraries of mass spectra. With beam instruments, it should be noted that the sensitivity required for recording the spectrum depends on the efficiency of the ion source, the transmission through the analyzer, and, most particularly, the dwell time of the ions. The dwell time per mass is given by the width of the mass scan (e.g. 50–550 Da) and the scan/acquisition rate of the chromatogram (e.g. 500 ms). From this, a scan rate of 1000 u/s is calculated, which is a "dwell time" of just 1 ms/u. Effective scan rates of modern quadrupole instruments exceed 10 000 u/s with maximum scan rates up to 20 000 u/s. Each mass from the selected mass range is measured only once during a scan over a short period (Figure 2.251). All other ions from the substance formed in parallel in the ion source are not detected during the mass scan and get lost (quadrupole as mass filter),

$$\text{dwell time} = \frac{m_2 - m_1}{\text{scan rate}} \text{ [ms]},$$

with
m_1 being the start mass of scan,
m_2 being the end mass of scan.

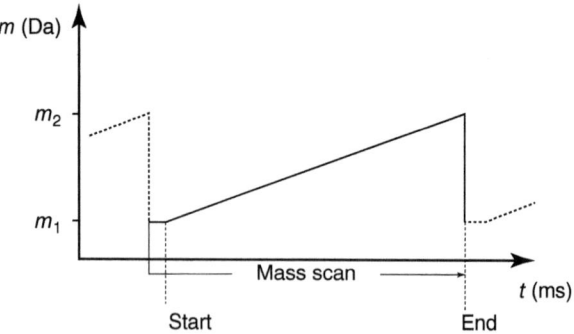

Figure 2.251 Scan function of the quadrupole analyzer: each mass between the start of the scan (m_1) and the end (m_2) is only registered once during the scan.

Typical full scan sensitivities for most compounds with benchtop quadrupole systems are in the mid to low pg region. Extending the scan time increases the dwell time; hence, the sensitivity of these systems for full scan operation. However, there is an upper limit for a useful scan time determined by the minimum required number of data points across a chromatographic peak. In practice, for coupling with capillary GC, scan rates of 0.2–0.5 s are used. For quantitative determinations, it should be ensured that the scan rate across the chromatographic peak is adequate in order to determine the area and height correctly (see also Section 3.3.1). The SIM mode is chosen in targeted analyses to increase the sensitivity and scan rates of quadrupole systems for this reason.

2.3.4.2 Recording Individual Masses (SIM)

In quadrupole or magnetic sector mass spectrometers (beam instruments), the achievable detection limit in the full scan mode is insufficient for trace analysis because the analyzer can only use short dwell times per ion during the scan. Additional sensitivity is achieved by sharing the same scan time on a few selected ions by individual mass recording (SIM) (Table 2.54 and Figure 2.252).

At the same time, a higher scan rate at improved sensitivity can be chosen so that chromatographic peaks can be plotted more precisely, and fast GC applications are

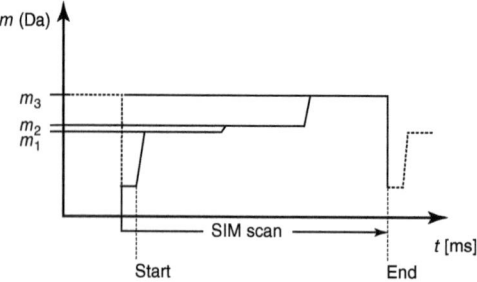

Figure 2.252 SIM scan with a quadrupole analyzer: the total scan time is divided here into the three individual masses m_1, m_2 and m_3 with correspondingly long dwell times.

Table 2.54 Dwell times per ion and relative sensitivity in SIM analysis (for beam instruments) at constant scan rates.

Number of SIM ions	Total scan time[a] [ms]	Total voltage setting time[b] [ms]	Effective dwell time per ion[c] [ms]	Relative sensitivity[d] [%]
1	500	2	498	100
2	500	4	248	50
3	500	6	165	33
4	500	8	123	25
10	500	20	48	10
20	500	40	23	5
30	500	60	15	3
40	500	80	11	3
50	500	100	8	2
For comparison in full scan				
500	500	2	1	1

a) The total scan time is determined by the necessary scan rate of the chromatogram and is held constant.
b) Total voltage setting times are necessary in order to adjust the mass filter for the subsequent SIM mass acquisition. The actual times necessary can vary slightly depending on the type of instrument.
c) Duty cycle/ion.
d) The relative sensitivity is directly proportional to the effective dwell time per ion.

supported. The SIM technique is used exclusively for quantifying data on known target compounds, especially in trace analysis.

The mode of operation of a GC-MS system as a mass-selective detector in the SIM mode requires the selection of individual ions (fragments, molecular ions) so that the desired analytes can be detected selectively (targeted approach), and if possible not affected by the sample matrix. Other compounds contained in the sample besides those chosen for analysis remain undetected as long as not the same masses are generated. Thus, the matrix present in trace analysis can be masked out, as are analytes whose appearance is not expected in the sample.

In the choice of masses for detection, it is assumed that for three selective signals in the fragmentation pattern per substance, a secure basis for a yes/no decision can be found (Sphon, 1978). Identification of substances by comparison with spectral libraries is no longer possible. The relative intensities (ion ratios) of the selected ions serve as quality criteria (qualifiers) (1 ion ⇒ no criterion, 2 ions ⇒ 1 criterion and 3 ions ⇒ 3 criteria!). Confirmation of positive results, and statistically of negative results as well, is required by international directives either by full scan, MS/MS, ion ratios, or HRMS. Planning an analysis in the SIM mode first requires a standard run in the full scan mode to determine both the retention times and the mass signals necessary for the SIM detection (Tables 2.55 and 2.56). SIM analyses for large sets of analytes are significantly facilitated by using the retention timed SIM (t-SIM) mode several modern GC-MS instruments offer.

Table 2.55 Characteristic ions (m/z values) for selected (poly)aromatic hydrocarbons (PAHs) and their alkyl derivatives (in the elution sequence for methylsilicone phases).

Component name	m/z
Benzene-d_6	92, 94
Benzene	77, 78
Toluene-d_8	98, 100
Toluene	91, 92
Ethylbenzene	91, 106
Dimethylbenzene	91, 106
Methylethylbenzene	105, 120
Trimethylbenzene	105, 120
Diethylbenzene	105, 119, 134
Naphthalene-d_8	136
Naphthalene	128
Methylnaphthalene	141, 142
Azulene	128
Acenaphthene	154
Biphenyl	154
Dimethylnaphthalene	141, 155, 156
Acenaphthene-d_{10}	162, 164
Acenaphthene	152
Dibenzofuran	139, 168
Dibenzodioxin	184
Fluorene	165, 166
Dihydroanthracene	178, 179, 180
Phenanthrene-d_{10}	188
Phenanthrene	178
Anthracene	178
Methylphenanthrene	191, 192
Methylanthracene	191, 192
Phenylnaphthalene	204
Dimethylphenanthrene	191, 206
Fluoranthene	202
Pyrene	202

Table 2.55 (Continued)

Component name	m/z
Methylfluoranthene	215, 216
Benzofluorene	215, 216
Phenylanthracene	252, 253, 254
Benzanthracene	228
Chrysene-d$_{12}$	240
Chrysene	228
Methylchrysene	242
Dimethylbenz[a]anthracene	239, 241, 256
Benzo[b]fluoranthene	252
Benzo[j]fluoranthene	252
Benzo[k]fluoranthene	252
Benzo[e]pyrene	252
Benzo[a]pyrene	252
Perylene-d$_{12}$	264
Perylene	252
Methylcholanthrene	268
Diphenylanthracene	330
Indeno[1,2,3-cd]pyrene	276
Dibenzanthracene	278
Benzo[b]chrysene	278
Benzo[g,h,i]perylene	276
Anthanthrene	276
Dibenzo[a,l]pyrene	302
Coronene	300
Dibenzo[a,i]pyrene	302
Dibenzo[a,h]pyrene	302
Rubicene	326
Hexaphene	328
Benzo[a]coronene	350

Table 2.56 Recommended SIM ions and relative intensities for major pesticides and some of their derivatives (Thier and Kirchhoff, 1992).

Compound	Molar mass	1	2	3	4	5	6
Acephate	183	43 (100)	44 (88)	136 (80)	94 (58)	47 (56)	95 (32)
Alaclor	269	45 (100)	188 (23)	160 (18)	77 (7)	146 (6)	224 (6)
Aldicarb	190	41 (100)	86 (89)	58 (85)	85 (61)	87 (50)	44 (50)
Aldrin	362	66 (100)	91 (50)	79 (47)	263 (42)	65 (35)	101 (34)
Allethrin	302	123 (100)	79 (40)	43 (32)	81 (31)	91 (29)	136 (27)
Atrazine	215	43 (100)	58 (84)	44 (75)	200 (69)	68 (43)	215 (40)
Azinphos-methyl	317	77 (100)	160 (77)	132 (67)	44 (30)	105 (29)	104 (27)
Barban	257	51 (100)	153 (76)	87 (66)	222 (44)	52 (43)	63 (43)
Benzazolin methyl ester	257	170 (100)	134 (75)	198 (74)	257 (73)	172 (40)	200 (31)
Bendiocarb	223	151 (100)	126 (58)	166 (48)	51 (19)	58 (18)	43 (17)
Bromacil	260	205 (100)	207 (75)	42 (25)	70 (16)	206 (16)	162 (12)
Bromacil N-methyl derivative	274	219 (100)	221 (68)	41 (45)	188 (41)	190 (40)	56 (37)
Bromophos	364	331 (100)	125 (91)	329 (80)	79 (57)	109 (53)	93 (45)
Bromophos-ethyl	392	97 (100)	65 (35)	303 (32)	125 (28)	359 (27)	109 (27)
Bromoxynil methyl ether	289	289 (100)	88 (77)	276 (67)	289 (55)	293 (53)	248 (50)
Captafol	347	79 (100)	80 (42)	77 (28)	78 (19)	151 (17)	51 (13)
Captan	299	79 (100)	80 (61)	77 (56)	44 (44)	78 (37)	149 (34)
Carbaryl	201	144 (100)	115 (82)	116 (48)	57 (31)	58 (20)	63 (20)
Carbendazim	191	159 (100)	191 (57)	103 (38)	104 (37)	52 (32)	51 (29)
Carbetamid	236	119 (100)	72 (54)	91 (44)	45 (38)	64 (37)	74 (29)
Carbofuran	221	164 (100)	149 (70)	41 (27)	58 (25)	131 (25)	122 (25)
Chlorbromuron	292	61 (100)	46 (24)	62 (11)	63 (10)	60 (9)	124 (8)
Chlorbufam	223	53 (100)	127 (20)	51 (13)	164 (13)	223 (13)	70 (10)
cis-Chlordane	406	373 (100)	375 (84)	377 (46)	371 (39)	44 (36)	109 (36)
trans-Chlordane	406	373 (100)	375 (93)	377 (53)	371 (47)	272 (36)	237 (30)
Chlorfenprop-methyl	232	125 (100)	165 (64)	75 (46)	196 (43)	51 (43)	101 (37)
Chlorfenvinphos	35	81 (100)	267 (73)	109 (55)	269 (47)	323 (26)	91 (23)
Chloridazon	221	77 (100)	221 (60)	88 (37)	220 (35)	51 (26)	105 (24)
Chloroneb	206	191 (100)	193 (61)	206 (60)	53 (57)	208 (39)	141 (35)
Chlorotoluron	212	72 (100)	44 (29)	167 (28)	132 (25)	45 (20)	77 (11)
3-Chloro-4-methylaniline (GC degradation product of Chlorotoluron)	141	141 (100)	140 (37)	106 (68)	142 (36)	143 (28)	77 (25)

Table 2.56 (Continued)

Compound	Molar mass	Main fragment m/z (intensities) 1	2	3	4	5	6
Chloroxuron	290	72 (100)	245 (37)	44 (31)	75 (21)	45 (19)	63 (16)
Chloropropham	213	43 (100)	127 (49)	41 (35)	45 (20)	44 (18)	129 (16)
Chlorpyrifos	349	97 (100)	195 (59)	199 (53)	65 (27)	47 (23)	314 (21)
Chlorthal-dimethyl	330	301 (100)	299 (81)	303 (47)	332 (29)	142 (26)	221 (24)
Chlorthiamid	205	170 (100)	60 (61)	171 (50)	172 (49)	205 (35)	173 (29)
Cinerin I	316	123 (100)	43 (35)	93 (33)	121 (27)	81 (27)	150 (27)
Cinerin II	360	107 (100)	93 (57)	121 (53)	91 (50)	149 (35)	105 (33)
Cyanazine	240	44 (100)	43 (60)	68 (60)	212 (48)	41 (47)	42 (34)
Cypermethrin	415	163 (100)	181 (79)	165 (68)	91 (41)	77 (33)	51 (29)
2,4-DB methyl ester	262	101 (100)	59 (95)	41 (39)	162 (36)	69 (28)	63 (25)
Dalapon	142	43 (100)	61 (81)	62 (67)	97 (59)	45 (59)	44 (47)
Dazomet	162	162 (100)	42 (87)	89 (79)	44 (73)	76 (59)	43 (53)
Demetron-S-methyl	230	88 (100)	60 (50)	109 (24)	142 (17)	79 (14)	47 (11)
Desmetryn	213	213 (100)	57 (67)	58 (66)	198 (58)	82 (44)	171 (39)
Dialifos	393	208 (100)	210 (31)	76 (20)	173 (17)	209 (12)	357 (10)
Di-allate	269	43 (100)	86 (62)	41 (38)	44 (25)	42 (24)	70 (19)
Diazinon	304	137 (100)	179 (74)	152 (65)	93 (47)	153 (42)	199 (39)
Dicamba methyl ester	234	203 (100)	205 (60)	234 (27)	188 (26)	97 (21)	201 (20)
Dichlobenil	171	171 (100)	173 (62)	100 (31)	136 (24)	75 (24)	50 (19)
Dichlofenthion	314	97 (100)	279 (92)	223 (90)	109 (67)	162 (53)	251 (46)
Dichlofluanid	332	123 (100)	92 (33)	224 (29)	167 (27)	63 (23)	77 (22)
2,4-D isooctyl ester	332	43 (100)	57 (98)	41 (76)	55 (54)	71 (41)	69 (27)
2,4-D methyl ester	234	199 (100)	45 (97)	175 (94)	145 (70)	111 (69)	109 (68)
Dichlorprop isooctyl ester	346	43 (100)	57 (83)	41 (61)	71 (48)	55 (47)	162 (41)
Dichlorprop methyl ester	248	162 (10)	164 (80)	59 (62)	189 (56)	63 (39)	191 (35)
Dichlorvos	220	109 (100)	185 (18)	79 (17)	187 (6)	145 (6)	47 (5)
Dicofol	368	139 (100)	111 (39)	141 (33)	75 (18)	83 (17)	251 (16)
o,p'-DDT	352	235 (100)	237 (59)	165 (33)	236 (16)	199 (12)	75 (12)
p,p'-DDT	352	235 (100)	237 (58)	165 (37)	236 (16)	75 (12)	239 (11)
Dieldrin	378	79 (100)	82 (32)	81 (30)	263 (17)	77 (17)	108 (14)
Dimethirimol methyl ether	223	180 (100)	223 (23)	181 (10)	224 (3)	42 (2)	109 (2)
Dimethoate	229	87 (100)	93 (76)	125 (56)	58 (40)	47 (39)	63 (33)
DNOC methyl ether	212	182 (100)	165 (74)	89 (69)	90 (57)	212 (48)	51 (47)

(Continued)

Table 2.56 (Continued)

Compound	Molar mass	\multicolumn{6}{c}{Main fragment m/z (intensities)}

Compound	Molar mass	1	2	3	4	5	6
Dinoterb methyl ether	254	239 (100)	209 (41)	43 (36)	91 (35)	77 (33)	254 (33)
Dioxacarb	223	121 (100)	122 (62)	166 (46)	165 (42)	73 (35)	45 (31)
Diphenamid	239	72 (100)	167 (86)	165 (42)	239 (21)	152 (17)	168 (14)
Disulfoton	274	88 (100)	89 (43)	61 (40)	60 (39)	97 (36)	65 (23)
Diuron	232	72 (100)	44 (34)	73 (25)	42 (20)	232 (19)	187 (13)
Dodine	227	43 (100)	73 (80)	59 (52)	55 (47)	7 (46)	100 (46)
Endosulfan	404	195 (100)	36 (95)	237 (91)	41 (89)	24 (79)	75 (78)
Endrin	378	67 (100)	81 (67)	263 (59)	36 (58)	79 (47)	82 (41)
Ethiofencarb	225	107 (100)	69 (48)	77 (29)	41 (26)	81 (21)	45 (17)
Ethirimol	209	166 (100)	209 (17)	167 (14)	96 (12)	194 (4)	55 (2)
Ethirimol methyl ether	223	180 (100)	223 (23)	85 (14)	181 (12)	55 (10)	96 (9)
Etrimfos	292	125 (100)	292 (91)	181 (90)	47 (84)	153 (84)	56 (73)
Fenarimol	330	139 (100)	107 (95)	111 (40)	219 (39)	141 (33)	251 (31)
Fenitrothion	277	125 (100)	109 (92)	79 (62)	47 (57)	63 (44)	93 (40)
Fenoprop isooctyl ester	380	57 (100)	43 (94)	41 (85)	196 (63)	71 (60)	198 (59)
Fenoprop methyl ester\f "compound"	282	196 (100)	198 (89)	59 (82)	55 (36)	87 (34)	223 (31)
Fenuron\f "compound"	164	72 (100)	164 (27)	119 (24)	91 (22)	42 (14)	44 (11)
Flamprop-isopropyl	363	105 (100)	77 (44)	276 (21)	106 (18)	278 (7)	51 (5)
Flamprop-methyl	335	105 (100)	77 (46)	276 (20)	106 (14)	230 (12)	44 (11)
Formothion	257	93 (100)	125 (89)	126 (68)	42 (49)	47 (48)	87 (40)
Heptachlor	370	100 (100)	272 (81)	274 (42)	237 (33)	102 (33)	
Iodofenphos	412	125 (100)	377 (78)	47 (64)	79 (59)	93 (54)	109 (49)
Loxynil isooctyl ether	483	127 (100)	57 (96)	41 (34)	43 (33)	55 (26)	37 (16)
Loxynil methyl ether	385	385 (100)	243 (56)	370 (41)	127 (13)	386 (10)	88 (9)
Isoproturon	206	146 (100)	72 (54)	44 (35)	128 (29)	45 (28)	161 (25)
Jasmolin I	330	123 (100)	43 (52)	55 (34)	93 (25)	91 (24)	81 (23)
Jasmolin II	374	107 (100)	91 (69)	135 (69)	93 (67)	55 (66)	121 (58)
Lenacil	234	153 (100)	154 (20)	110 (15)	109 (15)	152 (13)	136 (10)
Lenacil N-methyl derivative	248	167 (100)	166 (45)	168 (12)	165 (12)	124 (9)	123 (6)
Lindane	288	181 (100)	183 (97)	109 (89)	219 (86)	111 (75)	217 (68)
Linuron	248	61 (100)	187 (43)	189 (29)	124 (28)	46 (28)	44 (23)

Table 2.56 (Continued)

Compound	Molar mass	1	2	3	4	5	6
MCPB isooctyl ester	340	87 (100)	57 (81)	43 (62)	71 (45)	41 (42)	69 (29)
MCPB methyl ester	242	101 (100)	59 (70)	77 (40)	107 (25)	41 (22)	142 (20)
Malathion	330	125 (100)	93 (96)	127 (75)	173 (55)	158 (37)	99 (35)
Mecoprop isooctyl ester	326	43 (100)	57 (94)	169 (77)	41 (70)	142 (69)	55 (52)
Mecoprop methyl ester	228	169 (100)	143 (79)	59 (58)	141 (57)	228 (54)	107 (50)
Metamitron	202	104 (100)	202 (66)	42 (42)	174 (35)	77 (24)	103 (19)
Methabenzthiazuron	221	164 (100)	136 (73)	135 (69)	163 (42)	69 (30)	58 (25)
Methazole	260	44 (100)	161 (44)	124 (36)	187 (31)	159 (24)	163 (23)
Methidathion	302	85 (100)	145 (90)	93 (32)	125 (22)	47 (21)	58 (20)
Methiocarb	225	168 (100)	153 (84)	45 (40)	109 (37)	91 (31)	58 (21)
Methomyl	162	44 (100)	58 (81)	105 (69)	45 (59)	42 (55)	47 (52)
Metobromuron	258	61 (100)	46 (43)	60 (15)	91 (13)	258 (13)	170 (12)
Metoxuron	228	72 (100)	44 (27)	183 (23)	228 (22)	45 (21)	73 (15)
Metribuzin	214	198 (100)	41 (78)	57 (54)	43 (39)	47 (38)	74 (36)
Mevinphos	224	127 (100)	192 (30)	109 (27)	67 (20)	43 (8)	193 (7)
Monocrotophos	223	127 (100)	67 (25)	97 (23)	109 (14)	58 (14)	192 (13)
Monolinuron	214	61 (100)	126 (63)	153 (42)	214 (34)	46 (29)	125 (25)
Napropamide	271	72 (100)	100 (81)	128 (62)	44 (55)	115 (41)	127 (36)
Nicotine	162	84 (100)	133 (21)	42 (18)	162 (17)	161 (15)	105 (9)
Nitrofen	283	283 (100)	285 (67)	202 (55)	50 (55)	139 (37)	63 (37)
Nuarimol	314	107 (100)	235 (91)	203 (85)	139 (60)	123 (46)	95 (35)
Omethoat	213	110 (100)	156 (83)	79 (39)	109 (32)	58 (30)	47 (21)
Oxadiazon	344	43 (100)	175 (92)	57 (84)	177 (60)	42 (35)	258 (22)
Parathion	291	97 (100)	109 (90)	291 (57)	139 (47)	125 (41)	137 (39)
Parathion-methyl	263	109 (100)	125 (80)	263 (56)	79 (26)	63 (18)	93 (18)
Pendimethalin	281	252 (100)	43 (53)	57 (43)	41 (41)	281 (37)	253 (34)
Permethrin	390	183 (100)	163 (100)	165 (25)	44 (15)	184 (15)	91 (13)
Phenmedipham	300	133 (100)	104 (52)	132 (34)	91 (34)	165 (31)	44 (27)
Phosalone	367	182 (100)	121 (48)	97 (36)	184 (32)	154 (24)	111 (24)
Pirimicarb	238	72 (100)	166 (85)	42 (63)	44 (44)	43 (24)	238 (23)
Pirimiphos-ethyl	333	168 (100)	318 (94)	152 (88)	304 (79)	180 (73)	42 (71)
Pirimiphos-methyl	305	290 (100)	276 (93)	125 (69)	305 (53)	233 (44)	42 (41)
Propachlor	211	120 (100)	77 (66)	93 (36)	43 (35)	51 (30)	41 (27)

(Continued)

Table 2.56 (Continued)

		Main fragment m/z (intensities)					
Compound	Molar mass	1	2	3	4	5	6
Propanil	217	161 (100)	163 (70)	57 (64)	217 (16)	165 (11)	219 (9)
Propham	179	43 (100)	93 (88)	41 (42)	120 (24)	65 (24)	137 (23)
Propoxur	209	110 (100)	152 (47)	43 (28)	58 (27)	41 (21)	111 (20)
Pyrethrin I	328	123 (100)	43 (62)	91 (58)	81 (47)	105 (45)	55 (43)
Pyrethrin II	372	91 (100)	133 (70)	161 (55)	117 (48)	107 (47)	160 (43)
Quintozene	293	142 (100)	237 (96)	44 (75)	214 (67)	107 (62)	212 (61)
Resmethrin	338	123 (100)	171 (67)	128 (52)	143 (49)	81 (38)	91 (28)
Simazine	201	201 (100)	44 (96)	186 (72)	68 (63)	173 (57)	96 (40)
Tecnazene	259	203 (100)	201 (69)	108 (69)	215 (60)	44 (57)	213 (51)
Terbacil	216	160 (100)	161 (99)	117 (69)	42 (45)	41 (41)	162 (37)
Terbacil N-methyl derivative	230	56 (100)	174 (79)	175 (31)	57 (24)	176 (23)	41 (20)
Tetrachlorvinphos	364	109 (100)	329 (48)	331 (42)	79 (20)	333 (14)	93 (9)
Tetrasul	322	252 (100)	254 (67)	324 (51)	108 (49)	75 (40)	322 (40)
Thiabendazole	201	201 (100)	174 (72)	63 (12)	202 (11)	64 (11)	65 (9)
Thiofanox	218	57 (100)	42 (75)	68 (39)	61 (38)	55 (34)	47 (33)
Thiometon	246	88 (100)	60 (63)	125 (56)	61 (52)	47 (49)	93 (47)
Thiophanat-methyl	342	44 (100)	73 (97)	159 (89)	191 (80)	86 (72)	150 (71)
Thiram	240	88 (100)	42 (25)	44 (20)	208 (18)	73 (15)	45 (10)
Tri-allate	303	43 (100)	86 (73)	41 (43)	42 (31)	70 (23)	44 (21)
Trichlorfon	256	109 (100)	79 (34)	47 (26)	44 (20)	185 (17)	80 (8)
Tridemorph	297	128 (100)	43 (26)	42 (18)	44 (13)	129 (11)	55 (5)
Trietazine	229	200 (100)	43 (81)	186 (52)	229 (52)	214 (50)	42 (48)
Trifluralin	335	43 (100)	264 (33)	306 (32)	57 (7)	42 (6)	290 (5)
Vamidothion	287	87 (100)	58 (47)	44 (40)	61 (29)	59 (26)	60 (25)
Vinclozolin	285	54 (100)	53 (93)	43 (82)	124 (65)	212 (63)	187 (61)

The data refer to EI ionization at 70 eV. The relative intensities can depend on tuning and the type of mass spectrometer used. For confirmation, mass spectra should be consulted which were run under identical instrumental conditions.

As the SIM analysis is targeted to the detection of certain analyte ions in small mass ranges and the comparability of a complete spectrum for a library search is not required, a special SIM tuning of the mass analyzer with emphasis on the improved ion transmission in the upper mass range is recommended. The standard tuning aims to produce a balanced spectrum which corresponds to the data in a reference list for the reference substance FC43 (perfluorotributylamine, PFTBA), in order

> **SIM Setup**
>
> 1) Choice of column and oven temperature program for optimal GC separation, paying particular attention to analytes with similar fragmentation patterns.
> 2) Full scan analysis of an average substance concentration to determine the selective ions based on the full spectrum (SIM masses, min. 3 ions/component). More SIM ions are required with challenging matrix conditions. Chose SIM masses at high mass with a high S/N ratio.
> 3) Determination of the retention times of the individual components.
> 4) Set the data acquisition interval (time window) for the individual SIM descriptors, or use the retention t-SIM technique with a symmetrical acquisition window centered to the compound retention time.
> 5) Test analysis of a low standard, or better a typical matrix spike, for optimization of SIM masses, and if necessary, the separation conditions.
> 6) Tune the quadrupole analyzer toward high mass sensitivity.
> 7) SIM ions not impaired by matrix are used for quantitation.
> 8) Calibration and quantification are based on single or the sum of mass traces. The setup can change depending on the matrix.

to guarantee good comparability of the spectra run with those in the library. Certain U.S. EPA methods require the source tuning using BFB (4-bromofluorobenzene) or DFTPP (decafluorotriphenylphosphine) for compliance with predefined mass intensities as set in the operating procedure. In the optimization of the ion source, special attention should be paid to the mass range involved in SIM data acquisition. The source and lens potentials should then be selected manually so that a nearby fragment of the reference compound or an ion produced by column bleed (GC temperature about 200 °C) can be detected with the highest intensity but good resolution. In this way, a significant additional increase in sensitivity can be achieved with quadrupole analyzers for the SIM mode. The same consideration applies for SRM with MS/MS detection (refer to Section 2.3.4.4).

The chromatogram in Figure 2.253 shows the single quadrupole SIM run of a PCB standard mixture. In this case, two masses are chosen as SIM masses for each PCB chlorination degree. The switching points of the individual descriptors are visible as steps in the base line. The different base line heights arise as a result of the different contributions of the chemical noise to these signals. To control the elution, the substance signals can be represented as peaks in the expected retention time windows (Figure 2.254). A deviation from the calibrated retention time (Figure 2.254, right segment with the masses m/z 499.8 and 497.8) leads to a shift of the peak from the middle to the edge of the window and should be a reason for further checking. Qualifiers like isotope intensities or typical fragmentation patterns should be checked based on the relative intensities of the selected SIM masses by the chromatography data system (Figure 2.255).

374 | 2 Fundamentals

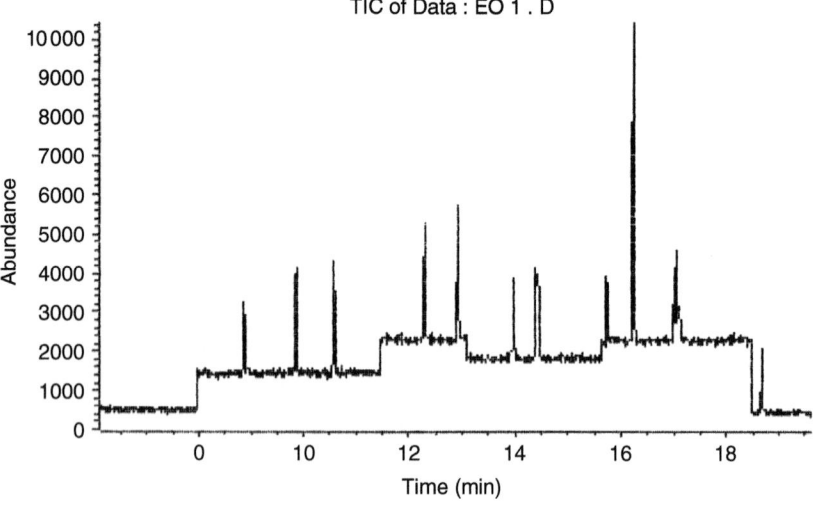

Figure 2.253 Example of a typical PCB analysis in the SIM mode. The steps in the base line show the switching points of the SIM descriptors.

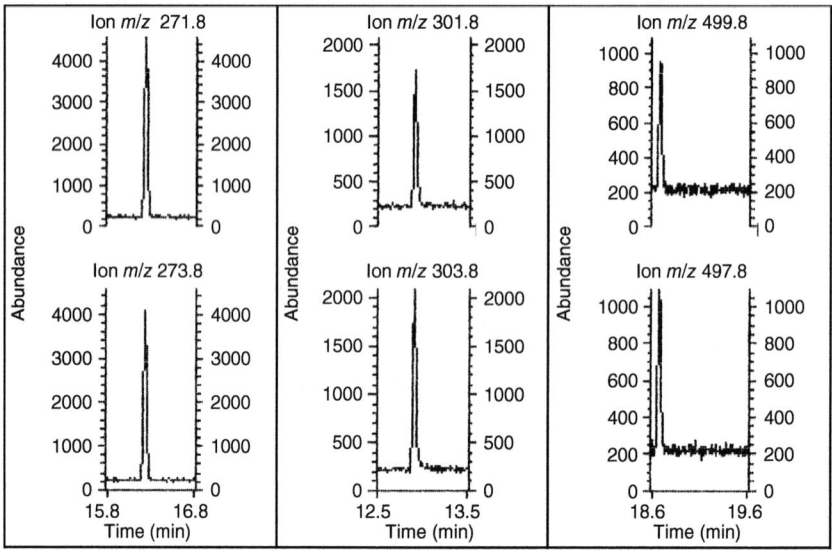

Figure 2.254 Evaluation of the PCB analysis from Figure 2.253 by display of the peaks at the selective mass traces (see text).

2.3.4.3 High Resolution Accurate Mass SIM Data Acquisition

Target compound analysis using high mass resolution on a magnetic sector mass spectrometer, for example, for PCDD/PCDFs, pesticides, persistent organic pollutants (POPs), or pharmaceutical residues, are typically performed by monitoring the compound-specific accurate mass ions at the expected retention time. High

Figure 2.255 Evaluation of isotope patterns from an SIM analysis of hexachlorobiphenyl by comparison of relative intensities (shown as a bar graph spectrum).

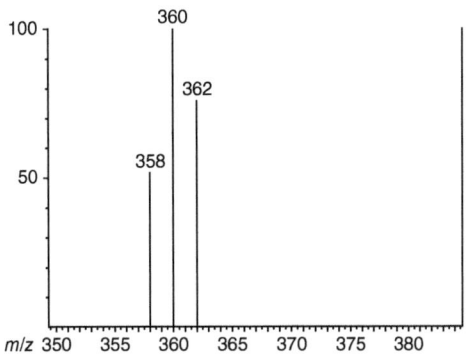

resolution GC-MS target compound applications on a magnetic sector MS benefit from a unique technical feature referred to as *the lock-mass technique* for performing SIM analyses. The lock-mass technique provides ease of use, combined with a maximum quantitative precision and certainty in analyte confirmation.

The basic equation for sector mass spectrometers,

$$m/z = c \cdot \frac{B^2}{V}$$

with

c = the instrument constant,
B = magnetic field strength,
V = acceleration voltage,

shows that accurate mass scans are feasible by either the acceleration voltage V with a constant calibrated magnetic field B or vice versa. For SIM data acquisition, the fixed magnet setting with a variable acceleration voltage is typically used. The mass calibration for SIM data acquisition follows a special procedure during the data acquisition in the form of a scan inherent mass calibration. This internal mass calibration is performed during SIM analysis in every scan before acquisition of the target compound intensities. For most precise mass assignment, two reference masses are checked, one below and one above the target masses. This exact mass calibration is referenced as the "lock-plus-cali mass" technique or in the short "lock mode". It provides optimum mass accuracy for peak detection even in difficult chromatographic situations. The scan-to-scan mass calibration provides the best confidence for the acquired analytical data and basically is the accepted feature for the requirement of HRGC-HRMS as a confirmation method (see also U.S. EPA Method 1613B 1994).

The internal mass calibration process is performed during acquisition in the background without being noticed by the operator. It provides superior stability especially for high sample throughput with extended runtimes. A reference compound is leaked continuously from the reference inlet system during the GC run into the ion source. Typically, perfluorotributylamine (FC43) is used as a reference compound in HRGC-HRMS for e.g. dioxin analysis. Other reference compounds may be used to suit individual experimental conditions.

The exact ion masses of the reference compound are used in the SIM acquisition windows for internal calibration. For the lock-and-cali-mass technique, two ions of the reference substance are individually selected for each SIM window, one mass close, but below the analyte target mass, and the second, which is slightly above the analyte or internal standard target masses. Although both reference masses are used for the internal calibration, it became common practice to name the lower reference mass as the "lock mass" and the upper reference mass as the "calibration mass". During the SIM scan, the mass spectrometer sets the magnetic field strength at the start of each SIM window (locking) and then performing the mass calibration using the lock and calibration masses followed by the acquisition of the target and internal standard mass intensities.

Lock-Mass Technique At the start of each SIM retention time window, the magnet is automatically set to one mass below the lowest mass found in the SIM descriptor. The magnet is parked, or "locked", and remains with this setting throughout the entire SIM window. All analyzer jumps to the calibration and target compound masses are done by fast electrical jumps of the acceleration voltage.

Internal Mass Calibration The lock mass (L) as shown in Figure 2.256 ① is scanned in a small mass window starting below the mass peak by slowly decreasing the ion source acceleration voltage. The mass resolution of the lock-mass peak is calculated and written to the data file. Using the lock-mass setting, the second reference mass is used for building the SIM mass calibration. The calibration mass is checked by an electrical jump. A fine adjustment of the electrical calibration is made based on this measurement ②. The electrical "jump" at ③ is very fast and takes only a few

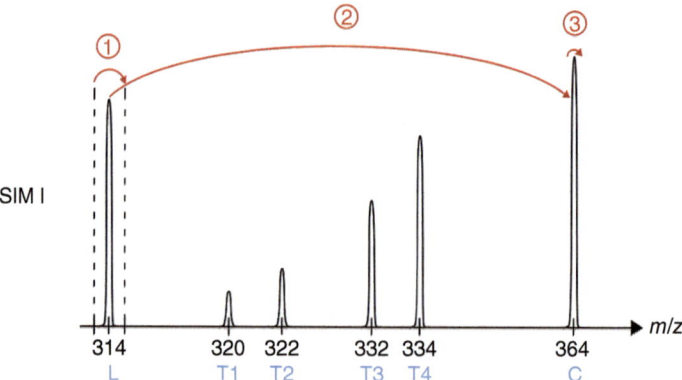

Figure 2.256 Mass detection scheme in HRMS SIM calibration, in this case for TCDD detection. The arrows show the sequence of measurement in the mass calibration steps (the magnet is "locked" in this example at m/z 313). ① Magnet locking and "lock mass" sweep, mass calibration, and resolution determination. ② Electrical jump to calibration mass. ③ Calibration mass sweep and mass calibration. L from FC43 lock mass (L), m/z 313.983364. C from FC43 calibration mass (C), m/z 363.980170. T1, T2 from native TCDD analyte target masses m/z 319.895992, 321.893042. T3, T4 from ^{13}C-TCDD internal standard masses m/z 331.936250, 333.933300.

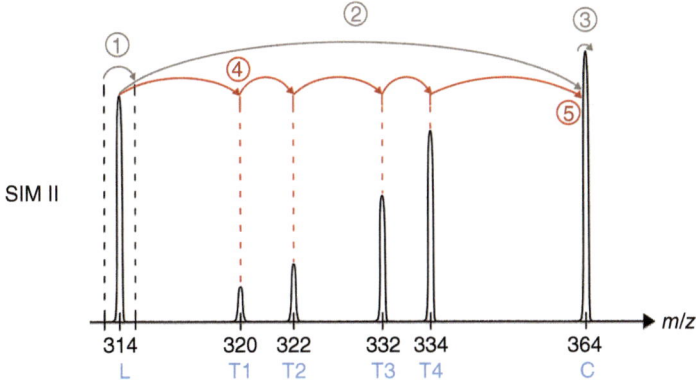

Figure 2.257 Mass detection scheme in HRMS SIM data acquisition. The arrows show the sequence of measurement in the target compound and internal standard data acquisition (legend for the mass scale see Figure 2.256). ④ Consecutive electrical jumps to target and internal standard masses. ⑤ Electrical jump to calibration mass, mass calibration.

milliseconds. The dwell times for the sufficiently intense reference ions are short. The resulting electrical calibration is used for the following SIM data acquisition of the analyte signal.

Data Acquisition With the updated and exact mass calibration settings, the analyzer sets the acceleration voltage to the masses of the target ions. The intensity of each ion is measured based on a preset dwell time, as pointed out in Figure 2.257 ④. The dwell times to measure the analyte target ion intensities are significantly longer than the lock or calibration mass ions' dwell times. This is done to achieve the optimum detection sensitivity for each analyte ion. The exact positioning on the top of the target ion mass peak allows for higher dwell times, significantly increased sensitivity, and higher S/N values compared to formerly known sweep-scan techniques still used in older technology HRMS systems. It is important to note that the lock-plus-cali mass technique extends the dynamic range significantly into the lower concentration range.

Advantages of the Lock-Plus-Cali Mass Technique The "lock-plus-cali mass" calibration technique provides extremely stable conditions for data acquisitions of long sample sequences even over days, for example, over the weekend, and always includes the resolution performance documentation for quality control documented in the data file.

The electrical jumps of the acceleration voltage are very fast, deliver the required number of data points to define even fast GC peaks for integration, and provide an excellent instrument duty cycle. Any outside influences from incidental background ions, long-term drift, or minute electronic fluctuations are taken care of and do not influence the result. Both the lock and cali masses are monitored in parallel during the run, providing an excellent confirmation and documentation of the system stability for data certainty. Together with the constant resolution monitoring, this

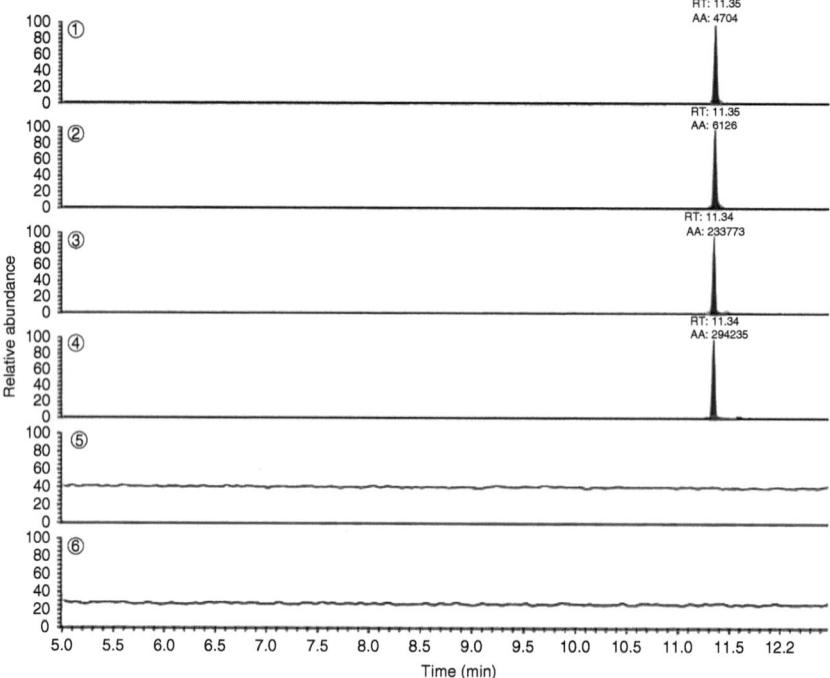

Figure 2.258 Resulting mass chromatograms of a TCDD standard solution at 100 fg/μL (DB-5MS, 60 m × 0.25 mm × 0.1 μm). ① Ratio mass of 2,3,7,8-TCDD (native) m/z 319.8960 ② Quan mass of 2,3,7,8-TCDD (native) m/z 321.8930 ③ Ratio mass of 2,3,7,8-13C12-TCDD (ISTD) m/z 331.9362 ④ Quan mass of 2,3,7,8-13C12-TCDD (ISTD) m/z 333.9333 ⑤ Lock mass of FC43 m/z 313.983364 ⑥ Cali mass of FC43 m/z 363.980170

technique provides the traceability in SIM data analysis especially required for the highly regulated dioxin analysis.

Other acquisition techniques have been formerly used employing just one lock-mass position. This technique requires a separate pre-run electrical mass calibration and does not allow a scan internal correction of the mass position which may arise due to long-term temperature drifts of the analyzer during data acquisition. Consequently, with the one mass lock techniques, the mass jumps are less precise with increasing run times. Deviations from the peak top position when acquiring data at the peak slope result in less sensitivity, less reproducibility, and poor isotope ratio confirmation.

The "lock-plus-cali mass" technique has proven to be superior in achieving lower limit of quantitations (LOQs) and higher S/N values in high resolution SIM. Figure 2.258 shows the typical chromatogram display in dioxin analysis with the TCDD target masses as well as the ^{13}C internal standard masses. In addition, the continuously monitored FC43 lock and cali masses are displayed as constant mass traces. Both traces are of valuable diagnostic use and confirm the correct measurement of the target compounds.

2.3 Mass Spectrometry

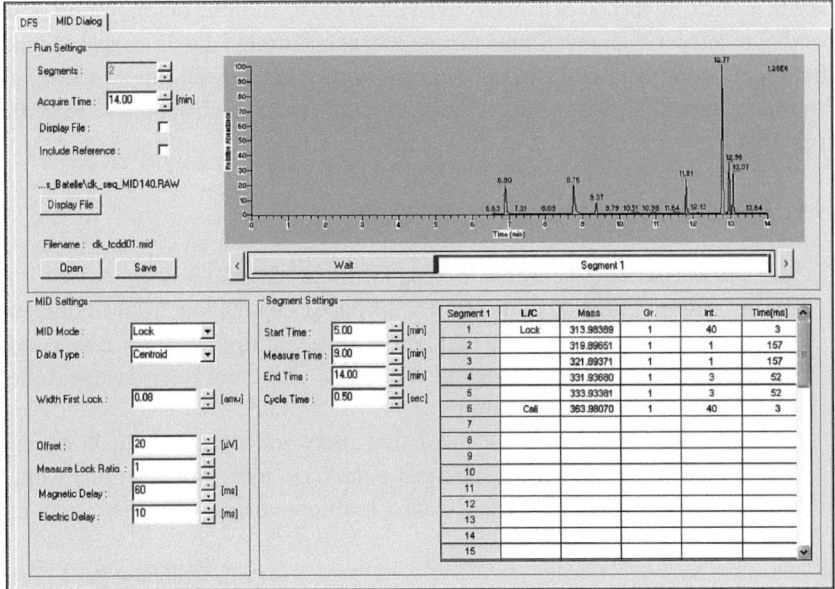

Figure 2.259 SIM/MID editor with sample chromatogram (top), target mass list, and duty cycles for the highlighted "Segment 1" retention time window (bottom right).

Setup of the SIM Descriptor The SIM descriptor for the data acquisition contains all the information required by the HRMS mass analyzer. Included in each descriptor is the retention time information for switching between different target ions, the exact mass calibration, the target masses to be acquired with the corresponding dwell times, long for analytes, short for lock and cali masses.

A sample chromatogram facilitates the setting of the retention time windows, as shown Figure 2.259. The sample chromatogram is used to optimize the GC component separation and set the SIM windows before data analysis (often called "window finder"), and usually consists of a medium concentration standard mix.

Optimized SIM Cycle Time In order to provide a representative and reproducible GC peak integration, the total SIM cycle time on the chromatographic time scale should allow for the acquisition of 8–10 data points over a chromatographic peak (Figure 2.259). The cycle time has a direct influence on the available measurement time for each ion (dwell time). If the SIM cycle time is too short, the sensitivity of the instrument is compromised; too high values lead to a poor GC peak definition.

Note the different dwell time settings in Figure 2.259 for the lock and calibration masses (3 ms each), the labeled internal standards (52 ms each), and the native compounds (157 ms each). This shifts most of the available dwell time and sensitivity in favor for the low concentrated native compounds of the sample. This differential acquisition time setup is possible due to the higher intensity of the lock and cali masses and higher concentration of the labeled standards that only need short acquisition times. It guarantees optimum sensitivity and quantitative precision at lowest analyte levels. In case higher PCB concentrations need to be analysed in the same run as dioxins special care needs to be taken to the number of ions monitored in parallel to keep a high dwell time for the low concentrated dioxin congers high, so that higher concentrated PCBs are acquired with shorter dwell times. A similar differential acquisition time setup is available with triple quadrupole GC-MS/MS instruments.

2.3.4.4 MS/MS – Tandem Mass Spectrometry

> Can atomic particles be stored in a cage without material walls?
> This question is already quite old. The physicist Lichtenberg from Göttingen wrote in his notebook at the end of the 18th century: "I think it is a sad situation that in the whole area of chemistry we cannot freely suspend the individual components of matter.
> This situation lasted until 1953. At that time, we succeeded, in Bonn, in freely suspending electrically charged atoms, i.e. ions, and electrons using high frequency electric fields, so-called multipole fields. We called such an arrangement an ion cage".
>
> Prof. Wolfgang Paul 1991
> 1913–1993, German physicist, who co-developed the non-magnetic quadrupole mass filter, shared the Nobel Prize in physics in 1989 for this work with Prof. Hans Georg Dehmelt

As part of the further development of instrumental techniques in mass spectrometry, the tandem mass spectrometry (MS/MS) analysis has become the method of choice for routine trace analysis of target compounds in complex matrices. Most of the current applications involve the determination of substances in the parts per billion and parts per trillion ranges in samples of urine, blood, and animal or plant tissues, food, or in many environmental analyses. In addition to trace level quantitation, the determination of molecular structures is still an important area of application for GC-MS/MS analysis (McLafferty, 1983; Busch et al., 1988).

The information content of the GC-MS/MS technique was evaluated in 1983 by Richard A. Yost (University of Florida, co-inventor of the MS/MS technique with Christie G. Enke (Yost and Enke, 1978). On the basis of the theoretical task of detecting one of the five million substances catalogued at that time by the Chemical Abstracts Service, a minimum information content of 23 bits ($\log_2(5 \times 10^6)$) was required for the result of the chosen analysis procedure, and the MS procedures available were evaluated accordingly. The calculation showed that capillary GC-MS/MS can give 1000 times more information than the traditional GC-MS methods (Table 2.57)!

As an analytical background to the use of the GC-MS/MS technique in residue analysis, it should be noted that in general the S/N ratio for analytes increases with the number of analytical steps (Figure 2.260). But, all clean-up steps lower the potential signal intensity by losses of the analyte. The sequence of wet chemical or instrumental sample preparation steps can easily lead to the situation whereby,

Table 2.57 Information content of mass spectroscopic techniques.

Technique	P	Factor
MS[a]	$1.2 \cdot 10^4$	0.002
Packed GC-MS[b]	$7.8 \cdot 10^5$	0.12
Capillary GC-MS[c]	$6.6 \cdot 10^6$	1
MS/MS	$1.2 \cdot 10^7$	2
Packed GC-MS/MS	$7.8 \cdot 10^8$	118
Capillary GC-MS/MS	$6.6 \cdot 10^9$	1000

a) MS: 1000 Da, unit mass resolution, maximum intensity 2^{12}.
b) Packed GC: $2 \cdot 10^3$ theoretical plates, 30 min separating time.
c) Capillary GC: $1 \cdot 10^5$ theoretical plates, 60 min separating time.
(Adapted from Kaiser, 1978; Yost, 1983).

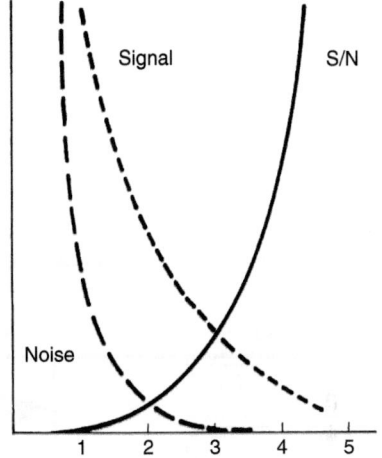

Figure 2.260 Relationship between signal, noise and the number of analytical steps (after Cooks and Busch, 1982).

because of processing losses, a substance can no longer be detected. From this consideration, the first separation step (MS1) in a GC-MS/MS system can be regarded as a mass-specific clean-up in the analysis of extracts containing matrix. After the subsequent induced fragmentation of selected ions in a collision cell, an analyte is identified using the characteristic mass spectrum of the product ions, or, it is quantified using structure selective fragment ions for target compound determination in difficult matrices.

Soft ionization techniques instead of the high energetic EI are the preferred ionization processes for MS/MS analysis. Although fragmentation in the EI mode of GC-MS is desirable for substance identification, frequently, only low selectivity and sensitivity are achieved in complex matrices. Reduced voltage ionization in the EI mode is a viable solution available in many MS systems. This low eV ionization finds application for instance in dioxin analysis for sensitivity increase (see Section 4.27). Also, soft ionization techniques, such as CI, concentrate the ion flux to a few intense

MS/MS Tandem Mass Spectrometry

- Ionization of the sample (EI, CI, and other ionization methods).
- Selection of a precursor ion.
- Fragmentation to product ions (collision induced dissociation, CID).
- Optimization of the collision energy.
- Mass analysis of product ions (product ion scan).
- Detection as a complete product ion spectrum (full product ion scan), or pre-selected individual masses (SRM/MRM).

ions which can form a good starting point for MS/MS detection. For this reason, the high performance liquid chromatography (HPLC) coupling techniques, APCI, and ESI are also in the forefront of the development and extension of MS/MS analysis. The use of the GC-MS/MS technique with PCI and NCI ionization can provide specific advantages for trace analysis. In case the EI ionization is indispensable for multi-methods like pesticides, an optimized analyzer tuning in favor of the higher target masses is performed (high mass tune).

Tandem MS operation comprises several consecutive processes as shown for a triple quadrupole MS in Figure 2.261. In the ion source, all substances eluting from the GC column (analytes, co-eluents, matrix, column bleed, etc.) get ionized. From the resulting mixture of ions, the precursor ion of interest with a particular m/z value is selected in the first step (Q1). This ion can in principle be formed from different molecules, and even with different empirical formulae, they can be of the

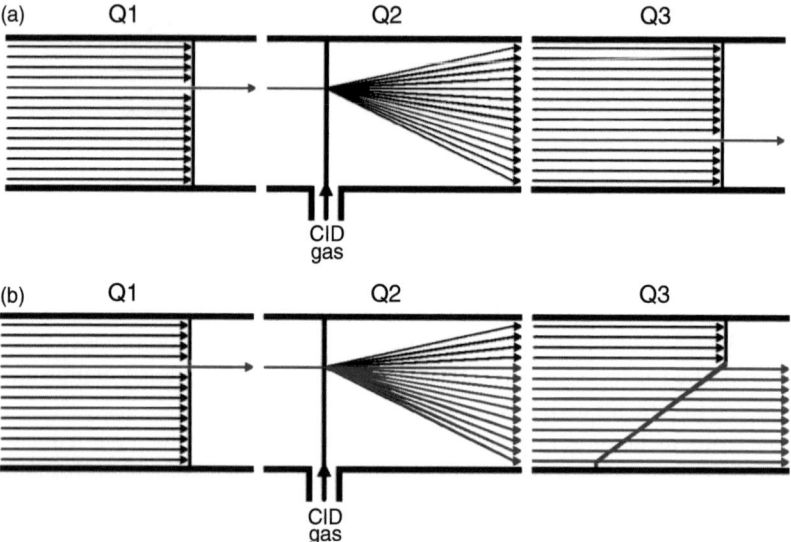

Figure 2.261 Modes of operation of a triple quadrupole mass spectrometer. Q_1 = first mass separating quad, mass selection of the precursor ion. Q_2 = collision cell, CID process. Q_3 = second mass separating quad, SRM detection (a) or product ion scan (b).

same nominal *m/z* value but differ in structure. The Q1 step is identical to SIM analysis in single quadrupole GC-MS, which, for the reasons mentioned above, would exhibit a low S/N, and cannot rule out false-positive signals. The fragmentation of the precursor ions to product ions occurs in a collision cell Q2 by collisions of the selected ions with neutral gas molecules (CID or CAD, collision induced/activated decomposition). In this collision process, the kinetic energy of the precursor ions is converted into internal energy, which leads to a structure-specific fragmentation by cleavage or rearrangement of bonds, and the loss of neutral particles. A mixture of product ions with lower *m/z* values is formed and leaves the collision cell toward Q3. Following this fragmentation step, the second mass spectrometric separation in Q3 is necessary for the mass analysis of the product ions. The spectrum of the product ions is finally detected as the product ion or MS/MS spectrum of the selected precursor ion. In the case of a targeted SRM analysis, Q3 selects the structure specific product ions for a selective detection.

The efficiency of the CID process is of critical importance for the use of GC-MS/MS in residue analysis. In beam instruments, optimization of the collision energy via the entry lens potential of Q2 and the collision gas pressure is necessary. The optimization is limited by scattering effects at high chamber pressures and subsequent fragmentation of product ions. Square instead of round rods are typically used for the enclosed quadrupole collision cell devices (Q2). Hexapole or octapole rod systems have been employed also in the past. The collision cell is operated in an RF only made without mass separating but ion focusing and acceleration capabilities. Helium, nitrogen, argon, and xenon at pressures of about 10^{-3}–10^{-4} Torr are used as collision gases (Johnson and Yost, 1985). Heavier collision gases increase the yield of product ions. The collision energies in quadrupole MS/MS are typically in the range of 5–50 eV (magnetic sector >1 keV) and are controlled by a preceding lens stack accelerating the precursor ions to the set energy level (see Figure 2.262). Cross talk from

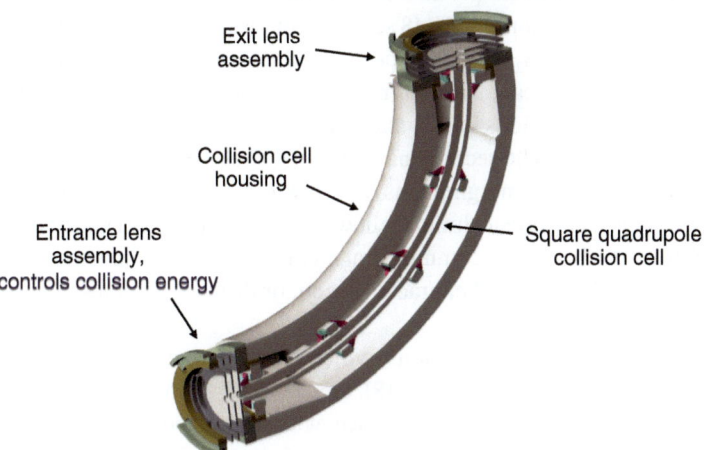

Figure 2.262 Curved collision cell of a triple quadrupole system (90°) for reduction of nonspecific noise by eliminating neutral particles and photons from the ion flight path (TSQ Quantum series, courtesy Thermo Fisher Scientific).

different precursors to the same product ion is effectively eliminated by an active cleaning step of the collision cell between the scans (adds to the interscan time).

When is MS/MS Used?

In quantification:
- All types of target compound analyses.
- The sample matrix contributes significantly to the chemical background noise in SIM.
- Low S/N values, difficult peak integration, and low precision data in SIM.
- Quantification with the highest sensitivity in difficult matrices is necessary.
- The SIM analysis requires additional confirmation.
- Co-elution with isobaric impurities occurs.
- Short sample clean-up extracts of multi-methods such as QuEChERS.

In qualitative analysis:
- Unknown compounds need to be identified by generating precursor fragmentation spectra.
- Structure elucidation of unknown compounds.
- Similar/homologeous compounds identification by neutral loss scan.

Regarding existing analysis procedures, for target compound quantitation, high speed and flexibility in the selection of precursor ions are necessary. For the analysis with internal standards (e.g. deuterated standards) and multi-component methods, the fast switch between multiple precursor ions is necessary to achieve a sufficiently high data rate for a reliable peak definition of all coeluting compounds, in particular for quantitation.

Besides recording a spectrum for product ions (product ion scan), or of an individual mass in SRM, two additional MS/MS scan techniques give valuable analysis data. By linking the scans in MS_2 and MS_1, specific and targeted analysis routes are possible. In the precursor ion scan, the first mass analyzer (MS_1) is scanned over a preselected mass range. All the ions in this mass range reach the collision chamber and form product ions (CID). The second mass analyzer (MS_2) is held constant for a specific fragment mass. Only emerging ions of MS_1 which form the selected fragment are recorded. This recording technique allows the identification of substances of related structure, which lead to common fragments in the mass spectrometer used, e.g. for biomarkers in crude oil characterization, or drug metabolites (Noble, 1995).

In the neutral loss scan, all precursor ions, which lose a particular neutral particle (that otherwise cannot be detected in MS), are detected. Both mass analyzers scan, but with a constant selected mass difference, which corresponds to the mass of the neutral particle lost in the collision cell. This analysis technique is particularly meaningful if molecules contain the same functional groups (e.g. metabolites as acids, glucuronides, or sulfates). In this way, it is possible to identify the starting ions which are characterized by the loss of a common structural element. Both

MS/MS scan techniques can be used for substance-class specific detection in triple quadrupole systems.

When evaluating MS/MS spectra, it should be noted that independent of the type of analyzer, no isotope intensities appear in the product ion spectrum. During selection of the precursor ion for the CID process, naturally occurring isotope mixtures are separated and isolated. The formation of product ions is usually achieved by the loss of common neutral species. The interpretation of these spectra is generally straightforward and less complex compared with EI spectra. When comparing product ion spectra of different instruments, the acquisition parameter used must be taken into account. Especially with beam instruments, the relative intensities of the spectra can reflect the recording parameters.

MS/MS Scan Techniques

Scan mode in		Result	Application
MS1	MS2		
Single ion	Scan	Product ion spectrum (MS/MS spectrum)	**Identification and confirmation** of compounds, structure determination, create libraries of product ion spectra for comparison.
Single ion	Single ion	Individual intensities of product ions (SRM/MRM)	Structure selective, **highly selective, and highly sensitive target compound quantitation** with complex matrices, e.g. pesticides, dioxins, etc.
Scan	Single ion	Precursor masses of certain fragments (precursor ion scan)	Specific analysis of compounds (classes of substance) with **common structural features**, screening, for example, crude oil biomarker.
Scan	Scan-NL	Precursor ions, which undergo loss of neutral particles with NL Da (neutral loss scan)	Specific analysis of compounds (classes of substance) with **common functional groups/structural features**, for example, loss of COCl from PCDD/PCDF, or CO_2 from acids (NL = neutral loss).

Structure Selective Detection Using MS/MS Transitions Triple quadrupole mass spectrometers besides structure elucidation find their major application in quantitation of target compounds in difficult matrices. The increased selectivity when observing specific transitions from a precursor ion to a structural related product ion provides highly confident analyses with excellent LOQs even in matrix samples. In the

SRM mode, an intense precursor ion from the spectrum of the target compound is selected in the first quadrupole Q1, fragmented in the collision cell, and monitored on a selective product ion for quantitation, see the principle given graphically in Figure 2.261. The high selectivity of the SRM method is controlled by the mass resolution of the first quadrupole Q1. While round rod quadrupoles in Q1 and Q3 work at unit mass resolution, hyperbolic quadrupoles typically are set to 0.7 Da peak width FWHM as a standard and are operated in the highly resolved H-SRM mode for maximum selectivity even at a narrow peak width of 0.4 Da FWHM, also refer to Section 2.3.2.1.

Analogue to the SIM analysis mode used with single quadrupole instrumentation, the SRM mode omits full spectral information for substance confirmation. A structure selective characteristic of the assay is given by the mass difference of the transition monitored. Typically, two independent transitions together with the chromatographic retention time provide the positive confirmation of the occurrence of a particular compound. State-of-the-art triple quadrupole instrumentation provide transition times as low as 0.5–2 ms, offering the potential for screening of a large number of compounds in a given chromatographic window, for example, for multi-component pesticide analysis.

Setting up MRM Methods for Quantitation The setup of the MS/MS method (SRM) for target compound quantitation can be a very time-consuming process if hundreds of compounds are involved. Each compound needs to be treated individually to define the optimum detection conditions. For each compound, one or more specific precursor ions are selected in the first quadrupole Q1 and fragmented in the collision cell Q2 using the optimum collision energy. Finally, from the dissociation products of the CID process, the structure-specific product ion is detected with the third quadrupole Q3. The SRM MS/MS process is graphically displayed in Figure 2.263. By switching the transitions by retention time, an MRM method is created. Using the timed-SRM mode, each of the programmed SRM transitions is activated only in a short time window around the compound retention time. This way, complex SRM methods, for example, for pesticides analysis screening of many hundreds of target compounds, can be established with optimum sensitivity for each analyte.

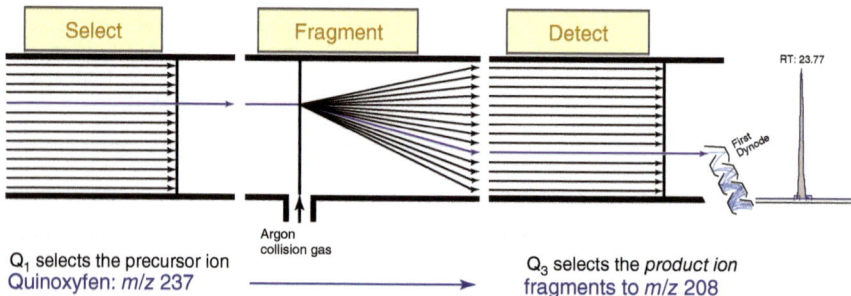

Figure 2.263 SRM process for MS/MS target compound detection, shown quinoxyfen with precursor *m/z* 237 fragmented into the detected product ion *m/z* 208.

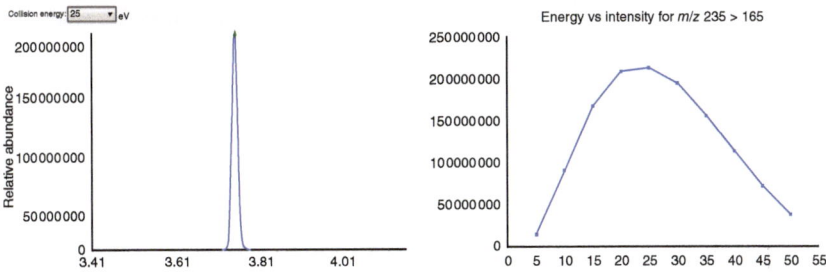

Figure 2.264 AutoSRM optimization of the collision energy between 5 and 50 V with optimum 25 V.

The setup of SRM methods requires the following steps for each compound, typically analyzing one or more standard mixes:

1) Get the compound retention time (a first standard run in full scan).
2) Get the full scan spectrum from first run.
   - Note the most intense ions as precursor ions.
   - One precursor for each planned SRM transition.
3) Define the number of SRM transitions per compound:
   - used for quantitation
   - and ion ratio confirmation.
4) Get the product ion spectra from each precursor ion (second run).
   - Decide on argon or nitrogen as collision gas.
   - Use a medium collision energy.
   - Note the most intense product ion for quantitation.
   - Note the other intense product ions for ion ratio confirmation.
5) Optimize the collision energies for the selected SRMs (third and more runs).
   - Start at 5 V and ramp up to 50 V in steps of 5 V (or/and refer to publications).
   - One injection per voltage setting is required if the instrument cannot change the energy during one scan). Figure 2.264 shows the automated AutoSRM process with stepped collision energies of 5 V, gaining a maximum product ion intensity at 25 V.
   - Use the collision energy of the highest product ion intensity for the method.

This laborious procedure is necessary only once for the initial setup, or if new compounds need to be added. In many cases, already proven compound data bases are available by several GC-MS/MS vendors easing up the task, for example, for pesticides with many hundreds of the most analyzed compounds. Another approach is the AutoSRM procedure, which organizes the above steps automatically and comes up with the ready-to-use acquisition method. Optionally, the acquired product ion spectra can be added to libraries for additional use.

Data Dependent Data Acquisition The advanced electronic capabilities of modern triple quadrupole instruments offer additional features for delivering compound-specific information even during SRM quantitation analyses. Depending on the

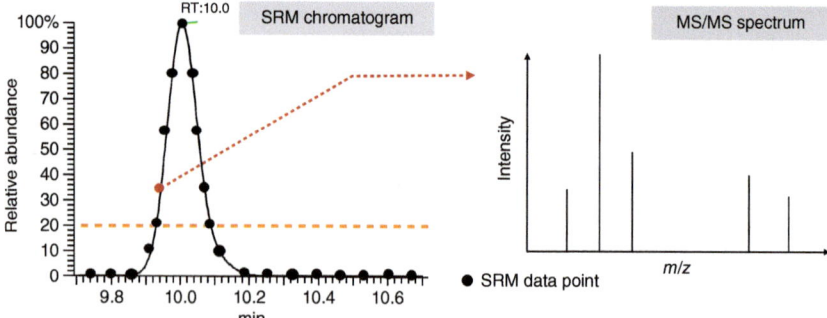

Figure 2.265 Data-dependent data acquisition scheme. The first SRM scan intensity above a user-defined threshold (dotted line) is acquired as the MS/MS product ion spectrum (on the right).

quality of a scan currently being acquired, the mode of the data acquisition for a following scan (data point) can be switched to acquire a full product ion spectrum. The level of product ion intensity that triggers the switch of the acquisition mode is set method and compound specific. As a result, one data point in the substance peak is used to generate and acquire the compound product ion spectrum (see Figure 2.265). This spectrum provides the structural information about the detected substance and is available for library search and compound confirmation. Using the data-dependent acquisition mode, the final data file contains both quantitative and qualitative information.

For the generation of the MS/MS product ion spectrum, the applied collision energy has a substantial impact on the information content of the spectrum, represented by the occurrence and intensities of fragment and precursor ions. The maximum information on the compound structure is achieved by variation of the collision energy during the product ion scan. High collision energies lead to the generation of low mass fragments, while lower-collision energies provide the favored main fragments and still some visible intensity of the precursor ion. The instrumental approach to this solution is the variation of the collision energy during the scan from a high energy to a low energy level by ramping down the ion acceleration voltage using the lens stack located at the entrance of the collision cell (see Figure 2.262). The resulting product ion spectrum consequently delivers the full available information about the structural characteristics of an unknown or the compound to be confirmed (see Figure 2.266).

2.3.5 Mass Calibration

A calibration of the mass scale is necessary to operate a GC-MS system. The calibration converts applied voltages controlling the analyzer or flight times into m/z values. For the calibration of the mass scale, a mass spectrum of a known chemical compound is used, where both the m/z values of the generated ions and their intensities are known and stored in the data system in the form of a reference table.

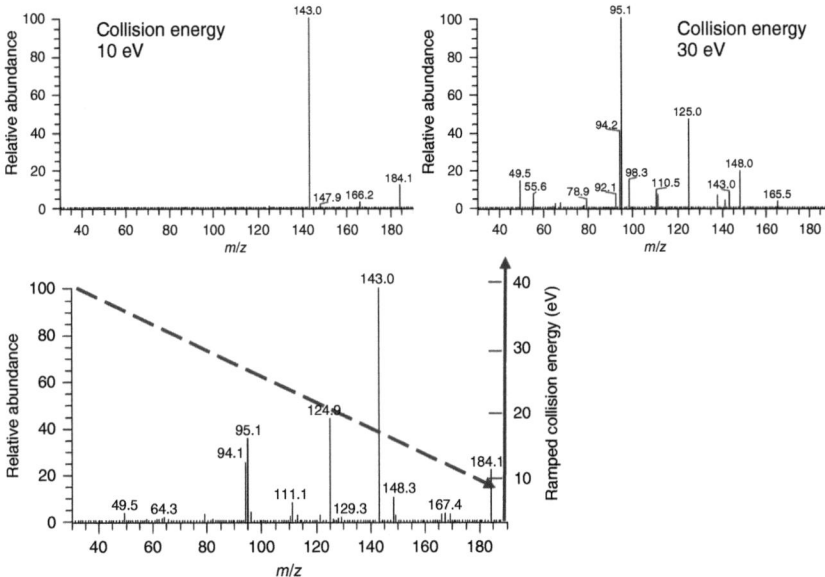

Figure 2.266 Decreasing collision energy ramp in the MS/MS mode provides the information rich product ion spectrum.

With modern GC-MS systems, performing an up-to-date mass calibration is generally the final process in a tuning or autotune program. This is preceded by a series of necessary adjustments and optimizations of the ion source, beam focusing, and mass resolution settings, which affect the position of the signals on the mass scale. Tuning the lens potentials affects the transmission in individual mass areas. In particular with beam instruments, focusing must be adapted to the intensities of the reference substances in order to obtain the intensity pattern of the reference spectrum for comparable mass spectra. The m/z values contained in a stored reference table are identified by the calibration program in the spectrum of the reference compound measured. The relevant centroid of the reference peak is calculated and correlated with the operation of the analyzer. Using the stored reference table, a precise calibration function for the whole mass range of the instrument can be calculated. The data are plotted graphically and are available to the user for assessment and documentation. For quadrupole and ion trap instruments, the calibration graph is linear, whereas with magnetic sector instruments, the graph is exponential (Figure 2.267). With TOF MS the mass vs. time values form a quadratic relationship.

Perfluorinated compounds are generally used as calibration standards (Table 2.58). Despite of their high molecular masses, the volatility of these compounds is sufficient to allow a controllable leak current into the ion source. In addition, fluorine has a negative mass defect ($^{19}F = 18.9984022$) so that the fragments of these standards are below the corresponding nominal mass and can easily be separated from a potential background of hydrocarbons with positive mass defects (also refer to Section 2.3.2.1). The requirements of the reference substance are determined by the type of analyzer. Quadrupole and ion trap instruments are calibrated with

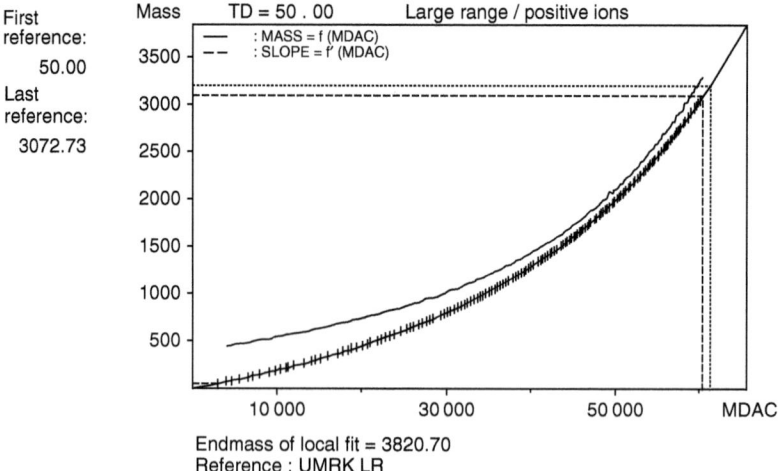

Figure 2.267 Exponential mass calibration of a magnetic sector analyzer.

Table 2.58 Calibration substances and their areas of use.

Calibrant name	Formula	M	m/z max.	Magnetic sector [m/z]	Orbitrap/ Quadrupole/ Ion trap [m/z]
FC43[a]	$C_{12}F_{27}N$	671	614	<620	>1000
FC5311	$C_{14}F_{24}$	624	624	<620	>1000
PFK[b]	—	—	1017	<1000	>1000
(C_7)[c]	$C_{24}F_{45}N_3$	1185	1185	800–1200	—
(C_9)[d]	$C_{30}F_{57}N_3$	1485	1485	800–1500	—
Fomblin[e]	$(OCF(CF_3)CF_2)_x\text{-}(OCF_2)_y$	—	—	<2500	—
Ultramark1621[f]	$C_{40}H_{19}O_6N_3P_3F_{68}$	2120	2120	<2120	—
CsI[g]	—	—	15 981	high mass range	—

a) Perfluorotributylamine.
b) Perfluorokerosene.
c) Perfluorotriheptylazine.
d) Perfluorotrinonyltriazine.
e) Poly(perfluoropropylene oxide) (also used as diffusion pump oil).
f) Fluorinated phosphazine.
g) Cesium iodide.

perfluorotributylamine (FC43, PFTBA independent of the available MS mass range. Figure 2.268 shows the complete mass spectrum of FC43, Tables 2.59 and 2.60)

HRAM instruments like magnetic sector or Orbitrap MS calibrate the accurate mass scale as well "on the fly" (scan-to-scan) with an "internal mass calibration" using a continuously leaking reference, e.g. FC43, or in GC-MS simply by referencing

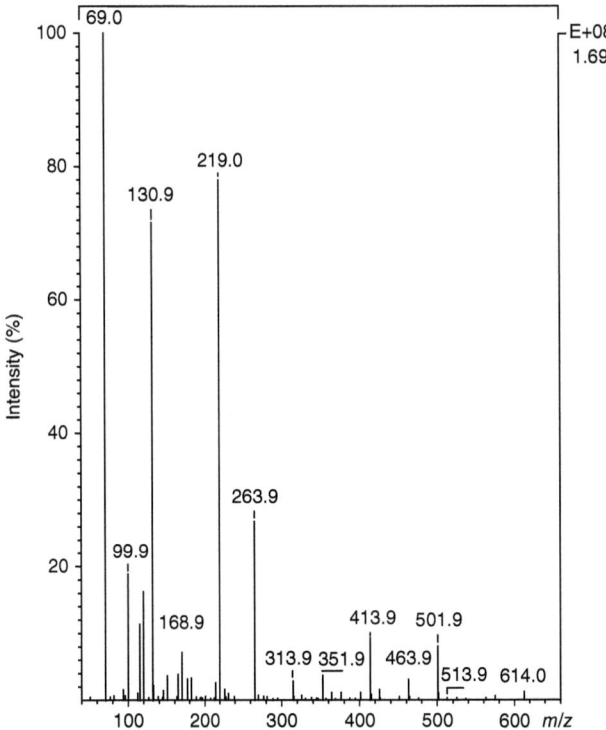

Figure 2.268 FC43/PFTBA spectrum (Finnigan TSQ 700, EI ionization, Q3).

selected siloxane compounds of the GC column bleed. A magnetic sector field analyzer can then be positioned on the exact mass of the target analyte ion relative to the measured centroid of the known reference. At the beginning of the next scan, the exact position of the centroid of the reference mass is determined again and is used as a new internal basis for the next scan (see Section 2.3.4.3).

For the compliance of a series of U.S. EPA methods and other regulations, the targeted source tuning according to the specific manufacturers' guidelines is required, see Table 2.61 (Eichelberger et al., 1975). This, for instance, refers to *U.S. EPA methods* 501.3, 524.2, 8260B, and CLP-SOW for the determination of volatiles using 1-bromo-4-fluorobenzene (BFB) and the U.S. EPA methods 625, 1625, 8250, and 8270 for base/neutrals/acids or semi-volatiles referring to DFTPP as tuning compounds providing consistent and instrument independent ion ratio profiles for quantitation.

The precision of the mass scan and the linearity of the calibration allow the line to be extrapolated beyond the highest fragment which can be determined, which is m/z 614 for FC43. For magnetic sector instruments, the use of perfluorokerosene (PFK) has proved successful besides FC43. It is particularly suitable for magnetic field calibration as it gives signals at regular intervals up to over m/z 1000. Also, perfluorophenanthrene (FC5311) is frequently used as an alternative calibration compound

Table 2.59 Reference table FC43/PFTBA (EI, intensities >1%, quadrupole instrument; Thermo Electron, 1999).

Exact mass [Da]	Intensity [%]	Formula
68.9947	100.0	CF_3^+
92.9947	2.0	$C_3F_3^+$
99.9931	19.0	$C_2F_4^+$
113.9961	11.0	$C_2NF_4^+$
118.9915	16.0	$C_2F_5^+$
130.9915	72.0	$C_3F_5^+$
149.9899	4.0	$C_3F_6^+$
168.9883	7.0	$C_3F_7^+$
175.9929	3.0	$C_4NF_6^+$
180.9883	3.0	$C_4F_7^+$
213.9898	2.0	$C_4NF_8^+$
218.9851	78.0	$C_4F_9^+$
225.9898	2.0	$C_5NF_8^+$
263.9866	27.0	$C_5NF_{10}^+$
313.9834	3.0	$C_6NF_{12}^+$
351.9802	4.0	$C_6NF_{14}^+$
363.9802	1.0	$C_7NF_{14}^+$
375.9802	1.0	$C_8NF_{14}^+$
401.9770	2.0	$C_7NF_{16}^+$
413.9770	9.0	$C_8NF_{16}^+$
425.9770	1.0	$C_9NF_{16}^+$
463.9738	3.0	$C_9NF_{18}^+$
501.9706	8.0	$C_9NF_{20}^+$
613.9642	2.0	$C_{12}NF_{24}^+$

to FC43, which covers about the same mass range (for reference tables of PFK and FC5311, see Hübschmann, 2015). Higher mass ranges can be calibrated using perfluorinated alkyltriazines, Ultramark, or cesium iodide. Reference tables frequently also take into account the masses from the lower mass range, such as He, N_2, O_2, Ar, or CO_2, which always form part of the background spectrum.

The calculated reference masses in the given tables are based on the following values for isotopic masses: ^1H 1.0078250321 Da, ^4He 4.0026032497 Da, ^{12}C 12.0000000000 Da, ^{14}N 14.0030740052 Da, ^{16}O 15.9949146221 Da, ^{19}F 18.9984032000 Da, and ^{40}Ar 39.9623831230 Da. The mass of the electron 0.00054857991 Da was taken into account for the calculation of the ionic masses (Audi and Wapstra, 1995; Mohr and Taylor, 1999).

Table 2.60 Reference table FC43/PFTBA (EI, intensities >0.1%, high resolution magnetic sector instrument).

Exact mass [Da]	Intensity [%]	Formula
4.00206	—	He^+
14.01510	—	CH_2^+
18.01002	—	H_2O^+
28.00560	—	N_2^+
30.99786	—	CF
31.98928	—	O_2^+
39.96184	—	Ar^+
43.98928	—	CO_2^+
49.99626	3.0	CF_2^+
68.99466	50.7	CF_3^+
75.99933	1.0	$C_2NF_2^+$
80.99466	1.2	$C_2F_3^+$
92.99466	1.0	$C_3F_3^+$
99.99306	3.9	$C_2F_4^+$
113.99614	3.5	$C_2NF_4^+$
118.99147	5.9	$C_2F_5^+$
130.99147	49.8	$C_3F_5^+$
149.98987	2.1	$C_3F_6^+$
168.98827	5.1	$C_3F_7^+$
175.99295	1.3	$C_4NF_6^+$
180.98827	1.3	$C_4F_7^+$
199.98668	0.2	$C_4F_8^+$
213.98975	2.5	$C_4NF_8^+$
218.98508	100.0	$C_4F_9^+$
225.98975	1.4	$C_5NF_8^+$
230.98508	1.1	$C_5F_9^+$
242.98508	0.0	$C_6F_9^+$
263.98656	37.6	$C_5NF_{10}^+$
275.98656	0.2	$C_6NF_{10}^+$
280.98189	0.1	$C_6F_{11}^+$
294.98496	0.2	$C_6NF_{11}^+$
313.98336	2.7	$C_6NF_{12}^+$
325.98336	0.3	$C_7NF_{12}^+$
344.98177	0.0	$C_7NF_{13}^+$
351.98017	1.7	$C_6NF_{14}^+$
363.98017	1.0	$C_7NF_{14}^+$

(Continued)

Table 2.60 (Continued)

Exact mass [Da]	Intensity [%]	Formula
375.98017	1.2	$C_8NF_{14}^+$
401.97698	2.0	$C_7NF_{16}^+$
413.97698	10.8	$C_8NF_{16}^+$
425.97698	4.0	$C_9NF_{16}^+$
451.97378	0.6	$C_8NF_{18}^+$
463.97378	7.3	$C_9NF_{18}^+$
475.97378	0.0	$C_{10}NF_{18}^+$
501.97059	22.7	$C_9NF_{20}^+$
513.97059	0.1	$C_{10}NF_{20}^+$
525.97059	0.1	$C_{11}NF_{20}^+$
551.96740	0.0	$C_{10}NF_{22}^+$
563.96740	0.1	$C_{11}NF_{22}^+$
575.96740	1.3	$C_{12}NF_{22}^+$
613.96420	3.6	$C_{12}NF_{24}^+$

To record high resolution data in manual work, peak matching was employed in the past (Webb et al., 2004). At a given magnetic field strength, one or two reference peaks and an ion of the substance being analyzed are alternately shown on an oscilloscope screen. By changing the acceleration voltage (and the electric field coupled to it), the peaks are superimposed. From the known mass and voltage difference, the exact mass of the substance peak is determined. This process is an area of the solid probe technique, as here the substance signal can be held constant over a longer period.

The validity period of the calibration depends on the type of instrument and can last for a time of up to several days or weeks. All tuning parameters, in particular the adjustment and the increasing contamination of the ion source, affect the calibration as described above. Also, the analyzer scan speed has a strong impact on the mass calibration with magnetic sector instruments. Special attention should also be paid to a constant temperature of the ion source. Regular mass calibration using analysis conditions is recommended to comply with the lab internal QA/QC documentation procedures.

The carrier gas flow setting of the GC can also show effect on the position of the mass calibration. Ion sources with small volumes and also ion trap instruments with internal ionization can show a significant drift of several tenths of a mass unit if the carrier gas flow rate is significantly changed by a temperature program. Calibration at an average elution temperature, the use of an open split, or equipping the gas chromatograph with electronic pressure programming for analysis at constant flow is imperative in this case. Severe contamination of the ion source or the ion optics also affects the mass calibration. However, reduced transmission of such an instrument should force cleaning to be carried out in good time.

Table 2.61 Ion abundance criteria for the U.S. EPA BFB and DFTPP target tuning.

Compound	m/z	Ion abundance criteria (relative abundance)	
BFB	50	15–40%	of m/z 95
	75	30–60%[a]	of m/z 95
	95	100%	base peak
	96	5–9%	of m/z 95
	173	<2%	of m/z 174
	174	>50%	of m/z 95
	175	5–9%	of m/z 174
	176	>95% but <101%	of m/z 174
	177	5–9%	of m/z 176
DFTPP	51	30–60%	of m/z 198
	68	<2%	of m/z 69
	70	<2%	of m/z 69
	127	40–60%	of m/z 198
	197	<1%	of m/z 198
	198	100%	base peak
	199	5–9%	of m/z 198
	275	10–30%	of m/z 198
	365	>1%	of m/z 198
	441	present	of m/z 443
	442	>40%	of m/z 198
	443	17–23%	of m/z 442

a) Thirty to eighty percent of m/z 95 for U.S. EPA 524.2.

For analyses with differing scan rates using magnetic instruments, calibrations are carried out at the different rates which are required for the subsequent measurements. For scan rates which differ significantly, a mass drift can otherwise occur between calibration and measurement. Fast scanning quadrupole instruments consider the different ion flight times of low and high mass ions in the calibration algorithms.

Depending on the type of instrument, a new mass calibration is required if the ionization process is changed. While for ion trap instruments switching from EI to CI ionization is possible without alterations to the analyzer, with other types, switching of the ion source or changing the ion volume is required. For an optimized CI reaction, a lower source temperature is frequently used compared with the EI mode. After these changes have been made, a new mass calibration is necessary. This involves running a CI spectrum of the reference substance and applying a CI reference table. The perfluorinated reference substances can be used for both PCI and

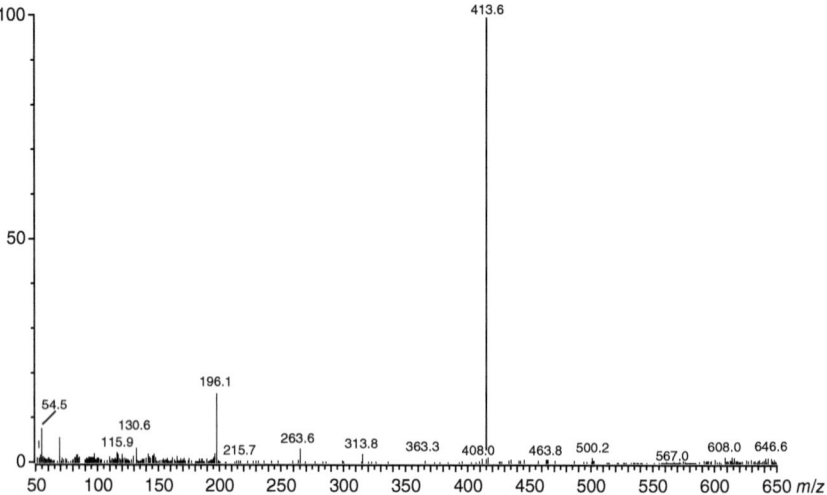

Figure 2.269 FC43/PFTBA spectrum in the PCI mode (CI gas methane, Finnigan GCQ).

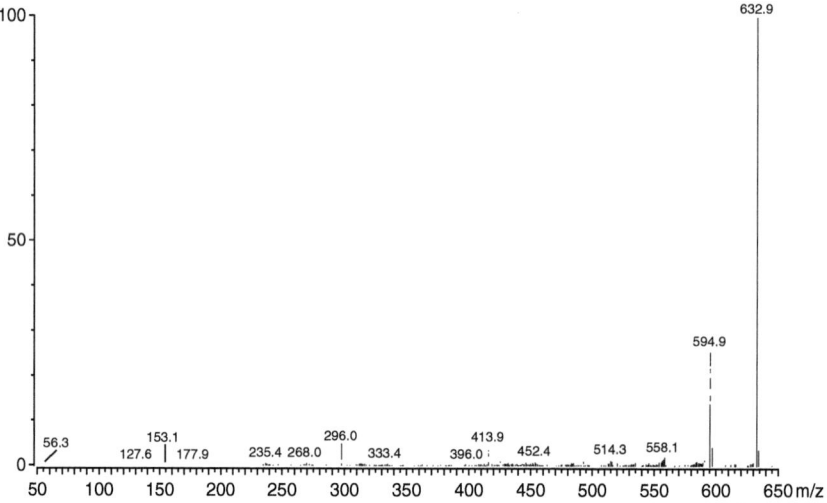

Figure 2.270 FC43/PFTBA spectrum in the NCI mode (CI gas methane, Finnigan GCQ).

NCI (see Figures 2.269 and 2.270). The intensities given in the CI reference tables can also be used to optimize the flow adjustment of the reagent gas.

For reasons of quality control, mass calibrations should be carried out regularly and should be documented with a printout, as shown in Figure 2.267.

2.3.6 Vacuum Systems

The carrier gas for GC-MS is mostly either helium or hydrogen. It is well known that the use of hydrogen significantly improves the performance of the GC, lowers the elution temperatures of compounds, and permits shorter analysis times due to higher flow rates. The more favorable van Deemter curve for hydrogen accounts for these improvements in analytical performance. As far as MS is concerned, when hydrogen is used, the mass spectrometer requires a higher vacuum capacity than for helium and thus a more powerful pumping system. The type of analyzer and the pumping system determine the advantages and disadvantages of these carrier gases.

For the turbomolecular pumps, in short term "turbopump", the important performance data is the compression ratio. The compression ratio describes the ratio of the outlet pressure (forepump) of a particular gas to the inlet pressure, the MS analyzer vacuum. Turbopumps give a completely background-free high vacuum and exhibit excellent start-up properties, which is important for benchtop instruments or those for mobile use. The use of helium lowers the performance of the pump compared to the specified nitrogen performance. Furthermore, the compression ratio decreases about two orders of magnitude for hydrogen. The reason for the much lower performance with hydrogen is its lower molecular mass and the high diffusion rate of hydrogen (Freeman, 1985). Often turbopumps comprise an integrated "Holweck" pumping stage. The Holweck stage is a molecular pump stage with a rotor of screw-shaped pump channels in the stator. It increases the permissible fore vacuum pressure level markedly compared to a regular turbopump.

A turbomolecular pump essentially consists of the rotor and a stator (Figure 2.271). Rotating and stationary discs are arranged alternately. All the discs have diagonal channels, whereby the channels on the rotor disc are arranged so that they mirror the positions of the channels of the stator discs. Each channel of the disc forms an elementary molecular pump. All the channels on the disc are arranged in parallel. A rotor disc together with a stator disc forms a single pump stage, which produces a certain compression. The pumping process is such that a gas molecule which meets the rotor acquires a velocity component in the direction of rotation of the rotor in addition to its existing velocity. The final velocity and the direction in which the molecule continues to move are determined from the vector sum of the two velocity components (Figure 2.272). The thermal motion of a molecule, which is initially undirected, is converted into directed motion when the molecule enters

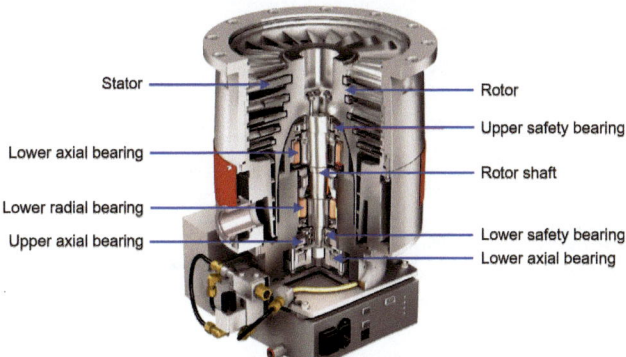

Figure 2.271 Cut-away of a magnetically levitated turbopump with emergency bearings for use in mass spectrometry (Pfeiffer Vacuum, 2024).

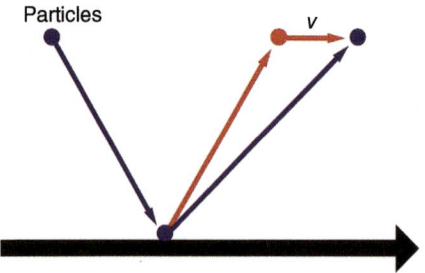

Figure 2.272 Principle of the molecular pump (Pfeiffer Vacuum, 2003).

the pump. An individual pumping step produces a compression of only about 30. Several consecutive pumping steps, which reinforce each other's action, lead to high compression rates.

The reduction in the performance of the pump when using hydrogen leads to a measurable increase of the pressure in the analyzer. In the case of beam instruments, quadrupoles, or magnetic sector MS, residual gas particles cause collisions of analyte ions with on their flight path through the analyzer. The transmission and thus the sensitivity of the instrument are reduced (Figure 2.273). The mean free path L of an ion is calculated according to

$$L = p^{-1} \cdot 5 \cdot 10^{-3} (\text{cm}). \tag{2.30}$$

The effect can be compensated for by using higher performing pumps with a higher compression ratio. Differentially pumped systems for the source and analyzer are typically in use with MS/MS, Orbitrap, TOF, and magnetic sector instruments.

Modern magnetically levitated turbopumps feature a rotor held in suspension by electromagnets with no mechanical contact between the rotor and outer casing. In the event of power failure, power is supplied internally by the rotational energy of the rotor itself until the rotor lands securely on the emergency bearings. Magnetic bearing turbopumps are maintenance free throughout the entire service life and do not require lubricants, which eliminates the risk of hydrocarbon contamination in mass spectrometry (Pfeiffer, 2024).

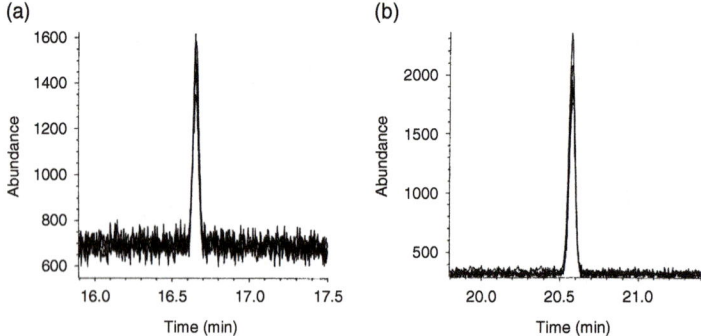

Figure 2.273 Effect of the carrier gas on the signal/noise ratio in the quadrupole GC-MS (Schulz, 1987). PCB 101, 50 pg, SIM m/z 256, 326, 328. (a) Carrier gas hydrogen, S/N 4 : 1. (b) Carrier gas helium, S/N 19 : 1.

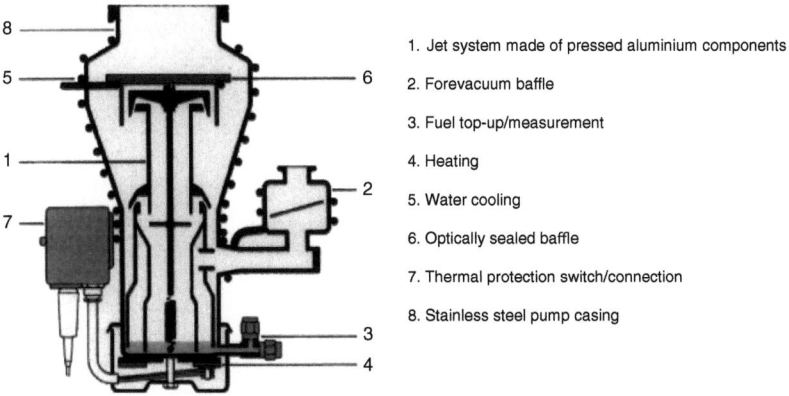

1. Jet system made of pressed aluminium components
2. Forevacuum baffle
3. Fuel top-up/measurement
4. Heating
5. Water cooling
6. Optically sealed baffle
7. Thermal protection switch/connection
8. Stainless steel pump casing

Figure 2.274 Cross section of a classical oil diffusion pump (Pfeiffer Vacuum, 2003).

When hydrogen is used as the carrier gas in GC, the use of oil diffusion pumps is of advantage, although outdated today for benchtop GC-MS instruments. The pump capacity of diffusion pumps is largely independent of the molecular mass and is therefore very suitable for hydrogen and helium. Oil diffusion pumps have long service lives and are economical as no moving parts are involved. The pump operates using oil as a propellant, which is evaporated on a heating plate (Figure 2.274). The propellant vapor is forced downward via several stages of baffles downward and back into a fluid reservoir. Gas molecules diffuse into the propellant stream and are conveyed deeper into the pump. They are finally evacuated by a forepump. Perfluoropolyethers (e.g. Fomblin) or polyphenyl ethers (e.g. Santovac S) are used as diffusion pump fluids in MS. However, the favorable operation of the pump results in disadvantages for the operation of the mass spectrometer. Because an electrical heating plate is used, the diffusion pump starts sluggishly and can only be vented again after cooling. Baffles closing the pump on top help for MS maintenance. While older models require water cooling, modern diffusion pumps for GC-MS instruments are air cooled by a fan so that heating and cool-down times of 30 min and longer are unavoidable. The final vacuum of diffusion pumps is limited by the vapor pressure of the oil used, reaching between 10^{-8} and 10^{-9} Torr. Oil vapor can lead to a permanent background in the mass spectrometer, which makes the use of a liquid nitrogen-cooled baffle necessary, depending on the instrument design. The detection of negative ions can be affected by the use of fluorinated polymer oils.

Both turbomolecular pumps and oil diffusion pumps require a mechanical forepump (rough pump) as the compression is not sufficient to work against atmospheric pressure. Rotary vane pumps are generally used as forepumps (Figure 2.275). Mineral oils are used as the operating fluid. Oil vapors from the rotary vane pump passing into the vacuum tubing to the turbo pump are visible in the mass spectrometer as a hydrocarbon background, in particular, on frequent venting of the system. Special devices for separation or removal are necessary. As an alternative for a hydrocarbon-free forevacuum, spiromolecular pumps can be used. These pumps can be run "dry" without the use of oil. Their function involves

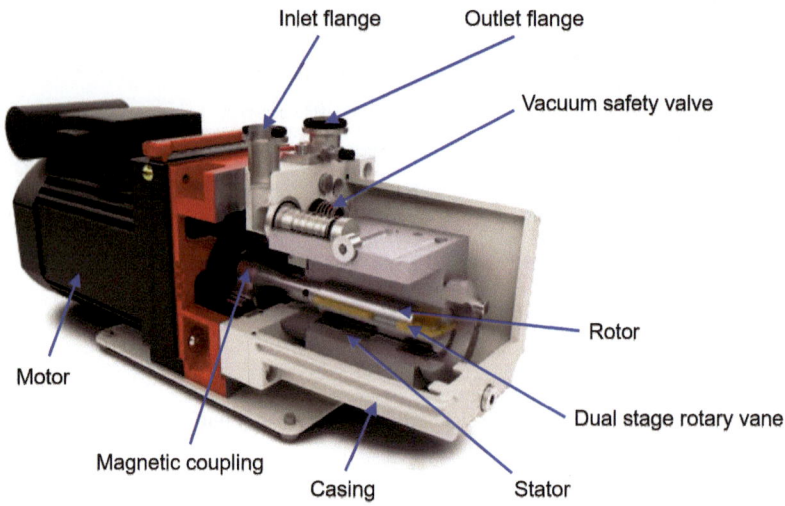

Figure 2.275 Cross section of a two-stage rotary vane vacuum pump (Pfeiffer Vacuum, 2003).

centrifugal acceleration of the gas molecules through several pumping steps like the way it is in a turbomolecular pump. For the start-up, an integrated membrane pump is used. In particular, for mobile use of GC-MS systems and in other cases where systems are frequently disconnected from the main electricity, spiromolecular pumps have proven useful.

References

Section 2.3 Mass Spectrometry

Busch, K.L. (2001) Units of mass spectrometry. *Spectroscopy*, **16** (11), 28–31, https://www2.chemistry.msu.edu/courses/cem832/units_of_mass.pdf.

McLafferty, F.W. and Michnowicz, J.A. (1992) State-of-the-art GC/MS. *Chemtech.*, **22** (3), 182–189.

Section 2.3.1 Ionization

Alam, M.S., Stark, C., and Harrison, R.M. (2016) Using variable ionization energy time-of-flight mass spectrometry with comprehensive GC×GC to identify isomeric species. *Anal. Chem.*, **88** (8), 4211–4220, https://doi.org/10.1021/acs.analchem.5b03122.

Allgood, C., Lin, Y., Ma, Y.-C., and Munson, B. (1990) Benzene as a selective chemical ionization reagent gas. *Org. Mass Spectrom.*, **25** (10), 497–502, https://doi.org/10.1002/oms.1210251003.

Ardenne, M., Steinfelder, K., and Tümmler, R. (1971) *Elektronenanlagerungsmassen spektrometrie organischer Substanzen*, Springer-Verlag, Berlin.

Aue, D.H. and Bowers, M.T. (1979) Stability of positive ions from equilibrium gas-phase basicity measurements, in *Gas Phase Ion Chemistry*, Vol. 2 (ed. M.T. Bowers), Academic Press, New York.

Ayala-Cabrera, J.F., Montero, L., Meckelmann, S.W., Uteschil, F., and Schmitz, O.J. (2023) Review on atmospheric pressure ionization sources for gas chromatography-mass spectrometry. Part I: Current ion source developments and improvements in ionization strategies. *Anal. Chim. Acta*, **1238**, 340353, https://doi.org/10.1016/j.aca.2022.340353.

Buser, H.-R. and Müller, M. (1994) Isomer- and enantiomer-selective analyses of toxaphene components using chiral high resolution gas chromatography and detection by mass spectrometry/mass spectrometry. *Environ. Sci. Technol.*, **28**, 119–128.

Class, T.J. (1991) Determination of pyrethroids and their degradation products in indoor air and on surfaces by HRGC-ECD and HRGC-MS (NCI), in *12th International Symposium on Capillary Chromatography, Riva del Garda, May 1991* (ed. P. Sandra), Huethig.

Crow, F.W., Bjorseth, A., Knapp, K.T., and Bennett, R. (1981) Determination of polyhalogenated hydrocarbons by glass capillary gas chromatography-negative ion chemical ionization mass spectrometry. *Anal. Chem.*, **53**, 619, http://dx.doi.org/10.1021/ac00227a014.

Bartmess, J. and McIver, R. (1979) The gas phase acidity scale, in *Gas Phase Ion Chemistry*, Vol. 2 (ed. M.T. Bowers), Academic Press, New York.

Brodbelt, J., Liou, C.-C., and Donovan, T. (1991) Selective adduct formation by dimethyl ether chemical ionization in a quadrupole ion trap mass spectrometer and a conventional ion source. *Anal. Chem.*, **63** (5), 1205–1209.

Budzikiewicz, H. (1981) Massenspektrometrie negativer Ionen. *Angew. Chem.*, **93**, 635–649.

Budzikiewicz, H. and Schäfer, M. (2012) *Massenspektrometrie*, Wiley-VCH Publishers, Weinheim, Germany, https://www.wiley-vch.de/de/fachgebiete/naturwissenschaften/chemie-11ch/analytische-chemie-11ch1/spektroskopie-11ch15/massenspektrometrie-978-3-527-32911-3.

Chernetsova, E.S., Revelsky, A.I., Revelsky, I.A., Mikhasenko, I.A., and Sobolevsky, T.G. (2002) Determination of polychlorinated dibenzo-p-dioxins, dibenzofurans, and biphenyls by gas chromatography/mass spectrometry in the negative chemical ionization mode with different reagent gases. *Mass Spectrom. Rev.*, **21**, 373–387.

DePuy, C.H., Grabowski, J.J., and Bierbaum, V.M. (1982) Chemical reactions of anions in the gas phase. *Science*, **218**, 955–960.

Dougherty, R.C. (1981) Negative chemical ionization mass spectrometry. *Anal. Chem.*, **53**, 625A–636A.

Fang, J. et al. (2019) Performance of atmospheric pressure gas chromatography-tandem mass spectrometry for the analysis of organochlorine pesticides in human serum. *Anal. Bioanal. Chem.*, **411** (18), 4185–4191, https://doi.org/10.1007/s00216-019-01822-1.

Frigerio, A. (1974) *Essential Aspects of Mass Spectrometry*, John Wiley & Sons Inc., Spectrum Publishers, New York.

Fürst, P. (1994) Personal communication.

Gross, J.H. (2004) *Mass Spectrometry*, Springer, Heidelberg.

Grimme, S. (2013) Towards first principles calculation of electron impact mass spectra of molecules. *Angew. Chem. Int. Ed.*, **52**, 6306–6312.

Hainzl, D., Burhenne, J., and Parlar, H. (1993) Isolation and characterization of environmental relevant single toxaphene components. *Chemosphere*, **27** (10), 1857–1863, https://doi.org/10.1016/0045-6535(93)90381-E.

Hainzl, D., Burhenne, J., and Parlar, H. (1994) Isolierung von Einzelsubstanzen für die Toxaphenanalytik. *GIT Fachz. Lab.*, **4**, 285–294.

Harrison, A.G. (1992) *Chemical Ionization Mass Spectrometry*, CRC Press, Boca Raton, FL.

Horning, E.C., Horning, M.G., Carroll, D.I., Dzidic, I., and Stillwell, R.N. (1973) New picogram detection system based on a mass spectrometer with an external ionization source at atmospheric pressure. *Anal. Chem.*, **45**, 936–943, https://doi.org/10.1021/ac60328a035.

Horning, E.C., Caroll, D.I., Dzidic, I., and Stillwell, R.N. (1981) Negative ion atmospheric pressure ionization mass spectrometry and the electron capture detector, in *Electron Capture, J. Chrom. Library*, Vol. 20 (eds A. Zlatkis and C.F. Poole), Elsevier, Amsterdam.

Howe, I., Williams, D.H., and Bowen, R. (1981) *Mass Spectrometry*, 2nd edn, McGraw Hill, New York.

Hunt, D.F., Stafford, G.C., Crow, F., and Russel, J. (1976) Pulsed positive negative ion chemical ionization mass spectrometry. *Anal. Chem.*, **48**, 2098–2105.

Jones, R., Hall, K., Douce, D., Dunstan, J., Rosnack, K., and Ladak, A. (2018) Confirmation of dioxins and dioxin-like substances at sub-femtogram levels using atmospheric pressure gas chromatography (APGC) MS/MS. Poster at the 66th ASMS Conference, June 3–7, 2018, San Diego, CA, USA. Waters Corp.

Knapp, D.R. (1979) *Handbook of Analytical Derivatization Reactions*, Wiley-Interscience, ISBN: 978-0-471-03469-8.

Landrock, A., Richter, H., and Merten, H. (1995) Water CI$^+$, a new selective and highly sensitive method for the detection of environmental components using ion trap mass spectrometers. *Fres. J. Anal. Chem.*, **351**, 536–543.

McGregor, L. and Barden, D. (2014) A new outlook on soft ionization for GC-MS. *Spectroscopy Online*, **12** (1), https://www.spectroscopyonline.com/view/new-outlook-soft-ionization-gc-ms.

McGuire, J.M. and Munson, B. (1985) Comparison of isopentane and isobutane as chemical ionization reagent gases. *Anal. Chem.*, **57** (3), 680–683, https://doi.org/10.1021/ac00280a024.

McLafferty, F.W. and Turecek, F. (1993) *Interpretation of Mass Spectra*, 4th edn, University Science Books, Mill Valley, CA.

Munson, M.S.B. and Field, F.H. (1966a) Chemical ionization mass spectrometry. I. General introduction. *J. Am. Chem. Soc.*, **12**, 2621–2630, https://doi.org/10.1021/ja00964a001.

Munson, M.S.B. and Field, F.H. (1966b) Chemical ionization mass spectrometry. II. Esters. *J. Am. Chem. Soc.*, **88** (19), 4337–4345, https://doi.org/10.1021/ja00971a007.

Pizzutti, I.R. et al. (2012) A multi-residue method for pesticides analysis in green coffee beans using gas chromatography-negative chemical ionization mass spectrometry in selective ion monitoring mode. *J. Chrom. A*, **1251**, 16–26.

Smit, A.L.C. and Field, F.H. (1977) Gaseous anion chemistry. Formation and reaction of OH⁻; reactions of anions with N_2O; OH⁻ negative chemical ionization. *J. Am. Chem. Soc.*, **99**, 6471–6483, https://doi.org/10.1021/ja00462a001.

Spiteller, M. and Spiteller, G. (1973) *Massenspektrensammlung von Lösungsmitteln, Verunreinigungen. Säulenbelegmaterialien und einfachen aliphatischen Verbindungen*, Springer, Wien.

Stan, H.-J. and Kellner, G. (1981) Analysis of organophosphoric pesticide residues in food by GC-MS using positive and negative chemical ionisation, in *Recent Developments in Food Analysis* (eds W. Baltes, P.B. Czedik-Eysenberg, and W. Pfannhauser), Verlag Chemie, Weinheim.

Stout, S.J. and Steller, W.A. (1984) Application of gas chromatography negative ion chemical ionization mass spectrometry in confirmatory procedures for pesticide residues. *Biomed. Mass Spectrom.*, **11**, 207–210.

Ten Dam, G. et al. (2016) The performance of atmospheric pressure gas chromatography-tandem mass spectrometry compared to gas chromatography-high resolution mass spectrometry for the analysis of polychlorinated dioxins and polychlorinated biphenyls in food and feed samples. *J. Chrom. A*, **1477** (December), 76–90, https://doi.org/10.1016/j.chroma.2016.11.035.

Theobald, F. and Hübschmann, H.-J. (2007) High Sensitive MID Detection Method for Toxaphenes by High Resolution GC-MS. Application Note AN30128, Thermo Fisher Scientific, Bremen.

van Bavel, B. et al. (2015) Atmospheric-pressure chemical ionization tandem mass spectrometry (APGC/MS/MS) an alternative to high resolution mass spectrometry (HRGC/HRMS) for the determination of dioxins. *Anal Chem.*, **87** (17), 9047–9053. doi: 10.1021/acs.analchem.5b02264.

Waters (2023) APGC - No Compromise Atmospheric Pressure Ionization GC-MS. Product brochure, April 2023 720004755EN GJ-PDF, Waters Corporation, Milford, MA, USA.

Yang, Y., Lin, M., Tang, J., Ma, S., and Yu, Y. (2020) Derivatization gas chromatography negative chemical ionization mass spectrometry for the analysis of trace organic pollutants and their metabolites in human biological samples. *Anal. Bioanal. Chem.*, **412** (25), 6679–6690, https://doi.org/10.1039/d1ra01185e.

Yost, R.A. (1988) Analytical applications of ion trap mass spectrometry. *Spectra*, **11** (2).

Section 2.3.2 Mass Analysis

Abel, R.J. (2014) Identification of unknown compounds from quadrupole GC-MS data using Cerno Bioscience MassWorks™. *Can. Soc. Forensic Sci. J.*, **47** (2), 74–98, https://doi.org/10.1080/00085030.2014.904605.

Agilent (2011b) Method for the Determination of Dioxins in Foodstuffs and Animal Feed. Application Note, 5990-7689EN, Agilent Technologies Inc.

Agilent (2011a) Time-of-Flight Mass Spectrometry. Technical Overview, 5990-9207EN, Agilent Technologies, Inc.

Agilent (2017) The Agilent Dioxins in Feed and Food Analyzer. Report, 5991-5471EN, Agilent Technologies Inc.

Baumeister, T.U.H., Ueberschaar, N., and Pohnert, G. (2018) Gas-phase chemistry in the GC Orbitrap mass spectrometer. *J. Am. Soc. Mass Spectrom.*, **30** (4), 573–580, https://doi.org/10.1007/s13361-018-2117-5.

Belarbi, S. *et al.* (2021) Comparison of new approach of GC-HRMS (Q-Orbitrap) to GC-MS/MS (triple-quadrupole) in analyzing the pesticide residues and contaminants in complex food matrices. *Food Chem.*, **359** (October), 129932, https://doi.org/10.1016/j.foodchem.2021.129932.

Bendix Corporation (1960) Bendix MA-2 Time-of-Flight Mass Spectrometer 1960-1969. Photographs from the Bendix Time-of-Flight Mass Spectrometer Collection, Box 1. Science History Institute, Philadelphia, https://digital.sciencehistory.org/works/mg74qm60n.

Borman, S. (1998) A Brief History in Mass Spectrometry, May 26, 1998, https://masspec.scripps.edu/learn/ms/history/mass-spectrometery-instrumentation.html (accessed 26 April 2024).

Brunnée, C. (1987) The ideal mass analyzer: fact or fiction? *Int. J. Mass Spectrom. Ion Proc.*, **76**, 121–237.

Burnum-Johnson, K.E. *et al.* (2019) Ion mobility spectrometry and the omics: distinguishing isomers, molecular classes and contaminant ions in complex samples. *Trends Anal. Chem.*, **116**, 292–299, https://doi.org/10.1016/j.trac.2019.04.022.

Cojocariu, C., Abalos, M., Abad Holgado, E., and Silcock, P. (2014) Meeting the European Commission Performance Criteria for the Use of Triple Quadrupole GC-MS/MS as a Confirmatory Method for PCDD/Fs in Food and Feed Samples. Application Note 10380, Thermo Fisher Scientific, Runcorn, UK.

Cojocariu, C., Roberts, D., and Silcock, P. (2016) The Power of High Resolution Accurate Mass Using Orbitrap Based GC-MS. White paper 10456, Thermo Fisher Scientific, Runcorn, UK.

Cotter, R.J. (1992) Time-of-flight mass spectrometry for the structural analysis of biological molecules. *Anal. Chem.*, **64**, 1027A–1039A, https://doi.org/10.1021/ac00045a726.

Cotter, R.J. (1997) *Time-of-Flight Mass Spectrometry*, American Chemical Society, Washington, DC, USA.

Dawson, P.H. and Whetten, N.R. (1968) Mass spectrometry using radio-frequency quadrupole fields. *J. Vac. Sci. Technol.*, **5**, 1.

De Vooght-Johnson, R. (2019) GC-orbitrap combination can give spectral alteration. *Wiley Anal. Sci.*, **16**, 9–11.

Denisov, E., Damoc, E., Lange, O., and Makarov, A. (2012) Orbitrap mass spectrometry with resolving powers above 1,000,000. *Int. J. Mass Spectrom.*, **325–327** (July), 80–85, https://doi.org/10.1016/j.ijms.2012.06.009.

Dimandja, J.M. (2003) A new tool for the optimized analysis of complex volatile mixtures: comprehensive two-dimensional gas chromatography/time-of-flight mass spectrometry. *Am. Lab.*, **2**, 42–53.

Dindal, A., Thompson, E., Strozier, E., and Billets, S. (2011) Application of GC-HRMS and GC × GC-TOFMS to aid in the understanding of a dioxin assays performance for soil and sediment samples. *Environ. Sci. Technol.*, **45** (24), 10501–10508, https://doi.org/10.1021/es202149k.

Dodds, J.N., May, J.C., and McLean, J.A. (2017) Correlating resolving power, resolution and collision cross section: unifying cross platform assessment of separation efficiency in ion mobility spectrometry. *Anal Chem.*, **89** (22), 12176–12184, https://doi.org/10.1021/acs.analchem.7b02827.

Eljarrat, E. and Barcelo, D. (2002) Congener-specific determination of dioxins and related compounds by gas chromatography coupled to LRMS, HRMS, MS/MS and TOFMS. *J. Mass Spectrom.*, **37** (11), 1105–1117, https://doi.org/10.1002/jms.373.

European Commission (2006) Commission regulation No. 1883. *Off. J. Eur. Union*, **L364**, 32–43.

European Commission (2012) Commission regulation No. 252. *Off. J. Eur. Union*, **L84**, 1–22.

European Commission (2014) Commission regulation No. 589. *Off. J. Eur. Union*, **L164**, 18–40.

Fabrellas, B., Sanz, P., Abad, E., Rivera, J., and Larrazábal, D. (2004) Analysis of dioxins and furans in environmental samples by GC-ion trap MS/MS. *Chemosphere*, **55** (11), 1469–1475, https://doi.org/10.1016/j.chemosphere.2004.01.039.

Fürst, P. and Bernsmann, T. (2010) personal communication.

Focant, J.F. et al. (2005) Comprehensive two-dimensional gas chromatography with isotope dilution time-of-flight mass spectrometry for the measurement of dioxins and polychlorinated biphenyls in foodstuffs: comparison with other methods. *J. Chrom. A*, **1086** (1–2), 45–60, https://doi.org/10.1016/j.chroma.2005.05.090.

Focant, J.F. (2012) Dioxin food crises and new POPs: challenges in analysis. *Anal. Bioanal. Chem.*, **403**, 2797–2800, https://doi.org/10.1007/s00216-012-5801-5.

Gall, L.N., Golikov, Y.K., Aleksandrov, M.L., Pechalina, Y.E., and Holin, N.A. (1986) USSR Inventor's Certificate 1247973.

GAS (2014) Headspace Analysis of Tea. Application Note, G.A.S. Gesellschaft für analytische Sensorsysteme mbH, Dortmund, Germany, https://www.gas-dortmund.de (accessed 1 May 2024).

GAS (2019a) IMS Working Principle. Web page, G.A.S. Gesellschaft für analytische Sensorsysteme mbH, Dortmund, Germany, https://www.gas-dortmund.de (accessed 1 May 2024).

GAS (2019b) Ultra Trace Headspace Analysis. Application Note, G.A.S. Gesellschaft für analytische Sensorsysteme mbH, Dortmund, Germany, https://www.gas-dortmund.de (accessed 1 May 2024).

Gerhardt, N., Birkenmeier, M., Sanders, D., Rohn, S., and Weller, P. (2017) Resolution-optimized headspace gas chromatography-ion mobility spectrometry (HS-GC-IMS) for non-targeted olive oil profiling. *Anal. Bioanal. Chem.*, **409** (16), 3933–3942, https://doi.org/10.1007/s00216-017-0338-2.

Gu, M. and Wang, Y. (2008) Calibration stability for formula determination on a single quadrupole GC-MS system. *Curr. Trends Mass Spectrom.*, **5** (May), 25–29.

Guilhaus, M. (1995) Review of TOF-MS. *J. Mass Spectrom.*, **30**, 1519.

Gummersbach, J. (2010) Private communication, Thermo Fisher Scientific, GC-MS Applications Laboratory, Dreieich, Germany.

Groessl, M., Graf, S., and Knochenmuss, R. (2015) High resolution ion mobility-mass spectrometry for separation and identification of isomeric lipids. *Analyst*, **140** (20), 6904–6911, https://doi.org/10.1039/c5an00838g.

Hädener, M., Kamrath, M.Z., Weinmann, W., and Groessl, M. (2018) High resolution ion mobility spectrometry for rapid cannabis potency testing. *Anal. Chem.*, **90** (15), 8764–8768, https://doi.org/10.1021/acs.analchem.8b02180.

Hecht, E.S., Scigelova, M., Eliuk, S., and Makarov, A. (2019) Fundamentals and advances of orbitrap mass spectrometry, in *Encyclopedia of Analytical Chemistry*, John Wiley & Sons, Ltd, https://doi.org/10.1002/9780470027318.a9309.pub2.

Jemal, M. and Ouyang, Z. (2003) Enhanced resolution triple-quadrupole mass spectrometry for fast quantitative bioanalysis using liquid chromatography/tandem mass spectrometry: investigations of parameters that affect ruggedness. *Rapid Commun. Mass Spectrom.*, **17**, 24–38.

Jiang, W. and Robinson, R.A.S. (2013) Ion mobility-mass spectrometry, in *Encyclopedia of Analytical Chemistry*, John Wiley & Sons, Ltd, https://doi.org/10.1002/9780470027318.a9292.

Jones, P.D. and Giesy, J.P. (2017) Quantification of dioxins by GC-Orbitrap MS. *Organohalogen Comp.*, **79**, 589–592.

Kendrick, E. (1963) A mass scale based on CH2 = 14.00000 for high resolution mass spectrometry of organic compounds. *Anal. Chem.*, **35** (13), 2146–2154. doi: 10.1021/ac60206a048.

Koppenaal, D.W. et al. (2005) MS detectors. *Anal. Chem.*, **77**, 418A–427A.

Kotz, A. et al. (2009) GC-MS/MS determination of PCDD/Fs and PCBs in feed and food - comparison with GC-HRMS. *Organohalogen Compd.*, **73**, 1–4.

Kotz, A. et al. (2012) Analytical criteria for use of MS/MS for determination of dioxins and dioxin-like PCBs in feed and food. *Organohalogen Compd.*, **74**, 156–159.

Lanucara, F., Holman, S.W., Gray, C.J., and Eyers, C.E. (2014) The power of ion mobility-mass spectrometry for structural characterization and the study of conformational dynamics. *Nat Chem.*, **6** (4), 281–294, https://doi.org/10.1038/nchem.1889.

Lapthorn, C., Pullen, F., and Chowdhry, B.Z. (2013) Ion mobility spectrometry-mass spectrometry (IMS-MS) of small molecules: separating and assigning structures to ions. *Mass Spectrom. Rev.*, **32** (1), 43–71, https://doi.org/10.1002/mas.21349.

Liu, Y.-M., Akervik, K., and Maljers, L. (2006) Optimized high resolution SRM quantitative analysis using a calibration correction method on a triple quadrupole system. Poster Presentation, TP08, #115, 54th ASMS Conference, May 28–June 1, 2006, Seattle, WA, USA.

Makarov, A. (1999a) The Orbitrap: a novel high performance electrostatic trap. Proceedings of the 47th Conference on Mass Spectrometry and Allied Topics. 47th

ASMS conference on Mass Spectrometry and Allied Topics, Dallas, Texas, USA, June 13–17, 1999.

Makarov A.A. (1999b) Mass Spectrometer, U.S. Patent 5,886,346.

Makarov, A. (2000) Electrostatic axially harmonic orbital trapping: a high performance technique of mass analysis. *Anal. Chem.*, **72** (6), 1156–1162.

Malavia, J., Santos, F.J., and Galceran, M.T. (2008) Comparison of gas chromatography-ion trap tandem mass spectrometry systems for the determination of polychlorinated dibenzo-*p*-dioxins, dibenzofurans and dioxin-like polychlorinated biphenyls. *J. Chromatogr. A*, **1186**, 302–311.

Mamyrin, B.A., Karataev, V.I., Shmikk, D.V., and Zagulin, V.A. (1973) The mass-reflectron, a new nonmagnetic time-of-flight mass spectrometer with high resolution. *Sov. Phys. JETP*, **37** (1), 45–48.

Mamyrin, B.A. (2001) Time-of-flight mass spectrometry (concepts, achievements, and prospects). *Int. J. Mass Spectrom.*, **206** (3), https://doi.org/10.1016/S1387-3806(00)00392-4.

McDaniel, E.W. and Martin, D.W. (1970) Mass Spectrometric Studies of Mobilities, Diffusion, and Reactions of Ions in Gases. Accession Number: AD0711083, Georgia Institute of Technology, https://apps.dtic.mil/sti/citations/AD0711083.

McClenathan, D. and Ray, S.J. (2004) Plasma source TOFMS. *Anal. Chem.*, **76** (9), 159A–166A.

McLafferty, F.W. and Turecek, F. (1993) *Interpretation of Mass Spectra*, 4th edn, University Science Books, Mill Valley, CA.

Meruva, N.K., Sellers, K.W., Brewer, W.E., Goode, S.R., and Morgan, S.L. (2000) Comparisons of Chromatographic Performance and Data Quality using Fast Gas Chromatography. Paper 1397, Pittcon 2000, New Orleans, LA, 17 March 2000.

Miller, P.E. and Denton, M.B. (1986) The quadrupole mass filter: basic operating concepts. *J. Chem. Educ.*, **63**, 617–622.

Miyagawa, H. (2016) A New Gold Standard for Dioxin Analysis. Shimadzu Corporation, presented at the 36th Dioxin Conference 2016, Florence, Italy.

Mol, H.G.J., Tienstra, M., and Zomer, P. (2016) Evaluation of gas chromatography – electron ionization – full scan high resolution orbitrap mass spectrometry for pesticide residue analysis. *Anal. Chim. Acta*, **935** (September), 161–172, https://doi.org/10.1016/j.aca.2016.06.017.

Moura, P.C., Vassilenko, V., Fernandes, J.M., and Santos, P.H. (2020) Indoor and outdoor air profiling with GC-IMS, in *Technological Innovation for Life Improvement. DoCEIS 2020. IFIP Adv. Information Communication Technol.*, Vol. **577** (eds L. Camarinha-Matos, N. Farhadi, F. Lopes, and H. Pereira), Springer, Cham, https://doi.org/10.1007/978-3-030-45124-0_43.

Münster, H. and Taylor, L. (2009) Mass Resolution and Resolving Power. Technical Note 30175, Thermo Fisher Scientific, Bremen.

Murray, K.K. (2022) Resolution and resolving power in mass spectrometry. *J. Am. Soc. Mass Spectrom.*, **33** (12), 2342–2347, https://doi.org/10.1255/ejms.727.

Parastar, H. and Weller, P. (2024) Towards greener volatilomics: is GC-IMS the New Swiss army knife of gas phase analysis? *TrAC - Trends Anal. Chem.*, **170** (June 2023), 117438, https://doi.org/10.1016/j.trac.2023.117438.

Paul, W. and Steinwedel, H. (1953) Ein neues Massenspektrometer ohne Magnetfeld. *Z. Naturforsch.*, **8a**, 448–450.

Paul, W. and Steinwedel, H. (1956) Apparat zur Trennung von geladenen Teilchen mit unterschiedlicher spezifischer Ladung. German Patent 944 900 (U.S. Patent 2 939 952 v. 7. 6. 1960).

Paul, W., Reinhard, H.P., and von Zahn, U. (1958) Das elektrische Massenfilter als Massenspektrometer und Isotopentrenner. *Z. Phys.*, **152**, 143–182.

Peterson, A., Quarmby, S.T., McAllister, G.C., and Coon, J.J. (2009) Implementation of an EI/CI Interface on a hybrid Orbitrap system for ultra high resolution GC-MS. Presentation at the 57th ASMS Conference, May 31-June 4, 2009, Philadelphia, PA, USA.

Peterson, A.C., McAlister, G.C., Quarmby, S.T., Griep-Raming, J., and Coon, J.J. (2010) Development and characterization of a GC-enabled QLT-orbitrap for high resolution and high mass accuracy GC-MS. *Anal. Chem.*, **82** (20), 8618–8628.

Peterson, A.C. et al. (2012) A Novel Applications-Grade, Bench-Top GC/Quadrupole-Orbitrap Mass Spectrometer for High Resolution and High Mass Accuracy GC/MS. Poster TP 685, 60th ASMS Conference, Vancouver, Canada.

Sammarco, G. et al. (2023) A geographical origin assessment of Italian hazelnuts: gas chromatography-ion mobility spectrometry coupled with multivariate statistical analysis and data fusion approach. *Food Res. Int.*, **171** (May), 113085, https://doi.org/10.1016/j.foodres.2023.113085.

Skoczyńska, E., Korytár, P., and De Boer, J. (2008) Maximizing chromatographic information from environmental extracts by GC × GC-ToF-MS. *Env. Sci. Technol.*, **42** (17), 6611–6618, https://doi.org/10.1021/es703229t.

Slayback, J.R.B. and Taylor, P.A. (1983) Analysis of 2.3.7,8-TCDD and 2,3,7,8-TCDF in environmental matrices using GC-MS/MS techniques. *Spectra*, **9** (4), 18–24.

Sleno, L. (2012) The use of mass defect in modern mass spectrometry. *J. Mass Spectrom.*, **47** (2), 226–236, https://doi.org/10.1002/jms.2953.

Takakura, M., Lehardy, T., and March, P. (2017) Analysis of Dioxins in Foods and Feeds Using GC-MS/MS. Technical Report C146-E376, Shimadzu Corporation.

TOFWERK (2024) *What Is High Resolution Ion Mobility Spectrometry?* Web page, https://www.tofwerk.com/high-resolution-ion-mobility-spectrometry/ (accessed 2 May 2024).

Ubukata, M. et al. (2015) Non-targeted analysis of electronics waste by comprehensive two-dimensional gas chromatography combined with high resolution mass spectrometry: using accurate mass information and mass defect analysis to explore the data. *J. Chrom. A*, **1395**, 152–159, https://doi.org/10.1016/j.chroma.2015.03.050.

Urban, J., Afseth, N.K., and Štys, D. (2014) Fundamental definitions and confusions in mass spectrometry about mass assignment, centroiding and resolution. *TrAC - Trends Anal. Chem.*, **53**, 126–136, https://doi.org/10.1016/j.trac.2013.07.010.

Verenchikov, A.N., Kirillov, S.N., Vorobyev, A.V. et al. (2025) Multi reflecting TOF MS approaching resolution of 1,000,000 in a wide mass range. *Int. J. Mass Spectrom.*, **508** (November 2024), 117395, https://doi.org/10.1016/j.ijms.2024.117395.

Wang, Y. and Gu, M. (2010) The concept of spectral accuracy for MS. *Anal. Chem.*, **82**, 7055–7062.

Warner, N.A., Benedetti, P., Zheng, X., Roberts, D., and Kutscher, D. (2023) Orbitrap Exploris GC Mass Spectrometer Performance-Based Analysis of Polychlorinated Dibenzo-p-Dioxins/Furans (PCDD/F). Technical Note 001984, Thermo Fisher Scientific, Bremen, Germany.

Webb, K., Bristow, T., Sargent, M., and Stein, B. (2004) *Methodology for Accurate Mass Measurement of Small Molecules, Best Practice Guide*, LGC Limited, Teddington, UK, ISBN: 0-948926-22-8.

Wells, G., Prest, H., and Russ, C.W.I.V. (2011) Why use signal-to-noise as a measure of MS performance when it is often meaningless? *Curr. Trends Mass Spectrom.*, **9**, 28–33.

Wiley, W.C. and MacLaren, I.H. (1955) Time-of-flight mass spectrometer with improved resolution. *Rev. Sci. Instrum.*, **26**, 1150, https://doi.org/10.1063/1.1715212.

Xian, F., Hendrickson, C.L., and Marshall, A.G. (2012) High resolution mass spectrometry. *Anal. Chem.*, **84**, 708–719, https://doi.org/10.1021/ac203191t.

Zacometti, C. et al. (2024) Authenticity assessment of ground black pepper by combining headspace gas-chromatography ion mobility spectrometry and machine learning. *Food Res. Int.* **179** (January), 114023, Contents, https://doi.org/10.1016/j.foodres.2024.114023.

Section 2.3.3 Isotope Ratio Monitoring GC-MS

Boato, G. (1960) Isotope Fractionation Processes in Nature. Summer Course on Nuclear Geology, Pisa, Italy.

Bradley, D. (2002) Tracking cocaine to its roots. *Today's Chem. Work*, **5**, 15–16.

Brand, A. (2004) in *Handbook of Stable Isotope Analytical Techniques*, Vol. 1 (ed. P.A. de Groot), Elsevier, Amsterdam, pp. 835–856.

Brand, W.A., Tegtmeyer, A.R., and Hilkert, A. (1994) Compound-specific isotope analysis: extending toward $^{15}N^{14}N$ and $^{18}O^{16}O$. *Org. Geochem.*, **21**, 585, https://doi.org/10.1016/0146-6380(94)90004-3.

Craig, H. (1957) Isotopic standards for carbon and oxygen and correction factors for mass-spectrometric analysis of carbon dioxide. *Geochim. Cosmochim. Acta*, **12**, 133–149, https://doi.org/10.1016/0016-7037(57)90024-8.

Ehleringer, J.R., Casale, J.F., Lott, M.J., and Ford, V.L. (2000) Tracing the geographical origin of cocaine. *Nature*, **408**, 311–312, https://doi.org/10.1038/35042680.

Ehleringer, J.R. and Cerling, T.E. (2002) in *The Earth System: Biological and Ecological Dimensions of Global Environmental Change*, Encyclopaedia of Global Environmental Change, Vol. 2 (eds H.A. Mooney and J.G. Canadell), John Wiley & Sons, Ltd, Chichester, pp. 544–550.

Fry, B. (2006) *Stable Isotope Ecology*, Springer, New York.

Groening, M. (2004) in *Handbook of Stable Isotope Analytical Techniques*, Vol. 1 (ed. P.A. de Groot), Elsevier, Amsterdam, pp. 875–906.

Heuer, K., Brand, W.A., Hilkert, A.W., Juchelka, D., Mosandl, A., and Podebred, F. (1998) *Z. Lebensm. Unters. Forsch.*, **206**, 230.

Hilkert, A., Douthitt, C.B., Schlüter, H.J., and Brand, W.A. (1999) Isotope ratio monitoring gas chromatography/mass spectrometry of D/H by high temperature conversion isotope ratio mass spectrometry. *Rapid Commun. Mass Spectrom.*, **13**,

1226–1230, https://doi.org/10.1002/(SICI)1097-0231(19990715)13:13<1226::AID-RCM575>3.0.CO;2-9.

Junk, G. and Svec, H.J. (1958) The absolute abundance of the nitrogen isotopes in the atmosphere and compressed gas from various sources. *Geochim. Cosmochim. Acta*, **14** (3), 234–243.

Lock, C.M. and Meier-Augenstein, W. (2008) Investigation of isotopic linkage between precursor and product in the synthesis of a high explosive. *Forensic Sci. Int.*, **179** (2–3), 157–162, https://doi.org/10.1016/j.forsciint.2008.05.015.

Matthews, D.E. and Hayes, J.M. (1978) Isotope-ratio-monitoring gas chromatography-mass spectrometry. *Anal. Chem.*, **50**, 1465, https://doi.org/10.1021/ac50033a022.

Matucha, M., Jockisch, W., Verner, P., and Anders, G. (1991) Isotope effect in gas-liquid chromatography of labelled compounds. *J. Chromatogr.*, **588**, 251–258, https://doi.org/10.1016/0021-9673(91)85030-J.

Meyer-Augenstein, W. (1997) The chromatographic side of isotope ratio mass spectrometry: pitfalls and answers. *LCGC Int.*, **10**, 17–25.

Rautenschlein, M., Habfast, K., and Brand, W. (1990) in *Stable Isotopes in Paediatric, Nutritional and Metabolic Research* (eds T.E. Chapman, R. Berger, D.J. Reijngoud, and A. Okken), Intercept Ltd, Andover, pp. 133–148.

Rosman, K.J.R. and Taylor, P.D.P. (1998) Isotopic compositions of the elements 1997. *Pure Appl. Chem.*, **70** (1), 217–235.

Rossmann, A. (2001) Determination of stable isotope ratios in food analysis. *Food Rev. Int.*, **17** (3), 347–381.

Sano, M. et al. (1976) A new technique for the detection of metabolites labelled by the isotope 13C using mass fragmentography. *Biomed. Mass Spectrom.*, **3**, 1–3.

Schulz, J. (1987) *Nachweis und Quantifizieren von PCB mit dem Massenselektiven Detektor*. LaborPraxis, **6**, 648–667.

Sharp, Z. (2007) *Principles of Stable Isotope Geochemistry*, Pearson Prentice Hall, Upper Saddle River, NJ, https://doi.org/10.25844/h9q1-0p82.

St. Jean, G. (2002) Basic principles of stable isotope geochemistry. Manuscript of the 9th Canadian CF-IRMS Workshop, Short Course, August 25, 2002.

de Vries, J.J. (2000) Natural abundance of the stable isotopes of C, O and H, in *Environmental Isotopes in the Hydrological Cycle*, Vol. 1: Introduction - Theory, Methods, Review (ed. W.G. Mook), IAEA, Vienna, www.iaea.org (accessed 2 May 2024).

Section 2.3.4 Acquisition Techniques in GC-MS

Busch, K.L., Glish, G.L., and McLuckey, S.A. (1988) *Mass Spectrometry/Mass Spectrometry: Techniques and Applications of Tandem Mass Spectrometry*, VCH Publishers, New York.

Cooks, R.G. and Busch, K.L. (1982) Counting molecules by desorption ionization and mass spectrometry/mass spectrometry. *J. Chem. Educ.*, **59** (11), 926–933.

Thier, H.P. and Kirchhoff, J. (eds) (1992) *DFG Deutsche Forschungsgemeinschaft, Manual of Pesticide Residue Analysis*, Vol. II, Wiley-VCH Verlag GmbH, Weinheim, pp. 25–28.

Johnson, J.V. and Yost, R.A. (1985) Tandem mass spectrometry for trace analysis. *Anal. Chem.*, **57**, 758A, *https://doi.org/10.1021/ac00284a001*.

Kaiser, H. (1978) Foundations for the critical discussion of analytical methods. *Spectrochim. Acta Part B*, **33 b**, 551.

McLafferty, F.E. (ed.) (1983) *Tandem Mass Spectrometry*, John Wiley & Sons, Inc., New York.

Noble, D. (1995) MS/MS flexes its muscles. *Anal. Chem.*, **67**, 265A–269A.

Paul, W. (1991) A Cage for Atomic Particles - A Basis for Precision Measurements in Navigation, Geophysics and Chemistry. *Frankfurter Allgemeine Zeitung*, Wednesday 15th December 1993 (291) N4.

Sphon, J.A. (1978) Use of mass spectrometry for confirmation of animal drug residues. *J. Assoc. Off. Anal. Chem.*, **61** (5), 1247–1252, *https://doi.org/https://doi.org/10.1093/jaoac/61.5.1247*.

Thermo Electron (1999) Perfluorotributylamine (PFTBA, FC43) Reference Table. Data Sheet PS30040_E, Thermo Electron Corporation.

U.S. EPA (1994) *Method 1613 - Tetra-Through Octa-Chlorinated Dioxins and Furans by Isotope Dilution HRGC-HRMS–Revision B*, United States Environmental Protection Agency, Washington, DC, USA.

Yost, R.A. and Enke, C.G. (1978) Selected ion fragmentation with a tandem quadrupole mass spectrometer. *J. Am. Chem. Soc.*, **100** (7), 2274. doi: 10.1021/ja00475a072.

Yost, R.A. (1983) MS/MS: tandem mass spectrometry. *Spectra*, **9** (4), 3–6.

Section 2.3.5 Mass Calibration

Audi, G. and Wapstra, A.H. (1995) The 1995 update to the atomic mass evaluation. *Nucl. Phys. A*, **595**, 409–480.

Eichelberger, J.W., Harris, L.E., and Budde, W.L. (1975) DFTPP tuning. *Anal. Chem.*, **47**, 995.

Hübschmann, H.-J. (2015) *Handbook of GC-MS: Fundamentals and Applications*, 3rd edn, Wiley-VCH Verlag GmbH & Co. KGaA, Weinheim, Germany, ISBN: 978-3-527-33474-2.

Mohr, P.J. and Taylor, B.N. (1999) CODATA recommended values of the fundamental physical constants: 1998. *J. Phys. Chem. Ref. Data*, **28** (6), 1713–1852.

Webb, K., Bristow, T., Sargent, M., and Stein, B. (2004) *Methodology for Accurate Mass Measurement of Small Molecules*. Best Practice Guide, LGC Limited, Teddington, UK.

Section 2.3.6 Vacuum Systems

Freeman, J.A. (1985) How to Select High Vacuum Pumps, Microelectronic Manufacturing and Testing, October, 1985 (Balzers reprint 1985).

Pfeiffer Vacuum (2003) Working with Turbopumps. Technical Report PT 0053 PE.

Pfeiffer Vacuum (2024) Turbopumps. Web presentation, *https://www.pfeiffer-vacuum.com* (accessed 5 May 2024).

3
Evaluation of GC-MS Analyses

3.1 Display of Chromatograms

Chromatograms obtained by gas chromatography–mass spectrometry (GC-MS) are plots of the signal intensity against the retention time, as with classical GC detectors. Nevertheless, there are considerable differences between the two types of chromatograms arising from the fact that data from GC-MS analyses are in three dimensions. Figure 3.1 shows a section of the chromatogram of the total ion current (TIC) in the analysis of volatile halogenated hydrocarbons. The retention time axis also shows the number of continually registered mass spectra (scan number). The mass axis is drawn above the time axis at an angle. The elution of each individual substance can be detected by evaluating the mass spectra using a "maximizing masses peak finder" program and can be shown by a marker. Each substance-specific ion shows a local maximum at these positions, which can be determined by the peak finder. The mass spectra of all the analytes detected are shown in a three-dimensional representation for the purposes of screening. For further evaluation, the spectra can be examined individually.

3.1.1 Total Ion Current Chromatograms

The intensity axis in GC-MS analysis is shown as a TIC chromatogram or as a calculated ion chromatogram (extracted/reconstructed ion chromatogram, XIC/RIC). The intensity scale may be given in absolute values, but a percentage scale is more frequently used. Both terms describe the mode of representation characteristic of the recording technique. At constant scan rates, the mass spectrometer plots spectra over the preselected mass range and thus gives a three-dimensional data field arising from the retention time, mass scale, and intensity. A signal parameter equivalent to a flame ionization detector (FID) is not directly available. Magnetic sector mass spectrometer were initially equipped with a TIC detector directly at the ion source until the end of the 1970s! A total signal intensity comparable to the FID signal at a particular point in the scan can, however, be calculated from the sum of the intensities of all the ions at this point. All the ion intensities of a mass spectrum are added together by the chromatography data system (CDS) and stored as a total intensity value (TIC)

Handbook of GC-MS: Fundamentals and Applications, Fourth Edition. Hans-Joachim Hübschmann.
© 2025 WILEY-VCH GmbH. Published 2025 by WILEY-VCH GmbH.

414 | *3 Evaluation of GC-MS Analyses*

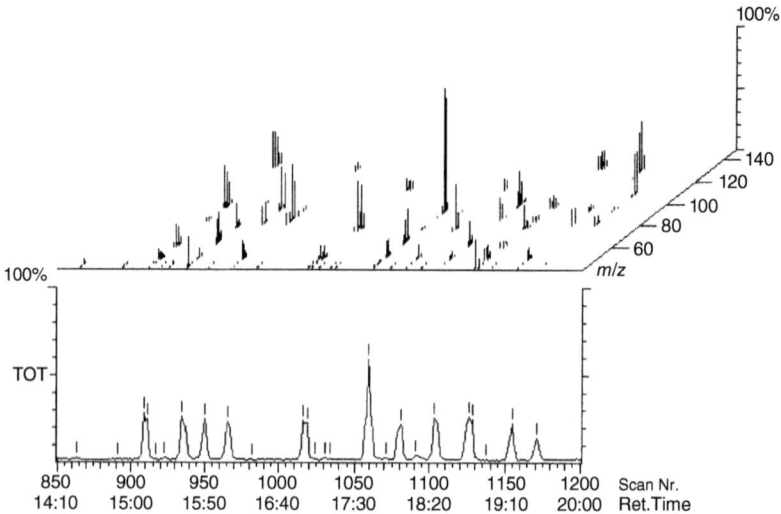

Figure 3.1 Three-dimensional data field of a GC-MS analysis showing retention time, signal intensity, and mass axis.

together with the spectrum. The TIC chromatogram, hence called 'reconstructed ion chromatogram' (RIC), thus constructed is therefore dependent on the scan range used for data acquisition. When making comparisons, it is essential to take the data acquisition conditions into consideration (Barwick *et al.*, 2006).

Selected ion monitoring (SIM), multiple ion detection (MID), or selected reaction monitoring (SRM)/multiple reaction monitoring (MRM) analyses give chromatograms in the same way but no mass spectrum is retrievable. The TIC in these acquisition modes is composed of the intensities of the selected ions. The analysis runs that switch the detection masses at fixed retention times show jumps in the base line (see Figure 2.245).

The appearance of a GC-MS chromatogram (TIC/XIC/RIC) showing the peak intensities is therefore strongly dependent on the mass range chosen for display and that of the data acquisition. The repeated GC-MS analysis of one particular sample employing mass scans of different widths leads to peaks of different heights above the base line of the TIC. The starting mass of the scan has a significant effect here. The result of a low start mass of the scan is a more or less strong recording of unspecific background which manifests itself in a higher or lower base line of the TIC chromatogram. Peaks of the same concentration are therefore shown with different signal/noise (S/N) ratios in the TIC at different scan ranges. In spite of differing representations of the substance peaks, the detection limit of the GC-MS system naturally does not change. Particularly in trace analysis, the concentration of the analytes is usually of the same order of magnitude or even below that of the chemical noise (matrix) in spite of good sample processing so that the TIC cannot represent the elution of these analytes. Only the use of selective information from an extracted mass chromatogram (XIC) (see Section 3.1.2) brings the substance peak sought on screen for further evaluation.

In the case of data acquisition using selected individual masses in SIM/MID or SRM/MRM, only the intensity changes of the masses selected by the data acquisition parameter are shown. During data acquisition, only those signals (ion intensities) that correspond to the acquisition mass range selected by the user are recorded from the TIC. The greater part of the ion current generated by the ion source is therefore not detected using the SIM/MID or SRM/MRM techniques (see Section 2.3.4). Only substances that give signals of the masses selected for acquisition as fragment or molecular ions are shown as peaks. A mass spectrum for the purpose of checking identity is therefore not available. For confirmation, a mass spectrum should be acquired using an alternating full scan/SIM mode or the check of the RI in a subsequent analysis. An exception is the acquisition of the product ion spectrum in MS/MS. This product ion spectrum, formerly also called the "daughter" spectrum, contains the full fragmentation information and can be stored and used in library searches. The retention time and the relative intensities of two or three specific ion intensities are used as qualifying features. In trace analysis, unambiguous detection is not possible by just using one ion signal. Positive results of an SIM/MID analysis basically require additional confirmation by a mass spectrum. SRM/MRM analyses offer the recording of the MS/MS product ion spectrum for confirmation or require the monitoring of multiple transitions as additional qualifiers.

3.1.2 Mass Chromatograms

The three-dimensional data field of GC-MS analyses in the full scan mode does not only allow the determination of the total ion intensity at a point in the scan. The individual analytes can be displayed selectively by the intensities of selected ions (m/z values) from the TIC and plotted as an intensity/time trace (EIC or XIC, extracted ion chromatogram, also called "mass chromatogram"). A meaningful assessment of S/N ratios of certain substance peaks can only be carried out using mass chromatograms of substance-specific ions or the sum of them (fragment/molecular ions).

The evaluation of these mass chromatograms allows the exact determination of the detection limit (limit of detection [LOD]) using the S/N ratio of the substance-specific ion produced by a compound. With the SIM/MID mode, this ion would be detected exclusively, but a complete mass spectrum for confirmation would not be available. In the case of complex chromatograms of real samples, mass chromatograms offer the key to the isolation of co-eluting components so that they can be integrated perfectly and quantified. Using the MS/MS mode, often there is no baseline noise detected due to the structural selectivity of the collision-induced fragmentation. In such cases, but also with a regular baseline, the statistical determination of the LOD by the precision of repeated measurements, the statistical instrument detection limit (IDL) is the preferred choice (Wells et al., 2011).

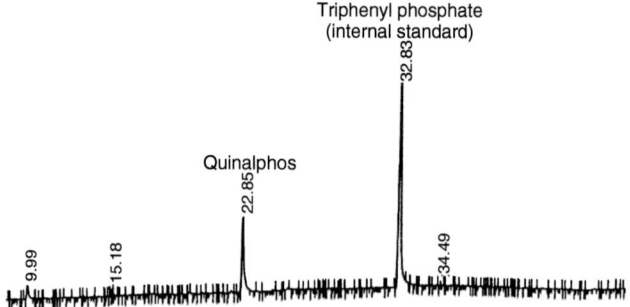

Figure 3.2 Analysis of a lemon extract using the NPD as detector. The chromatogram shows the elution of a pesticide as well as the internal standard.

As an example, in the analysis of lemons for pesticide residues, a co-elution situation was discovered by data acquisition in the full scan mode and was evaluated using mass chromatograms.

The routine testing with GC-MS in full scan mode gives a trace that differs from that using an element-specific NPD (nitrogen-phosphorous detector) (Figure 3.2). Many different peaks appear by MS detection in the retention region of quinalphos (Figure 3.3) while the NPD chromatogram shows a very clean chromatogram. The quinalphos peak has a shoulder on the left side and is closely followed by another less intense component. In the mass chromatogram of the characteristic individual masses (XIC of fragment ions), it can be deduced from the TIC that another eluting substance is present (Figure 3.4). Unlike NPD detection, with GC-MS analysis, it becomes clear after evaluating the mass chromatograms and mass spectra that the co-eluting substance is chlorfenvinphos.

In routine analysis, this evaluation is carried out for target compound analysis by the data system. If the information on the retention time of an analyte, the mass spectrum, the selective quantifying mass, and a valid calibration are supplied, a chromatogram can be evaluated in a very short time for a large number of components (Figure 3.5, see also Section 3.2).

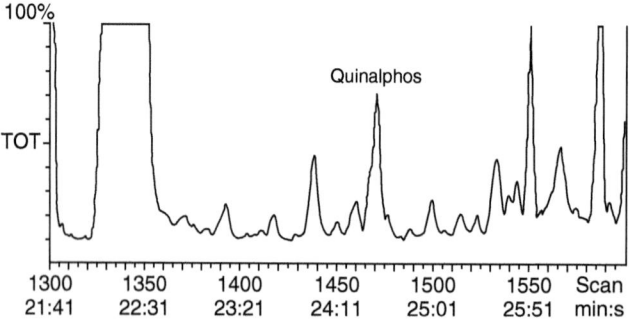

Figure 3.3 Pesticides analysis in a lemon extract by full scan GC-MS. The total ion current clearly shows the questionable peak with a shoulder. A co-eluting second substance gives rise to the left shoulder.

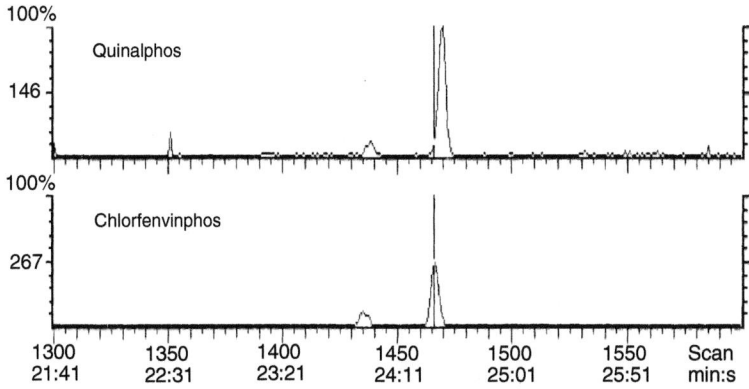

Figure 3.4 Mass chromatograms for the specific masses show the co-elution of quinalphos (m/z 146) and chlorfenvinphos (m/z 267). The retention time range is identical with that in Figure 3.3. The selective plot of the mass signals (XIC) confirms the coelution of both substances.

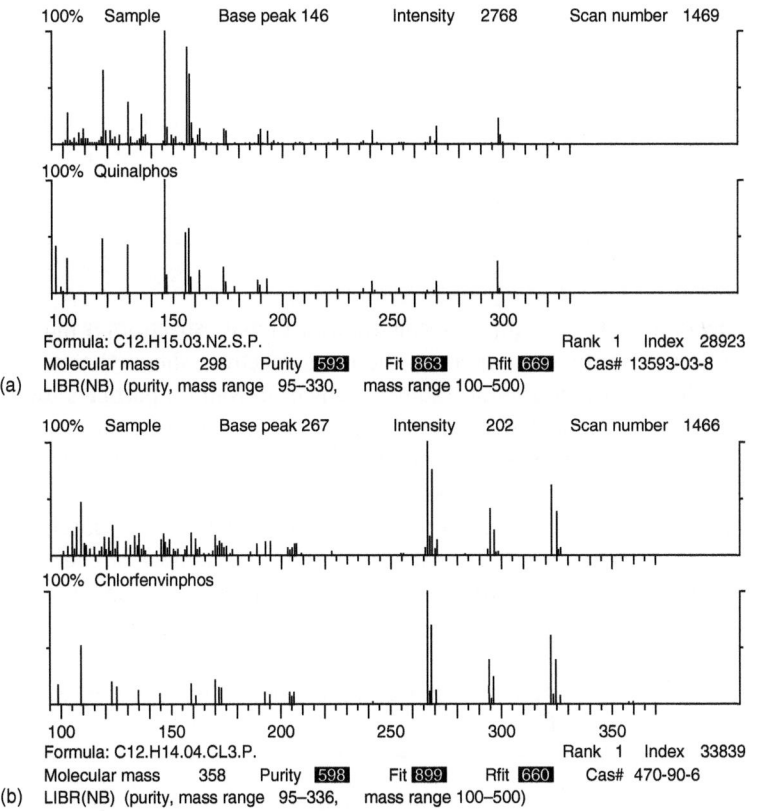

Figure 3.5 The phosphoric acid esters are identified by a library comparison after extraction of the spectra by a background subtraction. (a) Quinalphos is confirmed by comparison with the NBS library (FIT value 863). (b) Chlorfenvinphos is confirmed by a comparison with the NBS library (FIT value 899). (The former NBS is the predecessor of the current NIST mass spectral library).

3.2 Substance Identification

3.2.1 Reading Mass Spectra

First, what does a mass spectrum mean? In the graphical representation, the mass-to-charge ratio m/z of ions measured is plotted along the horizontal line. As ions with unit charge are generally involved in GC-MS with a few exceptions (e.g. PAHs), this axis is generally taken as the mass scale and gives the molar mass of an ion (see Figure 3.6). The intensity scale shows the frequency of occurrence of an ion under the chosen ionization conditions. The scale is usually given both in percentages relative to the base peak (100% intensity) or in measured intensity values (counts). Neutral particles lost in the fragmentation process or rearrangement of a molecular ion M^+ cannot be detected by the analyzer. The mass of such neutral particles is deduced from the difference between the fragment ions and the molecular ion (or precursor fragments) as illustrated in Figure 3.7.

In electron ionization (EI) at an ionization energy of 70 eV, the extent of the fragmentation reactions observed for most organic compounds is independent of the design of the ion source. For building up libraries of spectra, the comparability of the mass spectra produced is thus ensured. All commercially available libraries of mass spectra are acquired under these standard conditions and allow the fragmentation pattern of an unknown substance to be compared with the spectra available in the library (see Section 3.2.5).

Introduction

Learning how to identify a simple molecule from its EI mass spectrum is much easier than from other types of spectra. The mass spectrum shows the mass of the molecule and the masses of fragment ions of it. Neutral particles lost during fragmentation are coded by the difference to the molecular ion. Thus, the chemist does not have to learn anything new – the approach is similar to an arithmetic brain-teaser. Try one and see.

In the bar graph form of a spectrum as shown in Figure 3.6, the abscissa indicates the mass (actually m/z, the ratio of mass to the number of charges on the ions employed) and the ordinate indicates the relative intensity. If you need a hint, remember the atomic masses of hydrogen and oxygen are 1 and 16, respectively.

Prof. McLafferty, Interpretation of Mass Spectra (McLafferty and Turecek, 1993)

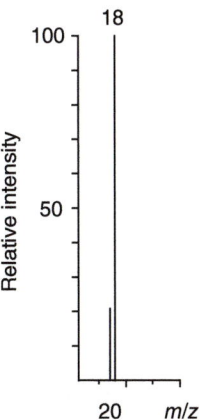

Figure 3.6 McLafferty's unknown spectrum.

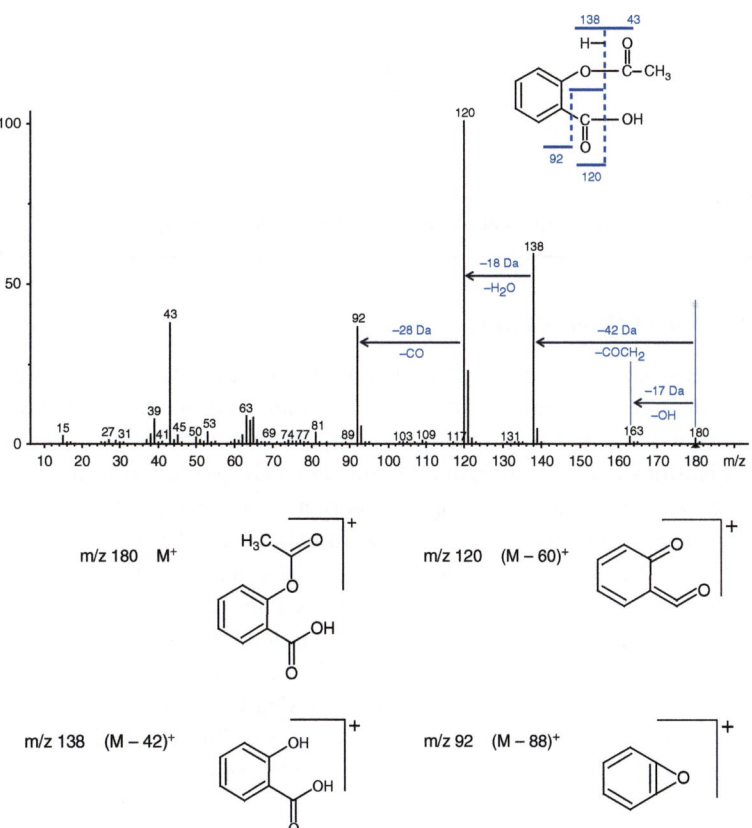

Figure 3.7 A typical line spectrum in organic mass spectrometry with a molecular ion, fragment ions and the loss of neutral particles (acetylsalicylic acid, $C_9H_8O_4$, M 180).

Types of Ions in Mass Spectrometry

- *Molecular ion:*
  The unfragmented positively or negatively charged ion M⁺ or M⁻ with a mass equal to the molecular mass and of a radical nature because of the unpaired electron.
- *Cationized/anionized molecular ion:*
  Ions associated with the molecular mass that are formed through CI, for example, as (M+H)⁺, (M−H)⁺, or (M−H)⁻ and are not radicals. The term quasi-molecular ion is deprecated.
- *Adduct ions:*
  Ions that are formed through the addition of charged species, for example, $(M + NH_4)^+$ through CI with ammonia as the CI gas.
- *Fragment ions:*
  Ions that are formed by the cleavage of one or more bonds.
- *Rearrangement products:*
  Ions that are formed following bond cleavage and migration of an atom (see McLafferty rearrangement, see 3.2.6.2).
- *Metastable ions:*
  Ions (m_1) that lose neutral species (m_2) during the flight time through the analyzer and are detected on a magnetic sector instrument with mass $m^* = (m_2)^2/m_1$, not detected in quadrupol or ion trap MS.
- *Base ion:*
  This ion has the highest intensity (100%) *aka* base peak in a mass spectrum.

3.2.2 Extraction of Mass Spectra

One of the great strengths of MS is the generation of direct information about an eluting component and not the measure of a derived indirect effect like absorption. The careful extraction of the substance-specific signals from the chromatogram is critical. For identification or confirmation of individual GC peaks, recording mass spectra that are as complete as possible is an important basic prerequisite.

By plotting mass chromatograms, co-elution situations can be discovered, as shown in Section 3.1.2. The mass chromatograms of selected ions give important information via their signal-maximizing behavior. Only when peak maxima of different ions (*m/z* values) show up at exactly the same retention time, it can be assumed that the fragments observed originate from a single substance, that is, from the same chemical structure. The only exception is the ideal simultaneous co-elution of compounds. If peak maxima with different retention times are shown by various ions, it must be assumed that there are co-eluting components (see Figures 3.8 and 3.9).

3.2.2.1 Manual Spectrum Subtraction

By subtraction of the background or the co-elution spectra before or behind a questionable GC peak, the mass spectrum of the substance sought is extracted from the

3.2 Substance Identification

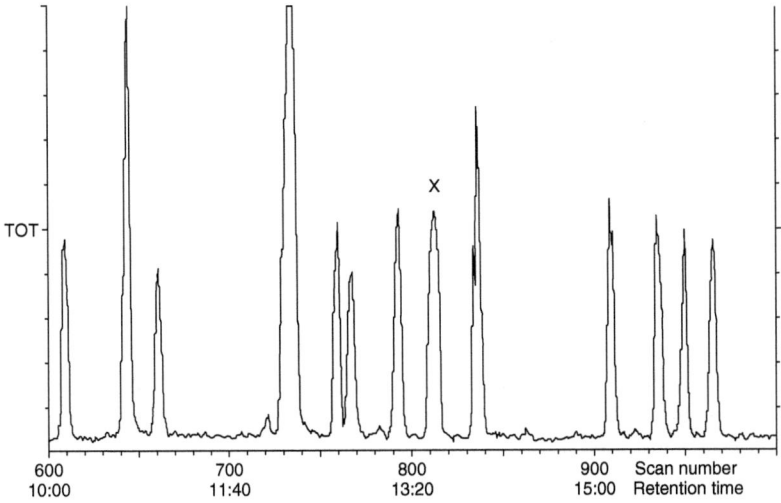

Figure 3.8 Chromatogram of an analysis of volatile halogenated hydrocarbons by purge and trap GC-MS. The component marked with X has a larger peak width than the neighboring compounds.

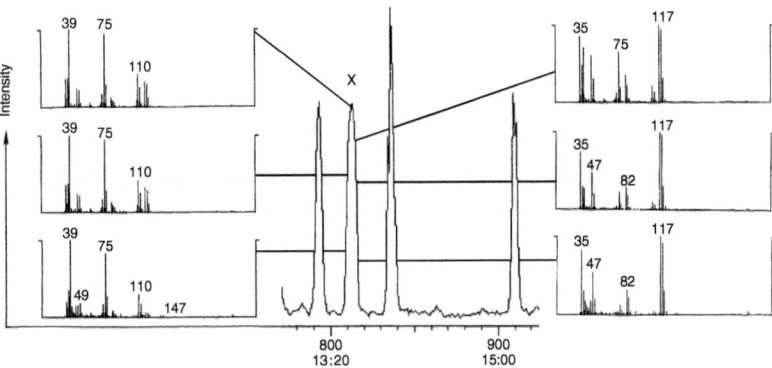

Figure 3.9 Continuous plotting of the mass spectra in the peak marked X in Figure 3.8.

chromatogram as free as possible from other signals. All substances coeluting with an unknown substance including the matrix components and column bleed are described in this context as chemical background. The differentiation between the substance signals and the background and its elimination from the substance spectrum is of particular importance for successful spectroscopic comparison in a library search (Mallard and Reed, 2012). In the example of the GC-MS analysis of lemons for pesticides described earlier, this procedure is used to determine the identity of the active substances.

The possibilities for the subtraction of mass spectra are shown in the following real example of the analysis of volatile halogenated hydrocarbons by purge and trap-GC-MS. Figure 3.8 shows part of a TIC chromatogram. The peak marked with X shows a larger width than that of the neighboring components. On closer inspection of the individual spectra in the peak, it can be seen that in the rising slope of the peak ions with m/z 39, 75, 110, and 112 dominate. As the elution of the peak continues, other ion signals appear. The ions with m/z 35, 37, 82, 84, 117, 119, and 121 appear in increased strength, while the previously dominant signals decrease. Figure 3.9 shows this situation using the continuing presentation of individual mass spectra in this GC peak.

From the individual mass spectra in Figure 3.9, it can be recognized that some signals obviously belong together. Figure 3.10 shows the mass chromatograms of the ions m/z 110/112 and m/z 117/119 (as a sum in each case) above the TIC from the detected mass range of m/z 33–260. The mass chromatograms show an intense GC peak at the questionable retention time in each case. The peak maxima are not superimposed and are slightly shifted toward each other. This is an important indication of the co-elution of two components (Figure 3.11).

If other ions are included in this first mass analysis, it can be concluded that the fragments belong together from their common maximizing masses behavior (Biller and Biemann, 1974).

After the individual mass signals have been assigned to the two components, the extraction of the spectrum for each compound can be performed. Figure 3.12 shows the division of the peak into the front peak slope A and the back peak slope B. With the manual background subtraction function contained in all data systems, the spectra in the shaded areas A and B are added and subtracted from one another.

The subtraction of the areas A and B results in clean spectra of the co-eluting analytes. In Figure 3.13 the subtraction A − B shows the spectrum of the first component, which is shown to be 1,3-dichloropropene (Figure 3.14) from a library comparison. The reverse procedure, that is, the subtraction B − A, gives the identity of the second component as tetrachloromethane (Figures 3.15 and 3.16).

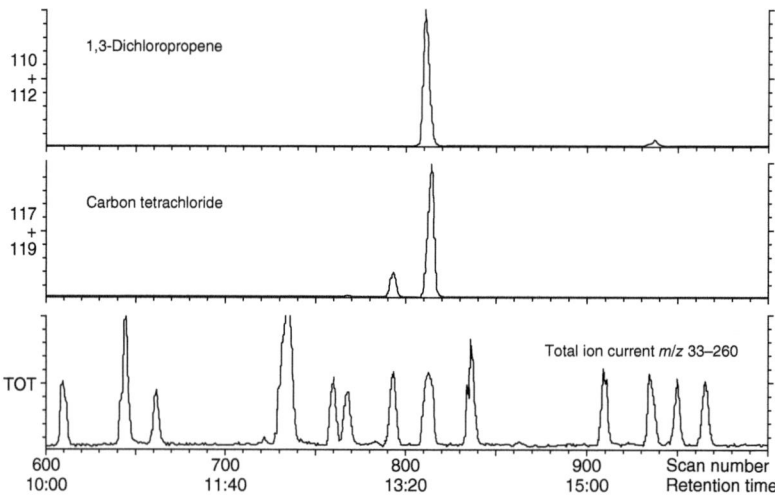

Figure 3.10 Mass chromatograms for m/z 110/112 and m/z 117/119 shown above a total ion current chromatogram.

3.2 Substance Identification | 423

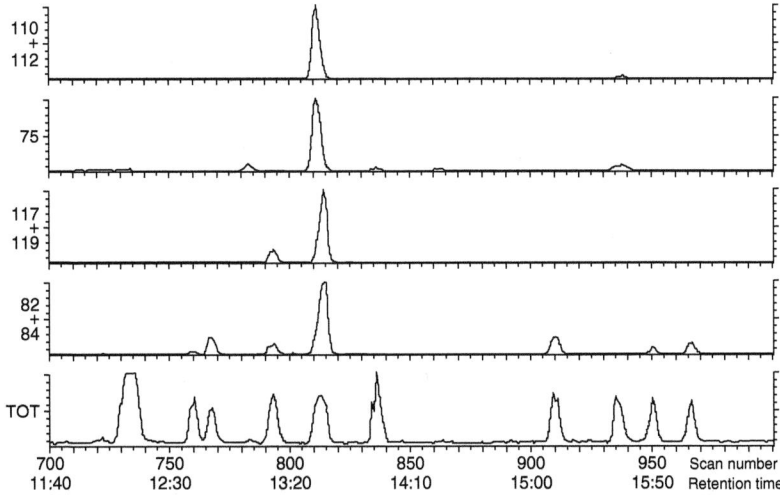

Figure 3.11 Analysis of a co-elution situation by inclusion of additional fragment ions.

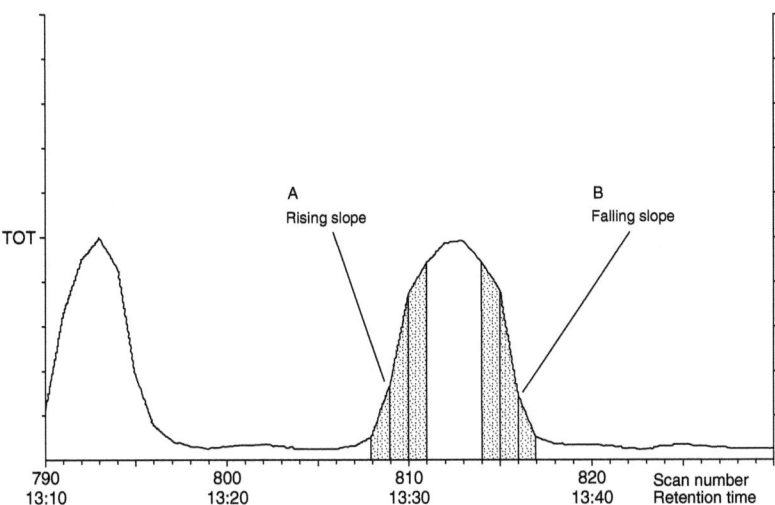

Figure 3.12 Plot of a peak with selected areas for spectra subtraction.

Further frequent use of spectrum subtraction allows the removal of background signals caused by the matrix or column bleed. Figure 3.17 shows the elution of a minor component from the analysis of volatile halogenated hydrocarbons, which elutes in the region where column bleed begins. For background subtraction, the spectra in the peak and from the region of increasing column bleed are subtracted from one another.

The result of the background subtraction is shown in Figure 3.18. While the spectrum clearly shows column bleed with m/z 73, 207 and a weak CO_2 background at m/z 44, the resulting substance spectrum is free from the signals

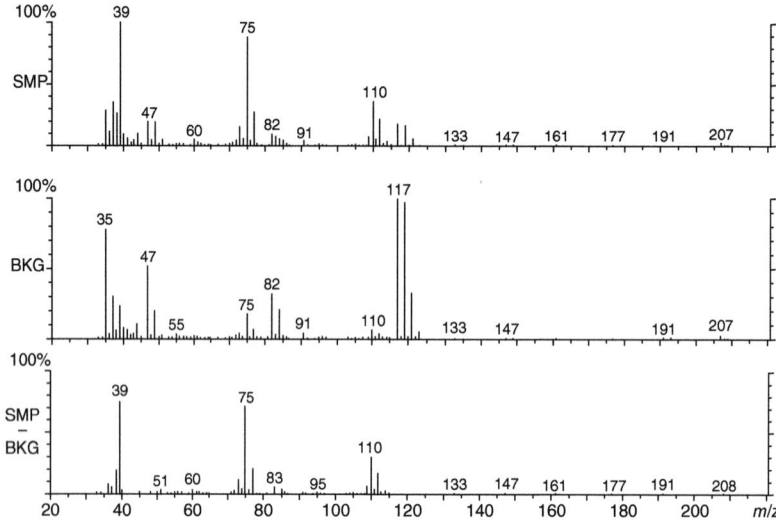

Figure 3.13 Spectral subtraction of the areas A – B in Figure 3.12: SMP = spectra of the rising peak slope (A) (sample), BKG = spectra of the falling peak slope (B) (background), SMP – BKG = resulting spectrum of the component eluting first.

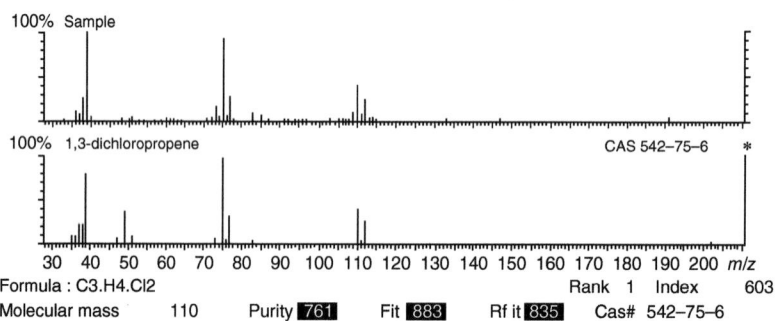

Figure 3.14 Identification of the first component by library comparison.

of the interfering chemical background after the subtraction. This clean spectrum can then be used for a library search in which it can be confirmed as 1,2-dibromo-3-chloropropane.

In the subtraction of mass spectra, it should generally be noted that in certain cases, primary substance signals can also be reduced. In these cases, it is necessary to choose another background area. If changes in the substance spectrum cannot be prevented in this way, the library search should be carried out with a small proportion of chemical noise. In some of the library search programs, there is also the possibility of editing the spectrum before the start of the search. In critical cases, this option should also be followed to remove known interfering signals resulting from the chemical noise from the substance spectrum.

3.2 Substance Identification | 425

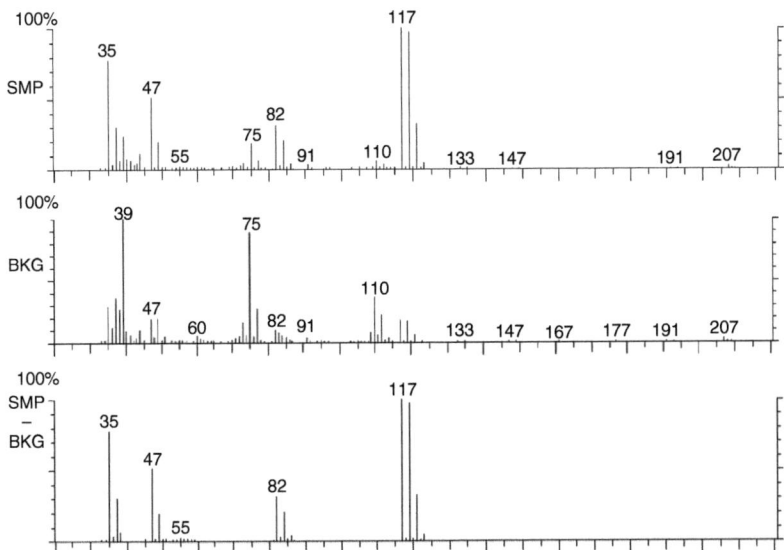

Figure 3.15 Spectral subtraction of the areas B – A in Figure 3.12: SMP = spectra of the falling peak slope (B) (sample), BKG = spectra of the rising peak slope (A) (background), SMP – BKG = resulting spectrum of the component eluting second.

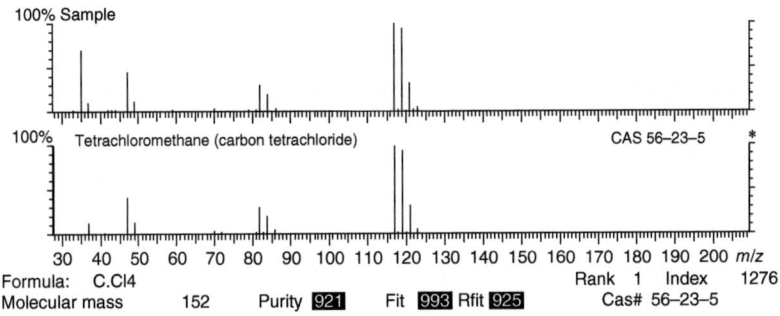

Figure 3.16 Identification of the second component by library comparison.

3.2.2.2 Deconvolution of Mass Spectra

The advancements in full scan sensitivity and its use, especially Orbitrap and time-of-flight (TOF) MS instrumentation as well as the increased application of fast and two-dimensional GC methods, created a strong demand for post-acquisition deconvolution methods (Dimandja, 2004; Shao et al., 2004). A deconvolution example of a complex elution situation is shown in Figure 3.19. The extraction of pure spectra from compounds co-eluting with other analytes or interferences with conventional background subtraction methods of GC-MS data systems as discussed in Section 3.2.2 is of limited use and can hardly recognize the transient dependence of ion intensities of multiple compounds eluting close together. Many professional programs for peak deconvolution and enhanced

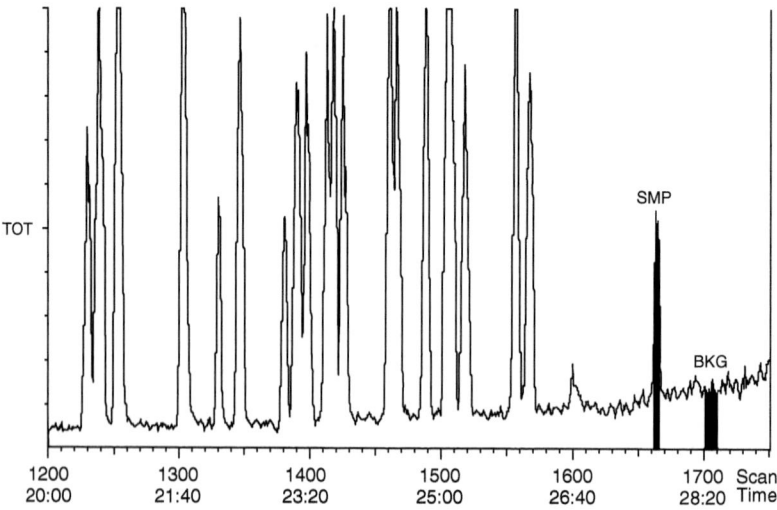

Figure 3.17 The total ion current chromatogram of an analysis of volatile halogenated hydrocarbons shows the elution of a minor component at the beginning of column bleed (the areas of background subtraction are shown in black).

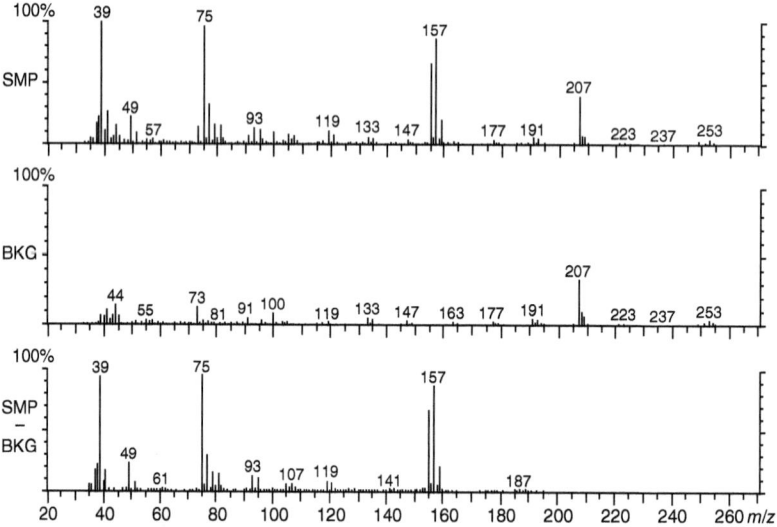

Figure 3.18 Result of background subtraction from Figure 3.17: SMP = spectra from the substance peak (sample), BKG = spectra from the background (column bleed), SMP − BKG = resulting substance spectrum.

spectrum extraction are available integrated or as add-on to chromatography data systems.

AMDIS is a widely used free program for automated mass spectra deconvolution and identification. AMDIS was developed at the National Institute of Standards and Technology (NIST) with the support of the Special Weapons Agency of the

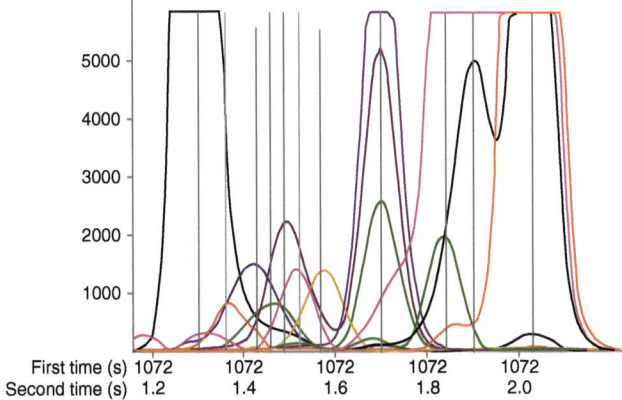

Figure 3.19 Deconvolution of overlapping peaks in GC × GC/TOF-MS. Every vertical line indicates the peak of an identified component. (Dimandja, 2004; reprinted with permission from Analytical Chemistry. Copyright 2004 by American Chemical Society.)

Department of Defense, for the critical task of verifying the Chemical Weapons Convention ratified by the United States Senate in 1997. In order to meet the requirements for this purpose, AMDIS was tested against more than 30 000 GC-MS data files accumulated by the U.S. EPA Contract Laboratory Program without a single false positive for the target set of known chemical warfare agents. After two years of development and extensive testing, it has been made available to the general analytical chemistry community for download from the Internet.

The AMDIS program analyses the individual ion signals and extracts and identifies the spectrum of each component in a mixture analyzed by GC-MS (see as well Section 4.16). The software comprises an integrated set of procedures for first extracting the pure component spectra from the chromatogram and then identifying the compound by a reference library (Figures 3.20–3.25).

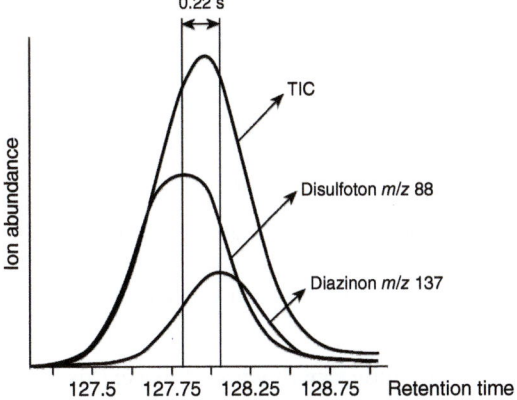

Figure 3.20 Deconvolution example illustrating coeluting disulfoton and diazinon.

428 | 3 Evaluation of GC-MS Analyses

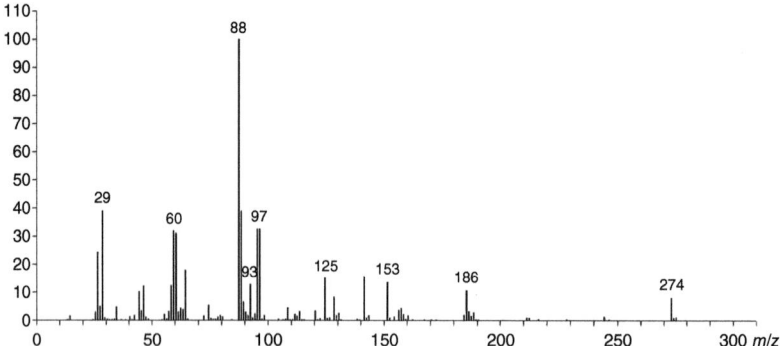

Figure 3.21 Disulfoton, spectrum of the pure compound, NIST# 52689.

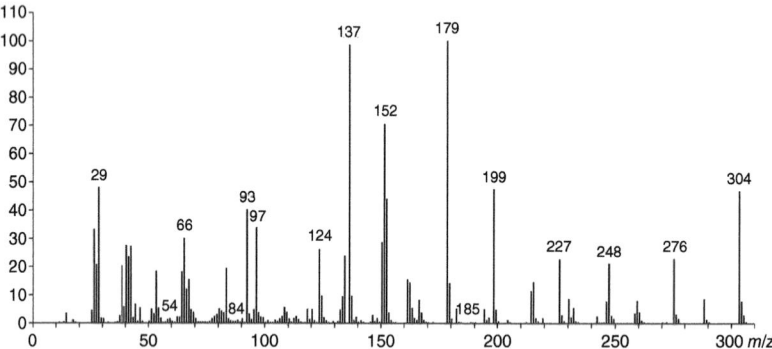

Figure 3.22 Diazinone, spectrum of the pure compound, NIST# 147231.

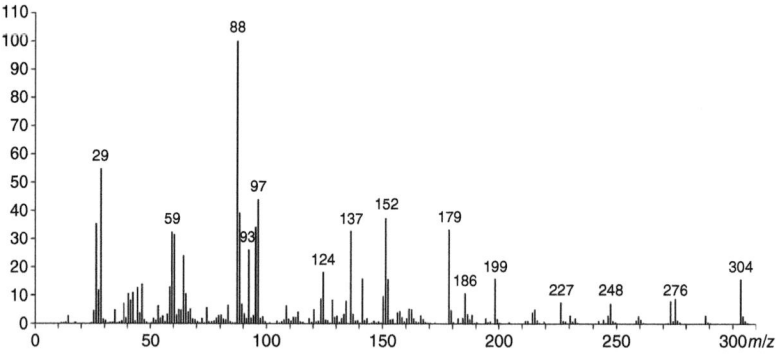

Figure 3.23 Co-elution spectrum 3 : 1 (at the peak maximum of disulfoton in Figure 3.20).

The overall deconvolution process involves four sequential steps for spectrum refinement and identification:

1) Noise analysis by a complete analysis of noise signals with the use of this information for component perception. A correction for baseline drift is done for each component in case the chromatogram does not have a flat baseline.

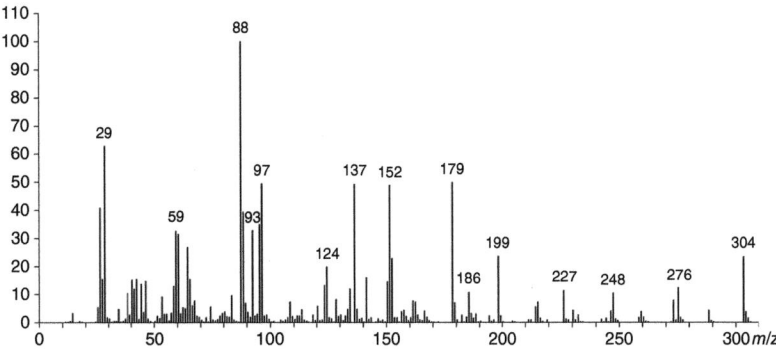

Figure 3.24 Co-elution spectrum 2 : 1 (at the peak maximum of diazinone in Figure 3.20).

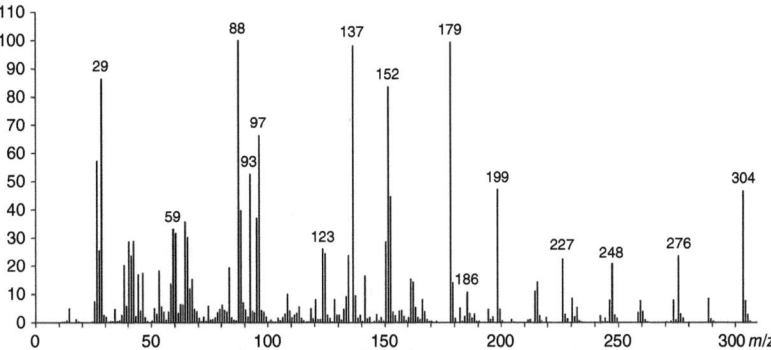

Figure 3.25 Co-elution spectrum 1 : 1 (at the right peak side in Figure 3.20).

2) Component perception identifies the location of each of the eluted components on the retention time scale by investigating the maximizing masses elution peak profile.
3) True spectral "deconvolution" of the data. Even if there is no available constant background for subtraction, AMDIS extracts clean spectra. The extraction of closely co-eluting components is possible even for analytes that peak within a single scan of each other in a wide range of each component's concentration.
4) Library search for compound identification to match each deconvoluted spectrum to a reference library spectrum.

AMDIS includes uncertainties in the deconvolution, purity, and retention indices in the match factor. The final match factor is a measure of both the quality of the match and of the confidence in the identification.

AMDIS can operate as a "black box" chemical identifier, displaying all identifications that meet a user-selectable degree of confidence (also refer to application 4.16). Identification can be aided by internal standards and retention times. Also employed can be retention index (RI) data when identifying target compounds and internal and external standards are maintained in separate libraries. AMDIS reads GC-MS raw data files in the formats of the leading GC-MS manufacturer or is already integrated in instrument data systems.

With its unique deconvolution algorithms, AMDIS has proven its capabilities for the efficient removal of overlapping interferences in many GC-MS applications (Mallard and Reed, 1997). The deconvolution process is independent of the type of mass analyzer and scan rate used to resolve overlapping peaks for substance identification as well as multi-component residue analysis (Dimandja, 2004; Zhang et al., 2006). Without time-consuming manual data evaluation, AMDIS provides sensitive compound information even with complex background present (Halket et al., 1999).

AMDIS

AMDIS has been designed to reconstruct "pure component" spectra from complex GC-MS chromatograms even when components are present at trace levels. For this purpose, observed chromatographic behavior is used along with a range of noise-reduction methods. AMDIS works as well with specialized libraries (environmental, flavor and fragrance, and drugs and toxins) which can be linked to the search with the NIST Library. AMDIS has a range of additional features, including the ability to search the entire NIST Library with any of the spectra extracted from the original data file. It can also employ RI information when identifying target compounds and can make use of internal and external standards maintained in separate libraries. Also MS/MS product ion spectra search features and a mass spectra interpreter tool are available. A history list of selected performance standards is also maintained.

As of version 2.73 (April 25, 2017), AMDIS reads data files in the following formats. The NetCDF format is supported by many manufacturers as an export file format.

Agilent Files	mzXML/mzData Files
Agilent MS Engine Files	NetCDF Files
Bruker Files	PerkinElmer Files
Finnigan GCQ Files	Shimadzu MS Files
Finnigan INCOS Files	Schrader/GCMate Files
Finnigan ITDS Files	Varian MS Files
INFICON Files	Varian SMS Files
JEOL/Schrader Files	Varian XMS Files
Kratos Mach3 Files	Viking Files
MassLynx NT Files	Xcalibur Raw Files
Micromass Files	

3.2.3 The Retention Index

If the chromatographic conditions are kept constant, the retention times of the compounds remain the same. All identification concepts using classical detectors function on this basis. The retention times of compounds, however, can change through aging of the column and more particularly through differing matrix effects between samples.

The measurement of the retention times relative to a co-injected standard can help to overcome these influences. Fixed retention indices (RI) are assigned to these standards. An analyte is bracketed in an RI system with the RI values of standards eluting before and after it. It is assumed small variances in the retention times affect both the analyte and the standards so that the RI values calculated remain constant (Deutsche Forschungsgemeinschaft, 1982; Onuska and Karasek, 1984).

The first RI system to become widely used was developed by Kovats in 1958 (Kovats, 1958; Zenkevich, 2002). In this system, a series of n-alkanes is used as the standard. Each n-alkane is assigned the value of the number of carbon atoms multiplied by 100 as the RI (pentane 500, hexane 600, heptane 700, etc.). For isothermal operations, the RI values for other substances are calculated as follows:

$$\text{Kovats index RI} = 100 \cdot c + 100 \frac{\log (t'_R)_x - \log (t'_R)_c}{\log (t'_R)_{c+1} - \log (t'_R)_c} \quad (3.1)$$

The t'_R values give the retention times of the standards and the substance corrected for the dead time t_0 ($t'_R = t_R - t_0$). As the dead time is constant in the cases considered, uncorrected retention times are mostly used. The determination of the Kovats indices (Figure 3.26) can be carried out very precisely, and on comparison, the Kovats indices are reproducible within ±10 RI units between different laboratories. In libraries of mass spectra, the RIs are also given, for instance, in the NIST libraries, the terpene library by Adams, or the toxicology library by Pfleger/Maurer/Weber (see Section 3.2.4).

Figure 3.26 Determination of the Kovats index for a substance X by interpolation between two n-alkanes (Schomburg, 1987).

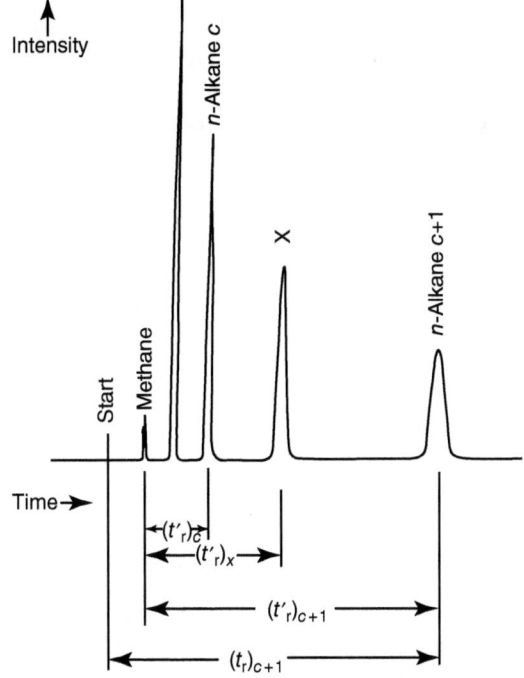

Figure 3.27 n-Alkylbis(trifluoromethyl)phosphine sulfides (M series with n = 6, 8, 10, ≤20).

```
           FPD
    SCD     ↓    PID
      ↘     S    ↙
            ‖
        CF₃-P-(CH₂)ₙ₋₁-CH₃
            |
            CF₃
      ↗           ↑
    ECD          FID
```

On working with linear temperature programs, a prediction and a generalization for use were introduced, whereby direct retention times are applied instead of the logarithmic terms used by Kovats (Van Den Dool and Kratz, 1963; Bemgard et al., 1994; Bianchi et al., 2007):

$$\text{Modified Kovats index RI} = 100 \cdot c + 100 \frac{(t'_R)_x - (t'_R)_c}{(t'_R)_{c+1} - (t'_R)_c} \quad (3.2)$$

A weakness of RI systems using alkanes lies in the fact that not all analytes are affected by variances in the measuring system to the same extent and that also selective detectors like ECD or NPD are used. For these special purposes, a homologous series of substances which are closely related have been developed (Hall et al., 1986; Lipinski and Stan, 1989; Kostiainen and Nokelainen, 1990). For use in trace analysis in environmental chemistry, particularly for the analysis of pesticides and chemical weapons, the homologous M-series (Figure 3.27) of n-alkylbis(trifluoromethyl)phosphine sulfides were developed (Manninen et al., 1987).

The molecule of the M-series contains several heteroatoms that respond to the selective detectors ECD, NPD, FPD, SCD, AED, and PID and also give a good response in FID and MS. The typical detection limits are with ECD about 1 pg and FID about 300 pg (Figure 3.28). In the mass spectrometer, all components of the M-series show intense characteristic ions at M-69 and M-101 and a typical fragment

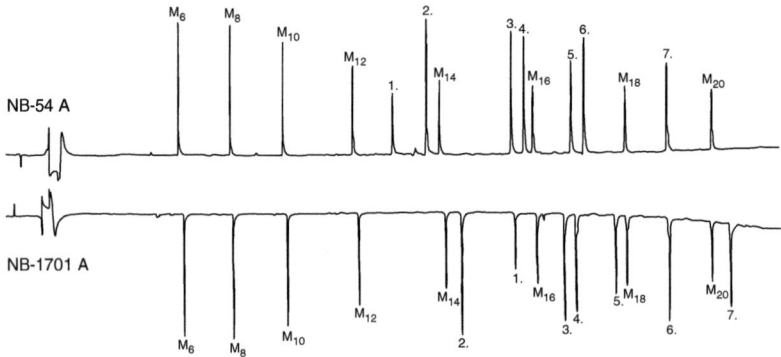

Figure 3.28 Chromatograms of the M-series and of pesticides (phosphoric acid esters) on columns of different polarities (HNU/Nordion). Carrier gas He, detector NPD, program: 50°C (2 min), 150°C (20°C/min), 270°C (6°C/min). Components: M series M₆, M₈ to M₂₀, 1. dimethoate, 2. diazinon, 3. fenthion, 4. trichloronate, 5. bromophos-ethyl, 6. ditalimphos, and 7. carbophenothion.

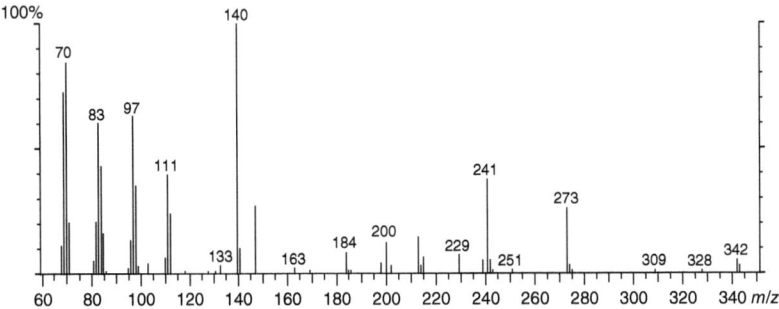

Figure 3.29 M-series: EI Mass Spectrum of the component M10 (HNU).

at m/z 147 (Figure 3.29). The M-series is versatile with positive (PCI) and negative chemical ionization (NCI) as well.

The use of RIs despite, or perhaps because of, the wide use of GC-MS systems is now becoming more important again because of the outstanding stability of fused silica capillaries and the good reproducibility of gas chromatographs. The chromatography data systems are increasingly prepared for the processing of RIs for compound identification. Main GC-MS applications including a RI evaluation are found in the food and flavor industry with polar and non-polar separations on two-column systems (see Section 4.16).

Retention indices can also be used with the NIST AMDIS software (see also Sections 3.2.2.2 and 3.2.5) (Mallard and Reed, 1997). More than 491 790 RI values for 180 610 compounds are listed in the latest 2023 version of the NIST library. RIs are used in the AMDIS library search in combination with MS to mainly provide the identification of isomers of similar spectral fragmentation, but show a difference in their retention behavior. If there is a chromatogram of a mix of hydrocarbons available using the same GC conditions, AMDIS interpolates between bracketing hydrocarbons to determine analyte's RI (Little, 2022). In addition, there are artificial intelligence (AI)-estimated RIs for all the compounds in the EI library (Sparkman, 2023).

If the RI of a compound is not known, it can be estimated empirical from considerations of the elements and partial structures present in the molecule (Weber and Weber, 1992; Katritzky et al., 1994) (Tables 3.1 and 3.2). A first approximation can already be made using the empirical formula of an analyte. This is particularly valuable for assessing suggestions from the GC-MS library search because, besides good correspondence to a spectrum, plausibility regarding the retention behavior can be tested. According to Weber, the values determined give a correct estimation, also using structural characteristics within 10% (Tables 3.2 and 3.3).

Polar groups with hydrogen bonding increase the boiling point of a compound and are thus responsible for stronger retention. For the second and every additional polar group, the RI increases by 150 units. Branches in the molecule increase the volatility. For each quaternary carbon atom present in a t-butyl group, the RI is reduced by 100 units. Values can be estimated with higher precision from RIs for known structures by calculating structure elements according to Tables 3.1 and 3.2.

Table 3.1 Contributions for the determination of the retention index from the empirical formula (Weber and Weber, 1992).

Element	Index contribution
H, F	0
C, N, O	100
Si in Si(CH$_3$)$_3$	0
P, S, Cl	200
Br	300
J	400

Table 3.2 Retention behavior of structural isomers (Weber and Weber, 1992).

Alkyl branches	Tertiary < secondary < n-alkyl
Disubstituted aromatics	Ortho < meta, para

3.2.4 Libraries of Mass Spectra

In EI (70 eV), a large number of fragmentation reactions take place with organic compounds. These are independent of the manufacturer's design of the ion source. The tuning of the ion source has a greater effect on the characteristics of a mass spectrum, which leads to a particularly wide adjustment range, especially in the case of quadrupole analyzers. The relative intensities of the higher and lower mass ranges can easily be reversed. In the early days of the use of quadrupole instruments, this possibility was highly criticized by those using the established magnetic sector instruments. The problem is resolved in so far as both the manual and the automatic tuning of the instruments are aimed at giving the intensities of a reference compound, in contrast to a manual SIM tuning which aims to high sensitivity within a specific mass range. Perfluorotributylamine (PFTBA, FC43) is used as the reference substance in all GC-MS systems. Other influences on the mass spectrum in GC-MS systems are caused by sharp changes in the substance concentration during a slow mass scan on beam instruments. On running spectra over a large mass range (e.g. in the case of methyl stearate with a scan of 50–350 units in 1 s), sharp fast GC peaks lead to a mismatch of intensities (skewing) between the front and back slopes of the peak. This effect can only be counteracted using faster scan rates, which, however, result in lower full scan sensitivity. In practice, standardized spectra must be used for these systems in order to calculate the compensation (for background subtraction see Section 3.2.2). Ion trap mass spectrometers do not show this skewing of intensities because there is parallel storage and detection of all the ions formed. A mass spectrum should therefore not be regarded as naturally constant, but as the result of a complex process.

In practice, such observed variations affect the relative intensities of groups of ions in the mass spectrum. The fragmentation processes themselves are not affected (the same fragments are found with all GC-MS instruments) nor are the

3.2 Substance Identification

Table 3.3 Comparison of calculated retention indices with empirically determined values (Weber and Weber, 1992).

Substance	Calculated RI	Determined RI
Atrazine	1500	1716
Parathion-ethyl	2000	1970
Triadimefon	2000	1979

Examples of retention index calculations

	Element	Number	Contribution	Total
Parathion-ethyl		$C_{10}H_{14}NO_5PS$		
	C	10	100	1000
	H	14	—	—
	N	1	100	100
	O	5	100	500
	P	1	200	200
	S	1	200	200
			Sum	2000
Atrazine		$C_8H_{14}ClN_5$		
	C	8	100	800
	H	14	—	—
	Cl	1	200	200
	N	5	100	500
			Sum	1500
Triadimefon		$C_{14}H_{14}ClN_3O_2$		
	C	14	100	1400
	H	14	—	—
	Cl	1	200	200
	N	3	100	300
	O	2	100	200
	tert-C	1	−100	−100
			Sum	2000

isotope ratios that result from natural distribution (except the small retention time differences during elution which is compensated for by summation of spectra as of Section 3.2.2).

The comparability of the mass spectra produced is thus ensured for building up libraries of mass spectra. All commercially available library spectra were acquired in EI under standard conditions of 70 eV ionization energy and allowed the comparison of the fragmentation pattern of an unknown substance with those available from the library. For the large universal libraries, it should be assumed that many of the spectra were initially not run with GC-MS systems. Still, many reference spectra are acquired using a solid probe inlet. For example, a former reference spectrum of Aroclor 1260 (a mixture of polychlorinated biphenyls [PCBs] with a 60% degree of chlorination) can only be explained in this way. Information on the inlet system used can be found in such library entries.

EI spectra are particularly informative because the seen fragmentation patterns reveal structural details. The same applies to product ion spectra from collision-induced fragmentation with MS/MS instruments. The introduction of substructure libraries (MS/MS product ion spectra) and accurate mass spectra libraries is ongoing (NIST). The commercially available libraries can be divided into general extensive collections and special task-related collections with a defined range of applications.

3.2.4.1 Universal Mass Spectral Libraries

NIST/EPA/NIH Mass Spectral Library The "Electron Ionization Library Component NIST/EPA/NIH Mass Spectral Library with the NIST GC Retention Index Database" is probably the most popular and most widely distributed library for GC-MS instruments. The 2023 edition has been largely expanded by the number of EI mass spectra and Kovats RIs for polar and non-polar columns (Babushok et al., 2007). Over the last decade, over 10 000 spectra have been added each year. In addition, the NIST Chemistry WebBook provides access to data compiled and distributed by NIST under the Standard Reference Data Program (NIST, 2023a). The NIST MS Search Program is integrated in many commercial instrumental GC-MS software suites.

Extensive spectra evaluation and quality control have been involved in the NIST database. Each spectrum was critically examined by experienced mass spectrometrists, and each chemical structure has been examined for correctness and consistency, using both human and computer methods (Ausloos et al., 1999). Spectra of stereoisomers have been intercompared, chemical names have been examined by experts, and IUPAC names provided. CAS registry numbers have been verified.

The NIST "Tandem Mass Spectral Library" of MS/MS spectra covers the product ion mass spectra of ions generated by LC-MS electrospray ionization and fragmented by collisional dissociation. The 2023 MS/MS mass spectra library contains 2 374 064 spectra of 399 267 precursor ions from 51 501 chemical compounds. Spectra have been acquired on ion trap, QTOF, magnetic sector, and triple quadrupole instruments. Spectra for the latter instrument classes have been acquired over a wide range of collision energies to ensure matching regardless of instrument settings.

Also, when available, high mass accuracy spectra are stored. Listed compounds include human and plant metabolites, pollutants, contaminants, drugs, and a wide range of other "societally important" compounds.

The contents of the NIST 2023 Mass Spectral Libraries are specified as follows (NIST, 2023b, 2023c; SIS, 2023):

Library description	Library contents
NIST/EPA/NIH mass spectral library (EI)	
Main EI MS library	394 054 spectra
Compounds	347 100 compounds
Replicates	46 954 spectra
Retention indices	346 757 Kovats RI values
Compounds	82 868 compounds
Tandem (ESI MS/MS) library	
MS/MS Library 2023	2 374 064 Spectra
Precursor ions	399 267 ions
Compounds	51 501 compounds

Wiley Registry of Mass Spectral Data The Wiley Registry of Mass Spectral Data has been published in its 2023 release (Wiley, 2023a). It is the largest and most comprehensive mass spectral library made commercially available in the most common MS software formats and compatible with most manufacturer data systems. The collection covers a wide range of applications such as environmental, forensics, toxicology, metabolomics, pharmaceutical, biotech, food/cosmetics, defense, homeland security, and many more.

The 2023 release of the Wiley Registry contains:

Library description	Library contents
Wiley Registry of Mass Spectral Data 2023	
GC-MS library	873 300 spectra
MS/MS spectra	841 100 compounds
Searchable structures	399 000 precursor ions
Unique compounds	741 000 compounds
Replicates	201 000 spectra
Retention indices	738 400 Kovats RI values
Wiley Registry/NIST 2023 combined library	
EI mass spectra	1 180 800 spectra
Unique compounds	950 200 compounds
Searchable structures	1 148 600 compounds
Chemical names, synonyms	>4 000 000 compounds

The Wiley database records include searchable details when available for a record such as:

- Chemical Structure
- Chemical Name
- Exact Mass
- Calculated Kovats RI values
- Splash IDs
- Wiley ID
- Formula
- InChI/InChIKey
- Molecular Mass
- Quality Index (QI)
- Digital Object Identifier (DOI)

Also available is the combination of the large Wiley Registry with the current NIST database with the Wiley Registry/NIST 2023 combined library. It provides a comprehensive coverage with over 3 million spectra including EI and MS-MS data, 1 148 600 searchable structures, 950 200 unique compounds, replicate spectra, and over 4 million chemical names and synonyms. A broad range of compounds is applied for targeted and non-targeted analyses. Also, it includes world patents from the United States, China, Japan, and Europe. The spectra contain searchable additional information such as physical properties, structures, and more to narrow results even further. The library is available in the most common manufacturer formats and includes the NIST 2023 Software MS Search, AMDIS, and MS Interpreter. The overlap to the NIST 2023 library is reported to be 130 000 spectra only (NIST, 2023b; Wiley, 2023b).

3.2.4.2 Application Libraries of Mass Spectra

Mass Spectra of Designer Drugs This mass spectrum collection edited by Peter Rösner covers the entire range of designer drugs with a most recent update in 2024 (Roesner et al., 2007). It is the first database featuring systematic structures in depth, carefully compiled by the mass spectral experts at the Regional Departments of Criminal Investigation in Kiel, Hamburg, and the Federal Criminal Laboratory in Wiesbaden, Germany, and other partners worldwide. This database in the 2024 edition includes 35 094 mass spectra of 26 712 unique chemical compounds like designer drugs and medicinal drugs with detailed information and chemical structures. Chemical warfare agents are added due to the recent interest in homeland security. Over 500 new, unique compounds have been added to classification groups such as fentanyls (3180 spectra), various opioids (371 spectra), synthetic cannabinoids (1997 spectra) and many more. Measured Kovats indices of 22 057 compounds are included.

All data have been taken from both legal and underground literature, providing the most comprehensive picture of these compounds available worldwide. Even highly potential hallucinogens like the Bromo-Dragon Fly are covered. Due to the evolving market of designer drugs, this collection is updated annually. The library is available in the most common instrumentation manufacturer formats (Wiley, 2024).

Mass Spectra of Volatiles in Food This mass spectral database is dedicated to the application areas of the food and flavor industries and was selected and quality controlled by the mass spectral experts at the Central Institute of Nutrition and Food Research in the Netherlands. The collection in its second edition includes 4182 reference mass spectra and structures with 3927 unique compounds. It covers the whole range of volatile compounds in food. Apart from the large number of natural, nature-identical and artificial flavors and aromas, there are, among others, food additives and solvents, pesticides, and veterinary pharmaceutical compounds, which are frequently found as residues. Derivatives of non-volatile compounds such as sugars or polyhydroxyphenols are also included. The database is available in its second edition (Central Institute of Nutrition and Food Research, 2003; Wiley-VCH ISBN 978-0-471-64825-3).

Flavors and Fragrances of Natural and Synthetic Compounds The library of Flavors and Fragrances of Natural and Synthetic Compounds (FFNSC), edited by Prof. Luigi Mondello, contains 3462 mass spectra, and searchable chemical structures in its third edition, linear RI data, measured Kovats RI on polar and non-polar columns of compounds of interest for the flavors and fragrances industry as well as research applications (Mondello, 2015).

The library is available in the data formats of all leading GC-MS manufacturers (Wiley, ISBN: 978-1-119-06984-3).

Adams Essential Oil Components Library The Adams Essential Oil mass spectral library is available in its fourth edition (Adams, 1989, 2007). This comprehensive collection of mass spectra and retention times of common components in plant essential oils covers 2205 compounds, each including:

RT: retention time on DB-5 capillary column
AI: arithmetic RI
KI: Kovat's RI
CAS#: chemical abstracts service number
MF: molecular formula
FW: formula mass
MSD LIB#: entry number in library
CN: chemical name

Also included in it is a list of synonyms and the source of compounds used for the spectrum. If the compound occurs in nature, two additional sources for the compound (concentration at % oil, plant name, literature reference) are included.

The former print version Allured Books, ISBN 1932633219, is out of print. Diablo Analytical offers the e-book for download https://diabloanalytical.com/ms-software/essentialoilcomponentsbygcms/. The NIST has licensed the spectra contained in the Adam's Library and has included them in the NIST 23 Mass Spectral Library.

Mass Spectral and GC Data of Drugs, Poisons, Pesticides, Pollutants and Their Metabolites
This specialized collection known as "Maurer Meyer Pfleger Weber library" (MMPW) is the reference in forensic analysis, occupational toxicology, food, and environmental analysis and contains data obtained from clinical samples over the course of more than 30 years. Prof. Karl Pfleger is the former and Prof. Hans H. Maurer is the current head of the Clinical Toxicology Laboratory at the clinical campus of Saarland University in Homburg, Germany. Together with Armin Weber, they have developed this unique and most comprehensive toxicological database (Pfleger et al., 2023).

The sixth edition of 2023 includes 10 948 mass spectra and GC retention indices from 10 055 unique compounds with 10 653 structures in 175 categories, including psychoactive substances, almost all relevant therapeutic drugs, and over 7800 of their metabolites. It covers substances from simple analgesics to designer drugs, and from pesticides and pollutants to chemical warfare agents, including metabolites to allow the identification of the mother substance. The database records include the following details when available for the record:

- Data of nearly all the new drugs relevant to clinical and forensic toxicology, doping control, food chemistry, and so on.
- Nearly complete coverage of trimethylsilylated, perfluoroacylated, perfluoroalkylated, and methylated compounds.
- Sections on sample preparation and GC-MS methods.
- Chemical name, formula, molecular mass, exact mass, and structure.
- InChI/InChI key.
- Classifications.

Mass Spectra of Physiologically Active Substances Including Drugs, Steroid Hormones, and Endocrine Disruptors This mass spectral database is focused on doping control, endocrinology, and clinical toxicology. It has been complied by the Institute of Biochemistry, German Sport University Cologne and covers 4182 mass spectra and structures from 3927 unique compounds of drugs, steroid hormones, including their trimethylsilyl-, O-methoxyoxime- and acetal derivatives, endocrine disruptors, and β-2-agonists compounds. The library includes chemical names, molecular formulas, and synonyms (Parr et al., 2011).

The database is available as part of the Wiley KnowItAll Mass Spectral Library subscription in multiple MS data system formats like ACD, Agilent Chemstation, NIST, Finnigan GCQ, SSQ, TSQ, ICIS, ITS40, Magnum, INCOS, PE Turbomass, Shimadzu QP-5000, Thermo Scientific Xcalibur, Thermo Galactic SpectralID, Varian Saturn, VG Labbase, Masslab, Waters Masslynx.

Mass Spectra of Pesticides The collection edited by Rolf Kühnle contains 1238 mass spectra of mass spectra of pesticides such as insecticides, acaroids, nematicides, fungicides, rodenticides, mollucicides, and repellents. The mass spectra are unreduced. Chemical structure, CAS #, synonym and systematic name, molecular mass, molecular formula, and experimental conditions complete the data record (Wiley-VCH ISBN 978-3-527-32488-0).

Available database formats: Agilent Chemstation; NIST MSSEARCH; Finnigan GCQ, SSQ, TSQ, ICIS, INCOS, Iontrap, ITS40, Magnum; INCOS; PE Turbomass; Shimadzu QP-5000; Thermo Galactic SpectralID; Varian Saturn; VG Labbase, Masslab; Waters Masslynx; Xcalibur (Kühnle, 2009).

SWGDRUG MS Library The scientific working group for the analysis of seized drugs (SWGDRUG) has compiled a mass spectral library containing drugs and drug-related compounds. Pure drug spectra, plus GC breakdown products and pure metabolite standards have been edited into this compilation of 3500 mass spectra. SWGDRUG is open to accepting contributions from the forensic community to further extend the library. The collection will be updated on a regular basis.

All spectra were run on Agilent quadrupole GC-MS instruments tuned against PFTBA (FC43). The current version of the full spectra library was last updated as the SWGDRUG MS Library Version 3.13 of June 30, 2023, available for download at www.swgdrug.org (SWGDRUG, 2023). The mass spectral database and the list of entries are available for download from the internet in the formats for NIST, JCAMP, and the data systems of Agilent, Shimadzu, and Thermo Fisher Scientific. The library is supported by the NIST MSSEARCH program.

The Fiehn Library The Fiehn Library "FiehnLib" is a mass spectral and RI library for comprehensive metabolic profiling. The library comprises over 1212 EI mass spectra and retention indices for quadrupole and TOF GC-MS for over 1000 primary metabolites below 550 Da, covering lipids, amino acids, fatty acids, amines, alcohols, sugars, amino-sugars, sugar alcohols, sugar acids, organic phosphates, hydroxyl acids, aromatics, purines, and sterols, included as methoximated and trimethylsilylated derivatives, that are currently screened by the Fiehn Laboratory (Kind et al., 2009). The FiehnLib libraries comprised 68% more compounds and twice as many spectra with higher spectral diversity than the public Golm Metabolite Database. The compound list is continually extended. In addition, there is the open source BinBase metabolomic database (https://sourceforge.net/projects/binbase/) that currently comprises 5598 unique mass spectra from 47 plant, animal, and microorganism species.

The Fiehn mass spectral library is commercially available for GC-quadrupole mass spectrometers from Agilent Technologies and for GC-TOF mass spectrometers from Leco Corporation. For details see https://fiehnlab.ucdavis.edu/projects/fiehnlib.

The Golm Metabolome Database The Golm Metabolome Database (GMD) provides public access to custom mass spectra libraries, metabolite profiling experiments, and other necessary information related to the field of metabolomics. The main goal is the representation of an exchange platform for experimental research activities and bioinformatics to develop and improve metabolomics by multidisciplinary cooperation. The GMD comprises mass spectra and retention time indices of pure reference substances, silylized derivatives, and frequently observed mass spectral tags (MST: mass spectrum linked to chromatographic retention) of yet unidentified metabolites (Hummel et al., 2013; Wikipedia, 2024).

GMD is maintained by the GMD Consortium, a joint venture of researchers from various research areas having different research interests. The consortium is based on an interdisciplinary cooperation to joining expertise from these research interests. For further details see http://gmd.mpimp-golm.mpg.de/.

The Lipid Library The Archives of Mass Spectra by W.W. Christie comprise in total c. 2232 mass spectra in archives of methyl esters, 3-pyridylcarbinol esters, 4,4′-dimethyloxazolines, pyrrolidides, alkyl esters, and other lipids (Christie and Han, 2010; Christie, 2021). They are made available on the web for study but without interpretation for the following compound groups:

- Methyl esters of fatty acids (FAMEs)
- Picolinyl esters
- DMOX (4,4-dimethyloxazoline) derivatives
- Pyrrolidine derivatives
- Miscellaneous fatty acid derivatives, lipids, artifacts, and so on.

All the mass spectra illustrated on these pages were obtained by EI at an ionization potential of 70 eV on quadrupole mass spectrometers. The AOCS website also offers the Bibliography of Mass Spectra with lists of references mainly concerning the use of MS for structural analysis of fatty acids, and comprehensive tutorials on MS of fatty acid derivatives (Christie, 2021).

Mass Spectra of Geochemicals, Petrochemicals, and Biomarkers This database is focused on organic, geochemical, and petrochemical applications (De Leeuw, 2004) and comprises:

- 1093 mass spectra of well-defined compounds.
- 1087 chemical structures.
- Information including mass spectra, chemical structure, chemical name, molecular formula, molecular mass (nominal mass), base peak, reference, and measurement condition.
- Chemical structures elucidated, if necessary, by a variety of techniques including NMR spectroscopy and single crystal X-ray structure analysis (Wiley, 2020).

Pyrolysis-GC/MS Data Book of Synthetic Polymers This compilation of pyrograms, thermograms, and mass spectral data by Shin Tsuge of the Nagoya University, Hajime Ohtani, Nagoya Institute of Technology, and Chuichi Watanabe, Frontier Laboratories Ltd. in Japan comprises the major pyrolyzates for 163 typical polymer samples with detailed peak assignments, tables, and thermograms for each polymer. Also, the compilation contains pyrograms of 33 condensation polymers through reactive pyrolysis in the presence of tetramethyl ammonium hydroxide (TMAH) as a derivatizing agent. The appendix refers to monographs and reviews for Pyrolysis-GC (Tsuge et al., 2011).

Chemical Concepts Library of Mass Spectra The Mass Spectra Chemical Concepts Fourth Edition comprises mass spectra of more than 40 000 compounds. The main part of this mass spectra reference library comprises the Industrial Chemicals Collection of Prof. Dieter Henneberg, Max-Planck-Institut für Kohlenforschung, Mülheim, Germany. Also, universities and institutes such as ETH, Zürich, Switzerland and ISAS, Dortmund, Germany, have contributed their research spectra to this collection. Before being included in the library, the data pass consistency and quality checks were performed at the Max-Planck-Institute (Publisher: Wiley-VCH; 4th edition, February 4, 1998, ISBN-13 978-0471440369). It was included in the release of the NIST 2011 library.

Additional information included with the mass spectra is as follows:

- Chemical structure
- Chemical name
- Molecular formula
- Molecular mass (nominal mass)
- Base peak
- Reference
- Measurement conditions

Alexander Yarkov – Mass Spectra of Organic Compounds The new specialized data collection contains 37 055 mass spectra of physiologically active organic compounds. The data resulted from quality control in combinatorial synthesis and covered a wide range of compound classes.

Additional information included with the mass spectra is as follows (Yarkov, 2004):

- Chemical structure
- Chemical name
- Molecular formula
- Molecular mass (nominal mass)
- Base peak
- Reference
- Measurement conditions

3.2.5 Library Search Programs

> "*Library searching has limitations and can be dangerous in novice hands.*"
>
> Wendy A. Warr, Ph.D., active in the Division of Chemical
> Information of the ACS, serves on the Advisory Board
> of ACS Software, and is on the Editorial Board
> of the Journal of Cheminformatics (Warr, 1993a).

In general, it is expected that the identity of an unknown compound will be found in a library search procedure. However, it is necessary to consider the results of a search procedure from the aspect of similarity between the reference and the unknown spectrum (Stein, 1994). Other information for confirmation of identity, such as sample preparation procedure, retention time and index (RI) on polar and non-polar column, and other spectroscopic data, should always be consulted.

Examples of critical cases are different compounds that have very similar spectra (isomers), the same compounds with different spectra (measuring conditions, reactivity, and decomposition) or the fact that a substance being searched for is not in the library at all, but similar spectra are suggested. In particular, the limited scope of the libraries and the quality of the unknown spectrum must be taken into account (Sparkman, 1996).

A good overview of the total number of known and well-characterized substances can be found at the Chemical Abstract Service (CAS) statistics. A CAS Registry Number is a unique and unambiguous identifier for a specific substance that allows clear communication and links together all available data and research about that substance. The majority of the registered substances are of organic nature and hence applicable to mass spectrometric analysis. A strong growth in registered compounds is evident having 60 million substances recorded in 2011, and 70 million in 2012, and over 90 million in 2014. In 2024, only ten years later, CAS reported that the registry has listed more than 279 million organic substances, alloys, coordination compounds, minerals, mixtures, polymers, and salts disclosed in publications since the early 1800s. See https://www.cas.org/support/documentation/chemical-substances/faqs.

A look at the molecular mass distribution gives some background on the limited compound coverage of the available GC-MS libraries and reveals the large potential for small molecule analysis to be run by GC-MS or LC-MS, depending on the polarity of the compounds. Ninety-nine percent of all CAS-registered substances are in the molecular mass range of up to 1000 Da as shown for the 2009 status in Figure 3.30 (Little et al., 2011, 2013), a typical mass range capable for most of the modern GC-MS and LC-MS instrumentation today. For unknown compounds detected, the search in large spectra libraries will stay as an important step in compound identification, but the manual skills for spectra interpretation will prevail (McLafferty and Turecek, 1993).

Today, the library search program provided by the NIST is undoubtably of the widest distribution. It was introduced in 1995 together with the NIST/U.S. Environmental Protection Agency (U.S. EPA)/National Institutes of Health (NIH) Mass Spectral Database for use on the Microsoft Windows™ platform. Certain algorithm components of the formerly widely used INCOS library search program build a basis of the NIST search procedure (Davies, 1978; Warr, 1993a, 1993b; Stein, 1994, 1999). Today, the data systems of the leading GC-MS manufacturer provide a direct data link and integrate by this way the NIST library search into the proprietary data processing software. Mass spectra can also be imported from a text file. Probability-based components later have significantly improved the recent versions of the NIST search (Stein, 1994) and have increasingly replaced the independent probability-based match (PBM) search program (Stauffer et al., 1985a, 1985b). The SISCOM procedure (Search for Identical and Similar Compounds)

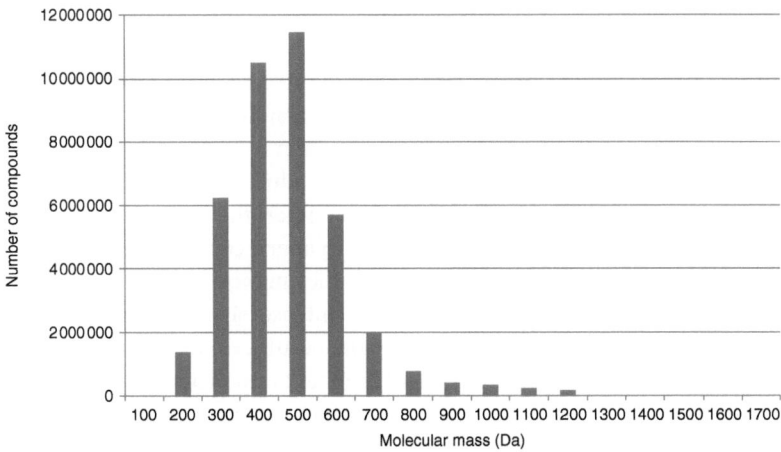

Figure 3.30 Molecular mass distribution of CAS registered compounds (Little et al., 2011, 2013).

developed by Damen, Henneberg and Weimann stands out on account of its excellent performance for a structure-related interpretation of mass spectra (Damen et al., 1978; Henneberg and Weimann, 1984; Henneberg et al., 1993; MassLib, 2014). The procedures for the determination of similarity between spectra are based on the classical spectrum interpretation considerations (Neudert et al., 1987). The NIST and PBM procedures aim to give suggestions of possible substances to explain an unknown spectrum. The PBM and SISCOM search algorithms are described in detail in the previous editions of this Handbook (Hübschmann, 2015). Other search procedures, such as the earlier Biemann search using the two largest peaks in every 14 m/z interval, have been replaced by newer developments and broadening of the algorithms by the manufacturers of spectrometers (Heller, 1999).

3.2.5.1 The NIST Search Procedure

At the beginning of the 1970s, the INCOS company presented a search procedure that operated both on the principles of pattern recognition and with the components of classical interpretation techniques, which could reliably process data from different types of mass spectrometer (Davies, 1978). The newest development in computer-supported library searches, which is the further development of the INCOS procedure, has been presented by Steven Stein (NIST) through targeted optimization of the weighting and combination with probability values (Stein, 1994, 1999; Stein and Scott, 1994). The early years of GC-MS were characterized by the rapid development of quadrupole instruments, which were ideal for coupling with gas chromatographs, because of their fast scan rates, which were high compared with the magnetic sector instruments of that time. But at that time, the available spectral libraries had been drawn up from spectra acquired on magnetic sector instruments.

From the beginning, the INCOS procedure was able to take into account the relatively low intensities of the higher masses in spectra run on quadrupole systems,

3 Evaluation of GC-MS Analyses

besides the typical high mass intensity magnetic sector spectra. The INCOS search has remained virtually unchanged since the 1970s and became the standard on Finnigan GC-MS systems. The search was known for its high hit probability, even with mass spectra with a high proportion of matrix noise obtained in residue analysis and its independence from the type of instrument.

After a significance weighting (square root of the product of mass and intensity), data reduction by a noise filter, and a redundancy filter, the extensive reference database is searched for suitable candidates for a pattern comparison in a rapid pre-search. The NIST search uses all masses of the unknown spectrum in this pre-search (former INCOS only the eight most intense signals). The intensity ratios are not yet considered at this point, only the occurrence of the mass signal. Reference spectra, which only contain a small number or no matching masses, are excluded from the list of possible candidates and are not further processed.

The main search is the critical step in the search algorithm, in which the candidates selected in the pre-search are compared with the unknown spectrum and arranged in a prioritized list of suggestions. Of critical importance for the tolerance for different types of mass spectrometer and differing conditions of data acquisition (and thus for the high hit rate) is a process known as *local normalization*.

Local normalization introduces an important component into the search procedure, which is comparable to the visual comparison of two patterns (Figure 3.31). Individual clusters of ions and isotope patterns are compared with one another in a local mass window. The central mass of such a window from the reference spectrum is compared with the intensity of this mass in the unknown spectrum in order to assess the matching of the line pattern to the left and right. In this way, the nearby region of each mass signal is examined and, for example, the matching of isotope patterns of e.g. C, Cl, Br, S, Si, and cleavage reactions are assessed.

The advantage of this procedure lies in the fact that deviating relative intensities caused by spectral skewing, or the type of data acquisition do not influence the result of the search. A variance in the relative signal intensities in a mass spectrum can be caused by varying the choice of spectra from the rising or falling slopes of the peak in

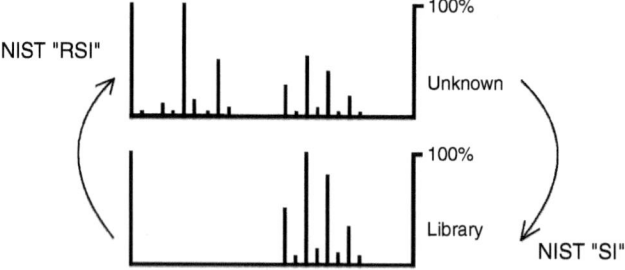

Figure 3.31 Diagram showing local normalization and the principle of match and reverse match calculation. Reverse match value RSI high: All mass signals of the library spectrum are present in the unknown spectrum and the isotope pattern also fits after "local" normalization of the intensities. Match value SI low: The unknown spectrum has more mass signals than the library spectrum. Only some of the masses from the unknown spectrum are present in the library spectrum.

the case of quadrupole and magnetic sector instruments (skewing) and by changes in the tuning parameters of the ion source, or its increasing contamination. Furthermore, local normalization has a positive effect on spectra with a high proportion of noise (trace analysis, chemical background). Local normalization is the major reason why spectral libraries require only one mass spectrum per substance entry.

Two values are determined for spectral comparison as a result of the main search. The reverse match value (NIST "RSI", former INCOS "FIT"), simply translated as the fit value, gives a measure of how well the reference spectrum is represented with its masses in the unknown spectrum (reverse search procedure, ignoring all peaks that are in the sample spectrum but not the reference spectrum). The forward-looking mode of searching, whereby the presence of the unknown spectrum in the reference spectrum is examined (forward search procedure, all peaks of the sample spectrum are compared), is expressed as the match value (NIST "SI", former INCOS reverse fit "RFIT"). The combination of the two values gives information on the purity of the unknown spectrum (Figure 3.31). If the reverse match value is high (NIST "RSI") and the match value is lower (NIST "SI"), it can be assumed that the spectrum measured contains considerably more mass signals than the reference spectrum used for comparison. It would be necessary to evaluate whether a co-eluate, chemical noise, the presence of a homologous substance, or another reason is responsible for the appearance of the additional lines. This is typically done by displaying mass chromatograms and performing background subtractions.

All the candidates found in the pre-search are processed in the main search as described. As a result, a sorted list according to match "SI" and reverse match "RSI" values is displayed. Sorting can be selected by "SI" or "RSI" values (Table 3.4). The initial sorting according to reverse match is recommended because with this value, the best estimation of the possible identity is achieved. Further sorting according to the match values gives additional solutions, which generally supplement the further steps toward identification with valuable information on partial structures or identifying a particular class of compound.

The maximum achievable match value for identical spectra is set to 999. As of general experience, values greater than 900 indicate an excellent match. Good matches to consider give values greater than 800. Values below 800 are considered fair, and values below 600 are considered very poor. Unknown spectra with a large number of peaks tend to yield lower match factors than spectra with fewer peaks. Fewer peaks in a spectrum tend to increase the RSI match factors, as can be experienced with

Table 3.4 Results of the NIST spectral comparison.

SI	RSI	Prob.	Assessment
High	High	High	Identification or that of an isomer very probable
Low	High	Low	Identification possible, but homologues, co-elution, or noise present
High	Low	Low	Possibly an incomplete spectrum

Sorting of the suggestions should first be carried out according to RSI and then according to SI values.

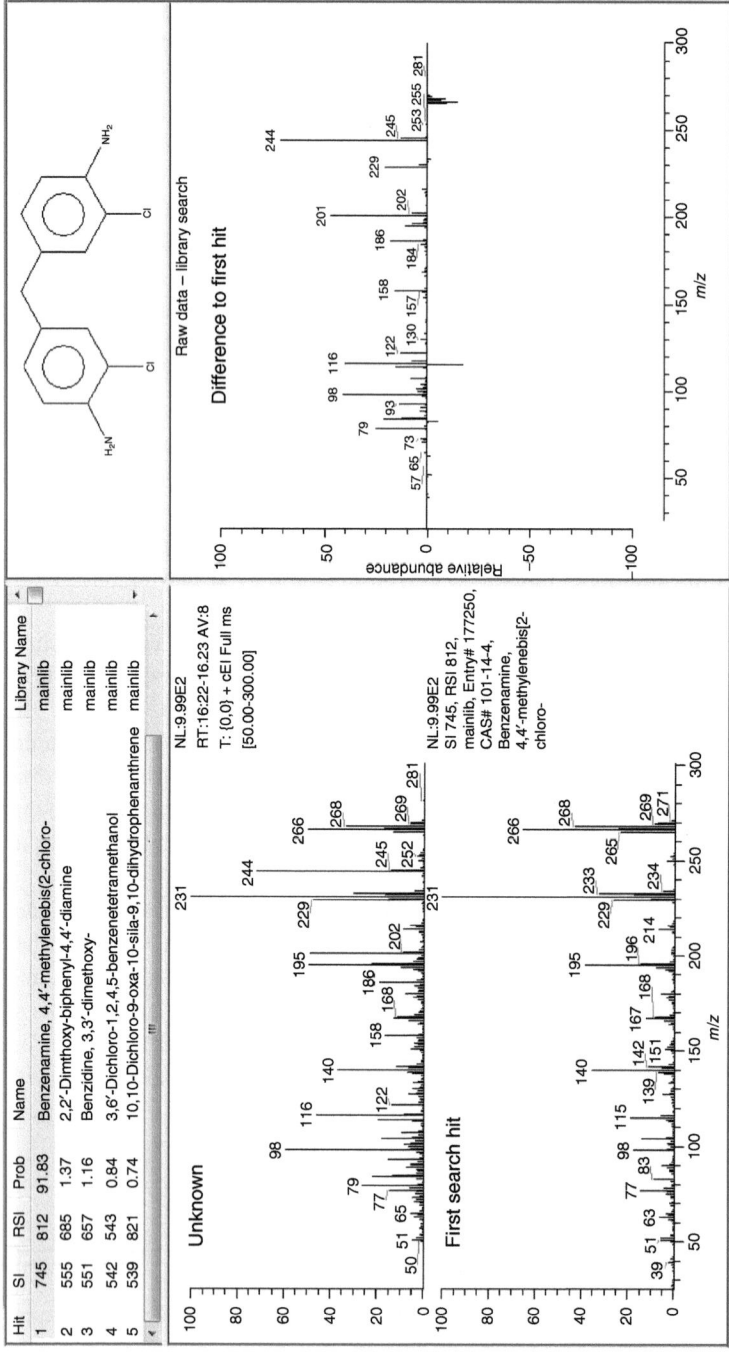

Figure 3.32 NIST library search result of a co-elution spectrum. The hit list upper left is sorted by SI match value. The RSI match value with 812 is high, SI value 745 low due to additional masses in the unknown spectrum (upper left) which are not in the library reference (bottom left). The probability with 91.8% is high, indicating a unique spectrum with significant difference to the next search hit. The difference spectrum (upper right) can be retrieved for search by spectrum subtraction in the chromatogram.

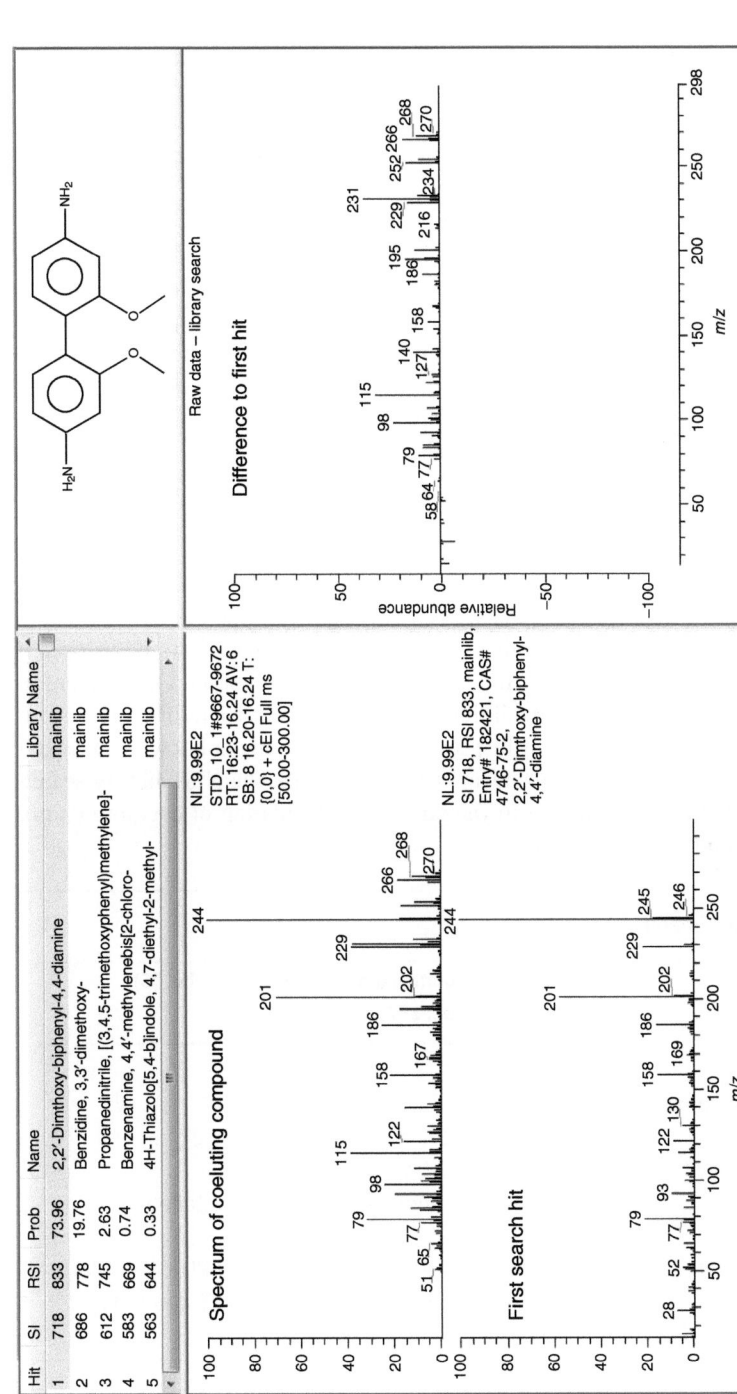

Figure 3.33 NIST library search of the difference spectrum of Figure 3.31 after careful background subtraction. The RSI match value 833 is high, SI value with 718 is low due to other fragments from background ions, the probability value of 74% is high due to the unique spectrum of the first hit. The difference spectrum shows signals of the coeluting compound of Figure 3.31.

polyaromatic hydrocarbon (PAH) spectra. On the contrary, spectra with many ions, especially those with background or matrix interferences from highly sensitive full scan measurements of real-life samples, tend to lower SI match values, compared to clean spectra, due to the additional chemical interferences (Figure 3.32). A spectrum subtraction has to be done manually in the chromatogram (Figure 3.33). NIST does not offer the function of searching difference spectra to the components of the hit list.

An additional probability value is given with the NIST hit list. The probability value (Prob.%) for a certain compound hit is derived as a relative probability from the spectral differences between adjacent hits in the list. Several high values indicate high spectra similarities of the hits, low values indicate different spectral patterns (see also Figure 3.34i and g). The probability calculation requires the search of the NIST database with the set of replicate spectra. Another separate probability factor ("InLib") is provided as a measure of the probability that the unknown compound is contained in the searched databases (Stein, 2011).

In the 1990s, Steven Stein from NIST took the former INCOS approach and extended it to the most common situations when unknown compounds are not present in the library. Structurally similar compounds can appear in the NIST library search hit list. The identity search presumes that the unknown compound is represented in the reference database, as it is designed to find the exact match to the unknown spectrum. The extended search mode for similarity should be used if it becomes apparent that the spectrum of the unknown compound is not present in the library. Stein also added probabilities to the hit list that give information about common substructures which may be present or absent in the unknown compound (Stein, 1994, 1995). Based on this advanced performance, the NIST library search is recommended as the first step in the structural elucidation of compounds not found in reference libraries (Stein, 2011).

INCOS Library Search

Principle	Pattern recognition (after Joel Karnovsky, INCOS) (Sokolow et al., 1978)
Course	
1. Significance weighting	$\sqrt{m\,I}$
2. Noise filter	Window ± 50 Da, ≥40 masses
3. Redundancy filter	Window ± 7 Da, ≥6 masses
4. Pre-search	8 masses + molecular mass (NIST today 16 masses)
5. Main search	Local normalization FIT, RFIT and PUR* calculation (*PUR as normalized product of FIT and RFIT)
6. Sorting and display	

Advantages

- high hit probability
- secure identification even with spectra with high noise levels
- search is independent of scan times and type of instrument because of local normalization
- only one spectrum per substance necessary in the library
- very fast
- available significantly advanced for a variety of data systems in the form of NIST algorithm.

Limitations

- manual difference calculation necessary for co-eluates
- with spectra with many equally distributed fragments.

3.2.6 Interpretation of Mass Spectra

There are no hard and fast procedures for fully interpreting a mass spectrum. Unlike spectra obtained using other spectroscopic procedures, such as UV, IR, NMR, or fluorescence, mass spectra do not show uptake or emission of energy by the compound (i.e. the intact molecule), but reflect the qualitative and quantitative analysis of the processes accompanying ionization (charge localization, fragment formation, rearrangements, chemical reactions) by the measurement of the resulting ionic species (see Section 2.3.2). The time factor and the energy required for ionization (electron beam, temperature, pressure) also play a role. With all other spectroscopic methods, the intensity of interaction with certain functional groups or other structural elements is well determined. In MS, however, the appearance of certain characteristics of a structure in the mass spectrum always depends on the total structure and elemental composition of the compound. The failure of expected signals to appear generally does not prove in MS that certain structural elements are not present, only positive signals and differences between signals count. It is also true that a mass spectrum cannot be associated with a particular chemical structure without additional information.

Nevertheless, following the below scheme is recommended for deciphering information coded in a mass spectrum. (Certain users accuse experienced mass spectrometrists of having a criminological feel for the subject – and they are justified!) In this spectroscopic discipline, the experience of a frequently investigated class of substances is rapidly built up. New groups of substances usually require new methods of solution. It is therefore of high importance that other parameters relating to the substance besides MS (spectrum and accurate mass), such as UV, IR, NMR spectra, solubility, elution temperature, acidic or basic clean-up, synthesis reaction equations, or those of conversion processes, should be incorporated into the interpretation of the spectrum.

452 | *3 Evaluation of GC-MS Analyses*

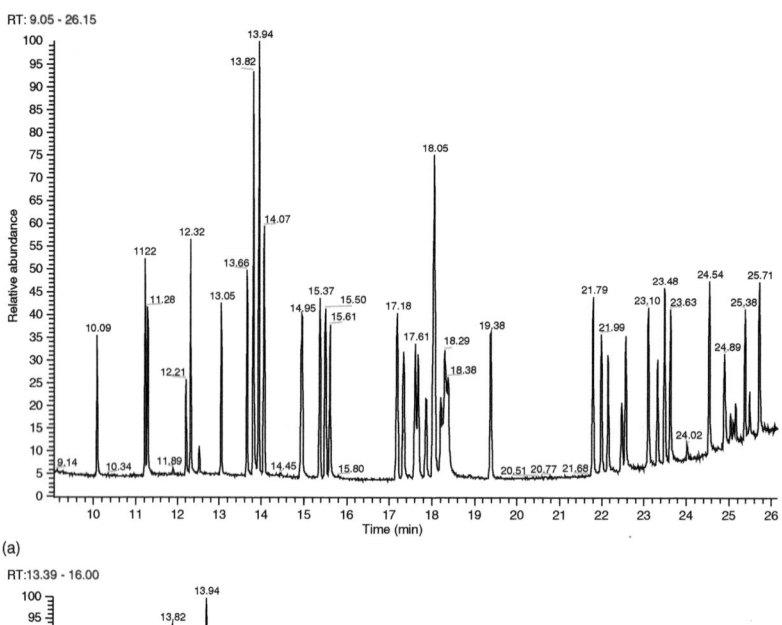

(a)

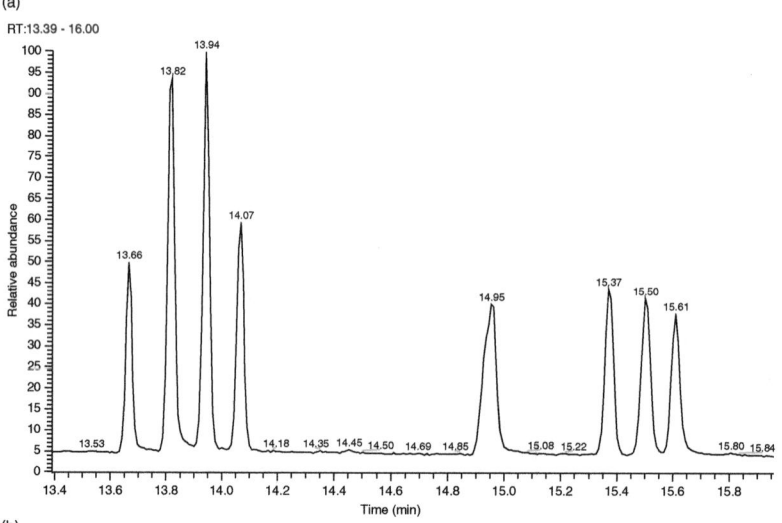

(b)

Figure 3.34 The graphics (a) to (i) illustrate the sequence of a peak analysis for coeluting compounds using mass chromatograms, spectrum background subtraction and library search. (a) Full scan total ion chromatogram (TIC) of a pesticide mixture. (b) Chromatogram detail, peak at RT 14.95 min shows unsymmetrical peak shape. (c) Mass spectrum of peak at RT 14.95 min. (d) Background spectrum before peak at RT 14.95 min (*m/z* 77, 91, 105, 121 indicate alkyl-phenylen, *m/z* 207, 281 polysiloxane components of the stationary phase of the used Rtx-5ms column). (e) Peak analysis using mass chromatograms. (f) Corrected mass spectrum at RT 14.93 min (left peak slope subtracted by right peak slope). (g) Library search result (NIST), first hit quinoclamine with spectrum comparison, top unknown of Figure 3.34f, bottom first hit. (h) Corrected mass spectrum at RT 14.96 min (right peak slope subtracted by left slope and column background of Figure 3.34d. (i) Library search result (NIST), first hit cyanazine with spectrum comparison, top unknown of Figure 3.34h, bottom first hit. Note: *m/z* 172 in the unknown spectrum got lost due to the manual subtraction process.

3.2 Substance Identification

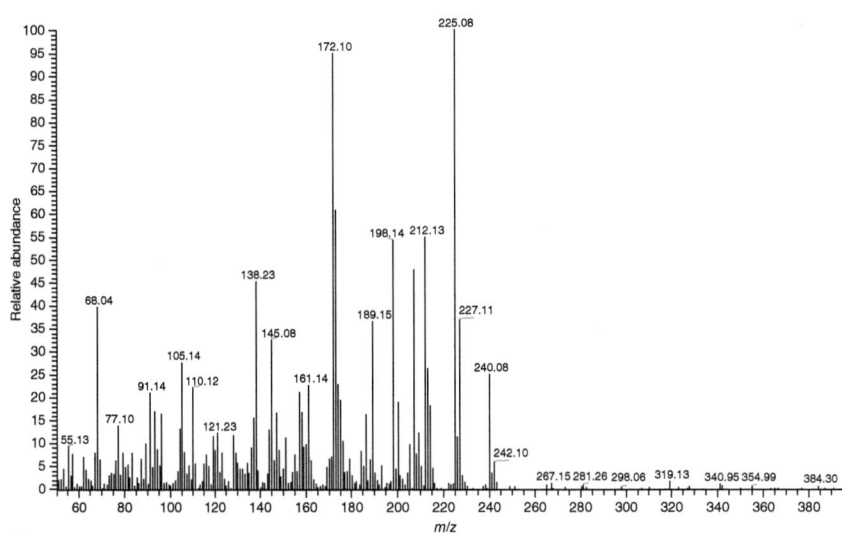

(c)

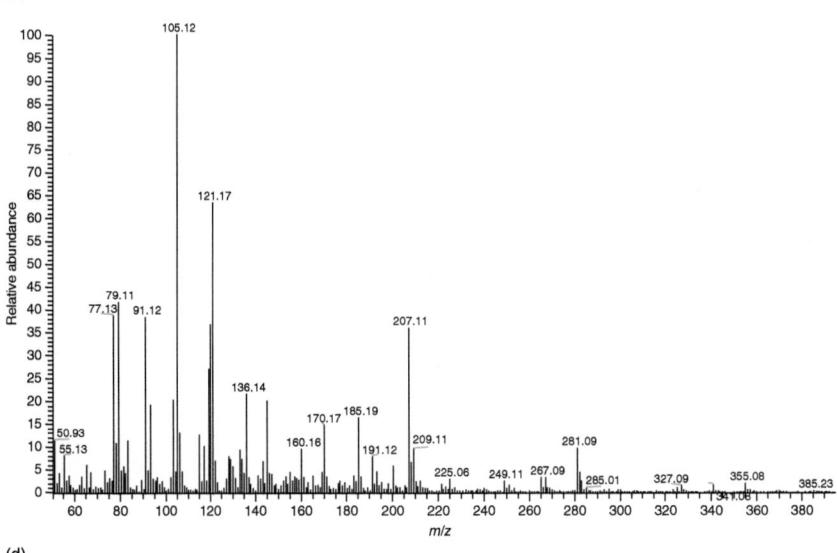

(d)

Figure 3.34 (Continued)

454 | *3 Evaluation of GC-MS Analyses*

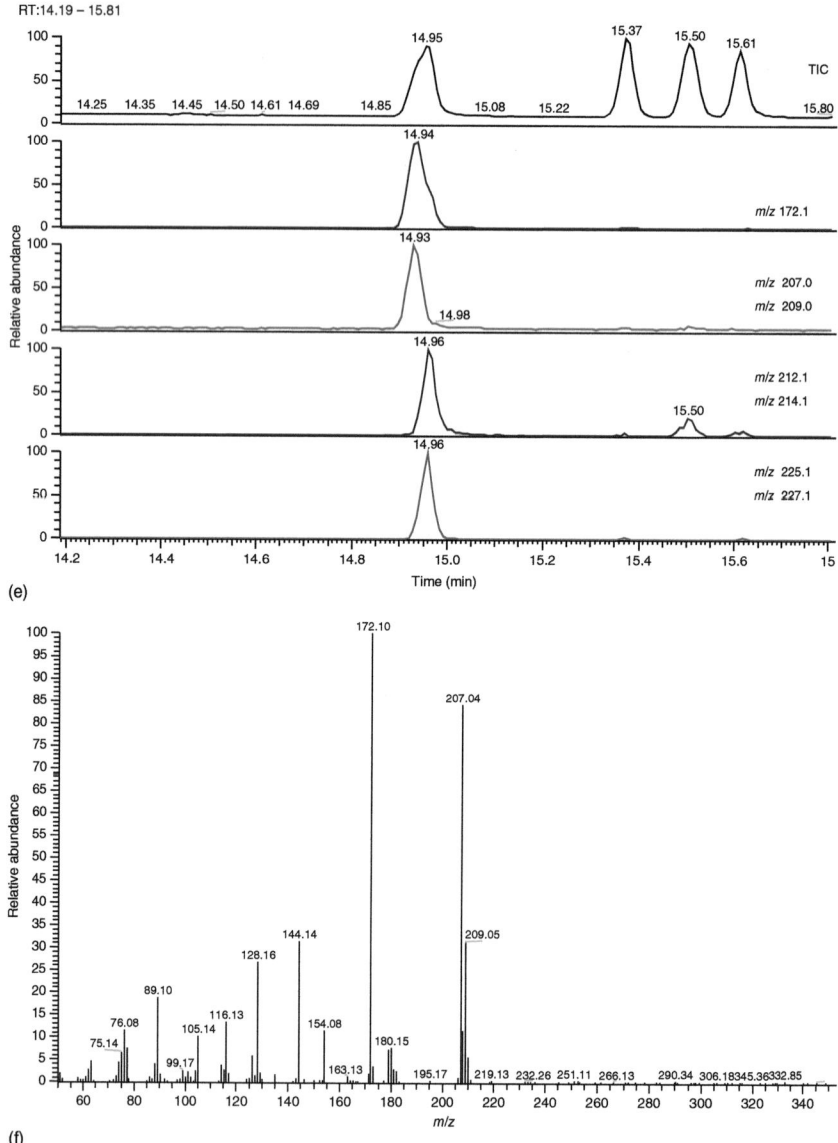

Figure 3.34 *(Continued)*

3.2 Substance Identification | 455

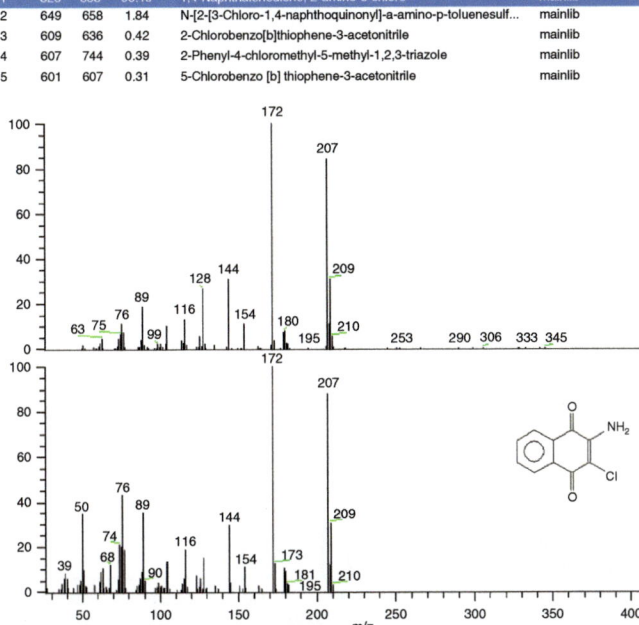

(g)

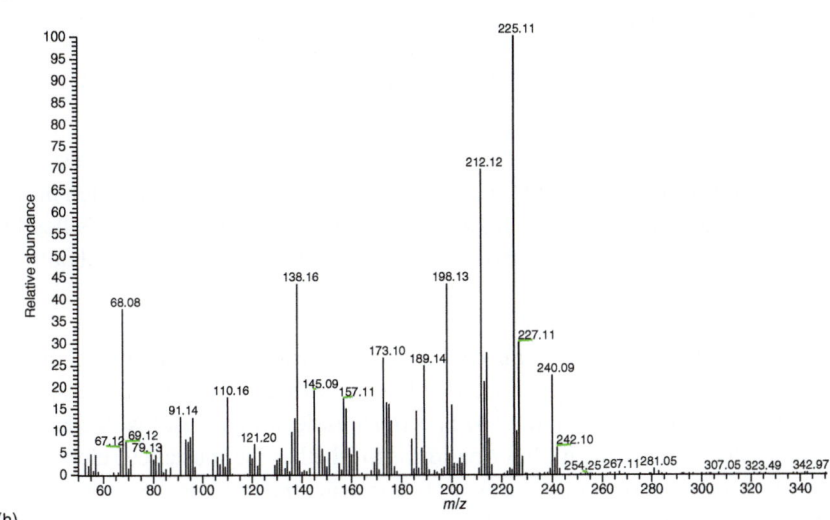

(h)

Figure 3.34 (Continued)

3 Evaluation of GC-MS Analyses

Hit	SI	RSI	Prob	Name	Library Name
1	850	862	96.47	Cyanazine	mainlib
2	543	548	0.96	Oxazolidin-2-one, 3-[2-(5,7-dichloro-2-methyl-3-indol...	mainlib
3	521	523	0.38	2,3-Dicyano-2-(2-oxo-cyclododecyl)-succinonitrile	mainlib
4	505	585	0.21	Cyprazine	mainlib
5	499	549	0.17	Cyanamide,N-[4-chloro-6-(dimethylamion)-1,3,5-tria...	mainlib

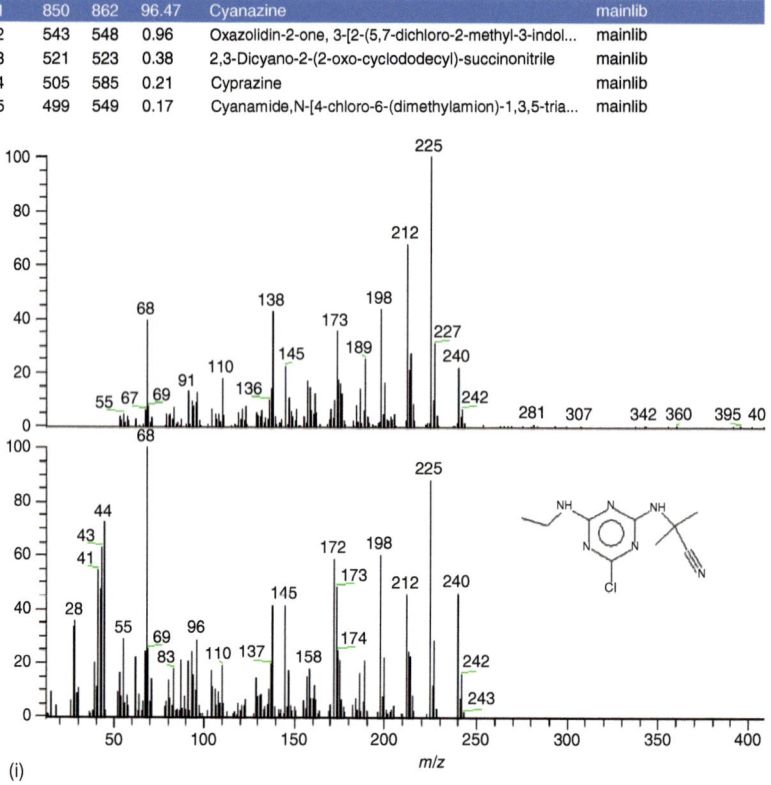

Figure 3.34 *(Continued)*

The procedure shown in the following scheme has proved to be effective:

1) *Spectrum display*
   Does the mass spectrum originate from a single substance or are there signals which appear not to belong to it? Can a deconvolution or the subtraction of the background give a clearer representation? Is the spectrum possibly falsified by the subtraction? What information do the mass chromatograms give about the most significant ions?

2) *Library search*
   As all GC-MS systems are connected to powerful computer systems, each interpretation process for EI spectra should begin with a search through available spectral libraries (see Section 3.2.4). Spectral libraries are an inestimable source of knowledge, which can give information as to whether the substance belongs to a particular class or on the appearance of clear structural features, even when identification seems improbable. Careful use of the database spares time and gives important suggestions.

3) *Molecular ion*
   Which signal could be that of the molecular ion? Is an $M^{+/-}$, $(M+H)^+$ or an $(M-H)^{+/-}$ present? Which signals are considered to be noise or chemical

background? Mass chromatograms can also help here. If the molecular mass is known, for example, from soft ionization results, the library search should be carried out again limited to the estimated molecular mass.

4) *Isotope pattern*

Is there an obvious isotope pattern, for example, for chlorine, bromine, silicon, or sulfur? The molecular ion shows all elements with stable isotopes in the compound. Is it possible to find out the maximum number of carbon atoms? This is a typical limitation with residue analysis as usually it is not possible to detect the ^{13}C signal intensity with certainty. Also, a noticeable absence of isotope signals, particularly with individual fragments, can be important for identifying the presence of phosphorus, fluorine, iodine, arsenic, and other monoisotopic elements. Only the molecular or cationized molecular ions give complete information on all isotopes in the elemental formula.

5) *Nitrogen rule*

Is nitrogen present? The even atomic mass 14 of nitrogen with three binding sites allows this simple rule: An uneven molecular mass indicates an uneven number of nitrogen atoms in the empirical formula.

6) *Fragmentation pattern*

Which information does the fragmentation pattern give? Are there pairs of fragments, the sum of which give the molecular mass? Which fragments could be formed from an α-cleavage? Here the use of a table giving details of neutral losses from molecular ions $(M-X)^+$ listed in Table 3.5 is advisable.

7) *Key fragments*

What can be said about characteristic fragments in the lower mass region? Is there information on aromatic building blocks or even ions formed through rearrangements (McLafferty, retro-Diels-Alder)? Also, the application of MS/MS with the registration of product ion spectra from major ions is a successful route for structure elucidation. Here tables with plausible explanations of fragment ions are helpful (see Table 3.5).

8) *Structure postulate*

Bringing together such information rapidly leads to a rough interpretation, which initially gives a partial structure, and finally, it postulates the molecular structure. Test this postulated structure for plausibility: which fragmentation pattern would the proposed structure give? Is the proposed substance available as a reference and does this correspond to the unknown spectrum? Does the rentention behavior (RI) fit?

Within this interpretation scheme, spectroscopic comparison through library searching is placed at the beginning of the interpretation. Confirmation of a suggestion from a library search is achieved through an assessment using the scheme mentioned earlier. Also, some compound drawing programs can assist in the calculation of potential fragment ions, e.g. the free MSC Pro™ calculator (Quadtech Associates, 2024). Interpretation has never finished simply with the print-out of the list of suggestions!

Steps to the Interpretation of Mass Spectra
1) Library search!
2) Only one substance?
3) Molecular ion?
4) Isotope pattern?
5) Nitrogen?
6) Fragmentation pattern?
7) Fragments? MS/MS product ion spectra?
8) Reference spectrum?

Table 3.5 Mass correlations to explain cleavage reactions (M–X)+ and key fragments X+.

m/z	Fragment X+	M+ – X	Explanations
12	C		
13	CH		
14	CH_2, N, N_2^{2+}		
15	CH_3	M+ – 15	Nonspecific, CH_3 at high intensity
16	O, NH_2, O_2^{2+}, CH_4	M+ – 16	Rarely CH_4, (but frequently R+-CH_4 in alkyl fragments), O from N-oxides and nitro compounds, NH_2 from anilines
17	OH, NH_3	M+ – 17	Nonspecific O-indication, NH_3 from primary amines
18	H_2O, NH_4	M+ – 18	Nonspecific O-indication, strong for many alcohols, some acids, ethers and lactones
19	H_3O, F	M+ – 19	F-indication
20	HF, Ar^{2+}, CH_2CN^{2+}	M+ – 20	F-indication
21	$C_2H_2O^{2+}$		(rarely)
22	CO_2^{2+}		
23	Na		(rarely)
24	C_2		
25	C_2H	M+ – 25	Rarely with a terminal C≡CH group
26	C_2H_2, CN	M+ – 26	From purely aromatic compounds, rarely from cyanides
27	C_2H_3, HCN	M+ – 27	CN from cyanides, C_2H_3 from terminal vinyl groups and some ethyl esters
28	C_2H_4, N_2, CO	M+ – 28	CO from aromatically bonded O, ethylene through RDA from cyclohexenes, by H-migration from alkyl groups, nonspecific from alicyclic compounds
29	C_2H_5, CHO	M+ – 29	Aromatically bonded O, nonspecific with hydrocarbons
30	C_2H_6, H_2NCH_2, NO, CH_2O, BF (N-fragment)	M+ – 30	CH_2O from cyclic ethers and aromatic methyl ethers, NO from nitro compounds and nitro esters

Table 3.5 (Continued)

m/z	Fragment X+	M+ − X	Explanations
31	CH$_3$O, CH$_2$OH, CH$_3$NH$_2$, CF, (O-fragment)	M+ − 31	Methyl esters, methyl ethers, alcohols
32	O$_2$, CH$_3$OH, S	M+ − 32	Methyl esters, some sulfides and methyl ethers
33	CH$_3$OH$_2$, SH, CH$_2$F	M+ − 33	SH nonspecific S-indication, M+-18-15 nonspecific O-indication, strong with alcohols
34	SH$_2$ (S-fragment)	M+ − 34	Nonspecific S-indication, strong with thiols
35	^{35}Cl, SH$_3$	M+ − 35	Chlorides, nitrophenyl-compounds (M+-17-18)
36	HCl, C$_3$	M+ − 36	Chlorides
37	^{37}Cl, C$_3$H		
38	H^{37}Cl, C$_3$H$_2$		
39	C$_3$H$_3$	M+ − 39	Weak with aromatic hydrocarbons
40	Ar, C$_3$H$_4$	M+ − 40	Rarely with CH$_2$CN
41	C$_3$H$_5$, CH$_3$CN	M+ − 41	C$_3$H$_5$ from alicyclic compounds, CH$_3$CN from aromatic N-methyl and o-C-methyl heterocycles
42	CH$_2$=C=O, C$_3$H$_6$, C$_2$H$_4$N	M+ − 42	Nonspecific with aliphatic and alicyclic systems, strong through RDA from cyclohexenes, by rearrangement from α-, β-cyclo-hexenones, enol and enamine acetates
43	CH$_3$CO, C$_3$H$_7$, C$_2$H$_4$N, CONH	M+ − 43	Acetyl, propyl, aromatic methyl ethers (M+-15-28), nonspecific with aliphatic and alicyclic systems
44	CO$_2$, CH$_3$NHCH$_2$ (N-fragment), CH$_2$CHOH	M+ − 44	CO$_2$ from acids, esters, butane from aliphatic hydrocarbons
45	C$_2$H$_5$O, HCS (Sulfides)	M+ − 45	Ethyl esters, ethyl ethers, lactones, acids, CO$_2$H from some esters; CH$_3$NHCH$_3$ from dimethylamines
46	C$_2$H$_5$OH, NO$_2$	M+ − 46	Ethyl esters, rarely acids, nitro compounds, n-alkanols (M+-18-28)
47	CH$_3$S (S-fragment), C^{35}Cl, C$_2$H$_5$OH$_2$, CH(OH)$_2$		
48	CH$_3$SH, CH^{35}Cl		
49	C^{37}Cl, CH$_2$ ^{35}Cl		
50	C$_4$H$_2$, CH ^{35}Cl		
51	C$_4$H$_3$ (Aromatic fragment)		
52	C$_4$H$_4$, CH$_3$ ^{37}Cl (Aromatic fragment)		
53	C$_4$H$_5$		
54	⟋⟍, C$_2$H$_4$CN	M+ − 54	Cyclohexene (RDA)
55	C$_4$H$_7$, C$_2$H$_3$CO	M+ − 55	C$_4$H$_7$ from alicyclic systems and butyl esters
56	C$_4$H$_8$, C$_2$H$_4$CO	M+ − 56	Nonspecific with alkanes and alicyclic systems

(Continued)

Table 3.5 (Continued)

m/z	Fragment X+	M+ − X	Explanations
57	C_4H_9, C_2H_5CO, C_3H_2F	M+ − 57	Nonspecific with alkanes and alicyclic systems
58	CH_3COHCH_2, $C_2H_5CHNH_2$, $C_2H_6NCH_2$	M+ − 58	C_3H_6O from α-methylaldehydes and acetonides
59	C_2H_6COH, $C_2H_5OCH_2$, CO_2CH_3, CH_3CONH_2	M+ − 59	Methyl esters
60	$CH_2CO_2H_2$, CH_2ONO	M+ − 60	O-acetates (M+-AcOH), methyl esters (M+-CH$_3$OH-CO)
61	$CH_3CO_2H_2$, C_2H_4SH		
62	$HOCH_2CH_2OH$	M+ − 62	Ethylene ketals
63	C_5H_3		
64	SO_2	M+ − 64	SO_2 cleavage from sulfonic acids
65	C_5H_5		
67	(furyl)		
68	, C_4H_4O, C_3H_6CN		
69	C_5H_9, C_3H_5CO, CF_3, C_3HO_2 (1,3-dioxyaromatics)		
70	C_5H_{10}, (pyrrolyl)		
71	C_5H_{11}, C_4H_7CO, (furanyl)		
72	$C_4H_{10}N$, $C_3N_7NHCH_2$, $C_2H_5COHCH_2$		
73	$CO_2C_2H_5$, $C_3H_7OCH_2$, $CH_2CO_2CH_3$, C_4H_8OH (O-fragments)		
74	$CH_2=COHOCH_3$, $CH_3CH=COHOH$		
75	$C_2H_5CO_2H_2$, $C_2H_5SCH_2$, $CH_3OCHOCH_3$, (dimethyl acetates)		
76	C_6H_4		
77	C_6H_5		
78	C_6H_6		
79	C_6H_7, ^{79}Br		
80	C_6H_8, $H^{79}Br$, (pyridyl), (pyrrolyl)CH_2, CH_3S_2H		
81	C_6H_9, ^{81}Br, (furanyl)CH_2		
82	C_6H_{10}, $H^{81}Br$		
83	C_6H_{11}, C_4H_7CO		
84	(pyridyl), (N-methylpyrrolyl)		
85	C_6H_{13}, C_4H_9CO		
86	$C_3H_7COH=CH_2$		
87	$CO_2C_3H_7$, $CH_2CO_2C_2H_5$, $CH_2CH_2CO_2CH_3$, (O-fragments)		
88	$CH_2=COHOC_2H_5$, $CH_3CH=COHOCH_3$		

Table 3.5 (Continued)

m/z	Fragment X+	m/z	Fragment X+
91	⌬–CH₂, ⌬–Cl, n-Alkyl chlorides		
92	pyridine–CH₂, benzyl=CH₂ with –H		
93	CH₂⁷⁹Br, aniline-NH		
94	CH₃⁷⁹Br, phenol fragment, pyrrole–CO	120	benzoyl-type C=O
95	CH₂⁸¹Br, furyl–CO	121	HO-C₆H₄-CO, CH₃O-C₆H₄-CH₂
96	C₅H₁₀CN, CH₃⁸¹Br	127	J
97	C₇H₁₃, thiophene–CH₂	128	HJ, naphthalene
98	piperidine–CH₂	130	indole fragment
99	C₇H₁₅, (Ethylene ketals)	131	C₃F₅
104	C₂H₅CHONO₂, styrene	135	–Br (n-Alkyl bromides)
105	benzoyl–CO, phenyl–N=N, –C₂H₄	141	naphthyl–CH₂
111	thiophene–CO	142	quinoline–CH₂
115	indenyl–H	149	phthalic anhydride–OH
119	phenyl–C(CH₃)₂, tolyl–CO with CH₃	152	biphenylene

3.2.6.1 Isotope Patterns

For organic MS, only a few stable elements with noticeable isotope patterns are important, while in inorganic MS, there are many isotope patterns of metals, some of them very complex. In Table 3.6 it can be seen that the elements carbon, sulfur, chlorine, bromine, and silicon consist of naturally occurring stable, nonradioactive isotopes. The elements fluorine, phosphorus, and iodine are among the few monoisotopic elements in the periodic table (Rosman and Taylor, 1998).

Table 3.6 Exact masses and natural isotope abundance.

Element	Isotope	Nominal mass[a] [g/mol]	Exact mass [g/mol]	Abundance[b] [%]	Factors for calculating the isotope intensity[c] M+1	M+2
Hydrogen	^{1}H	1	1.007825	99.99		
	D, ^{2}H	2	2.014102	0.01		
Carbon	^{12}C	12	12.000000[d]	98.9		
	^{13}C	13	13.003354	1.1	1.1	0.006
Nitrogen	^{14}N	14	14.003074	99.6	0.4	
	^{15}N	15	15.000108	0.4	0.4	
Oxygen	^{16}O	16	15.994915	99.76		
	^{17}O	17	16.999133	0.04	0.04	
	^{18}O	18	17.999160	0.20		0.20
Fluorine[e]	F	19	18.998405	100		
Silicon	^{28}Si	28	27.976927	92.2		
	^{29}Si	29	28.976491	4.7	5.1	
	^{30}Si	30	29.973761	3.1		3.4
Phosphorus	P	31	30.973763	100		
Sulfur	^{32}S	32	31.972074	95.02		
	^{33}S	33	32.971461	0.76	0.8	
	^{34}S	34	33.976865	4.22		4.4
Chlorine	^{35}Cl	35	34.968855	75.77		
	^{37}Cl	37	36.965896	24.23		32.5
Bromine	^{79}Br	79	78.918348	50.5		
	^{81}Br	81	80.916344	49.5		98.0
Iodine	I	127	126.904352	100		

a) The calculation of the nominal mass of an empirical formula is carried out using the mass numbers of the most frequently occurring isotope, for example, for lindane $^{12}C_6\,^1H_6\,^{35}Cl_6$ the exact mass of M$^+$ is 287.860065 g/mol, the nominal mass is 288 g/mol.
b) The isotope abundance is a relative parameter. The abundances of an element add up to 100%.
c) In the isotope pattern of the ion the intensity of the first mass peak (M) is assumed to be 100%. The intensities of the isotope peaks (satellites) M+1 and M+2 are given by multiplying the factors with the number of atoms of an element in the ion:

    Example $C_{10}H_{22}$ M$^+$ 142   Intensity m/z 143: $10 \cdot 1.1 = 11\%$
                    C_6Cl_6 M$^+$ 282   Intensity m/z 284: $6 \cdot 32.5 = 195\%$
                    S_6 M$^+$ 192   Intensity m/z 194: $6 \cdot 4.4 = 26.4\%$

d) See the box "Why is ^{12}C the Official Reference Mass" (p 463).
e) Fluorine, sodium, aluminum, phosphorus, manganese, arsenic and iodine, for example, appear as monoisotopic elements in mass spectrometry.

Why is ^{12}C the Official Reference Mass for Atomic Mass Units?

"Before the 1970s, two conventions were used for determining relative atomic masses. Physicists related their mass spectrometric determinations to the mass of ^{16}O (i.e. ^{16}O has a mass of exactly 16 on the 'atomic mass unit' scale), the most abundant isotope of oxygen, and chemists used the weighted mass of all three isotopes of oxygen: ^{16}O, ^{17}O and ^{18}O. At an international congress devoted to the standardization of scientific masses and measures, the redoubtable A.O. Nier proposed a solution to these disparate conventions whose negative consequences were becoming serious. He suggested that the carbon-12 isotope (^{12}C) be the reference for the atomic mass unit (amu). By definition, its mass would be exactly 12 amu, a convention that would be acceptable to the physicists. In accordance with this convention, the average mass for oxygen (the weighted sum of the three naturally occurring isotopes) became 15.9994 amu, a number close enough to 16 to satisfy the chemists" (Sharp, 2017).

If isotope signals appear in mass spectra, there is the possibility that these elements can be recognized by their typical pattern, and the number of them in molecular and fragment ions can be determined from this pattern. With carbon, there are limitations to this procedure in trace analysis, because usually the quantity of substance available is often small for a sufficiently stable signal to be obtained. In order to allow conclusions to be drawn on the maximum number of carbon atoms, a larger quantity of substance is advisable. For this, the technique of individual mass registration (SIM, MID) with longer dwell times or accurate mass detection are particularly suitable for giving reproducible ion statistics. In the evaluation, in the case of carbon, only the maximum number of carbon atoms (isotope intensity/1.1%) can be calculated, as contributions from other elements must be taken into account.

Some elements, in particular the halogens chlorine and bromine, which are contained in many active substances, plastics, and other technical products, can be recognized by the typical isotope patterns. These easily recognized patterns are shown in Figures 3.35–3.42. The intensities shown are scaled to a unit ion stream of the isotope pattern. The lowering of the specific response of the compound as a function of, for example, the degree of chlorination, is illustrated. The simple occurrence of the elements shown as a series is used as a reference in each case. The relative intensities within an isotope pattern are given as a percentage in the caption underneath the isotope lines with the mass contribution to the molecular mass based on the most frequent occurrence in each case. Notice for chlorine and bromine the distance between the isotopes of two mass units. Provided that chlorine and bromine occur separately, the degree of chlorination or bromination can be easily determined by visual comparison of the variation in the intensities. For compounds that contain both chlorine and bromine the degree of substitution cannot be determined by comparison of the patterns alone. In these cases, the high atomic mass of bromine is of help (for calculation of RIs from the

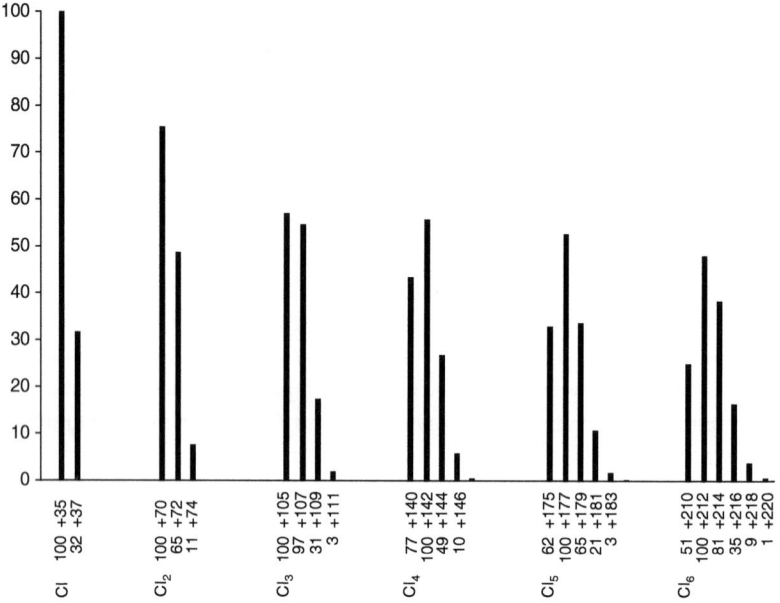

Figure 3.35 Isotope pattern of chlorine (Cl to Cl$_6$).

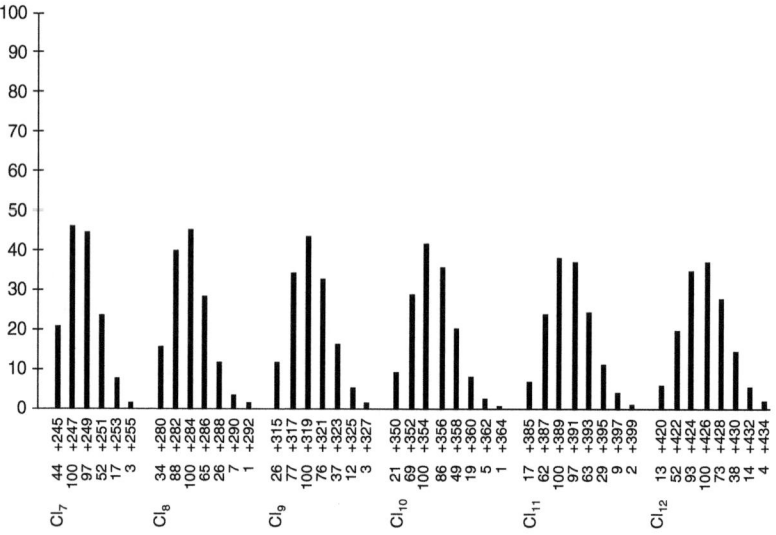

Figure 3.36 Isotope pattern of chlorine (Cl$_7$ to Cl$_{12}$).

empirical formula see Section 3.2.3). In GC-MS coupling, relatively high molecular masses (fragment ions) are detected but through the presence of bromine in a molecule at earlier retention times than can be expected by the molecular mass. In library searches, mixed isotope patterns of chlorine and bromine are reliably recognized.

3.2 Substance Identification | 465

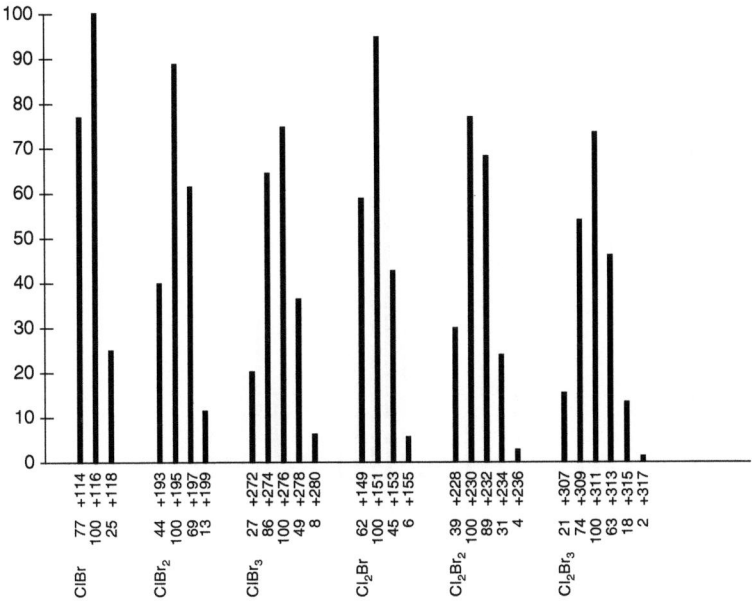

Figure 3.37 Isotope pattern of chlorine/bromine (ClBr to Cl$_2$Br$_3$).

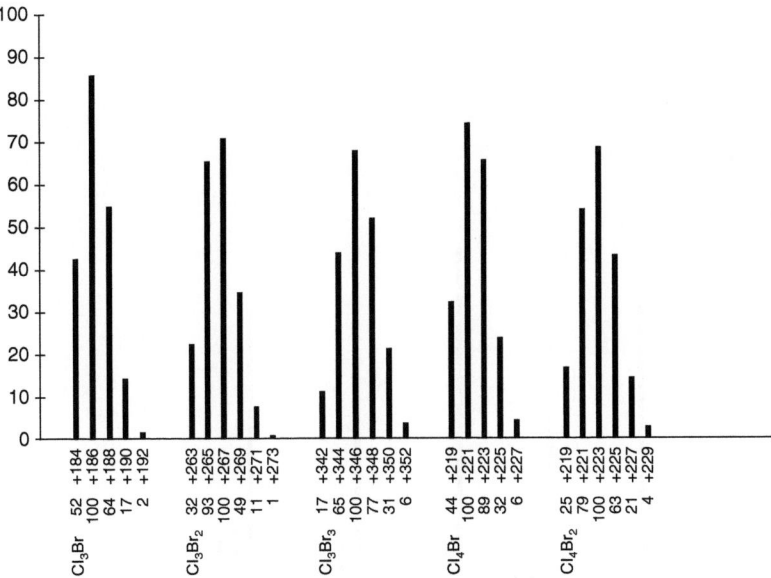

Figure 3.38 Isotope pattern of chlorine/bromine (Cl$_3$Br to Cl$_4$Br$_2$).

Silicon occurs frequently in trace analysis. Silicon gets into the analysis by derivatization (silylation), and frequently through clean-up (joint grease), and more frequently through bleeding from the septum or the column (septa of autosampler vials, silicone phases). The typical isotope pattern of all silicone masses

466 | *3 Evaluation of GC-MS Analyses*

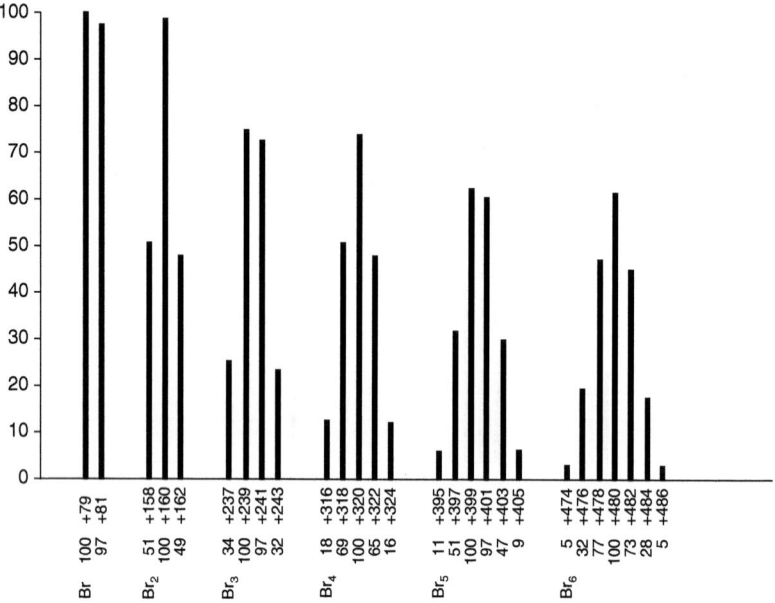

Figure 3.39 Isotope pattern of bromine (Br to Br$_6$).

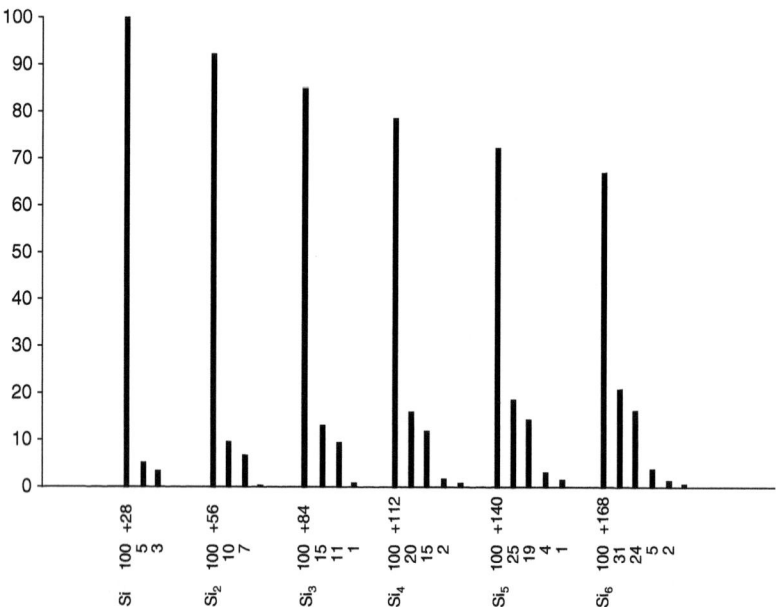

Figure 3.40 Isotope pattern of silicon (Si to Si$_6$).

can be recognized rapidly and excluded from further evaluation measures if from background (Figure 3.40).

With sulfur, the distance between the isotope peaks is also two mass units. However, when the proportion of sulfur in the molecules is low (e.g. in the case of VSCs or phosphoric acid esters), it is difficult to be sure of the presence of sulfur by the mass spectrum alone (Figure 3.41). Here a detailed investigation of the

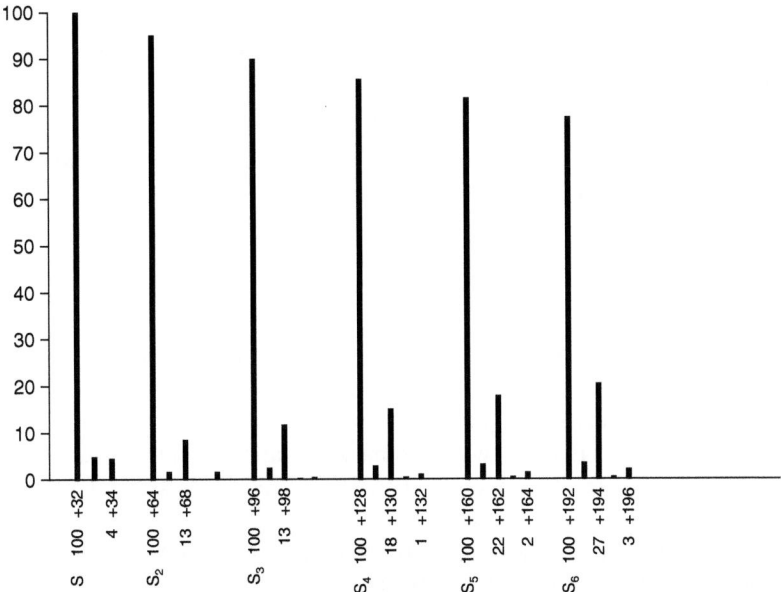

Figure 3.41 Isotope pattern of sulfur (S to S_6).

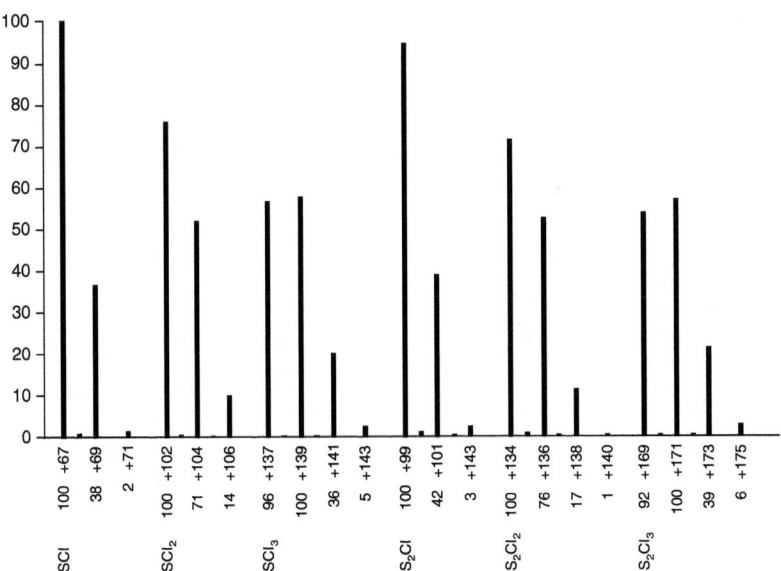

Figure 3.42 Isotope pattern of sulfur/chlorine (SCl to S_2Cl_3).

fragmentation is necessary. A parallel sulfur-specific detector (SCD) detects such components. Higher sulfur contents give clear information (see also Section 3.4). The combinations of sulfur and chlorine in which chlorine is clearly dominant are also shown (Figure 3.42). Differences are extremely difficult to identify visually and with library comparisons only when measurements are obtainable with representative ion statistics.

The theoretical pattern of isotopic distributions of ions can be calculated from the abundance in Table 3.6. A detailed description is provided by Jürgen Gross in his textbook (Gross, 2004).

3.2.6.2 Fragmentation and Rearrangement Reactions

> "Structure determination by mass spectrometry has been hampered by the relatively unpredictable possibilities of molecular rearrangement."
>
> Fred Warren McLafferty, 1959
> 1923–2021, Peter J. W. Debye Professor of Chemistry,
> Cornell University, Ithaka, New York, USA.

The starting point for ion fragmentation is the molecular ion in EI or the cationized molecular ion generated by soft ionization (Figure 3.43). All reactions follow the thermodynamic aim of achieving the most favorable energy balance possible. A large number of reactions that follow the primary ionization need to be described here. The basic mechanisms that are involved in the generation of spectra of organic compounds will be discussed briefly here. For in-depth discussions, the references cited should be consulted (Howe et al., 1981; Pretsch et al., 1990; McLafferty and Turecek, 1993; Budzikiewicz and Schäfer, 2012).

There are two possible mechanisms for the cleavage of carbon chains following ionization, known as *α-cleavage* and the *formation of carbenium ions*. The starting point after ionization is the localization of positive charge on electron-rich structures in the molecule.

α-Cleavage α-Cleavage takes place after ionization by loss of one nonbonding electron from a heteroatom (e.g. in amines, ethers, ketones, see Tables 3.7–3.9) or on the formation of an allylic or benzylic carbenium ion (from alkenes and alkylaromatics).

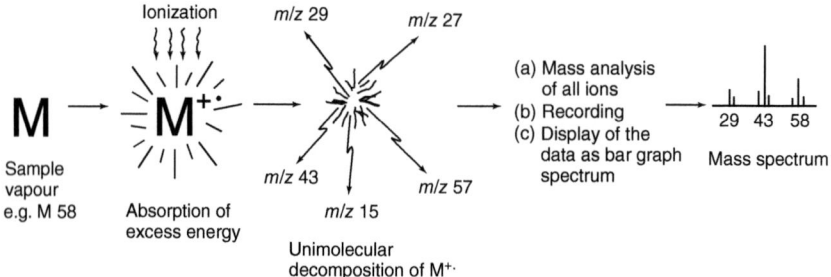

Figure 3.43 Principle of mass spectrometry: formation of molecular ions as the starting point of a fragmentation (after Frigerio, 1974).

Table 3.7 Characteristic ions from the α-cleavage of amines.

Amines

$R_1-\overset{+\cdot}{\underset{R_2}{N}}-CH_2-R \longrightarrow R_1-\overset{+}{\underset{R_2}{N}}=CH_2 + R^\bullet$

m/z	R_1	R_2
30	H	H
44	CH_3	H
58	CH_3	CH_3

Table 3.8 Characteristic ions from the α-cleavage of ethers.

Ethers

$R_1-\overset{+\cdot}{O}-CH_2-R \longrightarrow R_1-\overset{+}{O}=CH_2 + R^\bullet$

m/z	R_1
31	H
45	CH_3
59	C_2H_5
73	C_3H_7
87	C_4H_9

Table 3.9 Characteristic ions from the α-cleavage of ketones.

Ketones

$R_1-\overset{\overset{+\cdot}{O}}{C}-R \longrightarrow R_1-C\equiv\overset{+}{O} + R^\bullet$

m/z	R_1
29	H
43	CH_3
57	C_2H_5
71	C_3H_7
85	C_4H_9

Formation of Carbenium Ions Fragmentation involving the formation of carbenium ions takes place at a double bond in the case of aliphatic carbon chains (allylic carbenium ion) or at a branch. With alkylaromatics, side chains are cleaved giving a benzylic carbenium ion (benzyl cleavage), which dominates as the tropylium ion m/z 91 in many spectra of aromatics.

Alkenes

$$R-CH=CH-CH_2-R \longrightarrow R-\overset{+}{C}H-\overset{\cdot}{C}H-CH_2-R \longrightarrow R-\overset{+}{C}H=CH=CH_2 + R^{\bullet}$$

Branched carbon chains

$$R_3\overset{+}{C}\cdot R \longrightarrow R_3C^+ + R^{\bullet}$$

The formation of carbenium ions occurs preferentially at tertiary branches rather than secondary ones.

Alkylaromatics (benzyl cleavage)

[Reaction scheme: benzyl-CH₂-R → cyclohexadienyl-CH₂-R → cyclohexadienyl-CH₂ + R• → benzyl-CH₂⁺]

Loss of Neutral Particles The elimination of stable neutral particles is a common fragmentation reaction. These include for instance H_2O, CO, CO_2, NO, HCN, HCl, RCOOH, and alkenes (McLafferty, 1959). These reactions can usually be recognized from the corresponding ions and mass differences in the spectra. Eliminations are particularly likely to occur when the α-cleavage is not possible. MS/MS mass spectrometer providing the neutral loss scan mode allows a substance class-specific detection based on the elimination of neutrals.

Alcohols (loss of water)

$$[C_nH_{2n+1}OH]^{+\bullet} \longrightarrow [C_nH_{2n}]^{+\bullet} + H_2O$$

Carbonyl functions (loss of CO)

[Anthraquinone radical cation → fluorenone radical cation + CO]

Heterocycles (loss of HCN)

[Pyridine radical cation → $C_4H_4^{+\bullet}$ + HCN]

Retro-Diels-Alder (loss of alkenes; in short RDA)

[Cyclohexene radical cation → butadiene radical cation + CH₂=CH₂]

The McLafferty Rearrangement The "McLafferty rearrangement" involves the migration of an H-atom in a six-membered ring transition state (McLafferty, 1959). The following conditions must be fulfilled for the rearrangement to take place:

- The double bond C=X is C=C, C=O or C=N.
- There is a chain of three σ-bonds ending in a double bond.
- There is an H-atom in the γ-position relative to the double bond which can be abstracted by the element X of the double bond.

According to convention, the McLafferty rearrangement is classified as the loss of neutral particles (alkene elimination with a positive charge remaining on the fragment formed from the double bond, see Table 3.10).

Table 3.10 Characteristic ions formed in the McLafferty rearrangement.

m/z	R_1	Found
44	H	in aldehydes
60	OH	in organic acids
74	O-CH_3	in methyl esters

3.2.6.3 DMOX Derivatives for Location of Double Bond Positions

The location of double bonds in polyunsaturated fatty acids by GC-MS involves many different methodologies either by suitable derivatization or CI techniques (López and Grimalt, 2004; Jham et al., 2005; Christie and Han, 2010). Derivatization reactions include the reaction with DMOX (Fay and Richli, 1991; Dobson and Christie, 2002), dimethyl disulfide (DMDS) (Moss and Lambert-Fair, 1989), as well as methoxy and methoxybromo derivatives (Shantha and Kaimal, 1984) besides other known derivatives. CI allows specific reactions with C,C-double and triple bonds with suitable reagent gases, which in many cases permit a location of the sites of unsaturation in organic molecules (Budzikiewicz, 1985).

Methoxybromo derivatives of unsaturated fatty acids including conjugated acids yield simple mass spectra to locate the position of double bonds in these acids. Unlike other methods using methoxy derivatives, the methoxybromo derivatives yield fewer ions, the diagnostic peaks forming the most intense ions of the spectra with the characteristic appearance of fragments corresponding to $[CH_3(CH_2)_n CH(OMe)CH(Br)CH_2 2H]^+$.

The preparation of derivatives using 4,4-dimethyloxazoline (DMOX) is most widely applied for routine identification of fatty acids in unknown samples. A relatively mild reaction minimizes possible isomerization reactions. The DMOX

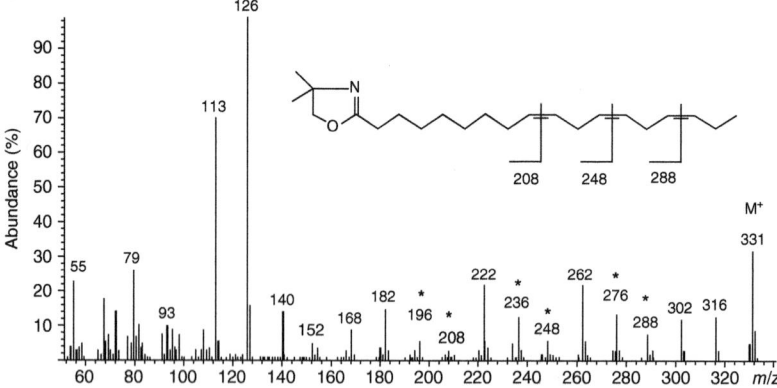

Figure 3.44 Spectrum of the DMOX derivative of 9,12,15-octadecatrieenoate (C18: 3, ω3). Masses indicated by * are the diagnostic ions with 12 Da mass distance for locating double bonds (Christie, 2021).

derivatives are comparable to FAMEs in volatility and hence in chromatographic resolution. When ionized under regular EI conditions, radical-induced cleavage processes give rise to mass spectra that are easy to interpret in terms of locating double bond positions in the hydrocarbon chain. The total number of carbons and the degree of unsaturation can be taken from the molecular ion information. Typical mass distances of 12 Da unveil double bonds corresponding to fragments containing n and $n-1$ carbons (see the mass peaks marked by * with the pairs m/z 196/208, 236/248, and 276/288 in Figure 3.44). Monoenes with double bonds between C7 and C15 follow these rules and exhibit intense allylic ions. Especially with an increasing degree of unsaturation and for conjugated double bonds, DMOX spectra are more informative compared to other derivatives. With double bonds closer to the carboxyl end, the spectra show at C4, C5, and C6 characteristic odd-numbered ions at m/z 139, 153, and 167, respectively. At C3, the base peak m/z 152, and at C2, the base peak m/z 110 are prominent ions (Christie, 2021).

3.2.7 Mass Spectroscopic Features of Selected Substance Classes

The compounds involved and detected by analytical GC-MS workflows like volatiles, PCBs, or pesticides are discussed with representative mass spectra in the following sections. These compounds do not necessarily belong to the same functional compound class with a chemical classification, for instance, of amines, ester, phenols, hydrocarbons, and others. Many standard references cover these characteristic fragmentation patterns of specific functional groups and compound classes in detail (Howe et al., 1981; McLafferty and Turecek, 1993; Budzikiewicz and Schäfer, 2012).

3.2.7.1 Volatile Halogenated Hydrocarbons

This group of compounds does not belong to a single class of compounds (Figures 3.45–3.54). In a single analysis, more than 60 aliphatic and aromatic

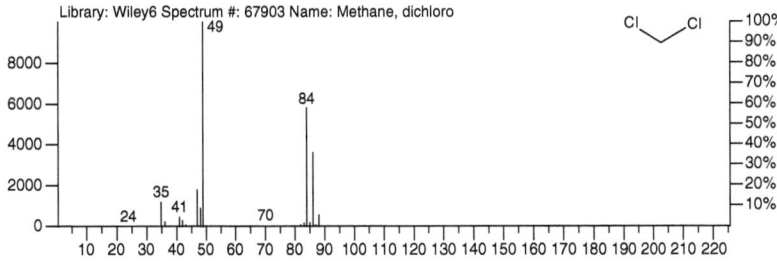

Figure 3.45 Dichloromethane (R30) CH$_2$Cl$_2$, M: 84, CAS Reg. No.: 75-09-02.

compounds (VVOCs, VOCs, Magic 60) are typically determined by headspace GC-MS (static/dynamic HS, purge and trap). More than 300 compounds were collected and identified from air samples (Ciccioli et al., 1993). A noticeable feature is the frequent appearance of chlorine and bromine isotope patterns in the mass spectra. With aliphatic compounds, molecular ions do not always appear. With increasing molecular size, the M$^+$ intensities decrease. Usually, the loss of Cl (and also Br, F) as a radical from the molecular ion occurs. Fluorine can be recognized as HF elimination from M$^+$ with a difference of 20 Da or as the CF fragment m/z 31. Bromine should be assumed from signals with significantly higher masses but relatively short retention times. For detection, it is a good recommendation to include the masses m/z 35/37 in the scan to guarantee ease of identification during library search.

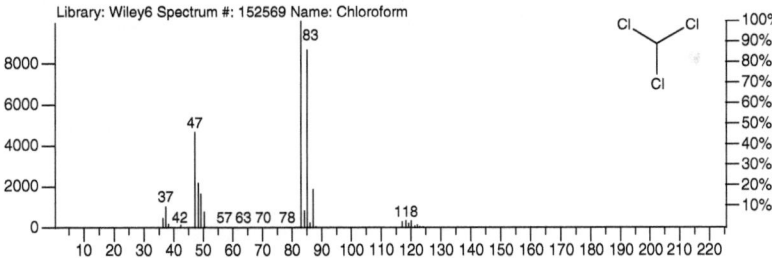

Figure 3.46 Chloroform CHCl$_3$, M: 118, CAS Reg. No.: 67-66-3.

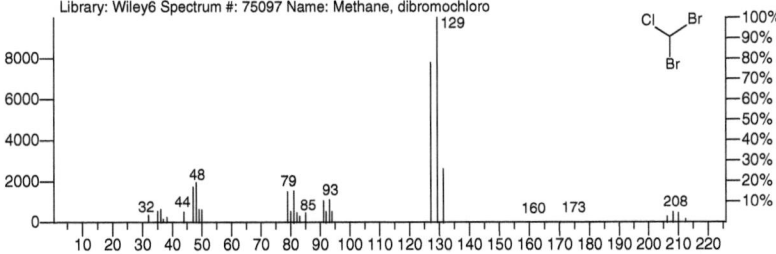

Figure 3.47 Dibromochloromethane CHBr$_2$Cl, M: 206, CAS Reg. No.: 124-48-1.

474 | *3 Evaluation of GC-MS Analyses*

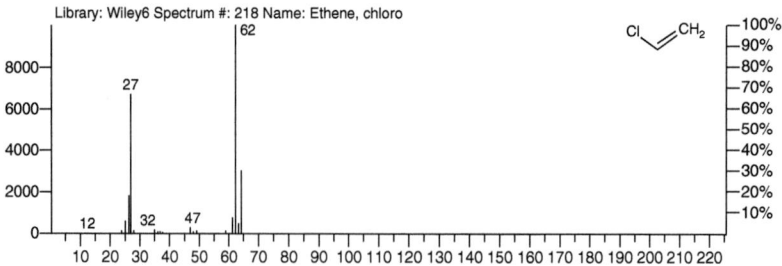

Figure 3.48 Vinyl chloride C$_2$H$_3$Cl, M: 62, CAS Reg. No.: 75-01-4.

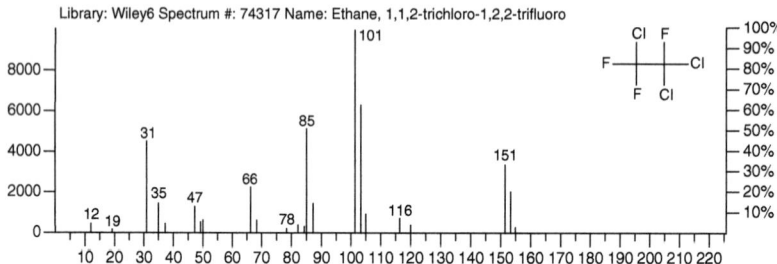

Figure 3.49 1,1,2-Trifluoro-1,2,2-trichloroethane (R113) C$_2$Cl$_3$F$_3$, M: 186, CAS Reg. No.: 76-13-1.

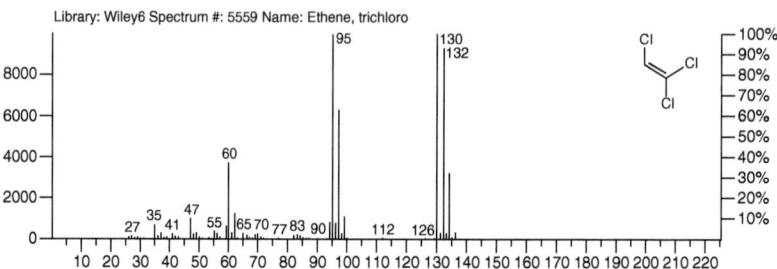

Figure 3.50 Trichloroethylene C$_2$HCl$_3$, M: 130, CAS Reg. No.: 79-01-6.

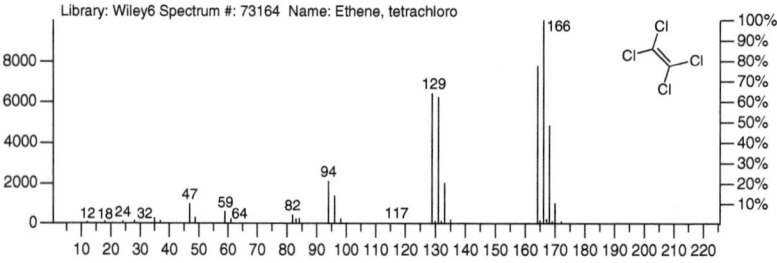

Figure 3.51 Tetrachloroethylene (Per) C$_2$Cl$_4$, M: 164, CAS Reg. No.: 127-18-4.

3.2 Substance Identification | 475

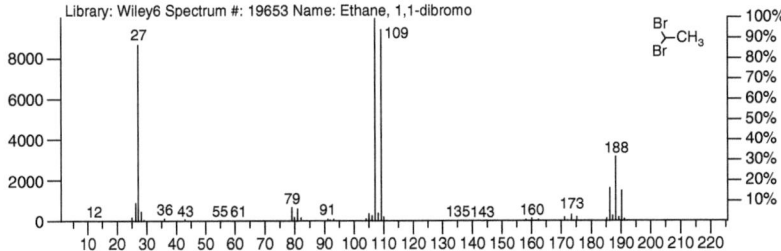

Figure 3.52 1,1-Dibromoethane $C_2H_4Br_2$, M: 186, CAS Reg. No.: 557-91-5.

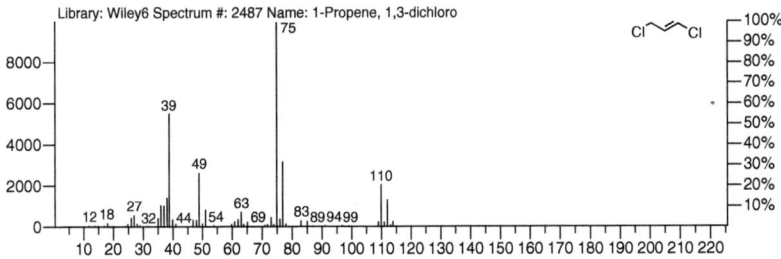

Figure 3.53 Dichloropropene $C_3H_4Cl_2$, M: 110, CAS Reg. No.: 542-75-6.

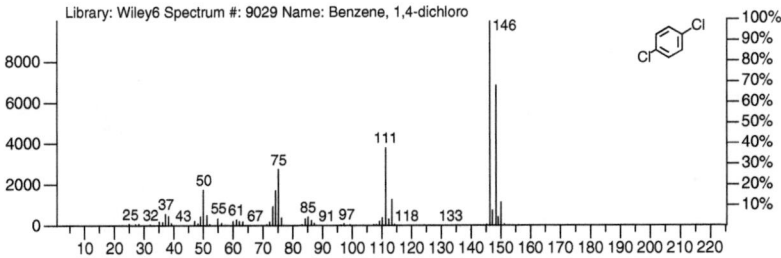

Figure 3.54 p-Dichlorobenzene $C_6H_4Cl_2$, M: 146, CAS Reg. No.: 106-46-7.

Aromatic halogenated hydrocarbons generally show an intense molecular ion. There is successive radical cleavage of chlorine. In the lower mass range, the characteristic aromatic fragments appear with lower intensity.

3.2.7.2 Benzene/Toluene/Ethylbenzene/Xylenes (BTEX, Alkylaromatics)

Alkylaromatics form very stable molecular ions which can be detected with very high sensitivity (Figures 3.55–3.63). The tropylium ion occurs at m/z 91 as the base peak, which is, for example, responsible for the uneven base peak in the toluene spectrum (M-92). The fragmentation of the aromatic skeleton leads to typical series of ions with m/z 38–40, 50–52, 63–67, and 77–79 ("aromatic rubble"). Ethylbenzene and the xylenes cannot be differentiated from their spectra because they are isomers. In these cases, the retention times of the components are more meaningful.

With alkyl side chains, the problem of isomerism must be considered. For di-methylnaphthalene, for example, there are 10 isomers! Consider RIs for

476 | *3 Evaluation of GC-MS Analyses*

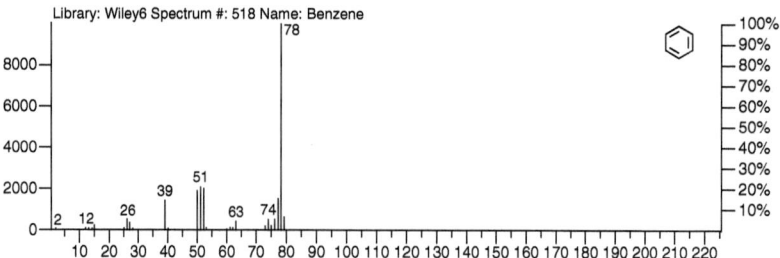

Figure 3.55 Benzene C$_6$H$_6$, M: 78, CAS Reg. No.: 71-43-2.

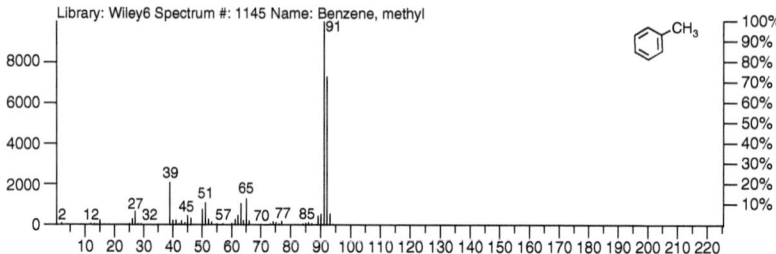

Figure 3.56 Toluene C$_7$H$_8$, M: 92, CAS Reg. No.: 108-88-3.

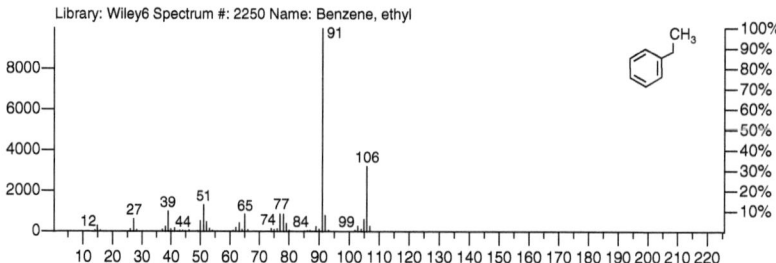

Figure 3.57 Ethylbenzene C$_8$H$_{10}$, M: 106, CAS Reg. No.: 100-41-4.

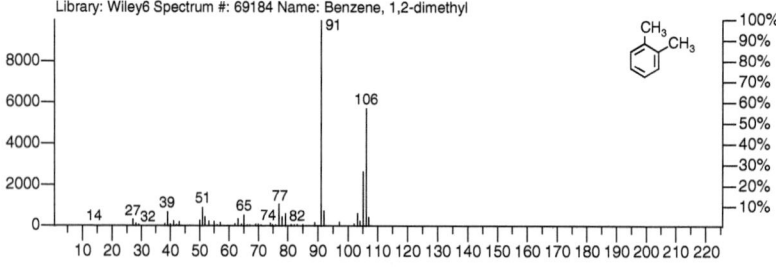

Figure 3.58 *o*-Xylene C$_8$H$_{10}$, M: 106, CAS Reg. No.: 95-47-6.

3.2 Substance Identification | 477

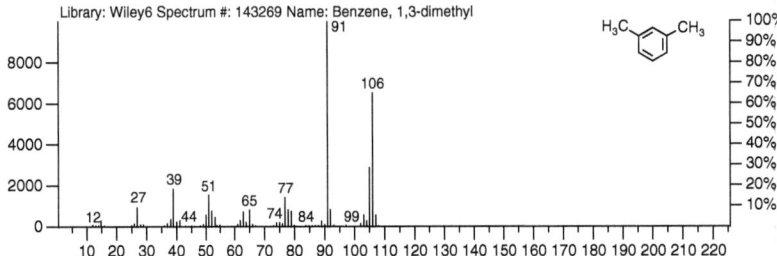

Figure 3.59 m-Xylene C$_8$H$_{10}$, M: 106, CAS Reg. No.: 108-38-3.

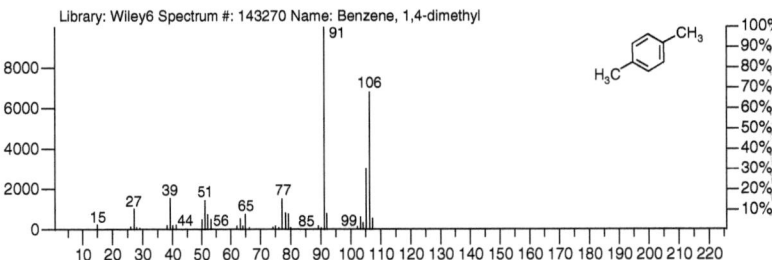

Figure 3.60 p-Xylene C$_8$H$_{10}$, M: 106, CAS Reg. No.: 106-42-3.

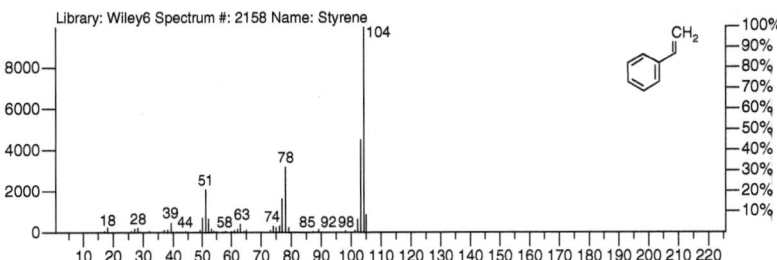

Figure 3.61 Styrene C$_8$H$_8$, M: 104, CAS Reg. No.: 100-42-5.

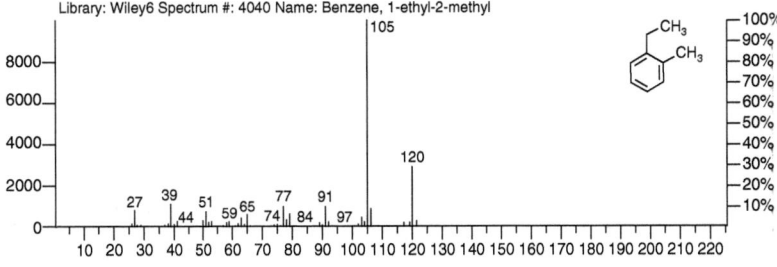

Figure 3.62 1-Ethyl-2-methylbenzene C$_9$H$_{12}$, M: 120, CAS Reg. No.: 611-14-3.

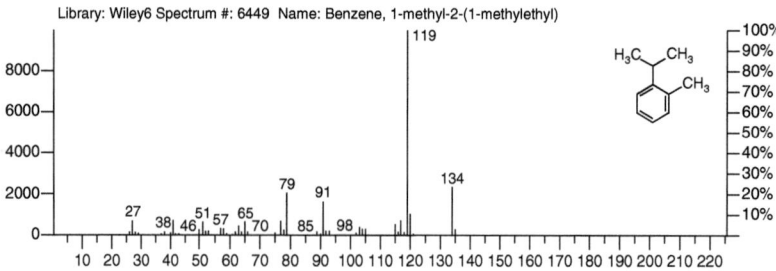

Figure 3.63 1-Methyl-2-isopropylbenzene $C_{10}H_{14}$, M: 134, CAS Reg. No.: 1074-17-5.

identification. Alkylaromatics fragment through benzyl cleavage. From a propyl side chain onwards benzyl cleavage can take place with H-transfer to the aromatic ring (to C_8H_{10}, m/z 106) or without transfer (to C_8H_9, m/z 105) depending on the steric or electronic conditions.

3.2.7.3 Polyaromatic Hydrocarbons

PAHs form very stable molecular ions (Figures 3.64–3.72). They can be recognized easily from a "half mass" signal caused by doubly charged molecular ions as $m/2z = 1/2\,m/z$. Be aware of heterocycles with odd numbers of nitrogen, the doubly charged ion appears on two masses in LRMS, see acridine in Figure 3.70. Masses in the range m/z 100 to m/z 320 should be scanned for PAH full scan analysis, if not SIM or SRM is used. In this range, all polycondensed aromatics from naphthalene to coronene (including all 16 U.S. EPA components) can be determined and a possible

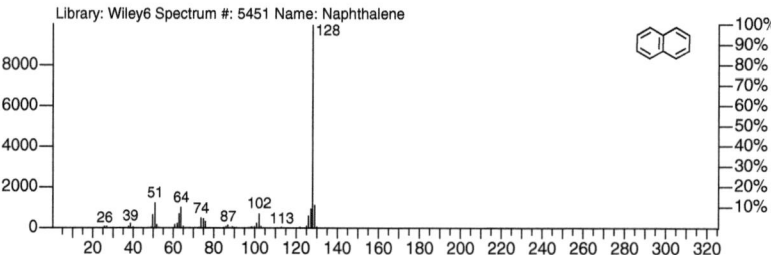

Figure 3.64 Naphthalene $C_{10}H_8$, M: 128, CAS Reg. No.: 91-20-3.

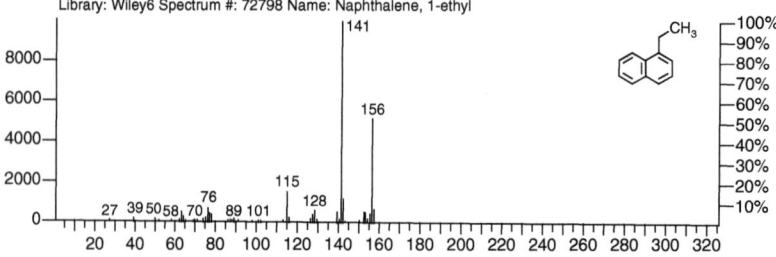

Figure 3.65 1-Ethylnaphthalene $C_{12}H_{12}$, M: 156, CAS Reg. No.: 1127-76-0.

3.2 Substance Identification | 479

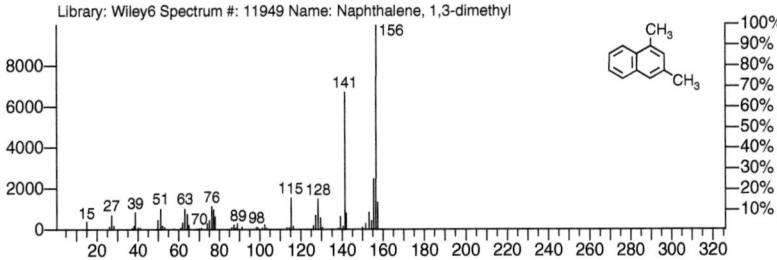

Figure 3.66 1,3-Dimethylnaphthalene $C_{12}H_{12}$, M: 156, CAS Reg. No.: 575-41-7.

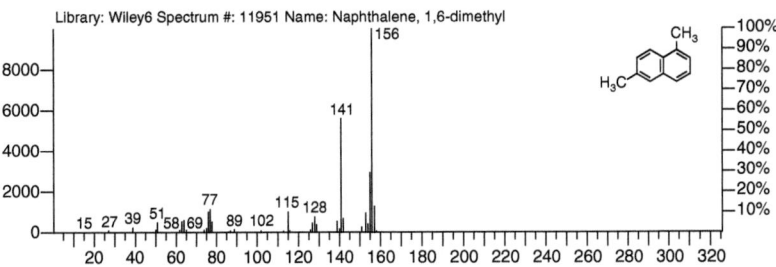

Figure 3.67 1,6-Dimethylnaphthalene $C_{12}H_{12}$, M: 156, CAS Reg. No.: 575-43-9.

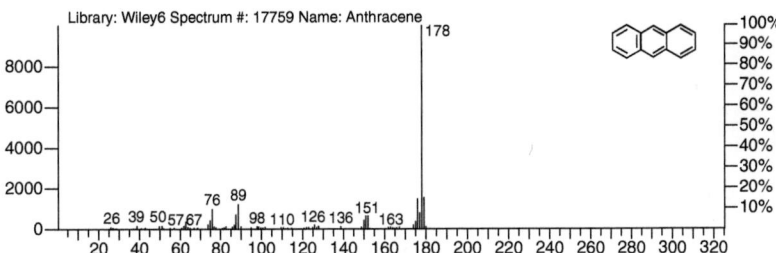

Figure 3.68 Anthracene $C_{14}H_{10}$, M: 178, CAS Reg. No.: 120-12-7.

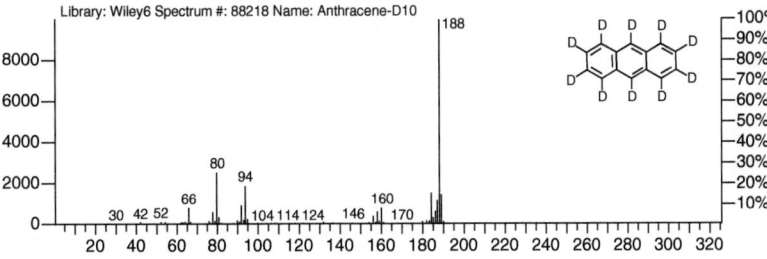

Figure 3.69 Anthracene-d10 $C_{14}D_{10}$, M: 188, CAS Reg. No.: 1719-06-8.

480 | *3 Evaluation of GC-MS Analyses*

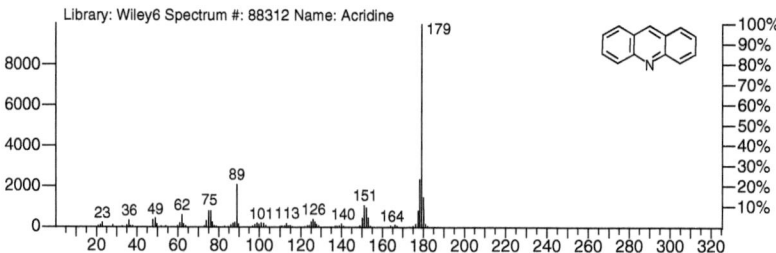

Figure 3.70 Acridine C$_{13}$H$_9$N, M: 179, CAS Reg. No.: 260-94-6.

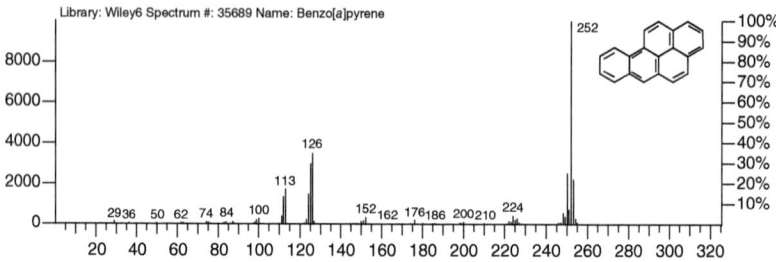

Figure 3.71 Benzo[a]pyrene C$_{20}$H$_{12}$, M: 252, CAS Reg. No.: 50-32-8.

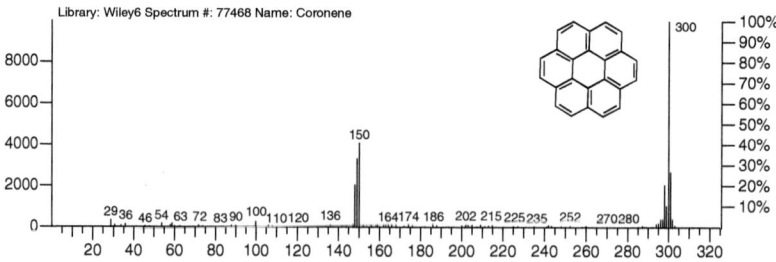

Figure 3.72 Coronene C$_{24}$H$_{12}$, M: 300, CAS Reg. No.: 191-07-1.

matrix background of hydrocarbons with aliphatic character can be almost completely excluded from detection.

3.2.7.4 Phenols

In MS, phenolic substances are determined by their aromatic character (Figures 3.73–3.78). Depending on the side chains, intense molecular ions and less intense fragments appear. In GC-MS, phenols are usually chromatographed as their methyl esters or acetates. Phenols especially halogenated phenols are acidic and tend to severe peak tailing. In trace analysis, chlorinated and brominated phenols are the most important and can be recognized by their clear isotope patterns. The loss of CO (M-28) gives a less intense signal but is a clear indication of the presence of phenols. Halogenated phenols clearly show the loss of HCl (M-36) and HBr (M-80) in their spectra. With phenols, isomers are also best recognized from their RIs rather than their mass spectra.

3.2 Substance Identification | 481

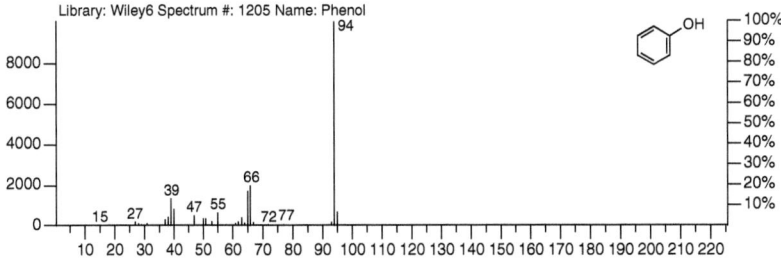

Figure 3.73 Phenol C$_6$H$_6$O, M: 94, CAS Reg. No.: 108-95-2.

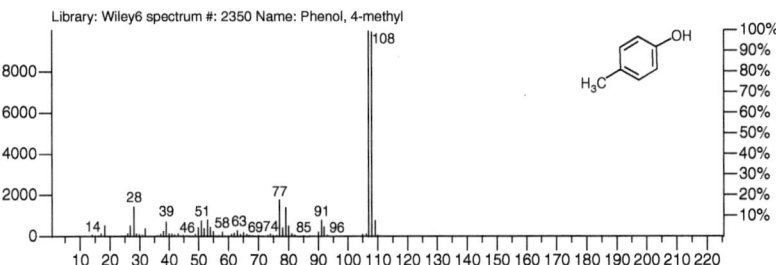

Figure 3.74 p-Cresol C$_7$H$_8$O, M: 108, CAS Reg. No.: 106-44-5.

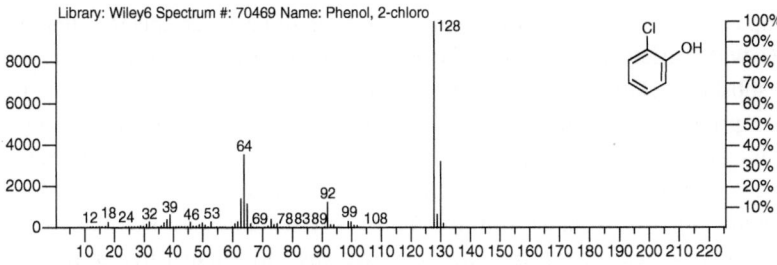

Figure 3.75 o-Chlorophenol C$_6$H$_5$ClO, M: 128, CAS Reg. No.: 95-57-8.

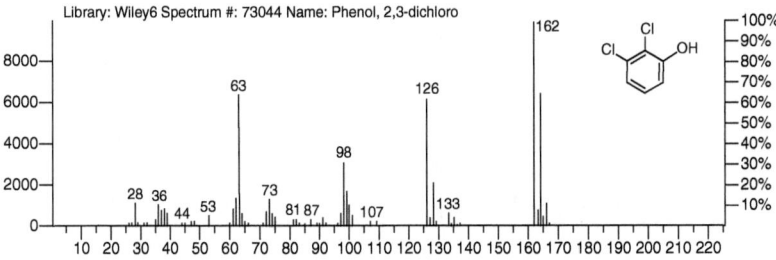

Figure 3.76 2,3-Dichlorophenol C$_6$H$_4$Cl$_2$O, M: 162, CAS Reg. No.: 576-24-9.

482 | 3 Evaluation of GC-MS Analyses

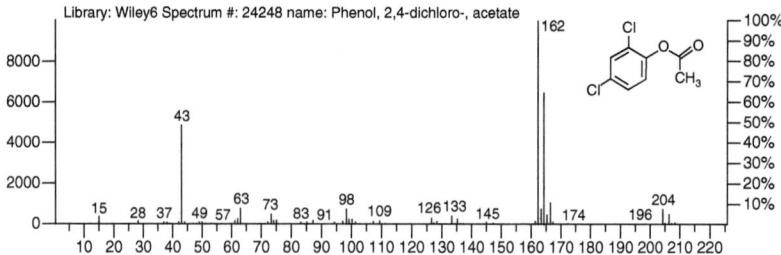

Figure 3.77 2,4-Dichlorophenyl acetate $C_8H_6Cl_2O_2$, M: 204, CAS Reg. No.: 6341-97-5.

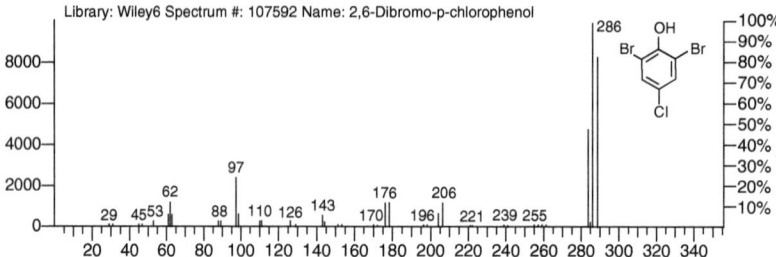

Figure 3.78 2,6-Dibromo-4-chlorophenol $C_6H_3Br_2ClO$, M: 284.

3.2.7.5 Pesticides

The common use as plant protection agents forms the basis of the classification of these compounds (Figures 3.79–3.110). In a collection of pesticide spectra, there is therefore a wide variety of compound classes, which are covered to some extent by the other substance classes described here. Even when considering what appears to be a single group, such as phosphoric acid esters, it is virtually impossible to give generalizations on fragmentation. Usually, only stable compounds with aromatic character form intense molecular ions. In other cases, molecular ions are of low intensity and cannot be isolated from the matrix in trace analyses. For many pesticides (phosphoric acid esters, triazines, phenylureas, chlorinated hydrocarbons, etc.), the use of PCI is advantageous for confirming identities or allowing selective detection by NCI.

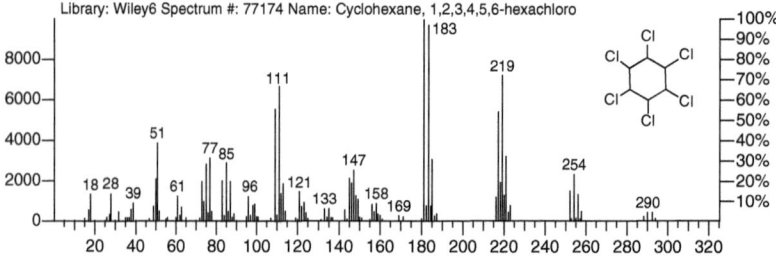

Figure 3.79 Lindane $C_6H_6Cl_6$, M: 288, CAS Reg. No.: 58-89-9 (isomers with similar spectra).

3.2 Substance Identification | 483

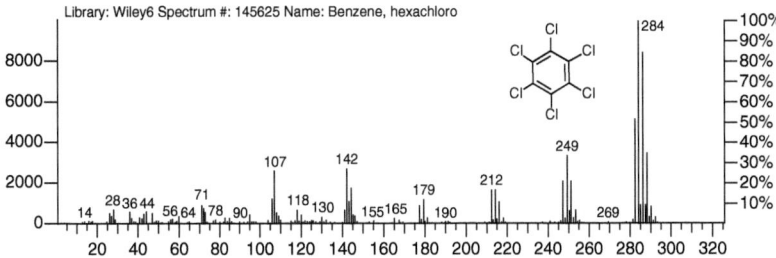

Figure 3.80 Hexachlorobenzene (HCB) C_6Cl_6, M: 282, CAS Reg. No.: 118-74-1.

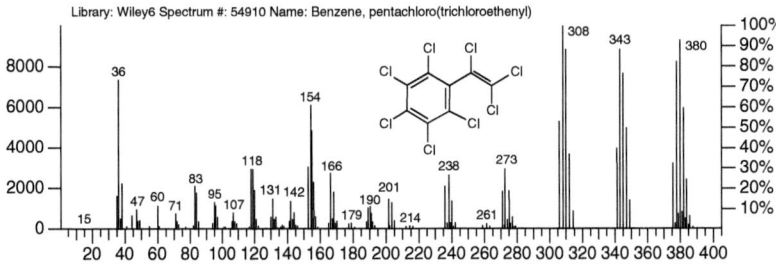

Figure 3.81 Octachlorostyrene C_8Cl_8, M: 376, CAS Reg. No.: 29082-74-4.

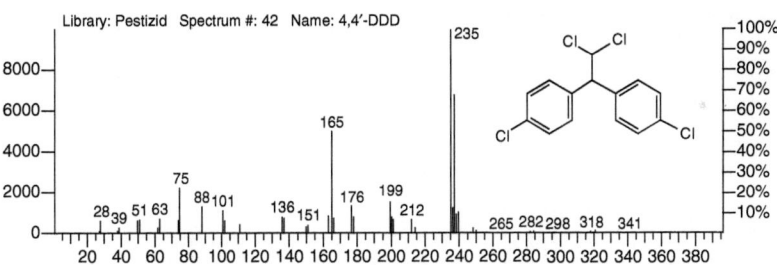

Figure 3.82 4,4′-DDD $C_{14}H_{10}Cl_4$, M: 318, CAS Reg. No.: 72-54-8.

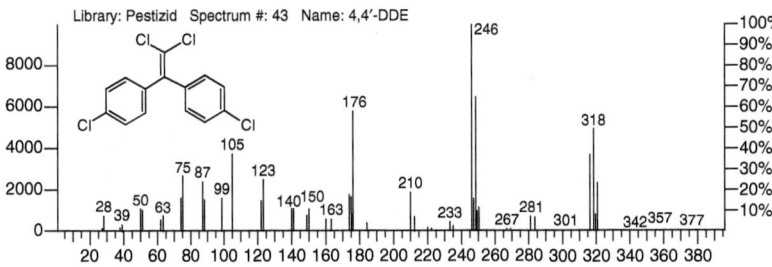

Figure 3.83 4,4′-DDE $C_{14}H_8Cl_4$, M: 316, CAS Reg. No.: 72-55-9.

484 | *3 Evaluation of GC-MS Analyses*

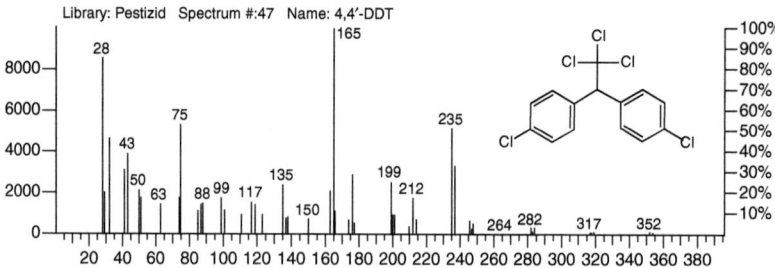

Figure 3.84 4,4′-DDT C$_{14}$H$_9$Cl$_5$, M: 352, CAS Reg. No.: 50-29-3.

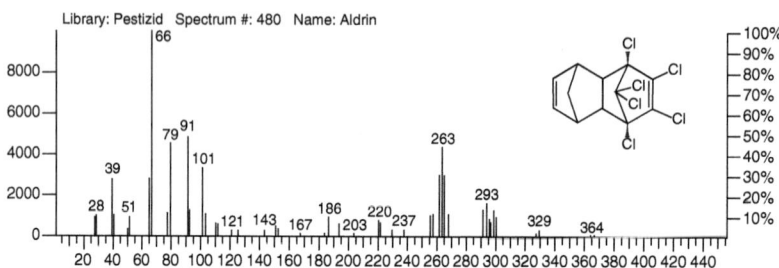

Figure 3.85 Aldrin C$_{12}$H$_8$Cl$_6$, M: 362, CAS Reg. No.: 309-00-2.

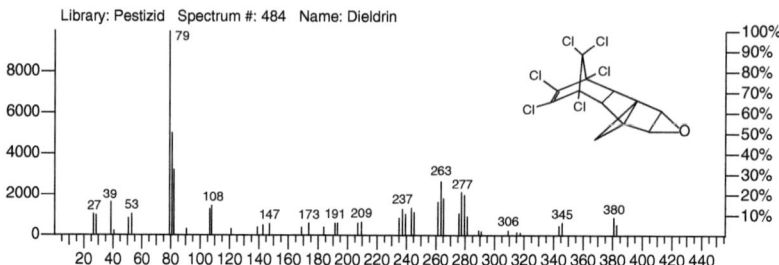

Figure 3.86 Dieldrin C$_{12}$H$_8$Cl$_6$O, M: 378, CAS Reg. No.: 60-57-1.

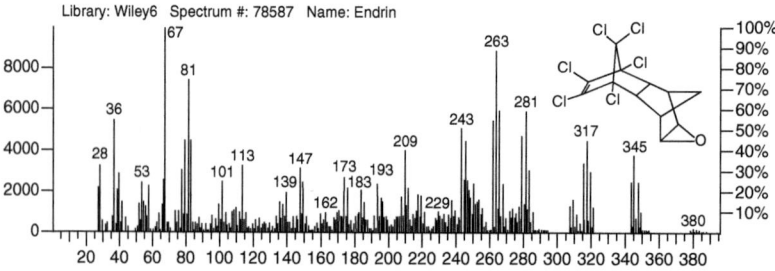

Figure 3.87 Endrin C$_{12}$H$_8$Cl$_6$O, M: 378, CAS Reg. No.: 72-20-8.

3.2 Substance Identification | 485

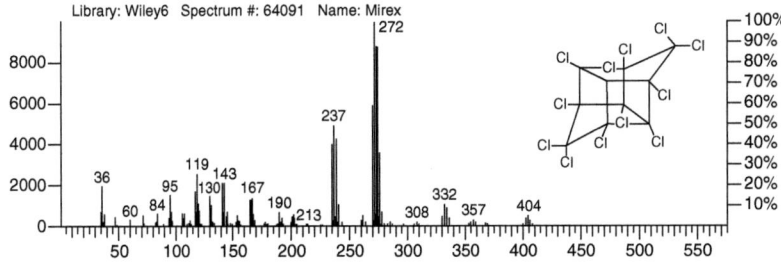

Figure 3.88 Mirex $C_{10}Cl_{12}$, M: 540, CAS Reg. No.: 2385-85-5.

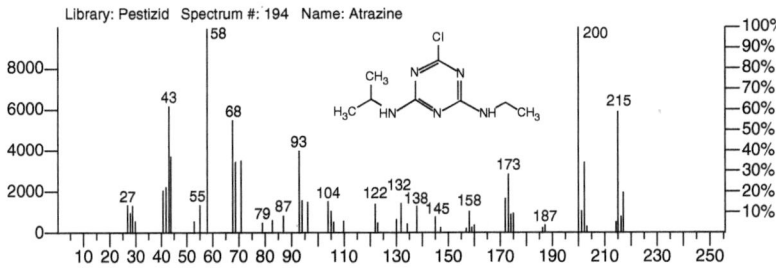

Figure 3.89 Atrazine $C_8H_{14}ClN_5$, M: 215, CAS Reg. No.: 1912-24-9.

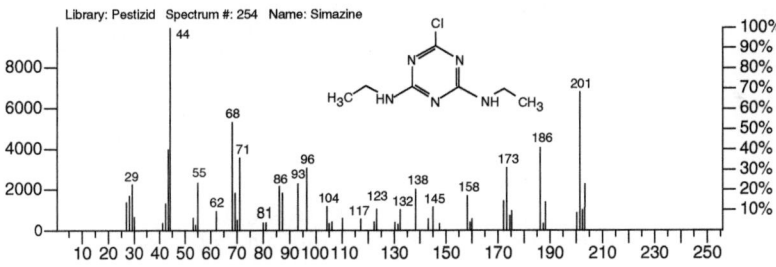

Figure 3.90 Simazine $C_7H_{12}ClN_5$, M: 210, CAS Reg. No.: 122-34-9.

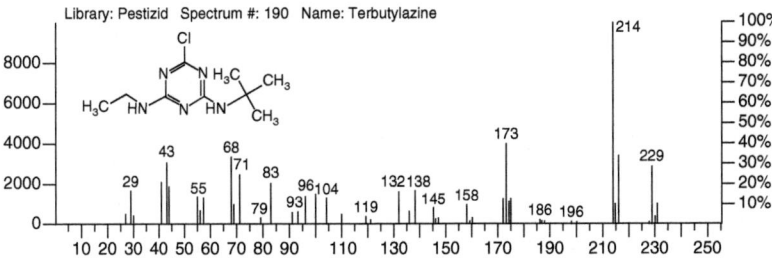

Figure 3.91 Terbutylazine $C_9H_{16}ClN_5$, M: 229, CAS Reg. No.: 5915-41-3.

486 | *3 Evaluation of GC-MS Analyses*

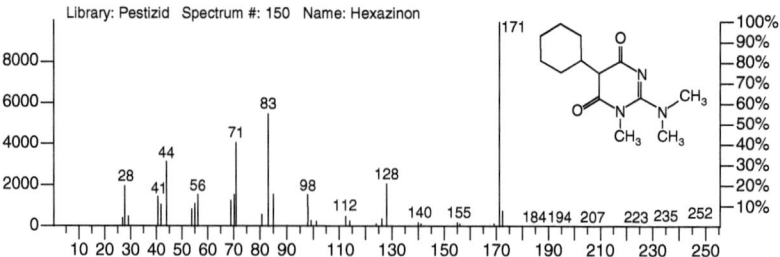

Figure 3.92 Hexazinon $C_{12}H_{20}N_4O_2$, M: 252, CAS Reg. No.: 51235-04-2.

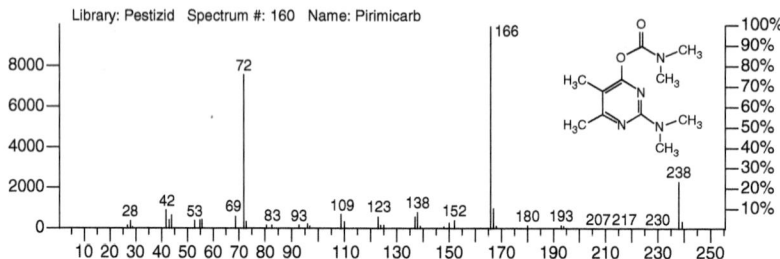

Figure 3.93 Pirimicarb $C_{11}H_{18}N_4O_2$, M: 238, CAS Reg. No.: 23103-98-2.

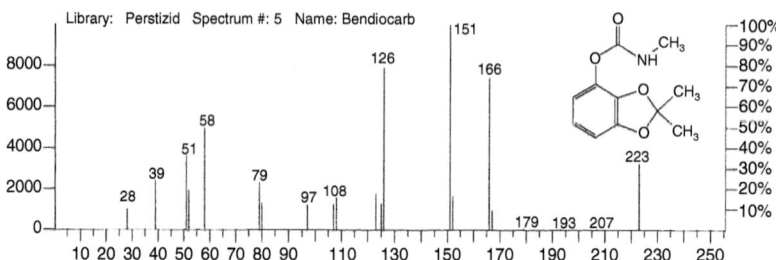

Figure 3.94 Bendiocarb $C_{11}H_{13}NO_4$, M: 223, CAS Reg. No.: 22781-23-3.

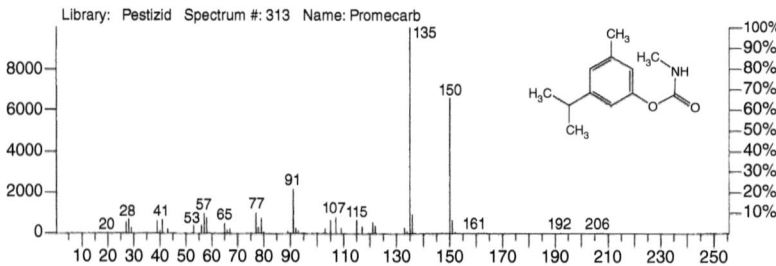

Figure 3.95 Promecarb $C_{12}H_{17}NO_2$, M: 207, CAS Reg. No.: 2631-37-0.

3.2 Substance Identification | 487

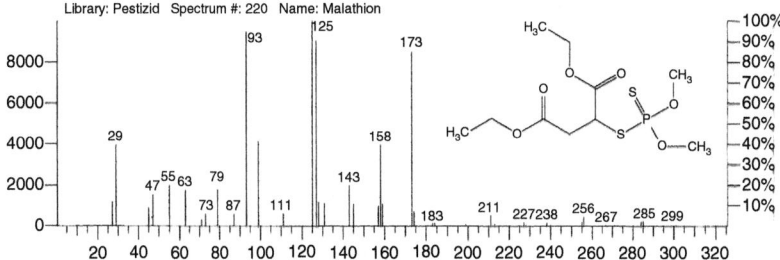

Figure 3.96 Malathion $C_{10}H_{19}O_6PS_2$, M: 330, CAS Reg. No.: 121-75-5.

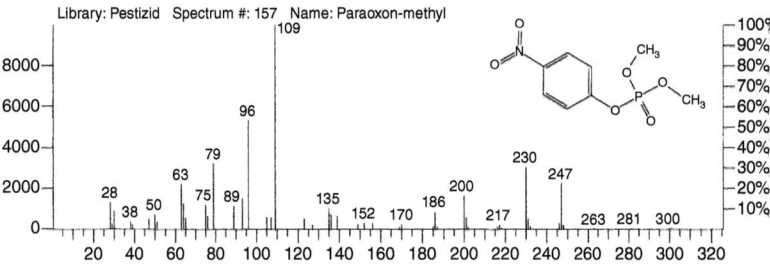

Figure 3.97 Paraoxon-methyl $C_8H_{10}NO_6P$, M: 247, CAS Reg. No.: 950-35-6.

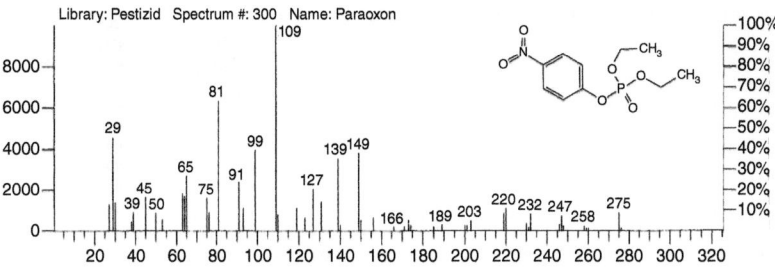

Figure 3.98 Paraoxon(-ethyl) $C_{10}H_{14}NO_6P$, M: 275, CAS Reg. No.: 311-45-5.

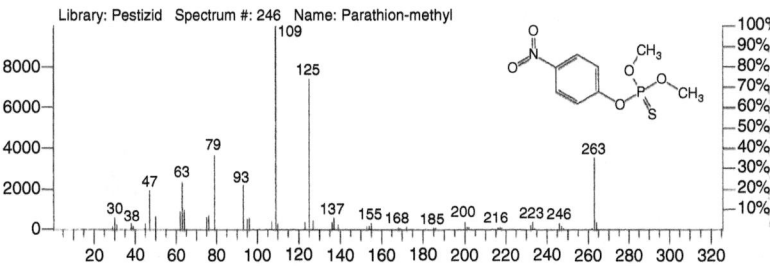

Figure 3.99 Parathion-methyl $C_8H_{10}NO_5PS$, M: 263, CAS Reg. No.: 298-00-0.

488 | 3 Evaluation of GC-MS Analyses

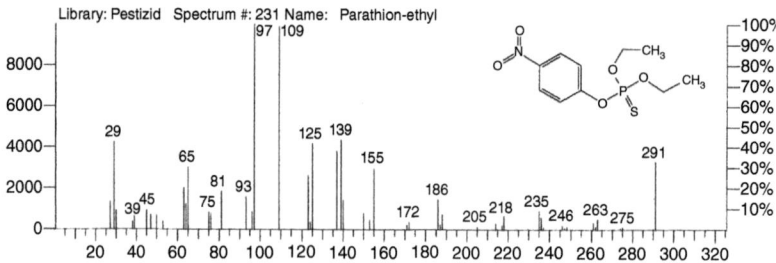

Figure 3.100 Parathion(-ethyl) C$_{10}$H$_{14}$NO$_5$PS, M: 291, CAS Reg. No.: 56-38-2.

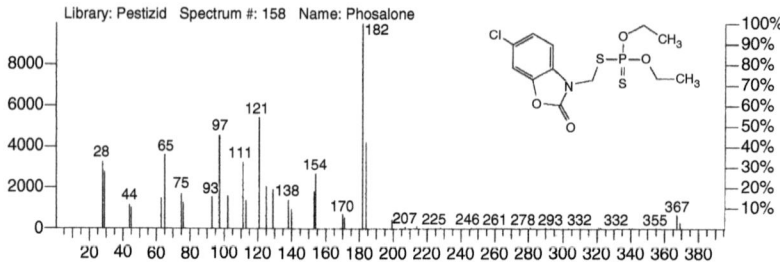

Figure 3.101 Phosalone C$_{12}$H$_{15}$ClNO$_4$PS$_2$, M: 367, CAS Reg. No.: 2310-17-0.

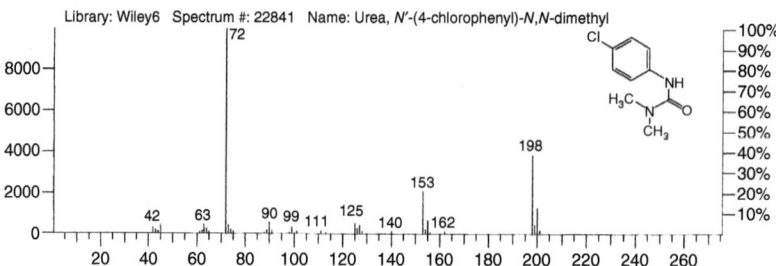

Figure 3.102 Monuron C$_9$H$_{11}$ClN$_2$O, M: 198, CAS Reg. No.: 150-68-5.

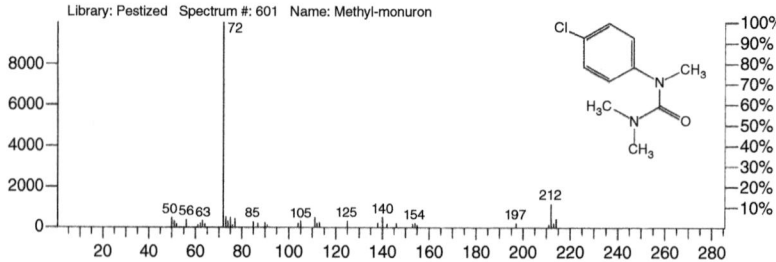

Figure 3.103 Methyl-monuron C$_{10}$H$_{13}$ClN$_2$O, M: 212.

3.2 Substance Identification | 489

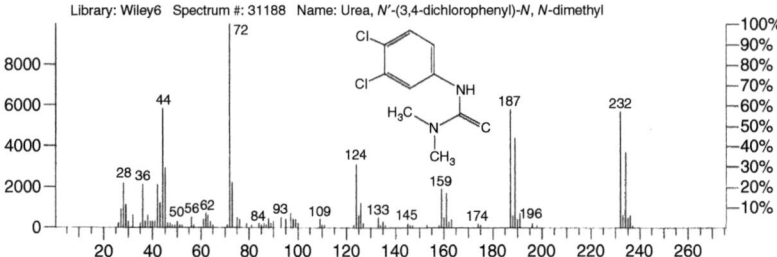

Figure 3.104 Diuron $C_9H_{10}Cl_2N_2O$, M: 232, CAS Reg. No.: 330-54-1.

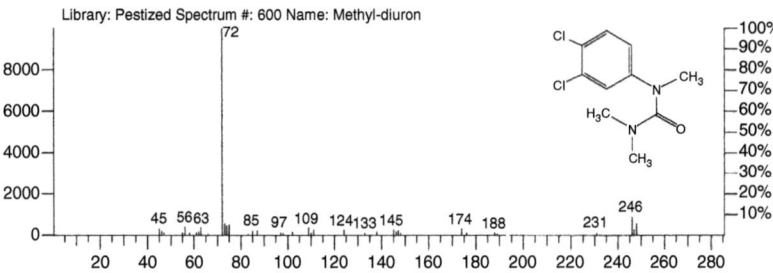

Figure 3.105 Methyl-diuron $C_{10}H_{12}Cl_2N_2O$, M: 246.

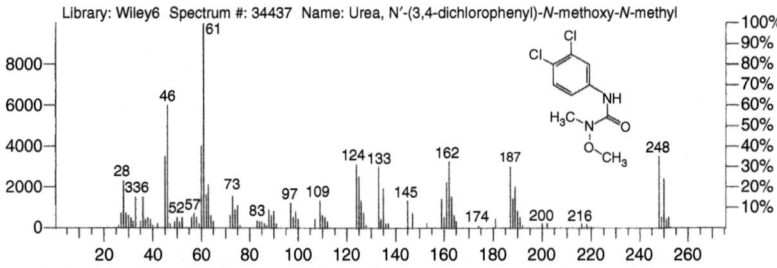

Figure 3.106 Linuron $C_9H_{10}Cl_2N_2O_2$, M: 248, CAS Reg. No.: 330-55-2.

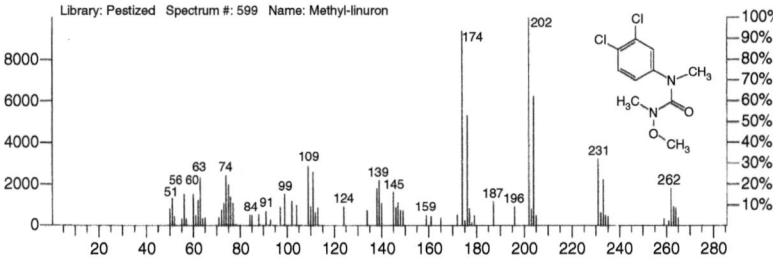

Figure 3.107 Methyl-linuron $C_{10}H_{12}Cl_2N_2O_2$, M: 262.

3 Evaluation of GC-MS Analyses

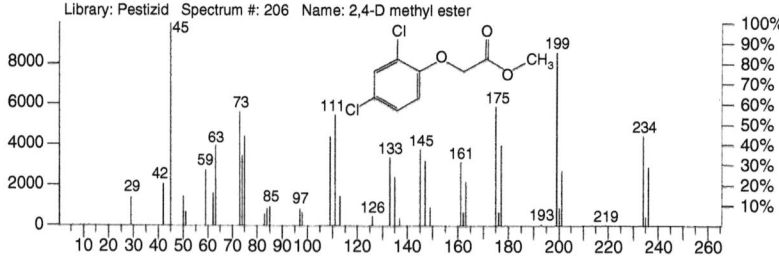

Figure 3.108 2,4-D methylester C$_9$H$_8$Cl$_2$O$_3$, M: 234, CAS Reg. No.: 1928-38-7.

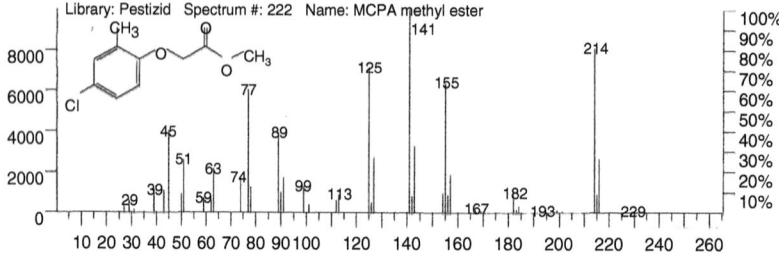

Figure 3.109 MCPA methylester C$_{10}$H$_{11}$ClO$_3$, M: 214, CAS Reg. No.: 2436-73-9.

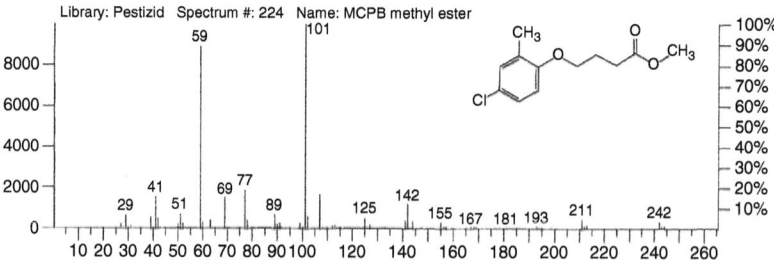

Figure 3.110 MCPB methylester C$_{12}$H$_{15}$ClO$_3$, M: 242, CAS Reg. No.: 57153-18-1.

Chlorinated Hydrocarbons In this group of organochlorine pesticides (OCPs), there is a large number of different types of compounds (Figures 3.79–3.88). In the example of lindane (γ-HCH, BHC) and hexachlorobenzene (HCB), the difference between compounds with saturated and aromatic character can clearly be seen (molecular ion, fragmentation).

The high proportion of chlorine in these analytes appears with intense and characteristic isotope patterns. With the nonaromatic compounds (polycyclic polychlorinated alkanes made by Diels-Alder reactions, e.g. dieldrin, aldrin), spectra with large numbers of lines are formed through extensive fragmentation of the molecule. These compounds can be readily analyzed using negative CI, whereby the fragmentation is prevented.

Triazines Triazine herbicides are substitution products of 1,3,5-triazines and thus belong to a single series of substances (Figures 3.89–3.91). Hexazinon is also determined together with the triazine group in analysis (Figure 3.92). Without exception the EI spectra of triazines show many fragment ions and usually also contain the molecular ion with varying intensity. The high degree of fragmentation is responsible for the low specific response of triazines in trace analysis. All triazine analyses can be confirmed and quantified readily by positive CI (e.g. with NH$_3$ as the reagent gas).

Carbamates The highly polar carbamate pesticides cannot always be analyzed by GC-MS (Figures 3.93–3.95). The low thermal stability leads to decomposition even in the injector. Substances with definite aromatic character, however, form stable intense molecular ions.

Phosphoric Acid Esters This large group of organophosphorous pesticides (OPPs) does not exhibit uniform fragmentation behavior (Figures 3.96–3.101). In trace analysis, the detection of molecular ions is usually difficult, except in the case of aromatic compounds (e.g. parathion). The high degree of fragmentation frequently extends into the area of matrix noise. Because of this, it is more difficult to detect individual compounds, as in full scan analysis, low starting masses must be used. Positive CI is suitable for phosphoric acid esters because generally a strong CI reaction can be expected because of the large number of functional groups. In NCI the occurrence of the molecular anion is dependent on the structure of the molecule and its ability to stabilize the negative ion by charge delocalization (Stan and Kellner, 1982). The presence of the halogens Cl or Br can only be determined with certainty from the (cationized) molecular ions. Phosphoric acid esters are also used as highly toxic chemical warfare agents (Tabun, Sarin and Soman, see Section 3.2.7.10).

The occurrence of fragments belonging to a particular group in the spectra of phosphoric acid esters has been intensively investigated (Table 3.11). Phosphoric acid esters can be subdivided as follows with R the short methyl- or ethyl-, and Z the compound determining side chains (Stan et al., 1977, Stan, 1981):

Table 3.11 Fragments typical of various groups from phosphoric acid ester (PAE) pesticides (Stan et al., 1977).

Group		R	m/z 93	m/z 97	m/z 109	m/z 121	m/z 125
Dithio-PAE	Ia	CH$_3$	+	–	–	–	+
	Ib	C$_2$H$_5$	+	+	–	+	+
Thiono-PAE	IIa	CH$_3$	+	–	+	–	+
	IIb	C$_2$H$_5$	+	+	+	+	+
Thiol-PAE	IIIa	CH$_3$	(+)	–	+	–	+
	IIIb	C$_2$H$_5$	–	+	+	+	–
PAE	IVa	CH$_3$	(+)	–	+	–	–
	IVb	C$_2$H$_5$	+/–	–	+	–	–

I	Dithiophosphoric acid esters	$(RO)_2\text{-P(S)-S-Z}$
II	Thionophosphoric acid esters	$(RO)_2\text{-P(S)-O-Z}$
III	Thiophosphoric acid esters	$(RO)_2\text{-P(O)-S-Z}$
IV	Phosphoric acid esters	$(RO)_2\text{-P(O)-O-Z}$

Phenylureas GC-MS can be used for the determination of phenylureas only after the derivatization of the active substances (Färber and Schöler, 1991). HPLC or HPLC-MS are currently the most suitable analytical methods because of the thermal lability and polarity of these compounds. A spectrum with few lines is obtained in GC-MS analyses, which is dominated by the dimethylisocyanate ion m/z 72, which is specific to the group. The molecular ion region is of higher specificity but usually lower intensity (Figures 3.102–3.106). The phenylureas are rendered more suitable for GC by methylation of the azide hydrogen (e.g. with trimethylsulfonium hydroxide [TMSH] in the PTV [programmed temperature vaporizer] injector after Färber). The mass spectra of the methyl derivatives correspond to those of the parent substances, except that the molecular ions are 14 masses higher with the same fragmentation pattern (Figures 3.103, 3.105, and 3.107).

Phenoxyalkylcarboxylic Acids The free acids cannot be analyzed by GC-MS in the case of trace analysis. They are determined using the methyl ester (Figures 3.108–3.110). If the aromatic character predominates, intense molecular ions occur in the upper mass range. Increasing the length of the side chains significantly reduces the intensity of the molecular ion and leads to signals in the lower mass range. It should be noted that the presence of Cl or Br can be determined with certainty only from the (cationized) molecular ion. With EI the molecular ion fragments thereby losing a Cl radical. Because of this, the isotope signals of the fragments cannot be evaluated conclusively (see MCPB methyl ester Figure 3.110). A final confirmation can be achieved through CI.

3.2.7.6 Polychlorinated Biphenyls

The spectra of PCBs show similar features independent of the degree of chlorination. The spectra of compounds with different degrees of chlorination are shown in Figures 3.111–3.120. As they are all derived from the same aromatic backbone, their molecular ions are strongly pronounced. The degree of chlorination can clearly be

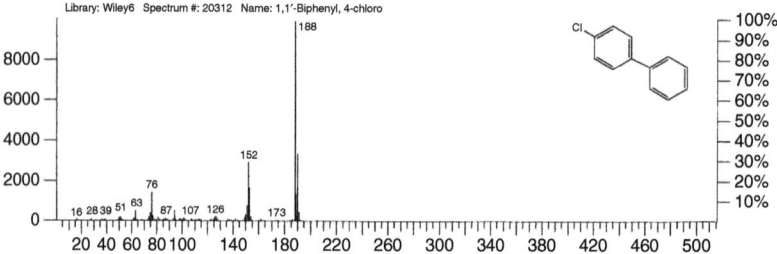

Figure 3.111 Monochlorobiphenyl $C_{12}H_9Cl$, M: 188.

3.2 Substance Identification | 493

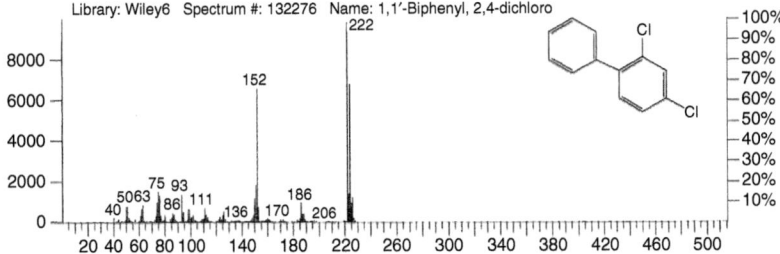

Figure 3.112 Dichlorobiphenyl $C_{12}H_8Cl_2$, M: 222.

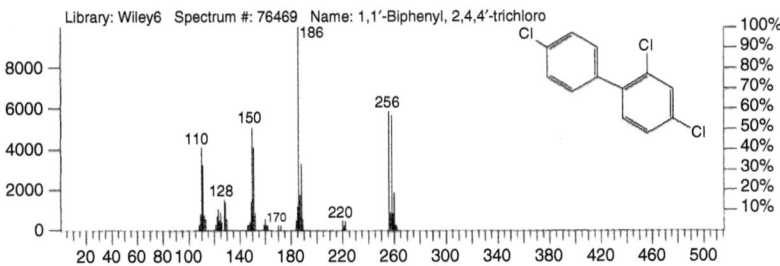

Figure 3.113 Trichlorobiphenyl (e.g. PCB 28, 31) $C_{12}H_7Cl_3$, M: 256.

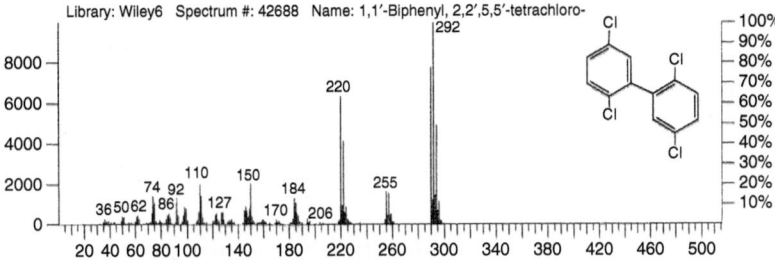

Figure 3.114 Tetrachlorobiphenyl (e.g. PCB 52) $C_{12}H_6Cl_4$, M: 290.

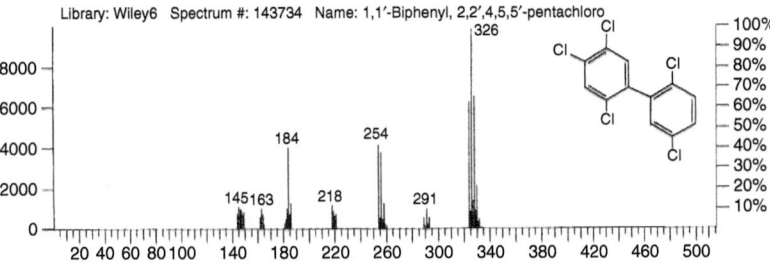

Figure 3.115 Pentachlorobiphenyl (e.g. PCB 101, 118) $C_{12}H_5Cl_5$, M: 324.

494 | *3 Evaluation of GC-MS Analyses*

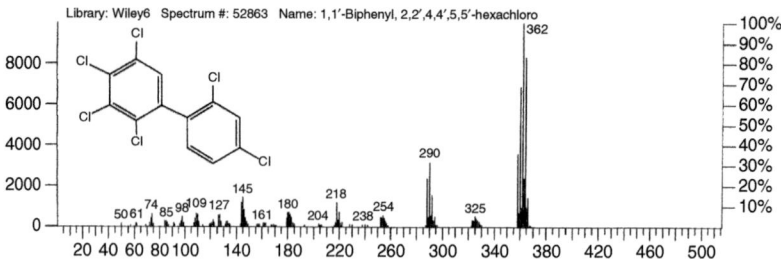

Figure 3.116 Hexachlorobiphenyl (e.g. PCB 138, 153) $C_{12}H_4Cl_6$, M: 358.

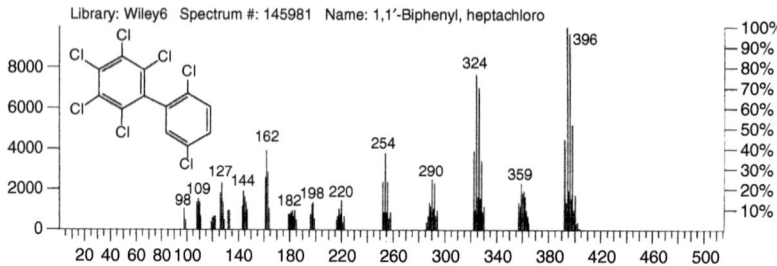

Figure 3.117 Heptachlorobiphenyl (e.g. PCB 180) $C_{12}H_3Cl_7$, M: 392.

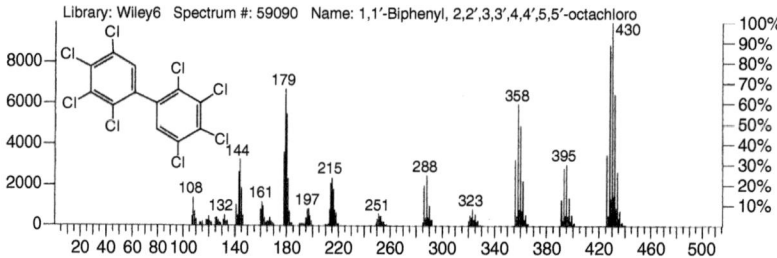

Figure 3.118 Octachlorobiphenyl $C_{12}H_2Cl_8$, M: 426.

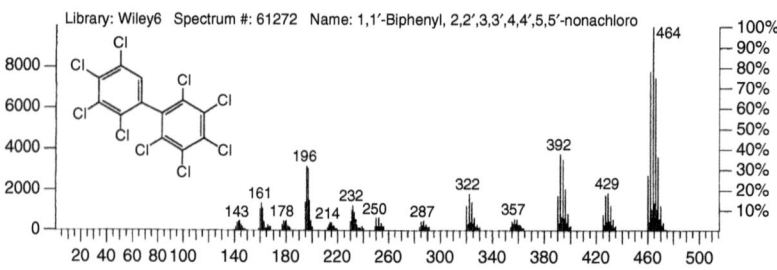

Figure 3.119 Nonachlorobiphenyl $C_{12}HCl_9$, M: 460.

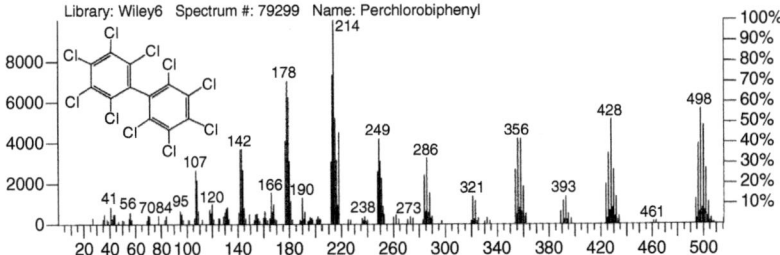

Figure 3.120 Decachlorobiphenyl (PCB 209) $C_{12}Cl_{10}$, M: 494.

Table 3.12 Indicator PCB congeners for quantitation (indicate the presence or absence of PCB mixtures in a sample).

PCB No.	Structure (Cl substitution)	
28	2,4,4'	Cl_3-PCB
52	2,2',5,5'	Cl_4-PCB
101	2,2',4,5,5'	Cl_5-PCB
118	2,3',4,4',5	Cl_5-PCB
138	2,2',3,4,4',5	Cl_6-PCB
153	2,2',4,4',5,5'	Cl_6-PCB
180	2,2',3,4,4',5,5'	Cl_7-PCB
209	2,2',3,3',4,4',5,5',6,6'	Cl_{10}-PCB[a]

a) used as internal standard.

determined from the isotope pattern. Fragmentation involves successive loss of Cl radicals and, in the lower mass range, degradation of the basic skeleton. For data acquisition in full scan, the mass range above m/z 100 or 150 is required, so that detection of PCBs is usually possible above an accompanying matrix background. SIM or SRM analysis is applied to achieve a high selectivity. Individual isomers with a particular degree of chlorination have almost identical mass spectra. They can be differentiated based on their retention times and therefore good gas chromatographic separation is a prerequisite for the determination of PCBs (see Figure 2.102). The isomers 31 and 28 (systematic numbering after Ballschmiter and Zell, 1980) are used as resolution criteria (Table 3.12).

3.2.7.7 Polychlorinated Dioxins/Furans (PCDDs/PCDFs)

The persistence of this class of substance in the environment parallels their mass spectroscopic stability. As they are aromatic, the polychlorinated and -brominated dioxins and furans give good molecular ion intensities with pronounced isotope patterns and only low fragmentation (Figure 3.121 and Figure 3.122). Internal standards with 6- and 12-fold isotopic labeling (^{13}C) are used

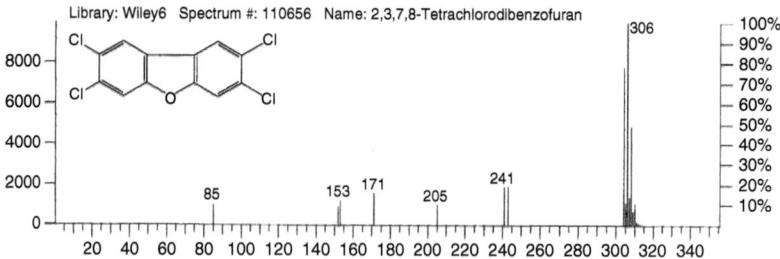

Figure 3.121 2,3,7,8-Tetrachlorodibenzofuran (2,3,7,8-TCDF) $C_{12}H_4Cl_4O$, M: 304, CAS Reg. No.: 51207-31-9.

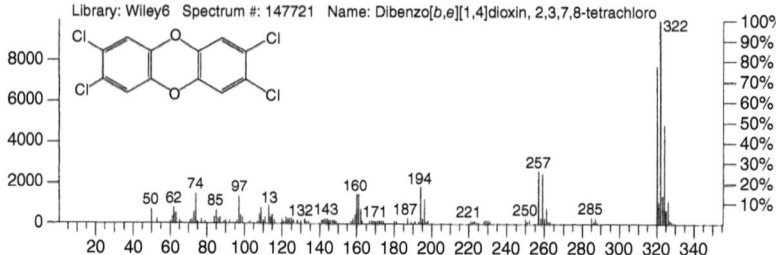

Figure 3.122 2,3,7,8-Tetrachlorodibenzodioxin (2,3,7,8-TCDD) $C_{12}H_4Cl_4O_2$, M: 320, CAS Reg. No.: 1746-01-6.

as internal standards for quantitation. High levels of labeling are necessary in order to obtain mass signals for the standard above the native Cl isotope pattern.

3.2.7.8 Drugs

Various classes of active substances are assigned to the group of drugs. The fragmentation of amphetamines is of particular interest because their EI spectra are dominated by α-cleavage (Figures 3.123–3.125). In this case, the mass scan must be started at a correspondingly low mass to determine the ions m/z 44 and 58. The drugs of the morphine group (morphine, heroin, codeine, and cocaine) give stable molecular ions and can be recognized with certainty on the basis of their fragmentation pattern (Figures 3.126–3.129). The selective detection and confirmation of the identity of drugs is also possible in complex matrices using PCI with ammonia or isobutane as the CI gas.

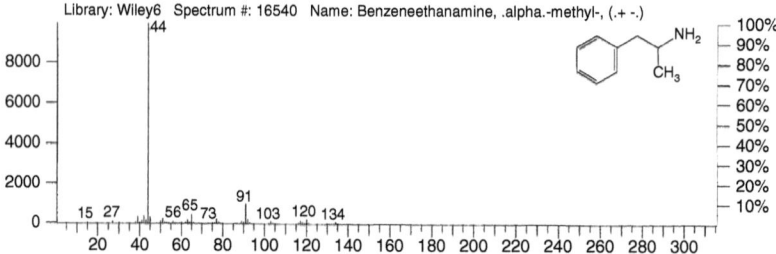

Figure 3.123 Amphetamine $C_9H_{13}N$, M: 135, CAS Reg. No.: 300-62-9.

3.2 Substance Identification | 497

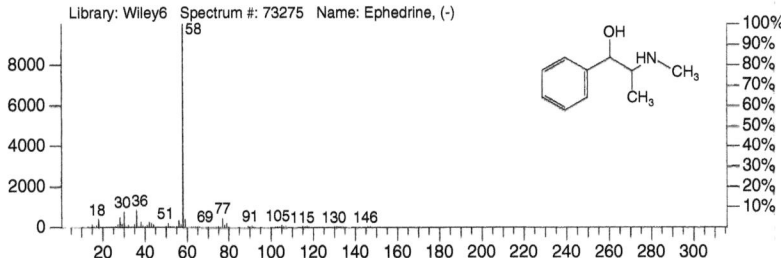

Figure 3.124 Ephedrine $C_{10}H_{15}NO$, M: 165, CAS Reg. No.: 299-42-3.

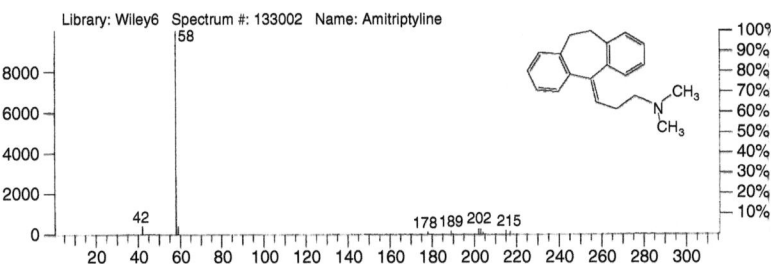

Figure 3.125 Amitriptyline $C_{20}H_{23}N$, M: 277, CAS Reg. No.: 50-48-6.

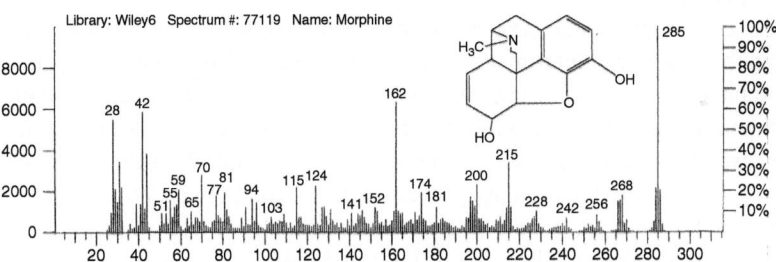

Figure 3.126 Morphine $C_{17}H_{19}NO_3$, M: 285, CAS Reg. No.: 57-27-2.

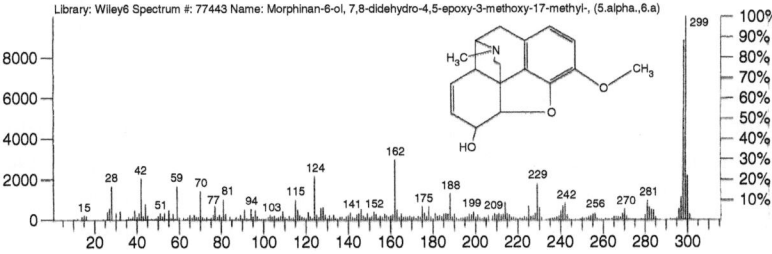

Figure 3.127 Codeine $C_{18}H_{21}NO_3$, M: 299, CAS Reg. No.: 76-57-3.

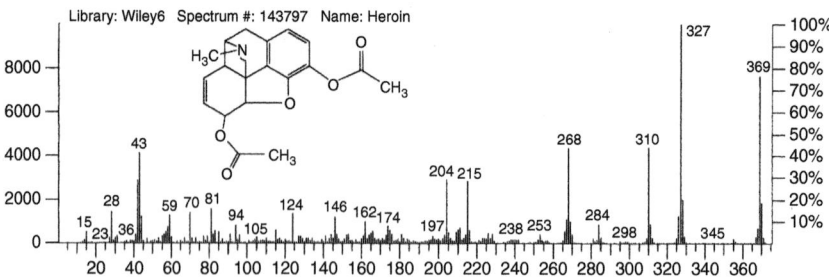

Figure 3.128 Heroin $C_{21}H_{23}NO_5$, M: 369, CAS Reg. No.: 561-27-3.

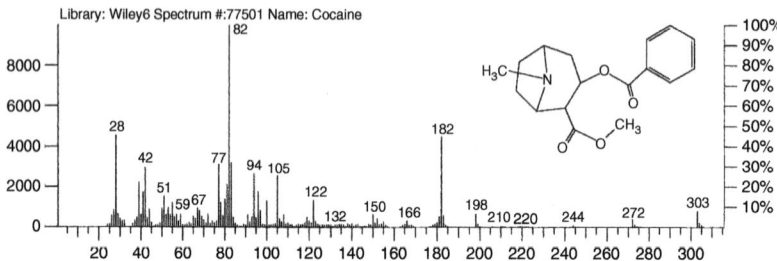

Figure 3.129 Cocaine $C_{17}H_{21}NO_4$, M: 303, CAS Reg. No.: 50-36-2.

3.2.7.9 Explosives

All explosives contain a high proportion of oxygen in the form of nitro groups. GC-MS analysis of the nonaromatic compounds (hexogen, octogen, nitropenta, etc.) is problematic because of decomposition in the injector. However, aromatic nitro compounds and their metabolites (aromatic amines) can be detected with high sensitivity because of their stability (Figures 3.130–3.145). With EI, molecular ions usually appear at low intensity because nitro compounds eliminate NO (M-30). o-Nitrotoluenes are an exception because a stable ion is formed after the loss of an OH radical (M-17) because of the proximity of the two groups (ortho effect) (Figures 3.130 and 3.131). If the scan range includes the mass m/z 30, the mass chromatogram can indicate the general elution of the nitro compounds. In this area, CI is useful for confirming and quantifying results. In particular, water has proved to be a useful CI gas in ion trap systems for residue analysis of explosives.

3.2.7.10 Chemical Warfare Agents

The identification of chemical warfare agents is important for testing disarmament measures and for checking disused military sites. Here also, there is no single chemical class of substances. Volatile phosphoric acid esters and organoarsenic compounds belong to this group (Figures 3.146–3.151). PCI is the method of choice for identification and confirmation of identity.

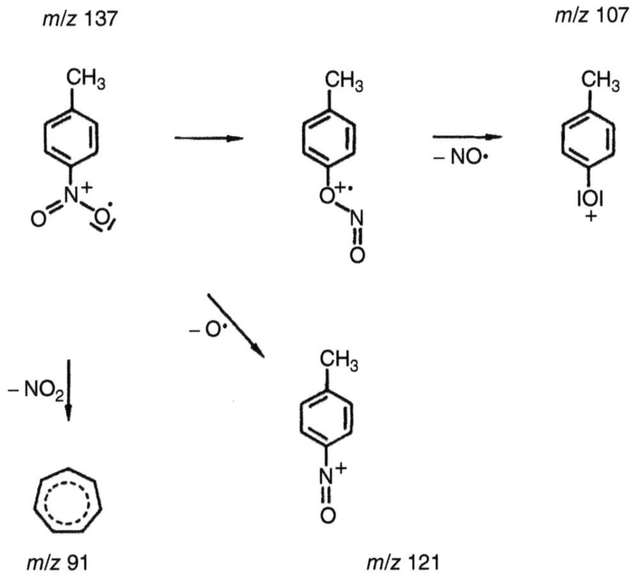

3.2 Substance Identification | 499

Figure 3.130 Fragmentation pathways of 4-nitrotoluene.

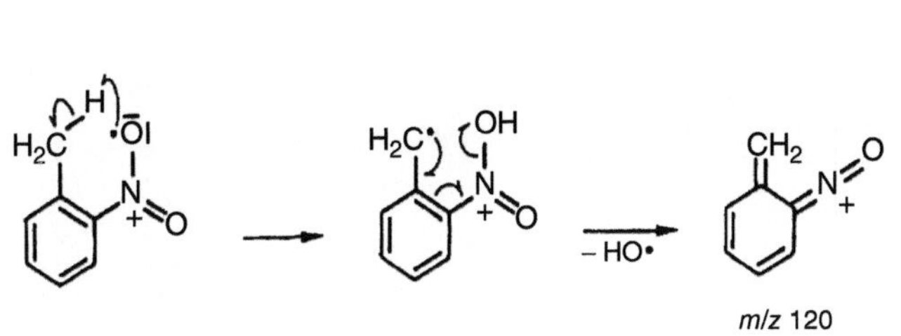

Figure 3.131 Fragmentation of 2-nitrotoluene (ortho effect).

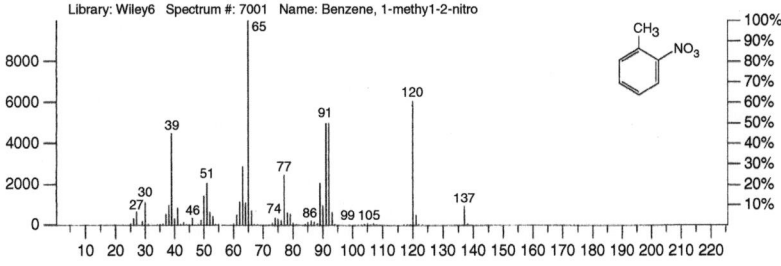

Figure 3.132 2-Nitrotoluene $C_7H_7NO_2$, M: 137, CAS Reg. No.: 88-72-2.

3 Evaluation of GC-MS Analyses

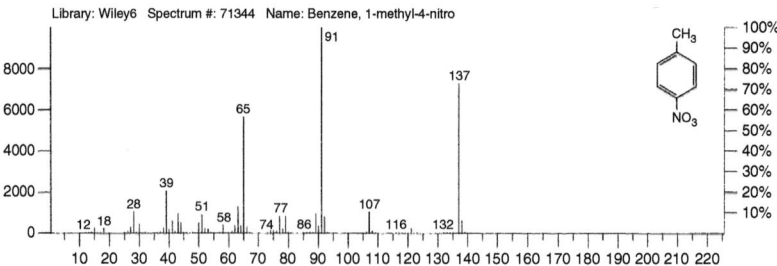

Figure 3.133 4-Nitrotoluene $C_7H_7NO_2$, M: 137, CAS Reg. No.: 99-99-0.

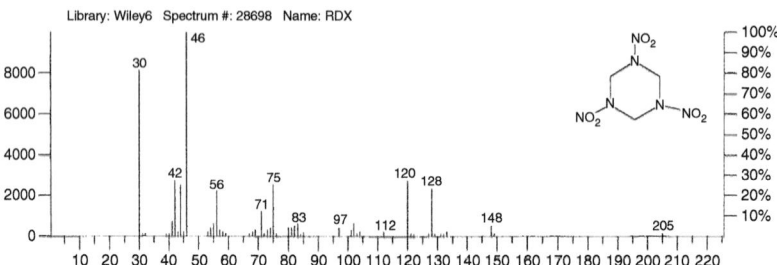

Figure 3.134 Hexogen (RDX) $C_3H_6N_6O_6$, M: 222, CAS Reg. No.: 121-82-4.

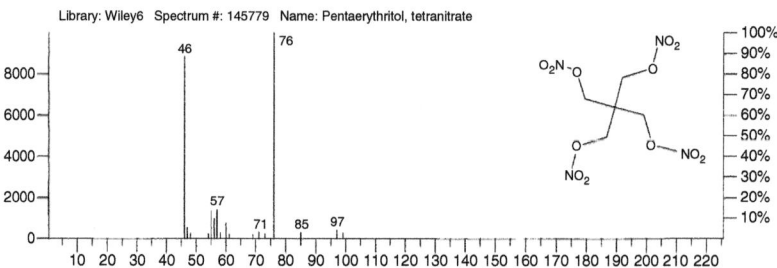

Figure 3.135 Nitropenta (PETN) $C_5H_8N_4O_{12}$, M: 316, CAS Reg. No.: 78-11-5.

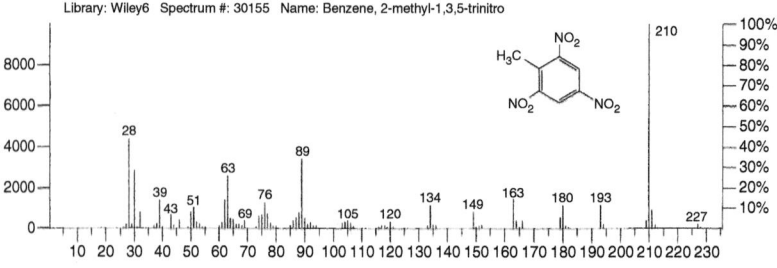

Figure 3.136 Trinitrotoluene (TNT) $C_7H_5N_3O_6$, M: 227, CAS Reg. No.: 118-96-7.

3.2 Substance Identification | 501

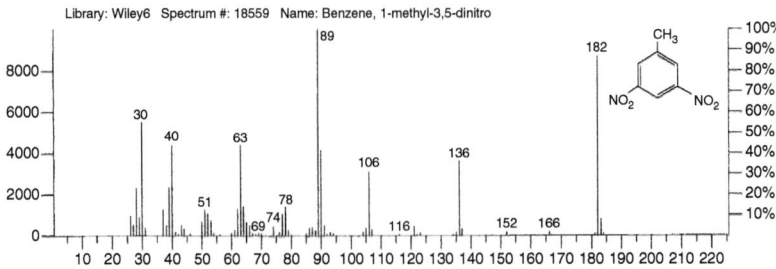

Figure 3.137 3,5-Dinitrotoluene (3,5-DNT) $C_7H_6N_2O_4$, M: 182, CAS Reg. No.: 618-85-9.

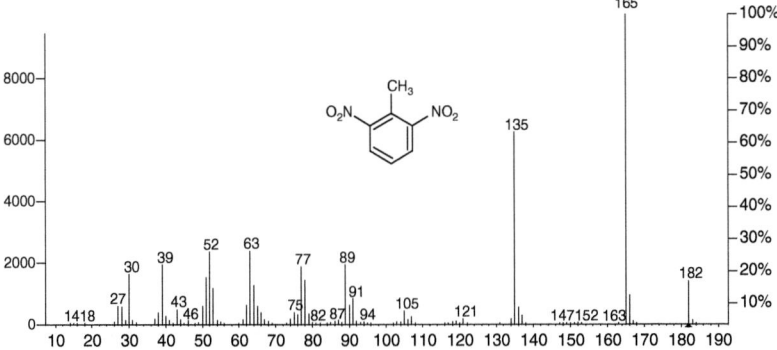

Figure 3.138 2,6-Dinitrotoluene (2,6-DNT) $C_7H_6N_2O_4$, M: 182, CAS Reg. No.: 606-20-2.

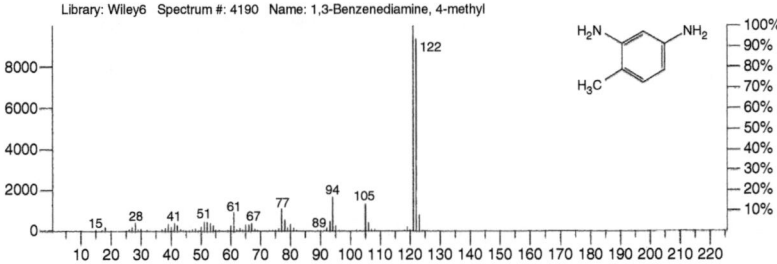

Figure 3.139 2,4-Diaminotoluene (2,4-DAT) $C_7H_{10}N_2$, M: 122, CAS Reg. No.: 95-80-7 (Note: 2,4-DAT and 2,6-DAT form a critical pair during GC separation).

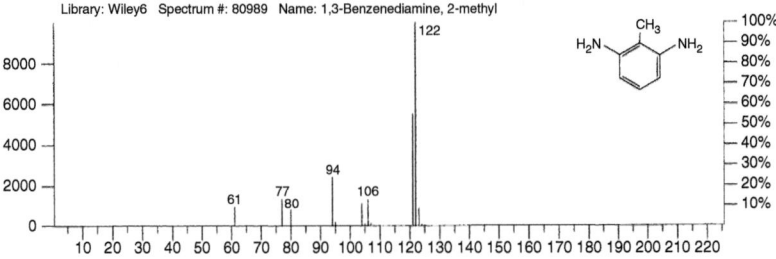

Figure 3.140 2,6-Diaminotoluene (2,6-DAT) $C_7H_{10}N_2$, M: 122, CAS Reg. No.: 823-40-5.

502 | *3 Evaluation of GC-MS Analyses*

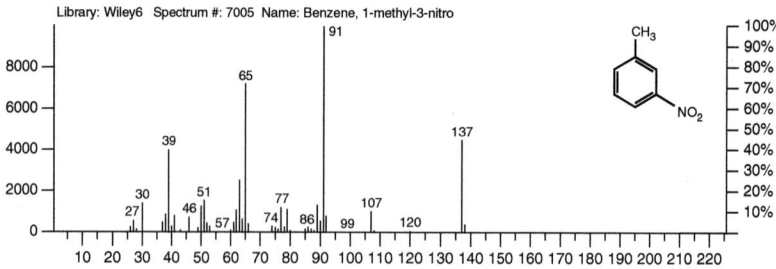

Figure 3.141 (Mono-)3-nitrotoluene (3-MNT) $C_7H_7NO_2$, M: 137, CAS Reg. No.: 99-08-1.

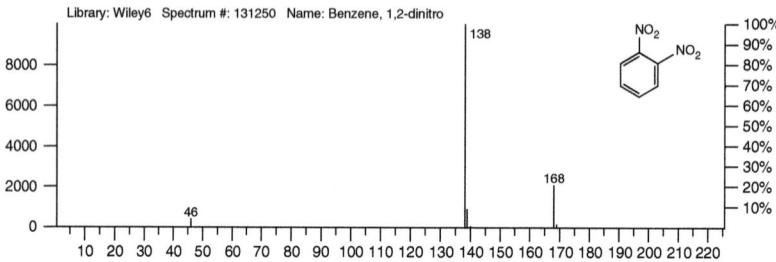

Figure 3.142 1,2-Dinitrobenzene (1,2-DNB) $C_6H_4N_2O_4$, M: 168, CAS Reg. No.: 528-29-0.

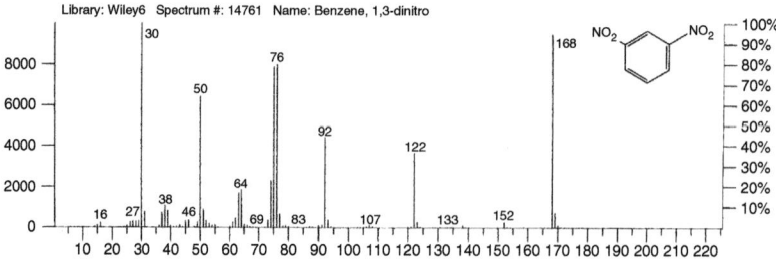

Figure 3.143 1,3-Dinitrobenzene (1,3-DNB) $C_6H_4N_2O_4$, M: 168, CAS Reg. No.: 99-65-0.

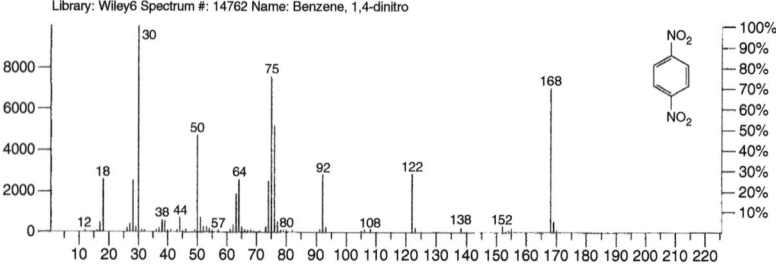

Figure 3.144 1,4-Dinitrobenzene (1,4-DNB) $C_6H_4N_2O_4$, M: 168. CAS Reg. No.: 100-25-4.

3.2 Substance Identification | 503

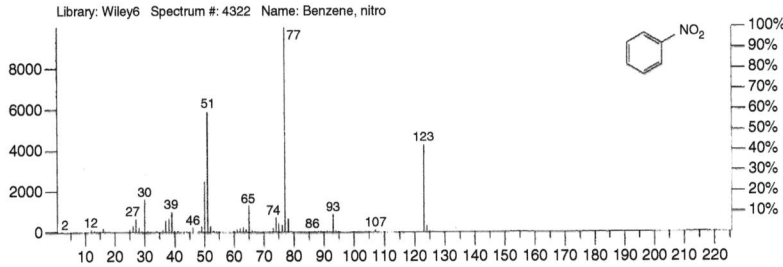

Figure 3.145 (Mono-)nitrobenzene (MNB) $C_6H_5NO_2$, M: 123, CAS Reg. No.: 98-95-3.

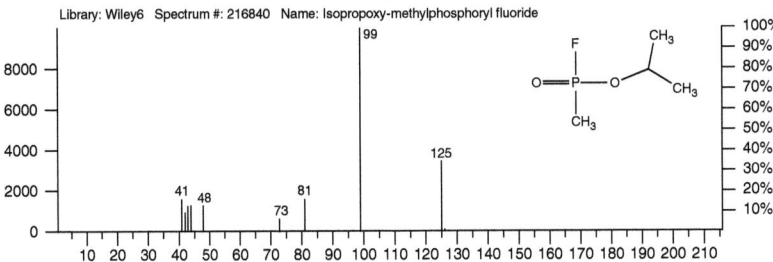

Figure 3.146 Sarin $C_4H_{10}FO_2P$, M: 140, CAS Reg. No.: 107-44-8.

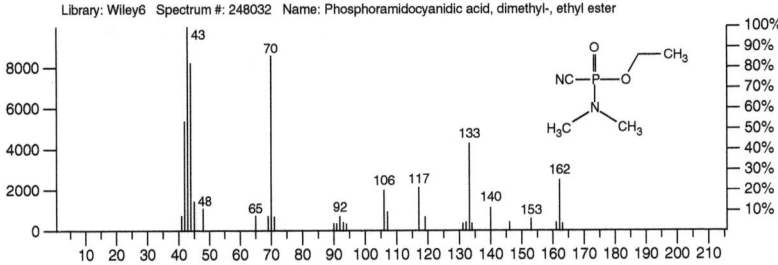

Figure 3.147 Tabun $C_5H_{11}N_2O_2P$, M: 162, CAS Reg. No.: 77-81-6.

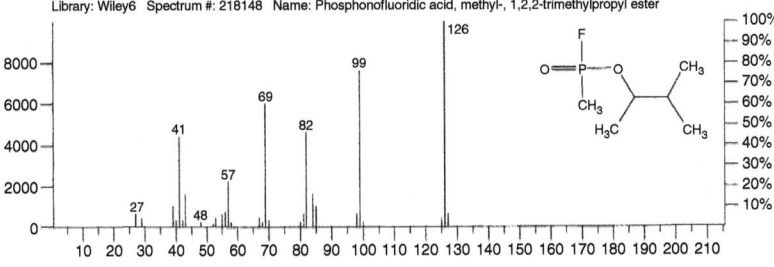

Figure 3.148 Soman $C_7H_{16}FO_2P$, M: 182, CAS Reg. No.: 96-64-0.

3 Evaluation of GC-MS Analyses

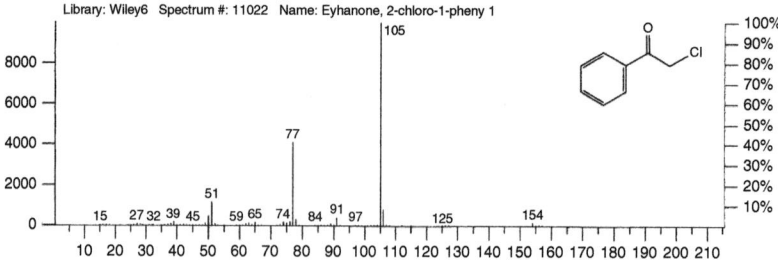

Figure 3.149 Chloroacetophenone (CN) C_2H_7ClO, M: 154, CAS Reg. No.: 532-27-4.

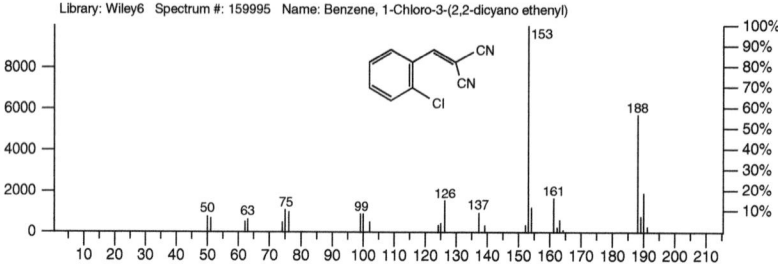

Figure 3.150 o-Chlorobenzylidenemalnonitrile (CS) $C_{10}H_5ClN_2$, M: 188, CAS Reg. No.: 2698-41-1.

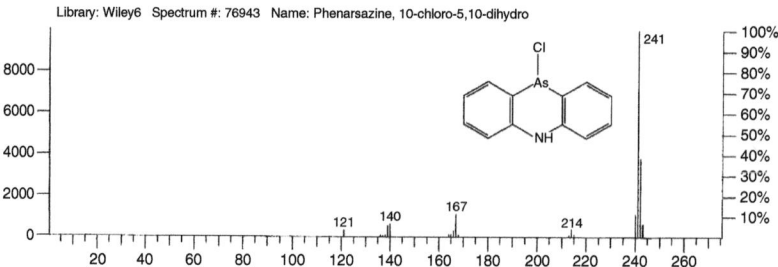

Figure 3.151 Adamsite (DM) $C_{12}H_9AsClN$, M: 277, CAS Reg. No.: 578-94-9.

3.2.7.11 Brominated Flame Retardants (BFRs)

Highly brominated compounds are used as flameproofing agents (Figures 3.152 and 3.153). The polybrominated biphenyls (PBBs) and polybrominated diphenylether (PBDE), which are mostly used, have molecular masses of up to 1000 (decabromobiphenyl M 950). BFRs have recently become of interest because burning materials containing PBB and PBDE can lead to the formation of polybrominated dibenzodioxins and furans. Decabromodiphenylether (PBDE 209) is thermolabile and tends to disintegrate in the GC injector and during separation on the column. A short column with higher flows for low elution temperatures is required for sensitive analysis. The spectra of PBB and PBDE are characterized by the symmetrical isotope pattern of the bromine and the high stability of the aromatic molecular ion. Flameproofing agents based on brominated alkyl phosphates have a greater tendency to fragment.

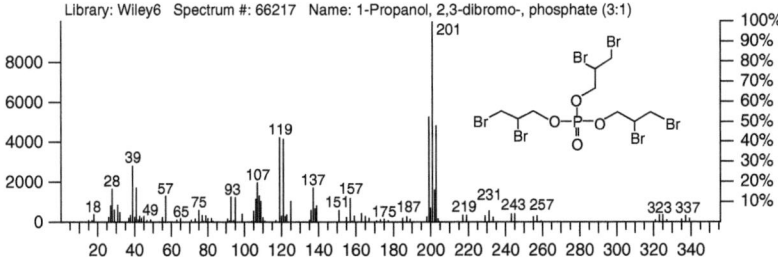

Figure 3.152 Bromkal P67-6HP (Tris) $C_9H_{15}Br_6O_4P$, M: 692, CAS Reg. No.: 126-72-7.

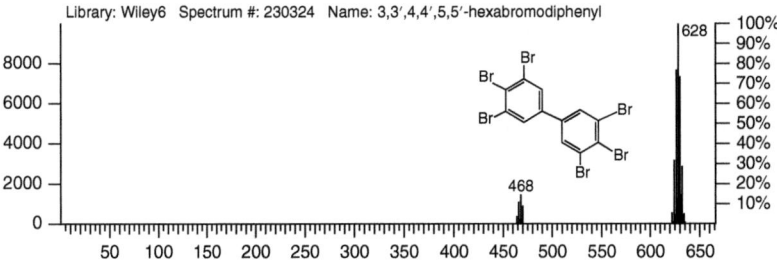

Figure 3.153 Hexabromobiphenyl (HBB) $C_{12}H_4Br_6$, M: 622.

3.3 Quantitation

It is important to point out that accuracy assessment is a continuous process, which should be implemented in the routine work as a part of the QA/QC set-up of the laboratory.

Richard Boqué 2008
Universitat Rovira i Virgili, Faculty of Chemistry,
Tarragona, Spain

Besides the identification of components of a mixture, the use of GC-MS systems to determine the concentration of target compounds according to legal requirements is a primary application in governmental, industrial as well as private control labs. The need to determine quantitatively an increasing number of components in complicated matrices in ever smaller concentrations makes the use of GC-MS systems in routine analysis the economic choice. Gone are the days when only positive results from classical GC were confirmed by GC-MS. In many areas of trace analysis, the development of targeted multicomponent methods has become possible only through the selectivity of detection and the reliability of identification of GC-MS and GC-MS/MS systems. The development of GC-MS data systems has therefore been successful in recent years, particularly in areas where the use of the mass spectrometric substance information coincides with the integration of chromatographic peaks from mass selective detection techniques and thus increases the certainty of quantitation. Compared with chromatography data systems for stand-alone GC and HPLC, there are therefore differences and additional features which arise from the use of the mass spectrometer as the detector for quantification.

3.3.1 Acquisition Rate

Mass spectrometers do not continuously record the substance stream arriving in the detector (as, for example, with FID, ECD and other classical detectors). The chromatogram is comprised of a series of measurement points which are represented by mass spectra. The scan rate chosen by the user establishes the time interval between the data points. The maximum possible scan rate depends on the scan speed of the analyzer. It is determined by the width of the mass range or the SIM/SRM window times to be acquired and the necessity of achieving an analytical detection capacity of the instrument which is as high as possible. For quantitative measurements, full scan rates below 1.0 s/scan, and SIM/SRM window acquisition times significantly shorter are chosen. Compared with a sharp concentration change in the slope of a GC peak, these scan rates can be slow.

In the integration of GC-MS chromatograms, the choice of scan rate is particularly important for the determination of the peak area for quantitation. With the scan rate of 1s/scan, which is sometimes used, the peak area of rapidly eluting components cannot be determined reliably as too few data points for the correct plotting of the GC peak are recorded. The definition of the top of the peak in the calculation of height and area cannot be carried out correctly. With too few data points, a low precision in the determination of the peak area and height results, also small peaks may be missed (Figure 3.154). A slow scan with sharp peaks will distort the relative intensities as the compound concentration in the ion source changes rapidly, important signals in the lower or upper mass range may be missed (spectrum skewing). A minimum number of 7 data points describes the peak profile well. A range of 8–12 data points is recommended to achieve a reliable peak area and height determination with low RSD%, as shown in Figure 3.155 for fast chromatography. More data points do not deliver more precision but reduce the response as the dwell time for the quantitation ions drops significantly. Also, the deconvolution of chromatograms benefits from a higher acquisition rate, if the sensitivity allows. Figure 3.155 also indicates that the area calculation delivers better precision than using the peak height with lower scan rates.

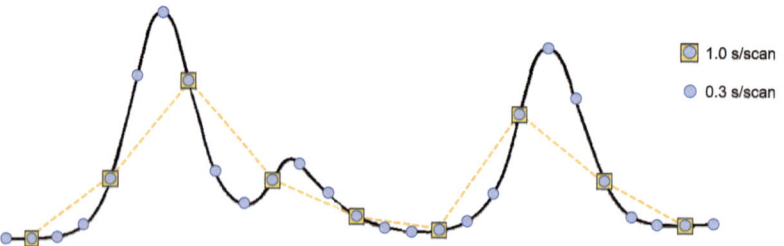

Figure 3.154 Comparison of an actual chromatogram (black trace) with one reconstructed from data points at a low scan rate of 1 s/scan (dotted line). For the peak width of c. 3 s a scan rate of min. 0.3 s/scan is required (blue data points) (after Chang, 1985).

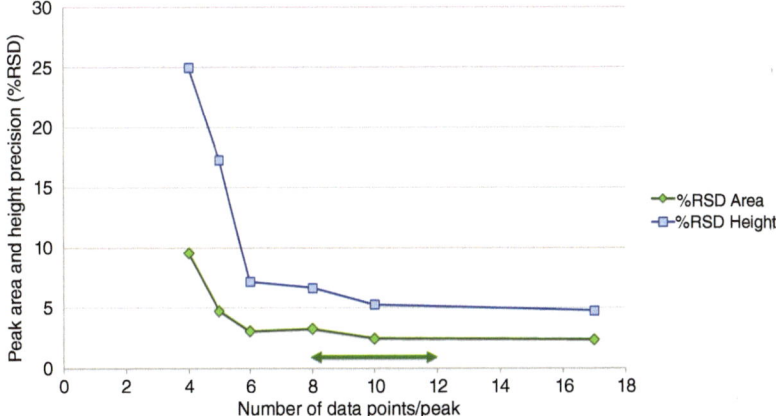

Figure 3.155 Precision of the peak area determination depending on the scan rate (after Dallüge et al., 2002). The green bar represents the recommended working range.

For the determination of area values in quantitative analyses, special attention should be paid to the recording parameters. To determine the optimal scan rate and sampling frequency, the base peak width of sharp early eluting components should be investigated. Some valuable information is available from systematic investigations on the impact of the GC-MS sampling frequency on the precision of the peak area and peak in full scan and SIM acquisition and deal with dwell time optimization for sensitivity and S/N (Kirchner et al., 2005). In particular for fast GC, the discussion on the minimum number of data points is required for reliable peak area determination (Dallüge et al., 2002). It should be noted that the overall sensitivity is significantly decreasing with increased scan and acquisition rate due to a reduction of the measurement time per ion, defined as the dwell time:

$$\text{max. scan time [s]} = \frac{\text{base peak width [s]}}{\text{min. \# data points}}$$

$$\frac{\text{dwell time [s]}}{\text{ion}} = \frac{\text{scan time [s]}}{\text{mass range}}$$

The big advantage of SIM as well as the comparable SRM/MRM is the significant increase of dwell time and sensitivity for the selected ions on the cost of delivering a complete mass spectrum.

3.3.2 Decision Limit

The question of when a substance can be said to be detected cannot be answered differently for quantitative GC-MS compared to all other chromatographic systems.

In the basic adjustment of mass spectrometers, unlike classical GC detectors, the zero point is adjusted correctly (electrometer zero) to ensure the exact plot of isotope patterns. For this, the adjustment is chosen in such a way that the minimum noise of the electronics is determined which can then be removed during data acquisition by software filters. Besides electrical noise, there is also chemical noise (matrix, column bleed, leaks, etc.), particularly in trace analysis, often clearly visible in the chromatograms.

The decision as to whether a substance has been detected or not is a qualitative decision usually assessed in the *signal domain*. It is established that the decision limit is such that the smallest detectable signal from the substance can be clearly differentiated from a blank value. The answer is found in the GC-MS chromatogram with the determination of the S/N ratio.

In the measurement of a substance-free sample (blank sample), an average signal is obtained which corresponds to a so-called blank value. Multiple measurements confirm this blank value statistically and give its precision with the standard deviation. The determination of the baseline is carried out in the immediate vicinity of the substance peak, whereby the noise widths before and after the peak are taken into consideration. The average noise and the signal intensity can also be determined manually from a printout. Also, suitable S/N algorithms are available in data systems. A substance can be said to be detected if the substance signal exceeds a certain multiple of the noise width (standard deviation). This value is chosen arbitrarily as two, three, or higher depending on the laboratory SOP on the application. It is commonly used as the deciding criterion in most routine evaluations of GC-MS data systems, as further illustrated in Figure 3.156.

In a *statistically* defined definition, it is specified that the substance is considered to be detected if the measured signal exceeds a certain multiple of the standard deviation. The decision limit takes into account the measurement uncertainty, for instance, also above a legal threshold limit. It is then the smallest detectable signal that can be unambiguously distinguished from a blank value. The European Commission defines the "decision limit (CC α) as the limit at and above which it can

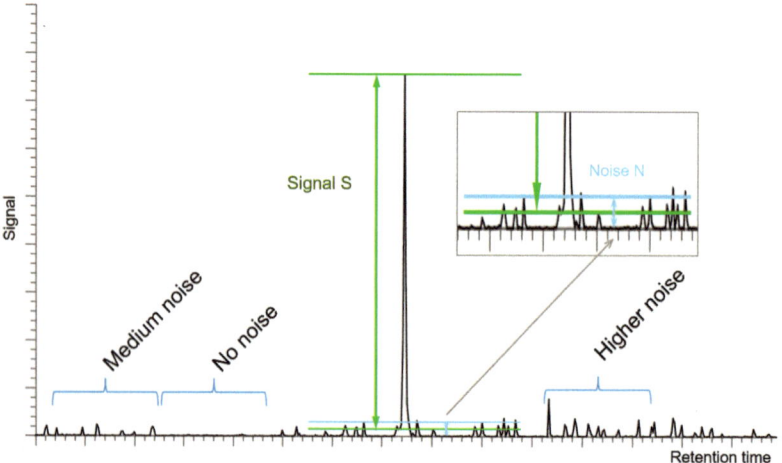

Figure 3.156 Graphical S/N determination with choice of different background noise windows.

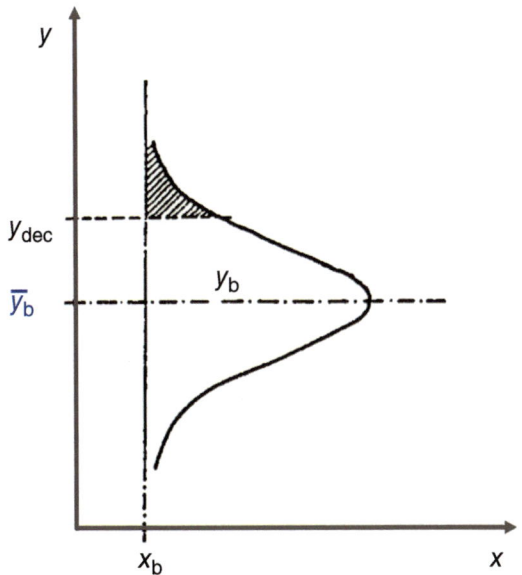

Figure 3.157 Statistical interpretation of the decision limit. y_b = measurement distribution of the blank samples, y_{dec} = decision limit (x substance domain, y signal domain, Ebel and Dorner, 1987).

be concluded with an error probability of α that a sample is non-compliant", and "the detection capability (CC β) as the smallest content of the substance that may be detected, identified and/or quantified in a sample with an error probability of β" (Commission Decision, 2002/657/EC).

With for instance a 99% certainty the decision limit is the lowest concentration of the analyte that can be detected in the sample with a 1% chance of a false negative decision (Figure 3.157). The decision needs multiple measurements of the blank value and a user-defined acceptable error probability. In practice, the uncertainty of the result may be caused by the injection, instrument variances, or matrix effects (Shukla et al., 2023).

3.3.3 Detection Limit

How does the decision limit relate to the LOD? Usually, both terms are used synonymously, which is incorrect! The LOD of a method is never given in counts or parts of a scale but is given a dimension, such as pg/μL, ng/L, or ppb, in any case the LOD is expressed in the *substance domain*. The transfer from the signal domain of the decision limit into the substance domain of the LOD is accomplished by a valid calibration function (see Figure 3.158). A method detection limit (MDL) is accordingly defined as the detection limit considering the complete method with all steps of the sample prep and analysis (U.S. EPA, 2023).

For practical reasons, the signal domain is used to calculate the LOD from a single run displayed on screen or via printouts as illustrated in Figure 3.156. In a chromatogram of a low concentration standard, the width of the noise band is determined, and from the average noise level the peak height is taken (Wells et al., 2023). This can simply be done in centimeters or counts, or a suitable software routine

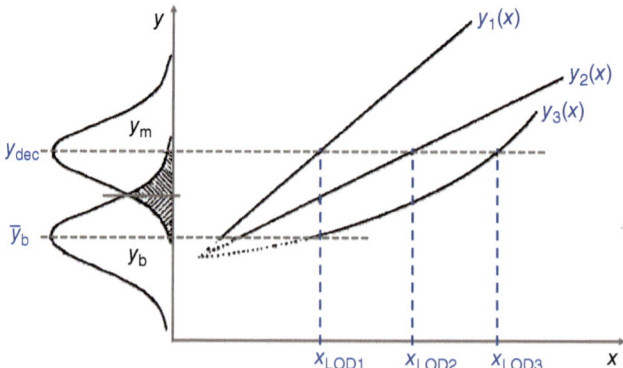

Figure 3.158 Definition of the limit of detection (x_{LOD}) from the decision limit y_{dec} as of Figure 3.156. The limit of detection (LOD) is defined as that quantity of analyte, concentration or content x_{LOD} which is given for the decision limit y_{dec} using the calibration function. Calibration functions with different sensitivities (slopes) lead to different limits of detection at the same signal height (x substance domain, y signal domain; Ebel and Dorner, 1987).

used. The result is an S/N ratio for a certain peak and concentration injected, often the analyte of interest. But what is a representative background? With high selectivity detection, it is increasingly difficult to choose a suitable background range. Especially with MS/MS instruments and HRMS accurate mass acquisitions, a continuous noise band is not accessible anymore due to the high selectivity of these analyzer types and the very noise-free digital signal processing. With a noise band tending to zero, the S/N calculation becomes increasingly meaningless. Here, the LOD can only be regarded as a qualitative parameter from just one measurement. The injection variation and the uncertainty of the calibration function is not taken into account. If used, an S/N ratio of 3 is often the practical criterion for an analyte LOD.

The *statistically* determined LOD is based on the results of a series of measurements at low concentration level. The detection limit is the upper limit (decision limit) of the confidence interval of the measurement signals for $x = 0$ as explained in Figure 3.157. The variance of peak areas from consecutive measurements shows all "noise" influences for each injection, and the overall precision of the measurement is expressed in the relative standard deviation (%RSD) (Glaser et al., 1981).

A theoretical instrument detection limit (IDL) can be calculated from the peak variance of a series of measurements. The IDL, describing the performance of an individual instrument, is calculated with the analyte in solvent. Similarly, a detection limit for a complete method (MDL) is calculated with the analyte dissolved in an average sample matrix and includes all steps of sample preparation and measurement (U.S. EPA, 2023; Agilent, 2024). The concentration of the test solution should be in the range of 5 to 10 close to the expected LOD. For good statistics, the number of replicate injections should be at least 7, although 8–10 is preferred (Sheehan and Yost, 2015). For IDL calculation, the Student *t*-distribution is used to calculate the 99% confidence factor t_α, taken from a Student *t*-distribution table for $n-1$ measurements:

$$\text{IDL} = t_\alpha \times \%\text{RSD} \times \text{Amount injected (in solvent*)}$$

with

IDL instrument detection limit.

t_α confidence factor for the desired level of 99% from the Student t-distribution for $n-1$ measurements.

RSD relative standard deviation in % from n consecutive measurements.

*Using a spiked matrix sample an MDL is calculated similarly.

According to the example shown in Table 3.13, the calculated IDL is given with 1.8 fg of OFN. This result is based on the estimate that this 1.8 fg substance signal can be statistically distinguished from the mean value of a series of blank measurements. With higher precision of the measurements, hence lower RSDs, the IDL value tends to be lower. Yet, it is not specified at which concentration the measurements are carried out. This calculated value does not specify a real injection of this very low concentration and has yet to be documented on this instrument.

Table 3.13 Example of consecutive measurements of 10 fg OFN, peak area and height determination on m/z 272.

Injection #	Response of OFN ion m/z 272 Peak area	Peak height	Moving RSD[a] Peak area	Peak height
1	2074	2629	—	—
2	1859	2250	—	—
3	1553	1953	—	—
4	1640	1948	—	—
5	1690	2024	—	—
6	1792	2221	—	—
7	1559	2028	—	—
8	1531	1869	11.0%	11.6%
9	1738	2151	7.2%	6.7%
10	1566	2041	6.0%	5.6%
11	1814	2242	6.6%	6.3%
12	1871	2579	7.7%	10.0%
		Lowest RSD	6.0%	
		t_α (of $n-1$ runs)	2.998	
		IDL[b]	1.8 fg	

a) A moving RSD from the 8 runs before is calculated, the lowest RSD taken for IDL calculation.
b) Calculated as IDL = 2.998 × 0.06 × 10 fg

3.3.4 Limit of Quantitation

The limit of quantitation (LOQ) of a method is often given as a quantity of substance or concentration in the *substance domain* as well. As a rule of thumb, in chromatographic practice, a factor of two to three of the noise band is usually regarded as the LOD and a factor of five to ten as the LOQ.

In a statistical definition, the LOQ takes into account the quantitative calibration with its slope and thus also the uncertainty (error consideration) of the measurements (Ebel and Kamm, 1983). The LOQ is defined as the smallest amount or concentration of an analyte that can be quantified. It is the lower limiting concentration x_{LOQ} at a user-defined statistical error probability which can be quantitatively determined. It differs significantly from zero, the blank value (Montag, 1982; Ebel and Kamm, 1983; ISO 11843, 2019). All parameters of the calibration and the confidence area are included in the definition of the LOD (Figure 3.158).

The International Committee on Harmonization of analytical procedures (ICH) discusses LOQ approaches for visual evaluation, by S/N, and statistically on standard deviation and slope of the calibration function. In the statistical approach, the LOD is calculated as

$$LOD = 3.3\sigma/S$$

and the limit of quantification LOQ as

$$LOQ = 10\sigma/S$$

Here σ is the standard deviation of the measurements of the lowest point of the calibration curve in the range of the expected LOQ. S is the slope of the calibration curve for the analyte (ICH, 2005) (Figure 3.159). The calculation of the LOQ

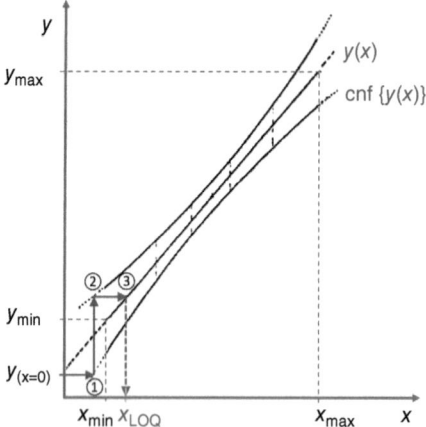

Figure 3.159 Statistical definition of the limit of quantitation. ① Measured blank value, analyte concentration = 0 (signal domain). ② The upper limit of the confidence interval determines the limit of quantitation. This signal can be differentiated significantly from zero, taking the standard deviation into account. ③ Limit of quantitation x_{LOQ} in the substance domain is taken from the calibration function. (x substance domain, y signal domain; Ebel and Dorner, 1987; DIN 38402-1:2011-09).

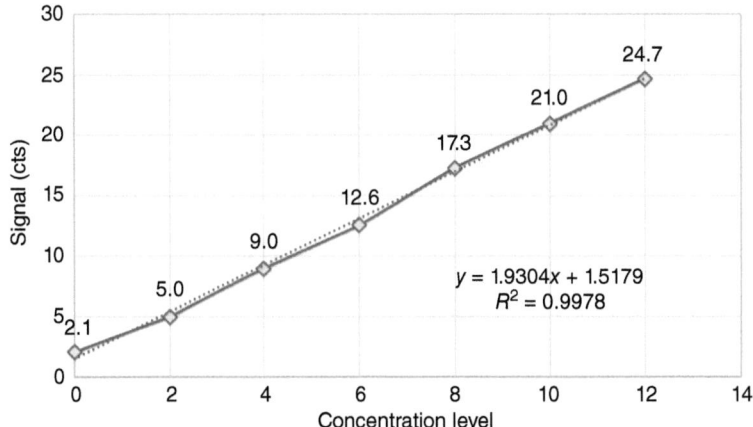

Figure 3.160 Calibration curve example with trendline and regression function.

Table 3.14 Regression statistics received from data analysis of Figure 3.160.

Regression Statistics	
Calibration slope	1.9304
Intercept	1.5179
R^2	0.9978
Standard error	0.224
Calibration points	7

according to ICH guidelines can be done with a few steps in MS Excel with the Data Analysis ToolPak installed. An example is provided with the below potential calibration curve displayed in MS Excel (Figure 3.160; Table 3.14) (Dolan 2021). These data lead to the following LOQ calculation:

$$\text{LOQ} = 10\sigma/S = 10 \times 0.2240/1.9304 = 1.2 \text{ ng/mL}$$

In the European Union the "Guidance Document on the Estimation of LOD and LOQ for Measurements in the Field of Contaminants in Feed and Food" provides the details of LOD and LOQ calculation and focuses on the analysis of PAHs, heavy metals, mycotoxins, and "dioxins" (i.e. PCDDs, PCDFs, and PCBs) (Wenzl et al., 2016). In particular, for congener-based LOQs, as is the case for "dioxins," the LOQ is defined as the concentration of the analyte which produces a response with an S/N ratio of 3 : 1 for the less intensive raw data signal of two different ion traces (usually the quantitation and the confirmation ion). In case of absence of noise, as it can be the case for triple quadrupole or Orbitrap instruments, the LOQ is calculated from the lowest concentration of the calibration curve that gives an acceptable precision of ≤30% RSD calculated for all points on the calibration curve.

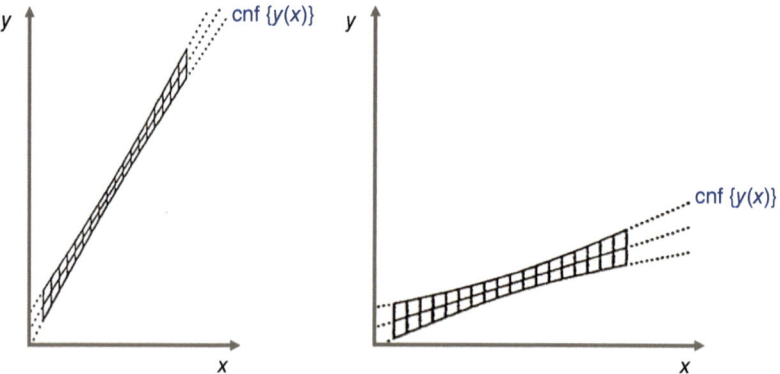

Figure 3.161 Height of the confidence intervals for calibration curves with the same distribution of the measured values but with different slopes (x substance domain, y signal domain; Ebel and Dorner, 1987).

3.3.5 Sensitivity

The term *sensitivity* is frequently used to describe the quality of a residue analysis method or the current state of a GC-MS instrument. It is often used incorrectly as a synonym for the lowest possible LOD or LOQ. A sensitive analysis procedure, however, exhibits a large change in signal with a small change in substance concentration. Consequently, the International Union of Pure and Applied Chemistry (IUPAC) defines sensitivity as the slope of the calibration curve (i.e. the plot of signal vs. amount or concentration of an analyte). The sensitivity of a procedure thus describes the slope of the linear calibration function (see Section 3.3.6). High sensitivity provides a narrow confidence interval for the measured points of the calibration function (Figure 3.161). Due to the narrower confidence intervals analysis procedures achieve lower LOQs, as it is illustrated in Figure 3.159. The confidence intervals cnf{y(x)} are only defined for a specific y-value and always only in the y-direction. The dotted lines, in particular the calibration line, can be calculated but are not metrologically proven.

3.3.6 The Calibration Function

The calibration function is built from the measurement of known concentrations. Fixed response factors are used only to a limited degree in GC-MS because the intensity of the signal depends on the operating parameters of the mass spectrometer. Venting, restarting, or tuning the MS requires an update of the quantitative calibration. The calibration function generally describes the dependence of the signal on the substance concentration. In the case of linear dependence, the regression calculation gives a straight line for the calibration function with the general linear equation

$$f(x) = a_0 + ax \tag{3.3}$$

which comprises the blank value a_0 and the sensitivity a.

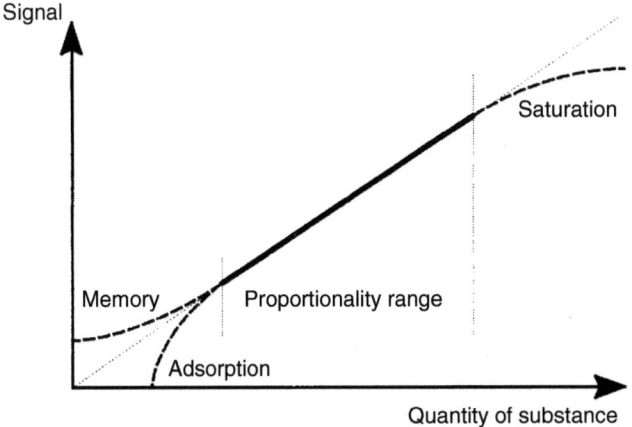

Figure 3.162 Dependance of the signal intensity with the quantity of substance.

The calibration function is defined exclusively within the working range given by the experimental calibration from the lowest to the highest calibration level. GC-MS systems achieve very low LODs so that a dense collection of calibration points near the lowest calibration point achieves a higher precision at the low end, and a potential nonlinear area can be described. A non-linear area of the calibration can be caused by unavoidable active sites (residual activities) in the system and a "swallow up" of small but constant quantity of substance. Such a calibration function tends to approach the x-axis before reaching the origin. Small memory effects turn the lower calibration end towards the y-axis. In the upper concentration range an increasing saturation of the detector can be observed. The signal hardly increases at all with increasing substance concentration. The calibration function turns to asymptotic as illustrated in Figure 3.162.

The best fit of the calibration curve to the measured points is determined by a regression calculation. The plot of the data should first be inspected for possible outliers from faulty standard preparations or measurements with a high leverage on the curve (Rousseeuw and Leroy, 1987). From the least squares calculation, the regression coefficient gives information on the quality of the curve fit. Linear regressions are not always suitable for the best fit in GC-MS analysis; quadratic regressions frequently give better results. Especially in the area of trace analysis, the type of fit within the calibration can change through nonlinear effects described earlier. In such cases it is helpful to limit the regression calculation close to the calibration level that is needed for the concentration in the samples (local linearization, next three to four data points). If individual data systems do not allow this, a point-to-point calibration can be carried out instead, provided an acceptable density of data points. Usually, the manufacturers CDS provide a flexible regression calculation, otherwise MS Excel with the Data Analysis ToolPak installed, or similar programs can be used.

Important aspects regarding the optimization of a calibration can be derived from the facts discussed (Prichard and Barwick, 2003):

- One-point calibrations have no statistical confirmation and can therefore only be used as an estimation. With FID detection it is only useful with confirmed response values.
- Multipoint calibrations must cover the expected concentration range. For regulated methods, a factor of 10 below the regulated concentration (maximum residue level, MRL) is required.
- A dense distribution of calibration points near the LOQ improves the regression fit (unlike the equidistant position of the calibration level).
- Multiple measurements at an individual calibration level define the confidence interval. Continuous calibration procedures replace older values and keep track of potential changes.
- If the measured data at the blank value obey a normal distribution, the blank value can be included in the experimental calibration function. Usually in chromatography, this is not the case.
- Extrapolation beyond the experimentally measured points is not justified. Calibration functions do not have to pass through the origin.

3.3.7 Quantitation and Standardization

In order to determine the substance concentration in an unknown analysis sample, the peak areas or heights of the sample are calculated using the calibration function, and the results are given in terms of quantity or concentration. Concerning the discussion whether to use peak area or peak height, we observe the peak area determination as the prevailing measure of response, although there is the word "the peak height is the safer measurand" (Kuss, 2009). Also, a peak width-based response calculation is proposed (Kadjo et al., 2017). In reality, peak shapes are not perfectly symmetrical, showing tailing or interferences that need to be recognized (Folley, 1987). In any way, the chosen method needs to be validated and its fit for purpose confirmed (Eurachem, 2014). The chromatography data systems allow options, also to process the sample amount as weight, dilution, or concentration steps in order to represent the amount of the original sample.

The minimum required performance limits (MRPL) with definitions for quantitation in GC-MS and LC-MS are laid down in the European Council Directive 96/23/EC concerning the performance of analytical methods and the interpretation of results (Commission Decision, 2002).

For GC-MS, the methods of external or internal standard calibration or standard addition are used in quantification procedures (Funk et al., 2007). Preparing calibration standard solutions is critical in human hands. Automated procedures are recommended in quantitative assays to avoid human error and bias, and improve accuracy and reproducibility (Zhang et al., 2014; Wiest, 2020; Hübschmann, 2021; Banerjee and Hübschmann, 2022).

3.3.7.1 External Standardization

External standardization corresponds to the classical calibration procedure. The substance to be quantified is used to prepare standard solutions with a known concentration. Measurements are made with standard solutions of different concentrations (calibration levels). For calibration, the peak areas or heights determined are plotted against the concentrations of the different calibration levels.

With external standardization, standard deviations below 10% are obtained. The causes of a data variation are sample preparation, injection errors, and small changes in the mass spectrometer. Absolute measurement values from the analysis runs are used to build the external standardization. The calibration function shows all the effects, which can cause a variation in response. Various factors can contribute to this, for example, slightly different ionization efficiency through geometric changes in the filament, slightly varying transmission due to contamination of the lens systems with matrix, or a contribution from the multiplier on signal production. These factors can be compensated for by internal standardization (Guichon and Guillemin, 1988; Naes and Isakson, 1992).

3.3.7.2 Internal Standardization

The principle of the internal standard is based on the calculation of relative values, which are determined within the same analysis. One or more additional substances are introduced as a fixed reference parameter, the concentration of which is kept constant in the standard solutions and is always added to the analysis sample at the same concentration. For the calculation, the peak area values (or peak heights) of the substance being analyzed relative to the peak area (or height) of the internal standard are used. In this way, potential volume errors and variations in the function of the instrument are compensated for, and quantitative determinations of the highest precision are achieved (see also ISO 5725-6, 1994). Standard deviations of less than 5% can be achieved with internal standardization.

The time at which the internal standard is added during the sample preparation depends on the method requirement. For example, the internal standard can be added to the sample at an early stage (surrogate standard added before sample preparation) to simplify and control the clean-up. The addition of different standards at different stages of the clean-up allows the efficiency of individual clean-up steps to be monitored. Addition of the quantitation standard to the extract directly before the measurement serves to calculate the recoveries in the sample processing steps. This so-called syringe standard can be added very precisely by many autosamplers in the sandwich mode from a separate internal standard vial right before injection.

The choice of the internal standard is particularly important also in GC-MS. Basically, the internal standard should behave as far as possible in the same way as the substance being analyzed. Unlike classical GC detectors, the GC-MS procedure offers the unique possibility of using isotopically labeled, but nonradioactive analogs of the substances being analyzed. Deuterated or ^{13}C carbon standards are frequently used for this purpose as they fulfill the requirement of comparable behavior during clean-up and analysis to the greatest degree. The degree of labeling should always be sufficiently high to avoid interference with the natural isotope intensities of the unlabeled substance. Note, deuterated internal standards elute slightly before the native compound, depending on the degree of labeling, and must be integrated on the selective mass chromatograms, well visible in Figure 3.163. See also the following section on "Isotope Dilution."

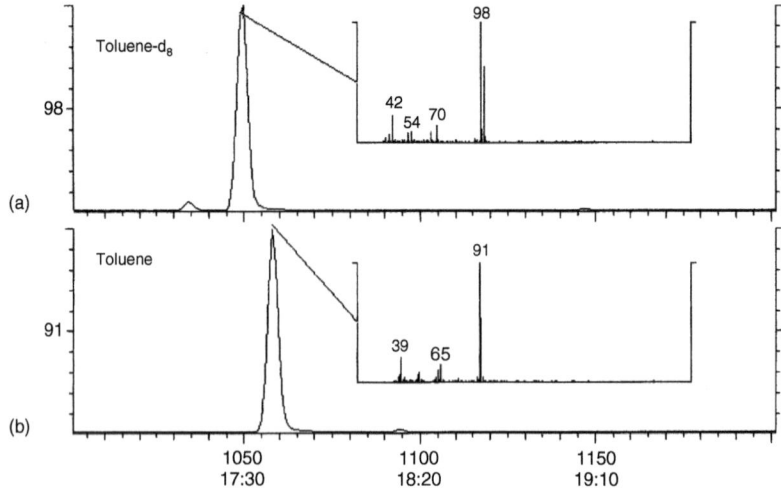

Figure 3.163 Elution of toluene-d_8 as the internal standard using purge and trap GC-MS (mass chromatograms), (a) Toluene-d_8, m/z 98 and (b) toluene m/z 91.

Requirements for Internal Standards in GC-MS Analysis

- The standard itself must not be present in the original sample.
- The internal standard chosen must be stable to clean-up and analysis and as inactive as possible.
- As far as possible the properties of the standard should be comparable to those of the analyte regarding sample preparation and analysis; therefore, isotopically labeled standards should ideally be used.
- The retention behavior in the GC should be adjusted so that elution occurs in the same section of the program (isothermal or heating ramp).
- The use of several standards allows them to be used as retention time standards.
- The retention behavior of the internal standard should be adjusted to ensure that overlap with matrix peaks or other components to be determined is avoided and faultless integration is possible.
- The ionization and fragmentation behavior (mass spectrometric response) should be comparable.
- The choice of the quantifying mass of the standard should be in the same mass range and exclude interference by the matrix or other components.
- Common naming convention: Internal standard (ISTD) is added to the sample extract immediately before analysis. Surrogate standard (SS) is added to the sample before sample preparation/extraction. The ratio is used for a "recovery" calculation.

In the preparation of standard solutions for calibration, the concentration of the internal standards must be kept constant at all calibration levels. Pipetting the same

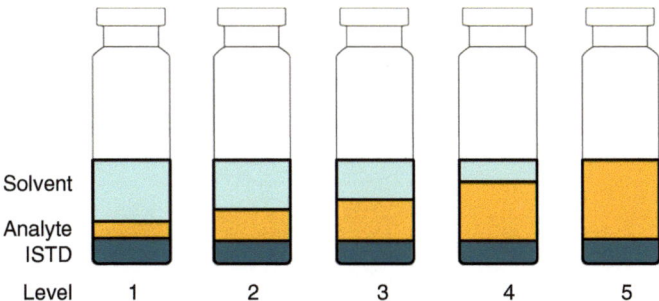

Figure 3.164 Preparation of solutions for calibration with internal standards: c(ISTD) = constant, c(analytes) = variable, total volume = constant, filled up with solvent, calculate concentrations on the basis of the total volumes.

volumes of the internal standard followed by different volumes of a mixed analyte standard stock solution and then making up with solvent to the same fixed volume in the sample vial has proved to be a good method (Figure 3.164). The preparation of the internal standard calibration vials can be performed in routine analysis by program control of a robotic liquid autosampler as well. Keeping the mixed standard and the internal standard vials available on an autosampler platform simplifies many error-prone manual steps.

The internal standard calibration curves plot the area ratio of the analyte relative to the area of the internal standard against the concentration in the sample (Figure 3.165). The parameter determined relative to the internal standard is thus independent of deviations in the injection volume and possible variations in the performance of the detector, as all these influences affect the analyte and the internal standard to the same extent. The sample extracts receive the same amount of internal standard before measurements. To calculate the analysis results, the ratio of the peak area of the analyte to that of the internal standard is determined, and the concentration is calculated using the calibration function. An example of the analytical setup is shown in Table 3.15. In this VOC analysis, two types of internal standards are added, the just discussed internal standard (ISTD), two compounds early and late eluting, and a "surrogate standard" (SS). The difference is the point of addition to the sample. The "surrogate" is a different internal standard compound added to the sample prior to extraction and allows the determination of the recovery with all sample preparation influences. The ISTD is added at the end of the sample preparation to the extract just before analysis. The ratio of ISTD to SS defines the recovery as a measure of the performance of the complete workflow of extraction and sample preparation.

Isotope Dilution The term "isotope dilution quantitation" refers to internal standard quantification methods, providing the highest accuracy of a quantitative method. As internal standard compounds, the most similar analogs of the native analytes labeled with stable isotopes are administered. Widely used are ^{13}C-labeled or deuterated compounds, for example $^{13}C_{12}$-TCDD or d_{12}-benzo[a]pyrene. The chemical and physical behavior of the native and labeled compound is almost identical

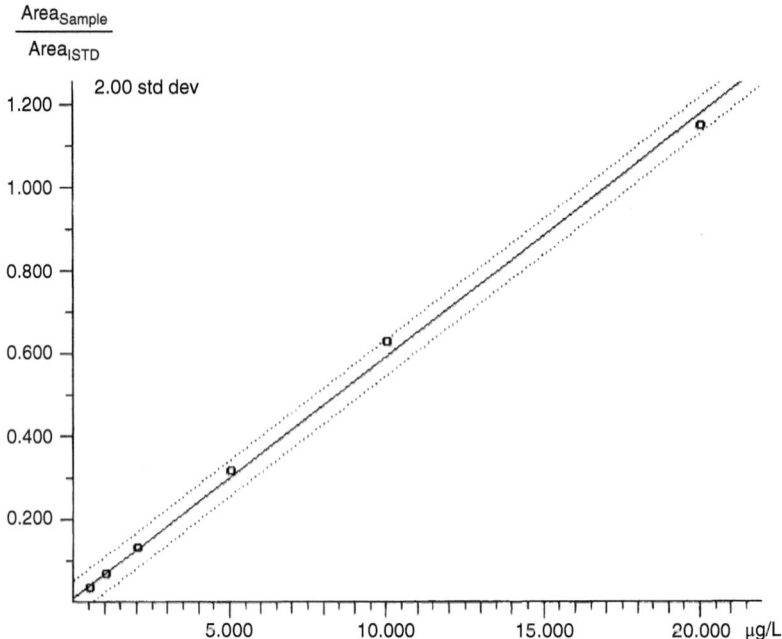

Figure 3.165 Calibration function using the internal standard procedure. Plot: area of the sample/area of the internal standard (ISTD, dimensionless) against concentration in the sample. The dotted lines indicate the confidence interval.

as ideally required by the concept of internal standardization. It is assumed that unspecific binding sites in the sample extract or the chromatographic system are covered by the labeled and the native analyte in the same way delivering improved recovery (carrier effect) and best trueness of the results.

The term *isotope dilution* is borrowed from isotope ratio MS for quantitative organic analysis. The basic definition asserts that it is not necessary to carry out a standard curve by different standard dilutions. Instead, a known quantity of a rare isotope is added as a spike to each sample. The measurement of the isotope ratio of the resulting mix compared to the known isotope ratio of the spike and natural abundances allows the calculation of the concentration of the unknown in the individual sample. In organic quantitation methods, typically the average relative response of standards in different concentrations (native vs. labeled ISTD) is used as the basis for quantitative calibration.

A typical example for extensive use of the isotope dilution method is the quantification of PCDDs and PCDFs. Sixfold and 12-fold labeled dioxins and furans in the required chlorination degrees are used as recovery and internal standards during sample preparation and GC-MS analysis, as described in the widely applied U.S. EPA 1613 method. The nearly identical compound characteristics during clean-up, the chromatographic process, and detection guarantee the most reliable

Table 3.15 Analytical setup and results of a VOC analysis in drinking water using the ISTD procedure (Types: I: internal standard, S: surrogate standard, A: analyte, *: set concentrations).

Number	Substance	Type	Scan #	Retention time	Me	Calculated concentration	Unit
1	Fluorobenzene ISTD1	I	680	17:00	BB	*5.000	g/L
2	1,2-Dichlorobenzene ISTD2	I	1050	26:15	BB	*5.000	g/L
3	Bromofluorobenzene SS1	S	943	23:34	BB	4.789	g/L
9	Trichlorofluoromethane	A	424	10:36	BB	0.045	g/L
11	Dichloromethane	A	522	13:03	BB	0.518	g/L
17	Chloroform	A	630	15:45	BB	8.714	g/L
18	1,1,1-Trichloroethane	A	641	16:01	BB	0.071	g/L
21	Benzene	A	664	16:36	BB	0.056	g/L
22	Carbon tetrachloride	A	651	16:17	BB	0.023	g/L
24	Trichloroethylene	A	702	17:33	BV	0.060	g/L
26	Dichlorobromomethane	A	732	18:18	BB	7.415	g/L
29	Toluene	A	779	19:28	BB	0.167	g/L
32	Dibromochloromethane	A	829	20:44	BB	4.860	g/L
34	Tetrachloroethylene	A	813	20:19	BV	0.058	g/L
35	Chlorobenzene	A	868	21:42	BB	0.032	g/L
37	Ethylbenzene	A	874	21:51	BV	0.037	g/L
38	meta, para-Xylene	A	881	22:02	VB	0.054	g/L
39	Bromoform	A	923	23:05	BB	0.635	g/L
45	Bromobenzene	A	954	23:51	BB	0.034	g/L
51	1,2,4-Trimethylbenzene	A	998	24:57	BB	0.084	g/L
55	Isopropyltoluene	A	1019	25:28	BB	0.057	g/L
59	Naphthalene	A	1182	29:33	BB	0.155	g/L
61	Hexachlorobutadiene	A	1176	29:24	BV	0.126	g/L

quantitation results based on the sample individual recovery values in one analysis run (see details in the application section on dioxin analysis 4.27).

Also for the clean-up of biological material, the carrier effect can be exploited through the addition of internal standards. The standard added at comparatively high concentrations can cover up active sites in the matrix and thus improve the extraction of the substance being analyzed. The result is a improved extraction recovery from active matrices for the native compound for a reliable representation of the given content in the sample.

In chromatography, the ^{2}H and ^{13}C labeled standards have a slightly shorter retention time than the native analytes. This retention time difference, though small, is always visible in the mass chromatogram, which is used for selective integration of the individual components (see also Figure 2.238).

3.3.7.3 Standard Addition

Matrix effects frequently lead to varying extraction yields. The standard addition method is an important tool in the confirmation of critical levels to compensate for matrix effects. The headspace and purge and trap techniques are affected in particular but it is also common in pesticide and similar types of analysis with legal backgrounds on MRL levels. With the standard addition method, the calibration is analogous to the external standardization described earlier, but involves the addition of known quantities of the analyte to be determined to one and the same sample (Miller, 1992; Basilicata et al., 2005).

Two types of matrix effects can be distinguished which either shift the calibration independent of the analyte concentration ("translational" effect) or change the slope ("rotational" effect) (Thompson, 2009). The "rotational" matrix effect is proportional to the analyte concentration and can be compensated by the standard addition method (Figure 3.166).

For the standard addition method, several sample vials are prepared, usually for six measurements. The vials are prepared with constant quantities of the sample material and by addition of increasing amounts of the standard solution. One sample is left neat as it is, that is, no standard is added. The other spiked samples serve the requirement for the confirmation of a linear calibration. A better precision of the calibration slope is achieved, and less of the standard dilution work required, if only one level of standard addition is prepared with multiple measurements of both, the native sample and the spiked vials. The concentration of the added analyte should be at least five times higher than the estimated concentration of the analyte.

The analysis results are calculated by plotting the peak areas against the quantity added. The concentration in the native sample can be read off as the absolute value at the intercept of the extrapolated calibration function with the x axis as indicated in Figures 3.167 and 3.168.

Despite of the good results obtained with this procedure, in practice, there is a major disadvantage with the manual preparation. Today, robotic x,y,z-autosampler allow the handling of the standard addition within an automated workflow instead. A further caveat of the standard addition procedure arises from the statistical aspect. The linear extrapolation of the calibration function is carried out assuming its validity for this area also. For routine analysis of certain analytes, it is required to confirm

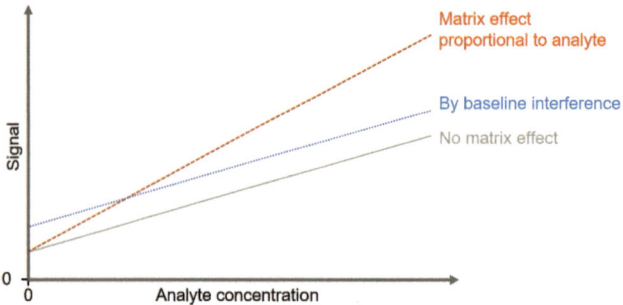

Figure 3.166 Matrix effects in standard addition quantitation.

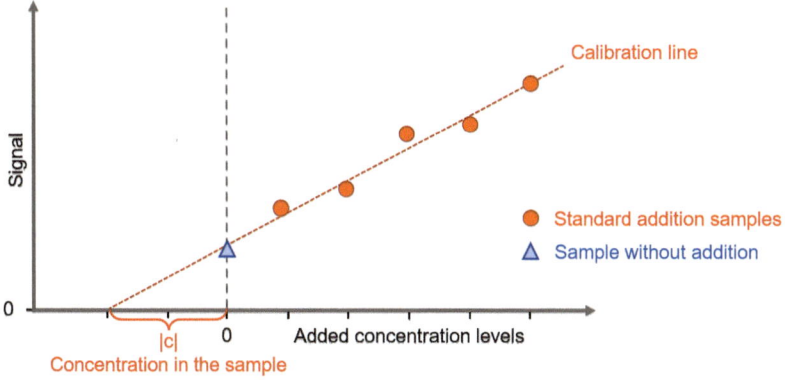

Figure 3.167 Standard addition calibration (orange dots) extrapolated for an unknown sample (blue triangle).

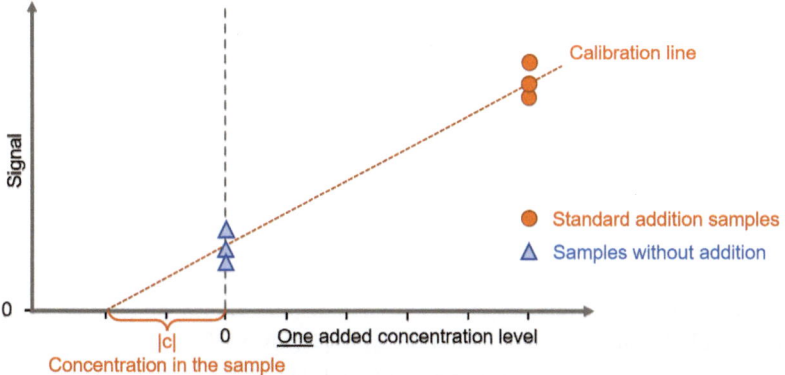

Figure 3.168 Optimized standard addition procedure with one spiked level (orange dots).

the linearity separately. A possible nonlinear deviation in this lower concentration range would give rise to considerable errors.

3.4 Frequently Occurring Impurities

With increasing analyte concentration during sample preparation and increasing sensitivity of GC-MS systems, the question of what are the necessary instrument and laboratory conditions for trace analysis is becoming more important. Signals from a permanent background or impurities can reduce S/N levels, interfere with peak integration, or are diagnostic for the instrument status.

One root cause of background in the mass spectrometer can be leaks on the GC or the MS side. Carrier gas leaks on the GC should never be checked using a "liquid leak detector", which is soap, and seeking for bubbles, although advertised. A long-lasting hydrocarbon background like fatty acid methyl esters will appear soon (Table 3.16). The problem is that a leak on the GC side works in both directions. While carrier

Table 3.16 Mass signals of frequently occurring contaminants.

m/z values	Possible cause
18/19, 28 (14), 32, 40, 44	H_2O, N_2, O_2, Ar, CO_2 from small leaks of the MS. m/z 19 from moisture after source maintenance. m/z 14 as doubly charged N_2 from a larger leak. Ar at m/z 40 may be high with an ICP instrument in the lab. With m/z 44 CO_2 is visible in the background.
31	Methanol
43, 58	Acetone
149, 167, 279	Phthalate plasticizers, various derivatives
129, 185, 259, 329	Tri-n-butyl acetyl citrate plasticizer
99, 155, 211, 266	Tributyl phosphate plasticizer
91, 165, 198, 261, 368	Tricresyl phosphate plasticizer
108, 183, 262	Triphenylphosphine (synthesis by-product)
51, 77, 183, 201, 277	Triphenylphosphine oxide (synthesis by-product)
41, 55, 69, 83, …	Hydrocarbon background series (forepump oil, greasy fingers, tap grease, etc.), always 14 masses apart, see Figure 3.173
43, 57, 71, 85, …	
99, 113, 127, 141, …	
285, 299, 313, 327, …	
339, 353, 367, 381, …	
407, 421, 435, 449, …	
409, 423, 437, 451, …	
64, 96, 128, 160, 192, 224, 256	Sulfur (as S_8), see Figure 3.175
205, 220	Antioxidant 2,6-di-t-butyl-4-methylphenol (BHT, Ionol) and isomers (technical mixture)
115, 141, 168, 260, 354, 446	Poly(phenyl ether) (diffusion pump oil)
262, 296, 298	Chlorophenyl ether (impurity in diffusion pump oil)
43, 59, 73, 87, 89, 101, 103, 117, 133	Poly(ethylene glycol) (all Carbowax phases)
73, 147, 207, 221, 281, 355, 429	Silicone rubber (all silicone phases), see Figure 3.169
133, 207, 281, 355, 429	Silicone grease, see Figure 3.169
233, 235	Rhenium oxide ReO^- (from the filament in NCl)
217, 219	Rhenium oxide ReO^-
250, 252	Rhenium oxide $HReO^-$

gas goes out, the soap, as well as atmospheric oxygen, can and will go in. In high purity carrier gas, the partial pressure of oxygen is low, so there is enough energy to penetrate against the leak. Oxygen in the GC column will deteriorate the stationary phase and cause the shown silicon background (Figure 3.169). Capillary columns can develop a high background noise by column bleed at elevated temperatures, in

3.4 Frequently Occurring Impurities | 525

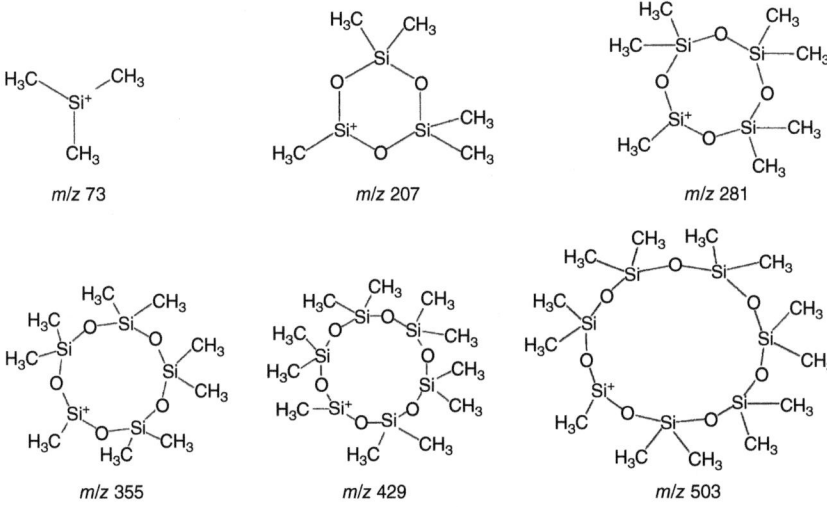

Figure 3.169 Structures of the most important signals resulting from column bleed of silicone phases (Spiteller and Spiteller, 1973).

particular non-bonded and polar phases. The only way to check for leaks on the GC fittings is by using a leak detector, available for He and H$_2$ to detect the lowest leaks.

Leaks on the MS side deal with the system vacuum. The MS itself is the leak detector. Course leaks are indicated by the vacuum readings. A 10^{-5}–10^{-6} Torr range is typical for a quadrupole GC-MS system at a He carrier gas flow of about 1.0 mL/min. Even at that level leaks may interfere with measurements and should be checked. The manual tune is recommended to check for the ions of water m/z 18, 19, air m/z 28, 32, or argon m/z 40 and CO$_2$ m/z 44. Argon is usually about 1% in ambient air. If there is an ICP-MS in the room, the ratio is significantly higher.

After ion source maintenance usually the water signal is high, m/z 19 is visible. It should be down to zero after equilibration. A leak-free MS shows the nitrogen signal m/z 28 similar or lower than water m/z 18, with almost no m/z 19. In larger air leaks the air signals dominate.

Often the transfer line fitting may cause leaks. Vespel/graphite ferrules used here can shrink with oven temperature programs and need a regular check. For investigation of the location of a leak, it is helpful to take the column off and cap the transfer line. The vacuum reading should get to the 10^{-6}–10^{-7} Torr level. Leak checks on the MS side, including the transfer line fitting can be done by a gentle stream of Ar if available, but also solvents like iso-propanol or methanol work well with a quick jump in the pressure reading and masses m/z 31 or 45. Also, electronic duster sprays are often recommended with ions of 1,1-difluoroethane m/z 51, 65, 1,1,1-trifluoroethane m/z 69, or 1,1,1,2-tetrafluoroethane m/z 69, 83, but volatile fluor compounds should be avoided in favor of a green analytical chemistry.

Often, there are software-based performance checks of air and water available under the tuning section which compare the air background to the set calibration gas level.

Typical ions from frequently occurring backgrounds are listed in Table 3.16. These can stem from "cleaned" glass apparatus (e.g. rotary evaporators and pipettes), through contact with polymers (e.g. cartridges from solid phase extraction, septa, solvent, and sample bottles), from the solvents themselves (e.g. stabilizers), or even from the laboratory surroundings (e.g. dust, solvent vapors, etc.). There are also sources of interfering signals in the GC-MS system itself. These are often arising from the GC injector. They range from septum bleed or decomposing septa to impurities that have become deposited in the split vent and which get into the measuring system on subsequent injections. Also, the closures of autosampler vials should be considered as a common source of contamination. Sources of contamination in the mass spectrometer are also known. Among these are background signals arising from pump oil (rotary vane pump, diffusion pump) and degassing products from sealing materials (e.g. Teflon) and from ceramic parts after solvent cleaning.

For trace analysis, the carrier gas used (helium and hydrogen) should be of the highest available purity from the beginning of installation of the instrument. Contamination from the gas supply tubes (e.g. cleaned up with hydrocarbons "for ECD operation") leads to ongoing interference and can be removed only at great expense. With central gas supply, a gas purification (irreversible binding of organic contaminants to getter materials) directly at the entry to the GC-MS system is recommended for trace analysis. Leaks in the vacuum system, in the GC system, and in the carrier gas supply always lead to secondary effects and should therefore be carefully eliminated. Under no circumstances should plastic tubing be used for the carrier and auxiliary gas supply to GC-MS systems (Table 3.17).

Some typical interfering components frequently occurring in residue analysis are listed together with their spectra in Figures 3.170–3.186. They are listed in order of

Table 3.17 Diffusion of oxygen through various tubing materials (Air Products).

Line material	Contamination by O_2 [ppm]
Copper	0
Stainless steel	0
Kel-F	0.6
Neoprene	6.9
Polyethylene (PE)	11
Teflon (PTFE)	13
Rubber	40

Note: measured in Ar 6.0 at a tubing diameter of 6 mm and 1 m length with a flow rate of 5 L/h.

3.4 Frequently Occurring Impurities | 527

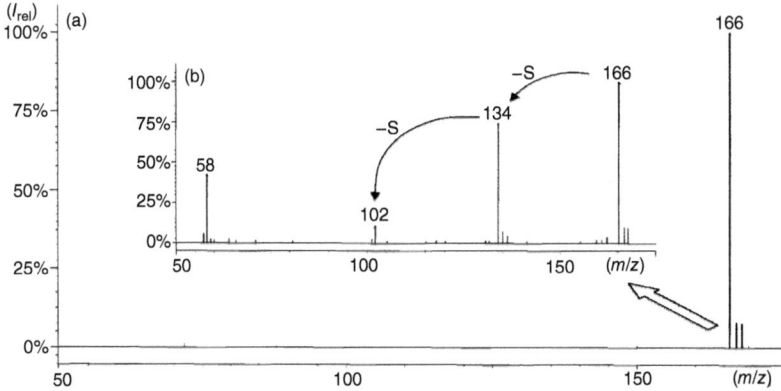

Figure 3.170 (a) GC-NCI-MS spectrum of a contaminant from butyl rubber septa. (b) Confirmation of two S atoms by GC-MS/MS.

Figure 3.171 Structure of the artefact 2-benzothiazolyl-N,N-dimethyldithiocarbamate.

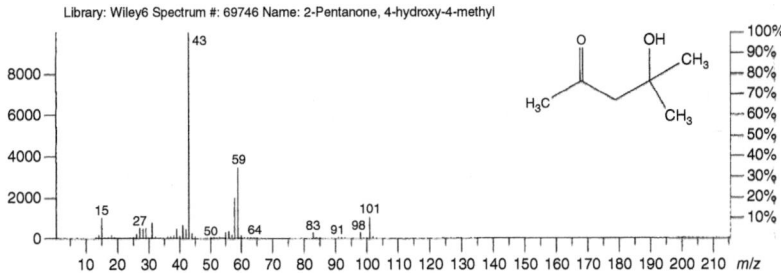

Figure 3.172 Diacetone alcohol $C_6H_{12}O_2$, M: 116, CAS: 123-42-2. Occurrence: acetone dimer, forms from the solvent under basic conditions.

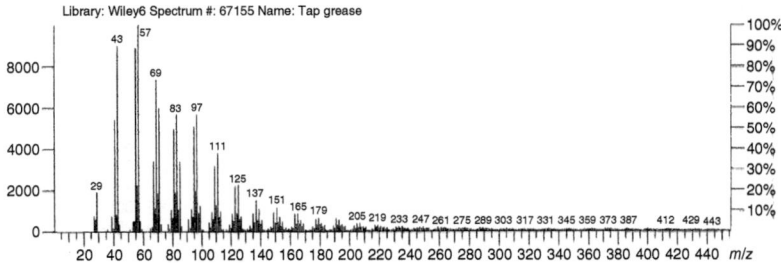

Figure 3.173 Hydrocarbon background. Occurrence: greasy fingers from maintenance work on the analyzer, the ion source or after changing the column, also from background of forepump oil (rotatory vane pumps), joint grease.

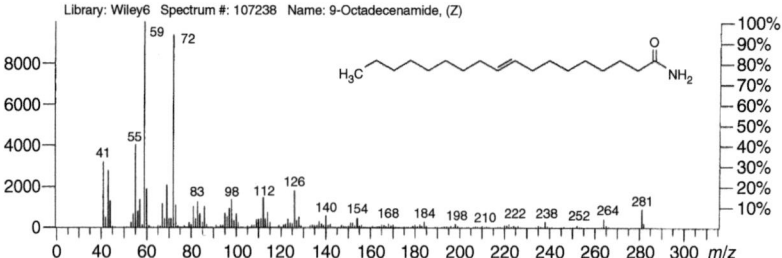

Figure 3.174 Oleic acid amide, oleamide C$_{18}$H$_{35}$NO, M: 281, CAS 301-02-0. Occurrence: lubricant for plastic sheeting. Erucamide C$_{22}$H$_{43}$NO, M: 337, CAS 112-84-5, occurs just as frequently.

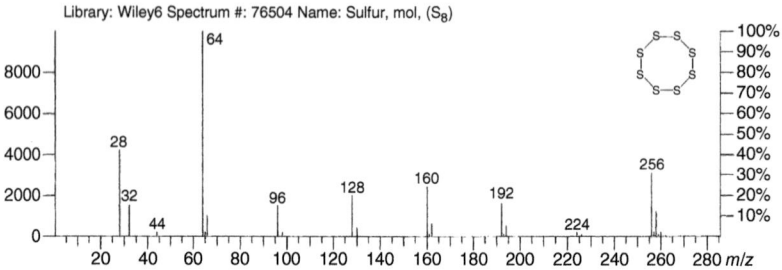

Figure 3.175 Molecular sulfur S$_8$, M: 256, CAS: 10544-50-0. Occurrence: from soil samples (microbiological decomposition), impurities from rubber objects, and by-product in the synthesis of thio-compounds.

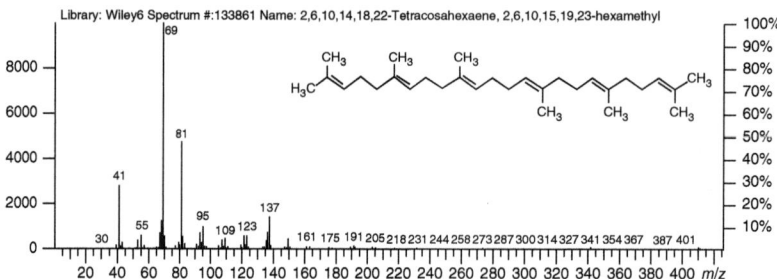

Figure 3.176 Squalene C$_{30}$H$_{50}$, M: 410, CAS: 7683-64-9. Occurrence: stationary phase in gas chromatography.

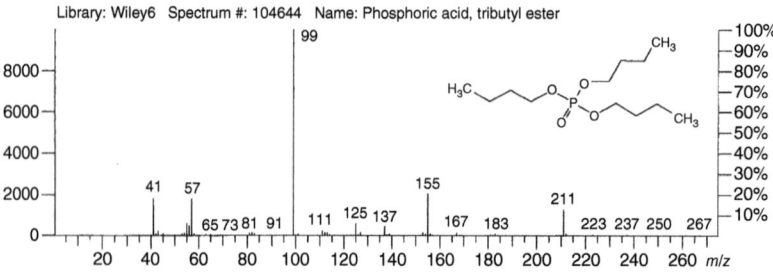

Figure 3.177 Tributyl phosphate C$_{12}$H$_{27}$O$_4$P, M: 266, CAS: 126-73-8. Occurrence: widely used plasticizer.

3.4 Frequently Occurring Impurities | 529

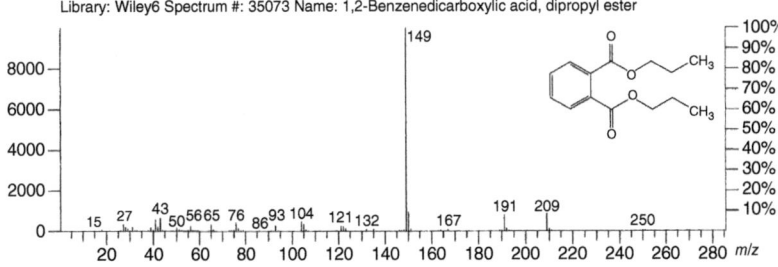

Figure 3.178 Propyl phthalate $C_{14}H_{18}O_4$, M: 250, CAS: 131-16-8. Occurrence: plasticizer, widely used in many plastics (also SPE cartridges, bottle tops, etc.), typical mass fragment of all dialkyl phthalates m/z 149.

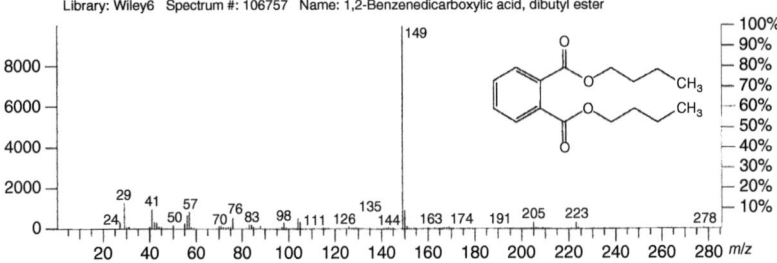

Figure 3.179 Dibutyl phthalate $C_{16}H_{22}O_4$, M: 278, CAS: 84-74-2.

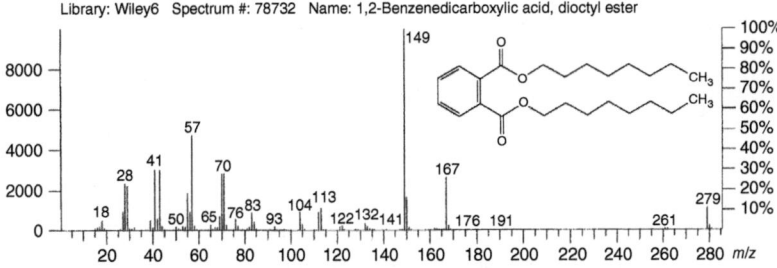

Figure 3.180 Dioctyl phthalate $C_{24}H_{38}O_4$, M: 390, CAS: 117-84-0.

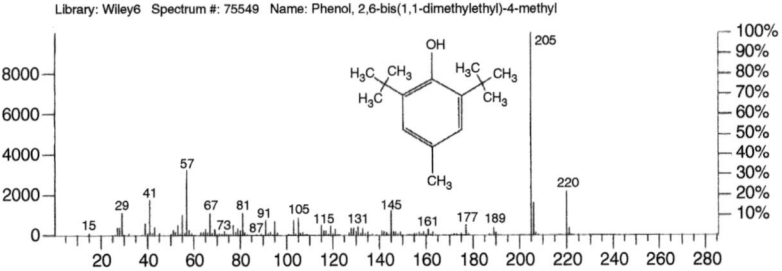

Figure 3.181 Ionol (BHT) $C_{15}H_{24}O$, M: 220, CAS: 128-37-0. Occurrence: antioxidant in plastics, stabilizer (radical scavenger) for ethers, THF, dioxan, technical mixture of isomers.

530 | *3 Evaluation of GC-MS Analyses*

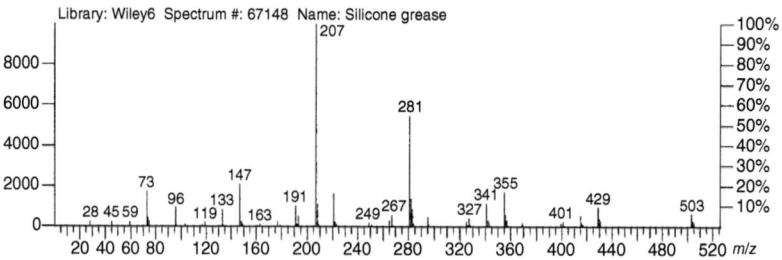

Figure 3.182 Silicones, silicone grease. Occurrence: typical column bleed from many silicone phases, from septum caps of sample bottles, injectors, and so on.

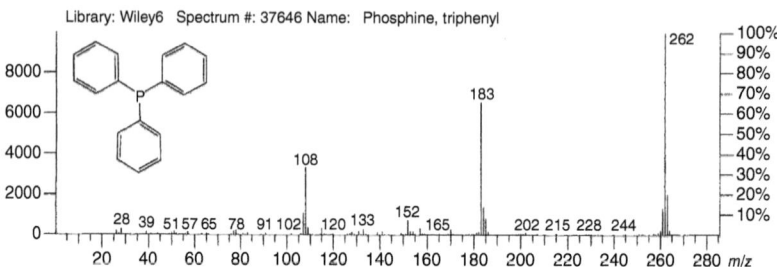

Figure 3.183 Triphenylphosphine $C_{18}H_{15}P$, M: 262, CAS: 603-35-0. Occurrence: catalyst for syntheses.

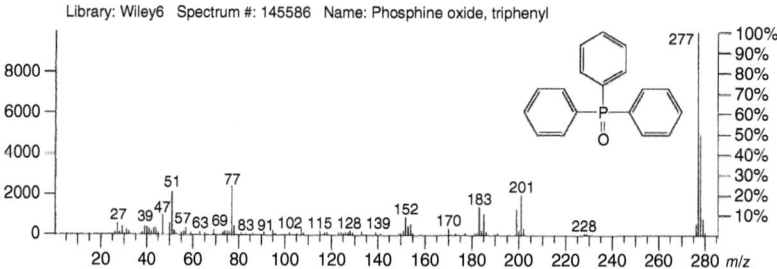

Figure 3.184 Triphenylphosphine oxide $C_{18}H_{15}OP$, M: 278, CAS: 791-28-6. Occurrence: forms from the catalyst in syntheses (e.g. the Wittig reaction).

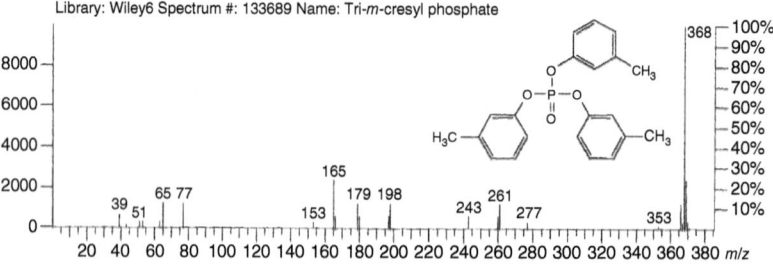

Figure 3.185 Tri-m-cresyl phosphate $C_{21}H_{21}O_4P$, M: 368, CAS: 563-04-2. Occurrence: plasticizer (e.g. in PVC (polyvinylchloride), nitrocellulose, etc.).

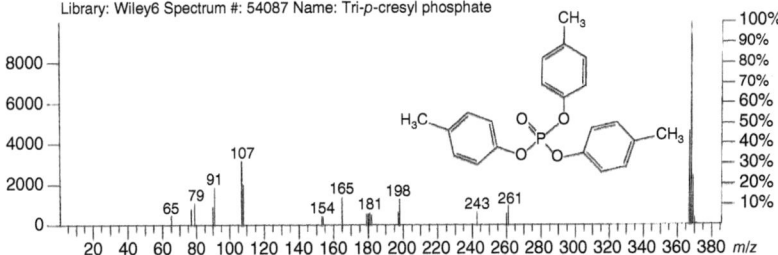

Figure 3.186 Tri-p-cresyl phosphate $C_{21}H_{21}O_4P$, M: 368, CAS: 78-32-0. Occurrence: plasticizer (e.g. in PVC, nitrocellulose, etc.).

the mass of the base peak. In GC-NCI/MS an intense contamination can be observed exhibiting m/z 166 with a significant isotope cluster. This compound was found to migrate from butyl rubber septa of autosampler vials into the sample and can be explained as artifacts from vulcanization facilitators (Figures 3.170 and 3.171) (Kapp and Vetter, 2006).

References

Section 3.1 Display of Chromatograms

Barwick, V., Langley, J., Mallet, T., Stein, B., and Webb, K. (2006) *Best Practice Guide for Generating Mass Spectra*, LGC, Teddington, UK, ISBN: 978-0-948926-24-2.

Wells, G., Prest, H., and Russ IV, C.W. (2011) Why use signal-to-noise as a measure of MS performance when it is often meaningless? *Curr. Trends Mass Spectrom.*, **May**, 28–33.

Section 3.2.2 Extraction of Mass Spectra

Biller, J.E. and Biemann, K. (1974) Reconstructed mass spectra, a novel approach for the utilization of gas chromatograph-mass spectrometer data. *Anal. Lett.*, **7**, 515–528. doi: 10.1080/00032717408058783.

Dimandja, J.M.D. (2004) GC × GC. *Anal. Chem.*, **76** (9), 167A–174A.

Halket, J.M., Przyborowska, A., Stein, S.E., Mallard, W.G., Down, S., and Chalmers, R. (1999) Deconvolution gas chromatography/mass spectrometry of urinary organic acids – potential for pattern recognition and automated identification of metabolic disorders. *Rapid Commun. Mass Spectrom.*, **13**, 279–284. doi: 10.1002/(SICI)1097-0231(19990228)13:4<279::AID-RCM478>3.0.CO;2-I.

Mallard, W.G. and Reed, J. (1997) *AMDIS Users Guide*, U.S. Department of Commerce, Technology Administration, National Institute of Standards and Technology (NIST), Standard Reference Data Program, Gaithersburg, MD, https://chemdata.nist.gov/mass-spc/amdis/docs/amdis.pdf (accessed 6 May 2024).

Mallard, G. (2012) AMDIS – an introduction to extracting high quality spectra from complex GC/MS data. *Sep. Sci.*, **19**.

Shao, X., Wang, G., Wang, S., and Su, Q. (2004) Extraction of mass spectra and chromatographic profiles from overlapping GC-MS signals with background. *Anal. Chem.*, **76**, 5143–5148. doi: 10.1021/ac035521u.

Zhang, W., Wu, P., and Li, C. (2006) Study of automated mass spectral deconvolution and identification system (AMDIS) in pesticide residue analysis. *Rapid Commun. Mass Spectrom.*, **20**, 1563–1568. doi: 10.1002/rcm.2473.

Section 3.2.3 The Retention Index

Bemgard, A., Colmsjö, A., and Wrangskog, K. (1994) Prediction of temperature-programmed retention indexes for polynuclear aromatic hydrocarbons in gas chromatography. *Anal. Chem.*, **66**, 4288–4294. doi: 10.1021/ac00095a027.

Bianchi, F., Careri, M., Mangia, A., and Musci, M. (2007) Retention indices in the analysis of food volatiles in temperature-programmed gas chromatography: database creation and evaluation. *J. Sep. Sci.*, **30**, 563–572. doi: 10.1002/jssc.200600393.

Deutsche Forschungsgemeinschaft (1982) *Gaschromatographische Retentionsindizes toxikologisch relevanter Verbindungen auf SE-30 oder OV-1, Mitteilung 1 der Kommission für Klinisch-toxikologische Analytik*, Verlag Chemie, Weinheim.

Hall, G.L., Whitehead, W.E., Mourer, C.R., and Shibamoto, T. (1986) A new gas chromatographic retention index for pesticides and related compounds. *HRC & CC*, **9**, 266–271, http://doi.wiley.com/10.1002/jhrc.1240090503.

Katritzky, A.L., Ignatchenko, E.S., Barcock, R.A., Lobanov, V.S., and Karelson, M. (1994) Prediction of gas chromatographic retention times and response factors using a general quantitative structure property relationship treatment. *Anal. Chem.*, **66**, 1799–1807. doi: 10.1021/ac00083a005.

Kostiainen, R. and Nokelainen, S. (1990) Use of M-series retention index standards in the identification of trichothecenes by electron impact mass spectrometry. *J. Chromatogr.*, **513**, 31–37. doi: 10.1016/S0021-9673(01)89421-8.

Kovats, E. (1958) Gas chromatographische Charakterisierung organischer Verbindungen. *Helv. Chim. Acta*, **41**, 1915–1932.

Lipinski, J. and Stan, H.-J. (1989) Compilation of retention data for 270 pesticides on three different capillary columns, in *10th International Symposium Capillary Chromatography, Riva del Garda, May 1989* (ed. P. Sandra), Huethig, Heidelberg, pp. 597–611.

Little, J. (2022) Creating and Using Retention Indices in NIST Software. Web presentation, *https://littlemsandsailing.com/2022/12/part-vi-creating-and-using-retention-indices-in-nist-software/* (accessed 6 May 2024).

Mallard, W.G. and Reed, J. (1997) *AMDIS Users Guide*, U.S. Department of Commerce, Technology Administration, National Institute of Standards and Technology (NIST), Standard Reference Data Program, Gaithersburg, MD, *https://chemdata.nist.gov/mass-spc/amdis/docs/amdis.pdf* (accessed 6 May 2024).

Manninen, A. et al. (1987) Gas chromatographic properties of the M-series of universal retention index standards and their application to pesticide analysis. *J. Chromatogr.*, **394**, 465–471.

Onuska, F.I. and Karasek, F.W. (1984) The retention index system, in *Open Tubular Column Gas Chromatography in Environmental Sciences*, Springer, Boston, MA. doi: 10.1007/978-1-4684-4688-3_6.

Schomburg, G. (1987) *Gaschromatographie*, 2nd edn, VCH Publishers, Weinheim, Germany.

Sparkman, O.D. (2023) NIST 23: the largest increases in compound coverage for the tandem and NIST/EPA/NIH EI libraries since NIST became curator. *Separation Science*, Web site, Aug 3, 2023, https://www.sepscience.com/nist-23/ (accessed 6 May 2024).

Van Den Dool, H. and Kratz, P.D. (1963) A generalization of the retention index system including linear temperature programmed gas-liquid partition chromatography. *J. Chrom.*, **11**, 463–471.

Weber, E. and Weber, R. (1992) *Methodik und Applikationen in der Kapillargaschromatographie*. Buch der Umweltanalytik, Band **4**, GIT Verlag, Darmstadt, Germany.

Zenkevich, I.G. (2002) Kovats' retention index system, in *Encyclopedia of Chromatography* (ed. J. Cazes), Marcel Dekker Inc., New York, NY, USA.

Section 3.2.4 Libraries of Mass Spectra

Adams, R.P. (1989) *Identification of Essential Oils by Ion Trap Mass Spectrometry*, Academic Press, San Diego, CA, USA.

Adams, R.P. (2007) *Identification of Essential Oil Components by Gas Chromatography/Mass Spectrometry*, 4th edn, Allured Books, Carol Stream, IL, USA.

Ausloos, P. et al. (1999) The critical evaluation of a comprehensive mass spectral library. *J. Am. Soc. Mass Spectrom.*, **10**, 287–299. doi: 10.1016/S1044-0305(98)00159-7.

Babushok, V.I. et al. (2007) Development of a database of gas chromatographic retention properties of organic compounds. *J. Chrom. A*, **1157** (1–2), 414–421. doi: 10.1016/j.chroma.2007.05.044.

Central Institute of Nutrition and Food Research (2003) *Mass Spectra of Volatiles in Food (SpecData)*, 2nd edn, John Wiley & Sons, Inc., ISBN: 978-0-471-64825-3, November 2003.

Christie, W.W. and Han, X. (2010) *Lipid Analysis - Isolation, Separation, Identification and Lipidomic Analysis*, 4th edn, 446 pages, Elsevier, ISBN: 9780081014653.

Christie, W.W. (2021), in *The AOCS Lipid Library - Mass Spectrometry of Fatty Acid Derivatives* (ed. A. Marangoni), https://lipidlibrary.aocs.org/ (accessed 8 May 2024).

De Leeuw, J.W. (2004) *Mass Spectra of Geochemicals, Petrochemicals and Biomarkers (SpecData)*, John Wiley & Sons, Ltd, August 2004, ISBN: 978-0-471-64798-0.

Hummel, J., Strehmel, N., Bölling, C., Schmidt, S., Walther, D., and Kopka, J. (2013) Mass spectral search and analysis using the Golm metabolome database, in *The Handbook of Plant Metabolomics*, Wiley-VCH Verlag GmbH, Weinheim, Germany, pp. 321–343, http://dx.doi.org/10.1002/9783527669882.ch18.

Kind, T. et al. (2009) FiehnLib - mass spectral and retention index libraries for metabolomics based on quadrupole and time-of-flight gas chromatography/mass spectrometry. *Anal. Chem.*, **81** (24), 10038–10048. doi: 10.1021/ac9019522.

Kühnle, R. (2009) *Mass Spectra of Pesticides*, John Wiley & Sons, Inc., ISBN: 978-3-527-32488-0.

MassLib (2014) MassLib's Searches. MassLib Home Page, MSP Kofel, Zollikofen, Switzerland, *http://masslib.com/Home.html* (accessed 20 June 2024).

Mondello, L. (2015) *FFNSC3 - Mass Spectra of Flavors and Fragrances of Natural and Synthetic Compounds*, 3rd edn, Wiley, ISBN: 978-1-119-06984-3.

NIST (2023a) *NIST Chemistry WebBook - NIST Standard Reference Database Number 69*. Web site, last update 2023, National Institute of Standards and Technology, Gaithersburg, MD, *https://webbook.nist.gov/chemistry/https://doi.org/10.18434/T4D303*.

NIST (2023b) NIST/EPA/NIH EI-MS LIBRARY 2023 Release. Web site, last update 2023, National Institute of Standards and Technology, Gaithersburg, MD, *https://chemdata.nist.gov/dokuwiki/lib/exe/fetch.php?media=chemdata:asms2023:asms2023_nist23_features.pdf*.

NIST (2023c) *Electron Ionization Library Component of the NIST/EPA/NIH Mass Spectral Library and NIST GC Retention Index Database*. Web site, updated August 11, 2023, National Institute of Standards and Technology, Gaithersburg, MD, USA, *https://www.nist.gov/programs-projects/electron-ionization-library-component-nistepanih-mass-spectral-library-and-nist-gc* (accessed 6 May 2024).

Parr, M.K., Opfermann, G., Schänzer, W., and Makin, H.L.J. (2011) *Mass Spectra of Physiologically Active Substances: Including Drugs, Steroid Hormones, and Endocrine Disruptors*. Data sheet, Wiley-VCH, Weinheim, Germany, ISBN: 978-3-527-33080-5.

Pfleger, K., Maurer, H.H., and Weber, A. (2023) *Mass Spectral and GC Data of Drugs, Poisons, Pesticides, Pollutants and Their Metabolites*, 6th edn, John Wiley & Sons, Inc., ISBN: 978-3-527-35286-9.

Roesner, P., Junge, T., Westphal, F., Fritschi, G., and Tenczer, J. (2007) *Mass Spectra of Designer Drugs*, 1st edn, Wiley-VCH, Weinheim, Germany, 2087 pages (Wiley-VCH Verlag GmbH & Co. KGaA), ISBN-13: 978-3527307982, *https://www.designer-drugs.de/about.pl* (accessed 8 May 2024).

SIS (2023) The NIST 23 Mass Spectral Library & Search Software (NIST 2023/2020/2017/EPA/NIH). Web site, Scientific Instrument Services, *https://www.sisweb.com/software/ms/nist.htm* (accessed 6 May 2024).

Tsuge, S., Ohtani, H., and Watanabe, C. (2011) *Pyrolysis – GC/MS Data Book of Synthetic Polymers*, Elsevier, ISBN: 9780444538925.

SWGDRUG (2023) SWGDRUG MS Library, *https://swgdrug.org/ms.htm* (accessed 6 May 2024).

Wikipedia (2024) Golm Metabolome Database, *https://en.wikipedia.org/wiki/Golm_Metabolome_Database* (accessed 8 May 2024).

Wiley (2020) *Mass Spectra of Geochemicals, Petrochemicals and Biomarkers*. Data sheet, John Wiley & Sons, Inc., *https://sciencesolutions.wiley.com/solutions/technique/gc-ms/mass-spectra-of-geochemicals-petrochemicals-and-biomarkers-specdata/*.

Wiley (2023a) *Wiley Registry of Mass Spectral Library 2023*. Data sheet, John Wiley & Sons Inc., *https://sciencesolutions.wiley.com/news-wiley-releases-wiley-registry-nist-mass-spectral-library-2023/* (accessed 6 May 2024).

Wiley (2023b) *Wiley Registry/NIST Mass Spectral Library 2023*. Data sheet, John Wiley & Sons Inc., https://sciencesolutions.wiley.com/solutions/technique/gc-ms/wiley-registry-of-mass-spectral-data/ (accessed 6 May 2024).

Wiley (2024) *Mass Spectra of Designer Drugs 2024*, ISBN: 9783527353903.

Yarkov, A. (2004) *Mass Spectra of Organic Compounds (SpecInfo)*, John Wiley & Sons, Inc., ISBN: 978-0-47166-773-5.

Section 3.2.5 Library Search Programs

Damen, H., Henneberg, D., and Weimann, B. (1978) Siscom - a new library search system for mass spectra. *Anal. Chim. Acta*, **103** (4), 289–302, https://doi.org/10.1016/S0003-2670(01)83095-6.

Davies, A.N. (1993) Mass spectrometric data systems. *Spectrosc. Eur.*, **5**, 34–38.

Heller, S.R. (1999) The history of the NIST/EPA/NIH mass spectral database. *Today's Chemist at Work*, **8** (2), 49–50, https://citeseerx.ist.psu.edu/document?repid=rep1&type=pdf&doi=0c5cbd5f9b29184f74e8db94a6d43fa706bedd83.

Henneberg, D. and Weimann, B. (1984) Search for identical and similar compounds in mass spectral data bases. *Spectra*, 11–14.

Henneberg, D., Weimann, B., and Zalfen, U. (1993) Computer-aided interpretation of mass spectra using databases with spectra and structures. I. Structure searches. *J. Mass Spectrom.*, **28** (3), 198–206, https://doi.org/10.1002/oms.1210280311.

Hübschmann, H.-J. (2015) *Handbook of GC-MS*, 3rd edn, Wiley-VCH, Weinheim, Germany, ISBN: 978-3-527-33474-2.

Little, J.L., Cleven, C.D., and Brown, S.D. (2011) Identification of 'known unknowns' utilizing Accurate Mass Data and Chemical Abstracts Service Databases. *J. Am. Soc. Mass Spectrom.*, **22** (2), 348–359. doi: 10.1007/s13361-010-0034-3.

Little, J.L., Cleven, C.D., Howard, A.S., and Yu, K. (2013) Identifying 'known unknowns' in commercial products by mass spectrometry. *LCGC N. Am.*, **31**, 114–125.

McLafferty, F.W. and Turecek, F. (1993) *Interpretation of Mass Spectra*, 4th edn, University Science Books, Mill Valley, CA. doi: 10.1002/bms.1200230614.

Neudert, B., Bremser, W., and Wagner, H. (1987) Multidimensional computer evaluation of mass spectra. *Org. Mass Spectrom.*, **22**, 321–329. doi: 10.1002/oms.1210220604.

Sokolow, S., Karnovsky, J., and Gustafson, P. (1978) The Finnigan Library Search Program. Application Report, Finnigan MAT.

Sparkman, D. (1996) Evaluating electron ionization mass spectral library search results. *J. Am. Soc. Mass Spectrom.*, **7**, 313–318. doi: 10.1016/1044-0305(95)00705-9.

Stauffer, D.B., McLafferty, F.W., Ellis, R.D., and Peterson, D.W. (1985a) Adding forward searching capabilities to a reverse search algorithm for unknown mass spectra. *Anal. Chem.*, **57**, 771–773. doi: 10.1021/ac00280a045.

Stauffer, D.B., McLafferty, F.W., Ellis, R.D., and Peterson, D.W. (1985b) Probability-based-matching algorithm with forward searching capabilities for matching unknown mass spectra of mixtures. *Anal. Chem.*, **57**, 1056–1060. doi: 10.1021/ac00283a021.

Stein, S.E. (1994) Estimating probabilities of correct identification from results of mass spectral library search. *J. Am. Soc. Mass Spectrom.*, **5**, 316–323.

Stein, S.E. and Scott, D.R. (1994) Optimization and testing of mass spectral library search algorithms for compound identification. *J. Am. Soc. Mass Spectrom.*, **5**, 859–866.

Stein, S.E. (1995) Chemical substructure identification by mass spectral library searching. *J. Am. Soc. Mass Spectrom.*, **6**, 644–655.

Stein, S.E. (1999) An integrated method for spectrum extraction and compound identification from gas chromatography/mass spectrometry data. *J. Am. Soc. Mass Spectrom.*, **10**, 770–781.

Stein, S. (2011) *NIST Standard Reference Database 1A, User's Guide*, U.S. Department of Commerce, National Institute of Standards and Technology, Standard Reference Data Program, Gaithersburg, MD, USA.

Warr, W.A. (1993a) Computer-assisted structure elucidation. Part 1. Library search and spectral data collections. *Anal. Chem.*, **65**, 1045A–1050A.

Warr, W.A. (1993b) Computer-assisted structure elucidation. Part 2: Indirect database approaches and established systems. *Anal. Chem.*, **65** (24), 1087A–1095A.

Section 3.2.6 Interpretation of Mass Spectra

Budzikiewicz, H. (1985) Structure elucidation by ion-molecule reactions in the gas phase: the location of C, C-double and triple bonds. *Fresenius J. Anal. Chem.*, **321** (2), 150–158.

Budzikiewicz, H. and Schäfer, M. (2012) *Massenspektrometrie*, 6th edn, Wiley-VCH Verlag GmbH, Weinheim, Germany, ISBN: 978-3-527-32911-3.

Christie, W.W. (2001) A practical guide to the analysis of conjugated linoleic acid. *Inform*, **12** (2), 147–152.

Christie, W.W. and Han, X. (2010) *Lipid Analysis - Isolation, Separation, Identification and Lipidomic Analysis*, 4th edn, Elsevier, 446 pages, ISBN: 9780081014653.

Christie, W.W. (2021), in *The AOCS Lipid Library - Mass Spectrometry of Fatty Acid Derivatives* (ed. A. Marangoni), https://lipidlibrary.aocs.org/ (accessed 8 May 2024).

Dobson, G. and Christie, W.W. (2002) Mass spectrometry of fatty acid derivatives. *Eur. J. Lipid Sci. Technol.*, **104**, 36–43, https://doi.org/10.1002/1438-9312(200201)104:1<36::AID-EJLT111136>3.0.CO;2-W.

Fay, L. and Richli, U. (1991) Location of double bonds in polyunsaturated fatty acids by gas chromatography-mass spectrometry after 4,4-dimethyloxazoline derivatization. *J. Chromatogr.*, **541**, 89–98. doi: 10.1016/S0021-9673(01)95986-2.

Frigerio, A. (1974) *Essential Aspects of Mass Spectrometry*, John Wiley & Sons Inc., Spectrum Publishers, New York.

Gross, J.H. (2004) *Mass Spectrometry - A Textbook*, Springer-Verlag, Berlin, Heidelberg, Germany.

Howe, I., Bowen, R., and Williams, D.H. (1981) *Mass Spectrometry: Principles and Applications*, McGraw-Hill, New York, ISBN: 978-0070705692.

Jham, G.N., Attygalleb, A.B., and Meinwalda, J. (2005) Location of double bonds in diene and triene acetates by partial reduction followed by methylthiolation. *J. Chromatogr. A*, **1077** (1), 57–67. doi: 10.1016/j.chroma.2005.01.073.

López, J.F. and Grimalt, J.O. (2004) Phenyl- and cyclopentylimino derivatization for double bond location in unsaturated C (37)-C (40) alkenones by GC-MS. *J. Am. Soc. Mass Spectrom.*, **15** (8), 1161–1172. doi: 10.1016/j.jasms.2004.04.024.

McLafferty, F.W. (1959) Mass spectrometric analysis. Molecular rearrangements. *Anal. Chem.*, **31** (1), 82–87. doi: 10.1021/ac60145a015.

McLafferty, F.W. and Turecek, F. (1993) *Interpretation of Mass Spectra*, 4th edn, University Science Books, Mill Valley, CA, USA.

Moss, C.W. and Lambert-Fair, M.A. (1989) Location of double bonds in monounsaturated fatty acids of Campylobacter cryaerophila with dimethyl disulfide derivatives and combined gas chromatography-mass spectrometry. *J. Clin. Microbiol.*, **27** (7), 1467–1470. doi: 10.1128/jcm.27.7.1467-1470.1989.

Pretsch, E., Clerc, T., Seibl, J., and Simon, W. (1990) *Tabellen zur Strukturaufklärung organischer Verbindungen mit spektrometrischen Methoden*, 3rd edn, 1st corr. reprint edn, Springer, Berlin, ISBN: 3540158952.

Quadtech Associates (2024) *Mass Spec Calculator Pro*, https://quadtechassociates.com/mscd01.html (accessed 21 June 2024).

Rosman, K.J.R. and Taylor, P.D.P. (1998) Isotopic compositions of the elements 1997. *Pure Appl. Chem.*, **70** (1), 217–235.

Shantha, N.C. and Kaimal, T.N.B. (1984) Mass spectrometric location of double bonds in unsaturated fatty acids including conjugated acids as their methoxybromo derivatives. *Lipids*, **19** (12), 971–974. doi: 10.1007/BF02534736.

Sharp, Z. (2017) *Principles of Stable Isotope Geochemistry*, 2nd edn, open access, https://digitalrepository.unm.edu/unm_oer/1/?fref=gc&dti=175833885799280 edn, *https://doi.org/10.25844/h9q1-0p82*.

Section 3.2.7 Mass Spectroscopic Features of Selected Substance Classes

Ballschmiter, K. and Zell, M. (1980) Analysis of polychlorinated biphenyls (PCB) by glass capillary gas chromatography - composition of technical aroclor- and clophen-PCB mixtures. *Fres. Z. Anal. Chem.*, **302** (1), 20–31. doi: 10.1007/BF00469758.

Budzikiewicz, H. and Schäfer, M. (2012) *Massenspektrometrie*, 6th edn, Wiley-VCH Verlag GmbH, Weinheim, Germany, ISBN: 978-3-527-32911-3.

Ciccioli, P., Brancaleoni, E., Cecinato, A., and Frattoni, M. (1993) A method for the selective identification of volatile organic compounds (VOC) in air by HRGC-MS, in *15th International Symposium Capillary Chromatography, Riva del Garda, May 1993* (ed. P. Sandra), Huethig, Heidelberg, pp. 1029–1042.

Färber, H. and Schöler, F. (1991) Gaschromatographische Bestimmung von Harnstoffherbiziden in Wasser nach Methylierung mit Trimethylaniliniumhydroxid oder Trimethylsulfoniumhydroxid. *Vom Wasser*, **77**, 249–262.

Howe, I., Bowen, R., and Williams, D.H. (1981) *Mass Spectrometry: Principles and Applications*, McGraw-Hill, New York, ISBN: 978-0070705692.

McLafferty, F.W. and Turecek, F. (1993) *Interpretation of Mass Spectra*, 4th edn, University Science Books, Mill Valley, CA, USA.

Stan, H.-J., Abraham, B., Jung, J., Kellert, M., and Steinland, K. (1977) Nachweis von Organophosphorinsecticiden durch Gas-Chromatographie-Massenspektroskopie. *Fresenius Z. Anal. Chem.*, **287**, 271–285. doi: 10.1002/cber.18950280434.

Stan, H.-J. and Kellner, G. (1982) Negative chemical ionization mass spectrometry. *Biomed. Mass Spectrom.*, **9** (11), 483–492. doi: 10.1002/bms.1200091106.

Section 3.3 Quantitation

Agilent (2024) *An Introduction to Using Instrument Detection Limit as a Performance Metric (IDL)*. Web site, *https://www.agilent.com/en/support/liquid-chromatography-mass-spectrometry-lc-ms/instrument-detection-limit* (accessed 2 May 2024).

Banerjee, K. and Hübschmann, H.-J. (2022) Automation in pesticide residue analysis in foods: a step toward smarter laboratories and green chemistry. *Agri. Sci. Technol.*, **2** (3), 426–429. doi: 10.1021/acsagscitech.2c00126.

Basilicata, P., Miraglia, N., Pieri, M., Acampora, A., Soleo, L., and Sannolo, N. (2005) Application of the standard addition approach for the quantification of urinary benzene. *J. Chrom. B*, **818** (2), 293–299. doi: 10.1016/j.jchromb.2005.01.013.

Boqué, R., Marato, A., and Vander Heyden, Y. (2008) Assessment of accuracy in chromatographic analysis. *LCGC Eur.*, **21**, 264–267.

Chang, C. (1985) Parallel mass spectrometry for high performance GC and LC detection. *Int. Lab.*, **5**, 58–68.

Commission Decision (2002) of 12 August, implementing council directive 96/23/EC concerning the performance of analytical methods and the interpretation of results. *Off. J. European Communities*, **L221** (8), 8–36. doi: 10.1017/CBO9781107415324.004.

Commission Decision 2002/657/EC *of 12 August 2002 implementing Council Directive 96/23/EC concerning the performance of analytical methods and the interpretation of results*, *https://eur-lex.europa.eu* (accessed 10 June 2024).

Dallüge, J., Vreuls, R.J., van Iperen, D.J., van Rijn, M., and Brinkman, U.A.T. (2002) Resistively heated gas chromatography coupled to quadrupole mass spectrometry. *J. Sep. Sci.*, **25** (9), 608–614. doi: 10.1002/1615-9314(20020601)25:9<608::AID-JSSC608>3.0.CO;2-R.

DIN 38402 (2011) German Standard Methods for the Examination of Water, Waste Water and Sludge – General Information (group A) – Part 1: Recording of Analysis Results (A 1). DIN 38402-1:2011-09, DIN Media, *https://dx.doi.org/10.31030/1804070*.

Dolan, J.W. (2021) Chromatographic measurements, Part 5: Determining LOD and LOQ based on the calibration curve. *Sep. Sci.*, **126**, www.sepscience.com.

Eurachem (2014) Eurachem Guide: *The Fitness for Purpose of Analytical Methods – A Laboratory Guide to Method Validation and Related Topics* (eds B. Magnusson and U. Örnemark), 2nd edn, Vol. 379. doi: 10.1016/S0014-2999(99)00500-2.

Ebel, S. and Kamm, K. (1983) Statistische Definition der Bestimmungsgrenze. *Z. Anal. Chem.*, **316**, 382–385. doi: 10.1007/BF00487800.

Ebel, S. and Dorner, W. (1987) *Jahrbuch Chemielabor 1987*, VCH Publishers, Weinheim, Germany.

Folley, J.P. (1987) Systematic errors in the measurement of peak area and peak height for overlapping peaks. *J. Chrom. A*, **384** (C), 301–313. doi: 10.1016/S0021-9673(01) 94679-5.

Funk, W., Dammann, V., and Donnevert, G. (2007) *Quality Assurance in Analytical Chemistry*, 2nd edn, Wiley-VCH Verlag GmbH, Weinheim, Germany, ISBN: 978-3-527-31114-9.

Guichon, G. and Guillemin, C.L. (1988) *Quantitative Gas Chromatography for Laboratory Analyses and On-Line Process Control*, Elsevier, Amsterdam, Oxford, New York, Tokyo, eBook ISBN: 9780080858470.

Hübschmann, H.-J. (2021) *Automated Sample Preparation – Solutions for GC-MS and LC-MS*, Wiley-VCH Verlag GmbH & Co. KGaA, Weinheim, Germany.

ICH Expert Working Group (2005) Validation of analytical procedures: text and methodology Q2(R1), in *The Textbook of Pharmaceutical Medicine*, pp. 1–13, https://database.ich.org/sites/default/files/Q2%28R1%29Guideline.pdf.

ISO 11843 (2019) *ISO 11843-6:2019 – Capability of Detection*. Technical Committee ISO/TC 69/SC 6, International Organization for Standardization, Geneva.

ISO 5725-6:1994 (1994) *Accuracy (Trueness and Precision) of Measurement Methods and Results*. Parts 1-4, International Organization for Standardization, Geneva.

Kadjo, A.F., Dasgupta, P.K., Su, J., Liu, S., and Kraiczek, K.G. (2017) Width based quantitation of chromatographic peaks: principles and principal characteristics. *Anal. Chem.*, **89** (7), 3884–3892. doi: 10.1021/acs.analchem.6b04857.

Kirchner, M., Matisova, E., Hrouzkova, S., and de Zeeuw, J. (2005) Possibilities and limitations of quadrupole mass spectrometric detector in fast gas chromatography. *J. Chrom. A*, **1090**, 126–132. doi: 10.1016/j.chroma.2005.06.090.

Kuss, H.-J. (2009) Interpretation of chromatograms, in *Quantification in LC and GC: A Practical Guide to Good Chromatographic Data* (eds H.-J. Kuss and S. Kromidas), WILEY-VCH Verlag GmbH, Weinheim, Germany, pp. 153–166.

Glaser, J., Foerst, D.L., McKee, G.D., Quave, S.A., and Budde, W.L. (1981) Trace analyses for wastewaters. *Env. Sci. Techn.*, **15** (12), 1426–1435. doi: 10.1021/es00094a002.

Miller, J.N. (1992) VI: The method of standard additions. *Spectroscopy Europe*, **4** (6), 26–27.

Montag, A. (1982) Beitrag zur Ermittlung der Nachweis- und Bestimmungsgrenze analytischer Meßverfahren. *Fresenius Z. Anal. Chem.*, **312**, 96–100.

Naes, T. and Isakson, T. (1992) The importance of outlier detection in spectroscopy. *Spectrosc. Eur.*, **4** (4), 32–33. doi: 10.1255/nirn.136.

Prichard, L. and Barwick, V. (2003) Preparation of Calibration Curves – A Guide to Best Practice. LGC/VAM/2003/032, LGC Limited, https://doi.org/LGC/VAM/2003/032.

Rousseeuw, P.J. and Leroy, A.M. (1987) *Robust Regression and Outlier Detection*, John Wiley & Sons, New York, ISBN: 0471-85233-3.

Sheehan, T.L. and Yost, R.A. (2015) What's the most meaningful standard for mass spectrometry: instrument detection limit or signal-to-noise ratio? *Spectroscopy*, October.

Shukla, S.S., Pandey, R.K., Gidwani, B., and Kalyani, G. (2023) Errors and uncertainties in calibration, in *Pharmaceutical Calibration, Validation and Qualification: A*

Comprehensive Approach, Springer Nature Singapore Pte. Ltd., pp. 17–27, *https://link.springer.com/chapter/10.1007/978-981-19-9002-1_2*.

Thompson, M. (2009) Standard Additions: Myth and Reality. AMC Technical Briefs 3, Royal Society of Chemistry, *https://www.rsc.org/images/myth-reality-technical-brief-37_tcm18-214868.pdf*.

U.S. EPA (2023) Appendix B to Part 136 – Definition and Procedure for the Determination of the Method Detection Limit-Revision 2. Federal Register, Water Programs, Part 136-Guidelines Establishing Test Procedures for The Analysis of Pollutants: 385–388, *https://www.govinfo.gov/content/pkg/CFR-2023-title40-vol25/pdf/CFR-2023-title40-vol25-part136-appB.pdf*

Wells, G., Prest, H., and Russ, C.W. (2023) Signal, Noise, and Detection Limits in Mass Spectrometry. Technical Note, 5990-7651EN, Agilent Technologies, Inc..

Wenzl, T., Haedrich, J., Schaechtele, A., Robouch, P., and Stroka, J. (2016) *Guidance Document on the Estimation of LOD and LOQ for Measurements in the Field of Contaminants in Feed and Food*. EUR 28099, Publications Office of the European Union, Luxembourg, ISBN: 978-92-79-61768-3. doi: 10.2787/8931.

Wiest, L. (2020) Tips for Preparing Calibration Curve Standards and Avoiding Sources of Error. Technical Guide, Restek Corporation, Bellefonte, PA, USA, *https://www.restek.com/Technical-Resources/Technical-Library/General-Interest/gen_GNAR3169-UNV*.

Zhang, X. Chongtian, Y., Liang, L., and Hübschmann, H-J. (2014) Determination of BTEX in Cigarette Filter Fibers by GC-MS with Automated Calibration using the TriPlus RSH Autosampler Application Note 10399, Thermo Fisher Scientific, Shanghai, China.

Section 3.4 Frequently Occurring Impurities

Kapp, T. and Vetter, W. (2006) Migration von Additiven aus Buthylgummihaltigen Verschlüssen von Probegläschen. *Lebensmittelchem.*, **60**, 152.

Spiteller, M. and Spiteller, G. (1973) *Massenspektrensammlung von Lösemitteln, Verunreinigungen, Säulenbelegmaterialien und einfachen aliphatischen Verbindungen*, Springer, Wien.

4

Applications

The applications presented in this chapter have been chosen in order to describe typical areas of the wide usage of gas chromatography–mass spectrometry (GC-MS), such as air, water, soil, foodstuffs, the environment, waste materials, drugs, or pharmaceutical products. Special emphasis has been put on current and reproducible examples that give successful templates including often integrated and automated sample preparation for routine laboratories. Different MS detection methods with full scan, selected ion monitoring (SIM), MS/MS techniques with selected reaction monitoring (SRM) target compound analysis and high resolution accurate mass detection (HRAM) are included. The described instrumentation is exemplified and can be replaced by equivalent technologies and brands. The selection cannot be representative of the use of modern GC-MS but shows workflows from main areas. In addition, in special areas of application, such as the analysis of isotope-specific measuring procedures and the isotope dilution method for dioxin analysis are described.

Some applications are compiled from the references cited and are documented with various graphics and tables. The analysis conditions are described in full with comments to allow adaptation of the methods. For published methods, the sources are given for each section. References to the cited and related literature are provided.

4.1 Air Analysis According to U.S. EPA Method TO-14

Keywords: air analysis; SUMMA canister; EPA method TO-14; TO-15; thermodesorption; volatile halogenated hydrocarbons; VOCs; cryofocusing; Nafion water removal; thick film column; SIM; ion trap; full scan

4.1.1 Introduction

The U.S. Environmental Protection Agency describes in method TO-14 a process for sampling and analysis of volatile organic compounds (VOCs) in the atmosphere. It is based on the collection of air samples in passivated stainless steel canisters (SUMMA canisters) (U.S. EPA 1999). The organic components are separated by GC and determined using conventional GC detectors or by mass spectrometry (Figure 4.1). The use of mass spectrometers in the targeted selected ion monitoring

Handbook of GC-MS: Fundamentals and Applications, Fourth Edition. Hans-Joachim Hübschmann.
© 2025 WILEY-VCH GmbH. Published 2025 by WILEY-VCH GmbH.

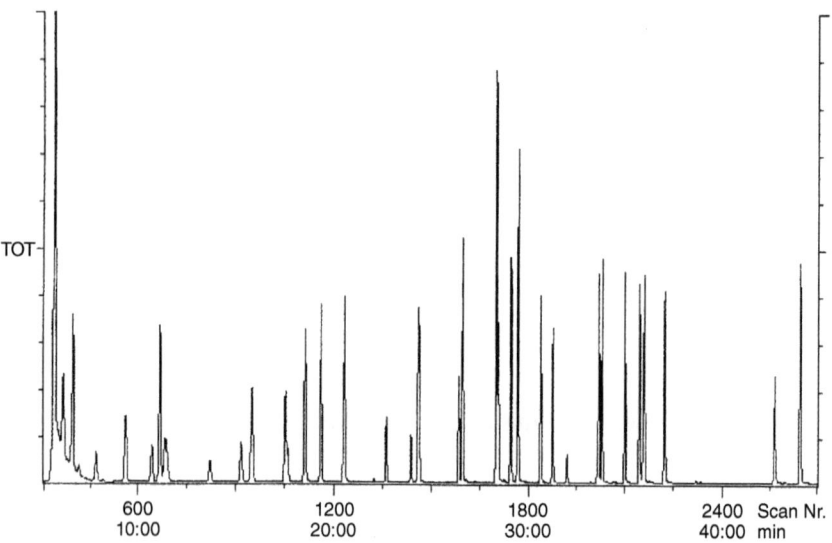

Figure 4.1 A typical chromatogram of a 10 ppbv VOC standard.

(SIM) or ion trap full scan mode allows the direct positive detection of individual components (Madden, 1994).

For a SIM analysis, the mass spectrometer is programmed in such a way that a certain number of compounds in a defined retention time (RT) range are detected. These SIM segments are switched at programmed RTs so that a list of target compounds can be worked through. Alternatively, a timed-SIM with a short acquisition window centered to the compound RT can be used. In the full scan mode, the mass spectrometer works as a universal detector acquiring complete mass spectra for analyte identification and confirmation (Pleil et al., 1991).

SUMMA canisters

The analysis of VOCs is frequently carried out by adsorption on to suitable materials; Tenax™ is mainly used (see Section 2.1.6). The limits of this adsorption method lie in the adsorption efficiency, which is dependent upon the compound, the breakthrough of the sample at higher air concentrations, the impossibility of multiple measurements on a sample, and the possible formation of artefacts by contact with catalytic iron. Stainless steel canisters, whose inner surfaces have been passivated by the SUMMA process, do not exhibit these limitations. This passivation process involves polishing the inner surface and applying a Cr/Ni oxide layer. Containers treated in this way have been used successfully for the collection and storage of air samples. As the Cr/Ni oxide coating can degrade over time Silonite™ ceramic coating for reduced risk of iron exposure and improved durability was introduced. Purification and handling of canisters and the sampling apparatus must be carried out carefully, however, because of possible contamination problems.

For analysis, a cryoconcentrator with a three-way valve system can be used for concentration. The sample can be applied by two routes without having to alter the screw joints on the tubing. Usually, the inlet is via control of a mass flow regulator to a cryofocusing unit. The direct measurement of the sample volume provides very precise data. A Nafion™ drier is used to dry the air only for TO-14 nonpolar compounds (Schnute and McMillan, 1993). Another means of sample injection is the loop injection. The sample is drawn through a 5 mL sample loop directly into the cryoconcentrator. This method is suitable for higher concentrated samples as only a small quantity of sample is required. In this case, the Nafion drier is avoided.

The Nafion drier is a system for removal of water from the air sample, which uses a semipermeable membrane (see Figure 2.243; Section 2.3.3.10 Water Removal). Nonpolar compounds pass the membrane unaffected, while polar ones, such as water, are held by the membrane and diffuse outward. The outer side of the membrane is dried in countercurrent by a dry gas stream, and the water thus separated is removed from the system. The Nafion drier is recommended by the TO-14 method to prevent blockage of the cryofocusing unit by the formation of ice crystals.

In this application, the GC is coupled to the mass spectrometer by an open split interface. A restrictor limits the carrier gas flow. Open coupling was chosen because the sensitivity of the ion trap GC-MS makes the concentration of large quantities of air superfluous. Open coupling dilutes remaining moisture, which may be contained in the sample, to an acceptable level so that cryofocusing can be used without additional drying.

The U.S. EPA guidance document "Compendium of Methods for the Determination of Toxic Organic Compounds in Ambient Air" includes two similar canister methods, TO-14A and TO-15. The difference is mainly the list of compounds (TO-14A for nonpolar and TO-15 for nonpolar and polar compounds) and the management of humidity to avoid losses of the polar fraction.

4.1.2 Analysis Conditions

Sample material			Air
Sample concentration	Sampling type		SUMMA canister
	Concentrator		Grasby Nutech model 3550A cryoconcentrator with a 354A cryofocusing unit, Nafion drier
	Autosampler		Nutech 3600 16-position sampler
GC method	Column	Type	J&W DB-5
		Dimensions	60 m length × 0.25 mm ID × 1.0 µm film thickness
			The thick film columns DB-1 or DB-5 guarantee chromatographic separation even at start temperatures just above room temperature if cryofocusing is used.
	Pre-column		2 m, 0.53 mm ID, deactivated
	Carrier gas		Helium

	Flow		Constant flow, 1 mL/min
	Sample injection		The sample is drawn from the SUMMA sampling canister through the Nafion drier and reaches the cryoconcentrator cooled to −160 °C. It is then heated rapidly to transfer the sample to the cryofocusing unit of the GC (Figure 4.2).
	Cryofocusing		Liquid N_2, −190 °C sample focusing, heated to 150 °C for injection
	Oven program	Start	35 °C, 6 min
		Ramp 1	8 °C/min
		Final temperature	200 °C
	Transferline	Temperature	250 °C
	Open split interface		SGE type GMCC/90, mounted in the transfer line to the MS restrictor capillary 0.05 mm ID, adjusted to 2.5% transmission
MS method	System		Finnigan MAGNUM
	Analyzer type		Ion trap MS with internal ionization
	Ionization		EI
	Electron energy		70 eV
	Ion trap	Temperature	250 °C
	Acquisition	Mode	Full scan
	Mass range		35–300 Da
	Scan rate		1 s/scan
	Resolution		Nominal mass
Calibration	External standard		Based on 1 L samples from a SUMMA canister
	Calibration range		0.1 and 20 ppbv (Figure 4.3)
	Data points		6

4.1.3 Limit of Detection

For the U.S. EPA method TO-14A (Table 4.1 and Figure 4.4), a limit of detection of 0.1 ppbv is required. This requirement is achieved using an ion trap or single quadrupole MS system in full scan mode even at the high split rate of the interface.

4.1.4 Results

With this method, data acquisition in the full scan mode allows mass spectra to be acquired for subsequent identification through library searching even at the required

4.1 Air Analysis According to U.S. EPA Method TO-14

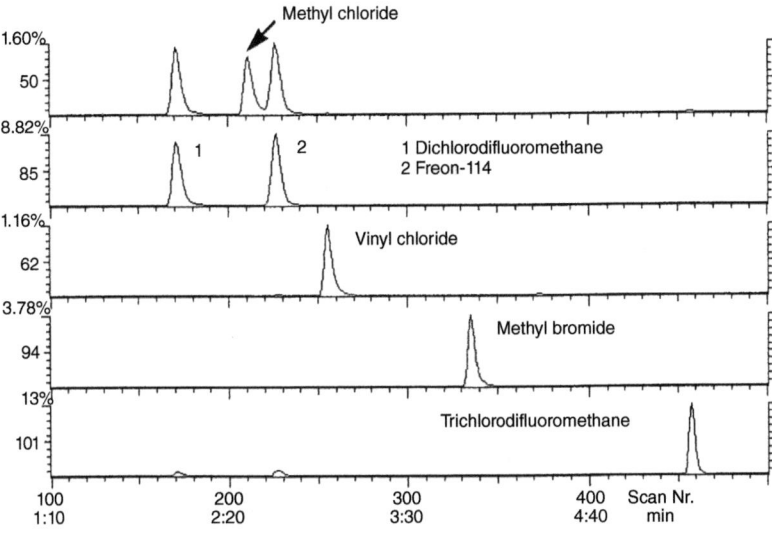

Figure 4.2 Elution of gaseous VOCs following injection using cryofocusing.

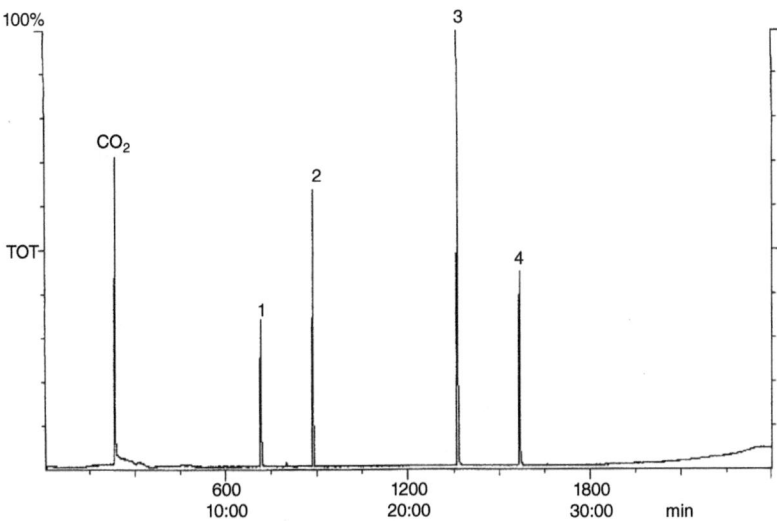

Figure 4.3 Continuous flushing of the gas lines ensures contamination-free analyses. 1. bromochloromethane, 2. 1,4-difluorobenzene, 3. chlorobenzene-d_5 as an internal standard, and 4. bromofluorobenzene.

limit of detection of 0.1 ppbv. Figures 4.5 and 4.6 show examples of the spectra library search and the identification of dichlorobenzene at 20.0 and 0.1 ppbv with high FIT and PURity values. With the procedure described, both the compounds required according to TO-14A and other unexpected components in the critical concentration range can be identified.

Table 4.1 List of U.S. EPA TO-14A compounds with targeted SIM masses and LOD of the ion trap MS full scan detection.

Compound	SIM (m/z)	LOD (ion trap) (ppbv)
Dichlorodifluoromethane	85, 87	0.01
Chloromethane	50, 52	0.01
Freon-114	85, 135, 87	0.02
Vinyl chloride	62, 27, 64	0.01
Bromomethane	94, 96	0.01
Chloroethane	64, 29, 27	0.08
Trichlorofluoromethane	101, 103	0.01
1,1-Dichloroethene	49, 84, 86	0.02
Methylene chloride	49, 84, 86	0.01
Chloropropane	42, 29, 27	0.02
Freon-113	151, 101, 103	0.01
1,1-Dichloroethane	63, 27, 65	0.01
cis-1,2-Dichloroethene	61, 96, 98	0.01
Chloroform	83, 85, 47	0.01
1,1,1-Trichloroethane	97, 99, 61	0.01
1,2-Dichloroethane	62, 27, 64	0.01
Benzene	78, 77, 50	0.01
Tetrachloromethane	117, 119	0.01
1,2-Dichloropropane	63, 41, 62	0.02
Trichloroethene	130, 132, 95	0.02
cis-1,3-Dichloropropene	75, 39, 77	0.01
trans-1,3-Dichloropropene	75, 39, 77	0.01
Toluene	91, 92	0.01
1,1,2-Trichloroethane	97, 83, 61	0.01
1,2-Dibromoethane	107, 109, 27	0.01
Tetrachloroethene	166, 164, 131	0.01
Chlorobenzene	112, 77, 114	0.01
Ethylbenzene	91, 106	0.01
m/p-Xylene	91, 106	0.01
Styrene	104, 78, 103	0.02
o-Xylene	91, 106	0.01
1,1,2,2-Tetrachloroethane	83, 85	0.09
4-Ethyltoluene	105, 120	0.02
1,3,5-Trimethylbenzene	105, 120	0.02

4.1 Air Analysis According to U.S. EPA Method TO-14 | 547

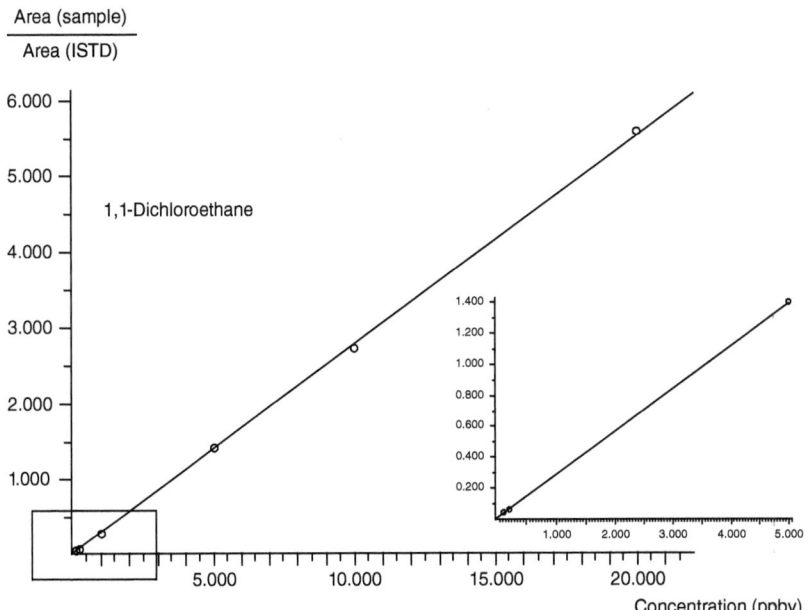

Figure 4.4 Six-point calibration from 0.1 to 20 ppbv. The lower region is shown magnified (ppbv = parts per billion in volume).

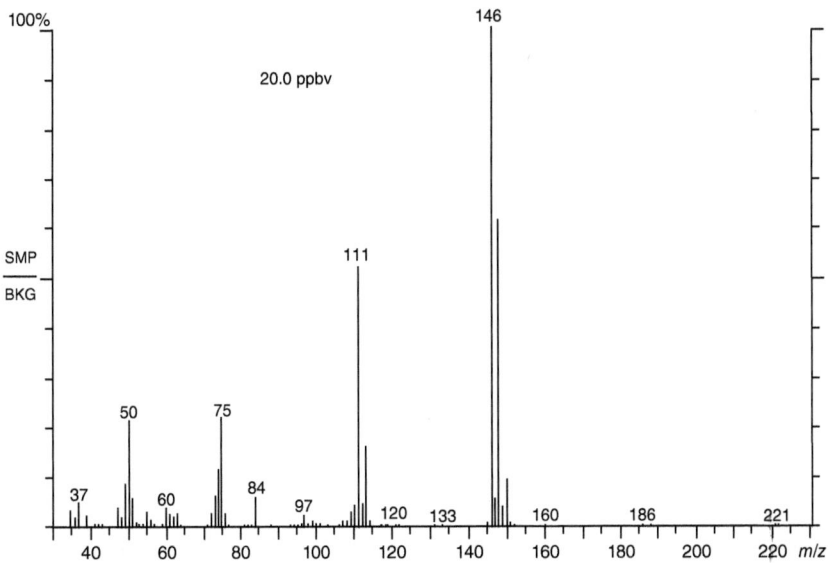

Figure 4.5 Spectrum and result of the library search (NIST) for dichlorobenzene at a concentration of 20 ppbv.

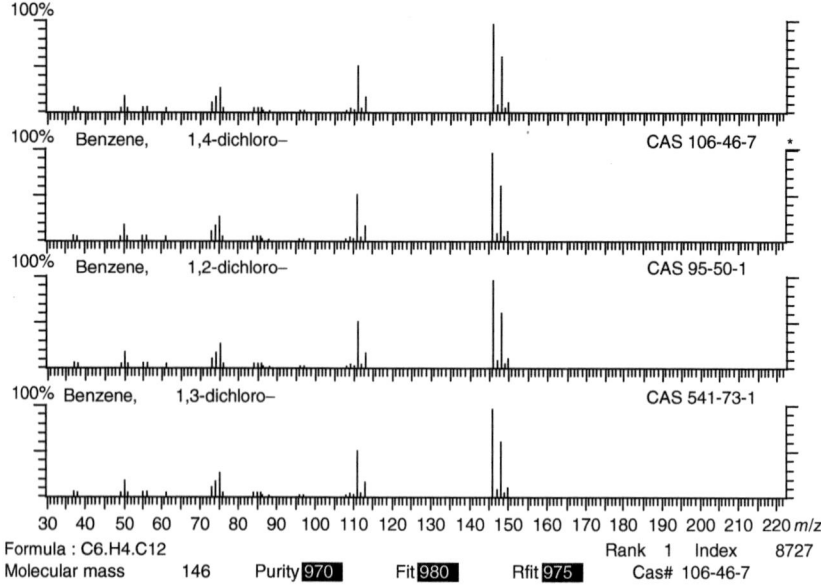

Figure 4.5 (Continued)

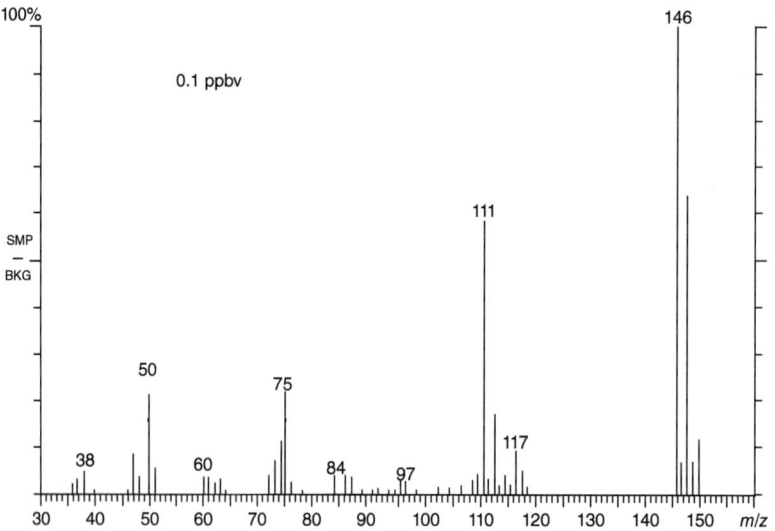

Figure 4.6 Spectrum and result of the library search (NIST) for dichlorobenzene at a concentration of 0.1 ppbv.

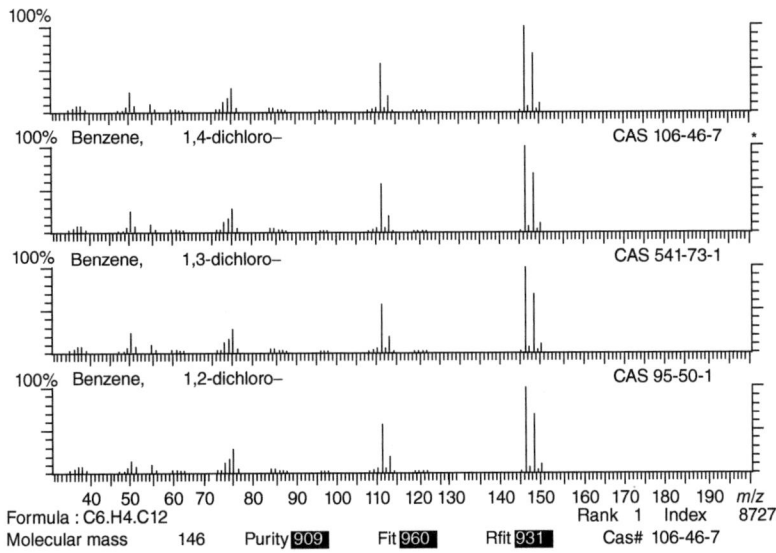

Figure 4.6 (Continued)

4.2 BTEX in Surface Water as of U.S. EPA Method 8260

Keywords: EPA Method 8260; environmental analysis; surface water; wastewater; BTEX; purge and trap; single quadrupole MS; BFB tuning

4.2.1 Introduction

The U.S. EPA method 8260 is used to determine VOCs by purge and trap (P&T) GC-MS in a wide variety of solid and aqueous sample matrices (U.S. EPA, 2006). The method covers more than 100 VOC compounds and is applied for the trace determination of the benzene, toluene, ethylbenzene, and isomeric xylenes (BTEX) components.

BTEX compounds can be released to surface water streams from, for example, oil spills or industrial processes and get monitored on low levels. During the production of gasoline products, oil refineries generate a waste stream that contains petroleum by-products such as BTEX. These volatile, monoaromatic hydrocarbons can be toxic to receiving water bodies.

4.2.2 Sample Preparation

Before analysis, samples containing high levels of BTEX in the ppm range were diluted with organic-free water. The applied P&T unit automatically spikes the internal standards (ISTD) and surrogates into a 5 mL sample.

For calibration, a solution of 250 ng/µL is prepared in P&T grade methanol and 1 µL is spiked into 5 mL of organic-free water at a final concentration of 50 µg/L. A calibration curve is generated from 20 to 200 µg/L.

Standards are prepared in organic-free water. Each standard gets transferred into a 40 mL sampler vial and loaded onto the P&T autosampler. A 5 mL aliquot is transferred automatically from the 40 mL vial into a fritted sparger for analysis.

4.2.3 Experimental Conditions

The GC was configured for P&T analysis by installing the transfer line of the P&T unit with a dedicated adapter to the SSL injector (Butler, 2013). The selected internal trap #10 containing three different trapping materials of increasing strength is recommended for the EPA methods 524.2, 624, and 8260 with MS detection (OI Analytical 2010).

4.2.4 Analysis Conditions

Sample material			Water, surface water, wastewater
Purge and trap	System		O.I Analytical Eclipse model 4660
	Autosampler		O.I Analytical model 4551A
	Sample	Volume	5 mL
		Purge temperature	40 °C
		Purge flow	40 mL/min
		Purge time	11 min
	Trap	Type, material	#10 (Tenax, silica gel, carbon molecular sieve)
		Water management	Purge 110 °C, bake 240 °C
		Desorb preheat	180 °C
		Desorb time	30 s
	Bake	Rinse cycles	Two times
		Temperature, cycle time	210 °C, 10 min
GC method	System		Thermo Scientific™ TRACE 1300 GC
	Column	Type	TG-VMS
		Dimensions	20 m length × 0.18 mm ID × 1.0 µm film thickness
	Carrier gas		Helium
	Flow		Constant flow, 1 mL/min
	SSL injector	Injection mode	Split
		Split flow	30 mL/min
		Injection temperature	200 °C
		Injection volume	1 µL

	Oven program	Start	45 °C, 4.5 min
		Ramp 1	8 °C/min to 100 °C
		Ramp 2	25 °C/min to 230 °C
		Final temperature	230 °C, 2 min
	Transferline	Temperature	200 °C
			SGE type GMCC/90, mounted in the transfer line to the MS restrictor capillary 0.05 mm ID, adjusted to 2.5% transmission
MS method	System		Thermo Scientific ISQ™
	Analyzer type		Single quadrupole MS
	Ionization		EI
	Electron energy		70 eV
	Ion source	Temperature	230 °C
	Acquisition	Mode	Full scan
	Mass range		35–350 Da
	Scan rate		0.2 s (5 spectra/s)
	Resolution	Setting	Normal (0.7 Da)
Calibration	Internal standard	Compounds	Deuterium- and fluorine-labeled analogues
		Range	20–200 µg/L

The analyte separation was carried out using a narrow-bore thick film column installed to a regular split/splitless inlet with a narrow ID inlet liner. The list of compounds analyzed is given in Table 4.2 with RTs.

For compliance with the U.S. EPA 8260 method requirements on the MS tune conditions, a 1 µL injection of 25 ng/µL 4-bromofluorobenzene (BFB) solution was injected at the beginning of each analysis shift for a continuous performance documentation.

4.2.5 Results

Before sample analysis, the tuning of the MS was checked by using the autotune function and running the BFB check sample. The used single quadrupole MS met the tuning criteria for BFB listed in the method (Figure 4.7).

The entire runtime of one analysis cycle on the 20 m column was 16 min, as shown for the 200 µg/L BTEX standard in Figure 4.8. The analyzed surface water sample shown in Figure 4.9 contained 0.62 ppm benzene. Method detection limits (MDL) were generated by running replicate samples at 0.1 µg/L BTEX concentration (see Tables 4.2 and 4.3).

4 Applications

Table 4.2 List of BTEX compounds and standards analyzed with achieved precision and MDL data.

Retention time (min)	Compound	% RSD calibration	MDL (µg/L)	% RSD at MDL
4.24	Dibromofluoromethane (surrogate)	2.13	—	2.32
4.79	Benzene	6.16	0.008	3.15
4.99	1,2-Dichloroethane-d4 (surrogate)	2.65	—	2.13
5.41	Fluorobenzene (ISTD)	9.38	—	4.51
7.86	Toluene-d8 (surrogate)	1.00	—	1.30
7.94	Toluene	5.67	0.010	3.48
10.31	Chlorobenzene-d5 (ISTD)	11.06	—	4.78
10.46	Ethyl benzene	5.98	0.009	3.46
10.72	m- and p-Xylene	6.49	0.024	4.70
11.41	o-Xylene	7.12	0.015	5.85
12.23	BFB (surrogate)	1.18	—	1.93
13.39	1,4-Dichlorobenzene-d4 (ISTD)	10.26	—	4.12
	Average values	4.27	0.013	3.48

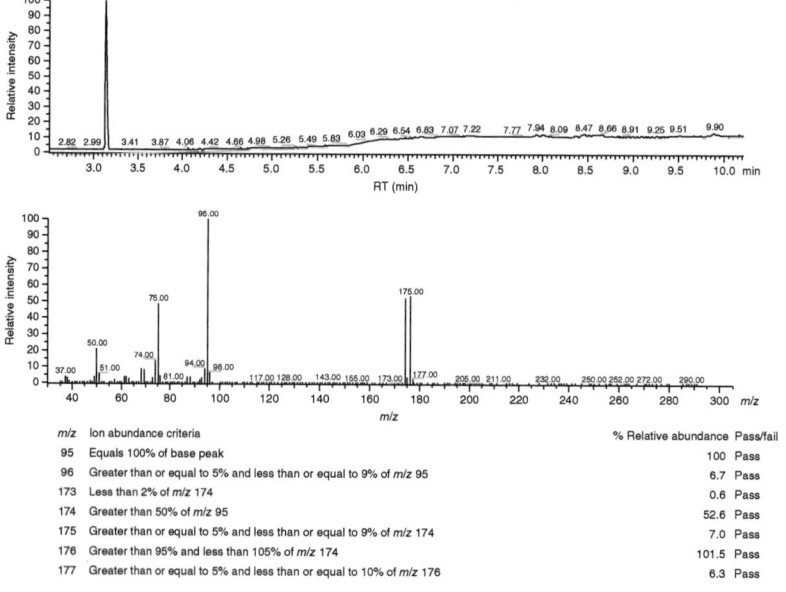

Figure 4.7 BFB tune criteria and report.

4.2 BTEX in Surface Water as of U.S. EPA Method 8260 | 553

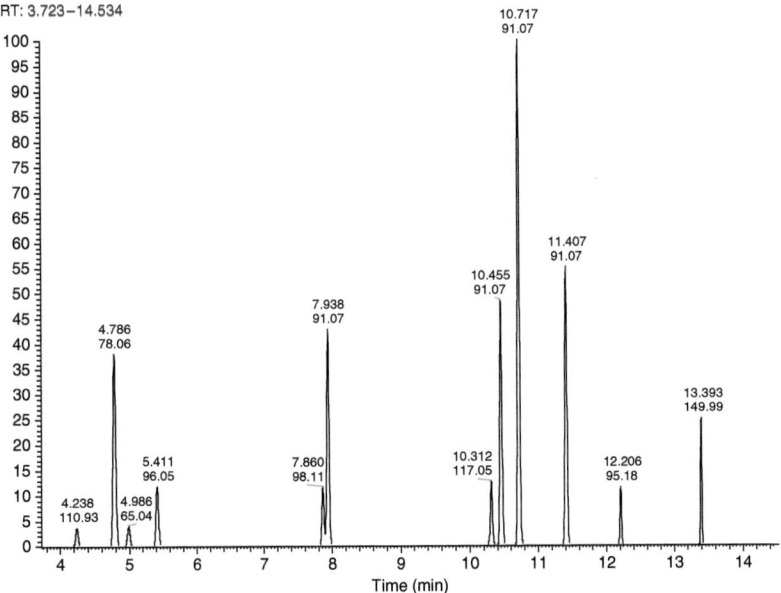

Figure 4.8 Typical BTEX chromatogram, 200 µg/L standard, compounds, see Table 4.2.

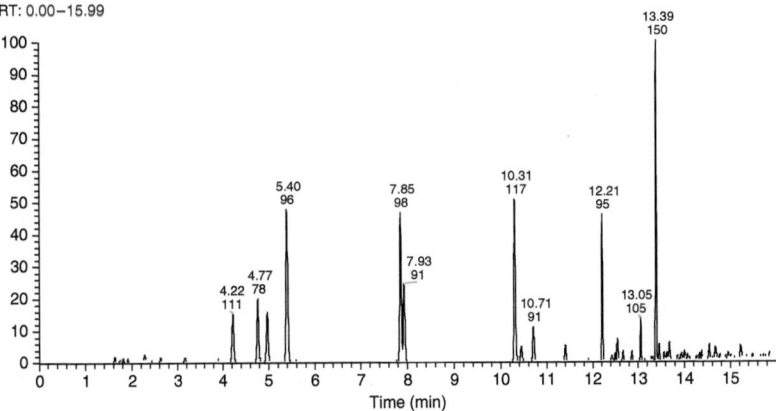

Figure 4.9 Analysis of a surface water sample.

4.2.6 Conclusions

The applied mass spectrometer met the quality control criteria of the U.S. EPA method 8260 and generated a typical spectrum for BFB within the method criteria limits.

The BTEX calibration curve from 20 to 200 µg/L in water showed a wide linear range and good precision with an average relative standard deviation (RSD) of 4.3%.

Table 4.3 Results of the surface water sample of Figure 4.9.

Compound	Sample (ppm)
Benzene	0.62
Toluene	1.163
Ethylbenzene	0.29
o-Xylene	0.32
m- and p-Xylene	0.73

The average MDL for water produced from eight replicates of 0.1 µg/L spiked organic-free water was 0.013 µg/L, showing very good sensitivity and precision with an average standard deviation of better than 4% RSD (Butler, 2013).

4.3 Volatile Priority Pollutants

Keywords: headspace; VOCs; water; soil; quantification; internal standard; single quadrupole MS; automation

4.3.1 Introduction

Many analytical techniques are used for the quantification of VOCs in water and soil, including liquid/liquid microextraction, solid phase microextraction (SPME), and P&T. Automated static headspace analysis (HS) offers the advantages of simplicity and robustness especially when a large sample throughput is required (CTC, 2022). A typical application of the static headspace method described here is the analysis of surface and wastewater (see Figure 4.10). Other techniques such as P&T GC-MS are better designed for the ultra trace analysis of drinking water. The water samples were carefully transferred to 20 mL headspace vials containing sodium sulfate, together with an ISTD (e.g. toluene-d_8), and capped immediately to prevent losses of the highly volatile VOC compounds. Using the sample agitation feature, the headspace equilibrium is reached very quickly, allowing all sampling operations to take place during the GC runtime (Belouschek et al., 1992).

The GC was operated in accordance with the "whole column trapping technique" using a standard medium bore capillary column with a 1 µm stationary phase and liquid CO_2 as the coolant. The analytes were trapped in the column inlet at subambient temperature, without the need for a dedicated cold trapping device. Trapping of the VOCs helped to maintain optimum chromatographic efficiency by focusing the analytes at the column inlet. High sensitivity was achieved by injecting a relatively high volume of headspace using a low split flow. The injection speed was controlled in order to prevent the injector from overflowing. The MS was operated in the electron ionization (EI) mode and acquisition of full mass spectra, thereby enabling both target analysis and identification of unknowns during a single

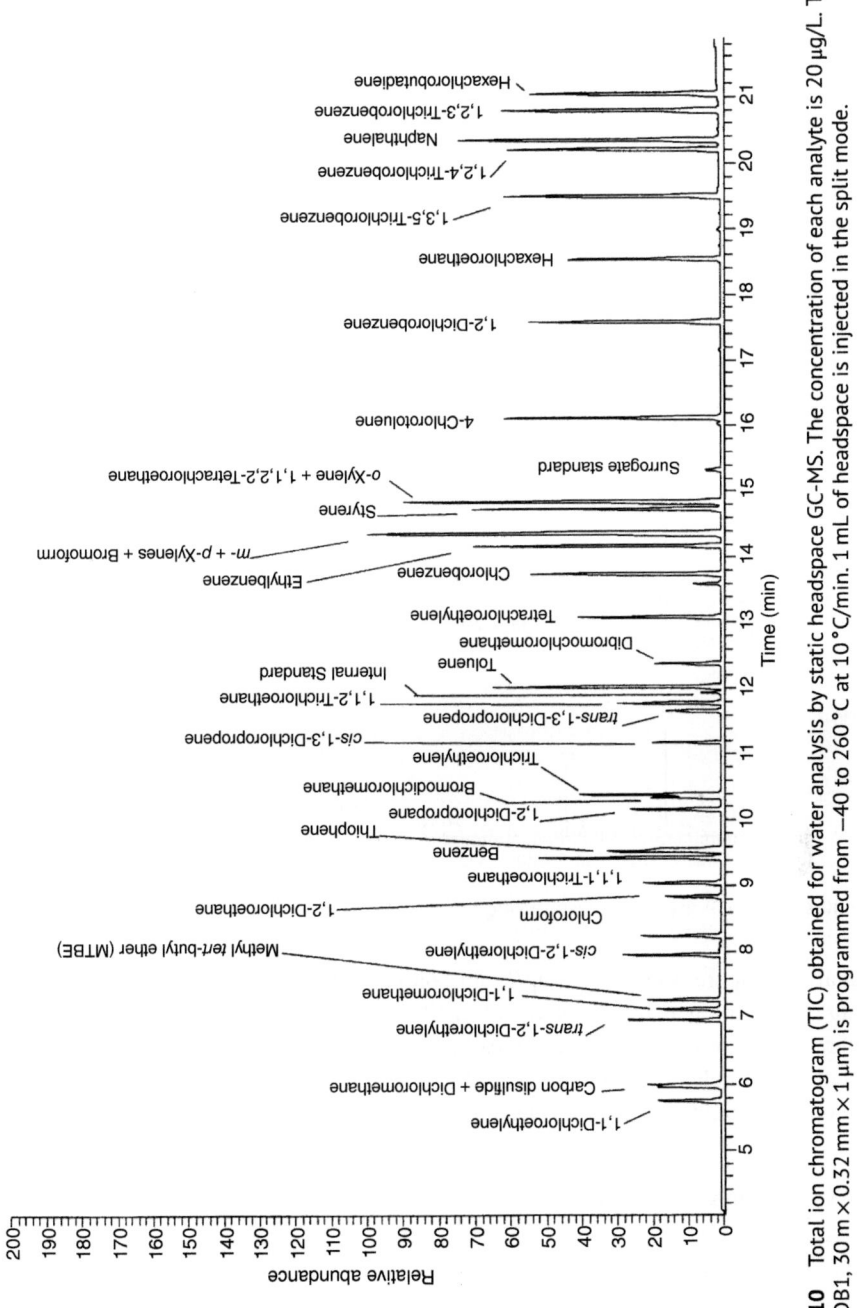

Figure 4.10 Total ion chromatogram (TIC) obtained for water analysis by static headspace GC-MS. The concentration of each analyte is 20 µg/L. The column (DB1, 30 m × 0.32 mm × 1 µm) is programmed from −40 to 260 °C at 10 °C/min. 1 mL of headspace is injected in the split mode.

556 | 4 Applications

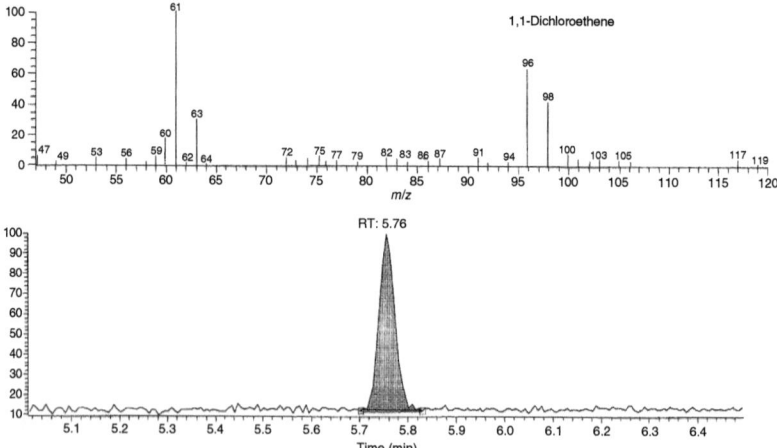

Figure 4.11 Mass spectrum and chromatogram obtained for 1,1-dichloroethylene at 0.5 µg/L.

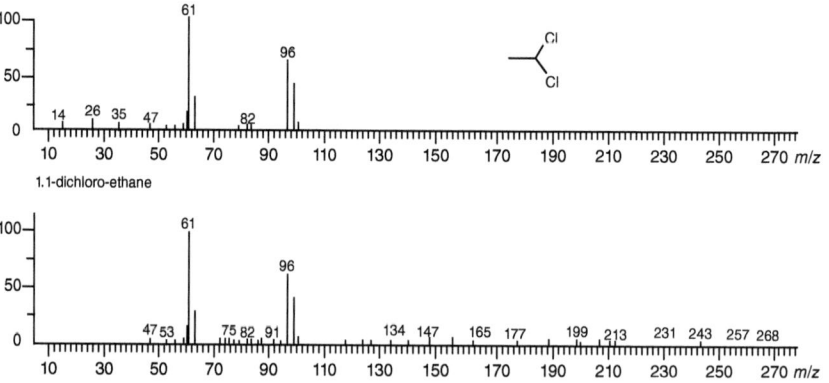

Figure 4.12 Library search result using the NIST library.

analysis (Figure 4.11). At 0.5 µg/L, all compounds were identified by automated library searching, using the National Institute of Standards and Technology (NIST) library (Figure 4.12). The quantification was based on an ISTD method, using a deuterated analogue of toluene. A surrogate standard was also added to the samples in order to control the long-term stability.

4.3.2 Analysis Conditions

Sample material			Water, soil
Sample preparation	Static headspace	System	CE Instruments HS2000
		Incubation	40 °C, 10 min
		Sample volume	1 mL

GC method	System		Finnigan TRACE MS
	Column	Type	DB-1
		Dimensions	30 m length × 0.32 mm ID × 1.0 μm film thickness
	Cryofocusing		Liquid CO_2 oven cooling
	Carrier gas		Helium
	Flow		Constant flow, 1 mL/min
	SSL injector	Injection mode	Splitless
		Injection temperature	200 °C
		Injection volume	1 μL
	Split	Closed until	1 min
		Open	1 min to end of run
	Oven program	Start	−40 °C
		Ramp 1	10 °C/min to 260 °C
		Final temperature	260 °C
	Transferline	Temperature	260 °C
MS method	System		Finnigan TRACE MS
	Analyzer type		Single quadrupole MS
	Ionization		EI
	Electron energy		70 eV
	Ion source	Temperature	200 °C
	Acquisition	Mode	Full scan
	Mass range		45–270 Da
	Scan rate		0.4 s/scan
Calibration	External standard	Range	0.1–100 μg/L
	Data points		6

4.3.3 Results

The technique of static headspace GC-MS offers significant benefits for laboratories tasked with VOC analyses. The virtual elimination of sample carry-over, thanks to a programmable temperature cleaning cycle of the syringe and needle heaters in the headspace autosampler, obviates the need for running blank samples between specimen samples.

The wide linear dynamic range of the quadrupole mass spectrometer in full scan mode permits target compounds to be accurately quantified over a concentration range of at least three decades. Full scan operation enables unknown peaks to be detected and identified. Even after a period of several weeks of unattended operation, highly sensitive and reproducible results are achieved.

Table 4.4 Substance list. LODs are given for the static headspace method in water.

Substance name	% RSD	Precision (R^2)	LOD (µg/L)
1,1-Dichloromethane	0.552	0.9998	0.010
Carbon disulfide	0.945	0.9990	0.013
Dichloromethane	1.150	0.9992	0.018
trans-1,2-Dichloroethane	0.698	0.9997	0.012
MTBE	0.697	0.9993	0.013
cis-1,2-Dichloroethane	1.318	0.9963	0.022
Chloroform	0.578	0.9996	0.013
1,2-Dichloroethane	0.362	0.9989	0.011
1,1,1-Trichloroethane	0.747	0.9998	0.014
Benzene	0.568	0.9994	0.016
Thiophene	0.793	0.9992	0.012
Carbon tetrachloride	0.385	0.9996	0.008
Bromodichloromethane	0.397	0.9995	0.010
Trichloroethane	0.585	0.9994	0.008
cis-1,3-Dichloropropene	0.909	0.9933	0.019
trans-1,3-Dichloropropene	1.017	0.9883	0.026
1,1,2-Trichloroethane	0.772	0.9989	0.015
Toluene	0.409	0.9997	0.016
Dibromochloromethane	0.225	0.9994	0.006
Tetrachloroethane	0.432	0.9997	0.006
Chlorobenzene	0.792	0.9990	0.007
Ethylbenzene	0.365	0.9996	0.011
Bromoform	0.397	0.9987	0.013
m- + p-Xylenes	0.219	0.9999	0.011
Styrene	0.833	0.9998	0.013
o-Xylene	0.193	0.9993	0.005
1,1,2,2-Tetrachloroethane	0.663	0.9983	0.013
4-Chlorotoluene	0.178	0.9996	0.007
1,2-Dichlorobenzene	0.605	0.9996	0.006
Hexachloroethane	0.496	0.9996	0.007
1,3,5-Trichlorobenzene	2.032	0.9936	0.021
1,2,4-Trichlorobenzene	2.550	0.9912	0.026
Naphthalene	1.216	0.9950	0.016
1,2,3-Trichlorobenzene	2.825	0.9933	0.027
Hexachlorobutadiene	1.569	0.9933	0.020

The precision study, based on 72 replicate injections of the low standard (0.5 µg/L), represents analyses acquired over 3 weeks of continuous operation (see Table 4.4). The figures were derived from multiple calibration curves spanning three orders of magnitude (0.1–100 µg/L) and accumulated over three days (Figure 4.13). The calculated limits of detection (LODs) for each target compound lie in the low parts per billion range, making this technique suitable for analyses of VOCs in water or other materials like soil or sediments (Figure 4.14). The described system yields very high stability and sensitivity. The % RSDs and LOD were based on analyses performed

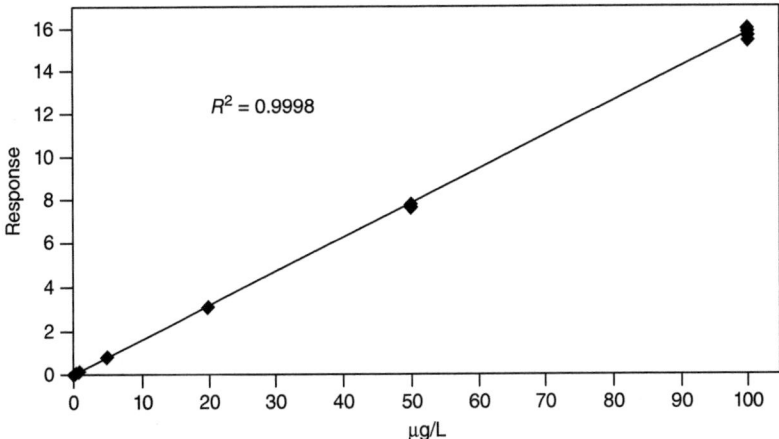

Figure 4.13 Calibration curve obtained for 1.1-dichloroethylene over the 0.1–100 µg/L concentration range. The graph is based on nine calibrations (consecutive calibration runs were repeated over 3 days).

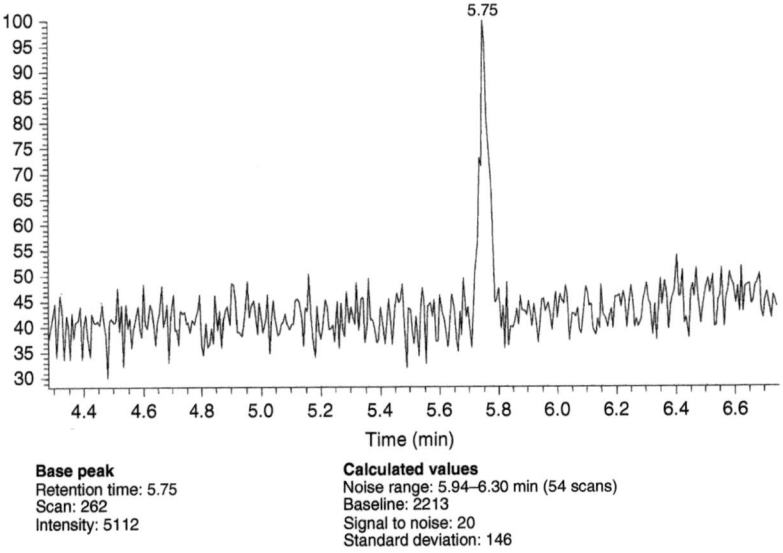

Base peak
Retention time: 5.75
Scan: 262
Intensity: 5112

Calculated values
Noise range: 5.94–6.30 min (54 scans)
Baseline: 2213
Signal to noise: 20
Standard deviation: 146

Figure 4.14 Automated S/N calculation for the 1,1-dichloroethylene peak at 0.1 µg/L.

at 0.5 µg/L (n = 72, over 3 weeks of continuous operation). The detection limits are calculated according to LOD = 3σ. The precision of the calibration with R^2 values obtained is described in Table 4.4.

4.4 MAGIC 60 – VOC Analysis

Keywords: purge and trap; water; soil; volatile halogenated hydrocarbons; VOCs, BTEX; drinking water; U.S. EPA 8260, 8240, 524.2; ion trap MS; internal standards

4.4.1 Introduction

The method described in this application for the quantification of VOCs in water and soil is taken from the U.S. EPA methods 8260, 8240, and 524.2. These approximately 60 substances are referred to by the term *MAGIC 60*. The method also allows the determination of compounds in solid materials, such as soil samples, and is generally applicable for a broad spectrum of organic compounds, which have sufficiently high volatility and low solubility in water for them to be determined effectively by the P&T procedure. Vinyl chloride can also be determined with certainty by this method (Figures 4.15 and 4.16).

Gastight syringes (5 and 25 mL) with open/shut valves have been shown to be useful for the preparation of water samples. The syringe is carefully filled with the sample from the plunger, the plunger is inserted, and the volume is adjusted. The standard solutions are added through the valve with a microliter syringe. The syringe is then connected to the P&T unit via the open/shut valves, and the prepared sample is transferred to a needle sparger or frit sparger vessel.

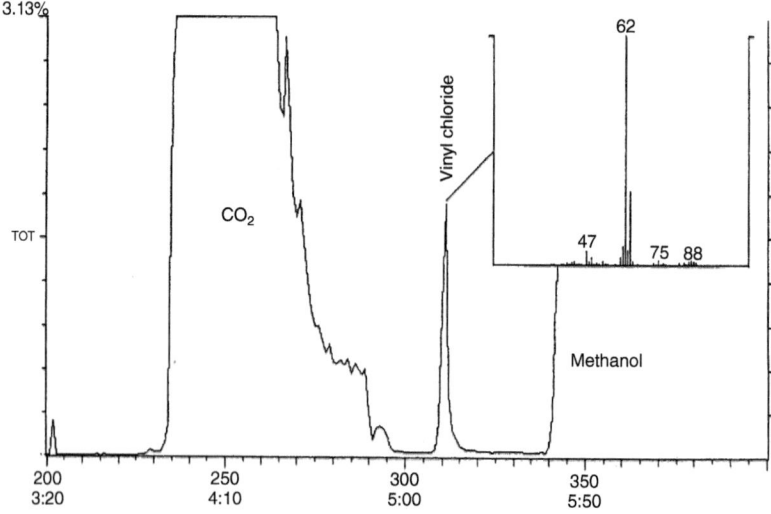

Figure 4.15 Elution of vinyl chloride (50 µg/L) between CO_2 and methanol (DB-624, for analysis parameters, see text). The methanol peak can be avoided by using polyethylene glycol as the solvent for the preparation of standard solutions (Figure 4.16).

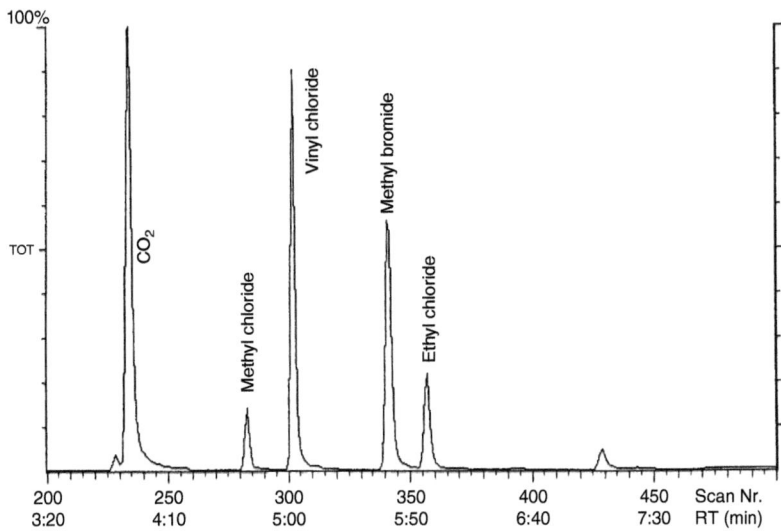

Figure 4.16 Elution of vinyl chloride (50 µg/L) from a DB-VRX column (for analysis parameters, see text) using polyethylene glycol as the solvent.

The analysis of soil samples is similar to that of water, but, depending on the expected concentration of volatile halogenated hydrocarbons in the sample, a different preparation procedure is chosen. For samples where a concentration of less than 1 mg/kg is expected, 5 g of the sample is placed directly in a 5 mL needle sparger and treated with 5 mL reagent water. Solutions of the ISTDs are added with a microliter syringe. At higher concentrations, 4 g of the sample is weighed into a 10 mL vessel, which can be closed with a Teflon-coated septum (e.g. headspace vial), and is treated with 9.9 mL methanol and 0.1 mL of the surrogate standard solution. After the solid phase has settled, 5–100 µL of the methanol phase is taken up, depending on the concentration, in a prepared 5 mL (25 mL) syringe together with reagent water. After addition of the ISTD, the sample is injected into the P&T apparatus (needle sparger). An ion trap GC-MS unit in the full scan mode is used for detection.

4.4.2 Analysis Conditions

Sample material	Drinking water, soil
Sample concentration System	Tekmar purge and trap concentrator LSC 2000
Autosampler	Tekmar ALS 2016
	With 25 mL frit sparge glass vessels for water, needle sparge vessels for soil samples
Trap	Supelco Vocarb 3000

	Sample	Volume	5 mL
		Standby temperature	Room temperature for water and soil
	Purge	Sample preheat time	2.50 min
		Sample temperature	40 °C
		Pre-purge	0 min
		Purge duration	12 min
		Purge flow	40 mL/min
		Dry purge	0 min
	Desorb	Preheat	255 °C for Vocarb 3000
		Desorb temperature	260 °C
		Desorb time	4 min
	Water management	MCM Desorb	0 °C
		Trap bake	20 min at 260 °C
		MCM bake	90 °C
	Instrument	Bake gas bypass (BGB)	On after 120 s
		Mount	110 °C
		Valve	110 °C
		Transfer line	110 °C
		Connection to GC	LSC 2000, linked into the carrier gas supply of the injector
	Cleaning cycle	Sample	Blank DI water
		Purge time	5 min
		Desorption	1 min
		Bake mode	5 min
		Valves, transfer line	200 °C
		All other parameters remain unchanged	
GC method	System		Finnigan MAGNUM
	Analyzer type		Ion trap with internal ionization
	Column	Type	DB-VRX, DB-624, or Rtx-624
		Dimensions	60 m length × 0.32 mm ID × 1.8 µm film thickness
	Carrier gas		Helium
	Flow		1 mL/min, constant pressure
	Injector	Injection mode	Split
		Base temperature	200 °C
		Transfer rate	Isothermal

	Split	Open	Until end of run
		Split flow	20 mL/min
		Purge flow	5 mL/min
	Oven program	Start	40 °C, 5 min
		Ramp 1	7 °C/min to 180 °C
		Ramp 2	15 °C/min to 220 °C
		Final temperature	220 °C, 5 min
	Transferline	Temperature	250 °C
MS method	System		Finnigan MAGNUM
	Analyzer type		Ion trap with internal ionization
	Ionization		EI
	Electron energy		70 eV
	Ion source	Temperature	250 °C
	Mass range		33–260 Da
	Scan rate		1 s
	Resolution	Setting	Normal (0.7 Da)
Calibration	Internal standards		Toluene-d8, fluorobenzene, 4-bromofluorobenzene, 1,2-dichloroethane-d4
			Solvent for all standard solutions is methanol

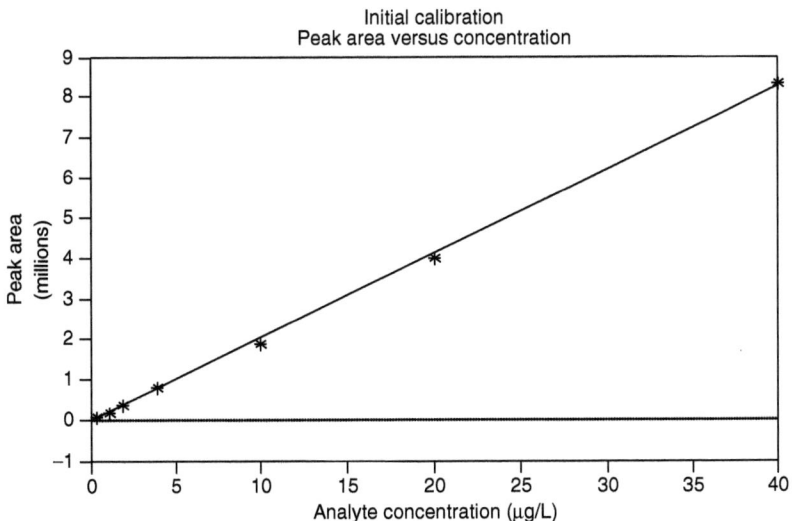

Figure 4.17 Calibration function for cis-1,2-dichloroethylene based on m/z 61, 96, and 98. Correlation factor R^2 0.9991 and relative standard deviation (RSD) 4.96%.

Table 4.5 MAGIC 60-substance list with details of quantitation masses, CAS (Chemical Abstracts Service) numbers, limits of detection (LODs), and limits of quantitation (LOQs), arranged alphabetically.

Compound	m/z	CAS No.	LOD (µg/L)	LOQ (µg/L)
Benzene	77, 78	71-43-2	0.05	0.1
Bromobenzene	77, 156, 158	108-86-1	0.1	0.2
Bromochloromethane	49, 128, 130	74-97-5	0.05	0.1
Bromodichloromethane	83, 85, 127	75-27-4	0.05	0.1
Methyl bromide	94, 96	74-96-4	n.b.	n.b.
Bromoform	173, 175, 252	95-25-2	0.05	0.1
n-Butylbenzene	91, 134	104-51-8	0.05	0.1
sec-Butylbenzene	105, 134	135-98-8	0.05	0.1
t-Butylbenzene	91, 119	98-06-6	0.05	0.1
Chlorobenzene	77, 112, 114	108-90-7	0.1	0.2
Ethyl chloride	64, 66	75-00-3	n.a.	n.a.
Methyl chloride	50, 52	74-87-3	n.a.	n.a.
Chloroform	83, 85	67-66-3	0.05	0.1
2-Chlorotoluene	91, 126	95-49-8	0.05	0.1
4-Chlorotoluene	91, 126	106-43-4	0.05	0.1
Dibromochloromethane	127, 129	124-48-1	0.05	0.1
1,2-Dibromo-3-chloropropane	75, 155, 157	96-12-8	0.25	0.5
Dibromomethane	93, 95, 174	74-95-3	0.1	0.2
1,2-Dibromoethane	107, 109	106-93-4	0.1	0.2
1,2-Dichlorobenzene	111, 146	95-50-1	0.15	0.2
1,3-Dichlorobenzene	111, 146	541-73-1	0.1	0.2
1,4-Dichlorobenzene	111, 146	106-46-7	0.1	0.2
1,1-Dichloroethane	63, 112	75-34-3	0.05	0.1
1,2-Dichloroethane	62, 98	107-06-2	0.1	0.2
1,1-Dichloroethylene	61, 63, 96	75-35-4	0.05	0.1
cis-1,2-Dichloroethylene	61, 96, 98	156-59-4	0.05	0.1
trans-1,2-Dichloroethylene	61, 96, 98	156-60-5	0.05	0.1
Dichlorodifluoromethane	85, 87	75-71-8	0.1	0.2
Dichloromethane	49, 84, 86	75-09-2	0.5	0.5
1,2-Dichloropropane	63, 112	78-87-5	0.1	0.2
1,3-Dichloropropane	76, 78	142-28-9	0.05	0.1
2,2-Dichloropropane	77, 97	590-20-7	0.05	0.1
1,1-Dichloropropylene	75, 110, 112	563-68-6	0.05	0.1

Table 4.5 (Continued)

Compound	m/z	CAS No.	LOD (µg/L)	LOQ (µg/L)
cis-1,3-Dichloropropylene	75, 110, 112	10061-01-5	0.05	0.1
trans-1,3-Dichloropropylene	75, 110, 112	10061-01-5	0.05	0.1
Ethylbenzene	91, 106	100-41-4	0.05	0.1
Hexachlorobutadiene	225, 260	87-68-3	0.15	0.2
Isopropylbenzene	105, 120	48-82-8	0.1	0.2
4-Isopropyltoluene	91, 119, 134	99-87-6	0.05	0.1
Naphthalene	128	91-20-3	0.5	0.5
Styrene	78, 104	100-42-5	0.1	0.2
1,1,1,2-Tetrachloroethane	131, 133	620-30-6	0.1	0.2
1,1,2,2-Tetrachloroethane	83, 85, 131	79-34-5	0.1	0.2
Tetrachloroethylene	129, 166, 168	127-18-4	0.1	0.2
Carbon tetrachloride	117, 119	56-23-5	0.05	0.1
Toluene	91, 92	108-88-3	*	*
1,2,4-Trichlorobenzene	180, 182	120-82-1	0.3	0.3
1,2,3-Trichlorobenzene	180, 182	87-61-6	0.4	0.4
1,1,1-Trichloroethane	61, 97, 99	71-55-6	0.05	0.1
1,1,2-Trichloroethane	83, 85, 97	79-020-5	0.1	0.2
Trichloroethylene	95, 130, 132	79-01-6	0.05	0.1
Trichlorofluoromethane	101, 103	75-69-4	0.05	0.1
1,2,3-Trichloropropane	75, 77	96-18-4	0.1	0.2
1,2,4-Trimethylbenzene	105, 120	95-63-6	0.1	0.2
1,3,5-Trimethylbenzene	105, 120	108-67-8	0.1	0.2
Vinyl chloride	62, 64	75-01-4	n.b.	n.b.
m-Xylene	91, 106	108-38-3	0.1	0.2
o-Xylene	91, 106	95-47-6	0.1	0.2
p-Xylene	91, 106	95-47-6	0.1	0.2
Internal standards				
4-Bromofluorobenzene	95, 174	—	—	—
1-Chloro-2-bromopropane	77, 79	—	—	—
1,2-Dichlorobenzene-d_4	115, 150, 152	—	—	—
1,2-Dichloroethane-d_4	65, 102	—	—	—
Fluorobenzene	77, 96	—	—	—
Toluene-d_8	70, 98, 100	—	—	—

566 | 4 Applications

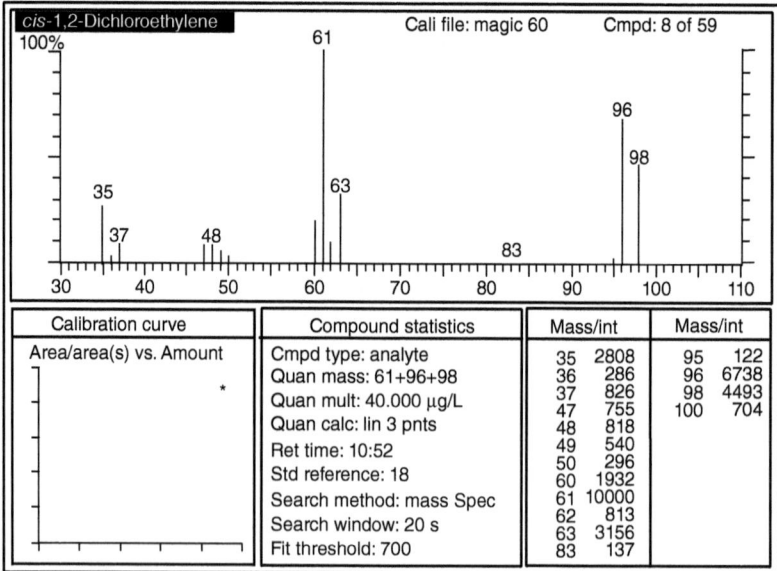

Figure 4.18 MAGIC 60: Standard chromatogram for a water sample. All analytes 20 µg/L, internal standard 2 µg/L.

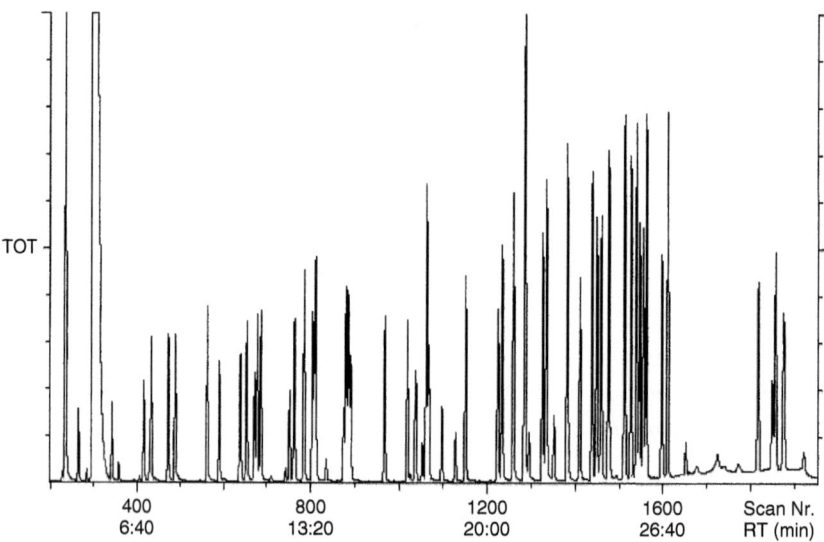

Figure 4.19 MAGIC 60: standard chromatogram of a soil sample. All analytes 100 µg/kg, internal standard 10 µg/kg.

4.4 MAGIC 60 – VOC Analysis | 567

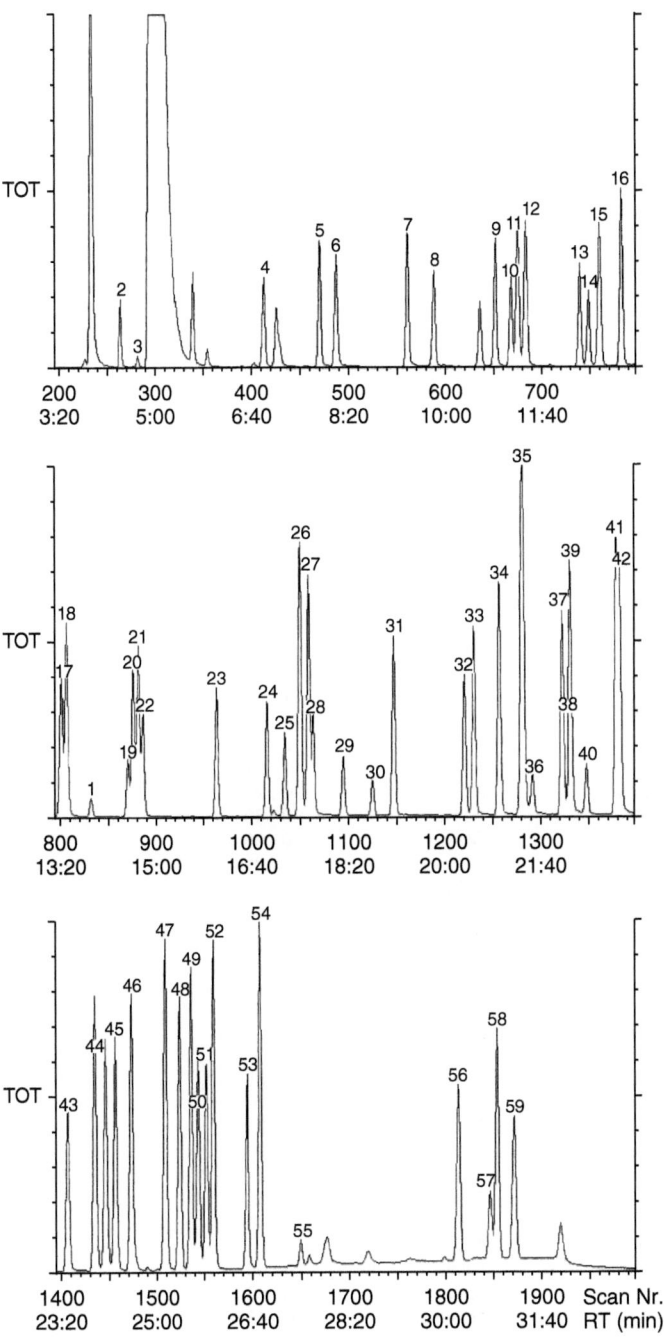

Figure 4.20 Data system entry for cis-1,2-dichloroethylene from the calibration file.

No.	Compound name	Status	Type	RT (min)
1	Benzene	Active	I	13 : 52
2	Fluorobenzene	Active	A	4 : 24
3	Dichlorodifluoromethane	Active	A	4 : 42
4	Chloromethane	Active	A	6 : 54
5	Trichlorofluoromethane	Active	A	7 : 50
6	1,1–Dichloroethylene	Active	A	8 : 07
7	Methylene chloride	Active	A	9 : 20
8	trans–1,2–Dichloroethylene	Active	A	9 : 48
9	1,1–Dichloroethylene	Active	A	10 : 52
10	cis–1,2–Dichloroethylene	Active	A	11 : 09
11	Bromochloromethane	Active	A	11 : 15
12	Chloroform	Active	A	11 : 24
13	2,2–Dichloropropane	Active	A	12 : 20
14	1,2–Dichloroethane-d4	Active	A	12 : 29
15	1,2–Dichloroethane	Active	A	12 : 40
16	1,1,1–Trichloroethane	Active	A	13 : 02
17	1,1,–Dichloropropylene	Active	A	13 : 22
18	Carbon tetrachloride	Active	A	13 : 27
19	Dibromomethane	Active	A	14 : 31
20	1,2–Dichloropropane	Active	A	14 : 36
21	Trichloroethylene	Active	A	14 : 42
22	Bromodichloromethane	Active	A	14 : 47
23	cis-1,3–Dichloropropylene	Active	A	16 : 04
24	trans-1,3–Dichloropropyle	Active	A	16 : 56
25	1,1,2–Trichloroethane	Active	A	17 : 14
26	Toluene-d8	Active	A	17 : 30
27	Toluene	Active	A	17 : 38
28	1,3–Dichloropropane	Active	A	17 : 44
29	Dibromochloromethane	Active	A	18 : 15
30	1,2–Dibromoethane	Active	A	18 : 44
31	Tetrachloroethylene	Active	A	19 : 07
32	1,1,1,2–Tetrachloroethane	Active	A	20 : 21
33	Chlorobenzene	Active	A	20 : 30
34	Ethylbenzene	Active	A	20 : 56
35	m,p–Xylene	Active	A	21 : 21
36	Bromoform	Active	A	21 : 32
37	Styrene	Active	A	22 : 02
38	1,1,2,2–Tetrachloroethane	Active	A	22 : 09
39	o–Xylene	Active	A	22 : 11
40	1,2,3–Trichorpropane	Active	A	22 : 28
41	Isopropylbenzene	Active	A	22 : 58
42	4–Bromofluorobenzene	Active	A	23 : 02
43	Bromobenzene	Active	A	23 : 27
44	2–Chlorotoluene	Active	A	23 : 55
45	4–Chlorotoluene	Active	A	24 : 16
46	1,3,5–Trimethylbenzene	Active	A	24 : 34
47	tert–Butylbenzene	Active	A	25 : 09
48	1,2,4–Trimethylbenzene	Active	A	25 : 24
49	sec –Butylbenzene	Active	A	25 : 37
50	1,3–Dichlorobenzene	Active	A	25 : 44
51	1,4–Dichlorobenzene	Active	A	25 : 53
52	4–Isopropyltoluene	Active	A	26 : 00
53	1,2–Dichlorobenzene	Active	A	26 : 35
54	n–Butylbenzene	Active	A	26 : 48
55	1,2–Dibromo–3–Chloropropa	Active	A	27 : 29
56	1,2,4–Trichlorobenzene	Active	A	30 : 13
57	Naphthalene	Active	A	30 : 46
58	Hexachlorobutadiene	Active	A	30 : 53
59	1,2,3–Trichlorobenzene	Active	A	31 : 12

Figure 4.20 (Continued)

4.4.3 Results

For quantitation, a calibration with eight steps from 0.1 to 40 µg/L is built (Figure 4.17). The P&T procedure gives very good linearity over this range. A standard analysis of the compounds listed in Table 4.5 in water is shown in Figure 4.18. The search and identification of individual analytes in the chromatograms (Figure 4.19) are carried out automatically based on reference data, such as the RT and mass spectrum, which are stored in a calibration file (Figure 4.20).

4.5 irm-GC-MS of Volatile Organic Compounds

Keywords: compound specific isotope analysis; CSIA; VOCs; purge and trap; isotope ratio MS; sources of contamination; degradation pathways

4.5.1 Introduction

The compound specific isotope analysis (CSIA) is successfully used in the assessment of *in situ* remediation of contaminated environments, identification of pollutant degradation pathways, or the verification of contaminant sources. In these types of studies, the sensitivity of isotope ratio monitoring GC-MS (irm-GC-MS) is often a limiting factor since concentrations of organic contaminants in groundwater are very often in the low µg/L range. Hence, in order to be able to routinely use irm-GC-MS techniques in environmental studies, efficient extraction and pre-concentration procedures are required.

While P&T, providing the lowest MDLs for VOCs, is a routinely used extraction method for the trace level quantification, the online coupling with irm-GC-MS has rarely been reported (Zwank and Berg, 2003; Schmidt et al., 2004). Since the P&T procedure includes various phase transition steps that may shift the isotopic signature of the analytes due to phase transitions (evaporation, sorption, and condensation), the P&T method parameters of purge time, desorption time, and injection temperature have been carefully evaluated for the determination of the $\delta^{13}C$-values. The compound specific isotope ratios of 10 different VOCs ranging from the unpolar benzene to the polar methyl t-butyl ether (MTBE) as listed in Table 4.6 were determined (Zwank and Berg, 2003, 2013).

Aqueous solutions of the target analytes were obtained by spiking aliquots of methanolic stock solutions into tap water for method setup and parameter optimization. The isotopic signatures of all the compounds relative to Vienna Pee Dee Belemnite (VPDB) were obtained using CO_2 that was calibrated against referenced CO_2.

4.5.2 Analysis Conditions

Sample material			Groundwater
Sample concentration		System	Tekmar purge and trap concentrator LSC 3100
		Autosampler	Tekmar AQUATek 70
			Aqueous samples were filled into 40 mL vials without headspace
			25 mL automatically transferred to the frit sparge glass vessel
		Trap	Supelco Vocarb 3000
	Sample	Volume	25 mL
		Standby temperature	Room temperature
	Purge	Sample temperature	Room temperature
		Pre-purge	0 min
		Purge duration	30 min
		Purge flow	40 mL/min, N_2
		Dry purge	0 min
	Desorb	Preheat	none
		Desorb temperature	250 °C
		Desorb time	1 min
GC method	System		Thermo Scientific TRACE GC Ultra
	Column	Type	Rtx-VMS
		Dimensions	60 m length × 0.32 mm ID × 1.8 μm film thickness, coupled online to the pre-column of the GC system
	Pre-column		0.5 m length × 0.53 mm ID, deactivated
	Cryofocusing		−120 °C, liquid N_2
	Carrier gas		Helium
	Flow		180 kPa, constant pressure
	Oven program	Start	40 °C, 2 min
		Ramp 1	2 °C/min to 50 °C, 4 min
		Ramp 2	8 °C/min to 100 °C
		Ramp 3	40 °C/min to 210 °C
		Final temperature	210 °C, 3.5 min
	Interface	Type	Thermo Scientific Combustion interface GC-C III
		Temperature	940 °C

MS method	System	Thermo Scientific DELTAplus XL
	Analyzer type	Magnetic sector MS
	Ionization	EI
	Scan mode	Simultaneous ion detection
	Masses	m/z 44, 45, 46

4.5.3 Results

P&T allowed MDLs ranging from 0.25 to 5.0 µg/L, depending on the analyte, which corresponds to the highest sensitivity of CSIA for volatile compounds, reported so far. These results were due to both the high sample volume (25 mL) and the high extraction efficiency (up to 80%) of P&T for the analytes. Thus, P&T-irm-GC-MS allows determining compound-specific stable isotope signatures of contaminant concentrations found in groundwater. As Figure 4.21 and Table 4.6 show, P&T allows highly reproducible CSIA measurements.

The averaged MDLs for the analytes correspond to 0.4 + 0.1 nmol C expressed as the absolute amount of carbon injected on-column. As can be seen from Figure 4.22, the P&T method yields very clean chromatograms with very sharp peaks due to cryofocusing. The detection limit could even be lowered.

Since the validated extraction technique shows little if any carbon isotopic fractionations due to the high extraction efficiency, it is also applicable for CSIA of D/H ratios, which require 10–20 times higher analyte concentrations than $\delta^{13}C$ analysis. P&T pre-concentration methods could also be used to lower analyte concentrations needed for $\delta^{15}N$ and are expected to work also for $\delta^{18}O$ analysis.

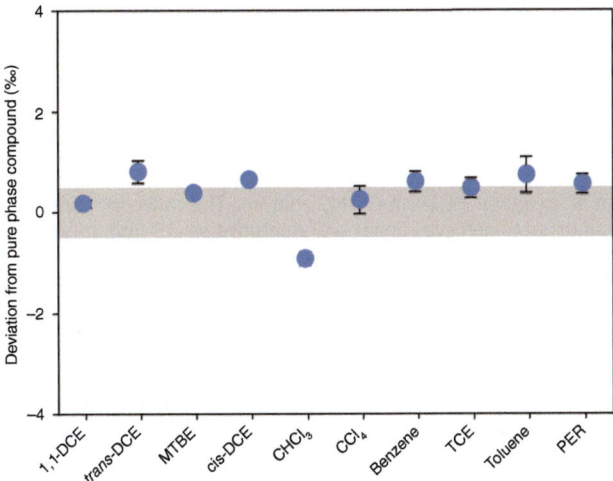

Figure 4.21 Reproducibility and accuracy of P&T-irm-GC-MS. Plotted are the differences from the pure liquid standards measured by EA-IRMS and the horizontal bars correspond to a ^{13}C-measurement within a +0.5‰ interval of the EA-IRMS measurements allowing a direct comparison between the different compounds. (Zwank and Berg, 2013)

Table 4.6 Compounds in order of GC elution, method-detection limits (MDL) in water, accuracy, and reproducibility of purge and trap extraction coupled to irm-GC-MS compared to the elemental analyzer (EA) technique.

Compound	MDL (µg/L)	δ^{13}C EA-IRMS VPDB (‰)	δ^{13}C P&T-irm-GC-MS VPDB (‰)
1,1-Dichloroethylene	3.6	−29.25 + 0.14	−29.07 + 0.08
trans-1,2-Dichloroethylene	1.5	−26.42 + 0.17	−25.61 + 0.22
Methyl-t-butyl ether	0.63	−28.13 + 0.15	−27.75 + 0.09
cis-1,2-Dichloroethylene	1.1	−26.61 + 0.06	−25.96 + 0.07
Chloroform	2.3	−45.30 + 0.19	−46.22 + 0.14
Tetrachloromethane	5.0	−38.62 + 0.01	−38.37 + 0.27
Benzene	0.3	−27.88 + 0.20	−27.27 + 0.20
Trichloroethylene	1.4	−26.59 + 0.08	−26.11 + 0.20
Toluene	0.25	−27.90 + 0.24a	−27.16 + 0.35
Tetrachloroethylene	2.2	−27.32 + 0.14	−26.76 + 0.19

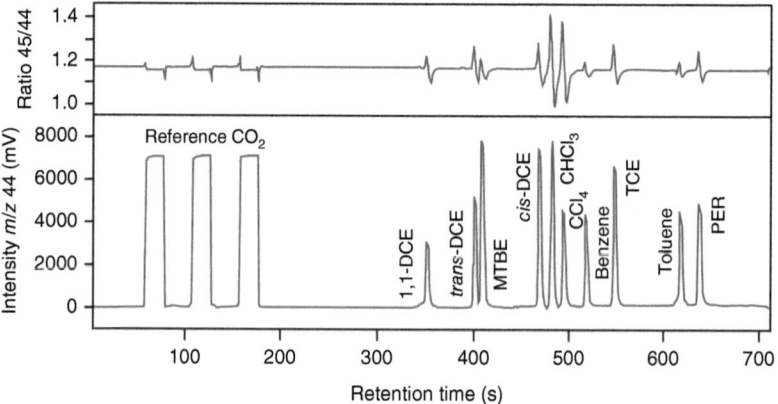

Figure 4.22 Chromatogram of an online P&T-irm-GC-MS analysis. The concentrations of the different analytes were adjusted to achieve similar signal intensities (1.7 µg/L toluene to 32 µg/L CCl$_4$). The three first peaks correspond to the reference CO$_2$ gas pulses.

4.6 Organotin Compounds in Water

Keywords: food safety; water; European Union Water Framework Directive; triple quadrupole MS/MS; timed-SRM; *in situ* derivatization; LLE

4.6.1 Introduction

The monitoring of organotin compounds is regulated by the European Union Water Framework Directive (Commission Directive 2014/101/EU). This includes the mono-, di-, tri-, tetra-butyl, and triphenyl tin compounds. Tributyltin compounds are considered to be most hazardous. Studies have reported adverse health effects causing kidney and central nervous disorders in humans (Ostrakhovitch and Cherian, 2007; OSPAR Commission, 2011). Organotins are used for a variety of applications. Tributyl and triphenyltin are well known as antifouling agents on underwater structures and a fungicide in crop protection. Mono- and dibutyltin are used as stabilizers in plastics and catalysts in soft foam production (European Food Safety Authority, 2004; OSPAR Commission, 2011).

Organotin compounds are lipophilic and get accumulated in adipose tissue. The toxicity of these compounds at low concentrations drives the requirement for accurate and sensitive analytical methods for their detection, quantification, and research for less-toxic replacements (Takeuchi et al., 2000; Morabito and Quevauviller, 2002). The low concentration with ppb to ppt levels in water samples creates a particular analytical challenge for the sample preparation and a very selective and sensitive detection in matrix samples (Butler and Phillips, 2009).

The extraction of organotin compounds from water samples involves the analyte ethylation with tetraethyl borate as an *in situ* derivatization step, followed by liquid extraction with pentane and subsequent extract concentration. Also, SPME methods from the sample headspace after derivatization are in place (Coscollà et al., 2014). Gas chromatography-mass spectrometry/mass spectrometry (GC-MS/MS) detection is preferred exploiting the high MS/MS selectivity (de Dobbeleer et al., 2011).

4.6.2 Analysis Conditions

Sample material			Sea water, surface water
Sample preparation	Sample		400 mL water
			Adjust pH to 5 using a 1 M acetic acid/sodium acetate buffer
	Derivatization		Ethylation, add a 2% w/v sodium tetraethyl borate solution in 0.1 M NaOH
	Liquid extraction		Shake with pentane, 10 min
			Concentrate organic phase to 400 µL
GC method	System		Thermo Scientific TRACE GC Ultra
	Injector		Temperature programmable injector (PTV)
	Column	Type	TG-5MS, 5% phenyl silicone
		Dimensions	30 m length × 0.25 mm ID × 0.25 µm film thickness
	Guard column		DMTPS deactivated, 2 m × 0.53 mm ID pre-column
	Carrier gas		Helium

		Flow	Constant flow, 1.4 mL/min
	PTV injector	Injection mode	Splitless
		Inlet liner	Straight glass liner, SurfaSil treated
		Injection volume	3 μL
		Base temperature	50 °C, 0.1 min
		Transfer rate	8 °C/s
		Transfer temperature	280 °C, 1 min
		Cleaning step	350 °C, 11 min
	Split	Closed until	2 min
		Open	2 min to end of run
		Split flow	50 mL/min
	Oven program	Start	45 °C, 2 min
		Ramp 1	55 °C/min to 175 °C
		Ramp 2	35 °C/min to 300 °C
		Final temperature	300 °C, 2 min
	Transferline	Temperature	300 °C
MS method	System		Thermo Scientific TSQ Quantum XLS
	Analyzer type		Triple quadrupole MS
	Ionization		EI, 70 eV
	Ion source	Temperature	250 °C
	Acquisition	Mode	Timed-SRM
			2 SRMs per analyte
	Collision gas		Ar, 1.0 mTorr
	Scan rate		0.2 s (5 spectra/s)
	Resolution	Setting	0.7 Da Q1 and Q3
Calibration	Internal standards		Tripropyltin, monoheptyltin, diheptyltin
	Calibration		Single-point, 10 ng/L spiked to a water sample

4.6.3 Experimental Conditions

Special attention is required for an inert GC inlet system to achieve a tailing-free peak shape and high compound response. For this purpose, the PTV injector as a temperature programmable injection system was chosen, with a special liner deactivation using the SurfaSil treatment for high inertness (Thermo Fisher Scientific, 2008).

The deteriorating impact of the sample matrix on the chromatographic performance was reduced using a guard column (pre-column) connected in front of the analytical column. The pre-column allows a quick preventive maintenance without replacing the analytical column while keeping the compound RTs constant.

The GC-MS/MS detection method described uses a timed-SRM setup with three transitions for each analyte for a confirming ion ratio confirmation. The t-SRM window is typically set from 30 to 60 s width for all compounds. Peaks were positively identified using the product ion ratios for each compound. The compliance with the calibrated compound ion ratios of standards is checked by the processing software for the peaks identified in samples (Figures 4.23, 4.24, and Table 4.7).

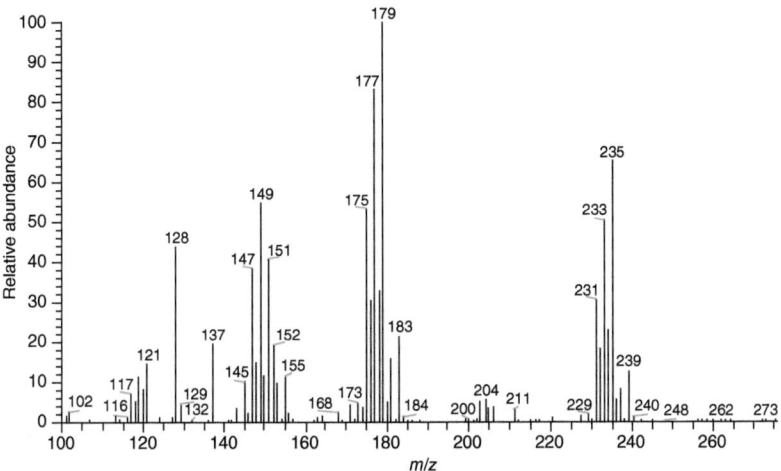

Figure 4.23 Full scan spectrum of monobutyltin (MBT) with the clearly visible Sn isotope pattern.

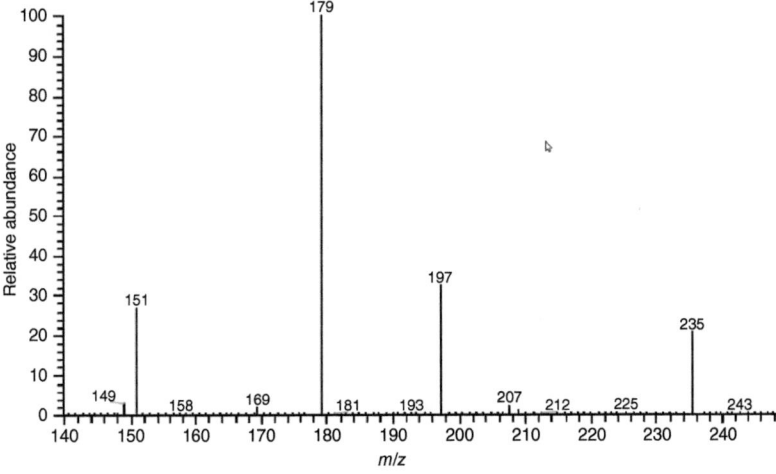

Figure 4.24 Product ion spectrum of monobutyltin (MBT) precursor ion m/z 235.

Table 4.7 t-SRM acquisition method for organotin compounds and S/N achieved at 0.05 ng/L.

Compound name	Precursor ion (m/z)	Product ion (m/z)	Collision energy (V)	Retention time (min)	S/N at 0.05 ng/L
Monobutyltin	235.08	150.98	6	7.94	
	233.08	176.95	6	7.94	
	235.08	178.95	6	7.94	103
Tripropyltin, ISTD	249.08	164.91	8	8.89	
	245.08	160.91	8	8.89	
	247.08	162.91	8	8.89	
Tetrapropyltin, ISTD	249.08	164.91	8	8.89	
	245.08	160.91	8	8.89	
	247.08	162.91	8	8.89	
Dibutyltin	261.03	205.03	8	9.42	
	263.03	150.98	8	9.42	
	263.03	207.03	8	9.42	47
Monoheptyltin, ISTD	178.95	150.98	8	10.55	
	275.07	176.95	8	10.55	
	277.07	178.95	8	10.55	
Tributyltin	287.09	174.94	8	11.21	
	291.08	235.08	8	11.21	
	289.09	176.95	8	11.21	784
Monooctyltin	287.09	174.94	8	11.21	
	289.09	176.95	8	11.21	2458
	291.08	178.95	8	11.21	
Tetrabutyltin	287.09	174.94	8	11.21	
	289.09	176.95	8	11.21	1651
	291.08	235.08	8	11.74	
Diheptyltin, ISTD	275.07	176.95	8	13.71	
	277.07	178.95	8	13.71	
	245.08	145.98	8	13.71	
Dioctyltin	261.03	148.98	8	14.94	80
	263.03	150.98	8	14.94	
	375.17	263.03	8	14.94	
Triphenyltin	196.94	119.90	18	16.21	
	349.15	194.98	22	16.21	
	351.15	196.94	22	16.21	923
Tricyclohexyltin	233.08	150.98	8	16.28	187
	287.09	205.03	8	16.28	
	315.04	150.98	8	16.28	

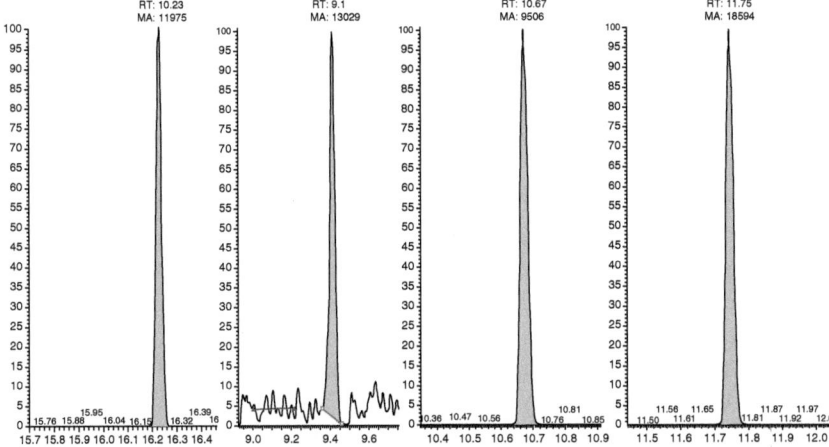

Figure 4.25 Mono-, di-, tri-, and tetra-butyltin signals from a spiked water sample at 0.05 ng/L (0.2 pg on column).

4.6.4 Results

The result shown in Figure 4.25 demonstrates the method capability to reach 0.05 ng/L of organotins in the water sample and below. Calculations were performed using tripropyltin as the ISTD.

The described method demonstrates the strength of GC-MS/MS analysis to comply and exceed the allowable average of 0.2 ng/L of the European Water Framework Directive for the detection and quantification of organotin compounds. The used t-SRM functionality gives the instrument the ability to automatically determine optimum SRM dwell times for high sensitivity detection, even allowing for partially overlapping SRM transitions.

4.7 Analysis of Dithiocarbamate Pesticides

Keywords: food safety; dithiocarbamate fungicides; DTCs; hydrolysis; thiram; single quadrupole MS; SIM

4.7.1 Introduction

Dithiocarbamate (DTC) fungicides are used in agriculture as fungicides. They are nonsystemic and typically remain at the site of application. DTCs are characterized by a broad spectrum of activity, low acute mammal toxicity, and low production costs (Crnogorac and Schwack, 2009).

DTCs are not stable and cannot be extracted or analyzed directly. Contact with acidic plant juices degrades DTCs rapidly, and they decompose into carbon disulfide (CS_2) and the respective amine. DTCs cannot be extracted by organic solvents from

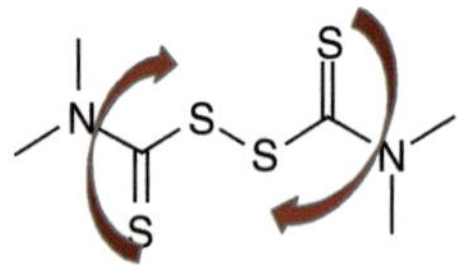

Figure 4.26 Thiram – 1 mol of thiram generates 2 mol of CS_2.

homogenized plant samples as it is the Quick-Easy-Cheap-Efficient-Rugged-and-Safe (QuEChERS) standard procedure in pesticide residue analyses. The described method is a nonspecific DTC sum method that does not distinguish between the different species of DTCs in the sample. Interferences are known from natural precursors, for example, from crops or brassica, that can produce CS_2 as well during hydrolysis (Reynolds, 2006; Crnogorac and Schwack, 2009).

DTCs can be quantitatively converted to CS_2 by reaction with tin(II)chloride in aqueous HCl (1 : 1) in a closed bottle at 80 °C. The CS_2 gas produced is dissolved into iso-octane and measured by GC-MS. The analysis of DTCs for this application follows the acid-hydrolysis method using $SnCl_2$/HCl (Reynolds, 2006). For method validation of the DTC pesticides, thiram (99.5% purity) is used as the representative DTC compound considering its simple structure (1 mol of thiram = 2 mol of CS_2, i.e. 1 mg of thiram theoretically generates 0.6333 mg CS_2, 1 mL of 100 ppm thiram in 25 g of grapes = 2.5 ppm of CS_2), see Figure 4.26. The DTC residues were estimated by the analysis of CS_2 as the DTC hydrolysis products by GC-MS.

4.7.2 Analysis Conditions

Sample material		Fruits, vegetables, spices
Sample preparation		$SnCl_2$/HCl acid-hydrolysis method (Crnogorac and Schwack, 2009)
		The homogenized sample (25 g) was taken in a 250 mL glass bottle, 75 mL of the reaction mixture was added, followed by 25 mL iso-octane. The bottle was closed immediately and placed in a water bath at 80 °C for 1 h with intermittent shaking and inverting the bottle after every 20 min. After cooling the bottle to <20 °C by ice water, a 1–2 mL aliquot of the upper iso-octane layer gets transferred into a micro centrifuge tube, and centrifuged at 5000 rpm for 5 min at 10 °C. The supernatant was transferred into GC autosampler vial.
GC method	System	Thermo Scientific TRACE GC Ultra
	Autosampler	Thermo Scientific TriPlus RSH

	Column 1	Type	TraceGOLD TG-624
		Dimensions	30 m length × 0.32 mm ID × 1.8 µm film thickness
	Column 2	Type	TraceGOLD TG-5MS
		Dimensions	30 m length × 0.25 mm ID × 0.25 µm film thickness
	Carrier gas		Helium
	Flow		Constant flow, 1 mL/min
	PTV injector	Injection mode	PTV-LVI
		Injection volume	4 µL
		Base temperature	40 °C, 0.1 min at 100 kPa
		Evaporation	10 °C/s to 80 °C, 0.3 min at 200 kPa
		Transfer temperature	10 °C/s to 110 °C
		Transfer time	0.5 min at 200 kPa
		Cleaning phase	14.5 °C/s to 290 °C
		Final temperature	290 °C
	Solvent vent	Open until	0.17 min
	Split	Closed until	4 min
		Open	4 min to end of run
		Split flow	20 mL/min
	Oven program	Start	40 °C, 5 min
		Ramp 1	40 °C/min to 200 °C
		Final temperature	200 °C, 5 min
	Transferline	Temperature	295 °C
			Direct coupling
MS method	System		Thermo Scientific ITQ 900
	Analyzer type		Ion trap MS with external ionization
	Ionization		EI, 70 eV
	Ion source	Temperature	200 °C
	Acquisition	Mode	SIM
	Mass range	SIM mode	m/z 76, 78
	Scan rate		0.5 s/scan
Calibration	External standardization		CS_2, thiram
	Data points		6
	Calibration range		0.04–1.3 µg/mL

4.7.3 Sample Preparation

The published $SnCl_2$/HCl acid-hydrolysis method was employed for sample preparation (CRL, 2009). The described method follows the established methods applied in the EU reference laboratories and European commercial testing laboratories for CS_2 analysis. From the homogenized sample, 25 g is taken in a 250 mL glass bottle, and 75 mL of the reaction mixture is added, followed by 25 mL iso-octane. The bottle is closed gastight immediately and placed in a water bath at 80 °C for 1 hour with intermittent shaking and inverting the bottle after every 20 min.

After cooling the bottle to <20 °C by ice water, a 1–2 mL aliquot of the upper iso-octane layer is transferred into a micro centrifuge tube and centrifuged at 5000 rpm for 5 min at 10 °C. The supernatant is then transferred into GC vials, and the residues of DTCs are estimated by determining the CS_2 concentration by GC-MS. The sample preparation procedure depending on the type of food used takes c. 1–2 hours.

4.7.4 Preparation of Standard Solutions and Reaction Mixture

For method validation, thiram (99.5% purity) was used as the representative DTC compound considering its simple structure (1 mol of thiram = 2 mol of CS_2).

4.7.4.1 Carbon Disulfide Standard Solution

A stock solution of CS_2 (2000 µg/mL) was prepared by accurately pipetting 79 µL of CS_2 into a volumetric flask (certified A class, 50 mL) containing approximately 45 mL of iso-octane and made up to 50 mL with iso-octane. The CS_2 stock solution was kept in refrigerator at −20 °C and used within 2 days of preparation. CS_2 working standard solutions of 200 and 20 µg/mL concentrations (10 mL each) were prepared by the serial dilution of stock solution with iso-octane.

4.7.4.2 Standard Solution of Thiram

An amount of 10 (±0.05) mg of thiram was weighed into a 10 mL volumetric flask (certified A class) and dissolved in ethyl acetate up to the mark to get a stock solution of 1000 µg/mL. A 100 µg/mL thiram working standard was prepared from stock solution by dilution.

4.7.4.3 Preparation of Reaction Mixture

An amount of 30 g of tin(II)chloride was accurately weighed in the 1000 mL volumetric flask (certified A class) to which 1000 mL of concentrated HCl (35%) was added. Then the solution was gradually added to 1000 mL water with continuous stirring to get clear solution.

4.7.4.4 Calibration Standards

Calibration standard solutions of CS_2 at six different concentration levels (0.04, 0.08, 0.16, 0.32, 0.64, and 1.3 µg/mL) were prepared by appropriate dilutions of 20 µg/mL CS_2 working standard in iso-octane.

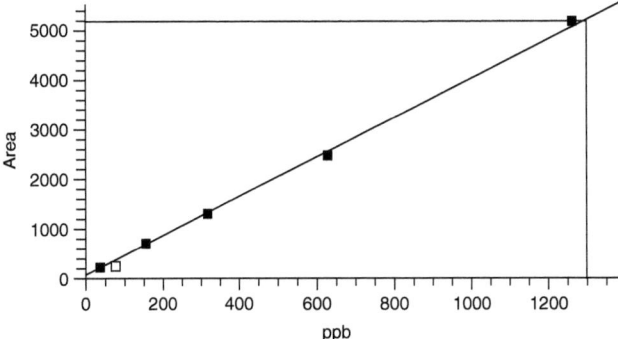

Figure 4.27 Calibration curve, range 0.04–1.300 µg/mL thiram matrix spike, $R^2 = 0.9990$.

Matrix-matched standards at the same concentrations were prepared by spiking the iso-octane extract of fresh control grapes, potato, tomato, green chilli, and eggplant (all organically grown) using the following formula derived from the above conversion of thiram to CS_2:

$$\text{Spike quantity} = \frac{\text{concentration to be achieved} \times \text{weight of the sample}}{0.6333 \times \text{concentration of the stock solution}}$$

Before the preparation of matrix-matched standards, the control samples were carefully monitored for the absence of DTCs (in terms of CS_2) (Figure 4.27).

4.7.5 Experimental Conditions

Two GC columns of different polarities, stationary phases, and film thicknesses have been evaluated. The first column was a medium polar cyanopropylphenyl phase (6% cyanopropylphenyl / 94% dimethylpolysiloxane, 30 m × 0.32 mm ID, 1.8 µm film thickness, e.g. TraceGOLD TG-624), and the second column was a low polar 5%-phenyl stationary phase (5% diphenyl / 95% dimethylpolysiloxane, 30 m × 0.25 mm ID, 0.25 µm film thickness, e.g. TraceGOLD TG-5MS). The TG-624 column type is a mid-polarity column ideally suited for the analysis of volatile analytes, whereas the TG-5MS column is more commonly used especially for pesticide analysis and mostly available with all laboratories. Both columns were thus tested for the applicability of the method. Either column can be used for DTC analysis (Figure 4.28).

4.7.6 Sample Measurements

A typical GC-MS batch consisted of matrix-matched calibration standards, samples, one matrix blank, and one recovery sample for performance check after a set of every six samples.

The data acquisition was carried out in the SIM mode with compound-specific ions m/z 76 and 78 (the ^{34}S isotope, ion ratio 10 : 1) for a selective identification of CS_2 (Figure 4.29).

582 | *4 Applications*

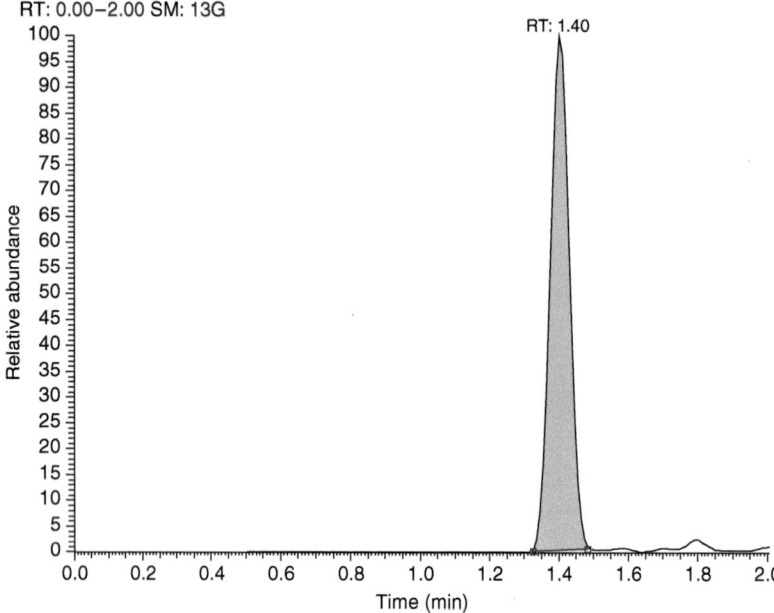

Figure 4.28 CS$_2$ chromatogram, 5 ppb matrix spike calibration.

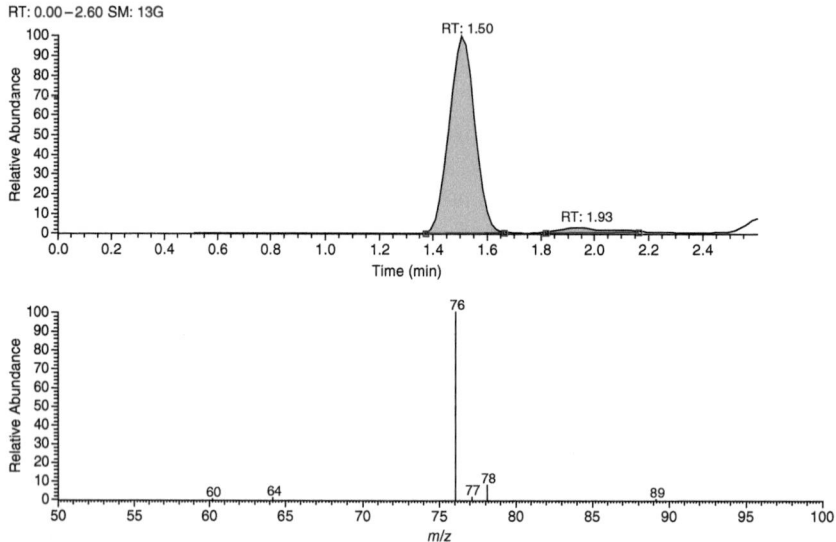

Figure 4.29 Chili sample analysis with confirming CS$_2$ ion ratio *m/z* 76/78 c. 100 : 10.

Table 4.8 Recoveries from different food commodities.

Spike level (ppb)	Grapes (%)	Chilli (%)	Potato (%)	Eggplant (%)	Tomato (%)
1300	96 (±4)	81 (±10)	90 (±9)	90 (±5)	81 (±4)
160	94 (±10)	80 (±13)	94 (±10)	92 (±8)	85 (±10)
40	104 (±15)	79 (±9)	104 (±15)	86 (±10)	96 (±15)

4.7.7 Results

4.7.7.1 Sensitivity
The sensitivity of the method was evaluated in terms of the LOD and LOQ which were 0.005 and 0.04 µg/mL, respectively. The LOD is the concentration at which the signal-to-noise ratio (S/N) for the quantifier ion is >3, whereas LOQ is the concentration for which the S/N is >10.

4.7.7.2 Recovery
The recovery experiments were carried out on fresh untreated potatoes, tomatoes, eggplants, green chillies, and grapes by fortifying 25 g of the samples with thiram solution at 0.04, 0.16, and 1.30 µg/g levels in six replicates. The control samples of each of the tested commodities were obtained from an organic farm near Pune, India, and screened for the absence of DTC residues before spiking. The spiked samples were extracted using sample preparation method as described above. The quantitation of the residues was performed using matrix-matched standards (Table 4.8).

4.7.7.3 Accuracy
The precision of repeatability was determined by three analysts preparing six samples each on a single day. The intermediate precision was determined by the same analysts with six samples each on six different days. The method accuracy was determined with 0.04 mg/kg.

4.7.8 General Guidelines for DTC Analysis

The analysis of cruciferous crops, including brassica samples, may not be unequivocal because they contain naturally occurring compounds that may generate CS_2.

It is necessary to avoid the use of rubber material (natural/synthetic), for example, gloves, when performing DTC analyses as they contain DTCs, and this could lead to contamination problems. Silicone rubber and polythene do not contain DTCs.

Samples, other than fresh foodstuffs, will be comminuted by cryogenic milling. Fresh samples should be subsampled prior to extraction by removing segments from fresh samples following current Codex Alimentarius guidelines.

The samples should be analyzed within 4 weeks of cryogenic milling. If the storage of fresh produce is necessary, it should be in a cool place (<−10 °C) keeping condensation at minimum (Amvrazi, 2005).

4.7.9 Conclusions

A reliable routine method for the analysis of DTCs with high precision in different vegetable and fruit commodities has been described (Dasgupta et al., 2013). The extraction uses a $SnCl_2$/HCl acid-hydrolysis with iso-octane as the solvent to form CS_2 which finally gets quantified by GC-MS. The recovery from different food commodities has been shown to be very high with 79–104%. The method allows a wide calibration range of 0.04–1.30 µg/mL thiram. The LOQ has been determined as 0.04 µg/mL.

The GC injection method and column separation have been optimized for the injection of 4 µL of extract, using GC columns of standard film and dimensions, typically used for other types of residue analyses as well, so that a column change to a specific column for CS_2 determination only is not required.

4.8 Pesticides Multi-Method by Single Quadrupole MS

Keywords: multi-residue analysis; pesticides; fruits; vegetables; SPE clean-up; single quadrupole MS

4.8.1 Introduction

The efficient monitoring of pesticide residues and its degradation products requires fast and comprehensive multi-methods. The referenced method fulfils the requirement for monitoring of a wide scope of pesticide compounds and degradation products in fruit and vegetable samples using single quadrupole GC-MS.

The described ISTD method is suitable for the quantification of more than 200 components. The LLE extraction process with acetonitrile (MeCN) similar to QuEChERS including a salting-out step is followed by a clean-up with solid phase extraction (SPE). The compound detection is done by SIM using the RT and selected ion abundance ratios as qualifiers for identification. Two runs are required to determine the complete set of pesticides given in Table 4.9 according to the required SIM acquisition windows described as FV-1 and FV-2.

The quantitation using aldrin as an ISTD is based on blank matrix spikes to compensate for matrix effects. In addition, the published method also includes the parallel determination of carbamates by high performance liquid chromatography (HPLC) with fluorescence detection (FD) (Health and Welfare Canada, 2000; Fillion et al., 2000).

Table 4.9 List of compounds included in the multi-method with retention times, target and qualifier ions (Q), and ratios (Q/tgt).

Compound	Method	RT (min)	Target	Q1	Q2	Q3	Q/tgt	Q2/tgt	Q3/tgt
Dichlorvos-naled	FV-1	7.24	185	109	220	—	4.61	0.19	—
EPTC	FV-1	7.89	189	132	—	—	1.40	—	—
Bendiocarb degr.	FV-2	7.99	166	151	126	—	1.86	1.82	—
Allidochlor	FV-2	8.28	138	132	173	—	1.12	0.14	—
Methamidophos	FV-2	8.88	141	94	95	—	2.31	1.60	—
Butylate	FV-1	9.06	217	174	—	—	5.20	—	—
Promecarb degr.	FV-1	9.09	135	150	—	—	0.35	—	—
Chlorthiamid degr.	FV-2	9.36	171	173	136	—	0.64	0.18	—
Dichlobenil	FV-1	9.37	171	173	—	—	0.62	—	—
Dichlormid	FV-1	9.51	168	166	—	—	1.53	—	—
Vernolate	FV-2	9.59	86	128	—	—	1.14	—	—
Pebulate	FV-2	9.97	128	161	203	—	0.12	0.06	—
Aminocarb degr.	FV-1	10.06	150	151	136	—	1.27	0.79	—
Etridiazole	FV-2	10.12	211	183	—	—	0.97	—	—
Chlormephos	FV-1	10.46	234	121	—	—	3.19	—	—
Nitrapyrin	FV-2	10.85	194	196	198	—	0.97	0.31	—
Mexacarbate degr.	FV-2	11.18	165	150	134	—	0.81	0.34	—
Bufencarb degr.	FV-1	11.24	121	122	107	—	0.37	0.22	—
cis-Mevinphos	FV-1	11.82	127	192	164	109	0.20	0.06	0.23
Propham	FV-2	11.85	179	137	—	—	1.28	—	—
trans-Mevinphos	FV-1	12.17	127	192	164	109	0.22	0.06	0.23
Chloroneb	FV-2	12.43	191	193	206	—	0.64	0.53	—
o-Phenylphenol	FV-2	13.57	170	169	141	115	0.66	0.29	0.21
Tecnazene	FV-1	14.96	261	215	—	—	1.54	—	—
Cycloate	FV-2	15.09	154	83	215	—	2.26	0.07	—
Captafol degr.	FV-2	15.56	151	79	80	—	1.31	0.73	—
Captan degr.	FV-1	15.60	151	79	80	—	1.26	0.71	—
Heptenophos	FV-2	15.89	124	215	250	—	0.09	0.07	—
Acephate	FV-1	15.89	136	94	—	—	0.54	—	—
Demeton-S	FV-1	16.04	171	88	143	—	6.27	0.55	—
HCB	FV-1	16.51	284	286	282	—	0.77	0.56	—
Ethoprophos	FV-2	16.98	158	242	139	—	0.08	0.47	—
Diphenylamine	FV-1	17.09	169	167	168	—	0.34	0.63	—

(Continued)

Table 4.9 (Continued)

Compound	Method	RT (min)	Target	Q1	Q2	Q3	Q/tgt	Q2/tgt	Q3/tgt
Di-allate 1	FV-1	17.46	234	236	—	—	0.36	—	—
Chlordimeform	FV-2	17.67	196	181	—	—	0.75	—	—
Propachlor	FV-1	17.68	120	176	211	—	0.23	0.05	—
Demeton-S-methyl	FV-2	18.08	88	109	142	—	0.26	0.17	—
Di-allate 2	FV-1	18.48	234	236	—	—	0.36	—	—
Ethalfluralin	FV-1	18.61	276	316	—	—	0.50	—	—
Phorate	FV-1	18.86	260	231	—	—	0.73	—	—
Trifluralin	FV-1	19.27	306	264	—	—	1.18	—	—
Sulfallate	FV-2	19.31	188	116	148	—	0.09	0.06	—
Chlorpropham	FV-2	19.39	213	127	—	—	3.21	—	—
Benfluralin	FV-1	19.45	292	264	—	—	0.23	—	—
Sulfotep	FV-1	19.54	322	202	—	—	1.15	—	—
α-HCH	FV-1	19.83	219	183	—	—	1.16	—	—
Bendiocarb	FV-2	20.74	151	166	223	—	0.41	0.08	—
Quintozene	FV-1	20.82	295	237	—	—	1.79	—	—
Promecarb degr.	FV-1	20.93	135	150	—	—	0.62	—	—
Omethoate	FV-1	21.15	156	110	—	—	1.10	—	—
Terbufos	FV-2	21.71	231	153	—	—	0.38	—	—
Demeton-O	FV-1	21.74	88	114	170	—	0.15	0.12	—
Desethylatrazine	FV-1	21.86	172	174	—	—	0.31	—	—
Clomazone	FV-2	21.92	204	125	—	—	2.14	—	—
Prometon	FV-2	22.01	225	210	168	—	1.44	1.17	—
Tri-allate	FV-1	22.32	268	270	—	—	0.67	—	—
Fonofos	FV-1	22.40	246	109	—	—	4.50	—	—
Diazinon	FV-1	22.67	304	179	—	—	3.58	—	—
Terbumeton	FV-2	22.77	210	169	225	—	0.94	0.25	—
Dicrotophos	FV-1	22.77	127	193	—	—	0.07	—	—
Lindane	FV-2	22.98	219	183	—	—	1.28	—	—
Dioxathion	FV-1	23.18	125	153	270	—	0.29	0.35	—
Disulfoton	FV-2	23.60	142	274	—	—	0.29	—	—
Profluralin	FV-1	23.68	318	330	—	—	0.30	—	—
Dicloran	FV-2	23.78	206	176	—	—	1.37	—	—
Propazine	FV-1	24.04	229	214	—	—	1.85	—	—
Atrazine	FV-2	24.12	215	200	—	—	1.88	—	—

Ions monitored (m/z); Abundance ratio of qualifier ion/target

Table 4.9 (Continued)

Compound	Method	RT (min)	Target	Q1	Q2	Q3	Q/tgt	Q2/tgt	Q3/tgt
Etrimfos	FV-2	24.14	292	277	—	—	0.44	—	—
Simazine	FV-1	24.21	201	186	—	—	0.70	—	—
Heptachlor	FV-1	24.50	272	274	—	—	0.79	—	—
Chlorbufam	FV-1	24.59	223	164	—	—	1.00	—	—
Schradan	FV-2	24.59	199	135	—	—	1.74	—	—
Aminocarb	FV-1	24.65	151	208	150	—	0.12	0.73	—
Terbuthylazine	FV-1	24.78	214	173	—	—	0.55	—	—
Monolinuron	FV-2	24.80	214	126	—	—	1.43	—	—
Secbumeton	FV-2	25.33	196	210	225	—	0.20	0.14	—
Dichlofenthion	FV-1	25.82	279	223	—	—	1.36	—	—
Cyanophos	FV-1	26.09	243	125	—	—	1.34	—	—
Isazofos	FV-2	26.10	161	257	313	—	0.28	0.04	—
Mexacarbate	FV-2	26.28	165	222	150	—	0.21	0.86	—
Propyzamide	FV-1	26.45	173	175	—	—	0.63	—	—
Aldrin ISTD	FV-1	26.48	263	265	—	—	0.66	—	—
Pirimicarb	FV-2	26.50	166	238	—	—	0.16	—	—
Dimethoate	FV-1	26.60	87	229	143	—	0.04	0.09	—
Monocrotophos	FV-2	26.68	127	192	109	—	0.09	0.11	—
Chlorpyrifos-methyl	FV-1	26.93	286	288	—	—	0.65	—	—
Fluchloralin	FV-2	27.56	306	326	264	—	0.82	0.83	—
Fenchlorphos	FV-1	27.75	285	287	—	—	0.68	—	—
Desmetryn	FV-1	27.91	198	213	—	—	1.49	—	—
Dinitramine	FV-2	28.00	305	307	261	—	0.37	0.41	—
Dimetachlor	FV-2	28.23	210	134	197	—	8.78	3.13	—
Chlorothalonil	FV-1	28.75	266	264	—	—	0.78	—	—
Alachlor	FV-1	28.80	188	160	—	—	1.27	—	—
Prometryn	FV-1	29.22	241	226	—	—	0.68	—	—
Metobromuron	FV-2	29.23	258	61	—	—	16.26	—	—
Cyprazine	FV-2	29.33	212	227	229	—	0.89	0.12	—
Ametryn	FV-2	29.34	227	212	—	—	1.12	—	—
Simetryn	FV-1	29.45	170	155	—	—	0.93	—	—
Pirimiphos-methyl	FV-1	29.53	290	305	—	—	0.52	—	—
Vinclozolin	FV-1	29.55	285	287	—	—	0.62	—	—
Thiobencarb	FV-2	29.79	100	257	125	—	0.09	0.25	—

(Continued)

Table 4.9 (Continued)

Compound	Method	RT (min)	Target	Q1	Q2	Q3	Q/tgt	Q2/tgt	Q3/tgt
Metribuzin	FV-2	29.82	198	199	—	—	0.29	—	—
β-HCH	FV-1	29.90	219	183	—	—	1.17	—	—
Terbutryn	FV-1	30.21	226	241	—	—	0.51	—	—
Metalaxyl	FV-2	30.37	206	249	—	—	0.46	—	—
Parathionmethyl	FV-2	30.42	263	125	—	—	2.15	—	—
Chlorpyrifos	FV-1	30.61	314	199	—	—	3.18	—	—
Aspon	FV-1	30.97	211	253	—	—	0.27	—	—
Dicofol	FV-1	31.14	250	139	—	—	5.88	—	—
Oxychlordane	FV-2	31.21	115	185	149	—	0.62	0.51	—
Malaoxon	FV-1	31.25	268	195	—	—	1.85	—	—
Chlorthal-dimethyl	FV-2	31.26	301	299	—	—	0.81	—	—
Phosphamidon	FV-1	31.31	264	193	—	—	0.28	—	—
δ-HCH	FV-2	31.35	183	219	217	—	0.86	0.68	—
Metolachlor	FV-1	31.63	238	162	—	—	2.14	—	—
Terbacil	FV-2	31.91	160	161	216	—	1.33	0.02	—
Fenthion	FV-1	32.01	278	169	—	—	0.33	—	—
Bromophos	FV-1	32.15	331	329	—	—	0.77	—	—
Dichlofluanid	FV-1	32.39	226	123	—	—	5.26	—	—
Fenitrothion	FV-1	32.46	277	260	—	—	0.59	—	—
Pirimiphos-ethyl	FV-2	32.67	333	304	—	—	1.26	—	—
Malathion	FV-1	32.68	158	125	—	—	2.23	—	—
Paraoxon	FV-1	32.71	275	109	—	—	11.06	—	—
Heptachlor epoxide	FV-2	32.76	237	183	217	—	2.02	1.21	—
Nitrothal-isopropyl	FV-2	33.21	236	212	—	—	0.75	—	—
Butralin	FV-2	33.42	266	250	—	—	0.14	—	—
Ethofumesate	FV-1	33.48	161	286	207	—	0.25	1.04	—
Triadimefon	FV-1	33.99	208	210	—	—	0.30	—	—
Parathion	FV-2	34.05	291	139	—	—	1.20	—	—
Isopropalin	FV-1	34.15	280	238	—	—	0.60	—	—
Pendimethalin	FV-2	34.34	252	281	—	—	0.08	—	—
Fenson	FV-2	34.36	268	141	—	—	4.41	—	—
Linuron	FV-2	34.40	248	160	250	—	1.44	0.62	—
α-Endosulfan	FV-1	34.42	277	339	243	—	0.57	1.37	—
Chlorthiamid	FV-2	34.45	205	170	—	—	2.36	—	—
Chlorbenside	FV-1	34.54	125	268	—	—	0.07	—	—

Ions monitored (m/z) / Abundance ratio of qualifier ion/target

Table 4.9 (Continued)

Compound	Method	RT (min)	Target	Q1	Q2	Q3	Q/tgt	Q2/tgt	Q3/tgt
Allethrin	FV-2	34.82	123	136	—	—	0.28	—	—
Chlorfenvinphos	FV-1	34.97	323	267	—	—	2.51	—	—
trans-Chlordane	FV-1	35.22	373	375	—	—	0.93	—	—
Bromophos-ethyl	FV-1	35.35	359	303	—	—	1.31	—	—
Chlorthion	FV-2	35.56	297	125	—	—	4.38	—	—
Quinalphos	FV-1	35.59	146	298	—	—	0.09	—	—
Propanil	FV-2	35.69	161	163	—	—	0.66	—	—
Diphenamid	FV-2	35.71	167	239	—	—	0.16	—	—
Crufomate	FV-2	35.79	256	182	—	—	0.83	—	—
cis-Chlordane	FV-1	35.91	373	375	—	—	0.93	—	—
Isofenphos	FV-1	35.92	213	255	—	—	0.29	—	—
Metazachlor	FV-2	36.11	209	133	—	—	1.82	—	—
Phenthoate	FV-2	36.19	246	274	—	—	3.83	—	—
Chlorfenvinphos	FV-1	36.26	323	267	—	—	2.51	—	—
Penconazole	FV-1	36.33	248	159	—	—	1.85	—	—
Tolylfluanid	FV-1	36.61	238	137	—	—	3.56	—	—
p,p'-DDE	FV-1	37.06	318	246	—	—	1.75	—	—
Folpet	FV-2	37.21	260	262	—	—	0.69	—	—
Prothiofos	FV-2	37.50	309	267	—	—	1.34	—	—
Dieldrin	FV-1	37.55	277	263	—	—	1.47	—	—
Butachlor	FV-1	37.75	176	160	—	—	0.84	—	—
Chlorflurecol-methyl	FV-2	37.77	215	217	152	—	0.32	0.44	—
Captan	FV-1	38.05	149	79	—	—	6.28	—	—
Iodofenphos	FV-1	38.08	377	379	—	—	0.44	—	—
Tribufos	FV-2	38.16	169	202	—	—	0.50	—	—
Chlozolinate	FV-2	38.23	259	331	—	—	0.43	—	—
Crotoxyphos	FV-2	38.56	193	127	—	—	3.65	—	—
Methidathion	FV-1	38.70	145	85	—	—	0.75	—	—
Tetrachlorvinphos	FV-1	38.83	329	331	—	—	0.95	—	—
Chlorbromuron	FV-2	38.84	61	294	—	—	0.06	—	—
Procymidone	FV-1	38.85	283	285	—	—	0.69	—	—
Flumetralin	FV-2	38.98	143	157	—	—	0.14	—	—
Endrin	FV-1	39.09	263	281	—	—	0.44	—	—
Bromacil	FV-2	39.17	205	207	—	—	0.97	—	—
Flurochloridone 1	FV-2	39.19	311	187	—	—	2.48	—	—

(Continued)

Table 4.9 (Continued)

Compound	Method	RT (min)	Target	Q1	Q2	Q3	Q/tgt	Q2/tgt	Q3/tgt
Triadimenol	FV-2	39.22	112	168	—	—	0.40	—	—
Profenofos	FV-1	39.28	339	337	—	—	1.06	—	—
DDD, o,p'	FV-1	39.67	235	237	—	—	0.64	—	—
Flurochloridone 2	FV-2	39.67	311	187	—	—	1.85	—	—
Ethylan	FV-2	39.78	223	165	—	—	0.10	—	—
Cyanazine	FV-1	40.04	225	240	—	—	0.49	—	—
Chlorfenson	FV-1	40.09	302	175	—	—	6.83	—	—
TCMTB	FV-2	40.40	238	180	—	—	6.23	—	—
DDT, o,p'	FV-1	40.71	235	237	—	—	0.63	—	—
Oxadiazon	FV-2	40.89	258	175	—	—	3.48	—	—
Carbetamide	FV-2	41.06	119	120	236	—	0.22	0.04	—
Tetrasul	FV-2	41.13	252	324	—	—	0.52	—	—
Imazalil	FV-1	41.15	215	173	217	—	0.98	0.61	—
Aramite 1	FV-1	41.17	185	319	—	—	0.18	—	—
Fenamiphos	FV-2	41.48	303	217	—	—	0.59	—	—
Erbon	FV-1	42.08	169	171	—	—	0.64	—	—
Aramite 2	FV-1	42.13	185	319	—	—	0.08	—	—
Methoprotryne	FV-2	42.18	256	213	—	—	0.37	—	—
Chloropropylate	FV-1	42.24	251	139	253	—	1.14	0.63	—
Methyl trithion	FV-2	42.40	157	314	—	—	0.10	—	—
Nitrofen	FV-1	42.42	283	202	—	—	0.73	—	—
Chlorobenzilate	FV-2	42.55	251	139	—	—	1.26	—	—
Carboxin	FV-2	42.81	143	235	—	—	0.29	—	—
Flamprop-methyl	FV-1	42.89	105	77	276	—	0.29	0.04	—
Bupirimate	FV-2	43.17	273	316	208	—	0.16	0.94	—
β-Endosulfan	FV-1	43.36	241	237	—	—	1.06	—	—
DDD, p,p	FV-2	43.62	235	237	—	—	0.64	—	—
Oxyfluorfen	FV-2	43.82	252	300	—	—	0.27	—	—
Chlorthiophos	FV-1	43.92	325	360	—	—	0.46	—	—
Ethion	FV-1	44.42	231	153	384	—	0.76	0.04	—
Etaconazole 1	FV-1	44.42	245	173	—	—	1.51	—	—
Sulprofos	FV-2	44.47	322	156	140	—	1.83	2.14	—
Etaconazole 2	FV-1	44.60	245	173	—	—	1.70	—	—
Flamprop-isopropyl	FV-2	44.85	105	276	363	—	0.07	0.01	—
DDT, p,p'	FV-1	44.96	235	237	—	—	0.64	—	—

Headers: Ions monitored (m/z); Abundance ratio of qualifier ion/target

Table 4.9 (Continued)

Compound	Method	RT (min)	Target	Q1	Q2	Q3	Q/tgt	Q2/tgt	Q3/tgt
Carbophenothion	FV-1	45.17	157	342	121	—	0.13	0.54	—
Fluorodifen	FV-2	45.19	190	328	162	—	—	0.15	—
Myclobutanil	FV-2	46.02	179	288	—	—	0.06	—	—
Benalaxyl	FV-1	46.15	148	206	325	—	0.16	0.02	—
Edifenphos	FV-1	46.84	173	310	201	—	0.27	0.27	—
Propiconazole 1	FV-2	47.25	259	261	—	—	0.61	—	—
Fensulfothion	FV-2	47.52	293	308	—	—	0.16	—	—
Mirex	FV-1	47.61	272	274	237	—	0.80	0.62	—
Propargite	FV-2	47.63	135	350	150	201	0.02	0.12	0.03
Propiconazole 2	FV-2	47.63	259	261	—	—	0.64	—	—
Diclofop-methyl	FV-1	47.64	253	340	281	—	0.40	0.34	—
Propetamphos	FV-2	48.22	124	208	—	—	2.24	—	—
Triazophos	FV-2	48.40	162	161	—	—	1.54	—	—
Benodanil	FV-1	48.92	231	323	203	—	0.17	0.21	—
Nuarimol	FV-1	49.15	314	235	203	—	3.39	3.03	—
Bifenthrin	FV-2	49.19	181	165	166	—	0.27	0.27	—
Endosulfan sulfate	FV-1	49.43	272	387	—	—	0.27	—	—
Bromopropylate	FV-2	50.26	341	183	—	—	1.11	—	—
Oxadixyl	FV-2	50.29	163	132	278	—	0.79	0.06	—
Methoxychlor	FV-1	50.41	227	228	—	—	0.16	—	—
Benzoylprop-ethyl	FV-2	50.84	105	77	292	—	0.26	0.04	—
Tetramethrin 1	FV-2	50.89	164	123	—	—	0.28	—	—
Fenpropathrin	FV-1	51.21	181	265	—	—	0.33	—	—
Tetramethrin 2	FV-2	51.25	164	123	—	—	0.29	—	—
Leptophos	FV-2	51.51	171	377	—	—	0.29	—	—
EPN	FV-1	51.70	169	157	—	—	2.46	—	—
Norflurazon	FV-2	51.87	303	145	—	—	2.48	—	—
Hexazinone	FV-1	52.08	171	128	—	—	0.13	v	—
Phosmet	FV-1	52.16	160	161	317	—	0.12	0.02	—
Iprodione	FV-2	52.30	314	316	187	—	0.64	1.31	—
Tetradifon	FV-1	52.47	229	356	—	—	0.40	—	—
Bifenox	FV-2	52.65	341	173	—	—	0.54	—	—
Oxycarboxin	FV-2	53.16	175	267	—	—	0.26	—	—
Phosalone	FV-1	53.43	182	367	—	—	0.08	—	—
Chloridazon	FV-2	53.48	221	220	—	—	0.51	—	—

(Continued)

Table 4.9 (Continued)

Compound	Method	RT (min)	Target	Q1	Q2	Q3	Q/tgt	Q2/tgt	Q3/tgt
Azinphos-methyl	FV-2	53.49	160	132	—	—	0.79	—	—
Nitralin	FV-2	53.59	274	316	—	—	0.95	—	—
cis-Permethrin	FV-1	53.69	183	163	165	—	0.22	0.18	—
Fenarimol	FV-2	53.79	219	139	—	—	2.18	—	—
trans-Permethrin	FV-1	54.05	183	163	165	—	0.30	0.23	—
Pyrazophos	FV-1	54.06	232	221	373	—	3.77	0.20	—
Azinphos-ethyl	FV-1	54.37	160	132	—	—	1.18	—	—
Dialifos	FV-2	54.40	210	208	—	—	3.13	—	—
Cyfluthrin 1	FV-2	55.79	226	206	—	—	1.08	—	—
Prochloraz	FV-2	55.80	180	308	—	—	0.33	—	—
Coumaphos	FV-2	56.02	362	210	—	—	1.20	—	—
Cypermethrin 1	FV-1	56.03	181	163	—	—	1.59	—	—
Cyfluthrin 4	FV-2	56.33	226	206	—	—	1.24	—	—
Cypermethrin 4	FV-1	56.58	181	163	—	—	1.91	—	—
Fenvalerate 1	FV-2	57.67	167	225	419		0.44	0.06	—
Fenvalerate 2	FV-2	58.16	167	225	419		0.44	0.06	—
Deltamethrin	FV-2	59.50	181	251	—	—	0.51	—	—

degr. = degradation product
(Fillion et al., 2000/Oxford University Press.)

4.8.2 Analysis Conditions

Sample material		Fruits, vegetables
Sample preparation	Liquid/liquid extraction	Homogenize 50 g sample with 100 mL MeCN for 5 min, add 10 g NaSO$_4$, homogenize 5 min
	SPE clean-up	Condition a C18 SPE tube with 2 mL MeCN extract, transfer 15 mL MeCN extract onto the cartridge and elute 13 mL, add NaSO$_4$ to 15 mL, shake and centrifuge evaporate 10 mL aliquot (equivalent of 5 g sample) to 0.5 mL transfer extract to the carbon SPE tube, elute with 20 mL MeCN/toluene (3 + 1), evaporate, exchange solvent by adding 2 × 10 mL acetone, add 50 µL ISTD (aldrin 1.0 ng/µL), add acetone to 2.5 mL, transfer 0.5 mL for analysis

GC method	System		Agilent 5890 Series
	Autosampler		Agilent liquid autosampler 7673A
	Column	Type	DB-1701
		Dimensions	30 m length × 0.25 mm ID × 0.15 µm film thickness
	Pre-column		30 cm of the same type is used as a retention gap
	Carrier gas		Helium
	Flow		Constant flow
	SSL injector	Injection mode	Splitless
		Injection temperature	250 °C
		Injection volume	2 µL
	Oven program	Start	70 °C, 2 min
		Ramp 1	25 °C/min to 130 °C
		Ramp 2	2 °C/min to 220 °C
		Ramp 3	10 °C/min to 280 °C
		Final temperature	280 °C, 6.6 min
MS method	System		Agilent MSD 5972A
	Analyzer type		Single quadrupole MS
	Ionization		EI
	Acquisition	Mode	SIM
	Scan rate		0.2 s (5 spectra/s)
Calibration	Internal standard		Aldrin, 1.0 ng/µL

4.8.3 Results

For most of the components, the achieved recovery has been between 70% and 120%. For mirex, S-ethyl-dipropylthiocarbamate (EPTC), butylate, hexachlorobenzene (HCB), folpet, oxycarboxin, and chlorthiamid with recoveries below 50%, the method is considered a screening procedure; for individual recoveries and discussion, see the original reference (Fillion et al., 2000). The published LODs range between 0.02 and 1.0 mg/kg with 80% of the compounds having LODs below 0.04 mg/kg.

Additional compounds have been tested but not included to the method. Benzoximate gave four peaks; chloroxuron, metoxuron, and oxydemeton-methyl gave three peaks of degradation products; fluazinam and flualinate showed many degradation peaks and poor sensitivity; diclone showed poor sensitivity; ditalimfos was not recovered, and vamidothion degraded in solution.

Data analyses have been done by macro-driven automation with spreadsheet report and printouts of the selected ion chromatograms. It is reported that this method typically allows a sample throughput of 42 samples including blanks and spikes per week with results for priority samples within 1 day.

Limitations of the method do not arise from the sensitivity of a single quadrupole MS, which is comparable to a triple quadrupole instrument, but limited from the impaired target ion selectivity of the SIM detection with matrix samples. The extract clean-up is of utmost importance in single quadrupole GC-MS for low real-life LOQs.

4.9 QuEChERSER Analysis of Pesticides

Keywords: pesticides; multi-residue analysis; QuEChERS; QuEChERSER; triple quadrupole MS/MS; structure related selectivity; fast GC; LPGC-MS; quantification; µSPE clean-up; automation

4.9.1 Introduction

The analysis of biological materials, such as plant and animal matter, for pesticide residues by GC-MS has traditionally been difficult due to the very complex sample matrix. The advent of the QuEChERS sample preparation has revolutionized the GC-MS and liquid chromatography-mass spectrometry (LC-MS) analysis not only for pesticides but also for many environmental contaminants (Anastassiades et al., 2003a; Lehotay et al., 2010a; Anastassiades, 2024). QuEChERS is based on the buffered liquid/liquid portioning and extraction using acetonitrile (MeCN) with a dispersive SPE clean-up. The extracts can be used directly for GC-MS or LC-MS injection. Recent developments introduced the automated extraction and an improved clean-up using micro-SPE (µSPE) on industry standard x,y,z-robotic samplers for improved clean-up efficiency (Michlig and Lehotay, 2022). Test portions applied for reproducible quantification went down to 2 g or less after cryohomogenization allowing a seamless automation of the complete workflow (Lehotay et al., 2019; Chong and Huebschmann, 2023).

With the increasing number of pesticides in the assay, it turned out that samples of different pH values are affecting the recovery of some of the pesticides. The further method development led to the U.S. and European standardization:

USA: AOAC Official Method 2007.01 Pesticide Residues in Foods by Acetonitrile Extraction and Partitioning with Magnesium Sulfate (AOAC, 2007).

Europe: EN 15662 Version 2007-10-24 Foods of Plant Origin – Determination of Pesticide Residues Using GC-MS and/or LC-MS/MS Following Acetonitrile Extraction/Partitioning and Clean-up by Dispersive SPE (QuEChERS method) (EN 15662, 2018).

For basic pesticides in low pH matrices, the AOAC method with the acetate buffer of pH 4.8 gives a higher and more consistent recovery, e.g. for chlorothalonil,

Table 4.10 QuEChERS method comparison as initial published, U.S. and EU standard. The differences are highlighted.

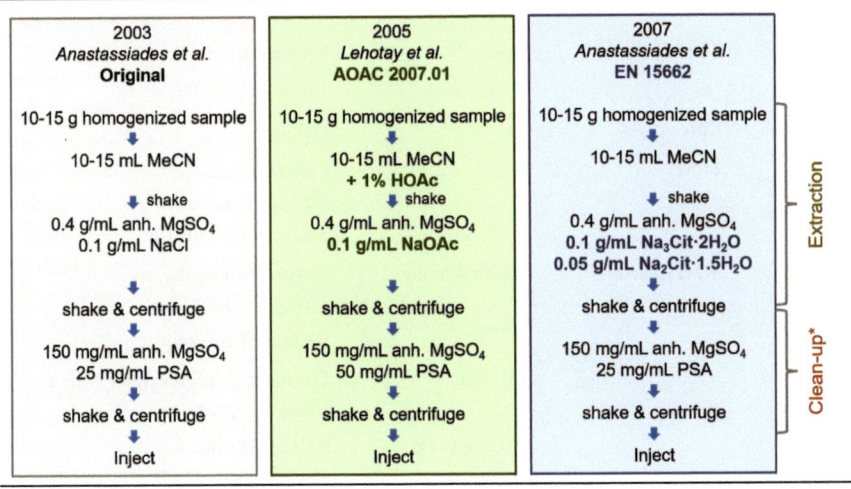

The manual clean-up steps involve optional 50 mg C18 and 2.5–7.5 mg GCB depending on the matrix.

folpet, tolylfluanid, or pymetrozine (Lehotay et al., 2010b). In Europe, Michelangelo Anastassiades chose a citrate buffer with a pH of 5–5.5 for the locally prevailing pesticides where most pesticides, labile under acidic or alkaline conditions, are sufficiently stabilized (Anastassiades et al., 2003b). A comparison of the method variations is outlined in Table 4.10. Commodity specific modifications of the QuEChERS method are made available on the QuEChERS website, e.g. for avocado (Anastassiades, 2024).

4.9.2 Analysis Conditions

Sample material			Food, fruits, avocado, vegetables, herbs, tea, fish, meat, soil, etc.
Sample preparation	Liquid/liquid extraction	QuEChERS, AOAC Official Method 2007.01	10–15 g of cryohomogenized sample Automated clean-up by a micro-SPE (µSPE) cartridge, sorbent mix: 45 mg anh. MgSO$_4$, primary secondary amine (PSA), C18, CarbonX, ratio of 20/12/12/1, w/w/w/w
GC method	System		Agilent 7890A GC
	x,y,z-Robot		GERSTEL MPS3, or PAL RTC
	Guard column	Type	Uncoated restrictor, Restek
		Dimensions	5 m × 0.18 mm ID

	Column	Type	Phenomenex ZB-5MSi
		Dimensions	15 m × 0.53 mm ID × 1 µm film thickness
	Transfer line	Type	Uncoated, Restek
		Dimensions	1 m × 0.53 mm ID
	Carrier gas		Helium
	Flow		Constant flow, 2.25 mL/min, 3 min 1.50 mL/min, 10 min
	MMI injector	Injection mode	Solvent vent mode, 0.31 min vent closed, He flow 50 mL/min
		Inlet liner	Deactivated splitless liner with a glass wool plug
		MMI program	80 °C, 0.31 min, 320 °C/min to 420 °C, 420 °C to end of run
		Injection volume	3 µL extract + 1 µL analyte protectants + 1 µL air plug
		Pressure pulse	40 psi (276 kPa), 0.75 min
	Oven program	Start	80 °C, 1 min
		Ramp 1	45 °C/min to 320 °C
		Final temperature	320 °C, 10 min
	Transferline	Temperature	280 °C
MS method	System		Agilent 7010 MS
	Analyzer type		Triple quadrupole
	Ionization		EI, 70 eV
	Ion source	Temperature	320 °C
	Acquisition	Mode	Dynamic SRM
			Quantifier and qualifier per analyte
	Collision gas		N_2, 1.5 mL/min
	Quench gas		He, 2.25 mL/min
Calibration	External standard	Range	blank, 5–150 ng/mL, 1/10th for PCBs, matrix matched
	Data points		7
Analyte protectants		MeCN/water solution with 1.1% formic acid, containing	Ethylglycerol 25 mg/mL, gulonolactone 2.5 mg/mL, d-sorbitol 2.5 mg/mL, shikimic acid 1.25 mg/mL (Mastovska et al., 2005)

4.9.3 Analysis

After QuEChERS extraction, the raw extracts were automatically cleaned up by µSPE on an x,y,z-robotic sampler equipped with an automated tool change for online GC-MS injection (Figure 4.30). The clean-up workflow used in this application is described in Table 4.11 (Lehotay et al., 2016). Important to note, for improved sample throughputs, the described clean-up procedure only takes few minutes and is executed in the so-called *"prep ahead"* mode during an ongoing GC-MS analysis, being ready for injection with the expected *GC Ready* signal.

A fast low pressure GC-MS (LPGC-MS) analysis was used for the quantification of the compounds (Lehotay et al., 2020). LPGC-MS saves time with only little loss of separation efficiency vs. standard GC-MS, but it provides many important advantages for multi-compound pesticide analyses. The peaks are taller in LPGC. Because of the short analysis time, less peak broadening and tailing is observed; hence, a greater sensitivity can be achieved. Especially those analytes benefit that normally elute at the end of the chromatogram. A 2.5-fold enhancement in signal is common in LPGC-MS. Another aspect is that the greater sample capacity of the thicker film allows larger injection volumes compared to conventional GC columns, which supports lower levels that can be detected. In addition, there is more robustness for sensitive compounds as the fast separation elutes the analytes at lower oven temperatures, and less column bleed provides improved S/N ratios (Lehotay et al., 2020).

The MS/MS detection used with dedicated quantifier and qualifier ions introduces a structure-related selectivity. While single quadrupole and time-of-flight (TOF) MS cannot distinguish coeluting isobaric compounds (ions) with different structures, e.g. pesticide vs. matrix, the triple quadrupole analyzer fragments such isobaric precursor ions in the collision cell and detects only the specific fragments

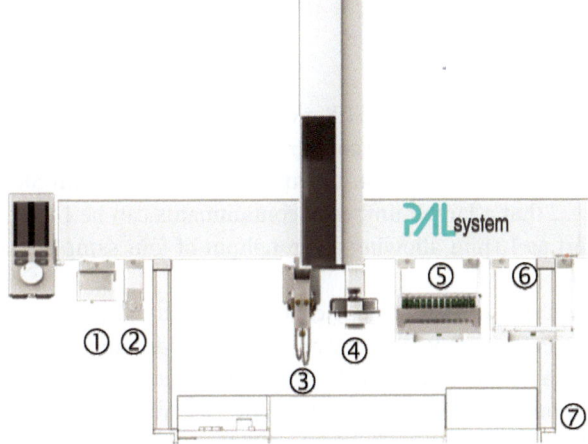

Figure 4.30 Configuration of a PAL x,y,z-robotic sampler for the µSPE clean-up of QuEChERS extracts. ① Tool park station; ② Standard wash station; ③ Fast wash station; ④ Solvent module; ⑤ µSPE clean-up tray holder; ⑥ Tray holder for additional sample vials (optional); ⑦ Mounting kit for GC.

Table 4.11 Workflow for µSPE clean-up of QuEChERS raw extracts (8 min total prep time).

Step	Description
1	Select the µSPE tool with a 1000 µL syringe
2	Wash the 1 mL syringe with MeCN (2 × 500 µL)
3	Pull up the 300 µL raw extract from the Sample Rack into the 1000 µL syringe
4	Transfer the µSPE cartridge by syringe needle transport to the Elution Rack
5	Elute extract through the µSPE cartridge at 2 µL/s
6	Discard the cartridge into waste receptacle
7	Wash the 1000 µL syringe with 1/1/1 MeCN/MeOH/water (2 × 500 µL)
8	Wash the 1000 µL syringe with MeCN (4 × 500 µL)
9	Switch to the 100 µL syringe and wash with MeCN (2 × 50 µL)
10	Add 25 µL MeCN to the eluate collection vial (with glass insert) in rack 2
11	Add 25 µL AP + QC solution to the eluate collection vial (with glass insert) in rack 2
12	Mix the solvents in the collection vial with syringe aspiration/dispense
13	Wash the 100 µL syringe with 1/1/1 MeCN/MeOH/water (5 × 50 µL)
14	Wash the 100 µL syringe with MeCN (3 × 50 µL)
15	Switch to the 1000 µL syringe to prepare the next sample

of the target structure. This explains the substantial improvement in selectivity using triple quadrupole acquisition methods. For quantification in pesticides multi-compound analysis, the selected reaction monitoring (SRM) mode is applied eliminating the unspecific chemical background for a low level detection. The acquisition rate of triple quadrupole instruments is high and accommodates well the fast LPGC-MS setup.

4.9.4 Results

What is more than QuEChERS? The QuEChERSER mega-method for the analysis of pesticides as it was coined by Steven J. Lehotay combines the standardized QuEChERS extraction with the automated µSPE clean-up, and fast LPGC-MS with SRM detection. It is demonstrated that a large number of contaminants can be cleaned up and analyzed in as short as 10 min allowing a throughput of four samples the hour (Sapozhnikova, 2018).

The automated workflow using µSPE cartridge clean-up for fatty and non-fatty foods yielded excellent results for nearly all the tested 252 pesticides and environmental contaminants. An example is shown for the compound acephate spiked at 10 ng/mL in a pork and salmon sample in Figure 4.31. Recoveries are reported in the range of 80–120% in 91% of the cases. The results obtained with septumless µSPE cartridges showed that 93.8% had RSDs <10%, and only 1.4% of them gave RSD >20% (Michlig and Lehotay, 2022).

The application of the QuEChERSER mega-method reaches far beyond pesticides. It also covers in the same setup the analysis of veterinary drugs, and contaminants

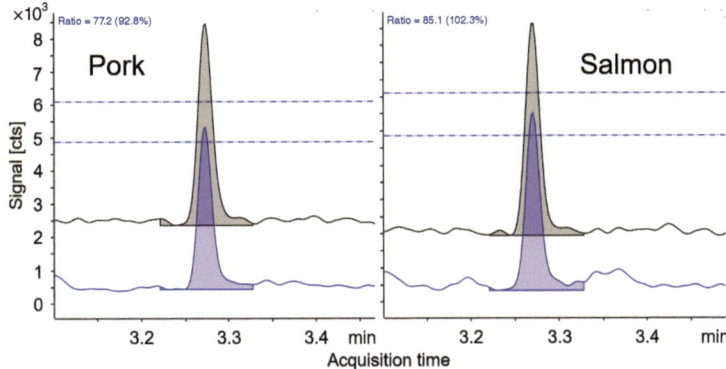

Figure 4.31 Summation integration of the quantifier (grey) and qualifier (purple) ion from 10 ng/mL spikes of acephate in fatty matrices pork and salmon. The horizontal dotted lines indicate the ±10% window of acceptable ion ratio for confirmation. (as of Lehotay et al., 2016/Springer Nature/CC BY 4.0.)

like polychlorinated biphenyls (PCBs), polyaromatic hydrocarbons (PAHs), or polybrominated diphenylethers (PBDEs) in a wide variety of food, even fatty foods, and environmental samples. With this described high potential, the QuEChERSER mega-method addresses the global challenges in food safety analyses.

4.10 Pesticide Analysis with Ethyl Acetate Extraction and Automated Micro-SPE Clean-up

Keywords: pesticides; food; multi-residue analysis; ethyl acetate extraction; automated sample preparation; µSPE clean-up; triple quadrupole MS/MS; confirmation

4.10.1 Introduction

The rising number of samples and demand for swift reporting in multi-residue and multi-matrix pesticide analyses in routine food control require a uniform and capable extraction, and most desired a consistent clean-up procedure. These requirements call for a high degree of automation to replace the traditional multi-step extract clean-up using different sorbent material mixes, and the traditional fat removal from high lipid containing foods, before GC-MS analysis.

Currently, the prevailing extraction methods for routine pesticides analysis are the QuEChERS method using buffered acetonitrile as a solvent (Anastassiades et al., 2003a) and the so-called "Swedish ethyl acetate method" (SweEt) with ethyl acetate as the extraction medium (Ekroth, 2011). The ethyl acetate extraction is based on the works of A. Andersson and B. Ohlin (Andersson and Ohlin, 1989) and was further refined into a multi-residue multi-matrix method (Mol et al., 2007).

Ethyl acetate is especially suitable for the extraction of high sugar commodities since sugar has limited solubility in ethyl acetate. It is also reported that polar compounds such as captan, captafol, folpet, endrine, and iprodione are extracted with high recoveries. Higher coextractives are observed requiring thorough clean-up (Andersson and Palsheden, 1991). In addition, there is a more than 3-fold cost advantage and less toxicological concern with the use of ethyl acetate compared to acetonitrile (Banerjee et al., 2007; Hildmann et al., 2015; Schürmann et al., 2023).

4.10.2 Analysis Conditions

Sample material			Several types of food matrices
Sample preparation			"SweEt" ethyl acetate extraction and µSPE clean-up
GC method	System		Thermo Scientific TRACE GC 1310
	x,y,z-Robot		Thermo Scientific TriPlus RSH smart
	Column	Type	DB-5 ms Ultra Inert
		Dimensions	15 m length × 0.25 mm ID × 0.25 µm film thickness
	Carrier gas		Helium 6.0
	Flow		Constant pressure 70 kPa
	PTV injector	Injection mode	Solvent vent mode
		Liner	Baffled liner without glass wool
		Split flow	30 mL/min
		Injection temperature	55 °C
		Injection volume	3 µL
		Temp ramp	2.5 °C/s to 330 °C for 12 min
	Split	Open	During injection
		Closed	After 6 s, for 3 min
		Open	3 min to end of run
	Oven program	Start	55 °C, 2 min
		Ramp 1	20 °C/min to 165 °C
		Ramp 2	3 °C/min to 205 °C
		Ramp 3	10 °C/min, 3 min
	Transferline	Temperature	290 °C
MS method	System		Thermo Scientific TSQ 9610
	Analyzer type		Triple quadrupole MS
	Ionization		EI, 70 eV
	Ion source	Temperature	220 °C
	Acquisition	Mode	Timed-SRM, 425 transitions for 213 compounds, and TPP as ISTD

4.10.3 Ethyl Acetate Extraction

The samples analyzed were iceberg lettuce, avocadoes, raspberries, ground paprika spice, whole egg, and liver. 10 g test portions of the cryomilled sample were used, for egg 5 g, and for paprika spice and liver 2 g each. The paprika spice powder was soaked with 10 mL of water prior to extraction. The sample portions were transferred into 50 mL tubes with 6 g $MgSO_4$ and 1.5 g sodium acetate. Ten milliliters of ethyl acetate (EtOAc) with 1% acetic acid (HOAc) and the procedural standard triphenyl phosphate (TPP) were added. After shaking for 5 min, the samples were centrifuged. One milliliter of the raw extract was transferred to 2 mL autosampler vials for the subsequent automated µSPE clean-up with online GC injection.

4.10.3.1 Micro-SPE Clean-up

The automation of a clean-up procedure for pesticides using µSPE cartridges was first investigated by Bruce Morris and Richard Schriner in 2014 by replacing the manual dispersive SPE (dSPE) by miniaturized SPE cartridges (Morris and Schriner, 2015). An optimized sorbent mix for GC-MS analysis was found with 45 mg of a mixture of primary secondary amine (PSA), C18EC, CarbonX, and $MgSO_4$ sorbents, as specified in Table 4.12. While µSPE applications typically use acetonitrile, this described analysis procedure uses ethyl acetate as the extraction solvent that is applied to the automated µSPE clean-up of the raw extracts. The customized clean-up procedure prepared for the ethyl acetate extracts is illustrated in Figure 4.32. Due to the installation of the x,y,z-robot on top of the GC, the injection follows online right after clean-up. The *"prep ahead"* mode of the robot allows the clean-up of a next raw extract during the GC-MS analysis run so that all extracts are treated on the same timeline to avoid decomposition during and after contact with the sorbent material improving reproducibility of the recovery of the target analytes.

4.10.4 Results

With the described µSPE clean-up workflow, samples such as eggs, avocados, or liver, which would have formerly been processed with a customized clean-up, can be run in routine without the need of, e.g. a separate freeze-out of fats or a time-consuming GPC separation. Known critical matrices, like spices with a high

Table 4.12 Sorbent material composition used for the µSPE cartridge.

Sorbent	Bedmass	Units	%
PSA	12	mg	27
C18EC	12	mg	27
CarbonX	1	mg	2
$MgSO_4$	20	mg	44
Total	45	mg	100

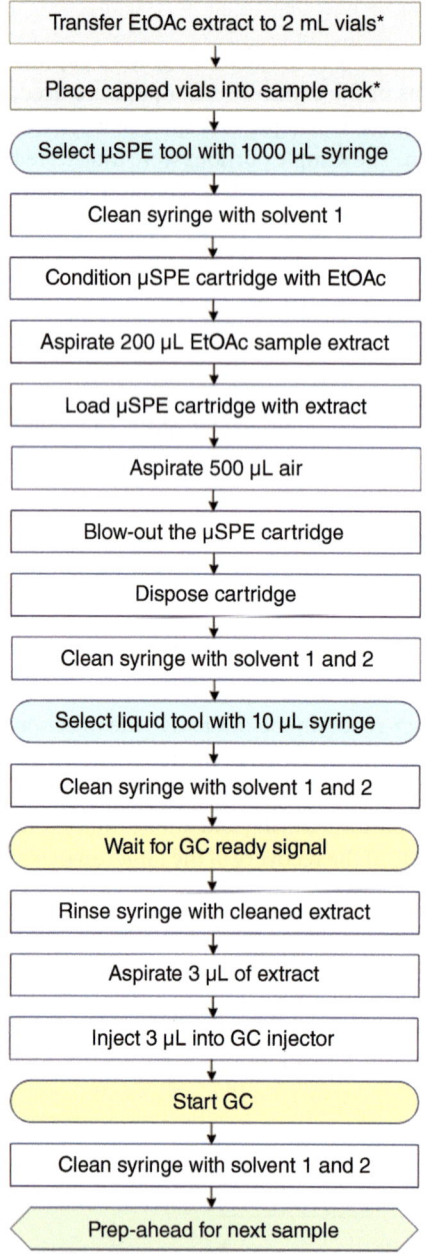

Figure 4.32 Stepwise workflow of the µSPE clean-up for EtOAc pesticide extracts.

content of essential oils, and unfamiliar matrices are treated using the described workflow without alteration.

The validation criteria with a recovery of 70–120% and RSD ≤20% at a concentration level of 0.1 mg/kg were fulfilled in lettuce, raspberries, liver, and eggs for 207, 195, 182, and 177 of 212 compounds. Folpet and captan, two typical but difficult GC analytes, were successfully analyzed. The comparison of the automated µSPE workflow to the earlier manual method using a dispersive SPE clean-up with optimized sorbent mixes for particular food commodities showed very good compliance. The described workflow allowed in routine to use only one type of µSPE cartridge for all matrices. No further optimization of the sorbent mix for different food types was required. This demonstrates that the method is fit for purpose in routine operation with such mixed sample series.

A big advantage for a routine laboratory is the time saved compared to the previous approach with manual dSPE. Even more a time-consuming GPC purification for high lipid matrix containing samples is not necessary and can be avoided.

As a result of the applied online µSPE raw extract clean-up, significantly less matrix was injected to the GC. A liner exchange is preventively performed only once a week after about 100 sample runs, reducing system downtime significantly. Even at that time of change, the liner still appeared to be clean without visible residues. The low matrix burden after the online µSPE raw extract clean-up also shows up with the extended lifetime of the GC column in use. The column got clipped about half a meter only after six months of use and more than 2600 sample analyses run on the system.

The described µSPE workflow is in routine operation and shows high reliability and is also applied for unattended overnight runs, releasing time from earlier manual workload to be used for other duties such as data evaluation and reporting (Schürmann et al., 2023).

4.11 Multi-Residue Pesticides Analysis in Ayurvedic Churna

Keywords: pesticides; multi-residue analysis; food safety; traditional herbal medicine; spices; liquid/liquid extraction; timed-SRM; triple quadrupole MS/MS

4.11.1 Introduction

Ayurveda is a Sanskrit term, made up of the words "ayus" and "veda" meaning life and science, together translating to "science of life". A blend of several herbs and spices makes up the powdered mixture known as "churna". Depending on its intended use for medicinal, beauty, or culinary purpose, the recipe varies. Avipittakara "churna" is a traditional ayurvedic formula used widely and almost daily to control vitiated pitta dosha, remove heat in the digestive system, and control indigestion, constipation, vomiting, and anorexia. A major analytical challenge for these types of samples is mainly the addition of multiple herbs with sugar and the natural color of herbs (Narayanaswamy, 1981; Lohar, 2008).

The dried leaves result in highly complex extracts from the sample preparation due to the rich content of active ingredients, essential oils, and the typical high boiling natural polymer compounds. Owing to the use of pesticides in the fresh herbs, the "churna" may contain residual pesticides. Analysis of pesticide residues in ayurvedic churna is governmentally controlled. Strict quality parameters have been implemented to preserve the quality and efficacy of these "churnas".

4.11.2 Analysis Conditions

Sample material			Ayurvedic churna, herbs, spices
Sample preparation	Liquid/liquid extraction		15 g powdered sample
			15 mL acetonitrile containing 1% acetic acid
			Vortex with 3 g $MgSO_4$, 1.5 g NaOAc, 10 g NaCl, centrifuge
			Shake 1 mL supernatant with 200 mg PSA and 10 mg GCB, centrifuge
			Inject from supernatant
GC method	System		Thermo Scientific™ TRACE 1300 GC
	Autosampler		Thermo Scientific™ TriPlus RSH
	Column	Type	TraceGOLD™ TG-5SilMS
		Dimensions	30 m length × 0.25 mm ID × 0.25 µm film thickness
	Carrier gas		Helium
	Flow		Constant flow, 1.2 mL/min
	PTV injector	Inlet liner	Baffled, Siltek™ deactivated
		Injection mode	Splitless
		Injection volume	3 µL
		Base temperature	87 °C, 0.3 min
		Transfer rate	14.5 °C/s
		Transfer temperature	285 °C
		Transfer time	2.5 min
		Cleaning ramp	14.5 °C/s
		Cleaning phase	20 min
		Final temperature	290 °C
	Split	Closed until	3 min
		Open	3 min to end of run
		Split flow	30 mL/min
		Purge flow	5 mL/min
	Oven program	Start	70 °C, 2 min
		Ramp	10 °C/min to 285 °C
		Final temperature	285 °C, 8.5 min
	Transferline	Temperature	285 °C
			Direct coupling

MS method	System		Thermo Scientific TSQ 8000
	Analyzer type		Triple quadrupole MS
	Ionization		EI
	Electron energy		70 eV
	Ion source	Temperature	230 °C
	Acquisition	Mode	Timed-SRM
			2 SRMs per analyte
	Collision gas		Ar, 1.5 mL/min
	SRM method		See Table 4.15
	Scan rate		0.2 s (5 spectra/s)
	Resolution	Setting	Normal (0.7 Da)
Calibration	External standard	Range	2.5–50 µg/L
	Data points		5

4.11.3 Sample Preparation

The sample preparation involves the extraction of a 15 g powdered sample of Avipittakara "churna" with 15 mL acetonitrile containing 1% acetic acid in the presence of 3 g $MgSO_4$, 1.5 g sodium acetate, and 10 g NaCl. One microliter of the supernatant was collected after centrifugation, and a dSPE clean-up was performed using 200 mg PSA and 10 mg graphitized carbon black (GCB). The extract then centrifuged at 10 000 rpm for 5 min. Three microliter of the supernatant was injected to GC-MS. For recovery and validation studies, 15 g of a confirmed blank "churna" matrix is fortified with appropriate quantities of a pesticide standard mixture.

4.11.4 Experimental Conditions

The analytical method comprises a triple quadrupole GC-MS/MS system with a liquid autosampler. The gas chromatograph is equipped with a temperature programmable PTV injector for splitless injection. The separation is performed using a 5% phenyl phase column of standard dimensions. The Thermo Scientific TraceFinder™ software with compound database of pesticides was used for method setup and data processing.

For all pesticide compounds, two SRM transitions have been chosen for the acquisition method. The first transition was used for quantitation and the second transition for confirmation by checking the ion intensity ratio during data processing, see Table 4.13.

Table 4.13 SRM transitions for quantitation and confirmation, optimized collision energy (CE), and the precision of the calibration (R^2) of matrix spike standards.

No.	Compound name	RT (min)	Quantitation (m/z)	CE (V)	Confirmation (m/z)	CE (V)	R^2
1.	Diflubenzuron (degr. i-cyanat)	5.24	153.02 > 90.01	20	153.02 > 125.01	20	0.9969
2.	Diflubenzuron (degr. aniline)	5.75	127.01 > 65.01	30	127.01 > 100.01	30	0.9949
3.	Methamidophos	5.87	141.00 > 95.00	10	141.00 > 126.00	5	0.9930
4.	Dichlorphos (DDVP)	5.94	184.95 > 92.98	17	219.95 > 184.95	10	0.9960
5.	Dichlobenil	6.82	135.97 > 99.98	10	170.96 > 135.97	15	0.9960
6.	Mevinphos	7.39	127.03 > 109.02	10	192.04 > 127.03	12	0.9964
7.	Acephate	7.50	136.01 > 42.00	10	136.01 > 94.01	15	0.9904
8.	Dichloraniline, 3,5-	7.61	160.98 > 89.99	25	160.98 > 98.99	25	0.9989
9.	Molinate (Ordram)	8.58	126.07 > 55.03	10	187.10 > 126.07	10	0.9941
10.	TEPP	8.60	263.06 > 179.04	15	263.06 > 235.06	5	0.9946
11.	Omethoate	9.00	110.01 > 79.01	15	156.02 > 110.01	10	0.9969
12.	Fenobucarb	9.11	121.07 > 77.05	15	150.09 > 121.07	10	0.9977
13.	Propoxur	9.13	110.06 > 64.03	10	152.08 > 110.06	10	0.9981
14.	Propachlor	9.16	176.06 > 120.04	10	196.07 > 120.04	10	0.9980
15.	Ethoprophos	9.38	158.00 > 80.90	15	158.00 > 114.00	5	0.9949
16.	Trifluralin	9.58	264.09 > 160.05	15	306.10 > 264.09	15	0.9944
17.	Chlorpropham	9.62	213.00 > 127.00	5	213.00 > 171.00	5	0.9981
18.	Benfluralin	9.63	292.10 > 160.05	21	292.10 > 264.09	10	0.9923
19.	Sulfotep	9.70	322.02 > 202.01	15	322.02 > 294.02	10	0.9943
20.	Bendiocarb	9.72	166.06 > 151.06	15	166.06 > 166.06	15	0.9996
21.	Monocrotophos	9.80	127.03 > 95.03	20	127.03 > 109.03	25	0.9971
22.	Methabenzthiazuron	9.82	164.05 > 136.04	12	164.05 > 164.05	10	0.9974
23.	HCH, alpha	10.15	180.91 > 144.93	15	218.89 > 182.91	15	0.9970
24.	Metamitron	10.36	202.09 > 174.07	5	202.09 > 186.08	10	0.9969
25.	Atrazine	10.54	215.09 > 173.08	10	215.09 > 200.09	10	0.9945
26.	Pencycuron	10.62	125.05 > 89.04	12	180.07 > 125.05	12	0.9914
27.	Dioxathion	10.72	125.00 > 97.00	15	125.00 > 141.00	15	0.9936
28.	HCH, beta	10.73	180.91 > 144.93	15	218.89 > 182.91	15	0.9933
29.	Propetamphos	10.74	236.07 > 166.05	15	236.07 > 194.06	5	0.9918
30.	HCH, gamma (Lindane)	10.81	180.91 > 144.93	15	218.89 > 180.91	5	0.9939
31.	Terbuthylazine	10.84	214.10 > 132.06	10	229.11 > 173.08	10	0.9935
32.	Diazinon	10.88	137.05 > 84.03	10	304.10 > 179.06	15	0.9987

Table 4.13 (Continued)

No.	Compound name	RT (min)	Quantitation (m/z)	CE (V)	Confirmation (m/z)	CE (V)	R^2
33.	Propyzamide	10.93	173.01 > 145.01	15	175.02 > 147.01	15	0.9939
34.	Fluchloralin	10.95	264.04 > 206.03	10	306.05 > 264.04	10	0.9967
35.	Pyroquilon	11.07	173.08 > 130.06	20	173.08 > 145.07	20	0.9974
36.	Pyrimethanil	11.11	198.11 > 158.09	30	198.11 > 183.10	15	0.9953
37.	Tefluthrin	11.16	177.02 > 127.02	20	197.03 > 141.02	15	0.9991
38.	Etrimfos	11.29	292.06 > 153.03	10	292.06 > 181.04	10	0.9935
39.	Pirimicarb	11.50	166.10 > 96.06	10	238.14 > 166.10	15	0.9937
40.	HCH, delta	11.54	180.91 > 144.93	15	204.07 > 91.03	15	0.9949
41.	Iprobenfos	11.54	204.07 > 122.04	15	218.89 > 182.91	15	0.9997
42.	Formothion	11.74	126.00 > 93.00	8	172.00 > 93.00	5	0.9982
43.	Phosphamidon II	11.83	227.05 > 127.03	15	264.06 > 193.04	15	0.9977
44.	Dichlofenthion	11.90	222.98 > 204.98	10	278.97 > 222.98	15	0.9946
45.	Dimethachlor	11.94	197.08 > 148.06	10	199.08 > 148.06	10	0.9992
46.	Dimethenamid	11.95	230.06 > 154.04	10	232.06 > 154.04	10	0.9953
47.	Propazine	12.02	214.09 > 172.08	12	214.09 > 214.09	10	0.9970
48.	Propanil	12.06	217.01 > 161.00	10	219.01 > 163.00	10	0.9934
49.	Malaoxon	12.07	127.02 > 99.02	10	127.02 > 109.02	20	0.9978
50.	Chlorpyrifos-methyl	12.08	124.96 > 78.97	10	285.91 > 92.97	20	0.9945
51.	Metribuzin	12.13	198.08 > 82.03	20	198.08 > 110.05	20	0.9997
52.	Spiroxamine I	12.15	100.09 > 58.05	15	100.09 > 72.06	15	0.9909
53.	Vinclozolin	12.16	212.00 > 172.00	15	285.00 > 212.00	15	0.9957
54.	Carbofuran, 3-hydroxy	12.21	137.06 > 81.03	18	180.08 > 137.06	15	0.9974
55.	Parathion-methyl	12.22	263.00 > 109.00	15	263.00 > 246.00	15	0.9966
56.	Alachlor	12.23	161.07 > 146.06	12	188.08 > 160.07	10	0.9997
57.	Tolclofos-methyl	12.25	264.96 > 92.99	20	264.96 > 249.96	15	0.9932
58.	Propisochlor	12.31	162.08 > 144.07	10	223.11 > 147.07	10	0.9983
59.	Metalaxyl	12.37	249.13 > 190.10	10	249.13 > 249.13	5	0.9911
60.	Carbaryl	12.41	144.06 > 115.05	20	144.06 > 116.05	20	0.9919
61.	Fuberidazol	12.41	183.80 > 156.10	10	183.80 > 183.10	20	0.9902
62.	Fenchlorphos (Ronnel)	12.47	284.91 > 269.92	13	286.91 > 271.91	20	0.9994
63.	Prosulfocarb	12.63	100.00 > 72.00	10	128.00 > 43.10	5	0.9938
64.	Pirimiphos-methyl	12.66	290.09 > 233.07	10	305.10 > 290.09	15	0.9911
65.	Spiroxamine II	12.75	100.09 > 58.05	15	100.09 > 72.06	15	0.9916
66.	Ethofumesate	12.80	207.08 > 161.06	10	277.02 > 109.01	8	0.9907

(Continued)

Table 4.13 (Continued)

No.	Compound name	RT (min)	Quantitation (m/z)	CE (V)	Confirmation (m/z)	CE (V)	R^2
67.	Fenitrothion Confirming 1	12.80	277.02 > 260.02	10	286.11 > 207.08	12	0.9997
68.	Methiocarb	12.84	168.06 > 109.04	15	168.06 > 153.06	15	0.9971
69.	Malathion	12.92	127.01 > 99.01	10	173.02 > 127.01	10	0.9951
70.	Dichlofluanid	12.95	223.97 > 122.99	15	225.97 > 122.99	15	0.9971
71.	Phorate sulfone	13.01	153.00 > 125.00	5	199.00 > 143.00	10	0.9942
72.	Dipropetryn	13.02	241.90 > 149.80	20	254.90 > 180.30	20	0.9906
73.	Chlorpyrifos (-ethyl)	13.12	198.96 > 170.96	15	313.93 > 285.94	12	0.9995
74.	Fenthionoxon	13.22	277.80 > 109.10	25	329.60 > 298.90	10	0.9927
75.	Chlorthal-dimethyl (DCPA)	13.24	300.91 > 300.91	15	331.90 > 300.91	15	0.9986
76.	Flufenacet	13.26	211.04 > 123.02	10	211.04 > 183.03	10	0.9959
77.	Endosulfan I (alpha)	13.43	240.89 > 205.91	20	264.88 > 192.91	22	0.9942
78.	Imazethapyr	13.49	201.90 > 133.00	15	252.00 > 145.90	20	0.9944
79.	Butralin	13.50	266.14 > 190.10	15	266.14 > 220.11	15	0.9996
80.	Pirimiphos (-ethyl)	13.54	304.12 > 168.06	15	333.13 > 318.12	15	0.9992
81.	Pendimethalin	13.86	252.12 > 162.08	12	252.12 > 191.09	12	0.9912
82.	Fipronil	13.87	212.97 > 177.98	16	366.95 > 212.97	25	0.9938
83.	Cyprodinil	13.91	224.13 > 208.12	20	225.13 > 210.12	18	0.9959
84.	Metazachlor	13.92	133.05 > 117.04	20	209.07 > 132.05	12	0.9939
85.	Penconazole	14.01	248.06 > 157.04	25	248.06 > 192.04	15	0.9977
86.	Tolylfluanid	14.05	137.05 > 91.03	20	238.09 > 137.05	15	0.9922
87.	Chlorfenvinphos-Z	14.05	266.98 > 158.99	15	322.97 > 266.98	15	0.9904
88.	Allethrin	14.06	123.08 > 81.05	10	136.08 > 93.06	10	0.9923
89.	Mecarbam	14.09	226.04 > 198.03	5	329.05 > 160.03	10	0.9979
90.	Phenthoate	14.18	146.01 > 118.01	10	274.03 > 246.02	10	0.9951
91.	Mephosfolan	14.20	196.02 > 140.02	15	196.02 > 168.02	10	0.9973
92.	Quinalphos	14.21	146.03 > 118.02	15	157.03 > 129.02	13	0.9943
93.	Triflumizole	14.31	179.04 > 144.04	15	206.05 > 179.04	15	0.9925
94.	Procymidone	14.31	283.02 > 96.01	15	283.02 > 255.02	10	0.9983
95.	Bromophos-ethyl	14.50	358.89 > 302.91	20	358.89 > 330.90	10	0.9985
96.	Methidathion	14.60	124.98 > 98.99	22	144.98 > 84.99	10	0.9945
97.	Chlordane, alpha (cis)	14.62	372.81 > 265.87	18	374.81 > 267.87	15	0.9967
98.	DDE, o,p	14.63	245.95 > 175.97	25	317.94 > 245.95	20	0.9946
99.	Sulfallate	14.68	188.02 > 132.02	22	188.02 > 160.02	16	0.9945

Table 4.13 (Continued)

No.	Compound name	RT (min)	Quantitation (m/z)	CE (V)	Confirmation (m/z)	CE (V)	R^2
100.	Paclobutrazol	14.72	236.10 > 125.06	15	236.10 > 167.07	15	0.9926
101.	Disulfoton sulfone	14.74	213.01 > 125.01	10	213.01 > 153.01	5	0.9912
102.	Picoxystrobin	14.77	303.09 > 157.04	20	335.09 > 303.09	10	0.9937
103.	Endosulfan II (beta)	14.88	271.88 > 236.89	18	338.85 > 265.88	15	0.9973
104.	Mepanipyrim	14.89	222.11 > 207.10	15	223.11 > 208.10	15	0.9965
105.	Chlordane, gamma (trans)	14.89	372.81 > 265.87	18	374.81 > 267.87	15	0.9991
106.	Flutriafol	14.97	123.04 > 75.03	15	219.07 > 123.04	15	0.9915
107.	Napropamide	15.00	128.07 > 72.04	10	271.16 > 128.07	5	0.9972
108.	Flutolanil	15.03	173.06 > 145.05	15	173.06 > 173.06	15	0.9988
109.	Pretilachlor	15.13	162.09 > 147.08	15	216.05 > 174.04	20	0.9935
110.	Hexaconazole, confirming 1	15.13	231.06 > 175.04	10	262.14 > 202.11	15	0.9962
111.	Isoprothiolane	15.14	290.06 > 118.03	15	290.06 > 204.05	15	0.9961
112.	Profenofos	15.21	138.98 > 96.98	8	338.94 > 268.95	20	0.9939
113.	Oxadiazon	15.26	258.05 > 175.04	10	304.06 > 260.05	10	0.9927
114.	DDE, p,p	15.32	245.95 > 175.97	25	317.94 > 245.95	20	0.9964
115.	Myclobutanil	15.40	179.07 > 125.05	15	179.07 > 152.06	15	0.9912
116.	Buprofezin	15.43	172.09 > 57.03	10	249.13 > 193.10	10	0.9906
117.	Kresoxim-methyl	15.44	206.09 > 116.05	15	206.09 > 131.06	15	0.9921
118.	DDT, o,p′	15.47	234.94 > 164.96	15	234.97 > 164.98	20	0.9935
119.	DDT, o,p′, confirming 1	15.47	236.94 > 164.96	20	236.97 > 164.98	20	0.9963
120.	Aramite-1-1	15.48	185.06 > 63.02	15	319.10 > 185.06	15	0.9959
121.	Aramite-2	15.69	185.06 > 63.02	15	319.10 > 185.06	15	0.9971
122.	Carpropamid	15.78	139.00 > 103.10	10	222.00 > 125.00	18	0.9982
123.	Cyproconazole	15.79	222.09 > 125.05	20	224.09 > 127.05	20	0.9989
124.	Nitrofen	15.85	201.99 > 138.99	21	282.98 > 252.98	15	0.9997
125.	Chlorobenzilate	15.98	251.02 > 139.01	20	253.03 > 141.01	15	0.9978
126.	Oxadiargyl	15.99	149.90 > 122.90	15	285.00 > 255.00	14	0.9963
127.	Fenthion sulfoxide	16.05	279.01 > 153.01	15	294.02 > 279.01	8	0.9958
128.	Diniconazole	16.11	268.06 > 232.05	15	270.06 > 234.05	15	0.9949
129.	Ethion	16.12	230.99 > 202.99	15	383.99 > 230.99	10	0.9973
130.	Oxadixyl	16.16	132.06 > 117.05	15	163.07 > 132.06	10	0.9985
131.	DDT, p,p′	16.20	234.94 > 164.96	20	234.94 > 164.96	20	0.9979
132.	DDD, p,p′	16.20	234.97 > 164.98	20	236.97 > 164.98	20	0.9959

(Continued)

Table 4.13 (Continued)

No.	Compound name	RT (min)	Quantitation (m/z)	CE (V)	Confirmation (m/z)	CE (V)	R^2
133.	Chlorthiophos1	16.20	324.96 > 268.97	15	324.96 > 296.97	10	0.9969
134.	Imiprothrin	16.36	123.00 > 81.00	5	324.90 > 269.20	14	0.9967
135.	Mepronil	16.45	269.14 > 119.06	10	269.14 > 210.11	10	0.9945
136.	Triazophos	16.46	161.03 > 134.03	10	257.05 > 162.03	10	0.9936
137.	Ofurace	16.58	186.05 > 158.05	10	232.07 > 186.05	10	0.9973
138.	Carfentrazone-ethyl	16.59	330.03 > 310.03	20	340.03 > 312.03	10	0.9919
139.	Benalaxyl	16.63	234.12 > 174.09	10	266.14 > 148.08	10	0.9951
140.	Trifloxystrobin	16.65	116.04 > 89.03	15	190.06 > 130.04	10	0.9962
141.	Propiconazole, peak 1	16.77	259.02 > 69.01	20	259.02 > 173.02	20	0.9989
142.	Edifenphos	16.78	173.01 > 109.01	15	310.03 > 173.01	10	0.9904
143.	Quinoxyfen	16.84	272.00 > 237.00	20	307.00 > 272.00	10	0.9982
144.	Endosulfan sulfate	16.85	271.88 > 236.89	15	273.88 > 238.89	15	0.9929
145.	Clodinafop-propargyl	16.87	349.05 > 238.04	15	349.05 > 266.04	15	0.9991
146.	Fluopicolide	16.90	208.80 > 182.00	20	261.00 > 175.00	24	0.9988
147.	Hexazinone	17.02	171.00 > 71.00	10	171.00 > 85.00	10	0.9998
148.	Propargite	17.16	135.06 > 107.05	15	350.16 > 201.09	10	0.9991
149.	Diflufenican	17.21	266.05 > 246.05	10	394.07 > 266.05	10	0.9981
150.	Triphenyl phosphate (TPP)	17.26	325.07 > 169.04	25	326.07 > 325.07	10	0.9995
151.	Iprodione	17.65	187.02 > 124.01	20	187.02 > 159.02	40	0.9979
152.	Bifenthrin	17.77	181.05 > 153.05	6	181.05 > 166.05	15	0.9922
153.	Picolinafen	17.90	376.08 > 238.05	15	376.08 > 239.05	15	0.9981
154.	Bromopropylate	17.91	184.98 > 156.98	20	342.96 > 184.98	20	0.9967
155.	Fenoxycarb	17.93	186.08 > 186.08	10	255.11 > 186.08	10	0.9933
156.	Fenpropathrin	18.01	181.09 > 152.07	23	265.13 > 210.10	15	0.9956
157.	Fenamidone	18.10	238.08 > 237.08	20	268.09 > 180.06	20	0.9994
158.	Tebufenpyrad	18.11	276.13 > 171.08	15	333.16 > 276.13	10	0.9997
159.	Fenazaquin	18.23	145.08 > 117.07	15	160.09 > 117.07	20	0.9951
160.	Imazalil	18.25	173.03 > 145.02	20	215.04 > 173.03	15	0.9954
161.	Furathiocarb	18.27	163.07 > 107.04	10	325.13 > 194.08	10	0.9989
162.	Flurtamone	18.38	199.06 > 157.05	20	333.10 > 120.04	15	0.9945
163.	Tetradifon	18.46	226.93 > 198.94	18	353.88 > 158.95	15	0.9973
164.	Phosalone	18.54	181.99 > 111.00	15	181.99 > 138.00	10	0.9985
165.	Triticonazole	18.57	217.09 > 182.07	10	235.10 > 217.09	10	0.9945
166.	Pyriproxyfen	18.68	136.06 > 78.03	15	136.06 > 96.04	15	0.9941

Table 4.13 (Continued)

No.	Compound name	RT (min)	Quantitation (m/z)	CE (V)	Confirmation (m/z)	CE (V)	R^2
167.	Cyhalofop butyl	18.70	256.10 > 120.05	10	256.10 > 256.10	10	0.9969
168.	Tralkoxydim	18.80	137.00 > 57.20	10	181.04 > 152.03	23	0.9995
169.	Cyhalothrin, lambda	18.80	197.04 > 141.03	15	234.90 > 217.20	15	0.9997
170.	Lactofen	18.83	344.04 > 223.02	15	344.04 > 300.03	15	0.9975
171.	Benfuracarb	19.03	164.08 > 149.07	10	190.09 > 144.07	10	0.9975
172.	Pyrazophos	19.05	221.05 > 193.04	10	232.05 > 204.05	10	0.9930
173.	Fenarimol	19.15	139.01 > 111.01	15	219.02 > 107.01	15	0.9993
174.	Azinphos-ethyl	19.20	132.01 > 77.01	20	160.02 > 132.01	5	0.9944
175.	Fenoxaprop-*P*	19.41	288.03 > 260.03	10	361.04 > 288.03	10	0.9998
176.	Bitertanol1	19.59	170.09 > 115.06	25	170.09 > 141.07	20	0.9993
177.	Permethrin, peak 1	19.68	183.04 > 165.03	15	183.04 > 168.03	15	0.9973
178.	Bitertanol2	19.71	170.09 > 115.06	25	170.09 > 141.07	20	0.9993
179.	Permethrin, peak 2	19.81	183.04 > 165.03	15	183.04 > 168.03	15	0.9909
180.	Prochloraz	19.88	180.01 > 138.01	15	310.03 > 268.02	10	0.9932
181.	Cafenstrole	20.21	100.04 > 72.03	15	188.08 > 119.05	15	0.9991
182.	Cyfluthrin, peak 1	20.26	163.02 > 91.01	12	163.02 > 127.02	10	0.9915
183.	Fenbuconazole	20.34	129.04 > 102.03	15	198.07 > 129.04	10	0.9996
184.	Cypermethrin I	20.65	163.03 > 127.02	10	181.03 > 152.03	25	0.9996
185.	Boscalid (Nicobifen)	20.84	342.03 > 140.01	15	344.03 > 142.01	15	0.9977
186.	Flucythrinate, peak 1	20.85	199.07 > 107.04	22	199.07 > 157.06	10	0.9958
187.	Quizalofop-Ethyl	20.92	299.07 > 255.06	20	372.09 > 299.07	15	0.9969
188.	Etofenprox	21.08	163.09 > 107.06	16	163.09 > 135.07	10	0.9987
189.	Flucythrinate, peak 2	21.12	199.07 > 107.04	22	199.07 > 157.06	10	0.9989
190.	Fenvalerate, peak 1	21.94	167.05 > 125.04	10	419.13 > 225.07	10	0.9978
191.	Fluvalinate, peak 1	22.09	250.06 > 200.05	20	252.06 > 200.05	20	0.9973
192.	Pyraclostrobin	22.17	132.03 > 77.02	15	325.08 > 132.03	20	0.9936
193.	Fluvalinate, peak 2	22.20	250.06 > 200.05	20	252.06 > 200.05	20	0.9977
194.	Fenvalerate, peak 2	22.28	167.05 > 125.04	10	419.13 > 225.07	10	0.9996
195.	Difenoconazole, peak 1	22.76	323.05 > 265.04	15	325.05 > 267.04	20	0.9995
196.	Indoxacarb	22.95	203.03 > 106.01	20	203.03 > 134.02	20	0.9996
197.	Deltamethrin II	23.28	252.99 > 93.00	18	252.99 > 173.99	18	0.9987
198.	Azoxystrobin	23.63	344.10 > 329.10	20	388.11 > 345.10	15	0.9991
199.	Dimethomorph-1	23.91	301.10 > 165.05	10	387.12 > 301.10	12	0.9992
200.	Dimethomorph-2	24.60	301.10 > 165.05	10	387.12 > 301.10	12	0.9990

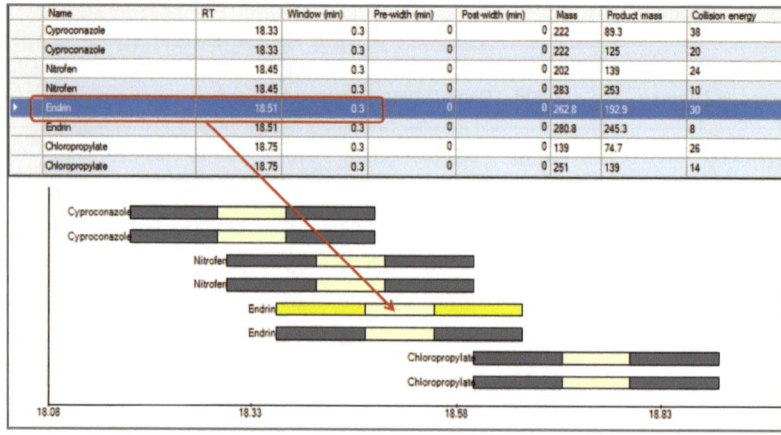

Figure 4.33 Principle of the timed-SRM acquisition method in the retention time range of 10.29–11.04 min. The white center parts of the individual compound acquisition windows stand for the pesticide peak width centered to the compound retention time, the grey areas before and after the peak indicate the width of the SRM acquisition window. Yellow highlighted the beta HCH t-SRM window.

The RTs were synchronized between data processing of QC (quality control) standard samples with the acquisition method for the timed-SRM protocol (Figure 4.33) in order to update all compound RTs to keep track of potential RT variations due to a matrix impact from real-life samples.

4.11.5 Results

All the 200 compounds included in the described method showed a precise calibration with good calibration correlation coefficients of >0.99 in the concentration range of 2.5–50 ng/g, as shown in Figure 4.34. The obtained recoveries from spiked blank churna samples are reported high within 70–120% with <20% associated RSDs.

The method setup as describe above was applied for samples bought from the regional market. The analysis of market samples indicated a contamination with chlorpyriphos at 2.3 µg/kg and kresoxim at 2.7 µg/kg (Figure 4.35) (Surwade et al., 2013).

4.11.6 Conclusions

The described method offers the routine screening for a large number of pesticides in herbs, spices, and similar products using a quick and fast sample preparation with a relatively short total analysis time of c. 28 min. This method can be utilized for detection and confirmation of trace amounts of pesticides in difficult matrices such as herbal churnas or traditional medicine preparations.

A good linearity, specificity, recovery, and repeatability of the method were established with minimal sample preparation time. GC-MS/MS as the detection method provided a very high selectivity for the sensitive detection and quantitation

4.11 Multi-Residue Pesticides Analysis in Ayurvedic Churna | 613

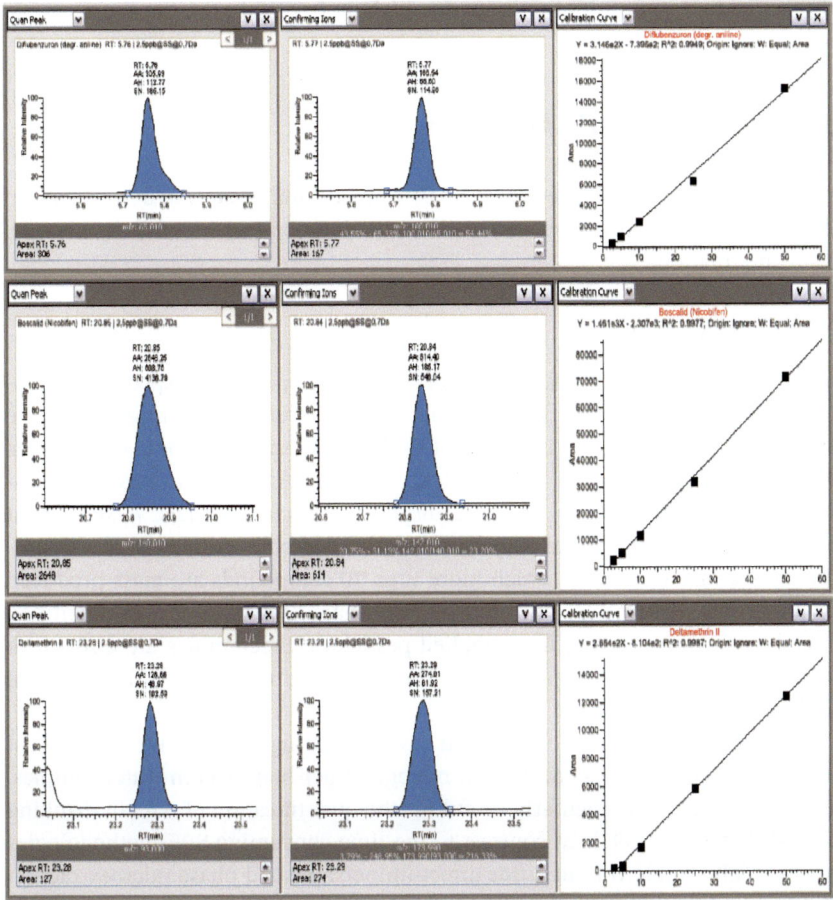

Figure 4.34 Selected pesticide chromatograms of diflubenzuron (degr. aniline), boscalid (nicobifen), and deltamethrin II at 2.5 ng/g with quantitation peaks, confirming ions and calibration curves.

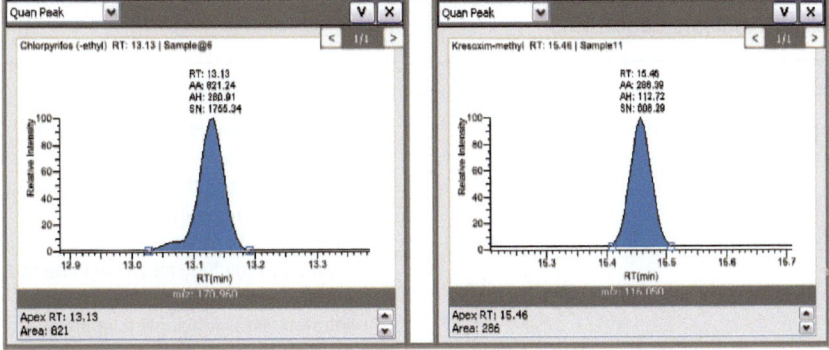

Figure 4.35 Traces of chlorpyrifos ethyl and kresoxim methyl were detected at 2.3 and 2.7 µg/kg, respectively, in regional market samples.

of the pesticides even from samples with a high matrix load from the short sample preparation.

4.12 Polar Aromatic Amines by SPME

Keywords: nitroaromatic compounds; NAC; aromatic amines; *in situ* derivatization; iodination; DI-SPME; drinking water; EU regulation; ion trap MS; full scan confirmation, automation.

4.12.1 Introduction

Aromatic amines (AAs) are known to react with DNA, posing a suspected carcinogenic risk. Six AAs are classified as carcinogenic or probably carcinogenic by the International Agency for the Research of Cancer. The source of these compounds is manifold. The major use is in the polyurethane production with possible emissions during the production, use, and disposal of such materials. A significant contribution to environmental distribution is the microbial reduction of nitroaromatic compounds (NACs) of which more than 70 compounds are mass produced with more than 1000 mt/yr. Until today, more than 30 AAs have been identified as metabolites, also deriving from applied pesticides. Free amines are difficult to extract from sample matrices and due to their basic nature are difficult to analyze by GC directly.

The described quantification method provides a highly sensitive and robust method for monitoring AAs at the low nanogram per liter level in water samples. The SPME method is particularly useful for the quantification of AAs in drinking water (Müller et al., 1997). In contrast to the time-consuming SPE methodologies requiring high sample volumes, this direct-immersion solid phase microextraction (DI-SPME) method utilizes *in situ* derivatization of the polar compounds in aqueous solution (Zimmermann et al., 2004; Pan et al., 1997) forming iodinated apolar derivatives. The automated SPME GC-MS procedure allows the routine quantification and full scan confirmation of large series of water samples.

4.12.2 Analysis Conditions

Sample material	Drinking water, groundwater
Sample preparation *In situ* derivatization	10 mL water sample, acidified 0.2 mL of hydroiodic acid, shaken with 0.5 mL NaNO$_3$ solution (10 g/L), 20 min, add 1 mL amidosulfonic acid solution (50 g/L), shake for 45 min, heat the solution to 100 °C, 5 min, cool to room temperature, add 0.25 mL NaSO$_3$ solution (sat.), adjust pH to approx. 8 with 0.25 mL K$_2$HPO$_4$ (0.25 mol/L) and 0.4 mL NaOH (5 mol/L), fill solution into 13 mL crimp top vials without headspace for analysis.

4.12 Polar Aromatic Amines by SPME

	Extraction	SPME	Supelco PDMS/DVB fiber, 65 μm
		Immersion	30 min into aqueous sample while shaking, use "prep ahead" mode during GC runtime
	Desorption	Desorption time	5 min
		Temperature program	50 °C initial temperature
			250 °C final temperature after 1 min, splitless injection
		Split ratio	100 : 1, open after 3 min
GC method	System		Varian 3800
	Injector		Varian split/splitless, temperature programmable 1079 type
	Inlet liner		Siltek deactivated SPME liner, 0.8 mm ID
	Column	Type	Rtx-CLPesticides
		Dimensions	30 m length × 0.25 mm ID × 0.25 μm film thickness with a 1.5 m retention gap and transfer capillary
	Carrier gas		Helium
	Flow		Constant flow, 2 mL/min
	Split	Closed until	3 min
		Open	3 min to end of run
		Split flow	200 mL/min
	Oven program	Start	40 °C, 3 min
		Ramp 1	15 °C/min to 130 °C
		Ramp 2	30 °C/min to 160 °C
		Isothermal	160 °C, 5 min
		Ramp 3	30 °C/min to 200 °C
		Final temperature	20 °C/min to 250 °C
	Transferline	Temperature	280 °C
MS method	System		Varian Saturn 2000
	Analyzer type		Ion trap MS with internal ionization
	Ionization		EI, emission current 40 μA,
	Electron energy		70 eV
	Ion trap	Temperature	200 °C
	Acquisition	Mode	Full scan
	Mass range		60–450 Da
	Scan rate		0.35 s/scan
Calibration	Internal standard	Range	Aniline-d_5, 1 mg/L

4.12.3 Results

DI-SPME has been successfully applied for the extraction of nonpolar components from aqueous samples. The *in situ* derivatization step lowers the polarity of the polar amines in an easy "one pot" reaction with extraction efficiencies of the aromatic iodine derivatives of better than 95%, with the only exception of 2-amino-4,6-dinitrotoluene (2A4,6DNT) with only 77%. Spike solutions have been prepared with water and may not be diluted in alcohols as the alcohol component will react as nucleophile (Figure 4.36).

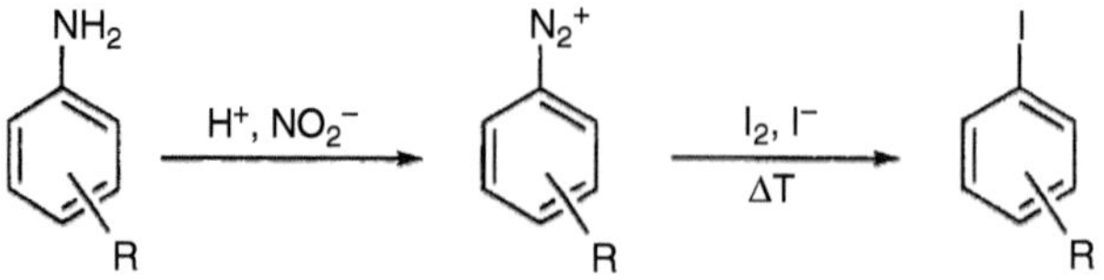

Figure 4.36 Derivatization reaction by diazotation followed by iodination.

SPME allows the complete transfer of the fiber accumulated analyte amount to the GC-MS analysis; also small sample sizes can be analyzed with the given performance. The thermal stress to the fiber has been minimized by the temperature profile during injection, resulting in the repeated use of one fiber bundle of up to 80 times.

For optimization of the chromatographic conditions, a standard sample diluted in ethyl acetate was injected using identical conditions. Figure 4.37 shows the

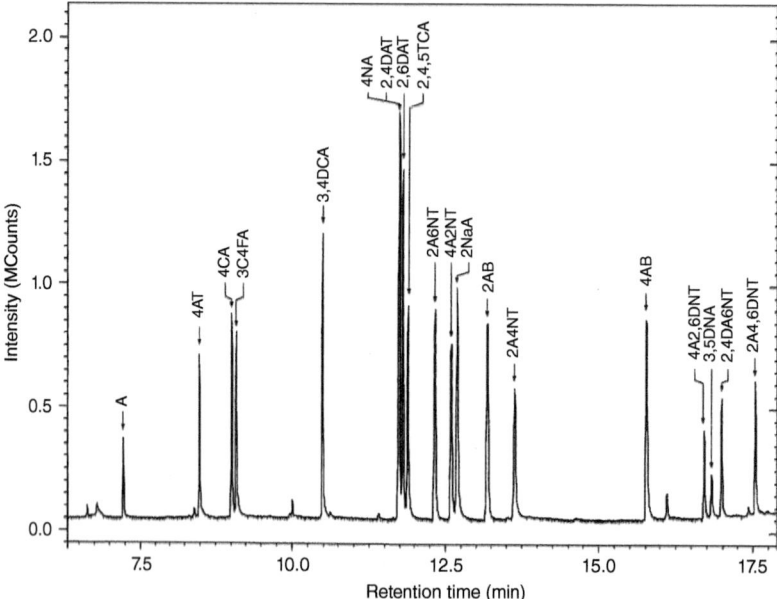

Figure 4.37 Standard chromatogram of the derivatized aromatic amines (abbreviations see Table 4.16). (Reproduced from Zimmermann et al., 2004/with permission of American Chemical Society.)

Table 4.14 List of aromatic amine compounds[a].

Compound	Acronym	Derivative	RT (min)	M	Detected ions[a] (m/z)	LOD (ng/L)	RSD % at 0.5 µg/L
Aniline	A	Iodobenzene	8.36	204	**204**, 77, 127	4	6.0
4-Aminotoluene	4AT	4-Iodotoluene	9.54	218	**218**, 91, 65	12	4.3
4-Chloroaniline	4CA	1-Chloro-4-iodobenzene	9.98	238	**238**, 111, 75	2	3.8
3,4-Dichloroaniline	3,4DCA	1,2-Dichloro-4-iodobenzene	11.55	272	**272**, 145, 109	3	6.7
2,4,5-Trichloroaniline	2,4,5TCA	1,2,4-Trichloro-5-iodobenzene	13.41	306	**306**, 179, 143	6	11
3-Chloro-4-fluoroaniline	3C4FA	2-Chloro-4-iodo-1-fluorobenzene	10.04	256	**256**, 129, 109	3	4.3
2,4-Diaminotoluene	2,4DAT	2,4-Diiodotoluene	13.22	344	**344**, 217, 90	13	11
2,6-Diaminotoluene	2,6DAT	2,6-Diiodotoluene	13.31	344	**344**, 217, 90	7	9.2
2-Naphthylamine	2NaA	2-Iodonaphthaline	14.59	254	**254**, 127, 74	11	7.4
2-Aminobiphenyl	2ABP	2-Iodobiphenyl	15.20	280	**280**, 152, 127	5	14
4-Aminobiphenyl	4ABP	4-Iodobiphenyl	16.98	280	**280**, 152, 127	9	11
4-Nitroaniline	4NA	1-Iodo-4-nitrobenzene	13.22	249	**249**, 219, 203	5	20
2-Amino-4-nitrotoluene	2A4NT	2-Iodo-4-nitrotoluene	15.69	263	**263**, 90, 105	8	16
2-Amino-6-nitrotoluene	2A6NT	2-Iodo-6-nitrotoluene	14.05	263	**246**, 89, 119	2	14
4-Amino-2-nitrotoluene	4A2NT	4-Iodo-2-nitrotoluene	14.44	263	**246**, 89, 119	3	16
2-Amino-4,6-dinitrotoluene	2A4,6DNT	2-Iodo-4,6 dinitrotoluene	18.49	308	**291**, 164, 89	38	20
4-Amino-2,6-dinitrotoluene	4A2,6DNT	4-Iodo-2,6-dinitrotoluene	17.76	308	**291**, 89, 63	27	13
2,4-Diamino-6-nitrotoluene	2,4DA6NT	2,4-Diiodo-6-nitrotoluene	18.00	389	**372**, 344, 216	30	16
Aniline-d₅ (ISTD)	A-d₅	Iodobenzene-d₅	8.35	209	**209**, 82, 127	—	—

a) Quantitation mass in bold.
(Reproduced from Zimmermann et al., 2004/with permission of American Chemical Society.)

separation of a standard mixture with almost baseline separation of all components. The MS was operated in the full scan mode to allow the detection of additional compounds for non-targeted analysis.

An overview of the components included in the described method is given in Table 4.14. Listed with the names and used abbreviations of the compounds are the chromatographic RTs, molecular mass of the iodine derivative, and its characteristic mass peaks. The base peaks used for compound selective quantification have been typically the molecular ion. The observed $(M-17)^+$ base peaks are due to the ortho effect fragmentation (see also Section 3.2.7.9). With the iodinated derivatives, the quantitation mass has been shifted by 111 Da per amino group to the higher mass range. This effect allows the very sensitive detection of the derivatives with high S/N values as unspecific background typically appears in the lower mass range.

The calibration range has been selected in accordance with the European regulation for toxic organic pollutants in drinking water, with the regulated level of 0.1 µg/L for all target compounds. The calibration using aniline-d$_5$ as an ISTD is linear from 0.05 µg/L over two orders of magnitude. A saturation of the SPME fiber has not been observed in this range. LODs have been calculated between 2 and

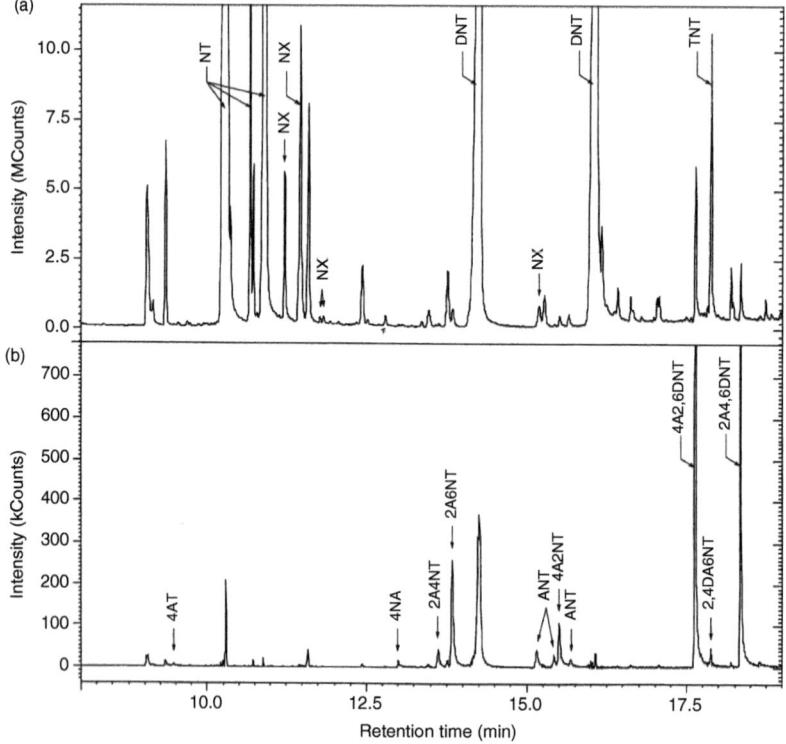

Figure 4.38 Analysis of a contaminated groundwater from the area of a former ammunition plant. (Reproduced from Zimmermann et al., 2004/with permission of American Chemical Society.)

13 ng/L except for 2A4,6DNT, 4A2,6DNT, and 2,4DA6NT with 27–38 ng/L, meeting excellently the EU regulatory levels.

Real water samples have been analyzed from different sources and demonstrated the applicability of the method for wastewater, groundwater, surface water, and drinking water. Figure 4.38 shows the analysis of a contaminated groundwater sample from the area of a former ammunition plant. Repeated analyses gave RSDs in the range of 3–10% (Zimmermann et al., 2004).

4.13 Phthalates in Liquors

Keywords: food safety; spirits; plasticizer; phthalic acid ester; dinonyl-phthalate; DNP; single quadrupole MS

4.13.1 Introduction

Phthalates (phthalic acid esters, PAEs) have widespread use in the polymer industry as plasticizers and softeners to increase the elasticity of polymer materials. They are chemically inert, have high density, low to medium volatility, and a high solubility in organic solvents, and they are easily released during the ageing of the polymer materials. Phthalates had been reported as functional solvents in the aromatic, essential oil, and even beverage industries. Phthalate plasticizers can migrate from plastic containers or closures into soft drinks and alcoholic beverages.

Phthalate residues in food and beverages are regulated internationally. In China, the Ministry of Health issued a public notice on June 1 2011 that phthalate esters are prohibited as nonfood substances for use in food. PAEs are introduced into the food chain primarily through food packaging materials. Alcoholic beverages in plastic containers are of potential risk as well since the ethanol provides a very good solubility for leaching PAEs into the beverages from the plastic contact materials. The contamination risk increases with liquors having a high ethanol content.

4.13.2 Analysis Conditions

Sample material			Spirits, Baijiu, alcoholic beverages
GC Method	System		Thermo Scientific TRACE 1310 GC
	Autosampler		AS 1310 liquid autosampler
	Column	Type	Thermo Scientific TRACE TR-35MS
		Dimensions	30 m length × 0.25 mm ID × 0.25 µm film thickness
	Carrier gas		Helium
	Flow		Constant flow, 1 mL/min

	SSL injector	Injection mode	Splitless
		Injection temperature	280 °C
		Injection volume	1 µL
	Oven program	Start	80 °C, 1 min
		Ramp 1	10 °C/min to 280 °C
		Final temperature	280 °C, 10 min
	Transferline	Temperature	280 °C
			Direct coupling
MS method	System		Thermo Scientific ISQ
	Analyzer type		Single quadrupole MS
	Ionization		EI
	Electron energy		70 eV
	Ion source	Temperature	280 °C
	Acquisition	Mode	Full scan
	Mass range		50–350 Da
	Scan rate		0.2 s (5 spectra/s)
	Resolution	Setting	Normal (0.7 Da)
Calibration	External standard	Range	0.10–4.00 mg/L, matrix spike

4.13.3 Sample Preparation

The beverage sample used for this application was a white spirit. Chinese "white wine" (baijiu, 白酒) contains 30–60 vol% of ethanol. As phthalate esters are highly soluble in ethanol, the extraction of phthalate esters using n-hexane as a solvent is less effective (Dongliang, 2010). The removal of the major part of ethanol from the liquor before n-hexane extraction is necessary to avoid low recoveries. For optimization of the extraction procedure and recovery determination, one liquor sample was spiked with 4 mg/L concentration of a commercial phthalate standard.

An accurate amount of the 5.0 mL sample was transferred in a glass centrifuge tube and heated in a boiling water bath to remove the ethanol (Dongliang, 2010). The heating time depends on the alcoholic strength of the spirit sample. Usually, the tube gets removed from the water bath with a residual volume of 2–3 mL. After cooling to room temperature, 2.0 mL of n-hexane was added. The glass tube was then shaken for extraction and left standing 5 min for phase separation. The supernatant was transferred to autosampler vials for analysis.

For the determination of the recoveries, the PAE standard solution was added to the sample to obtain a spiked solution at the 0.80 mg/L concentration level. The results were compared with and without ethanol removal, shown in Table 4.15. After the removal of ethanol before the extraction with n-hexane, good and consistent recoveries of the phthalate compounds in the range of 89–112% were obtained.

Table 4.15 Comparison of recovery of phthalates from liquor without and with prior removal of ethanol before extraction.

Compound	CAS #	Acronym	Without ethanol removal recovery (%)	With ethanol removal recovery (%)
Dimethyl phthalate	131-11-3	DMP	60.0	102.0
Diethyl phthalate	84-66-2	DEP	35.4	107.0
Di-isobutyl phthalate	84-69-5	DIBP	99.5	94.4
Di-n-butyl phthalate	84-74-2	DBP	106.0	104.0
Di-(4-methyl-2-pentyl) phthalate	146-50-9	DMPP	99.7	95.1
Di-(2-methoxy)-ethyl phthalate	117-82-8	DMEP	3.4	88.8
Diamyl phthalate	131-18-0	DPP	109.0	108.0
Di-(2-ethoxy)-ethyl phthalate	605-54-9	DEEP	13.6	103.0
Dihexyl phthalate	685-15-50-4	DHP	104.0	101.0
Butylbenzyl phthalate	85-68-7	BBP	88.4	108.0
Di-(2-ethylhexyl) phthalate	117-81-7	DEHP	106.0	108.0
Di-(2-butoxy)-ethyl phthalate	117-83-9	DBEP	83.1	104.0
Dicyclohexyl phthalate	84-61-7	DCHP	94.8	102.0
Di-n-octyl phthalate	117-84-0	DNOP	103.0	106.0
Diphenyl phthalate	84-62-8	DPhP	77.1	112.0
Dinonyl phthalate	84-76-4	DNP	110.0	109.0

4.13.4 Experimental Conditions

All measurements have been carried out using a single quadrupole GC-MS system equipped with split/splitless injector and liquid autosampler (Lv et al., 2013a).

4.13.5 Sample Measurements

The elution order of the phthalate compounds was determined by analyzing a standard mixture at medium concentration. The analyses had been run with full scan data acquisition. The spectra observed were compared with the NIST database for identification and RT determination. Although the full scan chromatogram gives high background and includes the elution of many other compounds dissolved in the spirit, the selective mass traces of the major phthalate ions allow a very good selectivity for a reliable peak area integration as shown in Figure 4.39.

The compound quantification was performed by selecting the most intense and unique ions of the PAE compounds providing selective mass chromatograms for individual peak integration. Finally, eight commercial liquor samples were prepared by the described sample preparation method for determining possible contamination by phthalate esters.

622 | *4 Applications*

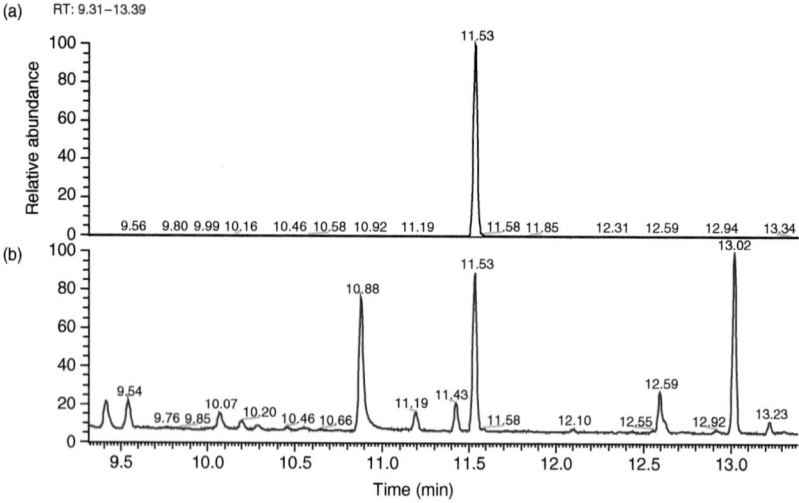

Figure 4.39 Dimethyl-phthalate chromatograms from a spiked sample with the selective mass chromatogram *m/z* 163 (a) and the full scan trace (b) allowing the interference-free peak area integration of the PAE compound.

4.13.6 Results

The full scan detection of three selected PAE components dimethyl-phthalate, di-isobutyl-phthalate, and di-(2-ethylhexyl) phthalate are shown as examples in Figures 4.40–4.42. The mass spectra are compared with the NIST library to confirm the compound identity.

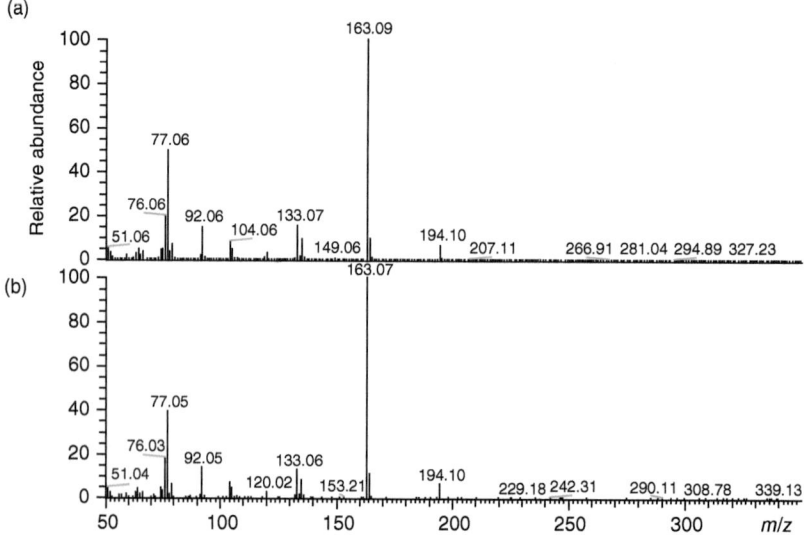

Figure 4.40 Dimethyl-phthalate spectra from standard (a) and sample (b).

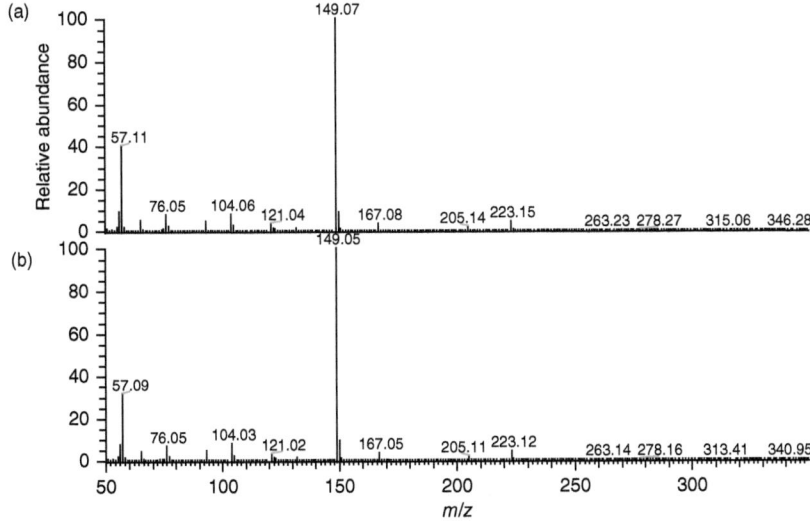

Figure 4.41 Di-isobutyl-phthalate spectra from standard (a) and sample (b).

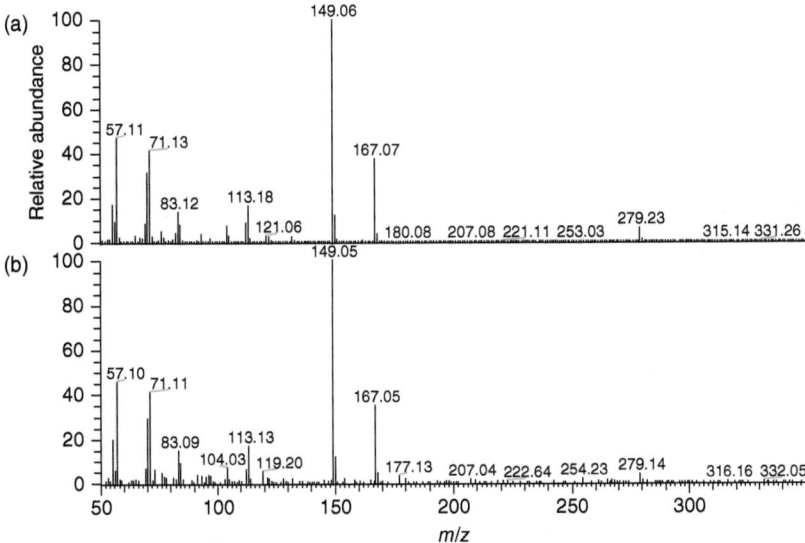

Figure 4.42 Di-(2-ethylhexyl) phthalate spectra from standard (a) and sample (b).

4.13.7 Quantitation

A series of matrix-spiked samples with five different concentrations were prepared in the range of 0.10–4.00 mg/L of the standard solution. The samples were injected in sequence from low to high concentration. The peak areas were calculated for the calibration curve with linear regression with very good precision and R^2 value of 0.999 for all PAE compounds. The results for 15 phthalate esters show a very good

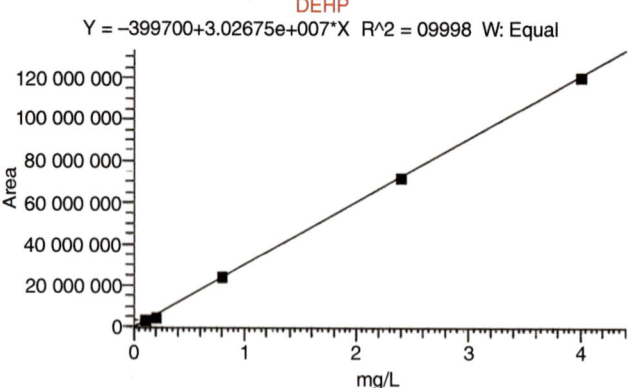

Figure 4.43 The linear calibration curve of DEHP with R^2 0.9998.

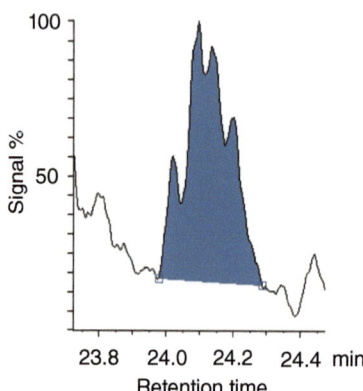

Figure 4.44 Quantitation peaks of the unresolved DNP isomers over a set retention time range.

linear relationship in the full calibration range of 0.1–4.0 mg/L. As an example, the quantitative calibration of the anticipated cancerogenic di-(2-ethylhexyl) phthalate (DEHP) (NTP, 2021) in excellent precision with R^2 0.9998 is shown in Figure 4.43.

The dinonyl-phthalates (DNPs) create a special analytical challenge. In contrast to calibration mixtures, the DNPs in use typically consist of a technical mixture of C9-isomers. Hence, the response of DNPs in real-life samples is distributed over several isomers with slightly different RTs. The integration of such unresolved DNP peaks needs to be performed over a wider but constant retention-time range from the data processing software, as shown in Figure 4.44. A linear calibration range for DNP of 0.4–4.0 mg/L could be achieved.

4.13.8 Sensitivity

The determination of the LOD and LOQ was based on the characteristic extracted ion mass chromatograms with a peak $S/N \geq 3$ for LOD, and $S/N \geq 10$ for LOQ, as given in Table 4.16 for the individual phthalate compounds.

Table 4.16 Phthalate quantification – Linear dynamic range with limit of detection (LOD) and limit of quantification (LOQ), average R^2 0.9990.

Compound name	Retention time (min)	Quantitation ion (m/z)	Linear range (mg/L)	Correlation coefficient R^2	LOD (µg/L)	LOQ (µg/L)
DMP	11.53	163	0.1–4.0	0.9994	0.1	0.3
DEP	13.02	149	0.1–4.0	0.9999	0.1	0.3
DIBP	15.64	149	0.1–4.0	0.9981	0.1	0.3
DBP	16.72	149	0.1–4.0	0.9986	0.1	0.3
DMPP	17.33/17.36	149	0.1–4.0	0.9993	0.2	0.6
DMEP	17.74	59	0.1–4.0	0.9984	0.2	0.6
DPP	18.43	149	0.1–4.0	0.9996	0.1	0.3
DEEP	18.59	72	0.1–4.0	0.9996	0.1	0.3
DHP	20.02	149	0.1–4.0	0.9990	0.1	0.3
BBP	20.94	149	0.1–4.0	0.9998	0.2	0.6
DEHP	21.37	149	0.1–4.0	0.9969	0.2	0.6
DBEP	21.45	149	0.1–4.0	0.9993	0.5	1.5
DCHP	22.50	149	0.1–4.0	0.9985	0.2	0.6
DOP	23.43	149	0.1–4.0	0.9998	0.5	1.5
DPhP	23.70	225	0.1–4.0	0.9988	0.2	0.6
DNP	24.0–24.4	149	0.4–4.0	0.9983	50.0	150.0

4.13.9 Method Precision and Recovery

The liquor samples were spiked by two low concentration levels at 0.1 and 0.3 mg/L and measured five times at each level. The results show that the average recovery even at trace level was 83.2–110%, and the achieved RSD range ($n = 5$) was 1.3–8.4%. The recovery and precision results are shown in Table 4.17.

Eight samples of commercially available liquor brands were analyzed using the described method. The concentrations of phthalate ester residues found are shown in Table 4.18. The samples tested showed that the reasonably anticipated cancerogenic DEHP was found in low concentration in all samples, and DIBP, DBP, and DEHP in many of the analyzed liquor samples.

4.13.10 Conclusions

The described application follows the China regulation GB/T 21911-2008 for the determining of phthalates in food (Standardization Administration of China, 2008). The used sample preparation procedure was optimized from GB/T 21911-2008 with the ethanol removal from liquor beverages followed by an n-hexane extraction and GC-MS detection. The method is sensitive, rapid, and accurate and covers a wide linear concentration range to meet the need for trace level detection of phthalate ester contaminations in alcoholic beverages.

Table 4.17 Method recovery and precision data at trace level (average recovery 103%).

Compound name	Spike level 0.1 mg/L Recovery %	RSD %	Spike level 0.3 mg/L Recovery %	RSD %
DMP	95.0	5.4	99.0	4.7
DEP	103.0	5.5	108.0	2.2
DIBP	101.0	2.0	101.0	3.2
DBP	107.0	6.6	101.0	1.3
DMPP	105.0	3.3	107.0	5.7
DMEP	86.3	5.3	83.2	3.4
DPP	109.0	6.0	104.0	1.6
DEEP	103.0	4.1	104.0	3.2
DHP	104.0	4.6	109.0	3.7
BBP	110.0	3.6	103.0	3.7
DEHP	102.0	1.4	105.0	4.1
DBEP	104.0	5.0	108.0	4.6
DCHP	103.0	4.1	103.0	3.6
DOP	105.0	5.8	104.0	2.6
DPhP	108.0	4.2	109.0	1.8
DNP	107.0	8.4	101.0	5.4

Table 4.18 The phthalate ester concentration in eight commercial baijiu samples (mg/L). DMPP, DMEP, DPP, DEEP, DHP, BBP, DBEP, DCHP, DOP, DPhP, and DNP were not detected in any sample.

Compound	Sample 1	Sample 2	Sample 3	Sample 4	Sample 5	Sample 6	Sample 7	Sample 8
DMP	ND	0.303	ND	ND	0.005	ND	ND	0.025
DEP	ND	ND	ND	ND	0.011	ND	ND	ND
DIBP	ND	1.526	ND	1.373	0.106	ND	ND	ND
DBP	ND	1.024	0.045	0.656	0.133	ND	0.469	0.064
DEHP	0.086	0.029	0.010	0.236	0.014	0.006	0.017	0.016

Note: ND, not detected.

4.14 Natural Spice Ingredients Capsaicin, Piperine, Thymol, and Cinnamaldehyde

Keywords: food safety; product safety; spices; active ingredients; traditional Chinese medicine, TCM; personal defense products; pepper spray; triple quadrupole MS/MS; analyte protectants

4.14.1 Introduction

Pungent spices are common ingredients for food preparations in all cooking traditions. Spices have been used as well for a long time in the traditional Chinese medicine (TCM). Beyond that, there is a modern use of the active ingredients of spices in a variety of personal defense and law enforcement products, such as pepper spray, due to their immediate physiological irritation effects.

Piperine

Capsaicin

Thymol

α-Methyl-t-Cinnamaldehyde

4.14.2 Analysis Conditions

Sample material	Spices, fruits, medical applications, personal defense products
Sample preparation	Dried material
	Solvent extraction with acetonitrile, methanol, or ethanol
	Filtration, concentration

GC method	System		Thermo Scientific TRACE 1300 GC
	Autosampler		Thermo Scientific TriPlus RSH
	Column	Type	Rtx-5Sil MS
		Dimensions	15 m length × 0.25 mm ID × 0.25 µm film thickness
	Carrier gas		Helium
	Flow		Constant flow, 1.2 mL/min
	SSL injector	Injection mode	Splitless
		Injection temperature	300 °C
		Injection volume	1 µL
	Split	Closed until	1 min
		Open	1 min to end of run
		Split flow	50 mL
		Gas saver	2 min, 20 mL/min
		Purge flow	5 mL/min
	Oven program	Start	50 °C, 2 min
		Ramp 1	20 °C/min to 300 °C, 2 min
	Transferline	Temperature	280 °C
			Direct coupling
MS method	System		Thermo Scientific TSQ 8000
	Analyzer type		Triple quadrupole MS
	Ionization		EI
	Electron energy		70 eV
	Ion source	Temperature	250 °C
	Acquisition	Mode	Timed-SRM
	Mass range		See the SRM table (Table 4.20)
	Scan rate		200 ms
	Resolution	Setting	Normal (0.7 Da)
Calibration			External standardization
	Calibration range		10–1000 ppb

4.14.3 Experimental Conditions

This application describes the analysis of extracts from spices by GC-MS/MS as a highly selective tool for the quantitative trace determination of natural active spice ingredients, i.e. capsaicin, piperine, thymol, and cinnamaldehyde as a flavoring component (Table 4.19).

4.14 Natural Spice Ingredients Capsaicin, Piperine, Thymol, and Cinnamaldehyde

Table 4.19 SRM data acquisition method for the spice analysis by GC-MS/MS.

Compound name	CAS number	RT (min)	Precursor ion (m/z)	Product ions (m/z)	Collision energy (V)	Peak width (min)
Thymol	89-83-8	6.24	135.1, 150.1	91.1, 135.1	15, 10	5
α-Methyl-*trans*-cinnamaldehyde	101-39-3	6.51	145.1	91.1, 115.1	25, 20	5
Capsaicin	404-86-4	12.64	137.0	94.0, 122.0	20, 15	5
Dihydrocapsaicin	19408-84-5	12.89	137.0	94.0, 122.0	20, 15	5
Piperine	94-62-2	14.08	200.8, 285.0	115.1, 172.7	20, 10	5

4.14.4 Sample Measurements

The active compounds capsaicin and dihydrocapsaicin elute from the GC column with only a short RT difference. A good separation free from peak tailing is necessary for the reliable peak integration at low concentration levels. It was found with different types of GC columns that the quality of the column deactivation, age of the column, and matrix deposits have a detrimental effect on the capsaicin and dihydrocapsaicin peak shape and quantitative reproducibility. Also, piperine and cinnamaldehyde were affected, while thymol always showed symmetrical peak shapes, apparently being unaffected by the increasingly active column conditions.

An analyte protectant was co-injected with the extract of active analytes to preserve inert conditions with the inlet liner and analytical column for high

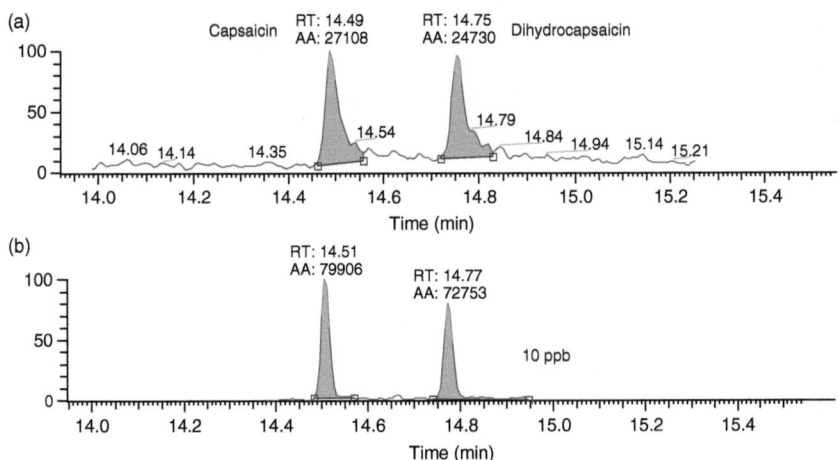

Figure 4.45 GC peak shapes of capsaicin and dihydrocapsaicin, (a) without and (b) with analyte protectant, both runs are at 10 ppb concentration using a 30 m 5%-phenyl column.

630 | *4 Applications*

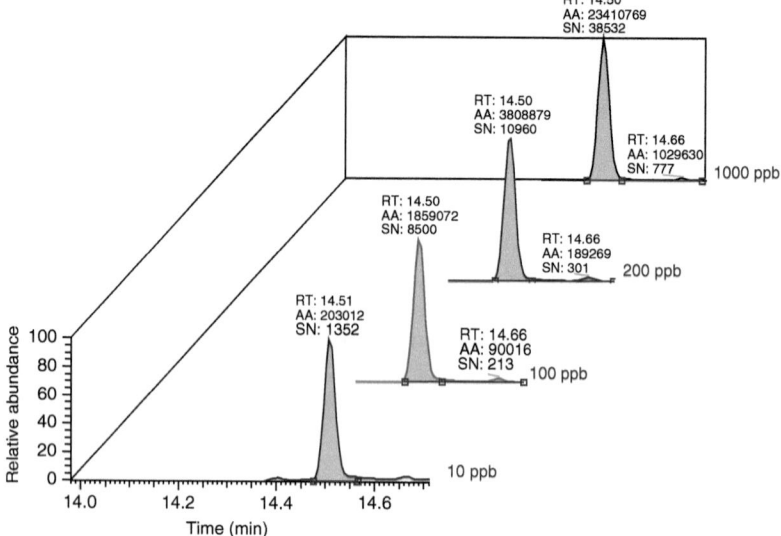

Figure 4.46 Capsaicin calibration peaks 10–1000 ppb.

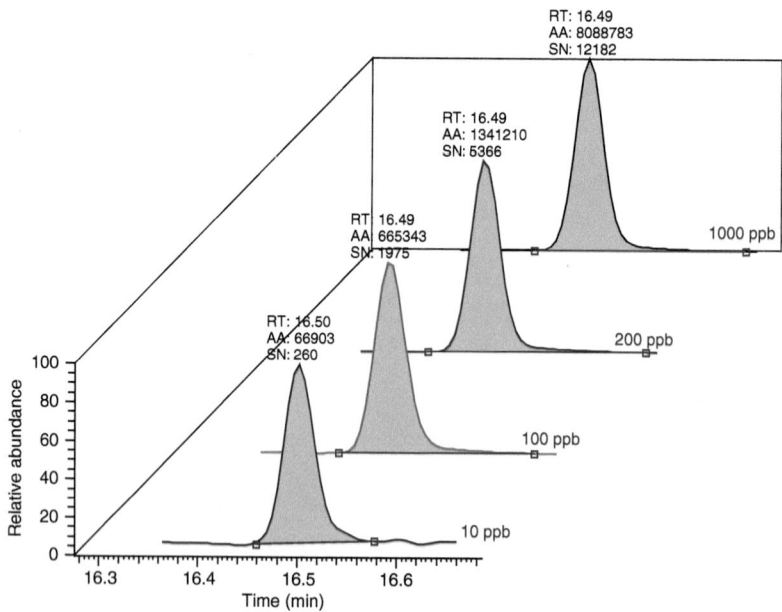

Figure 4.47 Piperine calibration peaks 10–1000 ppb.

quantitative precision and reproducible results. In this case, a concentration of 2 ppm of sorbitol was added to the extracts in the shown experiments.

4.14.5 Results

With the co-injection of sorbitol as an analyte protectant, symmetrical peak shapes for all described compounds could be achieved, including the critical pair capsaicin and dihydrocapsaicin, as shown in Figure 4.45. It is important to note here that it is not only the immediately visible peak shape and peak separation that is positively affected, but due to a significantly reduced tailing, the integrated peak areas and S/N ratios are exceptionally increased. The individual peaks for selected compounds capsaicin and piperine of the calibration runs up to 1000 ppb, normalized to 100% each, are given in Figures 4.46 and 4.47. The linear quantitative calibrations with a zoom into the low concentration range of 10–200 ppb are shown in Figure 4.48.

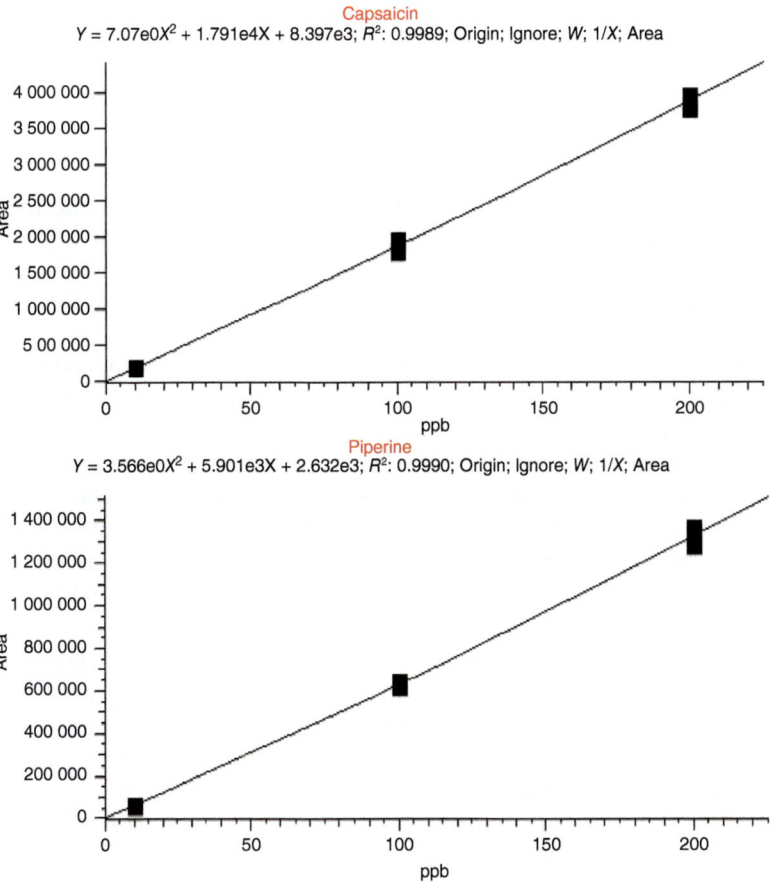

Figure 4.48 Quantitative calibrations on low concentration side for capsaicin and piperine.

Table 4.20 Precision of a spiked spice sample analysis.

Compound name	Day 1(area cts)	Day 2(area cts)	Day 3(area cts)	RSD(%)
Thymol	96.513	94.128	91.462	2.7
α-Methyl-*trans*-cinnamaldehyde	97.665	93.579	92.918	2.7
Capsaicin	100.669	105.363	99.392	3.1
Dihydrocapsaicin	102.752	103.852	101.089	1.4
Piperine	104.307	106.685	103.274	1.7

4.14.6 Conclusions

The reproducibility for a series of measurements using a spiked real-life spice sample with the application of the abovementioned analyte protectant was determined on three consecutive days. The precision of the area results has been calculated as RSD %. The peak area data in Table 4.20 indicate the good precision of the spiked spice sample analysis at a low level spike of below 10 ppb. The reproducibility over three days for all compounds tested is in the range of 1–3%.

4.15 Geosmin and Methylisoborneol in Drinking Water

Keywords: drinking water; off-odors; SPME arrow; automation: quantification; single quadrupole MS

4.15.1 Introduction

Most complaints about the quality of drinking water found relate to odor and taste. 1,2,7,7-tetramethyl-2-norborneol (geosmin) and 2-methylisoborneol (2-MIB) are compounds that can be found in drinking water produced by blue–green algae (cyanobacteria) and actinomycete bacteria and cause musty, earthy odors. Although these compounds have not been shown to be of health concern in public water supplies, the caused odors require removal, and thus, concentrations of geosmin and 2-MIB are monitored routinely in areas where they occur. The odor threshold for these compounds is very low, and humans can typically detect them in drinking water at 30 and 10 ng/L (ppt) for geosmin and 2-MIB.

Past analytical techniques to determine concentrations of these off-odor compounds used elaborate closed-loop stripping (CLS) and P&T sample preparation that provided the required sensitivity but have been either time consuming (CLS) or required a significant technical effort (P&T) (Preti et al., 1993). SPME has shown in many applications to be a very versatile and sensitive sampling strategy with the potential of high sensitivity and full automation (Chang et al., 2007; Bristow et al., 2019; Kaziur et al., 2019). This application describes the automated SPME technique from the sample headspace for the sensitive analysis of these compounds with a high sample throughput (Table 4.21).

4.15 Geosmin and Methylisoborneol in Drinking Water

Table 4.21 Geosmin and 2-MIB compound characteristics.

Compound name	Chemical structure	M (Da)	CAS #
Geosmin $C_{12}H_{22}O$		182.3	23333-91-7
2-Methyisoborneol (2-MIB) $C_{11}H_{20}O$		168.3	2371-42-8
2-Isopropyl-3-methoxypyrazine (IPMP, ISTD) $C_3H_{12}N_2O$		152.2	25773-40-4

4.15.2 Analysis Conditions

Sample material			Drinking water 5 mL/20 mL vial
Sample concentration	SPME Arrow	Sorbent type	CTC DVB/CAR/PDMS, 1.1 mm OD
		SPME Arrow conditioning	250 °C/10 min
		Salting out	1.5 g NaCl/20 mL vial
		Heatexer temperature	60 °C
		Extraction time	60 min
GC method	System		Agilent 8890 GC
	x,y,z-Robot		CTC PAL RTC
	Column	Type	Agilent DB-5MS UI
		Dimensions	30 m length × 0.25 mm ID × 0.25 µm film thickness
	Carrier gas		Helium 5.0
	Flow		Constant flow, 1.0 mL/min
	SSL injector	Injection mode	Splitless, 2 min
		Temperature	250 °C
	Split	Closed until	2 min
		Split flow	10 mL/min
		Purge flow	5 mL/min

	Oven program	Start	60 °C, 2.0 min
		Ramp 1	10 °C/min to 270 °C
		Final temperature	270 °C, 2.0 min
	Transferline	Temperature	280 °C
MS method	System		Agilent 5977C GC/MSD
	Analyzer type		Single quadrupole MS
	Ionization		EI, 70 eV
	Ion source	Temperature	250 °C
	Acquisition	Mode	SIM
			2-MIB: m/z 95,107,108
			Geosmin: m/z 111,112,125
			ISTD: m/z 137, 152
Calibration	Internal standard		Isopropylmethoxypyrazine (IPMP)
	Range		1–100 ng/L (ppt)
	Data points		5

SPME fiber sorbents: DVB – divinylbenzene; CAR – Carboxen; PDMS – polydimethylsiloxane.

4.15.3 SPME Method

Before first use, the SPME arrow was conditioned at 270 °C for 1 h. All standards were dissolved in methanol as a solvent. Approx. 1.5 g sodium chloride was added to empty 20 mL screw cap vials, and then 5 mL of the water sample plus 5 µL of the internal standard (ISTD) solution. For headspace solid phase microextraction (HS-SPME), a sampling temperature of 60 °C and a time of 30 min were found to be optimal. For analysis, the SPME arrow is moved after extraction automatically by a PAL RTC x,y,z-robot to the injection port and is thermally desorbed at 250 °C for 2 min. During the desorption time, the injector split is closed. A series of standard dilutions with concentrations of 1, 10, 20, 50, and 100 ng/L were prepared from the stock solution.

4.15.4 Results

The full scan mass spectra of the target compounds 2-MIB and geosmin are shown in Figures 4.49 and 4.50, respectively. The standard solution for the 10 ng/L level of 2-MIB and geosmin was measured five times in the SIM mode to obtain the intraday reproducibility data of this method. The area precision of 2-MIB and geosmin repeat measurements was 1.65% and 3.7% RSD ($n = 5$), respectively (Figure 4.51).

The calibration was generated calculating the response relative to the ISTD isopropylmethoxypyrazine (IPMP). The calibration curve of 2-MIB and geosmin

4.15 Geosmin and Methylisoborneol in Drinking Water | 635

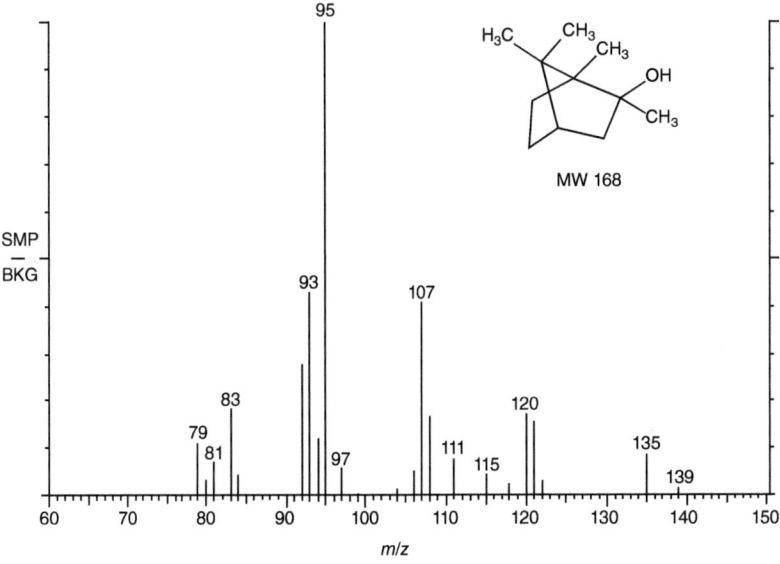

Figure 4.49 Mass spectrum of 2-MIB, the molecular ion M+ is not detected.

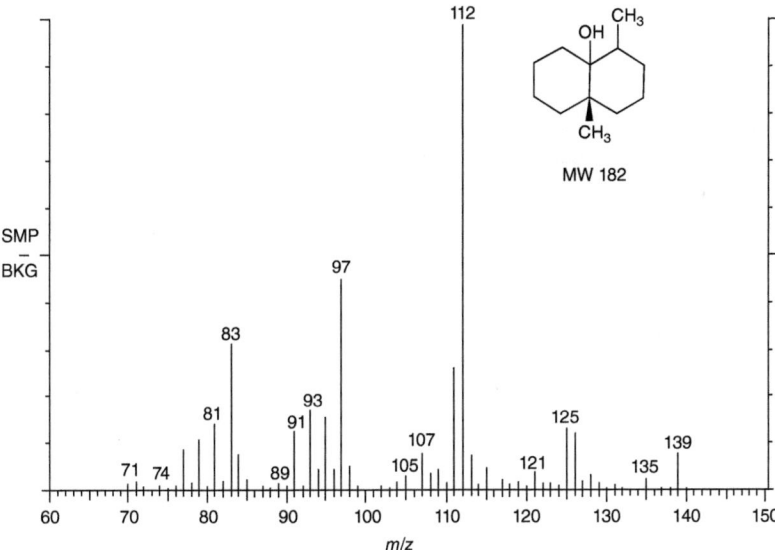

Figure 4.50 Mass spectrum of geosmin, the molecular ion M+ is not detected.

is obtained with very good precision and linearity with the linear correlation coefficient of 0.999 and better for both compounds shown in Figures 4.52 and 4.53.

The MDLs of the described HS-SPME method attained for 2-MIB and geosmin were 0.37 ng/L for 2-MIB and 0.22 ng/L for geosmin, calculated from 8 consecutive

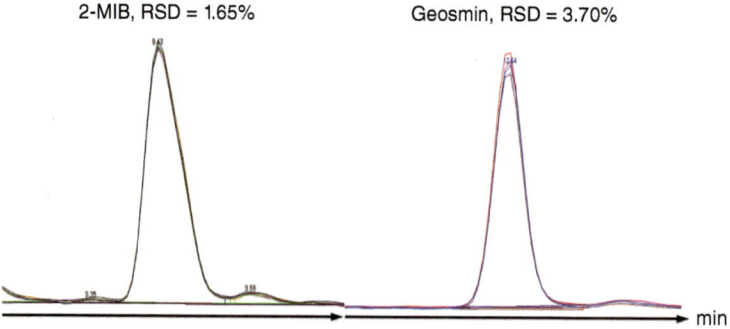

Figure 4.51 Intraday reproducibility data of 2-MIB and geosmin (10 ng/L, n = 5).

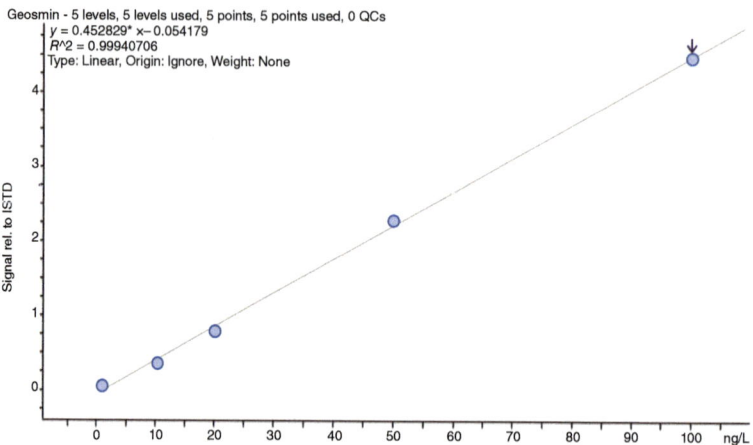

Figure 4.52 Relative response calibration for geosmin, in the range of 1–100 ng/L with precision $R^2 = 0.9994$.

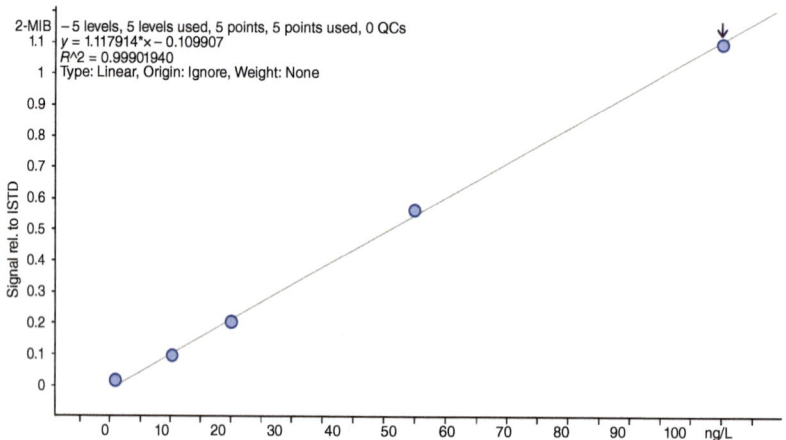

Figure 4.53 Relative response calibration for 2-MIB in the range of 1–100 ng/L with precision $R^2 = 0.9990$.

runs of a 2 ng/L dilution. At this very low concentration level, S/N 4.6 is achieved for 2-MIB and S/N 6.0 for geosmin. MDLs of less than 2 ng/L (2 ppb), which is less than the concentrations typically perceived by the human nose, can be achieved by the described method.

SPME as a green solventless sample preparation technique includes extraction, concentration, and sample GC injection of the analytes in a single procedure. Automation of this sample preparation process offers a valuable tool to greatly extend the sample throughput. SPME has gained widespread acceptance as the technique of preference for the analysis of such malodorous compounds.

4.16 Flavor and Fragrance Profiling by Dual-Column GC-MS

Keywords: essential oils; perfume; cosmetics; beverages; aroma profiling; quality control; dual column separation; retention index calibration; coeluting compounds; mass spectral library; AMDIS deconvolution; single quadrupole MS

4.16.1 Introduction

Essential oil, perfume, and flavoring samples are concentrated liquids of complex mixtures of hundreds of individual volatile compounds from natural and synthetic sources. The characterization of flavored food, detergent, and cosmetic samples is a time-consuming task with the identification of the compounds for product quality, new product development, or adulteration (David and Doro, 2023). In addition to mass spectral identification, the chromatographic retention information is indispensable for analysis: "Given reproducible retention times, the unequivocal identification of essential oil components is possible by retention time/mass spectrometry/computer searching algorithms" (Adams, 2012).

An efficient solution to provide detailed sample insights is the automated processing of chromatograms based on the deconvolution of the mass spectral data together with retention index information contained as additional information in library entries. A tabular output listing the compounds portion in the whole is typical for understanding the composition of a flavor mixture. The solution for flavor and fragrance profiling describes a turn-key solution based on an automated process of the GC-MS analysis and the employment of individual compound databases tailored to the laboratory requirements.

The identification of essential oil components via full scan mass spectra is often limited due to the subtle differences in mass spectra, coelutions, or the similarity of spectra of isomeric compounds. MS tuning, source conditions, and the type of mass analyzer may vary from those used for generating common library spectra. The compound retention information on polar and nonpolar columns complement the spectral information with the important influence of the molecular structure. Important here is the different retention behavior of isomeric compounds which are significant for the perception of flavors and fragrances. While RTs can vary even on columns of the same type from different manufacturers, also on measurement conditions, actual length or age, the calculation of retention indices (RIs) compensate such individual conditions well (Bianchi et al., 2007; Babushok, 2015). Known examples are the similar mass spectra of the stereoisomers cis-β-ocimene (CAS 3338-55-4) and trans-β-ocimene (CAS 3779-61-1), well separated on a 5%-phenyl column with RTs of 8:96 and 9:42 min, respectively, or the isomeric compounds iso-caryophyllene (CAS 118-65-0) and β-caryophyllene (CAS 87-44-5) with 24:95 and 25:26 min RTs.

4.16.2 Analysis Conditions

Sample material			Perfume, cosmetics, beverages, neat standard mixes
Sample preparation	LLE; dilution		DCM
GC method	System		Agilent 8890GC
	Autosampler		AS7673A
	Nonpolar column	Type	5% phenyl type MEGA5 MS
		Dimensions	60 m × 0.25 mm × 0.25 μm
	Polar column	Type	Carbowax type, MEGAWAX MS
		Dimensions	60 m × 0.25 mm × 0.25 μm
	Carrier gas		Helium 5.0
	Flow		Constant flow @ 30 cm/s, 1.4 mL/min
	SSL injector	Injection mode	Split
		Temperature	250 °C
		Liner volume	870 μL
	Split	Open	100 mL/min
	Oven program	Start	40 °C for 2 min
		Ramp 1	5 °C/min
		Final temperature	250 °C
	Transferline	Temperature	250 °C
MS method	System		Agilent MS5977B
	Analyzer type		Single quadrupole
	Ionization		EI, 70 eV
	Ion source	Temperature	230 °C
	Acquisition	Mode	Full scan
	Range		35–400 Da

4.16.3 Sample Preparation

The sample preparation depends on the type of sample to be analyzed. Standard mixes of pure substances as well as perfumes were diluted with a solvent.

A solid/liquid or the liquid/liquid extraction can be used for beverages and aqueous cosmetics. Other cosmetics were evaporated at mild temperature using the full evaporation of a small sample. The outgassed volatiles get trapped on Tenax with subsequent thermal desorption (TD). Diluted samples and extracts were placed in 2 mL vials on the autosampler for analysis according to a sequence table. Also, SPME and static and dynamic HS are used for sample preparation and injection to GC-MS.

4.16.4 Experimental Setup

The two capillary columns of different polarities, a nonpolar 5% phenyl, and a polar wax type column are installed preferably on two instruments for parallel analyses. If a column installation is planned for just one GC-MS unit, the columns get installed on separate injectors and fed together via a Vespel/graphite two-hole ferrule on the transfer line into the ion source of the MS (Nakasuji and Hiramatsu, 2023). Also, a dual column adapter is versatile as it allows the independent installation and positioning of each of the both columns (SIM, 2024).

In a dual column installation, the oven temperature program and transfer line temperature, the maximum permitted temperature (continuous and ramped), of the polar column need to be taken into account. During analysis, only the column selected for injection gets the analytical carrier gas flow and the other column a small standby flow only. This keeps the carrier gas inflow to the ion source in the allowable and optimum range. Two GC programs are prepared with the appropriate flow settings to be used for an alternating injection to each of the both injectors.

Two separate datafiles are acquired and processed. The RIs are generated from the elution profile of a mix of alkanes on both column types. The calibration of RIs in the range of C8 (RI 800) to C32 (RI 3200) is shown in Figure 4.54. A chromatogram with the addition the same alkane mix to a fragrance standard is shown

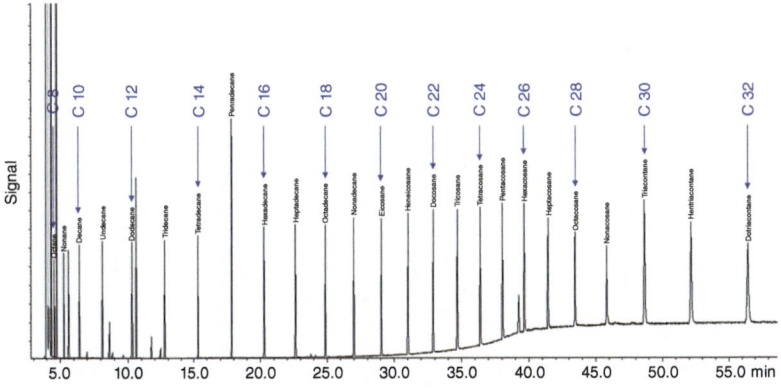

Figure 4.54 Retention index calibration C8 (RI 800) to C32 (RI 3200), even C numbers blue labeled.

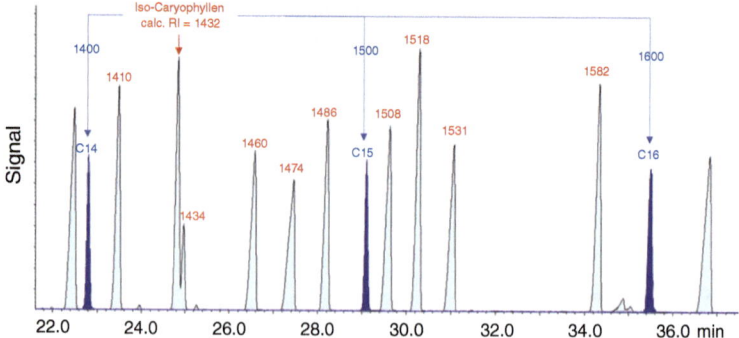

Figure 4.55 RI calculation between adjacent alkanes, e.g. iso-caryophyllene (red RI calculated from blue calibrants). Fragrance peaks from a synthetic mixture with signs of column overload.

in Figure 4.55. It illustrates the calculation of RIs of the individual compounds eluting between the alkane standards; in this case, for iso-caryophyllene, an RI of 1432 is calculated. The AMDIS analysis of an unknown sample based on the combination of full scan mass spectra retrieved from a spectral library and the calibrated RIs provides the identification of the eluting compounds. For easy review, the identified peaks are automatically labeled in the chromatograms with the compound names of the library spectra (Figure 4.60).

A detail of the deconvolution of coeluting compounds is illustrated in Figure 4.56. The total ion chromatogram (TIC) and the individual selective mass traces are

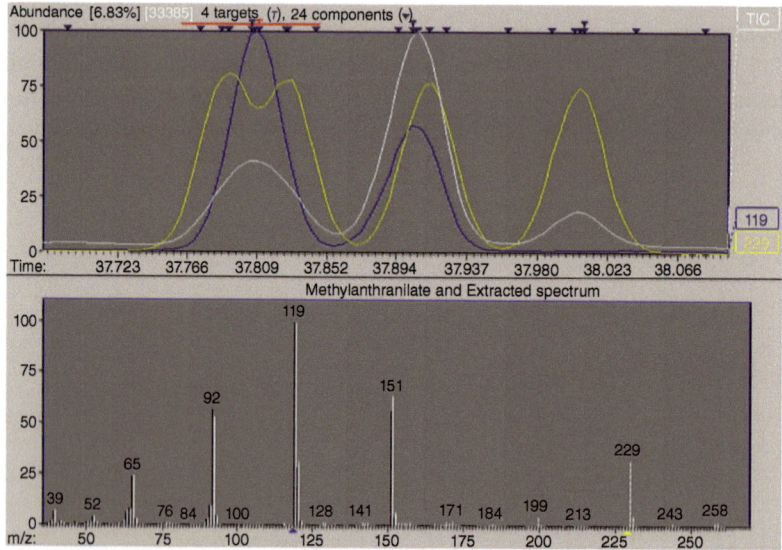

Figure 4.56 AMDIS deconvolution displays of methylanthranilate identification from a coeluted peak. Top chromatogram: white TIC, blue mass trace m/z 119 of methylanthranilate, yellow a coeluting substance. Bottom deconvoluted spectrum: white lines extracted, black library spectrum, dotted from the coeluting compound).

Table 4.22 Deconvolution report detail: 125 targets identified in total, methylanthranilate listed at Rt 37.810 min with the calibrated RI 2248, 0.41% area.

125 target(s).

Name	Rt	RI-Exp.	Area	% Area	CAS	Formula	Ri-RILib	Scan
GALAXOLIDE	37.805	2248	22,086,299	0.38	999000-33-8	C18H26O	-3.40	8808
METHYL ANTHRANILATE	37.810	2248	23,890,360	0.41	000134-20-3	C8H9NO2	-20.10	8809
Vertofix	37.904	2254	81,901,168	1.40	999001-14-8	C17H26O	-2.90	8831
GALAXOLIDE	38.008	2259	14,601,147	0.25	999000-34-9	C18H26O	-4.20	8856
Myristicin	38.224	2271	4,200,299	0.07	000607-91-0	C11H12O3	0.00	8906
TRASEOLIDE	38.372	2279	1,906,976	0.03	068140-48-7	C18H26O	-3.20	8941
hedione (IMP)	38.471	2285	6,708,920	0.11	999008-78-5	C13H22O3	10.30	8964
HEDIONE, (Z)-	38.674	2296	475,411,149	8.11	024851-98-7	C13H22O3	-6.30	9011
GALAXOLIDE	39.305	2364	182,156,184	3.11	999000-32-7	C18H26O	14.70	9158

displayed in different colors. The extracted mass spectrum and the spectrum of the library are displayed below, indicating not belonging signals as dotted lines. A final report is provided for each chromatogram (Table 4.22). Besides the list of compounds identified in the analysis, also the relative area is determined, allowing a quantitative estimate of the relative quantity. This 100% report delivers the estimated composition of the sample with major and minor compounds (Soulier, 2024).

4.16.5 Results

The efficiency of the deconvolution algorithm and the required high detection dynamic of the mass spectrometer to detect and to identify the presence of flavor relevant trace components underneath a large solvent peak is shown in Figure 4.57. In this case, terpineol and fenchol are identified in a very large peak of propylene glycol eluting in the middle of the chromatogram. Once the components are identified by AMDIS, it is easy to plot reconstructed signals specific to each

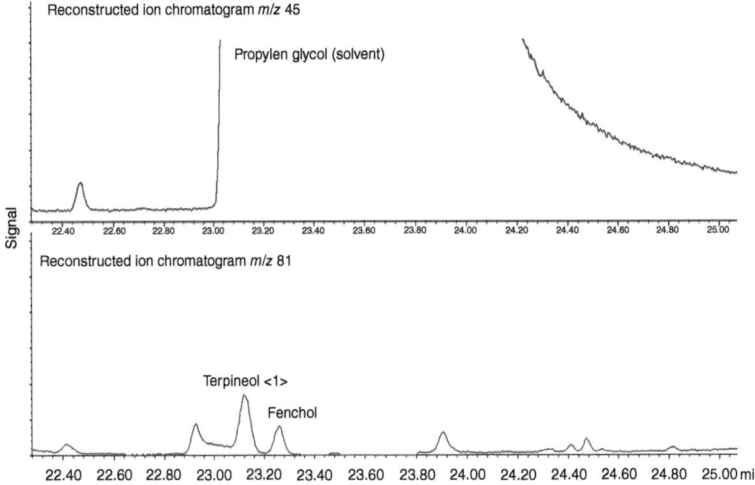

Figure 4.57 Minor compound detection and identification of terpineol <1> and fenchol underneath a large solvent peak of propylene glycol.

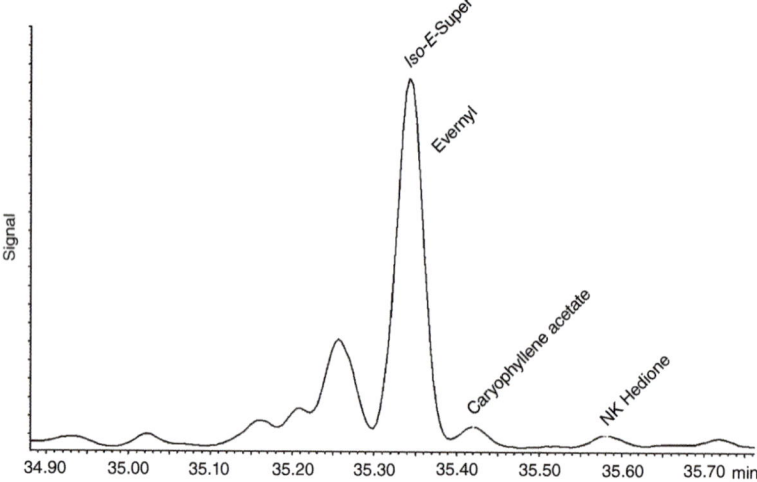

Figure 4.58 Coelution in a fragrance mixture is shown with the compound Iso-E-Super.

compound, m/z 45 for propylene glycol, and m/z 81 for terpineol <1> and fenchol. This feature allows a better understanding of the mixture, and it also permits the correct identification of the coeluted components by a manual subtraction of the signals.

An often-observed situation of coelution in a fragrance mixture is shown with the elution of the compound Iso-E-Super (Figure 4.58). Although the substance peak looks symmetrical, the component Evernyl (CAS# 4707-47-5) is perfectly hidden in the right slope. The AMDIS deconvolution algorithm analyzes the common and differing signals, extract, and library searches the spectra (Little, 2024). The identification of the compounds is performed by both spectral comparison and the calculation of the experimental retention index compared to the theoretical index stored along with the spectrum. An RI deviation of only −0.5 is reported for the compound Evernyl (Figure 4.59). Four scans' difference between the two peaks

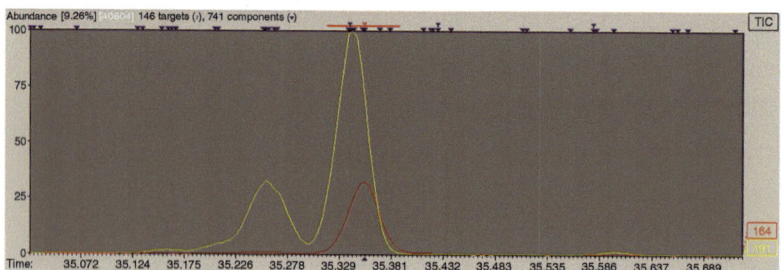

Figure 4.59 AMDIS coelution identification by selective mass traces m/z 164 of Evernyl and m/z 191 of Iso-E-Super.

4.16 Flavor and Fragrance Profiling by Dual-Column GC-MS | 643

Table 4.23 AMDIS Deconvolution report with Rt, RI, area, CAS#, and RI difference to library.

Name	Rt	RI-Exp.	Area	% Area	CAS	Ri-RILib
ISO E SUPER	34.440	1679	9908966	0.610	999001-62-6	1.10
ISO E SUPER	34.654	1688	127951714	7.876	999001-61-5	1.50
HEXYL SALICYLATE	34.719	1691	26532081	1.633	006259-76-3	-4.60
METHYL DIHYDROJASMONATE...	34.737	1692	31389168	1.932	002630-39-9	1.20
ISO E SUPER	35.339	1719	29413213	1.811	999001-58-0	0.80
EVERNYL	35.354	1720	4434223	0.273	004707-47-5	-0.50
CARYOPHYLLENE ACETATE	35.426	1723	1712930	0.105	057082-24-3	5.90
NK hedione	35.580	1731	771215	0.047	999008-77-4	1.00

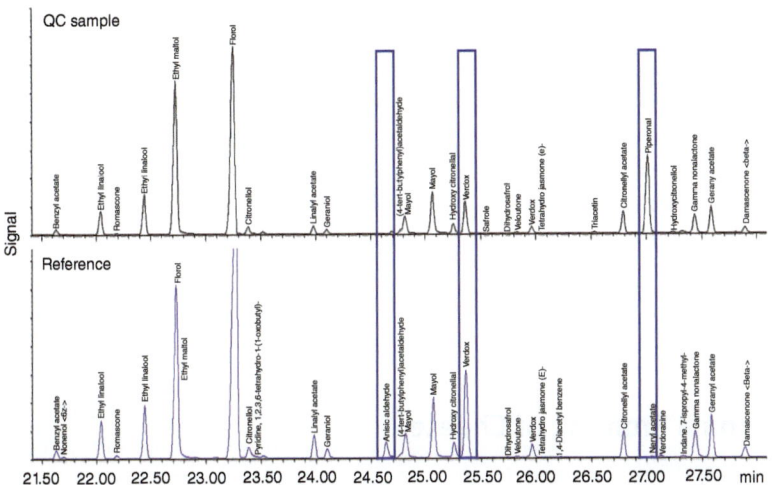

Figure 4.60 Quality control with visual support of the control and suspect sample chromatograms.

is enough to compute almost pure spectra by correct spectral subtractions. The RT difference between the two peaks is less than 900 ms. The apex scan number Iso-E-Super is 8232, Rt = 35.3394 min, and of Evernyl is 8236, Rt = 35.3544 min (Table 4.23).

The same process is also applied to quality control analyses in order to provide a visual support with the reference for qualitative and quantitative differences. Chromatograms with labeled peaks offer an easy reading of the data for comparison, for example, in Figure 4.60 with the missing anisic aldehyde (CAS#123-11-5) at 24.64 min and the presence of piperonal (CAS# 120-57-0) at 27:00 min in the sample chromatogram trace. Verdox (CAS# 88-41-5) at 25:36min is found in different concentrations. The combination of the deconvolution results along with the traditional GC-MS software permits a fast and easy understanding of the composition of a fragrance mixture.

4.16.6 Conclusions

The compound RT information is calibrated by an alkane mix on two different column types of different polarities. The use of AMDIS allows an efficient management of coeluting compounds by RIs and the extraction of deconvoluted mass spectra for library search. The described solution is in use for generating flavor and fragrance product compositions. A 100% report of the composition can be generated based on the AMDIS identification and integration of deconvoluted peak areas. Getting the integration information from the deconvoluted signals brings a particular benefit in terms of time saving when working with complex matrices and particularly when coelutions occur. In this case, the regular integration algorithms of the chromatography data systems (CDS) cannot differentiate individual quantities of mixed signals.

The described concept can also be used for the analysis of target compounds through dedicated libraries of, for instance, allergens, pesticides, and other marker compounds particularly for the detection of trace molecules which are often hidden behind intense compound peaks. The automated processing of the RI information, mass spectra, and response information allows the rapid determination of the sample composition providing a list of identified compounds with relative quantitation.

4.17 Aroma Profiling of Cheese

Keywords: food quality; aroma profiling; off-odors; cheese; thermal extraction; dynamic headspace; TOF MS

4.17.1 Introduction

The aroma impression is important for consumers in spontaneous grocery food checking. Aroma profiling on a professional aspect is a key component in the industrial control amongst others of constant quality, shelf life, composition, or maturity. In the case of cheese, a complex blend of low and high level compounds combines to create their distinctive odors, the subject of further analytical investigation.

The nature of cheese varies considerably, and the subtle differences in aroma, flavor, and texture all add to its appeal. Various factors are involved in the nature of the final product, including the type of animal, its diet, processing of the milk, the cheese culture, and the ageing conditions. Manufacturers spend a great deal of effort in ensuring the products are of a consistently high quality.

The extraction of the diversity of aroma components from cheese at the typical temperature of consumption, trapping on sorbent tubes followed by TD to a GC-MS system, allows the evaluation of the aroma profile and the identification and quantitation of individual compounds.

4.17.2 Analysis Conditions

Sample material			Food, cheese
Sample preparation			
	Extraction	Thermal extractor	Markes Ltd. Micro-Chamber
		Purge gas	Nitrogen
		Trap	Quartz wool – Tenax TA – Carbograph 5TD
		Sample volume	5 g cheese, grated
		Chamber temperature	40 °C
		Equilibration time	20 min
		Purge flow	50 mL/min
		Sampling time	20 min
	Thermo desorber		Markes Ltd. UNITY 2
		Flow path temperature	160 °C
		Transfer line	200 °C
		Focusing trap	Material emissions
		Dry purge	2 min, 20 mL/min flow to split
		Primary (tube) first stage	150 °C for 5 min
		Desorption	50 mL/min trap flow, no split flow
		Second stage	300 °C for 5 min
			50 mL/min trap flow, no split flow
		Secondary (trap)	Trap low: 30 °C, trap high: 300 °C
		Desorption	Heating rate: 24 °C/s
		Hold time	5.0 min, split flow
			50 mL/min (high split)
			5 mL/min (low split)
		Pre-trap fire purge	2 min, 50 mL/min trap flow
			50 mL/min split flow (high split)
			5 mL/min split flow (low split)
		Overall TD split	51 : 1 (high split), 6 : 1 (low split)
GC method	System		Thermo Scientific TRACE GC Ultra
	Autosampler		Markes Ltd. TD-100

	Column	Type	HP-Innowax
		Dimensions	30 m length × 0.25 mm ID × 0.25 µm film thickness
	Carrier gas		Helium
	Flow		Constant flow, 1 mL/min (initial $P = 6.4$ psi)
	SSL injector	Injection mode	Splitless
		Injection temperature	160 °C
	Oven program	Start	40 °C, 2.0 min
		Ramp 1	5 °C/min to 180 °C
		Ramp 2	20 °C/min to 260 °C
		Final temperature	260 °C, 6.0 min
	Transferline	Temperature	265 °C
MS method	System		ALMSCO International BenchTOF-dx
	Analyzer type		Time-of-flight MS with axial ion acceleration
	Ionization		EI
	Ion source	Temperature	260 °C
	Acquisition	Mode	Full scan
	Mass range		33–350 Da
	Scan rate		2 Hz

4.17.3 Sample Preparation

Samples of the following cheeses were analyzed:

- full-fat Welsh Cheddar (extra-mature),
- low fat Cheddar (mild),
- Comté (a hard cheese made with unpasteurized cow's milk),
- Emmental (a medium-hard cheese), and
- Channel Island Brie (a soft cheese).

The cheese samples were cut and placed in equal amounts into inert-coated heated extraction chambers for release and trapping of the volatile compounds. Nitrogen as an inert gas is passed through the chambers at a constant flow rate while sweeping the headspace vapors onto individual sorbent packed tubes. The used micro-chamber thermal extractor allows up to six samples to be extracted in parallel under identical conditions for comparable results.

4.17.4 Experimental Conditions

4.17.4.1 Dynamic Headspace Sampling

The sample chamber is set to a temperature of 40 °C for pre-heating and extraction. This temperature was chosen for a good representation of the released VOCs of the cheeses as a mock-up of the typical mouth feeling. In this dynamic headspace sampling step, the volatile headspace is flushed from the cheese samples by using the inert gas nitrogen for collection onto the sorbent packed tubes. After collection, the tubes were transferred to a TD unit.

4.17.4.2 Thermal Desorption

TD concentrates the collected cheese aroma constituents into a narrow elution band for GC-MS analysis. Each sorbent tube of the individual cheese samples was thermally desorbed by evaporating them inverse to the sampling direction into the GC carrier gas stream using an automated thermal desorber.

During the automated desorption process, the released components were transferred to a Peltier-cooled smaller internal "focusing" trap integrated within the TD system. After completion of the slower primary sampling tube desorption, the focusing trap was very quickly desorbed by heating it rapidly in reverse flow for injection of the compounds into the GC column. This two-stage desorption process optimizes the concentration enhancement and produces narrow chromatographic peaks for optimum separation and sensitivity.

The choice for the INNOWax column with a polyethylene glycol (PEG) stationary phase was taken due to the high polarity and high upper temperature limits to provide optimum separation for the containing polar aroma compounds expected to be present in these samples.

The MS detection has been carried out using the BenchTOF-dx time-of-flight mass spectrometer for its sensitivity in the full scan acquisition mode. This allows recording of the complete mass spectra at similar levels to that of quadrupole instruments run in the SIM mode. The full scan mass spectra then can be identified by matching against commercial spectra libraries NIST and Wiley.

4.17.5 Sample Measurements

The chromatograms shown in Figure 4.61 are taken from four types of cheeses, collected to individual sorbent tubes simultaneously from adjacent chambers. The intensities are normalized to the chromatogram base peaks. All samples had been analyzed with identical TD-GC-MS conditions. The pattern of the collected volatile compounds can be compared easily by overlaying the chromatograms. Even without a quantitative calibration, the peak areas give a good approximation of the relative abundances of the individual components, estimating comparable response factors for these organic compounds of a similar structure.

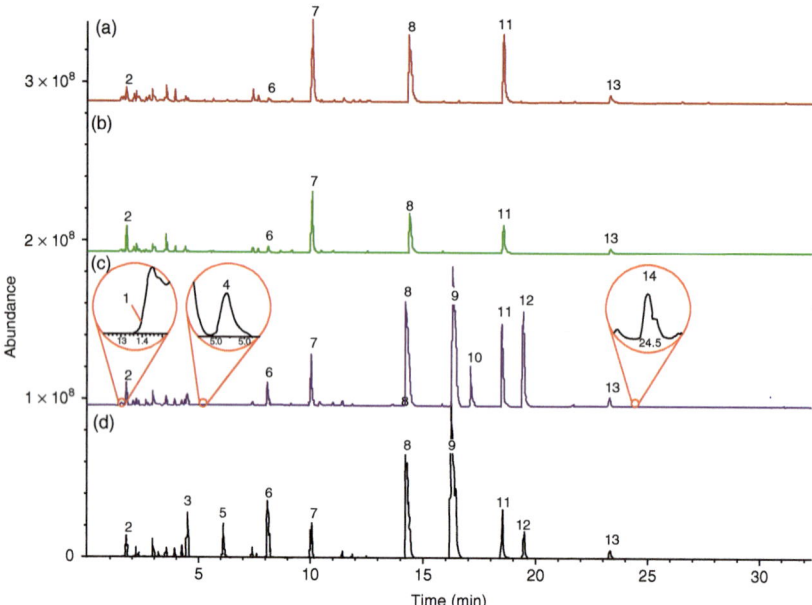

Figure 4.61 Analysis of the VOC profiles of (a) full-fat (extra-mature) Cheddar, (b) low fat Cheddar, (c) Comté, and (d) Emmental, with major compounds and some sulfur compounds labeled (insets). Compounds: (1) hydrogen sulfide; (2) carbon disulfide; (3) propan-1-ol (4) dimethyl disulfide; (5) pentan-2-ol; (6) 3-methylbutan-1-ol; (7) 3-hydroxybutan-2-one (acetoin); (8) acetic acid; (9) propanoic acid; (10) 2-methylpropanoic acid; (11) butanoic acid; (12) 3-methylbutanoic acid; (13) hexanoic acid; (14) dimethyl sulfone.

4.17.6 Results

Similar aroma profiles had been acquired of the low fat and full-fat Cheddar cheese with the main differences being in peak intensities.

Looking at the individual compounds identified, the presence of VSCs with carbon disulfide is seen in all four samples, along with trace levels of hydrogen sulfide, dimethyl disulfide, and dimethyl sulfone in the Comte.

Fatty acids, both straight-chain and branched, are key contributors to the aromas of various cheeses. The presence of large amounts of propanoic acid and 3-methylbutanoic acid had been found in the odor profiles of the Emmental and Comte, with the added presence of 2-methylpropanoic acid in the latter.

The presence of 3-methylbutan-1-ol has been identified in all four samples. This compound, which is known to confer a pleasant aroma of fresh cheese, was most abundant in the Comte and Emmental. In the Emmental, relatively large amounts of propan-1-ol and pentan-2-ol have been determined. The respective "sweet" and "fresh" notes that have been reported for these components may contribute to the distinctive aroma of this cheese.

4.17.7 Conclusions

Thermal extraction of foods in combination with TD of the collected volatile headspace constituents, in this reported case of different cheeses varieties, is a viable method for the reproducible aroma characterization.

The described technology was used to evaluate the aroma profile from a range of cheese samples. Besides the chromatographic pattern, both the identification and relative quantitation of desirable compounds as well as off-odors can be performed. The method also enables the identification of changes in the composition over time, providing useful data relating to shelf life and product safety (Markes, 2010, 2012).

4.18 Allergens

Keywords: product safety; consumer protection; toys; fragrances; allergens; ion trap MS/MS; EU regulation; liquid/liquid extraction; triple quadrupole MS/MS

4.18.1 Introduction

Thousands of fragrance ingredients are used in perfumes and perfumed consumer goods such as cosmetics, detergents, fabric softeners, and other household products to give them a specific, usually pleasant smell. They can sometimes cause skin irritations or allergic reactions (Scientific Committee on Consumer Safety, 2012). The safety of children toys is regulated by the EU Directive 2009/48/EC, also including a list of allergens that shall not be contained. The presence of traces of these fragrances shall be allowed provided that such a presence is technically unavoidable under good manufacturing practice (European Commission, 2009, 2023).

The referenced publication provides the general methodology for the determination of 48 out of the 66 restricted fragrance allergens by the EU regulation extracted from children toys. Eight of the remaining 18 fragrance allergens are natural extracts that are not amenable to a GC method, and another 10 are reported to be at that time commercially unavailable (Lv et al., 2013b). The analysis of allergens from indoor air uses glass cartridges with sorbent material (e.g. Amberlite XAD-2) for collection with subsequent LLE solvent elution (Balducci et al., 2022).

4.18.2 Analysis Conditions

Sample material	Plastics, plush and paper toys, clays
Sample preparation Liquid extraction	Dissolution of the plastic material, methanol liquid extraction
	Liquid extraction with acetone (clays, plush, paper toys)

	SPE	Material	ENVI-Carb cartridges
		Conditioning	5 mL methanol
		Sample application	Flow 3 mL/min, effluent collected
		Elution	With 15 mL dichloromethane, effluent collected
		Concentration	Combined effluents to 5 mL volume
		Filter	0.45 µm polytetrafluoroethylene (PTFE) membrane prior to injection
GC method	System		Varian 450
	Autosampler		Varian CP-8400
	Column	Type	HP-1MS
		Dimensions	50 m length × 0.20 mm ID × 0.50 µm film thickness
	Carrier gas		Helium, 99.999%
	Flow		Constant flow, 0.7 mL/min
	SSL injector	Injection mode	Splitless
		Injection temperature	280 °C
		Injection volume	1 µL
	Split	Closed until	1 min
		Open	1 min to end of run
	Oven program	Start	50 °C, 1 min
		Ramp 1	5 °C/min to 155 °C, 6 min
		Ramp 2	3 °C/min to 260 °C
		Final temperature	260 °C
	Transferline	Temperature	280 °C
			Direct coupling
MS method	System		Varian 240
	Analyzer type		Ion trap MS with internal ionization
	Ionization		EI
	Filament/multiplier delay time		8 min
	Electron energy		70 eV
	Ion source	Temperature	280 °C
	Manifold temperature		50 °C
	Ion trap temperature		220 °C
	Acquisition	Mode	MS/MS, resonant waveform CID
	Precursor ion width		3–5 Da
Calibration	External standard		Target compounds spiked into blank samples
	Range		0.005 – 100 mg/kg

4.18.3 Sample Preparation

4.18.3.1 Extraction

The extraction method uses a complete dissolution of plastic toys and then the liquid extraction by an immiscible polar solvent, in this case methanol. Clays, paper, and other not dissolvable materials are extracted directly after chopping into small pieces of less than 2 mm size. Cryomilling to preserve the volatile fragrances is recommended.

Plastic toys: 1 g of the sample is dissolved with 10 mL of an appropriate solvent, for instance, acetone for acrylonitrile butadiene styrene (ABS), dichloromethane (DCM) for polystyrene (PS), and tetrahydrofuran (THF) for polyvinylchloride (PVC) in an ultrasonic bath. 10 mL of methanol is added to the solution and shaken. The methanol phase is centrifuged, applied to SPE clean-up, and concentrated.

Play clays: 1 g of the sample is extracted with 10 mL acetone and methanol in an ultrasonic bath. The combined solutions get centrifuged, applied to SPE clean-up, and concentrated.

Plush and paper toys: The samples are cut into pieces of less than 5 mm size and extracted with 20 mL of acetone in an ultrasonic bath. Prior to GC-MS analysis, the extract is filtered using a 0.45 µm PTFE filter.

For clean-up and concentration, an SPE step is used. The eluate gets concentrated to 5 mL volume for injection to GC-MS.

4.18.4 Experimental Conditions

The MS/MS mode of an ion trap MS was chosen at that time as the detection method. Triple quadrupole MS/MS systems are applied as the current alternative. The single ion detection (SIM) has been compared using matrix samples but did not deliver the required compound selectivity for trace level quantification in the matrix.

For MS/MS method development, the individual standards are analyzed in the full scan mode (40–400 Da) for selection of the precursor ions. Two MS/MS transitions are used for each compound, one for quantitation and the second as a qualifier for confirmation. The ion trap MS/MS excitation voltage was optimized for the precursor ions of each individual compound for the best isolation efficiencies and maximum response of the characteristic product ions. The complete setup for MS/MS acquisition is shown in Table 4.24. For data acquisition, the chromatogram was divided into 13 RT segments with the parallel detection of the assigned compounds.

4.18.5 Results

The separation of the 48 fragrance allergens could be achieved with good performance using a 50 m unpolar thick film column. The chromatogram of the total ion current (TIC) of a standard compound mix is shown in Figure 4.62, with a zoom into the congested peak region in the center of the chromatogram in Figure 4.63.

Table 4.24 Fragrance allergens with MS/MS acquisition and quantitation details.

No.	Compound name	EU Reg.	Retention time (min)	CAS Number	Quantitation Transition m/z > m/z	Excitation voltage m/z	Excitation isol. wind (Da)	Qualifier transition m/z > m/z	Excitation voltage* (V)	isol. wind. (Da)	Linear ranges (mg/L)	Correlation coefficient R^2
1	Ethyl acrylate	a)	8.17	140-88-5	99 > 77	0.4	4	99 > 81	0.4	4	0.2–50	0.9990
2	Methyl-*trans*-2-butenoate	a)	9.97	623-43-8	85 > 57	0.3	3	100 > 69	0.4	3	0.05–20	0.9995
3	5-Methyl-2,3-hexanedione	a)	12.30	13706-86-0	85 > 57	0.5	3	85 > 41	0.5	3	0.02–10	0.9996
4	*trans*-2-Heptenal	a)	16.34	18829-55-5	83 > 55	0.4	4	95 > 79	0.4	5	0.05–10	0.9993
5	*trans*-2-Hexenal-dimethyl-acetal	a)	17.58	18318-83-7	113 > 71	0.6	3	113 > 97	0.6	3	0.2–20	0.9968
6	Benzyl alcohol	a)	18.93	100-51-6	108 > 79	0.6	3	91 > 65	0.4	4	0.005–5	0.9995
7	d-Limonene	b)	19.65	5989-27-5	107 > 91	0.6	3	136 > 94	0.5	3	0.05–10	0.9992
8	Dimethyl citraconate	b)	20.53	617-54-9	127 > 99	0.4	4	127 > 69	0.4	4	0.005–5	0.9993
9	*trans*-2-Hexenal-diethyl-acetal	a)	21.33	67746-30-9	127 > 85	0.7	3	127 > 98	0.8	3	0.5–10	0.9943
10	Linalool	b)	21.51	78-70-6	93 > 77	0.6	4	121 > 93	0.4	3	0.02–10	0.9994
11	Benzyl cyanide	a)	21.87	140-29-4	117 > 90	0.4	4	90 > 63	0.9	4	0.002–5	0.9985
12	Diethyl maleate	a)	22.93	141-05-9	127 > 99	0.4	4	127 > 82	0.5	4	0.01–10	0.9995
13	Methyl heptine carbonate	b)	24.29	111-12-6	123 > 93	0.7	4	123 > 67	0.5	4	0.05–20	0.9996
14	4-Methoxyphenol	a)	24.70	150-76-5	124 > 109	0.6	4	109 > 81	0.6	3	0.01–20	0.9988
15	Citronellol	b)	25.98	106-22-9	95 > 67	0.4	4	128 > 81	0.7	3	0.05–20	0.9991
16a	Citral, isomer 1	a)	26.41	5392-40-5	137 > 95	0.5	3	137 > 109	0.4	3	0.2–20	0.9996
16b	Citral, isomer 2	a)	27.62	5392-40-5	137 > 95	0.5	3	137 > 109	0.4	3	0.2–20	0.9996
17	Geraniol	a)	27.06	106-24-1	93 > 65	1.2	4	123 > 81	0.7	5	0.5–20	0.9986
18	Cinnamal	a)	27.36	104-55-2	131 > 103	0.7	4	131 > 77	0.8	4	0.01–20	0.9999

			CAS									
19	4-Ethoxy-phenol	a)	27.71	622-62-8	138	>	110	0.6	3	110 > 82	0.6	3	0.02–20	0.9995
20	Anisyl alcohol	b)	27.82	105-13-5	138	>	109	0.5	4	109 > 94	0.7	5	0.02–20	0.9990
21	Hydroxy-citronellal	a)	28.10	107-75-5	95	>	67	0.4	4	121 > 93	0.7	3	0.5–50	0.9991
22	4-tert-Butylphenol	a)	28.66	98-54-4	135	>	107	0.6	3	107 > 77	0.6	3	0.005–5	0.9995
23	Cinnamyl alcohol	a)	29.12	104-54-1	92	>	65	1.0	4	115 > 89	0.7	4	0.1–20	0.9986
24	4-Phenyl-3-buten-2-one	a)	31.64	122-57-6	103	>	77	0.5	4	131 > 103	0.4	3	0.02–10	0.9996
25	Eugenol	a)	31.91	97-53-0	164	>	149	0.6	3	164 > 131	0.7	3	0.2–20	0.9992
26	Dihydrocoumarin	a)	32.51	119-84-6	120	>	91	0.6	4	148 > 120	0.8	4	0.005–5	0.9991
27	Coumarin	a)	35.17	91-64-5	118	>	90	0.4	3	146 > 118	0.8	3	0.005–5	0.9987
28	Isoeugenol	a)	36.23	97-54-1	164	>	149	0.6	4	164 > 131	0.7	4	0.2–20	0.9985
29	2,4-Dihydroxy-3-methylbenzaldehyde	a)	36.83	6248-20-0	151	>	67	1.2	4	151 > 95	1.2	4	0.2–10	0.9768
30	alpha-iso-Methylionone	b)	38.54	127-51-5	135	>	79	0.8	4	135 > 107	0.4	4	0.01–10	0.9999
31	Lilial	b)	39.95	80-54-6	189	>	131	1.0	3	147 > 129	0.8	4	0.01–20	0.9999
32a	Pseudoionone, isomer 1	a)	40.16	141-10-6	109	>	79	0.6	4	149 > 93	0.7	4	0.2–20	0.9995
32b	Pseudoionone, isomer 2	a)	42.34	141-10-6	109	>	79	0.6	4	149 > 93	0.7	4	0.2–20	0.9995
33	6-Methylcoumarin	a)	40.88	92-48-8	160	>	132	0.4	4	132 > 103	1.6	4	0.02–20	0.9993
34	7-Methylcoumarin	a)	40.88	2445-83-2	160	>	132	0.4	4	132 > 103	1.6	4	0.02–20	0.9993
35	Diphenylamine	a)	43.62	122-39-4	169	>	140	3.2	3	169 > 115	3.2	3	0.005–5	0.9998
36	4-(p-Methoxyphenyl)-3-butene-2-one	a)	44.50	943-88-4	161	>	133	0.6	3	176 > 145	0.6	4	0.02–10	0.9983

(Continued)

Table 4.24 (Continued)

No.	Compound name	EU Reg.	Retention time (min)	CAS Number	Quantitation Transition m/z m/z	Excitation voltage m/z	isol. wind (Da)	Qualifier transition m/z m/z	Excitation voltage* (V)	isol. wind. (Da)	Linear ranges (mg/L)	Correlation coefficient R^2
37	Amylcinnamal	a)	45.09	122-40-7	203 > 145	1.2	4	203 > 129	1.6	4	0.05–20	0.9991
38	Lyral	a)	45.47	31906-04-4	136 > 107	0.4	3	136 > 79	0.7	3	0.5–50	0.9996
39	Amylcinnamyl alcohol	a)	46.57	101-85-9	133 > 115	0.7	4	187 > 130	0.7	3	0.5–50	0.9997
40a	Farnesol, isomer 1	b)	47.28	4602-84-0	93 > 77	0.4	4	107 > 91	0.5	4	1.0–50	0.9963
40b	Farnesol, isomer 2	b)	47.81	4602-84-0	93 > 77	0.4	4	107 > 91	0.5	4	1.0–50	0.9963
40c	Farnesol, isomer 3	b)	48.24	4602-84-0	93 > 77	0.4	4	107 > 91	0.5	4	1.0–50	0.9963
41	7-Methoxycoumarin	a)	47.37	531-59-9	148 > 133	0.7	3	176 > 148	0.8	3	0.05–20	0.9989
42	1-(p-Methoxyphenyl)-1-penten-3-one	a)	48.64	104-27-8	161 > 133	0.6	4	190 > 161	1.4	4	0.01–10	0.9997
43	Hexyl-cinnamaldehyde	b)	49.08	101-86-0	145 > 117	0.7	4	129 > 102	1.7	5	0.05–20	0.9999
44	Benzyl benzoate	b)	49.46	120-51-4	194 > 165	1.7	3	105 > 77	0.6	3	0.02–20	0.9992
45	4-tert-Butyl-3-methoxy-2,6-dinitrotoluene	a)	52.23	83-66-9	253 > 219	0.6	5	253 > 121	1.3	5	0.2–50	0.9985
46	Benzyl salicylate	a)	53.47	118-58-1	91 > 65	0.4	3	228 > 210	0.8	3	0.2–10	0.9850
47	7-Ethoxy-4-methylcoumarin	a)	56.21	87-05-8	204 > 148	1.9	4	148 > 91	1.1	4	0.1–10	0.9988
48	Benzylcinnamate	b)	60.61	103-41-3	131 > 103	0.7	3	192 > 115	1.1	4	0.05–20	0.9984

a) Banned compound as of EU Directive 2009/48/EC, technically unavoidable traces may not exceed 100 mg/kg.
b) Requires declaration if concentration exceeds 100 mg/kg.
* Related to ion trap MS.

(Lv et al., 2013b, reproduced with permission of John Wiley & Sons).

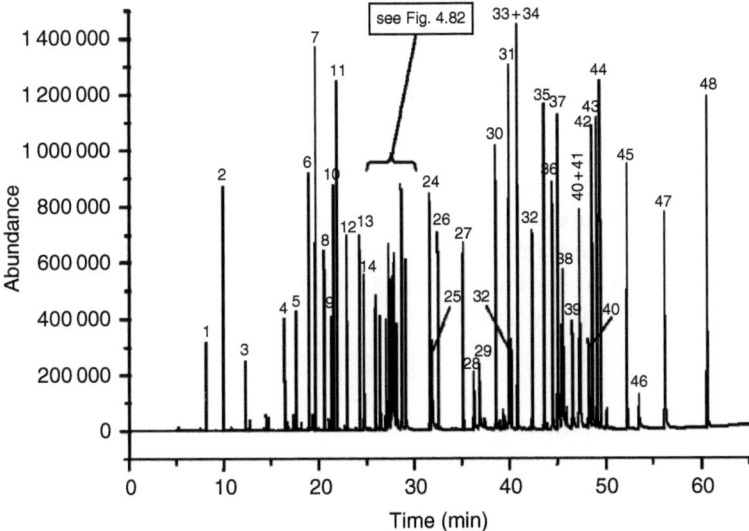

Figure 4.62 Separation of the 48 fragrance allergens, standard at 40 mg/L, peak numbers as of Table 4.28. (Lv et al., 2013b, reproduced with permission of John Wiley & Sons.)

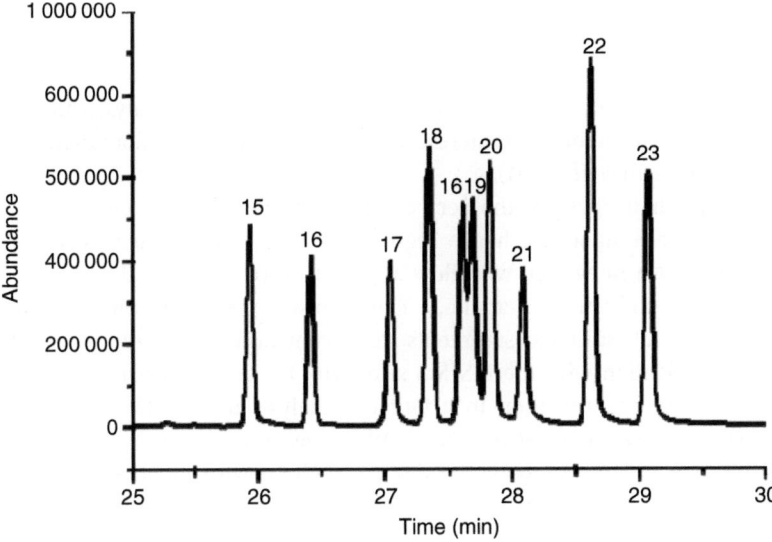

Figure 4.63 Compound separation detail of Figure 4.81. (Lv et al., 2013b, reproduced with permission of John Wiley & Sons.)

The LOD and LOQ values of the method have been determined based on the attained S/N ratio with three, respectively 10 times the S/N. The achieved LODs in toy materials ranged from 0.005 to 5.0 mg/kg and the LOQs from 0.02 to 20 mg/kg, being significantly lower than the required limits in the EU regulation of 100 mg/kg. The obtained recoveries ranged from 79.5% to 109.1%. Commercial toys were analyzed in order to demonstrate the applicability of the method (Lv et al., 2013b).

4.18.6 Conclusions

The described method provides a general approach for the accurate and effective determination of 48 fragrance allergens in toy samples and similar consumer goods. The MS/MS acquisition mode instead of SIM allows the identification and quantitation of the fragrance allergens at levels compliant with the current EU regulations in the complex matrix of the toy samples.

4.19 Metabolite Profiling

Keywords: metabolomics; phenotype; genotype; SPME; single quadrupole MS; triple quadrupole MS/MS; AMDIS; deconvolution; biomarker; derivatization; methoximation; silylation; library search; quantification

4.19.1 Introduction

Metabolite profiling and identification is the qualitative and quantitative study of biotransformations. The analysis of metabolites is highly complex due to the chemical diversity of such small biological molecules. Metabolite target analyses may be restricted to specific metabolites of interest, which can be selectively monitored and quantified (Weckwerth et al., 2004).

The analytical challenges are the very complex chromatographic coelution of a huge number of compounds, hence requiring a mass spectral deconvolution for identification. An integrated workflow approach of two analytical phases for discovery and quantification is applied. This two-stage concept combines the advantages of the full scan measurements for deconvolution and library search compound identification with the MS/MS selectivity for chemical structures and allows for absolute quantification in a targeted analysis using SRM, as shown in Figure 4.64 (Hübschmann et al., 2012). With the described 2-tier workflow, scientists are able to conduct both the discovery phase with identification and the quantification analysis with a single instrument using the triple quadrupole GC-MS/MS technology.

4.19.1.1 Workflow Phase I: Discovery

The discovery phase provides the identification of as many metabolites as possible by GC-MS analysis in the full scan mode. This first phase is dominated by the deconvolution of the chromatograms to extract the full and representative mass spectra as well as the compound RT for subsequent identification. Figure 4.65 shows the elution of the different metabolite compound classes using the described analytical method (Fragner et al., 2010).

4.19 Metabolite Profiling | 657

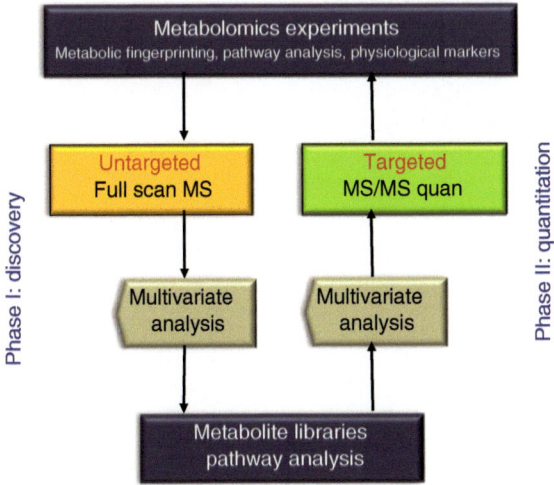

Figure 4.64 Metabolite profiling two-stage GC-MS/MS Workflow.

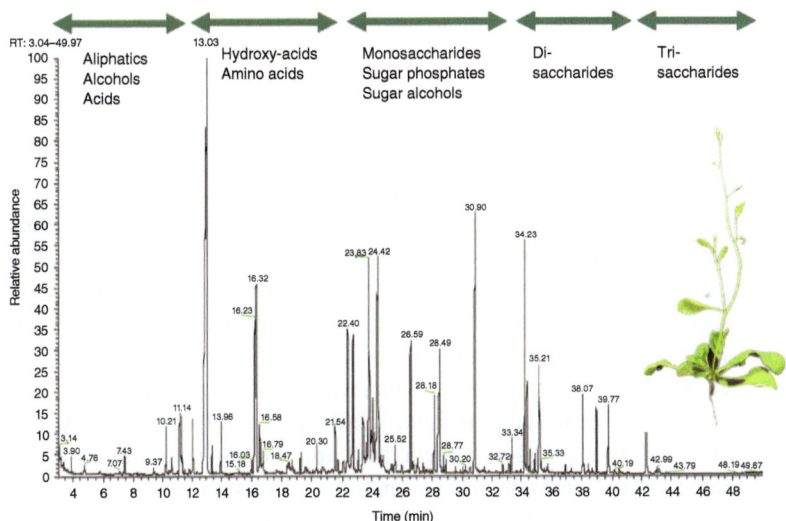

Figure 4.65 General metabolite profile with the elution of the different metabolite compound classes (Arabidopsis Thaliana).

The deconvolution step is greatly facilitated by using the AMDIS program and the NIST mass spectral library search program (Mallard and Reed, 1997). Unique compound mass spectra are extracted by the analysis of all the transient ion signals allocating ion masses and relative intensities for each of the eluting compounds.

The identification of metabolites is based on the characteristic EI fragmentation patterns as well as on RT. The mass spectrum identification is facilitated by searching large databases with NIST, Wiley, or dedicated collections of mass spectra like

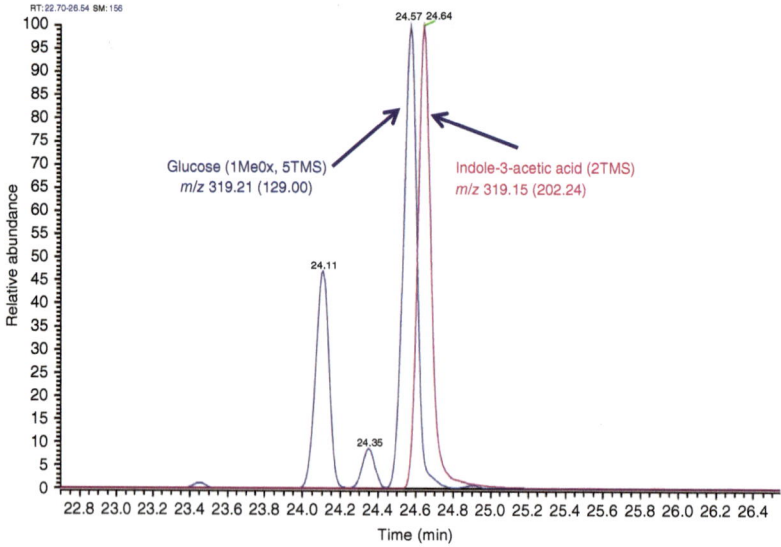

Figure 4.66 Separation of coeluting compounds with same precursor ion by MS/MS, indole-3-acetic acid (IAA): m/z 319.15 > 202.24, glucose: m/z 319.21 > 129.00.

the Fiehn metabolite spectrum library, or already available in-house metabolite databases, which are linked seamlessly into the search procedure.

4.19.1.2 Workflow Phase II: Targeted Quantitation

For the targeted quantification phase, suitable MS/MS transitions for each compound are applied using the identical chromatography setup and the RTs defined in phase I. Using a triple quadrupole MS, the selected precursor ions get fragmented to structure specific product ions in the collision cell. With the applied triple quadrupole MS, the AutoSRM optimization procedure can be used at this stage for a large number of the selected metabolites. The integration of the detected product ion peaks provides the selective quantification of all target metabolites. Figure 4.66 shows the chromatograms of indole-3-acetic acid and glucose compounds at the level of 50 pmol injected amount. Both compounds although coeluting can be integrated independently from each other due to the different SRM transitions used, to m/z 129.00 for glucose and to m/z 202.24 for indole-3-acetic acid.

4.19.2 Analysis Conditions

Sample material		Seeds, grains, plant material
Sample preparation	SPME extraction	Fiber type DVB/CAR/PDMS
		4 g of rice into a 20 mL headspace vial, capped
		Incubation temperature 80 °C, 30 min
		Extraction time 30 min

		Derivatization	Liquid extraction with methanol/water/DCM
			Dried extracts derivatized with BSTFA or MSTFA
GC method	System		Thermo Scientific TRACE 1310 GC
	Injector		Split/splitless, SPME desorption time 5 min
	Column	Type	TG-5MS
		Dimensions	30 m length × 0.25 mm ID × 0.25 µm film thickness
	Carrier gas		Helium
	Flow		Constant flow, 1.2 mL/min
	SSL injector	Injection mode	Split 1 : 10
		Inlet liner	Split liner with glass wool
		Injection temperature	285 °C
		Injection volume	1 µL
	Oven program	Start	60 °C, 4 min
		Ramp 1	8 °C/min to 170 °C
		Ramp 2	4 °C/min to 300 °C
		Final temperature	300 °C, 15 min
	Transferline	Temperature	285 °C
MS method	System		Thermo Scientific TSQ 8000
	Analyzer type		Triple quadrupole MS
	Ionization		EI, 70 eV
	Acquisition	Mode	Full scan, timed-SRM
			SRM, 2 SRMs per compound
	Collision gas		Ar, 1.5 mL/min
	Mass range		29–350 Da (SPME), 44–600 Da (derivatized extracts)
	Scan rate		0.2 s (5 spectra/s)
	Resolution	Setting	Normal (0.7 Da)

4.19.3 Sample Preparation

For the analysis of volatile compounds, the sample material, for example, rice or leaves, is weighed in equal amounts into headspace vials and capped. For the analysis of the extracted metabolites, plant material (leaves) gets homogenized under liquid nitrogen. About 50 mg is applied to extraction with a water/chloroform/methanol mixture to extract water soluble metabolites. The polar phase is dried in a vacuum centrifuge. Next, a two-step derivatization is applied: First,

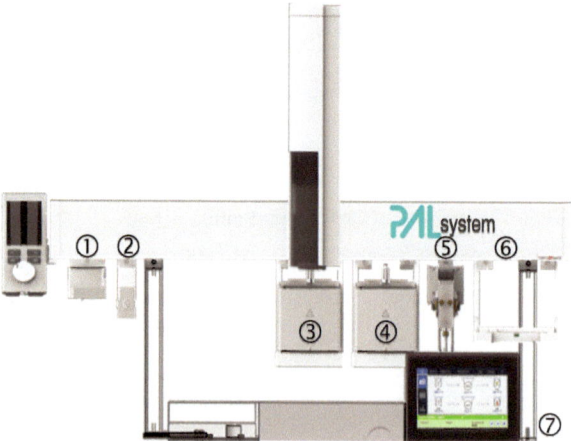

Figure 4.67 PAL RTC x,y,z robotic sampling system in configuration for two different incubation temperatures. 1 Tool park station for sample prep and GC injection tool. 2 Standard wash station for reagents and standards. 3 Incubator 1 30 °C (optional). 4 Incubator 2 37 °C. 5 Syringe fast wash station. 6 Tray holder for sample vials. 7 Mounting on GC top.

a methoxyamination (methoxyamine hydrochloride in pyridine) is used to suppress keto-enol tautomerism, followed by a regular silylation using BSTFA (N,O-bis(trimethylsilyl)trifluoroacetamide) or MSTFA (N-methyl-N-trimethylsilyl-trifluoroacetamide) to derivatize polar functional groups. The final derivatization volume is 100 μL. Standards get dissolved in methanol or water, diluted into various concentrations, dried, and derivatized using the same procedure (Weckwerth et al., 2004; Fragner et al., 2010). Recent developments using industry standard x,y,z robotic samplers allow the automated two-step derivatization of samples for large sample series (Chong and Huebschmann, 2020). A PAL RTC x,y,z-robot in the used configuration for, if wanted, the initially proposed two different reaction temperatures of 30 °C, and the regular 37 °C is shown in Figure 4.67 mounted on top of a GC-MS unit for online injection. The automated workflow derivatizes all samples on the same timeline in a so-called *"prep ahead"* mode during a previous run, ready for injection upon a next *GC Ready* signal (Figure 4.68). The timely derivatization identical for all samples without waiting times secures the optimum reproducibility for large sample series.

4.19.4 Results

The HS-SPME measurements presenting the flavor profile had been carried out to distinguish different varieties of rice. The achieved full scan chromatograms in Figure 4.69 show a significantly different pattern of the collected volatile components. Based on the recorded complete mass spectra, the AMDIS deconvolution with peak identification can be performed. Selected compounds that had been identified with their relative signal intensity to the first sample are given in Table 4.25.

4.19 Metabolite Profiling | 661

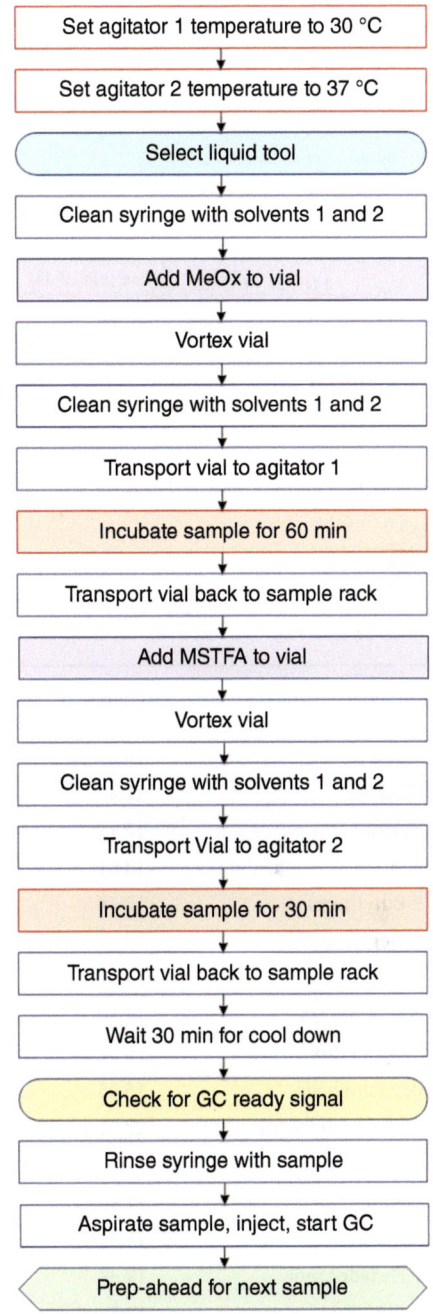

Figure 4.68 Automated workflow for the two-step methoximation and silylation of large samples series operating in the "prep ahead" mode.

4 Applications

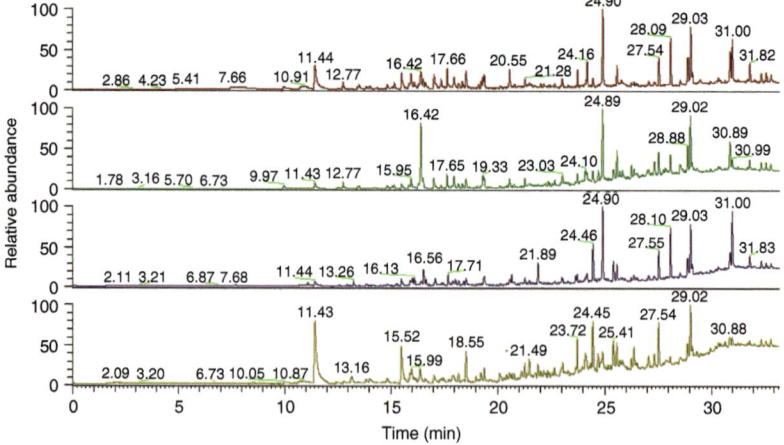

Figure 4.69 Phenotype analysis by SPME GC-MS (four different rice varieties).

Table 4.25 Selected compounds detected by HS-SPME in samples of four different rice varieties with relative intensity to sample 1.

Compound name	RT (min)	Sample 1	Sample 2	Sample 3	Sample 4
Toluene	10.03	+++	++	+	+
Hexanal	11.42	+++	+	+	++
Xylene	12.77	+++	++	+	+
Styrene	13.51	+++	++	+	+
2-Butylfuran	13.14	+++	+	+	++
2-Heptanone	13.83	+++	+	+	++
Heptanal	14.01	+++	+	+	++
2-Pentylfuran	15.50	+++	+	+	++
1-Octen-3-ol	15.91	+++	+	++	++
Benzaldehyde	15.95	+++	++	+	+
2-Ethyl-1-hexanol	17.05	+++	++	+	+
Undecane	17.36	+++	++	++	+
Phenol	17.66	+++	++	++	+
Nonanal	18.54	+++	+	+	++
1-Hexadecanol	19.33	+++	++	+	−
Dodecane	19.39	+++	++	+	+
1-Methylene-1H-indene	20.56	+++	++	++	+
2-Butyl-2-octenal	23.71	+++	+	+	++
4-Octadecylmorpholine	24.90	+++	++	++	+

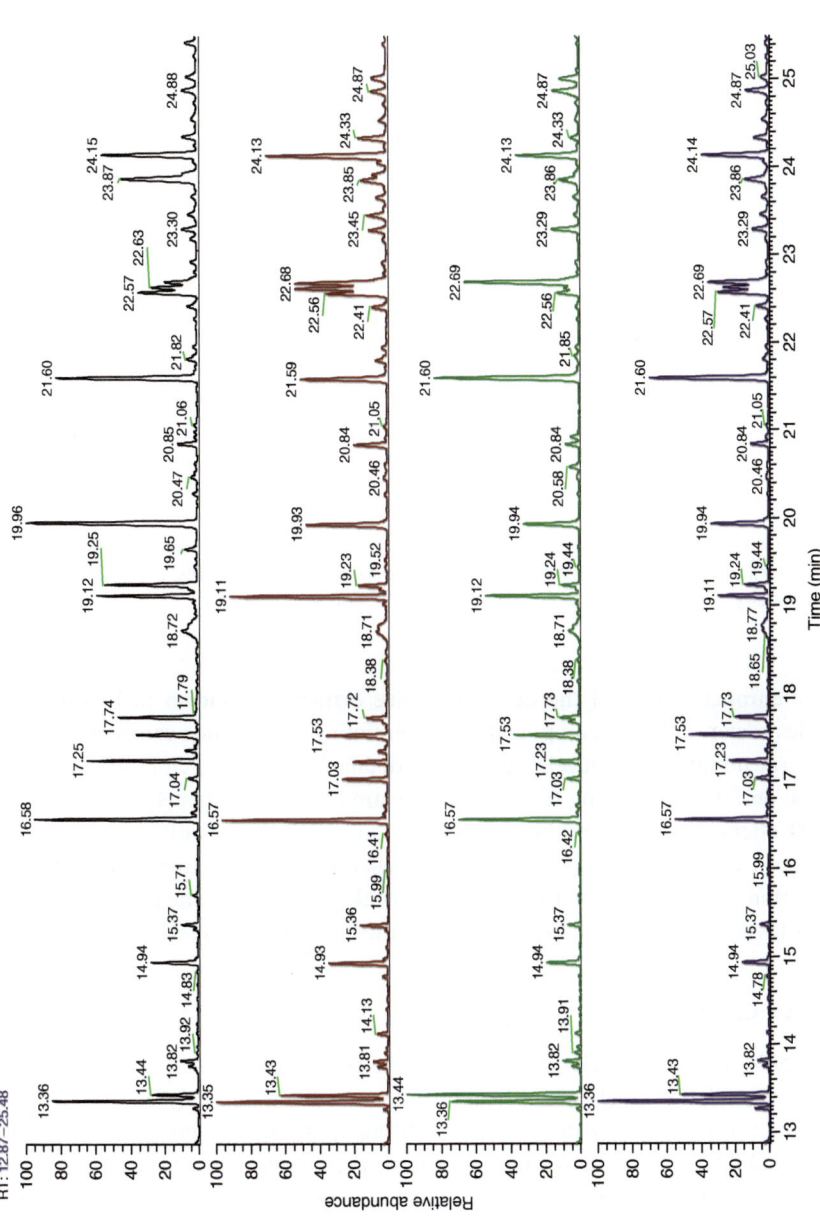

Figure 4.70 Extract analyses after BSTFA derivatization (four difference rice varieties).

4 Applications

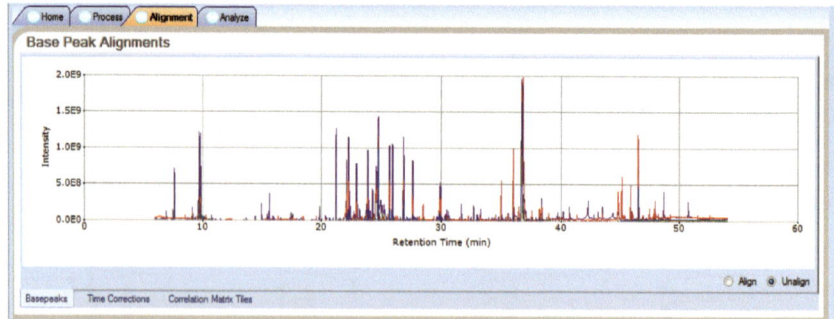

Figure 4.71 Retention time alignment of the sample chromatogram (SIEVE software).

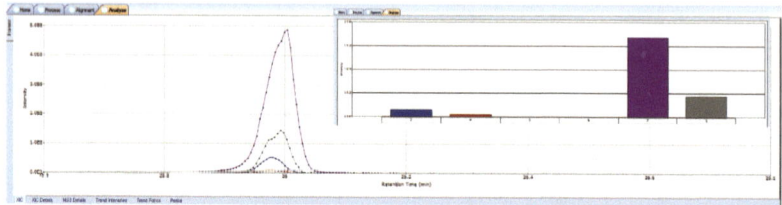

Figure 4.72 A discriminator compound identified in four of six samples, three times with low intensity, one sample with high intensity (SIEVE software).

The chromatograms of the derivatized rice extracts are shown in Figure 4.70. Besides major common components differentiating compounds can be identified on a large number at low level intensities. A detailed data analysis for differentiators can be performed by using the SIEVE software (Thermo Fisher Scientific). After the RT alignment of the chromatograms in Figure 4.71, the quantitative analysis of the deconvoluted peaks is performed and presented in tabular form and for visual inspection in Figure 4.72 with the overlaid peak profiles from the chromatograms under investigation, as well as bar graphs of the peak areas as quantitative results.

4.19.5 Conclusion

The ability to use a single instrument when performing both the screening for discovery and the target quantification of metabolites not only accelerates the profiling process, thereby increasing the throughput, but also provides optimized reproducibility in the fragmentation pattern, peak intensities, and mass accuracies. In addition, compounds which typically cannot be separated through gas chromatography can be separated using compound-specific SRM transitions in a single run for individual quantification. The high sensitivity, dynamic range, and selectivity of the triple quadrupole method enable scientists to effectively perform metabolite profiling.

GC-MS/MS provides both metabolomic workflow phases on one instrument platform:

Phase I: Discovery phase analysis

- Marker identification in the full scan mode
- Fast full scan analysis with deconvolution
- Access to the largest mass spectral know-how bases
- Reference library building

Phase II: Target compound quantification

- Selective and accurate compound quantification using the MS/MS mode
- High dynamic range for interesting metabolite biomarkers
- Complex mixture analysis using the compound-specific SRM mode
- Coeluting compounds get separated and individually quantified by SRM.

Large sample series as typical in metabolite profiling can be processed automatically using industry standard x,z,y-robots for highest reproducibility and comparability of the achieved data for large data analyses.

4.20 Extractables and Leachables

Keywords: product safety; pharmaceutical products; quality control; leachables; extractables; solid/liquid extraction; purge and trap; derivatization; AMDIS deconvolution; unknown identification; Mass Frontier spectrum interpretation; single quadrupole MS

4.20.1 Introduction

Leachables in pharmaceutical products are "trace amounts of chemicals originating from containers, medical devices or process equipment that end up as contaminants in medicinal products resulting in exposure to patients" (Moffat, 2011).

The U.S. Food and Drug Administration (FDA) defines extractables as compounds that can be extracted from a container material when in the presence of a solvent, and leachables as those compounds that leach into the drug product formulation as a result of direct contact with the formulation (Lewis, 2011).

The European Agency for the evaluation of Medical Products Guideline (European Medicines Agency, EMA, 2022) states that "it should be determined whether any of the extractables are also leachables present in the formulation at the end of the shelf life of the product or to the point equilibrium is reached if sooner". A guideline on the assessment and control of extractables and leachables (E&L) is proposed by the International Council for Harmonisation of Technical Requirements for Pharmaceuticals for Human Use (ICH) (ICH, 2020).

The identification of leachables, and attribution to the contact component from which they originate, is important because such species may react with the drug product or formulation ingredients, compromise the efficacy of the drug product, or interfere with dosage consistency, which finally may pose a negative health effect.

GC-MS analysis is mostly applied for volatile components using HS, or after a solvent extraction step for semi-volatile compounds. This application describes the analysis of a polymer plunger material using different extraction techniques, derivatization, and HS analysis by single quadrupole GC-MS. A parallel classical flame ionization detector (FID) detection channel was configured which shows the chromatogram similarity and can be used for a future simplified GC-FID routine QC method.

4.20.2 Analysis Conditions

Sample material			Elastomer material, polymers
Sample preparation	Purge and trap		Tekmar LSC 2000
		Sparge vessel	Frit sparge, 25 mL volume
		Purge gas	Nitrogen
		Trap	Tenax
		Sample volume	5 mL
		Sample temperature	40 °C
		Standby	30 °C
		Sample preheat	2.50 min
		Purge duration	10 min
		Purge flow	40 mL/min
		Desorb preheat	175 °C
		Desorb temperature	180 °C
		Desorb time	4 min
		MCM desorb	5 °C
		Bake	10 min at 200 °C
		MCM bake	90 °C
		BGB	Off
		Mount	60 °C
		Valve	220 °C
		Transfer line	200 °C
		Connection to GC	LSC 2000, in line with the carrier gas supply of the injector
GC method	System		Thermo Scientific TRACE 1300 GC
	Autosampler		Thermo Scientific TriPlus RSH
	Column	Type	TG-5MS
		Dimensions	30 m length × 0.25 mm ID × 0.25 μm film thickness

	Transfer capillaries	To FID	0.2 m × 0.2 mm ID
		To MS	2.0 m × 0.15 mm ID
	Carrier gas		Helium
	Flow		Constant pressure, 125 kPa
	Injection		Autosampler Thermo Scientific TriPlus RSH and manual
	SSL injector	Inlet liner	4 mm ID with glass wool
		Injection mode	Splitless
		Injection temperature	320 °C
		Injection volume	1 µL of liquid extracts
			1 mL headspace volume
	Split	Closed until	1 min for liquid extracts
		Open	For headspace injections
		Open	1 min to end of run
		Split flow	20 mL/min
		Purge flow	5 mL/min
	Oven program liquid injections	Start	40 °C, 1 min
			8 °C/min to 325 °C
		Final temperature	325 °C, 14 min
	Oven program headspace injections	Start	30 °C, 3 min
		Ramp 1	8 °C/min to 280 °C
		Final temperature	280 °C, 10 min
	Transferline	Temperature	300 °C
			Direct coupling
Detector	FID	Temperature	300 °C
		Air	350 mL/min
		Hydrogen	35 mL/min
		Nitrogen	40 mL/min
MS method	System		Thermo Scientific ISQ
	Analyzer type		Single quadrupole MS
	Ionization		EI, 70 eV
	Ion source	Temperature	220 °C
		Emission current	50 µA
	Acquisition	Mode	Full scan
	Mass range		25–700 Da
	Scan rate		4 scans/s (250 ms/Scan)
	Resolution	Setting	Normal (0.7 Da)

4.20.3 Sample Preparation

The elastomeric plunger material was examined in different ways. The volatiles were determined via direct static HS analysis. For the headspace injection, 10 plungers had been placed in a 20 mL headspace vial.

The extractables of the sample were analyzed by preparing different liquid extracts using three extraction procedures and derivatization:

1) Aqueous extraction of the elastomer material, followed by a dichloromethane (DCM) extraction of the aqueous phase, no derivatization.
2) The above DCM extract has been derivatized using BSTFA (Kumirska et al., 2013).
3) Isopropanol (IPA) extraction, no derivatization.

4.20.4 Experimental Conditions

The analyses were performed using GC with dual detection by FID and single quadrupole MS. The dual detection was accomplished by using a SilFlow™ microfluidic connection device, also allowing a no-vent option for easy column change without a venting of the mass spectrometer. The GC was equipped with an autosampler for both liquid and headspace injections.

4.20.5 Data Processing and Results

The mass spectrometer detection in the full scan mode was chosen for identification of the unknown compounds. The parallel FID detection was checked for compliance with the MS TIC for future fast quality control purposes.

The chromatogram of the static HS analysis is shown in Figure 4.73. The analyses of the liquid extractions is shown in Figures 4.74–4.76, all of them with the MS TIC and FID traces. All chromatograms demonstrate the parallel detection with FID plus full scan MS in very good agreement of the eluted compound pattern.

AMDIS was used for a deconvolution of the complex chromatograms extracting the "clean" background and coelution corrected single compound mass spectra. The NIST library was used for unknown search and spectral comparison. AMDIS associates the found RT and mass spectrum for an improved identification. All results can optionally be transferred to MS Excel™ for further investigation and documentation.

Mass spectra not satisfactorily identified by the library search were analyzed in the fragmentation pattern by using Mass Frontier resulting in structural proposals (Thermo Fisher Scientific, 2011).

4.20.6 AMDIS Chromatogram Deconvolution

The AMDIS deconvolution program works in three steps executed in the background of the processing (Mallard and Reed, 1997):

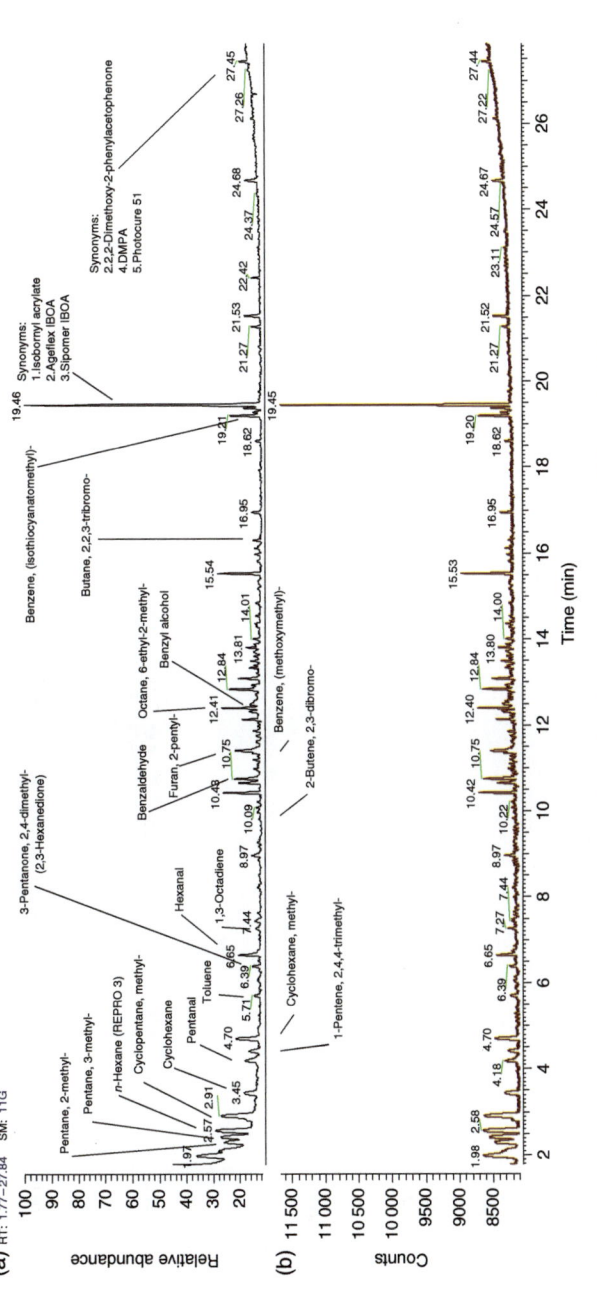

Figure 4.73 Chromatogram of the elastomeric plunger material by headspace analysis, (a) MS TIC and (b) FID.

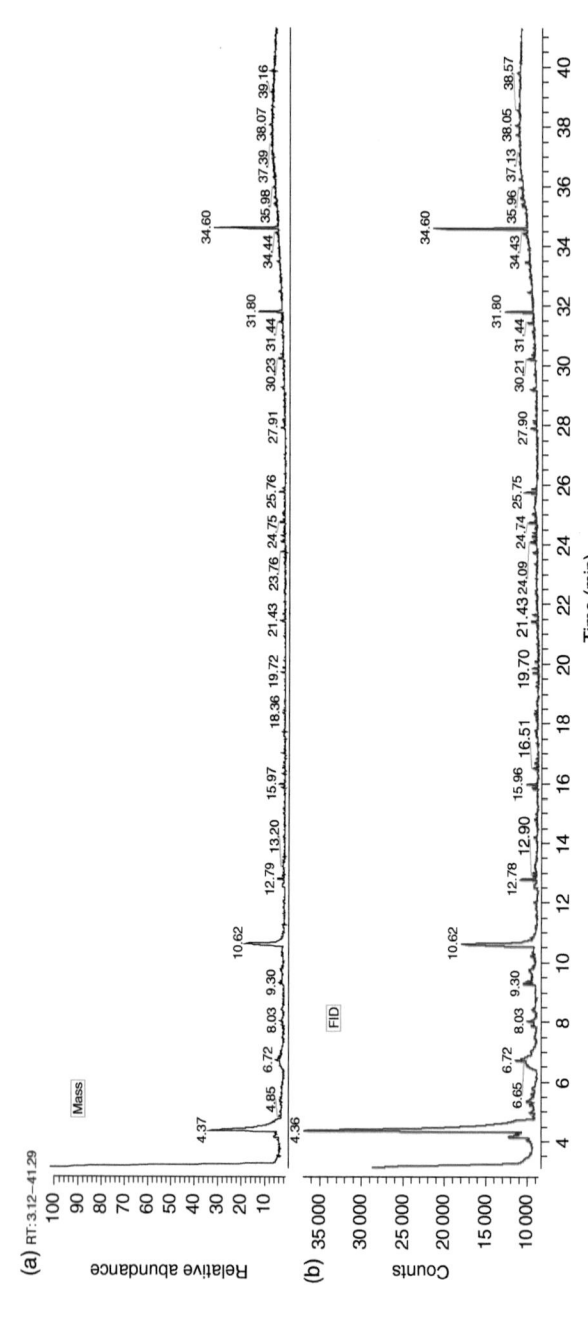

Figure 4.74 Chromatogram of the liquid DCM extract, (a) MS TIC and (b) FID.

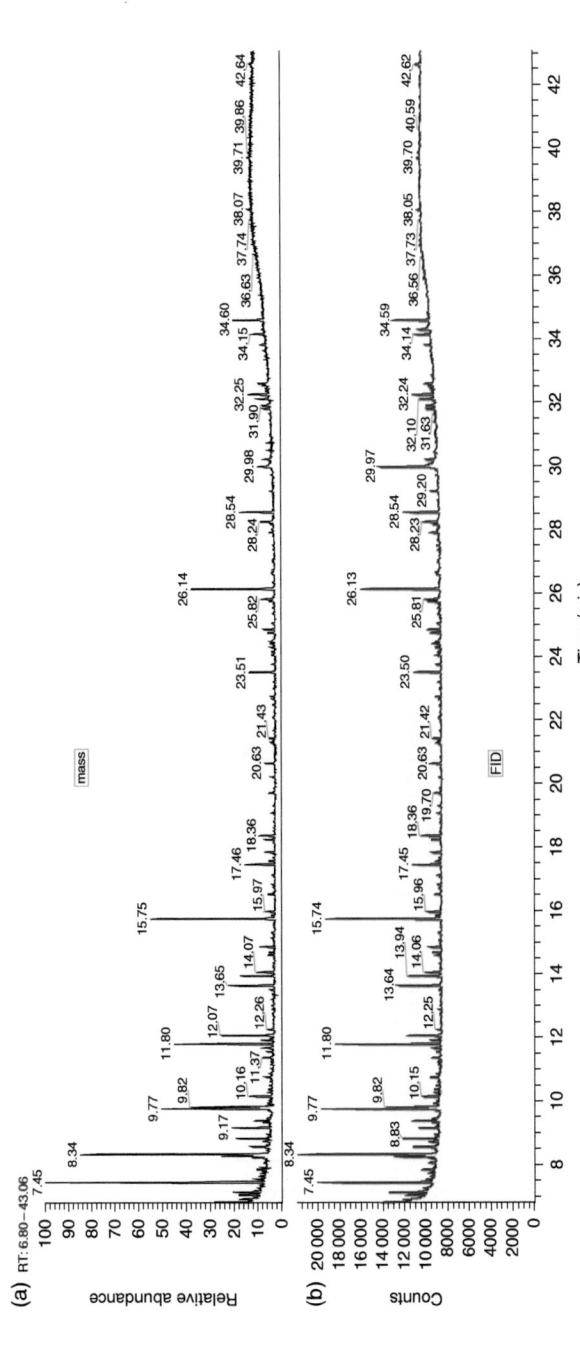

Figure 4.75 Chromatogram of the liquid DCM extract, derivatized with BSTFA, (a) MS TIC and (b) FID.

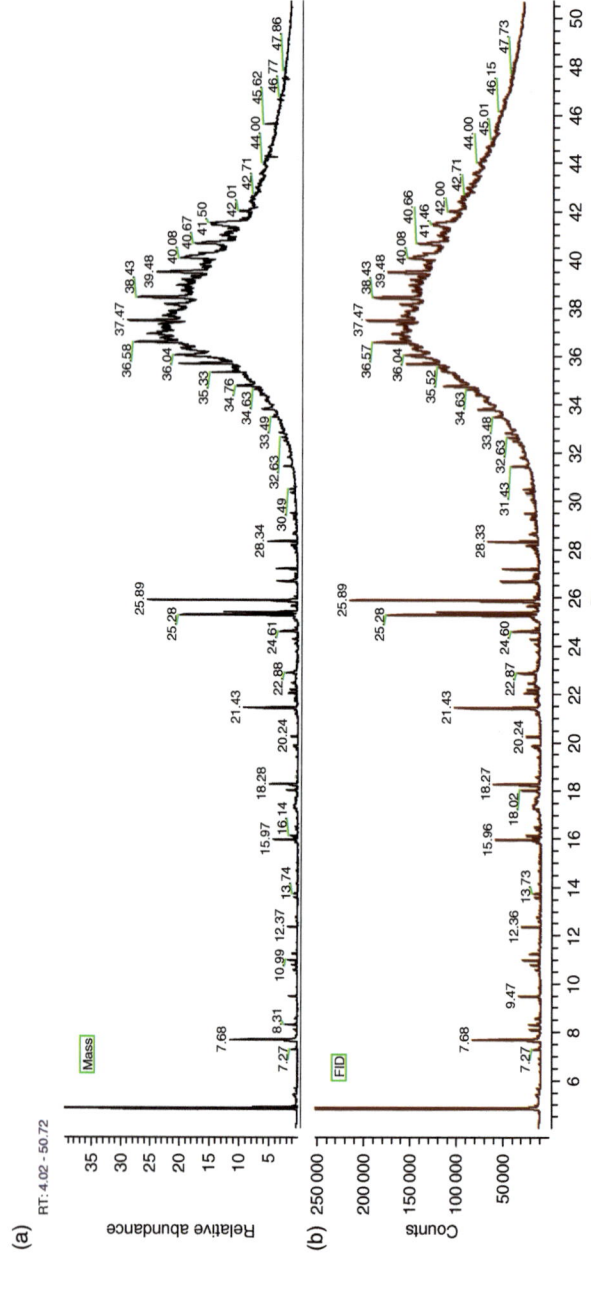

Figure 4.76 Chromatogram of the liquid IPA extract, (a) MS TIC and (b) FID.

Table 4.26 Compounds identified by static headspace analysis (AMDIS).

Name	RT (min)	Reverse FIT value	Area (cts)	Area (%)	Height (cts)	Height (%)
Pentane, 2-methyl-	2.28	85	1917	4.7	368	2.4
Pentane, 3-methyl-	2.40	87	2308	5.6	442	2.8
Cyclopentane, methyl-	2.93	79	2847	7.0	515	3.3
Cyclohexane, methyl-	4.72	82	1333	3.3	346	2.2
2,3-Hexanedione	6.40	83	238	0.6	106	0.7
Hexanal	6.65	82	825	2.0	300	1.9
Propanal, 2,2-dimethyl-	10.43	88	950	2.3	538	3.5
Octane, 2,2,6-trimethyl-	10.66	88	445	1.1	259	1.7
Benzaldehyde	10.75	92	1031	2.5	461	3.0
Octane, 2,6,6-trimethyl-	13.34	79	164	0.4	115	0.7
2-Propenoic acid	19.46	95	6465	15.8	3438	22.1
Ethanone, 2,2-dimethyl	27.44	82	217	0.5	125	0.8

Step 1: AMDIS analyzes the chromatogram. It counts the number of eluted compounds based on a minimum of ions that show a common RT maximum (maximizing masses peak finder). The corresponding mass spectrum is extracted and cleared from potential contributions from baseline and coeluting compound mass intensities.

Step 2: AMDIS checks if target compounds from a user library are present by matching simultaneously the RT (or retention index if available) and mass spectrum.

Step 3: All detected compound spectra are compared with the spectra of the linked libraries allowing a filter with different criteria.

Tables 4.26–4.28 summarize the identified compounds of the described extraction experiments after AMDIS deconvolution with compound name, CAS numbers as well as the peak quality parameters with RT, the measured peak width, and tailing information. The result of the mass spectrum library comparison is given for each compound. The "Reverse FIT" column informs about the match quality in % of the spectrum pattern of the proposed library entry with the unknown spectrum (reverse search).

4.20.7 Mass Frontier Spectrum Interpretation

Some of the acquired spectra are not included in the commercial libraries, while some matches show structural similarities. The Mass Frontier software analyzes the unknown mass spectrum and associates fragmentation pathways and ion structures with the unknown spectral pattern calculated from the included knowledge base of known fragmentation rules. Two examples of spectrum interpretation of

Table 4.27 Compounds identified by liquid IPA extraction (AMDIS).

Name	RT (min)	Reverse FIT value	Area (cts)	Area (%)	Height (cts)	Height (%)
Isopropyl alcohol	4.85	78	1 737 928	31.0	626 333	28.9
Tricyclo[3.1.0.0(2,4)]hex-3-ene-3-carbonitrile	8.52	78	11 506	0.2	5 183	0.2
Benzyl alcohol	9.48	94	43 713	0.8	23 090	1.1
Benzene, (bromomethyl)-	10.80	90	23 050	0.4	10 571	0.5
Benzyl isopentyl ether	10.99	89	31 562	0.6	18 724	0.9
Benzyl isocyanate	11.25	94	21 095	0.4	10 942	0.5
Heptanoic acid, propyl ester	11.41	86	3 680	0.1	2 310	0.1
Decanone-2	12.60	79	3 716	0.1	2 230	0.1
Dodecane	12.77	92	5 686	0.1	3 521	0.2
Butane, 1,2,2-tribromo-	13.12	79	1 611	0.0	1 096	0.1
Tridecane	14.69	81	3 736	0.1	2 055	0.1
Benzene, (isothiocyanatomethyl)-	15.97	91	114 760	2.1	47 845	2.2
2-Dodecanone	16.40	84	7 773	0.1	3 757	0.2
Tetradecane	16.52	94	11 436	0.2	5 989	0.3
1-Bromo-3-(2-bromoethyl)heptane	18.02	65	57 416	1.0	19 643	0.9
1-Bromo-3-(2-bromoethyl)heptane	18.31	66	143 599	2.6	50 332	2.3
Pentadecane, 3-methyl-	19.39	88	5 468	0.1	2 763	0.1
2-Tetradecanone	19.82	89	16 697	0.3	7 514	0.4
Hexadecane	19.87	90	18 647	0.3	8 612	0.4
N-Benzylidenebenzylamine	22.32	94	17 660	0.3	5 632	0.3
Tetradecanenitrile	24.29	81	17 783	0.3	8 962	0.4
Hexadecanoic acid, methyl ester	24.61	97	61 160	1.1	28 869	1.3
Phthalic acid, butyl cyclobutyl ester	25.06	88	10 308	0.2	4 055	0.2
Decane, 5,6-bis(2,2-dimethylpropylidene)-, (E,Z)-	25.29	72	403 320	7.2	163 977	7.6
Isopropyl palmitate	25.89	76	572 407	10.2	204 258	9.4
Oleanitrile	26.72	80	90 096	1.6	41 750	1.9
Heptadecanenitrile	26.97	76	8 891	0.2	4 527	0.2
Octadecanoic acid, methyl ester	27.19	92	73 043	1.3	38 750	1.8
Isopropyl stearate	28.34	90	99 362	1.8	53 712	2.5
Diisooctylphthalate	31.81	82	22 826	0.4	6 778	0.3

4.20 Extractables and Leachables

Table 4.28 Compounds identified by solid/liquid extraction with DCM (AMDIS).

Name	RT (min)	Reverse FIT value	Area (cts)	Area (%)	Height (cts)	Height (%)
Toluene	4.17	96	17 526	3.7	2 914	3.2
Benzene, 1-fluoro-4-methyl-	4.30	88	6 223	1.3	2 039	2.2
Benzene, 1-fluoro-2-methyl-	4.37	93	244 339	51.0	27 516	29.7
Benzyl alcohol	7.89	90	800	0.2	307	0.3
Benzylalcohol	9.38	88	3 632	0.8	1 081	1.2
Cyclopropyl carbinol	10.63	77	55 086	11.5	9 086	9.8
Decanal	12.79	88	4 070	0.9	2 030	2.2
Diethylphthalate	19.72	92	1 195	0.3	682	0.7
Dibutyl phthalate	25.05	90	1 253	0.3	559	0.6
Ethylhexyl phthalate	31.81	93	6 673	1.4	3 399	3.7

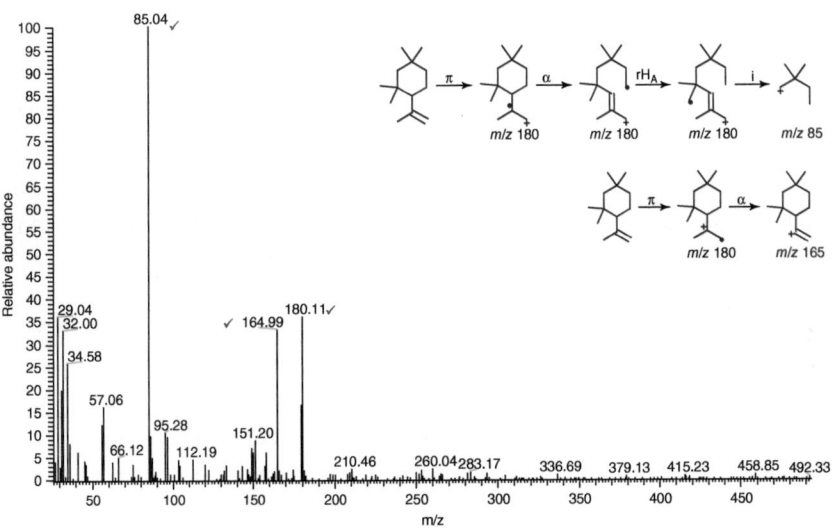

Figure 4.77 DCM extract, unknown peak at RT 12.68 min.

unknown compound spectra are given in Figure 4.77 from the DCM extract and in Figure 4.78 from the IPA extract. The generated proposals of the Mass Frontier expert system show a good plausibility and explain the mass spectrum pattern well (Table 4.29).

4.20.8 Conclusions

The parallel detection using full scan MS and FID shows very good compliance in the detected compound pattern. After identification of the typical major components

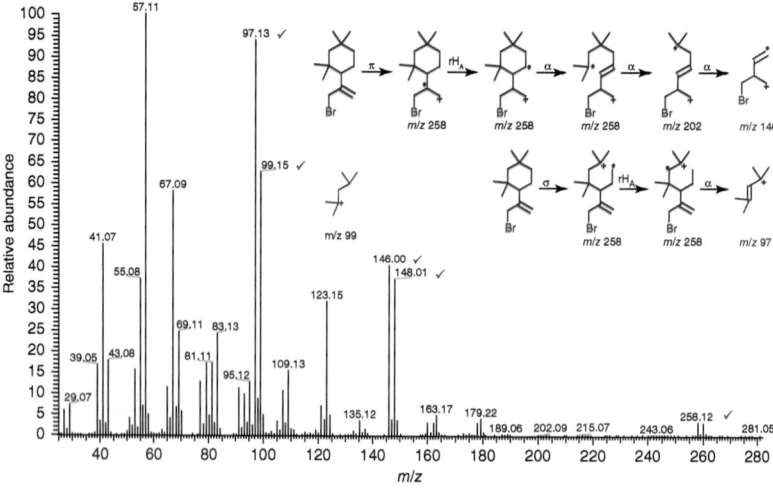

Figure 4.78 IPA extract, unknown peak at RT 18.31 min.

Table 4.29 Compounds identified by solid/liquid extraction with DCM and BSTFA derivatization (AMDIS).

Name	RT (min)	Reverse FIT value	Area (cts)	Area (%)	Height (cts)	Height (%)
Disiloxane, hexamethyl-	6.83	91	1 426	0.5	1 588	1.0
Dimethyl sulfone	7.14	92	5 078	1.7	3 467	2.1
Trifluoromethyl-*bis*-(trimethylsilyl) methyl ketone	7.45	94	14 274	4.8	10 694	6.6
Octane, 4-ethyl-	7.85	94	1 932	0.6	1 158	0.7
1,2-Bis(trimethylsiloxy)ethane	8.34	93	20 903	7.0	12 757	7.8
Cyclopropane, 1-heptyl-2-methyl-	8.58	87	3 614	1.2	2 127	1.3
Silane, (cyclohexyloxy)trimethyl-	8.83	89	4 655	1.6	3 102	1.9
Tetrasiloxane, decamethyl-	9.34	86	3 844	1.3	2 285	1.4
Silane, (1-cyclohexen-1-yloxy)trimethyl-	9.77	90	18 016	6.0	10 881	6.7
Propanoic acid, 2-[(trimethylsilyl)oxy]-, trimethylsilyl ester	9.82	95	8 888	3.0	4 868	3.0
Glycolic acid	10.17	88	3 744	1.3	1 516	0.9
Silane, trimethyl(phenylmethoxy)-	11.80	94	16 044	5.3	9 758	6.0
3,6,9-Trioxa-2-silaundecane, 2,2-dimethyl-	11.93	89	1 501	0.5	974	0.6
Benzoic acid trimethylsilyl ester	13.65	94	7 195	2.4	4 126	2.5
Octanoic acid, trimethylsilyl ester	13.95	87	5 175	1.7	3 037	1.9
Octane, 2,4,6-trimethyl-	14.70	85	1 338	0.5	790	0.5
Butanedioic acid, *bis*(trimethylsilyl) ester	14.86	89	1 936	0.6	1 089	0.7
Nonanoic acid, trimethylsilyl ester	15.75	91	19 185	6.4	10 747	6.6

Table 4.29 (Continued)

Name	RT (min)	Reverse FIT value	Area (cts)	Area (%)	Height (cts)	Height (%)
Benzene, (isothiocyanatomethyl)-	15.97	86	2 663	0.9	1 295	0.8
Decanoic acid	17.46	90	5 016	1.7	2 737	1.7
Lauric acid TMS	20.63	91	1 867	0.6	1 006	0.6
Tetradecanoic acid, trimethylsilyl ester	23.51	86	4 824	1.6	2 576	1.6
Phthalic acid, butyl cyclobutyl ester	25.05	89	655	0.2	348	0.2
Hexadecanoic acid, trimethylsilyl ester	26.14	89	14 483	4.8	7 421	4.6
Octadecanoic acid, trimethylsilyl ester	28.55	86	6 429	2.1	3 375	2.1
Phthalic acid, ethyl hexyl ester	31.81	89	1 724	0.6	869	0.5
4-Methyl-2,4-*bis*(4'-trimethylsilyl-oxyphenyl)pentene-1	33.43	91	652	0.2	380	0.2

using the mass spectrometer, the routine quality control for such compounds can be run reliably using a simplified GC-FID method.

The deconvolution using the AMDIS software allows a precise isolation of the mass spectra from coeluting compounds. The possibility to use an individual library of target compounds, combining RT and mass spectra, makes it a powerful tool for analytical control.

For unknown mass spectra, the Mass Frontier software is a unique tool for spectrum interpretation. Structure proposals and fragmentation pathways are provided for mass spectra allowing a deeper sample and unknown elucidation.

4.21 Volatiles in Car Interiors

Keywords: automobile industry; car interior materials; outgassing; VOCs; SVOCs; FOG; VDA 278; ISO 11890-2; tube absorption; thermal desorption; single quadrupole MS

4.21.1 Introduction

The smell of new cars is often the enchanting scent of a long planned and eagerly awaited delight. This typical scent is prepared intentionally but also contributed by a variety of interior polymer materials of the new automobiles. Owners become increasingly concerned about the air quality inside of a new car as the emitted chemicals are supposed to create a potential health risk.

Studies show that the indoor air of new vehicles carries a high concentration of VOCs released from the vehicle interiors. The total volatile organic compound (TVOC) concentration within the interior of a minivan was determined as high as

7500 µg/m³ of inside air on the second day after delivery, which is approximately two orders of magnitude higher than regular outdoor TVOC concentrations. Over 60 chemicals had been identified in this study inside the interior of vehicles released from different materials such as carpets, pedals, seat covers, door linings, and so on (Grabbs et al., 2000).

The European so-called "solvents emissions" directive provides the currently most stringent regulations on the release of odors from VOCs or fogging of the interior windscreen by semi-volatile emissions (SVOCs, semi-volatile organic compounds, FOG) with the maximum limit values for vehicle refinishing products (European Commission, 2004, 2010).

The standards for the sampling and gas chromatographic testing of paints, varnishes, and related products used in the automobile industry are set by the ISO norm 11890-2 in the new revision from 2020 (ISO, 2020). Other methods ISO 11890-1 and ASTM D2369 uses gravimetric weight difference methods (ASTM, 2024). In the United States, the California Standard Section 01350 specification (CDPH Standard method) is the relevant standard for evaluating and restricting VOC emissions in indoor air, which is also applied to volatile emissions in automobiles (California Department of Public Health, 2017). In China, the overall automotive interior air quality is regulated by the GB Standards for Interior Air Quality of Vehicles GB/T 27630-2011 (Guidelines for Air Quality Assessment in Cars, 2011) and GB 8410-2006 for flammability.

On the background of these international regulations, the organic materials used for automobile interiors need to be screened at the manufacturer and the raw material suppliers for VOC and SVOC release to ensure the air quality inside the car. The reference method for the determination of VOCs and FOG in automotive interior materials is the VDA 278 standard (or GMW 15634) using a TD GC-MS method (VDA 278, 2016; GM Engineering Standards, 2020). VDA stands for the German quality management system for the automobile industry (Verband der Automobilindustrie, Germany). The VDA 278 is part of the delivery specs of the car manufacturers Daimler, BMW, Audi, Porsche, and Volkswagen. GM/Opel uses the corresponding GM Engineering Standards GMW 15634. The VDA 278 analysis procedure serves for the determination of emissions from nonmetallic materials which are used for interior parts in motor vehicles like textiles, carpets, adhesives, scaling compounds, foam materials, leathers, plastic parts, foils, lacquers, or combinations of different materials (VDA 278, 2016). It provides semiquantitative values of the emission from these materials of VOCs and the semi-volatile condensable substances (SVOCs, FOG). The term "fog" is used here as these less volatile substances can condense at ambient temperatures and contribute to the fogging of the windshield. The suggested analysis method uses TD GC-MS for both VOCs and SVOC/FOG analysis.

In a first step, the VOC analysis is determined with a thermal desorption of the sample material at 90 °C for 30 min. The emitted compounds are analyzed and calibrated using a toluene standard. The VOC concentration is expressed as a toluene equivalent. The SVOC analysis (FOG) is run from the same sample in a second desorption at 120 °C for 60 min. The emitted substances are calibrated using a C16 alkane standard. The FOG result is expressed as a hexadecane equivalent (VDA 278, 2016).

4.21.2 Analysis Conditions

Sample material			All vehicle interior materials, polymer materials, leather, rubber
Sample preparation	Thermo desorber	Type	Markes Ltd. UNITY 2
		Desorption tubes	Empty glass tube ID 4 mm, length 9 cm, with a glass wool plug for calibration
		Carrier gas	Helium
		Transfer line	200 °C
		Focusing trap	Material emissions
	Desorption	Pre-purge	1 min
		Desorption temperature/time	90 °C for 30 min (VOCs), 120 °C for 60 min (SVOCs), 300 °C for 10 min (calibration, standards)
		Desorption flow	50 mL/min trap flow, splitless
	Cold trap focusing	Initial temperature	−30 °C
		Heating rate	100 °C/s
		Desorption temperature	300 °C
		Hold time	3 min (VOCs), 5 min (SVOCs), 10 min (standards)
		Pre-purge	3 min, bypass 20 mL/min
		Cold trap desorption flow	30 mL/min trap flow
		Split flow	20 mL/min
GC method	System		Thermo Scientific TRACE 1300 GC
	Column	Type	TG-5MS, 5% phenyl methyl siloxane
		Dimensions	60 m length × 0.25 mm ID × 0.25 µm film thickness
	Carrier gas		Helium
	Flow		Constant flow, 1.3 mL/min
	SSL injector	Injection mode	Splitless
		Injection temperature	280 °C
	Oven program	Start	40 °C, 2.0 min
		Ramp 1	20 °C/min to 80 °C, 2 min
		Ramp 2	10 °C/min to 160 °C, 5 min
		Ramp 3	20 °C/min to 320 °C

		Final temperature	310 °C, 15.0 min
	Transferline	Temperature	280 °C
MS method	System		Thermo Scientific ISQ
	Analyzer type		Single quadrupole MS
	Ionization		EI, 70 eV
	Ion source	Temperature	280 °C
	Acquisition	Mode	Full scan
	Mass range		33–350 Da
	Scan rate		300 ms/scan

4.21.3 Sample Measurements

4.21.3.1 Sample Preparation

Samples are taken directly into a glass desorption tube. Specific sampling requirements apply according to the investigated materials. For ABS, PVC, leather, and other plastic parts, about 30 ± 5 mg are used, cut into pieces of about 4 cm length and 3 mm width.

4.21.3.2 VOC testing

A standard series with concentrations of 10, 50, 100, 200, 500, and 1000 ng/μL in methanol is prepared as the working standards used for calibration. The calibration solutions are applied directly into Tenax-filled desorption tubes and analyzed using the above-described method for TD-GC-MS measurements. The chromatogram of the VOC compounds as the TIC is shown in Figure 4.79. In Table 4.30, the BTEX

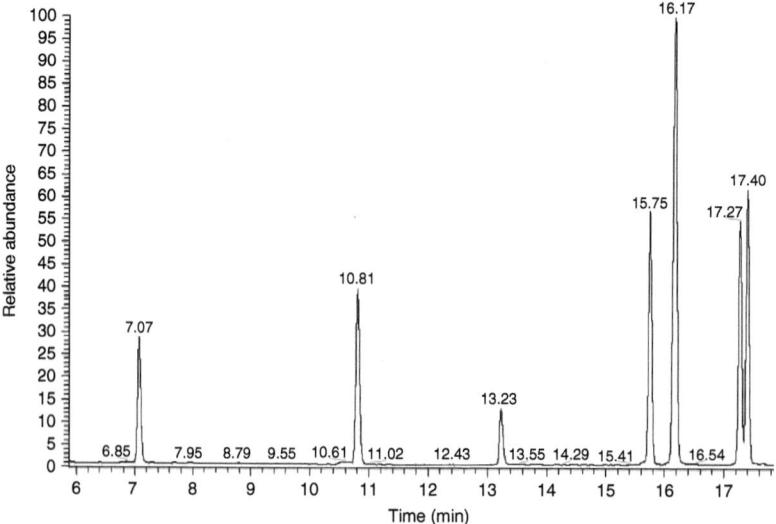

Figure 4.79 TD-GC-MS TIC of the volatile organic standard.

Table 4.30 Volatile organic compounds (VOCs) of the BTEX test with retention times and quantitative precision.

Compounds	Retention time (min)	Quantitation ion (m/z)	Linearity R^2
Benzene	7.07	78	0.9991
Toluene	10.81	91	0.9999
Ethylbenzene	15.75	91	0.9990
p/m-Xylene	16.17	91	0.9998
Styrene	17.27	91	0.9993
o-Xylene	17.40	91	0.9994

compounds analyzed are listed with RTs and the specific ions used for selective quantification, also showing the resulting R^2 values giving the precision of the quantitative calibration.

4.21.3.3 SVOC Testing

The SVOC testing provides the semi-volatile emission value (TVOC) of the material analyzed. It is determined according to VDA 278 by the peak area integration of the sample chromatogram after desorption at the higher temperature of 120 °C. The result is calculated and expressed as the area comparison with a 100 ng toluene analysis as standard. The calibration solution dissolved in methanol has been applied directly into Tenax-filled desorption tubes and analyzed using the above-described method for TD-GC-MS measurements.

The TVOC value of a sample is determined by the integration of the chromatographic peak area between C6 and C16 with 100 ng total integrated area toluene peak comparison calculated. The two standards hexane and n-hexadecane determine the RT position of the C6 and C16 peaks (see Figure 4.80). For the test samples, the peak area between the calibrated RTs of n-hexane and n-hexadecane is determined as a total peak area. Representative chromatograms of a leather and a sponge sample are shown with the TIC in Figures 4.81 and 4.82.

The TVOC concentration of the sample is calculated according to the following formula:

$$C_s = \frac{100 \times A}{A_T \times m_s}$$

with

A: Sample, C6 – C16 chromatographic total peak area integration (area cts),
A_T: Toluene reference, 100 ng injection, chromatographic peak area integration (area cts/100 ng),
m_s: Sample volume (mg), and
C_s: TVOC concentration (FOG value) in the sample (ng/mg).

682 | 4 Applications

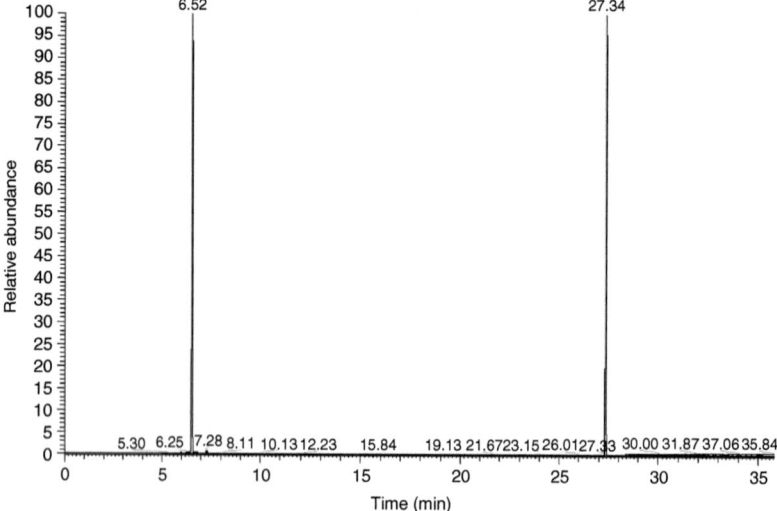

Figure 4.80 C6 and C16 retention times determining the total ion current (TIC) integration limits.

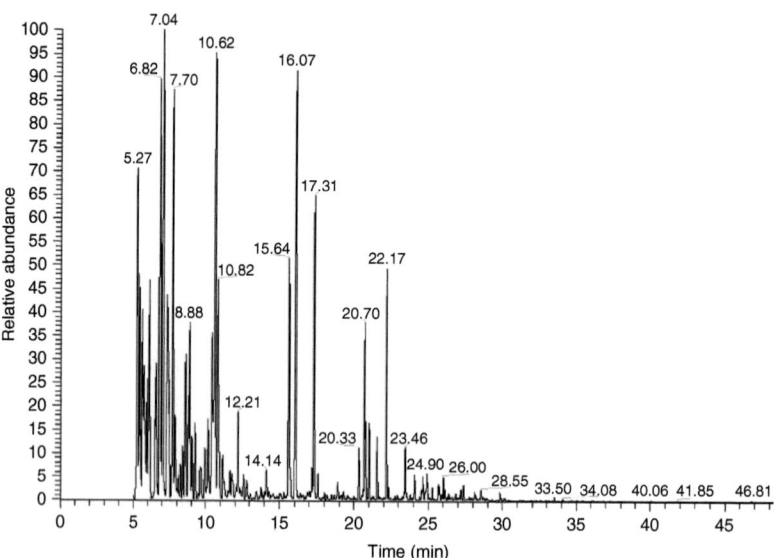

Figure 4.81 A leather sample TD-GC-MS analysis (TIC).

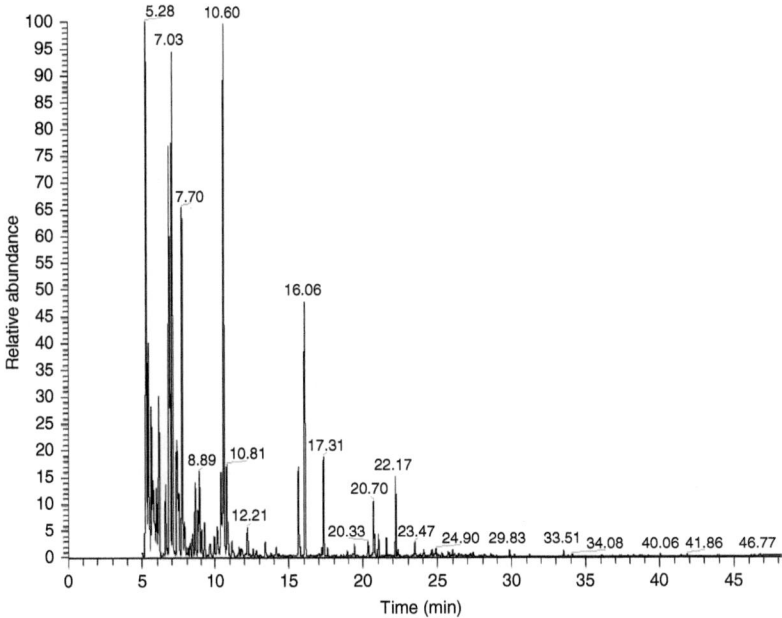

Figure 4.82 A sponge sample TD-GC-MS analysis (TIC).

4.21.4 Conclusion

This application demonstrates the standardized analysis of VOCs and SVOCs in automotive interior materials using TD GC-MS with a single quadrupole MS. The described analytical method follows the internationally recognized method VDA 278 for the analysis of volatiles for the automotive industry. The TD sample preparation using glass tubes is easy to handle; the measurements run fully automated providing high sensitivity, precision, and a wide linear range. This method setup provides the proven analytical solution for testing automotive interior materials for VOC and SVOC/FOG data.

4.22 Azo Dyes in Leather and Textiles

Keywords: product safety; textiles; leather; azo dyes; cancerogenic amines; quantification; single quadrupole MS; library search

4.22.1 Introduction

Azo dyes are compounds characterized by their vivid colors and provide excellent coloring properties. They are important and widely used as coloring agents in the textile and leather industries. The risk in the use of azo dyes arises mainly from biotransformation by various skin bacteria that release aromatic amines (AAs)

Table 4.31 Amine compounds included in the azo dye analysis method. CAS# is the CAS Registry Number, RT is the expected retention time, and Quan is the quantitation ion.

Amine compound	CAS #	RT (min)	Quan (m/z)	Comment
Aniline	62-53-3	5.42	93	a)
o-Toluidine	95-53-4	6.43	106	
2,4-Xylidine	95-68-1	7.35	121	a)
2,6-Xylidine	87-62-7	7.39	121	a)
d-Naphthalene	1146-65-2	7.50	136	ISTD
2-Methoxyaniline	90-04-0	7.61	108	
p-Chloroaniline	106-47-8	7.97	127	
m-Anisidine	536-90-3	8.40	123	a)
p-Cresidine	120-71-8	8.50	122	
2,4,5-Trimethylaniline	137-17-7	8.57	120	
4-Chlor-o-toluidine	95-69-2	8.90	106	
1,4-Phenylenediamine (1,4-Benzendiamine)	106-50-3	8.91	108	a),b)
2,4-Toluenediamine	95-80-7	10.09	122	b)
2,4-Diaminoanisole	615-05-4	10.91	123	a)
2,4,5-Trichloroaniline	636-30-6	11.08	195	ISTD
2-Napthylamine	91-59-8	11.44	143	
5-Nitro-o-toluidine	99-55-8	11.85	152	
4-Aminodiphenyl	92-67-1	12.66	169	
p-Aminoazobenzene	60-09-3	14.36	92	
4,4-Oxydianiline	101-80-4	14.62	200	
4,4-Diaminodiphenylmethane	101-77-9	14.66	198	
Benzidine	92-87-5	14.71	184	
o-Aminoazobenzene	2835-58-7	15.01	106	a)
3,3-Dimethyl-4,4-diaminodiphenylmethane	838-88-0	15.31	226	
3,3'-Dimethylbenzidine	119-93-7	15.47	212	
4,4'-Thiodianiline	139-65-1	16.04	216	
4,4-Methylene-*bis*-2-chloroaniline	101-14-4	16.23	231	
3,3'-Dimethoxybenzidine	119-90-4	16.24	244	
3,3-Dichlorobenzidine	91-94-1	16.26	252	

a) Additional compounds included in the assay, not part of EN ISO 17234-1.
b) Compounds are unstable and tend to deteriorate at temperatures above 20°C.

which might be dermally absorbed. The AAs can be created by reductive cleavage of the azo group (Brüschweiler and Merlot, 2017). Due to the toxicity, carcinogenicity, and potential mutagenicity of thus formed AAs, the use of certain azo dyes as textile and leather colorants and the exposure of consumers using the textile and leather colored with azo compounds cause a serious health concern (European Commission, 1999; European Parliament and the Council of the European Union 2006).

The EU Commission classified 22 amines as proven or suspected human carcinogens. Azo dyes that, by reductive cleavage of one or more azo groups, may release one or more of these AAs in detectable concentrations above 30 ppm in the finished articles or in the dyed parts thereof may not be used in textile and leather articles which may come into direct and prolonged contact with the human skin or oral cavity (European Commission, 2002). The EU Directive 2002/61/EC has banned the use of dangerous azo colorants, placing textiles and leather articles colored with such substances on the market, and requested the development of a validated analytical methodology for control. The described application is compliant with the requirements of the EN ISO 14362-1 standard procedure for the analysis of certain azo dyes in cotton and silk textiles (also refer to EN ISO 14362-2 and -3) (EN ISO 14362-1, 2017; EN ISO 14362-2, 2003; EN ISO 14362-3, 2017). Following appropriate sample preparation methods, the described analytical setup can be used for the analysis of azo dyes in leather and synthetic fabrics as well.

The MS was set to run in the full scan mode, providing the complete mass spectra of the detected compounds for identification, confirmation, and quantification. The complete list of azo compounds and ISTDs is given in Table 4.31 with RTs for the given analytical conditions and the masses for the selective quantification (Purwanto and Chen 2013).

4.22.2 Analysis Conditions

Sample material	Textiles, leather
Sample preparation	Textiles made of cellulose and protein fibers, for example, cotton, viscose, wool, or silk make the azo dyes accessible to a reducing agent. The EN ISO 17234-1 standard method for the analysis of such textiles is based on the chemical reduction of azo dyes followed by SPE with ethyl acetate.
	Synthetic fibers like polyester, polyamide, polypropylene, acrylic, or polyurethane materials require prior extraction of the azo dyes and is described in the EN 14362-2 standard method.
	The analysis of leather samples follows the EN ISO 17234 standard method.

			The azo group of the most azo dyes can be reduced in the presence of sodium dithionite ($Na_2S_2O_4$) under mild conditions (pH = 6, T = 70 °C), resulting in the cleavage of the diazo group and forming two aromatic amines as the reaction products. The amines are extracted by liquid/liquid extraction with methyl t-butyl ether (MTBE), concentrated, adjusted to a certain volume with MTBE, and then analyzed by GC-MS. The quantitation is performed with the internal standards d-naphthalene and 2,4,5-trichloroaniline. In the EN ISO 17234-1 standard method, the directly reduced amines are isolated by SPE.
GC method	System		Thermo Scientific TRACE 1300 GC
	Carrier gas		Helium
	Flow		Constant flow, 1.0 mL/min
	Column		TG-35MS or equivalent polarity column
			30 m length × 0.25 mm ID × 0.25 µm film thickness
	Injection		Autosampler, AS 1310
	SSL injector	Injection mode	Splitless
		Injection temperature	200 °C
		Injection volume	1.0 µL
	Split	Closed	1 min
		Open	1 min to end of run
		Split flow	50 mL/min
		Gas Saver	10 mL/min at 3 min
	Oven program	Start	60 °C, 1 min
		Ramp 1	15 °C/min to 200 °C
		Ramp 2	25 °C/min to 310 °C
		Final temperature	310 °C, 5 min
	Transferline	Temperature	295 °C
			Direct coupling
MS method	System		Thermo Scientific ISQ
	Analyzer type		Single quadrupole MS
	Ionization		EI, 70 eV
	Ion source temperature		220 °C
	Acquisition mode	Full scan	50–350 Da
	Scan rate		0.075 s/scan
Calibration	Internal standards		d-Naphthalene, 2,4,5-trichloroaniline
	Calibration range		Representing 5, 30, and 100 ppm in the textile

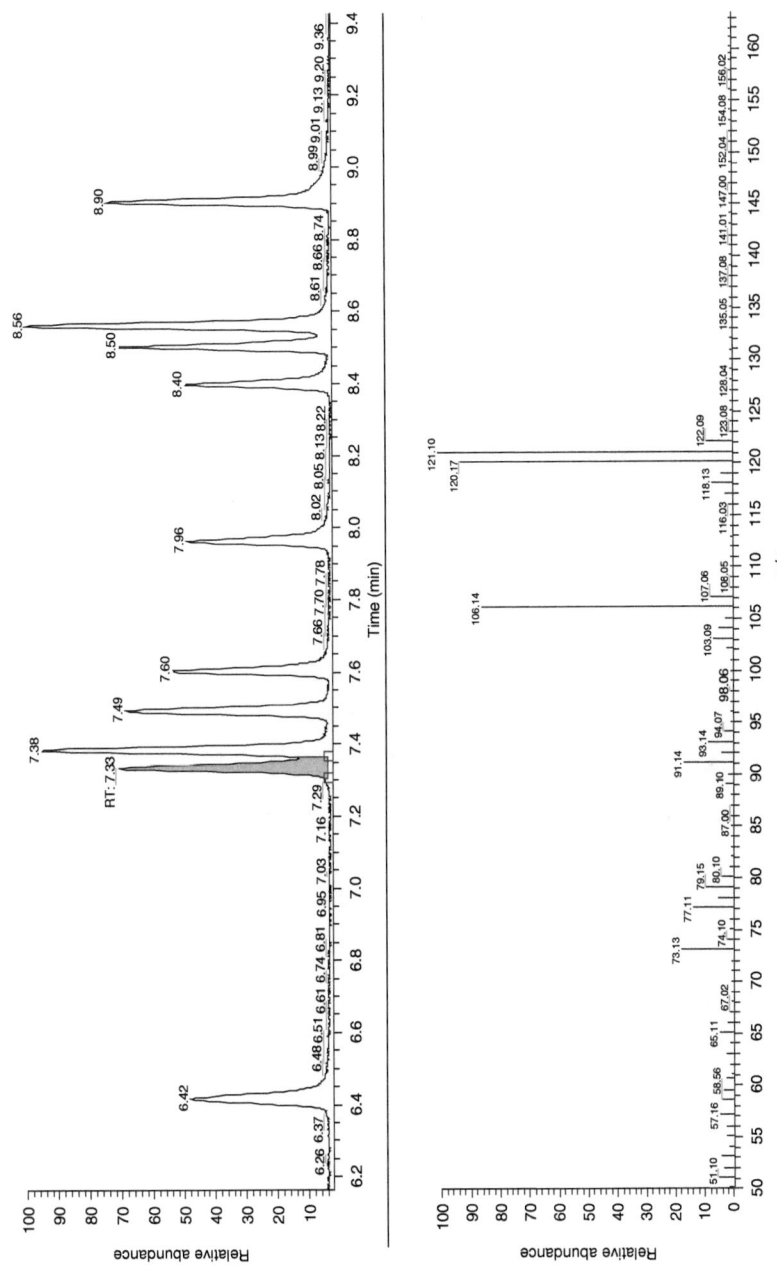

Figure 4.83 Chromatographic separation of 2,4- and 2,6-xylidine at retention times 7:33 and 7:38 min, with mass spectrum of the integrated 2,4-xylidine peak below.

4.22.3 Results

The scan rate of the MS run in the full scan mode has been set to a very fast scan speed of only 75 ms/scan. This allows a very high chromatographic resolution especially of unresolved chromatographic peaks in the applied complex mixture of amines. Short analysis cycle times can be achieved for an increased sample throughput.

Isomeric amine compounds need to be separated on the chromatographic time scale for individual component quantification. The mass spectra of isomers are very similar and typically do not offer unique quantification ions for the independent quantitation. This is the case for the isomeric xylidine compounds 2,4-xylidine and 2,6-xylidine. The applied method separated both compounds very well with 7:33 and 7:38 min RTs, see Figure 4.83.

All other target amine analytes produce distinct mass spectra with unique ions available for selective quantification, even if the chromatographic peaks in the TIC are not resolved from each other. In these situations, when several analytes coelute, individual mass spectra can still be isolated and identified by a library search. The intensities of the unique ions allow the interference-free quantification based on the separated mass traces. In case of the coelution of the three compounds 4,4-methylene-bis-2-chloroaniline (RT 16:23 min), 3,3′-dimethoxybenzidine (RT 16:24 min), and 3,3-dichlorobenzidine (RT 16:26 min) shown in Figure 4.84,

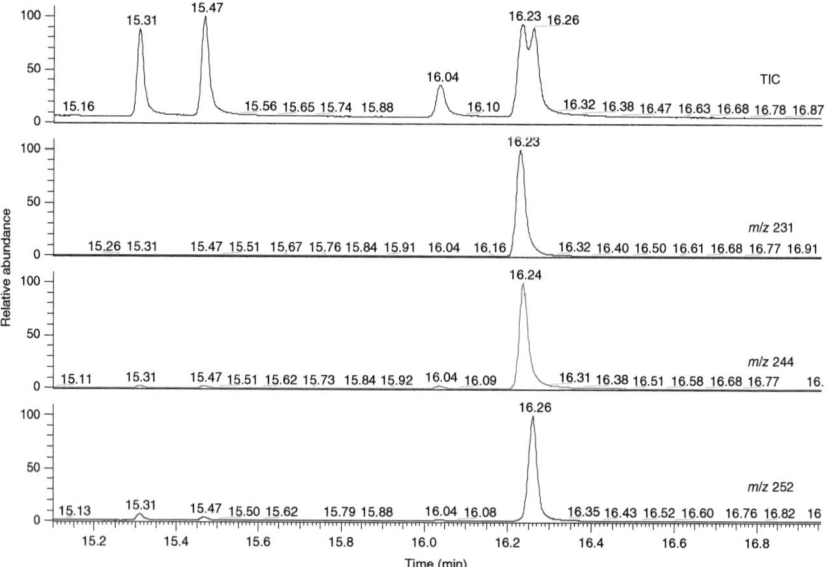

Figure 4.84 Coelution of three different amines, separated by individual mass traces.

4.22 Azo Dyes in Leather and Textiles | 689

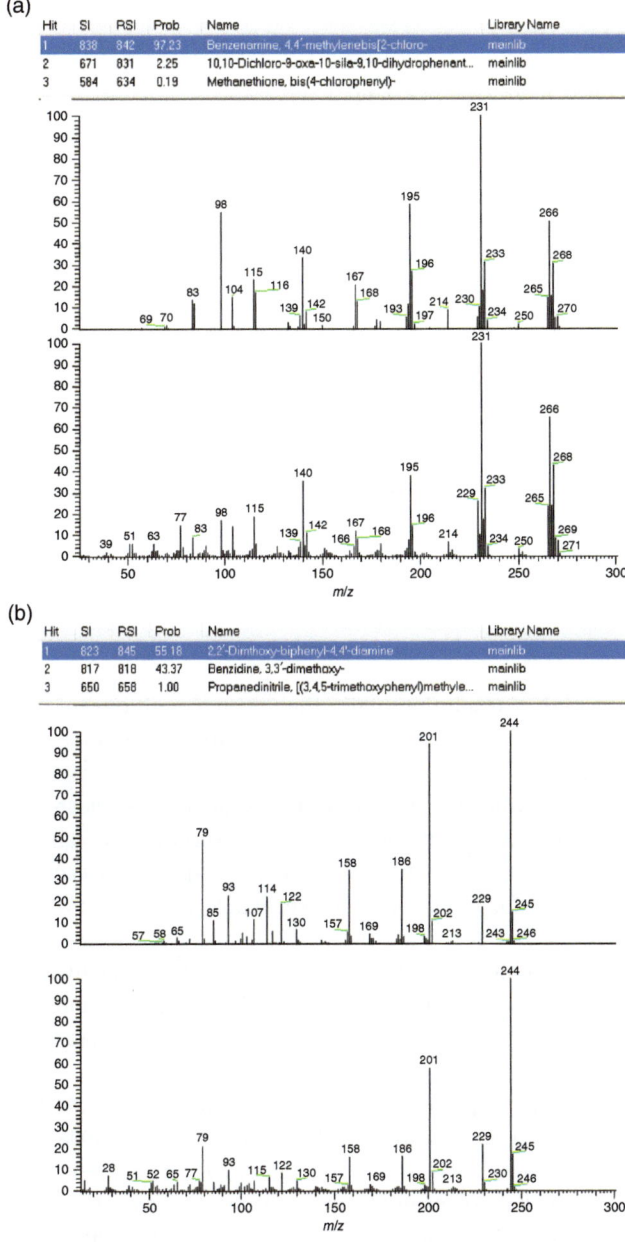

Figure 4.85 Acquired mass spectra and library search results for the coeluting compounds of Figure 4.86 top acquired spectrum from data file in Figure 4.86 bottom spectrum from NIST library: (a) 4,4-methylene-bis-2-chloroaniline (CAS 101-14-4), (b) dimethoxybenzidine (CAS 119-90-4), and (c) 3,3-dichlorobenzidine (CAS 91-94-1).

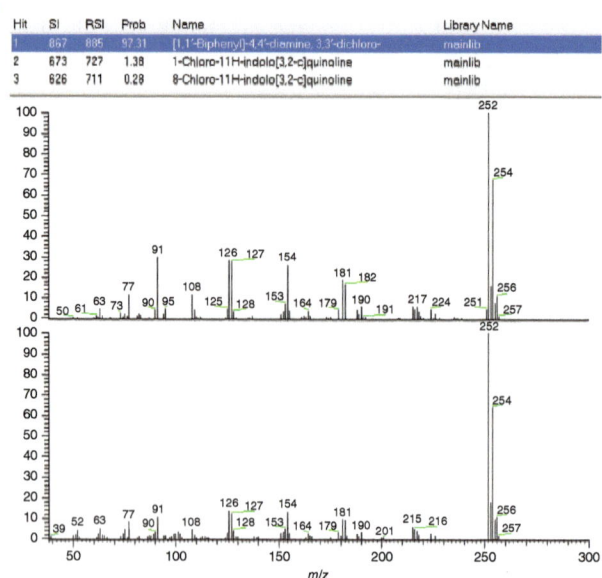

Figure 4.85 (Continued)

only a partial chromatographic resolution with the given conditions of the method is achieved, but due to specific fragment ions, a safe peak integration and quantification is accomplished.

Although coeluting the quantification is performed separately by means of individual fragment ions. The fast scanning of the mass spectrometer allows the detection of small RT differences so that even the clean spectra of each of the compounds can be isolated for library search, see Figure 4.85a–c. All three compounds are safely identified by searching the NIST library. The spectra have been taken from a spiked calibration file at the 3 ppm level. The high sensitivity of the mass spectrometer provides even at the required low detection level and fast scanning speed very complete mass spectra for a solid compound confirmation in parallel to the quantification.

The quantitation was done by using the two ISTDs d-naphthalene RT 7:50 min and 2,4,5-trichloroaniline RT 11:08 min. A linear calibration with $R^2 > 0.9998$ was created using standards over three levels at 0.5, 3.0, and 10.0 ppm, representing 5, 30, and 100 ppm in the textile. An example of a routine QC report for the decision level of 30 ppm is shown in Figure 4.86. The list of compounds and the chromatogram of the QC sample are shown. All compounds were detected, integrated, and passed the confirmation check based on the compound mass spectrum.

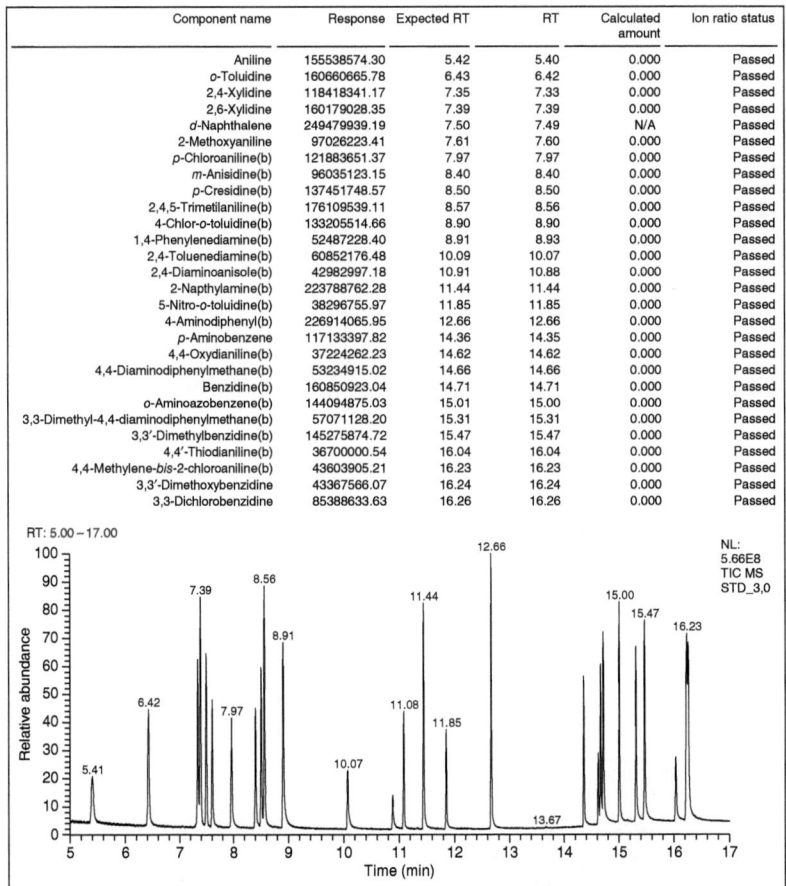

Figure 4.86 Report of a routine QC sample at the decision level of 30 ppm.

4.23 Fast GC of 16 Priority PAHs

Keywords: PAHs; benzo[a]pyrene; EU regulation; fast GC; magnetic sector MS; HRMS; food; meat; flavors; smoke; flavoring

4.23.1 Introduction

In 2002, the European Scientific Committee on Food reasoned that a list of 15 PAH compounds may be regarded as potentially genotoxic and carcinogenic to humans and therefore represent a priority group in the assessment of the risk of long-term adverse health effects following dietary intake of PAHs (Figures 4.87

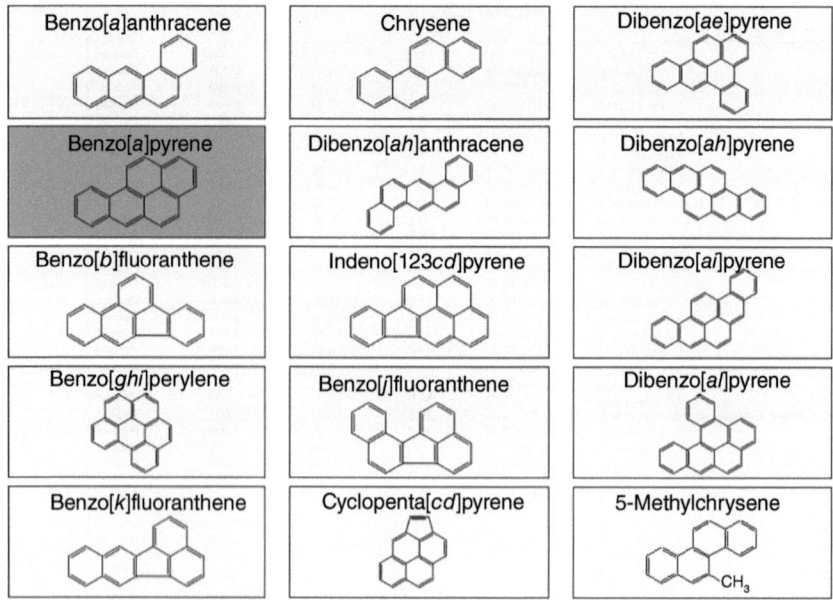

Figure 4.87 15 PAH priority compounds classified by the European Scientific Committee on Food. (European Commission, 2005).

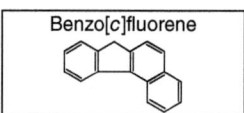

Figure 4.88 Additional PAH priority compound to be monitored. (Adapted from JECFA, 2005.)

and 4.88). In 2008, the European Food Safety Agency (EFSA) expert panel published a new scientific opinion about polycyclic aromatic hydrocarbons in food (EFSA, 2008). Benzo(a)pyrene is no longer accepted as the most suitable indicator of PAH in food.

In this opinion, EFSA concluded that benzo(a)pyrene is not a suitable marker for the occurrence of PAHs in food. A system of four specific substances (PAH4) with the sum of benzo(a)pyrene, benz(a)anthracene, benzo(b)fluoranthene, and chrysene should be used. A set of eight substances (PAH8) with benzo(a)pyrene, chrysene, benz(a)anthracene, benzo(b)fluoranthene, benzo(k)fluoranthene, benzo(g,h,i) perylene, dibenz(a,h)anthracene, and indeno(1,2,3-c,d)pyrene is suggested alternatively. In the following, the European Commission amended the regulation No 1881/2006 with the Commission Regulation No 835/2011 on

August 19 2011 (European Commission, 2011a). New maximum levels for the sum of PAH4 were set in force.

During GC-MS method development, it turned out that single quadrupole MS instrumentation could not provide the required selectivity and sensitivity at the low decision level for the list of 16 PAH for a reliable detection in real-life samples. High resolution and accurate mass MS (HRAM) with its particularly high mass selectivity provided the requested robustness and safety in the quantitative determination (Kleinhenz et al., 2006; Ziegenhals and Jira, 2006, 2007).

The quantitation was done using the isotope dilution technique by the addition of isotope-labeled and fluorinated standards before extraction, as well as for the determination of the response factors of all PAHs (Ziegenhals et al., 2008). Recovery values have been determined by the addition of three deuterium-labeled compounds (Table 4.32).

Table 4.32 Exact masses of PAHs and labeled internal standards.

PAH native	Acronym	Exact mass native (Da)	PAH ISTD labeled	Exact mass labeled (Da)
Benzo[c]fluorine	BcL	216.0934	5-F-BcL	234.0839
Benzo[a]anthracene	BaA	228.0934	$^{13}C_6$-BaA	234.1135
Chrysene	CHR	228.0934	$^{13}C_6$-CHR	234.1135
Cyclopenta[cd]pyrene	CPP	226.0777	—	—
5-Methylchrysene	5MC	242.1090	d_3-5MC	245.1278
Benzo[b]fluoranthene	BbF	252.0934	$^{13}C_6$-BbF	258.1135
Benzo[j]fluoranthene	BjF	252.0934	—	—
Benzo[k]fluoranthene	BkF	252.0934	$^{13}C_6$-BkF	258.1135
Benzo[a]pyrene	BaP	252.0934	$^{13}C_4$-BaP	256.1068
Indeno[123cd]pyrene	IcP	276.0934	d12-IcP	288.1687
Dibenzo[ah]anthracene	DhA	278.1090	d_{14}-DhA	292.1969
Benzo[ghi]perylene	BgP	276.0934	$^{13}C_{12}$-BgP	288.1336
Dibenzo[al]pyrene	DlP	302.1090	13-F-DlP	320.0996
Dibenzo[ae]pyrene	DeP	302.1090	$^{13}C_6$-DeP	308.1291
Dibenzo[ai]pyrene	DiP	302.1090	$^{13}C_{12}$-DiP	314.1493
Dibenzo[ah]pyrene	DhP	302.1090	—	—
PAH recovery standards deuterated				
D_{12}-Benzo[a]anthracene			d_{12}-BaA	240.1687
D_{12}-Benzo[a]pyrene			d_{12}-BaP	264.1687
D_{12}-Benzo[ghi]perylene			d_{12}-BgP	288.1687

4.23.2 Analysis Conditions

Sample material			Food, meat, meat products, herbs, spices
Sample preparation	Extraction		Pressurized solvent extraction (ASE)
	Clean-up		Size exclusion chromatography (SEC)
			Solid phase extraction (SPE)
GC method	System		Thermo Scientific TRACE GC Ultra
	Autosampler		Thermo Scientific TriPlus RSH
	Column	Type	TR-50MS
		Dimensions	10 m length × 0.1 mm ID × 0.1 µm film thickness
	Carrier gas		Helium
	Flow		Constant flow, 0.6 mL/min
	SSL injector	Injection mode	Splitless, hot needle
		Injection temperature	320 °C
		Injection volume	1.5 µL
	Split	Closed until	1 min
		Open	1 min to end of run
	Oven program	Start	140 °C, 1 min
		Ramp 1	10 °C/min to 240 °C
		Ramp 2	5 °C/min to 270 °C
		Ramp 3	30 °C/min to 280 °C
		Ramp 4	4 °C/min to 290 °C
		Ramp 5	30 °C/min to 315 °C
		Ramp 6	3 °C/min to 330 °C
		Final temperature	330 °C, 0 min
	Transferline	Temperature	300 °C
MS method	System		Thermo Scientific DFS
	Analyzer type		Magnetic sector high resolution MS
	Ionization		EI, 45 eV
	Ion source	Temperature	280 °C
	Acquisition	Mode	SIM
			2 ions per analyte (see Table 4.34)
	Scan rate		0.7 s
	Resolution	Setting	R 8000 at 5% peak height, R 16000 FWHM
Calibration	Internal standard	Range	^{13}C-labeled internal standards

4.23.3 Results

Although an initial use of a 50% phenyl capillary column of 60 m length × 0.25 mm ID × 0.25 µm film thickness at constant pressure provided the required chromatographic resolution of the various isomers, the RT of more than 90 min turned out to be not appropriate for a control method with high productivity. The application of the fast GC column technology reduced the required retention by more than 75% to only 25 min maintaining the necessary chromatographic resolution (see Figure 4.89). The critical separation components are shown in Figures 4.90 and 4.91. For all components, the fast GC method provides a robust peak separation and quantitative peak integration.

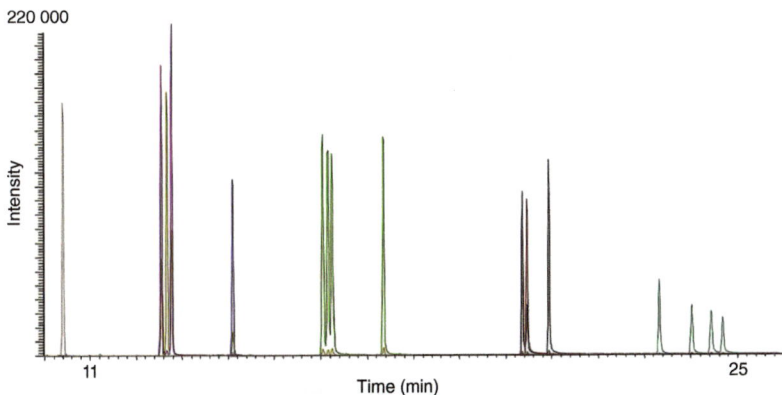

Figure 4.89 Fast-GC-HRMS separation of 16 EC priority PAHs in only 25 min. (Data acquisition SIM descriptor as of Table 4.33.)

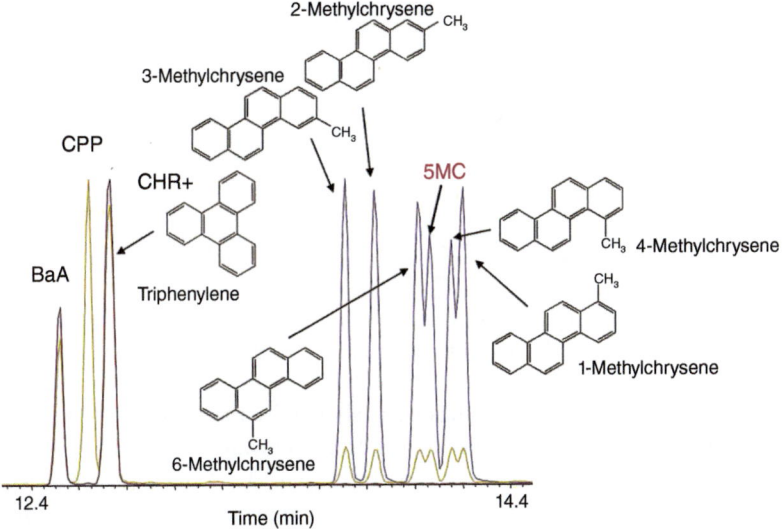

Figure 4.90 Detail of isomer separation from Figure 4.89. Elution sequence first peak cluster BaA, CPP, CHR, second peak cluster 3MC, 2MC, 6MC, 5MC, 4MC, and 1MC.

696 | 4 Applications

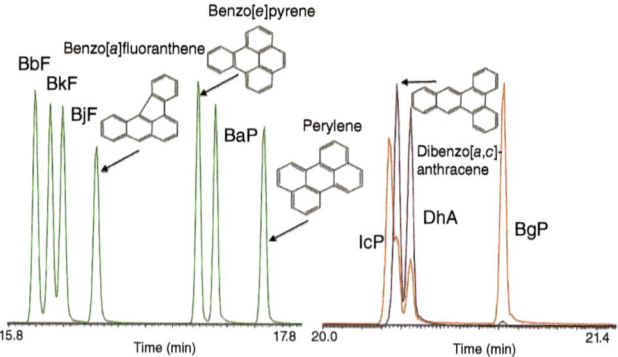

Figure 4.91 Detail of isomer separation from Figure 4.89. Elution sequence left peak cluster 15.8–17.8 min is BbF, BkF, BjF, BaF, BeP, BaP, and PER and right peak cluster 20.0–21.4 min is IcP, DcA, DhA, and BgP.

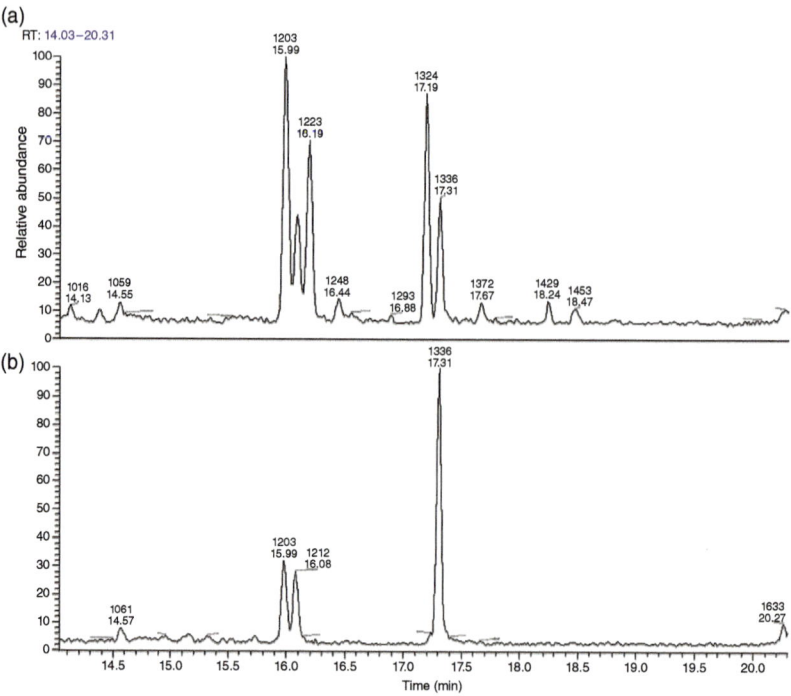

Figure 4.92 Benzo[a]pyrene determination (RT 17:31 min) in caraway seeds at a level of 0.02 µg/kg (a. native PAH and b. ^{13}C-BaP, elution sequence BbF, BkF, BjF, BeP, and BaP).

Applicability for different matrices has been shown for many critical matrices. Figure 4.92 shows the analysis of the extract from caraway seeds with a determined concentration of benzo[a]pyrene of 0.02 µg/kg. An LOD of 0.005 µg/kg and an LOQ of 0.015 µg/kg can be estimated for the analysis of spices, when the sample weight is 1–1.5 g. The recovery values achieved with the described sample preparation has been between 50% and 120%.

Table 4.33 SIM descriptor for PAH fast-GC/HRMS data acquisition.

RT (min)	Exact mass (Da)	Function	Dwell time (ms)
8:50	216.09375	Native	82
	218.98508	Lock	2
	226.07830	Native	82
	228.09383	Native	82
	234.08450	Native	82
	234.11400	Native	82
	240.16920	Native	82
	263.98656	Cali	6
13:00	218.98508	Lock	2
	242.10960	Native	82
	245.12840	Native	82
	252.09390	Native	82
	256.10730	Native	82
	258.11400	Native	82
	263.98656	Cali	6
	264.16920	Native	82
19:00	263.98656	Lock	2
	276.09390	Native	74
	278.10960	Native	74
	288.13410	Native	74
	292.19740	Native	74
	313.98340	Cali	6
22:00	263.98656	Lock	2
	302.10960	Native	120
	308.12970	Native	120
	313.98340	Cali	6
	314.13980	Native	120
	320.10010	Native	120

4.24 Environmental Contaminants in Fish

Keywords: food safety; micro pollutants; environmental analysis; liquid/liquid extraction; POPs; OCPs; PCBs; PAHs; BFRs; fish feed; fish farming; triple quadrupole MS/MS

4.24.1 Introduction

Fish with high fat contents is an important source of long-chain n-3 polyunsaturated fatty acids (PUFAs) in the human diet, in particular for docosahexaenoic acid (DHA) and eicosapentaenoic acid (EPA) (EFSA, 2005). On the other side, a wide range of environmental contaminants have been reported to be accumulated in fish that can pose a potential human health hazard (Nadal et al., 2015). Major contaminations include the different organochlorine pesticides (OCPs), PAHs, or the widely distributed persistent organic pollutants (POPs) with the PCBs and the flame retardant compound class of the PBDEs. Also farmed fish can significantly contribute to dietary exposure to various contaminants due to the use of land-sourced fish feed (Nadal et al., 2015; Focker et al., 2022). A multi-method published by Kamila Kalachova for efficient control of fish and fish feed for various groups of contaminants from PCBs, OCPs, brominated flame retardants, and PAHs is outlined in the following analytical procedure (Kalachova et al., 2013).

4.24.2 Analysis conditions

Sample material			Fish, fish feed
Sample preparation	Liquid extraction		10 g fish tissue, homogenized (1 g for fish feed)
			Add recovery ISTDs BDE 37 (3,4,4′-triBDE) and $^{13}C_{12}$-PCB 77 (3,3′,4,4′-tetraCB)
			Mix with 5 mL distilled water (14 mL distilled water for fish feed)
			Shake with 10 mL ethyl acetate
			Add 4 g anhydr. $MgSO_4$, 2 g NaCl, shake, centrifuge
			5 mL organic layer, dried under nitrogen
			Redissolve in 1 mL hexane
			Apply to clean-up
	Clean-up	Silica minicolumn	1 g sorbent (for up to 0.1 g fat)
			5 g sorbent (for up to 0.8 g fat)
			Elute with 20 mL hexane-DCM (3 : 1, v/v)
			Evaporate solvents
			Redissolve in 0.5 mL i-octane
			Add quantitation ISTDs
GC method		System	Thermo Scientific TRACE GC Ultra
		Autosampler	Thermo Scientific TriPlus RSH

	Column	Type	Rxi-17Sil MS
		Dimensions	30 m length × 0.25 mm ID × 0.25 µm film thickness
	Carrier gas		Helium
	Flow		Constant flow, 1.3 mL/min
	PTV injector	Injection mode	Splitless, 2 min
		Injection volume	1 µL
		Base temperature	95 °C, 0.05 min
		Transfer rate	14.5 °C/s
		Transfer temperature	200 °C
		Transfer time	1 min
		Cleaning phase	4.5 °C/s to 320 °C
		Final temperature	320 °C, 3 min
		Inlet liner	Baffled liner, 2 mm ID, Siltek™ deactivated
	Split	Closed until	2 min
		Open	2 min until end of run
	Oven program	Start	80 °C, 2 min
		Ramp 1	30 °C/min to 240 °C
		Ramp 2	10 °C/min to 340 °C
		Final temperature	340 °C, 20 min
	Transferline	Temperature	320 °C
			Direct coupling
MS method	System		Thermo Scientific TSQ Quantum GC
	Analyzer type		Triple quadrupole MS
	Ionization		EI, 70 eV
	Ion source	Temperature	270 °C
		Emission current	50 µA
	Acquisition	Mode	Timed-SRM
			2 SRMs per analyte
	Collision gas		Ar, 1.5 mL/min
	Scan rate		0.3 s
	Resolution	Setting	0.7 Da for Q1, Q3
Calibration	Internal standard		Recovery and quantitation ISTDs
	Calibration range		0.05–100 ng/mL

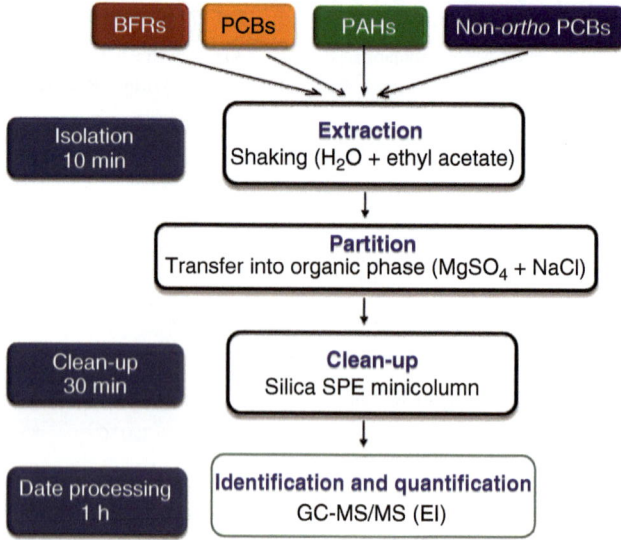

Figure 4.93 Multi-class sample preparation workflow.

4.24.3 Sample Preparation

The multi-class sample preparation for PCBs, OCPs, BFRs, and PAHs is based on the liquid extraction of the hydrophobic target analytes from an aqueous sample suspension into ethyl acetate. The clean-up step of the ethyl acetate phase provides the removal of the co-extracted lipids on a silica/Na_2SO_4 SPE minicolumn, see Figure 4.93 (Kalachova et al., 2011).

Ten grams of fish tissue is homogenized and spiked with 10 ng of the internal recovery standards BDE 37 (brominated diphenyl ether) and a ^{13}C-labeled PCB 77, mixed with 5 mL distilled water, and shaken with 10 mL ethyl acetate for 1 min. Four gram anhydrous $MgSO_4$ and 2 g of NaCl are added to the mixture and shaken for another 1 min, and centrifuged. Finally, 5 mL of the organic layer is withdrawn and dried under nitrogen. The residue is dissolved in 1 mL of n-hexane and purified using a silica minicolumn.

As internal quantitation standards, ^{13}C-labeled BDE 209 (50 ng/mL), ^{13}C-labeled PCB 101 (40 ng/mL), and a 16-EPA-PAH mix as ^{13}C-labeled compounds (2 ng/mL) are spiked to the final extract.

4.24.4 Sample Measurements

Special attention in this multi-component setup is necessary to the GC inlet system. A temperature programmable injector (PTV) is used, offering a cleaning step at an elevated temperature after each extract injection. The choice of liner and deactivation turned out to be critical for the long-term robustness. This is caused by the short sample preparation generating extracts with a high matrix load, as shown in Figure 4.94.

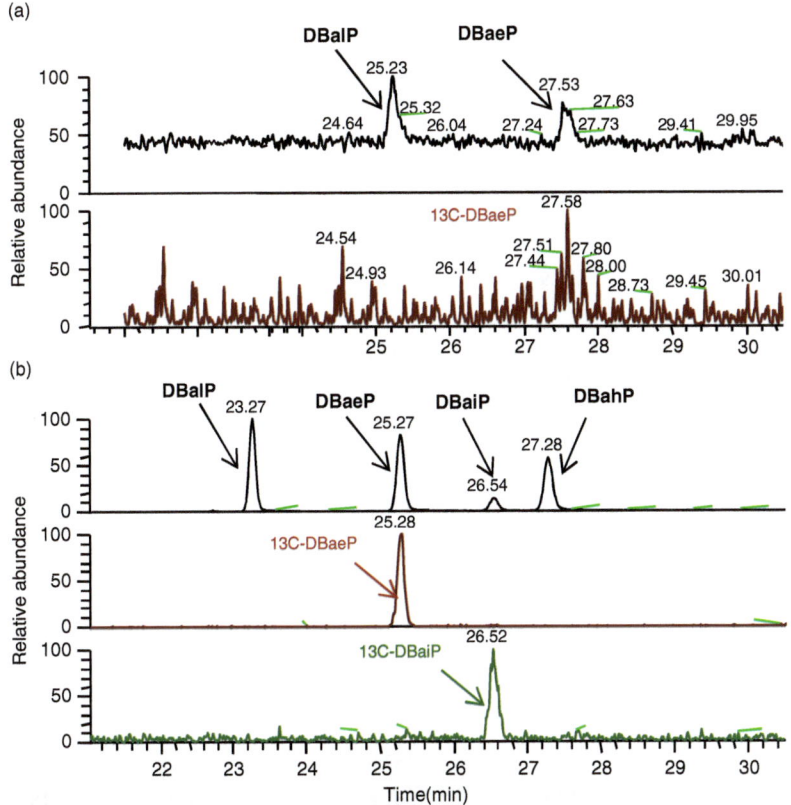

Figure 4.94 Impact of inlet liner choice on GC performance after months of operation, (a) Silcosteel liner and (b) Siltek deactivated baffle liner.

A mid-polarity phase GC column dedicated for the separation of PAH components was selected (Thomas et al., 2010). The GC oven temperature program was optimized to obtain the best separation for all target analytes BFRs, PCBs, OCPs, and PAHs. Particular attention is required for the coeluting isomeric PAH compounds (i) benzo[b]fluoranthene (BbFA), benzo[j]fluoranthene (BjFA), and benzo[k]fluoranthene (BkFA); (ii) indeno[1,2,3-cd]pyrene (IP), benzo[ghi]perylene (BghiP), and dibenzo[ah]anthracene (DBahA); and (iii) benz[a]anthracene (BaA), chrysene (CHR), cyclopenta[cd]pyrene (CPP) which use the same SRM transition for detection (for the acronyms used, see Table 4.35). Each target compound was detected using two MS/MS transitions to fulfil the EU SANCO identification criteria (European Commission, 2021). The quantitation ions, confirmation transitions, and collision energies had been optimized for each analyte, as given in Tables 4.34–4.38.

The chosen target analytes are present in fish typically in a wide concentration range. Trace levels of dl-PCBs or PAHs can occur together with high concentrations of major PCB congeners. For this reason, the standard solutions for the quantitative calibration covers three orders of magnitude with a concentration range from 0.05 to 100 ng/mL.

Table 4.34 SRM transitions of target BFRs.

Analyte	M (g/mol)	RT (min)	Precursor ion (m/z)	Production (m/z)	CE (V)
BDE 28	406.9	10.5	405.8	245.9	15
			407.9	247.9	18
BDE 37	406.9	10.9	405.8	245.9	15
			407.9	247.9	18
PBT	486.6	10.9	406.7	246.7	20
			485.5	324.8	30
PBEB	500.7	11.1	499.7	485.1	10
			501.5	487.0	15
BDE 49	485.8	12.0	485.8	325.8	18
			487.8	327.8	18
BDE 47	485.8	12.3	485.8	325.8	18
			487.8	327.8	18
HBB	551.5	12.5	551.6	470.6	25
			551.6	389.6	30
BDE 66	485.8	12.6	485.8	325.8	18
			487.8	327.8	18
BDE 77	485.8	13.0	485.8	325.8	18
			487.8	327.8	18
BDE 100	564.7	13.8	561.8	401.8	18
			565.8	405.8	18
BDE 99	564.7	14.2	561.8	401.8	18
			565.8	405.8	18
BDE 85	564.7	15.4	561.8	401.8	18
			565.8	405.8	18
BDE 154	643.6	15.5	641.7	481.7	18
			645.7	485.7	18
BDE 153	643.6	16.3	641.7	481.7	18
			645.7	485.7	18
BDE 183	722.5	20.2	721.8	561.8	20
			723.7	563.8	20
BTBPE	687.6	21.0	356.7	277.4	15
			356.7	328.4	15

Table 4.35 Compound acronyms used.

Acronym	Compound name
PBDE	Polybrominated diphenyl ether
BDE	Brominated diphenyl ether
HBB	Hexabromobenzene
PBT	Pentabromotoluene
PBEB	Pentabromoethylbenzene
BTBPE	bis(2,4,6-Tribromophenoxy) ethane
OBIND	Octabromo-1-phenyl-1,3,3-trimethylindane
DBDPE	Decabromodiphenyl ethane
cis-HEPO	cis-Heptachloroepoxide
trans-HEPO	trans-Heptachloroepoxide
AC	Acenaphthene
ACL	Acenaphthylene
AN	Anthracene
BaA	Benz[a]anthracene
BaP	Benzo[a]pyrene
BbFA	Benzo[b]fluoranthene
BcFL	Benzo[c]fluorene
BeP	Benzo[e]pyrene
BjFA	Benzo[j]fluoranthene
BkFA	Benzo[k]fluoranthene
BghiP	Benzo[ghi]perylene
CHR	Chrysene
CPP	Cyclopenta[cd]pyrene
DBahA	Dibenz[ah]anthracene
DBaeP	Dibenzo[ae]pyrene
DBahP	Dibenzo[ah]pyrene
DBaiP	Dibenzo[ai]pyrene
DBalP	Dibenzo[al]pyrene
dl-PCBs	Dioxin-like PCBs
FA	Fluoranthene
FL	Fluorene
IP	Indeno[1,2,3-cd]pyrene
NA	Naphthalene
PHE	Phenanthrene
PY	Pyrene
TRI	Triphenylene

(Continued)

Table 4.35 (Continued)

Acronym	Compound name
1MC	1-Methylchrysene
1MN	1-Methylnaphthalene
1MPH	1-Methylphenanthrene
1MP	1-Methylpyrene
2MA	2-Methylanthracene
2MN	2-Methylnaphthalene
3MC	3-Methylchrysene
5MC	5-Methylchrysene

Table 4.36 SRM transitions of target PCBs.

Analyte	M (g/mol)	RT (min)	Precursor ion (m/z)	Product ion (m/z)	CE (V)
PCB 28	257.5	8.2	255.9	219.9	20
			257.9	150.9	35
PCB 52	291.9	8.4	291.9	221.9	22
			291.9	256.9	15
PCB 101	324.6	9.2	323.8	253.8	30
			325.8	290.8	14
$^{13}C_{12}$-PCB 101	336.6	9.2	338.8	265.8	30
			336.8	302.8	14
PCB 81	291.9	9.7	291.9	221.9	22
			291.9	256.9	15
PCB 77	291.9	9.9	291.9	221.9	22
			291.9	256.9	15
$^{13}C_{12}$-PCB 77	303.9	9.9	302.9	233.9	22
			302.9	268.9	15
PCB 123	324.6	10.0	323.8	253.8	30
			325.8	290.8	14
PCB 118	324.6	10.1	323.8	253.8	30
			325.8	290.8	14
PCB 153	360.9	10.2	357.8	287.9	25
			359.8	289.9	25
PCB 114	324.6	10.3	323.8	253.8	30
			325.8	290.8	14

Table 4.36 (Continued)

Analyte	M (g/mol)	RT (min)	Precursor ion (m/z)	Product ion (m/z)	CE (V)
PCB 105	324.6	10.6	323.8	253.8	30
			325.8	290.8	14
PCB 138	360.9	10.8	357.8	287.9	25
			359.8	289.9	25
PCB 126	324.6	10.9	323.8	253.8	30
			325.8	290.8	14
PCB 167	360.9	11.0	357.8	287.9	25
			359.8	289.9	25
PCB 156	360.9	11.5	357.8	287.9	25
			359.8	289.9	25
PCB 157	360.9	11.6	357.8	287.9	25
			359.8	289.9	25
PCB 180	395.3	11.6	391.8	321.8	25
			393.8	323.8	25
PCB 169	360.9	12.0	357.8	287.9	25
			359.8	289.9	25
PCB 189	395.3	12.5	391.8	321.8	25
			393.8	323.8	25

For quality assurance, the standard reference materials such as Lake Michigan fish tissue, SRM 1947, mussel tissue, SRM 1974b, fresh fish from the retail market, and fish feed were used. The fish feed used was composed of fish meal (48.8%), fish oil (5.7%), wheat (17.4%), wheat by-products (8.8%), soya (13.4%), and other components (5.9%). Typical chromatograms of the target contaminants in retail market samples are shown in Figures 4.95–4.97.

4.24.5 Results

The described sample preparation method provides recoveries for all target analytes in the range from 70% to 121%, except for the OCPs dieldrin, endrin, endosulfan sulfate, α- and β-endosulfan, and the volatile PAHs acenaphthene (AC), acenaphthylene (ACL), phenanthrene (PHE), naphthalene (NA), 1-methylnaphthalene (1MN), and 2-methylnaphthalene (2MN). For these analytes, recoveries close to zero are reported. The overall recoveries and the RSDs for both matrices fish and fish feed were in the ranges for PCBs 74–119% (RSD 1–19%), OCPs 72–120% (RSD 3–20%), BFRs 73–116% (RSD 3–19%), and PAHs 70–119% (RSD 1–20%), see also Table 4.39.

Table 4.37 SRM transitions of target OCPs.

Analyte	M (g/mol)	RT (min)	Precursor ion (m/z)	Product ion (m/z)	CE (V)
HCB	284.8	7.4	248.8	213.9	20
			283.8	213.9	20
HCH-α	290.8	7.5	216.9	180.9	15
			218.9	182.9	15
HCH-γ	290.8	7.9	216.9	180.9	15
			218.9	182.9	15
HCH-β	290.8	8.0	216.9	180.9	15
			218.9	182.9	15
Heptachlor	373.3	8.1	271.9	236.9	15
			273.9	238.9	12
Aldrin	364.9	8.5	262.9	192.9	32
			262.9	227.9	32
HEPO-cis	389.3	9.0	352.8	262.9	15
			354.8	264.9	15
HEPO-trans	389.3	9.1	288.9	218.9	15
			352.8	252.9	15
Chlordane-trans	409.8	9.3	276.9	203.9	16
			372.8	265.9	15
DDE, o,p'	318.0	9.3	246.0	176.0	25
			317.9	245.9	20
Chlordane-cis	409.8	9.4	372.8	265.9	18
			409.8	374.8	5
Endosulfan-α	406.9	9.5	240.9	205.9	20
			264.9	192.9	22
DDE, p,p'	318.0	9.6	246.0	176.0	25
			317.9	245.9	20
Dieldrin	380.9	9.9	262.9	192.9	26
			262.9	227.9	5
DDD, o,p'	320.0	10.0	235.0	165.0	20
			237.0	165.0	20
Endrin	380.9	10.3	262.9	190.9	25
			280.9	244.9	12
DDD, p,p'	320.0	10.3	235.0	165.0	20
			237.0	165.0	20
DDT, o,p'	354.5	10.4	234.9	165.0	15
			236.9	165.0	20

Table 4.37 (Continued)

Analyte	M (g/mol)	RT (min)	Precursor ion (m/z)	Product ion (m/z)	CE (V)
Endosulfan-β	406.9	10.7	240.9	205.9	20
			271.9	236.9	18
DDT, p,p'	354.5	10.8	234.9	165.0	20
			236.9	165.0	20
Endosulfan sulfate	422.9	11.3	271.9	236.9	15
			273.9	238.9	15

Table 4.38 SRM transitions of target PAHs.

Analyte	M (g/mol)	RT (min)	Precursor ion (m/z)	Product ion (m/z)	CE (V)
$^{13}C_6$-NA	134.2	5.0	134.0	83.0	30
			134.0	107.0	20
NA	128.2	5.0	128.0	77.0	30
			128.0	102.0	20
1MN	142.2	5.6	141.0	115.0	15
			142.0	115.0	25
2MN	142.2	5.7	141.0	115.0	15
			142.0	115.0	25
$^{13}C_6$-ACL	158.2	6.5	158.0	130.0	30
			158.0	156.0	20
ACL	152.2	6.5	152.0	102.0	30
			152.0	126.0	20
$^{13}C_6$-AC	160.2	6.6	159.0	158.0	20
			160.0	159.0	20
AC	154.2	6.6	153.0	126.0	40
			153.0	151.0	40
$^{13}C_6$-FL	172.2	7.1	171.0	145.0	30
			171.0	169.0	30
FL	166.2	7.1	165.0	139.0	30
			165.0	163.0	30
$^{13}C_6$-PHE	184.2	8.1	184.0	156.0	30
			184.0	182.0	30
PHE	178.2	8.1	178.0	152.0	20
			178.0	176.0	20

(Continued)

Table 4.38 (Continued)

Analyte	M (g/mol)	RT (min)	Precursor ion (m/z)	Product ion (m/z)	CE (V)
$^{13}C_6$-AN	184.2	8.1	184.0	156.0	30
			184.0	182.0	30
AN	178.2	8.1	178.0	152.0	20
			178.0	176.0	20
1MPH	192.3	8.6	192.0	165.0	30
			192.0	189.0	30
2MA	192.3	8.8	192.0	165.0	30
			192.0	189.0	30
$^{13}C_6$-FA	208.3	9.7	208.0	206.0	30
			208.0	180.0	30
FA	202.3	9.7	202.0	176.0	30
			202.0	200.0	30
$^{13}C_3$-PY	205.3	10.1	205.0	203.0	30
			205.0	204.0	30
PY	202.3	10.1	202.0	176.0	30
			202.0	200.0	30
BcFL	216.3	10.7	216.0	189.0	30
			216.0	215.0	20
1MP	216.3	11.0	216.0	189.0	30
			216.0	215.0	20
$^{13}C_6$-BaA	234.3	12.3	234.0	208.0	30
			234.0	232.0	30
BaA	228.3	12.3	228.0	202.0	30
			228.0	226.0	30
$^{13}C_6$-CHR	234.3	12.5	234.0	208.0	30
			234.0	232.0	30
CPP	226.3	12.5	226.0	200.0	30
			226.0	224.0	30
CHR	228.3	12.5	228.0	202.0	30
			228.0	226.0	30
1MC	242.3	13.1	242.0	226.0	30
			242.0	240.0	30
5MC	242.3	13.3	242.0	226.0	30
			242.0	240.0	30
3MC	242.3	13.4	242.0	226.0	30
			242.0	240.0	30
$^{13}C_6$-BbFA	258.3	14.6	258.0	230.0	30
			258.0	256.0	30

Table 4.38 (Continued)

Analyte	M (g/mol)	RT (min)	Precursor ion (m/z)	Product ion (m/z)	CE (V)
BbFA	252.3	14.6	252.0	226.0	30
			252.0	250.0	30
$^{13}C_6$-BkFA	258.3	14.7	258.0	230.0	30
			258.0	256.0	30
BkFA	252.3	14.7	252.0	226.0	30
			252.0	250.0	30
BjFA	252.3	14.8	252.0	226.0	30
			252.0	250.0	30
$^{13}C_4$-BaP	256.3	15.7	256.0	230.0	30
			256.0	256.0	30
BaP	252.3	15.7	252.0	226.0	30
			252.0	250.0	30
$^{13}C_6$-IP	282.3	19.3	282.0	254.0	30
			282.0	280.0	30
IP	276.3	19.3	276.0	248.0	40
			276.0	274.0	40
$^{13}C_6$-DBahA	284.3	19.3	284.0	256.0	30
			284.0	282.0	30
DBahA	278.3	19.3	278.0	252.0	30
			278.0	276.0	30
13C$_{12}$-BghiP	288.3	20.8	288.0	260.0	30
			288.0	286.0	30
BghiP	276.3	20.8	276.0	248.0	40
			276.0	274.0	40
DBalP	302.4	27.9	302.0	276.0	40
			302.0	300.0	30
$^{13}C_6$-DBaeP	308.4	30.7	308.0	282.0	40
			308.0	306.0	30
DBaeP	302.4	30.7	302.0	276.0	40
			302.0	300.0	30
$^{13}C_{12}$-DBaiP	314.4	32.5	314.0	288.0	40
			314.0	312.0	30
DBaiP	302.4	32.5	302.0	276.0	40
			302.0	300.0	30
DBahP	302.4	33.5	302.0	276.0	40
			302.0	300.0	30

710 | *4 Applications*

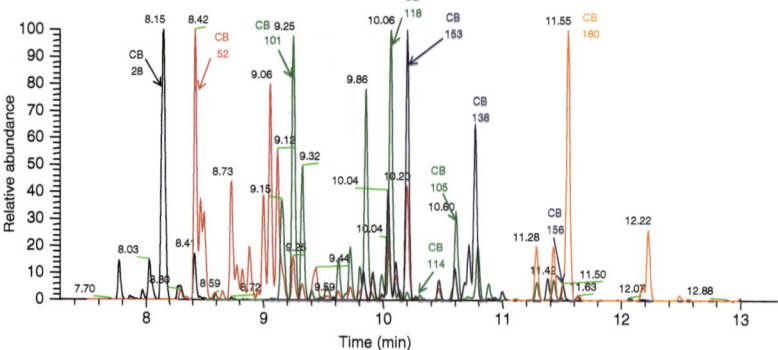

Figure 4.95 Herring fresh fish from the Baltic Sea (PCBs).

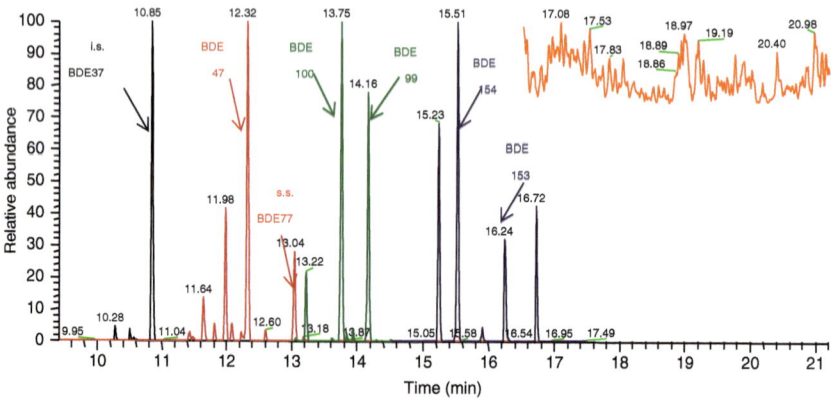

Figure 4.96 Herring fresh fish from the Baltic Sea (PBDEs).

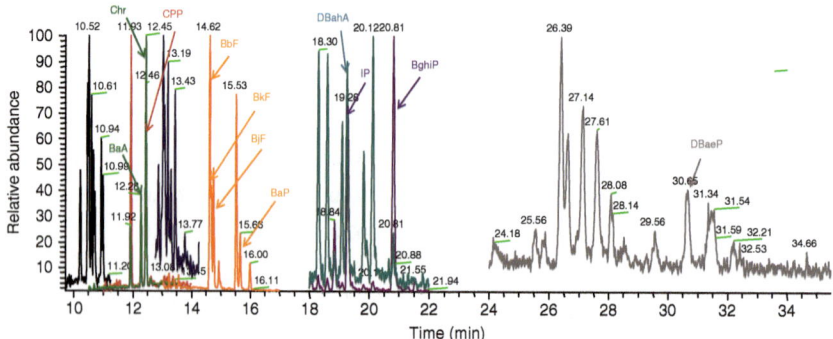

Figure 4.97 Bivalves fresh from Northeast Atlantic Ocean (PAHs).

Table 4.39 Method validation results.

n = 6	PAHs	PCBs	PBDEs
Recovery (%)	76–90	90–115	92–116
RSD (%)	5–14	2–11	9–16
LOQ (µg/kg)	0.025–0.5	0.005–0.01	0.025–0.5
Linearity R^2	0.993–1	0.993–0.999	0.991–0.999

For the quantitative calibrations, a precision with R^2 values higher than 0.99 is reported for all calibration ranges tested. The quantification of PAHs was highly influenced by matrix effects that are explained by active sites of the injector inlet liner, as can be seen in Figure 4.94. Injections' effects have been minimized using a deactivated inlet liner. The result calculation corrected recoveries with the internal syringe standard.

The achieved method quantitation limits (MQLs) allow the determination of dl-PCBs and the major PCBs well below the maximum limits set by the EU legislation. Also, for BaP and the sum of BaP, BaA, BbFA, and CHR, low MQLs of 0.025 and 0.1 µg/kg could be achieved. The reported MQLs were defined by $S/N > 6$ for the quantitative transition and $S/N > 3$ for the confirmation transition (European Commission, 2021).

4.24.6 Method Limitations

The labile highly brominated compounds BDE 196, 197, 203, 206, 207, and 209, OBIND (octabromo-1-phenyl-1,3,3-trimethylindane), and decabromodiphenyl ethane (DBDPE) cannot be included in the method due to an unsatisfactory sensitivity. For the trace analysis of these compounds, a dedicated shorter capillary column is required for a separate analysis to avoid compound degradation (Krumwiede et al., 2005).

With the chosen analytical column, a coelution of several PCBs is observed. PCBs 28/31, 84/101, 138/163, and 118/123 are not baseline separated. Coelution also occurs for the PAHs CHR/TRI (triphenylene) which are detected on the same SRM transition. These coeluting compounds need to be reported as a sum value.

The described method cannot be used for the OCPs dieldrin, endrin, endosulfan sulfate, and α- and β-endosulfan, as well as the PAHs AC, ACL, PHE, NA, 1MN, and 2MN. These volatile PAHs get partially lost during the evaporation step in the clean-up.

4.24.7 Conclusions

This method describes the trace determination of multi-class environmental contaminants in fish and fish feed. The analytical performance meets the EU SANCO criteria for the control of food and feed contaminants with high recoveries and good repeatability.

A total number of 73 target compounds comprising 18 PCBs, 16 OCPs, 14 BFRs, and 25 PAHs can be determined using a common liquid extraction sample preparation with a short silica minicolumn clean-up. Very low MQLs are achieved for all target analytes in the range from 0.005 to 1 µg/kg for fish muscle tissue and from 0.05 to 10 µg/kg for fish feed. The recoveries in both matrices are in the range of 70–120%.

The analytical relevance of those compounds which were excluded from the method in real fish or fish feed samples is seen of minor significance with respect to their frequency of presence and contamination levels. The volatile PAHs are typical air pollutants, and their presence in aquatic organisms indicates acute environmental contamination. Also, the occurrence of the five excluded OCPs dieldrin, endrin, endosulfan sulfate, and α- and β-endosulfan in fish is not seen as a significant source of human exposure (Kalachova et al., 2013).

4.25 Fast GC of PCBs

Keywords: Fast GC; PCBs; dl-PCBs; WHO-PCBs; sewage; sludge; single quadrupole MS

4.25.1 Introduction

The PCBs with its 209 possible congeners, although out of production since years, are still today subject to be monitored in food and environmental analysis (Lorenzi et al., 2020; UN, 2024; U.S. EPA, 2003). PCBs found widespread use as dielectric and heat-transferring fluids in power transformers, hydraulic fluids, as well as plasticizers, solvents, and flame retardants due to their stable molecular structure, which made them part of the POPs. In particular, the coplanar "dioxin-like" dl-PCBs, or non-ortho-substituted PCBs, are of increasing analytical importance because of their toxicity comparable to 2,3,7,8-TCDD (tetrachlorodibenzodioxin) with a significant contribution to the sample toxicity equivalent (TEQ) value.

12 PCB congeners are coplanar non-ortho substituted. They exhibit toxicological properties similar to dioxins, called the dl-PCBs or WHO-PCBs congeners (Table 4.40). Another group, the "non-dioxin-like" ndl-PCBs do not exhibit dioxin-like toxicity but are of different toxicological profile (Table 4.41). A set of six "indicator" PCBs are considered as appropriate marker for the occurrence and human exposure to a total of ndl-PCBs. Due to the toxicological properties of the dl-PCBs congeners, toxicity equivalent factors (TEF values) are applied in the final

Table 4.40 "Dioxin-like" dl-PCBs or "WHO-PCBs" congeners.

PCB type	Ballschmiter number
Tetra Cl-PCB	77, 81
Penta Cl-PCB	105, 114, 118, 123, 126
Hexa Cl-PCB	156, 157, 167, 169
Hepta Cl-PCB	189

Table 4.41 "Non-dioxin-like" ndl-PCBs *aka* "marker" or "indicator" PCBs.

PCB type	Ballschmiter number
Tri Cl-PCB	28
Tetra Cl-PCB	52
Penta Cl-PCB	101
Hexa Cl-PCB	138, 153
Hepta Cl-PCB	180

TEQ calculation, comprising the concentrations for PCDD/F congeners as well (see Section 4.26) (European Commission, 2011b, 2022).

The nomenclature of PCBs by systematic numbering follows the recommendation by Karlheinz Ballschmiter of the University in Ulm, Germany (Ballschmiter and Zell, 1980).

The GC separation of these PCB congeners was based for a long time on low and intermediate-polarity phases, such as 5% phenylsilicone (typically DB-5MS, TR-5MS, etc.). Problems with co-elution, poor detection at low levels, and the need for better separation of the toxic coplanar PCBs demand columns with increased selectivity.

The HT8 carborane phase provides good selectivity for the above dl-PCBs as well as the indicator congeners and separates them from most of the potential co-eluting congeners (Frame et al., 1996; Matsumara et al., 2002; SGE, 2004, 2005). The analysis is usually performed using a 50 m and 0.25 mm ID capillary column. The analysis run takes up to 60 min.

The analysis time for PCBs can be significantly reduced with fast GC conditions by using a short 10 m thin film column with 0.1 mm ID. The HT8 FAST PCB column has been successfully used as an excellent screening capillary column for MS detection (Gummersbach, 2011).

The unique selectivity of the HT8 column is attributed to the presence of the carborane unit having an affinity toward chlorinated biphenyls with the least number of ortho-substitutions. Fewer substitutions in the ortho position increase the freedom of rotation, allowing the chlorinated biphenyl moiety to have greater interaction with the carborane unit (de Boer et al., 1992). This phenomenon causes non-ortho-substituted PCBs to have increased elution times compared to their ortho-substituted congeners. This allows detection and quantification of the important congeners used to monitor PCB occurrence and distribution in the environment.

4.25.2 Analysis Conditions

Sample material			Sewage sludge
GC method	System		Thermo Scientific TRACE GC Ultra
	Injector		Split/splitless
	Column	Type	HT-8
		Dimensions	10 m length × 0.1 mm ID × 0.1 µm film thickness
	Carrier gas		Helium
	Flow		Constant flow, 0.6 mL/min
	SSL injector	Injection mode	Splitless
		Inj. Temperature	250 °C
		Inj. Volume	1 µL
	Split	Closed until	1 min
		Open	1 min to end of run
	Oven program	Start	90 °C, 1 min
		Ramp 1	30 °C/min to 220 °C
		Ramp 2	15 °C/min to 330 °C
		Final temperature	300 °C, 5 min
	Transferline	Temperature	280 °C
MS method	System		Thermo Scientific DSQ II
	Analyzer type		Single quadrupole MS
	Ionization		EI
	Electron energy		70 eV
	Ion source	Temperature	280 °C
	Acquisition	Mode	Full scan
	Mass range		100–500 Da
	Scan rate		0.1 s (10 spectra/s)

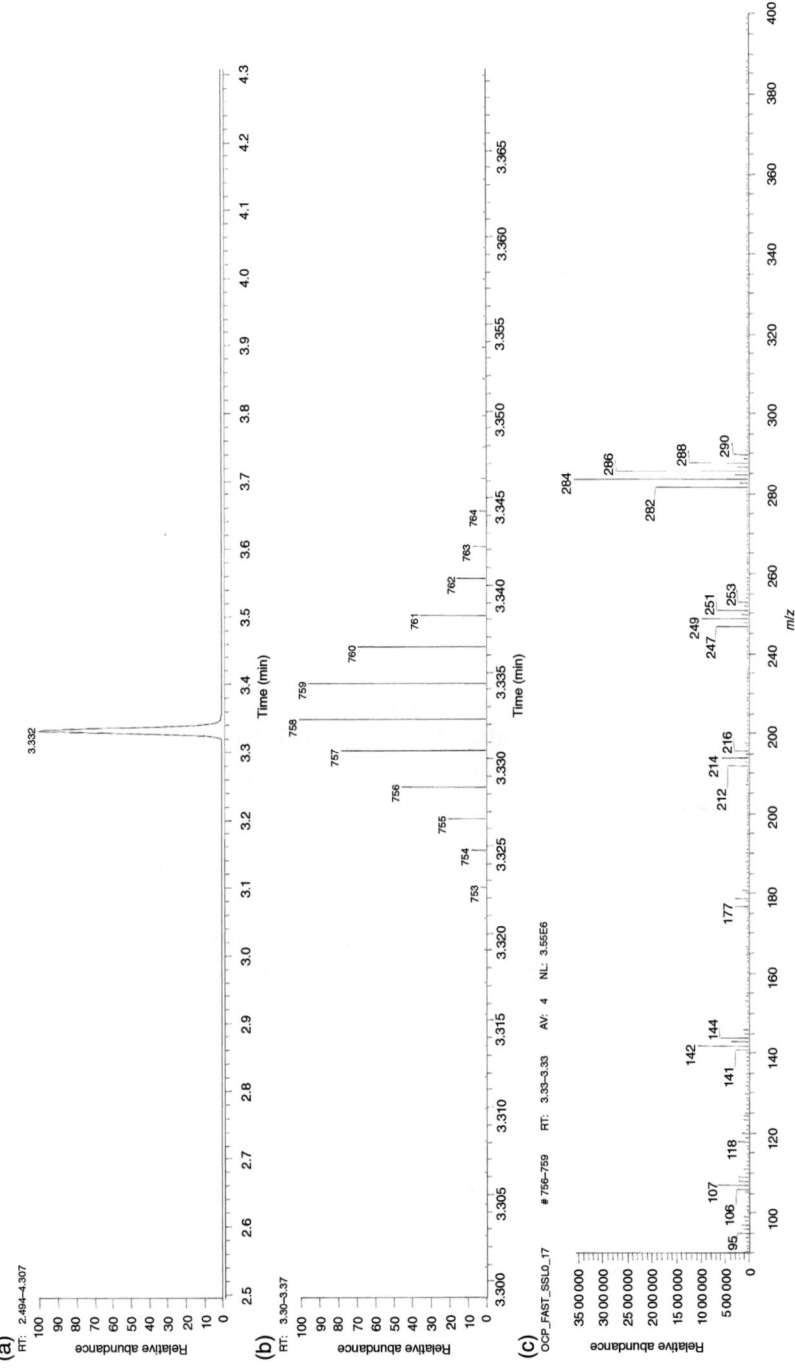

Figure 4.98 Hexachlorobenzene (HCB) spectrum in fast GC mode. Chromatographic peak (a); number of data points over the fast GC peak (b) and full scan spectrum (c).

716 | 4 Applications

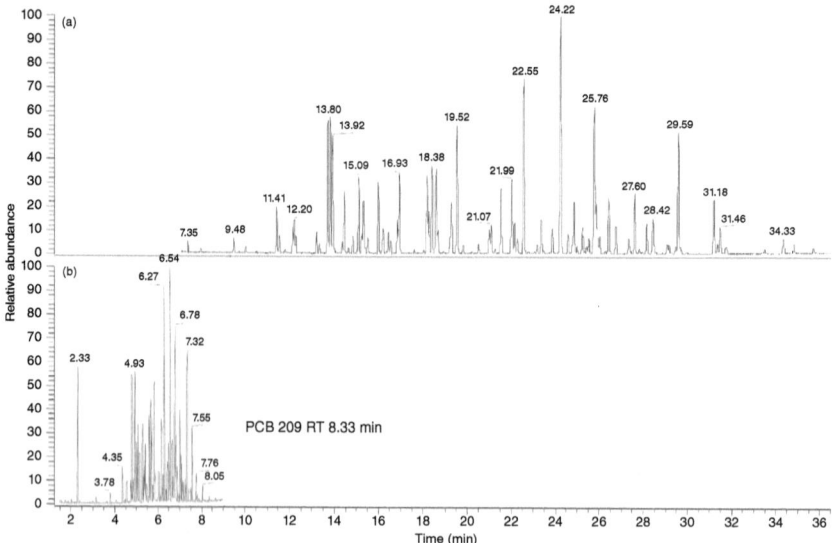

Figure 4.99 Aroclor Mix A30, A40, A60: Comparison of normal and Fast GC: (a) 30 m column length, 0.25 µm ID, (b) 10 m column length, 0.1 µm ID.

4.25.3 Results

The Fast GC separation of PCBs on the HT8 phase maintains the excellent congener separation but within a total analysis time of only 9 min including the ISTD decachlorobiphenyl (PCB 209). Laboratories with a high sample throughput benefit in particular from the increased productivity.

The very high scan speed of the employed MS detector in the full scan mass mode allows the acquisition of sufficient data points over the fast peak. An HCB peak with data points and mass spectrum is provided with Figure 4.98. Figure 4.99 compares the "normal" GC run with Fast GC for an Aroclor standards mix, completing the run in less than 10 min. The detailed chromatogram comparison in Figure 4.100 of the test mix shows a very comparable separation of the critical pairs, for example, congeners 31/28 (about 80%) and 163/138 (about 50%). The achieved separation power shows that the FAST PCB capillary column in combination with a very fast scanning quadrupole mass spectrometer is an excellent choice for the screening and quantification for PCBs.

Further potential for increased detection sensitivity could be achieved by running the MS in the SIM detection mode. The analysis of a contaminated sewage sludge sample acquired in the full scan mode given in Figure 4.101 shows the well-known mass chromatogram patterns of the PCB chlorination degrees, but in the very short analysis runtime of less than 10 min including the elution of the internal quantification standard PCB 209.

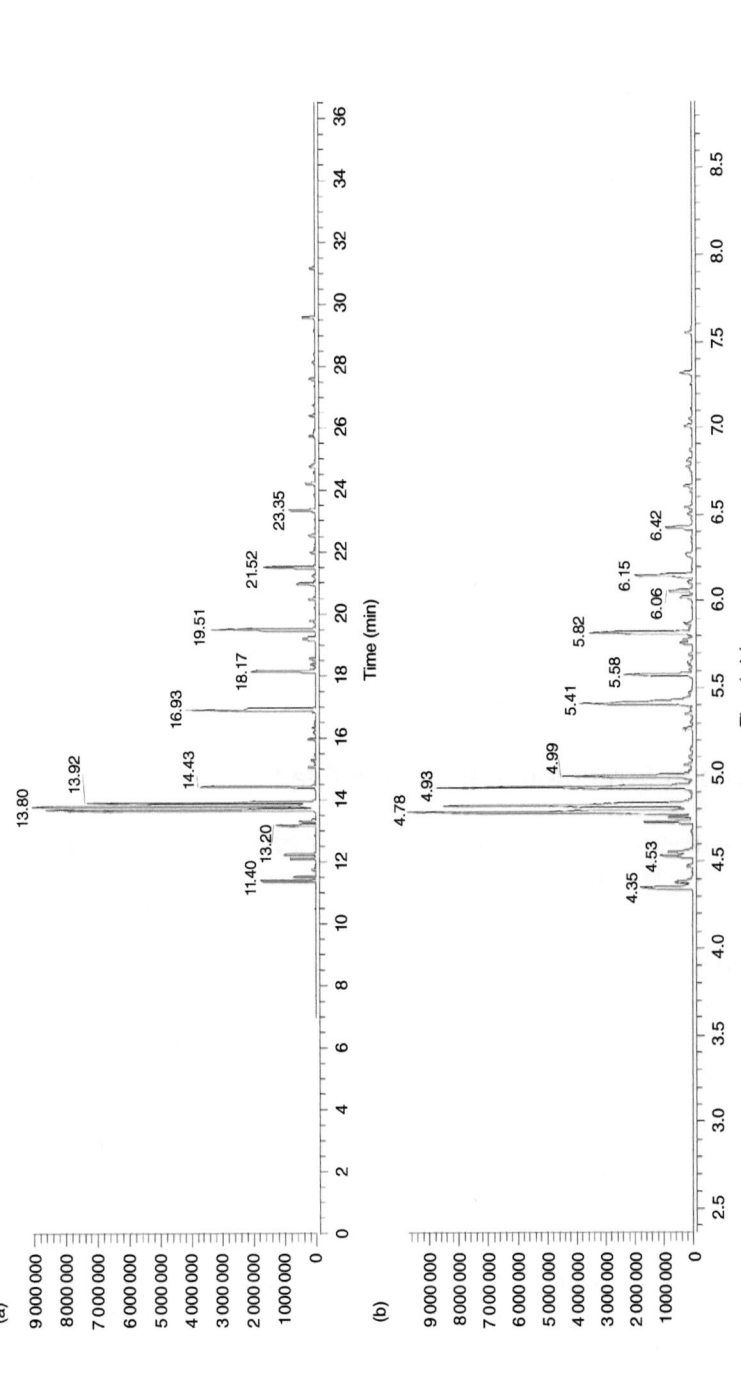

Figure 4.100 Aroclor Mix A30, A40 and A60: separation power comparison of (a) normal chromatography with 30 m column length to (b) Fast GC with 10 m column length.

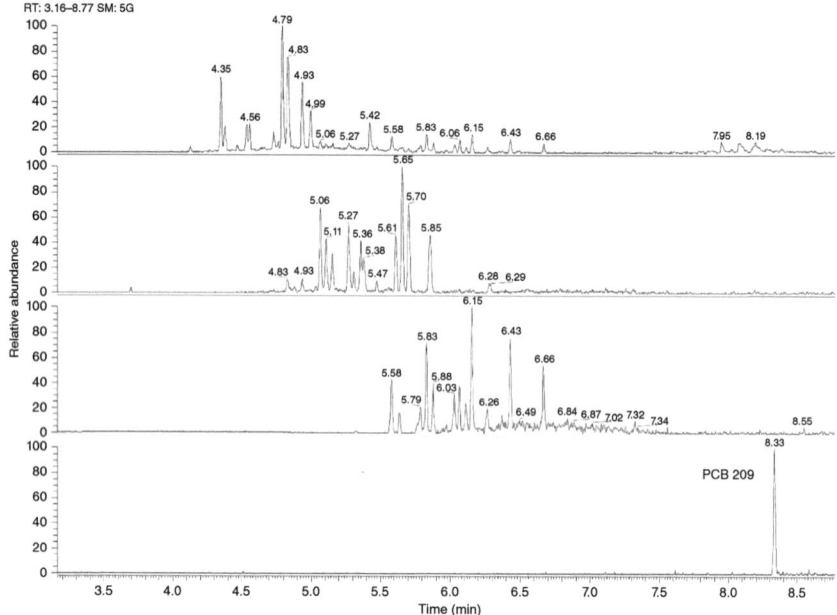

Figure 4.101 Fast GC-MS analysis of a sewage sludge sample with the extracted mass chromatograms of tri- (*m/z* 256), tetra- (*m/z* 293), penta- (*m/z* 326) and deca-chlorobiphenyl (*m/z* 498, ISTD).

4.26 Dioxin Screening in Food and Feed

Keywords: food safety; animal feed; POPs; PCDDs; PCDFs; dl-PCBs; EU regulations; triple quadrupole MS/MS; GC-HRMS; ion ratio confirmation

4.26.1 Introduction

The monitoring of food and feed stuff for the presence of polychlorinated dibenzo-*p*-dioxins (PCDDs), polychlorinated dibenzofurans (PCDFs), and the dioxin-like polychlorinated biphenyls (dl-PCBs) can be performed with screening and confirmatory methods.

The methods of sampling and analysis for the official control of levels of PCDD/PCDFs and dl-PCBs for food and feeding stuffs are defined by European Commission Regulations. Screening methods comprise GC-MS, GC-MS/MS, or bioanalytical methods. Confirmatory methods are defined as gas chromatography/high resolution mass spectrometry (GC-HRMS) (U.S. EPA, 1994a) and in Europe as well by GC-MS/MS methods (Commission Regulation (EU) 2017/771) with criteria as outlined in Table 4.42. The control of the regulated maximum level by screening methods should allow a cost-effective high sample throughput with the goal to avoid false negative results. Here, GC-MS methods as well as bioanalytical methods are in use. The confirmation of results in samples with significant levels requires

Table 4.42 Criteria for screening and confirmatory methods (Commission Regulation (EU) 2017/771).

	Screening with bioanalytical or physicochemical methods	Confirmatory methods
False-compliant rate	<5%	
Trueness		−20 to +20%
Repeatability (RSD$_r$)	<20%	
Within-laboratory reproducibility (RSD$_R$)	<25%	<15%

a confirmatory method. The quality criteria for the screening and confirmatory methods in the EU regulations are oriented toward high reproducibility and a low false compliant rate.

For GC-MS methods, the method performance needs to be demonstrated to cover the level of interest with an acceptable precision (coefficient of variation, CV). The required LOQ needs to be at 20% of the level of interest, which requires in absolute amounts detectable quantities of PCDD/PCDFs in the upper femtogram range, non-ortho-PCBs in the low picogram range, and other dl-PCBs in the nanogram range. The limit of quantitation of an individual congener is defined in this context as the concentration of an analyte in the extract of a sample which produces an instrumental response on two different ions with a S/N ratio of 3 : 1 for the less sensitive signal.

Also, GC-MS screening methods require the control of recoveries by the addition of ^{13}C-labeled ISTDs at the first sample preparation step. Screening methods require at least one congener as an ISTD for any homologous group or mass spectrometric recording function. In confirmatory methods, all ^{13}C-labeled ISTDs shall be used. The range of recoveries shall cover 30–140% for screening methods and 60–120% for confirmatory methods.

For the clean-up and fractionation as well as the chromatographic separation of isomers, the same requirements apply for screening and confirmation.

Due to the high performance of modern triple quadrupole mass spectrometers for PCDD/PCDF and dl-PCB analysis (Kotz *et al.*, 2012; Fürst and Bernsmann, 2014), further amendments of the criteria for confirmatory methods for inclusion of MS/MS (tandem mass spectrometry) have been set in effect with the European Commission Regulation (EU) No 709/2014, amended by the Commission Regulation (EU) 2017/771: "Technical progress and developments have shown that the use of gas chromatography/tandem mass spectrometry (GC-MS/MS) should be allowed for use as a confirmatory method for checking compliance with the maximum level, in addition to gas chromatography/high resolution mass spectrometry (GC-HRMS)" (European Commission, 2014, 2017):

- The calculation of LOQ needs to be based on the S/N or lowest concentration point on the calibration curve.

- The data acquisition requires the monitoring of at least two specific precursor ions, each with one specific corresponding transition product ion for all labeled and unlabeled analytes (comparable to GC-HRMS).
- The mass resolution setting for each quadrupole needs to be set equal to or better than unit mass resolution in order to minimize possible interferences on the analytes of interest.
- The maximum permitted tolerance of relative ion intensities needs to be demonstrated to be in the range of ±15% for selected SRM product ions in comparison to calculated or measured values (average from calibration standards).

4.26.2 Analysis Conditions

Sample material			Food, feeding stuff
GC method	System		Thermo Scientific TRACE GC Ultra
	Autosampler		Thermo Scientific TriPlus RSH
	Column	Type	TR-Dioxin, TG-5MS, or DB-5MS
		Dimensions	60 m length × 0.25 mm ID × 0.25 μm film thickness
	Carrier gas		Helium
	Flow		Constant flow, 1 mL/min
	PTV injector	Injection mode	Splitless
		Injection volume	5 μL, extract in toluene
		Injection speed	5 μL/s
		Base temperature	100 °C
		Transfer rate	13.3 °C/s
		Transfer temperature	340 °C
		Final temperature	340 °C
	Split	Closed until	1 min
		Open	1 min to end of run
		Flow	20 mL/min
	Oven program	Start	120 °C
		Ramp 1	17 °C/min to 250 °C
		Ramp 2	2.5 °C/min to 285 °C
		Final temperature	285 °C, 13 min
	Transferline	Temperature	280 °C
			Direct coupling

MS method	System		Thermo Scientific TSQ Quantum XLS Ultra
	Analyzer type		Triple quadrupole MS
	Ionization		EI, 40 eV
	Ion source	Temperature	250 °C
		Emission current	50 µA
	Acquisition	Mode	Timed-SRM
			2 SRMs per analyte
	Collision gas		Ar, 1.5 mTorr
	Collision energy		22 V (all congeners)
	Resolution	Setting	Normal (0.7 Da) for Q1 and Q3
Calibration	Internal standard		Isotope dilution method, ^{13}C-labeled standards
	Calibration range		0.0125–10.0 pg/µL, congener dependent
	Data points		5

4.26.3 Sample Preparation

The extraction, clean-up and fractionation process for indicator, mono-ortho and non-ortho PCBs and PCDD/Fs used for food and feed samples were performed according Figure 4.102 for GC-MS/MS and GC-HRMS. The identical extracts were measured on both instrument types for performance evaluation.

4.26.4 Experimental Conditions

For GC-MS/MS analysis, a GC equipped with PTV allows the injection of larger extract volumes with solvent split as described. Regular split/splitless injectors are used with 1–2 µL injections.

Special attention is required concerning the analytical column choice, carrier gas flow, and the setup of the GC temperature program. The U.S. EPA method 1613 sets certain requirements for the TCDD and TCDF (tetrachlorodibenzofuran) congener separation (U.S. EPA 1994b). The valley between 2,3,7,8-TCDD and the other tetra-dioxin isomers with m/z 319.8965 and between 2,3,7,8-TCDF and the other tetra-furan isomers at m/z 303.9016 shall not exceed 25%, as shown in Figure 4.103 using a 60 m TR-5MS column type. The column length of 60 m is also required to separate the 1,2,3,4,7,8-HxCDF and 1,2,3,6,7,8-HxCDF isomers with better than 25% valley.

Other known interferences observed on the most used 5% phenyl columns can be caused by the content of PCBs in the sample. PCB 87 can interfere with PCB

722 | 4 Applications

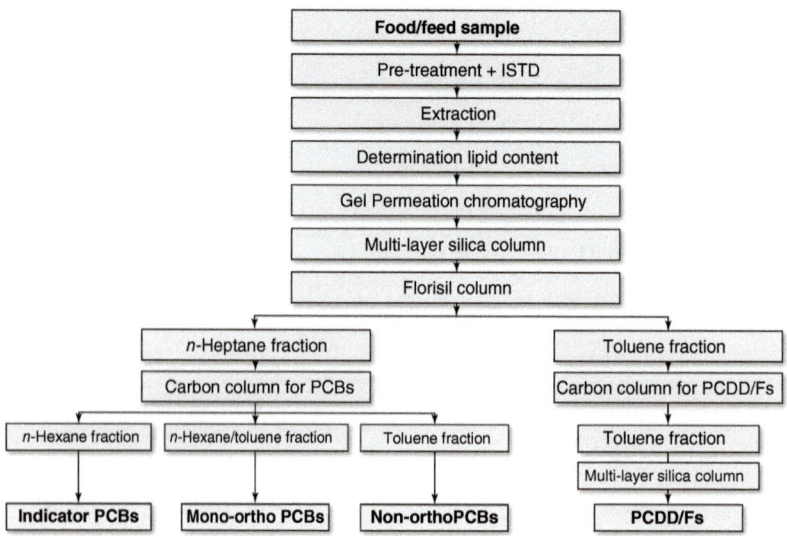

Figure 4.102 Extraction and clean-up process for food and feed samples. (Kotz *et al.*, 2011/European Union.)

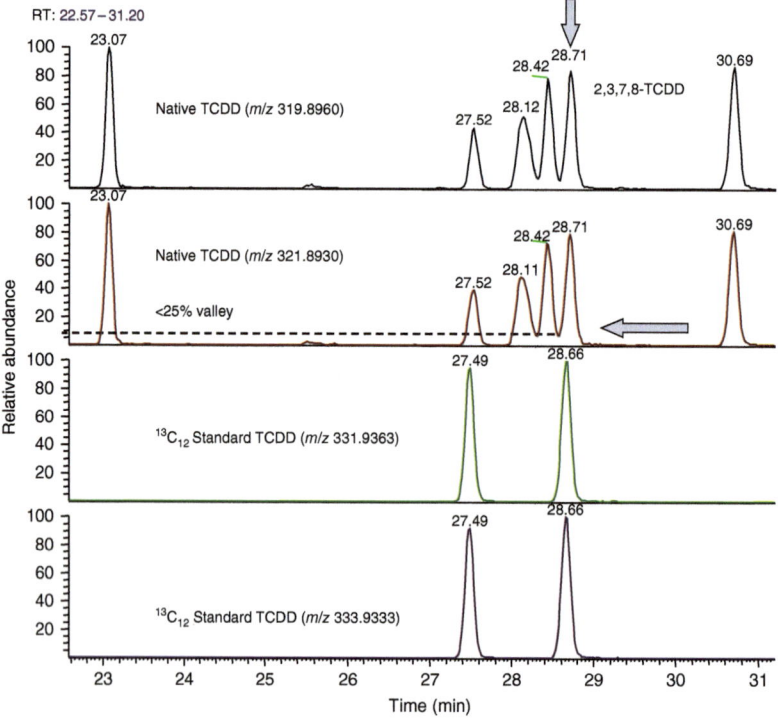

Figure 4.103 Separation performance of <25% valley between 2,3,7,8-TCDD and the other tetra-dioxin congeners as of U.S. EPA 1613 requirement.

Table 4.43 Separation potential of the 17 2,3,7,8-substituted congeners with the 2022 WHO TEF values on different GC columns. Highlighted the current combination of nonpolar and polar columns for the separation of all WHO TEF congeners.

PCDDs/PCDFs	2022 WHO TEF	DB-5, HP-5MS, Rtx-5MS, Equity-5	CP-Sil 8 CB/MS, VF-5ms	DB-5MS, ZB-5MS	VF-Xms	ZB-SUMS	Rtx-Dioxin2	DB-XLB	BPX-DXN	DB-225	RH-12ms	SP-2331
				non-polar, low polar						polar		
2,3,7,8-TCDD	1	++	+-	++	++	++	++	+-	++	+-	--	+-
1,2,3,7,8-PnCDD	0.4	++	+-	+-	++	--	--	--	++	--	--	--
1,2,3,4,7,8-HxCDD	0.09	++	++	++	++	++	++	++	++	++	--	++
1,2,3,6,7,8-HxCDD	0.07	++	++	++	++	++	++	+-	++	++	++	++
1,2,3,7,8,9-HxCDD	0.05	--	+-	+-	++	++	++	++	++	++	++	++
1,2,3,4,6,7,8-HpCDD	0.05	++	++	++	++	++	++	++	++	++	++	++
OCDD	0.001	++	++	++	++	++	++	++	++	++	++	++
2,3,7,8-TCDF	0.07	--	+-	+-	++	++	++	++	++	++	++	+-
1,2,3,7,8-PnCDF	0.01	++	++	++	++	++	++	++	++	--	++	--
2,3,4,7,8-PnCDF	0.1	--	--	--	--	--	--	--	++	++	++	++
1,2,3,4,7,8-HxCDF	0.3	--	++	++	++	++	++	++	++	++	++	--
1,2,3,6,7,8-HxCDF	0.09	++	++	++	++	++	++	++	++	--	--	++
2,3,4,6,7,8-HxCDF	0.1	+-	--	--	+-	--	--	+-	--	--	++	++
1,2,3,7,8,9-HxCDF	0.2	++	--	--	+-	--	--	++	--	++	--	++
1,2,3,4,6,7,8-HpCDF	0.02	++	++	++	++	++	++	++	++	++	++	++
1,2,3,4,7,8,9-HpCDF	0.1	++	++	++	++	++	++	++	++	++	++	++
OCDF	0.002	++	++	++	++	++	++	++	++	++	++	++

with + + baseline separation or at least 10% valley; + − quantifiable result, corresponds to at least 50% valley; − − interference.
(Adapted from Fishman et al., 2007, 2011; Mamoon, 2013; Theobald, 2024.)

81, and PCBs 129 and 178 with PCB 126. Also, PCBs 77 and 110 can interfere on the column and need be separated in the clean-up step. Other interferences are reported to occur on the masses of the HxCBs 169 and 1,2,3,7,8-PeCDD which are very similar and coelute on 30 and 60 m DB-5MS. Other newly developed column phases which are specialized for the dioxin/PCB congener separation should be considered for improvements of particular congener coelutions. For instance, the BPX-DXN column is able to separate between HxCB 169 and 1,2,3,7,8-PeCDD, but at lower sensitivity of the 60 m BPX-DXN compared to a 30 m DB-5MS column. The Rtx-Dioxin2 and TG-Dioxin columns offer a different selectivity and separate well 2,3,7,8-TCDD and 2,3,7,8-TCDF from other congeners, improving the 2,3,4,7,8-PeCDF and 1,2,3,7,8-PeCDF separation. The separation on a VF-Xms column has shown the potential to separate all the 17 toxic 2,3,7,8-substituted dioxins and furans (Fishman et al., 2011; Mamoon, 2013). About limitations on congener separation and the most suitable column choice, also refer to Section 4.27 and Table 4.43.

All target congeners are monitored using two precursor ions with each one product ion. As precursor ions usually the intense M^+ ions are used. For the PCDD/PCDFs, the product ions typically generated by COCl loss are selected, see Table 4.44. With PCBs, the elimination of chlorine is observed, see Table 4.45.

4 Applications

Table 4.44 SRM transitions for tetra to octa PCDFs/PCDDs and the labeled internal standards.

PCDD/PCDF type	Precursor ion 1 (m/z)	Product ion 1 (m/z)	Precursor ion 2 (m/z)	Product ion 2 (m/z)
TCDF	303.90	240.94	305.90	242.94
$^{13}C_{12}$-TCDF ISTD	315.94	251.97	317.94	253.97
TCDD	319.90	256.90	321.89	258.89
$^{13}C_{12}$-TCDD ISTD	331.94	267.97	333.93	269.97
PeCDF	339.86	276.90	341.86	278.89
$^{13}C_{12}$-PeCDF ISTD	351.90	287.93	353.90	289.93
PeCDD	355.85	292.85	357.85	294.85
$^{13}C_{12}$-PeCDD ISTD	367.90	303.90	369.89	305.89
HeCDF	371.82	308.86	373.82	310.86
$^{13}C_{12}$-HeCDF ISTD	383.86	319.90	385.86	321.89
HeCDD	387.82	324.82	389.82	326.82
$^{13}C_{12}$-HeCDD ISTD	399.86	335.86	401.86	337.86
HpCDF	407.78	344.82	409.78	346.82
$^{13}C_{12}$-HpCDF ISTD	419.82	355.86	421.82	357.85
HpCDD	423.78	360.78	425.77	362.77
$^{13}C_{12}$-HpCDD ISTD	435.82	371.82	437.81	373.81
OCDF	441.76	378.80	443.76	380.79
$^{13}C_{12}$-OCDF ISTD	453.78	389.82	455.78	391.81
OCDD	457.74	394.74	459.74	396.74
$^{13}C_{12}$-OCDD ISTD	469.78	405.78	471.78	407.78

For calibration, reference standards in the concentrations as given in Table 4.46 were applied. The chromatogram peaks of the congeners are displayed in Figure 4.104. For checking of the performance of the GC-MS/MS system in the low concentration range, additionally 1 : 2 and 1 : 5 dilutions of the lowest calibration point were measured. The calculated WHO-PCDD/PCDF-TEQ for these calibrations are based on the analysis equivalent of 3 g of fat and an injection of 5 µL out of 20 µL of the concentrated final extract volume, ranged between 0.12 and 24 pg/g fat.

4.26.5 Results

One important criterion for quality control in the identification of the measured PCDD/PCDF congeners is the ion ratio between the two monitored product ions for each congener. For quality control, the ion abundance ratios can be compared

Table 4.45 SRM transitions for PCBs and the labeled internal standards.

PCB type	Precursor ion 1 (m/z)	Product ion 1 (m/z)	Precursor ion 2 (m/z)	Product ion 2 (m/z)
MoCB	188.04	153.04	190.04	153.04
$^{13}C_{12}$-MoCB ISTD	200.08	165.10	202.08	165.10
DiCB	222.00	152.06	224.00	152.06
$^{13}C_{12}$-DiCB ISTD	234.04	164.10	236.04	164.10
TrCB	255.96	186.02	257.96	186.02
$^{13}C_{12}$-TrCB ISTD	268.00	198.02	270.00	198.02
TeCB	289.92	219.98	291.92	219.98
$^{13}C_{12}$-TeCB ISTD	301.96	232.02	303.96	232.02
PeCB	323.90	253.95	325.90	255.95
$^{13}C_{12}$-PeCB ISTD	335.92	265.99	337.92	267.99
HxCB	357.80	287.90	359.80	289.95
$^{13}C_{12}$-HxCB ISTD	369.90	299.51	371.90	301.95
HpCB	391.80	321.90	393.80	323.90
$^{13}C_{12}$-HpCB ISTD	403.80	333.90	405.80	335.90
OcCB	427.80	357.80	429.80	357.80
$^{13}C_{12}$-OcCB ISTD	439.80	369.90	441.80	369.90
NoCB	461.70	391.80	463.70	393.80
$^{13}C_{12}$-NoCB ISTD	473.80	403.80	475.80	405.80
DeCB	495.70	425.80	497.70	427.80
$^{13}C_{12}$-DeCB ISTD	507.70	437.80	509.70	439.80

with calculated or measured values, as demonstrated in Figure 4.105. The ratios are comparable with the measured ratios if identical collision energy and collision gas pressure are applied for both transitions. The measured ion abundance ratios in the calibration run match the calculated theoretical values within the QC limits of ±15%. The calculated ratio depends on the relative abundance of the two selected precursor ions of the molecular ion and the probability of the loss of the two chlorine isotopes with the possible leaving groups $CO^{35}Cl$ or $CO^{37}Cl$, leading to the formation of two product ions of different masses. Figures 4.106 and 4.107 show the principle for TCDD. The ion abundance ratio 1.04 for TCDD between the quantitation ion and confirming ion for TCDD is calculated by

$$\frac{\text{Relative abundance (quantitation ion)} \times \text{Probability (quantitation ion)}}{\text{Relative abundance(confirming ion)} \times \text{Probability(confirming ion)}} = 1.04$$

The results of triple quadrupole GC-MS/MS were compared with magnetic sector GC-HRMS measurements. The deviation of the GC-MS/MS results from

4 Applications

Table 4.46 Concentrations of individual congeners in calibration solutions Cal 1 to Cal 5.

	Cal 1 (pg/µL)	Cal 2 (pg/µL)	Cal 3 (pg/µL)	Cal 4 (pg/µL)	Cal 5 (pg/µL)
2,3,7,8-TCDD	0.0125	0.025	0.05	0.2	0.5
1,2,3,7,8-PeCDD	0.0250	0.050	0.10	0.4	1.0
1,2,3,4,7,8-HxCDD	0.0250	0.050	0.10	0.4	1.0
1,2,3,6,7,8-HxCDD	0.0625	0.125	0.25	1.0	2.5
1,2,3,7,8,9-HxCDD	0.0250	0.050	0.10	0.4	1.0
1,2,3,4,6,7,8-HpCDD	0.1250	0.250	0.50	2.0	5.0
OCDD	0.2500	0.500	1.00	4.0	10.0
2,3,7,8-TCDF	0.0125	0.025	0.05	0.2	0.5
1,2,3,7,8-PeCDF	0.0125	0.025	0.05	0.2	0.5
2,3,4,7,8-PeCDF	0.0625	0.125	0.25	1.0	2.5
1,2,3,4,7,8-HxCDF	0.0250	0.050	0.10	0.4	1.0
1,2,3,6,7,8-HxCDF	0.0250	0.050	0.10	0.4	1.0
1,2,3,7,8,9-HxCDF	0.0125	0.025	0.05	0.2	0.5
2,3,4,6,7,8-HxCDF	0.0125	0.025	0.05	0.2	0.5
1,2,3,4,6,7,8-HpCDF	0.0250	0.050	0.10	0.4	1.0
1,2,3,4,7,8,9-HpCDF	0.0125	0.025	0.05	0.2	0.5

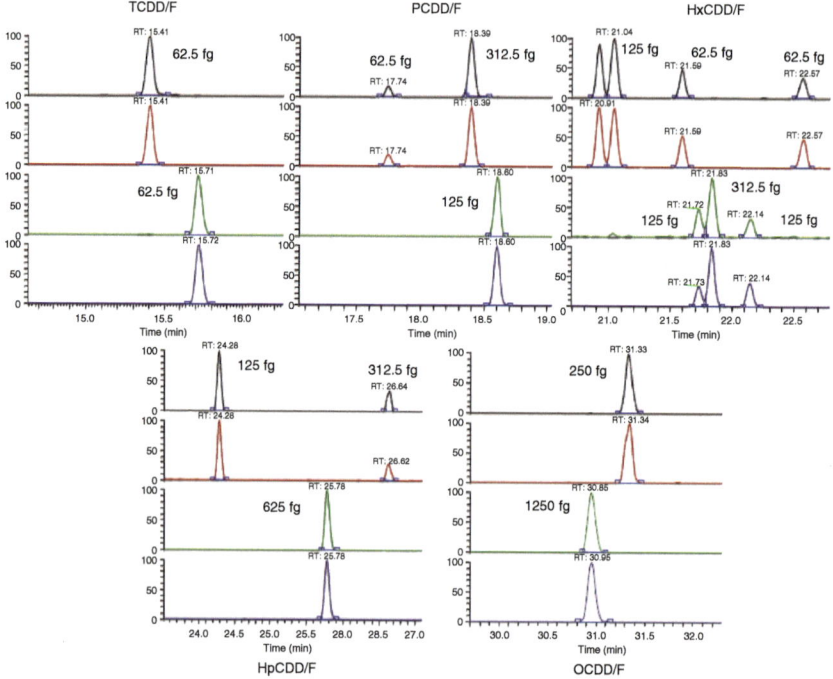

Figure 4.104 Lowest level calibrations peaks (Cal 1) for tetra- to octa-CDD/F congeners, top traces PCDDs, bottom PCDFs with injected amounts.

4.26 Dioxin Screening in Food and Feed | 727

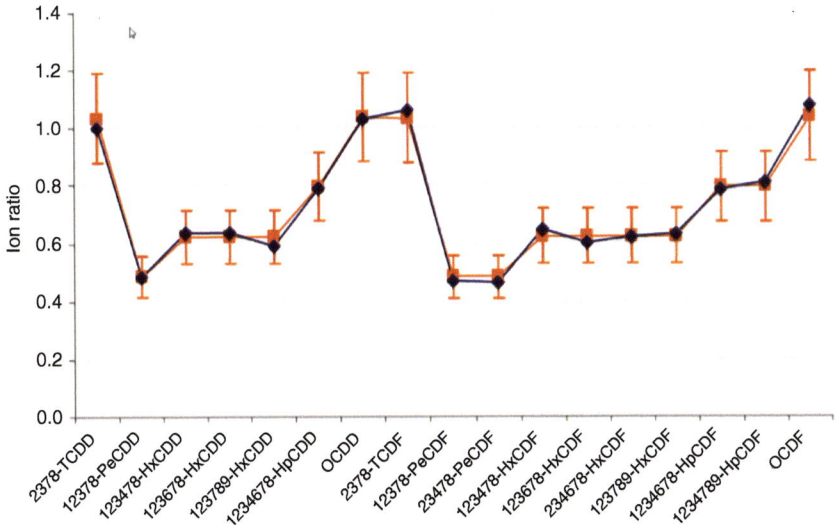

Figure 4.105 Comparison of the calculated (square) and measured (diamond) ion ratio on a TSQ Quantum XLS Ultra GC-MS/MS system.

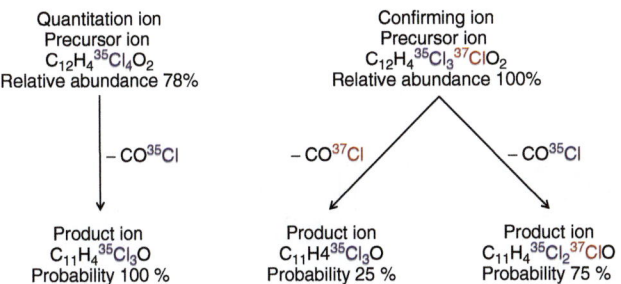

Figure 4.106 Calculation of theoretical ion abundance ratio for TCDD.

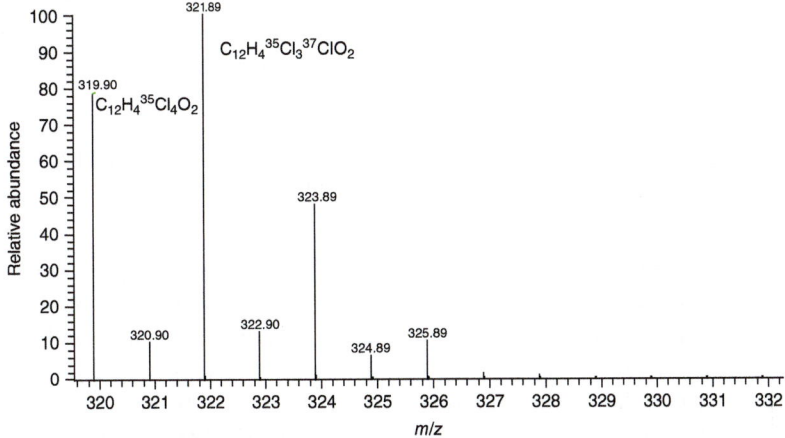

Figure 4.107 TCDD molecular ion cluster with m/z 319.90 $C_{12}H_4{}^{35}Cl_4O_2$ and m/z 321.89 $C_{12}H_4{}^{35}Cl_3{}^{37}ClO_2$ due to the statistical distribution of the chlorine isotopes.

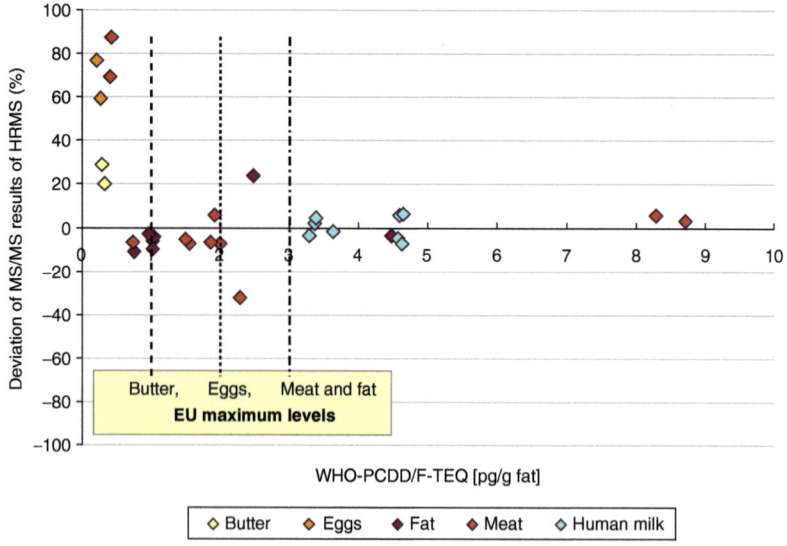

Figure 4.108 Compliance of GC-MS/MS with the required EU PCDD/PCDF maximum levels for butter, eggs, meat and fat in pg/g fat and % difference to the GC-HRMS method. (Kotz et al., 2011/European Union.)

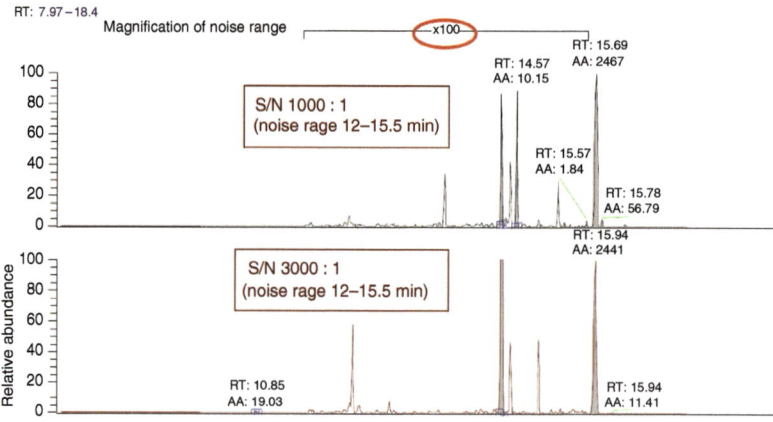

Figure 4.109 2,3,7,8-TCDD detection at RT 15.69 min in a spiked fat sample by GC-MS/MS, determined as 62.5 fg, top quantitation, bottom ion ratio transition.

GC-HRMS for different food samples and human milk, covering a concentration range between 0.1 and 10 pg WHO-PCDD/PCDF-TEQ/g fat, is below 20% for most of the samples. The applicable maximum levels for butter, eggs, meat, and fat together with difference GC-MS/MS vs. GC-HRMS are shown in Figure 4.108. For all sample types, the maximum levels can be controlled by triple quadrupole GC-MS/MS with good compliance compared to the GC-HRMS method. As examples, a spiked fat sample is shown in Figure 4.109 at a concentration of

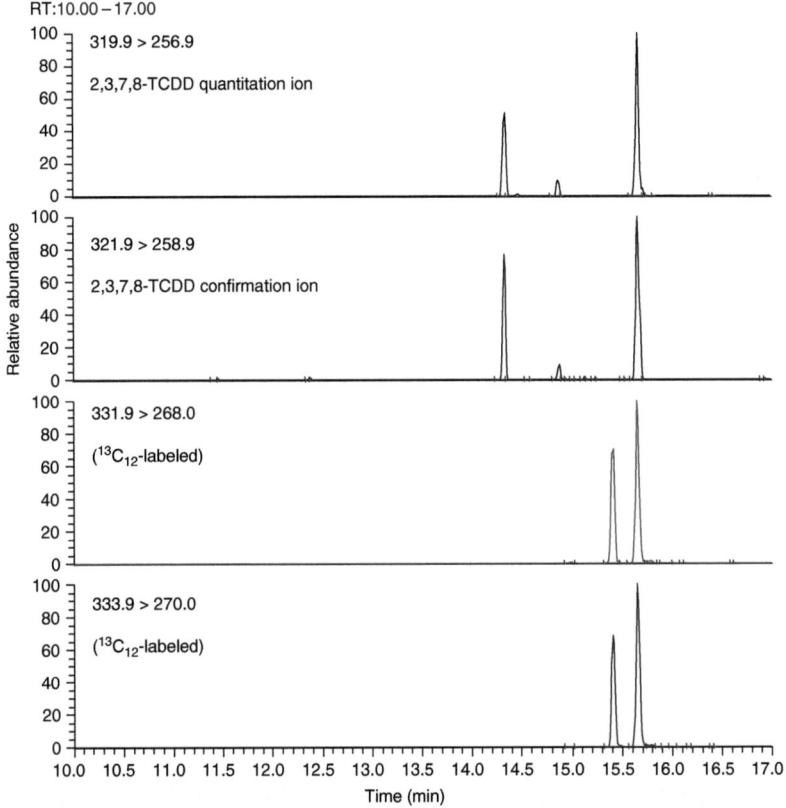

Figure 4.110 2,3,7,8-TCDD mass traces of a contaminated real fat sample by GC-MS/MS, determined as 150 fg, from top quantitation, ion ratio confirmation, bottom: labeled internal standard transitions.

62.5 fg 2,3,7,8-TCDD, and a real-life sample of a contaminated fat sample in Figure 4.110.

4.26.6 Conclusions

Latest developments in triple quadrupole technology make GC-MS/MS applicable for routine analysis of PCDD/PCDFs and dioxin-like PCBs in food and feed at the levels of regulatory importance. The triple quadrupole GC-MS/MS systems provide a cost-efficient alternative analysis technique for the determination of PCDD/PCDFs and dioxin-like PCBs in food and feed in Europe. Outside of Europe as of the status in 2024, GC-MS/MS is applicable for in-house screening assays, requiring confirmation of positive results by GC-HRMS as described in Chapter 4.27. The sample preparation and clean-up workflow as well as the chromatographic separation are kept unchanged as of the standard procedures established in GC-HRMS protocols.

A good correlation between the results of GC-MS/MS and GC-HRMS could be observed for the concentration range above 0.5 pg WHO-PCDD/PCDF-TEQ/g fat for food and human milk samples. For fish and feed samples, acceptable deviations were observed in the concentration range considerably below the established legal limits (Kotz et al., 2011).

4.27 Confirmation Analysis of Dioxins and Dioxin-like PCBs

Keywords: dioxin; PCDD; PCDF; dl-PCBs; TEFs; TEQ calculation; confirmation; GC-HRMS; magnetic sector MS; accurate mass; mass resolution; U.S. EPA Method 1613

4.27.1 Introduction

More than 90% of human exposure to dioxins and dioxin-like substances is through food, mainly meat and dairy products, fish, and shellfish, and hence leads to continuous control (WHO, 2024). "Dioxin analyses" have been performed since the mid-1970s as millions of broilers died in the United States from a chicken feed contamination later identified as chlorodibenzo-p-dioxins (Firestone, 1973). The need for monitoring is still up to date due to natural processes but also due to food and feed contamination scandals in the past. A low but permanent background contamination level is entering the food chain. But, with the gratefully observed

Table 4.47 Number of PCDD/PCDF and PCB isomers for analysis.

Chlorination degree	PCDDs	PCDFs	PCBs
Mono	2	4	3
Di	10	16	12
Tri	14	28	24
Tetra	22	38	42
Penta	14	28	46
Hexa	10	16	42
Hepta	2	4	24
Octa	1	1	12
Nona	—	—	3
Deca	—	—	1
Total	**75**	**135**	**209**
Total of 2,3,7,8 Cl config.	7	10	12[a]

a) PCBs with ≤ 1 Cl ortho-substitution (non/mono-ortho) are coplanar and exhibit "dioxin"-like toxicity.

Table 4.48 Summary of 1998, 2005, and new 2022 WHO TEF values.

Compound	1998 WHO TEF	2005 WHO TEF[a]	2022 WHO-TEF[b]
Chlorinated dibenzo-p-dioxins (PCDDs)			
2,3,7,8-TCDD	1	1	1
1,2,3,7,8-PeCDD	1	1	**0.4**
1,2,3,4,7,8-HxCDD	0.1	0.1	**0.09**
1.2.3.6.7.8-HxCDD	0.1	0.1	**0.07**
1.2.3.7.8.9-HxCDD	0.1	0.1	**0.05**
1,2,3,4,6,7,8-HpCDD	0.01	0.01	**0.05**
OCDD	0.0001	0.0003	**0.001**
Chlorinated dibenzofurans (PCDFs)			
2,3,7,8-TCDF	0.1	0.1	**0.07**
1,2,3,7,8-PeCDF	0.05	0.03	**0.01**
2,3,4,7,8-PeCDF	0.5	0.3	**0.1**
1,2,3,4,7,8-HxCDF	0.1	0.1	**0.3**
1.2.3.6.7.8-HxCDF	0.1	0.1	**0.09**
1.2.3.7.8.9-HxCDF	0.1	0.1	**0.2**
2,3,4,6,7,8-HxCDF	0.1	0.1	**0.1**
1.2.3.4.6.7.8-HpCDF	0.01	0.01	**0.02**
1.2.3.6.7.8.9-HpCDF	0.01	0.01	**0.1**
OCDF	0.0001	0.0003	**0.002**
Non-ortho-substituted chlorinated biphenyls (PCBs)			
3,3′,4,4′-tetraCB (PCB 77)	0.0001	0.0001	**0.0003**
3,4,4′,5-tetraCB (PCB 81)	0.0001	0.0003	**0.006**
3,3′,4,4′,5-pentaCB (PCB 126)	0.1	0.1	**0.05**
3,3′,4,4′,5,5′-hexaCB (PCB 169)	0.01	0.03	**0.005**
Mono-ortho-substituted chlorinated biphenyls (PCBs)			
2,3,3′,4,4′-pentaCB (PCB 105)	0.0001	0.00003	0.00003
2,3,4,4′,5-pentaCB (PCB 114)	0.0005	0.00003	0.00003
2,3′,4,4′,5-pentaCB (PCB 118)	0.0001	0.00003	0.00003
2′,3,4,4′,5-pentaCB (PCB 123)	0.0001	0.00003	0.00003
2,3,3′,4,4′,5-hexaCB (PCB 156)	0.0005	0.00003	0.00003
2,3,3′,4,4′,5′-hexaCB (PCB 157)	0.0005	0.00003	0.00003
2,3′,4,4′,5,5′-hexaCB (PCB 167)	0.00001	0.00003	0.00003
2,3,3′,4,4′,5,5′-heptaCB (PCB 189)	0.0001	0.00003	0.00003

a) van den Berg *et al.* (2006).
b) Bold values indicate the updated TEF value (DeVito *et al.*, 2024)

lower dioxin levels in food, feed, and tissues, more demanding LOD, selectivity, sensitivity, and QC checks are required to trace their presence at the expected further decreasing levels (Aylward and Hays, 2002). Studies document the declines in exposure and body burden in the United States (Lorber, 2002; Hays and Aylward, 2003; Turner et al., 2006, CDC, 2009, 2014; Hernández et al., 2020). As the dioxin and dl-PCB concentrations are related to the low fat level in blood, significantly improved sensitivities of the analytical instrumentation applied are required to further establish an efficient control at further decreasing levels, especially in the younger population. With children and toddlers, the small available blood sample sizes become the limiting factor (Turner et al., 2004).

The complexity of analysis is caused by the large number of occurring isomers of the dioxins and dioxin-like compounds. The PCDD/PCDFs with 2,3,7,8 Cl substitution, and the PCBs with chlorine at meta positions and at one or none of the ortho positions on the biphenyl ring (non/mono-ortho) are coplanar and exhibit "dioxin"-like toxicity. As of Table 4.47, the number of compounds to be analyzed is 75 PCDDs, 135 PCDFs, and 209 PCBs, in total 419 chlorinated compounds of which 29 compounds must be quantified using the WHO TEF as of Table 4.48. Additional potentially toxic brominated or mixed chlorinated/brominated compounds are not taken into account at this point.

4.27.2 Analysis Conditions

Sample material			Food, meat eggs, milk, blood, and more
GC method	System		Thermo Scientific TRACE GC Ultra
	Injector		Split/splitless
	Column	Type	TG-5MS
		Dimensions	60 m length × 0.25 mm ID × 0.25 µm film thickness
	Carrier gas		Helium
	Flow		Constant flow, 1.3 mL/min
	SSL injector	Injection mode	Splitless
		Injection temperature	260 °C
		Injection volume	2 µL
	Split	Closed until	1 min
		Open	1 min to end of run
		Gas saver	2 min, 10 mL/min
	Oven program	Start	120 °C, 3 min
		Ramp 1	19 °C/min to 210 °C
		Ramp 2	3 °C/min to 275 °C
		Final temperature	275 °C, 12 min
	Transferline	Temperature	280 °C

MS method	System		Thermo Fisher Scientific DFS
	Analyzer type		Double focusing magnetic sector MS
	Ionization		EI
	Electron energy		45 eV
	Ion source	Temperature	270 °C
	Resolution	Setting	R 10 000 (at 5% peak height)
Calibration	Internal standard Range		^{13}C-labeled congeners

4.27.2.1 Chromatographic Analysis

After extraction and a standard clean-up, the concentrated extract is injected for GC-HRMS analysis. The sample preparation workflow is described in Section 4.26. Despite many years of development, there is currently not one type of GC column available for the separation of all compounds of interest. The analytical challenge is the interference free peak integration of the congeners listed in the WHO TEF document in the presence of a huge number of nontoxic congeners in complex chromatograms. Missing peak assignment or faulty peak integration can lead to significant errors and overestimation in the result calculation. Quantitative results of a sample expressed as TEQs are calculated by the sum of the products of compound concentration times TEF value:

$$\text{TEQ [pg/g]} = \sum^{i} c_i \text{ (PCCDs, PCDFs, dl-PCBs) [pg/g]} \times \text{TEF}_i$$

with c_i being the concentration of the individual compounds listed in Table 4.48.

For a quantitative analysis with the calculation of the TEQ values, a summary of the WHO 2005 and the revised WHO 2022 TEF values are given in Table 4.48 (van den Berg et al., 1998, 2006; DeVito et al., 2024). The coelution of nontoxic congeners with the toxic compounds of the WHO TEF list must be avoided for the correct calculation of the sample total TEQ value from the concentrations of the individual compounds with TEF values.

In this challenge, the choice of analytical column is crucial. The same discussion applies to dioxin analyses using a triple quadrupole MS/MS system. As an example, in Figures 4.112 and 4.113, the mass traces of four hexafurans (HxCDFs) are shown. As the congeners cannot be mass separated, the column separation is required. The problematic congeners are 2,3,4,6,7,8-HxCDF at RT 41.5 min and 1,2,3,7,8,9-HxCDF at RT 43.5 min both with highly contributing 2022 TEF values of 0.1 and 0.2, as shown in Table 4.48.

The HxCDF mass traces of the two selected columns show the 2,3,4,6,7,8-HxCDF peak at RT 41.5 min partially separated from a coeluting nontoxic congener. This mandatory separation cannot be achieved on the often-used DB5MS type columns. For the 1,2,3,7,8,9-HxCDF at RT 43.5 min, a partial separation from the interfering congener is possible on the VM-Xms column type, shown in Figure 4.113. This partial separation is also possible on the DB5MS type columns. The Rtx-Dioxin2 column shows just one peak with the coeluting congener and could lead to overestimation (Figure 4.112).

The programs of modern automated dioxin sample preparation systems are commonly designed in such a way that the result of the sample purification delivers a fraction with dioxins together with the four non-ortho PCBs and a second fraction with the non-coplanar PCBs. Having the four non-ortho PCBs with the congener numbers 77, 81, 126, and 169 in the dioxin fraction can cause interferences with dioxin peaks. Especially, the HxCB congener 169 is problematic since this congener interferes on the typical measured mass traces for PnCDD and usually both compounds cannot be separated chromatographically. The similar RTs of the interference cause the failure of the ion ratio verification which checks the ratio of the areas of the two measured PnCDD masses. In such cases, the ratio verification indicates ion ratios out of the allowed limits. The theoretical mass traces are displayed with the complete isotope pattern in unit mass resolution, and with the HRMS signal in Figure 4.114. At R 10 000, the signals still cannot be distinguished. A solution to overcome the problem is to modify the acquisition method by selecting a different and unaffected mass for the quantitation of PnCDD. For instance, m/z 353.8576 could be used.

In general, for dioxin analysis, columns of 60 m length with 0.25 mm ID and 0.25 µm film thickness are applied. A comprehensive investigation of the available GC column options for PCDD/F compound separation demonstrated in detail that two columns, nonpolar and polar, are required for an interference-free quantitation (Fishman et al., 2007). In a later investigation, it was shown that the separation on a VF-Xms column has the potential to serve as a single column type for the separation of all 17 2,3,7,8-substituted dioxins and furans (Fishman et al., 2011; Mamoon, 2013). A practical limitation with the else required polar columns arises from their intrinsic higher column background at high elution temperatures. A column combination often used can be the nonpolar DB-5MS or ZB-5MS with the polar SP-2331 column. In Japan, many laboratories use the combination of BPX-DXN (60 m × 0.25 mm × 0.25 µm) and the locally preferred RH-12ms (60 m × 0.25 mm × 0.25 µm). Table 4.43 summarizes the current data of the PCDD and PCDF separation potential with columns used in dioxin analysis, together with the 2022 TEF values of the regulated congeners. With a dual GC configuration on the Thermo Fisher Scientific DFS magnetic sector HRMS system both columns are permanently installed and connected to the HRMS ion source. Both columns are served by a robotic sampler according to the workflow. The carrier gas flow of the inactive column is then reduced to a minimum.

GC-HRMS Method Setup The described GC-HRMS isotope dilution method for the quantification of PCDDs, PCDFs, and dl-PCBs, commonly referred to as "dioxins", follows the U.S. EPA 1613 Rev. B method (U.S. EPA, 1994a, 1998a,b). The SIM scheme on the accurate target masses typically uses isotope ratio qualifiers besides the specific RT for all native dioxin/furan congeners, as well as for their specific ^{13}C-labeled ISTDs, one quantification mass and one ratio mass. The analytical setup for the high resolution GC-MS is given with the SIM descriptor, as shown in Table 4.49 listing the exact masses of PCDD/PCDFs including the ^{13}C-labeled standards. A typical SIM setup for the data acquisition of the individual groups of

Table 4.49 Accurate mass SIM set up for PCDD and PCDF analysis in SIM lock-and-cali mode (width first lock: 0.3 Da and voltage settling time delays: 10 ms).

SIM window no. (time window)	Reference masses (FC43) m/z lock mass (L), cali mass (C)	Target masses m/z native (n) ^{13}C internal standard (is)	SIM cycle time, intensity, dwell time (ms)
1. Tetra-PCDD/PCDF (9.00–19.93 min)	313.98336 (L), 363.98017 (C)	303.90088(n), 305.89813(n), 315.94133(is), 317.93838(is), 319.89651(n), 321.89371(n), 331.93680(is), 333.93381(is)	0.75 s (L/C: 30, 4 ms; n: 1, 137 ms; is: 7, 19 ms)
2. Penta-PCDD/PCDF (19.93–23.52 min)	313.98336 (L), 363.98017(C)	339.85889(n), 341.85620(n), 351.89941(is), 353.85702(n), 353.89646(is!), 355.85400(n), 365.89728(is), 367.89433(is)	0.80 s (L/C: 30, 4 ms; n: 1, 147 ms; is: 7, 21 ms)
3. Hexa-PCDD/PCDF (23.52–26.98 min)	375.97974 (L), 413.97698 (C)	371.82300(n), 373.82007(n), 385.86044(is), 387.85749(is), 389.81494(n), 391.81215(n), 401.85535(is), 403.85240(is)	0.80 s (L/C: 30, 4 ms; n: 1, 147 ms; is: 7, 21 ms)
4. Hepta-PCDD/PCDF (26.98–32.06 min)	413.97698 (L), 463.97378 (C)	407.78101(n), 409.77826(n), 419.82147(is), 421.81852(is), 423.77588(n), 425.77317(n), 435.81638(is), 437.81343(is)	0.90 s (L/C: 35, 4 ms; n: 1, 169 ms; is: 7, 24 ms)
5. Octa-PCDD/PCDF (32.06–36.00 min)	425.97681 (L), 463.97378 (C)	441.74219(n), 443.73929(n), (453.78250(is)), (455.77955(is)), 457.73706(n), 459.73420(n), 469.77741(is), 471.77446(is)	0.95 s (L/C: 40, 4 ms; n: 1, 183 ms; is: 7, 22 ms)

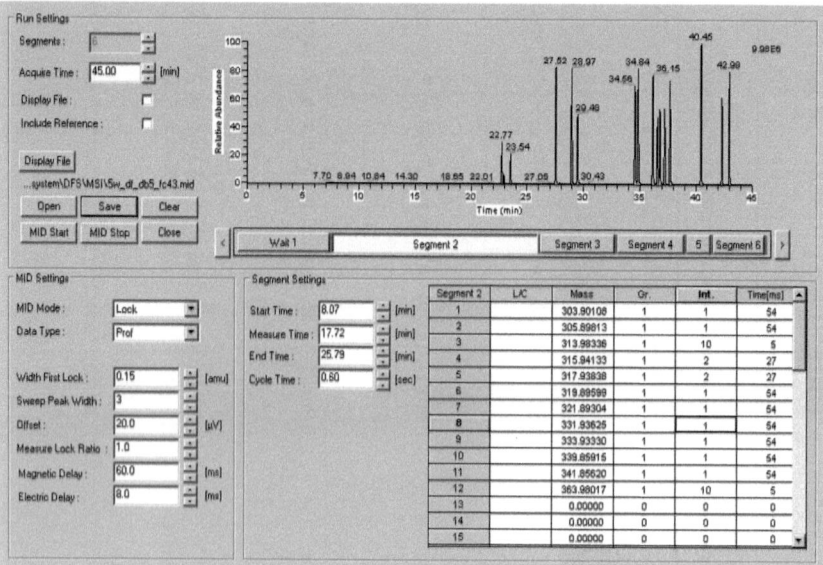

Figure 4.111 Typical SIM setup with the individual acquisition segments (retention time windows) for each eluting congener group of different chlorination degree.

chlorinated compounds is given in Figure 4.111. The mass spectrometer is operated at a mass resolution power of equal or better than 10 000 as required by the U.S. EPA 1613B method. The effective resolution achieved during the analysis of real-life samples has to be constantly monitored on the reference masses and documented in the data file for each SIM segment.

4.27.3 Results

The comparison of a GC separation using the same real-life sample with the HxCDF mass traces is shown in Figures 4.112 and 4.113 acquired using the parameters from Table 4.49 (Theobald, 2024). The chromatograms show the different

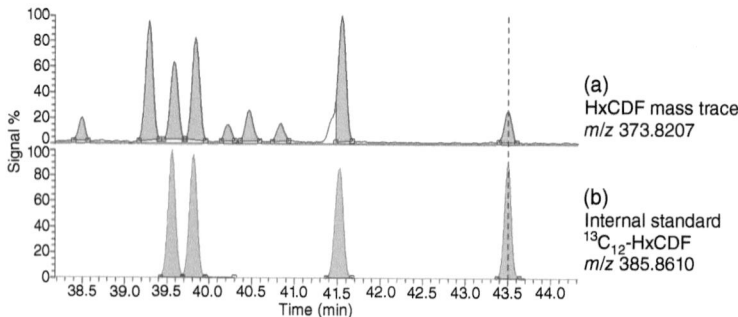

Figure 4.112 Mass chromatograms of the HxCDF compounds eluting from a Rtx-Dioxin2 column and labeled internal standards (60 m length × 0.25 mm ID × 0.25 μm film).

4.27 Confirmation Analysis of Dioxins and Dioxin-like PCBs | 737

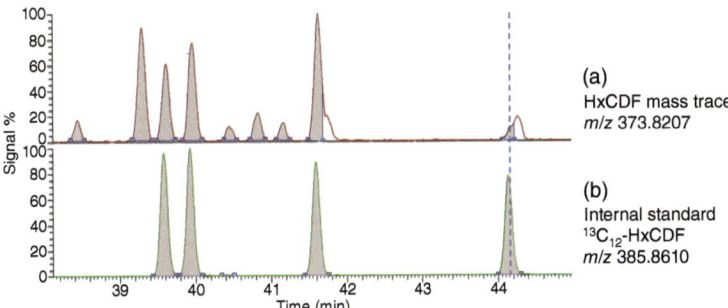

Figure 4.113 Mass chromatograms of the HxCDF compounds eluting from a VF-Xms column and labeled internal standards (60 m length × 0.25 mm ID × 0.25 μm film).

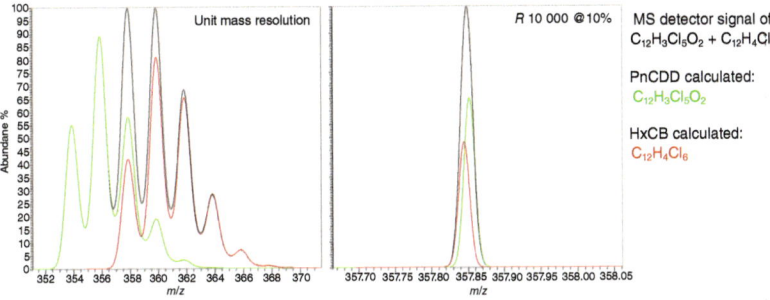

Figure 4.114 The problematic of interfering congener masses from HxCB 169 (red) on PnCDD (green) on m/z 357.8511. The detector signal trace is shown in black. Left: unit mass resolution full isotope pattern. Right: HRMS signal at resolution power R 10 000 (Theobald, 2024).

separation characteristics of both columns. The dotted line shows the elution of 1,2,3,7,8,9-HxCDF with a TEF of 0.2 and the nontoxic 1,2,3,4,8,9-HxCDF. A partial separation and manual peak integration is possible, as shown in Figure 4.113. The overlapping of the ion masses of a PnCDD with a HxCB is shown in Figure 4.114 for both unit mass and high mass resolution (R 10 000). The choice of a suitable quan ion can avoid the potential overestimation of the dioxin content.

The stability of confirmation ratios (relative areas of quantification and ratio masses) for all dioxins/furans in repeated injections of 17 fg/μL were evaluated for a blood pool sample (Figure 4.115) (Krumwiede, 2006a). All 2,3,7,8-TCDD results of a sample measurement series over several days of the blood sample gave excellent results within the required ±15% window at the lowest detection levels and provided the confirmation ion ratios in compliance with U.S. EPA 1613 requirements.

For the routine check and confirmation of the analytical performance for both, triple quadrupole and HRMS instruments, a dioxin standard of different TCDD congeners aka "Step Standard" in decreasing concentration, eluting well separated on any of the column types in use, was introduced (Krumwiede, 2006b). The concentration profile ranges from 2 to 100 fg including a labeled ISTD and allows the continuous monitoring of peak profiles and sensitivity (Table 4.50 and Figure 4.116).

738 | 4 Applications

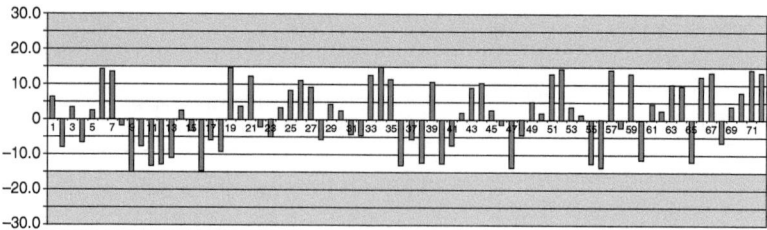

Figure 4.115 Confirmation of TCDD HRMS ion ratios m/z 319.8960/321.8930 in % for repeated injections of a blood sample extract at 17 fg/µL, ±15% window complies with the U.S. EPA 1613 method.

Table 4.50 The so-called dioxin "Step Standard" with six different native TCDD congeners at different concentrations of 2–100 fg/µL, eluting at different retention time as demonstrated in Figure 4.116.

1,3,6,8-TCDD	2 fg/µL
1,3,7,9-TCDD	5 fg/µL
1,3,7,8-TCDD	10 fg/µL
1,4,7,8-TCDD	25 fg/µL
1,2,3,4-TCDD	50 fg/µL
2,3,7,8-TCDD	100 fg/µL
2,3,7,8-$^{13}C_{12}$ TCDD (ISTD)	100 fg/µL

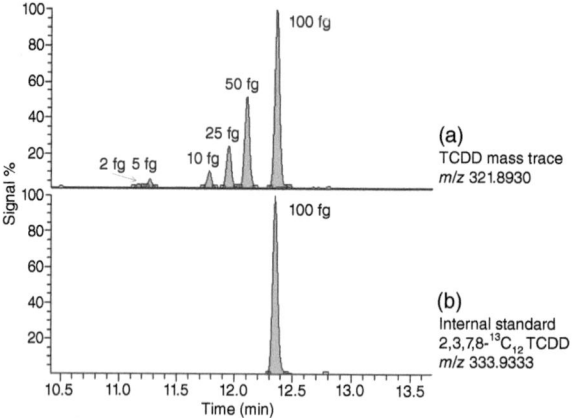

Figure 4.116 Chromatogram of the dioxin "Step Standard" for continuous monitoring of instrument performance measured on GC-HRMS.

4.28 U.S. EPA 1614 Brominated Flame Retardants PBDEs

Keywords: BFRs; PBDE; deca-BDE; PTV OCI mode; degradation; U.S. EPA Method 1614; magnetic sector MS; HRMS

4.28.1 Introduction

Polybrominated diphenyl ethers (PBDEs) are within the brominated flame retardants (BFRs) the most important and most widely used flame proofing compounds in a range of different industrial and consumer products and stepwise restricted between 2003 and 2010 (ECHA, 2023). Although PBDEs have been banned or restricted in the EU and U.S., they are still persistent in the environment. PBDEs can leach into the air, water, soil, food, and feed (EFSA, 2024), moving them into the focus of legislation resulting in a ban for certain BPDE congeners (European Parliament, 2003a). The EU directive 2003/11/EC prohibits the use of penta-BDE and octa-BDE for the member states of the European community (European Parliament, 2003b). Maximum values were recently set (European Parliament, 2022). The European Food Safety Authority (EFSA) is updating its scientific opinion on BFRs in food, taking into account new occurrence data and scientific information. The work is scheduled for completion by 2025 (EFSA, 2024).

As a result, BDEs have received rising interest in recent years by the analytical community (Martinez *et al.*, 2021). The by far most efficient analysis technique is high resolution GC-MS using the isotope dilution technique according to U.S. EPA 1614 for highest precision quantification with highest significance (Krumwiede, 2006c).

4.28.2 Analysis Conditions

Sample material			Industrial and consumer products
GC method	System		Thermo Scientific TRACE GC Ultra
	Column	Type	TG-5MS
		Dimensions	15 m length × 0.25 mm ID × 0.1 µm film thickness
	Carrier gas		Helium
	Flow		Constant flow, 1.0 mL/min
	SSL injector	Injection mode	Splitless
		Injection temperature	280 °C
		Injection volume	2 µL

	Split	Closed until	1 min
		Open	1 min to end of run
		Slit flow	50 mL/min
	Oven program	Start	120 °C, 2 min
		Ramp 1	15 °C/min to 230 °C
		Ramp 2	5 °C/min to 270 °C
		Ramp 3	10 °C/min to 330 °C
		Final temperature	330 °C, 5 min
	Transferline	Temperature	280 °C
MS method	System		Thermo Scientific DFS
	Analyzer type		Double focusing magnetic sector MS
	Ionization		EI, 40 eV
	Ion source	Temperature	270 °C
	Acquisition	Mode	Selected ion detection (SIM) (Table 4.51)
	Mass resolution		R 10 000 (at 5% peak height)

4.28.3 Results

The full scan results proved that for all bromination degrees, either the molecular ion or the fragment ion showing the loss of 2 Br atoms is the most abundant ion (see Figure 4.117). The change of the most intense ion from M^+ to $[M-2Br]^+$ is typically

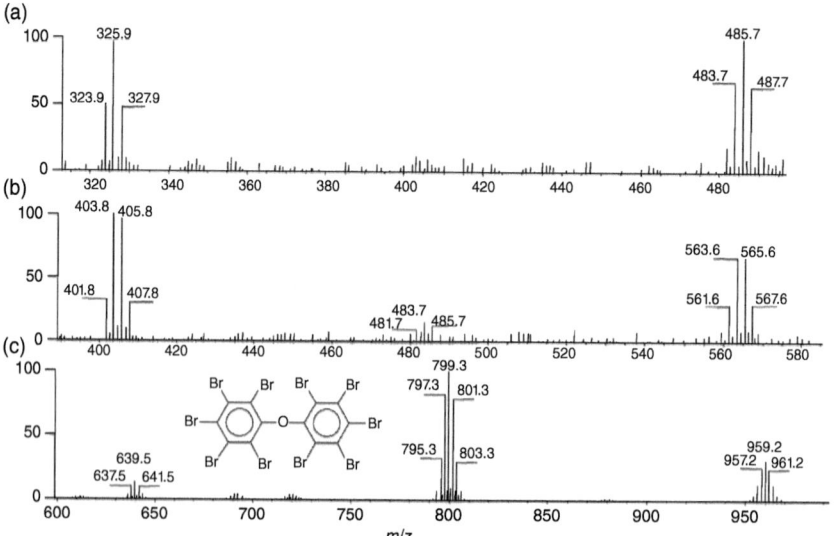

Figure 4.117 Mass spectra of PBDE congeners: (a) tetra-BDE, (b) penta-BDE, and (c) deca-BDE and the molecular structure of deca-BDE.

observed with higher bromination degrees and starting to become significant already with penta- and hexa-BDEs, depending on the ion source conditions. The loss of Br is temperature dependent and varies with GC elution and ion source temperatures. Therefore, with different instrument conditions, the transition of the most abundant ion from M^+ to $[M-2Br]^+$ might be shifted to tetra/penta- or hexa/hepta-BDE. In general, the relative intensity $[M-2Br]^+/M^+$ is increasing with the degree of bromination. For deca-BDE, the intensity gain when using the $[M-2Br]^+$ mass peak for the SIM analysis is at least a factor of 4 compared to M^+ (Krumwiede and Huebschmann, 2007, 2008). For the SIM setup the exact masses and cycle times as of Table 4.51 were used. For a list of exact masses including relative intensities, see Tables 4.52 and 4.53.

Mass spectrometer tuning parameters similar to those typically used for dioxin/PCB analysis were found to give optimum sensitivity for BDE analysis as well. A slightly higher ion source temperature of 270 °C is recommended, taking the high boiling characteristics of the BFRs into account (although highly brominated compounds appear at lower elution temperatures as the same molecular mass hydrocarbons). The use of PFK (perfluorokerosene) as an internal mass reference for accurate mass measurements is mandatory because lock-and-cali masses in the high mass range are needed (e.g. for deca-BDE). Autotuning for gaining highest sensitivity was carried out on the PFK mass m/z 480.9688.

Table 4.51 Accurate mass SIM setup: SIM lock-and-cali mode (target masses in brackets: optional second ratio mass for native PBDE).

SIM window no.	Reference masses (PFK) L = lock mass, C = cali mass	Target masses (second ratio mass native)	SIM cycle time (s)
1. Tri-BDE	392.9753 (L), 430.9723 (C)	(403.8041), 405.8021, 407.8001, 417.8424, 419.8403	0.55
2. Tetra-BDE	480.9688 (L), 492.9691 (C)	(483.7126), 485.7106, 487.7085, 495.7529, 497.7508	0.55
3. Penta-BDE	554.9644 (L), 592.9627 (C)	(561.6231), 563.6211, 565.6190, 575.6613, 577.6593	0.60
4. Hexa-BDE	480.9688 (L), 504.9691 (C)	481.6976, 483.6956, (485.6937), 493.7372, 495.7352	0.60
5. Hepta-BDE	554.9644 (L), 592.9627 (C)	(559.6082), 561.6062, 563.6042, 573.6457, 575.6436	0.70
6. Deca-BDE	754.9531 (L), 766.9531 (C)	797.3349, 799.3329, (801.3308), 809.3752, 811.3731	0.90

Table 4.52 PBDE exact mass references (1): for natives and internal standards (molecular ions).

#Br	Native Exact mass M⁺ (Da)	Native Relative intensity (%)	$^{13}C_{12}$ Standard Exact mass M⁺ (Da)	$^{13}C_{12}$ Standard Relative intensity (%)	$^{13}C_6$ Standard Exact mass M⁺ (Da)	$^{13}C_6$ Standard Relative intensity (%)
Br₁	247.98313	100.0	260.02339	100.0	254.00326	100.0
	249.98108	97.3	262.02134	97.3	256.00121	97.3
Br₂	325.89364	51.4	337.93390	51.4	331.91377	51.4
	327.89159	100.0	339.93185	100.0	333.91172	100.0
	329.88955	48.6	341.92981	48.6	335.90968	48.6
Br₃	403.80415	34.3	415.84441	34.3	409.82428	34.3
	405.80211	100.0	417.84237	100.0	411.82224	100.0
	407.80006	97.3	419.84032	97.3	413.82019	97.3
	409.79801	31.5	421.83827	31.5	415.81814	31.5
Br₄	481.71467	17.6	493.75492	17.6	487.73480	17.6
	483.71262	68.5	495.75288	68.5	489.73275	68.5
	485.71057	100.0	497.75083	100.0	491.73070	100.0
	487.70853	64.9	499.74878	64.9	493.72866	64.9
	489.70648	15.8	501.74674	15.8	495.72661	15.8
Br₅	559.62518	10.6	571.66544	10.6	565.64531	10.6
	561.62313	51.4	573.66339	51.4	567.64326	51.4
	563.62109	100.0	575.66134	100.0	569.64122	100.0
	565.61904	97.3	577.65930	97.3	571.63917	97.3
	567.61699	47.3	579.65725	47.3	573.63712	47.3
	569.61495	9.2	581.65520	9.2	575.63508	9.2
Br₆	637.53569	5.4	649.57595	5.4	643.55582	5.4
	639.53365	31.7	651.57390	31.7	645.55377	31.7
	641.53160	77.1	653.57186	77.1	647.55173	77.1
	643.52955	100.0	655.56981	100.0	649.54968	100.0
	645.52751	73.0	657.56776	73.0	651.54763	73.0
	647.52546	28.4	659.56572	28.4	653.54559	28.4
	649.52341	4.6	661.56367	4.6	655.54354	4.6
Br₇	715.44620	3.1	727.48646	3.1	721.46633	3.1
	717.44416	21.1	729.48442	21.1	723.46429	21.1
	719.44211	61.7	731.48237	61.7	725.46224	61.7
	721.44006	100.0	733.48032	100.0	727.46019	100.0
	723.43802	97.3	735.47828	97.3	729.45815	97.3
	725.43597	56.8	737.47623	56.8	731.45610	56.8
	727.43392	18.4	739.47418	18.4	733.45405	18.4
	729.43188	2.6	741.47214	2.6	735.45201	2.6

Table 4.52 (Continued)

	Native		$^{13}C_{12}$ Standard		$^{13}C_6$ Standard	
#Br	Exact mass M+ (Da)	Relative intensity (%)	Exact mass M+ (Da)	Relative intensity (%)	Exact mass M+ (Da)	Relative intensity (%)
Br$_8$	793.35672	1.6	805.39697	1.6	799.37685	1.6
	795.35467	12.4	807.39493	12.4	801.37480	12.4
	797.35262	42.3	809.39288	42.3	803.37275	42.3
	799.35058	82.2	811.39084	82.2	805.37071	82.2
	801.34853	100.0	813.38879	100.0	807.36866	100.0
	803.34648	77.8	815.38674	77.8	809.36661	77.8
	805.34444	37.9	817.38470	37.8	811.36457	37.8
	807.34239	10.5	819.38265	10.5	813.36252	10.5
	—	—	—	—	815.36047	1.3
Br$_9$	871.26723	0.9	883.30749	0.9	877.28736	0.9
	873.26518	7.8	885.30544	7.8	879.28531	7.8
	875.26314	30.2	887.30339	30.2	881.28327	30.1
	877.26109	68.5	889.30135	68.5	883.28122	68.5
	879.25904	100.0	891.29930	100.0	885.27917	100.0
	881.25700	97.3	893.29725	97.3	887.27713	97.3
	883.25495	63.1	895.29521	63.1	889.27508	63.1
	885.25290	26.3	897.29316	26.3	891.27303	26.3
	887.25086	6.4	899.29111	6.4	893.27099	6.4
Br$_{10}$	949.17774	0.5	961.21800	0.5	955.19787	0.5
	951.17570	4.4	963.21595	4.4	957.19582	4.4
	953.17365	19.4	965.21391	19.4	959.19378	19.4
	955.17160	50.3	967.21186	50.3	961.19173	50.3
	957.16956	85.7	969.20981	85.7	963.18968	85.7
	959.16751	100.0	971.20777	100.0	965.18764	100.0
	961.16546	81.1	973.20572	81.1	967.18559	81.1
	963.16342	45.1	975.20367	45.1	969.18354	45.1
	965.16137	16.4	977.20163	16.4	971.18150	16.4
	967.15932	3.6	979.19958	3.6	973.17945	3.6

The calculated reference masses are based on the following values for isotopic masses: ^{1}H 1.0078250321 Da, ^{12}C 12.0000000000 Da, ^{13}C 13.0033548378 Da, ^{16}O 15.9949146221 Da, ^{79}Br 78.9183376 Da and ^{81}Br 80.9162910 Da. All listed masses refer to singly positively charged ions. Masses for isotope peaks have been calculated for a resolving power of 10 000 (10% valley definition). The mass of the electron (0.000548579911 Da) was taken into account for the calculation of the ionic masses.
(Adapted from Audi and Wapstra, 1995; Peter and Taylor, 1999.)

Table 4.53 PBDE exact mass references (2): for natives and internal standards (M-2Br ions).

#Br of M	Ratio (%) M⁺/(M-2Br)⁺	Native Exact mass (Da) (M-2Br)⁺	Native Relative intensity (%)	$^{13}C_{12}$ Standard Exact mass (Da) (M-2Br)⁺	$^{13}C_{12}$ Standard Relative intensity (%)	$^{13}C_{6}$ Standard Exact mass (Da) (M-2Br)⁺	$^{13}C_{6}$ Standard Relative intensity (%)
Br₁	100	n/a		n/a		n/a	—
Br₂	100	168.056966	100.0	180.097224	100.0	174.077095	100.0
Br₃	100	245.967479	89.8	258.007737	100.0	251.987608	95.8
		247.965432	100.0	260.005690	97.7	253.985561	100.0
Br₄	100	323.877991	47.5	335.918249	51.0	329.898120	49.4
		325.875945	100.0	337.916203	100.0	331.896074	100.0
		327.873898	54.4	339.914156	48.5	333.894027	51.4
Br₅	85	401.788504	31.0	413.828762	33.9	407.808633	33.0
		403.786457	96.4	415.826715	100.0	409.806586	99.6
		405.784411	100.0	417.824669	97.4	411.804540	100.0
		407.782364	35.3	419.822622	31.4	413.802493	34.2
Br₆	60	479.699017	16.0	491.739275	17.4	485.719146	16.7
		481.696970	65.7	493.737228	68.2	487.717099	66.9
		483.694923	100.0	495.735181	100.0	489.715052	100.0
		485.692877	69.5	497.733135	64.7	491.713006	66.9
		487.690830	17.9	499.731088	15.7	493.710959	17.0
Br₇	55	557.609529	2.8	569.649787	3.0	563.629658	2.9
		559.607483	19.8	571.647741	21.0	565.627612	20.5
		561.605436	59.7	573.645694	61.7	567.625565	60.5
		563.603389	99.7	575.643647	100.0	569.623518	100.0
		565.601343	100.0	577.641601	97.4	571.621472	98.5
		567.599296	60.9	579.639554	56.9	573.619425	59.0
Br₈	50	635.520042	4.8	647.560300	5.3	641.540171	5.1
		637.517995	29.5	649.558253	31.6	643.538124	30.6
		639.515949	74.2	651.556207	77.2	645.536078	75.5
		641.513902	100.0	653.554160	100.0	647.534031	100.0
		643.511855	75.6	655.552113	73.1	649.531984	74.6
		645.509809	31.2	657.550067	28.4	651.529938	29.8
		647.507762	5.1	659.548020	4.6	653.527891	5.0

Table 4.53 (Continued)

		Native		¹³C₁₂ Standard		¹³C₆ Standard	
#Br of M	Ratio (%) M⁺/(M-2Br)⁺	Exact mass (Da) (M-2Br)⁺	Relative intensity (%)	Exact mass (Da) (M-2Br)⁺	Relative intensity (%)	Exact mass (Da) (M-2Br)⁺	Relative intensity (%)
Br₉	40	713.430554	2.8	725.470812	3.0	719.450683	2.9
		715.428508	19.8	727.468766	21.0	721.448637	20.5
		717.426461	59.7	729.466719	61.7	723.446590	60.5
		719.424414	99.7	731.464673	100.0	725.444544	100.0
		721.422368	100.0	733.462626	97.3	727.442497	98.5
		723.420321	60.9	735.460579	56.9	729.440450	58.9
		725.418275	20.7	737.458533	18.4	731.438404	19.6
		727.416228	2.8	739.456486	2.6	733.436357	2.8
Br₁₀	25	791.341067	1.4	803.381325	1.6	797.361196	1.5
		793.339020	11.5	805.379278	12.3	799.359149	11.9
		795.336974	40.1	807.377232	42.2	801.357103	41.3
		797.334927	80.5	809.375185	82.1	803.355056	81.2
		799.332880	100.0	811.373138	100.0	805.353009	100.0
		801.330834	80.5	813.371092	77.8	807.350963	79.0
		803.328787	40.8	815.369045	37.8	809.348916	39.4
		805.326741	11.9	817.366999	10.5	811.346870	11.2
		807.324694	1.4	819.364952	1.3	813.344823	1.4

The calculated reference masses are based on the following values for isotopic masses: ^{1}H 1.0078250321 Da, ^{12}C 12.0000000000 Da, ^{13}C 13.0033548378 Da, ^{16}O 15.9949146221 Da, ^{79}Br 78.9183376 Da, and ^{81}Br 80.9162910 Da. All listed masses refer to singly positively charged ions. Masses for isotope peaks have been calculated for a resolving power of 10 000 (10% valley definition). The mass of the electron (0.000548579911 Da) was taken into account for the calculation of the ionic masses.
The given ratio M⁺/(M-2Br)⁺ provides typical values, depending on actual ion source conditions.
(Adapted from Audi and Wapstra, 1995; Peter and Taylor, 1999.)

All congeners in the employed BDE standard can be separated on a 15 m column (see Figure 4.118). Similar to dioxins, the BDE congeners are separated on unpolar columns groupwise in the order of their bromination degree. The use of a short 10–15 m column with a thin film is recommended to analyze the thermolabile deca-BDE more efficiently.

The LOQs achieved compare to those known for dioxin and PCB analysis. They can also be achieved for the analysis of the far higher boiling BDEs in the low femtogram range (see Figure 4.119). The quantitation linearity proved to fulfil highest standards, as shown in Figure 4.120.

746 | 4 Applications

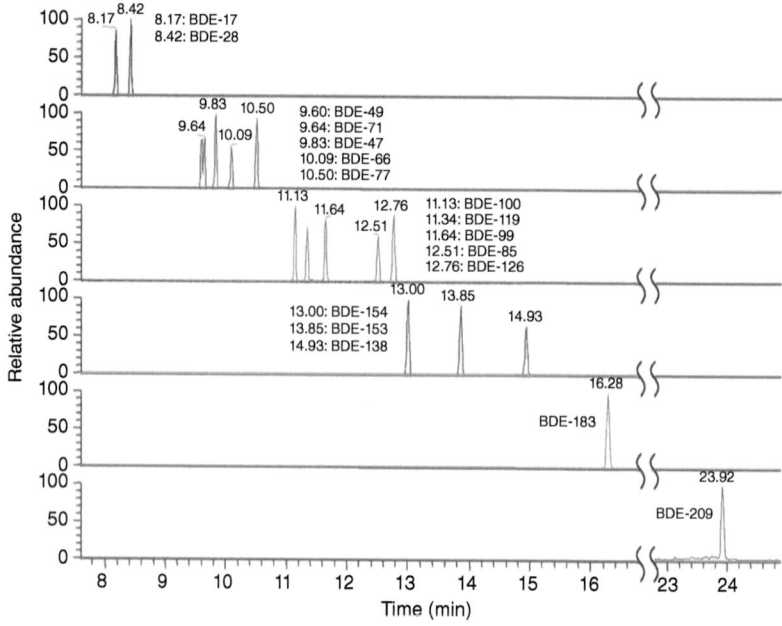

Figure 4.118 Mass chromatograms of tri- to hepta- and deca-BDE showing the separation according to the bromination degree on a 5% phenyl phase column, length 15 m.

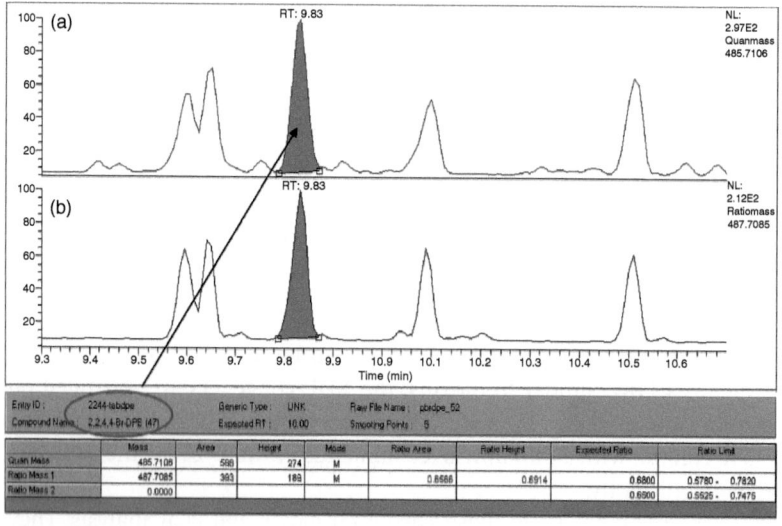

Figure 4.119 25 fg BDE 47 (tetra-BDE): (a) quantitation mass and (b) ratio mass.

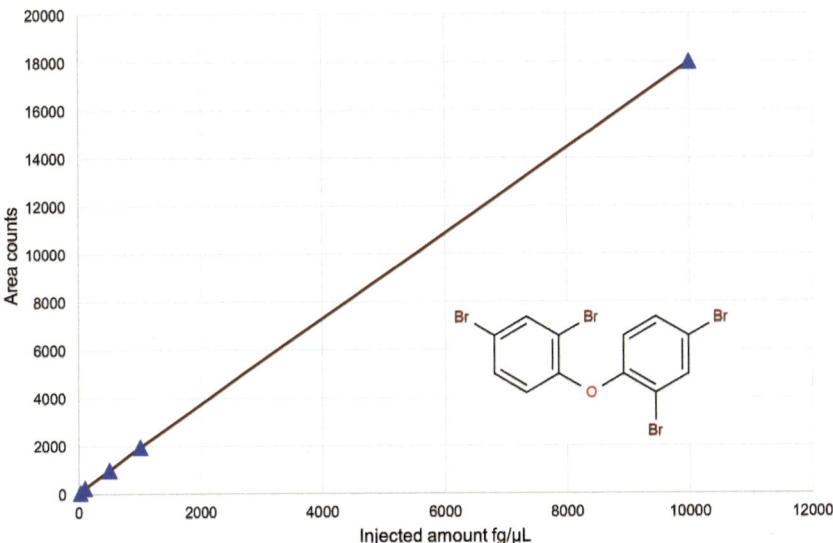

Figure 4.120 Quantitative calibration showing linearity for 2,2′,4,4′-tetrabromodiphenyl ether (BDE-47) in the range of 25–10 000 fg/μL, R^2 0.9994.

4.29 PBB Analysis by SPME

Keywords: PBBs; PBDEs; HS-SPME; ion trap MS/MS; triple quadrupole MS/MS; quantification

4.29.1 Introduction

Polybrominated biphenyls (PBBs) are with the PBDEs (see Section 4.28) among the most used flame retardants with a widely distributed contamination in the environment causing a high risk by accumulation in food and feedstuff. Although typically the highly brominated congeners are applied in flame retardants, the lower brominated species are found more often in environmental samples (de Boer et al., 2001).

According to the focus of potential accumulation from environmental sources, this application covers the congener range of up to the hexabrominated species of both PBBs and PBDEs. A fast and reliable automated online extraction method for BFR from water samples is described. The solventless SPME is applied for the trace determination of BFRs. In contrast to conventional methods using the Soxhlet extraction with an extended extract clean-up, this method uses the fast and straightforward extraction of the analytes from the sample headspace. This is possible due to a high water vapor volatility of the PBBs and PBDEs and the low partition coefficient of these BFR compounds in aqueous media. The SPME method can be integrated using robotic x,y,z-samplers, is fast, prevents the chromatographic system from the typical matrix background load of Soxhlet-extracted samples, and provides clean chromatograms (de Boer *et al.*, 2001). Tap water and urban wastewater has been analyzed with high recoveries (Polo *et al.*, 2004).

The mass spectrometric detection in this application is achieved by MS/MS using an ion trap mass spectrometer providing excellent selectivity and sensitivity for trace level quantification in environmental water samples. Alternatively, triple quadrupole MS/MS systems can be employed using a similar setup. Highest selectivity and lowest determination levels have been achieved also by using a high resolution mass spectrometer (de Boer *et al.*, 2001; Krumwiede, 2006b).

4.29.2 Analysis Conditions

Sample material			Environmental samples, water
Sample preparation	SPME		10 mL of a filtered aqueous sample is filled in 22 mL headspace vials, sealed with a Teflon-faced septum, equilibrated at 100 °C for 5 min
	Extraction	Fiber	Supelco PDMS fiber, 65 µm
		Time	30 min, stir bar agitation while sampling
	Desorption	Temperature	300 °C
GC method	System		Varian 3800
	Injector		Split/splitless
	Column	Type	CP-Sil 8 CB
		Dimensions	25 m length × 0.25 mm ID × 0.25 µm film thickness
	Carrier gas		Helium
	Flow		Constant flow, 1.2 mL/min
	SSL injector	Injection mode	Splitless
		Injection temperature	300 °C
	Split	Closed until	2 min
		Open	2 min to end of run
		Split flow	50 mL/min
	Oven program	Start	60 °C, 2 min
		Ramp 1	30 °C/min to 250 °C
		Ramp 2	5 °C/min to 280 °C
		Final temperature	280 °C, 8 min
	Transferline	Temperature	280 °C

MS method	System		Varian Saturn 2000
	Analyzer type		Ion trap MS with internal ionization (with MS/MS waveboard)
	Ionization		EI, 70 eV
	Ion trap	Temperature	250 °C
	Acquisition	Mode	Full scan, MS/MS resonant waveform
	Mass range		50–650 Da
	Scan rate		1 s (1 spectrum/s)
Calibration	External standard	Range	0.12–500 pg mL

4.29.3 Results

Despite of the high molecular mass of the BFRs, the investigated compounds have been extracted with higher recoveries from the sample headspace than from direct immersion into the aqueous media. This is due to an the high water vapor volatility of the brominated aromatics, which is known from PCBs as well. Using headspace SPME, the transfer of high boiling matrix components to the chromatographic system is prevented usually causing high background levels during MS detection. At the same time, the useful lifetime of the fiber is significantly extended compared to DI-SPME. Extractions are supported by stir bar agitation in the headspace vial. The method parameters are based on a comprehensive and systematic optimization study (Polo *et al.*, 2004).

The described method shows excellent sensitivity, linearity, and quantitative precision up to the hexabromo congeners. The detection limits are in the low picogram per liter range (7.5–190 pg/L) with a calibration range by spiking tap water between 120 fg/mL and 500 pg/mL, as given in Table 4.54. The achieved

Table 4.54 Linearity and LODs for the HS-SPME analysis of PBBS and PBDEs.

Compounds	Concentration range (pg/mL)	Correlation factor R^2	LOD S/N 3 (pg/L)
BDE-3	1.00–498	0.9977	190
PBB-15	1.01–503	1.0000	9.0
PBB-49	0.95–476	1.0000	7.5
BDE-47	0.41–205	0.9998	20
BDE-100	0.12– 60	1.0000	60
BDE-99	0.41–205	0.9999	47
BDE-154	0.34– 17	1.0000	150
BDE-153	0.23– 12	0.9995	100

Table 4.55 Repeatability of the HS-SPME analysis at two different concentration levels (n = 3).

Compounds	Concentration 1 (pg/mL)	Repeatability RSD %	Concentration 2 (pg/mL)	Repeatability RSD %
BDE-3	10.0	4.4	199	6.3
PBB-15	10.1	3.8	201	12
PBB-49	9.5	2.8	190	1.7
BDE-47	4.1	17	82	1.2
BDE-100	1.2	15	24	10
BDE-99	4.1	20	82	1.2
BDE-154	0.34	24	6.8	9.3
BDE-153	0.23	26	4.6	8.8

Table 4.56 HS-SPME recoveries of BFRs from real-life water samples.

Compounds	Tap water spiked conc. (pg/mL)	Recovery (%)	Effluent water spiked conc. (pg/mL)	Recovery (%)	Influent water[a] spiked conc. (pg/mL)	Recovery RSD (%)
BDE-3	1.00	99 ± 1	10.0	100 ± 4	199	106 ± 10
PBB-15	1.01	97 ± 4	10.1	92 ± 2	201	90 ± 6
PBB-49	0.95	90 ± 21	9.5	94 ± 1	190	93 ± 5
BDE-47	0.41	91 ± 8	4.1	97 ± 7	82	87 ± 7
BDE-100	0.12	100 ± 4	1.2	83 ± 17	24	95 ± 4
BDE-99	0.41	87 ± 19	4.1	90 ± 12	82	92 ± 8
BDE-154	0.034	n.d.	0.34	100 ± 25	6.8	74 ± 11
BDE-153	0.023	n.d.	0.23	117 ± 13	4.6	82 ± 11

n.d., not determined.
a) Chromatogram see Figure 4.121.

correlation values are excellent in the range of 0.9977–1.0. Method precision and LOD were determined with triplicate measurements as given in Table 4.55. The recovery experiments were performed with three different types of blank matrix samples including tap water and effluent and influent wastewaters from an urban sewage plant. The results are given in Table 4.56 and illustrated using the SRM detection mode in Figure 4.121.

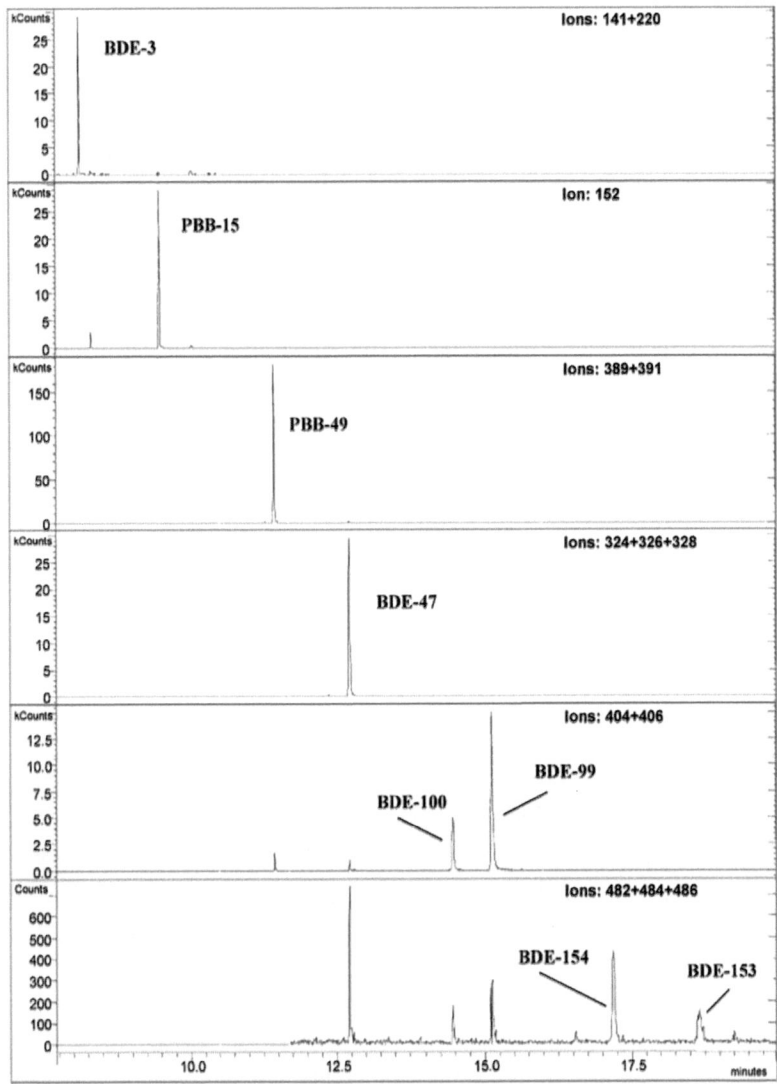

Figure 4.121 Analysis of a spiked influent wastewater sample as of Table 4.56. (Reproduced from Polo et al., 2004/with permission of American Chemical Society.)

4.30 THC-A in Urine by NCI

Keywords: THC-A; cannabis; Marinol™; Dronabinol; forensic analysis; metabolites; SPE; derivatization; negative chemical ionization; single quadrupole MS

4.30.1 Introduction

It is a well-documented fact that the widespread use of marijuana products commonly named like grass, weed, dope, hashish, bhang, pot, shit, or ganja continues to make Cannabis Sativa one of the most used drug plants of our time (Kapusta et al., 2006; Beck et al., 2007; Jones et al., 2009). Furthermore, recent research has led to the use of THC (tetrahydrocannabinol), which is the main psychoactive ingredient of the plant, as an FDA-approved therapeutically drug under the brand name Marinol® in the USA, marketed under the generic international non-proprietary name Dronabinol in other countries (Levin and Kleber, 2008). Also, the legality of cannabis products for medical and recreational use varies by country (Wikipedia, 2024).

Urine is the matrix of choice in epidemiological studies dealing with addictive drugs in physiological or forensic research, workplace drug testing, and doping control (Musshoff and Madea, 2006). This application was developed at the Robert Koch-Institute in Berlin, Germany, and presents a highly reliable routine method for an easy and effective sample pre-treatment with high recovery. The selective quantitative determination of the THC acid metabolite (THC-A, THC-COOH) was achieved with a mixed derivatization using pentafluoropropanol (PFPOH) and hexafluorobutyric acid (HFBA) derivative in the low pico- to middle femtogram-range using the negative chemical ionization (NCI) mode (Melchert et al., 2009).

4.30.2 Analysis Conditions

Sample material			Urine
Sample preparation	SPE		
		Material	Extrelut™ NT1
		Sample application	Wait 5 min
		Elution	1.4 mL iso-octane
			2.2 mL iso-octane
		Concentration	Evaporation close to dryness
	Derivatization	Reagents	50 µL pentafluoropropanol
			80 µL heptafluorobutyric acid anhydride
		Reaction	65 °C for 20 min
			Remaining reagents evaporated
		Solvent	200 µL toluene
GC method	System		Thermo Scientific TRACE GC Ultra
	Autosampler		CTC PAL x,y,z-robot
	Column	Type	Restek Rtx 5Sil MS with 10 m Integra-Guard
		Dimensions	30 m length × 0.25 mm ID × 0.25 µm film thickness

	Carrier gas		Helium
	Flow		Constant flow, 30 cm/s
	PTV injector	Injection mode	Splitless
		Injection volume	1 μL
		Base temperature	90 °C
		Transfer rate	10 °C/s
		Transfer temperature	200 °C
	Split	Closed until	1 min
		Open	1 min to end of run
		Purge flow	5 mL/min
	Oven program	Start	70 °C, 1 min
		Ramp 1	40 °C/min to 200 °C
		Ramp 2	20 °C/min to 265 °C
		Final temperature	265 °C, 3.5 min
	Transferline	Temperature	285 °C
			Direct coupling
MS method	System		Thermo Scientific DSQ II
	Analyzer type		Single quadrupole MS
	Ionization	EI	For method development
		CI	For routine quantification
	CI Gas		Methane, 1.2 mL/min
	Electron energy		70 eV
	Ion source	Temperature	230 °C
	Acquisition	Mode	SIM
	Mass range		m/z 197, 474, and 483
Calibration	Internal standard		THC-COOH-d9-PFPOH/HFBA
	Calibration range		1–10 pg, 25–500 pg
			THC-COOH-d9-HFBA

4.30.3 Sample Preparation

4.30.3.1 Hydrolysis

Because the THC metabolite THC-COOH is excreted in urine in conjugated form, a hydrolysis step is necessary to receive the free THC-COOH before extraction. Hydrolysis is done for a 1 mL urine sample with 100 μL 12 N KOH solution after the addition of 20 μL of the ISTD, equivalent to 100 ng of deuterated THC-COOH. Hydrolysis is achieved by heating the mixture in a crimp-cap septum closed N 11-1 vial to 60 °C for

20 min. After cooling to room temperature, 350 µL of acetic acid anhydride is added to set pH < 4 for the reaction mixture.

4.30.3.2 Extraction
The hydrolyzed sample is transferred quantitatively on an Extrelut™ NT1 extraction column. After waiting for 5 min, the sample is first extracted with 4 mL iso-octane and then with additional 2 mL iso-octane. The combined extracts are collected in glass vials (volume 10 mL) with a conical bottom, evaporated to dryness, and concentrated in the conical part of the vial.

4.30.3.3 Derivatization
The extracted THC-COOH is transferred completely into N 11 glass vials. As derivatization agents, 50 µL pentafluoropropanol and 80 µL heptafluorobutyric acid anhydride are added and the reaction mixture is heated to 65 °C for 20 min in the crimp-closed vial. The remaining derivatization reagents are then completely evaporated, and the remainder is carefully transferred to conical autosampler microvials (ND 8, 1.1 mL) using a microliter syringe and toluene as a solvent so that an end volume of 200 µL is achieved for the injection of 1 µL aliquots of the derivatized extract.

4.30.4 Experimental Conditions
A single quadrupole GC-MS system equipped with a liquid autosampler was used. The GC-MS measurements have been done in the EI mode for method development and the NCI mode for routine quantification.

4.30.5 Sample Measurements

4.30.5.1 Reproducibility of Retention Times
Under the described GC conditions, the reproducibility of the RTs for the THC-COOH-derivatives was <±0.5% within one day and <±3% within a total working time of 6 months, covering more than 1000 sample injections.

4.30.5.2 Limit of Detection and Limit of Quantification
For measurements of THC-COOH as PFPOH/HFBA-derivatives from spiked urine samples in the NCI mode, a LOD value of 300 fg with S/N > 3 could be achieved. A LOQ value of 1 pg with S/N > 10 was determined.

4.30.5.3 Recovery and Calibration
For urine samples spiked with THC-COOH between 15 and 65 ng/mL, recovery rates of 85–95% were found. The fragment ions m/z 483 (deuterated derivative) and m/z 474 (native derivative) have been used for quantification.

For standard analytical work, the ISTD calibration method was used in a range of 25–500 pg for the THC-COOH-PFPOH/HFBA derivatives, as shown in Figure 4.122.

4.30 THC-A in Urine by NCI

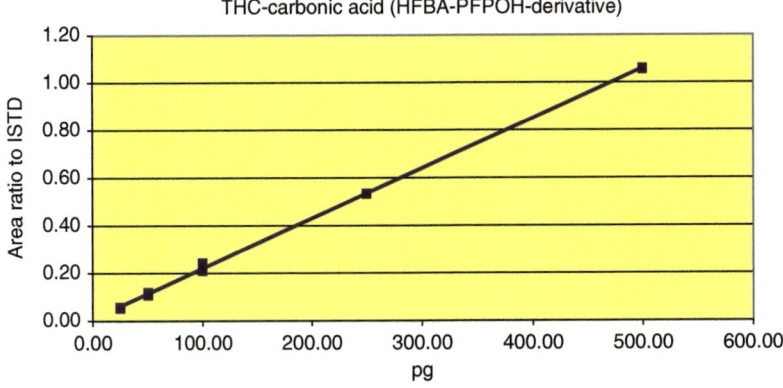

Figure 4.122 Internal standard calibration for the THC-COOH-PFPOH-HFBA derivative, range 25–500 pg, R^2 0.9989.

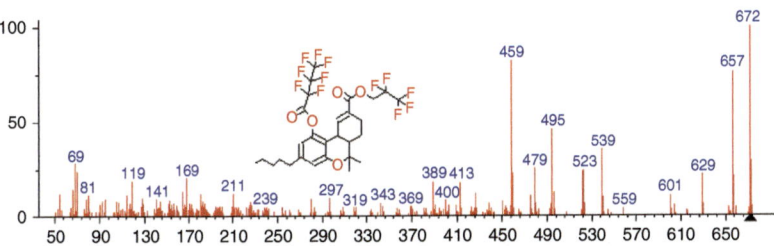

Figure 4.123 EI mass spectrum of the THC-COOH-PFPOH-HFBA derivative.

For urine samples of low level exposure samples, the calibration was done in the 1–10 pg range.

4.30.6 Results

4.30.6.1 Mass Spectra and GC Separation

Figure 4.123 shows the mass spectrum of the THC-COOH-PFPOH/-HFBA derivative measured in the EI mode. Figure 4.124 shows the mass spectrum in the NCI mode. Both spectra had been submitted for inclusion into the NIST mass spectral library. The following Figure 4.125 presents the mass chromatogram of the native and deuterated THC-COOH derivative from a hydrolyzed, extracted, and derivatized urine sample. The three traces show the TIC trace (top), the mass trace m/z 474 for the native THC-COOH derivative (middle), and m/z 483 for the ISTD d_9-THC-COOH derivative (bottom). Although the ion intensity of m/z 474 appears to be low in the NCI spectrum of Figure 4.124, the higher mass delivers a less matrix interference, leading to higher S/N values in urine samples than the use of the lower mass fragment m/z 197.

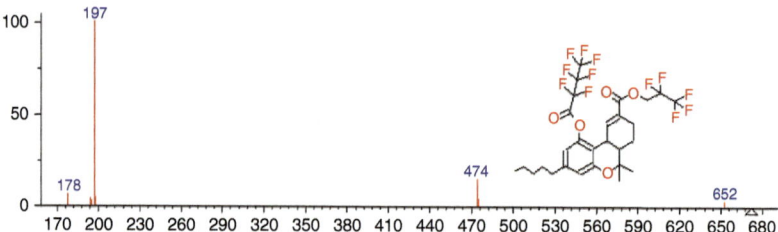

Figure 4.124 NCI mass spectrum of the THC-COOH-PFPOH-HFBA derivative.

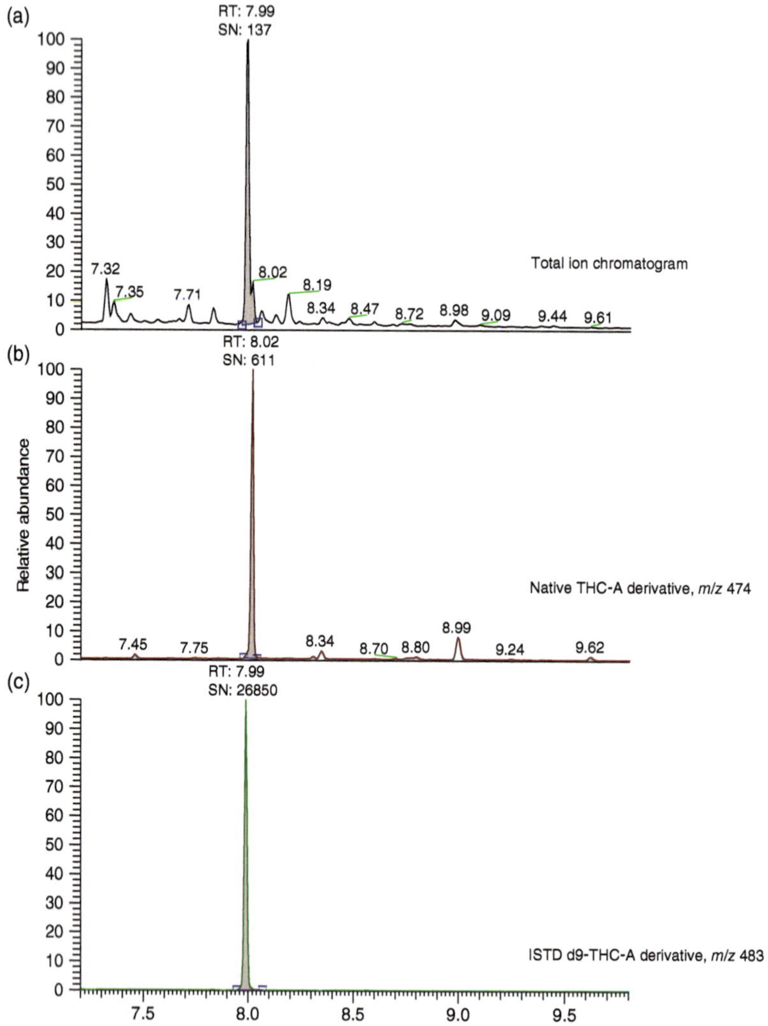

Figure 4.125 Chromatograms of (a) the total ion, (b) the native compound, and (c) the deuterated THC-COOH derivative from a hydrolyzed, extracted and derivatized urine sample.

4.30.7 Quality Control Samples

Multiple measurements ($n = 10$) of the Medidrug BTM U-Screen Level 2 quality control urine samples with a target value of 65 ng/mL have been used for verification. The results have been well within the expected range (±4%). CEDIA control urine samples of the 'THC 25 Control Set' "low" and "high" (Cat.-Nr. 1661086) were measured as an additional control. The expected target ranges for the "low" samples (18.75 µg/L) and for the "high" samples (31.25 µg/L) could be verified with ±6% for the "low" and ±4% for the "high" urine samples. Spiked native urine samples of cannabis naive persons did show that the abovementioned recovery rates between 85% and 95% could be achieved continuously. In none of the analyzed urines of cannabis naive persons as well as in commercial drug-free urine control samples, THC-COOH was detected.

4.30.8 Conclusions

THC-COOH, as the main metabolite of THC in urine, can be detected for many days after the last consumption of cannabis products. The THC-COOH concentration depends not only on the amount of drugs used but also on the THC-content of the consumed cannabis product. Measurements with the method described above have shown that none of the commonly prescribed or over-the counter (OTC) drugs interfered with the quantification of THC-carbonic acid in urine samples. This is due to the effective Extrelut extraction procedure. In addition, the derivatization of the isolated THC-COOH leads to very selective and "clean" chromatograms. The NCI mass spectral fragmentation pattern is unique to the THC-COOH-PFPOH/HFBA derivative. The high resolution power of capillary GC, coupled with mass spectrometry in the NCI mode, can safely overcome interferences that are often observed in immunological methods from present drugs.

4.31 Comprehensive Drug Screening and Quantitation

Keywords: emergency toxicology; poisoning; full scan screening; targeted; non-targeted analysis; LLE; one-point calibration; single quadrupole MS

4.31.1 Introduction

The analytical tasks in emergency toxicology laboratories are manifold and comprise, for instance, the support for poisonings, detoxication, or diagnosis. The described analytical workflow covers the full scan screening, spectrum identification, and a fast one-point calibration for quantitation (Maurer, 2007, 2012; Meyer et al., 2014).

4.31.2 Analysis Conditions

Sample material			Plasma
Sample preparation	Liquid/liquid extraction		1 mL plasma
			Add ISTD 100 µL of methanolic trimipramine-d3 (1.0 mg/L)
			Add 2.0 mL saturated aqueous Na_2SO_4 solution
			Add diethyl ether, ethyl acetate (1 : 1; v/v, 5 mL, shaking 30 s)
			Centrifuge
			Evaporate organic layer to dryness
			Add to aqueous layer 0.5 mL of aqueous sodium hydroxide (1 M)
			Extract again with a solvent mixture
			Evaporate organic layer to dryness
			Reconstitute residue in 100 µL of methanol
			Transfer into an autosampler vial for injection
GC method	System		Agilent 6890 GC
	Column	Type	HP-1
		Dimensions	12 m length × 0.2 mm ID × 0.33 µm film thickness
	Carrier gas		Helium
	Flow		Constant flow, 1 mL/min
	SSL injector	Injection mode	Splitless
		Injection temperature	200 °C
		Injection volume	1 µL
	Oven program	Start	100 °C
		Ramp 1	30 °C/min to 310 °C
		Final temperature	310 °C, 5 min
	Transferline	Temperature	280 °C
			Direct coupling
MS method	System		Agilent 5973 MSD
	Analyzer type		Single quadrupole MS
	Ionization		EI
	Electron energy		70 eV

	Ion source Temperature	220 °C
	Acquisition Mode	Full scan
	Mass range	50–550 Da
	Scan rate	1 scan/min
Calibration	Standard	One-point calibrator (OPC), 0.5–10.0 mg/L
	Internal standard (ISTD)	Trimipramine-d3, 1.0 mg/L
	Data points	1 per compound

4.31.3 Sample Preparation

The liquid/liquid extraction (LLE) is the standard extraction method in clinical and forensic toxicology for a broad range of drugs (Maurer et al., 2023). The available sample volume allows the application of the extracts for GC-MS and LC-MS analysis. For the sample preparation, 1 mL of plasma is mixed with ISTD, 100 µL of methanolic trimipramine-d3 (1.0 mg/L), and 2.0 mL of saturated aqueous sodium sulfate solution. The LLE is performed with a mixture of diethyl ether and ethyl acetate (1 : 1; v/v, 5 mL, 30 s). The organic layer is taken after centrifugation and evaporated to dryness at 60 °C. The remaining aqueous layer is then basified with 0.5 mL of aqueous sodium hydroxide (1 M) and extracted again. The organic layer is transferred to the same flask and evaporated to dryness. The dry residue is reconstituted in 100 µL of methanol and transferred into autosampler vials for GC injection (Meyer et al., 2014).

4.31.4 Experimental Conditions

The extracts are analyzed using a single quadrupole GC-MS system. As an analytical column, a 12 m short nonpolar capillary column with 0.2 mm ID and a thick film of 0.33 µm has been used. One microliter of the extract was injected in splitless mode.

For the GC-MS quality control check, a methanolic solution of typical drugs is injected routinely. The ratio of peak areas between morphine and codeine (both at 50 ng/µL concentration) should be at least 1 : 10 as minimum QC requirement. Otherwise, the system requires preventive maintenance.

4.31.5 Sample Measurements

The set of drugs comprising the described screening solution shown in Table 4.57 was based on the experience of the most often quantified compounds by GC-MS in the referenced laboratory and selected for a fast and reliable quantification in emergency toxicology (Meyer et al., 2014).

The total number of compounds has been split in two one-point calibrator groups (OPC A and B, shown in Table 4.57) to avoid chromatographic overlapping.

Table 4.57 Drugs, quantifier ions, one-point calibrator group, therapeutic and toxic plasma reference concentrations, and plasma concentrations in OPCs and QC low and high levels.

Drug name	Quantifier ion (m/z)	OPC group	Therapeutic plasma conc. (mg/L)	Toxic plasma conc. higher than (mg/L)	OPC (mg/L)	QC low (mg/L)	QC high (mg/L)
Amitriptyline	58	A	0.05–0.2	0.5	1	0.2	4
Biperiden	98	A	0.005–0.1	0.08	1	0.1	4
Bromazepam	236	B	0.08–0.17	0.3	1	0.5	4
Citalopram	58	B	0.02–0.2	0.5	1	0.2	4
Clobazam	300	B	0.1–0.4	0.5	1	0.4	4
Clomipramine	269	B	0.1–0.2	0.4	1	0.2	4
Clozapine	243	B	0.35–0.6	0.8	1	0.6	4
Codeine	299	A	0.01–0.05	0.3	1	0.1	4
Diazepam	256	A	0.1–1	1.5	1	1	4
Diphenhydramine	58	A	0.1–1.0	1	1	1	4
Dihydrocodeine	301	B	0.03–0.25	1	1	0.25	4
Diltiazem	58	B	0.05–0.4	0.8	1	0.4	4
Doxepin	58	A	0.02–0.15	0.5	1	0.3	4
Doxylamine	58	A	0.05–0.2	1	1	0.5	4
Imipramine	58	B	0.05–0.15	0.4	1	0.2	4
Lamotrigine	185	B	4–10	15	5	10	30
Levetiracetam	126	A	10–37	400	10	40	400
Levomepromazine	185	A	0.02–0.1	0.4	1	0.2	4
Melperone	112	B	0.05–1	1	1	1	4
Methadone	72	A	0.05–0.8	0.5	0.5	1	4
Metoclopramide	86	B	0.04–0.15	0.2	1	0.1	4
Midazolam	310	B	0.08–0.25	1	0.5	0.25	4
Mianserin	193	B	0.015–0.07	0.5	1	0.1	4
Mirtazapine	195	A	0.04–0.3	1	1	0.4	4
Moclobemide	100	B	0.3–1	5	1	1	4
Nordazepam	242	A	0.2–0.8	1.5	1	1	4
Olanzapine	242	B	0.02–0.08	0.2	2	0.1	4
Perazine	70	B	0.02–0.35	0.5	0.5	0.3	4
Pethidine	247	A	0.1–0.8	1	1	1	4
Phenprocoumon	251	B	1–3	5	5	5	10
Promethazine	72	A	0.1–0.4	1	1	0.25	4
Prothipendyl	58	A	0.05–0.2	0.5	1	0.2	4
Quetiapine	210	B	0.025–0.9	1.8	2	1	4
Sertraline	274	A	0.05–0.25	0.3	1	0.2	4

Table 4.57 (Continued)

Drug name	Quantifier ion (m/z)	OPC group	Therapeutic plasma conc. (mg/L)	Toxic plasma conc. higher than (mg/L)	Plasma concentration OPC (mg/L)	QC low (mg/L)	QC high (mg/L)
Temazepam	271	A	0.3–0.9	1	1	1	4
Tramadol	58	A	0.01–0.25	1	1	0.5	4
Venlafaxine	58	A	0.2–0.75	1	1	0.5	4
Verapamil	303	B	0.05–0.35	0.9	1	0.4	4
Zolpidem	235	A	0.08–0.3	0.5	1	0.3	4
Zotepine	58	A	0.01–0.12	0.2	1	0.1	4

(Meyer et al., 2014, reproduced with permission of John Wiley & Sons.)

The identification is based on the complete mass spectra using AMDIS (Meyer et al., 2010). The compound quantification is performed on the specific ion chromatograms for each compound using the quantifier ions given in Table 4.57, in ratio to the applied ISTD. The qualifier ions had been selected according to selectivity and sensitivity, but could show potential interferences from matrix or contaminants at high concentrations: tramadol on m/z 58 with palmitic acid (alternatively use m/z 263 at lower sensitivity), methadone on m/z 72 with stearic acid, and nordazepam on m/z 242 with diisoctylphthalate.

4.31.6 Results

The recovery for the drug compounds screened was achieved for the most analytes between 72% and 97%. Low recovery exceptions are reported for levetiracetam 23%, phenprocoumon 37%, and olanzapine with 66%.

The linearity of the quantitation was tested for the practical application of a fast OPC approach with levels in the practical range from high therapeutic up to toxic plasma concentrations, typically comprising 0.1, 0.5, 2, 4, and 6 mg/L. Exceptions are the highly dosed drugs such as levetiracetam, lamotrigine, phenprocoumon, or quetiapine (see Table 4.57).

The proof of method applicability has been achieved by analyzing proficiency test samples separately validated by reference methods. See the typical reconstructed compound specific mass chromatograms in Figure 4.126.

4.31.7 Conclusions

The described GC-MS screening method allows the fast, accurate, and reliable quantification of drugs relevant to emergency toxicology in the range of therapeutic to toxic concentrations.

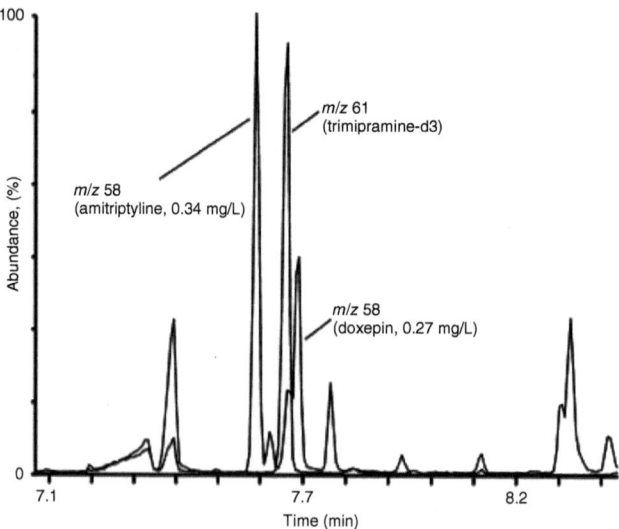

Figure 4.126 Mass chromatograms from a plasma proficiency test for quantitation: amitriptyline (m/z 58, 0.34 mg/L), doxepin (m/z 58, 0.27 mg/L), and ISTD timipramine-d3 (m/z 61). (Meyer et al., 2014, reproduced with permission of John Wiley & Sons.)

4.32 Drugs of Abuse

Keywords: drug screening; heroin; codeine; morphine; hydrolysis; derivatization; ion trap MS; triple quadrupole MS; forensic analysis;

4.32.1 Introduction

Detection of drug taking by the investigation of blood and urine samples is one of the tasks of a forensic toxicology laboratory in workplace drug testing. The routine analysis of drug screening can be carried out by TLC, HPLC, or immunological methods. Positive results require confirmation. Determination using GC-MS is recognized as a reference method (Braithwaite et al., 1995; Simpson et al., 1997). Triple quadrupole MS/MS methods provide the increased selectivity for a list of target compounds (Weller et al., 2000; Smith et al., 2007). Full scan methods are preferred over SIM or SRM acquisition for comprehensive screening procedures because of the higher specificity and universality (Maurer et al., 2023).

Heroin (3,6-diacetylmorphine) is usually not determined directly (Musshoff et al., 2004). The unambiguous detection of heroin consumption is carried out by determining 6-MAM, which is formed from heroin as a metabolite (Figure 4.127). If morphine is difficult or impossible to detect when clean-up is carried out without hydrolysis, then the latter should be used. It should be noted that nonspecific hydrolases can also effect the degradation of 6-MAM to morphine. If the substitute

Figure 4.127 Opiate metabolism (* determination by GC-MS).

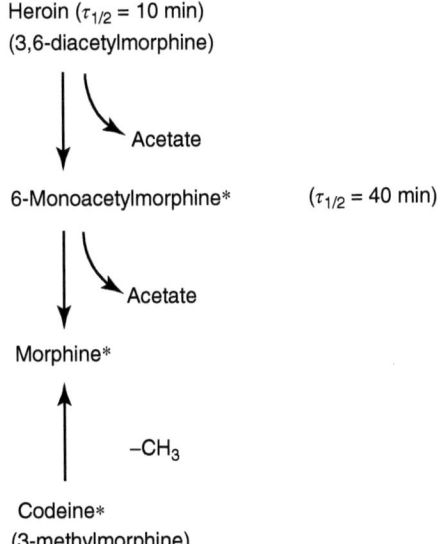

drugs methadone and dihydrocodeine are present, these are also determined using the procedure described. As blood from corpses also has to be processed during routine operations to some extent, a comparatively time-consuming extraction including a re-extraction is carried out. For the gas chromatography of morphine derivatives, derivatization of the extract by silylation, acetylation, or pentafluoropropionylation, for example, is basically necessary (Maurer, 1990, 1993). Mass spectrometric detection using NCI is possible by using fluorinated derivatives. In the procedure described, silylation with MSTFA and detection in the EI mode has been chosen (Donike, 1969).

4.32.2 Analysis Conditions

Sample material		Urine, blood
Sample preparation	Serum, blood	1 mL sample, add internal standard (e.g. 100 ng morphine-d_3 or codeine-d_3)
		Dilute to 7 mL with phosphate buffer (pH 6)
	SPE extraction	Condition with 2 mL methanol and 2 mL phosphate buffer pH 6 (Bond-Elut-Certify 130 mg)
		Elute with 2 × 1 mL chloroform/isopropanol/25% ammonia (70 : 30 : 4)
	Derivatization	With 50 µL MSTFA at 80 °C, 30 min

		Urine	Hydrolyze 2 mL sample with 20 μL enzyme solution (ß-glucuronidase), 60 min, 60 °C
			Adjust sample pH to 8–9 with 0.1 N sodium carbonate solution
		SPE extraction	Activate a C18-cartridge with methanol, condition with 0.1 N sodium carbonate solution
			Apply the sample and wash the cartridge with 0.1 N sodium carbonate solution
			Elute with 1 mL acetone/chloroform (50 : 50)
			Evaporate eluate to dryness
		Derivatization	With MSTFA at 80 °C, 30 min
GC method	System		Finnigan GCQ GC
	Injector		Split/splitless
	Column	Type	DB-1
		Dimensions	30 m length × 0.25 mm ID × 0.25 μm film thickness
	Carrier gas		Helium
	Flow		constant flow, 40 cm/s
	SSL injector	Injection mode	Splitless
		Injection temperature	275 °C
		Injection volume	1 μL
	Split	Closed until	0.1 min
		Open	0.1 min to end of run
	Oven program	Start	100 °C
		Ramp 1	10 °C/min to 310 °C
		Final temperature	310 °C
	Transferline	Temperature	280 °C
MS method	System		Finnigan GCQ
	Analyzer type		Ion trap MS with internal ionization
	Ionization		EI
	Acquisition	Mode	Full scan
	Mass range		70–440 Da
	Scan rate		0.5 s (2 spectra/s)

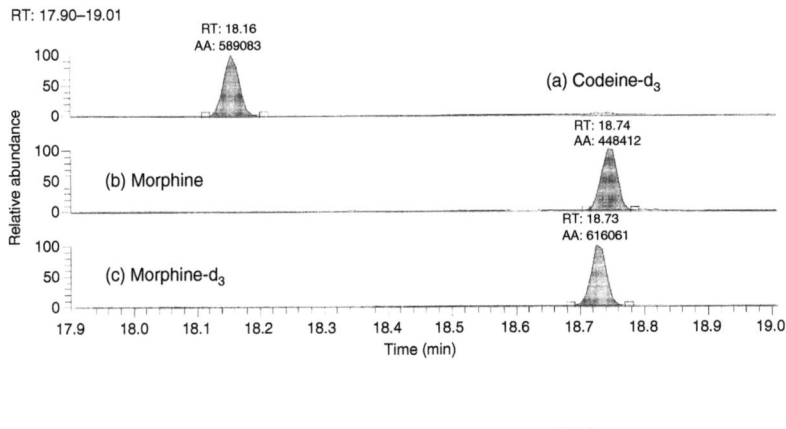

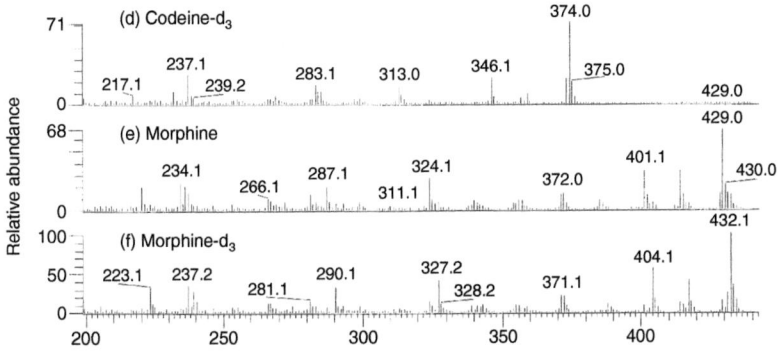

Figure 4.128 Analysis of an authentic serum sample fortified with 100 ng morphine-d₃ and 100 ng codeine-d₃. Above: (a) Mass trace for codeine-d₃ (m/z 374), (b) mass trace for morphine (m/z 429), and (c) Mass trace for morphine-d₃ (m/z 432). Below: (d) spectrum of codeine-d₃, € spectrum of morphine, and (f) spectrum of morphine-d₃.

4.32.3 Results

The chromatogram and mass spectra after clean-up of a typical serum sample are shown in Figure 4.128. The mass traces for codeine, morphine-d₃ and morphine, and the TIC are shown. For morphine, a concentration of 60 ng/L is calculated with reference to morphine-d₃ using a calibration curve. The resulting mass spectra allow identification, for example, on comparison with relevant toxicologically orientated spectral libraries. The clean-up and analysis methods can be used for other basic drugs, such as amphetamine derivatives, methadone, and cocaine and their metabolites.

A list of morphine derivatives and synthetic opiates, which have been found by the procedure described after clean-up of serum samples, is shown in Table 4.58. For the corresponding trimethylsilyl derivatives, the specific search masses are given which allow identification in combination with the retention index.

Table 4.58 Selective masses and RI data of the opiates as TMS derivatives.

Opiate	Retention index	m/z values
Levorphanol-TMS	2 188	329, 314
Pentazocine-TMS	2 262	357, 342
Levallorphan-TMS	2 318	355, 328
Dihydrocodeine-TMS	2 365	373, 315
Codeine-TMS	2 445	371, 343
Hydrocodone-TMS	2 447	371, 356
Dihydromorphine-TMS	2 459	431, 416
Hydromorphone-TMS	2 499	357
Oxycodone-TMS	2 503	459, 444
Morphine-bis-TMS	2 513	429, 414
Norcodeine-bis-TMS	2 524	429
Monoacetylmorphine-TMS	2 563	399, 340
Nalorphine-bis-TMS	2 656	455, 440

Note: All TMS derivatives show an intense signal for the trimethylsilyl fragment at m/z 73, which does not appear when the scan is begun at m/z 100 and is unnecessary for substance confirmation. The RIs were determined on DB-1, 30 m × 0.25 mm × 0.25 μm with an n-alkane mixture of up to C_{32} (Weller, 1990).

4.33 Structure Elucidation by CI and MS/MS

Keywords: structure elucidation; compound identification; chemical ionization; molecular ion; ion trap MS/MS; amines; Mass Frontier software; fragmentation pathways

4.33.1 Introduction

Structural elucidation is important in any synthetic work to check the identity of final products and by-products. Mass spectrometry in this context is the tool of choice for the confirmation of expected compound structures or the identification of new components found, especially after a chromatographic separation of mixtures.

The mass spectra for structure elucidation are typically generated by EI providing rich information from the fragmentation to the structure of an unknown compound. The reliable interpretation requires within the first steps the allocation of the molecular ion of the compound in the spectrum, giving access to the possible fragmentation pathways from the chemical structure of the molecule. Many compound classes only show a molecular ion of low intensity, or even do not show the molecular ion in the spectrum at all. A final structure confirmation using the EI spectrum only is not concluding in these cases. The vast majority of members of the large compound classes of hydrocarbons, alcohols, ketones, acids, esters, or amines are known to not reveal the molecular information in their respective EI mass spectra.

Soft ionization, for instance, the chemical ionization (CI) mode, in combination with EI can be a solution for a routine structure elucidation. Most useful is the positive chemical ionization (PCI) using different protonating reagent gases like methane, iso-butane, or ammonia. PCI often leads to the formation of an abundant cationized molecular ion $(M+H)^+$ which then can be also used for MS/MS fragmentation revealing the required structural details. Only for special compound classes with strong electronegative groups like nitrates or halogens, the NCI is used.

This example of structural elucidation deals with the analysis of long-chain alkyl amides applying the reagent gases methane and ammonia. MS/MS fragmentation of the molecular precursor ion is used for structural confirmation. The data system-supported interpretation using the Mass Frontier program is used to confirm the fragmentation pathways.

4.33.2 Analysis Conditions

Sample material			Industrial amines, cationic surfactant products
Sample preparation			Aliquot weighed out, dissolved in isopropanol, diluted in acetone to final concentration of about 2000 µg/mL
GC method	System		Thermo Scientific TRACE GC Ultra
	Autosampler		Thermo Scientific AS 3000 II
	Column	Type	TG5-Amine
		Dimensions	30 m length × 0.25 mm ID × 0.50 µm film thickness
	Carrier gas		Helium
	Flow		Constant flow, 1 mL/min
	PTV injector	Injection mode	Splitless
		Injection volume	1 µL
		Base temperature	50 °C
		Transfer rate	14 °C/s
		Transfer temperature	300 °C
		Transfer time	10 min
	Split	Closed until	0.05 min
		Open	0.05 min to end of run
		Gas saver	2 min, 10 mL/min
		Purge flow	5 mL/min

	Surge	Pressure, time	120 kPa, 2.55 min
	Oven Program	Start	40 °C, 2.5 min
		Ramp	30 °C/min to 300 °C
		Final temperature	300 °C, 15 min
MS method	System		Thermo Scientific™ ITQ
	Analyzer type		Ion trap MS with external ionization
	Ionization		EI/PCI
	CI Gas		Methane, ammonia, 1 mL/min at 10 psi
	Electron energy		70 eV
	Ion source	Temperature	200 °C
		Buffer gas	0.3 mL He/min
		Emission current	250 µA
	Acquisition	Mode	Full scan
		Mass range	40–450 Da
	Acquisition	Mode	MS/MS
		Precursor ion	m/z 200.3
		Product ions	m/z 72.1, 85.0, 116.1, 130.2
		Coll. energies	0.54, 1.07, 1.93 (ACE optimized)
		Scan time	0.17 s (6 spectra/s)

4.33.3 Experimental Conditions

As a model compound representing the general MS behavior of alkyl amides, the compound N,N-dimethyldecanamide, CAS No. 14433-76-2, is used for the development of the methodology. The GC injection and separation parameters were optimized for the separation of the amides on the applied amine column.

4.33.4 Sample Measurements

First, the regular 70 eV EI mass spectrum of the amide has been acquired (Vale et al. 2011). The search of the EI spectrum against the NIST library gave a match for "N,N-dimethyldecanamide", which has a molecular mass of 199 g/mol, see Figure 4.129. The abundance of the molecular ion m/z 199 with c. 2% is very low.

PCI with methane and ammonia was used to form the molecular ion, which in the next steps was fragmented for structural elucidation by MS/MS. The Mass Frontier software was then used to predict the theoretical fragments and fragmentation pathways of the detected product ions to confirm their identity.

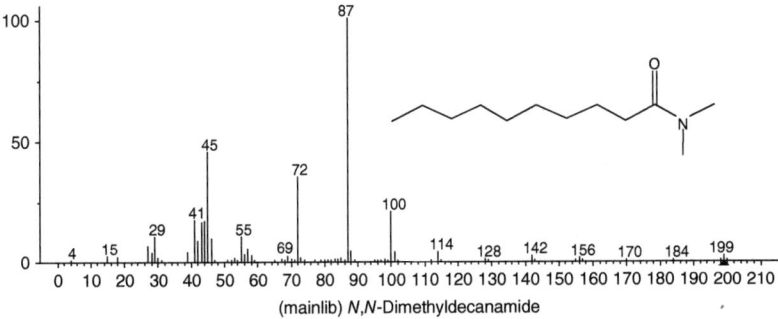

Figure 4.129 NIST library entry "N,N-dimethyldecanamide", M 199.

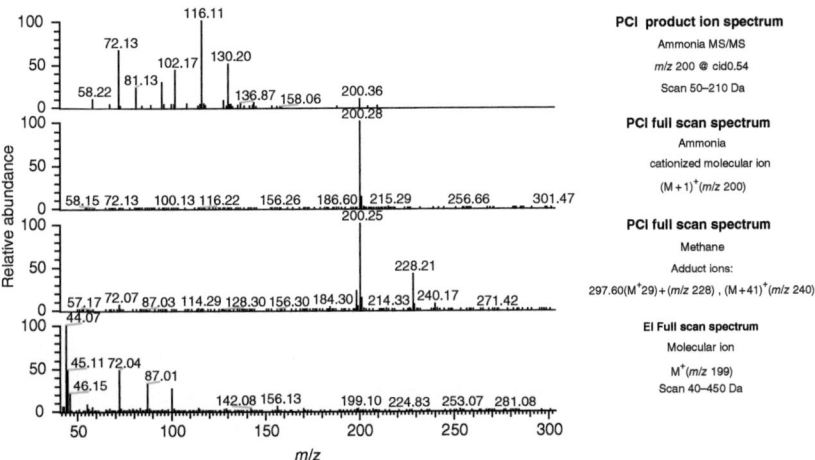

Figure 4.130 PCI MS/MS product ion and full scan spectra with ammonia and methane, EI full scan spectrum.

4.33.5 Results

The chromatographic separation was optimized by a cold injection on the PTV and a capillary column specialized for the analysis of amines.

The ITQ GC-MS/MS system provided the required structure elucidation and identification of N,N-dimethyldecanamide, see Figure 4.130. By choosing the proper reagent gas, PCI techniques can selectively protonate molecules and provide a high intensity of the cationized molecular ion for the MS/MS process. The use of methane as a reagent provides the confirmation of the molecular mass 199 with the protonated molecular ion m/z 200 and the expected adduct formation of $(M+29)^+$ to m/z 228 and $(M+41)^+$ to m/z 240.

Ammonia PCI does not form adducts and provides the cationized molecular ion $(M+1)^+$ ion unfragmented with high intensity due to its high proton affinity of 854.2 kJ/mol (see Section 2.3.1.3). The high intensity cationized molecular ion is the ideal precursor for MS/MS. The product ion spectrum with the fragment ion

770 | 4 Applications

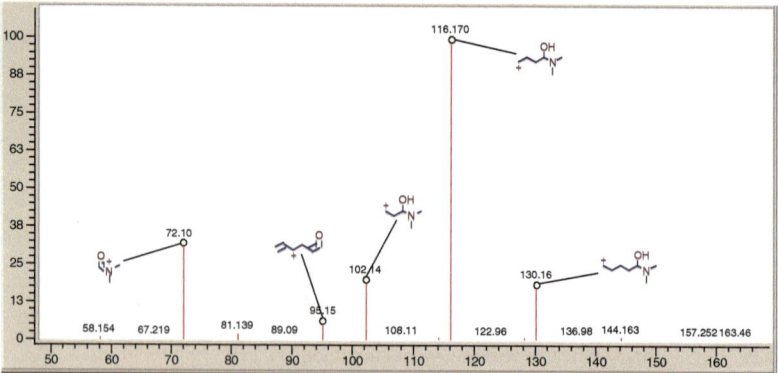

Figure 4.131 Structure annotated product ion spectrum (Mass Frontier).

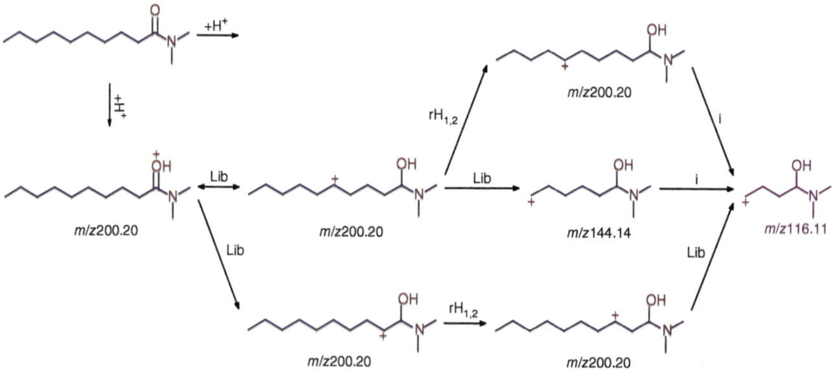

Figure 4.132 Product ion m/z 116 pathway confirmation (Mass Frontier).

structures predicted by the Mass Frontier software for the protonated molecular ion is given with Figure 4.131. The mechanism for confirmation of the m/z 116 ion by Mass Frontier is shown as a fragmentation pathway in Figure 4.132. In conclusion, the generation of the molecular ion by PCI and Mass Frontier interpretation assisted in the final structural identification of N,N-dimethyldecanamide.

References

Section 4.1 Air Analysis According to U.S. EPA Method TO-14

U.S. Environmental Protection Agency (U.S. EPA) (1999) Compendium method TO-14A – determination of volatile organic compounds (VOCs) in ambient air using specially prepared canisters with subsequent analysis B, in *Compendium of Methods for the Determination of Toxic Organic Compounds in Ambient Air*, 2nd edn, January, https://www.epa.gov/sites/default/files/2019-11/documents/to-14ar.pdf.

Madden, A.L. (1994) Analysis of Air Samples for the Polar and Non-Polar VOCs Using a Modified Method TO-14. Tekmar Application Report Vol. 5.3, Cincinnati, OH.

Pleil, J.D., Vossler, T.L., McClenny, W.A., and Oliver, K.D. (1991) Optimizing sensitivity of SIM mode of GC/MS analysis for EPA's TO-14 air toxics method. *J. Air Waste Manage. Assoc.*, **41** (3), 287–293. doi: 10.1080/10473289.1991.10466845.

Schnute, B. and J. McMillan (1993) TO-14 Air Analysis Using the Finnigan MAT Magnum TM Air System. Application Report 230, Finnigan MAT, San Jose, CA.

Section 4.2 BTEX in Surface Water as of U.S. EPA Method 8260

Butler, J. (2013) Analysis of BTEX in Wastewater by ISQ GC-MS, Thermo Fisher Scientific, Austin, TX, personal communication.

U.S. EPA (2006) Method 8260D (SW-846): Volatile Organic Compounds by Gas Chromatography/Mass Spectrometry (GC/MS). Revision 3, Washington, DC.

OI Analytical (2010) Proper Trap Selection for the OI Analytical Model 4560 and 4660 Purge-and-Trap Sample Concentrators. Application Note 12861111, OI Analytical, College Station, TX.

Section 4.3 Volatile Priority Pollutants

Belouschek, P., Brand, H., and Lönz, P. (1992) Bestimmung von chlorierten Kohlenwasserstoffen mit kombinierter Headspace- und GC-MS-Technik. *Vom Wasser*, **79**, 3–8.

CTC (2022) Static and Dynamic Headspace Analysis – The PAL Compendium. White Paper, CTC Analytics AG, Zwingen, Switzerland, *https://www.palsystem.com/en/* (accessed 14 May 2024)

Section 4.5 irm-GC/MS of Volatile Organic Compounds

Schmidt, T.C., Zwank, L., Elsner, M., Berg, M., Meckenstock, R.U., and Haderlein, S.B. (2004) Compound-specific stable isotope analysis of organic contaminants in natural environments: a critical review of the state of the art, prospects, and future challenges. *Anal. Bioanal. Chem.*, **378**, 283–300.

Zwank, L. and Berg, M. (2013) Enhanced Method Detection Limits for irm-GC-MS of Volatile Organic Compounds Using Purge and Trap Extraction. Application Note 30053, Thermo Electron.

Zwank, L. and Berg, M. (2003) Compound specific carbon isotope analysis of volatile organic compounds in the low microgram per liter range. *Anal. Chem.*, **75**, 5575–5583. doi: 10.1021/ac034230i.

Section 4.6 Organotin Compounds in Water

Butler, J. and Phillips, E. (2009) Analysis of Organotins by LVI GC-MS SIM. Application Note 10305, Thermo Fisher Scientific, Austin, TX.

Coscollà, C., Navarro-Olivares, S., Martí, P., and Yusà, V. (2014) Application of the experimental design of experiments (DoE) for the determination of organotin compounds in water samples using HS-SPME and GC-MS/MS. *Talanta*, **119**, 544–552. doi: 10.1016/j.talanta.2013.11.052.

de Dobbeleer, I., Huebschmann, H.-J., Mayer, A., and Gummersbach, J. (2011) The Determination of Organotins in Water Using Triple Quadrupole GC-MS/MS. LCGC – The Application Notebook 8, August 02.

Commission Directive (2014) Commission Directive 2014/101/EU of 30 October 2014 amending Directive 2000/60/EC of the European Parliament and of the Council establishing a framework for Community action in the field of water policy. *Off. J. Europ. Union*, **L311**, 32–35.

European Food Safety Authority (2004) Opinion of the Scientific Panel on contaminants in the food chain [CONTAM] to assess the health risks to consumers associated with exposure to organotins in foodstuffs. *EFSA J.*, **102**, 1–119. https://doi.org/10.2903/j.efsa.2004.102.

Morabito, R. and Quevauviller, P. (2002) Performances of spectroscopic methods for tributyltin (TBT) determination in the 10 years of the EU-SM&T organotin programme. *Spectrosc. Eur.*, **4**, 18.

OSPAR Commission (2011) *Background Document on Organic Tin Compounds*, OSPAR Commission Publication Number: 535/2011, ISBN: 978-1-907390-76-0.

Ostrakhovitch, E. and Cherian, M. (2007) Tin, in *Handbook on the Toxicology of Metals*, 3rd edn (eds G.F. Nordberg, B.A. Fowler, M. Nordberg, and L. Friberg), Elsevier, pp. 839–854.

Takeuchi, M., Mizuishi, K., and Hobo, T. (2000) Determination of organotin compounds in environmental samples. *Anal. Sci.*, **16**, 349.

Thermo Fisher Scientific (2008) *AquaSil™ and SurfaSil™ Siliconizing Fluids. Instructions for Use*, Thermo Fisher Scientific Inc., USA.

Section 4.7 Analysis of Dithiocarbamate Pesticides

Amvrazi, E.G. (2005) *Fate of pesticide residues on raw agricultural crops after postharvest storage and food processing to edible portions*, in *Pesticides – Formulations, Effects, Fate* (ed. M. Stoytcheva), IntechOpen, ISBN: 978-953-307-532-7.

CRL (2009) Analysis of Dithiocarbamate Residues in Foods of Plant Origin Involving Cleavage Into Carbon Disulfide, Partitioning Into Isooctane and Determinative Analysis by GC-ECD, Version 2. Community Reference Laboratories for Residues of Pesticides, CVUA Stuttgart, Fellbach, Germany, https://www.eurl-pesticides.eu//library/docs/srm/meth_DithiocarbamatesCs2_EurlSrm.PDF (accessed 15 May 2024).

Crnogorac, G. and Schwack, W. (2009) Residue analysis of dithiocarbamate fungicides. *Trends Anal. Chem.*, **28** (1), 40–50.

Dasgupta, S., Mujawar, S., Banerjee, K., and Huebschmann, H.-J. (2013) Analysis of Dithiocarbamate Pesticides by GC-MS. Application Note AN10333, Thermo Fisher Scientific, Austin, TX, USA.

Reynolds, S. (2006) Analysis of dithiocarbamates. Presentation at the SELAMAT Workshop, Bangkok, Thailand, July 5–7, 2006, Central Science Laboratory, York, UK (today FERA), *https://www.selamat.net/upload_mm/8/4/e/b6d4cb03-4786-4997-853f-103fd3686589_a13.pdf* (accessed 15 May 2024)

Section 4.8 Multi-Method for Pesticides by Single Quadrupole MS

Fillion, J., Sauvé, F., and Selwyn, J. (2000) Multiresidue method for the determination of residues of 251 pesticides in fruits and vegetables by gas chromatography/mass spectrometry and liquid chromatography with fluorescence detection. *J. AOAC Int.*, **83**, 698–713.

Health and Welfare Canada (2000) *Analytical Methods for Pesticide Residues in Food*, Health Canada, Ottawa.

Section 4.9 QuEChERSER Analysis of Pesticides

Anastassiades, M., Lehotay, S.J., Stajnbaher, D., and Schenk, F.J. (2003a) Fast and easy multiresidue method employing acetonitrile extraction/partitioning and 'dispersive solid phase extraction' for the determination of pesticide residues in produce. *J. AOAC Int.*, **86** (2), 412–431. doi: 10.1093/jaoac/86.2.412.

Anastassiades, M. (2024) Home of the QuEChERS Method. Web site, Chemisches und Veterinäruntersuchungsamt Stuttgart (CVUA), Fellbach, Germany, *https://www.quechers.eu/* (accessed 16 May 2024)

AOAC (2007) Pesticide Residues in Foods by Acetonitrile Extraction and Partitioning with Magnesium Sulfate, *Method Number 2007.01*, AOAC, *https://academic.oup.com/aoac-publications* (accessed 16 May 2024)

Chong, C.M. and Huebschmann, H.-J. (2023) Fully automated QuEChERS extraction and clean-up of organophosphate pesticides in orange juice. *J. N. Food Sci. Tech.*, **4** (2), 1–8, *https://unisciencepub.com/jnfst-volume-4-issue-2-year-2023/*.

EN 15662 (2018) Foods of Plant Origin – Multimethod for the Determination of Pesticide Residues Using GC- and LC-Based Analysis Following Acetonitrile Extraction/Partitioning and Clean-up by Dispersive SPE – Modular QuEChERS method. DIN EN 15662:2018, German Institute for Standardisation (Deutsches Institut für Normung).

Lehotay, S.J., Han, L., and Sapozhnikova, Y. (2016) Automated mini-column solid phase extraction clean-up for high throughput analysis of chemical contaminants in foods by low pressure gas chromatography-tandem mass spectrometry. *Chromatographia*, **79**, 1113–1130. doi: 10.1007/s10337-016-3116-y.

Lehotay, S.J., Anastassiades, M., and Majors, R.E. (2010a) The QuEChERS revolution. *LCGC. Europe*, **9**, September, 418–429.

Lehotay, S.J. et al. (2010b) Comparison of QuEChERS sample preparation methods for the analysis of pesticide residues in fruits and vegetables. *J. Chromatogr. A*, **2017** (16), 2548–2560. doi: 10.1016/j.chroma.2010.01.044.

Lehotay, S.J., Michlig, N., and Lightfield, A.R. (2019) Assessment of test portion sizes after sample comminution with liquid nitrogen in the high throughput analysis of pesticide. *J. Agri. Food Chem*, **68** (5), 1468–1479. doi: 10.1021/acs.jafc.9b07685.

Lehotay, S.J., De Zeeuw, J., Sapozhnikova, Y., Michlig, N., Rousova Hepner, J., and Konschnik, J.D. (2020) There is no time to waste: low pressure gas chromatography-mass spectrometry (LPGC-MS) is a proven solution for fast, sensitive, and robust GC-MS analysis. *LCGC N. Am.*, **38** (8), 457–466.

Mastovska, K., Lehotay, S.J., and Anastassiades, M. (2005) Combination of analyte protectants to overcome matrix effects in routine GC analysis of pesticide. *Anal. Chem.*, **77** (24), 8129–8137. doi: 10.1021/AC0515576.

Michlig, N. and Lehotay, S.J. (2022) Evaluation of a septumless mini-cartridge for automated solid phase extraction clean-up in gas chromatographic analysis of more than 250 pesticides and environmental contaminants in fatty and nonfatty foods. *J. Chrom. A*, **1685**, 463596. doi: 10.1016/j.chroma.2022.463596.

Sapozhnikova, Y. (2018) High throughput analytical method for 265 pesticides and environmental contaminants in meats and poultry by fast low pressure gas chromatography and ultrahigh performance liquid chromatography tandem mass spectrometry. *J. Chrom. A*, **1572**, 203–211. doi: 10.1016/j.chroma.2018.08.025.

Section 4.10 Pesticide Analysis with Ethyl Acetate Extraction and Automated Micro-SPE Clean-up

Anastassiades, M., Lehotay, S.J., Stajnbaher, D., and Schenck, F.J. (2003b) A fast and easy multiresidue method employing acetonitrile extraction/partitioning with dispersive solid phase extraction for the determination of pesticide residues in produce. *J. AOAC Int.*, **86**, 412.

Andersson, A. and Ohlin, B. (1989) A capillary gas chromatographic multiresidue method for determination of pesticides in fruits and vegetables. *Vaar Foeda*, Suppl. (Sweden), 79–109 *https://agris.fao.org/agris-search/search.do?recordID=SE8611088*.

Andersson, A. and Pålsheden, II. (1991) Comparison of the efficiency of different GLC multi-residue methods on crops containing pesticide residues. *Fres. J. Anal. Chem.*, **339** (6), 365–367. doi: 10.1007/BF00322349.

Banerjee, K. et al. (2007) Validation and uncertainty analysis of a multi-residue method for pesticides in grapes using ethyl acetate extraction and liquid chromatography-tandem mass spectrometry. *J. Chromatogr A*, **1173** (1–2), 98–109. doi: 10.1016/j.chroma.2007.10.013.

Ekroth, S. (2011) Simplified Analysis of Pesticide Residues in Food Using the Swedish Ethyl Acetate Method (SweEt). National Food Administration (NFA) Sweden, *http://www.laprw2011.fq.edu.uy/pdf/Lunes/Susanne%20Ekroth..pdf* (accessed 17 May 2024)

Hildmann, F., Gottert, C., Frenzel, T., Kempe, G., and Speer, K. (2015) Pesticide residues in chicken eggs – a sample preparation methodology for analysis by gas and liquid chromatography/tandem mass spectrometry. *J. Chrom. A*, **1403**, 1–20. doi: 10.1016/j.chroma.2015.05.024.

Mol, H.G.J., Rooseboom, A., van Dam, R., Roding, M., Arondeus, K., and Sunarto, S. (2007) Modification and re-validation of the ethyl acetate-based multi-residue method for pesticides in produce. *Anal. Bioanal. Chem.*, **389** (6), 1715–1754. doi: 10.1007/s00216-007-1357-1.

Morris, B.D. and Schriner, R.B. (2015) Development of an automated column solid phase extraction clean-up of QuEChERS extracts, using a zirconia-based sorbent, for pesticide residue analyses by LC-MS/MS. *J. Agric. Food Chem.*, **63**, 5107–5119. doi: 10.1021/jf505539e.

Schürmann, A., Crüzer, C., Duss, V., Kämpf, R., Preiswerk, T., and Huebschmann, H.-J. (2023) Automated micro solid phase extraction clean-up and gas chromatography-tandem mass spectrometry analysis of pesticides in foods extracted with ethyl acetate. *Anal. Bioanal. Chem.*, **416**, 689–700. doi: 10.1007/s00216-023-05027-5.

Section 4.11 Multi-Residue Pesticides Analysis in Ayurvedic Churna

Lohar, D.R. (2008) *Protocol for Testing Guideline for Ayurvedic, Siddha and Unani Medicines*, Government of India, Department of AYUSH, Ministry of Health & Family Welfare, Pharmacopoeial Laboratory for Indian Medicines, Ghaziabad.

Narayanaswamy, V. (1981) Origin and development of ayurveda (a brief history). *Anc. Sci. Life*, **1** (1), 1–7.

Surwade, M., Kumar, S.T., Karkhanis, A., Kumar, M., Dasgupta, S. and Hübschmann, H.-J. (2013) Analysis of Multiresidue Pesticides Present in Ayurvedic Churna by GC-MS/MS. Application Note AN 10361, Thermo Fisher Scientific, Austin, TX, USA.

Section 4.12 Determination of Polar Aromatic Amines by SPME

Müller, L., Fattore, E., and Benfenati, E. (1997) Determination of aromatic amines by solid phase microextraction and gas chromatography-mass spectrometry in water samples. *J. Chrom. A*, **791** (1–2), 221–230. doi: 10.1016/S0021-9673(97)00795-4.

Pan, L. and Pawliszyn, J. (1997) Derivatization/solid phase microextraction: new approach to polar analytes. *J. Anal. Chem.*, **69**, 196–205. doi: 10.1021/ac9606362.

Pan, L., Chong, J.M., and Pawliszyn, J. (1997) Determination of amines in air and water using derivatization combined with solid phase microextraction. *J. Chrom. A*, **773** (1–2), 249–260. doi: 10.1016/S0021-9673(97)00179-9.

Zimmermann, T., Ensigner, W.J., and Schmidt, T.C. (2004) In situ derivatisation/solid phase microextraction: determination of polar aromatic amines. *Anal. Chem.*, **76**, 1028–1038. doi: 10.1021/ac035098p.

Section 4.13 Phthalates in Liquors

Dongliang, S. (2010) Determination of phthalate ester residues in white spirit by GC-MS. *Chem. Anal. Meter.*, **19** (6), 33–35.

Lv, J., Liang, L., and Huebschmann, H.-J. (2013a) Determination of Phthalates in Liquor Beverages by Single Quadrupole GC-MS. Application Note AN10339, Thermo Fisher Scientific, Austin, TX, USA.

NTP (2021) *15th Report on Carcinogens – Di(2-Ethylhexyl) Phthalate*, National Toxicology Program, U.S. Department of Health and Human Services, https://doi.org/10.22427/NTP-OTHER-1003.

Standardization Administration of China (2008) *Determination of Phthalates in Food*, Standardization Administration of China. China method GB/T 21911-2008.

Section 4.15 Geosmin and Methylisoborneol in Drinking Water

Bristow, R.L., Young, I.S., Pemberton, A., Williams, J., and Maher, S. (2019) An extensive review of the extraction techniques and detection methods for the taste and odor compound geosmin (trans-1, 10-dimethyl-trans-9-decalol) in water. *TrAC Trends Anal. Chem.*, **110** (January), 233–248. doi: 10.1016/J.TRAC.2018.10.032.

Chang, J., Biniakewitz, R., and Harkey, G. (2007) Determination of Geosmin and 2-MIB in Drinking Water by SPME-PTV-GC-MS. Application Note AN10213, Thermo Fisher Scientific, San Jose, CA.

Kaziur, W., Salemi, A., Jochmann, M.A., and Schmidt, T.C. (2019) Automated determination of picogram-per-liter level of water taste and odor compounds using solid phase microextraction arrow coupled with gas chromatography-mass spectrometry. *Anal. Bioanal. Chem.*, **411**, 2653–2662. doi: 10.1007/s00216-019-01711-7.

Preti, G., Gittelman, T.S., Staudte, P.B., and Luitweiler, P. (1993) Letting the nose lead the way – malodorous components in drinking water. *Anal. Chem.*, **65**, 699A–702A.

Section 4.16 Flavor and Fragrance Profiling by Dual Column GC-MS

Adams, R.P. (2012) *Identification of Essential Oil Components by Gas Chromatography – Mass Spectroscopy*, 4th edn, Allured Publishing Corporation, Carol Stream, IL, USA.

Babushok, V.I. (2015) Chromatographic retention indices in identification of chemical compounds. *TrAC – Trends Anal. Chem.*, **69** (June), 98–104. doi: 10.1016/j.trac.2015.04.001.

Bianchi, F., Careri, M., Mangia, A., and Musci, M. (2007) Retention indices in the analysis of food aroma volatile compounds in temperature-programmed gas chromatography: database creation and evaluation of precision and robustness. *J. Sep. Sci.*, **30** (4), 563–572. doi: 10.1002/jssc.200600393.

David, O.R.P. and Doro, F. (2023) Industrial fragrance chemistry: a brief historical perspective. *Europ. J. Org. Chem.*, **26** (44), e202300900. doi: 10.1002/ejoc.202300900.

Little, J. (2024) Part 4: Processing GC-MS Data with AMDIS. Educational Video, *https://www.youtube.com/watch?v=wKoy-vdMEtU* (accessed 2 June 2024)

Nakasuji, Y. and Hiramatsu T. (2023) Quantitative Analysis of 57 Fragrance Allergens in Cosmetics Using Twin Line MS System. Application News 01-00526-EN, Shimadzu Corporation.

SIM (2024) Dual Column Adapters, *https://sim-gmbh.de/en/products/accessoires/127-gc-accessories/714-dual-column-adapters.html* (accessed 2 June 2024).

Soulier, C. (2024) Technical Consulting, Provence-Alpes-Côte d'Azur, France, personal communication, *https://www.linkedin.com/in/christian-soulier-6bb12583* (accessed 13 November 2024)

Section 4.17 Aroma Profiling of Cheese

Markes (2010) Food Decomposition Analysis Using the Micro-Chamber/Thermal Extractor and TD-GC-MS. Application Note TDTS 95, Markes Int. Ltd., Bridgend, UK.

Markes (2012) Rapid Aroma Profiling of Cheese Using a Micro-Chamber/Thermal Extractor with TD-GC-MS Analysis. Application Note TDTS 101, Markes Int. Ltd., Bridgend, UK.

Section 4.18 48 Allergens

Balducci, C., Cerasa, M., Avino, P., Ceci, P., Bacaloni, A., and Garofalo, M. (2022) Analytical determination of allergenic fragrances in indoor air. *Separations*, **9** (4), 1–12. doi: 10.3390/separations9040099.

European Commission (2009) Directive 2009/48/EC of the European Parliament and of the Council of 18 June 2009 on the safety of toys. *Off. J. Eur. Union*, **L170**, 1–37.

Scientific Committee on Consumer Safety (2012) Opinion on fragrance allergens in cosmetic products. Scientific Committee on Consumer Safety, The SCCS adopted this opinion at its 15th plenary meeting of June 26–27, 2012, SCCS/1459/11, ISSN: 1831-4767.

European Commission (2023) COMMISSION REGULATION (EU) 2023/1545 of 26 July 2023 Amending Regulation (EC) No 1223/2009 of the European Parliament and of the Council as regards labelling of fragrance allergens in cosmetic products. *Off. J. Eur. Union*, **L188** 27.7.2023: L188/1–L188/23.

Lv, Q. et al. (2013b) Determination of 48 fragrance allergens in toys using GC with ion trap MS/MS. *J. Sep. Sci.*, **36**, 3534–3549. doi: 10.1002/jssc.201300586.

Section 4.19 Metabolite Profiling of Natural Volatiles and Extracts

Chong, C.M. and H.-J. Huebschmann (2020) Metabolite profiling by automated methoximation and silylation. Poster at the 16th (online) Metabolomics Conference, July 6–10, 2020, Shanghai, China 2020, CTC Analytics Pte. Ltd., Singapore, https://doi.org/10.13140/RG.2.2.29085.03045.

Fragner, L., Weckwerth, W., and Hübschmann, H.-J. (2010) Metabolomics Strategies Using GC-MS/MS Technology. Application Note AN 51999, Thermo Fisher Scientific, Austin, TX, USA.

Hübschmann, H.-J., Fragner, L., Weckwerth, W., and Cardona, D. (2012) Metabolomics strategies using GC-MS/MS technology. Poster at the 8th International Conference of the Metabolomics Society, Washington, DC, USA, June 25–28, 2012.

Mallard, W.G. and Reed, J. (1997) *AMDIS – User Guide*, U.S, Department of Commerce, Technology Administration, National Institute of Standards and Technology (NIST), Standard Reference Data Program, Gaithersburg, MD, USA.

Weckwerth, W., Wenzel, K., and Fiehn, O. (2004) Process for the integrated extraction identification and quantification of metabolites, proteins, and RNA to reveal their co-regulation in biochemical networks. *Proteomics*, **4** (1), 78–83. doi: 10.1002/pmic.200200500.

Section 4.20 Identification of Extractables and Leachables

EMA (2022) *Quality Documentation for Medicinal Products When Used with a Medical Device – Scientific Guideline*, European Medicines Agency, Amsterdam, The Netherlands. ICH (2020).

ICH Q3E: Guideline for Extractables and Leachables (E&L). International Council for Harmonisation of Technical Requirements for Pharmaceuticals for Human Use, July 10, 2020. *https://database.ich.org/sites/default/files/ICH_Q3E_ConceptPaper_2020_0710.pdf* (accessed 20 October 2024).

Kumirska, J. et al. (2013) Chemometric optimization of derivatization reactions prior to gas chromatography-mass spectrometry analysis. *J. Chrom. A*, **1296**, 164–178. doi: 10.1016/j.chroma.2013.04.079.

Lewis, D.B. (2011) Current FDA perspective on leachable impurities in parenteral and ophthalmic drug products. Presentation at the AAPS Workshop on Pharmaceutical Stability, Washington, DC, 2011.

Mallard, W.G. and Reed, J. (1997) *AMDIS – User Guide, U.S*, Department of Commerce, Technology Administration, National Institute of Standards and Technology (NIST), Standard Reference Data Program, Gaithersburg, MD, USA.

Moffat, F. (2011) Extractables and leachables in pharma – a serious issue. Solvias Whitepaper. Solvias AG, Kaiseraugst.

Thermo Fisher Scientific (2011) *Mass Frontier Version 7.0. User Guide*, Thermo Fisher Scientific, San Jose, CA. USA.

Section 4.21 Volatiles in Car Interior Materials

ASTM (2024) D2369 – 20 Standard Test Method for Volatile Content of Coatings. ASTM International, West Conshohocken, PA, *https://doi.org/10.1520/D2369-20, https://www.astm.org/d2369-20.html* (accessed 21 May 2024).

California Department of Public Health (2017) Standard Method for the Testing and Evaluation of Volatile Organic Chemical Emissions from Indoor Sources Using Environmental Chambers (California Specification 01350), Version 1.2, January 2017.

European Commission (2004) Directive 2004/42/CE of the European Parliament and of the Council of 21 April 2004 on the limitation of emissions of volatile organic compounds due to the use of organic solvents in certain paints and varnishes and vehicle refinishing products and amending Directive 1999/13/EC. *Offi. J. Europ. Union*, **L143**, 87–96.

European Commission (2010) Commission Directive 2010/79/EU of 19 November 2010 on the adaptation to technical progress of Annex III to Directive 2004/42/EC of the European Parliament and of the Council on the limitation of emissions of volatile organic compounds. *Off. J. Eur. Union*, **L304**, 18.

GM Engineering Standards GM15634 (2020) *Determination of Volatile and Semi-Volatile Organic Compounds from Vehicle Interior Materials*, 1 July 2020, General Motors Corporation, Detroit, MI, USA.

Guidelines for Air Quality Assessment in Cars (2011) GB/T 27630-2011, China Ministry of Ecology and Environment, *https://english.mee.gov.cn/Resources/standards/Air_*

Environment/quality_standard1/201605/t20160511_337499.shtml (accessed 14 November 2024)

Grabbs, J., Corsi, R., and Torres, V. (2000) Volatile organic compounds in new automobiles: screening assessment. *J. Environ. Eng.*, **126** (10), 974–977.

ISO (2020) *Paints and varnishes – determination of volatile organic compounds (VOC) and/or semi volatile organic compounds (SVOC) content*, in *Part 2: Gas-Chromatographic Method*, ISO 11890-2:2020(en).

VDA 278 (2016) *Thermal Desorption Analysis of Organic Emissions for the Characterization of Non-Metallic Materials for Automobiles* (Version 05/2016), VDA Verband der Automobilindustrie, Germany, https://webshop.vda.de/VDA/de/vda-278-05-2016 (accessed 20 May 2024).

Section 4.22 Analysis of Azo Dyes in Leather and Textiles

Brüschweiler, B.J. and Merlot, C. (2017) Azo dyes in clothing textiles can be cleaved into a series of mutagenic aromatic amines which are not regulated yet. *Regul. Toxicol. Pharmacol.*, **88**, 214–226. doi: 10.1016/j.yrtph.2017.06.012.

European Commission (2002) Directive 2002/61/EC of the European Parliament and of the Council of 19 July 2002 amending for the nineteenth time Council Directive 76/769/EEC relating to restrictions on the marketing and use of certain dangerous substances and preparations (azocolorants) of 19 July 2002 amending for the nineteenth time Council Directive 76/769/EEC relating to restrictions on the marketing and use of certain dangerous substances and preparations (azocolorants). *Off. J. Europ. Union*, **L243**, 0015–0018.

European Parliament and the Council of the European Union (2006) Corrigendum to Regulation (EC) No 1907/2006 of the European Parliament and of the Council of 18 December 2006 concerning the Registration, Evaluation, Authorisation and Restriction of Chemicals (REACH). *Offi. J. Europ. Union*, **L136**, L 136/3–L 136/280.

European Commission (1999) Opinion on risk of cancer caused by textiles and leather goods coloured with azo-dyes. Health and Consumers Scientific Committees, Expressed at the 7th CSTEE Plenary Meeting, Brussels, January 18, 1999. https://ec.europa.eu/health/scientific_committees/environmental_risks/opinions/sctee/sct_out27_en.htm (accessed 20 May 2024).

European Commission (2005) Commission Recommendation of 4 February 2005 on the further investigation into the level of polycyclic aromatic hydrocarbons in certain. *Off. J.*, **L34**, 43–45.

EN ISO 14362-1 (2017) Textiles – methods for determination of certain aromatic amines derived from azo colorants, in *Detection of the Use of Certain Azo Colorants Accessible With and Without Extracting the Fibres*, EN ISO 14362-1:2017, ISBN 978 0 580 85045 5.

EN ISO 14362-2 (2003) Textiles – methods for determination of certain aromatic amines derived from azo colorants, in *Part 2: Detection of the Use of Certain Azo Colorants Accessible by Extraction the Fibres*, EN ISO 14362-2:2003.

EN ISO 14362-3 (2017) Textiles – methods for determination of certain aromatic amines derived from azo colorants, in *Part 3: Detection of the Use of Certain Azo Colorants,*

Which May Release 4-Aminoazobenzene (Supplementary to ISO 14362-1), EN ISO 14362-3:2017.

Purwanto, A. and Chen, A. (2013) Detection, Identification and Quantitation of Azo Dyes in Leather and Textiles by GC-MS. Application Note AN10329, Thermo Fisher Scientific, San Jose, CA, USA.

Section 4.23 Fast GC of 16 Priority PAHs

Kleinhenz, S., Jira, W., and Schwind, K.-H. (2006) Dioxin and polychlorinated biphenyl analysis: automation and improvement of clean-up established by example of spices. *Mol. Nutr. Food Res.*, **50** (4–5), 362–367.

EFSA (2008) Polycyclic aromatic hydrocarbons in food – scientific opinion of the panel on contaminants in the food chain. *EFSA J.*, **724**, 1–114. doi: 10.2903/j.efsa.2008.724.

European Commission (2011a) Commission Regulation (EU) No 835/2011 of 19 August 2011 of 19 August 2011 amending Regulation (EC) No 1881/2006 as regards maximum levels for polycyclic aromatic hydrocarbons in foodstuffs. *Offi. J. Europ. Union* L 215/4 (835), 4–8.

JECFA (2005) Summary and conclusion of the Joint FAO/WHO Expert Committee on food additives. Sixty-Fourth Meeting, Rome, February 8–17, 2005, JECFA/64/SC.

Ziegenhals, K. and Jira, W. (2006) *Bestimmung der von der EU als prioritär eingestuften polyzyklischen aromatischen Kohlenwasserstoffe (PAK) in Lebensmitteln*, Kulmbach Kolloquium September, 2006.

Ziegenhals, K. and Jira, W. (2007) High sensitive PAH method to comply with the new EU directives. Presentation at the European High Resolution GC-MS Users Meeting, Venice, Italy, March 23–24, 2007.

Ziegenhals, K., Speer, K., Hübschmann, H.-J., and Jira, W. (2008) Fast GC/HRMS to quantify the EU priority PAH. *J. Sep. Sci.*, **31** (10), 1779–1786.

Section 4.24 Environmental Contaminants in Fish

EFSA (2005) Opinion of the scientific panel on contaminants in the food chain on a request from the European parliament related to the safety assessment of wild and farmed fish. *EFSA J.*, **236**, 1–118.

European Commission (2021) Method Validation Procedures for Pesticide Residues Analysis in Food and Feed. SANTE 11312/2021. Directorate-General for Health and Food Safety, *https://food.ec.europa.eu/system/files/2023-11/pesticides_mrl_guidelines_wrkdoc_2021-11312.pdf* (accessed 20 May 2024)

Focker, M. et al. (2022) Review of food safety hazards in circular food systems in Europe. *Food Res. Int.*, **158** (June), 111505. doi: 10.1016/j.foodres.2022.111505.

Kalachova, K., Pulkrabova, J., Drabova, L., Cajka, T., Kocourek, V., and Hajslova, J. (2011) Simplified and rapid determination of polychlorinated biphenyls, polybrominated diphenyl ethers and polycyclic aromatic hydrocarbons in fish and shrimps integrated into a single method. *Anal. Chim. Acta*, **707**, 84–91. doi: 10.1016/j.aca.2011.09.016.

Kalachova, K., Pulkrabova, J., Cajka, T., Drabova, L., Stupak, M., and Hajslova, J. (2013) Gas chromatography-triple quadrupole tandem mass spectrometry: a powerful tool for the (ultra)trace analysis of multiclass environmental contaminants in fish and fish feed. *Anal. Bioanal. Chem.*, **405**, 7803–7815. doi: 10.1007/s00216-013-7000-4.

Krumwiede, D., Griep-Raming, J., and Muenster, H. (2005) Comparative studies of PTV on-column like injection for improved sensitivity in GC-MS analysis of thermolabile high boiling brominated flame retardants. Poster, ASMS Conference, 2005, Thermo Electron, Bremen, Germany.

Nadal, M., Marquès, M., Mari, M., and Domingo, J.L. (2015) Climate change and environmental concentrations of POPs: a review. *Env. Res.*, **143** (November), 177–185. doi: 10.1016/j.envres.2015.10.012.

Thomas, J. et al. (2010) *Analysis of Polyaromatic Hydrocarbons Using Next Generation Highly Selective Stationary Phases*. Product Information, Restek Corporation, Bellefonte, PA, USA.

Section 4.25 Fast GC of PCBs

Ballschmiter, K. and Zell, M. (1980) Analysis of polychlorinated biphenyls (PCB) by glass capillary gas chromatography – composition of technical aroclor- and clophen-PCB mixtures. *Fres. Zeit. Anal. Chemie*, **302** (1), 20–31. doi: 10.1007/BF00469758.

de Boer, J., Dao, Q., and van Dortmond, R. (1992) Retention times of fifty one chlorobiphenyl congeners on seven narrow bore capillary columns coated with different stationary phases. *J. High Resolut. Chromatogr.*, **15**, 249–255. doi: 10.1002/jhrc.1240150409.

Frame, G.M., Cochran, J.W., and Bøwadt, S.S. (1996) Complete PCB congener distributions for 17 aroclor mixtures determined by 3 HRGC systems optimized for comprehensive, quantitative, congener-specific analysis. *J. High Res. Chrom.*, **19** (12), 657–668. doi: 10.1002/jhrc.1240191202.

European Commission (2011b) Commission Regulation (EU) No 1259/2011 of 2 December 2011 amending regulation (EC) No 1881/2006 as regards maximum levels for dioxins, dioxin-like PCBs and non dioxin-like PCBs in foodstuffs. *Off. J. Europ. Union*, **L320**, 18–23.

European Commission (2022) Commission Regulation (EU) 2022/2002 of 21 October 2022 amending regulation (EC) No 1881/2006 as regards maximum levels of dioxins and dioxin-like PCBs in certain foodstuffs. *Off. J. Europ. Union*, **L274**, 64–66.

Gummersbach, J. (2011) *Thermo Fisher Scientific*, Application Laboratory, Dreieich, Germany, personal communication.

Lorenzi, V. et al. (2020) PCDD/Fs, DL-PCBs, and NDL-PCBs in dairy cows: carryover in milk from a controlled feeding study. *J. Agri. Food Chem.*, **68** (7), 2201–2213. doi: 10.1021/acs.jafc.9b08180.

Matsumara, C., Tsurukawa, M., Nakano, T., Ezaki, T., and Ohashi, M. (2002) Elution orders of all 209 PCBs congeners on capillary column 'HT8-PCB'. *J. Enviro. Chem.*, **12** (4), 855–865.

SGE (2004) Elution Orders of All 209 PCBs Congeners on Capillary Column 'HT8-PCB'. Technical Report, TA-0122-C. SGE International Pty. Ltd.

SGE (2005) HT8: the perfect PCB column. Product information. Publication No. AP-0040-C Rev:04 08/05. SGE International Pty. Ltd.

U.S. EPA (2003) Table of PCB Species by Congener Number, *https://www.epa.gov/sites/default/files/2015-09/documents/congenertable.pdf* (accessed 22 May 2024).

UN (2024) PCBs – a Forgotten Legacy? *Web site*. UN Environment Program, *https://www.unep.org/topics/chemicals-and-pollution-action/pollution-and-health/persistent-organic-pollutants-pops/pcbs* (accessed 20 May 2024).

Section 4.26 Dioxin Screening in Food and Feed

European Commission (2014) Commission Regulation (EU) No 709/2014 of 20 June 2014 amending regulation (EC) No 152/2009 as regards the determination of the levels of dioxins and polychlorinated biphenyls. *Off. J. European Comm*, **L 188**, 1–18.

European Commission (2017) Commission Regulation (EU) 2017/771 of 3 May 2017 amending regulation (EC) No 152/2009 as regards the methods for the determination of the levels of dioxins and polychlorinated biphenyls. *Off. J. European Comm*, **L 115**, 22–42.

Fürst, P. and Bernsmann T. (2014) A new evaluation of triple qudrupole GC/MS for dioxin analysis in food samples using a novel source technology. Presentation, Dioxin Conference Madrid 2014, Chemical and Veterinary Analytical Institute Münsterland-Emscher-Lippe (CVUA-MEL), Münster, Germany.

Kotz, A. et al. (2011) GC-MS/MS determination of PCDD/Fs and PCBs in feed and food–comparison with GC-HRMS. *Organohalogen Compd.*, **73**, 688–691.

Kotz, A. et al. (2012) Analytical criteria for use of MS/MS for determination of dioxins and dioxin-like PCBs in feed and food. *Organohalogen Compd.*, **74**, 156–159.

U.S. EPA (1994a) Tetra- through octa-chlorinated dioxins and furans by isotope dilution HRGC-HRMS. *Method 1613, Rev. B*, U.S. Environmental Protection Agency, Washington, DC, USA

Section 4.27 Confirmation Analysis of Dioxins and Dioxin-like PCBs

Aylward, L.L. and Hays, S.M. (2002) Temporal trends in human TCDD body burden: Decreases over three decades and implications for exposure levels. *J. Expo. Anal. Environ. Epidemiol.*, **12**, 319–328.

CDC (2009) Fourth Report on Human Exposure to Environmental Chemicals U.S. Department of Health and Human Services, Centers for Disease Control and Prevention, Atlanta, GA, *http://www.cdc.gov/exposurereport/* (accessed 22 May 2024)

CDC (2014) Fourth Report on Human Exposure to Environmental Chemicals, Updated Tables. U.S. Department of Health and Human Services, Centers for Disease Control and Prevention, Atlanta, GA, USA, *http://www.cdc.gov/exposurereport/* (accessed 22 May 2024)

DeVito, M. et al. (2024) The 2022 World Health Organization reevaluation of human and mammalian toxic equivalency factors for polychlorinated dioxins, dibenzofurans and biphenyls. *Reg. Tox. Pharmacol.*, **146** (July 2023), 105525. doi: 10.1016/j.yrtph.2023.105525.

Firestone, D. (1973) Etiology of chick edema disease. *Environ. Health Perspect.*, **5**, 59–66.

Fishman, V.N., Martin, G.D., and Lamparski, L.L. (2007) Comparison of a variety of gas chromatographic columns with different polarities for the separation of chlorinated dibenzo-p-dioxins and dibenzofurans by high resolution mass spectrometry. *J. Chromatogr. A*, **1139**, 285–300. doi: 10.1016/j.chroma.2006.11.025.

Fishman, V.N., Martin, G.D., and Wilken, M. (2011) Retention time profiling of all 136 tetra- through octa- chlorinated dibenzo-p-dioxins and dibenzofurans on a variety of si-arylene gas chromatographic stationary phases. *Chemosphere*, **84** (7), 913–922. doi: 10.1016/j.chemosphere.2011.06.012.

Hays, S.M. and Aylward, L.L. (2003) Dioxin risks in perspective: past, present and future. *Regul. Toxicol. Pharmacol.*, **37** (2), 202–217.

Hernández, C.S. et al. (2020) Biomonitoring of polychlorinated dibenzo-p-dioxins (PCDDs), polychlorinated dibenzofurans (PCDFs) and dioxin-like polychlorinated biphenyls (Dl-PCBs) in human milk: exposure and risk assessment for lactating mothers and breastfed children from Spain. *Sci. Total Env.*, **744** (November), 140710. doi: 10.1016/j.scitotenv.2020.140710.

Krumwiede, D. and Hübschmann, H.-J. (2006a) Confirmation of Low Level Dioxins and Furans in Dirty Matrix Samples using High Resolution GC-MS. Application Note AN30112, Thermo Fisher Scientific, Bremen.

Krumwiede, D. (2006b) *'Step Standard' for Continuous Monitoring of Instrument Performance*. Personal information.

Lorber, M. (2002) A pharmacokinetic model for estimating exposure of Americans to dioxin-like compounds in the past, present and future. *Sci. Total Environ.*, **288**, 81–95.

Mamoon, H.A. (2013). Evaluation of the separation capacity of different GC columns for tetra- to octachlorinated PCDD/Fs. Master Thesis. Department of Chemistry, Umeå University, Sweden.

Theobald, F. (2024) Analysis of POPs compounds using magnetic sector HRMS instrumentation. Training documentation, POPsLab, Valencia, Spain, private communication.

Turner, W. et al. (2004) The phantom menace – determination of the true method detection limit (MDL) for background levels of PCDDs, PCDFs and cPCBs in human serum by high resolution mass spectrometry. *Organohalogen Compd.*, **66**, 264–271.

Turner, W.E. et al. (2006) Instrumental approaches for improving the detection limit for selected PCDD congeners in samples from the general U.S. population as background levels continue to decline. Poster at the 26th International Symposium on Halogenated Persistent Organic Pollutants, Oslo, Norway, August 21–25, 2006.

U.S. EPA (1994b) Tetra-Through Octa-Chlorinated Dioxins and Furans by Isotope Dilution HRGC-HRMS, U.S. Environmental Protection Agency Office of Water Engineering and Analysis Division, Washington, DC, U.S. EPA Method 1613 Rev.B, October 1994.

U.S. EPA (1998a) Polychlorinated dibenzo-p-dioxins and polychlorinated dibenzofurans by high resolution gas chromatography-low resolution mass spectrometry (HRGC-LRMS). *U.S. EPA Meth. 8280B* Rev. 2, January 1998.

U.S. EPA (1998b) Polychlorinated diebenzo-p-dioxins and polychlorinated dibenzofurans by high resolution gas chromatography-high resolution mass spectrometry (HRGC-HRMS). *U.S. EPA Meth. 8290B* Rev. 1, January 1998.

van den Berg, M.L. et al. (1998) Toxic equivalent factors (TEFs) for PCBs, PCDDs, PCDFs for humans and wildlife. *Environ. Health Perspect.*, **106** (12), 775–792.

van den Berg, M. et al. (2006) The 2005 World Health Organization re-evaluation of human and mammalian toxic equivalency factors for dioxins and dioxin-like compounds, 2005 WHO re-evaluation of TEFs. *Tox. Sci.*, **93** (2), 223–241.

WHO (2024) Dioxins and Their Effects on Human Health. *Fact sheet, https://www.who.int/news-room/fact-sheets* (accessed 22 May 2024).

Section 4.28 U.S. EPA 1614 Brominated Flame Retardants PBDEs

Audi, G. and Wapstra, A.H. (1995) The 1995 update to the atomic mass evaluation. *Nuclear Phys. A*, **595**, 409–480.

ECHA (2023) *Regulatory Strategy for Flame Retardants*, European Chemicals Agency, Helsinki, Finland. doi: 10.2823/854233.

EFSA (2024) *Brominated Flame Retardants, https://www.efsa.europa.eu/en/topics/topic/brominated-flame-retardants* (accessed 20 May 2024)

European Parliament (2003a) Directive 2002/95/EC of the European Parliament and of the Council of the European Union, January 27, 2003. *J. Eur. Union*, **L37**, 19–23.

European Parliament (2003b) Directive 2003/11/EC of The European Parliament and of The Council of 6 February 2003 amending for the 24th time Council Directive 76/769/EEC relating to restrictions on the marketing and use of certain dangerous substances and preparations (pentabromodiphenyl ether, octabromodiphenyl ether). *J. Eur. Union*. L 42, 45–46.

European Parliament (2022) Regulation (EU) 2022/2400 of the European Parliament and of the Council of 23 November 2022 amending annexes IV and V to regulation (EU) 2019/1021 on persistent organic pollutants. *Off. J. Europ. Union*, **L317**, 24–31.

Krumwiede, D. and Hübschmann, H.-J. (2006c) Trace analysis of brominated flame retardants with high resolution mass spectrometry. *LCGC Europe*, **19** (7), 45–46.

Krumwiede, D. and Hübschmann, H.-J. (2007) High resolution GC-MS as a viable solution for conducting environmental analyses. *Peak*, **11**, 7–15.

Krumwiede, D. and Hübschmann, H.-J. (2008) DFS – Analysis of Brominated Flame Retardants with High Resolution Mass Spectrometry. Application Note 30098, Thermo Fisher Scientific, Bremen, Germany.

Martinez, G., Niu, J., Takser, L., Bellenger, J.P., and Zhu, J. (2021) A review on the analytical procedures of halogenated flame retardants by gas chromatography coupled with single quadrupole mass spectrometry and their levels in human samples. *Env. Poll.*, **285**, 117476. doi: 10.1016/j.envpol.2021.117476.

Peter, J. and Taylor, B.N. (1999) CODATA recommended values of the fundamental physical constants. *J. Phys. Chem. Ref. Data*, **28** (6), 1713–1852 and references cited therein.

Section 4.29 SPME Analysis of PBBs

de Boer, J., Allchin, C., Law, R.J., Zegers, B., and Boon, J.P. (2001) Methods for the analysis of polybrominated diphenylethers in sediments and biota. *Trends Anal. Chem.*, **10**, 591–599. doi: 10.1016/S0165-9936(01)00097-8.

Krumwiede, D. and Hübschmann, H.-J. (2006c) Analysis of Brominated Flame Retardants by High Resolution GC-MS. Application Note 30098, Thermo Fisher Scientific, Bremen, Germany.

Polo, M., Gomez-Noya, G., Quintana, J.B., Llompart, M., Garcia-Jares, C., and Cela, R. (2004) Development of a solid phase microextraction gas chromatography/tandem mass spectrometry method for polybrominated diphenyl ethers and polybrominated biphenyls in water samples. *Anal. Chem.*, **76**, 1054–1062. doi: 10.1021/ac030292x.

Section 4.30 THC-A in Urine by NCI

Beck, F., Legleye, S., and Spilka, S. (2007) The use of cannabis by adolescents and young adults: comparison of European consumption practices. *Sante Pub.*, **19**, 481–488.

Jones, A.W., Kugelberg, F.C., Holmgren, A., and Ahlner, J. (2009). Five-year update on the occurrence of alcohol and other drugs in blood samples from drivers killed in road-traffic crashes in Sweden. *Forensic Sci. Int.*, **186**, 56–62.

Kapusta, N.D. et al. (2006) Epidemiology of substance use in a representative sample of 18-year-old males. *Alcohol.* **41** (2), 188–192. doi: 10.1093/alcalc/agh251.

Levin, F.R. and Kleber, H.D. (2008) Use of dronabinol for cannabis dependence: two case reports and review. *Am. J. Addict.* **17**, 161–164. doi: 10.1080/10550490701861177.

Melchert, H.-U., Hübschmann, H.-J., and Pabel, E. (2009) Analytik der THC-Carbonsäure – Spezifische Detektion und hochsensitive Quantifizierung im Harn durch NCI-GC-MS. *Labo*, **1**, 8–12.

Musshoff, F. and Madea, B. (2006). Review of biologic matrices (urine, blood, hair) as indicators of recent or ongoing cannabis use. *Ther. Drug Monit.* **28** (2), 155–163. doi: 10.1097/01.ftd.0000197091.07807.22.

Wikipedia (2024) Legality of Cannabis, *https://en.wikipedia.org/wiki/Legality_of_cannabis* (accessed 27 December 2024).

Section 4.31 Comprehensive Drug Screening and Quantitation

Maurer, H.H. (2007) Demands on scientific studies in clinical toxicology. *Forensic Sci. Int.*, **165**, 194.

Maurer, H.H. (2012) How can analytical diagnostics in clinical toxicology be successfully performed today? *Ther. Drug Monit.*, **34**, 561.

Maurer, H.H., Meyer, M., Pfleger, K., and Weber, A. (2023) *GC-MS Library of Drugs, Poisons, and Their Metabolites*, Wiley-VCH, Weinheim, Germany.

Meyer, M.R., Peters, F.T., and Maurer, H.H. (2010) Automated mass spectral deconvolution and identification system for GC-MS screening for drugs, poisons and metabolites in urine. *Clin. Chem.*, **56**, 575.

Meyer, G.M.J., Weber, A.A., and Maurer, H.H. (2014) Development and validation of a fast and simple multi-analyte procedure for quantification of 40 drugs relevant to emergency toxicology using GC-MS and one-point calibration. *Drug Test. Anal.*, **6** (5), 472–481. doi: 10.1002/dta.1555.

Section 4.32 Screening for Drugs of Abuse

Braithwaite, A., Jarvie, D.R., Minty, P.S., Simpson, D., and Widdop, B. (1995) Screening for drugs of abuse. I: Opiates, amphetamines and cocaine. *J. Clin. Path.*, **32**, 123–153. doi: 10.1177/000456329503200203.

Donike, M. (1969) N-methyl-N-trimethylsilyl-trifluoracetamid ein neues Silylierungsmittel aus der Reihe der silylierten Amide. *J. Chromatogr.*, **42**, 103–104.

Maurer, H.H. (1990) Identifizierung unbekannter Giftstoffe und ihrer Metaboliten in biologischem Material. *GIT*, (Suppl.1), 3–10.

Maurer, H.H. (1993) GC-MS Contra Immunoassay? *Symposium Aktuelle Aspekte des Drogennachweises, Mosbach, Germany, April 15, 1993, Abstracts.*

Maurer, H.H., Meyer, M., Pfleger, K., and Weber, A. (2023) *GC-MS Library of Drugs, Poisons, and Their Metabolites*, Wiley-VCH, Weinheim, Germany.

Musshoff, F., Trafkowski, J., and Madea, B. (2004) Validated assay for the determination of markers of illicit heroin in urine samples for the control of patients in a heroin prescription program. *J. Chromatogr. B*, **811**, 47–52.

Simpson, D. *et al.* (1997) Screening for drugs of abuse (II): cannabinoids, lysergic acid diethylamide, buprenorphine, methadone, barbiturates, benzodiazepines and other drugs. *Ann. Clin. Biochem.*, **34**, 460–510. doi: 10.1177/000456329703400502 087.

Smith, M., Vorce, S.P., Holler, J.M., Shimomura, E., Magluilo, J., and Jacobs, A.J. (2007) Modern instrumental methods in forensic toxicology. *J. Anal. Tox.*, **31** (5), 237–253 and 8A–9A.

Weller, J.P., Wolf, M., and Szidat, S. (2000) Enhanced selectivity in the determination of Δ9-tetrahydrocannabinol and two major metabolites in serum using ion trap GC-MS/MS. *J. Anal. Toxicol.*, **24**, 1–6.

Section 4.33 Structural Elucidation by Chemical Ionization and MS/MS

Vale, G., Butler, J., Herbold, R., Hübschmann, H.-J., and O'Brian, P. (2011) Structural Elucidation of Alkylamines in Process Streams by PCI Ammonia GC-MS/MS. Application Note AN52255, Thermo Fisher Scientific, Austin, TX, USA.

Weller, J.P. and Wolf, M. (1990) Nachweis von Morphinderivaten im Blut Mittels Ion Trap-Detektor. Application Note No. 76 (AR 7/1047), Finnigan MAT GmbH, Bremen, Germany.

Glossary

This glossary of chromatographic and mass spectrometric terms and commonly used abbreviations also contains selected terms of recommendations for nomenclature, definitions, and acronyms in mass spectrometry of the IUPAC Analytical Chemistry Division.

A

α	Separation factor of two adjacent peaks; $\alpha = k_2/k_1$.
α-Cleavage	Homolytic cleavage where the bond fission occurs between at the atom adjacent to the atom at the apparent charge site and an atom removed from the apparent charge site by two bonds.
A_P	Peak area.
a-Ion	Fragment ion containing the N-terminus formed upon dissociation of a protonated peptide at a backbone C—C bond.
AC	Alternating current.
Accelerating voltage	Electrical potential used to impart translational energy to ions in a mass spectrometer.
Accelerator mass spectrometry (AMS)	Mass spectrometry technique in which atoms extracted from a sample are ionized, accelerated to megaelectron volt energies, and separated according to their momentum, charge, and energy.
Acceptable quality range	The interval between specified upper and lower limits of a sequence of values within which the values are considered to be satisfactory.
Acceptable value	An observed or corrected value that falls within the acceptable range.
Accreditation	A formal recognition that a laboratory is competent to carry out specific tests or specific types of tests.

Handbook of GC-MS: Fundamentals and Applications, Fourth Edition. Hans-Joachim Hübschmann.
© 2025 WILEY-VCH GmbH. Published 2025 by WILEY-VCH GmbH.

Accuracy	The closeness of agreement between a test result and the accepted reference value, determined by determining trueness and precision.
Accurate mass	Experimentally determined mass of an ion. It can be used to determine an elemental formula. Note: Accurate mass and exact mass are not synonymous. The former refers to a measured mass and the latter to a calculated mass.
ACN	Acetonitrile.
Adiabatic ionization	Process whereby an electron is removed from an atom, an ion, or a molecule in its lowest energy state to produce an ion in its lowest energy state.
Adduct ion	Ion formed by the interaction of an ion with one or more atoms or molecules to form an ion containing all the constituent atoms of the precursor ion as well as the additional atoms from the associated atoms or molecules.
AED	Atomic emission detector.
AFS	Amperes full scale.
AGC	Automatic gain control controls the variable ionization or ion storage time in ion trap mass spectrometers depending from the total ion current in the selected mass range by a quick pre-scan. This control provides the high inherent full scan sensitivity of ion trap mass spectrometers at low ion streams by storing ions until the full trap capacity is reached.
Alumina	A gas–solid adsorbent stationary phase with GC columns.
Aliquot	A subsample derived by a divisor that divides a sample into a number of equal parts and leaves no remainder; a subsample resulting from such a division. In analytical chemistry, the term *aliquot* is generally used to define any representative portion of the sample, regardless of whether a remainder is left or not.
AMDIS	Automated mass spectra deconvolution and identification system, analyzes the individual ion signals and extracts and identifies the deconvoluted spectrum of each component in co-eluting peaks analyzed by GC-MS. Compound identification is provided via library search and retention index, → deconvolution, → retention index.

Glossary

Analogue ion	Ions that have similar chemical valence, for example, the acetyl cation CH_3CO^+ and the thioacetyl cation CH_3CS^+.
Analyte	The substance that has to be detected, identified, and/or quantified and derivatives emerging during its analysis.
Analytical scan	The part of the ion trap MS scan function that produces the mass spectrum.
Analyte spike recovery	Recovery of an analyte spike added to a sample prior to sample preparation. Determination of analyte recovery is based on results provided by spiked and unspiked samples. Used to estimate matrix effects and sample preparation losses.
APCI	Atmospheric pressure chemical ionization, ionization mode for GC-MS coupling and LC-MS. The chemical ionization takes place using a nebulized liquid and corona discharge at atmospheric pressure before ions enter the vacuum of the MS.
APE	Atom percent excess, commonly used expression in tracer experiments employing labeled substances for the degree of enrichment above the natural isotope content: $APE = at.\% - at.\%_{nat}$ Deprecated term, replace SI conform with "atom fraction" expressed as percent.
APGC	GC coupling method to an APCI interface, $\rightarrow$ APCI.
Appearance energy (AE)	Minimum energy that must be imparted to an atom or a molecule to produce a detectable amount of a specified ion. In mass spectrometry, it is the voltage, which corresponds to the minimum electron energy necessary for the production of a given fragment ion. The term *appearance potential* (AP) is deprecated.
ARC	Automatic reaction control, variable ionization, and reaction time control in internal ionization ion trap mass spectrometers for chemical ionization. A build-up of CI ions in the ion trap is facilitated by the AGC control until the maximum capacity or preset maximum reaction time is reached. This results in the inherent high sensitivity of the ion trap analyzer for data acquisition in the full scan mode.

Aromagram	A chromatogram representative for the odor intensity of the eluting compounds, → GC-olfactometry.
Array detector	Detector comprising several ion collection elements, arranged in a line or grid where each element is an individual detector.
Associative ion/molecule reaction	Reaction of an ion with a neutral species in which the reactants combine to form a single ion.
Associative ionization	Ionization process in which two atoms or molecules, one or both of which is in an excited state, react to form a single positive ion and an electron.
Atmospheric pressure chemical ionization	→ APCI
Atmospheric pressure ionization (API)	Ionization process in which ions are formed in the gas phase at atmospheric pressure before the generated ions enter the vacuum of the MS.
Atmospheric pressure photoionization (APPI)	Atmospheric pressure photo ionization in which the reactant ions are generated by photoionization before they enter the vacuum of the MS.
Atom% or at.%	Unit commonly used for the expression of isotope ratios, for example, tracer experiments. $$\text{at. \%} = \frac{n}{n_i} \cdot 100 = \frac{1}{(1+1/R)} \cdot 100\,(\%)$$ with n = number of isotope atoms; $n_i = R = {}^{13}C/{}^{12}C$.
Autodetachment	Formation of a neutral species when a negative ion in a discrete state with an energy greater than the detachment threshold loses an electron spontaneously without further interaction with an energy source.
Autoionization	Formation of an ion when an atom or molecule in a discrete state with an internal energy greater than the ionization threshold loses an electron spontaneously without further interaction with an energy source.
Average mass	Mass of an ion or a molecule calculated using the average mass of each element weighted for its natural isotopic abundance.

B

β	Phase ratio. The ratio of mobile to stationary-phase volumes. Thicker stationary-phase films yield longer retention times and higher peak capacities. For open-tubular columns, $\beta = V_G/V_L \sim d_c/4d_f$.
β-Cleavage	Homolytic cleavage where the bond fission occurs between at an atom removed from the apparent charge site atom by two bonds and an atom adjacent to that atom and removed from the apparent charge site by three bonds.
b-Ion	Fragment ion containing the N-terminus formed upon dissociation of a protonated peptide at a backbone C—N bond.
Backflush	Occurs when compounds (often high boiling matrix) in a → pre-column or at the end of a chromatogram are flushed from the column to vent or to another column by flow reversal.
Bake out	Generally, a thermal cleaning step in various applications: Gas chromatography – the process of removing contaminants from a column by operation at elevated temperatures, which should not exceed a column's maximum operating temperature. Purge and trap methodology – the purification step of the adsorption trap. Mass spectrometry – the cleaning of the analyzer from dissolved gases and contaminants by heating the steel manifold for an extended time.
Balanced pressure injection	Headspace injection technique whereby the equilibrated sample vial is depressurized into the GC column.
Band broadening	Several processes that cause solute profiles to broaden as they migrate through a column.
Base peak (BP)	The peak in a mass spectrum that has the greatest intensity.
Beam instruments	Type of mass spectrometers in which beams of ions are continuously formed from an ion source, passed through ion optics and are resolved for mass analysis, typically magnetic sector and quadrupole instruments.
Benzyl cleavage	A fragmentation reaction of alkylaromatics forming the benzylic carbenium ion, which appears as the tropylium ion with m/z 91 in many spectra of aromatics.

Bias	The difference between the mean measured value and the true, accepted reference value.
Blank	1. Material (a sample or a portion or extract of a sample) known not to contain detectable levels of the analyte(s) sought. Also known as a *matrix blank* 2. A complete analysis conducted using the solvents and reagents only; in the absence of any sample (water may be substituted for the sample, to make the analysis realistic). Also known as a *reagent blank* or *procedural blank*.
Blank value	Many analysis methods require the determination of blank values in order to be able to compensate for nonspecific analyte/matrix interaction. A differentiation is made between reagent blank samples and sample blank samples.
Bleed	The loss of material from a column or septum caused by high temperature operation. Bleed can result in ghost peaks, baseline offset and noise.
Blank sample	An analyte-free sample or sample of a standard matrix processed to measure artefacts in the measurement (sampling and analysis) process, providing the blank value.
Bonded phase	A stationary phase that is chemically bonded to the inner column wall, → also cross-linked phase.
Bracketing calibration	Organization of a batch of determinations such that the detection system is calibrated immediately before and after the analysis of the samples. For example, calibrant 1, calibrant 2, sample 1, sample n, calibrant 1, calibrant 2.
BSIA	Bulk sample isotope analysis, the isotope ratio MS analysis of a sample after bulk conversion in a crucible of an elemental analyzer, contrary to → CSIA.
BSTFA	Silylating agent for derivatization reactions, bis(trimethylsilyl)trifluoroacetamide.
BTEX	Abbreviation for the analysis of the benzene, toluene, ethylbenzene, and xylene isomer group of aromatics, mostly by headspace sample analysis.
BTV	Breakthrough volume, for example, of an adsorption trap.

BTX	→ BTEX.
Buffer gas	Inert gas used for collisional deactivation of internally excited ions or of the translational energies of ions confined in an ion trap.

C

CAD	Collision activated decomposition in MS/MS measurements, → CID.
Calibrant	→ Calibration standard.
Calibration	The set of operations that establish, under specified conditions, the relationship between values indicated by a measuring instrument or measuring system, and the corresponding known values. The result of a calibration is sometimes expressed as a calibration factor or as a series of calibration factors in the form of a calibration curve or function.
Calibration check standard	A standard independently prepared (different source, different analyst) from the calibration standards and run after the original calibration to verify the original calibration. There is usually at least one calibration check standard per batch.
Calibration curve	Defines the relation between analyte concentration and analytical response (function). Normally, at least three to five appropriately placed calibration standards are needed to adequately define the curve. The curve should incorporate a low standard not exceeding 10 times the detection limit. Analytical response, where appropriate, is zeroed using a reagent blank. Either a linear or other curve fit, as appropriate, may be used. Standards and samples must have equivalent reagent backgrounds (e.g. matrix, solvent, acid content, etc.) at the point of analysis.
Calibration drift	The difference between the instrument response and a reference value after a period of operation without recalibration.
Calibration standard	A solution (or dilution) of the analyte (and internal standard, if used) or reference material used to calibrate an instrument.
Capacity factor	The k' value of a column describes the molar ratio of a substance in the stationary phase to that in the mobile phase from the relationship of the net retention time to the dead volume.

Carbosieve	Carbon molecular sieve used as an adsorption material for air analysis, → also VOCARB (Supelco).
Carbowax	Trademark of polyethyleneglycols, →PEG.
Carboxen	Carbon molecular sieve used as an adsorption material for air analysis, → also VOCARB (Supelco).
Carry over	Taking the analyte to the next analysis, also known as memory effect.
CAS No.	The unique registration number of a chemical compound or substance assigned by the Chemical Abstract Service, a division of the American Chemical Society. The intention is to make database searches more convenient as chemicals often have many names. Almost all molecule databases today allow searching by the CAS number. The CAS number usually is given in the substance entries of many library mass spectra.
Cationized molecule	An ion formed by the association of a cation with a molecule M, for example, $[M + Na]^+$ and $[M + K]^+$. The terms *quasi-molecular ion* and *pseudo-molecular ion* are deprecated.
CCM	A scan-by-scan calibration correction method used, for example, in high resolution selected reaction monitoring, → H-SRM.
CDEM	→ Continuous dynode electron multiplier.
CE	Coating efficiency. A metric for evaluating column quality. The minimum theoretical plate height divided by the observed plate height; CE = H_{min}/H.
Centroid	The calculated center of a mass peak acquired in the scan mode. The centroid value can be calculated precisely independent of the resolution power of the mass spectrometer in use. Values displayed in the spectrum with three or more digits are often misleadingly associated with the resolving power of the instrument. In LRMS, special care has to be taken as the centroid value gives the center of gravity of the mass peak composed of many compounds falling in the same wide mass window. In HRMS, centroid mass values are used to calculate a possible sum formula within a deviation of < 2 ppm mass precision.

Certification	A formal recognition that a laboratory is competent to carry out specific tests or specific types of tests.
Certified reference material (CRM)	A reference material having one or more property values that are certified by a technically valid procedure, accompanied by or traceable to a certificate or other documentation that is issued by a certifying authority.
Charge exchange ionization	Type of CI reaction in PCI, interaction of an ion with an atom or a molecule in which the charge on the ion is transferred to the neutral without the dissociation of either charge → transfer ionization.
Charge transfer reaction	Type of CI reaction in NCI, action of an ion with a neutral species in which some or all of the charge of the reactant ion is transferred to the neutral species.
Chemical ionization	Formation of a new ion in the gas phase by the reaction of a neutral analyte species with an ion. The process may involve the transfer of an electron, a proton, or other charged species between the reactants.
	Note 1: When a positive ion results from chemical ionization, the term may be used without qualification. When a negative ion results, the term *negative ion chemical ionization* should be used.
	Note 2: This term is not synonymous with chemi-ionization.
	Through chemical ionization usually a soft ionization is achieved providing information on the (cationized-)molecular mass of a substance. The selectivity of the reaction and extent of fragmentation are controlled by the choice of the reagent gas.
Chemi-ionization	Reaction of an atom or a molecule with an internally excited atom or molecule to form an ion.
	Note that this term is not synonymous with chemical ionization.
CI	→ Chemical ionization.
CID	Collision induced dissociation, leads in MS/MS to the formation of a product ion spectrum from a selected precursor ion.
C-ion	Fragment ion containing the N-terminus formed upon dissociation of a protonated peptide at a backbone N—C bond.

Clean-up	Generally, the sample preparation procedure involving the removal of the matrix for isolation and often concentration of analytes.
Cluster ion	Ion formed by a multicomponent atomic or molecular assembly of one or more ions with atoms or molecules, such as $[(H_2O)_n H]^+$, $[(NaCl)_n Na]^+$, or $[(H_3PO_3)_n HPO_3]^-$.
Co-chromatography	A procedure in which the extract prior to the chromatographic step(s) is divided into two parts. Part one is chromatographed as such. Part two is mixed with the standard analyte that is to be measured and chromatographed. The amount of added standard analyte has to be similar to the estimated amount of the analyte in the extract. This method is designed to improve the identification of an analyte, especially when no suitable internal standard can be used, → also standard addition method.
Cold injection	A GC injection that occurs at injector temperatures lower than the oven temperature, usually at or below the solvent boiling point. Requires a temperature programmable vaporizer, → PTV.
Cold trapping	A chromatographic technique for focusing volatile compounds at the beginning of the GC column by cooling a section of the column, for example, with ice or liquid CO_2 below the boiling point of the compounds or solvent, → also cryofocusing.
Collision gas	Inert gas used for collisional excitation in MS/MS measurements and ion/molecule reactions. Typically argon or nitrogen are used. The term *target gas* is deprecated, → CID.
Collision-induced dissociation (CID)	Dissociation of an ion after collisional excitation. The term *collisionally activated dissociation* (CAD) is deprecated, → CID.
Collision quadrupole	Transmission quadrupole to which an oscillating radio frequency potential is applied so as to focus a beam of ions through a collision gas with no m/z separation.
	Note: a collision quadrupole is often indicated by a lower case q as in QqQ denoting a quadrupole MS/MS instrument.
Collisional activation (CA)	→ Collisional excitation.

Collisional excitation	Reaction of an ion with a neutral species in which all or part of the translational energy of the collision is converted into internal energy of the ion, for example, for the → CID process in MS/MS.
Collision cell	A transmission hexapole, octapole, or square quadrupole rod device to which an oscillating radio frequency potential is applied that is filled with a collision gas at low pressure and used to generate collision induced dissociation (→ CID) of ions to form a product ion spectrum. The collision cell has no mass separating capabilities. Axial fields accelerate product ions to leave the collision cell for a fast switch of → SRM/MRM reactions.
Comminution	The process of reducing a solid sample to small fragments before extraction, e.g. by cryo milling.
Comprehensive GC (GC × GC)	Two-dimensional GC technique in which all compounds experience the selectivity of two columns connected in series by a modulation device, thereby generating a higher and orthogonal separation than that attainable with a single column.
Compressibility correction factor (j)	This factor compensates for the expansion of a carrier gas as it moves along the column from the entrance, at the inlet pressure (p_i), to the column exit, at the outlet pressure (p_o).
Concurrent solvent recondensation (CSR)	Large volume GC injection technique with splitless injectors; due to recondensation of solvent inside a → pre-column (retention gap). The resulting pressure difference caused by the volume contraction due to the recondensation in the pre-column speeds up significantly the sample vapor transfer from the injector allowing larger sample volumes to be injected in standard inlet liners with glass wool. Typical injection volumes range up to 10 µL without and up to 50 µL with a pre-column.
Confidence interval	The set of possible values within which the true value will reside with a stated probability (i.e. confidence level).
Confidence level	The probability, usually expressed as a percentage, that a confidence interval will include a specific population parameter; confidence levels usually range from 90% to 99%.

Confidence limit	Upper or lower boundary values delineating the confidence interval.
Confirmation	Confirmation is the combination of two or more analyses that are in agreement with each other (ideally, using methods of orthogonal selectivity), at least one of which meets identification criteria. It is impossible to confirm the complete absence of residues. Adoption of a "reporting limit" at the → LCL avoids the unjustifiably high cost of confirming the presence or absence, of residues at unnecessarily low levels. The nature and extent of confirmation required for a positive result depends upon importance of the result and the frequency with which similar residues are found. Assays based on classical detectors tend to demand confirmation because of their limited selectivity. Mass spectrometric techniques are often the most practical and least equivocal approach to confirmation. AQC procedures for confirmation should be rigorous.
Confirmatory method	A method that provides full or complementary information enabling the substance to be unequivocally identified and if necessary quantified at the level of interest.
Consecutive reaction monitoring (CRM)	MS^n experiment with three or more stages of m/z separation and in which a particular multistep reaction path is monitored, typical with ion trap mass analyzers.
Constant neutral loss scan	A scan procedure for MS/MS designed to monitor a selected neutral loss mass difference from precursor ions by detection of the corresponding product ions produced by metastable ion fragmentation or collision-induced dissociation. Synonymous terms are constant neutral mass loss scan and fixed neutral fragment scan.
Constant neutral loss spectrum	Spectrum of all precursor ions that have undergone an operator-selected m/z decrement, obtained using a constant neutral loss scan; → Constant neutral mass loss spectrum and fixed neutral mass loss spectrum.
Constant neutral mass gain scan	Scan procedure for a tandem mass spectrometer designed to produce a constant neutral mass gain spectrum of different precursor ions by detection of the corresponding product ions of ion/molecule reactions with a gas in a collision cell.

Constant neutral mass gain spectrum	Spectrum formed of all product ions produced by gain of a pre-selected neutral mass following ion/molecule reactions with the gas in a collision cell, obtained using a constant neutral mass gain scan.
Continuous dynode electron multiplier	An ion-to-electron detector in which the ion strikes the inner continuous resistance surface of the device and induces the production of secondary electrons that in turn impinge on the inner surfaces to produce more secondary electrons. This avalanche effect produces an increase in signal in the final measured current pulse, → also SEM.
Continuous injection	Process for the production of defined atmospheres for the calibration of thermodesorption tubes.
Control chart	A graph of measurements plotted over time or sequence of sampling, together with control limit(s) and, usually, a central line. Control charts may be used to monitor ongoing performance as assessed by method blanks, verification standards, control standards, spike recoveries, duplicates, and references samples.
Control limits	Specified boundaries on a control chart that, if exceeded, indicates a process is out of statistical control, and the process must be stopped, and corrective action taken before proceeding. These limits may be defined statistically or based on protocol requirements. Control limits may be assigned to method blanks, verification/control standards, spiked recoveries, duplicates, and reference samples.
Control sample	A sample with predetermined characteristics which undergoes sample processing identical to that carried out for test samples and that is used as a basis for comparison with test samples. Examples of control samples include reference materials, spiked test samples, and method blanks.
Conversion dynode	Surface that is held at high potential so that ions striking the surface produce electrons that are subsequently detected, → Post-acceleration detector.
CRM	Certified reference material, means a material that has a specified and documented analyte content assigned to it.

Glossary

Cross-linked phase	A stationary phase that uses chemically cross-linked polymer chains. The phase is not bonded to the column inner wall, → also bonded phase.
Cross-talk	MS/MS signal artifacts in an SRM transition from the previous SRM scan can potentially occur when fragment ions from one SRM transition remain in the collision cell, while a second SRM transition takes place, which can be the source of false positives when different SRM events have the same product ions formed from different precursor ions. Linear acceleration of the product ions inside of the collision cell prevents cross-talk effects.
Cryofocusing	Capillary GC injection technique for volatile compounds in headspace, purge and trap, or thermodesorption systems. Instead of an evaporation injector, online coupling to the sample injection is set up in such a way that a defined region of the column (or within an injector) is cooled by liquid CO_2 or liquid N_2 to focus a volatile sample. The chromatography starts with heating up of the focusing region.
CSIA	Compound specific isotope analysis, the isotope ratio MS analysis of individual compounds after chromatographic separation followed by the online conversion to simple gases, contrary to a → BSIA.
CSR	→ Concurrent solvent recondensation.
Cyclotron motion	Circular motion of a particle of charge q moving at velocity v in a magnetic flux density B that results from the Lorentz force $qv \times B$.

D

2DGC	Two-dimensional GC, abbreviation used for describing heart cutting as well as comprehensive GC×GC methods. → Comprehensive GC.
δ notation	Notation in units per mil (‰), commonly used in isotope ratio mass spectrometry to express isotope abundance differences as a ‰ deviation between the sample against an international standard, for example, for ^{13}C $$\delta^{13}C = \left[\left(\frac{R_{Sample}}{R_{Standard}}\right) - 1\right] \times 10^3 [‰]$$ with $R = {}^{13}C/{}^{12}C$ 0.01 equals 10^{-5} at.% (→ also atom%). The international standards are assigned a value of 0.0 ‰ on their respective δ scales.

d_c	Average column inner diameter.
d_f	Average stationary-phase film thickness.
D_G	Gaseous diffusion coefficient; for instance ~0.05 for hydrocarbons in helium carrier gas and 0.1 in hydrogen carrier gas.
D_L	Liquid–liquid diffusion coefficient; ~1×10^{-5} for hydrocarbons in silicones.
d_p	Average particle diameter.
Dalton, Da	Non-SI unit of mass (symbol Da) that is identical to the unified atomic mass unit, based since 1961 on the mass definition of ^{12}C = 12.00000 g/mol.
Daly detector	Detector consisting of a conversion dynode, scintillator and photomultiplier. A metal knob at high potential emits secondary electrons when ions impinge on the surface. The secondary electrons are accelerated onto the scintillator that produces light that is then detected by the photomultiplier detector.
DC	Direct current.
DEA	The U.S. Drug Enforcement Agency.
Dead time	The time taken for a substance that is not retarded in a GC column to reach the detector, for example, air, methane for silicone phases, to pass through a chromatographic column. The carrier gas velocity is calculated from the dead time and the column length, → also HETP.
Dead volume	Extra volume experienced by analytes as they pass through a chromatographic system. Excessive dead volume causes additional peak broadening.
Decision limit	The limit at and above which it can be concluded with an error probability that a sample is non-compliant.
Deconvolution	Coeluting compounds are analyzed based on the concept of common → maximizing mass signals at identical retention times. The process is used e.g. in → AMDIS for extracting corrected mass spectra from complex elution situations.
DEGS	Diethylene glycol succinate; used as a stationary phase.

Glossary

Delayed extraction (DE)	Application of the accelerating voltage pulse after a time delay in desorption/ionization from a surface. The extraction delay can produce energy focusing in a time-of-flight mass spectrometer.
Desorb preheat	Purge and trap technique step for heating up the adsorption trap. The effective desorption and transfer of analytes from the adsorption trap into the GC-MS system is effected by switching a six-way valve first after a set preheat temperature of approx. 5 °C below of the desorption temperature has been reached.
Desorption ionization (DI)	Formation of ions from a solid or liquid material by the rapid vaporization of that sample in the ion source.
Detection capability	The smallest content of the substance that may be detected, identified, and/or quantified in a sample with an error probability of β.
Detection limit	It describes the smallest detectable signal of the analyte that can be clearly differentiated from a blank sample (assessed in the signal domain, detection criterion). It is calculated as the upper limit of the distribution range of the blank value.
Diagnostic ion	MS term for ions that are highly characteristic for the compound measured, especially for ions whose formation reveals structural or compositional information. For instance, the phenyl cation in an electron ionization mass spectrum is a diagnostic ion for benzene and derivatives. Used typically, for extracted ion chromatograms, → XIC.
DI-SDME	Direct-immersion single drop microextraction, → SDME.
DI-SPME	Direct immersion solid phase microextraction, the SPME device is exposed directly into an aqueous (liquid) sample for collection of analytes.
Dimeric ion	An ion formed by ionization of a dimer or by the association of an ion with its neutral counterpart such as $[M_2]^{+\bullet}$ or $[M-H+M]^+$.
Direct injection	The sample enters an inlet and is transferred in its entirety into a column by carrier gas flow. No sample splitting or venting occurs during or after the injection.

Direct insertion probe	Device for introducing a single sample of a solid or liquid, usually contained in a quartz or other nonreactive sample holder (e.g. crucible), into a mass spectrometer ion source.
Dissociative ionization	Reaction of a gas-phase molecule in chemical ionization that results in its decomposition to form products, one of which is an ion.
Distonic ion	Radical cation or anion arising formally by ionization of diradicals or zwitterionic molecules (including ylides). In these ions, the charge site and the unpaired electron spin cannot be both formally located in the same atom or group of atoms as it can be with a conventional ion.
dl-PCB	Dioxin-like PCBs with WHO TEF values, → PCB.
DLLME	Dispersive liquid–liquid microextraction. Extraction technique using the rapid injection of the extraction solvent mixed with a cosolvent to disperse the extraction phase in fine droplets in an aqueous sample. Centrifugation separates the immiscible phases and the extraction solvent is taken for analysis.
DMCS	Dimethylchlorosilane; used for silanizing glass GC parts.
DOD	The U.S. Department of Defense.
Double-focusing mass spectrometer	Mass spectrometer that incorporates a magnetic and an electric sector connected in series in such a way that ions with the same m/z but with distributions in both the direction and the translational energy of their motion are brought to a focus point.
Dry purge	Purge and trap analysis step whereby moisture is removed from the trap by the carrier gas before desorption.
Duty cycle	Degree of the effective ion acquisition time of, for example, a mass spectrometer. Here, the duty cycle is determined by the sum of the ion dwell times relative to the total scan cycle time including jump and stabilization times of the analyzer voltages.
Dwell time	Effective ion acquisition time typically in milliseconds, maximized by using a selected ion monitoring scan (→ SIM, MID, SRM).

E

ECD
Electron-capture detection, a detector ionizes analytes by collision with metastable carrier gas molecules produced by ß-emission from a radioactive source such as ^{63}Ni. The electron capture detector is one of the most sensitive detectors, and it responds strongly to halogenated analytes and others with high electron-capture cross-sections (→ also electron capture dissociation). Relative ECD-response to hydrocarbons:

10^1	Esters, ethers.
10^2	Monochlorides, alcohols, ketones, amines.
10^3	Dichlorides, monobromides.
10^4	Trichlorides, anhydrides.
10^5–10^6	Polyhalogenated, mono-, diiodides.

Eddy diffusion
Multipath effect mainly in packed and PLOT column chromatography, the root cause of peak broadening through diffusion processes.

Effective plates
The number of effective theoretical plates in a column, taking the dead volume into consideration (→ also HETP).

Efficiency
The ability of a column to produce sharp, well-defined peaks. More-efficient columns have more theoretical plates (N) and smaller theoretical plate heights (H).

EI
Electron ionization with modern GC-MS instrument ionization usually takes place at an ionization energy of 70 eV. Positive ions are formed predominantly.

EIC
Extracted ion chromatogram. The display of selected masses on screen as a mass chromatogram, →XIC.

Einzel lens
Three element ion lens in which the first and third elements are held at the same voltage. Such a lens produces focusing without changing the translational energy of the particle.

ELCD
Electrolytic-conductivity detector, also Hall detector, gives a mass-flow dependent signal. The detector catalytically reacts halogen-containing analytes with hydrogen (reductive mode) to produce strong acid by-products that are dissolved in a working fluid. The acids dissociate and the detector measures increased electrolytic conductivity. Other operating modes modify the chemistry for response to nitrogen- or sulfur-containing substances.

Electrical sector	→ Double-focusing mass spectrometer.
Electron affinity, E_{EA}	Electron affinity of a species M is the minimum energy required for the process $M^{-\bullet} \to M + e^-$ where $M^{-\bullet}$ and M are in their ground rotational, vibrational, and electronic states and the electron has zero translational energy.
Electron attachment ionization	Ionization of a gaseous atom or molecule by attachment of an electron to form $M^{-\bullet}$ ions.
Electron capture	The capture of thermal electrons by electronegative compounds. This process forms the basis of the → ECD (electron capture detector) and is also made use of in negative chemical ionization (→ NCI, ECD-MS).
Electron capture dissociation (ECD)	Process in which multiply protonated molecules interact with low energy electrons. Capture of the electron leads the liberation of energy and a reduction in charge state of the ion with the production of the $[M + nH]^{(n-1)+}$ odd electron ion, which readily fragments.
Electron energy	Magnitude of the electron charge multiplied by the potential difference through which electrons are accelerated, for example, from a filament in order to effect → electron ionization.
Electron impact ionization	→ Electron ionization.
Electron ionization	Ionization of an atom or molecule by electrons that are typically accelerated to energies between 10 and 150 eV in order to remove one or more electrons from the molecule. The term *electron impact* is deprecated.
Electron volt, eV	Non-SI unit of energy (symbol eV) defined as the energy acquired by a particle containing one unit of charge through a potential difference of 1V. An electron volt is equal to $1.60217733(49) \times 10^{-19}$ J.
Electrospray ionization (ESI)	A process for LC-MS in which ionized species in the gas phase are produced from a solution via highly charged fine droplets, by means of spraying the solution from a narrow-bore needle tip at atmospheric pressure in the presence of a high electric field (1000–10 000 V potential). The term *ion spray* is deprecated.
Electrostatic energy analyzer (ESA)	A device consisting of conducting parallel plates, concentric cylinders or concentric spheres that separates charged particles according to their ratio of translational energy to charge by means of a voltage difference applied between the pair.

Elution temperature	The temperature of the GC oven at which an analyte reaches the detector.
Energy	The SI unit is the Joule (J) $1\,J = 1\,Nm = 1\,Ws$ In mass spectrometry the energy of ions is usually given in (eV) and is calculated from the elemental charge and the acceleration voltage in the ion source. Conversion factors: $1\,eV = 23.0\,kcal = 96.14\,kJ$ (with $1\,cal = 4.18\,J$).
$E/2$ mass spectrum	Mass spectrum obtained using a sector mass spectrometer in which the electric sector field E is set to half the value required to transmit the main ion-beam. This spectrum records the signal from doubly charged ions.
EPA	The Environmental Protection Agency.
ESI	Electrospray ionization, an LC-MS coupling technique, → Electrospray ionization.
ESTD	External standard, quantitation by using external standardization. The analyte itself is used for quantitative calibration as a clean standard or added to a blank standard matrix. The signal height for a known concentration of the analyte is used for the calibration procedure. The calibration runs are carried out separately (externally) from the analysis of the sample.
Even-electron ion	An ion containing no unpaired electrons in its ground electronic state.
EVOO	Extra virgin olive oil, often subject to adulteration, analyzed e.g. by LC-GC coupling.
Exact mass	Calculated mass of an ion or molecular formula containing a distinctive number of isotopes of each atom. In MS mass of the lightest isotope of each element is used to calculate e.g. the molecular mass M of a compound.
External ionization	Process for the production of ions for an ion storage mass spectrometer, for example, ion trap MS. The ionization takes place in an attached ion source and transferred to the analyzer.
Extracted ion chromatogram	Chromatogram display created by plotting the intensity of the signal of a selected m/z value or a range of masses as a function of retention time, → Reconstructed ion chromatogram (RIC).

F

F_a	The column outlet flow-rate corrected to room temperature and pressure; for example, the flow-rate as measured by a flow meter. F_a can be calculated from the average carrier gas linear velocity and the column dimensions.
F_s	The split-vent flow-rate, measured at room temperature and pressure.
False negative	A result wrongly indicating that the analyte concentration does <u>not</u> exceed a specified value.
False positive	A result wrongly indicating that the analyte concentration <u>does</u> exceed a specified value.
FAME	Fatty acid methyl ester.
Faraday cup	A conducting cup or chamber that intercepts a charged particle beam and is electrically connected to a current measuring device.
FC43	Perfluorotributylamine ($C_{12}F_{27}N$, PFTBA), a widely used reference substance for calibration of the mass scale, M 671.
FFAP	Free fatty-acid phase, a polar stationary phase for GC columns.
FID	Flame ionization detector, providing a mass flow dependent signal, the detector ionizes most classes of organic compounds. FID is a universal detection technique. Little or no response have noble gases, CO, CO_2, O_2, N_2, H_2O, CS_2, NO_x, NH_3, perhalogenated compounds, formic acid/aldehyde.
Field desorption (FD)	Formation of gas-phase ions in the presence of a high electric field from a material deposited on a solid surface. The term *field desorption/ionization* is deprecated.
Field-free region (FFR)	Section of a mass spectrometer in which there are no electric or magnetic fields.
Field ionization (FI)	Removal of electrons from any species, usually in the gas phase, by interaction with a high electric field.
Fixed product ion scan	In a sector instrument, either a high voltage scan or a linked scan at constant B^2/E. Both give a spectrum of all precursor ions that fragment to yield a pre-selected product ion. Note: The term *daughter ion* is deprecated.
Fortified sample	A sample enriched with a spike of a known amount of the analyte to be detected.

Forward library search	A procedure of comparing a mass spectrum of an unknown compound with a mass spectral library so that the unknown spectrum is compared with the library spectra considering only the mass signals and intensities of the unknown spectrum. Mass signals not belonging to the compound of interest, e.g. matrix, column bleed or coeluting compounds lead to lower match values, and require background subtraction, → reversed library search.
Fourier transform ion cyclotron resonance mass spectrometer (FT-ICR)	A mass spectrometer based on the principle of ion cyclotron resonance in which an ion in a magnetic field moves in a circular orbit at a frequency characteristic of its m/z value. Ions are coherently excited to a larger radius orbit using a pulse of radio frequency energy and their image charge is detected on receiver plates as a time domain signal. Fourier transformation of the time domain signal results in a frequency domain signal which is converted to a mass spectrum based in the inverse relationship between frequency and m/z. A characteristic of FT-ICR is the very high mass resolution power ($R > 10^6$) and mass precision (< 1 ppm).
FPD	Flame photometric detector, providing a mass flow dependent signal. The detector burns a hydrogen-rich flame where analytes are reduced and excited. Upon decay of the excited species light is emitted of characteristic wavelengths. The visible-range atomic emission spectrum is filtered through an interference filter and detected with a photomultiplier tube. Different interference filters can be selected for sulfur, tin or phosphorus emission lines. The flame photometric detector is sensitive and selective.
Fragment ion	A product ion that results from the dissociation of a precursor ion during ionization or MS/MS measurements. Note: The term *daughter ion* is deprecated.
Fringe field	Electric or magnetic field that extends from the edge of a sector, lens or other ion optics element.
Frit sparger	U-shaped tube for the purge and trap analysis of water samples with built-in frit for fine dispersion of the purge gas, mandatory for U.S. EPA VOC methods.
Fritless sparger	U-shaped tube without a frit for the purge and trap analysis of moderately foaming water or solid samples.

FS	Fused silica.
FSOT	Fused-silica open-tubular column.
Full scan	Acquisition mode for the recording of complete mass spectra over a specified mass range.
FWHM	*Full width at half peak maximum*, term used in the definition of the mass peak resolution for the measurement of the peak width at half peak height, → Mass resolving power.

G

GALP	Good automated laboratory practice.
Gas isotope ratio mass spectrometry	Common name for the area of isotope ratio mass spectrometry for the determination of the stable isotopes of H, N, C, O, S and Si. Compounds containing these elements can be quantitatively converted into simple gases for mass spectrometric isotope ratio analysis, that is, H_2, N_2, CO, CO_2, O_2, SO_2, SiF_4 fed by viscous flow or entrained into a continuous He flow into the ion source of a dedicated isotope ratio mass spectrometer.
Gauge	The measure to specify the outside diameter of needles and tube products, abbreviation G. Typical GC injection syringes are of 23G (0.642 mm/0.0225″) or 26G (0.464 mm/0.0185″) needle diameters.
GCB	Graphitized carbon black, → VOCARB (Supelco).
GC-O	Abbreviation used for → GC-olfactometry.
GC-olfactometry	Refers to the use of the human nose as a sensitive and selective detector for odor-active compounds. The aim of this technique is to determine the odor activity of volatile compounds in a sample extract, and assign a relative importance to each eluting compound.
GC×GC	→ Comprehensive GC.
Ghost peaks	Peaks from compounds not present in the original sample. Ghost peaks can be caused by septum bleed, analyte decomposition or carrier gas contamination.
GIRMS	→ Gas isotope ratio mass spectrometry.
Glass cap cross	→ Werkhoff splitter.

GLC	Gas–liquid chromatography, using this technique analytes partition between a gaseous mobile phase and a liquid stationary phase. The partition of the analytes between the stationary phase and the mobile phase (carrier gas) cause different retention times in the column. Most current GC stationary phases are immobilized liquid phases at the specified operation temperature of the column.
GLP	Good laboratory practice.
GLPC	Gas–liquid phase chromatography, → GLC; gas–liquid chromatography.
GPC	Gel permeation chromatography, gel chromatography, *aka* size exclusion chromatography (SEC), is a liquid chromatography used for sample preparation, for example, to remove lipid matrix components in pesticide analysis clean-up online or offline to GC-MS.
Gridless reflectron	A reflectron in which ions do not pass through grids in their deceleration and turnaround thereby avoiding ion loss due to collisions with the grid.
GSC	Gas–solid chromatography, this technique, analytes partition between a gaseous mobile phase and a solid stationary phase (→ PLOT columns). Selective interactions between the analytes and the solid phase cause different retention times in the column.
Guard column	Pre-column, → retention gap.

H

H	Height equivalent to one theoretical plate. The distance along the column occupied by one theoretical plate, $H = L/N$.
H_{meas}	Height equivalent to one theoretical plate as measured from a chromatogram. $$H_{meas} = \frac{L}{5.54 \left(\frac{t_R}{W_h}\right)^2}$$
H_{min}	Minimum theoretical plate height at the optimum linear velocity, ignoring stationary-phase contributions to band broadening. For open-tubular columns: $$H_{min} = \left(\frac{d_c}{2}\right) \sqrt{\frac{1+6k+11k^2}{3(1+k)^2}}$$
h_p	Peak amplitude.

H_{theor}	Theoretical plate height. For open-tubular columns (Golay equation): $H_{theor} = \left(\frac{2D_G}{\bar{u}}\right) + \bar{u}\left\{\left[\frac{(1+6k+11k^2)}{96(1+k)^2}\right]\left(\frac{d_c^2}{D_G}\right) + \left[\frac{2k}{3(1+k)^2}\right]\left(\frac{d_f^2}{D_L}\right)\right\}$
Hall detector	→ ELCD.
Hard ionization	Formation of gas-phase ions accompanied by extensive fragmentation typically → EI ionization is considered a hard ionization technique.
Headspace GC	Synonymous for the static headspace GC.
Headspace sweep	Technique in purge and trap analysis for treating foaming samples whereby the purge gas is passed only over the surface of the sample instead of through it.
Heartcut	Multidimensional GC-GC technique in which two or more partially resolved peak regions that are eluted from one column are directed onto another column of different polarity for improved separation.
Heterolysis	→ Heterolytic cleavage.
Heterolytic cleavage	Fragmentation of a molecule or ion in which both electrons forming the single bond that is broken remain on one of the atoms that were originally bonded, → Heterolysis.
He-PDPID	Helium pulsed discharge photoionization detector, → PDD.
HETP	Height equivalent to one theoretical plate; discontinued term for plate height (H). The dependence of the plate high value on the carrier gas velocity determined from the van Deemter curve is used to optimize chromatographic separation.
HFBA	Heptafluorobutyric anhydride, derivatization agent for preparing volatile heptafluorobutyrates. It is used for introducing halogens into compounds to increase the response in ECD or NCI detection.
High energy collision-induced dissociation	Collision-induced dissociation process wherein the projectile ion has translational energy higher than 1 keV.
High resolution MS	→ HRMS.
HLB	Hydrophilic–lipophilic balanced sorbent material used in →SPE and →SPME, a divinylbenzene N-vinylpyrrolidone copolymer.

HLB-WCX	HLB with weak cationic exchange sorbent.
HLB-WAX	HLB with weak anionic exchange sorbent.
Homolysis	→ Homolytic cleavage.
Homolytic cleavage	Fragmentation of an ion or molecule in which the electrons forming the single bond that is broken are shared between the two atoms that were originally bonded. For an odd electron ion, fragmentation results from one of a pair of electrons that form a bond between two atoms moving to form a pair with the odd electron on the atom at the apparent charge site. Fragmentation results in the formation of an even electron ion and a radical. This reaction involves the movement of a single electron and is represented by a single barbed arrow, → Homolysis.
HRGC	High resolution gas chromatography, the inherent meaning of this term is capillary gas chromatography using fused silica columns in contrast to packed column chromatography.
HRMS	High resolution mass spectrometry. The empirical formula of an ion is usually obtained through accurate mass determination as part of the structure elucidation. Meaningful accurate mass determinations require high mass resolution to separate close isobaric mass signals. In GC-MS target compound analysis, this term demands for a resolution power of better than 10 000 at 10% valley providing the accurate ion mass (originating from the U.S. EPA method 1613). Mass spectrometers providing resolution power of $R > 10\,000$ (at 5% peak height) or $R > 20\,000$ (at FWHP) are generally termed high mass resolution instruments.
HS	Headspace sampling, gas-phase sampling technique in which the sample is taken from an enclosed space above a solid or liquid sample after attaining equilibrium conditions.
HSGC	Headspace gas chromatography.
HSSE	Headspace sorptive extraction. Describes the →SPME or →SBSE sampling of the headspace above a liquid or solid sample, also known as →HS-SPME.
HS-SDME	Headspace single drop microextraction, →SDME.

H-SRM	Highly resolved selected reaction monitoring, MS/MS target compound scan technique using enhanced mass resolution at Q1 of a triple quadrupole analyzer for increased analyte selectivity.
HS-SPME	Headspace solid phase micro extraction, the SPME device is exposed to the headspace of a sample for collection of analytes.
HxCDDs	Hexachlorodibenzodioxin isomers.
Hybrid mass spectrometer	A mass spectrometer that combines *mass analyzers* of different types to perform tandem mass spectrometry.
Hydrogen/deuterium exchange	Exchange of hydrogen atoms with deuterium atoms in a molecule or preformed ion in solution prior to introduction into a mass spectrometer, or by reaction of an ion with a deuterated collision gas inside a mass spectrometer.

I

ICR-MS	Ion cyclotron resonance mass spectrometer, an ion storage mass spectrometer providing very high resolution power $R > 10^6$ and accurate mass measurements by measuring the cycle frequency of ions in a strong magnetic field followed by Fourier transformation.
IDL	Instrument detection limit, defined statistically as a measure for the potential instrument sensitivity at the 99% confidence level from the area precision (RSD%) of a series of measurements, applied in cases a → S/N cannot be calculated due to the absence of a suitable noise band in modern digitally filtered MS and MS/MS systems. The IDL does not inform about the lowest real life detectable concentration → LOD.
IL	Ionic liquid. A salt of ions in liquid state at ambient conditions. Due to its favorable properties of high chemical stability and low volatility ILs found multiple applications also for analytical purposes, e.g. → SPME.
INCOS	*Integrated computer systems*, the abbreviation refers to a former data system company and was used synonymously for the mass spectra library search in former Finnigan GC-MS data systems.

Indicator PCBs	Mixtures of PCBs are generally assessed on the basis of the analysis of the sum of the seven indicator PCBs #28, 52, 101, 118, 138, 153, and 180, → PCB.
Inductive cleavage	A heterolytic cleavage of an ion. For an odd electron ion, inductive cleavage results from the pair of electrons that forms a bond to the atom at the apparent charge site moving to that atom while the charge site moves to the adjacent atom. The movement of the electron pair is represented by a double-barbed arrow.
In-house validation	→ Single laboratory study.
In-source collision-induced dissociation	The dissociation of an ion as a result of collisional excitation during ion transfer from an atmospheric pressure ion source and the mass spectrometer vacuum. This process is similar to ion desolvation but uses higher collision energy.
Interference	A positive or negative response produced by a compound(s) other than the analyte, contributing to the response measured for the analyte or making integration of the analyte response less certain or accurate. Interference is also loosely referred to as *chemical noise* (as distinct from electronic noise, "flame noise", etc.). → Matrix effects are a subtle form of interference. Some forms of interference may be minimized by greater selectivity of the detector using HRMS or GC-MS/MS. If the interference cannot be eliminated or compensated, its effects may be acceptable if there is no significant impact on accuracy.
Interlaboratory study	The organization, performance and evaluation of tests on the same sample by two or more laboratories in accordance with predetermined conditions to determine testing performance. According to the purpose the study can be classified as collaborative study or proficiency study.
Internal Standard (ISTD)	A compound for quantitative reference not contained in the sample with chemical characteristics similar to those of the analyte. Isotopically labeled analogues of the target analytes are typically applied, for example, in dioxin analysis. The ISTD provides an analytical response that is distinct from the analyte and not subject to interference. Internal standards are used to adjust for variations in the analytical response due to

– matrix effects,
– sample preparation losses,
– final sample volume,
– variable injection volumes, or
– instrumental effects.

One or more ISTDs are added at different stages of the sample preparation process for instance to the sample immediately prior to sample preparation (→ surrogate standard) and/or the extract vial before injection (→ recovery standard). The ratio between surrogate and recovery standard is used to calculate the analyte recovery.

Ion	An atomic, molecular or radical species with an unbalanced electrical charge. The corresponding neutral species need not be stable.
Ion desolvation	The removal of solvent molecules clustered around a gas-phase ion by means of heating and/or collisions with gas molecules.
Ionic dissociation	The dissociation of an ion into another ion of lower mass and one or more neutral species or ions with a lower charge.
Ion injection	The transfer of ions formed in an external ion source to the analyzer of an ion storage mass spectrometer for mass analysis, e.g. the Orbitrap analyzer.
Ion/ion reaction	The reaction between two ions, typically of opposite polarity. The term *ion/ion reaction* is deprecated.
Ion mobility spectrometry (IMS)	Separation of ions according to their drift velocity through a buffer gas under the influence of an electric field.
Ion/molecule reaction	Reaction of an ion with a molecule. Note: the term *ion-molecule reaction* is deprecated because the hyphen suggests a single species that is both an ion and a molecule.
Ion/neutral complex	A particular type of transition state that lies between precursor and product ions on the reaction coordinate of some ion reactions.
Ion/neutral reaction	Reaction of an ion with a neutral atom or molecule.
Ion/neutral exchange reaction	Reaction of an ion with a neutral species to produce a different neutral species as the product.
Ion-pair formation	Reaction of a molecule to form both positive ion and negative ion fragments among the products.
Ion source	The section of the mass spectrometer where ions are produced.

Ion storage MS	Mass spectrometer equipped with internal or external ion generation and ion collection. The collection and analysis of ions take place discontinuously, for example, in quadrupole ion traps (LRMS) and ICR or → Orbitrap MS (HRMS).
Ion-to-photon detector	Detector in which ions strike a conversion dynode to produce electrons that in turn strike a phosphor layer and the resulting photons are detected by a photomultiplier, → Daly detector.
Ion trap (IT)	Device for spatially confining ions using electric and magnetic fields alone or in combination.
Ionization cross section	A measure of the probability that a given ionization process will occur when an atom or molecule interacts with a photon, electron, atom or molecule.
Ionization efficiency	Ratio of the number of ions formed to the number of molecules consumed in the ion source.
Ionizing collision	Reaction of an ion with a neutral species in which one or more electrons are removed from either the ion or neutral.
IRMS	→ Isotope ratio mass spectrometry.
Isomers	Two substances with the same molecular sum formula, same M, but different structural formula, or different spatial arrangement of the atoms (stereoisomers). Isomers differ chemically and physically. Mass spectra of isomers can often not be differentiated by library search, but due to different interactions with the stationary phase, they can be chromatographically separated. Also → IMS is capable of separating isomers.
Isotope	Although having the same nuclear charge (number of protons) most of the elements exist as atoms that have nuclides with varying numbers of neutrons, known as *isotopes*, from the Greek "*isos*" equal and "*tópos*" place, in the periodic table. They belong to the same chemical element but have different physical behavior, for example, masses. While in chemical synthesis the natural isotope composition is not taken into consideration (use of the average atomic mass), in mass spectrometry the distribution of the isotopes over the different masses is visible and the isotope pattern is assessed, for example, in dioxin analysis (molecular mass calculation based on the principle chlorine isotope). → IRMS determines isotope ratios as a result of fractionation processes highly precise.

Isotope dilution mass spectrometry	A quantitative mass spectrometry technique based on the measurement of the isotopic abundance of a nuclide after isotope dilution with the test portion. An isotopically enriched compound, for instance a deuterated or ^{13}C-labeled compound, is used as an internal standard.
Isotope effect	Alteration of either the equilibrium constant or the rate constant of a reaction if an atom in a reactant molecule is replaced by one of its isotopes, distinguished are kinetic isotope effect, equilibrium isotope effect, primary isotope effect, secondary isotope effect.
Isotope pattern	Characteristic intensity pattern in a mass spectrum derived from the different abundance of the isotopes of an element. From the isotope pattern of a signal in the mass spectrum conclusions can be drawn on the number of atoms of this element in the elemental formula. Important isotope patterns in organic analysis are shown by Cl, Br, Si, C and S. Organometallic compounds show characteristic patterns rich in lines. Molecular ions show the combined isotope pattern of all the elements of the chemical sum formula.
Isotope ratio	Ratio of the number of atoms of one isotope to the number of atoms of another isotope of the same chemical element in the same system. The number ratio, R, is the number obtained by counting a specified entity (usually molecules, atoms or ions) divided by the number of another specified entity of the same kind in the same system.
Isotope ratio mass spectrometry (IRMS)	The precise measurement of the relative quantity of the different isotopes of an element in a material using a magnetic sector mass spectrometer. Isotope ratio differences are typically measured against international standards and expressed as deviation from the standard on a delta per mille scale (‰).
Isotopologue ions	Ions that differ only in the isotopic composition of one or more of the constituent atoms. For example, $CH_4^{+\bullet}$ and $CH_3D^{+\bullet}$ or $^{10}BF_3$ and $^{11}BF_3$ or the ions forming an isotope cluster. The term *isotopologue* is a short form of isotopic homologue.
Isotopomeric ions	Isomeric ions having the same numbers of each isotopic atom but differing in their positions. Isotopomeric ions can be either configurational isomers in which two atomic isotopes exchange positions or isotopic stereoisomers. The term *isotopomer* is a shortening of isotopic isomer.

ISTD	→ Internal standard.
ITD	Ion-trap detector, a mass spectrometric (MS) detector that uses an ion-trap device to generate mass spectra.
ITEX	In-tube extraction, *aka* ITEX DHS, automated dynamic headspace extraction technique using a sorbent packed syringe needle, injection by thermal desorption in a GC injector, capable of replacing purge and trap instrumentation.

J

j	Mobile phase compressibility correction factor, a factor, applying to a homogeneously filled column of uniform diameter, that corrects for the compressibility of the mobile phase in the column, also called *Compressibility Correction Factor*. In liquid chromatography the compressibility of the mobile phase is negligible. In gas chromatography, the correction factor can be calculated as: $$j = \frac{3}{2}\frac{p^2-1}{p^3-1} = \frac{3}{2}\frac{(p_i/p_o)^2-1}{(p_i/p_o)^3-1}$$
Jet separator	Interface construction for the coupling of mass spectrometers to wide bore and packed chromatography columns. The quantity of the lighter carrier gas is reduced by dispersion. Loss of analyte cannot be prevented.

K

k	Retention factor, a measurement of the retention of a peak: $$k = \frac{(t_R - t_M)}{t_M}$$
K	Partition coefficient. The relative concentration of an analyte in the mobile and stationary phases: $$K = \beta \cdot k$$ For static and dynamic headspace methods the partition coefficient is defined as $$K = \frac{c_S}{c_G} = \frac{\text{concentration in the sample}}{\text{concentration in the gas phase}}$$
Kovats index	The most used retention index system in gas chromatography for describing the retention behavior of substances. The retention values are based on a standard mixture of alkanes: Alkane index = number of C-atoms × 100.

L

L	Column length.
Laboratory sample	A sample prepared for sending to a laboratory and intended for inspection or testing.
Laser ionization (LI)	Formation of ions through the interaction of photons from a laser with a material or with gas-phase ions or molecules.
Level of interest	The concentration of substance or analyte in a sample that is significant to determine its compliance with legislation.
Line spectrum	Representation of a mass spectrum as a series of vertical lines indicating the ion abundance intensities across the mass scale of m/z values.
Linear ion trap (LIT)	A two-dimensional → Paul ion trap in which ions are confined in the axial dimension by means of an electric field at the ends of the trap. Used for instance as collision cell in → Orbitrap MS.
Linear range (LR)	Also called *linear dynamic range*. The range of analyte concentration or amount in which the detector response per amount is constant within a specified percentage.
Linear velocity (u)	The speed at which the carrier gas moves through the column, usually expressed as the average carrier gas linear velocity (u_{avg}), → van Deemter plot.
Linked scan	A scan in a tandem mass spectrometer with two or more mass analyzers, e.g. a triple quadrupole or a sector mass spectrometer that incorporates at least one magnetic sector and one electric sector. The analyzers are scanned simultaneously so as to preserve a predetermined relationship between scan parameters to produce a product ion, precursor ion or constant neutral loss or gain spectrum.
Liquid phase	In GC, a stationary liquid layer coated or chemically bonded on the inner column wall (WCOT column) or on a support (packed, SCOT column) that selectively interacts with the analytes to produce different retention times.
LLOD	Lower limit of detection, also called *detection limit* or *limit of detection*, → LOD.

LOD	Limit of detection, the lowest concentration of a substance that can still be detected unambiguously (assessed in the signal domain). The value is obtained from the decision limit (smallest detectable signal) using the calibration function or the distribution range of the blank value, typically expressed in a → S/N ratio of 3 to 5, based on statistical reasoning.
LOQ	Limit of quantitation, unlike the limit of detection, the limit of quantitation is confirmed by the calibration function. The value gives the lower limiting concentration, which differs significantly from a blank value and can be determined unambiguously and quantitatively with a given precision. The LOQ value is therefore dependent on the largest statistical error, which can be tolerated in the results. Often expressed using the signal domain for the determination of the lowest quantifiable concentration providing a S/N ratio > 10, here ignoring the calibration function with a significantly higher S/N requirement than applied for the LOD.
Low energy collision-induced dissociation	A collision-induced dissociation process wherein the precursor ion has translational energy lower than 1 keV. This process typically requires multiple collisions for the transformation of kinetic into internal vibrational energy. The collisional excitation process is cumulative.
LPGC-MS	Low pressure GC-MS uses for speed of separation a short 0.53 or 0.32 mm ID analytical column that is inserted directly into the MS and a flow restrictor on the GC inlet side.
LRMS	Low resolution mass spectrometry, covers all MS analyzer technologies providing nominal mass resolution like quadrupoles, in contrast to → HRMS with magnetic sector, QTOF, or Orbitrap analyzer.
LVI	Large volume injection. GC injection using solvent volumes typically above 10 µL.
LVSI	Large volume splitless injection, using the → concurrent solvent recondensation effect (CSR).

M

μ scan	The shortest scan unit in ion trap mass spectrometers, depending on the pre-selected scan rate of the chromatogram, several μ scans are accumulated to form the stored and displayed mass spectrum.
Magic 60	A laboratory term referring to the c.60 analytes of the combined volatile halogenated hydrocarbon/BTEX determination (VOC), which are analyzed together in EPA methods, for example, by purge and trap GC-MS.
Magnetic sector	A device that produces a magnetic field perpendicular to a charged particle beam that deflects the beam to an extent that is proportional to the particle momentum per unit charge. For a monoenergetic beam, the deflection is proportional to m/z.
Magnetic sector MS	Single or double focusing mass spectrometer with magnetic (and electrostatic) analyzer for the spatial separation of ions on individual flight paths and focusing on the exit slit at the detector, double focusing mass spectrometer are employed for high resolution MS, → also HRMS.
MAM	Monoacetylmorphine.
MAOT	Maximum allowable operating temperature, highest continuous column operating temperature that will not damage a column, if the carrier gas is free of oxygen and other contaminants. Slightly higher temperatures are permissible for short periods of time during temperature programed operation or intermediate column bakeouts.
Mass accuracy	The deviation of the measured accurate mass from the calculated exact mass of an ion. It can be expressed as an absolute value in millidalton (mDa) or as a relative value in parts-per-million (ppm) error. The absolute accuracy is calculated as accurate mass − exact mass. Example: The experimentally measured mass $m/z = 239.15098$, the theoretical exact mass of the ion $m/z = 239.15028$. The absolute mass accuracy $(239.15098 − 239.15028) = 7.0$ mDa. In relative terms the mass accuracy is calculated as ((accurate mass − exact mass)/exact mass) × 10^6. The relative mass accuracy $(239.15098 − 239.15028)/239.15028 \times 10^6 = 2.9$ ppm.

Mass calibration	A means of determining m/z values in a scan from their times of detection relative to initiation of acquisition of a mass spectrum. Most commonly this is accomplished using a computer-based data system and a calibration file obtained from a mass spectrum of a compound that produces ions whose m/z values are known, → PFTBA, → FC43.
Mass defect	The difference between the (nominal) mass number and the exact monoisotopic mass of a molecule or atom.
Mass excess	The negative of the mass defect.
Mass filter	A quadrupole analyzer works as a mass filter. From the large number of different ion species formed in the ion source a quadrupole filters ions of specific m/z values only. A mass spectrum is acquired by cyclic ramping (scan) the transmission window of 1 Da width over the selected mass range, typically upwards (full scan). Switching the filter characteristics to pre-selected masses only ions of distinct m/z values are acquired (SIM, selected ion monitoring).
Mass gate	A set of plates or grid of wires in a time-of-flight mass spectrometer that is used to apply a pulsed electric field with the purpose of deflecting charged particles in a given m/z range.
Mass number	Sum of the number of protons and neutrons in an atom, molecule or ion, → Nucleon number.
Mass range	Range of m/z over which a mass spectrometer can detect ions or is operated to record a mass spectrum.
Mass resolution	Smallest mass difference Δm between two equal magnitude peaks so that the valley between them is a specified fraction of the peak height.
Mass resolving power	The measure R describing the mass resolution setting or capability of a mass analyzer, expressed as the observed mass divided by the difference between two masses that can be separated: $R = m/\Delta m$. The procedure by which Δm was obtained, e.g. at half peak height (FWHM), and the mass at which the measurement was made should be reported.

	For magnetic sector instruments another definition is used. The "10% valley", method measures the peak width at 5% peak height. The difference between the two definitions is a factor of 2 (i.e. 10 000 resolving power by the 10% valley method equals 20 000 resolving power by FWHM). Note: Mass resolving power is often confused or interchangeably used with → mass resolution.
Mass scale	A mass scale always implies the m/z scale (mass to charge value). Multiply charged ions with n charges appear in the spectrum at m/n, e.g. doubly charged coronene (M 300) with a signal at m/z 150.
Mass selective instability	A phenomenon observed in a Paul ion trap whereby an appropriate combination of oscillating electric fields applied to the body and the end-caps of the trap leads to unstable trajectories for ions within a particular range of m/z values and thus to their ejection from the trap for instance to an ion detector.
Mass spectral library	A collection of mass spectra of chemical compounds. Spectral libraries can carry additional compound information like CAS number, sum formula, exact molecular mass, Kovacz retention index, compound structure and a collection of synonym names. MS/MS and accurate mass libraries are the current developments.
Mass spectrometry/mass spectrometry (MS/MS)	The acquisition and study of the spectra of selected precursor ions, or of precursor ions of a selected neutral mass loss. Structure selective target analyte quantitations use → SRM/MRM data acquisition methods. MS/MS can be accomplished using beam instruments incorporating more than one analyzer (tandem mass spectrometry in space, triple quadrupole analyzer) or in trap instruments (tandem mass spectrometry in time). → Tandem mass spectrometry.
Mass spectrum	A plot of the relative abundances of ions as a function of the m/z values. The mass spectrum is the quantitative mass analysis of the ions generated from a chemical compound in the ion source resp. collision cell.
Matrix	The medium (e.g. food, soil, seawater) in which the analyte of interest may be contained.

Matrix duplicate	An intralaboratory split sample that is used to document the precision of a method in a given sample matrix.
Matrix effect	The under- or overestimation of analytical results due to response differences between a clean standard solution and a matrix spike calibration. The matrix can be standardized by using → analyte protectants, → analyte spike recovery, → interference.
	The response of some determination systems (e.g. GC-MS, LC-MS) to certain analytes maybe affected by the presence of co-extractives from the sample matrix. Partition in headspace analyses and SPME is also frequently affected by components present in the samples. These matrix effects derive from various physical and chemical processes and may be difficult or impossible to eliminate. More reliable calibration may be obtained with the → Matrix-matched calibration or the → Standard addition method when it is necessary to use techniques or equipment that are potentially prone to the effects.
Matrix-matched calibration	Calibration intended to compensate for matrix effects and acceptable interference, common use in pesticide residue analysis. The matrix blank (→ Blank) should be prepared for analysis of samples. In practice, the pesticide is added to a blank extract (or a blank sample for headspace analysis) of a matrix similar to that analyzed. The blank matrix used may differ from that of the samples if it is shown to compensate for the effects. However, for determination of residues approaching or exceeding the MRL, the same matrix (or standard addition) should be used. A matrix-matched calibration may compensate for matrix effects but does not eliminate the underlying cause. Because the underlying cause remains, the intensity of effect may differ from one matrix or sample to another, and also according to the "concentration" of matrix. Isotope dilution or standard addition should be considered where matrix effects are sample dependent.
Matrix spike	A confirmed blank sample spiked with a known concentration of target analyte(s). The spiking occurs prior to sample preparation and analysis. A matrix spike is used to document the bias of a method in a given sample matrix (→ recovery, → surrogate standard). The term is also used in case of quantitative calibrations using a confirmed

	blank sample to prepare the calibration curve instead of clean solvent to prevent systematic errors by the → matrix effect, → matrix spike calibration.
Matrix spike calibration	Calibration procedure using a confirmed blank matrix as basis for a quantitative calibration instead of clean solvents, → matrix effect.
Mass unit	The SI unit for the atomic mass m is given in kg. The additional unit used in chemistry is the atomic mass unit u defined as

$1\,u = 1.660 \times 10^{-27}$ kg

In mass spectrometry the Dalton [Da] is commonly used as a mass unit, in particular in life sciences applications, defined as 1/12 of the mass of the carbon isotope ^{12}C, → Dalton.

In earlier literature mass units are given also in amu (atomic mass units), and the mmu (millimass unit) used for 1/1000 amu (use deprecated by IUPAC). |
Mathieu stability diagram	A graphical representation expressed in terms of reduced coordinates that describes the stability of charged particle motion in a quadrupole mass filter or quadrupole ion trap mass spectrometer, based on an appropriate form of the Mathieu differential equation.
Mattauch–Herzog geometry	An arrangement for a double-focusing mass spectrometer in which a deflection of $n/(4\sqrt{2})$ radians in a radial electrostatic field is followed by a magnetic deflection of $n/2$ radians.
Maximizing masses peak finder	A routine method of data handling in GC-MS analysis. The change in each in individual ion intensity with time is analyzed. If several ions have a common peak maximum at the same retention time, the elution of an individual substance is recognized and noted even in case of a co-elution. The peak finder is used for automatic analysis of complex chromatograms, → also AMDIS, → deconvolution.
McLafferty rearrangement	A dissociation reaction triggered by transfer of a hydrogen atom via a six-member transition state to the formal radical/charge site from a carbon atom four atoms removed from the charge/radical site (the α-carbon); subsequent rearrangement of electron density leads to expulsion of an olefin molecule. This term was originally applied to ketone ions where the charge/radical site is the carbonyl oxygen, but it is now more widely applied.

MCP	→ Microchannel plate detector, typically found as ion detector in TOF instruments, → Array detector.
MDL	→ Method detection limit.
MDQ	Minimum detectable quantity, the amount of analyte that produces a signal twofold that of the noise level.
MEPS	Micro-extraction by packed sorbent, miniaturized SPE sample preparation method using a packed syringe needle, and GC injection via liquid desorption in the GC injector.
Method blank	An analyte-free sample to which all reagents are added in the same volumes or proportions as used in sample processing. The method blank must be carried through the complete sample preparation and analytical procedure. The method blank is used to assess contamination resulting from the analytical process.
Method detection limit (MDL)	The minimum amount of analyte that can be analyzed within specified statistical limits of precision and accuracy, including sample preparation. The lower detection limit (→ LOD, LLOD) when applying the complete method from sample preparation, cleanup and analysis.
Merlin seal	Alternative injector septum solution "Merlin Microseal" using a septum less seal formed like a duck bill and sealed by the inside carrier pressure and spring load.
Metrology	The science of measurement and its application.
Membrane inlet (MI)	A semi-permeable membrane separator that permits the passage of analytes directly from solutions or ambient air to the mass spectrometer ion source.
Metastable ion	An ion that is formed with internal energy higher than the threshold for dissociation but with a lifetime great enough to allow it to exit the ion source and enter the mass analyzer where it dissociates before detection.
MFX	Multi fiber exchange accessory, used for x,y,z-robotic sampling systems for the automated exchange of →SPME probes.
MHE	Multiple headspace extraction, a quantitation procedure used in static headspace involving multiple extraction and measurement from a single sample. The vial pressure is released between the consecutive measurements using a dedicated MHE device on autosampler platforms.

MHS-SPME	Multiple headspace-SPME, →MHE, →MSPME.
Microchannel plate (MCP)	A thin plate that contains a closely spaced array of channels that each act as a continuous dynode particle multiplier. A charged particle, fast neutral particle or photon striking the plate causes a cascade of secondary electrons that ultimately exits the opposite side of the plate.
Microchannel devices	The microchannel or microfluidic devices are used for flow switching in GC, consisting of laser cut metal sheets (shims) with thicknesses from 20 to 500 µm. The resulting channel dimensions are similar to the conventional fused silica capillary columns. Shim stacks are bonded together with top and bottom plates for column connections, having low thermal mass for fast GC oven ramp rates.
Microfluidic devices	→ Microchannel devices.
MID	Multiple ion detection, the recording of several individual ions (m/z values) for increased dwell time and detection sensitivity in contrast to full scan, → also SIM.
Mixture search	Search mode of the → PBM library search algorithm whereby a mixture is analyzed by forming and searching difference spectra between sample and library.
MNT	Mononitrotoluene isomers.
Modifier	The addition of organic solvents in small amounts to the extraction agent in SFE. The modifier can be added directly to the sample (and is effective only in the static extraction step) or continuously by using a second pump. Typically up to 10% modifier is added to CO_2.
Molar mass	Mass of 1 mol of a compound: 6.022 1415(10) × 10^{23} atoms or molecules. Note: The term *molecular weight* is deprecated because "weight" is the gravitational force on an object that varies with geographical location. Historically, the term has been used to denote the molar mass calculated using isotope averaged atomic masses for the constituent elements.
Molecular ion	The non-fragmented ion formed by the removal of one or more electrons to form a positive ion M^+ or the addition of one or more electrons to form a negative ion M^-. Fragmentations start from the molecular ion to reduce the internal energy transferred by the ionization process.

Molecular sieve	A stationary phase that retains analytes by molecular size interactions.
Molecular weight	The term molecular weight is deprecated. The sum of the atomic masses of all the atoms present in a molecule. The term *molecular mass* is commonly used although actually masses are involved. The average molecular mass is calculated taking the natural isotopic distribution of the elements into account (stoichiometry), → Molar mass. Molecular mass in MS: The calculation is carried out exclusively using the atomic masses of the principle isotopes instead of averaged atomic masses, → Nominal mass.
Monoisotopic mass	Exact mass of an ion or molecule calculated using the mass of the principle (usually the lightest) isotope of each element.
MQL	Minimum quantitation limit, also → LOQ (limit of quantitation).
MRL	Maximum residue level.
MRM	Multiple reaction monitoring, MS/MS scan to monitor selected product ions only, essentially the same experiment as selected reaction monitoring, → SRM.
MRPL	Minimum required performance limit, the minimum concentration of an analyte in a sample, which at least has to be detected and confirmed. It is used to harmonize the analytical performance of methods for substances for which no permitted limit has been established.
MS/MS^n, MS^n	This symbol refers to multistage MS/MS experiments designed to record product ion spectra where n is the number of product ion stages (fragment ions). For ion trap MS, sequential MS/MS experiments can be undertaken where $n > 2$, whereas for a triple quadrupole system $n = 2$, → Multiple-stage mass spectrometry.
MSPME	Multiple →SPME analysis, performed in multiple steps in both →DI-SPME or →HS-SPME (analog to →MHE analysis).
M-series	*n*-Alkyl-bis (trifluoromethyl) phosphine sulfides, a homologous series used to construct a retention time index system. The compounds can be used in gas chromatography with FID, ECD, ELCD, NPD, FPD, PID and MS detectors.

MSE	Multiple step enrichment. Enrichment from several SPME extractions from different samples on an intermediate trap, then thermal desorption, for improved sensitivity.
MSTFA	N-Methyl-N-trimethylsilyltrifluoroacetamide, a derivatization agent for silylation, typically used, for example, in anabolic steroid derivatization.
Multidimensional	Separations performed sequentially with two or more columns of different separation mechanism (orthogonal) in which peaks are selectively directed onto or removed from at least one of the columns by a timed valve system, → backflush, → heartcut, → pre-cut.
Multiphoton ionization (MPI)	Photoionization of an atom or molecule in which in two or more photons are absorbed.
m/z	Symbol m/z is used to denote the dimensionless quantity formed by dividing the mass of an ion in unified atomic mass units by its charge number (regardless of sign). The symbol is written in italicized lower case letters with no spaces.
	Note 1: The term *mass-to-charge-ratio* is deprecated. Mass-to-charge-ratio has been used for the abscissa of a mass spectrum, although the quantity measured is not the quotient of the ion's mass to its electric charge. The symbol m/z is recommended for the dimensionless quantity that is the independent variable in a mass spectrum.
	Note 2: The formerly proposed unit Thomson (Th) is deprecated.

N

η	Viscosity. The carrier gas viscosity increases with increasing temperature.
N	Number of theoretical plates: $$N = 5.54\left(\frac{t_R}{w_h}\right)^2 \sim 16\left(\frac{t_R}{w_b}\right)^2$$
N_{eff}	The number of effective plates. This term is an alternate measurement of theoretical plate height that compensates for the non-partitioning nature of an unretained peak: $$N = 16\left(\frac{t'_R}{w_b}\right)^2$$
N_{req}	The number of theoretical plates required to yield a particular resolution (R) at a specific peak separation (α) and retention factor (k): $$N_{req} = 16 \cdot R^2\left(\frac{\alpha}{\alpha-1}\right)^2\left(\frac{k+1}{k}\right)^2$$

Nafion dryer	Device for drying analytical gas streams using capillary membrane tubes of polar Nafion™ material with outer counter flow of dry carrier gas.
NBS	The U.S. National Bureau of Standards, now → NIST.
ndl-PCB	Non dioxin-like PCBs, → indicator PCBs, PCB.
Needle sparger	Glass vessel used in purge and trap analysis of solids or foaming samples. The purge gas is passed into the sample via a needle perforated on the side or by means of a headspace sweep.
Negative ion	An atomic or molecular species having a net negative electric charge.
Negative ion chemical ionization (NCI, NICI)	Chemical ionization that results in the formation of negative ions.
Neutral loss	Loss of an uncharged species from an ion during either a rearrangement process or direct dissociation, → neutral loss scan.
Neutral loss scan	MS/MS experiment for analyzing precursor ions which undergo a common loss of neutral particles of the same mass. Analytes with functional groups in common are detected. Q1 and Q3 of a triple quadrupole MS scan simultaneously with constant mass difference, → Linked scan.
Nier–Johnson geometry	Arrangement for a double-focusing mass spectrometer in which a deflection of $n/2$ radians in a radial electrostatic field analyzer is followed by a magnetic deflection of $n/3$ radians to focus the ions on a focal plane.
NIH	U.S. National Institute of Health.
NIST	U.S. National Institute of Standards and Technology, part of the U.S. Department of Commerce.
Nitrogen rule	An organic molecule containing the elements C, H, O, S, P, or a halogen has an odd nominal mass if it contains an odd number of nitrogen atoms.
Nominal mass	Mass of an ion or molecule calculated using the mass of principle, often most abundant isotope of each element rounded to the nearest integer value and equivalent to the sum of the mass numbers of all constituent atoms. According to this convention CH_4 and CD_4 have the same nominal mass!

Nominal mass resolution	The mass spectrometric resolution for the separation of mass signals with a uniform peak width of 1000 mDa (1 mass unit) in the quadrupole or ion trap analyzer, → LRMS. The resulting mass numbers are generally given as → nominal mass numbers or to one decimal place (unlike high resolution data).
NPD	Nitrogen–phosphorus detection, the nitrogen–phosphorus detector, providing a mass flow dependent signal, catalytically ionizes nitrogen or phosphorus-containing solutes on a heated rubidium or cesium surface in a reductive atmosphere. The nitrogen–phosphorus detector is highly selective and provides sensitivity that is better than that of a flame ionization detector.
NPLC	Normal phase liquid chromatography.
Number of theoretical plates	Describes the separating capacity of a column.

O

OCDD	Octachlorodibenzodioxin.
OCI	On-column injection, sample enters the column directly from the syringe without prior volatilization and does not contact other surfaces. On-column injection usually signifies cold injection for capillary columns.
Odd-electron ion	→ Radical ion.
Odd-electron rule	Odd-electron ions may dissociate to form either odd or even-electron ions, whereas even-electron ions generally form even-electron fragment ions.
Olfactometry	→ GC-olfactometry, GC-O.
On-column	Sample injection technique in GC whereby a diluted liquid extract is injected directly (without evaporation) on to a → pre-column or on to the analytical column itself, → OCI. The temperature of the injection site at the beginning of the column is controlled by the oven temperature.
Onium ion	A positively charged hypervalent ion of the non-metallic elements. Examples are the methonium ion CH_5^+, the hydrogenonium ion H_3^+ and the hydronium ion H_3O^+. Other examples are the oxonium, sulfonium, nitronium, diazonium, phosphonium, and halonium ions.

Orbitrap	The Kingdon trap analyzer, an ion trapping device that consists of an outer barrel-like electrode and a coaxial inner spindle-like electrode that form an electrostatic field with quadro-logarithmic potential distribution. The frequency of harmonic oscillations of the orbitally trapped ions along the axis of the electrostatic field is independent of the ion velocity and is inversely proportional to the square root of m/z so that the trap can be operated as a mass analyzer using image current detection and Fourier transformation of the time domain signal.
	The trademarked term *Orbitrap* has been used to describe a Kingdon trap used as a mass spectrometer.
Orthogonal extraction	Pulsed acceleration of ions perpendicular to their direction of travel into a time-of-flight mass spectrometer. Ions may be extracted from a directional ion source, drift tube or m/z separation stage.

P

P	Relative pressure across the column: $P = p_i/p_o$.
Δp	Pressure drop across the column: $\Delta p = p_i - p_o$.
P_i	Absolute inlet pressure.
P_o	Absolute outlet pressure.
PAHs	Polyaromatic hydrocarbons, the group of polycyclic hydrocarbons from naphthalene to and beyond coronene.
Partition coefficient	The partition coefficient K used in the headspace analysis technique (e.g. static HS, SPME) is given by the partition of a compound between the liquid and the gaseous phases c_{liqu}/c_{gas}. The partition coefficient is temperature dependent.
PAT	Purge and trap technique, P&T.
Paul ion trap	Ion trapping device that permits the ejection of ions with an m/z lower than a prescribed value and retention of those with higher mass. It depends on the application of radio frequency voltages between a ring electrode and two endcap electrodes to confine the ion motion to a cyclic path described by an appropriate form of the

Mathieu equation. The choice of these voltages determines the m/z below which ions are ejected. The term *cylindrical ion trap* is deprecated.

The name was given to the device developed by Prof. Paul, University of Bonn, Germany, and known as the *ion trap analyzer*. Prof. Paul received the Nobel prize in 1989 for his work on the QUISTOR development at the beginning of the 1950s.

PBB	Polybrominated biphenyls.
PBDE	Polybrominated diphenylether.
PBM	Probability based match, library search procedure and program for mass spectra developed by Prof. McLafferty.
PCBs	Polychlorinated biphenyls with a total number of 209 congeners.

Seven indicator PCBs of the non dioxin-like PCBs (ndl-PCBs) are routinely monitored for regulatory purposes, listed here with their 'Ballschmiter numbers':

tetra Cl-PCB: 28, 52
penta Cl-PCB: 101, 118
hexa Cl-PCB: 138, 153
hepta Cl-PCB: 180.

12 so-called dioxin-like PCBs (dl-PCBs) or WHO-PCBs are the coplanar non-ortho substituted congeners of a total number of 68 coplanar congeners of which 20 in total are non-ortho substituted:

tetra Cl-PCB: 77, 81
penta Cl-PCB: 105, 114, 118, 123, 126
hexa Cl-PCB: 156, 157, 167, 169
hepta Cl-PCB: 189.

PCI	Positive chemical ionization.
PDD	Pulsed discharge detector.
PDECD	Pulsed discharge electron capture detector with similar or better sensitivity (MDQs of 10^{-15} to 10^{-12} g) than radioactive source ECDs.
PDED	Pulsed discharge emission detector (He pulsed discharge emission).
Peak capacity	In quantitative GC: The amount of analyte that can be injected without a significant loss of column efficiency.

In multidimensional GC: The maximum number of peaks that can be resolved.

Peak (in mass spectrometry)	Localized region of relatively large ion signal in a mass spectrum. Although peaks are often associated with particular ions, the terms *peak* and *ion* should not be used interchangeably.
Peak intensity	Height or area of a peak in a mass spectrum.
Peak matching	Procedure for measuring the accurate mass of an ion using scanning mass spectrometers, in which the peak corresponding to the unknown ion and that for a reference ion of known m/z are displayed alternately on a display screen and caused to overlap by adjusting the acceleration voltage.
Peak overload	If too much of analyte is injected, its peak can be distorted into a triangular shape exceeding the column capacity.
PEEK	Polyether ether ketone, hard plasticizer-free polymer with the general structure (with $x=2$, $y=1$): PEEK is a colorless semicrystalline thermoplastic polymer with high chemical inertness and resistance used as a sealing and tubing material and for screw joints in HPLC, SFE, SFC and in high vacuum areas of MS, thermally stable up to 340 °C.
PEG	Polyethylene glycol.
Permitted limit	The maximum residue limit, maximum level or other maximum tolerance for substances established in legislation.
PFBA	Pentafluoropropionic anhydride, derivatization agent, it is used for introducing halogens to increase the response in ECD and NCI.
PFK	Perfluorokerosene, calibrant in mass spectrometry, widely used with magnetic sector MS instruments.
PFPD	Pulsed flame photometric detector, provides two simultaneous signals for S and P by measurement of a fluorescence/time profile in the range of 2–25 ms with about 5 Hz cycle time (after Prof. Aviv Amirav, University Tel Aviv, Israel).
PFTBA	Perfluorotributylamine, calibrant in mass spectrometry with M 671, widely used with quadrupole, ion trap and magnetic sector MS instruments, → FC43.

Phase ratio	In the headspace analysis technique, the phase ratio of V_{gas}/V_{liqu} gives the degree of filling the headspace bottle.
	In capillary GC the phase ratio describing the ratio of the internal volume of a column (volume of the mobile phase) to the volume of the stationary phase. High performance columns are characterized by high phase ratios. A table showing different combinations of internal diameters and film thicknesses can be used to optimize the choice of column with regard to analysis time, resolution and capacity.
Photodissociation	Process wherein the reactant ion is dissociated as a result of absorption of one or more photons.
Photoionization (PI)	Ionization of an atom or molecule by a photon, $M + h\nu \rightarrow M^{+\bullet} + e^-$. The term *photon impact* is deprecated.
PID	Photoionization detector, the photoionization detector ionizes analyte molecules with photons in the UV energy range, provides a concentration dependent signal. The photoionization detector is a selective detector that responds to aromatic compounds and olefins when operated in the 10.2 eV photon range, and it can respond to other materials with a more energetic light source.
PIL	Polymerized ionic liquid. The polymeric ionic liquids are the polymeric form of ionic liquids (→IL).
PIONA	Paraffins, isoparaffins, olefins, napthenes, and aromatic compounds.
PLOT	Porous-layer open-tubular column, a fused silica capillary column type with solid adsorbents coated onto the inner surface of the capillary tubing to provide gas-solid chromatographic retention behavior.
PONA	Paraffins, olefins, napthenes and aromatic compounds.
Porous polymer	A stationary-phase material that retains analytes by selective adsorption or by molecular size interaction.
Positive ion	Atomic or molecular species having a net positive electric charge.
Post-acceleration detector	Detector in which a high voltage is applied after m/z separation to accelerate the ions and produce an improved signal, → Conversion dynode.

Post-source decay (PSD)	Technique specific to reflectron time-of-flight mass spectrometers where product ions of metastable transitions or collision-induced dissociations generated in the flight tube prior to entering the reflectron are m/z separated to yield product ion spectra.
PPINICI	Pulsed positive ion negative ion chemical ionization, the alternating data acquisition of positive and negative ions formed during chemical ionization, patented by former Finnigan Corp., San Jose, CA, USA.
Precision	The closeness of agreement between independent test results obtained under stipulated (predetermined) conditions. The measure of precision usually is expressed in terms of imprecision and computed as standard deviation of the test result. Less precision is determined by a larger standard deviation.
Pre-column	Guard column, → retention gap, used for → large volume injections or → column backflushing.
Precursor ion	Ion that reacts to form structure related product ions. The reaction can be unimolecular dissociation, ion/molecule reaction, isomerization or change in charge state. The term *parent ion* is deprecated.
Precursor ion scan	MS/MS scan function or process that records a precursor ion spectrum. It is used to detect substances with related structures which give common fragments. The term *parent ion scan* is deprecated.
Precursor ion spectrum	Mass spectrum recorded in which the appropriate m/z separation function can be set to record the precursor ion or ions of selected product ions. The term *parent ion spectrum* is deprecated.
Pre-cut	Peaks at the beginning of a chromatogram are removed to vent or directed onto another column of different polarity or at a different temperature for improved resolution.
Pre-purge	A preliminary step in purge and trap analysis, the atmospheric oxygen is removed by the purge gas before the sample is heated to avoid side reactions.
Pre-scan	The step in the ion trap scan function before the analytical scan. During the pre-scan the variable ionization time or ion collection time is adjusted to fill the trap to its optimum capacity with ions.

	Pre-search	Part of library searching of mass spectra in which a small group of candidates is selected from the whole number of entries for detailed comparison and ranking.
	Press fit	Glass tube connectors for fused silica capillaries. The cross cut column end is simply pushed into the conical opening and seal is achieved with the external polyimide coating. Caution is necessary when applying high temperatures during oven ramping for weakening the polymer at the sealing site resulting in air leaks.
	Pressure units	The SI unit is given in Pascal: $1\,\text{Pa} = 1\,\text{Nm}^{-2}$ $10^5\,\text{Pa} = 10^5\,\text{Nm}^{-2} = 1\,\text{bar}$ Pressure values are often given in traditional units. In MS vacuum technologies pressures are given in (Torr) or (mTorr) and gas pressures of GC supplies frequently in (kPa), (bar) or (psi).

Pressure unit conversion table:

	Pa	bar	Torr	psi	at	atm
Pa	1	$1\cdot10^{-5}$	$7.5\cdot10^{-3}$	$1.45\cdot10^{-4}$	$1.02\cdot10^{-5}$	$9.87\cdot10^{-6}$
bar	$1\cdot10^5$	1	750	14.514	1.02	0.987
Torr	133	$1.33\cdot10^{-3}$	1	$1.94\cdot10^{-2}$	$1.36\cdot10^{-3}$	$1.32\cdot10^{-3}$
psi	$6.89\cdot10^3$	$6.89\cdot10^{-2}$	51.67	1	$7.03\cdot10^{-2}$	$6.80\cdot10^{-2}$
at	$9.81\cdot10^4$	0.981	736	14.224	1	0.968
atm	$1.0133\cdot10^5$	1.0133	760	14.706	1.033	1

at: technical atmosphere 1 kp/cm².
atm: physical atmosphere 1.033 kp/cm².
psi: pound per square inch.

	Primary reaction	Conversion of the reagent gas used for CI into the reagent ions by electron ionization, → also CI.
	Principal ion	Most abundant ion of an element or molecular isotope cluster, such as the $^{11}\text{B}^{79}\text{Br}_2^{81}\text{Br}^{+\bullet}$ ion of m/z 250 of the cluster of → isotopologue molecular ions of BBr₃. The term *principal ion* has also been used to describe ions that have been artificially isotopically enriched in one or more positions such as $\text{CH}_3\text{-}^{13}\text{CH}_3^{+\bullet}$ or $\text{CH}_2\text{D}_2^{+\bullet}$, but those are best defined as isotopologue ions.

Product ion	An ion formed as the product of a reaction involving a particular precursor ion. The reaction can be unimolecular dissociation to form fragment ions, an ion/molecule reaction or by → CID during → SRM/MRM measurements. The term *fragment ion* is deprecated. The term *daughter ion* is deprecated.
Product ion scan	Specific MS/MS scan function or process that records a product ion spectrum. The terms *fragment ion scan* and *daughter ion scan* are deprecated.
Product ion spectrum	Mass spectrum recorded from any spectrometer in which the appropriate m/z separation scan function is set to record the product ion or ions of selected precursor ions. The terms *fragment ion scan* and *daughter ion scan* are deprecated. Note: The term *MS/MS spectrum* is deprecated; a scan specific term, for example, *precursor ion spectrum* or *second-generation product ion spectrum* should be used.
Proficiency study	Analyzing the same sample, allowing laboratories to choose their own methods, provided these methods are used under routine conditions. The study has to be performed according to ISO guide 43-1 and 43-2 and can be used to assess the reproducibility of methods.
Proton abstraction	Type of NCI reaction, ionization is effected by transfer of a proton from the analyte (abstraction) to the reagent ion, for example, from analytes with phenolic OH groups.
Proton affinity	Proton affinity of a species M is the negative of the enthalpy change for the reaction $M + H^+ [M + H]^+$ at 298 K.
Protonated molecule	An ion formed by interaction of a molecule with a proton, and represented by the symbolism $[M + H]^+$. Note 1: The term *protonated molecular ion* is deprecated; this would correspond to a species carrying two charges. Note 2: The terms *pseudo-molecular ion* and *quasi-molecular ion* are deprecated; a specific term such as protonated molecule or a chemical description such as $[M + Na]^+$, $[M - H]^-$, and so on, should be used.

Protonation	Type of PCI reaction, ionization is effected by transfer of one or more protons to the substance molecule. Protonating reagent gases include methane, methanol, water, isobutene, and ammonia.
PTGC	Programmed-temperature GC, the column temperature changes in a controlled manner as peaks are eluted.
PTI	Programmed-temperature injection, a cold injection technique in which the inlet temperature is specifically programmed from the gas chromatograph.
PTV	Programmed-temperature vaporizer, a widely used inlet system, *aka* multi injector, designed to perform a temperature programmed injection, a cold injection system for direct liquid injection for split or splitless injection, solvent split technique and cryo-enrichment.
P&T	Purge-and-trap sampling, dynamic headspace procedure, a concentration technique for volatile solutes. The sample is purged with an inert gas that entrains volatile components onto an adsorptive trap. The trap is then heated to desorb trapped components into a GC column.
Pure search	A mode of the → PBM search procedure, which only uses the forward search capability of the library search, → also Forward library search.
PyGC	Pyrolysis GC, the sample is pyrolyzed (decomposed) in the inlet before GC analysis.
PyMS	→ Pyrolysis mass spectrometry.
Pyrogram	Chromatogram received from the separation of pyrolysis products.
Pyrolysis mass spectrometry (PyMS)	A mass spectrometry technique in which the sample is heated to the point of decomposition and the gaseous decomposition products are introduced into the ion source.

Q

QED	Quantification enhanced detection by data-dependent MS/MS, a QED scan on a triple quadrupole instrument delivers an information rich product ion mass spectrum that can be used to confirm the existence of compounds by an in-built MS/MS library while they are being quantified using the MRM scan mode.

QIT	Quadrupole ion trap, → Paul ion trap.
QMS	Quadrupole mass spectrometer, → Transmission → Quadrupole mass spectrometer.
QqQ	Triple quadrupole mass spectrometer. Note: The lower case q denotes the collision cell.
Quality assurance (QA)	An integrated system of activities involving quality planning, quality control, quality assessment, quality reporting and quality improvement to ensure that a product or service meets defined standards of quality with a stated level of confidence.
Quality control (QC)	The overall system of technical activities whose purpose is to measure and control the quality of a product or service so that it meets the needs of users. The aim is to provide quality that is satisfactory, adequate, dependable and economical. For analytical chemistry, QC is a set of procedures applied to an analytical methodology to demonstrate that the analysis is in control.
Quality control sample	A sample or standard used either singly or in replicate to monitor method performance characteristics.
Quasimolecular ion	An ion to which the molecular mass is assigned, which is formed by, for example, chemical ionization as $(M+H)^+$, $(M+NH_4)^+$, $(M-H)^+$, or $(M-H)^-$. The term *quasimolecular ion* is deprecated by the IUPAC to use "cationized molecule" instead.
QuEChERS	Quick Easy Cheap Effective Rugged Safe, acronym for a "fast and easy multi-residue method employing acetonitrile extraction/partitioning" and "dispersive solid phase extraction for the determination of pesticide residues in produce", www.quechers.com, M. Anastassiades, S.J. Lehotay, D. Stajnbaher and F.J. Schenck, J AOAC Int 86, (2003) 412.
QUISTOR	Quadrupole ion storage trap, → Paul ion trap.

R

r	Relative retention. For peak i relative to a standard peak s; $r = k_i/k_s$.

R	Resolution, the quality of separation of two peaks. In GC for two closely eluted peaks using the formula: $$R = (t_{R,2} - t_{R,1})/w_{b,2}$$ where the subscripts 1 and 2 refer to the first and second peaks. From N, k_2 and α: $$R = \left(\frac{\sqrt{N}}{4}\right)\left(\frac{\alpha-1}{\alpha}\right)\left(\frac{k_2}{k_2+1}\right)$$ where k_2 is the retention factor of the second peak. A resolution of 1.5 is said to be baseline resolution. R incorporates both efficiency and separation.
Radical ion	An ion, either a cation or anion, containing unpaired electrons in its ground state. The unpaired electron is denoted by a superscript dot alongside the superscript symbol for charge, such as for the molecular ion of a molecule M, that is, $M^{+\bullet}$. Radical ions with more than one charge and/or more than one unpaired electron are denoted such as $M^{(2+)(2\bullet)}$. Unless the positions of the unpaired electron and charge can be associated with specific atoms, superscript charge designation should be placed before the superscript dot designation.
Reagent gas cluster	The spectrum of ions formed from the reagent gas for chemical ionization, monitored to adjust the correct reagent gas pressure in the ion source.
Reagent ion	An ion produced in large excess in a chemical ionization source that reacts with neutral sample molecules to produce an ionized form of the molecule through an ion/molecule reaction.
Reagent ion capture	Type of CI reaction in PCI/NCI. The ionization of the analyte is achieved by an addition reaction of the reagent ion.
Recombination energy	Energy released when an electron is added to an ionized molecule or atom, that is, the energy involved in the reverse process to that referred to in the definition of ionization energy.
Recovery	The percentage of the true concentration of a substance recovered during the analytical procedure. It is determined during validation, → surrogate standard.

Reference ion	Stable ion whose structure is known with certainty. These ions are usually formed by direct ionization of a molecule of known structure, and are used to verify by comparison the structure of an unknown ion.
Reference material	A material of which one or several properties have been confirmed by a validated method, so that it can be used to calibrate an apparatus or to verify a method of measurement, → CRM.
Reference method	A sampling or measurement method that has been officially specified by an organization as meeting its data quality requirements.
Reflectron	Constituent of a time-of-flight mass spectrometer that uses a static electric field to reverse the direction of travel of the ions entering it. A reflectron improves mass resolution by assuring that ions of the same m/z but different translational kinetic energy arrive at the detector at the same time.
Relative response factor	Ratio of the analyte response to the response of the related internal (recovery) standard.
Repeatability	Precision under repeatability conditions, where independent test results are obtained with the same method on identical test items in the same laboratory by the same operator using the same equipment.
Reproducibility	Precision under reproducibility conditions, where test results are obtained with the same method on identical test items in different laboratories with different operators using different equipment.
RER	Reduced energy ramp, collision energy regime during the generation of a MS/MS product spectrum in which the applied collision energy is ramped down to produce a richer product ion spectrum for structure elucidation and confirmation, especially in the higher mass range part of the spectrum.
Residual gas analyzer (RGA)	Mass spectrometer used to measure the composition and pressure of gases in an evacuated chamber.
Resonance ion ejection	Mode of ion ejection in a → Paul ion trap that relies on an auxiliary radio frequency voltage that is applied to the endcap electrodes. The voltage is tuned to the secular frequency of a particular ion to eject it.

Response	The specific height of the detector signal, usually calculated as the ratio of the peak area or height to the quantity of the analyte.
Retention gap	A short piece of deactivated but uncoated column used as "→ pre-column" or "guard column" placed between the inlet and the analytical column. There is no stationary phase hence the name as there is no retention or separation of the analytes. Retention gaps are used for → LVI, → LC-GC coupling, or → backflushing and avoids column solvent flooding. It also retains non-volatile sample contaminants from on-column injection. Retention gaps can be replaced upon contamination without changing the analyte retention times. After evaporation of the solvent the analytes are focused at the beginning of the analytical column.
Retention index	A uniform system of retention classification according to a solute's relative location between a pair of homologous reference compounds on a specific column under specific conditions. It compares the time a compound is eluting from the column to the times of a set of standard compounds. The most common set of compounds used for retention indices are hydrocarbons typically from C5 to C30, separated in the order of boiling points, → Kovats index.
Retention time	The time that the compound is held up on the GC column. Usually, a → retention index is used as compound specific rather than a retention time which is method dependent.
Retention volume	The carrier gas volume required to elute a component.
Reverse library search	A procedure of comparing a mass spectrum of an unknown compound with a mass spectral library so that the unknown spectrum is compared in turn with the library spectra, considering only the m/z peaks of the library spectrum. This comparison procedure serves to expose mixed spectra or to search for a substance in a GC-MS chromatogram.
RF	Response factor; radio frequency.
RI	Retention index, for example, → Kovats index.

RIC	Reconstructed ion chromatogram. The total ion current (→ TIC) is calculated from the sum of the intensities of all of the acquired mass signals in a mass spectrum. This value is generally stored with the mass spectrum and shown as the RIC or TIC value in the mass spectrum.

Plotting the RIC or TIC value along the scan axis (retention time) gives the conventional chromatogram display (older magnetic sector instruments used a dedicated total ion current detector for this purpose). For this reason the shape of the chromatogram in GC-MS is dependent from the mass range scanned. |
Robustness	A measure of the capacity of an analytical method to obtain comparable and acceptable results when perturbed by small but deliberate variations in procedural parameters. It provides an indication of the method's suitability and reliability during normal use. During a robustness study the GC MS method parameters such as solvent amounts, carrier gas flow, heating ramps and detector settings are intentionally varied to study the effects on analytical results.
RPLC	Reversed phase liquid chromatography.
RSD	Relative standard deviation.
RT	Retention time.
RTIL	Room-temperature ionic liquids. Organic salts with melting points at or below room temperature.
Ruggedness	The susceptibility of an analytical method to changes in experimental conditions which can be expressed as a list of the sample materials, analytes, storage conditions, environmental and/or sample preparation conditions under which the method can be applied as presented or with specified minor modifications.

S

s	Split ratio. The ratio of the sample amount that is vented to the sample amount that enters the column during split injection. Higher split ratios place less sample on the column. s is usually measured as the ratio of total inlet flow to column flow; $s = (F_s + F_c)/F_c$.

Sandwich technique	A syringe injection technique in which a sample plug is placed after or between one or two solvent plugs in the syringe barrel, separated by a short air plug, to rinse the syringe needle with solvent at injection and obtain better sample transfer into the inlet, → also solvent flush.
SBSE	Stir bar sorptive extraction, a glass coated magnetic stir bar is coated with PDMS which serves as high capacity sorption phase, extraction by thermal desorption in a dedicated injector system.
SBSE/LD	Stir bar sorptive extraction with liquid desorption of the collected analytes instead of thermal desorption, → SBSE.
Scan function	The control of ion trap or quadrupole mass analyzers by changing the applied voltages in time, represented by a diagram of voltage (U) against time t (ms).
SCD	Sulfur chemiluminescence detection, a sulfur chemiluminescence detector responds to sulfur-containing compounds by generating and measuring the light from chemiluminescence.
SCOT	Support-coated open-tubular column, a capillary column in which stationary phase is coated onto a support material that is distributed over the column inner wall. A SCOT column generally has a higher peak capacity than a → WCOT column with the same average film thickness.
Screening method	Methods that are used to detect the presence of a substance or class of substances at the level of interest. These methods have the capability for a high sample throughput and are used to sift large numbers of samples for potential non-compliant results. They are specifically designed to avoid false compliant results.
SDME	Single drop microextraction. A small drop of immiscible solvent, typ. 1 to 3 µL, is suspended from the needle of a syringe in the headspace or a liquid sample to extract target analytes.
SEC	Size exclusion chromatography, *aka* → GPC.
Secondary reaction	The chemical reaction of analyte molecules with the reagent ions in a CI ion source to form stable product ions, → CI.

Sector field mass spectrometer	Mass spectrometer consisting of one or more magnetic sectors for m/z selection in a beam of ions. Such instruments may also have one or more electric sectors for energy focusing.
Selected ion monitoring (SIM)	Operation of a mass spectrometer for target compound analysis in which the abundances of several ions of specific m/z values are recorded rather than the entire mass spectrum.
Selected reaction monitoring (SRM)	MS/MS scan technique, whereby a product ion spectrum is produced by collision induced decomposition of a selected precursor ion, in which only one or more selected fragment ions are detected, synonymous use to → MRM. In MS/MS instruments the SRM technique provides the highest possible selectivity together with the highest possible sensitivity (→ dwell time) for target compound quantitation.
	Data acquired from specific product ions corresponding to m/z selected precursor ions recorded via two or more stages of mass spectrometry. Selected reaction monitoring for target compound analysis can be performed as tandem mass spectrometry in time or tandem mass spectrometry in space. The term *multiple reaction monitoring* is deprecated.
Selectivity	Selectivity is the recommended term in analytical chemistry to express the extent of interferences. It is the recommended term to express the extent to which a particular method can be used to determine analytes under given conditions in mixtures or matrices without interferences from other components of similar behavior. Selectivity can be graded (in contrast to → specificity).
	In chromatographic methods the fundamental ability of a stationary phase to retain substances selectively based upon their chemical characteristics, including vapor pressure and polarity.
	In mass spectrometry the capability of a method or instrument to distinguish small mass differences, for example, between target ions and non-target matrix substances.
Selectivity tuning	Several techniques for adjusting the selectivity of separations that involve more than one column or stationary phase type. Serially coupled columns and mixed-phase columns can be selectivity-tuned.

SEM	Secondary electron multiplier for high amplification factors, usually built from discrete dynodes which are electrically connected by resistors to provide a voltage ramp across the number of dynodes (contrary to a → CDEM).
Sensitivity	In quantitation the slope of the calibration function. Also used for the degree of detector response to a specified analyte amount per unit time or per unit volume.
Separation α	The degree of separation of two peaks in time, → α and R.
Septum	Silicone or other elastomeric material that isolates inlet carrier pressure from the atmosphere and permits syringe penetration for injection (alternative septum solution → Merlin seal).
Septum purge	The carrier gas swept across the injector internal septum face to a separate vent so that material emitted from the septum does not enter the inlet.
SFC	Supercritical fluid chromatography, uses a supercritical fluid as the mobile phase, can be coupled directly to mass spectrometry by using capillary columns.
SFE	Supercritical fluid extraction, as extraction medium mostly CO_2 is used, modifiers, for example, methanol or ethyl acetate, can be added to optimize matrix specific extraction efficiencies.
Silica gel	A solid adsorbent, a porous form of silicon dioxide with high affinity to water.
Significant figures	Those digits in a number that are known with certainty, plus the first uncertain digit. Example: three significant figures have 0.104, 1.04, 104, 1.04×10^4. The 1 and the middle 0 are certain, and the 4 is uncertain, but significant. Initial zeroes are never significant. Exponential number has no effect on the number of significant figures.
SIM	Selected ion monitoring, recording of individual pre-selected ion masses for target compound detection, as opposed to full scan, → also MID.
SIM descriptor	Data acquisition control setup for SIM analyses, contains the selected specific masses of the analytes (m/z values), the individual dwell times and the retention times for timely switching the detection to other analytes in the same run.

SIMDIS	Simulated distillation, a boiling-point separation technique that simulates physical distillation of petroleum products.
Single laboratory study	An analytical study involving a single laboratory using one method to analyze the same or different test materials under different conditions over justified long time intervals (in-house validation).
SIS	Selected ion storage, describes the SIM measuring technique in ion trap MS, → SIM, → also waveform ion isolation.
SISCOM	Search for identical and similar compounds, program and search procedure for mass spectra in libraries developed by H. Damen, D. Henneberg, B. Weimann, Max Planck Institute for Coal Research, Mühlheim, Germany (DOI: 10.1016/S0003-2670(01)83095-6).
Skewing	Reversal of the relative intensities in a mass spectrum caused by changing the substance concentration during a scan. Skewing occurs with beam instruments during slow mass scans in the steep rising or falling slopes of GC peaks.
SN	Separation number or Trennzahl (TZ). A measurement of the number of peaks that could be placed with baseline resolution between two sequential peaks, z and $z+1$, in a homologous series such as two hydrocarbons: $$SN = \frac{t_{R(z+1)} t_{R(z)}}{w_{h(z+1)} + w_{h(z)}} - 1$$
S/N	Signal-to-noise ratio, the ratio of the peak height to the noise level. Peak height is measured from the average noise level to peak top. Noise is measured as the width of the noise band (typically 4 σ), excluding known signals.
Soft ionization	Formation of gas-phase ions without extensive fragmentation, → chemical ionization.
Solutes	Chemical substances that can be separated by chromatography.
Solvent effect	Focusing the analytes to a narrow band at the beginning of a capillary column by means of condensation of the solvent and directed evaporation.
Solvent flooding	A source of peak-shape distortion caused by excessive solvent condensation inside the column during and after splitless or on-column injection.

Solvent flush	GC injection technique, also known as *sandwich technique*, whereby a liquid (solvent, derivatization agent) is first drawn up into the syringe, followed by an air plug to act as a barrier and finally the sample, → Sandwich technique.
Solvent flushing	A column rinsing technique that can remove non-volatile sample residue and partially restore column performance.
Solvent split	GC injection technique using the PTV injector so that larger quantities of diluted extracts can be applied from more than 2 µL up to a LC-GC coupling. The excess of solvent is evaporated through the split line while analytes are concentrated on a support like glass wool in the inlet liner. A large difference between the boiling points (volatility) of the solvent and analytes is required.
Space charge effect	Result of mutual repulsion of particles of like charge that limits the current in a charged-particle beam or packet and causes some ion motion in addition to that caused by external fields.
SPE	Solid phase extraction, an analyte concentration and sample clean-up technique.
Specificity	The term *specificity* is considered as an absolute term, and thus cannot be graded. The IUPAC states that specificity is the ultimate of selectivity. A method is specific for one analyte or not. This suggests that no component other than the analyte contributes to the result. Hardly any method is that specific and, in general, the term should be avoided, → Selectivity.
SPI	Septum equipped programmable injector, special design of a cold injection system, which can be used exclusively for total sample transfer.
Spiked sample	A sample prepared by adding the target analytes to a blank matrix sample. Used in an analytical method to determine the effect of the matrix on the recovery efficiency, → Matrix effect, → Recovery.
Split injection	The sample size to be analyzed is adjusted to suit capillary column capacity requirements by splitting off a major fraction of sample vapors in the inlet so that as little as 0.1% enters the column. The rest is vented through the split exit.

Splitless injection	The total sample transfer into the analytical column. During the first minutes of the injection process, the sample is not split and enters only the analytical column. The split valve is opened after transfer of 2 to 3 liner volumes to purge the remaining sample vapor from the inlet. As much as 99% of the sample enters the column.
SPME	Solid phase micro extraction, the solventless sample extraction technique that uses sorbent coated thin fused silica or steel rods for manual or automated extractions from liquid immersion ($\rightarrow$ DI-SPME) or headspace ($\rightarrow$ HS-SPME) for thermal desorption in the GC injector (or mobile phase solvent dissolution for LC applications).
Spray and trap	Extraction procedure for foaming aqueous liquids, the liquid stream is sprayed into a purge gas stream.
SRM	$\rightarrow$ Selected reaction monitoring.
SSL	Split-/splitless injector.
Stable ion	Ion with internal energy sufficiently low that it does not rearrange or dissociate prior to detection in a mass spectrometer.
Standard addition method	A procedure in which the test sample is divided in two (or more) test portions. One portion is analyzed as such. Known amounts of an analyte standard are added to the other test portions before analysis. The concetration of the standard added has to be between two and five times the estimated amount of the analyte in the sample. This procedure is designed to determine the content of an analyte in a sample, taking account of the recovery of analytes impaired by matrix effects. Often used as quantitative calibration method in headspace analysis for matrix independent calibrations.
Standard operating procedures (SOPs)	The established written and approved analytical procedures of a laboratory.
Stationary phase	Liquid or solid materials coated inside an analytical column that selectively retain analytes.
SUMMA canister	Passivated stainless steel canisters for air analysis (e.g. EPA method TO 14/15) for collection of samples and standardization. The inner surface is deactivated by a passivation procedure involving a Cr/Ni oxide layer. Alternative passivation is available by a Silonite™ ceramic coating.

Surrogate standard	The internal standard for quantification, which is added to a homogenized sample before starting the sample preparation. The ratio of the surrogate standard to other internal standards added during the course of the clean-up is used to calculate the recovery efficiency.
SVOCs	Semi-volatile compounds, in aqueous media polar compounds with high partition coefficients, typical boiling point ranges above 240 °C. Examples are with pesticides, phenols, or plasticizers.
SWIFT	Stored waveform inverse Fourier transformation, technique to create excitation waveforms for ions in FT-ICR mass spectrometer or → Paul ion traps. An excitation waveform in the time-domain is generated by taking the inverse Fourier transform of an appropriate frequency domain programmed excitation spectrum, in which the resonance frequencies of ions to be excited are included. This procedure may be used for selection of precursor ions in MS/MS experiments.
Systematic error	A consistent deviation in the results of sampling or analytical processes from the expected or known value. Such error can be caused by a faulty instrument operation or calibration, by human or methodological bias.

T

T_0	Room temperature.
T_c	Column temperature.
t_M	Retention time of an unretained compound. The time required for one column volume (V_G) of carrier gas to pass through a column.
t_R	Retention time. The time required for a peak to pass through a column.
t'_R	Adjusted retention time; $t'_R = t_R - t_M$.
Tandem-in-space	MS/MS analysis with beam mass spectrometers. Ion formation, selection, collision induced decomposition and acquisition of the product ion spectrum take place continuously in separate sections of the mass spectrometer, for example, triple quadrupole MS.

Tandem-in-time	MS/MS analysis with ion storage mass spectrometers. Ion formation, selection, collision-induced decomposition and acquisition of the product ion spectrum take place in the same location of the mass spectrometer but sequentially in time, for example, ion trap MS, ICR MS.
Tandem mass spectrometry	→ Mass spectrometry/mass spectrometry coupling (MS/MS).
Target compound	Analyte to be quantitatively determined, usually taken from a list of compounds from regulations or directives. SRM/MRM modes typically serve the highly selective target compound analysis.
TCA	Target compound analysis, multicomponent analysis, multimethod. The analytical strategy with the setup and data evaluation for the quantitative monitoring of a selected group of target compounds. For data evaluation the analytes are grouped together in a target compound list for analysis and are searched for using automated routines by spectrum comparison in a retention time window, identified and quantified when found.
TCD	Thermal-conductivity detection, a thermal conductivity detector measures the differential thermal conductivity of carrier and reference-gas flows. Solutes emerging from a column change the carrier gas thermal conductivity and produce a response. TCD is a universal detection technique with moderate sensitivity depending on the thermal conductivity of the analytes.
TCDD	Tetrachlorodibenzo-p-dioxin isomers.
TCDF	Tetrachlorodibenzofuran isomers.
Tenax	Non-polar synthetic polymer based on 2,6-diphenyl-p-phenyleneoxide used as an adsorption material for the concentration of VOC air samples or in purge and trap analysis. Tenax TA (trapping agent) is the standard material in use, Tenax GR contains graphite.
TE/GC-MS	Thermal extraction GC-MS using a thermal extraction unit for solid sample material as inlet system of the gas chromatograph.
TF-SPME	Thin-Film-Solid Phase Microextraction is a further development of →SPME with a large volume and surface area of the sorbent phase for increased sensitivity. The sorbent material is applied to a rectangular mesh which is placed for extraction into the headspace of a sample or immersed into a liquid sample. Analysis is performed by → Thermal desorption in a GC injector.
TFME	Thin-Film-Microextraction, → TF-SPME.
Theoretical plate	A hypothetical entity inside a column that exists by analogy to a multiple plate distillation column. As solutes migrate through a column, they partition between the stationary phase and the carrier gas. Although this process is continuous, chromatographers often visualize a step-wise model. One step corresponds approximately to a theoretical plate.

Thermal ionization (TI)	Ionization of a neutral species through contact with a high temperature surface.
TIC	Total ion current, sum of all acquired intensities in a mass spectrum, → also RIC.
Time lag focusing	Energy focusing in a time-of-flight mass spectrometer that is accomplished by introducing a time delay between the formation of the ions and the application of the accelerating voltage pulse. Ion formation may be in the gas phase or at a sample surface. Related term: *delayed extraction*.
Time-of-flight mass spectrometer (TOF-MS)	Instrument that separates ions by m/z in a field-free region after acceleration to a fixed kinetic energy.
TMCS	Trimethylchlorosilane, trimethylsilyl chloride, silylation agent (catalyst).
TMS derivative	Trimethylsilyl derivative, a chemical derivative to increase substance volatility to facilitate chromatographic application, typical diagnostic mass m/z 73.
TMSH	Trimethylsulfonium hydroxide, derivatization agent for esterification and methylation, for example, free fatty acids, phenylureas, and so on. TMSH was used successfully for the derivatization in the insert of the → PTV cold injection system.
Total ion current (TIC)	Sum of all the separate ion currents carried by the ions of different m/z contributing to a complete mass spectrum or in a specified m/z range of a mass spectrum.
Total ion current chromatogram	Chromatogram obtained by plotting the total ion current detected in each of a series of mass spectra covering the acquired mass range recorded as a function of retention time.
Total sample transfer	GC injection technique whereby the entire injection volume reaches the column with or without evaporation, → splitless injection.
Traceability	The property of a result of a measurement whereby it can be related to appropriate standards, generally international or national standards, through an unbroken chain of custody.
Traceability chain	Relation of measurement results for a routine test sample to the reference point via a sequence of calibrations. Uncertainties, present in all procedures and calibrators, are propagated to the final result.

Transmission	The ratio of the number of ions leaving a region of a mass spectrometer to the number entering another region, e.g. ratio of ions leaving the ion source vs. reaching the detector.
Transmission quadrupole mass spectrometer	A mass spectrometer analyzer that consists of four parallel rods whose opposing poles are electrically connected. The voltage applied to the rods is a superposition of a static potential and a sinusoidal radio frequency potential. The motion of an ion in the x and y dimensions is described by the Mathieu equation whose solutions show that ions in a particular m/z range can be transmitted by oscillation along the z-axis.
Triple quadrupole mass spectrometer	A tandem mass spectrometer comprising two transmission quadrupole mass spectrometers in series, with a non-resolving (RF-only) quadrupole between them to act as a → Collision cell.
Tropylium ion	The characteristic ion in the spectrum of alkylaromatics, $C_7H_7^+$ m/z 91, is formed by benzyl cleavage of the longest alky chain. The structure of the seven-membered ring (after internal rearrangement) is responsible for the high stability of the ion.
TRT	Temperature rise time, heating up period in pyrolysis until the set pyrolysis temperature is reached. A fast TRT is a quality feature of the pyrolyzer.
Trueness	The measurement trueness is the closeness of agreement between the average of a number of replicate measured quantity values and a reference quantity value.
TSD	Thermionic-specific detection, → NPD (nitrogen–phosphorus detection).
TSIM	N-Trimethylsilyl-imidazole, a potent derivatization agent for silylation.
TZ	Trennzahl, → Separation number.
$\bar{u}_{avrg}$	Average linear carrier gas velocity; $\bar{u}_{avrg} = L/t_M$.
u_o	Carrier gas velocity at the column outlet; $u_o = \bar{u}_{avrg}/j$.

U

u_{opt}	Optimum linear gas velocity. The carrier gas velocity corresponding to the $\bar{u}$ minimum theoretical plate height, ignoring stationary-phase contributions to band broadening: $$u_{opt} = 8\left(\frac{D_G}{d_c}\right)\sqrt{\frac{3(1+k)^2}{1+6k+11k^2}}$$
u	The unified atomic mass unit u, a non-SI unit of mass defined as one twelfth of the mass of one atom of ^{12}C in its ground state and equal to $1.6605402(10) \times 10^{-27}$ kg. The term *atomic mass unit* is deprecated, → Dalton.

Note: The term *atomic mass unit* (amu) is ambiguous as it has been used to denote atomic masses measured relative to a single atom of ^{16}O or to the isotope-averaged mass of an oxygen atom or to a single atom of ^{12}C. |
UAR	*Unknown analytical response*, this term refers to GC peaks in a multicomponent target compound analysis (→ TCA), which are not included in the list of target compounds.
Uncertainty	The measurement uncertainty is the non-negative parameter characterizing the dispersion of the quantity values being attributed to a measurand and provides a quantitative indication of the quality of a measurement result.
Unimolecular dissociation	Fragmentation reaction in which the molecularity is treated as one, irrespective of whether the dissociative state is that of a metastable ion produced in the ion source or results from collisional excitation of a stable ion.
Unstable ion	Ion with sufficient energy to dissociate within the ion source.
UTE, UTE%	Utilization of theoretical efficiency, → CE (coating efficiency).

V

V_G	The volume of carrier gas contained in a column. For open-tubular columns and ignoring the stationary-phase film thickness: (d_f), $V_G = L\left(d_c^2/4\right)$.
V_L	The volume of (liquid) stationary phase contained in a column.

Glossary

Validation	The process of validation involves agreeing on specified requirements for performance characteristics such as selectivity, measuring interval, trueness and precision that are adequate for the intended use of the measurement procedure, and then confirming, on the basis of objective evidence, that they are fulfilled (verification).
Validated method	A method that has been determined to meet certain performance criteria for sampling or measurement operations.
van Deemter plot	The van Deemter plot, according to the equation of the same name, shows a hyperbolic function with the optimum carrier gas velocities for maximum column separation efficiency, → HETP.
Verification	Confirmation by examination and provision of evidence that specified requirements have been met.
Viton	The brand name of synthetic rubber and fluoropolymer elastomer commonly used in O-rings and other molded or extruded goods. The name is trademarked by DuPont Performance Elastomers L.L.C., → http://en.wikipedia.org/wiki/Viton.
VOCs	Volatile organic compounds, a group of analytes with low partition coefficient in aqueous media, determined by static or dynamic HS methods. Boiling point range from c. 50 to 100 °C to c. 240 to 260 °C. Examples are solvents, flavors, halogenated hydrocarbons, or →BTEX.
VOCARB	Non-polar adsorbent filling for concentration of air samples, and in purge and trap analysis, multilayer filling based on graphitized carbon black (Carboxen) and carbon molecular sieves (Carbosieve) (Supelco brand).
VOCOL	Volatile organic compounds column, Sigma-Aldrich.
VVOCs	Very volatile organic compounds with boiling points up to 50–100 °C. Examples are the organic gases methane, propane, mercaptans, and monochloromethane.

W

w_b	The peak width at its base, measured in seconds. For a Gaussian peak, $w_b = 1.596\,(A_p/h_p)$.
w_h	The peak width at half height, measured in seconds. For a Gaussian peak, $w_h = 0.940\,(A_p/h_p)$.
Waveform ion isolation	Ion storage technique for the ion trap analyzer. By using resonance frequencies the ion trap can exclude ions of several m/z values from storage (e.g. from the matrix) and collect selectively determined pre-selected analyte ions, also SIS.
WBOT	Wide-bore open-tubular column, open-tubular (capillary) column with a nominal inner diameter of 530 µm.
WCOT	Wall-coated open-tubular column, a capillary column in which the stationary phase is coated on the column wall.
Werkhoff splitter	Adjustable flow splitter for sample injection into two capillary columns, or a split located after the analytical column, also known as the *glass cap cross divider*.
Wiswesser line notation	Code for chemical structures using an alphanumerical system, for example, for Lindane, L6TJ AG BG CG DG EG FG *GAMMA, contained e.g. in the Wiley library of mass spectra.
Within-laboratory reproducibility	Precision obtained in the same laboratory under stipulated (predetermined) conditions (concerning, e.g. method, test materials, operators, environment) over justified long-time intervals.
Working standard	A general term used to describe traceable standard dilutions for daily routine use produced from the stock standard, which are used, for example, to spike for recovery determination or to prepare calibration standards.

X

XAD	Synthetic polymer (resin) used as an adsorption material in air or purge and trap analysis as well as in the clean-up for chlorinated compounds, for example, dioxins, furans, PCBs.
x-Ion	Fragment ion containing the C-terminus formed upon dissociation of a protonated peptide at a backbone C—C bond.

XIC	Extracted ion chromatogram. The display of chromatograms of selected masses on screen →EIC.

Y

y-Ion	Fragment ion containing the C-terminus formed upon dissociation of a protonated peptide at a backbone C—N bond.

Z

z-Ion	Fragment ion containing the C-terminus formed upon dissociation of a protonated peptide at a backbone N—C bond.

Further Reading

Barwick, V.J. (ed.) (2023) *Eurachem Guide: Terminology in Analytical Measurement – Introduction to VIM 3*, 2nd edn, ISBN: 978-0-948926-40-2, www.eurachem.org.

Coplen, T.B. (2008) Provisional recommendations. *Chem. Int., IUPAC*, **30** (3). doi: 10.1515/ci.2008.30.3.19.

Coplen, T.B. (2011) Guidelines and recommended terms for expression of stable isotope-ratio and gas-ratio measurement results. *Rapid Commun. Mass Spectrom.*, **25**, 2538–2560.

Hinshaw, J.V. (1992) A compendium of GC terms and techniques. *LCGC*, **10** (7), 516–522.

Irish National Accreditation Board (INAB) (2016) *Guide to Method Validation for Quantitative Analysis in Chemical Testing Laboratories*, ISBN: 0865426848.

IUPAC (1993) Nomenclature for chromatography. *Pure Appl. Chem.*, **65**, 819–872.

IUPAC Task Group MS Terms et al. (2013) Definitions of terms relating to mass spectrometry (IUPAC Recommendations 2013). *Pure Appl. Chem.*, **85** (7), 1515–1609. doi: 10.1351/PAC-REC-06-04-06.

Sadek, P.C. (2004) *Illustrated Pocket Dictionary of Chromatography*, John Wiley & Sons Inc., Hoboken, NJ, USA.

SANCO (2012) Method Validation and Quality Control Procedures for Pesticide Residue Analysis in Food and Feed. Document No. SANCO/12495/2011, implemented by 01/01/2012.

Schröder, E. (1991) *Massenspektroskopie*, Springer, Berlin.

Vessman, J. et al. (2001) Selectivity in analytical chemistry (IUPAC Recommendations 2001). *Pure Appl. Chem.*, **73** (8), 1381–1386.

Author Index

a
Adams, R.   19
Andersson, A.   599
Arthur, C.L.   27
Assadi, Y.   17
Aston, F.W.   5, 344

b
Ballschmiter, K.   495, 713
Bainbridge, K.T.   5
Bartky, W.   5
Berijani, S.   17
Beynon, J.H.   6
Biemann, K.   6
Boeker, P.   xx, 237

c
Christie, W.W.   442
Coon, J.   7
Cotter, R.J.   333
Cremer, E.   6

d
Damen, H.   445
Dandenau, R.D.   6
Dawson, P.H.   319
de Zeeuw, J.   233
Deans, D.R.   241, 272
Dehmelt, H.G.   7
Dempster, A.J.   5
Desty, D.H.   6, 128
Donike, M.   159

e
Ehleringer, J.R.   345
Enke, C.G.   6, 380
Ettre, L.S.   128

f
Field, F.H.   6, 296
Finnigan, B.   6

g
Gall, L.N.   329
Gamble, K.R.   25
Giacobbo, H.   92
Gohlke, R.S.   5
Golay, M.J.E.   6
Golikov, Y.   329
Good, T.   19
Grimme, S.   293
Grob, K.   3, 128, 136, 168
Gross, J.H.   468
Guzowski, J.P.   34, 56

h
Halasz, I.   128
Hayes, J.M.   345
Henneberg, D.   443, 445
Herzog, R.F.K.   5
Hipple, J.A.   5
Hustrulid, A.   5

j
James, A.T.   6

Handbook of GC-MS: Fundamentals and Applications, Fourth Edition. Hans-Joachim Hübschmann.
© 2025 WILEY-VCH GmbH. Published 2025 by WILEY-VCH GmbH.

Johnson, W.H. 6
Jordan, E.B. 5

k

Kovats, E. 225, 431
Kratz, P.D. 225

l

Lehotay, S.J. 1, 11, 13, 598
Lovelock, J.E. 6, 254

m

Majors, R.E. 99
Makarov, A. 7, 329
Mamyrin, B.A. 333
Markelov, M. 34, 56
Martin, A.J.P. 5, 216
Matthews, D.E. 345
Mattauch, J. 5
Maurer, H.H. 440
McDaniel, E.W. 336
McLafferty, F.W. 6, 291
McLaren, I.H. 6, 333
Michnowicz, J.A. 291
Mondello, L. 439
Morris, B.D. 601
Munson, B. 6, 296

n

Nier, A.O.C. 6, 344, 462

o

Ohlin, B. 599
Ohtani, H. 442

p

Paul, W. 6, 7
Pawliszyn, J. 27
Peterson, A. 7
Pfleger, K. 440
Poy, F. 128

r

Ramsay, N.F. 7
Rezaee, M. 17

s

Sander, R. 50
Sandra, P. 87
Schomburg, G. 128, 168
Schriner, R.B. 601
Shoulders, K. 6
Simmons, M.C. 239
Simon, W. 92
Smith, L.G. 5
Snyder, L.R. 239
Soddy, F. 344
Stein, B. 445, 450
Stephens, W.E. 5, 333
Synge, R.L.M. 5, 216

t

Telepchak, M.J. 19
Thomson 344
Tsuge, S. 442
Tsvet, M.S. 5

u

Uthe, P.M. 6

v

van den Dool, H. 225
Vogt, W. 128
von Ardenne, M. 306
von Zahn, U. 6

w

Watanabe, C. 442
Weber, A. 440
Weimann, B. 445
Whetten, N.R. 319
Wiley, W.C. 6, 333

y

Yarkov, A. 443
Yost, R.A. 6, 380

z

Zerenner, E.H. 6

Subject Index

a

absorption   42, 47, 262, 267, 277
accelerated solvent extraction   101, 694
accurate mass   7, 323, 332, 339, 374, 436, 462, 693, 730, 741
acetate buffer   14, 573, 594
acetylation   40, 763
Achilles heel   128
acquisition rate   235, 246, 335, 506, 598
acrylate   30, 37, 44, 45
acrylonitrile butadiene styrene (ABS)   651
actinomycete bacteria   632
activated charcoal   68, 79, 84
active ingredients   603, 626
active sites   222, 515, 711
adduct ion   301, 312, 420, 769
adsorption   22, 23, 42, 59, 62, 65, 77–85, 141, 144, 150, 515, 542
adulteration   173, 338, 637
AED   251–254, 432
AFID   259
air analysis   77–79, 86, 541
air-water distribution ratio   48
alcoholic beverages   619
allergens   644, 649
Amberlite XAD-2   78, 84, 649
Amberlite XAD-4   78, 81
ambient BP   161–163
AMDIS   426–433, 438, 637–644, 657, 668, 673–677
ammunition plant   618, 619

analysis time   28, 35, 148, 152, 161, 194, 197, 200–204, 216, 221, 226–228, 235, 240, 597, 612, 713
analyte protectants   222, 596, 626, 629
animal feed   101, 718
antifouling agents   573
APCI   310, 331, 382
APGC   2, 196, 311
appearance potential   294
Aquasil   141
aqueous samples   17, 25, 38, 47, 53, 64, 74, 87, 549, 570, 616, 700, 748
Arabidopsis Thaliana   657
aroma profiling   637, 644
artificial intelligence (AI)   249, 433
ASE *see* accelerated solvent extraction
ASTM Method 3710   191
ASTM Method 5501   191
ASTM Method 5504   191
ASTM Method 8368   268
ASTM Method D 2887   191
ASTM Method D 5134   191
ASTM Method D 5623   191
ASTM Method D 6584   192
ASTM Method D2369   678
ASTM Method D5116-97   96
ASTM Method D6196   82
ASTM Method D7143-05   96
ASTM Method D7706   96
ASTM Method D8071   268
ASTM Method D8267   268
ASTM oxygenates methods   192
ASTM Standard Practice D-7210   102

Handbook of GC-MS: Fundamentals and Applications, Fourth Edition. Hans-Joachim Hübschmann.
© 2025 WILEY-VCH GmbH. Published 2025 by WILEY-VCH GmbH.

Subject Index

ASTM Standard Practice D-7567   102
asymmetric peaks   217
atmospheric pressure chemical ionization *see* APCI
atomic emission detector *see* AED
authentication   173
authenticity   338
automation   9–12, 15, 17, 18, 25–29, 31, 32, 34, 36, 39, 42, 46, 47, 49, 50, 56, 59, 64, 70, 73, 74, 84, 85, 87–89, 98–101, 129, 135, 143, 158, 162, 164, 169, 171–173, 227, 250, 265, 554, 594, 599, 614, 632, 647, 665, 683
automobile industry   677
automotive paint   89
AutoSRM   387, 658
autotune   389, 741
AutoTwister   87
avocado   595
axial diffusion   197, 238
ayurvedic churna   604
azo dyes   683

b

backflush   66, 128, 151–155, 164, 250, 273, 355
background subtraction   417, 423–426, 449, 452
background   67, 69, 75, 78, 152, 269, 307, 324, 384, 399, 414, 480, 508, 523–527, 668, 734
baffled liner   143, 159, 167, 600, 699
Baijiu   619
bake-out phase   70
beverages   34, 73, 77, 253, 265–266, 268, 338, 637
BFB tune criteria   552
BFB tuning   373, 391, 395, 549
BFRs   504, 697, 700, 739, 747
biodiesel   192
bioethanol   191
biological samples   101, 102
biomarker   385, 656
biomedical analysis   240

BKF *see* backflush
blood alcohol   54, 193
blood   732, 763
bottom-sensing   17, 18
breakthrough volume   19, 77
brewing industry   189
BSIA *see* bulk stable isotope analysis
BTEXTRAP   69
BTV *see* breakthrough volume
bulk stable isotope analysis   351
butyl rubber septa   73, 527, 531

c

C18 sorbent   14, 16, 25, 212, 601
calculation of gas flow rates   271
calibration curve   344, 512–515
calibration function   389, 509, 513
calibration substances   390
cancerogenic amines   683
cannabis   751
car interior materials   677
carbenium ion   469
Carbograph   79, 80
Carbograph 5TD   645
carbon wide range   30, 34, 37, 45, 48
Carbon WR *see* carbon wide range
carbonized molecular sieve   79, 80
CarbonX   601
Carbopack   43, 66, 68–70, 77, 79, 82
Carbopack C   79
Carbosieve   69, 77, 79
Carbosieve SIII   68, 80–82
Carbotrap   78–80, 82, 167
Carboxen   43–45, 66, 68–70, 77, 79, 80, 187
carrier effect   520, 521
carrier gas(es)   41, 71, 100, 132, 136, 141, 143, 174, 194, 204, 212, 258
  argon   196
  helium   64, 132–134, 149, 174, 195, 198–201, 211, 216, 229, 253, 258, 261, 263–267, 271, 396
  hydrogen   64, 132, 149, 194, 195, 198–201, 216, 237, 257, 265–267, 271, 396

nitrogen 64, 132–134, 196, 199, 229, 265–267, 271
purity 526
carrier gas flow 138, 139, 146, 160, 197–200, 205, 225, 227, 233, 249, 251, 253
  constant flow 136, 198, 199, 215, 229–232, 253
  constant pressure 174, 198, 199, 229, 231, 253, 270
  optimum flow conditions 198
carrier gas flow rate 216
carrier gas re-routing 76, 130
carrier gas regulation 129, 131
  back pressure regulation 131
  forward pressure regulation 130
  gas saver 133
  mass flow controller 131
carrier gas saving 132
carrier gas supply 171
carrier gas velocity 148, 174, 198, 199, 201, 202, 221, 224, 233, 238
carryover 47, 72, 75, 149
centrifugation 17, 18, 605, 759
centrifuge 651
cereals 12, 13
certified reference standards 82
cesium iodide 392
charcoal 68, 82
cheese 644
chemical ionization 6, 196, 295–313, 766, 767
  ion volumes 312
  negative see negative chemical ionization
  positive see positive chemical ionization
chemical noise 174, 373, 414, 424, 447, 508
chicken 12
China Method GB 8410-2006 678
China Method GB/T 19649-2005 102
China Method GB/T 21911-2008 625
China Method GB/T 27630-2011 678
China Ministry of Health 619
chiral column phases 183

chlorophyll 14, 16
choice of GC columns 173
chromatographic resolution 167, 201, 216–227, 239–245, 250, 353, 688, 695
chromatography parameters 212
Chromosorb 80, 82, 147, 163, 228
ChromSync 90
CI see chemical ionization
CI gases
  ammonia 300, 307, 767, 769
  argon 301
  benzene 301
  carbon monoxide 301
  charge exchange 301
  isobutane 300, 307
  methane 300, 306, 753, 767, 769
  methanol 300, 308
  nitric oxide 301
  nitrogen 301
  proton affinities 300
  water 300, 310, 498
CID 382, 383
  collision energy 382
collision gases 383
citrate buffer 595
classical detectors 9, 249, 251, 268, 430, 506
Clausius-Clapeyron equation 149
clean-up 9–14, 16, 17, 19, 22, 25, 26, 50, 70, 75, 98, 99, 102, 105, 151, 584, 594, 599
α-cleavage 468, 496
closed loop cooling 64, 167
closed-loop stripping (CLS) 632
CLP SOW Method OLM04.2 102
co-elution 244, 384, 416–429, 447, 713
Codex Alimentarius 583
coeluting compounds 243, 323, 335, 384, 597, 637, 640, 658, 665, 673, 677, 690, 701, 733
coffee 173
cold needle injection 147
collision energy 387
collision induced dissociation see CID

column bleed   134, 135, 167, 174–176, 203–205, 209, 210, 212, 233, 269, 391, 423, 525
column clipping   152, 155
column flow   130, 132, 133, 138–140, 146, 202, 265
column length   202, 204, 216, 221, 229, 231, 232, 241
column lifetime   155, 206
column storage   205
Commission Regulation (EEC) 2568/91   173
Commission Regulation (EU) 2017/771   718
compound specific isotope analysis   351, 569
comprehensive GC   199, 239, 242, 248
concentrated samples   148, 171
concentration dependent detectors   251
concurrent backflush   151–154
concurrent solvent evaporation   165, 172
concurrent solvent recondensation   16, 24, 131, 150, 151, 162, 164
conditioning station   41
confidence interval   344, 510, 513–515, 520
confirmatory methods   343, 718, 719
consumer products   739
contour plot   248
cooled injection system CIS   135
cosmetics   637
cost per sample   11, 85, 99, 600
cost reduction   28
cross-contamination   59, 74
CRSs see certified references standards
cryo-enrichment   168
cryo-milling   584, 601, 651
cryo-trap   355, 357
cryoconcentrator   543
cryofocusing   35, 64, 71, 72, 75, 87, 166, 167, 203, 246, 541, 543, 545, 571
cryohomogenization   594
CSIA see compound specific isotope analysis
CSR see concurrent solvent recondensation
cup liner   143
Curie temperature   92
CWR see carbon wide range
cyanobacteria   632
cyclo liners   143

d

dairy samples   73
data acquisition   322
  centroid mode   322
  profile mode   322
de Broglie wavelength   292
dead time   174, 197, 215, 224, 431
Deans switch   249, 272
decision limit   507–510
deconvolution   247, 336, 425, 506, 637, 656, 665
deep eutectic solvents   11
degradation   593, 739, 762
degradation pathways   569
Depot Area Air Monitoring System tubes   86
derivatization   10, 27, 32, 39, 40, 44, 56, 103, 164, 304, 471, 492, 656, 665, 751, 762
  acetylation   763
  diazotation   616
  DMOX   471
  ethylation   573
  in situ derivatization   572, 614
  iodination   614, 616
  methoximation   656, 661
  methoxyamination   660
  methoxybromination   471
  methylation   349, 492
  pentafluorpropionylation   763
  silylation   349, 656, 660, 661, 763
  TMSH   492
  two-step derivatization   660
desorption phase   66
detergents   649
DFTPP tuning   395
diatomaceous earth   104

Diesel 268
DiffLok end-cap 85
diffusion 81, 85, 134, 139, 142, 145, 197, 198, 223, 226, 250
diffusion pumps 399
diffusion-locking 85
DIN EN 12393-2 151
dioxin 159, 164, 190, 255, 295, 302–304, 311, 317, 323, 339, 343, 375–381, 385, 495, 504, 513, 520, 718–738
dioxin-like 712, 718, 729–737
direct column heating 227, 235
direct immersion sampling 27, 28, 34, 46
direct liquid introduction 168
direct MS coupling 233
discrimination 144, 147, 148, 157
disperser 17
dispersive SPE 11, 15, 594, 601–605
dissociative electron capture 254
DLLME 17–19
doping control 752
double focusing MS 316
drinking water 62, 64, 88, 263, 265, 560, 614, 619, 632, 748
Dronabinol 751
drug analysis 27, 34, 40, 141, 189, 208, 267, 339, 345, 496
drug screening 762
dry purge phase 66, 69, 75, 79, 562, 570, 645
dSPE *see* dispersive SPE
dual detection 668
dual-column adapter 639
dual-column separation 637
dwell time 379, 507
dynamic headspace analysis 9, 10, 53, 62, 74, 76, 167, 338, 644, 647
  ITEX DHS 74, 76, 265
  purge and trap 62, 63, 67, 68, 70, 71, 76, 130
  spray-and-trap 63
dynamic range 158, 256, 265, 267, 335

e

ECD *see* electron capture detector
ECD-MS 302
Eddy diffusion 198, 215
edible oils 72, 173
eggs 601, 732
EIC *see* extracted ion chromatogram
ELCD *see* electrolytical conductivity detector
electrolytical conductivity detector 256
electron capture 10, 254, 261, 262
electron capture detector 10, 18, 66, 190, 250–252, 254–256, 260, 262, 432
electron mass 392
electronic components 98
electronic flow and pressure control 130
electronic pneumatic control 130
element-specific detectors 251
elemental formula 296, 329, 457
elution temperature 170, 203, 204, 209, 226, 233, 238, 741
emergency toxicology 757
EN ISO 16017 83
EN Method 14013 192
EN Method 14105 192
EN Method 14106 192
EN Method 14110 192
endocrine disruptors 88
enrichment 18, 19, 34, 35, 77, 79, 166
environmental analysis 1, 12, 26, 48, 62, 65, 72, 89, 98, 101, 206, 235, 240, 247, 253–255, 263, 265, 268, 549, 697, 748
environmental contaminants 235, 594, 598, 697–712
EPA Method 0030 82
EPA Method 0031 82
EPA Method 1613 190, 344, 520, 721, 730
EPA Method 1613B 375
EPA Method 1614 739
EPA Method 1624 79
EPA Method 1625 190
EPA Method 1994b 721
EPA Method 1P-1B 82
EPA Method 311 98
EPA Method 3445A 101
EPA Method 3540 102
EPA Method 3541 102

EPA Method 3550   102
EPA Method 502   263
EPA Method 502.2   189
EPA Method 503.1   263
EPA Method 504   190
EPA Method 505   190
EPA Method 507   191
EPA Method 508   190
EPA Method 524.2   189, 560
EPA Method 525.2   25
EPA Method 542.2   79
EPA Method 543   25
EPA Method 552   190
EPA Method 601   189
EPA Method 602   189, 263
EPA Method 608   190, 191
EPA Method 610   228, 229
EPA Method 624   79, 189, 550
EPA Method 625   190
EPA Method 8010   189
EPA Method 8020   189
EPA Method 8080   190
EPA Method 8081   190, 191
EPA Method 8081b   141
EPA Method 8082   190
EPA Method 8141A   191
EPA Method 8151   102, 190
EPA Method 8240   189, 560
EPA Method 8260   189, 549, 560
EPA Method 8270D   190
EPA Method 8270E   9
EPA Method 8275   98
EPA Method CLP   191
EPA Method SW-846   82
EPA Method TO-1   82
EPA Method TO-14   82, 87, 541, 546
EPA Method TO-15   87, 541, 543
EPA Method TO-17   82
EPA Method TO-2   82
EPA Methods 524.2   550
EPA SW-846 Test Method 3545A   102
epoxidized soy bean oil   173
equilibrium extraction   36, 38, 49–52, 70, 71
error probability   509

ESBO *see* epoxidized soy bean oil
essential oils   268, 602, 603, 637
EU Commission Regulation 278/2012   343
EU Commission Regulation No. 1883/2006   343
EU Commission Regulation No. 589/2014   343
EU Commission Regulations No. 252/2012   343
EU Directive 2002/61/EC   685
EU Directive 2003/11/EC   739
EU Directive 2009/48/EC   649
EU Regulation 1881/2006   692
EU Regulation 835/2011   692
EU SANCO identification criteria   701
European Council Directive 96/23/EC   516
European Union Water Framework Directive   572
evaporation   19, 20, 24, 32, 34, 100, 105, 138–141, 143, 144, 147–149, 151, 157, 159, 163, 165, 168, 169, 172, 754
exhaustive evaporation   70
exhaustive extraction   36, 40
exothermic reaction   13, 297, 306
explosives   193, 238, 339, 498
external standard   148, 429, 516, 522, 544, 557, 579, 596, 605, 620, 628, 650, 749, 806
extractables   16, 665–667
extractant   17
extracted ion chromatogram   415, 624
extraction time   38, 47, 88, 100
Extrelut   752

f

fabric softeners   649
FAMEs   180, 192, 204, 208–211, 257, 356, 442, 472
fast GC   131, 179, 204, 227, 228, 232, 233, 235–237, 239, 242, 272, 335, 337, 594, 597, 691, 712, 713
fat removal   105, 599

Subject Index

fatty matrices 13, 599
FD *see* fluorescence detector
FET *see* full evaporation technique
FF-TG-GC *see* flow-field thermal gradient GC
FID *see* flame-ionization detector
field sampling 27, 36, 74
film thickness 64, 70, 166, 174, 198, 200, 202, 204, 221, 224, 228, 232, 237
Firemaster 320
fish 12, 595, 698
fish farming 697
fish feed 697
flame-ionization detector 89, 132, 173, 211, 229–232, 241, 251–253, 255, 257–259, 261–263, 266, 268, 269, 432
flamephotometric detector 251, 253, 258, 265, 432
flameproofing agents 504
flavoring 628, 637, 691
flavors 27, 36, 53, 62, 77, 189, 206, 253, 257, 267, 268, 637, 691
flight time *see* ion flight time
flooded zone 170
flow optimization 197
flow path 131, 222, 273
flow switching 249, 271
flow-field thermal gradient GC 237
fluorescence detector 584
foaming samples 63, 65, 75
fog 677
Fomblin 399
food 11, 13, 25, 26, 46, 48, 50, 53, 57, 77, 89, 96, 101, 152, 173, 210, 240, 253, 254, 265, 267, 268, 338, 595, 599, 645, 691, 732
food container 46
food packaging materials 62, 72, 77
food quality 644
food safety 572, 577, 603, 619, 626, 697, 718
food safety analysis 1, 25, 26, 48, 173, 210, 235
forensic analysis 25, 26, 89, 268, 751

Fourier transformation 332
FPD *see* flamephotometric detector
fragment ion 90, 294, 328, 381, 416, 418, 420, 457, 462, 491, 690, 740, 754, 769
fragmentation 292–297, 300, 306–308, 311, 314, 339, 381–384, 415, 418, 434, 436, 468–472, 657
  neutral loss 384, 457, 470
fragmentation pathways 499, 673, 766–770
fragmentation pattern 365, 373, 419, 436, 457, 472, 492, 496, 668, 757
fragmentation rules 90, 468
fragrances 27, 44, 62, 189, 253, 257, 267, 268, 637, 649
frit liner 143
fruits 152, 578, 584, 595, 627
fuels 258
full evaporation technique 32, 56, 639
full scan screening 757
full scan spectra 195, 234, 240, 247, 332, 339, 363–365, 387, 769
full scan confirmation 614
fungicides 577

g

GAC *see* green analytical chemistry
gas-phase acidity scale 305
gasoline 549
GC column phases 175–193, 205–212
GC method
  isothermal phase 15, 131, 171, 226
  solvent peak 15, 53, 131, 163, 164, 169–171
GC method translator 196, 198
GC × GC 209, 223, 239–245, 335–339
  GC × GC/HRMS 247
  GC × GC/IRMS 247
  GC × GC/QMS 246
  modulator 242, 245, 246, 248
  GC × GC/TOF-MS 247
GC-ICP-MS 251
GC-IRMS 251
GC-LC Concordance 90

Subject Index

GC-O *see* olfactometry
GC-Q-TOF-MS   334, 335
GCB *see* graphitized carbon black
gel permeation chromatography   10, 99, 102, 601, 722
genotype   656
glass beads   64, 70, 167
glass cap cross divider   270
glass dome   249, 270
glass wool   138, 140, 141, 143, 144, 147, 150, 151, 157, 159, 163, 167
GMW 15634 standard   678
goose neck liner   139
GPC *see* gel permeation chromatography
grains   658
grapeseed oil   173
graphene   23, 48
graphite ferrules   273, 525
graphitized carbon black   14, 16, 69, 78–80, 605
green analysis   24
green analytical chemistry   9, 11, 17, 25, 27, 28, 98, 637
groundwater   614, 619
guard column   100, 205, 234, 269, 573, 595

h

Hagen–Poiseulle law   271
halfmil columns   256
Hall detector   256
hard discs   98
HayeSep D   80
height equivalent to a theoretical plate   174, 196–198, 201, 214, 216, 224
He-PDPID *see* PDID
headspace analysis   554
  dynamic headspace   9, 53, 56, 62, 76, 167, 639, 647
  multiple headspace extraction   40, 56
  pressure balanced injection   60
  sample loop   61
  static headspace   38, 49–52, 56, 59–75, 84, 148, 166, 554–558, 639, 673
  syringe injection   60
  vial pressurization   61
headspace liners   143
headspace sorptive extraction   87
heart cutting   239, 241, 242, 249, 272
heat of evaporation   149
helium ionization detector   252, 258, 261
Henry's law   50
herbicides   102, 191
herbs   595, 603, 694
HETP *see* height equivalent to a theoretical plate
HID, *see* helium ionization detector
high boiling compounds   3, 80, 93, 140, 145, 152, 154, 157, 159, 161, 170, 194, 225, 234, 273
high mass resolution   291, 323, 328, 332, 339, 374, 737
high mass tune   382
high resolution accurate mass   7, 10, 324, 339, 390, 541, 693
high resolution MS   240, 317, 322, 691, 694, 718, 739, 748
HLB *see* hydrophilic-lipophilic balance
HLB-WAX   48
HLB-WCX   48
H53 method   238
honey   12, 32
hops   268
hot wire detector   265
HPLC   34, 98, 99, 382, 492, 584
HRAM *see* high resolution accurate mass
HRMS *see* high resolution MS
HS-GC *see* headspace analysis
HS-SPME *see* SPME
HSSE *see* headspace sorptive extraction
human error   25
human nose detector   268
hydrocarbon background   324
hydrolysis   577, 753, 762
hydrophilic-lipophilic balance   25, 36, 48
HyperChrom GC   238
hyphenated techniques   9, 10

i

IDL *see* instrument detection limit
IDL calculation   511
IMS *see* ion mobility MS
in-cell clean-up   104
in-cell sample preparation   104
INCOS library search   444
incubation temperature   27, 38, 47, 50, 51, 53, 56
indoor air analysis   77, 96, 649, 677
information content   381
injection port septa   134
   average lifetime   135
   bleed and temperature-optimized (BTO)   135
   Merlin MicroSeal   135, 136
injection techniques   10, 129, 131, 138, 144
   cold needle with liquid band formation   138
   hot needle with thermospray   138, 144, 146
injection volume   60, 75, 131, 139, 145, 146, 149, 151, 162, 170
injector temperature gradient   135
inlet liner   41, 42, 138–140, 143, 152, 155, 158, 166, 171
inlet liner activity   140–142
inlet liner deactivation   140–142, 147, 150, 574, 700
instrument detection limit   344, 415, 510
internal diameter   159, 200–202, 204, 230
internal mass calibration   375, 390
internal standard   40, 53, 57, 59, 62, 74, 146, 148, 517
   isotopically labeled   247, 517, 519
interpretation of mass spectra   451
in-tube extraction   74–76
ion flight time   291, 314, 322, 335, 388, 395, 420
ion mobility MS   53, 336
ion ratio confirmation   718
ion source   10, 53, 195, 196, 202, 230, 233
ion source matrix effect   10, 16

ion trap MS   541, 560, 614, 762
ion trap MS/MS   649, 747, 766
ionic liquids   48, 209, 210
ionization   196, 258, 261–264, 292, 312, 359, 418, 468
   chemical ionization   295
   electron ionization   5, 292, 418, 436, 554
   energy   293
   ionization selectivity   297
   low voltage ionization   292, 331, 381
   potential (IP)   263, 292, 701
ions source vacuum   249
irm-GC-MS *see* isotope ratio mass spectrometry
ISO 11890-1   678
ISO 11890-2   677, 678
ISO 12219-3   96
ISO 14362-1 to-3   685
ISO/EN 16000-6/-9/-10/-11   96
isobaric interference   359
isomeric compounds   267, 433, 637, 688, 701, 721
   stereoisomers   638
isothermal phase   15, 131, 147, 151, 159, 160, 169, 171, 196–198, 204, 205, 224–226, 242, 431
isotope abundance   291, 346, 463
isotope dilution quantitation   517, 519, 520, 541, 693, 721, 734, 739
isotope effects   347, 349
isotope pattern   295, 375, 446, 457, 461–467, 473, 480, 490, 495, 496, 504, 507, 575, 734, 737
isotope ratio mass spectrometry (IRMS)   249, 344, 345, 569
   delta notation   346
   Faraday cup   359
   online combustion   355
   online reduction   355
   primary reference materials   362
isotope separation   353
isotopic signature   569
ISTD *see* internal standard

j

jet fuel 268
juices 12, 27, 577

k

kidney 12, 573
Kovats index 225, 431, 436–439

l

landfill gases 77
large volume injection 24, 25, 130, 155, 157, 162–165
larger sample volumes 163, 170
LC-GC analysis 10, 99, 100, 162, 165, 171–173, 200
LC-MS 12, 15, 16, 18, 25, 34, 87, 102, 311, 594
leachables 665
leak check 525
leak detector 134, 523, 525
leaks on GC side 48, 134, 136, 138, 273, 523–525
leaks on MS side 270, 508, 523–525
leather 679, 683
lettuce 601
library search 90, 195, 248, 267, 268, 556, 656, 683, 690, 768
  reverse search 673
life sciences analysis 268
limit of detection 2, 34, 238, 243, 344, 415, 509, 512, 514, 544, 546, 558–560, 564, 583, 593, 617, 624, 655, 696, 732, 749, 754
limit of quantitation 24, 39, 378, 385, 512–516, 564, 583, 594, 624, 655, 696, 711, 719, 745, 754
liner deactivation 140, 142, 147, 150, 273, 574
  AquaSil 141, 142, 574
  Siltek 604, 615, 699, 700
  SurfaSil 141, 142, 574
lipids 16, 173, 339, 441, 442, 700
lipophilic compounds 13
liquid band injection 144, 147
liquid CO_2 64, 70, 167, 245, 554
liquid desorption 87
liquid N_2 64, 70, 167, 245, 659
liquid/liquid extraction 9, 11, 15, 19, 23, 98, 572, 584, 603, 604, 620, 639, 649, 697, 700, 757
liver 601
LLE see liquid/liquid extraction
local normalization 446
location of double bonds 471
lock-mass technique 375
lock-plus-cali mass technique 375
LOD see limit of detection
loop type sampler 71
LOQ see limit of quantitation
low temperature HS 73
low pressure GC 200, 233, 597
LPGC see low pressure GC
LPGC-MS 234, 235, 594, 597
LVI see large volume injection
LVI-PTV 153, 162, 164

m

magnetic sector MS 246, 316, 327, 374, 390, 691, 734, 739
magnetic solid phase extraction 23
maintenance 13, 73, 75, 152, 155, 235, 264
makeup gas 257, 270
  Ar/10% methane 254
  nitrogen 252, 254
Marinol 751
mass calibration 321, 323, 375–379, 388–396
mass chromatogram 90, 160, 378, 415–422, 447, 452, 456, 498, 517, 521
mass defect 323, 324, 343, 389
mass flow controller 131
Mass Frontier software 665, 673, 766
mass resolution 314, 338, 720, 730
mass resolving power 314, 316
  10% valley definition 315
  FWHM definition 314
mass spectral libraries 90, 99, 637
  Adams essential oil library 439

Subject Index | 871

Chemical Concepts library 443
The Fiehn library 441, 658
Geochemicals, petrochemicals, and biomarkers database 442
The Golm metabolome database 441
Kühnle pesticides library 440
The lipid library 442
Maurer Meyer Pfleger Weber library (MMPW) 440
Mondello flavors and fragrances of natural and synthetic compounds (FFNSC) 439
NIST 657, 668, 768
NIST Tandem Mass Spectral Library 436
NIST/EPA/NIH Mass Spectral Library 436
Physiologically active substances of drugs, steroid hormones, and endocrine disruptors 440
Pyrolysis of synthetic polymers 442
Rösner designer drugs 438
SWGDRUG MS library 441
Volatiles in food 439
Wiley KnowItAllMass 440
Wiley Registry of Mass Spectral Data 437, 657
Yarkov library of organic compounds 443
mass flow dependent detectors 251
matrix effect 15, 16, 34, 40, 54, 522, 584
matrix-matched standards 581, 583
matrix solid phase dispersion 23
maximizing masses peak finder 413
maximum residue level 516
McLafferty rearrangement 471
MCSS *see* moving capillary stream switching
MDHS 72 82
MDL *see* method detection limit
meat 12, 595, 691, 732
medical applications 627
megabore columns 172, 200, 233
metabolites 751
metabolomics 27, 48, 77, 332, 656

metal columns 172, 200
metal ferrules 273
metastable ion 420
methanizer 257
method detection limit 25, 141, 321, 509, 511, 551–554, 571, 635
methylation 164
MHE 40, 53, 57, 59, 74
micro pollutants 697
micro-chamber 646
micro-SPE 595, 601
μSPE 9, 15, 19, 23–26, 235, 594–598, 600–603
microchannel devices 271
microextraction 10, 17, 27, 35, 48
microfluidic devices 271, 668
microplastic analysis 89, 94
microprep trap 251
microwave assisted extraction 100
milk 12, 36, 44, 732
miniaturization 9, 11, 19, 24, 27
minimum required performance limit 516
modulator 199, 223
  dual jet cryo modulator 245
  loop-based cryogenic-free flow modulators 245
moisture 77, 212, 266
moisture removal 64–67, 74, 75, 77, 104
molecular ion 293, 420, 456, 468, 766
molecular sieve 78–81
monoisotopic elements 462
MOSH/MOAH 173
moving capillary stream switching 249, 270
MRL *see* maximum residue level
MRPL *see* minimum required performance limit
MS Excel Data Analysis ToolPak 512, 515
MS/MS analysis 380, 415
  fragmentation 767
  neutral loss scan 384
  precursor ion scan 384
MSChromSearch 90

Subject Index

MSDP *see* matrix solid phase dispersion
MSPE *see* magnetic solid phase extraction
mulch film 94
multi-dimensional GC 239–241
multi-residue analysis 594, 603
multi-sorbent tubes 96
musk fragrances 44

n

Nafion dryer 87, 354, 541, 543
nanomaterial 11, 22, 23, 48
nanoplastics 89
narrowbore column 200, 233, 235
natural abundances 346, 347, 362, 520
natural gas 265
NCI *see* negative chemical ionization
needle penetration depth 147
negative chemical ionization 299, 301, 306, 433, 491, 751, 767
 charge transfer 304
 electron capture 302
 proton abstraction 304, 305
 reagent ion capture 306
negative ion mode 337, 338
neutral loss scan 470
$(NH_4)_2SO_4$ 54
NIOSH 2549 82
NIST Chemistry WebBook 436
NIST GC Retention Index Database 436
NIST library 556
NIST library search 444, 622
nitrogen 73, 253, 257, 259, 262–264, 266, 329
nitrogen rule 457
nitrogen/phosphorus detector 10, 251–253, 255, 259, 260, 416, 432
nominal mass resolution 291, 317, 326, 332
non-polar compounds 13, 69, 80, 81
non-polar hydrocarbons 209, 210
non-selective extractions 48
non-targeted analysis 4, 42, 47, 323, 332, 438, 618, 757
non-vaporizing injection 168

non-volatile samples 88
normal phase LC 99, 171
NPD *see* nitrogen/phosphorus detector
nuclear magnetic resonance (NMR) 293
number of theoretical plates N 197, 216, 224
nutrivolatilomics 77

o

O-FID 252, 257, 258
occupational health screening 62
OCI 131, 157, 171
odor threshold 632
odors 50, 257, 338, 678
off-line technique 10
off-odors 632, 644
olfactometry 249, 250, 268
olive oil 32, 173, 338
on-column cryofocusing 166
on-column injector 100, 130, 168
online analysis 9–11, 15, 18, 24, 25, 27, 29, 34, 39, 87, 94, 99, 163, 171–173, 198, 252
online coupling 10, 99
online SPE 23–25
open split interface 268, 351–354, 543
open-tubular columns 215
Orbitrap analyzer 323
Orbitrap MS 329, 390
ortho effect 498, 618
orthogonal separation 241
outgassing 62, 76, 94, 95, 167, 677
outlet splitter 265
outlier 515
oven temperature program 133, 151, 154, 155, 159, 160, 164, 170, 197, 198, 205, 220, 225–228, 235, 239, 242
oxygen 35, 46, 73–75, 77, 91, 102, 134, 136, 205, 212, 252, 253, 257, 259, 262, 264, 266
oxygen-specific detector 258

p

packaging materials 62, 72, 77, 256, 338, 619
packed columns 20, 66, 128, 168, 198, 214, 228–230, 256
P&T see dynamic headspace analysis; purge and trap
paprika 601
parallel MS detection 265, 266, 268
partial concurrent solvent evaporation 173
partition coefficient 17, 27, 38, 48–54, 70, 72, 74, 75, 212, 216, 219, 748
PBM library search 444
PCI see positive chemical ionization
positive chemical ionization 299, 306, 433, 491, 496, 498, 767
   adduct formation 301
   charge exchange 300
   hydride abstraction 300
   protonation 299
PDD see pulsed discharge detector
PDED see pulsed discharge detector
PDMS overcoated 28, 46
peak apex plot 243
peak area vs. height 506
peak broadening 71, 128, 171, 197, 198, 223, 226
peak capacity 225, 239–241, 243
peak matching 394
peak profile 217
peak symmetry 66, 222
peak tailing 222, 256, 631
peak width 217, 224
pepper spray 626
perfume 637
persistent organic pollutants 1, 12, 374, 697, 698, 712, 718
personal defense products 626
pesticides 11–16, 25–27, 48, 94, 99, 100, 102, 141, 146, 151, 152, 154, 159, 190, 191, 195, 205–208, 233–235, 251, 254, 255, 258, 323, 332, 374, 482, 584, 594, 599, 603
   basic pesticides 594

petrochemical analysis 240, 253, 254, 265, 266
PFE see pressurized fluid extraction
PFTBA 372, 390, 392, 393, 434
pH value 16, 38, 594
pH-sensitive analytes 16
pharmaceutical analysis 53, 159, 189, 206, 254, 266
pharmaceutical products 665
phase ratio β 52, 87, 203, 204, 221, 224
phenotype 656
photo ionization detector 262, 432
PID see photo ionization detector
PID lamp types 262
pigments 14, 16, 251, 712, 713, 730–734
pipette tip transport 26
planar analytes 16
plant material 658
plasma 758
plasticizer 20, 46, 100, 529, 619
PLE see pressurized liquid extraction
PLOT see porous-layer open tubular columns
poisoning 757
polar compounds 13, 16, 27, 34, 36, 39, 42, 69, 77, 79, 102, 141, 152, 159, 174, 203, 210, 212
polyethylene glycol 37, 46, 66, 181, 209
polymer analysis 89, 666
polymer materials 619, 677, 679
polystyrene (PS) 54, 651
polyvinylchloride (PVC) 57, 651
POPs see persistent organic pollutants
Porapak 80
pork 598
porous-layer open tubular columns 187
PPINICI see pulsed positive-ion negative-ion chemical ionization
pre-column 16, 99, 128, 131, 143, 152–155, 162, 164, 165, 169–173, 200, 249, 250, 270
precursor ion 321, 343, 382–388, 437, 575, 597, 629, 651, 658, 702–709, 720, 723–725, 727, 767
prep ahead mode 18, 597, 601, 660, 661

preparative GC 251
pressure corrected solvent BP 146, 149, 151, 158, 159, 169, 172, 226
pressure pulse 139, 140
pressure-balanced injection 71
pressurization 71
pressurized fluid extraction 101
pressurized liquid extraction 100–105
preventive maintenance 129, 135, 138, 141, 150
primary secondary amine 14, 16, 601–605
product ion spectrum 382, 385–389, 415, 575, 769, 770
product safety 626, 649, 665, 683
productivity 18, 44, 98, 103, 154, 198, 205, 210, 212, 226–228, 230
programmed temperature vaporizer (PTV) 15, 64, 70, 100, 131, 136, 154, 155, 157–159, 161–164, 167, 171
  cryofocusing mode 167
  heating rate 159
  large volume injection 24, 130, 153–158, 162–165, 579
  on-column injection mode 64, 100, 162, 165, 166, 171
  solvent elimination 165
  split injection 161
  splitless injection 159, 574, 605, 699, 720, 753
proteomics 332
PSA *see* primary secondary amine
PTV *see* programmed temperature vaporizer
pulsed discharge detector 261, 262
pulsed discharge HID (PDHID) *see* pulsed discharge detector
pulsed positive-ion negative-ion chemical ionization 299
pumping capacity 174, 195
purge and trap 549, 560, 569, 632, 665
Pyrofoil 93
pyrogram 88
pyrolysis 10, 88–91, 93–96, 167, 357
  Curie point pyrolysis 89, 92

foil pyrolysis 89, 90
furnace pyrolysis 89, 94
micro furnace pyrolysis 95

q

quadrupole analyzer 6, 174, 235, 240, 246, 310–312, 316–319, 321–327, 329, 331, 333–336, 363, 364, 389, 398, 434, 441, 442, 445, 447, 513, 525, 544, 557
  hyperbolic rods 6, 319–322, 386
  resolving power 319
  round rods 319, 321, 383, 386
qualifier 365, 373, 415, 584, 651, 734, 761
qualifier ion 585, 597, 599, 761
quality control 338, 637
quantification 40, 556, 594, 621, 632, 656, 683, 747
quantitation 47, 53, 56, 57, 62, 74, 148, 248
  isotope dilution 734, 739
  one-point calibration 757
quartz wool plug 141
quasimolecular ion 296, 420, 468
QuEChERS 11–16, 26, 151, 578, 584, 594
  AOAC method 2007.01 14, 594
  European Standard EN15662 14, 594
QuEChERSER 15, 594

r

radio frequency identification 86
rapeseed oil 173
raspberries 601
reagent gas cluster 297, 306–310
recondensation effect 16, 131, 148, 150, 151, 162
reconstructed ion chromatogram 413, 414, 641
residual gas analyzer 266
residual solvents 72, 189
residual water 73
resolving power (GC) 200, 216, 224
resolving power (MS) 291, 314–319, 322–326, 329, 335, 338, 340, 344, 743, 745

Subject Index

retention gap   100, 131, 151, 169–171, 200, 248, 593, 615
retention index   268, 429, 430–434, 436, 637, 673, 765
retro Diels Alder reaction   470, 490
retrospective analysis   333
reversed column flow   152
reversed Nier Johnson geometry   327
RFID *see* radio frequency identification
RIC *see* total ion chromatogram
rice   659, 663
robotic autosampler *see* x,y,z-robot
robotic x,y,z-sampler *see* x,y,z-robot
room-temperature ionic liquids *see* ionic liquids
rotary vane pumps   399
RTILs *see* ionic liquids
rubber   679
rule of thumb   71, 78, 139, 140, 146, 149, 169, 174, 324
100% report   641

S

S/N ratio   128, 158, 161, 174, 199, 226, 233, 243, 245, 297, 321, 344, 373, 378, 380, 384, 398, 415, 508–510, 512, 513, 559, 576, 583, 624, 631, 711, 717, 749
salmon   105, 598
salt addition *see* salting out
salting-out effect   11, 17, 38, 47, 54, 118, 584
sample capacity   10, 129, 138, 166, 200, 222, 235
sample collection   81
sample preparation   9–11, 20, 22–25, 27, 49, 73, 98, 99, 105, 129, 151, 163, 164, 171, 227, 599
  instrumental integrated   9
sample throughput   9, 10, 12, 13, 17, 28, 36, 50, 56, 86, 89, 163, 174, 227, 228, 234, 235, 242, 243, 632, 664
sandwich injection   146
SBSE *see* stir bar sorptive extraction
scan rate   335, 340, 363, 395, 506

scan speed   246, 394, 506, 688, 716
scavenging mode   19, 25
SCD *see* sulfur chemiluminescence detector
screening   96, 253, 310, 339, 385, 593, 664, 712, 718, 757, 762
sector field MS *see* magnetic sector MS
secureTD-Q technology   86
seeds   658
selected reaction monitoring   323, 598, 603, 656, 750
  AutoSRM   387
  data-dependent acquisition   388
  timed-SRM   572, 575, 659, 721
selectivity   15, 42, 44, 99, 209–212, 216, 218–220, 247, 251–253, 256–258, 260, 262–264, 267
selectivity tuning   239
sensitivity   1, 24, 27, 35, 41, 52, 53, 70, 88, 99, 167, 255, 292, 364, 507, 513
separation funnel   212
septum bleed   134, 137
septum purge   130–136, 149
sewage   712, 718
SFE *see* supercritical fluid extraction
shark fins   222
SI mass unit   291
SIEVE software   664
silica gel   22, 66, 68, 79, 105
silicon wafers   98
silicone grease   530
Silonite   87
SilTite   273
silylation   142, 349, 465, 656, 763
SIM mode   320, 323, 324, 340, 363, 364, 373, 374–379, 389, 415, 506
  retention timed SIM   365, 373, 542
SIMDIS *see* simulated distillation
simulated distillation   192
single quadrupole MS   549, 554, 577, 584, 619, 632, 637, 656, 665, 677, 683, 712, 751, 757
SISCOM library search   444
SLE *see* solid/liquid extraction
sludge   98, 712

smart technology 30
smoke 691
sniffing device 268
soft ionization 232, 295, 310, 381, 468, 767
soil 98, 554, 560, 595
solid phase extraction 9–11, 14, 15, 19, 20, 22–25, 27, 98, 100, 102, 584, 651, 700, 751, 763
  µSPE 9, 15, 19, 23, 25, 26, 594, 595, 598, 599
  C18 595
  dSPE 15, 601
  GCB 595
  online-SPE 25
  scavenging mode 25
  wash-and-elute mode 25
solid phase microextraction (SPME) 9, 10, 23, 27–32, 34, 36–45, 47, 48, 87, 88, 100, 135, 137, 138, 143, 203, 265, 573, 639
  arrow 30–32, 34, 36, 42, 48, 632
  arrow derivatization 32
  desorption 40
  devices 42
  DI-SPDE 34, 47
  DI-SPME 28, 38–40, 44, 46–48, 614
  fiber 28–32, 34, 37–39, 41, 42, 44–46, 48, 87, 137
  high capacity probe 28
  HiSorb probe 34
  HS-SPDE 34
  HS-SPME 27, 28, 32, 38, 48, 747
  injection 167
  injector temperature 40
  inlet liner diameter 41
  inlet liner 47, 143
  in-tube SPME 34
  MFX 42
  MHS-SPME 40
  multi-fiber exchange 42
  multiple injections 167
  multiple SPME extraction 40
  on-fiber derivatization 32, 39
  operation for GC-MS 36

rinsing 36, 46
sorbent materials 42
sorbent swelling 32, 46
SPDE 34
TFME 35
thin film SPME 35
TV-SPME 32
Vac-HS-SPME 32
solid samples 53, 72, 94, 96, 98, 100, 101
solid wastes 98
solid/liquid extraction 639
solvent displacement 84
solvent effect 146, 149, 150, 152, 159, 161, 170, 171, 173
solvent expansion volumes 139, 145
solvent vapor cloud 146, 151, 158
solvent vapor exit 100, 165, 171
solvent vapor volume 139, 146
sonication 17, 141
sorbent material 13, 19, 20, 25, 27, 30–32, 35, 38, 41, 48, 66, 76, 78, 87, 163
sorghum 99
sources of contamination 569
Soxhlet extraction 9, 100–102, 105, 747
soybean oil 173
SPE *see* solid phase extraction
spectrum skewing 336, 434, 506
spices 13, 578, 603, 626, 694
spirits 619
spiromolecular pumps 399
split at column end 270
split flow 130, 131, 133, 138, 163
split injection 138, 143, 148, 161
split ratio 148
split/splitless injector 16, 130, 131, 135, 144, 149, 150–155, 157, 158
splitless injection 139, 143, 148, 149, 159, 226
SPME *see* solid phase microextraction
SRM *see* selected reaction monitoring
SRSE *see* stir rod sorptive extraction
SS *see* surrogate standard
SSL injector *see* split/splitless injector
stabilizer 529

standard addition method 40, 47, 53, 148, 522
standard atmospheres 83
standard deviation 40, 53, 129, 215, 344, 508, 511, 512, 517
stationary phase 10, 19, 34, 134, 149, 170, 174, 175, 187, 197, 198, 200, 204, 205, 210, 212, 216, 220, 221, 223, 224
  polyethylene glycol 205
  polysilarylene 176, 179, 205
  siloxancarborane 205
stereoisomers 436
stir bar sorptive extraction 28, 87, 88, 167
  liquid desorption (SBSE/LD) 88
stir rod sorptive extraction 23
structure elucidation 295, 384, 385, 457, 766
structure related selectivity 594
SulfiCarb 80
sulfur chemiluminescence detector 265, 432
SUMMA canister 86, 541
sunflower oil 173
Supelcoport 163
supercritical fluid extraction 23, 100
surface water 549, 554, 619
surfactants 767
Surfasil 141
surge pressure 139, 140, 149
surrogate standard 517, 519, 561
SVE *see* solvent vapor exit valve
Swedish ethyl acetate method 15, 599
SweEt *see* Swedish ethyl acetate method
syringe needle 26, 76, 134, 137, 144, 147, 168
syringe needle transport 26, 598

t

tandem with FID 266
tapered liners 142
targeted analyses 4, 168, 240, 323, 332, 364, 372, 383, 438, 505, 541, 656, 757
TCD *see* thermal conductivity detector

TCM *see* traditional Chinese medicine
TD *see* thermal desorption
tea 13, 595
technical samples 101
TEF values 712, 730
  2005 WHO TEFs 733
  2022 WHO TEFs 723, 733, 734
temperature rise time 89–93
Tenax 64, 66–70, 78, 79, 82, 84, 96, 163, 167, 639, 680
Tenax GR 70, 77, 78, 81
Tenax TA 67, 77, 79, 81, 82, 645
TEQ *see* toxicity equivalents
textiles 683
thermal conductivity 266
thermal conductivity detector 196, 259, 266
thermal decomposition 73, 77, 158, 170
thermal degradation 157
thermal desorption 9, 36, 78, 79, 82–88, 94–96, 130, 131, 166, 167, 639, 647, 677
thermal desorption tubes 81, 83
  DiffLok end caps 85
  VOST stack sampling tubes 82
thermal electrons 253, 301, 304, 310
thermal extraction 10, 94, 96, 98, 644, 646
thermal gradient 237
thermodesorption 4, 35, 81, 256, 541
thermolabile compounds 88, 158, 226, 745
thick film columns 71, 202, 543, 551, 655, 759
thin film columns 202, 221, 230, 713, 745
3D contour plot 243
TIC *see* total ion current
time-of-flight MS 2, 5, 235, 240, 2455–248, 311, 316, 322, 333–339, 344, 389, 398, 425, 427, 441, 597, 646
timed SRM 386 572, 574, 600, 605, 612, 628, 659, 699, 721
TOF *see* time-of-flight MS
total ion current 303, 413, 426, 651, 682

total petroleum hydrocarbons 102, 191
total sample transfer 10, 139, 143, 148, 159, 161, 168
total volatile organic compounds 677, 681
toxicity equivalent factors *see* TEF values
toxicity equivalents 712, 724, 728, 733
toys 649
traceability 31, 82, 378
traditional Chinese medicine 626
transfer line 63, 64, 72, 77, 130, 132, 143, 166, 205, 268, 269
transfer time 149, 153, 154, 160
Trennzahl number 224
triple quadrupole MS 572, 594, 599, 603, 626, 651, 656, 697, 718, 747, 762
tropylium ion 469, 475
TRT *see* temperature rise time
tube absorption 677
tube tagging 86
turbomolecular pumps 397
TVOC value *see* total volatile organic compounds
Twister 87
two-hole ferrule 639

u

UFGC *see* ultra fast gas chromatography
UFM *see* ultra fast column module
ultra fast column module 235, 236
ultra fast gas chromatography 227, 230, 235–237
Ultramark 392
ultrasonic extraction 100, 651
UniCarb 80
unit mass resolution 291, 317, 319, 323, 326, 332, 340, 343, 381, 386, 720, 734, 737
unknown identification 249, 267, 292, 295, 296, 333, 384, 388, 418, 421, 436, 443–457, 471, 640, 665–677, 766
urine 34, 752
USE *see* ultrasonic extraction

v

vacuum-assisted 32, 34
vacuum ITEX 77
vacuum outlet 230
vacuum outlet GC 200, 230, 233
vacuum ultraviolet detector 252 253, 266–268
vacuum UV wavelength range 266
validation 86, 578, 580, 602, 605, 711
van Deemter 174, 197, 201, 214, 224
van Deemter curve 194, 196–199
vapor pressure 70, 78, 209
vapor volume calculator 139
VDA 278 98, 678–683
vegetables 152, 578, 584, 595
vehicle air analysis 96, 677–683
Vespel ferrules 273, 525, 639
veterinary drug analysis 12, 235
veterinary drugs 598
vial shaking 59
Vienna Pee Dee Belemnite 569
viscosity 59, 101, 194, 198, 209, 271
vITEX *see* vacuum ITEX
VOCARB 66, 68, 69, 79
volatile pesticides *see* pesticides
volatile polar compounds 48
volatilomics 338
volume contraction 149, 151
VPDB *see* Vienna Pee Dee Belemnite
VUV detector *see* vacuum ultraviolet detector
VUV spectra 267

w

wall-coated open tubular columns 175
wastewater 64, 265, 549, 554, 619, 748
water 11, 13, 15–17, 20, 25, 27, 28, 34, 35, 39, 44, 46–49, 52–54, 56, 62, 64, 65, 67, 69, 73, 74, 79, 84, 87, 139, 159, 257, 263, 266, 560, 572
   drinking water 35, 39, 64
water removal 66, 67, 104, 355–358
water vapor 53, 66, 74, 75, 261, 747, 749
waxes 206

weak anion exchange 48
weak cation exchange 48
wine 12, 189
workplace air monitoring 77
workplace drug testing 752

X

x,y,z-robot 10, 12, 15, 18, 32, 36, 40, 42, 76, 87, 88, 594, 601, 634, 665, 734, 752
XIC *see* mass chromatogram

Compound Index

a

AAs *see* aromatic amines
AC *see* acenaphthene
accelerated solvent extraction   101
acenaphthene   366, 703
acenaphthene-d$_{10}$ 366
acenaphthylene   703
acephate   368, 585, 598, 599, 606
acetic acid   648
acetoin   648
acetone   17, 44, 102, 142, 146, 151, 263, 266, 649, 767
acetonitrile   11, 13–17, 26, 45, 139, 146, 604
acetylene   263
acetylsalicylic acid   419
ACL *see* acenaphthylene
acridine   478, 480
adamsite   504
adipates   77
alachlor   587, 607
alaclor   368
alcohols   44, 48, 53, 66, 73, 80, 173, 207, 208
aldehydes   72, 80, 262
aldicarb   368
aldrin   368, 484, 490, 584, 587
aliphatic hydrocarbons   37
aliphatics   261
alkali flame ionization detector   259
alkanes   48, 93, 94, 140, 144, 159, 205, 224
alkanes, *N*-   431
alkyl amides   768
alkylaromatics   475
alkylbenzenes   79, 80
alkylbis(trifluoromethyl)phosphine sulfides, *N*-   432
allethrin   368, 589, 608
allidochlor   585
ametryn   587
amidosulfonic acid   614
amines   141, 181, 189, 207, 208, 256, 261, 262, 766
amino acids   193
aminoazobenzene, *o*-   684
aminoazobenzene, *p*-   684
aminobiphenyl, 2-   617
aminobiphenyl, 4-   617
aminocarb   585, 587
aminodiphenyl, 4-   684
aminopropyl   16
aminotoluene, 4-   617
amitriptyline   497, 760, 762
ammonia   189, 257, 260
amphetamines   496
amylcinnamal   654
amylcinnamyl alcohol   654
AN *see* anthracene
aniline   617, 684
aniline-d$_5$ 617
anisic aldehyde   643
anisidine, *m*-   684
anisyl alcohol   653
anthanthrene   367
anthracene   366, 479, 703

Handbook of GC-MS: Fundamentals and Applications, Fourth Edition. Hans-Joachim Hübschmann.
© 2025 WILEY-VCH GmbH. Published 2025 by WILEY-VCH GmbH.

Compound Index

anthracene-d10   479
aramite   609
aramite 1   590
aramite 2   590
argon   263, 264, 266
Aroclor   716
Aroclor 1260   436
aromatic amines   614, 616, 683
aromatic compounds   81, 206, 261
aromatic hydrocarbons   206
arsenic   253
aspon   588
atrazine   99, 368, 435, 485, 586, 606
azinphos-ethyl   592, 611
azinphos-methyl   368, 592
azoxystrobin   611
azulene   255, 366

b

BaA *see* benz[a]anthracene
BaP *see* benzo[a]pyrene
barban   368
base/neutrals and acids   102
basic compounds   159, 189
BbFA *see* benzo[b]fluoranthene
BBP *see* benzylbutylphthalate
BcFL *see* benzo[c]fluorene
BDE *see* brominated diphenyl ether
benalaxyl   591, 610
bendiocarb   368, 486, 585, 586, 606
benfluralin   586, 606
benfuracarb   611
benodanil   591
benz[a]anthracene   692, 701, 703
benzaldehyde   662, 673
benzanthracene   367
benzazolin methyl ester   368
benzendiamine, 1,4-   684
benzene   54, 69, 80, 84, 263, 266, 366, 476, 546, 549, 552, 558, 564, 569, 572, 681
benzene-d$_6$ 366
benzene, (bromomethyl)-   674
benzene, 1-fluoro-2-methyl-   675
benzene, 1-fluoro-4-methyl-   675
benzene, (isothiocyanatomethyl)-   674, 677
benzidine   684
benzo(b)fluoranthene   692
benzo(g,h,i)perylene   692
benzo(k)fluoranthene   692
benzo[a]coronene   367
benzo[a]pyrene   367, 480, 691, 703
benzo[a]pyrene   696
benzo[b]chrysene   367
benzo[b]fluoranthene   367, 701, 703
benzo[c]fluorene   703
benzo[e]pyrene   367, 703
benzo[g,h,i]perylene   367
benzo[ghi]perylene   701, 703
benzo[j]fluoranthene   367, 701, 703
benzo[k]fluoranthene   367, 701, 703
benzofluorene   367
benzoic acid trimethylsilyl ester   676
benzophenone   77
(Benzothiazol-2-ylthio)methyl thiocyanate (TCMTB)   590
benzoximate   593
benzoylprop-ethyl   591
benzyl alcohol   84, 652, 674, 675
benzyl benzoate   654
benzyl cyanide   652
benzyl isocyanate   674
benzyl isopentyl ether   674
benzyl salicylate   654
benzylalcohol   675
benzylbutylphthalate   46
benzylcinnamate   654
benzylidenebenzylamine, *N*-   674
BeP *see* benzo[e]pyrene
BFB *see* bromofluorobenzene
BghiP *see* benzo[ghi]perylene
bifenox   591
bifenthrin   591, 610
biperiden   760
biphenyl   366
bisphenols   77
bitertanol   611
BjFA *see* benzo[j]fluoranthene
BkFA *see* benzo[k]fluoranthene

boron 253
boscalid 611
bromacil 368, 589
bromacil N-methyl derivative 368
bromazepam 760
brominated alkyl phosphates 504
brominated diphenyl ether 703
brominated flame retardants 159, 698, 702, 739
bromine 253
Bromkal 505
Bromkal P 320
bromobenzene 564
bromochloromethane 545, 564
bromodichloromethane 38, 72, 558, 564
bromofluorobenzene, 4- (BFB) 373, 551, 552, 565
bromofluorobenzene 545
bromoform 38, 72, 79, 558, 564
bromomethane 546
bromophos 368, 588
bromophos-ethyl 368, 432, 589, 608
bromopropane, 1-chloro-2- 565
bromopropylate 591, 610
bromoxynil methyl ether 368
BSTFA 659, 660, 663, 668
BTBPE *see* pentabromoethylbenzene
BTEX 53, 66, 80–82, 84, 257, 262, 263, 475, 549, 553, 560
bufencarb 585
bupirimate 590
buprofezin 609
butachlor 589
butanal 72
butane, 1,2,2-tribromo- 674
butane 45
butanedioic acid, *bis*(trimethylsilyl) ester 676
butanoic acid 648
butanol 55, 83
butanol, N- 54
butanone, 2- 54
buten-2-one, 4-phenyl-3- 653
butene-2-one, 4- (*p*-methoxyphenyl)-3- 653

butralin 588, 608
butyl acetate, N- 54
butyl alcohol, N- 55
butylamine, N- 263
butylate 585, 593
butylbenzene, N- 564
butylbenzene, t- 564
butylbenzyl phthalate 621
butylfuran, 2- 662
butylphenol, 4-*tert*- 653
butyl phthalate, di-n- 621

C

cafenstrole 611
calcium chloride 54, 104
calcium sulfate 104
campesterol 173
cannabidiol 339
cannabinoids 438
capsaicin 628–630
captafol 13, 368, 585
captan 13, 368, 585, 589, 602
carbamates 491
carbaryl 368, 607
carbendazim 368
carbetamid 368
carbetamide 590
carbofuran 368
carbofuran, 3-hydroxy 607
carbohydrates 208
carbon dioxide 69, 77, 257–259, 262, 264, 266
carbon disulfide 73, 77, 84, 139, 558, 577, 648
carbon monoxide 77, 252, 257–259, 262, 266, 329
carbon tetrachloride 142, 257, 558, 565
carbon 253
carbon-13 253
carbonic acid 73, 75
carbonyl sulfide 338
carbophenothion 432, 591
Carbowax 46
carboxin 590
carfentrazone-ethyl 610

carpropamid 609
caryophyllene 640
caryophyllene, β- 638
chemical warfare agents 80, 491, 498
chlorbenside 588
chlorbromuron 368, 589
chlorbufam 368, 587
chlordane, alpha (cis) 608
chlordane, cis- 368, 589
chlordane, gamma (trans) 609
chlordane, trans- 368, 589
chlordimeform 586
chlorfenprop-methyl 368
chlorfenson 590
chlorfenvinphos 368, 416, 589
chlorfenvinphos-Z 608
chlorflurecol-methyl 589
chloridazon 368, 591
chlorinated hydrocarbons 80, 81
chlorinated pesticides 102
chlorine 253
chlormephos 585
chloroacetophenone (CN) 504
chloroaniline, 4- 617
chloroaniline, 4,4-methylene-bis-2- 684, 689
chloroaniline, p- 684
chlorobenzene 54, 546, 558, 564
chlorobenzene-d$_5$ 545, 552
chlorobenzilate 590, 609
chlorobenzylidenemalnonitrile, o- (CS) 504
chlorobutane, 2- 263
chlorodibromomethane 72
chloroethane 546
chloroethyl vinyl ether, 2- 79
chloroform 44, 72, 84, 142, 146, 266, 473, 546, 558, 564, 572
chloromethane 546
chloroneb 368, 585
chlorophenol 40
chlorophenol, o- 481
chlorophenol, 2,6-dibromo-4- 482
chloropropane 546
chloropropham 369

chloropropylate 590
chlorothalonil 16, 587, 594
chlorotoluene, 2- 564
chlorotoluene, 4- 558, 564
chlorotoluron 368
chloroxuron 369, 593
chlorpropham 586, 606
chlorpyrifos ethyl 608, 613
chlorpyrifos 311, 369, 588
chlorpyrifos-methyl 587, 607
chlorthal-dimethyl 369, 588, 608
chlorthiamid 369, 585, 588, 593
chlorthion 589
chlorthiophos 590, 610
chlozolinate 589
cholesterol 93, 94, 173
CHR see chrysene
chrysene 367, 692, 701, 703
chrysene-d$_{12}$ 367
C$_2$ hydrocarbons 81
cinerin I 369
cinerin II 369
cinnamal 652
cinnamaldehyde 628
cinnamaldehyde, α-methyl-trans- 629
cinnamyl alcohol 653
citalopram 760
citral, isomer 1 652
citral, isomer 2 652
citrate 77
citronellol 652
clobazam 760
clodinafop-propargyl 610
clomazone 586
clomipramine 760
clozapine 760
CO$_2$ see carbon dioxide
cocaine 496, 498
codeine 496, 497, 760, 762
codeine-d$_3$ 765
codeine-TMS 766
coronene 367, 478, 480
coumaphos 16, 592
coumarin 653
CPP see cyclopenta[cd]pyrene

cresidine, *p*- 684
cresol, *p*- 481
cresyl phosphate, tri-m- 530
cresyl phosphate, tri-*p*- 531
crotoxyphos 589
crufomate 589
CS$_2$ *see* carbon disulfide
cyanazine 369, 590
cyanophos 587
cycloate 585
cyclohexane 52, 54, 102, 146, 151
cyclohexane, methyl- 673
cyclopenta[*cd*]pyrene 701, 703
cyclopentane, methyl- 673
cyclopropane, 1-heptyl-2-methyl- 676
cyclopropyl carbinol 675
cyfluthrin 1 592
cyfluthrin 592, 611
cyhalofop butyl 611
cyhalothrin, lambda 611
cypermethrin 1 592
cypermethrin 4 592
cypermethrin I 611
cypermethrin 369
cyprazine 587
cyproconazole 609
cyprodinil 608

d

dacthal *see* dimethyl tetrachloroterephthalate
dalapon 369
dazomet 369
DBaeP *see* dibenzo[ae]pyrene
DBahA *see* dibenzo[ah]anthracene
DBahP *see* dibenzo[ah]pyrene
DBaiP *see* dibenzo[ai]pyrene
DBalP *see* dibenzo[al]pyrene
DBDPE *see* decabromodiphenyl ethane
DBEP *see* di-(2-butoxy)-ethyl phthalate
DBP *see* dibutylphthalate; di-n-butyl phthalate
DCHP *see* dicyclohexyl phthalate
DCM *see* dichloromethane

DCPA *see* dimethyl tetrachloroterephthalate
DDD *see* dichloro-diphenyl-dichloroethane
DDD, *o,p'* 590
DDD, p,p' 483, 609
DDD, *p,p* 590
DDE *see* dichloro-diphenyl-dichloroethylene
DDE, o,p 608
DDE, p,p 609
DDE, p,p' 483, 589
DDT *see* dichloro-diphenyl-trichloroethane
DDT, *o,p'* 369, 590
DDT, o,p' 609
DDT, *p,p'* 369, 590
DDT, p,p' 484, 609
DDVP *see* dichlorfos
deca-BDE 739
decabromobiphenyl 504
decabromodiphenyl ethane 703
decabromodiphenyl ether 203
decachlorobiphenyl (PCB 209) 495, 716
decachlorobornane 303
decafluorotriphenylphosphine 373
decanal 675
decane, 5,6-bis(2,2-dimethyl-propylidene)-, (*E,Z*)- 674
decanoic acid 677
decanone-2 674
DEEP *see* ethyl phthalate, di-(2-ethoxy)-
DEHP *see* ethyl phthalate, di-(2-butoxy)-
deltamethrin 154, 592
deltamethrin II 611
demeton-*o* 586
demeton-*S* 585
demeton-*S*-methyl 369, 586
DEP *see* diethyl phthalate
desethylatrazine 586
desmetryn 369, 587
deuterium 253
DFTPP 373, 391
DHA *see* docosahexaenoic acid
DHP *see* dihexyl phthalate

Compound Index

diacetone alcohol   527
diacetylmorphine, 3,6- *see* heroin
dialifos   369, 592
di-allate   369
di-allate 1   586
di-allate 2   586
diaminoanisole, 2,4-   684
diaminodiphenylmethane, 3,3-dimethyl-4,4-   684
diaminodiphenylmethane, 4,4-   684
diaminotoluene, 2,4- (2,4-DAT)   501, 617
diaminotoluene, 2,6- (2,6-DAT)   501, 617
diamyl phthalate   621, 625, 626
diazepam   760
diazinon   336, 369, 427, 432, 586, 606
dibenz[*a,h*]anthracene   692, 703
dibenzanthracene   367
dibenzo[*a,h*]pyrene   367
dibenzo[*a,i*]pyrene   367
dibenzo[*a,l*]pyrene   367
dibenzo[*ae*]pyrene   703
dibenzo[*ah*]anthracene   701
dibenzo[*ah*]pyrene   703
dibenzo[*ai*]pyrene   703
dibenzo[*al*]pyrene   703
dibenzodioxin   366
dibenzofuran   366
DIBP *see* di-isobutyl-phthalate
dibromo-3-chloropropane, 1,2-   424, 564
dibromochloromethane   473, 558, 564
dibromoethane, 1,1-   475
dibromoethane, 1,2-   546, 564
dibromofluoromethane   552
dibromomethane   564
dibutyl phthalate   529, 675
dibutylphthalate   46
dibutyltin   576
dicamba methyl ester   369
dichlobenil   369, 585, 606
dichlofenthion   369, 587, 607
dichlofluanid   369, 588, 608
dichloraniline, 3,5-   606
dichlormid   585
dichloroaniline, 3,4-   617
dichlorobenzene   547
dichlorobenzene, 1,2-d$_4$ 546, 552, 558, 564, 565
dichlorobenzene, 1,3-   564
dichlorobenzene, 1,4-d$_4$ 552, 564
dichlorobenzene, p-   475
dichlorobenzidine, 3,3-   684, 689
dichlorobiphenyl   493
dichlorodifluoromethane   546, 564
dichloro-diphenyl-dichloroethane   141
dichloro-diphenyl-trichloroethane   141
dichloro-diphenyl-dichloroethylene   340
dichloroethane, 1,1-   546, 564
dichloroethane, *cis*-1,2-   558
dichloroethane, *trans*-1,2-   558
dichloroethene, 1,1-   546
dichloroethene, *cis*-1,2-   546
dichloroethylene, 1,1-   556, 564, 572
dichloroethylcne, *cis*-1,2-   563, 564, 567, 572
dichloroethylene, *trans*-1,2-   564, 572
dichloromethane   54, 82, 84, 102, 139, 142, 146, 558, 564, 651
dichloromethane (R30)   473
dichloromethane, 1,1-   558
dichlorophenol, 2,3-   481
dichlorophenyl acetate, 2,4-   482
dichloropropane, 1,2-   546, 564
dichloropropane, 1,3-   564
dichloropropane, 2,2-   79, 564
dichloropropene, 1,3-   38, 422
dichloropropene   475
dichloropropene, *cis*-1,3-   546, 558
dichloropropene, *trans*-1,3-   546, 558
dichloropropylene, 1,1-   564
dichloropropylene, *cis*-1,3-   565
dichloropropylene, *trans*-1,3-   565
dichlorphos   606
dichlorprop isooctyl ester   369
dichlorprop methyl ester   369
dichlorvos (DDVP)   369, 606
dichlorvos   585
diclofop-methyl   591
diclone   593
dicloran   586

Compound Index

dicofol   369, 588
dicrotophos   586
dicyclohexyl phthalate   621, 625, 626
dieldrin   369, 484, 490, 589
diethyl maleate   652
diethyl phthalate   621, 625, 626
diethylamine   45
diethylbenzene   366
diethylether   146
diethylphthalate   675
difenoconazole   611
diflubenzuron   606
diflufenican   610
difluorobenzene, 1,4-   545
diheptyltin   576
dihexyl phthalate   621
dihydroanthracene   366
dihydrocapsaicin   629
dihydrocodeine   760, 763
dihydrocodeine-TMS   766
dihydrocoumarin   653
dihydromorphine-TMS   766
di-isobutyl-phthalate   621, 623, 625, 626
diisoctylphthalate   761
diisooctylphthalate   674
diisopropylamine   44
diltiazem   760
dimetachlor   587
dimethachlor   607
dimethenamid   607
dimethirimol methyl ether   369
dimethoate   369, 432, 587
dimethomorph-1   611
dimethomorph-2   611
dimethoxybenzidine, 3,3'-   684
dimethoxybenzidine   689
dimethyl citraconate   652
N,N-dimethyldecanamide   768, 769
dimethyl disulfide (DMDS)   471, 648
N,N-dimethylformamide   54, 263
dimethyl phthalate   621
dimethyl sulfide   338
dimethyl sulfone   648, 676
dimethyl tetracholro-terephthalate   608

dimethyl-, 3,6,9-trioxa-2-silaundecane, 2,2-   676
dimethylamine   45
dimethylbenz[a]anthracene   367
dimethylbenzene   366
dimethylbenzidine, 3,3'-   684
dimethyldichlorosilane   141
dimethylformamide   54
dimethylnaphthalene, 1,3-   479
dimethylnaphthalene, 1,6-   479
dimethylnaphthalene   366
dimethylphenanthrene   366
dimethyl-phthalate   621, 622, 625, 626
dimethyl sulfoxide   45
diniconazole   609
dinitramine   587
dinitrobenzene, 1,2- (1,2-DNB)   502
dinitrobenzene, 1,3- (1,3-DNB)   502
dinitrobenzene, 1,4- (1,4-DNB)   502
dinitrotoluene, 2,6- (2,6-DNT)   501
dinitrotoluene, 2-amino-4,6-   616, 617
dinitrotoluene, 3,5- (3,5-DNT)   501
dinitrotoluene, 4-amino-2,6-   617
dinitrotoluene, 4-*tert*-butyl-3-methoxy-2,6-   654
di-n-butyl phthalate   621
di-n-octyl phthalate   621, 622
dinonyl phthalate   619, 621, 624–626
dinoterb methyl ether   370
dioctyl phthalate   529, 625, 626
dioctyltin   576
dioxacarb   370
dioxane, 1,4-   35, 52, 54
dioxathion   586, 606
dioxin-like PCBs   703, 712, 713, 718, 730
dioxins   22, 100, 102, 103, 105, 159, 164, 190, 255, 323
diphenamid   370, 589
diphenhydramine   760
diphenyl phthalate   621, 625, 626
diphenylamine   585, 653
diphenylanthracene   367
dipropetryn   608
disiloxane, hexamethyl-   676
disulfoton sulfone   609

Compound Index

disulfoton   336, 370, 427, 586
ditalimfos   593
ditalimphos   432
dithiocarbamate   577
diuron   370, 489
divinylbenzene   30, 31, 34, 36, 42, 44, 46, 48, 81
dl-PCBs *see* dioxin-like PCBs
DMDCS *see* dimethyldichlorosilane
DMEP *see* ethyl phthalate, di-(2-methoxy)-
DMF *see* N,N-dimethylformamide
DMP *see* dimethyl phthalate
DMPP *see* phthalate, di-(4-methyl-2-pentyl)
DMSO *see* dimethyl sulfoxide
DNOC methyl ether   369
DNOP *see* octyl phthalate, di-n-
DNP *see* dinonyl phthalate
docosahexaenoic acid   698
dodecane   662, 674
dodecanone, 2-   674
dodine   370
DOP *see* dioctyl-phthalate
doxepin   760, 762
doxylamine   760
DPhP *see* diphenyl phthalate
DPP *see* diamyl phthalate
DTCs *see* dithiocarbamate
DVB *see* divinylbenzene

e

edifenphos   591, 610
eicosapentaenoic acid   698
endosulfan   370
endosulfan, α-   588
endosulfan, β-   590
endosulfan I (α)   608
endosulfan II (β)   609
endosulfan sulfate   591, 610
endrin aldehyde   141
endrin ketone   141
endrin   141, 370, 484, 589
EO *see* ethylene oxide
EPA *see* eicosapentaenoic acid

EPC   130, 132, 133, 198, 227, 241
ephedrine   497
EPN   591
EPTC   585, 593
erbon   590
erucamide   528
etaconazole 1   590
etaconazole 2   590
ethalfluralin   586
ethane   45, 263, 266
ethane, *bis*(2,4,6-tribromophenoxy)   703
ethane, bis(trimethylsiloxy), 1,2-   676
ethanol   37, 45, 52, 54, 56, 73, 139, 619
ethanone, 2,2-dimethyl   673
ethiofencarb   370
ethion   590, 609
ethirimol methyl ether   370
ethirimol   370
ethofumesate   588, 607
ethoprophos   585, 606
ethoxy-phenol, 4-   653
ethyl acetate   13, 15, 16, 26, 44, 54, 83, 139, 146, 151, 599, 700
ethyl acrylate   652
ethyl benzene   552
ethyl chloride   564
ethyl chloroformate   40
ethyl ether   44
ethyl phthalate, di-(2-butoxy)-   621
ethyl phthalate, di-(2-ethoxy)-   621
ethyl phthalate, di-(2-methoxy)-   621, 625, 626
ethylan   590
ethylbenzene   69, 84, 366, 476, 546, 549, 558, 565, 681
ethyl-dipropylthiocarbamate, S- *see* EPTC
ethylene glycol monophenyl ether   84
ethylene glycol   45
ethylene oxide   57, 80, 81
ethylene   263, 329
ethylhexyl phthalate   675
ethylnaphthalene, 1-   478
ethyltoluene, 4-   546
etofenprox   611

etridiazole 585
etrimfos 370, 587, 607
eugenol 653
Evernyl 642

f

FA *see* fluoranthene
FAMEs *see* fatty acid methyl esters
fatty acid methyl esters 192, 204, 208, 210, 211, 257
fatty acids 16, 36, 40, 70, 73, 207, 648
fatty acid TMS ester 159
FC43 372, 375, 390, 392, 393, 434
FC5311 391
fenamidone 610
fenamiphos 590
fenarimol 370, 592, 611
fenazaquin 610
fenbuconazole 611
fenchlorphos 587, 607
fenchol 641
fenitrothion 370, 588, 608
fenobucarb 606
fenoprop isooctyl ester 370
fenoprop methyl ester 370
fenoxaprop-*p* 611
fenoxycarb 610
fenpropathrin 591, 610
fenson 588
fensulfothion 591
fentanyls 438
fenthion sulfoxide 609
fenthion 432, 588
fenuron 370
fenthionoxon 608
fenvalerate 1 592
fenvalerate 2 592
fenvalerate 611
fipronil 608
FL *see* fluorene
flamprop-isopropyl 370, 590
flamprop-methyl 370, 590
flualinate 593
fluazinam 593
fluchloralin 587, 607

flucythrinate 611
flufenacet 608
flumetralin 589
fluopicolide 610
fluoranthene 366, 703
fluorene 366, 703
fluorine 253
fluoroaniline, 3-chloro-4- 617
fluorobenzene 552, 565
fluorodifen 591
flurochloridone 1 589
flurochloridone 2 590
flurtamone 610
flutolanil 609
flutriafol 609
fluvalinate 611
folpet 13, 589, 593, 595, 602
fonofos 586
formothion 370, 607
Freon R11 81
Freon-113 546
Freon-114 546
freons 255, 261
fuberidazol 607
furathiocarb 610

g

gasoline range organics 69
geosmin 632, 633, 635, 636
geraniol 652
germanium 253
glucose 658
glucuronidase, ß- 764
glycerol 44
glycolic acid 676

h

hallucinogens 438
halogenated compounds 263
halogenated hydrocarbons 62, 64–67, 70, 72, 80, 81, 159, 206, 255, 256, 338, 541
halogenated narcotics 80, 81
halogens 256
HBB *see* hexabromobenzene

Compound Index

HCB *see* hexachlorobenzene
HCH, γ- 606
HCH, α- 586, 606
HCH, β- 588, 606
HCH, δ- 588, 607
helium ionization detector 258
helium photoionization detector 261
helium 266
HEPO, *cis- see* heptachloroepoxide, *cis-*
HEPO, *trans- see* heptachloroepoxide, *trans-*
heptabromobiphenylene 303
heptachlor epoxide 588
heptachlor 370, 587
heptachlorobiphenyl (e.g. PCB 180) 494
heptachlorobiphenyl 340
heptachloroepoxide, *cis-* 703
heptachloroepoxide, *trans-* 703
heptadecanenitrile 674
heptafluorobutyric anhydride 304
heptanal 662
heptane1-bromo-3-(2-bromoethyl) 674
heptanoic acid, propyl ester 674
heptanone, 2- 662
heptenal, *trans-*2- 652
heptenophos 585
heroin 496, 498, 762
hexabromobenzene 703
hexabromobiphenyl 505
hexabromobiphenylene 303
hexachlorobenzene (HCB) 16, 483, 585, 593
hexachlorobiphenyl (e.g. PCB 138, 153) 494
hexachlorobutadiene 558, 565
hexachloroethane 558
hexaconazole 609
hexadecane 674
hexadecanoic acid, methyl ester 674
hexadecanoic acid, trimethylsilyl ester 677
hexadecanol, 1- 662
hexafluorobutyric acid 752
hexanal 662, 673
hexane 18, 25, 37, 45, 102, 139, 142

hexane, *N-* 146, 263, 620
hexane, n- 54
hexanedione, 2,3- 673
hexanedione, 5-methyl-2,3- 652
hexanoic acid 648
hexanol, 2-ethyl-1- 662
hexaphene 367
hexazinon 486, 490
hexazinone 591, 610
hexenal, 2- 72
hexenal-diethyl-acetal, *trans-*2- 652
hexenal-dimethyl-acetal, *trans-*2- 652
hexogen (RDX) 500
hexogen 498
hexyl-cinnamaldehyde 654
HFBA *see* hexafluorobutyric acid
HxCB 737
HxCDF, 1,2,3,4,7,8- 721
HxCDF, 1,2,3,4,8,9- 737
HxCDF, 1,2,3,6,7,8- 721
HxCDF, 1,2,3,7,8,9- 737
HxCDF 736, 737
H_2S *see* hydrogen sulfide
hydrocarbons 45, 59, 80, 134, 140, 157, 173, 255, 257, 258
hydrocodone-TMS 766
hydrogen sulfide 77, 338, 648
hydrogen 253, 259, 262, 266
hydroiodic acid 614
hydromorphone-TMS 766
hydroxybutan-2-one, 3- 648
hydroxy-citronellal 653
hydroxytetrachlorodibenzofuran 340

i

imazalil 590, 610
imazethapyr 608
imipramine 760
imiprothrin 610
indene, 1-methylene-1H- 662
indeno[1,2,3-*cd*]pyrene 367, 692, 701, 703
indicator PCBs 495
indole-3-acetic acid (IAA) 658
indoxacarb 611

Compound Index

iodine 253
iodobutane, 2- 263
iodofenphos 370, 589
ionol (BHT) 529
IPMP *see* methoxypyrazine 2-isopropyl-3-
iprobenfos 607
iprodione 332, 591, 610
iron 253
isazofos 587
isobutyraldehyde 263
isocaryophyllene 638
Iso-E-Super 642
isoeugenol 653
isofenphos 589
isomers, cis/trans 208
iso-octane 139, 580, 754
isooctyl ester 2,4-D, 369
iso-pentane 146
isopropalin 588
isopropanol 17, 37, 45, 668, 767
isopropyl alcohol 52, 54, 674
isopropyl palmitate 674
isopropyl stearate 674
isopropylbenzene, 1-methyl-2- 478
isopropylbenzene 565
isopropylmethoxypyrazine 634
isopropyltoluene, 4- 565
isoprothiolane 609
isoproturon 370

j

jasmolin I 370
jasmolin II 370

k

K$_2$CO$_3$ 54
ketones 80, 262
kresoxim-methyl 609, 613
krypton 261

l

lactofen 611
lamotrigine 760, 761
lauric acid TMS 677

lead 253
lenacil N-methyl derivative 370
lenacil 370
leptophos 591
levallorphan-TMS 766
levetiracetam 760, 761
levomepromazine 760
levorphanol-TMS 766
light hydrocarbons 18, 80
lilial 653
limonene, d- 652
linalool 652
lindane 370, 482, 586, 606
linuron 370, 489, 588
low pressure GC 233
loxynil isooctyl ether 370
loxynil methyl ether 370
lyral 654

m

2MA *see* methylanthracene, 2-
MAE *see* microwave assisted extraction
magnesium sulfate 13, 104
malaoxon 588, 607
malathion 371, 487, 588, 608
6-MAM *see* monoacetylmorphine, 6-
manganese 253
1MC *see* methylchrysene, 1-
3MC *see* methylchrysene, 3-
5MC *see* methylchrysene, 5-
MCPA methylester 490
MCPB isooctyl ester 371
MCPB methyl ester 371, 490, 492
mecarbam 608
MeCN *see* acetonitrile
mecoprop isooctyl ester 371
mecoprop methyl ester 371
melperone 760
mepanipyrim 609
mephosfolan 608
mepronil 610
mercaptanes 338
mercury 253
metalaxyl 588, 607
metamitron 371, 606

metazachlor   589, 608
methabenzthiazuron   371, 606
methadone   760, 761, 763
methamidophos   585, 606
methane   77, 263, 266
methanol   17, 25, 37, 44, 45, 66, 102, 142, 146, 651, 659
methazole   371
methidathion   371, 589, 608
methiocarb   371, 608
methomyl   371
methoprotryne   590
methoxyamine hydrochloride   660
methoxyaniline, 2-   684
methoxychlor   591
methoxycoumarin, 7-   653, 654
methoxyethanol, 2-   45
methoxyphenol, 4-   652
methoxypyrazine, 2-isopropyl-3-   633
methyl acetate   146
methyl bromide   79, 564
methyl chloride   564
methyl ester, 2,4-D   369, 490
methyl ester, 2,4-DB   369
methyl heptine carbonate   652
methyl isobutyl ketone   54
methyl isothiocyanate   263
methyl linolenate   293
methyl sterols   173
methyl trithion   590
methylamine   45
methylaniline, 3-chloro-4-   368
methylanthracene, 2-   704
methylanthracene   366
methylanthranilate   640
methylbenzaldehyde, 2,4-dihydroxy-3-   653
methylbenzene, 1-ethyl-2-   477
methylbutan-1-ol, 3-   648
methylbutanoic acid, 3-   648
methylcholanthrene   367
methylchrysene, 1-   704
methylchrysene, 3-   704
methylchrysene   367
methylchrysene, 5-   704

methylcoumarin, 6-   653
methylcoumarin, 7-ethoxy-4-   654
methyl-diuron   489
methylene chloride   546
methylethyl ketone   83
methylethylbenzene   366
methylethylketone   44
methylfluoranthene   367
methylionone, alpha-iso-   653
methylisoborneol, 2-   632, 633, 635, 636
methyl-linuron   489
methyl-monuron   488
methylnaphthalene   366
methylnaphthalene, 1-   704
methylnaphthalene, 2-   704
methylphenanthrene, 1-   704
methylphenanthrene   366
methylpropanoic acid, 2-   648
methylpyrene, 1-   704
methyl-t-butyl ether   48, 83, 146, 192, 572
methyl-*trans*-2-butenoate   652
metobromuron   371, 587
metoclopramide   760
metolachlor   588
metoxuron   371, 593
metribuzin   371, 588, 607
mevinphos   371, 606
mevinphos, *cis*-   585
mevinphos, *trans*-   585
mexacarbate   585, 587
$MgCl_2$ 54
$MgSO_4$ 14, 15, 54
mianserin   760
2-MIB *see* methylisoborneol   2-
midazolam   760
mirex   485, 591, 593
mirtazapine   760
1MN *see* methylnaphthalene, 1-
2MN *see* methylnaphthalene, 2-
moclobemide   760
molinate   606
(mono-)3-nitrotoluene (3-MNT)   502
(mono-)nitrobenzene (MNB)   503
monoacetylmorphine, 6-   762
monoacetylmorphine-TMS   766

Compound Index | 893

monoaromatic hydrocarbons   549
monobutyltin   575, 576
monochlorobiphenyl   492
monocrotophos   371, 587, 606
monoheptyltin   576
monolinuron   371, 587
monooctyltin   576
monuron   488
morphine   496, 497, 762, 765
morphine-bis-TMS   766
morphine-d₃   765
1MP *see* methylpyrene, 1-
1MPH *see* methylphenanthrene, 1-
M-series   432
MSTFA *see* trimethylsilyl-
　　trifluoroacetamide, *N*-methyl-*N*-
MTBE   558, 569
multimode injector   155, 158
multipath effect   214
multiple headspace extraction   40
multiple reaction monitoring   363, 598
multiresidue analysis   584, 599
myclobutanil   591, 609

n

NA *see* naphthalene
NAC *see* nitroaromatic compounds
NaCl   14, 37, 45, 46, 54
Na₂CO₃   53, 54
naled   585
nalorphine-bis-TMS   766
naphthalene   366, 478, 558, 565, 703
naphthalene, d-   684
naphthalene-d₈   366
naphthalenes   79
naphthoquinone, 1,4-   255
naphthylamine, 2-   617
napropamide   371, 609
napthylamine, 2-   684
Na₂SO₄   53, 54, 187
NH₄Cl   54
nickel   253
nicobifen   611
nicotine   371
nitralin   592

nitrapyrin   585
nitrate esters   260
nitroaniline, 4-   617
nitroaromatic compounds   614
nitrobenzene   255
nitrofen   371, 590, 609
nitrogen dioxide   77
nitrogen-15   253
nitropenta (PETN)   498, 500
nitrothal-isopropyl   588
nitrotoluene, 2-   499
nitrotoluene, 2,4-diamino-6-   617
nitrotoluene, 2-amino-4-   617
nitrotoluene, 2-amino-6-   617
nitrotoluene, 4-   499, 500
nitrotoluene, 4-amino-2-   617
nitrous oxides   77, 80, 255, 259, 260, 262
N₂O *see* nitrous oxide
N,O-bis(trimethylsilyl)trifluoroacetamide
　　see BSTFA
noble gases   262
nonachlorobiphenyl   340, 494
nonanal   662
nonane   72, 182, 209
nonanoic acid, trimethylsilyl ester   676
nontargeted analysis   757
nonvolatile matrix   138, 140, 141
norcodeine-bis-TMS   766
nordazepam   760, 761
norflurazon   591
nuarimol   371, 591

o

OBIND *see* octabromo-1-phenyl-1,3,3-
　　trimethylindane
ocimene, cis-β-   638
ocimene, trans-β-   638
OCPs   17, 191, 255, 256, 261, 490, 697, 698
octabromo-1-phenyl-1,3,3-
　　trimethylindane   703
octachlorobiphenyl   494
octachlorostyrene   483
octadecanoic acid, methyl ester   674

octadecanoic acid, trimethylsilyl ester 677
octadecylmorpholine, 4- 662
octafluoronaphthalene 302
octane, 2,2,6-trimethyl- 673
octane, 2,4,6-trimethyl- 676
octane, 2,6,6-trimethyl- 673
octane, 4-ethyl- 676
octanoic acid, trimethylsilyl ester 676
octanol 18, 48, 49
octen-3-ol, 1- 662
octenal, 2-butyl-2- 662
octogen 498
ofurace 610
olanzapine 760, 761
oleanitrile 674
oleic acid 218
oleic acid amide 528
oleic acids, cis/trans 211
omethoat 371
omethoate 586, 606
opiates 763
opioids 438
ordram 606
organic acids 16, 141
organic gases 64
organoarsenic compounds 498
organomercury compounds 253
organophosphorus compounds 77
organophosphorus pesticides (OPPs) 102, 491
organotin compounds 253, 572, 576
oxadiargyl 609
oxadiazon 371, 590, 609
oxadixyl 591, 609
oxycarboxin 591, 593
oxychlordane 588
oxycodone-TMS 766
oxydemeton-methyl 593
oxydianiline, 4,4- 684
oxyfluorfen 590
oxygenated compounds 80
oxygen-containing compounds 258
ozone 77

p

PA *see* polyacrylate
paclobutrazol 609
PAEs *see* phthalic acid esters
PAHs 17, 34, 39, 44, 79, 98, 159, 173, 190, 191, 206, 207, 228, 236, 239, 257, 262, 263, 291, 310, 478, 599, 691, 692, 697, 698, 710
palmitic acid 761
PAN *see* polyacrylonitrile
paraoxon(-ethyl) 487
paraoxon 588
paraoxon-methyl 487
parathion(-ethyl) 488
parathion 371, 588
parathion-ethyl 435
parathion-methyl 371, 487, 588, 607
PBBs *see* polybrominated biphenyls
PBDE 209 *see* decabromodiphenylether
PBDE *see* polybrominated diphenyl ether
PBEB *see* pentabromoethylbenzene
PBT *see* pentabromotoluene
PCB 209 *see* decachlorobiphenyl
PCBs 14, 22, 79, 98, 100, 102, 105, 159, 190, 191, 201, 204, 206–208, 251, 255, 256, 311, 323, 374, 436, 492, 599, 697, 698, 710, 712, 725
  coplanar PCBs 713
  dl-PCBs 712, 713
  indicator PCBs 712, 713
  ndl-PCBs 712, 713
  non-ortho-substituted PCBs 714
  WHO-PCBs 712, 713
PCDD *see* dioxins
PCDD/PCDF and PCB isomers 730
PCDDs 190, 374, 495, 718, 726
PCDFs 190, 374, 495, 718, 726, 730
PDMS *see* polydimethylsiloxane
pebulate 585
PeCDD, 1,2,3,7,8- 723
PeCDF, 2,3,4,7,8- 723
PEG *see* polyethylene glycol
penconazole 589, 608
pencycuron 606
pendimethalin 371, 588, 608

Compound Index

pentabromoethylbenzene 703
pentabromotoluene 703
pentachlorobiphenyl (e.g. PCB 101, 118) 493
pentachlorobiphenylene 340
pentachlorophenol 226, 227
pentadecane, 3-methyl- 674
pentafluorobenzyl bromide 40
pentafluoropropanol 752
pentan-2-ol 648
pentane 45, 84, 139, 573
pentane, n- 146
pentane, 2-methyl- 673
pentane, 3-methyl- 673
pentazocine-TMS 766
penten-3-one, 1-(p-methoxyphenyl)-1- 654
pentene-1,4-methyl-2,4-bis(4'-trimethylsilyl-oxyphenyl) 677
pentylfuran, 2- 662
perazine 760
perfluorinated alkyltriazines 392
perfluorobenzoyl chloride 304
perfluorocarbon 80
perfluorokerosene 391, 741
perfluorophenanthrene see FC5311
perfluoropolyethers 399
perfluorotributylamine see FC43
permanent gases 77, 80, 257, 266
permethrin 371, 611
permethrin, cis- 592
permethrin, trans- 592
perylene 367
perylene-d_{12} 367
pethidine 760
PFB-Br see pentafluorobenzyl bromide
PFK see perfluorokerosene
PFPOH see pentafluoropropanol
PFTBA see FC43
PHE see phenanthrene
phenanthrene 366, 703
phenanthrene-d_{10} 366
phenmedipham 371
phenol 48, 104, 226, 232, 263, 480, 481, 662

phenoxyalkylcarboxylic acids 492
phenprocoumon 760, 761
phenthoate 589, 608
phenylanthracene 367
phenylenediamine, 1,4- 684
phenylnaphthalene 366
phenylphenol, o- 585
phenylureas 492
phorate sulfone 608
phorate 586
phosalone 371, 488, 591, 610
phosmet 591
phosphamidon 588, 607
phosphorus 253
phthalate ester 17, 46, 77, 134
phthalate, di-(2-ethylhexyl) 621, 623
phthalate, di-(4-methyl-2-pentyl) 621, 625, 626
phthalic acid, butyl cyclobutyl ester 674, 677
phthalic acid esters 207, 619
phthalic acid, ethyl hexyl ester 677
phytane 237
phytosterols 173
picolinafen 610
picoxystrobin 609
piperine 628–630
piperonal 643
pirimicarb 371, 486, 587, 607
pirimiphos-ethyl 371, 588, 608
pirimiphos-methyl 371, 587, 607
PnCDD 737
polyacrylate 30, 37
polyacrylonitrile 48
polybrominated biphenyls 504, 747
polybrominated dibenzodioxins 504
polybrominated diphenyl ether 323, 504, 599, 698, 703, 710, 739, 740, 747
polybrominated diphenylethers 504, 599
polydimethylsiloxane 30, 34, 35, 42–46, 48, 69, 87
polyphenols 16
polyphenyl ethers 399
polyunsaturated fatty acids 471, 698
pregnane, 20-dimethylamino-5α- 294

Compound Index

pregnane, 5α- 294
pretilachlor 609
primary secondary amine 16, 601
pristane 236
prochloraz 592, 611
procymidone 589, 608
profenofos 590, 609
profluralin 586
promecarb 486, 585, 586
promethazine 760
prometon 586
prometryn 587
propachlor 371, 586, 606
propan-1-ol 648
propanal, 2,2-dimethyl- 673
propane 45
propanil 372, 589, 607
propanoic acid 648
propanoic acid, 2-[(trimethylsilyl)oxy]-, trimethylsilyl ester 676
propanol 45
propanol, 1- 44
propanol, 2- 37
propargite 591, 610
propazine 586, 607
propene 263
propenoic acid, 2- 673
propetamphos 591, 606
propham 372, 585
propiconazole 610
propiconazole 1 591
propiconazole 2 591
propisochlor 607
propoxur 372, 606
propyl phthalate 529
propylene glycol 641
propyzamide 587, 607
prosulfocarb 607
prothiofos 589
prothipendyl 760
pseudoionone 653
PUFAs *see*, polyunsaturated fatty acids
PVC *see* polyvinylchloride
PY *see* pyrene
pymetrozine 595

pyraclostrobin 611
pyrazophos 592, 611
pyrene 366, 703
pyrethrin I 372
pyrethrin II 372
pyridine 44
pyrimethanil 607
pyriproxyfen 610
pyroquilon 607

q

quetiapine 760, 761
quinalphos 416, 589, 608
quinoxyfen 610
quintozene 16, 372, 586
quizalofop-ethyl 611

r

resmethrin 372
ronnel 607
rubicene 367

s

Santovac S 399
sarin 491, 503
schradan 587
secbumeton 587
sec-butylbenzene 564
selenium 253
semivolatile organic compounds 9, 25, 27, 80, 81, 96, 190, 677
sertraline 760
SF$_6$ *see* sulfur hexafluoride
silane, (1-cyclohexen-1-yloxy)trimethyl- 676
silane, (cyclohexyloxy)trimethyl- 676
silane, trimethyl(phenylmethoxy)- 676
silicon 253
siloxanes 134
simazine 372, 485, 587
simetryn 587
sodium sulfate 104
sodium thiosulfate 39
soman 491, 503
sorbitol 631

Compound Index | 897

spiroxamine I 607
spiroxamine II 607
squalene 528
stanols 173
stanyl fatty acid esters 173
stearic acid 218, 761
sterols 16, 173
steryl fatty acid esters 173
stigmasterol 173
styrene 55, 80, 81, 84, 477, 546, 558, 565, 662, 681
sugars 16
sulfallate 586, 608
sulfotep 586, 606
sulfur compounds 189, 254, 258, 265
sulfur dioxide 77
sulfur hexafluoride 77, 255
sulfur 253, 528
sulfuric acid 105
sulprofos 590
SVOCs *see* semivolatile organic compounds

t

tabun 491, 503
TCDD, 2,3,7,8- 304, 324, 496, 712, 721, 722, 728, 729
TCDF, 2,3,7,8- 496, 721
TCMTB *see* (Benzothiazol-2-ylthio)methyl thiocyanate
tebufenpyrad 610
tecnazene 372, 585
tefluthrin 607
temazepam 761
TEPP *see* tetraethyl pyrophosphate
terbacil N-methyl derivative 372
terbacil 372, 588
terbufos 16, 586
terbumeton 586
terbuthylazine 587, 606
terbutryn 588
terbutylazine 485
terpenes 44, 207, 208
terpineol 641
tetrabutyltin 576

tetra-butyltin 577
tetrachlorobenzyltoluene 340
tetrachlorobiphenyl (e.g. PCB 52) 493
tetrachlorodibenzodioxin *see* TCDD, 2,3,7,8-
tetrachloroethane 54, 558
tetrachloroethane, 1,1,1,2- 565
tetrachloroethane, 1,1,2,2- 546, 558, 565
tetrachloroethene 54, 546
tetrachloroethylene 84, 565, 572
tetrachloroethylene (Per) 474
tetrachloromethane 422, 546, 572
tetrachloromethoxybiphenyl 340
tetrachlorvinphos 372, 589
tetradecane 674
tetradecanenitrile 674
tetradecanoic acid, trimethylsilyl ester 677
tetradecanone, 2- 674
tetradifon 591, 610
tetraethyl borate 573
tetraethyl pyrophosphate 606
tetrahydrocannabinol 339, 752
tetrahydrofuran 44, 651
tetramethrin 1 591
tetramethrin 2 591
tetramethylammonium hydroxide 422
tetramethyl-2-norborneol, 1,2,7,7- *see* geosmin
tetrapropyltin 576
tetrasiloxane, decamethyl- 676
tetrasul 372, 590
THC *see* tetrahydrocannabinol
THC-A 751, 756
THF *see* tetrahydrofuran
thiabendazole 16, 372
thiobencarb 587
thiodianiline, 4,4'- 684
thiofanox 372
thiometon 372
thiophanat-methyl 372
thiophene 558
thiram 372, 577, 578
thymol 628, 629
timipramine-d3 762

tin(II)chloride   578
tin   253
TMAH *see* tetramethylammonium hydroxide
TMSH *see* trimethylsulfonium hydroxide
tolclofos-methyl   295, 607
toluene   16, 18, 44, 53, 54, 69, 72, 82, 84, 102, 139, 142, 146, 263, 366, 476, 546, 549, 552, 558, 565, 572, 662, 675, 681
toluene-d_8 366, 518, 552, 565
toluenediamine, 2,4-   684
toluidine, 4-chor-*o*-   684
toluidine, 5-nitro-*o*-   684
toluidine, *o*-   684
tolylfluanid   589, 595, 608
toxaphen   303
TPHs *see* total petroleum hydrocarbons
TPP *see* triphenyl phosphate
tralkoxydim   611
tramadol   761
TRI *see* triphenylene
triadimefon   435, 588
triadimenol   590
tri-allate   372, 586
triazines   490
triazole pesticides   46
triazophos   591, 610
tribufos   589
tributyl phosphate   528
tributyltin   573, 576
trichlorfon   372
trichloroaniline, 2,4,5-   617, 684
trichlorobenzene   79
trichlorobenzene, 1,2,3-   558, 565
trichlorobenzene, 1,2,4-   558, 565
trichlorobenzene, 1,3,5-   558
trichlorobiphenyl (e.g. PCB 28, 31)   493
trichloroethane   558
trichloroethane, 1,1,1-   54, 82, 84, 546, 558, 565
trichloroethane, 1,1,2-   546, 558, 565
trichloroethane, 1,1,2-trifluoro-1,2,2- (R113)   474
trichloroethene   546

trichloroethylene   52, 84, 474, 565, 572
trichlorofluoromethane   546, 565
trichloromethane, 1,1,1-   54
trichloronate   432
trichloropropane, 1,2,3-   565
tricyclo[3.1.0.0(2,4)]hex-3-ene-3-carbonitrile   674
tricyclohexyltin   576
tridecane   674
tridemorph   372
trietazine   372
triethylamine   45
trifloxystrobin   610
triflumizole   608
trifluoromethyl-*bis*-(trimethylsilyl)methyl ketone   676
trifluralin   372, 586, 606
triglycerides   157, 173, 192, 206, 207
trimethylaniline, 2,4,5-   684
trimethylbenzene   366
trimethylbenzene, 1,2,4-   83, 565
trimethylbenzene, 1,3,5-   546, 565
trimethylsilyl-trifluoroacetamide, *N*-methyl-*N*-   659, 660, 763
trimethylsulfonium hydroxide   164
trinitrotoluene (TNT)   500
triphenyl phosphate   601, 610
triphenyl tin   573
triphenylene   703
triphenylphosphine   530
triphenylphosphine oxide   530
triphenyltin   576
tripropyltin   576
tris(dibromopropyl)phosphate   320
triterpenes   173
triticonazole   610

u

undecane   662
U-tube sparger   65, 75, 76

v

vamidothion   372, 593
vanadium   253
venlafaxine   761

verapamil   761
Verdox   643
vernolate   585
very volatile organic compounds (VVOCs)   65, 69, 77, 166, 202, 246
vinclozolin   372, 587, 607
vinyl chloride   81, 474, 546, 560, 565
vinylpyrrolidone, *N*-   48
VOCs *see* volatile organic compounds
volatile organic compounds   27, 44, 62, 65, 66, 77, 80, 81, 96, 166, 189, 190, 202, 208, 264, 338, 353, 473, 541, 545, 549, 554, 560, 569, 648, 677
volatile organic sulfur compounds   45
volatile sulfur compounds   189, 265, 338, 467, 648
vortexing   17
VOSs *see* volatile organic sulfur compounds
VSCs *see* volatile sulfur compounds
VVOCs *see* very volatile organic compounds
VX   80

X

xenon   261
xylene   52, 53, 82, 142, 549, 662
xylene, *m*-   84, 477, 546, 552, 558, 565, 681
xylene, *o*-   54, 84, 263, 476, 546, 552, 558, 565, 681
xylene, *p*-   84, 477, 546, 552, 558, 565, 681
xylidine, 2,4-   684, 687
xylidine, 2,6-   684, 687

Z

zolpidem   761
zotepine   761

MEAT
The Ultimate Cookbook

Meat

Copyright © 2021 by Appleseed Press Book Publishers LLC.

This is an officially licensed book by Cider Mill Press Book Publishers LLC.

All rights reserved under the Pan-American and International Copyright Conventions.
No part of this book may be reproduced in whole or in part, scanned, photocopied, recorded, distributed in any printed or electronic form, or reproduced in any manner whatsoever, or by any information storage and retrieval system now known or hereafter invented, without express written permission of the publisher, except in the case of brief quotations embodied in critical articles and reviews.
The scanning, uploading, and distribution of this book via the internet or via any other means without permission of the publisher is illegal and punishable by law. Please support authors' rights, and do not participate in or encourage piracy of copyrighted materials.

13-Digit ISBN: 978-1-64643-048-2
10-Digit ISBN: 1-64643-048-4

This book may be ordered by mail from the publisher. Please include $5.99 for postage and handling. Please support your local bookseller first!

Books published by Cider Mill Press Book Publishers are available at special discounts for bulk purchases in the United States by corporations, institutions, and other organizations. For more information, please contact the publisher.

Cider Mill Press Book Publishers
"Where good books are ready for press"
501 Nelson Place
Nashville, Tennessee 37214

cidermillpress.com

Typography: Adobe Garamond, Brandon Grotesque, Lastra, Sackers English Script
Front cover image: Grilled Rib Eye Steak, see page 57
Back cover image: Perfect Prime Rib, see page 73
Front endpaper image: Killer BBQ Spareribs, see page 183
Back endpaper image: Roasted Lamb Chops, see page 378

Printed in China

23 24 25 26 27 TYC 8 7 6 5 4

MEAT

The Ultimate Cookbook

KEITH SARASIN

CONTENTS

Introduction   9

Meat 101   15

Beef   37

Pork   159

Poultry   243

Lamb, Goat, Venison & Rabbit   377

Brines, Rubs, Stocks & Sauces   443

Industry Insiders   543

Sides, Salads & Accompaniments   611

Appendix   749

Index   787

The first four chapters of this book are divided into two sections: Prime Recipes and Go-To Recipes.

Prime Recipes celebrate the classic, special preparations for each protein.

Go-To Recipes feature equally delicious dishes that work for weeknight meals and use leftovers, guaranteeing that none of your meat goes to waste.

Unless otherwise noted, all salt called for in this book is kosher salt.

INTRODUCTION

A primal rush satiates part of our hunter-gatherer brain when we tuck into a thick, juicy steak. As we savor each succulent bite, a rush of dopamine overcomes our senses. Meat has been a part of human culture since humans figured out how to hunt, and it has been a salient element of American culture since before we became a nation. It has been the topic of political debates, ethical debates, and agricultural debates. Still, many people find themselves at backyard BBQs salivating over the richly scented smoke that permeates the air.

For the last eight years I have worked with countless farms across New England. I have helped farmers connect with restaurants and establish relationships with many amazing chefs. When I began this journey, I wanted to be able to trace my steak from the cow that grazed in the field to the moment it hit my plate. Like many people, I thought that this picture-perfect scene of a cow grazing in the warm sun was where my meat came from. Turns out, I was wrong.

The meat industry in the United States is a massive operation that has been championed and scrutinized since the earliest days of industrialization (see Upton Sinclair's *The Jungle*). Today, when we break down the farming, processing, and distribution of meat on a local level, we find incredible stories of hardworking farmers who have sacrificed massive amounts of time and money to deliver different options to consumers than they previously might have had. Some of these stories are included in this book, like Noah Bicchieri of Arkhive Farm in Chester, New Hampshire (see page 605). Bicchieri's dedication to raising American Wagyu beef and how he went from working in the financial sector to raising cows is an incredible story that showcases the ups and downs of local meat producers.

In this book I hope to not only showcase some amazing recipes but also to share with you how local farms can become a part of our lives. Through these men and women of the farming community I have found a community of my own. From late-night calls asking for help because a cow escaped to the sorrow of slaughter, the journey of our food system is an emotional one that teaches us to honor the sacrifices we all make.

THE SCOTTISH HIGHLANDER

"Hey buddies!"

The ground trembled as we hopped the fence to where Gary Bergeron's cows were grazing on a hot July day. Gary is the farmer and rancher at Rickety Ranch in New Hampshire. I had never been around a herd of cattle before and I wanted to get up close and personal with the entire process of beef making its way from farm to table.

"Hey buddies!" Gary shouts again.

Clouds moved across the sky, blocking out the midday sun. A herd of 30 to 40 cattle came

running down the hill, right toward us. Frozen in fear, I looked at Gary wide-eyed and managed to mutter a few broken words: "Umm, are we okay?" Gary glanced at me with a smirk and told me we were going to die.

The herd of cattle formed a circle around us. Imagining that I was going to be crushed by charging cows, my mind raced to the gladiatorial death rings of Rome. The air become charged with moos. The cows' massive horns clanked together as they shook their heads to ward off swarming flies. It was incredible to be in the middle of such majestic creatures. When you are surrounded like that, you don't think about your dinner plate. Rather, you feel extreme reverence, followed by a deep sense of guilt. This moment would send me on a path to rediscover the food I eat.

Over the years, I have spent many hours around livestock and my appreciation has continued to grow for all aspects of farming, but this book is about learning recipes and techniques I have picked up from years of being on the farm and in kitchens across America. Learning where your food comes from is an important part of any recipe, and I encourage everyone to support their local food systems. At the end of the day, we vote with our dollars.

From whole animal butchery to preservation methods, the journey to understand where our food originates and the best ways to utilize it is an ever-evolving quest. *Meat* represents many of the farmers, chefs, and craftspeople who have influenced and inspired me over the years. In this book you will find not only outstanding recipes, but also you will find a deep respect for the proteins we use.

Keith Sarasin

MEAT 101

A BRIEF HISTORY OF BUTCHERY

We have the French to thank for the word "butcher," which derives from the word *boucher*. But the French did not invent this craft. From the first time our earliest hunter-gatherer ancestors killed an animal to eat it, butchering has been an integral aspect of human culture. In 2010, bones from animals resembling a cow and goat were found in Ethiopia. These bones, over 3 million years old, showed evidence of cuts that indicated the animals had been skinned and that their marrow had been removed. This discovery has led anthropologists to revise, by 800,000 years, the estimate of how early in human history we started using stone tools and eating meat. Evidence from the last Ice Age of butchered giant sloths has been uncovered

in multiple locations around the world. Collections of the animal's rib bones show scrape marks, suggesting that a stone tool was used to separate the meat from the bone.

There are many visual and textual representations of standardized forms of butchery found in artifacts from ancient Egypt and Rome. This is a direct result of our ancestors learning how to domesticate animals, which went hand-in-hand with farming and becoming less and less reliant on being hunter-gatherers. Because this craft fed, literally, society, butchers were revered to the same degree as doctors and smiths.

Roman butchers are believed to have introduced their trade to the British. In London, the Worshipful Company of Butchers is the twenty-fourth oldest livery company in the city; there are references to Butchers' Hall "where craftsmen meet" dating to 975 CE, and the organization was granted its Royal Charter of Incorporation by King James I in 1605. The craft was codified under the guise of a guild, and the same thing began to happen across Europe. Meat was treated, rightfully so, as a valuable commodity, and the trained butchers knew how to get the most out of whatever they butchered.

As you might expect, recipes for some extraordinary dishes can be traced back to this time. In one of the manuscripts from the Harleian Collection at the British Library, there is a detailed description from circa 1450 for "Pecok Rosted":

Take a Pecok, brecke his necke, and kutte his throte. And fle him, the skyn and the fflethurs otgidre, and the hede still to the skyn of the nekke, And kepe the skyn and the fflethurs hole togiders; drawe him as an hen, And kepe the bone to the necke hole, and roste him, And set the bone of the necke aboue the broche, as he was wonte

to sitte a-lyve; And abowe the legges to the body, as he was wonte to sitte a-lyve, take him of, And lete him kele; And then wynde the skyn with the fethurs and the taile abought the body, And serue him for the as he were a-live; or else pull him dry, And roste him, and serue him as thou doest a henne.

The *Noble Boke off Cookry* includes a recipe for making capon into a meal for two by blowing up the skin. And from the Harleian Collection there is a recipe for spit-roasted "Cokyntryce"—"a mythical beast created from half a pig and half a cockerel—glazed with egg yolks, powdered ginger, and saffron and finished with parsley juice." And you thought a turducken was a bit over the top?

In North America, the earliest butchering tools discovered date back to 11,000 BCE and the Clovis culture, in what today is New Mexico. Plains Indians and other indigenous people of North America have been reverently butchering deer, elk, turkey, and bison for more than 2,000 years. Spanish explorers arriving in the sixteenth century introduced cattle to the New World.

With industrialization—from railways to refrigeration—agriculture and the transport of livestock developed, creating a demand for commercial butchering. Like so many industries that scaled up in this era, the craft was often replaced by automated processes. That history is well documented in countless books, the most famous of which, relating directly to butchery, is Upton Sinclair's *The Jungle*. As peo-

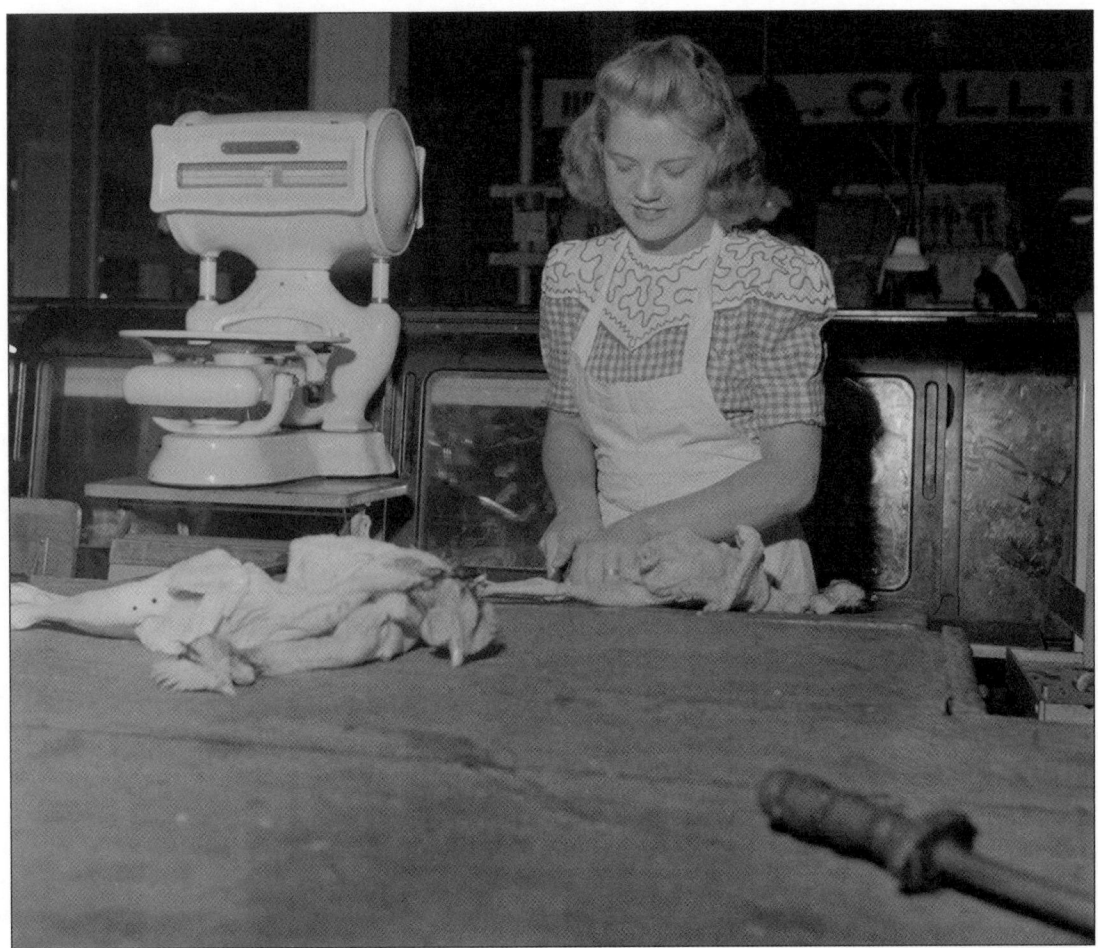

ple were exploited in the horrid working conditions of large slaughterhouses, the community butcher shop disappeared as shoppers became accustomed to buying prepackaged cuts of meat from supermarkets.

Today, there is still plenty of factory-farmed meat in supermarket meat departments, but more and more people have come to understand that while this meat is more affordable than what you find at a farmers market, it certainly isn't better. Factory-farmed meat does not have the best interests of the animals, the workers, the consumers, or the environment in mind.

Luckily, local farmers and butchers all across the United States are showing consumers the benefits of eating local, sustainably raised meat. And in doing so, these men and women have raised awareness about the sustainable tradition of this craft. Thanks to them, chefs like me can make food that not only tastes better but actually helps contribute to our local economies.

LEARNING TO IDENTIFY QUALITY MEAT

There you are at the local grocery store or staring at a package of steaks from your local farm. Should you get tri-tip or sirloin? How many people does a pound of beef feed? These are a few of the questions you are about to learn the answers to, along with learning how to pick out great cuts of meat.

VISUAL

The first thing I do when I am looking for a high-quality cut of meat is inspect the color and the marbling: the streaks of fat that separate the muscle. With beef, pork, and lamb you want to look for long streaks of white marbling dividing the muscle tissue.

SMELL

Quality meat should smell normal and have no rancid scent.

TENDERNESS

Quality meat should be tender and not overly firm. Aging meat can change this, but still look for something that isn't overly tough.

If you buy frozen meat, make sure there isn't any freezer burn as it will drastically affect the flavor of the meat.

PORTIONS

As a general rule, you can plan for half a pound per person. This accounts for fat loss and shrinkage as you cook the meat.

KNOW YOUR CUTS OF MEAT

Butchers and chefs know how to break down animals so that they become the recognizable cuts you buy in the market or find on your plate at a restaurant. But butchering a chicken to separate wings, thighs, and breasts is very different than isolating loins and ribs from a cow. Depending on what you are doing, there are certain tools for the job. For instance, if you are butchering a whole hog (see page 24) you are going to need different tools than you would for backyard grilling. Assuming that you are buying already butchered cuts of meat to use in the recipes that follow, here is my guide to understanding all things meat.

CUTS FROM A COW

Meat from cows is divided into four primal cuts, which are then broken down into food-service cuts, which are then further broken down into steaks, etc.

FOREQUARTER
BEEF CHUCK

Beef chuck comes from the front section of a cow. This is called the forequarter. The forequarter consists of the neck, shoulder blade, and upper arms. It is a very flavorful part of the cow but is known to be tough. Typical cuts from this section are stew meat, flat iron steak, roasts, and ground beef. You also get the classic 7-bone roast from this section of the cow.

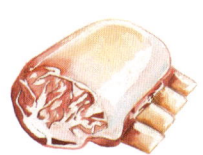

FOREQUARTER
RIBS

This section produces some really amazing cuts like the prime rib and rib eye. This section is classified as the center section of the rib, specifically the sixth through the twelfth rib.

FOREQUARTER
BEEF PLATE

The beef plate is often referred to as the short plate or the long plate depending on where it is taken from the rib. It comprises several cuts of meat, including the skirt steak and short ribs. There is a lot of connective tissue in the beef plate, which is why when cooking meat from this part of the cow we tend to focus on braising it, to make it more tender.

FOREQUARTER
BRISKET

Next to the beef plate is the brisket, located in the front section of the cow. Brisket is packed with flavor but must be cooked according to very specific guidelines for it to be tender.

FOREQUARTER
SHANK

Beef shank has lots of connective tissue and must be braised and cooked slowly to break it down. This is an incredible cut of meat that, with a little bit of time according to very specific guidelines, will melt in your mouth.

HINDQUARTER
SHORT LOIN

The short loin intersects the tenderloin, and it is the location of steaks like the porterhouse, T-bone, and strip steak (according to very specific guidelines). The short loin is about 16 inches long and yields eleven or more steaks. These steaks are prized and seared over high heat for a short amount of time.

GROUND BEEF FAT RATIOS EXPLAINED

When you look at a package of meat in the grocery store, you will see a percentage on the label. This ratio typically ranges from 73/27 to 96/4. A number like 80/20 indicates that the meat is 80 percent lean and 20 percent fat.

When it comes to different recipes, we look for different fat ratios. With the classic burger, we typically look for 80/20 as 20 percent fat helps keep the burger moist and juicy. When it comes to other recipes, like chili, the ratio can change to 90/10 because we are slow cooking the chilli with other ingredients that help break down the meat.

HINDQUARTER SIRLOIN

This section of the cow starts at the thirteenth rib and runs all the way back to the hip bone. This section is known for the top and bottom sirloin, tri-tip, and flap steaks.

HINDQUARTER BEEF FLANK

I love this section. Beef flank works great on the grill but can get very tough if it is overcooked. This section also works really well for taking on flavor through marinating. This cut tends to be cheaper than some of the other cuts, which makes it a great value.

HINDQUARTER TENDERLOIN

The tenderloin is the most prized section of the cow. It is where the filet mignon comes from. In an entire cow there are typically about six pounds of filet mignon. This cut is very tender, and packed with flavor. Because this section is so tender it is cooked over high heat for a short amount of time.

HINDQUARTER BEEF ROUND

The backside of the cow. This is the source of many roasts like the top round and bottom round roast. This section of the cow works great for braising.

MEAT 101 | 23

CUTS FROM A PIG

There are eight primal cuts when it comes to breaking down a pig.

PORK BUTT AKA BOSTON BUTT
This section of the pig is actually the shoulder. It includes the neck, shoulder blade, and part of the upper arm. This is where we get pulled pork. It is a tough cut that benefits from slow cooking.

PORK SIDE
This is where belly, bacon, and pancetta are from. You will find many recipes for this cut throughout this book. Typically this section is cured and smoked, but other techniques can be applied. This is one of the most prized sections of the pig.

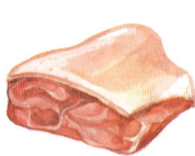

SHOULDER PICNIC
This section of the pig is located right below the Boston butt. This cut is typically where you get a lot of sausage from. Other times it is smoked or cured.

SPARERIB
Between the loin and the side, you have the sparerib. This section benefits from low and slow cooking and is very flavorful.

LOIN
The loin is where you get the tenderloin, baby back ribs, and also fatback. The loin can be cut into chops or cutlets, and the fatback can be added to sausage. A very versatile cut.

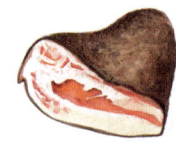

PORK JOWL
This is my favorite cut of the pig. Packed with flavor, the jowl (or cheek) benefits from braising. It is also used to make a cured meat known as guanciale. You can also use the jowl in place of a Boston butt for pulled pork.

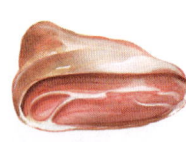

HAM
The ham of a pig comes from its back leg. Hams are typically cured, smoked, and baked. This is where you get amazing charcuterie like prosciutto and Serrano ham. Between the shank and the end of the ham you find a prized section called the ham hock. This is typically used to flavor bean soups and collard greens.

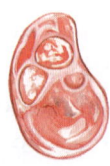

PORK FOOT
The foot, or trotters, is typically used for flavoring and making stocks. There is a ton of collagen in the trotter, and it makes for an unbelievable stock.

CUTS FROM A CHICKEN

When it comes to breaking down a chicken, you will find that it is not only an easy skill to learn; it is also a skill that will save you money. Sure, it's easy enough to buy legs and wings, but buying a whole bird is cheaper. Here's how you break down a whole chicken:*

1. Position the bird breast side up with wings faced away from you.

2. Bend the wing away from the body and cut through the skin in a half circle. This will reveal the joint. At the joint, cut through to free the wing from the body. Repeat with the other wing.

3. Pull the drumstick away from the body and make a slice between the drum and the breast to loosen the leg from the body.

4. Bend the leg down to reveal the joint. Bring the knife around the joint and cut to detach the leg from the body. Repeat with the other leg.

5. Flip the chicken over so the breast side is down. Using a pair of kitchen shears, cut along each side of the backbone to remove it (and reserve for making stock).

6. Place the chicken breast side up again and, with a chef's knife, remove the rib meat from both sides. Feel along the top of the breast where the breast bone resides. Run your knife along the side of the breast bone to slice the meat into the two breasts.

7. From here you can break down the chicken further by separating the drum from the thigh. To do this simply cut along the seam that separates the drum from the thigh.

8. You can also separate the wing tip from the drumette by doing the same thing with the wing if you choose.

*These same steps can be used to break down other poultry, such as turkey and duck.

MEAT 101 | 27

CUTS FROM A LAMB

Harvested from sheep are two types of red meat: lamb and mutton. Lamb is meat from animals that are under a year old; mutton comes from older sheep. Lamb is tender and milder in flavor; mutton tastes more like game.

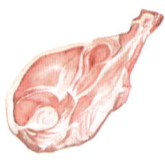

SHOULDER
The shoulder is full of flavor. Here is where we get the shoulder roast. This section benefits from slow cooking.

RUMP
The rump cut is from the backside of the lamb and is quite lean. It can dry out quickly, so it benefits from cooking quickly over high heat. It can also be cut into chops and grilled.

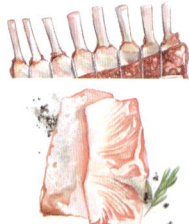

CHOP/RACK
The chop or rack is the prized section of lamb. Chops or cutlets are the most expensive cuts for lamb. They are taken from the ribs of the lamb and typically grilled. When the cutlets are left whole, we get a rack of lamb, which benefits from grilling or shorter cook times so the tender meat doesn't get overcooked.

LEG
The leg is another prized section of the lamb. It is packed with flavor and benefits from longer grilling, but also works great in longer braises for meat that is fall-off-the-bone tender. Be careful not to dry this cut out.

SHANK
The shank is a cheap cut that packs a lot of flavor. Like most tougher cuts, it benefits from longer cook times and works amazingly with a nice curry.

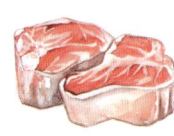

LOIN CHOP
The loin chop is like mini T-bones cut from the waist of the lamb. One side is the lamb loin and the other side is the filet. This cut is great for barbecuing or grilling.

NECK
The neck has a lot of bones, but that also means it has a lot of flavor! It is another cut that benefits from braising low and slow.

CUTS FROM A GOAT

Goat has long been popular in many parts of the world and the demand for this lean red meat has grown significantly in the United States.

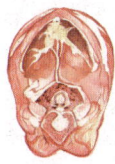

NECK
The neck is where you get cuts like the neck chops. The meat beneftis from braising, stewing, and roasting. The bones make for great stock as well.

SHOULDER
The main cuts from the shoulder are the bone-in shoulder, easy carve shoulder, forequarter rack, and forequarter chop. These cuts benefit from slow cooking methods such as braising, stewing, and roasting.

RACK
From the rack section you get great cuts like rack, ribs, and cultets. Ribs benefit from high heat and quick methods such as grilling, barbecuing, and pan-frying, but it can also benefit from cooking longer at lower temps as well.

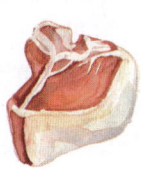

LOIN
In the loin you find loin chops, eye of loin, and tenderloin. Like all cuts of loin, they benefit from high heat and quick methods such as grilling, barbecuing, and pan-frying.

LEG
In the goat leg we find cuts like the bone-in leg, easy carve leg, boneless leg, mini roast, and leg steak. I find that the leg cuts work best with searing and roasting. For leg steaks, grilling, barbecuing, and pan-frying are the go-to methods.

HINDSHANK
The hindshank yields the shank and drumstick cuts. These cuts are tougher, so we tend to slow cook, braise, and stew them.

BREAST
The breast is part of the rib section. I typically slow cook this through smoking or roasting.

FORESHANK
Much like the hindshank, we get the shank and drumstick from this cut, and use slow cooking methods to tenderize the meat.

MEAT 101 | 31

CUTTING MEAT

You might have heard that you should slice meat against the grain. But what does that mean and why do we do it?

The grain of the meat refers to the direction in which the muscle fibers are running. It is not just the cut of meat that determines how tender it is; it is also how you slice it.

When cutting meat we want to first find the direction of the grain. Then we slice across the grain rather than parallel to it. So, if the grain of the meat was running vertical we would not want to slice vertical to the grain, we would slice horizontally, since we want to cut the fibers of the muscles, which shortens them. This makes the meat easier to chew.

THE IMPORTANCE OF LETTING MEAT REST

Simply put, letting meat rest after cooking allows the moisture to be reabsorbed. But there is also a second reason to let your meat rest. This is because of something we call "carry-over cooking." When you take meat off of heat, it continues to cook, as the heat inside the meat is a different temperature than the surface of the meat.

As meat cooks, the fibers of the muscles start to firm up and push out water. This water has to go somewhere, so it heads to the surface of the meat, where some of it evaporates, which is made even more visible if you slice the meat before resting it properly. You'll see more steam puff up from the meat, which means moisture is evaporating. Methods like roasting require time, and the water vapor needs time to be distributed.

When you cut meat right after it comes off of heat, that allows that water vapor to escape quickly and tends to dry out the meat. Letting the meat rest gives it time to redistribute the moisture that is has expelled.

CARVING & SERVING

Carving meat starts with proper tools. Having a sharp knife is the most important part of carving. Having a dull knife is not only useless; it is flat-out dangerous. Keep your knives sharp!

When it comes to carving, some of it is personal preference. For instance, when I carve a rib eye steak, I like to carve it differently if it is bone-in as opposed to boneless. For a bone-in rib eye, I like to carve along the bone, making sure to trim the bone and large pockets of fat. From there, I will cut the rib eye into thick strips about 1½-inches thick and plate it with the middle facing up.

When it comes to carving poultry, I will break down the bird into legs, wings, and breasts. From there I often like to slice the breast into ½-inch-thick slices that are manageable for people to eat. You can break down and carve as much or as little as you would like.

For tools, I use a boning knife for poultry and a 9-inch chef's knife for beef. These knives allow me to carve perfectly as needed. See page 753 for more on carving.

Now you know all the essentials for handling various types of meat. Let's get cooking!

BEEF

Prime rib, filet mignon, porterhouse—these are beef cuts that are more than food. They are symbols of luxury, and for good reasons. Cooked properly, the tender meat melts in your mouth, gilded by ribbons of fat. But there are plenty of other ways to use beef, whether letting it cook down like Barbacoa (see page 98) or using bone-flavored stock for Short Rib Ramen (see page 125), not to mention the wonderful flavors that can be achieved with the humbler ground beef. Explore all of the possibilities beef offers.

SEARED RIB EYE

YIELD: 2 SERVINGS / **ACTIVE TIME:** 10 MINUTES / **TOTAL TIME:** 1 HOUR

Rib eye is my favorite cut of meat, and this is how I cook it at home. The goal with cooking steak is to not over-season it, and not to overcook it.

1. Let the steak come to room temperature 40 minutes before you want to cook it.

2. Add canola oil to a cast-iron pan over high heat. As the pan gets hot, pat the steak dry and then salt both sides of the steak.

3. When the oil starts to smoke, add the steak. Cook for 4 to 5 minutes and flip.

4. Add 2 tablespoons of butter to the pan, tilt the pan slightly, and spoon the butter-oil mixture over the top of the steak. Repeat this 5 to 6 times.

5. Let the steak cook for 3 to 4 minutes, or until the desired doneness is achieved. Just before removing the steak from the pan, add the remaining butter and rosemary, tilt the pan slightly, and spoon the butter-oil mixture over the top of the steak. Repeat this 5 to 6 times.

6. Remove the steak from the pan and let rest on a cutting board for at least 15 minutes before serving.

INGREDIENTS:

- 1 LB. RIB EYE STEAK
- 2 TABLESPOONS CANOLA OIL
- 2 TEASPOONS SALT
- 4 TABLESPOONS UNSALTED BUTTER, DIVIDED
- 2 SPRIGS FRESH ROSEMARY

STEAK TEMPERATURE GUIDE

Very rare/rare	120°F to 125°F
Medium-rare	125°F to 130°F
Medium	135°F to 140°F
Medium-well	145°F to 150°F
Well-done	160°F and above

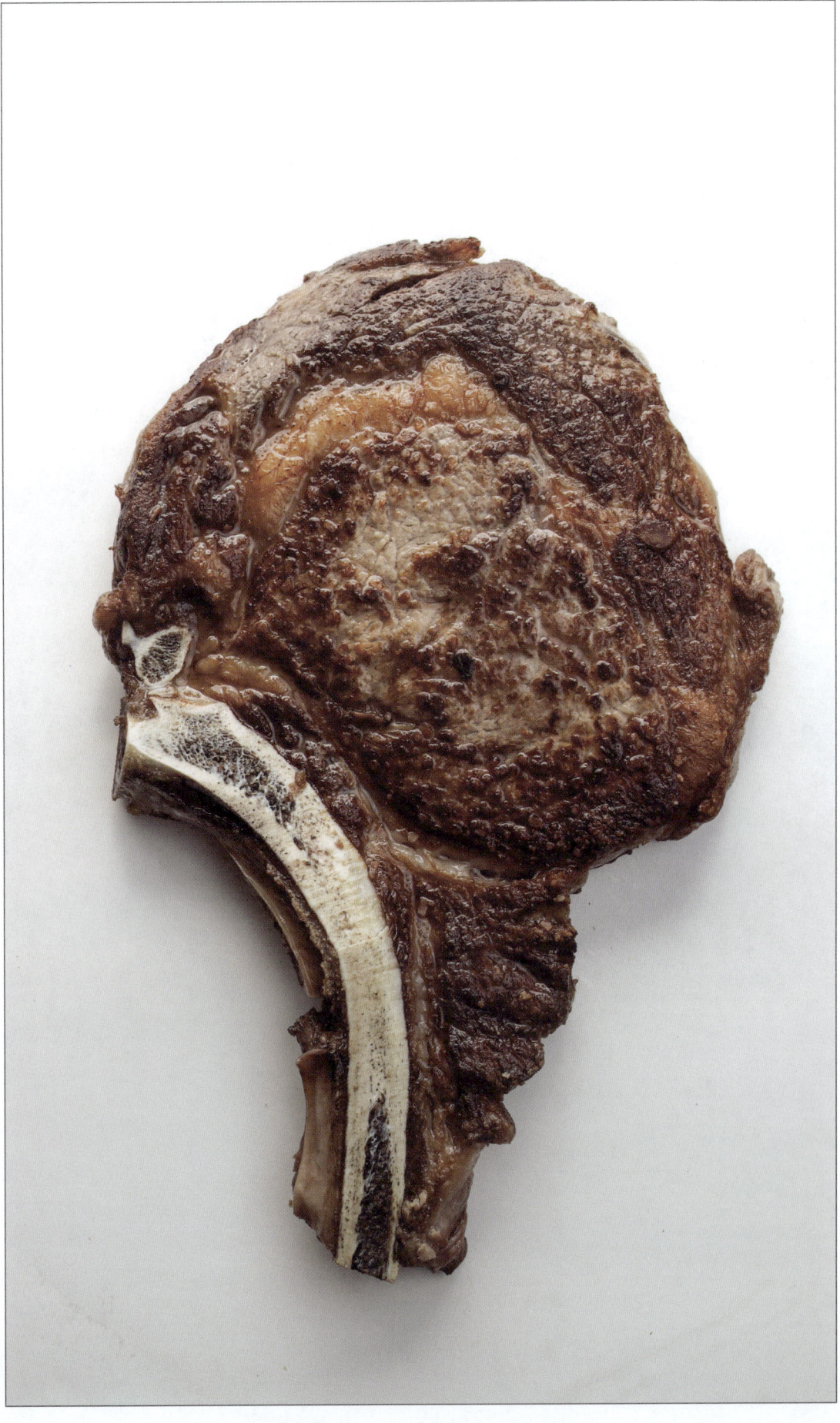

SOUS VIDE RIB EYE

YIELD: 2 SERVINGS / **ACTIVE TIME:** 15 MINUTES / **TOTAL TIME:** 1 HOUR AND 30 MINUTES

I love to sous vide rib eye steak for many reasons. One of the biggest is that you control the temperature perfectly while allowing the steak to seal in flavor. Once you try this method, you are going to love it too.

1. Preheat your sous vide cooker to 129.2°F.

2. Combine the steak, butter, and rosemary in a vacuum bag and place the bag in the sous vide cooker for at least 1 hour, or up to 3 hours. I typically remove it after an hour.

3. Remove the steak from the bag, being sure to save the liquid, and pat the steak dry.

4. Add canola oil to a cast-iron pan over high heat. As the pan gets hot, salt both sides of the steak.

3. When the oil starts to smoke, add the steak. Cook for 1 to 2 minutes.

4. Flip the steak and add the reserved liquid. Tilt the pan slightly and spoon the liquid over the top of the steak. Repeat this 5 to 6 times.

5. Remove steak from the pan and let rest on a cutting board for at least 15 minutes before serving.

INGREDIENTS:

- 1   LB. RIB EYE STEAK
- 4   TABLESPOONS UNSALTED BUTTER
- 2   SPRIGS FRESH ROSEMARY
- 2   TABLESPOONS CANOLA OIL
- 2   TEASPOONS SALT

SEARED NEW YORK STRIP STEAK

YIELD: 2 SERVINGS / **ACTIVE TIME:** 10 MINUTES / **TOTAL TIME:** 1 HOUR

The strip steak is a cut from the short loin of the cow. It is very tender because the muscle typically doesn't get much work. This cut is as good as many of the cuts you will get and often a bit cheaper as well.

1. Let the steak come to room temperature 40 minutes before you want to cook it.

2. Add canola oil to a cast-iron pan over high heat. As the pan gets hot, pat the steak dry and then salt both sides of the steak.

3. When the oil starts to smoke, add the steak. Cook for 4 to 5 minutes and flip.

4. Add 2 tablespoons of butter to the pan, tilt the pan slightly, and spoon the butter-oil mixture over the top of the steak. Repeat this 5 to 6 times.

5. Let the steak cook for 3 to 4 minutes, or until desired doneness is achieved. Just before removing steak from the pan, add the remaining butter and rosemary, tilt the pan slightly, and spoon the butter-oil mixture over the top of the steak. Repeat this 5 to 6 times.

6. Remove the steak from the pan and let rest on a cutting board for at least 15 minutes before serving.

INGREDIENTS:

- 1 LB. STRIP STEAK
- 2 TABLESPOONS CANOLA OIL
- 2 TEASPOONS SALT
- 4 TABLESPOONS UNSALTED BUTTER, DIVIDED
- 2 SPRIGS FRESH ROSEMARY

PORTERHOUSE

YIELD: 2 TO 3 SERVINGS / **ACTIVE TIME:** 20 MINUTES / **TOTAL TIME:** 1 HOUR AND 45 MINUTES

This is by far one of the most filling and flavorful steaks available.

INGREDIENTS:

- 2 PORTERHOUSE STEAKS, ABOUT 1½ INCHES THICK
- 4 TABLESPOONS OLIVE OIL
- COARSELY GROUND BLACK PEPPER
- SEA SALT

1. Rub both sides of the steaks with olive oil and let rest at room temperature for about 1 hour.

2. A half hour before cooking, prepare your gas or charcoal grill to medium-high heat.

3. When the grill is ready, about 400 to 450°F with the coals lightly covered with ash, season one side of the steaks with half of the coarsely ground pepper and sea salt. Place the seasoned sides of the steaks on the grill and cook for about 5 to 6 minutes, seasoning the tops of the steaks while waiting. When the steaks are charred, flip and cook for 4 to 5 more minutes for medium-rare, and 6 to 7 for medium. The steaks should feel slightly firm if poked in the center.

4. Remove the steaks from the grill and transfer to a large cutting board. Rest for 10 minutes before serving.

FILET MIGNON

YIELD: 2 TO 3 SERVINGS / **ACTIVE TIME:** 20 MINUTES / **TOTAL TIME:** 1 HOUR AND 30 MINUTES

The filet is one of the harder steaks to grill due to its thickness and relatively small surface area, proving it difficult to remain upright on the grate. I recommend using a seasoned cast-iron skillet over the grill for the initial searing and then transferring the filets to the oven.

INGREDIENTS:

- 1–2 FEET OF BUTCHER'S STRING
- 2 FILET MIGNON STEAKS, ABOUT 2 TO 2½ INCHES THICK
- 3 TABLESPOONS OLIVE OIL, DIVIDED
- COARSELY GROUND BLACK PEPPER
- SEA SALT

1. Tie the butcher's string tightly around each steak. Then rub both sides of the steaks with 2 tablespoons of the olive oil and let rest at room temperature for about 1 hour.

2. A half hour before cooking, place the cast-iron skillet on the grate and prepare your gas or charcoal grill to medium-high heat. Leave the grill covered while heating, as it will add a faint, smoky flavor to the skillet.

3. When the coals are ready, about 400°F with the coals lightly covered with ash, season one side of the steaks with half of the coarsely ground pepper and sea salt.

4. Spread the remaining tablespoon of olive oil in the skillet, and then place the steaks, seasoned sides down, into the cast-iron skillet. Wait 2 to 3 minutes until they are slightly charred, seasoning the uncooked sides of the steaks with the remaining pepper and sea salt while waiting. Turn the steaks and sear for another 2 to 3 minutes. Remove from the skillet and let rest, uncovered, for 30 minutes.

5. Preheat the oven to 400°F.

6. Put the steaks back into the cast-iron skillet and place in the oven. For medium-rare, cook for 11 to 13 minutes, and for medium, cook for 14 to 15.

7. Remove the steaks from the oven and transfer to a large cutting board. Let stand for 10 minutes. Remove the butcher's string from the steaks and serve warm.

BASIC FLANK STEAK

YIELD: 2 SERVINGS / **ACTIVE TIME:** 15 MINUTES / **TOTAL TIME:** 1 HOUR AND 15 MINUTES

Pay close attention to flank steak when cooking it. If it's overdone the meat will be chewy, but cooked properly and sliced thin it is a delicious and affordable cut.

1. Add olive oil, rosemary, and thyme to a bowl, mix well, and apply the rub to the steak. Let sit at room temperature for 1 hour.

2. A half hour before cooking, bring your grill to medium-high heat.

3. When the grill is ready, about 400 to 450°F with the coals lightly covered with ash, season one side of the steak with half of the coarsely ground pepper and sea salt. Place the seasoned side of the steak on the grill and cook for about 4 to 5 minutes, seasoning the uncooked side of the steak while waiting. When the steak seems charred, gently flip and cook for 4 to 5 more minutes for medium-rare and 6 more minutes for medium. The steak should feel slightly firm if poked in the center.

4. Remove the steak from the grill and transfer to a large cutting board. Let stand for 6 to 8 minutes. Slice the steak diagonally into long, thin slices.

INGREDIENTS:

- 1 FLANK STEAK, ABOUT 1 TO 1½ LBS.
- 2 TABLESPOONS OLIVE OIL
- 2 SPRIGS FRESH ROSEMARY, LEAVES REMOVED
- 2 SPRIGS FRESH THYME, LEAVES REMOVED
- COARSELY GROUND BLACK PEPPER
- SEA SALT

CHILE-RUBBED LONDON BROIL

YIELD: 2 TO 3 SERVINGS / **ACTIVE TIME:** 30 MINUTES / **TOTAL TIME:** 1 HOUR AND 30 MINUTES

The London broil is an economical steak for large family gatherings. Start with a tasty rub, and finish by slicing the meat diagonally into thin strips.

1. Combine the rub ingredients and mix thoroughly. Set aside.

2. Rub a very thin layer of olive oil on both sides of the steak and then generously apply the dry rub, firmly pressing it all around the steak. Let rest at room temperature for at least 1 hour.

3. A half hour before cooking, prepare your gas or charcoal grill to medium-high heat.

4. When the grill is ready, at about 400 to 450°F with the coals lightly covered with ash, place the steak on the grill. Cook until blood begins to rise from the top, about 4 to 5 minutes. When the steak is charred, flip and cook for another 3 to 4 minutes for medium-rare, and 5 to 6 for medium.

5. Remove the steak from the grill and transfer to a large cutting board. Let stand for 6 to 8 minutes. Slice the steak diagonally into long, thin slices. Serve warm.

INGREDIENTS:

- 1 LONDON BROIL STEAK, ABOUT ¾ TO 1 INCH THICK
- 1–2 TABLESPOONS OLIVE OIL

CHILE RUB

- 1 CUP ANCHO CHILE POWDER
- 2 TABLESPOONS PAPRIKA
- 1 TABLESPOON COARSELY GROUND BLACK PEPPER
- 1 TABLESPOON SEA SALT
- 2 TEASPOONS CUMIN
- 1 TEASPOON CAYENNE PEPPER
- 1 TEASPOON DRY MUSTARD
- 1 TEASPOON OREGANO

SOUS VIDE NEW YORK STRIP STEAK

YIELD: 2 SERVINGS / **ACTIVE TIME:** 15 MINUTES / **TOTAL TIME:** 1 HOUR AND 30 MINUTES

Why sous vide a NY Strip? Not only can you control the temperature perfectly, you also get an amazing infusion of flavor as the steak cooks in its own juices.

1. Preheat your sous vide cooker to 129.2°F.

2. Combine the steak, butter, and rosemary in a vacuum bag and place the bag in the sous vide cooker for at least 1 hour, or up to 3 hours. I typically remove it after an hour.

3. Remove the steak from the bag, being sure to save the liquid, and pat the steak dry.

4. Add canola oil to a cast-iron pan over high heat. As the pan gets hot, salt both sides of the steak.

5. When the oil starts to smoke, add the steak. Cook for 1 to 2 minutes.

6. Flip the steak and add the reserved liquid. Tilt the pan slightly and spoon the liquid over the top of the steak. Repeat this 5 to 6 times.

7. Remove steak from the pan and let rest on a cutting board for at least 15 minutes before serving.

INGREDIENTS:

- 1 LB. STRIP STEAK
- 4 TABLESPOONS UNSALTED BUTTER
- 2 SPRIGS FRESH ROSEMARY
- 2 TABLESPOONS CANOLA OIL
- 2 TEASPOONS SALT

GRILLED NEW YORK STRIP STEAK

YIELD: 2 SERVINGS / **ACTIVE TIME:** 10 MINUTES / **TOTAL TIME:** 1 HOUR

Grilling a strip steak gives it the deep char and smoky flavor that is almost impossible to reproduce in a kitchen.

1. Let the steak come to room temperature 40 minutes before you want to cook it.

2. Heat your grill to 500°F, or higher. I like to get the grill really hot. Don't be afraid to crank the heat.

3. Pat the steak dry and then generously salt both sides of the steak—there should be a thin layer of salt.

4. Once the grill is hot enough, place the steak on the grill and cook for 5 to 7 minutes before flipping.

5. After you flip the steak, brush the top with the butter and continue to cook uncovered for 3 to 5 minutes, or until desired doneness is reached.

6. Tent the steak in aluminum foil and let it rest for 10 minutes with a pat of butter on top.

INGREDIENTS:

- 1 LB. NEW YORK STRIP STEAK
- 2 TEASPOONS SALT
- 4 TABLESPOONS UNSALTED BUTTER

GRILLED RIB EYE STEAK

YIELD: 2 SERVINGS / **ACTIVE TIME:** 10 MINUTES / **TOTAL TIME:** 1 HOUR

The rib eye, from the rib section of the cow, is also known as the Delmonico steak. The marbling and tender nature of this cut make it prized, and often on the costly side. But it is well worth it. For this recipe it is grilled over high heat to get a perfect char crust.

INGREDIENTS:

- 1 LB. RIB EYE STEAK
- 2 TEASPOONS SALT
- 4 TABLESPOONS UNSALTED BUTTER

1. Let the steak come to room temperature 40 minutes before you want to cook it.

2. Heat your grill to 500°F, or higher. I like to get the grill really hot. Don't be afraid to crank the heat.

3. Pat dry the steak and then generously salt both sides of the steak—there should be a thin layer of salt.

4. Once the grill is hot enough, place the steak on the grill and cook for 5 to 7 minutes before flipping.

5. After you flip the steak, brush the top with the butter and continue to cook uncovered for 3 to 5 minutes, or until desired doneness is reached.

6. Tent the steak in aluminum foil and let it rest for 10 minutes with a pat of butter on top.

COFFEE-RUBBED SIRLOIN

YIELD: 2 SERVINGS / **ACTIVE TIME:** 10 MINUTES / **TOTAL TIME:** 1 HOUR

The sirloin is divided into other cuts: the top sirloin is the most sought after; the bottom sirloin is tougher than the top and much larger. In this recipe we use a coffee-and-salt rub to bring out some amazing flavor.

INGREDIENTS:

- 1 LB. SIRLOIN STEAK
- 2 TABLESPOONS FINELY GROUND MEDIUM ROAST COFFEE
- 2 TEASPOONS SALT
- 4 TABLESPOONS UNSALTED BUTTER

1. Let the steak come to room temperature 40 minutes before you want to cook it.

2. Heat your grill to 500°F, or higher. I like to get the grill really hot. Don't be afraid to crank the heat.

3. In a small bowl, mix together coffee and salt. Pat the steak dry and then generously season both sides of the steak with the rub—there should be a thin layer of seasoning.

4. Once the grill is hot enough, place the steak on the grill and cook for 4 to 6 minutes before flipping.

5. After you flip the steak, brush the top with the butter and continue to cook, uncovered, for 3 to 5 minutes, or until desired doneness is reached.

6. Tent the steak in aluminum foil and let it rest for 10 minutes with a pat of butter on top.

YANKEE SHORT RIBS
WITH ROASTED POTATOES & CARROTS

YIELD: 4 SERVINGS / **ACTIVE TIME:** 30 MINUTES / **TOTAL TIME:** 3 HOURS AND 30 MINUTES

When we create menus for The Farmers Dinner, we start by asking farms what they have too much of. When farms can't sell products, they aren't making money. This take on a classic New England dish came about thanks to a farm that had too much short rib. Short rib is an amazing cut from the brisket, chuck, or rib area of the cattle. It benefits from low and slow cooking.

1. Preheat the oven to 300°F.

2. In a Dutch oven, add oil over medium-high heat. As it warms up, salt the short ribs on all sides. Once oil is shimmering, sear the ribs on all sides until nicely browned, working in batches if necessary.

3. Once all the meat has been browned, add onions, carrots, potatoes, and stock to the Dutch oven, using a wooden spoon to scrape up any bits of meat off the bottom of the pan. Add bay leaves and short ribs.

4. Cover and cook in the oven until fork-tender, about 3½ hours. Once it is done, strain and reserve cooking liquid.

5. Pour reserved cooking liquid into a saucepan over medium-high heat, add rosemary, thyme, and red wine, and reduce until it starts to thicken slightly. Season to taste.

6. Serve meat and vegetables on a plate, drizzled with the sauce.

INGREDIENTS:

- 2 TABLESPOONS CANOLA OIL
- 4 LBS. BONE-IN OR BONELESS SHORT RIB
- 2 ONIONS, SLICED
- 4 CARROTS, PEELED AND DICED
- 4 POTATOES, DICED
- 2 QUARTS BEEF STOCK (SEE PAGE 475)
- 4 BAY LEAVES
- 2 SPRIGS FRESH ROSEMARY
- 2 SPRIGS FRESH THYME
- ½ CUP RED WINE
- SALT AND PEPPER, TO TASTE

GARLIC & THYME-RUBBED SIRLOIN

YIELD: 2 SERVINGS / **ACTIVE TIME:** 10 MINUTES / **TOTAL TIME:** 1 HOUR

With a simple garlic and thyme rub we elevate this sirloin steak to another level. Get ready, because this steak is no joke.

INGREDIENTS:

- 1 LB. SIRLOIN STEAK
- 6 GARLIC CLOVES, MINCED
- 2 TABLESPOONS FINELY CHOPPED FRESH THYME
- 2 TEASPOONS SALT
- 4 TABLESPOONS UNSALTED BUTTER

1. Let the steak come to room temperature 40 minutes before you want to cook it.

2. Heat your grill to 500°F, or higher. I like to get the grill really hot. Don't be afraid to crank the heat.

3. In a small bowl, mix together the garlic, thyme, and salt. Pat the steak dry and then generously season both sides of the steak with the garlic mixture—there should be a thin layer of seasoning.

4. Once the grill is hot enough, place the steak on the grill and cook for 4 to 6 minutes before flipping.

5. After you flip the steak, brush the top with the butter and continue to cook uncovered for 3 to 5 minutes, or until desired doneness is reached.

6. Tent the steak in aluminum foil and let it rest for 10 minutes with a pat of butter on top.

CHIPOTLE T-BONE STEAK

YIELD: 2 SERVINGS / **ACTIVE TIME:** 10 MINUTES / **TOTAL TIME:** 1 HOUR

The T-bone is a cut of beef that is from the short loin primal. It is then cut from the front portion of the larger loin primal.

1. Let the steak come to room temperature 40 minutes before you want to cook it.

2. Heat your grill to 500°F, or higher. I like to get the grill really hot. Don't be afraid to crank the heat.

3. In a small bowl, mix together the chipotle powder, salt, and olive oil. Pat the steak dry and then generously season both sides of the steak with the chipotle mixture—there should be a thin layer of seasoning.

4. Once the grill is hot enough, place the steak on the grill and cook for 4 to 6 minutes before flipping.

5. After you flip the steak, brush the top with the butter and continue to cook uncovered for 3 to 5 minutes, or until desired doneness is reached.

6. Tent the steak in aluminum foil and let it rest for 10 minutes with a pat of butter on top.

INGREDIENTS:

1	LB. T-BONE STEAK
1	TABLESPOON CHIPOTLE POWDER
2	TEASPOONS SALT
2	TABLESPOONS OLIVE OIL
4	TABLESPOONS UNSALTED BUTTER

GRILLED FILET

YIELD: 2 SERVINGS / **ACTIVE TIME:** 6 MINUTES / **TOTAL TIME:** 1 HOUR

The filet, aka filet mignon, is a cut taken from the smaller end of the tenderloin. This is the most tender cut of beef and often the most expensive.

1. Let the steak come to room temperature 40 minutes before you want to cook it.

2. Heat your grill to 500°F, or higher. I like to get the grill really hot. Don't be afraid to crank the heat.

3. Pat the steak dry and then generously salt both sides of the steak—there should be a thin layer of salt.

4. Once the grill is hot enough, place the steak on the grill and cook for 5 to 7 minutes before flipping.

5. After you flip the steak, brush the top with the butter and continue to cook, uncovered, for 3 to 5 minutes, or until desired doneness is reached.

6. Tent the steak in aluminum foil and let it rest for 10 minutes with a pat of butter on top.

INGREDIENTS:

1	LB. FILET MIGNON
2	TEASPOONS SALT
4	TABLESPOONS UNSALTED BUTTER

SEARED FILET MIGNON

YIELD: 2 SERVINGS / **ACTIVE TIME:** 6 MINUTES / **TOTAL TIME:** 1 HOUR

The key to getting a good sear on this filet is a well-seasoned and oiled cast-iron pan.

INGREDIENTS:

- 1 LB. FILET MIGNON
- 2 TABLESPOONS CANOLA OIL
- 2 TEASPOONS SALT
- 4 TABLESPOONS UNSALTED BUTTER, DIVIDED
- 2 SPRIGS FRESH ROSEMARY

1. Let the steak come to room temperature 40 minutes before you want to cook it.

2. Add canola oil to a cast-iron pan over high heat. As the pan gets hot, pat the steak dry and then salt both sides of the steak.

3. When the oil starts to smoke, add the steak. Cook for 4 to 5 minutes and flip.

4. Add 2 tablespoons of butter to the pan, tilt the pan slightly, and spoon the butter-oil mixture over the top of the steak. Repeat this 5 to 6 times.

5. Let the steak cook for 3 to 4 minutes, or until desired doneness is achieved. Just before removing steak from the pan, add the remaining butter and rosemary, tilt the pan slightly, and spoon the butter-oil mixture over the top of the steak. Repeat this 5 to 6 times.

6. Remove steak from the pan and let rest on a cutting board for at least 15 minutes before serving.

SEARED FILET MIGNON WITH PEPPERCORN CREAM SAUCE

YIELD: 2 SERVINGS / **ACTIVE TIME:** 6 MINUTES / **TOTAL TIME:** 1 HOUR

This peppercorn cream sauce gives a rich spicy flavor to the tender filet. Not only is this simple, you can wow your friends with the amazing presentation and taste.

1. Grind the peppercorns in a spice grinder for 2 to 3 seconds.

2. In a heavy-bottom saucepan over medium-high heat, add the peppercorns and toast for 1 to 2 minutes, until fragrant.

3. Add the cream and a pinch of salt and turn down the heat to medium-low. Stir this mixture for 2 to 3 minutes until it is well mixed. Remove from heat and set aside.

4. Add canola oil to a cast-iron pan over high heat. As the pan gets hot, pat dry the steak and then salt both sides of the steak.

5. When the oil starts to smoke, add the steak. Cook for 4 to 5 minutes and flip.

6. Add 2 tablespoons of butter to the pan, tilt the pan slightly, and spoon the butter-oil mixture over the top of the steak. Repeat this 5 to 6 times.

7. Let the steak cook for 3 to 4 minutes, or until desired doneness is achieved. Just before removing steak from the pan, add the remaining butter and rosemary, tilt the pan slightly, and spoon the butter-oil mixture over the top of the steak. Repeat this 5 to 6 times.

8. Remove steak from the pan and let rest on a cutting board for at least 15 minutes before serving. Serve with warmed cream sauce.

INGREDIENTS:

- 2 TABLESPOONS PEPPERCORNS
- 1 CUP HEAVY CREAM
- 2 TEASPOONS SALT
- 2 TABLESPOONS CANOLA OIL
- 1 LB. FILET MIGNON (BROUGHT TO ROOM TEMPERATURE 40 MINUTES BEFORE COOKING)
- 4 TABLESPOONS UNSALTED BUTTER, DIVIDED
- 2 SPRIGS FRESH ROSEMARY

PERFECT PRIME RIB

YIELD: 8 SERVINGS / **ACTIVE TIME:** 10 MINUTES / **TOTAL TIME:** 5 HOURS

Prime rib is absolutely worth the price. It is outstanding in texture and flavor and also makes one of the best sandwiches.

1. Take the roast out of the refrigerator and let it stand at room temperature for 2 to 3 hours.

2. Preheat the oven to 480°F and pat the roast dry with a paper towel.

3. In a small saucepan, melt the butter, then add the garlic and salt. Mix and then pour into a bowl and let cool slightly.

4. Place the meat in a roasting pan and rub the garlic butter all over the roast. Place the pan in the oven uncovered and cook for 20 to 25 minutes, until the roast has browned.

5. Once the roast has browned, reduce the temperature to 250°F and cover the roast with foil. Continue to cook for 1 to 1½ hours, until the internal temperature of the roast registers 118°F for medium-rare.

6. Allow the roast to rest for 20 minutes before cutting and serving. (During the rest, the meat's temperature will continue to rise to about 125°F for perfect medium-rare.)

INGREDIENTS:

- 6 LBS. PRIME RIB (4-BONE ROAST)
- 1 STICK UNSALTED BUTTER
- 1 HEAD GARLIC, MINCED
- 2 TEASPOONS SALT

FRENCHED PRIME RIB

YIELD: 6 TO 8 SERVINGS / **ACTIVE TIME:** 2 HOURS / **TOTAL TIME:** 6 HOURS

Frenching a cut of meat isn't too challenging and is normally done when serving to impress. Essentially, when you french a cut of meat, you trim the meat away from the upper portion of the bones so that when roasted, the bones flare out at the top in an elegant fashion. To french, all you need is a sharp carving knife and paring knife in order to remove the meat from the bones.

1. Remove the rib roast from the refrigerator and place on cooling racks over a large carving board, bone side down. In order to french the rib roast, you will need to cut off the meat that is on top of the bones. To do so, go about 2 inches down the ribs and using a carving knife, cut down through the meat until you reach the bone. Make a sharp cut and then cut up the bone so that the top of the meat can be peeled off. This top meat is very delicate and flavorful and works great in stews.

2. Stand the rib roast up and, starting with the left bone, cut down 1 to 2 inches along the bone, cut across to the next bone, and then cut back up so that you get a rectangular chunk of the meat to come apart from the space between the ribs. Do this to all the ribs, and then gently cut away the meat so that the ribs are left to stand openly and on their own. Using a paring knife, gently scrape away any bits of meat that still cling to the ribs. Set roast aside.

3. In a small bowl, combine the 2 tablespoons of coarse sea salt and coarsely ground black pepper. Using your hands, massage the seasoning into the rib roast so that the grains of salt and pepper cling to it firmly.

4. Next, in a small bowl, whisk together the extra-virgin olive oil and minced garlic, followed by the finely chopped thyme and rosemary. Brush the marinade over the rib roast, making sure that the majority of the finely chopped herbs are applied to the ends of the roast. Let the roast stand for about 30 minutes to 1 hour while you preheat the oven.

5. Preheat the oven to 450°F.

INGREDIENTS:

- 1   6-RIB RIB ROAST
- 2   TABLESPOONS COARSE SEA SALT
- 2   TABLESPOONS COARSELY GROUND BLACK PEPPER
- ¼   CUP EXTRA-VIRGIN OLIVE OIL
- 6   GARLIC CLOVES, MINCED
- 1   TABLESPOON CHOPPED FRESH THYME, PLUS 3 BUNCHES
- 1   TABLESPOON CHOPPED FRESH ROSEMARY, PLUS 3 BUNCHES

6. Using some butcher's twine, carefully tie the thyme and rosemary bunches into the spaces between the ribs so that they are firmly in place. Place the roast rib side down, meat side up on a rack in a large roasting pan. When the oven is ready, transfer the rib roast to the oven and cook for 15 minutes at 450°F so that the rib roast gets a strong initial searing. Reduce the heat to 325°F and continue to roast for about 3 to 4 more hours.

7. Toward the end of the recommended roasting time, use an instant-read thermometer to check the internal temperature of the meat. When it reads 125°F, pull the rib roast out for medium-rare.

8. Let the rib roast stand at room temperature for 10 minutes before carving, allowing it to properly store its juices and flavor.

CROWNED ROAST OF BEEF

YIELD: 12 SERVINGS / **ACTIVE TIME:** 2 HOURS / **TOTAL TIME:** 5 HOURS

Although crowning a roast of meat is traditionally done more often for roasts of pork and lamb, you can also crown a roast of beef. It's a little more challenging and requires you to do so on a larger cut of prime rib that is about 10 ribs. Built upon the fundamental method of frenching the meat, a crowned prime rib has the frenched ribs at the top, though it's sliced and bent into a circle so that the ribs, when positioned properly, resemble a crown. Butcher's twine is definitely required here, and the variation allows you to use the space between the roast for a stuffing—though if this is too tight a cavity, simply bake the stuffing in a baking dish for 1 hour until browned.

1. Remove the rib roast from the refrigerator and place it on cooling racks over a large carving board, bone side down. To begin, you will need to french the rib roast. First, cut the meat that covers the bones. To do so, go about 2 inches down the ribs and, using a sharp carving knife, cut through the meat until you reach the bone. Make a sharp cut and then cut up the bone so that the top of the meat can be peeled off.

2. Stand the rib roast up and, starting with the left bone, cut down 1 to 2 inches along the bone, across to the next bone, and then back up so that you get a rectangular chunk of the meat to come apart from the space between the ribs. Do this to all the ribs, and then gently cut away the meat so that the ribs are left to stand openly and on their own. Using a paring knife, gently scrape away any bits of meat that still cling to the ribs.

3. Bring the meat together into a circle, cut about ½ to 1 inch into the meat side of the rib roast between each bone. Make your cuts even and level. Stand the rib roast and, pushing back the ends of the roast, form it into a tight crown. Note that because it's fairly difficult to crown a roast of beef, you may need to cut deeper than 1 inch between the ribs so that it allows for more flexibility. Using butcher's twine, tie the crown tightly so that it'll remain in that position while roasting—you'll need to tie the roast around the bones themselves, and also around the equator of the roast. Set aside.

4. Mix the coarsely ground black pepper and sea salt in a small bowl. Using your hands, massage the seasoning into the rib roast.

5. In a small bowl whisk together the remaining ingredients. Using your hands, massage the paste into the rib roast. Let stand at room temperature for 30 minutes to 1 hour.

6. Preheat the oven to 450°F.

INGREDIENTS:

- 1 10-RIB RIB ROAST
- 3 TABLESPOONS COARSELY GROUND BLACK PEPPER
- 3 TABLESPOONS SEA SALT
- 1 CUP EXTRA-VIRGIN OLIVE OIL
- 8 GARLIC CLOVES, MINCED
- ⅓ CUP COARSELY CHOPPED FRESH THYME
- ⅓ CUP COARSELY CHOPPED FRESH ROSEMARY
- ¼ CUP FINELY CHOPPED FRESH SAGE

VARIATION:

- ½ CUP UNSALTED BUTTER
- 1 LB. WILD MUSHROOMS (PORCINI, SHIITAKE, CRIMINI), DICED
- 1 ROLL JIMMY DEAN HOT SAUSAGE
- 3 MEDIUM GARLIC CLOVES, FINELY CHOPPED
- 1 LARGE WHITE ONION, CHOPPED
- 4 CELERY STALKS, CHOPPED
- ¼ CUP FLAT-LEAF PARSLEY, COARSELY CHOPPED
- 1 TABLESPOON FRESH ROSEMARY, COARSELY CHOPPED

7. Place the standing rib roast in a large roasting pan on a large sheet of flat roasting racks. Cover the crown with aluminum foil so that it keeps the heat central. Roast at 450°F for 15 minutes so that the rib roast receives a nice initial searing. Lower the heat to 325°F and cook for another 2 to 3 hours, until the internal temperature of the meat reads 125°F for medium-rare. Baste the rib roast with its own juices every 30 minutes or so.

8. Remove the crown roast from the oven and place on a large serving piece. Let stand for 10 minutes before carving.

2	TEASPOONS FRESH THYME, COARSELY CHOPPED
½	CUP FRESH BABY SPINACH
3–4	CUPS CUBED PIECES DAY OLD BREAD
1	CUP CHICKEN STOCK (SEE PAGE 478)
1	LARGE EGG
	COARSELY GROUND BLACK PEPPER, TO TASTE
	SEA SALT, TO TASTE

CROWNED ROAST OF BEEF VARIATION

For a spinach and wild mushroom stuffing to go in the center of the prime rib, follow these instructions:

1. Set a medium cast-iron skillet over medium-high heat and add ¼ stick of butter. When hot, add the wild mushrooms to the skillet and cook until tender and browned, about 10 minutes. Remove from pan and set in a large bowl.

2. Next, add the sausage, garlic, white onion, and celery to the skillet and cook until brown, about 5 minutes. You want the sausage to break up into smaller pieces so that it'll mesh thoroughly with the cubed bread. Transfer the sausage mixture to the bowl of mushrooms.

3. Stir the remaining ingredients into the mushroom and sausage mixture and then set aside. When the crowned rib roast has about 1½ hours remaining, remove the aluminum foil from the crown of the rib roast and transfer the stuffing to its center. Season the top with the coarsely ground black pepper and sea salt. Transfer the rib roast back to the oven and cook the rib roast until the internal temperature reads 125°F and the stuffing is browned on top.

WOOD-FIRED PRIME RIB

YIELD: 6 TO 8 SERVINGS / **ACTIVE TIME:** 1 HOUR AND 30 MINUTES / **TOTAL TIME:** 4 HOURS

While grilling a rib roast over a wood fire, you'll want to be sure that the smoke does not directly rise up to the rib roast because that will give it too much of a smoky flavor. To combat this, you may want to set up the grill so that it features both direct and indirect heating. Simply stock your wood on one side of the fire pit and then place the rib roast toward the middle of the grilling rack. The area directly above the fire is your "direct" heating zone, and the area on the opposite side that is still hot, though it does not receive the same level of flame, will be your "indirect" zone.

INGREDIENTS:

- 1 6-RIB RIB ROAST
- 3 TABLESPOONS EXTRA-VIRGIN OLIVE OIL, DIVIDED
- 4 GARLIC CLOVES, MINCED
- 1 SMALL SHALLOT, FINELY CHOPPED
- 2 TABLESPOONS COARSELY GROUND BLACK PEPPER
- 2 TABLESPOONS COARSE SEA SALT
- 3 BUNCHES FRESH THYME
- 3 BUNCHES FRESH ROSEMARY

1. Rub the rib roast with 1 tablespoon of the extra-virgin olive oil and let rest at room temperature for 1 hour.

2. In a small bowl, combine the minced garlic and finely chopped shallot with the remaining extra-virgin olive oil. After the rib roast has rested at room temperature for about an hour, generously massage the meat with the garlic-shallot mixture so that it clings to the cap of the rib roast.

3. Season the rib roast generously with the coarsely ground black pepper and fresh sea salt. Take the bunches of thyme and rosemary and evenly distribute between the ribs. With butcher's twine, tie the bunches of herbs tightly around the ribs so that they will stay in place when you flip the meat on the grill. Once again, let the rib roast stand for 30 minutes while preparing the fire.

4. Prepare the wood fire to feature both direct and indirect heating at medium-low heat, about 325°F. You'll want to make sure that you have a strong foundational layer of coals so that you can easily maintain the heat and smoke as you grill your prime rib.

5. When the fire is ready, at about 325°F with the logs lightly covered with ash, place the rib roast toward the middle of the grill (in between both heating zones, away from the smoke) and sear each side, including the ends, for about 2 to 3 minutes each. Next, flip the rib roast so that the bone side is pressed against the rack, and then slowly roast for about 2 hours until the rib roast is charred and an instant thermometer reads 125°F.

6. Remove the rib roast from the grill and transfer to a large carving board. Let stand for 10 minutes before carving, allowing the meat to properly store its juices.

TIP: The choice of wood you use for your fire is always essential to the flavor of your meat. For a more medium level of flavor, go with a hickory or oak wood, giving your rib roast just the right amount of smoke flavor. If you'd like more of a kick to your rib roast, add a log or two of mesquite onto a fire that's already rooted with hickory or maple wood.

SMOKED PRIME RIB

YIELD: 6 TO 8 SERVINGS / **ACTIVE TIME:** 2 HOURS / **TOTAL TIME:** 6 HOURS

The key to a properly smoked rib roast lies in the fire. As with the Wood-Fired Prime Rib, the wood chips you chose will greatly influence the flavor of your meat. Be sure to soak your wood chips for about an hour before using on the grill, and after dispersing on top of the coals, cover the grill and align the air vent away from the fire so that the smoke will build underneath the lid and pillow over the meat. If you'd like more of a barbecued smoke flavor—perfect for summer evenings—mix 1 cup of mesquite wood chips with 2 cups of hickory or maple wood chips and then soak before throwing onto the flames.

---

1. Rub the rib roast with 1 tablespoon of the extra-virgin olive oil and let rest at room temperature for 1 hour.

2. In a small bowl, combine the minced garlic and finely chopped shallot with the 2 remaining tablespoons of extra-virgin olive oil. After the rib roast has rested at room temperature for about an hour, generously massage the meat with the garlic-shallot mixture so that it clings to the cap of the rib roast.

3. Season the rib roast generously with the coarsely ground black pepper and fresh sea salt. Take the bunches of thyme and rosemary and evenly distribute between the ribs. With butcher's twine, tie the bunches of herbs tightly around the ribs so that they will stay in place when you flip the meat on the grill. Let the rib roast stand for 30 minutes while preparing the fire.

4. Prepare the fire to feature both direct and indirect heat with an average low temperature of about 300°F. You'll want to make sure that you have a strong foundational layer of coals so that you can easily maintain the heat and smoke as you grill your prime rib. While you prepare your grill, add the 4 cups of hickory or maple wood chips to a bowl of warm water and set aside.

5. When the fire is ready, at about 300°F with the coals lightly covered with ash, place the rib roast over the direct heat of the grill and sear each side, including the ends, for about 2 to 3 minutes each. Transfer the rib roast over the direct heat and flip the rib roast to meat side up. Take a handful of wood chips and throw them over the flame. Cover the grill, aligning the air vent away from flame so that the smoke pillows around the rib roast, and begin slowly roasting for about 3 to 4 hours until the rib roast is charred and an instant thermometer reads 125°F. For the first 3 hours of the grilling process, distribute handfuls of the hickory or maple wood chips about every 30 minutes or so.

6. Remove the rib roast from the grill and transfer to a large carving board. Let stand for 10 minutes before carving, allowing the meat to properly store its juices.

INGREDIENTS:

1	6-RIB RIB ROAST
3	TABLESPOONS EXTRA-VIRGIN OLIVE OIL, DIVIDED
4	GARLIC CLOVES, MINCED
1	SMALL SHALLOT, FINELY CHOPPED
2	TABLESPOONS COARSELY GROUND BLACK PEPPER
2	TABLESPOONS COARSE SEA SALT
3	BUNCHES FRESH THYME
3	BUNCHES FRESH ROSEMARY
4	CUPS HICKORY OR MAPLE WOOD CHIPS

OVEN-ROASTED NEW YORK STRIP STEAK WITH WHIPPED POTATOES & SWISS CHARD

YIELD: 2 SERVINGS / **ACTIVE TIME:** 45 MINUTES / **TOTAL TIME:** 1 HOUR AND 30 MINUTES

As fall temperatures go from brisk to chilly, standing out by the grill doesn't seem quite as appealing. But that doesn't mean you have to say goodbye to the enchanting flavor of a quality steak.

1. Place the steak on a plate and let stand at room temperature for 20 to 30 minutes, as this will ensure that it cooks evenly. Preheat the oven to 475°F.

2. Place the potatoes in a saucepan, cover with cold water, and add the salt. Bring to a boil, reduce heat so that the potatoes simmer, and cook until tender, about 12 minutes. Strain and set aside.

3. Combine the cream and 6 tablespoons of the butter in a saucepan and heat until the butter is melted. Remove from heat, place the potatoes in the bowl of a stand mixer fitted with the whisk attachment, and add half of the cream-and-butter mixture. Whisk until the potatoes are smooth, adding more of the mixture as needed to achieve the desired consistency. Set aside.

4. Place a 12-inch cast-iron skillet over high heat and add the canola oil. Season the steak liberally with salt and black pepper. When the oil begins to smoke, carefully place the steak in the skillet and cook for 2 minutes on each side.

5. When both sides of the steak have been seared, use kitchen tongs to turn the steak onto the edge where the fat is. Hold the steak in place until the fat has rendered, about 2 minutes. Lay the steak flat in the pan and add the remaining butter, thyme, crushed garlic cloves, and shallot. Using a large spoon, baste the steak with the juices in the pan and cook for another minute. Remove from heat and place the steak on a wire rack resting in a baking sheet.

6. Place the steak in the oven and cook for 3 minutes. Remove and let stand for 8 to 10 minutes before slicing. The steak should be medium-rare, with an internal temperature of 145°F.

7. While the steak is resting, place the olive oil in a 12-inch cast-iron skillet and warm over medium-high heat. When the oil starts to smoke, add the Swiss chard and minced garlic and cook until the chard is wilted. Season with salt and pepper.

8. To serve, cut the steak into desired portions and plate alongside the potatoes and Swiss chard.

INGREDIENTS:

- 1 LB. NEW YORK STRIP STEAK
- 3 IDAHO POTATOES, PEELED AND CUT INTO 1-INCH PIECES
- 1 TABLESPOON SALT, PLUS MORE TO TASTE
- 1 CUP HEAVY CREAM
- 1 STICK UNSALTED BUTTER, DIVIDED
- 2 TABLESPOONS CANOLA OIL
- BLACK PEPPER, TO TASTE
- 3 SPRIGS FRESH THYME
- 3 GARLIC CLOVES, CRUSHED, PLUS 1 TABLESPOON MINCED GARLIC
- 1 SHALLOT, HALVED
- 2 TABLESPOONS OLIVE OIL
- ½ BUNCH SWISS CHARD, STEMMED AND CHOPPED

CARNE ASADA

YIELD: 4 SERVINGS / **ACTIVE TIME:** 15 MINUTES / **TOTAL TIME:** 6 HOURS AND 30 MINUTES

Carne Asada is grilled beef that is cut into thin strips and marinated. This recipe goes with the typical skirt steak, but you can also find preparations that use sirloin or rib steak.

1. Using a sharp knife, slice the steak across the grain into very thin strips. Place the strips in a large bowl.

2. In a separate bowl, combine the remainder of the ingredients, except the mezcal, and mix well. Pour the marinade over the steak and rub it into the meat with your hands. Cover and refrigerate for 6 hours.

3. When the steak is done marinating, let it come to room temperature. Add the mezcal and stir to incorporate.

4. Set a large cast-iron skillet over high heat. When the pan is ripping hot, let the excess marinade drip off the steak and evenly spread out the meat in the pan. Cook until the steak is well charred, about 5 minutes per side.

5. Rest the steak for 10 minutes before serving with tortillas and preferred taco condiments.

INGREDIENTS:

1	LB. SKIRT STEAK
2	TABLESPOONS OLIVE OIL
	JUICE OF 1 LIME
	SALT
1	TEASPOON CUMIN
1	TEASPOON SMOKED PAPRIKA
2	TEASPOONS ANCHO CHILE POWDER
1½	OZ. MEZCAL

TO SERVE

TORTILLAS (SEE PAGE 714)

BRISKET

YIELD: 10 SERVINGS / **ACTIVE TIME:** 15 MINUTES / **TOTAL TIME:** 9 HOURS

Brisket is the holy grail of BBQ. It is easy to mess up, but when you do it right, it can be life changing. First, be sure to buy the correct brisket cut. There are two separate cuts when it comes to brisket: the flat is one muscle and sliced with very little fat on it; the point is sliced with the fat. You want the point brisket, not the flat. Trimming brisket might seem scary, but it's pretty simple when you know that the goal is to trim away some of the fat to make the cook uniform but not so much that it dries out the meat.

1. Set the smoker to 225°F and preheat, with the lid closed, for 10 minutes. I like to use mesquite and hickory chips for brisket.

2. In a bowl, combine all the spices for the rub and then season the brisket on all sides with the mix.

3. Place brisket, fat side down, on grill grate and cook until it reaches an internal temperature of 160°F, around 3½ to 4 hours. When the brisket reaches temperature, remove from the grill.

4. Wrap meat in aluminum foil and add the stock to the foil packet. Return the brisket to the smoker and cook until it reaches an internal temperature of 205°F, about 3 hours. Once finished, remove the brisket from the grill and let rest for 15 minutes.

5. To serve, slice against the grain.

INGREDIENTS:

- 2 TABLESPOONS GARLIC POWDER
- 2 TABLESPOONS ONION POWDER
- 2 TABLESPOONS PAPRIKA
- 3 TABLESPOONS SALT
- 3 TABLESPOONS COARSELY GROUND BLACK PEPPER
- 1 TABLESPOON BROWN SUGAR
- 1 (15 LB.) BRISKET
- ½ CUP BEEF STOCK (SEE PAGE 475)

BURNT ENDS

YIELD: 5 SERVINGS / **ACTIVE TIME:** 10 MINUTES / **TOTAL TIME:** 6 TO 8 HOURS

Burnt ends are made from the point of the brisket. If you don't have the point of a brisket, only the flat, you can make a quick substitute with beef chuck.

1. Bring brisket up to room temperature.

2. In a bowl, combine the salt and pepper, mix well, and coat the meat.

3. Get your smoker to 275°F.

4. Add the brisket and close the lid. After 2½ hours, flip the meat. Continue to cook until a meat thermometer registers an internal temperature of 200°F.

5. Let the meat rest on a cutting board for about 15 minutes. Once cooled, slice meat into cubes and add the cubes to an aluminum pan. Toss in sauce and return the pan to the smoker for 1 to 1½ hours, then serve.

INGREDIENTS:

- 1 (5–6 LB.) BRISKET POINT OR CHUCK, TRIMMED
- ¼ CUP SALT
- ¼ CUP COARSELY GROUND BLACK PEPPER
- ½ CUP BBQ SAUCE OF YOUR CHOICE

BRAISED BEEF CHEEK

YIELD: 5 SERVINGS / **ACTIVE TIME:** 10 MINUTES / **TOTAL TIME:** 6 TO 8 HOURS

Beef cheek might sound different, but I can assure you that this recipe will shatter any doubts you might have.

1. Pat dry and trim any extra fat from the beef cheeks.

2. Add 2 tablespoons of olive oil to a frying pan over medium-high heat. When the oil begins to shimmer, sear half the beef cheeks on each side until nicely browned. Set aside seared beef cheeks on a plate and repeat with remaining cheeks.

3. Turn down the heat to medium and add garlic and onions. Sauté for 3 minutes until the onions are translucent. Add the celery and carrots, cook for an additional 3 minutes, and then remove the pan from the heat.

4. Pour the vegetable mixture into a slow cooker and place the beef cheeks on top.

5. Pour the wine into a pot over high heat, bring to a simmer, and then let simmer for 1 minute to cook out the alcohol.

6. Pour the wine into the slow cooker, then add all the remaining ingredients, starting with just a pinch of salt. Cook on low for 8 hours or on high for 6 hours, until the cheeks become fork-tender.

7. Remove the beef cheeks from the slow cooker and discard the bay leaves. Using an immersion blender, puree the braising liquid into a smooth sauce.

8. Transfer sauce to a large sauté pan and simmer over medium-high heat until the sauce turns a darker brown, 10 to 15 minutes.

9. Remove the pan from the heat, return beef cheeks to the sauce, cover, and keep warm until ready to serve.

INGREDIENTS:

- 3½ LBS. BEEF CHEEKS
- 4 TABLESPOONS OLIVE OIL, DIVIDED
- 5 GARLIC CLOVES, CHOPPED
- 2 SWEET ONIONS, DICED
- 1 CELERY STALK, DICED
- 1 CARROT, DICED
- 2 CUPS RED WINE
- SALT
- 1 CUP BEEF STOCK (SEE PAGE 475)
- 4 BAY LEAVES
- BLACK PEPPER

SLOW COOKER BARBACOA

YIELD: 6 TO 8 SERVINGS / **ACTIVE TIME:** 20 MINUTES / **TOTAL TIME:** 10 HOURS

Doing barbacoa in a slow cooker allows the flavors to meld together, and makes cleanup a breeze!

1. Add olive oil to a large skillet or Dutch oven over medium-high heat. When the oil is shimmering, sear the roast on each side until brown, about 4 to 5 minutes. Transfer to slow cooker.

2. In a food processor combine the remaining ingredients (except the ingredients for serving) and blend to a smooth consistency.

3. Pour sauce on top of roast. Cover and cook on low for 8 to 10 hours, or on high for 6 to 8 hours.

4. Carefully remove beef from the slow cooker. Shred beef and return to the cooker. Cook on high for an additional 10 minutes to absorb additional liquid.

5. Serve with warmed tortillas and desired toppings.

INGREDIENTS:

- 2 TABLESPOONS OLIVE OIL
- 3–5 LBS. BEEF CHUCK ROAST OR BEEF CHEEK
- ⅓ CUP FRESH ORANGE JUICE
- ¼ CUP FRESH LIME JUICE
- 1 CUP DARK BEER
- 3–4 CHIPOTLE PEPPERS IN ADOBO SAUCE, CHOPPED
- ¼ CUP LIGHT BROWN SUGAR
- 1 TEASPOON APPLE CIDER VINEGAR
- 2 TEASPOONS SALT
- 5 GARLIC CLOVES, MINCED
- 1 ONION, CHOPPED
- 1 TABLESPOON CUMIN
- 2 TEASPOONS FRESH OREGANO

TO SERVE

TORTILLAS (SEE PAGE 714)

CILANTRO

ONION

LIME WEDGES

ROAST BEEF AU JUS WITH VEGETABLES

YIELD: 6 SERVINGS / **ACTIVE TIME:** 25 MINUTES / **TOTAL TIME:** 1 HOUR AND 30 MINUTES

The bright sweetness provided by the beets makes this dish perfect for those days when winter is beating you down.

1. Preheat oven to 375°F. Place all of the ingredients except the olive oil, 2 tablespoons of the thyme, and the roast, in a 9x13-inch baking pan. Season with salt and pepper and stir to evenly coat the vegetables.

2. Liberally coat the roast on all sides with the olive oil. Season with salt, pepper, and the remaining thyme and place in the baking pan. Place the pan in the oven and cook until the interior of the roast reads 120°F on a digital thermometer, about 40 to 45 minutes. Remove the pan from the oven and place the roast upside down on a cutting board. Cover with foil and let rest for 15 to 20 minutes before carving.

3. Use a slotted spoon to transfer the vegetables to a serving dish. Ladle the juices in the pan into a separate bowl.

4. Cutting against the grain, carve the meat into 1- to 2-inch thick slices. Divide the vegetables between the plates and top with 2 or 3 slices of roast beef. Spoon the juices over the top and serve.

INGREDIENTS:

- 1 RED BEET, PEELED AND SLICED THIN
- 1 CARROT, MINCED
- 1 PARSNIP, MINCED
- 2 CELERY STALKS, DICED
- 1 YELLOW ONION, PEELED AND CUT INTO ½-INCH THICK SLICES
- 1 ANAHEIM PEPPER, SEEDED AND DICED
- 3 GARLIC CLOVES, MINCED
- 1 TABLESPOON DRIED SAGE
- 1 TABLESPOON GARLIC POWDER
- 1 CUP BEEF STOCK
- 1 CUP DRY RED WINE
- 2 TABLESPOONS WORCESTERSHIRE SAUCE
- 2 TABLESPOONS EXTRA-VIRGIN OLIVE OIL
- 4 TABLESPOONS FRESH THYME, MINCED
- 1 (2-LB.) EYE OF ROUND BEEF ROAST
- SALT AND PEPPER, TO TASTE

BASIC BURGERS

YIELD: 4 SERVINGS / **ACTIVE TIME:** 15 MINUTES / **TOTAL TIME:** 30 MINUTES

A good burger should taste like a burger, not just spices. I use white pepper because it brings the spice that you want with pepper but takes away a bit of black pepper's punch. This is the classic all-American burger.

1. Preheat your outdoor grill or cast-iron pan. If you don't have either, a frying pan will do. (And if using a pan, lightly grease it with neutral oil.)

2. In a bowl, combine all the ingredients until the mixture is well incorporated.

3. Divide the mixture into four equal parts. Form patties by rolling each piece into a ball in your hands and pressing it flat to form the patty. The patty should be about 1-inch thick.

4. Turn the grill to medium-high heat; if you're using a pan, set your burner on medium-high. Place patties on the grill or pan. Cover the patties and cook for 5 minutes.

5. Flip the patties and cover again, cooking for an additional 5 to 6 minutes until you have the desired doneness.

INGREDIENTS:

1	LB. GROUND BEEF (80/20 RECOMMENDED)
1	TEASPOON SALT
½	TEASPOON WHITE PEPPER
1	EGG WHITE
⅓	CUP BREAD CRUMBS

BACON CHEESEBURGERS

YIELD: 4 SERVINGS / **ACTIVE TIME:** 15 MINUTES / **TOTAL TIME:** 30 MINUTES

The bacon cheeseburger will never go out of style. The variations are endless, but for this recipe use thick-cut applewood-smoked bacon and American cheese.

1. Preheat your outdoor grill or cast-iron pan. If you don't have either, a frying pan will do. (And if using a pan, lightly grease it with neutral oil.)

2. In a bowl, combine all the ingredients (except the American cheese and bacon) until the mixture is well incorporated.

3. Divide the mixture into four equal parts. Form patties by rolling each piece into a ball in your hands and pressing it flat to form the patty. The patty should be about 1-inch thick.

4. Turn the grill to medium-high heat; if you're using a pan, set your burner on medium-high. Place patties on the grill or pan. Cover the patties and cook for 5 minutes.

5. Flip the patties and cover again, cooking for an additional 5 to 6 minutes. About 30 seconds before your burgers are done, top each one with a slice of cheese and cover until you have the desired doneness and the cheese is melted.

6. Place cooked bacon on top of the burgers and serve.

INGREDIENTS:

1	LB. GROUND BEEF (80/20 RECOMMENDED)
1	TEASPOON SALT
½	TEASPOON BLACK PEPPER
1	EGG WHITE
⅓	CUP BREAD CRUMBS
4	SLICES AMERICAN CHEESE
8	SLICES THICK-CUT BACON, COOKED

MEXICAN BURGERS WITH CILANTRO MAYONNAISE

YIELD: 4 SERVINGS / **ACTIVE TIME:** 15 MINUTES / **TOTAL TIME:** 30 MINUTES

Using traditional Mexican spices lends a deep flavor to this incredible burger.

1. Preheat your outdoor grill or cast-iron pan. If you don't have either, a frying pan will do. (And if using a pan, lightly grease it with neutral oil.)

2. In a bowl, combine the remaining ingredients.

3. Divide the mixture into four equal parts. Form patties by rolling each piece into a ball in your hands and pressing it flat to form the patty. The patty should be about 1-inch thick.

4. Turn the grill to medium-high heat; if you're using a pan, set your burner on medium-high. Place patties on the grill or pan. Cover the patties and cook for 5 minutes.

5. Flip the patties and cover again, cooking for an additional 5 to 6 minutes, until they are the desired level of doneness.

6. Top the burgers with the Cilantro Mayonnaise.

CILANTRO MAYONNAISE

1. In a food processor, blend all of the ingredients until smooth.

2. Transfer to a bowl, cover, and refrigerate for 1 hour.

INGREDIENTS:

- 1 LB. GROUND BEEF (80/20)
- 1 TEASPOON SALT
- ½ TEASPOON BLACK PEPPER
- 1 EGG WHITE
- ½ TEASPOON PAPRIKA
- ½ TEASPOON ONION POWDER
- ½ TEASPOON CUMIN
- 1 TABLESPOON DICED JALAPEÑO PEPPERS (OPTIONAL)
- ¼ CUP CORNMEAL

CILANTRO MAYONNAISE

- ½ BUNCH FRESH CILANTRO
- ½ CUP MAYONNAISE
- 1 TABLESPOON FRESH LIME JUICE
- PINCH OF SALT

BEEF & PORK BURGERS WITH CARAMELIZED ONION MAYO

YIELD: 4 SERVINGS / **ACTIVE TIME:** 15 MINUTES / **TOTAL TIME:** 30 MINUTES

From the richness of the beef and pork to the delicious sweetness of the mayo, this variation on a classic burger just might become a go-to recipe for all your BBQ parties.

1. Preheat your outdoor grill or cast-iron pan. If you don't have either, a frying pan will do. (And if using a pan, lightly grease it with neutral oil.)

2. In a bowl, combine all the ingredients until the mixture is well incorporated.

3. Divide the mixture into four equal parts. Form patties by rolling each piece into a ball in your hands and pressing it flat to form the patty. The patty should be about 1-inch thick.

4. Turn the grill to medium-high heat; if you're using a pan, set your burner on medium-high. Place patties on the grill or pan. Cover the patties and cook for 6 minutes.

5. Flip the patties and cover again, cooking for an additional 6 to 7 minutes, until you have the desired doneness.

INGREDIENTS:

- ½ LB. GROUND BEEF (80/20)
- ½ LB. GROUND PORK
- 1 TEASPOON SALT
- ½ TEASPOON BLACK PEPPER
- 1 EGG WHITE
- ⅓ CUP BREAD CRUMBS

CARAMELIZED ONION MAYO

- 1 TABLESPOON UNSALTED BUTTER
- 1 LARGE SWEET ONION, SLICED
- 1 TEASPOON SALT, PLUS A PINCH
- ½ CUP MAYONNAISE

CARAMELIZED ONION MAYO

1. Melt butter in a pan over medium heat.

2. Add onions and salt to the pan and cook, stirring occasionally, until they are brown and start to get dark brown.

3. Put onions in a bowl and allow them to cool. Once cool, place the onions and mayo, along with a pinch of salt, in a food processor and blend till smooth.

MUSHROOM & SWISS BURGERS

YIELD: 4 SERVINGS / **ACTIVE TIME:** 15 MINUTES / **TOTAL TIME:** 30 MINUTES

Mushrooms add a layer of umami—a deep savory element—to pretty much everything, including burgers, especially when sautéed.

INGREDIENTS:

- ½ LB. MUSHROOMS (BUTTON OR CREMINI WORK WELL)
- 1 TABLESPOON OLIVE OIL
- 1 TEASPOON SALT, PLUS MORE TO TASTE
- 1 LB. GROUND BEEF (80/20)
- ½ TEASPOON WHITE PEPPER
- 1 EGG WHITE
- ⅓ CUP BREAD CRUMBS
- 4 SLICES SWISS CHEESE

1. In a pan over medium heat, sauté mushrooms with olive oil and a pinch of salt until slightly brown. Remove from heat.

2. Preheat your outdoor grill or cast-iron pan. If you don't have either, a frying pan will do. (And if using a pan, lightly grease it with neutral oil.)

3. In a bowl, combine the remaining ingredients, except the cheese, until the mixture is well incorporated.

4. Divide the mixture into four equal parts. Form patties by rolling each piece into a ball in your hands and pressing it flat to form the patty. The patty should be about 1-inch thick.

5. Turn the grill to medium-high heat; if you're using a pan, set your burner on medium-high. Place patties on the grill or pan. Cover the patties and cook for 5 minutes.

6. Flip the patties and cover again, cooking for an additional 5 to 6 minutes. About 30 seconds before your burgers are done, top each one with a slice of cheese and cover until you have the desired doneness and the cheese is melted.

7. Place the mushroom on top of the burger and enjoy with your favorite condiments.

JALAPEÑO BURGERS WITH RED PEPPER JAM

YIELD: 4 SERVINGS / **ACTIVE TIME:** 15 MINUTES / **TOTAL TIME:** 30 MINUTES

Even if you don't like hot peppers, the sweetness of the red bell peppers, along with the sugar, add a friendly tang to this burger.

1. Preheat your outdoor grill or cast-iron pan. If you don't have either, a frying pan will do. (And if using a pan, lightly grease it with neutral oil.)

2. In a bowl, combine all the ingredients, except the cheese (if using) and jalapeños, until the mixture is well incorporated.

3. Divide the mixture into four equal parts. Form patties by rolling each piece into a ball in your hands and pressing it flat to form the patty. The patty should be about 1-inch thick.

4. Turn the grill to medium-high heat; if you're using a pan, set your burner on medium-high. Place patties on the grill or pan. Cover the patties and cook for 5 minutes.

5. Flip the patties and cover again, cooking for an additional 5 to 6 minutes. About 30 seconds before your burgers are done, top each one with a slice of cheese (if desired) and cover until the cheese has melted and you have the desired doneness. Top with jalapeños.

RED PEPPER JAM

1. In a small bowl, whisk together pectin and ¼ cup of sugar and set aside.

2. Seed and dice the peppers and pulse them in a food processor until finely blended.

3. In a saucepan, combine the pepper puree with the vinegar, red pepper flakes, salt, and the rest of the sugar and cook over medium-high heat. Once the mixture comes to a boil, let it continue to boil for about 4 minutes.

4. Add the pectin mixture and bring the contents of the saucepan back to a boil, stirring occasionally, until the mixture starts to thicken. Remove from heat and allow it to come to room temperature before refrigerating.

INGREDIENTS:

- 1 LB. GROUND BEEF (85/15)
- 1 TEASPOON SALT
- ½ TEASPOON WHITE PEPPER
- 1 EGG WHITE
- ⅓ CUP BREAD CRUMBS
- 4 SLICES CHEESE (OPTIONAL)
-   PICKLED JALAPEÑOS, TO TASTE

RED PEPPER JAM

- ¼ CUP PECTIN
- 1½ CUPS SUGAR, DIVIDED
- 3 RED PEPPERS
- ⅓ CUP APPLE CIDER VINEGAR
- 1 TEASPOON RED PEPPER FLAKES
- 1 TEASPOON SALT

TEXAS BBQ BURGERS

YIELD: 4 SERVINGS / **ACTIVE TIME:** 15 MINUTES / **TOTAL TIME:** 30 MINUTES

Texas flavor is delivered here thanks to the smoky bacon, tangy BBQ sauce, and pickled peppers.

1. Preheat your outdoor grill or cast-iron pan. If you don't have either, a frying pan will do. (And if using a pan, lightly grease it with neutral oil.)

2. In a bowl, combine the beef, salt, pepper, egg white, and bread crumbs, until the mixture is well incorporated.

3. Divide the mixture into four equal parts. Form patties by rolling each piece into a ball in your hands and pressing it flat to form the patty. The patty should be about 1-inch thick.

4. Turn the grill to medium-high heat; if you're using a pan, set your burner on medium-high. Place patties on the grill or pan. Cover the patties and cook for 5 minutes.

5. Flip the patties and cover again, cooking for an additional 5 to 6 minutes. About 30 seconds before your burgers are done, top each one with a slice of cheese and cover until the cheese has melted and you have the desired doneness. Top with BBQ sauce, bacon, and jalapeños (if desired).

INGREDIENTS:

- 1 LB. GROUND BEEF (85/15)
- 1 TEASPOON SALT
- ½ TEASPOON WHITE PEPPER
- 1 EGG WHITE
- ⅓ CUP BREAD CRUMBS
- 4 SLICES CHEESE
- ⅓ CUP BBQ SAUCE (SEE PAGES 497-505)
- ½ LB. BACON, COOKED
- PICKLED JALAPEÑOS, TO TASTE

BEEF STEW

YIELD: 5 SERVINGS / **ACTIVE TIME:** 10 MINUTES / **TOTAL TIME:** 2 HOURS

This recipe is a classic. Easy and absolutely delicious, you are going to want to make this all winter long.

1. In a bowl, combine flour, garlic powder, salt, and pepper and toss beef tips in this mixture.

2. Add olive oil to a large pot over medium-high heat. When the oil is shimmering, add the beef and onions and cook until browned, between 5 and 10 minutes.

3. Add wine and stock and scrape the sides and bottom of the pot with a wooden spatula or spoon to loosen any brown bits.

4. Stir in all the remaining ingredients, except for the cornstarch, water, and peas. Turn down heat to medium-low, cover, and simmer for an hour or so, or until beef is fork-tender.

5. In a bowl, mix cornstarch and water to create a slurry. Slowly add the slurry to the simmering stew and continue to cook until achieving the desired consistency.

6. Stir in peas and simmer for 5 minutes before serving.

INGREDIENTS:

- ½ CUP ALL-PURPOSE FLOUR
- 1 TEASPOON GARLIC POWDER
- SALT AND PEPPER, TO TASTE
- 2 LBS. BEEF TIPS
- 4 TABLESPOONS OLIVE OIL
- ONIONS
- 2 CUPS RED WINE
- 2 CUPS BEEF STOCK (SEE PAGE 475)
- 1 CUP CHOPPED CARROTS
- 1 CUP CHOPPED CELERY
- 4 BAY LEAVES
- 1 SPRIG FRESH ROSEMARY, CHOPPED
- 1 CUP CUBED POTATOES
- 3 TABLESPOONS CORNSTARCH
- 3 TABLESPOONS WATER
- 1 CUP PEAS

BEEF & VEGETABLE SOUP

YIELD: 12 SERVINGS / **ACTIVE TIME:** 20 MINUTES / **TOTAL TIME:** 3 HOURS

This soup reminds me of the classic canned soup many of us grew up on, but with less sodium and way tastier.

1. Pat dry the beef with paper towels and season with salt.

2. Add 1 tablespoon of olive oil to a large pot over medium-high heat. When the oil begins to shimmer add half the beef to the pot and brown for 2 minutes before turning and browning for an additional 2 minutes. Remove and set aside the browned beef. Add another ½ tablespoon of oil to the pot and repeat the browning with the remainder of the beef.

3. After all the meat is browned, add 1 tablespoon of oil to the pot, along with the onion, carrots, and celery. Sauté for 2 minutes then add garlic and cook for another minute.

4. Add the stock, tomatoes, browned beef, basil, oregano, and thyme and season with salt and pepper. Bring to a boil then reduce heat to low, cover, and simmer for 30 minutes.

5. Add potatoes and cook for another 20 minutes, or until the potatoes are soft.

6. Add green beans and simmer for another 15 minutes, or until the vegetables and beef are tender.

7. Add the corn and peas, simmer until heated through, about 5 minutes, season to taste, and serve.

INGREDIENTS:

- 1½ LBS. BEEF TIPS
- SALT AND BLACK PEPPER, TO TASTE
- 2½ TABLESPOONS OLIVE OIL, DIVIDED
- 1 ONION, DICED
- 1¼ CUPS CHOPPED CARROTS
- 1 CUP CHOPPED CELERY
- 1½ TABLESPOONS MINCED GARLIC
- 8 CUPS BEEF STOCK (SEE PAGE 475)
- 2 (14 OZ.) CANS DICED TOMATOES
- ¼ CUP CHOPPED BASIL
- 1 TEASPOON DRIED OREGANO
- ½ TEASPOON DRIED THYME
- 1 LB. RED OR YELLOW POTATOES, CHOPPED
- 1½ CUPS CHOPPED GREEN BEANS
- 1½ CUPS CORN KERNELS
- 1 CUP PEAS

CHICKPEA & BEEF CHILI

YIELD: 6 TO 8 SERVINGS / **ACTIVE TIME:** 15 MINUTES / **TOTAL TIME:** 2 HOURS AND 30 MINUTES

Chickpeas add a nice texture and really soak up the beefy goodness of this absolutely delicious chili.

INGREDIENTS:

- 1 TABLESPOON OLIVE OIL
- 1 ONION, DICED
- 5 GARLIC CLOVES, MINCED
- 1 TABLESPOON FRESH OREGANO
- 1 RED BELL PEPPER, DICED
- 1 TABLESPOON CUMIN
- 2 TEASPOONS CHILI POWDER
- 2 CUPS TURKEY STOCK (SEE PAGE 477)
- ½ LB. TOMATOES, CANNED OR FRESH
- 3 DRIED NEW MEXICAN CHILIES
- SALT, TO TASTE
- 1 (14 OZ.) CAN CHICKPEAS
- 2 LBS. GROUND BEEF
- 2 CUPS GRATED CHEDDAR CHEESE (OPTIONAL)
- 1 CUP SOUR CREAM (OPTIONAL)

1. Add oil to a Dutch oven and heat to medium-high heat. Once the oil is shimmering, add onion, garlic, oregano, pepper, cumin, and chili powder and cook for 5 minutes, stirring often.

2. Add stock, tomatoes, chilies, salt, chickpeas, and beef. Stir the mixture, cover, and reduce the heat so that the chili simmers, cooking for 2 hours and stirring occasionally.

3. To serve, scoop chili into a bowl and, if desired, top with the cheese and sour cream.

BEEF & BRAISED CABBAGE SOUP WITH HORSERADISH CREAM

YIELD: 6 SERVINGS / **ACTIVE TIME:** 30 MINUTES / **TOTAL TIME:** 2 HOURS AND 20 MINUTES

Consider bringing this to your next potluck dinner, as it's great with or without the steak. It also pairs well with game meats.

---

1. Preheat oven to 300°F.

2. In a mixing bowl, add the cabbage, onions, apple, brown sugar, garlic, nutmeg, caraway seeds, and vinegar with a ½ cup of the stock. Mix until well-combined.

3. Season with salt and pepper and transfer to a large, buttered casserole dish. Cover the pan and place it in the oven. Cook for 1½ hours, removing to stir the contents of the casserole dish.

4. Turn off the oven and open the oven door slightly.

5. When the dish has cooled slightly, remove it from the oven and set aside. Preheat oven to 450°F.

6. In a medium sauté pan, add the olive oil and warm over medium heat. Season the sirloin with salt and pepper and then add to pan. Cook until golden brown on both sides. Remove sirloin from the pan and set aside.

7. Spoon the cabbage mixture into a large saucepan. Add the remaining stock and bring to a boil. Reduce heat so that the soup simmers.

8. Place the sirloin in the oven and cook until it is the desired level of doneness.

9. Remove the sirloin from the oven and let it stand for 5 minutes. Ladle the soup into serving bowls. Thinly slice the steak and place it on top of the soup. Serve with Horseradish Cream and watercress.

HORSERADISH CREAM

1. Combine the horseradish, vinegar, mustard, and 4 tablespoons of the cream in a mixing bowl.

2. Lightly whip the remaining cream and then fold this into the horseradish mixture. Season to taste and refrigerate until ready to serve.

INGREDIENTS:

- 2 LBS. RED CABBAGE, CORE REMOVED, SHREDDED
- 2 ONIONS, PEELED AND FINELY SLICED
- 1 LARGE APPLE, PEELED, CORED, AND CHOPPED
- 3 TABLESPOONS BROWN SUGAR
- 2 GARLIC CLOVES, MINCED
- ¼ TEASPOON GRATED NUTMEG
- ½ TEASPOON CARAWAY SEEDS
- 3 TABLESPOONS APPLE CIDER VINEGAR
- 4 CUPS VEAL STOCK (SEE PAGE 476), DIVIDED
- SALT AND PEPPER, TO TASTE
- 2 TABLESPOONS EXTRA-VIRGIN OLIVE OIL
- 1½ LBS. SIRLOIN STEAK, FAT REMOVED
- WATERCRESS, FOR GARNISH

TO SERVE

HORSERADISH CREAM

HORSERADISH CREAM

- 2 TABLESPOONS FRESH HORSERADISH, PEELED AND GRATED
- 2 TEASPOONS WHITE WINE VINEGAR
- ½ TEASPOON DIJON MUSTARD
- 1 CUP HEAVY CREAM, DIVIDED
- SALT AND PEPPER, TO TASTE

BEEF, BARLEY & PORTOBELLO MUSHROOM SOUP

YIELD: 4 TO 6 SERVINGS / **ACTIVE TIME:** 20 MINUTES / **TOTAL TIME:** 2 HOURS

This simple and easy soup is even better the next day, so be sure to make enough. I like to enjoy it with a serving of warm, cheesy polenta.

1. In a large sauce pan, warm oil on medium-high heat.

2. Add the beef and cook for 5 minutes, or until evenly browned, remove with a slotted spoon and reserve.

3. Add the onion, celery, and carrots and cook for 5 minutes or until soft.

4. Add the red wine, garlic, and thyme and reduce by half.

5. Add the seared beef, the stock, and the barley and bring to a boil.

6. Reduce to a simmer, cover, and cook on low heat for 1½ hours.

7. Add the mushrooms and cook for 10 minutes or until the beef is very tender.

8. Season with salt and pepper and serve in warmed bowls.

INGREDIENTS:

- 1 TABLESPOON VEGETABLE OIL
- 1¾ LB. BEEF STEW MEAT, CUT INTO 1-INCH PIECES
- 1 ONION, CHOPPED
- 2 CELERY STALKS, CHOPPED
- 2 CARROTS, PEELED AND CHOPPED
- ½ CUP RED WINE
- 1 GARLIC CLOVE, MINCED
- 2 SPRIGS FRESH THYME, LEAVES REMOVED AND CHOPPED
- 8 CUPS BEEF OR VEAL STOCK (SEE PAGES 475-476)
- ¾ CUP PEARL BARLEY
- 1 LB. PORTOBELLO MUSHROOMS, SLICED
- SALT AND PEPPER, TO TASTE

FIVE-ALARM BEEF CHILI

YIELD: 6 TO 8 SERVINGS / **ACTIVE TIME:** 15 MINUTES / **TOTAL TIME:** 2 HOURS AND 30 MINUTES

If you want heat, this chili delivers it, thanks to the habanero and jalapeño peppers.

1. Add olive oil to a Dutch oven and heat to medium-high heat. Once the oil is shimmering, add onion, garlic, oregano, pepper, cumin, and chili powder and cook for 5 minutes, stirring often.

2. Add beer, tomatoes, all the chili peppers, beef, and salt. Stir, cover, and lower the heat to simmer for 2 hours, stirring occasionally.

3. To serve, scoop chili into a bowl and, if desired, top with the cheese and sour cream.

INGREDIENTS:

- 1 TABLESPOON OLIVE OIL
- 1 ONION, DICED
- 5 GARLIC CLOVES, MINCED
- 1 TABLESPOON FRESH OREGANO
- 1 RED BELL PEPPER, DICED
- 1 TABLESPOON CUMIN
- 2 TEASPOONS CHILI POWDER
- 2 CUPS BEER (WHEAT OR BELGIUM ALE)
- ½ LB. TOMATOES (CANNED OR FRESH)
- 3 DRIED NEW MEXICAN CHILIES
- 5 JALAPEÑOS, DICED
- 1 HABANERO, DICED
- 2 LBS. GROUND BEEF
- SALT, TO TASTE
- 2 CUPS GRATED CHEDDAR CHEESE (OPTIONAL)
- 1 CUP SOUR CREAM (OPTIONAL)

SHORT RIB RAMEN WITH BEEF BROTH

YIELD: 4 SERVINGS / **ACTIVE TIME:** 20 MINUTES / **TOTAL TIME:** 2 HOURS AND 30 MINUTES

Braised short rib and beef stock serve up a double dose of meaty richness in this outrageously comforting and hearty ramen. This can be topped with soft-boiled eggs, sliced scallions, cilantro, and toasted sesame seeds. You choose. Or, better yet, use them all!

1. Season the short ribs generously with salt.

2. Add the canola oil to a Dutch oven over medium-high heat. When it begins to shimmer, add the short ribs, working in batches if necessary, and brown on all sides, about 2 to 3 minutes per side. Remove and set aside.

3. Add the onion to the pot and sauté until the slices begin to brown, about 4 minutes. Stir in the garlic, ginger, and red pepper flakes and cook for about 1 minute.

4. Pour in the mirin to deglaze the pot, using a wooden spatula or spoon to scrape all the browned bits from the bottom of the pan. Cook until the mirin has reduced by half.

5. Add the soy sauce and stock and bring to a boil. Once boiling, return the short ribs to the pot and add the star anise. Cover, lower the heat to a simmer, and cook for 2 hours, or until the meat is fork-tender.

6. Remove the short ribs from the pot, place them on a cutting board, and shred into large chunks.

7. Add the water to the broth and continue to simmer. Taste and adjust seasoning if needed.

8. Bring a large saucepan full of water to a boil. Salt the water and cook the noodles according to package directions. Drain.

9. Serve by placing noodles and meat in a bowl and add broth, along with any additional toppings.

INGREDIENTS:

- 1½ LBS. BONELESS SHORT RIBS, CUT INTO 2-INCH PIECES
- SALT AND PEPPER, TO TASTE
- 1 TABLESPOON CANOLA OIL
- 1 ONION, SLICED
- 5 GARLIC CLOVES, MINCED
- 1 TEASPOON GRATED FRESH GINGER
- 1 TEASPOON RED PEPPER FLAKES, OR TO TASTE
- ⅓ CUP MIRIN
- ⅔ CUP SOY SAUCE
- 4 CUPS BEEF STOCK (SEE PAGE 475)
- 2 WHOLE STAR ANISE
- 2 CUPS WATER
- ½ LB. RAMEN NOODLES

BEEF KABOB

YIELD: 12 SERVINGS / **ACTIVE TIME:** 15 MINUTES / **TOTAL TIME:** 1 HOUR

For a tasty spin on the classic kabob, these beef kabobs are perfect!

1. Preheat the grill to high and soak skewers in water for 15 to 25 minutes.

2. In a large bowl, combine all of the remaining ingredients and mix with your hands. Form meat around the skewers, making an oblong sausage shape. Refrigerate the kabobs for 30 minutes.

3. Remove the skewers from the refrigerator and spray the grill with nonstick spray or oil. Grill the kabobs until they are charred on the outside, about 2 to 3 minutes, and rotate to cook evenly. The kabobs will be done in about 6 to 8 minutes.

4. Allow to cool for 5 minutes before removing the meat from the skewer and serving.

INGREDIENTS:

- 12 (6-INCH) WOODEN SKEWERS
- 1½ LBS. GROUND BEEF
- ½ CUP MINCED ONION
- 1 TEASPOON CHOPPED FRESH ROSEMARY
- 3 GARLIC CLOVES, MINCED
- 1 TEASPOON CUMIN
- ½ TEASPOON DRIED THYME
- SALT AND PEPPER, TO TASTE

MASALA BRAISED SHORT RIB SANDWICH WITH CILANTRO CHUTNEY

YIELD: 4 SERVINGS / **ACTIVE TIME:** 20 MINUTES / **TOTAL TIME:** 2 HOURS AND 30 MINUTES

This sandwich is a game changer. The use of many different Indian spices gives this sandwich a powerful kick of flavor that you are going to love.

1. Season the short ribs generously with salt.

2. Add the canola oil to a Dutch oven over medium-high heat. When it begins to shimmer, add the short ribs, working in batches if necessary, and brown on all sides, about 2 to 3 minutes per side. Remove and set aside.

3. Add the onion to the pot and sauté until the slices begin to brown, about 4 minutes. Stir in the garlic, ginger, and tomatoes. Cook about 1 minute then add the cumin, curry, garam masala, red pepper, and coriander powder.

4. Pour in the beef stock to deglaze the pot, using a wooden spatula or spoon to scrape all the browned bits from the bottom of the pan. Cook until the stock has reduced by half.

5. Add the short ribs back into the pot. Cover, lower the heat to a simmer, and cook for 2 hours, or until the meat is fork-tender.

6. When the meat is done let it rest for 20 minutes. Slice out the bone from the short rib, and assemble the sandwich, being sure to use plenty of Cilantro Chutney.

INGREDIENTS:

- 3 LBS. BONE-IN SHORT RIBS
- SALT AND FRESHLY GROUND PEPPER, TO TASTE
- 1 TABLESPOON CANOLA OIL
- 1 ONION, SLICED
- 5 GARLIC CLOVES, MINCED
- 1 TEASPOON GRATED FRESH GINGER
- 16 OZ. CANNED TOMATOES OR 2 LARGE FRESH TOMATOES, CHOPPED
- 1 TEASPOON CUMIN POWDER
- 1 TEASPOON CURRY POWDER
- 1 TEASPOON GARAM MASALA
- ½ TEASPOON RED PEPPER POWDER
- 1 TEASPOON CORIANDER POWDER
- 4 CUPS BEEF STOCK (SEE PAGE 475)
- 2 CUPS WATER
- 2 TABLESPOONS CILANTRO CHUTNEY (SEE PAGE 514)

HOT DOG WITH CURRY SPICED LENTILS

YIELD: 4 SERVINGS / **ACTIVE TIME:** 20 MINUTES / **TOTAL TIME:** 24 HOURS

These hot dogs are topped with a delicious lentil curry. The creaminess of the lentils with the snap of the hot dog works perfectly.

INGREDIENTS:

- ¾ CUP BLACK LENTILS
- ¼ CUP RED KIDNEY BEANS
- 1 TEASPOON SALT, PLUS MORE TO TASTE
- 3 TABLESPOONS UNSALTED BUTTER, DIVIDED
- 1 WHITE ONION, FINELY GRATED
- 2 TEASPOONS GINGER-GARLIC PASTE
- ½ CUP TOMATO SAUCE
- ¼ TEASPOON GARAM MASALA
- ½ TEASPOON RED CHILI POWDER
- ½ CUP WATER, PLUS MORE AS NEEDED
- ½ TEASPOON SUGAR
- ¼ CUP CREAM
- 4 HOT DOGS (SEE PAGE 131 FOR HOMEMADE)
- 4 HOT DOG BUNS

1. In a sieve, wash and rinse the lentils and beans. In a large bowl, combine the lentils and beans with 3 cups of water and soak overnight.

2. The next day, drain the lentils and beans and transfer them to a pressure cooker with 1 teaspoon salt and 3½ cups water. Pressure cook on medium-high heat for 20 minutes, then lower the heat to medium-low and cook for another 10 minutes. Let the pressure release naturally. The lentils and beans should be completely cooked and soft. Mash some of the mixture slightly, then set the pressure cooker to the lowest heat and let the mixture simmer as you prepare the rest of the recipe.

3. Add 2 tablespoons butter to a large pot over medium heat. When the butter begins to foam, add the onion and cook for 7 minutes, or until it turns a light golden brown. Add the ginger-garlic paste and cook for 1 to 2 minutes. Add the tomato sauce, mix well, and cook for 2 minutes.

4. Add in the lentils, beans, garam masala, red chili powder, and salt, to taste, and mix to combine. Add ½ cup water, stir, bring to a boil, and then set heat to low and simmer, uncovered, for 40 minutes, stirring often. Add more water, if necessary, to keep the sauce from burning.

5. After 40 minutes, add the remaining butter along with the sugar and cream. Mix well and simmer for 10 more minutes.

6. Once the cream is added, cook the hot dogs to your liking and serve on buns topped with the curry.

WAGYU HOT DOG

YIELD: 12 SERVINGS / **ACTIVE TIME:** 20 MINUTES / **TOTAL TIME:** 1 HOUR AND 30 MINUTES

This hot dog is a nod to Noah Bicchieri of Arkhive Farm and the American Wagyu he raises in New Hampshire (see page 605 for more about Noah). This recipe uses offcuts to make delicious hot dogs that showcase how great this beef is and how best to make sure not an ounce of it goes to waste.

1. In a food processor, puree the onion, garlic, coriander, mustard seed, and paprika. Add the pepper, egg white, sugar, salt, and milk and combine well. Set aside in a large bowl.

2. Using the fine blade of a food processor, grind the beef and fat cubes and then add the mixture to the bowl with the seasoning. Combine the mixture with your hands; wet your hands with cold water or oil to prevent the mixture from sticking to your hands. Refrigerate the mixture for 30 minutes and then puree it again. Set aside.

3. Rinse the casings under cool water to remove the salt and impurities. Place them in a bowl of cool water and soak it for 30 minutes. After soaking, rinse the casings again by slipping one end of the casings over the faucet nozzle; firmly holding them in place, turn on the cold water, gently at first, and then more forcefully. This will flush out any salt or impurities in the casings and help you spot any breaks or tears.

4. Put the casings in a bowl of water and add a splash of white vinegar. Vinegar softens the casings and makes it more transparent. Leave the casings in the water and vinegar solution until you are ready to use it. Rinse it well and drain before stuffing.

5. Using a sausage stuffer, fill the casings with the meat mixture and twist them off into 6-inch links. Boil the links, without separating them, in simmering water for 20 minutes.

6. Place the parboiled hot dogs in a bowl of ice water and chill. Remove, pat dry, and either cook to eat or store. The hot dogs will keep in the refrigerator for 1 week and in the freezer for a few months.

INGREDIENTS:

- ¼ CUP MINCED ONION
- 1 GARLIC CLOVE, FINELY CHOPPED
- 1 TEASPOON CORIANDER
- ½ TEASPOON GROUND MUSTARD SEED
- 1 TEASPOON SWEET PAPRIKA
- 1 TEASPOON FRESHLY FINE GROUND WHITE PEPPER
- 1 EGG WHITE
- 1½ TEASPOONS SUGAR
- 1 TEASPOON SALT
- ¼ CUP MILK
- 1¾ LBS. LEAN WAGYU, CUBED
- ¼ LB. WAGYU BEEF FAT, CUBED
- 4 FT. SHEEP CASINGS, OR SMALL HOG CASINGS (ABOUT 1½-INCH DIAMETER)
- 1 TABLESPOON WHITE VINEGAR

Continued...

CORN DOG

1. Soak the skewers in a bowl of water for 15 to 30 minutes.
2. In a bowl, combine cornmeal, flour, salt, sugar, and baking powder and mix well.
3. In a separate bowl, combine egg white and milk and whisk.
4. Add wet mixture to the dry mixture and stir well.
5. Skewer the hot dogs and roll in the batter until well coated.
6. Add oil to a Dutch oven and heat to 350°F. When the oil is at temperature, fry the corn dogs for about 3 minutes, or until golden brown.

CORN SALAD

1. Preheat the oven to 400°F.
2. Roast corn on a sheet pan until slightly golden in color.
3. In a bowl, combine the roasted corn with the remainder of the ingredients, and mix well.
4. Refrigerate for at least 3 hours before serving, but overnight is best.

CORN DOG

	WOODEN SKEWERS
1	CUP CORNMEAL
1	CUP ALL-PURPOSE FLOUR
2	TEASPOONS SALT
¼	CUP WHITE SUGAR
4	TEASPOONS BAKING POWDER
1	EGG WHITE
1	CUP MILK
4	CUPS CANOLA OIL, FOR FRYING
4	HOT DOGS (SEE PAGE 131 FOR HOMEMADE)

CORN SALAD

2	CUPS CORN KERNELS
2	TABLESPOONS UNSALTED BUTTER
1	JALAPEÑO PEPPER, DICED (PLUS MORE IF YOU LIKE MORE HEAT)
½	TEASPOON SALT
2	TABLESPOONS MAYONNAISE
2	TEASPOONS GARLIC POWDER
3	TABLESPOONS SOUR CREAM, OR MEXICAN CREMA
¼	TEASPOON CAYENNE PEPPER
¼	TEASPOON CHILI POWDER
2	TABLESPOONS FETA CHEESE
2	TABLESPOONS COTIJA CHEESE
2	TEASPOONS FRESH LIME JUICE
½	CUP CHOPPED CILANTRO
	SALT AND PEPPER, TO TASTE

COLA MEATBALLS

YIELD: 10 SERVINGS / **ACTIVE TIME:** 20 MINUTES / **TOTAL TIME:** 5 HOURS

Cola, in meatballs? Yes, cola in meatballs. If you doubt me, make this recipe.

1. Grease the slow cooker with cooking spray or lard.

2. In a bowl, combine all the ingredients, except the cola and salt. Using your hands, form the mixture into 1-inch meatballs.

3. Place uncooked meatballs in the slow cooker.

4. Pour cola over meatballs and cook on high for 3½ to 4½ hours. Season with salt before serving.

INGREDIENTS:

- 2 LBS. GROUND BEEF (80/20)
- ⅓ CUP ITALIAN BREAD CRUMBS
- ¼ CUP DICED ONION
- 1 TEASPOON ITALIAN SEASONING
- 1 EGG WHITE
- 1 TEASPOON SALT, PLUS MORE TO TASTE
- ¼ CUP FRESH CHOPPED PARSLEY
- 3 CUPS COLA

MEATBALLS WITH ARRABIATA

YIELD: 10 SERVINGS / **ACTIVE TIME:** 20 MINUTES / **TOTAL TIME:** 5 HOURS

This spicy arrabiata sauce is perfect for these meatballs. Let them cook low and slow and you are in for a real treat.

1. Grease the slow cooker with cooking spray or lard.

2. In a bowl, combine beef, bread crumbs, onion, 1 teaspoon Italian seasoning, egg white, and parsley. Using your hands, form the mixture into 1-inch meatballs.

3. Place uncooked meatballs in the slow cooker.

4. In a bowl, combine the remainder of the ingredients, pour the sauce over the meatballs, and cook on high for 3½ to 4½ hours.

INGREDIENTS:

- 2 LBS. GROUND BEEF (80/20)
- ⅓ CUP ITALIAN BREAD CRUMBS
- ¼ CUP DICED ONION
- 2 TEASPOONS ITALIAN SEASONING
- 1 EGG WHITE
- ¼ CUP CHOPPED FRESH PARSLEY
- 24 OZ. MARINARA SAUCE
- 2 GARLIC CLOVES
- 1 (14 OZ.) CAN CRUSHED TOMATOES
- 1 (28 OZ.) CAN DICED TOMATOES
- 2 TEASPOONS RED PEPPER FLAKES

CHEESE-STUFFED MEATBALLS

YIELD: 10 SERVINGS / **ACTIVE TIME:** 15 MINUTES / **TOTAL TIME:** 40 MINUTES

These cheddar-stuffed meatballs ooze cheesy goodness with every bite.

1. Preheat the oven to 400°F.

2. In a bowl, combine all the ingredients, except the cheese.

3. Using your hands, form the mixture into 1-inch meatballs. Be sure to leave some of the meat mixture in the bowl in order to seal in the cheese.

4. Using your thumb, make an indent in each meatball, fill the indent with cheese, and then cover with the reserved meat mixture.

5. Place the meatballs on a sheet pan and bake for 15 to 20 minutes.

INGREDIENTS:

- 1½ LBS. LEAN GROUND BEEF
- ⅓ CUP ITALIAN BREAD CRUMBS
- ¼ CUP DICED ONION
- 1 TEASPOON ITALIAN SEASONING
- 1 EGG WHITE
- ¼ CUP CHOPPED FRESH PARSLEY
- ¼ CUP GRATED CHEDDAR CHEESE

SWEET & SOUR MEATBALLS

YIELD: 10 TO 15 SERVINGS / **ACTIVE TIME:** 20 MINUTES / **TOTAL TIME:** 45 MINUTES

When I was young and just started cooking, I attended a potluck supper and was introduced to these meatballs. I loved them and asked for the recipe. The person who made them smiled and said, "Ketchup and grape jelly." I laughed and said, "Seriously, what's the recipe?" Turns out it wasn't a joke, and neither are these meatballs.

1. Preheat the oven to 350°F.

2. In a bowl, mix together the beef, onion, bread crumbs, eggs, and pepper until combined. Using your hands, form the mixture into 1-inch meatballs.

3. Bake for 15 to 20 minutes.

4. In a small saucepan over medium-low heat, combine ketchup and jelly. Stir until the jelly melts, then bring to a boil and let boil for 1 minute.

5. Toss meatballs in the sauce and simmer on low for 10 to 15 minutes (though I let them simmer longer to really absorb the sauce).

INGREDIENTS:

- 2 LBS. LEAN GROUND BEEF
- ½ SMALL ONION, DICED
- ½ CUP BREAD CRUMBS
- 2 EGGS
- 1 TEASPOON PEPPER
- 1½ CUPS KETCHUP
- ¾ CUP CONCORD GRAPE JELLY

AMERICAN CHOP SUEY

YIELD: 3 SERVINGS / **ACTIVE TIME:** 15 MINUTES / **TOTAL TIME:** 30 MINUTES

This is an incredibly simple recipe that my mom would make weekly. When I started my journey in the kitchen, I would look at recipes like this and scoff. As my career progressed, I realized that these are the soulful recipes you love the most. From my mother to you, here is my version of American Chop Suey.

1. Bring a large pot of water to boil. When it is boiling, throw in a couple pinches of salt and add the pasta.

2. As the pasta cooks, add the beef to a frying pan over medium-high heat. When the beef starts to brown, add the onions and sauté until the onions are translucent. Add the tomatoes and cook for another 3 minutes, stirring frequently.

3. Drain the cooked pasta, add it to the beef mixture, and stir. Season with salt and pepper and serve.

INGREDIENTS:

- SALT AND PEPPER, TO TASTE
- 1 LB. ELBOW MACARONI
- 1 LB. GROUND BEEF (80/20)
- 1 ONION, DICED
- 1 LB. DICED TOMATOES

ITALIAN MEATBALLS

YIELD: 10 MEATBALLS / **ACTIVE TIME:** 15 MINUTES / **TOTAL TIME:** 35 MINUTES

These Italian-style meatballs are my go-to when I make tomato sauce. They also freeze really well so feel free to make a bigger batch; you are going to need them.

1. Preheat the oven to 425°F.

2. In a bowl, combine bread crumbs and milk and allow the mixture to absorb for 15 minutes.

3. Mix together all of the ingredients until combined and using your hands, form the mixture into 1-inch meatballs.

4. Bake for 15 to 20 minutes.

INGREDIENTS:

- ½ CUP BREAD CRUMBS
- ½ CUP MILK
- 1 LB. LEAN GROUND BEEF
- 1 LB. GROUND PORK
- ½ SMALL ONION, DICED
- 2 EGGS
- 1 TEASPOON DRIED ROSEMARY
- 1 TEASPOON DRIED THYME
- 1 TEASPOON RED PEPPER FLAKES
- 1 TEASPOON PEPPER
- 2 TABLESPOONS GRATED PARMESAN CHEESE

SWEDISH MEATBALLS

YIELD: 8 SERVINGS / **ACTIVE TIME:** 20 MINUTES / **TOTAL TIME:** 40 MINUTES

This is a great dish for potluck suppers, and it's a lot easier to make than you think.

1. In a large bowl, combine the bread crumbs, milk, cream, egg, garlic, salt, peppers, and allspice. Once the milk has absorbed some, add the onion, meats, and parsley. Mix well with your hands to combine.

2. Using your hands, roll the meat mixture into about 20 small balls, or 12 larger balls.

3. In a saucepan over medium-high heat, add butter and oil. When the butter begins to foam, cook meatballs in batches, being sure not to crowd the pan. Cook for about 15 minutes, turning them every few minutes to brown all sides. Transfer to a warm plate and cover.

4. If making the Gravy, use the same pan that you used to cook the meatballs and melt ⅓ cup butter in the remaining juices.

5. Whisk in the flour until it turns brown, then add stock, cream, soy sauce, mustard, and the remaining butter. Bring to a simmer and season with salt and pepper. Continue to simmer until thickened to desired consistency.

INGREDIENTS:

- ⅓ CUP BREAD CRUMBS
- ½ CUP MILK
- ¼ CUP CREAM
- 1 LARGE EGG
- 1 GARLIC CLOVE, MINCED
- ⅓ TEASPOON SALT
- ¼ TEASPOON BLACK PEPPER
- ¼ TEASPOON GROUND WHITE PEPPER
- ¼ TEASPOON ALLSPICE
- ½ ONION, FINELY CHOPPED
- 1 LB. GROUND BEEF
- ½ LB. GROUND PORK
- 2 TABLESPOONS FINELY CHOPPED FRESH PARSLEY
- 2 TABLESPOONS UNSALTED BUTTER
- 2 TEASPOONS OLIVE OIL

GRAVY (OPTIONAL)

- ⅓ CUP UNSALTED BUTTER, PLUS 2 TABLESPOONS
- ¼ CUP ALL-PURPOSE FLOUR
- 1 CUP BEEF STOCK (SEE PAGE 475)
- 1 CUP HEAVY CREAM
- 2 TEASPOONS SOY SAUCE
- 1 TEASPOON DIJON MUSTARD
- SALT AND PEPPER, TO TASTE

KEFTA WITH WARM CHICKPEAS & SALAD

YIELD: 4 TO 6 SERVINGS / **ACTIVE TIME:** 30 TO 35 MINUTES / **TOTAL TIME:** 2 HOURS AND 10 MINUTES

Think of kefta as a Moroccan meatball, with the lemon zest lending a welcome brightness to these typically earthy elements.

INGREDIENTS:

FOR THE KEFTA
- 1 LB. GROUND BEEF (85/15 RECOMMENDED)
- 1 LB. GROUND LAMB
- ½ CUP MINCED WHITE ONION
- 2 GARLIC CLOVES, ROASTED AND MASHED
- ZEST OF 1 LEMON
- 1 CUP MINCED PARSLEY
- 2 TABLESPOONS MINT LEAVES
- 1 TEASPOON CINNAMON
- 2 TABLESPOONS CUMIN
- 1 TABLESPOON PAPRIKA
- 1 TEASPOON CORIANDER
- SALT AND PEPPER, TO TASTE
- 6 WOODEN SKEWERS
- ¼ CUP OLIVE OIL

FOR THE CHICKPEAS
- 1 (14 OZ.) CAN CHICKPEAS, DRAINED AND RINSED
- ½ ONION, DICED
- ½ CUP MINCED FRESH CILANTRO STEMS
- 2 TABLESPOONS OLIVE OIL
- JUICE OF 1 LEMON
- ¼ TEASPOON SAFFRON
- 1 TABLESPOON CUMIN
- 1 TEASPOON CINNAMON
- ½ TEASPOON RED PEPPER FLAKES
- SALT AND PEPPER, TO TASTE

FOR THE SALAD
- 1 LARGE TOMATO, CUT INTO ½-INCH-THICK SLICES
- 1 ENGLISH CUCUMBER, SEEDED AND CUT INTO ½-INCH-THICK SLICES
- ½ CUP CHOPPED FRESH PARSLEY
- JUICE OF ½ LEMON
- DOLLOP SOUR CREAM OR GREEK YOGURT
- SALT AND PEPPER, TO TASTE
- 2–3 TABLESPOONS PRESERVED LEMON RIND, MINCED (OPTIONAL)

1. To prepare the kefta, place all of the ingredients, except for the skewers and olive oil, in a bowl and stir until well combined. Cook a small bit of the mixture and taste. Adjust seasoning as necessary and then form the mixture into 18 ovals.

2. Place three meatballs on each skewer. Add the olive oil to a cast-iron Dutch oven and warm over medium-high heat. Working in batches, add three skewers to the pot and sear the kefta for 2 minutes on each side. Remove the skewers and set aside.

3. To prepare the chickpeas, place all of the ingredients in the Dutch oven and reduce the heat to medium. Cover and cook for 1 hour.

4. Remove lid, check the beans for doneness, and raise heat to medium-high. Cook for an additional 30 minutes or until approximately 85 percent of the liquid has evaporated.

5. Add the kefta skewers to the pot, cover, and remove it from heat. Let stand for 10 minutes so the kefta get cooked through.

6. To prepare the salad, place all of the ingredients in a small mixing bowl and stir until combined.

7. When the kefta are cooked through, remove the skewers and set aside. Place the chickpeas and salad on a plate, top with the kefta, and serve.

SHEPHERD'S PIE

YIELD: 4 TO 6 SERVINGS / **ACTIVE TIME:** 20 MINUTES / **TOTAL TIME:** 1 HOUR AND 20 MINUTES

This classic comfort dish is one I go back to all the time. Feel free to use lamb instead of beef.

1. Preheat the oven to 350°F.

2. Add oil and butter to a frying pan over medium-high heat. When the butter begins to foam, add the onion and sauté until slightly golden in color. Add the beef and continue to cook until brown.

3. Add the stock and reduce heat to medium-low. Allow the mixture to reduce for about 20 minutes. Add the corn and the thyme.

4. Pour the meat mixture into a baking dish and cover with the mashed potatoes. Bake for 40 minutes, or until the potatoes turn golden brown. Rest for 20 minutes before serving.

INGREDIENTS:

- 1 TABLESPOON CANOLA OIL
- 1 TABLESPOON UNSALTED BUTTER
- 1 SPRING ONION, DICED
- 1 LB. GROUND BEEF
- 1 CUP BEEF STOCK (SEE PAGE 475)
- 1 CUP CORN
- 1 TABLESPOON FRESH THYME
- 5 POTATOES, PEELED, COOKED, MASHED, AND SEASONED TO TASTE

BOLOGNESE

YIELD: 6 SERVINGS / **ACTIVE TIME:** 30 MINUTES / **TOTAL TIME:** 3 HOURS

Bolognese is like a hug. It's warm and comforting. This recipe has been my tried and true recipe for years; now it's yours.

1. Combine stocks and tomatoes in a large pot over medium heat and reduce the entire mixture by half. Set aside.

2. Add the oil and garlic to a heavy-bottom pot over medium heat and cook the garlic until it is fragrant, about 2 minutes. Remove the garlic and add the onion, carrot, and celery. Cook for 10 minutes to caramelize the vegetables. Add rosemary and cook for 5 more minutes.

3. Add the ground meats to the vegetable mixture and season with salt and black pepper.

4. Add the half-and-half and Parmesan rinds, bring to a simmer, and let reduce by three-quarters.

5. When it has reduced, add the wine and simmer until it reduces by half.

6. Add the tomato mixture and nutmeg to the meat and simmer until the Bolognese has thickened, about 2½ hours.

7. Before serving, add Parmesan. Serve over pasta with additional Parmesan.

INGREDIENTS:

- 2 CUPS CHICKEN STOCK (SEE PAGE 478)
- 2 CUPS VEAL STOCK OR BEEF STOCK (SEE PAGES 475-476)
- 2 LBS. PUREED SAN MARZANO TOMATOES
- ½ CUP OLIVE OIL
- 3 GARLIC CLOVES, MINCED
- 1 CUP DICED SWEET ONION
- 1 CUP DICED CARROT
- 1 CUP DICED CELERY
- 2 TABLESPOONS CHOPPED FRESH ROSEMARY
- 1 LB. GROUND VEAL
- 1 LB. GROUND BEEF
- 1 LB. GROUND PORK
- SALT AND PEPPER, TO TASTE
- 2 CUPS HALF-AND-HALF
- 2–3 PARMESAN CHEESE RINDS
- 1 CUP WHITE WINE
- 1 TEASPOON FRESHLY GRATED NUTMEG
- ½ CUP GRATED PARMESAN CHEESE, PLUS MORE FOR SERVING

SLOW COOKER CORNED BEEF

YIELD: 6 TO 8 SERVINGS / **ACTIVE TIME:** 15 MINUTES / **TOTAL TIME:** 9 HOURS

Corned beef is more than a St. Patrick's Day dish. It's an ideal potluck dish, though if you let a crowd of folks at it there won't be leftovers for sandwiches.

1. Place beef, fat side up, in the slow cooker. Add the remainder of the ingredients and cook on low for 9 hours.

2. After cooking the corned beef, heat the broiler to high. Put the meat on a sheet pan and broil it in the oven for 1 to 2 minutes, until it browns slightly.

3. Remove the corned beef from the oven, place it on a cutting board, cover with aluminum foil, and let rest for 10 minutes.

4. Slice to serve.

INGREDIENTS:

- 3 LBS. BEEF BRISKET
- 3 GARLIC CLOVES, CHOPPED
- 2 BAY LEAVES
- 2 TABLESPOONS SUGAR
- ½ TEASPOON BLACK PEPPER
- 2 TABLESPOONS APPLE CIDER VINEGAR
- 1½ CUPS WATER

OLD FASHIONED MEAT LOAF

YIELD: 4 TO 6 SERVINGS / **ACTIVE TIME:** 20 MINUTES / **TOTAL TIME:** 1 HOUR AND 30 MINUTES

Growing up, I was not a big fan of meat loaf. My mom would cook it until it was dry and I had to force myself to choke down as much as I could. I know that many people have similar experiences. Well, once you realize how good this dish can be, you'll understand why it is a classic.

1. In a small saucepan over medium-high heat, combine the ketchup, brown sugar, and 1 tablespoon Worcestershire sauce and bring to a boil. Let boil for 2 to 3 minutes and then set aside.

2. Preheat the oven to 400°F.

3. In a bowl, combine the remainder of the ingredients and mix well.

4. Fill a loaf pan with the meat mixture, pressing it down to make sure it fills all sides. Pour half the sauce over the meat.

5. Bake the meatloaf for 30 minutes, remove it from the oven to add the remaining sauce, and then bake it for 40 more minutes.

6. Let the cooked meatloaf rest for 15 minutes before serving.

INGREDIENTS:

- ½ CUP KETCHUP
- 2 TABLESPOONS BROWN SUGAR
- 2 TABLESPOONS WORCESTERSHIRE SAUCE, DIVIDED
- 2 EGGS
- 1 CUP BREAD CRUMBS
- 1 ONION, CHOPPED
- 2 LBS. GROUND BEEF (90/10)
- ½ CUP MILK
- 1 TEASPOON SALT
- 1 TEASPOON PEPPER
- 2 TEASPOONS MINCED GARLIC
- 1 TEASPOON GARLIC POWDER
- ½ TABLESPOON DIJON MUSTARD
- 1 TABLESPOON TOMATO PASTE

BBQ-GLAZED MEAT LOAF

YIELD: 4 TO 6 SERVINGS / **ACTIVE TIME:** 20 MINUTES / **TOTAL TIME:** 1 HOUR AND 30 MINUTES

This meat loaf is smoky and sweet with the addition of your favorite BBQ sauce.

1. Preheat the oven to 400°F.

2. In a bowl, combine all of the ingredients, except the BBQ sauce, and mix well.

3. Fill a loaf pan with the meat mixture, pressing it down to make sure it fills all sides. Pour half the BBQ sauce over the meat.

4. Bake the meat loaf for 30 minutes, remove it from the oven to add the second half of the sauce, and then bake it for 40 more minutes.

5. Let the cooked meat loaf rest for 15 minutes before serving.

INGREDIENTS:

- 2 EGGS
- 1 CUP BREAD CRUMBS
- 1 ONION, CHOPPED
- 2 LBS. GROUND BEEF (90/10)
- ½ CUP MILK
- 1 TEASPOON SALT
- 1 TEASPOON PEPPER
- 2 TEASPOONS MINCED GARLIC
- 1 TEASPOON GARLIC POWDER
- 1 TABLESPOON WORCESTERSHIRE SAUCE
- ½ TABLESPOON DIJON MUSTARD
- 1 TABLESPOON TOMATO PASTE
- 1 CUP BBQ SAUCE (SEE PAGES 497-505)

PORK

*P*ork has it all: unctuous, delicious fat; crispy, addictive skin; tender meat. It can be grilled, smoked, roasted, and braised, yielding wildly different results.

PORCHETTA

YIELD: 12 SERVINGS / **ACTIVE TIME:** 30 MINUTES / **TOTAL TIME:** 2 DAYS

Porchetta is basically rolled pork belly that is typically stuffed with pork loin, roasted to achieve a crisp exterior and super tender interior. If you have never had it, it's euphoric.

1. Two days before serving this recipe, lay the pork belly skin side down. Using a knife, score the belly's flesh in a diamond pattern and rub the herbs and garlic powder into the belly.

2. Flip the belly over and poke small holes in the flesh. Turn the belly back over and salt it.

3. Place the pork loin in the center of the belly and roll it up. Using kitchen twine, tie the roll every ½ inch.

4. Place the belly on a wire cooling rack with a drip pan under it and refrigerate for 2 days, uncovered. This allows the skin to dry out some and release any moisture. Occasionally pat the belly dry to get rid of extra moisture.

5. After 2 days, take the porchetta out of the fridge and allow it to come to room temp for 2 hours.

6. Preheat the oven to 480°F.

7. Roast the porchetta on a wire rack with the drip pan under it for 35 minutes, turning a couple of times to ensure even cooking.

8. Reduce the oven temperature to 300°F and continue to cook for an additional 1 to 2 hours, or until a meat thermometer reaches 145°F. Make sure that the skin is crisp. If it isn't, increase oven temperature to 500°F to ensure it crisps.

9. Let the porchetta rest for at least 15 minutes. Slice to serve.

INGREDIENTS:

- 5–6 LBS. PORK BELLY, SKIN ON
- 1 TABLESPOON MINCED FRESH ROSEMARY
- 1 TABLESPOON MINCED FRESH THYME
- 1 TABLESPOON MINCED FRESH SAGE
- 2 TEASPOONS GARLIC POWDER
- 1 TABLESPOON SALT
- 1 LB. PORK LOIN (CENTER CUT PREFERABLE)

PORK ROULADE WITH ORANGES, RAINBOW CHARD & WILD RICE

YIELD: 6 SERVINGS / **ACTIVE TIME:** 30 TO 45 MINUTES / **TOTAL TIME:** 1 HOUR AND 30 MINUTES

Roulades are a great way to pack flavor into a single piece of meat, as this floral, citrus-packed meal shows.

1. Preheat oven to 350°F. Season both sides of the butterflied pork loin with salt and pepper. Sprinkle the fennel seeds, celery seeds, and orange zest across the interior of the pork loin. Lay the leaves of chard across the interior and then roll the pork loin up until it is closed.

2. Cut a 3-foot section of kitchen twine and use it to tie the rolled-up pork loin closed. Drizzle the olive oil over the pork and rub it in. Place the pork roulade in a cast-iron skillet and cook over medium-high heat for 3 to 5 minutes on each side.

3. Place the remaining ingredients in a 9x13-inch baking pan. Place the seared pork loin in the center of the pan and top with the orange slices. Wrap the pan tightly with foil and cook for 45 minutes to 1 hour, until the center of the pork reaches 145°F and the rice is tender.

4. Remove the pan from the oven and place the pork loin on a cutting board. Cover it with foil and let rest for 15 minutes before removing the twine and carving. If the rice is not tender, cover the pan with foil and return to the oven until it is ready.

TIP: To butterfly the pork loin, you'll want to use a sharp knife and cut the pork loin lengthwise, taking care not to cut all the way through. Roll it out and repeat until the pork loin can lie flat and is ½- to 1-inch thick. If you don't think your knife skills are up to the task, you can always ask a butcher to do it for you.

INGREDIENTS:

- 1 (3–4 LB.) PORK LOIN, BUTTERFLIED ½-INCH THICK AND ROLLED OUT FLAT
- SALT AND PEPPER, TO TASTE
- 1 TEASPOON GROUND FENNEL SEEDS
- 1 TEASPOON GROUND CELERY SEEDS
- 2 ORANGES, ZESTED AND THEN SLICED INTO ¼-INCH-THICK ROUNDS
- 8 STALKS RAINBOW CHARD, STEMS REMOVED AND MINCED, LEAVES RESERVED
- 2 TABLESPOONS OLIVE OIL
- 2 GARLIC CLOVES, MINCED
- 1 FENNEL BULB, SLICED VERY THIN
- 1 CUP WILD RICE
- 2½ CUPS CHICKEN STOCK (SEE PAGE 478)
- 2 PLUM TOMATOES, MINCED

CLASSIC SMOKED RIBS

YIELD: 10 SERVINGS / **ACTIVE TIME:** 15 MINUTES / **TOTAL TIME:** 10 HOURS

Great barbecue is not a secret. It is all about time and temperature. Some of the best pit-masters in the world use nothing but salt and pepper. Find out why. These ribs don't need sauce. But if you want to sauce them, brush the ribs with the sauce of your choosing every 30 to 40 minutes for the last hour and a half before wrapping in foil.

INGREDIENTS:

- ¼ CUP SALT
- ½ CUP BLACK PEPPER
- 10 LBS. ST. LOUIS CUT RIBS

1. In a bowl, combine the salt and pepper and mix well. Rub the mixture all over the ribs. Refrigerate the ribs for 1 hour.

2. Bring the smoker to 265°F. Once it reaches temperature, add applewood chips and 1 cup of water to the steam tray. Place the ribs on the smoker and cook for 3 hours.

3. After the first hour, use a spray bottle to spritz the ends of the ribs with water. Do this again after the second hour of cooking.

4. Remove the ribs from the smoker and wrap them in aluminum foil. Cook them for another hour, until the meat shrinks from the bone; this typically takes about 4 to 6 hours.

TIP: Make sure to keep the steam tray full. You do not want to let the steam tray dry out.

APPLEWOOD SMOKED RIBS WITH MOLASSES BBQ SAUCE

YIELD: 10 SERVINGS / **ACTIVE TIME:** 15 MINUTES / **TOTAL TIME:** 5 HOURS

New England is often overlooked when it comes to BBQ. But there are some unbelievable BBQ joints across New England. Places like Ore Nell's in Kittery, Maine, and Andy Husbands's The Smoke Shop BBQ in Boston. Applewood is my favorite wood of choice when it comes to smoking pork. Not only do we have an abundance of apple trees across New England, it is an amazing flavor for pork. This recipe is a nod to colonial New England with the addition of the Molasses BBQ Sauce.

1. In a bowl, combine all of the dry ingredients together and mix well.

2. Massage the spice rub all over the ribs, then place ribs in the refrigerator for 1 hour.

3. Heat smoker to 250°F and then add applewood chips and 1 cup of juice (or vinegar) to the steam tray. For as long as the ribs are in cooking, be sure to keep the steam tray full; do not let the steam tray dry out.

4. Put the ribs in the smoker and baste them with the Molasses BBQ Sauce every 30 to 40 minutes. Cook the ribs until the meat shrinks from the bone, about 4 hours.

MOLASSES BBQ SAUCE

1. In a saucepan, combine all of the ingredients and bring to a boil. Reduce heat and simmer for 20 minutes, stirring periodically. Let cool and store.

INGREDIENTS:

- ¼ CUP SALT
- 2 TABLESPOONS LIGHT BROWN SUGAR
- 2 TABLESPOONS GARLIC POWDER
- 1 TABLESPOON ONION POWDER
- 1 TABLESPOON CHILI POWDER
- 1 TABLESPOON PAPRIKA
- 1 TABLESPOON CUMIN
- 10 LBS. ST. LOUIS CUT RIBS
- ½ GALLON APPLE JUICE OR APPLE CIDER

MOLASSES BBQ SAUCE

- ½ CUP KETCHUP
- ¼ CUP BROWN SUGAR
- 2 TABLESPOONS WHITE SUGAR
- 2 TABLESPOONS DIJON MUSTARD
- 3 TABLESPOONS APPLE CIDER VINEGAR
- 2 GARLIC CLOVES
- ¼ CUP BLACKSTRAP MOLASSES
- ¼ TEASPOON CLOVES
- ½ TEASPOON HOT SAUCE
- ¼ CUP HONEY

ROTISSERIE PIG ROAST

YIELD: 15+ SERVINGS / **ACTIVE TIME:** 1 HOUR / **TOTAL TIME:** 15 HOURS

Roasting a whole pig is a blast! This is how to do it. You'll need lots of time—the general formula is 1 hour of cooking per 10 pounds—and plenty of people around so you can share all of this delectable pork. It's worth noting that most processing houses will attach the pig to the spit. This process can be a pain, so let them do it for you.

1. Bring the rotisserie up to 250°F. In a bowl, combine the oil and paprika.

2. Spread out a plastic tablecloth and set the pig on it. Rub the pig all over with paprika mixture. If using a brine, inject pig with apple cider in the shoulders and ham.

3. Put foil pans under the pig with a little bit of water and/or apple juice—make sure these do not dry out when cooking the pig.

4. Put 1 meat thermometer into the ham and put another meat thermometer into the pit so temperature readings are always available.

5. Wrap the ears, snout, and tail with aluminum foil.

6. When the smoker reaches 250°F, put the pig on the rotisserie. Keep a squirt bottle full of water nearby to manage flare-ups. Keep an eye on the skin; if the skin breaks, hot fat can hit the coals and cause a serious fire.

7. Put 1 chunk of wood per grate of coal and check the temperature every 30 minutes without opening the lid. Cool or heat as necessary to maintain a temperature between 225 and 250°F. If running cold, open the intake valve at bottom and put a stone in the lid to allow air in; if that doesn't work, add more hot coals. If running hot, close off intake vents, but not all the way. Mist the pig with apple juice as needed.

8. When the ham reaches 185°F, about 8 hours in, coat the pig with BBQ sauce. The pig is done at this point, but let it cook for a final 15 minutes after applying the sauce.

9. After 15 minutes, carefully place the pig on a table covered in plastic tablecloths, wrap the pig in foil, and rest for 45 minutes before pulling off meat.

INGREDIENTS:

- 4 CUPS CANOLA OIL
- ½ CUP PAPRIKA
- 1 WHOLE PIG, ATTACHED TO ROTISSERIE
- SPRAY BOTTLE OF APPLE JUICE, FOR MISTING
- BBQ SAUCE (SEE PAGES 497-505)
- 2 CUPS APPLE CIDER VINEGAR, IF BRINING

CRISPY PORK BELLY WITH BALSAMIC GLAZE

YIELD: 5 SERVINGS / **ACTIVE TIME:** 15 MINUTES / **TOTAL TIME:** 28 HOURS

How do you make crispy pork belly better? Add a drizzle of balsamic glaze for a pop of flavor that cuts through the fattiness of the pork belly.

INGREDIENTS:

- 3 TABLESPOONS SALT
- 2 LBS. PORK BELLY, SKIN ON
- 1 CUP BALSAMIC VINEGAR
- ¼ CUP BROWN SUGAR

1. Rub 2 tablespoons salt all over the pork belly and then place on a rack over a sheet pan. Refrigerate, uncovered, for 24 hours.

2. The next day, pour boiling water over the skin of the pork belly. Leave it to cool and drain.

3. Preheat the oven to 300°F.

4. Pat the skin dry with paper towels, place the pork in a baking dish, and sprinkle with salt. Roast for 3 hours.

5. After 3 hours, remove the pork from the oven and turn up the oven to 500°F. When the oven is ready, cook the pork belly for another 10 or 15 minutes, or until the crackling is puffy and crisp. Transfer the meat to a cutting board.

6. In a saucepan, combine the vinegar and sugar and bring to a gentle boil. Reduce the heat to medium-low and simmer for 8 to 10 minutes, stirring often, until the mixture thickens. Remove from the heat and allow it to cool for 15 minutes.

7. To serve, slice the pork belly and drizzle with the glaze.

BONE-IN PORK ROAST

YIELD: 5 SERVINGS / **ACTIVE TIME:** 15 MINUTES / **TOTAL TIME:** 3 HOURS

This bone-in pork roast recipe is a classic.

1. Preheat the oven to 450°F.

2. Rinse the rack of pork under cold water and pat dry with paper towels. Place the rack fat side up in a roasting pan.

3. In a bowl, combine all of the spices and mix well.

4. Rub the mustard into the top of the roast and then cover the entire rack with the spice mixture. Add 1 cup of water to the bottom of the pan before roasting.

5. Place the pork in the oven and cook for 15 minutes. After 15 minutes, reduce the oven temperature to 325°F and continue to roast for 1½ to 2 hours, or until a meat thermometer inserted in the middle of the roast registers 145°F degrees for medium or 160°F for well done.

6. Remove the meat from the oven and transfer it to a cutting board. Rest for 20 minutes before slicing.

INGREDIENTS:

- 1 (7–8 LB.) BONE-IN, CENTER-CUT RACK PORK
- ¼ CUP SALT
- ¼ CUP PEPPER
- 1 TABLESPOON SMOKED PAPRIKA
- 1 TEASPOON GARLIC POWDER
- 1 TEASPOON ONION POWDER
- ¼ CUP DIJON MUSTARD

BASIC GRILLED ROAST PORK LOIN

YIELD: 5 TO 6 SERVINGS / **ACTIVE TIME:** 1 HOUR AND 15 MINUTES / **TOTAL TIME:** 2 HOURS

These basic ingredients bring out the natural juices and flavors. Just be sure to use a pan that can go from the oven to the grill.

INGREDIENTS:

- 5 TABLESPOONS OLIVE OIL
- 1–2 SPRIGS FRESH ROSEMARY
- 2¼ LBS. PORK LOIN
- SEA SALT, TO TASTE
- COARSELY GROUND BLACK PEPPER, TO TASTE

1. Fire up your grill and allow the coals to settle into a temperature of about 350°F. While the grill is heating, slowly sauté the olive oil and rosemary in a cast-iron or All-Clad style high heat-friendly pan. Be sure the pan is oven-and-grill friendly, as you will be placing this pan directly onto your grill.

2. After the oil and rosemary have been thoroughly heated and the flavors of the sprigs are infused throughout the oil (about 10 to 12 minutes), rub your pork loin with sea salt and fresh cracked pepper to your desired seasoning, and place the pork loin into the pan, turning it so the entire loin is covered with the heated oil.

3. Baste for 5 to 10 minutes at a medium heat until the loin begins to brown. Once your grill has reached the desired temperature, move the entire pan to your grill grate.

4. Cover your grill and allow the pork to cook for 45 minutes, turning and basting the pork occasionally so all sides are thoroughly browned from the heat of the hot pan.

5. At about 45 minutes, remove the pork loin from the pan and place directly on the grate. Continue to baste your pork loin using the infused oil from the pan, turning the loin evenly so the entire roast meets the heat side of your grill. Baste and turn for an additional 15 or so minutes or until the roast meets your desired temperature.

6. Remove from fire and let the loin rest for 10 to 12 minutes. Carve, season to taste, and serve with sides of your choice.

TIP: For any oven-to-grill recipe, avoid using any cookware that has a plastic, wood, or synthetic-type handle. It is best if the pan has rounded sides, high enough to prevent the oil from spilling or flaring up when basting.

GRILLED ROAST PINEAPPLE PORK LOIN

YIELD: 5 TO 6 SERVINGS / **ACTIVE TIME:** 1 HOUR AND 30 MINUTES / **TOTAL TIME:** 2 HOURS AND 15 MINUTES

Here we introduce more exotic flavors with a guest-worthy variation on grilled roast pork loin.

INGREDIENTS:

- 5 TABLESPOONS OLIVE OIL
- 1–2 SPRIGS FRESH ROSEMARY
- 1 CUP CRUSHED PINEAPPLE, DIVIDED
- ¼ CUP HONEY
- ¼ CUP WATER
- 1 TEASPOON GRATED GINGER
- 2¼ LBS. PORK LOIN
- SEA SALT, TO TASTE
- COARSELY GROUND BLACK PEPPER, TO TASTE

1. Fire up your grill and allow the coals to settle into a temperature of about 350°F.

2. While the grill is heating, slowly sauté the olive oil and rosemary in a cast-iron or All-Clad-style high heat-friendly pan. Be sure the pan is oven-and-grill friendly, as you will be placing this pan directly onto your grill.

3. After the oil and rosemary have been thoroughly heated and the flavors of the sprigs are infused throughout the oil, add in a ½ cup of crushed pineapple, honey, water, and ginger. Stir thoroughly, and bring the mixture to a soft boil.

4. Rub your pork loin with salt and pepper to your desired seasoning, and place the pork loin into the pan, turning it so the entire loin is covered with the basting sauce for 5 to 10 minutes at a medium heat until the loin begins to brown. Once your grill has reached the desired temperature, move the entire pan to your grill grate.

5. Cover your grill and allow the pork to cook for 45 minutes, turning and basting the pork occasionally so all sides are browned from the heat of the hot pan.

6. After about 45 minutes, remove the pork loin from the pan and place directly on the grate. Use the remaining ½ cup of crushed pineapple to baste the loin thoroughly, creating a golden-brown glaze as you turn the loin for another 15 minutes.

7. Remove from fire and let the loin rest for 10 to 12 minutes. Carve, season to taste, and serve with sides of your choice.

GRILLED ROAST PORK WITH ORANGE RIND

YIELD: 6 SERVINGS / **ACTIVE TIME:** 1 HOUR AND 30 MINUTES / **TOTAL TIME:** 2 HOURS AND 30 MINUTES

No bland meat here—orange, garlic, chili, and oregano come together beautifully in this grilled roast pork.

INGREDIENTS:

- 3 TABLESPOONS OLIVE OIL
- 1½ CUPS FRESH ORANGE JUICE WITHOUT PULP
- 1-2 TEASPOONS GRATED ORANGE RIND
- 1 GARLIC CLOVE, CHOPPED
- 2 BAY LEAVES
- 1 PINCH CHILI POWDER
- 1 PINCH DRIED OREGANO
- 2½ LBS. PORK LOIN
- SEA SALT, TO TASTE
- COARSELY GROUND BLACK PEPPER, TO TASTE

1. Fire up your grill and allow the coals to settle into a temperature of about 350°F.

2. Then, in a grill-friendly sauté pan, heat the olive oil and spread to cover the entire bottom of the pan. Once the oil begins to brown, gently pour in the fresh orange juice, adding the grated orange rind, chopped clove of garlic, two bay leaves, pinch of chili powder, and pinch of oregano and stir until the ingredients are heated and mixed thoroughly.

3. Rub your pork loin with sea salt and fresh cracked pepper to your desired taste, and place the pork loin into the pan.

4. Leave the pan on the grate and cover the grill, allowing the roast port to cook for about 1½ hours, basting often or at least every 15 to 20 minutes or so.

5. After 1 hour and 15 minutes, remove the roast from the pan and place directly on the grill for browning and searing. Continue basting and turning so the pork receives the direct heat evenly on all sides.

6. When the roast has browned and seared, remove it from the grill and allow it to sit for 10 minutes before carving.

BRAISED & GRILLED PORK WITH ROSEMARY

YIELD: 5 TO 6 SERVINGS / **ACTIVE TIME:** 45 MINUTES / **TOTAL TIME:** 1 HOUR AND 45 MINUTES

My favorite thing about this dish is the way the rosemary needles catch and burn over the grill, bringing a charred and sweet flavor to the roast.

1. To enhance the flavor and to prevent the rosemary from searing completely off during the cooking process, push the rosemary needles into the meat. This will help infuse the flavor throughout the pork. Leave a little bit of each rosemary sprig sticking out to catch and burn from the flame; this adds to the flavor.

2. Brush and coat the roast with ¼ cup olive oil.

3. Place the roast into a deep sauté pan that can withstand the direct heat of your grill. Place the remaining oil into the pan, turning and cooking the pork evenly on all sides until it reaches a lovely golden brown.

4. Add the garlic, onion, and remaining rosemary, and let the meat and seasoning cook together for about an hour. If you can control the temperature of your grill, bring the heat down so everything may simmer together for 1½ hours.

5. Just before the pork appears to be done, remove it from the pan and place it directly on the grill to sear off the rosemary sprigs and to gracefully char the exterior to your preference.

6. Remove the roast from the grill, and let it stand for 10 to 12 minutes. Slice thin before serving, and use some of the cooked juices as a light gravy if the pork happens to get slightly overdone.

INGREDIENTS:

- 2 SPRIGS FRESH ROSEMARY, STEMMED
- 2¼ LBS. BONELESS PORK LOIN
- ½ CUP OLIVE OIL
- 1 GARLIC CLOVE, CRUSHED
- ½ ONION, CHOPPED
- ¾ CUP WHITE WINE
- 1 TABLESPOON WHITE VINEGAR
- SEA SALT, TO TASTE
- COARSELY GROUND BLACK PEPPER, TO TASTE

KILLER BBQ SPARERIBS

YIELD: 6 TO 8 SERVINGS / **ACTIVE TIME:** 1 HOUR AND 45 MINUTES / **TOTAL TIME:** 4 TO 5 HOURS

Without approaching the task via a day long, low-heat smoking process, this recipe tackles ribs a little bit more conventionally (and much more simply). First, slow roast them in the oven at a low temperature of 200°F for about 3 hours. This allows the acids and seasonings to gently tenderize the meat, while the low heat loosens the meat from the bone so the cooked rib meat will pull away without fuss.

1. Preheat the oven to 325°F. Mix all of the ingredients except the ribs in a large sauce pan over low to medium heat, allowing the sugars to melt. Line the bottom the roasting pan with a thick layer of the BBQ sauce.

2. Place each rack of ribs into the roasting pan, layering them with a solid basting of the sauce so both sides of each rack of ribs are fully coated. Cover and allow the ribs to cook for 2½ to 3 hours. No need to turn or recoat the ribs during this process.

3. About 15 to 20 minutes before the ribs have finished cooking in your oven, fire up your grill to medium heat. A gas grill will work just fine, but there's nothing better than wood-grilled BBQ ribs, so consider your options carefully!

4. Use long tongs that will allow you to slide the tong the full length of the rack of ribs. This will help prevent the ribs from breaking off as the ribs will be soft and tender from their time in the oven.

5. Basting is perhaps the most important final step in preparing killer ribs. Continually baste the ribs to try to achieve a beautiful dark brown and black glazed surface. As soon as the flames char an edge of the meat, quickly baste over that area with a fresh coat of sauce and turn the ribs so the opposite side can be lightly and evenly charred by the fire as well. Unlike steaks on the grill, turn the ribs over and over, basting and turning each rack in order to achieve the best and most flavorful results. Do not worry about a little bit of blackening and charring; paint over all the charred areas with a fresh coat of BBQ sauce and the two flavors wed together beautifully.

6. As soon as the ribs reach the level of browning and blackening you desire, remove the ribs from the grill and place them onto a serving tray. Do not place them back into the roasting pan. Bring the ribs directly to the table and allow them to cool to touch before digging in.

TIP: The more layers of sauce, the richer the taste and the more gratifying the dining experience.

INGREDIENTS:

- 2–3 GARLIC CLOVES, SLICED EXTRA THIN
- 1 GARLIC CLOVE, CRUSHED OR MINCED
- 1 CUP LOCAL HONEY
- ⅓ CUP DARK MOLASSES
- ⅓ CUP LOCAL DARK MAPLE SYRUP
- 1½ TABLESPOONS PAPRIKA
- 1 TEASPOON SEA SALT
- 1½ TEASPOONS FRESH GROUND PEPPER
- 1 TABLESPOON ANCHO CHILE POWDER (ADD MORE IF YOU LIKE YOUR BBQ EXTRA SPICY)
- 2 TEASPOONS GROUND CUMIN
- ½ CUP APPLE CIDER VINEGAR (THE MORE YOU ADD, THE TANGIER THE FLAVOR)
- 1½ CUPS ORGANIC STRAINED TOMATOES
- 5–6 OZ. ORGANIC TOMATO PASTE (NO SUGAR ADDED)
- ¼ CUP CHILI SAUCE
- ¼ CUP WORCESTERSHIRE SAUCE
- 1½ TABLESPOONS FRESH SQUEEZED LEMON JUICE
- 5 TABLESPOONS CHOPPED ONIONS
- 1 TEASPOON MUSTARD POWDER
- ½ PINEAPPLE, CUBED (IF FRESH JUICE COLLECTS ON YOUR CUTTING BOARD, ADD THAT IN, TOO!)
- 4–5 LBS. BABY BACK PORK RIBS

CHAR SIU

YIELD: 4 SERVINGS / **ACTIVE TIME:** 15 MINUTES / **TOTAL TIME:** 25 HOURS

I love char siu, the popular Cantonese tender pork dish. If you're not familiar with it, you'll love the sweet and smoky flavor too once you make this recipe.

INGREDIENTS:

- 2 PIECES FERMENTED RED BEAN CURD
- 1 TABLESPOON MALTOSE OR HONEY
- 1 TABLESPOON SHAOXING WINE
- 1 TABLESPOON SOY SAUCE
- 1 TABLESPOON OYSTER SAUCE
- 1 TEASPOON DARK AND THICK SOY SAUCE
- 1 TEASPOON FIVE-SPICE POWDER
- ¼ TEASPOON WHITE PEPPER
- ½ CUP SUGAR
- 2 TABLESPOONS MINCED GARLIC
- 1 LB. SKINLESS PORK BELLY, CUT INTO 2 LONG STRIPS

1. In a large bowl, combine all of the ingredients, except the garlic and pork belly, and mix well.

2. In another large bowl, add the garlic and pork belly and coat the meat with half of the sauce. Refrigerate the meat, uncovered, for 24 hours. Reserve the remaining sauce in the refrigerator for the following day.

3. The next day, let the pork belly and reserved sauce come to room temperature and preheat the oven to 400°F.

4. Place the pork belly on a wire rack or in a pan lined with aluminum foil.

5. Roast the meat for 15 minutes, then remove it from the oven, turn the pork belly over, and brush some of the sauce on the pork. Continue to roast for another 15 minutes.

6. Set the oven to broil, and broil each side of the pork belly for about 1 minute, until each side becomes nicely charred; it is normal for it to turn a dark color. Remove from the oven and let rest for 15 minutes.

7. In a saucepan, simmer the remaining sauce over medium heat.

8. Slice the char siu into thin, bite-sized pieces and serve with steamed white rice and the warmed sauce.

GO-TO RECIPES

BASIC GRILLED DOUBLE-CUT RIB CHOPS

YIELD: 4 TO 6 SERVINGS / **ACTIVE TIME:** 25 MINUTES / **TOTAL TIME:** 45 MINUTES

I find pork delicious without any rubs, sauces, or marinades. Properly cooked, a great chop will burst with flavor. With this dish, there's no need to overpower the pork. Just sprinkle the chops with a hint of salt and fresh cracked pepper.

1. Preheat one side of your grill to 400°F. If possible, create a two-zone cooking area: one zone will be your hot zone, concentrating your coals or heat source beneath one side of your grilling area, while the second zone will be arranged on the opposite side of the grill with little to no coals or flame beneath this surface area. Use the hot side to sear your chops, and the cool side to allow your chops to cook through the radiating heat. This will help ensure that your chops do not overcook and dry out.

2. Place the chops directly over the hot zone and sear both sides of the chops until evenly browned, about 5 minutes per side. Keep a watchful eye on the grill during this stage as fat drippings can create flare-ups that will char rather than sear the meat.

3. Once the chops are seared golden brown, move them over to the cooler zone and let them cook more thoroughly and slowly. If using a meat probe thermometer, look for temperatures in the center of your cut from 135°F for medium rare and approximately 145°F for a tender and flavorful medium. Season with salt and pepper.

INGREDIENTS:

4–6 DOUBLE-CUT RIB CHOPS

SEA SALT, TO TASTE

FRESH CRACKED PEPPER, TO TASTE

JOWL BACON

YIELD: 12 SERVINGS / **ACTIVE TIME:** 5 MINUTES / **TOTAL TIME:** 7 DAYS

Pork jowl is a cut that many people might not know of, but it is one of the best cuts of meat. Today you are going to learn how to turn jowl into bacon! There is some math involved, but the effort is well worth it, trust me.

1. Weigh the jowl in grams. Then weigh out 2% of that weight of regular salt, and then 0.25% of that weight of Instacure. This should be about 4 tablespoons of regular salt and ¼ teaspoon of curing salt—but, please, for this to work, measure it precisely.

2. In a bowl, combine these salts with the brown sugar.

3. Rub the salt-and-sugar mixture into the jowl then put the jowl in a resealable plastic bag. Refrigerate for a week, rotating and flipping over the jowl daily.

4. When the jowl has been cured, remove it from the bag, give it a quick rinse, and pat dry. Set it on a rack over a sheet pan and refrigerate overnight.

5. The next day, bring the smoker to 165°F.

6. Take the jowl out of the refrigerator, set it in the smoker, and smoke it for 4 hours at 165°F.

7. Remove the jowl bacon and set it in the fridge to cool completely before storing.

INGREDIENTS:

- 1 PORK JOWL
- SALT, AS NEEDED
- INSTACURE NO. 1, AS NEEDED
- ¼ CUP BROWN SUGAR

CANDIED BACON

YIELD: 4 SERVINGS / **ACTIVE TIME:** 5 MINUTES / **TOTAL TIME:** 45 MINUTES

Sticky, sweet, and smoky, candied bacon is a delicious treat you didn't know you needed in your life.

1. Preheat the oven to 350°F.

2. Spread out the bacon strips on a sheet pan and coat both sides with brown sugar.

3. Cook for 35 to 40 minutes, until the sugar caramelizes the bacon and it's cooked through.

4. Let cool and enjoy.

INGREDIENTS:

- 1 LB. SMOKED BACON
- ½ CUP BROWN SUGAR

PORK PÂTÉ AKA GORTON

YIELD: 10 TO 15 SERVINGS / **ACTIVE TIME:** 20 MINUTES / **TOTAL TIME:** 4 HOURS

This recipe has a special place in my life. My grandmother prepared this French Canadian pâté, called gorton, for every holiday. Over the course of my career as a chef I have often found myself revisiting memorable dishes from my childhood. This is my version of my grandmother's classic dish. I hope when you make it you taste the same love she put into making this delectable spread.

1. Preheat oven to 300°F.

2. In a Dutch oven, combine all ingredients and cook for 3 to 4 hours, until the pork pulls apart easily with a fork. Discard the bay leaves and reserve the cooking liquid.

3. Place the pork on a cutting board and pull it apart with a fork until fully shredded.

4. Place the pork in a blender along with ½ to ¾ cup of the reserved cooking liquid and blend until smooth. You might need to add more liquid until it forms a smooth paste. Season with salt and pepper, to taste.

5. Once all the pork is blended to the desired consistency, spoon it into a jar, and cover it with the cooking liquid. Seal and refrigerate overnight before serving.

INGREDIENTS:

- 3–5 LBS. PORK BUTT
- 3 ONIONS, SLICED
- 2 TEASPOONS GROUND CLOVES
- 1 TABLESPOON SALT, PLUS MORE TO TASTE
- 4 BAY LEAVES
- 2 TEASPOONS BLACK PEPPER, PLUS MORE TO TASTE
- 1 TEASPOON FRESHLY GRATED NUTMEG

BACON, PEAR & ROSEMARY STUFFING

YIELD: 6 TO 8 SERVINGS / **ACTIVE TIME:** 15 MINUTES / **TOTAL TIME:** 1 HOUR

Stuffing is so versatile, and when you bring in the likes of smoky bacon, sweet pears, and the deep flavor of rosemary, you get a stuffing that won't stay on the table for long.

1. Preheat the oven to 350°F and put the bread into a greased 3-quart baking dish.

2. In a large saucepan over medium heat, cook the bacon until brown. With a slotted spoon, scoop out the bacon and drain on a paper towel–lined plate. Keep the bacon fat in the pan.

3. Add onion and celery to the bacon fat and cook until soft, about 5 minutes.

4. Add the remainder of the ingredients, bring to a boil, reduce the heat, and simmer for 10 minutes.

5. Pour the stock mixture over the bread in the baking dish and mix to incorporate all the ingredients. Cover the dish and bake for 30 minutes.

6. Uncover the dish and bake for an additional 25 to 30 minutes, or until the top of the mixture begins to brown.

INGREDIENTS:

- 13 CUPS BREAD, CUBED AND DRIED
- 1 LB. BACON, DICED
- 1 ONION, DICED
- 4 CELERY STALKS, CHOPPED
- 1 CUP TURKEY STOCK (SEE PAGE 477)
- 1 TABLESPOON SALT
- 1 TEASPOON PEPPER
- 1 TABLESPOON DRIED SAGE
- 3 SPRIGS FRESH ROSEMARY
- 4 PEARS, PEELED, CORED, AND SLICED

SAUSAGE, APPLE & PECAN STUFFING

YIELD: 6 TO 8 SERVINGS / **ACTIVE TIME:** 15 MINUTES / **TOTAL TIME:** 1 HOUR

This recipe delivers a sweet spin on a savory stuffing. The pecans complement the apples, and it's all rounded out with the beautiful flavor of smoky sausage.

1. Preheat the oven to 350°F.

2. In a large saucepan over medium heat, brown the sausage. Remove the sausage with a slotted spoon and set aside.

3. Cook the onion in the sausage fat until it starts to brown, about 10 minutes.

4. Add the stock to the pan along with the salt, pepper, sage, and browned sausage. Bring to a boil and then reduce heat and simmer for 10 minutes.

5. Pour the stock mixture over the bread in the baking dish.

6. Peel, core, and dice the apples and add them to the bread mixture, stirring to incorporate.

7. Top stuffing with pecans, cover the dish, and bake for 30 minutes.

8. Uncover the dish and bake for an additional 25 to 30 minutes, or until the top of the mixture begins to brown.

INGREDIENTS:

- 1 LB. SAUSAGE, CASINGS REMOVED
- 13 CUPS BREAD, CUBED AND DRIED
- 1 ONION, SLICED
- 2 CELERY STALKS, CHOPPED
- 1 CUP TURKEY STOCK (SEE PAGE 477)
- 1 TABLESPOON SALT
- 1 TEASPOON PEPPER
- 1 TABLESPOON DRIED SAGE
- 4 APPLES (GRANNY SMITH OR CORTLAND)
- 6 OZ. PECANS, CHOPPED

PORK GYOZA

YIELD: 10 SERVINGS / **ACTIVE TIME:** 10 MINUTES / **TOTAL TIME:** 40 MINUTES

These dumplings are filled with tender, flavored pork. They're great to serve at a party.

1. In a bowl, combine the pork, ginger, garlic, scallion, sesame oil, soy sauce, sugar, pepper, and salt and mix well.

2. Fill a small dish with water; this will be used to help seal the dumpling wrappers.

3. Place a wrapper in the palm of your hand. Use the index finger on your other hand to run a little water around the edges of the wrapper.

4. Place 3 teaspoons of pork mixture in the middle of the wrapper, fold it in half, bringing the edges almost together so it looks like a little taco shape. Press the bottom corner on one side together with your thumb and index finger and press together firmly. Now use the index finger and thumb to fold a small pleat over the thumb that is holding the seal closed. Move that thumb on top of the pleat. Repeat until fully sealed and then repeat until all of the dumplings have been filled.

5. Place dumplings in a steaming basket over simmering water, pleated edge pointing up, and steam for 5 minutes.

INGREDIENTS:

- ½ LB. GROUND PORK
- 1 TEASPOON GRATED GINGER
- 1 LARGE GARLIC CLOVE, MINCED
- 2 TABLESPOONS FINELY CHOPPED SCALLION
- ½ TEASPOON SESAME OIL
- 1 TEASPOON LIGHT SOY SAUCE
- ½ TEASPOON SUGAR
- ½ TEASPOON WHITE PEPPER
- ½ TEASPOON SALT
- 22 ROUND DUMPLING WRAPPERS
- 2 TABLESPOONS CANOLA OIL
- ⅓ CUP WATER

PORK BELLY & SWEET CORN CHOWDER

YIELD: 6 SERVINGS / ACTIVE TIME: 25 MINUTES / TOTAL TIME: 30 HOURS

This chowder plays off of the sweetness of in-season corn and the tender fattiness of pork belly.

INGREDIENTS:

- 1 LB. PORK BELLY
- 1 TABLESPOON SEA SALT
- 6 TABLESPOONS UNSALTED BUTTER
- 1 LARGE ONION, DICED
- 1 CUP DICED, COOKED BACON
- 2 CELERY STALKS, CHOPPED
- 2 GARLIC CLOVES, MINCED
- ¼ CUP ALL-PURPOSE FLOUR
- 3 CUPS FRESH CORN KERNELS
- 2 LARGE POTATOES
- 3 SPRIGS FRESH THYME
- ½ CUP CREAM
- ½ CUP WHOLE MILK
- SMOKED PAPRIKA, FOR GARNISH

1. The day before you plan on serving this recipe, rub the pork belly with salt, put the meat on a metal rack over a dish, and refrigerate overnight.

2. The next day, pour boiling water over the pork belly. Leave it to cool and drain.

3. Preheat the oven to 300°F.

4. Pat the skin dry, sprinkle with more salt, and roast for 3 hours. After 3 hours, remove the pork from the oven and raise the temperature to 500°F.

5. Using a knife, remove the skin from the pork to make cracklings. Once skin is removed, cover the pork with foil and rest.

6. To make cracklings, place the skin on a baking tray and cook it in the oven for 20 to 25 minutes, or until the skin is puffed and crisp. Set aside.

7. In a large saucepan over medium-high heat, melt butter then add onion, sautéing until slightly golden. Add bacon, celery, and garlic, stirring often for 5 minutes.

8. Add the flour to the saucepan and continue to stir, making sure to fully incorporate the flour into the mixture.

9. Add the corn, potatoes, thyme, cream, and milk and simmer until the potatoes are fork-tender. Season to taste.

10. To serve, ladle soup into a bowl, top with sliced pork belly, and garnish with paprika and the cracklings.

ROASTED CORN & RED PEPPER BISQUE WITH BACON

YIELD: 3 TO 5 SERVINGS / **ACTIVE TIME:** 30 MINUTES / **TOTAL TIME:** 40 MINUTES

At the height of summer in New England, sweet corn and peppers are bountiful, and they pair perfectly with smoky bacon in this divine bisque.

INGREDIENTS:

- ½ LB. BACON, DICED
- 3 CUPS FRESH CORN KERNELS
- 2 TABLESPOONS OLIVE OIL
- SALT AND PEPPER, TO TASTE
- 3 RED PEPPERS
- ½ CUP HEAVY CREAM
- ½ CUP MILK
- 1½ STICKS UNSALTED BUTTER, DIVIDED

1. Preheat the oven to 375°F.

2. Add bacon to the saucepan over medium heat and cook until brown. Using a slotted spoon, set bacon on a paper towel–lined plate.

3. On a baking sheet, spread out the corn in a thin layer, drizzle oil over it, sprinkle with salt, and bake for 12 minutes, or until the corn starts to darken in color.

4. Remove corn and turn the oven up to 425°F. Roast the peppers until they are charred on all sides. Let the peppers cool in a covered bowl. Once cool, remove the charred skin.

5. In a medium pot, combine corn, cream, milk, roasted pepper flesh, and half a stick of butter. Simmer over medium heat for 15 to 20 minutes, stirring to prevent the milk from scalding.

6. Remove the corn mixture from the heat and allow it to cool for 10 minutes.

7. Blend the corn mixture in a blender until smooth.

8. Warm the mixture if it cooled down too much; if not, serve hot. Top with the crumbled bacon.

BACON & PEA SOUP

YIELD: 6 SERVINGS / **ACTIVE TIME:** 30 MINUTES / **TOTAL TIME:** 1 HOUR

Bacon and peas are best friends, and with this soup you are going to find out why.

1. Add the bacon to a frying pan over medium heat and cook until browned. Remove the bacon and drain on a paper towel–lined plate.

2. Add butter to a saucepan over medium-low heat. When it begins to foam, add the onion and sauté until it turns soft and pale, about 6 minutes.

3. Add the peas and stock, bring to a boil, and then lower heat and simmer for 25 minutes.

4. Using an immersion blender, puree the soup and season to taste.

5. Serve in bowls topped with the cooked bacon.

INGREDIENTS:

- ½ LB. BACON, DICED
- 4 TABLESPOONS UNSALTED BUTTER
- 1 ONION, DICED
- 3 CUPS PEAS
- 4 CUPS CHICKEN STOCK (SEE PAGE 478)

BUTTERNUT SQUASH AND CHORIZO BISQUE

YIELD: 4 SERVINGS / **ACTIVE TIME:** 15 MINUTES / **TOTAL TIME:** 1 HOUR AND 30 MINUTES

Sweet and spicy is one of my favorite flavor pairings. The spiciness of the chorizo is really elevated in this smooth and sweet bisque.

1. Preheat the oven to 400°F.

2. Peel the squash, scoop out the seeds, and slice it. Toss squash and onion in oil and salt on a sheet pan and roast for 40 to 45 minutes, or until golden and fork-tender. Keep an eye on the onion to make sure it doesn't burn.

3. In a pot, combine the squash and onion with the remainder of the ingredients, except the butter, and bring to a boil. Once boiling, reduce heat and simmer for 30 minutes.

4. Turn off the burner. Using an immersion blender, blend the bisque in the pot until smooth. Add butter and stir until melted and fully incorporated in the bisque.

INGREDIENTS:

- 1 LARGE BUTTERNUT SQUASH
- 1 ONION, QUARTERED
- 2 TABLESPOONS VEGETABLE OIL
- SALT AND PEPPER, TO TASTE
- ½ LB. CHORIZO, CASING REMOVED
- 2 BAY LEAVES
- 1 CUP CREAM
- 1 CUP CHICKEN STOCK (SEE PAGE 478)
- 1 CUP MILK
- 4 TABLESPOONS UNSALTED BUTTER

JAMBALAYA WITH BACON & ANDOUILLE SAUSAGE

YIELD: 6 SERVINGS / **ACTIVE TIME:** 15 MINUTES / **TOTAL TIME:** 40 MINUTES

Jambalaya is one of my favorite dishes. I particularly like how andouille sausage and bacon add meaty and smoky flavors to this version of this Creole classic.

1. Add oil and bacon to a pot over medium-high heat and cook bacon until browned. Remove and drain on a paper towel–lined plate.

2. Sauté sliced andouille in the same pot as the bacon and cook until browned on both sides. Remove and set aside with the bacon.

3. Lower the heat to medium; add the butter, celery, carrot, peppers, onion, Creole seasoning, and bay leaves, and sauté for 8 minutes, or until the vegetables are soft. Add garlic and sauté for another 2 minutes.

4. Deglaze the pot with the beer and stir in tomato paste until it's coating all of the vegetables.

5. Add tomato, Worcestershire and hot sauce, and rice, stirring until everything is well incorporated.

6. Add the stock and pepper, increase heat to high, and when jambalaya starts to boil lower the heat and simmer, covered, for 15 minutes, stirring occasionally.

7. Return sausage and bacon to the pot, cover, and simmer for an additional 5 minutes.

INGREDIENTS:

- 2 TEASPOONS OLIVE OIL
- ½ LB. BACON, DICED
- 1 LB. ANDOUILLE SAUSAGE, CUT INTO ¾-INCH ROUNDS
- 1 TABLESPOON UNSALTED BUTTER
- 2 CELERY STALKS, CHOPPED
- 1 CARROT, PEELED AND DICED
- ½ GREEN PEPPER, CHOPPED
- ½ RED PEPPER, CHOPPED
- 1 ONION, CHOPPED
- 1 TABLESPOON CREOLE SEASONING
- 2 BAY LEAVES
- 5 GARLIC CLOVES, MINCED
- ½ CUP BEER
- 1 TABLESPOON TOMATO PASTE
- 1 LARGE TOMATO, CHOPPED
- 1 TEASPOON WORCESTERSHIRE SAUCE
- 2 TEASPOONS LOUISIANA HOT SAUCE
- 1 CUP RICE
- 3 CUPS CHICKEN STOCK (SEE PAGE 478)
- 1 TEASPOON PEPPER, DIVIDED

TWICE-COOKED PORK BELLY & BEAN SOUP WITH BROWN BREAD CROUTONS

YIELD: 6 TO 8 SERVINGS / **ACTIVE TIME:** 40 MINUTES / **TOTAL TIME:** 26 HOURS

Braising pork belly and pressing it overnight is the key to the intense amount of flavor locked in to this interpretation of a classic pork and bean soup.

PORK BELLY

1. The day before serving this recipe, preheat the oven to 300°F.

2. In a Dutch oven, combine all pork belly ingredients. Cover and cook for about 3½ hours, or until the pork belly starts to shred.

3. Place the pork belly skin side up in a shallow baking dish along with enough of the cooking liquid to cover half of the pork belly. Strain and reserve extra liquid.

4. Place parchment paper over the pork belly, top with another pan or dish that will press down on the pork belly, and refrigerate for 24 hours.

5. After 24 hours, take out the pork belly and slice to desired portion.

6. In a saucepan over medium heat, sear the portioned pork belly, starting skin side down. (Pork belly tends to pop quite a bit when searing, so be careful.)

CROUTONS

1. Preheat the oven to 400°F.

2. Cut the brown bread scraps into cubes and toss in oil and salt. Spread the cubes on a sheet pan and bake for 10 to 15 minutes, or until crisp.

SOUP

1. The day before serving this recipe, soak the beans overnight in a covered pot.

2. Rinse the soaked beans, add to a large pot, cover the beans with water, salt generously, and cook beans until tender, then drain.

3. In a large pot combine beans, tomato sauce, water, and stock, bring to a boil, and then simmer for 30 minutes.

4. Add brown sugar and apple cider vinegar and simmer for another 20 minutes. Season with salt and pepper, to taste.

5. To serve, ladle soup into a bowl and top with seared pork belly and croutons.

INGREDIENTS:

PORK BELLY
- 4 LBS. PORK BELLY
- 3 BAY LEAVES
- 3 CUPS APPLE CIDER
- 2 CINNAMON STICKS
- 2 TEASPOONS WHOLE CLOVES
- 1 ONION, SLICED
- 2 CUPS CHICKEN STOCK (SEE PAGE 478)
- 3 STAR ANISE
- 1 TABLESPOON PEPPERCORNS
- 2 CARROTS, PEELED AND CHOPPED
- 2 CELERY STALKS, CHOPPED

CROUTONS
- DAY-OLD BREAD
- 2 TABLESPOONS CANOLA OIL
- SALT AND PEPPER, TO TASTE

SOUP
- 1 LB. DRY NAVY BEANS
- SALT AND PEPPER, TO TASTE
- 1 QUART CHICKEN OR PORK STOCK (SEE PAGES 478 OR 481, RESPECTIVELY)
- 1 CUP TOMATO SAUCE
- 1 QUART WATER
- 2 TABLESPOONS BROWN SUGAR
- 2 TABLESPOONS APPLE CIDER VINEGAR

SAUSAGE-STUFFED PEPPERS

YIELD: 4 SERVINGS / **ACTIVE TIME:** 15 MINUTES / **TOTAL TIME:** 50 MINUTES

These stuffed peppers are self-contained meals that are ideal for a weeknight family dinner.

1. Cut off the top ½ inch of peppers and chop the tops into ¼ inch pieces. Discard stems and seeds.

2. Combine stock and rice in a saucepan over medium heat. Cover and cook until liquid is absorbed and rice is nearly tender, 13 to 15 minutes. Set aside.

3. Add sausage to a 12-inch nonstick skillet over medium-high heat. Break up meat into small pieces and cook until browned, 6 to 8 minutes. Using a slotted spoon, transfer sausage to a paper towel–lined plate.

4. Add onions and chopped peppers to the skillet and cook until browned, 8 to 10 minutes. Stir in garlic, oregano, ½ teaspoon salt, ¼ teaspoon pepper, and pepper flakes and cook until fragrant, about 30 seconds. Add tomatoes, bring to boil, and remove from heat. Stir basil into sauce and season to taste. If sauce cools down too much before serving, reheat to serve.

5. Preheat the oven to 350°F.

6. In a bowl, combine 1 cup sauce, sausage, 1 cup Parmesan, and rice.

7. Using a skewer, poke 4 holes in the bottom of each pepper. Fill each pepper with ¼ of the rice mixture and place upright on a sheet pan. Top peppers with remaining Parmesan. Cook in the oven for 25 minutes, or until peppers and rice are tender.

8. Transfer peppers to plate, spoon warm sauce over peppers, and serve.

INGREDIENTS:

- 4 BELL PEPPERS
- 1½ CUPS CHICKEN STOCK (SEE PAGE 478)
- ¾ CUP RICE
- ½ LB. HOT ITALIAN SAUSAGE, CASINGS REMOVED
- 2 ONIONS, DICED
- 6 GARLIC CLOVES, MINCED
- ½ TEASPOON DRIED OREGANO
- SALT AND PEPPER, TO TASTE
- ⅛ TEASPOON RED PEPPER FLAKES
- 1 (28 OZ.) CAN CRUSHED TOMATOES
- 2 TABLESPOONS CHOPPED FRESH BASIL
- 1¼ CUPS GRATED PARMESAN CHEESE

PULLED PORK

YIELD: 6 TO 8 SERVINGS / **ACTIVE TIME:** 40 MINUTES / **TOTAL TIME:** 6 HOURS

This moist, flavor-packed slow cooker pulled pork is my go-to when I don't want to fire up the smoker or get the oven going.

1. Preheat the oven to 300°F.

2. Salt the pork on all sides.

3. Add canola oil to a large pan over medium-high heat. Once the oil is shimmering, sear the pork on all sides until golden.

4. In a Dutch oven, combine the pork with the remainder of the ingredients. Cover and cook for 4 hours, or until the meat is fork-tender.

INGREDIENTS:

- 6–8 LBS. PORK SHOULDER
- 3 TABLESPOONS KOSHER SALT
- 2 TABLESPOONS CANOLA OIL
- 1 QUART CHICKEN STOCK (SEE PAGE 478)
- 2 TABLESPOONS PEPPERCORNS
- 1 TABLESPOON MUSTARD
- ¼ CUP BROWN SUGAR
- 2 TEASPOONS PAPRIKA
- 3 BAY LEAVES
- 1 LARGE ONION, DICED

BRAISED JOWL PULLED PORK

YIELD: 6 SERVINGS / **ACTIVE TIME:** 10 MINUTES / **TOTAL TIME:** 4 HOURS

This recipe for pulled pork uses jowl, a flavorful part of the pig that is typically cheaper than other cuts. People who try this will never know the difference.

1. Preheat the oven to 300°F.

2. Salt the pork on all sides.

3. In a Dutch oven over medium-high heat, sear the pork on all sides until golden.

4. Add the stock, onions, peppercorns, mustard, brown sugar, garlic, bay leaves, and paprika. Cover and bake for 4 hours, or until the meat is fork-tender.

INGREDIENTS:

- 3–4 LBS. PORK JOWL, TRIMMED OF ANY MEMBRANES
- SALT, TO TASTE
- 2 CUPS CHICKEN STOCK (SEE PAGE 478)
- 2 ONIONS, SLICED
- 2 TABLESPOONS PEPPERCORNS
- 1 TABLESPOON MUSTARD
- ¼ CUP BROWN SUGAR
- 3 GARLIC CLOVES, CHOPPED
- 3 BAY LEAVES
- 2 TEASPOONS PAPRIKA

ADOBO PULLED PORK

YIELD: 6 TO 8 SERVINGS / **ACTIVE TIME:** 40 MINUTES / **TOTAL TIME:** 5 HOURS

This spicy version of slow cooker pulled pork is packed with heat and flavor. The beautiful thing about this recipe is how easy it is.

1. Preheat the oven to 300°F.

2. Salt the pork on all sides.

3. Add canola oil to a large pan over medium-high heat. Once the oil is shimmering, sear the pork on all sides until golden.

4. In a Dutch oven, combine the pork with the remainder of the ingredients. Cover and cook for 4 hours, or until the meat is fork-tender.

INGREDIENTS:

- 6–8 LBS. PORK SHOULDER
- SALT, TO TASTE
- 2 TABLESPOONS CANOLA OIL
- 1 QUART CHICKEN STOCK (SEE PAGE 478)
- 1 (4 OZ.) CAN ADOBO CHILIES, WITH SAUCE
- 2 TABLESPOONS PEPPERCORNS
- ¼ CUP BROWN SUGAR
- 2 TEASPOONS PAPRIKA
- 3 BAY LEAVES
- 1 LARGE ONION, DICED

COLA-BRAISED PULLED PORK

YIELD: 6 TO 8 SERVINGS / **ACTIVE TIME:** 40 MINUTES / **TOTAL TIME:** 5 HOURS

Sweet cola takes pulled pork to the next level in this preparation.

1. Preheat the oven to 300°F.

2. Salt the pork on all sides.

3. Add canola oil to a large pan over medium-high heat. Once the oil is shimmering, sear the pork on all sides until golden.

4. In a Dutch oven, combine the pork with the remainder of the ingredients. Cover and cook for 4 hours, or until the meat is fork-tender.

INGREDIENTS:

- 6–8 LBS. PORK SHOULDER
- SALT, TO TASTE
- 2 TABLESPOONS CANOLA OIL
- 3 CUPS COLA
- 2 TABLESPOONS PEPPERCORNS
- 1 TABLESPOON MUSTARD
- ¼ CUP BROWN SUGAR
- 2 TEASPOONS PAPRIKA
- 3 BAY LEAVES
- 1 LARGE ONION, DICED

PULLED PORK WITH BLUE CHEESE POLENTA & ROASTED PEACH HOT SAUCE

YIELD: 6 TO 8 SERVINGS / **ACTIVE TIME:** 40 MINUTES / **TOTAL TIME:** 6 HOURS

Peach season in New England is a special time of year, and we spend countless hours preserving the season in a variety of ways. Overly ripe dropped, or "B" grade, peaches are used to create this sweet, sour, and spicy hot sauce.

PULLED PORK

1. Preheat the oven to 300°F.

2. Salt the pork on all sides.

3. Add canola oil to a large pan over medium-high heat. When oil begins to shimmer, sear the pork on all sides, until it is golden.

4. Place the pork in a Dutch oven along with onion, bay leaves, paprika, brown sugar, peppercorns, stock, and mustard. Cover and bake for 4 hours, or until the meat is fork-tender. When pork is done, let rest for 15 minutes.

5. To serve, spoon the polenta into a bowl, place some of the pulled pork on top of the polenta, and drizzle with hot sauce and additional crumbled blue cheese.

BLUE CHEESE POLENTA

1. In a large pot, combine cornmeal, stock, and water. Bring to a boil and then reduce to a simmer.

2. Keep the cornmeal simmering for 40 minutes to 1 hour, stirring often, until the polenta thickens. Once the polenta is thickened, add the butter and half the blue cheese and stir well. Season with salt and pepper.

ROASTED PEACH HOT SAUCE

1. Preheat the oven to 400°F.

2. Roast the peaches until they start to develop a darker color. (You can also grill the peaches until they begin to soften.)

3. In a saucepan over medium-low heat, combine peaches, vinegar, sugar, garlic, jalapeños, cayenne peppers, and lemon juice and bring to a simmer. Let simmer for 10 minutes.

4. Transfer this mixture to a blender and blend well. Strain and refrigerate.

INGREDIENTS:

PULLED PORK

- 6–8 LBS. PORK SHOULDER
- 2 TABLESPOONS KOSHER SALT
- 1 TABLESPOON CANOLA OIL
- 1 LARGE ONION, DICED
- 3 BAY LEAVES
- 2 TEASPOONS PAPRIKA
- ¼ CUP BROWN SUGAR
- 2 TABLESPOONS PEPPERCORNS
- 1 CUP CHICKEN STOCK (SEE PAGE 478)
- 1 TABLESPOON MUSTARD

BLUE CHEESE POLENTA

- 2 CUPS CORNMEAL
- 3 CUPS CHICKEN STOCK
- 2 CUPS WATER
- 4 TABLESPOONS UNSALTED BUTTER
- 1 CUP CRUMBLED BLUE CHEESE, PLUS MORE FOR GARNISH
- SALT AND PEPPER, TO TASTE

ROASTED PEACH HOT SAUCE

- 8 PEACHES, QUARTERED AND PITTED
- 2 CUPS APPLE CIDER VINEGAR
- ¾ CUP SUGAR
- 3 GARLIC CLOVES, CHOPPED
- 6 JALAPEÑO PEPPERS, SEEDED AND DICED
- 4 CAYENNE PEPPERS, SEEDED AND DICED
- ¼ CUP FRESH LEMON JUICE
- 1 STICK UNSALTED BUTTER

PORK & APPLE CASSEROLE

YIELD: 4 SERVINGS / **ACTIVE TIME:** 30 MINUTES / **TOTAL TIME:** 1 HOUR

Growing up, my mother would make her version of pork and apple casserole for special occasions. She used thin pork chops smothered in apples and baked for an hour. The result: overcooked pork and mushy apples. It was the love that she put in it that made her recipe so special to me. For this recipe, I channel her love into quality ingredients.

1. Preheat the oven to 325°F.

2. In a bowl, mix together apples, cinnamon, nutmeg, sugar, flour, and a pinch of salt. Add to a baking dish or Dutch oven, and then pour in apple cider.

3. In a small bowl, combine herbs with salt and pepper and then season the pork tenderloin, being sure to cover all sides.

4. Put the pork on top of the apple mixture, cover, and bake for 40 minutes, or until a meat thermometer registers 145°F.

5. Let the pork rest for at least 15 minutes. Serve sliced alongside the apples.

INGREDIENTS:

- 8 APPLES, SLICED
- 2 TEASPOONS CINNAMON
- 1 TEASPOON FRESHLY GRATED NURMEG
- ¼ CUP SUGAR
- ¼ CUP ALL-PURPOSE FLOUR
- SALT AND PEPPER, TO TASTE
- ¼ CUP APPLE CIDER
- 1½ LBS. PORK TENDERLOIN
- 2 TABLESPOONS CHOPPED FRESH ROSEMARY LEAVES
- 2 TABLESPOONS CHOPPED FRESH THYME LEAVES

BOILED DINNER

YIELD: 6 TO 8 SERVINGS / **ACTIVE TIME:** 30 MINUTES / **TOTAL TIME:** 2 HOURS AND 45 MINUTES

Another iconic New England dish is the boiled dinner. It is commonly done with either corned beef or ham. For this recipe I use a ham. The leftovers make an amazing sandwich, so be sure to save a little.

1. In a large pot, combine all of the ingredients and cover with water.

2. Bring to a boil. Once boiling, reduce the heat and simmer for 2½ hours.

3. Remove the vegetables and ham, tent with aluminum foil, and allow to cool for 15 minutes.

4. To serve, slice the ham and cut the vegetables to desired size. Whole-grain mustard and horseradish are ideal additions to this meal.

INGREDIENTS:

1	(6 LB.) HAM
1	HEAD CABBAGE, CUT INTO 6 WEDGES
6	CARROTS
6	POTATOES
3	BAY LEAVES
1	TABLESPOON PEPPERCORNS
1	ONION, QUARTERED

PORK BURGERS WITH PEPPER JACK & ARUGULA

YIELD: 4 SERVINGS / ACTIVE TIME: 15 MINUTES / TOTAL TIME: 30 MINUTES

Arugula's peppery flavor and the cheese's hint of spice work so well with this burger.

INGREDIENTS:

- 1 LB. GROUND PORK
- 1 TEASPOON SALT
- ½ TEASPOON BLACK PEPPER
- 1 EGG WHITE
- ⅓ CUP BREAD CRUMBS
- 4 SLICES PEPPER JACK CHEESE
- 3 OZ. ARUGULA

1. Preheat your outdoor grill or cast-iron pan. If you don't have either, a frying pan will do. (And if using a pan, lightly grease it with neutral oil.)

2. In a bowl, combine all the ingredients, except the cheese and arugula, until the mixture is well incorporated.

3. Divide the mixture into four equal parts. Form patties by rolling the mixture into a ball in your hands and pressing it flat to form the patty. The patty should be about 1-inch thick.

4. Turn the grill to medium-high heat; if you're using a pan, set your burner on medium-high. Place patties on the grill or pan. Cover the patties and cook for 6 minutes.

5. Flip the patties and cover again, cooking for an additional 6 to 7 minutes. About 30 seconds before your burgers are done, top each one with a slice of cheese and cover until the cheese has melted and you have the desired doneness.

6. Remove the burger from heat and top with arugula and your choice of condiments.

THICK-CUT PORK CHOPS WITH STONE FRUIT & BULGUR WHEAT

YIELD: 4 SERVINGS / **ACTIVE TIME:** 15 TO 25 MINUTES / **TOTAL TIME:** 40 MINUTES

Pork is the perfect vehicle for stone fruit, as its slightly salty flavor allows the fruit's natural sweetness to shine.

1. Place the stock in a cast-iron Dutch oven and bring to a boil. Place the bulgur wheat, salt, lemon zest, pepper, and olive oil in a bowl and pour the warmed stock over it. Cover tightly with plastic wrap and set aside for 30 minutes.

2. Preheat the oven to 375°F. Wipe out the pot, place 2 tablespoons of the vegetable oil in it, and warm over medium heat. When the oil starts to glisten, place the stone fruit flesh side down in the Dutch oven and sear for about 3 minutes per side. Add the turnip and the cipollini onions and cook for about 2 minutes per side. Remove the mixture from the pot and set aside.

3. Season the pork chops with salt and pepper. Add the remaining vegetable oil and then place the pork chops in the Dutch oven. Cook for about 5 minutes on each side, until a crust starts to form and the centers of the chops are 140 to 145°F. Remove the pork chops and set aside.

4. Chop the onions. Add the turnip and the onions to the bulgur wheat. Fluff with a fork and place the mixture in the Dutch oven. Arrange the fruit around the edge of the pot and place it in the oven. Cook for 5 to 7 minutes, until warmed through. Remove, serve with the pork chops, and garnish with the thyme.

INGREDIENTS:

- 2 CUPS CHICKEN STOCK (SEE PAGES 478)
- 1 CUP BULGUR WHEAT
- 1 TEASPOON SALT, PLUS MORE TO TASTE
- ZEST OF 1 LEMON
- ¼ TEASPOON CRACKED BLACK PEPPER, PLUS MORE TO TASTE
- 1 TABLESPOON OLIVE OIL
- ¼ CUP VEGETABLE OIL
- 4 PIECES PREFERRED STONE FRUIT, PITTED AND QUARTERED
- 1 TURNIP, PEELED AND MINCED
- 4-6 CIPOLLINI ONIONS, PEELED AND HALVED LENGTHWISE
- 4 (1-INCH THICK) PORK CHOPS
- 2 TABLESPOONS FRESH THYME LEAVES, FOR GARNISH

SAUSAGE RAGÙ

YIELD: 6 SERVINGS / **ACTIVE TIME:** 10 MINUTES / **TOTAL TIME:** 4 TO 6 HOURS

This recipe is close to my heart. As a kid, I loved it when my mother cooked this dish. It taught me that good food can take some time, but the result is very much worth it. I love to use hot Italian sausage for this but if you don't like heat, you can dial it back with sweet Italian sausage.

1. Add the sausage to a large pan over medium-high heat and sear until the sausage is cooked through. Set aside.

2. Add the oil and onion to a large pot over medium heat and sauté until the onion becomes translucent. Add the peppers and continue to cook for 3 to 4 minutes.

3. Add the tomatoes, basil, pepper flakes, Italian seasoning, and sausage, followed by a cup of water. Bring to a boil and then lower heat. Partly cover and simmer for 4 to 6 hours. Add water as needed so the sauce doesn't burn.

4. Serve with pasta of your choice.

INGREDIENTS:

- 2 LBS. ITALIAN SAUSAGE OF YOUR CHOOSING
- 2 TABLESPOONS OLIVE OIL
- 1 SWEET ONION, DICED
- 2 GREEN PEPPERS, DICED
- 2 (14 OZ.) CANS SAN MARZANO TOMATOES
- 1 CUP CHOPPED FRESH BASIL
- 2 TEASPOONS RED PEPPER FLAKES
- 1 TABLESPOON ITALIAN SEASONING
- 1 CUP WATER
- SALT AND PEPPER, TO TASTE

TO SERVE

1 LB. COOKED PASTA (SEE PAGES 725-743)

ASIAN MEATBALLS

YIELD: 10 SERVINGS / **ACTIVE TIME:** 15 MINUTES / **TOTAL TIME:** 35 MINUTES

These saucy and delicious meatballs are a perfect appetizer, and super easy to make.

1. Preheat the oven to 400°F.

2. In a large bowl, combine meat, 2 teaspoons sesame oil, panko, ½ teaspoon ginger, eggs, garlic, and scallions.

3. Using your hands, shape the meat into balls and place on a greased sheet pan. Bake for 10 to 12 minutes, or until meatballs are golden on the outside and cooked throughout.

4. As the meatballs are baking, whisk together the remaining ingredients until well blended.

5. Once meatballs have finished cooking, place them in a serving dish, pour the sauce over the meatballs, and gently stir them until coated.

HOISIN SAUCE

1. Add canola oil to a saucepan over medium heat. When the oil begins to shimmer add the garlic and cook for about 2 minutes, stirring often.

2. Add soy sauce, honey, vinegar, tahini, and sriracha and stir to incorporate. Cook until sauce is smooth, about 5 minutes. Set aside.

INGREDIENTS:

- 2 LBS. GROUND PORK
- 1 TABLESPOON SESAME OIL, DIVIDED
- ¾ CUP PANKO
- 1½ TEASPOONS GRATED GINGER, DIVIDED
- 2 EGGS
- 1 TABLESPOON MINCED GARLIC
- ½ CUP THINLY SLICED SCALLIONS
- ⅔ CUP HOISIN SAUCE (SEE RECIPE)
- ¼ CUP RICE VINEGAR
- 2 GARLIC CLOVES, MINCED
- 1½ TABLESPOONS SOY SAUCE

HOISIN SAUCE

- 4 GARLIC CLOVES, MINCED
- 2 TABLESPOONS CANOLA OIL
- ¼ CUP SOY SAUCE
- 3 TABLESPOONS HONEY
- 2 TABLESPOONS WHITE VINEGAR
- 2 TABLESPOONS TAHINI
- 2 TEASPOONS SRIRACHA

PORK TERRINE

YIELD: 10 TO 15 SERVINGS / **ACTIVE TIME:** 20 MINUTES / **TOTAL TIME:** 4 HOURS

This terrine, which is essentially a pâté, is hearty and homey.

1. Preheat the oven to 300°F.

2. In a Dutch oven, combine all ingredients and cook for 4 hours, or until the meat is fork-tender.

3. After the pork is done, let it cool and remove it from the Dutch oven. Discard the bay leaves and reserve the cooking liquid.

4. Once cooled, shred the pork and place it in a blender along with ½ to ¾ cup of the leftover cooking liquid and blend until a smooth paste forms. Season with salt and pepper, to taste.

5. Scoop the blended pork into a loaf pan that has been greased and lined with plastic wrap. Distribute the mixture evenly and pour some of the extra fat over it.

6. Refrigerate overnight before serving.

INGREDIENTS:

- 3–5 LBS. PORK BUTT
- 3 ONIONS, SLICED
- 2 TEASPOONS GROUND CLOVES
- 1 TABLESPOON SALT, PLUS MORE TO TASTE
- 4 BAY LEAVES
- 2 TEASPOONS BLACK PEPPER, PLUS MORE TO TASTE
- 1 TEASPOON FRESHLY GRATED NUTMEG

HEADCHEESE

YIELD: 8 TO 10 SERVINGS / **ACTIVE TIME:** 45 MINUTES / **TOTAL TIME:** 3 DAYS

Headcheese might strike you as next-level food, but it is the tasty essence of true nose-to-tail cooking. Ask a butcher or farmer for a pig head. Most are happy to sell it as it is not a popular part of the animal, but it is a flavor-packed part. Try this—you will love it!

INGREDIENTS:

- 1 GALLON MEAT BRINE (SEE PAGE 444)
- 1 PIG HEAD
- 4 TROTTERS AND/OR HAM HOCKS
- 3 ONIONS
- 4 CARROTS, CHOPPED
- 4 CELERY STALKS, CHOPPED
- 2 LEEKS, TRIMMED, RINSED WELL, AND CHOPPED
- 3 SPRIGS FRESH THYME
- 10 BAY LEAVES
- ¼ CUP PEPPERCORNS

1. Pour the brine into a large pot with a lid; place the pig head in the brine, cover, and refrigerate for 2 days.

2. After it has rested in the brine, place the head into a large pot with the remainder of the ingredients. Bring the mixture to a boil, reduce heat, and let simmer uncovered for 12 to 14 hours.

3. Once the pig head has been cooked, let the mixture cool down until it is lukewarm; this takes several hours.

4. Once cool, set the pig head on the counter, next to a large bowl. The meat will be fall-off-the-bone tender, so be careful moving the head.

5. Pick through the meat, placing the tender, fibrous pieces in the bowl; discard any hard and chewy pieces. Take your time with this process to make sure there are no hard bits.

6. As you pick through the meat, reduce the stock by maintaining a rolling boil. Reduce by at least half.

7. Put the meat into a greased loaf tin, compacting it well, and cover with reduced stock, about 1 to 2 cups. Cover and refrigerate overnight.

8. The next day the mixture will be solid. Slice to serve.

CHINESE EGGPLANT WITH SAUSAGE

YIELD: 4 SERVINGS / **ACTIVE TIME:** 15 MINUTES / **TOTAL TIME:** 1 HOUR

This recipe reminds me of one of my favorite dishes from the Baldwin Bar at Sichuan Garden in Woburn, Massachusetts. It is packed with flavor and I order it every time I go there. The addition of sausage gives this dish a depth and flavor you will love.

1. Place the eggplant slices in a bowl, add salt, cover with water, and let soak for 30 minutes. Drain slices on a paper towel–lined plate.

2. In a small bowl, combine soy sauces, chili garlic sauce, rice wine, vinegar, sugar, and stock and mix well. Set aside.

3. Add oil to a wok over medium-high heat. When the oil begins to shimmer, add the pork sausage and cook for about 3 minutes, using a spatula to break up the meat into small pieces.

4. Add the eggplant and red pepper, turn down the heat to medium-low, cover, and simmer for about 10 minutes, or until the eggplant is tender.

5. Add the sauce and mix it all together. Bring up the heat to medium-high and stir until the sauce coats everything.

6. Serve, garnishing with scallion and cilantro if desired.

INGREDIENTS:

- 2 CHINESE EGGPLANTS, SLICED ON A BIAS, ¾-INCH THICK
- 1 TABLESPOON SALT
- 1½ TABLESPOONS DARK SOY SAUCE
- 1½ TABLESPOONS LIGHT SOY SAUCE
- 1 TABLESPOON CHILI GARLIC SAUCE
- 1 TABLESPOON CHINESE RICE WINE (SHAOXING WINE)
- 1 TABLESPOON RICE VINEGAR
- ½ TEASPOON SUGAR
- ⅓ CUP CHICKEN STOCK (SEE PAGE 478)
- 1 TABLESPOON CANOLA OIL
- ½ LB. GROUND PORK SAUSAGE
- 1 RED PEPPER, CHOPPED
- 1 SCALLION, CHOPPED (OPTIONAL), FOR GARNISH
- CILANTRO, TO TASTE (OPTIONAL), FOR GARNISH

POULTRY

*C*hicken is the most popular meat in America, but to stop at chicken when we talk about poultry would be a disservice to your taste buds. From turkey to ducks and pheasant, poultry is more than just chicken. You are about to learn the secrets to making some amazing poultry dishes that will change the way you view poultry.

ROASTED WHOLE CHICKEN

YIELD: 4 SERVINGS / **ACTIVE TIME:** 10 MINUTES / **TOTAL TIME:** 1 HOUR

I have fond childhood memories of Sundays at my house and the smell of roasted chicken perfuming everything. This recipe always takes me back to those times. It is simple but teaches you the proper technique for roasting a whole chicken.

1. Preheat your oven to 450°F.

2. Rinse off the chicken, both inside and outside, then pat it dry with paper towels. This step is important, because it ensures that the chicken roasts and doesn't steam.

3. In a small bowl, combine at least a tablespoon each of salt and pepper. Apply the mixture to the bird, both inside and outside.

4. Truss the bird: place the chicken in front of you with the legs pointing at you; place the kitchen twine under the chicken, by the wings; loop the string over the wings to the front of the chicken, then back under to complete the loop; next, take the string and come over the top of the legs then loop back under the legs and back to the top of the legs. The string should now be at the backside of the chicken. Tie a knot—the chicken will shape up and look picture-perfect. Tie the legs together.

5. Place the chicken in a roasting pan and roast for 45 to 55 minutes, until a meat thermometer inserted in the breast registers 165°F.

6. When done, baste the chicken with the juices from the pan and let rest for about 15 minutes before carving and serving.

INGREDIENTS:

1   (3 LB.) CHICKEN

     SALT AND PEPPER, TO TASTE

SWEET & SMOKY ROASTED CHICKEN

YIELD: 6 SERVINGS / **ACTIVE TIME:** 10 MINUTES / **TOTAL TIME:** 25 HOURS

This dry brine combines sweet and smoky flavors for a tasty new way to enjoy chicken. This recipe prepares enough for 1 whole chicken.

1. Remove the innards from the chicken, reserving for gravy if you choose. Rinse the chicken under cold water and pat dry with paper towels.

2. In a bowl, combine all of the other ingredients and mix thoroughly.

3. Rub the mixture under the chicken skin, in the cavity, and all over the outside of the bird. Cover with foil and refrigerate for 24 hours.

4. Follow steps for Roasted Whole Chicken (see page 244).

INGREDIENTS:

- 1 (3–4 LB.) WHOLE CHICKEN
- 2 TABLESPOONS SALT
- 2 TABLESPOONS PEPPER
- ¼ CUP BROWN SUGAR
- 3 TABLESPOONS PAPRIKA
- 1 TABLESPOON GARLIC POWDER
- 1 TABLESPOON ONION POWDER
- 2 TEASPOONS LEMON ZEST

MARINATED CHICKEN WITH CHIPOTLE CAULIFLOWER

YIELD: 4 TO 5 SERVINGS / **ACTIVE TIME:** 1 HOUR AND 30 MINUTES / **TOTAL TIME:** 9 HOURS

For the marinade, be sure to let the bird soak for up to 6 hours—the longer the better. As for the chipotle cauliflower, if you would like to add a smoky flavor, throw some soaked wood chips over the coals and grill with the lid covered.

1. In a large bowl, combine the olive oil, onion, parsley, rosemary, garlic, and vinegar and mix thoroughly. Add the chicken skin side down into the marinade; keep in mind that the chicken will not be fully submerged. Let soak for 4 to 6 hours, turning the chicken with 1 hour remaining.

2. Remove the chicken from the marinade and season with pepper and salt. Let the chicken stand at room temperature for 30 minutes to 1 hour. A half-hour before grilling, prepare your gas or charcoal grill to medium heat.

3. While waiting, mix the cauliflower florets, olive oil, and lime juice in a medium bowl. Stir in the remaining ingredients. Transfer into a small frying pan. Set aside.

4. When the grill is ready, at about 400°F with the coals lightly covered with ash, place the chicken on the grill, skin side up. Cover the grill and cook for about 40 minutes. Before flipping, brush the top of the chicken with the remaining tablespoon of olive oil. Turn and cook for about 15 more minutes until the skin is crisp and a meat thermometer inserted into the thickest part of the thigh reads 165°F.

5. Remove the chicken, transfer to a large cutting board, and let stand for 15 minutes.

6. Position the aluminum pan of cauliflower on the grill and cover with lid. Cook for 8 to 9 minutes, until the florets are crisp with the chipotle powder. Remove from grill and serve alongside the chicken.

INGREDIENTS:

CHICKEN

- ½ CUP OLIVE OIL
- ½ SMALL WHITE ONION, FINELY CHOPPED
- ¼ CUP FLAT LEAF PARSLEY, FINELY CHOPPED
- 2 SPRIGS FRESH ROSEMARY, LEAVES REMOVED AND MINCED
- 2 GARLIC CLOVES, CRUSHED
- 2 TABLESPOONS RED WINE VINEGAR
- 1 (4–5 LB.) CHICKEN
- COARSELY GROUND BLACK PEPPER, TO TASTE
- SEA SALT, TO TASTE

CAULIFLOWER

- 2 LARGE HEADS CAULIFLOWER, CUT INTO FLORETS
- ¼ CUP OLIVE OIL
- ½ LIME, JUICED
- 3 GARLIC CLOVES, DICED
- 1 TABLESPOON CHIPOTLE POWDER
- 2 TEASPOONS PAPRIKA
- 2 TABLESPOONS BASIL LEAVES, SLICED
- COARSELY GROUND BLACK PEPPER, TO TASTE
- SEA SALT, TO TASTE

TRADITIONAL OVEN-ROASTED TURKEY

YIELD: 5 TO 10 SERVINGS / **ACTIVE TIME:** 30 MINUTES / **TOTAL TIME:** 3 TO 5 HOURS

This is my tried-and-true Thanksgiving turkey recipe. This method brings some simple aromatics to the bird without masking any of its succulent flavor.

INGREDIENTS:

- 1 ONION, HALVED
- 1 LEMON, HALVED
- 1 (10–20 LB.) WHOLE TURKEY, GIBLETS AND INNARDS REMOVED
- 6 FRESH SAGE LEAVES
- 5 TABLESPOONS UNSALTED BUTTER
- 2 TEASPOONS SALT

1. Preheat the oven to 450°F.

2. In a cast-iron skillet over high heat, sear the onion and lemon halves face down until they are dark but not black.

3. Remove the giblets from the cavity of the turkey. Save them for gravy (see page 471). Rinse the cavity and outside of the turkey with cold water. Pat dry the entire bird, including the cavity.

4. Place the seared onion and lemon along with sage leaves in the cavity of the turkey.

5. Slice a small slit in the skin of the breast, being careful not to puncture the meat. Using your index finger, carefully separate the skin from the meat. Once you have separated the skin, spread the butter in between the skin and meat.

6. Salt the outside of the turkey, making sure to rub the salt all over.

7. Set the turkey in a roasting pan and tent it with foil or a lid. Place the covered turkey in the oven for 30 minutes.

8. After 30 minutes, turn the oven down to 350°F. Cook for 2½ to 4½ hours, depending on the size of the bird. Cook until a meat thermometer registers 155°F when inserted in the breast.

9. Shortly before the breast reaches the desired temperature, uncover the turkey and turn the oven to 450°F once again to brown the skin. This should take around 10 minutes. Be sure to watch the bird so it doesn't overcook or burn.

TIP: I never stuff the bird. Stuffing a bird is not only risky because of pathogens, but also you run a risk of having dry meat. If you want the classic presentation of stuffing in the carcass, do it after the bird is done.

HOW TO THAW A TURKEY

Remember to keep the turkey inside its packaging while thawing.

The safest recommended method to properly thaw a turkey is over a few days in the refrigerator; it will take about 24 hours per 5 pounds of turkey to thaw.

The second method is thawing the turkey under cold water, which can take anywhere from 20 to 30 minutes per pound of turkey. With this method you need to be sure to rotate and flip the turkey and change out the water every 30 minutes to make sure that the whole turkey remains cold.

TRUSSING A TURKEY

One of the most important parts of cooking a whole turkey is learning to properly truss it. Trussing a whole bird helps it maintain shape and, more importantly, helps it cook evenly. This task may seem overwhelming, but it is very simple.

1. Cut a piece of kitchen twine 52 inches long.

2. Pat dry the bird and place it in front of you with the legs pointed toward you.

3. At the center point of the piece of twine, wrap the twine around the neck of the turkey and make a small knot.

4. Bring the twine under the turkey to the wings, tuck the wings under the turkey, and then wrap the twine around the wings.

5. Take the twine and pull it tightly under the breast near the cavity of the bird and make a tight knot.

6. Bring the legs together and tie them tight.

7. Cut any extra string.

HOW TO CARVE A TURKEY

Be sure to rest the turkey for 20 to 30 minutes after it has been removed from the oven before carving to ensure that the juices remain trapped in the meat and the turkey will stay moist. Reserve the pan drippings for your gravy.

1. Begin by removing any twine that you may have used to truss the turkey.

2. Carefully transfer the turkey to a carving board. Ensure that you have a stable surface to work on while carving and make sure that the knife you are using is sharpened.

3. Start by removing the thighs. Place your knife on the area of skin that joins the drumstick and the breast. Slice through that connective area until you reach the joint. From there, press down on the thigh meat until you hear a click where the joint has come out of the socket. Next, continue to use the knife to finish cutting through the bottom side of the thigh. Do this same process for both legs.

4. You may want to remove the wishbone before carving the breast meat, which makes it easier to carve, but it is not a necessary step. To do so, find the wishbone in the front of the breast and slide your fingers in and pull to remove.

5. To remove the breast meat, place your knife on either side of the breastplate and slice along the bone, keeping your knife as close to the bone as possible to make sure you get the maximum yield. While slicing down, use your opposite hand to help pull the meat away from the carcass. Repeat this step to the opposite side to remove the second breast. Lay each breast flat and locate the joint connecting the wing to the breast. Carve around that joint to remove and transfer to the serving platter.

6. Once the wings are removed, slice your turkey with the breast lying horizontally—be careful not to tear the skin off.

SPATCHCOCK TURKEY

YIELD: 5 TO 10 SERVINGS / **ACTIVE TIME:** 15 MINUTES / **TOTAL TIME:** 1 HOUR AND 30 MINUTES

Spatchcock is a technique that solves the biggest problem with cooking a turkey. Typically, the legs and the breast of a turkey have two different ideal temperatures. The spatchcock method solves this by allowing the bird to be flat, which means it cooks much more evenly. If you want to try this, a good pair of kitchen shears is a must.

INGREDIENTS:

- 2 TEASPOONS SALT
- 2 TEASPOONS PEPPER
- 1 TABLESPOON GARLIC POWDER
- 1 TABLESPOON ONION POWDER
- 1 (10-20 LB.) WHOLE TURKEY, GIBLETS AND INNARDS REMOVED
- 5 TABLESPOONS UNSALTED BUTTER, MELTED

1. Preheat the oven to 450°F. In a small bowl, mix together salt, pepper, garlic powder, and onion powder.

2. Place turkey breast side down on a cutting board. Using kitchen shears, cut out the backbone of the turkey.

3. Flip the turkey back over so the breast side is up and set it on a wire rack placed inside a roasting pan. Push down on the middle of the bird to flatten it as much as possible.

4. Rub melted butter all over the breast side of the turkey.

5. Rub the salt mixture all over the breast side of the turkey.

6. Cook until a meat thermometer registers 155°F when inserted in the breast and the legs are at least 165°F, typically about 1 hour and 15 minutes to 1 hour and 25 minutes.

SMOKED WHOLE TURKEY

YIELD: 5 TO 10 SERVINGS / **ACTIVE TIME:** 30 MINUTES / **TOTAL TIME:** 3½ TO 5 HOURS

Smoking a whole turkey might seem like a big task, but it is one of the tastiest ways to enjoy turkey. There are two major schools of thought when smoking turkey: low temperature (235 to 250°F) for about 30 minutes per pound or higher temperature, which tends to yield a crispier skin. If you want to go with the lower temperature, I recommend you finish the bird in an oven at 450°F for 7 to 10 minutes to crisp the skin. For wood, I like applewood; it gives the turkey a sweet, smoky flavor. Cherry and hickory also work.

1. Soak wood in water for 1 hour before using; this allows the smoke to dissipate slower.

2. Preheat the smoker to 300°F.

3. Pat dry the inside and outside of the turkey and rub the turkey with melted butter and salt, or use your favorite rub (see pages 458 to 467).

4. Pour 3 cups of water into the smoker's steam tray.

5. Place an aluminum drip pan on the grate below your bird to catch the drippings.

6. Set the turkey on a rack above your drip pan.

7. Add 2 chunks of wood, approximately 8 oz., to the smoker and close the smoker. Smoke for about 4 hours, making sure to add water to the steam pan and chunks of wood as needed. (After 2 or 3 hours I stop adding wood; at this point the smoke is sufficient and typically doesn't require more smoke for the rest of the cooking time.)

8. Cook until a meat thermometer registers 165°F when inserted in the breast. If the turkey starts to get too dark before it is done, wrap it loosely in aluminum foil to prevent further coloring.

9. Remove the turkey from the smoker, wrap it with aluminum foil, and let it rest for 20 minutes.

TIP: Use apple juice in place of the water to add a sweeter flavor to your turkey.

INGREDIENTS:

- APPLEWOOD CHIPS
- 1 (10 TO 20 LB.) WHOLE TURKEY, GIBLETS AND INNARDS REMOVED
- 5 TABLESPOONS UNSALTED BUTTER, MELTED
- 3 TABLESPOONS SALT

DEEP-FRIED TURKEY

YIELD: 5 TO 10 SERVINGS / **ACTIVE TIME:** 1 HOUR / **TOTAL TIME:** 2 HOURS AND 30 MINUTES

Deep-frying a turkey yields both incredibly crispy skin and juicy, tender meat. This method is quick but requires a lot of caution and cleanup.

1. Pat dry the turkey inside and out to get it as dry as possible. Salt both the cavity and outside of the turkey, or use your favorite rub (see pages 458 to 467 for options).

2. Heat the oil in your turkey fryer to 350°F. Make sure you leave plenty of room for the turkey so the oil doesn't spill over when you put the turkey in the fryer.

3. Place the turkey in the fryer basket and lower the basket into the fryer, fully submerging the turkey in the oil.

4. Cook for 3 minutes and 30 seconds per pound, being sure to keep the oil at 350°F.

5. Carefully remove the basket from the oil and drain the turkey. Insert a meat thermometer into the leg. If it registers 180°F, the turkey is done.

6. Drain the turkey on a paper towel–lined platter and let rest for at least 20 minutes.

7. Carve to serve.

INGREDIENTS:

- 1 (10–20 LB.) WHOLE TURKEY, GIBLETS AND INNARDS REMOVED
- 2 TEASPOONS SALT
- 3 GALLONS PEANUT OIL

Deep-frying a turkey is very dangerous. Please take every precaution if you decide to cook a turkey using this method. It is important to highlight a couple of critical precautions to start:

1. Only deep-fry a turkey outside.
2. Do not allow children or pets to be around the fryer at any time.
3. Use a proper turkey fryer and follow the instructions.
4. Never leave the fryer unattended.
5. Never try to fry a frozen turkey.

GRILLED WHOLE TURKEY

YIELD: 5 TO 10 SERVINGS / **ACTIVE TIME:** 15 MINUTES /
TOTAL TIME: 3 TO 4 HOURS (NOT INCLUDING BRINE; SEE PAGE 266 FOR MORE ABOUT BRINING)

Don't overlook grilling turkey on Thanksgiving. It adds a nice smoky flavor to the bird and frees up the oven for all the side dishes. Whenever I grill a whole turkey I make sure to use a brine and rub to give the turkey an extra kick of seasoning.

1. Pat dry the turkey inside and out. Massage the rub both inside the cavity and on the outside of the bird. Stuff the turkey with aromatics of your choosing.

2. Tuck the wing tips under the breast of the turkey and truss the legs together (see page 252).

3. Heat your grill on high for 10 minutes.

4. Place the turkey in a roasting pan with 2 cups of water. When it is time to grill the turkey, turn off one of the burners and place the pan containing the turkey over the burner that is still on. Close the grill and maintain a temperature of 450°F. Cook until a meat thermometer inserted in a leg registers 170°F.

5. Let rest for at least 20 minutes, then carve to serve.

INGREDIENTS:

- 1 (10–20 LB.) WHOLE TURKEY, GIBLETS AND INNARDS REMOVED
- FAVORITE RUB RECIPE (SEE PAGES 458 TO 467)
- AROMATICS OF YOUR CHOOSING, LIKE ONIONS, HERBS, AND LEMONS

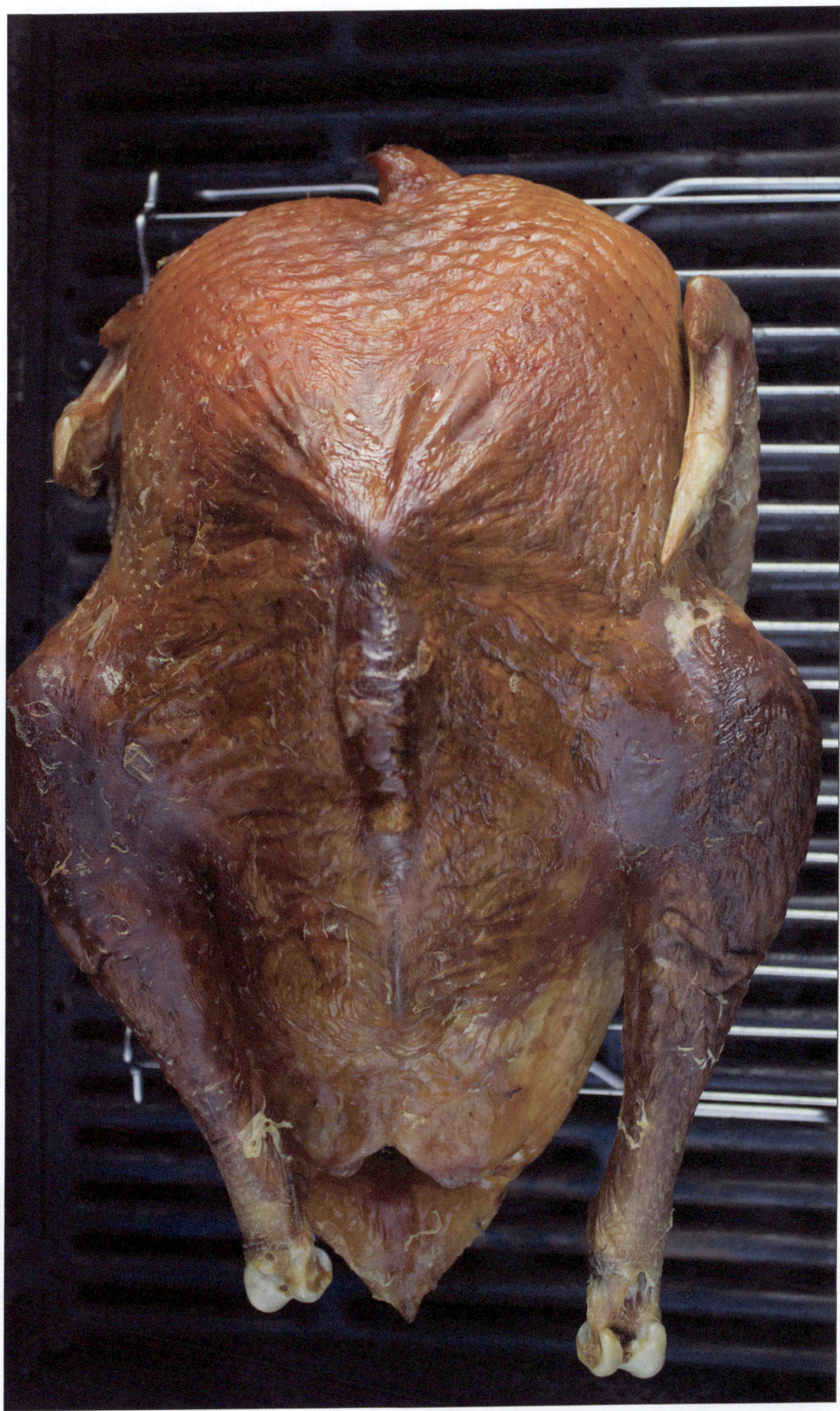

SPATCHCOCK GRILLED TURKEY

YIELD: 5 TO 10 SERVINGS / **ACTIVE TIME:** 25 MINUTES / **TOTAL TIME:** 8 TO 14 HOURS

So you need your turkey done fast? Enter the spatchcock grilled turkey. It takes about 90 minutes to cook a 12 to 14 pound bird using this method, so estimate about 8 minutes per pound.

1. Remove the backbone as you do for the spatchcock method (see page 255).

2. Massage the rub both inside the cavity and on the outside of the bird, then place the turkey on a wire rack, uncovered, and refrigerate for 6 to 12 hours.

3. Once the turkey is done air-drying in the refrigerator, heat your grill on high for 10 minutes.

4. Place the turkey in a roasting pan with 2 cups of water and cover with aluminum foil. When it is time to grill the turkey, turn off one of the burners and place the pan containing the turkey over the burner that is still on. Close the grill and maintain a temperature of 350°F.

5. Halfway through the cooking time, which depends on the size of the turkey, remove the foil and continue to cook until a meat thermometer inserted in the thickest part of a leg registers 165°F.

6. Let rest for at least 20 minutes, then carve to serve.

INGREDIENTS:

1 (10–20 LB.) WHOLE TURKEY, GIBLETS AND INNARDS REMOVED

FAVORITE RUB RECIPE (SEE PAGES 458 TO 467)

SOUS VIDE TURKEY

YIELD: 5 TO 10 SERVINGS / **ACTIVE TIME:** 50 MINUTES / **TOTAL TIME:** 24 HOURS

Sous vide is a method of cooking that translates to "under vacuum." It consists of cooking at a very precise temperature. Food is typically put into a bag or can and cooked underwater. This yields food that is juicy, tender, and cooked precisely.

This method solves one of the most difficult problems with cooking turkey. Typically, the legs should be cooked to 180°F and the breasts are done at 150°F. This discrepancy can lead to overcooked, dry breast meat. This is one of the reasons that turkey has a bad reputation and isn't more widely cooked beyond the holidays. Once you sous vide a turkey you won't ever look back. Trust me: you need to try this.

We are going to cook the legs and the breasts separately for this recipe. Make sure your turkey is broken down, and you will want to brine your turkey and have it ready to go the day before you want to eat it. Typical cooking time for the legs is 24 hours at 131°F. If you raise the temperature to 168°F, it will reduce the cooking time for the legs to 7 hours, but the method explained below is perfect and yields the best flavor.

---

1. The day before you want to serve this turkey, bring your immersion circulator to 150°F.

2. Coat the legs and breasts with salt, or your favorite rub, and then put the legs in one vacuum bag and the breasts in another vacuum bag.

3. Place the bag with the legs into the sous vide cooker, making sure to get all the air out as you submerge the bag in the water. Place the bag with the breasts in the refrigerator.

4. After 12 hours of cooking the legs, lower the temperature to 133°F and add the bag of turkey breasts to the water with the legs, making sure to get all the air out as you submerge the bag in the water. Cook for 12 more hours.

5. Thirty minutes before you are ready to eat, remove the bags from the sous vide cooker. Take out the turkey, pat dry the legs and breasts, and salt them.

6. Add oil to a large saucepan over medium-high heat. When the oil begins to shimmer, sear the turkey pieces, skin side down, for 3 to 5 minutes. Turn over the turkey and cook for an additional 3 minutes. Work in batches if necessary to avoid overcrowding the pan.

INGREDIENTS:

- 1 (10–20 LB.) WHOLE TURKEY, BROKEN DOWN INTO LEGS AND BREASTS
- SALT, TO TASTE
- FAVORITE RUB RECIPE (SEE PAGES 458 TO 467)
- 3 TABLESPOONS CANOLA OIL

BEER-BRINED TURKEY

YIELD: 10 SERVINGS / **ACTIVE TIME:** 45 MINUTES / **TOTAL TIME:** 24 HOURS

Hops, barley, and wheat infused into turkey. This wet beer brine is a perfect way to add a subtle flavor to turkey without overpowering it. I like to use a dark beer to bring out a rich flavor and beautiful color in the bird.

1. Remove the innards from the turkey, reserving for gravy if you choose. Rinse the turkey under cold water and pat dry with paper towels.

2. In a large pot, combine all of the remaining ingredients and stir to combine. Place the turkey in the brine, adding water if needed so that the turkey is fully submerged. Cover and refrigerate for 24 hours.

3. Follow steps for Traditional Oven-Roasted Turkey (see page 250).

INGREDIENTS:

- 1 (10–20 LB.) WHOLE TURKEY
- 3 (12 OZ.) BOTTLES OF DARK BEER
- ½ CUP SALT
- 1 CUP DARK BROWN SUGAR
- 4 BAY LEAVES
- 3 SPRIGS FRESH THYME
- 1 ONION, QUARTERED
- 3 SPRIGS FRESH ROSEMARY
- 1 TABLESPOON PEPPERCORNS

ROASTED WHOLE DUCK

YIELD: 5 SERVINGS / **ACTIVE TIME:** 10 MINUTES / **TOTAL TIME:** 3 HOURS

Roasted duck is a study in contrast—succulent, moist meat and crispy skin.

INGREDIENTS:

- 1 (5–6 LB.) WHOLE DUCK
- SALT, TO TASTE
- 1 LEMON, CHOPPED
- 4 GARLIC CLOVES, CHOPPED
- ¼ CUP HONEY, WARMED

1. Preheat the oven to 350°F.

2. Remove the giblets from the duck and set aside for stock (see page 484). Rinse the duck well and pat dry the inside and outside. Sprinkle salt on both the inside and outside of the duck.

3. Put lemon and garlic inside the cavity. The duck will have flapping skin on both ends. Fold that skin inward to hold the garlic and lemon inside. Tie up the duck legs with kitchen twine.

4. Place the duck breast side up in a baking dish and roast for 1 hour.

5. After the hour is up, flip the duck over and roast for 40 minutes.

6. Remove the baking pan from the oven and drain the extra duck fat into a container. Save the fat for roasting potatoes and vegetables.

7. Brush the duck with honey and place it back in the oven for another 40 minutes.

8. When the duck is done, remove it from the oven, transfer it to a cutting board, and let rest for 15 minutes before carving.

ROASTED PHEASANT

YIELD: 4 SERVINGS / **ACTIVE TIME:** 15 MINUTES / **TOTAL TIME:** 30 HOURS

Pheasant is like chicken in taste with a mild game flavor that makes it really delicious. This recipe is simple but takes a little time, so plan ahead.

INGREDIENTS:

- 7 CUPS WATER
- ½ CUP SALT, PLUS MORE TO TASTE
- 2 TABLESPOONS SUGAR
- 2 BAY LEAVES (OPTIONAL)
- 2 WHOLE PHEASANTS
- 2 TABLESPOONS OLIVE OIL OR ROOM-TEMPERATURE BUTTER
- 2 SMALL LEMON , CHOPPED

HERBS OF YOUR CHOOSING

BLACK PEPPER, TO TASTE

1. In a pot over medium-high heat, combine the water, salt, sugar, and bay leaves, if desired, and bring to a boil. Cover and let cool to room temperature. Submerge the pheasants in the brine and refrigerate for 6 hours.

2. After 6 hours, remove the birds from the brine and refrigerate, uncovered, for at least 12 hours, but ideally for 24 hours.

3. Bring pheasants to room temperature. Preheat the oven to 500°F.

4. Coat the pheasants with olive oil or butter, then season with salt. Stuff the cavity of each pheasant with a lemon and aromatic herbs, but do not pack the cavity too much.

5. Place the pheasants in a roasting pan and roast, uncovered, for 15 minutes.

6. Take the birds out of the oven and lower the temperature to 350°F; leave the oven door open for a couple minutes to help lower the temperature. Once the oven is down to 350°F, return the pheasants to the oven and cook for about 30 minutes, or until a meat thermometer inserted in a thigh registers 155 to 160°F.

7. Let rest for 10 minutes before removing lemon and herbs from the cavities and serving. Season with salt and pepper, to taste.

CORNISH GAME HENS WITH BABY BRUSSELS SPROUTS & CARAMELIZED ONIONS

YIELD: 4 SERVINGS / **ACTIVE TIME:** 20 TO 30 MINUTES / **TOTAL TIME:** 5 TO 9 HOURS

There is something so elegant about everyone getting their own bird. Brining will help the hens retain moisture, and keep them from getting dry if you overcook them, so it is well worth the additional time.

1. In a large stockpot, add the water, sugar, and salt and cook over medium-high heat until the sugar and salt dissolve. Add the remaining ingredients for the brine, remove from heat, and let cool to room temperature. Place the Cornish game hens in the brine, making sure they are completely covered, using a small bowl to weigh them down if necessary. Soak in the brine for 4 to 6 hours.

2. Preheat the oven to 375°F. Place the thyme, sage, rosemary, and minced garlic in a bowl and stir until combined.

3. Remove the Cornish game hens from the brine, pat dry, and rub with olive oil. Sprinkle with salt, pepper, and the herb-and-garlic mixture. Let the game hens stand at room temperature for approximately 30 minutes.

4. In a cast-iron Dutch oven, add 3 tablespoons of olive oil, the smashed garlic cloves, onion, and 1 tablespoon of salt. Cook over medium-high heat, reducing the heat if the onion starts to burn or dry out, until the onion is dark brown. Remove and set aside.

5. Place the Brussels sprouts, the remaining olive oil, and a pinch of salt in the Dutch oven. Stir until the brussels sprouts are evenly coated.

6. Spread the Brussels sprouts into a layer in the Dutch oven, cover with a layer of the garlic-and-onion mixture, and place the Cornish hens in the Dutch oven, breast side up. Put the half of a lemon in the center of the Dutch oven. Place the lemon quarters against the edge of the Dutch oven and between each hen. Place the Dutch oven in the oven and cook for 30 to 40 minutes.

7. Raise the heat to 400°F, remove the Dutch oven from the oven, and spin each hen 180°. Rub the top of each with the butter, return the Dutch oven to the oven, and cook for 20 minutes, or until the internal temperature of each hen is 165°F. Remove the Dutch oven from the oven and let stand for up to 30 minutes before serving.

TIP: If you plan on removing and discarding the skin after the hens are cooked, try cooking them breast side down, as this will ensure that they are even juicier.

INGREDIENTS:

FOR THE BRINE (OPTIONAL)
- 5 CUPS WATER
- 1 CUP SUGAR
- 1 CUP SALT
- 2 BAY LEAVES
- 1 TABLESPOON CRACKED BLACK PEPPER
- 1 TABLESPOON CORIANDER SEEDS, GROUND
- ½ TABLESPOON FENNEL SEEDS, GROUND
- ½ TABLESPOON CELERY SEEDS, GROUND

FOR THE HENS
- 4 (3 LB.) CORNISH GAME HENS
- 1½ TABLESPOONS MINCED FRESH THYME
- 1½ TABLESPOONS MINCED FRESH SAGE
- 1½ TABLESPOONS MINCED FRESH ROSEMARY
- 11 GARLIC CLOVES, 3 MINCED, 8 SMASHED
- 5 TABLESPOONS OLIVE OIL, PLUS MORE AS NEEDED
- 1 TABLESPOON SALT, PLUS MORE TO TASTE
- BLACK PEPPER, TO TASTE
- 1 LARGE WHITE ONION, SLICED
- 2 LBS. BABY BRUSSELS SPROUTS, HALVED
- 1 LEMON, QUARTERED, PLUS ½ OF A LEMON
- 4 TABLESPOONS UNSALTED BUTTER

GO-TO RECIPES

SWEET & STICKY CHICKEN WINGS

YIELD: 4 SERVINGS / **ACTIVE TIME:** 10 MINUTES / **TOTAL TIME:** 25 HOURS

Layered with sweet and savory Asian flavors, these wings don't last long.

1. Use paper towels to pat dry the chicken wings. Place the wings in a resealable plastic bag, add all of the remaining ingredients, except the sesame seeds, and shake to evenly coat the wings. Refrigerate the chicken for 24 hours.

2. Preheat the oven to 375°F.

3. Spread out the chicken wings on a sheet pan and bake for 25 to 30 minutes.

4. When the wings are done, let rest for 5 minutes, season to taste, garnish with sesame seeds, and serve.

INGREDIENTS:

- 4 LBS. CHICKEN WINGS, CUT INTO DRUMETTES AND FLATS
- ¼ CUP SOY SAUCE
- 2 TABLESPOONS MIRIN
- 3 TABLESPOONS OLIVE OIL
- 2 TABLESPOONS BROWN SUGAR
- 2 GARLIC CLOVES, MINCED
- 1 TABLESPOON MINCED GINGER
- 2 TABLESPOONS CHILI GARLIC SAUCE
- 2 TABLESPOONS HOISIN SAUCE
- SESAME SEEDS, FOR GARNISH

HABANERO, CALABRIAN CHILE & PINEAPPLE CHICKEN WINGS WITH CHIPOTLE BLUE CHEESE

YIELD: 2 SERVINGS / **ACTIVE TIME:** 10 MINUTES / **TOTAL TIME:** 40 MINUTES

Do you love spicy? This recipe is the best kind of spicy in how it delivers a sweet heat that really enhances the flavor of the chicken.

1. Preheat the oven to 325°F.

2. Toss the chicken wings in a bowl with 1 tablespoon canola oil and the salt. Place the chicken wings on a sheet pan lined with parchment paper and bake for 25 minutes, or until a meat thermometer inserted in a wing registers 160°F.

3. In a Dutch oven, heat the frying oil to 325°F.

4. In a small pot, combine the Habanero, Calabrian Chile & Pineapple Hot Sauce with the butter and bring to a simmer.

5. Gently add the chicken wings to the oil and cook until crispy, about 3 to 4 minutes.

6. In a medium-size bowl, toss the cooked chicken wings and hot sauce until fully coated. Garnish with cilantro, sesame seeds, and togarashi, if desired, and serve with blue cheese sauce.

CHIPOTLE BLUE CHEESE SAUCE

1. Place all of the ingredients in a food processor and pulse until smooth; if you prefer chunky blue cheese, mix the sauce by hand.

HABANERO, CALABRIAN CHILE & PINEAPPLE HOT SAUCE

1. In a pot over medium heat, toast the garlic and ginger for 1 to 2 minutes, until fragrant. Add the remainder of the ingredients, cover the pot, reduce heat to low, and cook for 30 minutes, or until the carrots are tender.

2. Let the mixture cool to room temperature, and then put in a food processor or blender and puree until very smooth. Season to taste.

INGREDIENTS:

- 12 CHICKEN WINGS
- 1 TABLESPOON CANOLA OIL, PLUS 3 QUARTS FOR FRYING
- 2 TEASPOONS SALT
- 1 CUP HABANERO, CALABRIAN CHILE & PINEAPPLE HOT SAUCE (SEE RECIPE)
- 1 TABLESPOON UNSALTED BUTTER
- CILANTRO, SESAME SEEDS, AND TOGARASHI, TO GARNISH (OPTIONAL)

CHIPOTLE BLUE CHEESE SAUCE

- 3 OZ. BLUE CHEESE, CRUMBLED
- 2 TABLESPOONS MINCED CHIPOTLE PEPPERS IN ADOBO
- 2 TABLESPOONS BUTTERMILK
- 4 TABLESPOONS SOUR CREAM
- ¼ CUP KEWPIE MAYO
- 1 TABLESPOON RICE WINE VINEGAR
- 2 TEASPOONS SUGAR
- SALT AND PEPPER, TO TASTE

HABANERO, CALABRIAN CHILE & PINEAPPLE HOT SAUCE

- 6 GARLIC CLOVES, MINCED
- 2 TABLESPOONS MINCED GINGER
- ½ LB. HABANERO CHILIES, STEMS REMOVED (IF YOU DON'T WANT A VERY SPICY SAUCE, USE ¼ LB.)
- 4 OZ. CALABRIAN CHILIES, STEMS REMOVED
- 1 LB. FRESH OR FROZEN PINEAPPLE, CHOPPED
- 2 CARROTS, CHOPPED
- 1½ CUPS RICE WINE VINEGAR
- ¼ CUP MAPLE SYRUP
- 2 TABLESPOONS SALT

MASALA WINGS

YIELD: 4 SERVINGS / **ACTIVE TIME:** 10 MINUTES / **TOTAL TIME:** 25 HOURS

This wing recipe uses Indian-inspired ingredients to deliver a ton of flavor packed into every bite.

1. Use paper towels to pat dry the chicken wings. Place the wings in a resealable plastic bag, add all of the remaining ingredients, and shake to evenly coat the wings. Refrigerate the chicken for 24 hours.

2. Preheat the oven to 375°F.

3. Spread out the chicken wings on a sheet pan and bake for 25 to 30 minutes.

4. When the wings are done, let rest for 5 minutes, season to taste, and serve.

INGREDIENTS:

- 4 LBS. CHICKEN WINGS, CUT INTO DRUMETTES AND FLATS
- 1 TABLESPOON CUMIN
- 1 TABLESPOON CORIANDER
- 1 TABLESPOON CURRY POWDER
- ½ TEASPOON TURMERIC
- 1 TEASPOON GARAM MASALA
- ½ TEASPOON RED CHILI POWDER, OR TO TASTE
- 3 TABLESPOONS OLIVE OIL
- SALT, TO TASTE

HOT & SPICY GRILLED WINGS

YIELD: 4 SERVINGS / **ACTIVE TIME:** 10 MINUTES / **TOTAL TIME:** 25 HOURS

This method for wings starts with a quick oven bake followed by a flavorful charring on the grill.

INGREDIENTS:

- 4 LBS. CHICKEN WINGS, CUT INTO DRUMETTES AND FLATS
- ¼ CUP HOT SAUCE, PLUS ½ CUP FOR COATING
- 3 TABLESPOONS OLIVE OIL
- SALT, TO TASTE
- 4 TABLESPOONS UNSALTED BUTTER, MELTED

1. Use paper towels to pat dry the chicken wings. Place the wings in a resealable plastic bag, add the hot sauce, olive oil, and salt, and shake to evenly coat the wings. Refrigerate the chicken for 24 hours.

2. Preheat the oven to 350°F.

3. Spread out the chicken wings on a sheet pan and bake for 15 minutes.

4. Preheat the grill to 450°F.

5. Once the wings are out of the oven, place them on the grill and close the lid. Cook for 5 to 7 minutes, or until they are nicely charred.

6. In a bowl, combine the butter and remaining hot sauce and toss the wings to thoroughly coat.

SOUS VIDE CAJUN CHICKEN WINGS

YIELD: 4 SERVINGS / **ACTIVE TIME:** 20 MINUTES / **TOTAL TIME:** 25 HOURS

This easy recipe delivers a ton of Cajun flavor and super tender wings.

1. Use paper towels to pat dry the chicken wings. Place the wings in a resealable plastic bag, add all of the remaining ingredients, and shake to evenly coat the wings. Refrigerate the chicken for 24 hours.

2. Set the sous vide to 140°F. Bag the chicken by following the vacuum seal directions or use a resealable bag, making sure all of the air is out of the bag when you seal it. You can do this by slowly lowering the bag of chicken into the water, using the water to push out air as you seal the zipper to close it. Cook for 45 minutes.

3. Remove the wings from the bag and pat dry with a paper towel. Preheat the oven to 450°F.

4. Spread out the chicken wings on a sheet pan and bake for 15 minutes, or until the wings develop a nice char. Let the wings rest for 5 minutes before serving.

INGREDIENTS:

- 4 LBS. CHICKEN WINGS, CUT INTO DRUMETTES AND FLATS
- ¼ CUP CAJUN SEASONING
- 3 TABLESPOONS OLIVE OIL
- ¼-½ TEASPOON SALT

SOUS VIDE TURKEY WINGS

YIELD: 4 SERVINGS / **ACTIVE TIME:** 20 MINUTES / **TOTAL TIME:** 25 HOURS

This easy method of cooking wings helps bring a ton of flavor to the wings while making them super tender.

1. Pat dry the wings with paper towels. In a resealable plastic bag, combine wings with the olive oil and salt. Toss to coat the wings and refrigerate for 24 hours.

2. The next day, set the sous vide to 140°F.

3. Bag the wings by following the vacuum seal directions or use a resealable plastic bag, making sure all the air is out of the bag by slowly lowering the bag of turkey into the water, using the water to push out air as you seal the zipper to close it.

4. Cook for 45 minutes and up to 1 hour. Once done, remove the turkey from the bag, pat dry with a paper towel, and place the wings on a sheet pan.

5. Preheat the oven to 450°F. Bake the wings for 15 minutes, or until the wings develop a nice char.

6. When the wings are done, toss in a large bowl with salt and pepper and let rest for 5 minutes before serving.

INGREDIENTS:

- 4 LBS. TURKEY WINGS, CUT INTO DRUMETTES AND FLATS
- 3 TABLESPOONS OLIVE OIL
- 1 TEASPOON SALT, PLUS MORE TO TASTE
- 1 TEASPOON FRESHLY GROUND BLACK PEPPER

SOUS VIDE LEMON-PEPPER TURKEY WINGS

YIELD: 4 SERVINGS / **ACTIVE TIME:** 20 MINUTES / **TOTAL TIME:** 25 HOURS

Lemon wings? It works. Trust me.

1. Pat dry the wings with paper towels. In a resealable plastic bag, combine wings with the olive oil, salt, lemon, and peppercorns. Toss to coat the wings and refrigerate for 24 hours.

2. The next day, set the sous vide to 140°F.

3. Bag the wings by following the vacuum seal directions or use a resealable plastic bag, making sure all the air is out of the bag by slowly lowering the bag of turkey into the water, using the water to push out air as you seal the zipper to close it.

4. Cook for 45 minutes and up to 1 hour. Once done, remove the turkey from the bag, pat dry with a paper towel, and place the wings on a sheet pan.

5. Preheat the oven to 450°F. Bake the wings for 15 minutes, or until the wings develop a nice char.

6. When the wings are done, toss in a large bowl with salt and lemon-pepper seasoning and let rest for 5 minutes before serving.

INGREDIENTS:

- 4 LBS. TURKEY WINGS, CUT INTO DRUMETTES AND FLATS
- 3 TABLESPOONS OLIVE OIL
- 1 TEASPOON SALT, PLUS MORE TO TASTE
- 1 LEMON, SLICED
- 1 TABLESPOON PEPPERCORNS
- 1 TABLESPOON LEMON-PEPPER SEASONING

SPICY GRILLED TURKEY WINGS

YIELD: 5 SERVINGS / **ACTIVE TIME:** 10 MINUTES / **TOTAL TIME:** 25 HOURS

If you like chicken wings, you'll love turkey wings. Start these in the oven and finish them on the grill for a nice smoky flavor.

INGREDIENTS:

- 4 LBS. TURKEY WINGS, CUT INTO DRUMETTES AND FLATS
- ¼ CUP HOT SAUCE, PLUS ½ CUP FOR COATING
- 3 TABLESPOONS OLIVE OIL
- SALT, TO TASTE
- 4 TABLESPOONS UNSALTED BUTTER, MELTED

1. Pat dry the wings with paper towels. In a resealable plastic bag, combine wings with the remaining ingredients, except the butter. Toss to coat the wings and refrigerate for 24 hours.

2. Preheat the oven to 350°F. Spread the turkey wings out on a sheet pan and bake for 20 minutes.

3. Preheat the grill to 450°F.

4. Once the wings are out of the oven, place them on the grill and close the lid. Cook for 6 to 8 minutes, until they start to char.

5. Place the wings in a bowl and toss in melted butter and remaining hot sauce.

SALT & PEPPER GRILLED TURKEY WINGS

YIELD: 5 SERVINGS / **ACTIVE TIME:** 10 MINUTES / **TOTAL TIME:** 40 MINUTES

Simple ingredients go a long way. These wings are tossed in simple salt and pepper at the end of their cooking. The flavor really comes through, and you are going to love it.

INGREDIENTS:

- 4 LBS. TURKEY WINGS, CUT INTO DRUMETTES AND FLATS
- 3 TABLESPOONS OLIVE OIL
- 1 TEASPOON SALT
- 1 TEASPOON FRESHLY GROUND BLACK PEPPER

1. Pat dry the wings with paper towels. In a bowl, combine wings with the olive oil and toss to coat.

2. Preheat the oven to 350°F. Spread out the turkey wings on a sheet pan and bake for 20 minutes.

3. Preheat the grill to 450°F.

4. Once the wings are out of the oven, place them on the grill and close the lid. Cook for 6 to 8 minutes, until they get a char.

5. Place the wings in a bowl and toss in the salt and pepper.

BBQ TURKEY WINGS

YIELD: 5 SERVINGS / **ACTIVE TIME:** 10 MINUTES / **TOTAL TIME:** 25 HOURS

These turkey wings are marinated, then placed in the oven and coated with rich BBQ sauce for a tangy and sweet wing that is irresistible.

1. Pat dry the wings with paper towels. In a resealable plastic bag, combine wings with the remaining ingredients, except the sauce and black pepper. Toss to coat the wings and refrigerate for 24 hours.

2. Preheat the oven to 350°F. Spread out the turkey wings on a sheet pan and bake for 20 minutes.

3. Place the wings in a bowl and toss in BBQ sauce and pepper.

INGREDIENTS:

- 4 LBS. TURKEY WINGS, CUT INTO DRUMETTES AND FLATS
- 1 TEASPOON SMOKED PAPRIKA
- 1 TABLESPOON BROWN SUGAR
- 1 TEASPOON CUMIN
- 3 TABLESPOONS OLIVE OIL
-   SALT, TO TASTE
- 1 CUP BBQ SAUCE (SEE PAGES 497 TO 505)
- ½ TEASPOON BLACK PEPPER

SOUS VIDE CHICKEN BREAST

YIELD: 4 SERVINGS / **ACTIVE TIME:** 5 MINUTES / **TOTAL TIME:** 1½ TO 4 HOURS

This method yields an incredibly moist and flavorful chicken breast. Make sure you try this!

INGREDIENTS:

- 2 **CHICKEN BREASTS (SKIN ON, BONELESS, OR BONE-IN)**
- **SALT, TO TASTE**
- **FRESH HERBS (OPTIONAL)**
- 1 **TABLESPOON CANOLA OIL**

1. Set your sous vide to 140°F.

2. Season the chicken with salt and herbs, if using.

3. Bag the chicken by following the vacuum seal directions or use a resealable plastic bag, making sure all the air is out of the bag when you seal in the chicken. You can do this by slowly lowering the bag of chicken into the water and using the water to push out air as you seal the zipper to close it. Cook for 1½ to 4 hours.

4. When done, remove chicken from the bag and pat dry with a paper towel.

5. Add oil to a fry pan over high heat. When the oil begins to shimmer, sear the chicken, skin side down, for 2 to 3 minutes, until the skin is crispy. Flip the chicken and cook for another 2 minutes.

6. Rest chicken for 5 minutes before slicing to serve.

SEARED CHICKEN BREAST

YIELD: 4 SERVINGS / **ACTIVE TIME:** 5 MINUTES / **TOTAL TIME:** 50 MINUTES

When done correctly, simple preparations can be sublime—this is the proper way to sear chicken breast.

1. Set the oven to 350°F.

2. Pat dry the chicken with a paper towel and season it with salt.

3. Add oil to an oven-safe frying pan over high heat. When the oil begins to shimmer, sear the chicken, skin side down, for 2 to 3 minutes, until the skin is crispy. Flip the chicken and cook for another 2 minutes.

4. Place the chicken in the oven and cook for 30 to 40 minutes, until a meat thermometer inserted in the thickest part of the breast registers 160°F.

5. Remove the chicken from the oven and rest in the pan; the temperature will rise to around 165°F.

INGREDIENTS:

- 2 **CHICKEN BREASTS (SKIN ON, BONELESS, OR BONE-IN)**
- **SALT, TO TASTE**
- 1 **TABLESPOON CANOLA OIL**

SMOKED CHICKEN THIGHS

YIELD: 6 SERVINGS / **ACTIVE TIME:** 10 MINUTES / **TOTAL TIME:** 4 HOURS

Time to break out the smoker. These smoked chicken thighs are simple to make, and this is an ideal recipe if you are looking to start learning how to smoke meats. I recommend using applewood or pecan chips for this preparation.

1. Bring the smoker to between 200 and 220°F.

2. In a small bowl, combine the paprika, chili powder, garlic powder, salt, and pepper and mix well.

3. Coat the chicken thighs with olive oil and then apply the seasoning rub.

4. Place the chicken on the smoker rack and add wood chips at the same time you add the meat. Smoke the chicken for 2 hours, rotating the thighs every 30 minutes to make sure they cook evenly.

5. Cook until a meat thermometer inserted in a thigh registers 165°F, about 3 to 4 hours total.

6. Rest cooked chicken for 10 minutes, season to taste, and serve.

INGREDIENTS:

- 2 TABLESPOONS SMOKED PAPRIKA
- 2 TABLESPOONS CHILI POWDER
- 1 TABLESPOON GARLIC POWDER
- 1 TABLESPOON SALT
- 1 TABLESPOON BLACK PEPPER
- 6 SKIN-ON CHICKEN THIGHS
- 2 TABLESPOONS OLIVE OIL

CHICKEN LEGS WITH POTATOES & FENNEL

YIELD: 6 SERVINGS / **ACTIVE TIME:** 30 MINUTES / **TOTAL TIME:** 1 HOUR AND 10 MINUTES

There's something so classic about these flavors that it's hard to resist the potato-and-fennel combo. This is a great comfort food for a cold day, as it's likely to stick with you for a while.

1. Place a cast-iron Dutch oven over medium-high heat and add the ⅓ cup of the olive oil. While the oil heats up, rub the chicken legs with the remaining oil and season with the salt and pepper. When the oil is hot, add half of the chicken, skin side down, and cook until the skin is golden brown and crusted. Remove, set aside, and repeat with the remaining chicken legs.

2. Preheat the oven to 400°F. Add the shallots and garlic to the Dutch oven and use a wooden spoon to scrape all of the browned bits from the bottom. Cook until the shallots and garlic darken, about 3 minutes.

3. Turn the heat up to high and add the remaining ingredients, except for the Chardonnay and the butter. Cook for about 15 minutes, stirring every few minutes.

4. Add the wine and the butter, stir, and then return the chicken to the pan, skin side up. Reduce the heat, cover, and cook until the potatoes are soft and the chicken is 155°F in the center, about 20 minutes. Remove the lid, transfer the Dutch oven to the oven, and cook until the chicken is 165°F in the center, about 10 minutes.

INGREDIENTS:

⅓	CUP OLIVE OIL, PLUS 2 TABLESPOONS
6	CHICKEN LEGS
1	TABLESPOON SALT
1	TABLESPOON GROUND BLACK PEPPER
½	CUP MINCED SHALLOTS
2	GARLIC CLOVES, MINCED
2	RED POTATOES, DICED
3-4	YELLOW POTATOES, DICED
3	FENNEL BULBS, DICED (RESERVE THE FRONDS FOR GARNISH)
1	TEASPOON CELERY SEEDS
1	TEASPOON FENNEL SEEDS
½	CUP SUN-DRIED TOMATOES
1	CUP CHARDONNAY
6	TABLESPOONS SALTED BUTTER

ROASTED CHICKEN THIGHS WITH TABBOULEH

YIELD: 4 SERVINGS / **ACTIVE TIME:** 25 MINUTES / **TOTAL TIME:** 1 HOUR

Boneless, skinless chicken makes prep a lot easier for many dishes, but you need both to stick around in this preparation. Searing the skin renders the fat and adds a tremendous amount of flavor, and cooking it on the bone ensures that the meat remains moist and tender.

1. Preheat the oven to 450°F. Place the olive oil in a cast-iron skillet and warm over medium-high heat. Sprinkle salt, pepper, the paprika, cumin, and ground fennel on the chicken thighs. When the oil starts to smoke, place the thighs skin side down in the pan and sear until brown.

2. Turn the thighs over and place the pan in the oven. Cook until the internal temperature is 165°F, about 16 minutes. Halfway through the cooking time, add the tomatoes, garlic, and shallot to the pan.

3. When the chicken is fully cooked, remove from the oven and transfer to a plate. Leave the vegetables in the pan, add the white wine, and place over high heat. Cook for 1 minute, while shaking the pan. Transfer the contents of the pan to a blender, puree until smooth, and season to taste. Set aside.

4. Prepare the tabbouleh. Place the bulgur, water, shallot, thyme, and salt in a saucepan and bring to a boil. Remove from heat, cover the pan with foil, and let sit until the bulgur has absorbed all the liquid. Fluff with a fork, remove the shallot and thyme, and add the remaining ingredients. Stir to combine and season with salt and pepper.

5. To serve, place some of the tabbouleh on each plate. Top with a chicken thigh and spoon some of the puree over it.

INGREDIENTS:

FOR THE CHICKEN THIGHS

- 2 TABLESPOONS OLIVE OIL
- SALT AND PEPPER, TO TASTE
- 2 TEASPOONS PAPRIKA
- 2 TEASPOONS CUMIN
- 2 TEASPOONS GROUND FENNEL SEEDS
- 4 CHICKEN THIGHS
- 1 CUP CHERRY TOMATOES
- 2 GARLIC CLOVES, CRUSHED
- 1 SHALLOT, SLICED
- ½ CUP WHITE WINE

FOR THE TABBOULEH

- 1 CUP BULGUR WHEAT
- 2 CUPS WATER
- 1 SHALLOT, HALVED
- 2 SPRIGS FRESH THYME
- 1 TABLESPOON SALT, PLUS MORE TO TASTE
- 1 TABLESPOON CHOPPED CILANTRO
- 1 TABLESPOON CHOPPED PARSLEY
- 2 TABLESPOONS CHOPPED SCALLIONS
- 1½ TABLESPOONS FRESH LIME JUICE
- ½ CUP DICED TOMATO
- ½ CUP DICED CUCUMBER
- ½ TEASPOON MINCED GARLIC
- 3 TABLESPOONS OLIVE OIL
- PEPPER, TO TASTE

KOREAN CHICKEN THIGHS WITH SWEET POTATO VERMICELLI

YIELD: 4 TO 6 SERVINGS / **ACTIVE TIME:** 45 MINUTES / **TOTAL TIME:** 3 HOURS AND 30 MINUTES

This is a Korean take on lo mein, the Chinese classic. The umami flavor of the sweet potato noodles, shiitake mushrooms, and cabbage is the perfect complement to the sweetness of the marinated chicken.

1. To prepare the marinade, place all of the ingredients in a blender and blend until smooth. Pour over the chicken thighs and let them marinate in the refrigerator for at least 2 hours.

2. Fill a large cast-iron Dutch oven with water and bring to a boil. Add the vermicelli and cook for about 6 minutes. Drain, rinse with cold water to keep them from sticking, and set aside.

3. Preheat the oven to 375°F. Remove the chicken from the refrigerator and place the pot back on the stove. Add the vegetable oil and warm over medium-high heat. Remove the chicken thighs from the marinade and place them skin side down in the Dutch oven. Reserve the marinade. Sear the chicken until a crust forms on the skin, about 5 to 7 minutes. Turn the chicken thighs over, add the reserved marinade, place the pot in the oven, and cook for about 15 to 20 minutes, until the centers of the chicken thighs reach 165°F.

4. Remove from the oven and set the chicken aside. Drain the Dutch oven and wipe it clean. Return it to the stove, add the cabbage, mushrooms, shallot, onion, garlic, scallion whites, and ginger, and cook for 8 minutes or until the cabbage is wilted.

5. Add the brown sugar, sesame oil, fish sauce, soy sauce, and rice vinegar to a small bowl and stir until combined. Add this sauce and the vermicelli to the pot, stir until the noodles are coated, and then return the chicken thighs to the Dutch oven. Top with the scallion greens and sesame seeds, return to the oven for 5 minutes, and serve.

INGREDIENTS:

FOR THE MARINADE

- 1 LEMONGRASS STALK, TENDER PART ONLY (THE BOTTOM HALF)
- 2 GARLIC CLOVES
- 1 TABLESPOON MINCED GINGER
- 1 SCALLION
- ¼ CUP BROWN SUGAR
- 2 TABLESPOONS CHILI PASTE
- 1 TABLESPOON SESAME OIL
- 1 TABLESPOON RICE VINEGAR
- 2 TABLESPOONS FISH SAUCE
- 1 TABLESPOON BLACK PEPPER

FOR THE CHICKEN THIGHS & VERMICELLI

- 4–6 CHICKEN THIGHS
- 10 OZ. SWEET POTATO VERMICELLI
- 2 TABLESPOONS VEGETABLE OIL
- ¼ HEAD NAPA CABBAGE, CHOPPED
- 4 OZ. SHIITAKE MUSHROOMS, SLICED THIN
- 1 SHALLOT, SLICED THIN
- 1 YELLOW ONION, SLICED THIN
- 2 GARLIC CLOVES, MINCED
- 2 SCALLIONS, CHOPPED AND GREENS RESERVED
- 2 TABLESPOONS MINCED GINGER
- ¼ CUP BROWN SUGAR
- 2 TABLESPOONS SESAME OIL
- 2 TABLESPOONS FISH SAUCE
- ¼ CUP SOY SAUCE
- ¼ CUP RICE VINEGAR
- ¼ CUP SESAME SEEDS

CONFIT DUCK LEGS

YIELD: 5 SERVINGS / **ACTIVE TIME:** 10 MINUTES / **TOTAL TIME:** 3 HOURS

Confit means to cook in fat. And confit duck legs are one of the all-time great preparations of this delicious bird.

1. Pat the duck dry with paper towels and then sprinkle with salt.

2. With the point of a knife or needle, gently poke the duck skin all around each leg. This will help release the fat as it renders. Let the legs rest at room temperature for at least 25 minutes.

3. Add the oil to an oven-safe pot and place the duck legs in the pot.

4. Place the uncovered pot in the oven at 285°F. Do not preheat the oven. We do this to slowly render out the duck fat.

5. After 1½ hours, check the duck. It should be under a layer of duck fat and the skin should be getting crisp. If the legs aren't browned and crispy, let the duck cook longer.

6. Once the skin is starting to crisp, turn the oven up to 375°F and let the duck cook for about 15 minutes.

7. Remove the pot from the oven, pull the duck from the fat, and let rest for 10 minutes.

INGREDIENTS:

- 5   **DUCK LEGS**
-     SALT, TO TASTE

SEARED DUCK BREAST

YIELD: 2 SERVINGS / **ACTIVE TIME:** 5 MINUTES / **TOTAL TIME:** 20 MINUTES

Searing a duck breast is not the same as searing chicken. It is pretty simple, though, and after reading this recipe, you'll know how to do it.

INGREDIENTS:

- 2 SKIN-ON DUCK BREASTS
- SALT, TO TASTE

1. Pat dry the duck with paper towels and then sprinkle with salt.

2. With a sharp knife, score the skin of the duck breast diagonally, making sure not to cut into the meat. Aim for a ⅛-inch-deep cut. Season with salt.

3. Place the breasts skin side down on a cold cast-iron pan and turn on the heat to medium-low. After 5 minutes the breast should begin to bubble a bit as the fat renders. Keep rendering the fat until the skin is golden and a meat thermometer inserted in the breast registers about 125°F.

4. Increase the heat to medium and brown the skin for another 2 minutes, then flip until the duck registers a temperature of 130°F.

5. Transfer the duck to a cutting board and rest for 5 minutes before serving.

CONFIT TURKEY LEGS

YIELD: 5 SERVINGS / **ACTIVE TIME:** 10 MINUTES / **TOTAL TIME:** 3 HOURS

These turkey legs are cooked in the fat they render, making for fall-off-the-bone tender turkey leg that is crispy and delicious.

INGREDIENTS:

- 5 TURKEY LEGS
- SALT, TO TASTE
- CANOLA OIL, AS NEEDED

1. Pat dry the turkey legs with paper towels and sprinkle with salt.

2. With the point of a knife or needle, gently poke the turkey skin on all the legs. This will help release the fat as it renders. Rest the legs at room temperature for at least 25 minutes.

2. Add oil to an oven-safe pot and place the legs in the pot.

3. Place the pot in the oven, uncovered, at 285°F. Do not preheat the oven. We do this so we can slowly render out the turkey fat.

4. After 1½ hours, check the turkey. It should be under a layer of turkey fat and the skin should be getting crisp. If it hasn't rendered, add a little bit of oil and cook longer.

5. Once the skin starts to crisp, turn the oven up to 375°F and let it cook for about 15 minutes.

6. Remove the pot from the oven, transfer the turkey to a cutting board, and rest for 10 minutes.

SEARED TURKEY BREAST

YIELD: 4 SERVINGS / **ACTIVE TIME:** 10 MINUTES / **TOTAL TIME:** 6 HOURS

This is my favorite method for searing turkey breast.

1. In a bowl, make the brine by combining 4 cups water with 2 tablespoons salt, sugar, peppercorns, allspice, and bay leaves and mixing well. Place the turkey in the brine for 2 to 6 hours.

2. Remove the turkey from the brine and pat dry with paper towels. Salt the turkey all over.

3. Preheat the oven to 350°F.

4. Add olive oil to a large oven-safe skillet over medium-high heat. When the oil begins to shimmer, place the turkey in the pan, skin side down. Cook for 6 to 8 minutes, or until the skin turns a deep golden color.

5. Flip and cook on the other side for 5 to 7 minutes.

6. Place the pan in the oven and cook for 15 to 20 minutes or until a meat thermometer inserted in each breast registers 165°F.

INGREDIENTS:

- 2 TABLESPOONS SALT, PLUS MORE TO TASTE
- ½ TABLESPOON SUGAR
- 1 TEASPOON PEPPERCORNS
- 1 TEASPOON ALLSPICE BERRIES
- 2 BAY LEAVES
- 2 SKIN-ON TURKEY BREASTS
- 3 TABLESPOONS OLIVE OIL

INDIAN CHICKEN & RICE SOUP

YIELD: 12 SERVINGS / **ACTIVE TIME:** 20 MINUTES / **TOTAL TIME:** 3 HOURS

Indian spices bring a nice heat and warmth for a tasty spin on chicken and rice soup.

1. In a large pot over medium-high heat, combine all of the ingredients except the rice and salt and pepper. Bring to a boil and then lower heat and simmer for at least 2 hours, until reduced by one-third.

2. Add rice 10 minutes before serving.

3. Season to taste and serve.

INGREDIENTS:

- 1 LB. COOKED CHICKEN, SHREDDED
- 2 QUARTS CHICKEN STOCK (SEE PAGE 478)
- 1 LB. POTATOES, DICED
- ½ LB. CARROTS, PEELED AND DICED
- 1 CUP DICED CELERY
- 2 LARGE ONIONS, SLICED
- 1 CUP DICED TOMATOES
- 1 TEASPOON CUMIN
- 1 TEASPOON CORIANDER
- 1 TEASPOON GARAM MASALA
- 4 BAY LEAVES
- 4 CUPS WATER
- 1 CUP RICE
- SALT AND PEPPER, TO TASTE

SMOKED CHICKEN NOODLE SOUP

YIELD: 12 SERVINGS / **ACTIVE TIME:** 20 MINUTES / **TOTAL TIME:** 3 HOURS

Put those leftover smoked chicken thighs (see page 295) to good use in this hearty soup.

1. In a large pot over medium-high heat, combine all of the ingredients except the pasta and salt and pepper. Bring to a boil and then lower heat and simmer for at least 2 hours, until reduced by one-third.

2. Add pasta 10 minutes before serving.

3. Remove the bay leaves and the thyme and rosemary stems, season to taste, and serve.

INGREDIENTS:

- 1 LB. SMOKED CHICKEN THIGHS, SHREDDED
- 2 QUARTS CHICKEN STOCK (SEE PAGE 478)
- 1 LB. POTATOES, DICED
- ½ LB. CARROTS, PEELED AND DICED
- 1 CUP DICED CELERY
- 2 LARGE ONIONS, SLICED
- 3 SPRIGS FRESH THYME
- 3 SPRIGS FRESH ROSEMARY
- 4 BAY LEAVES
- 4 CUPS WATER
- 2 CUPS ELBOW PASTA
-   SALT AND PEPPER, TO TASTE

CHICKEN & ROOT VEGETABLE STEW

YIELD: 12 SERVINGS / **ACTIVE TIME:** 20 MINUTES / **TOTAL TIME:** 2 HOURS AND 40 MINUTES

This hearty stew is the perfect vehicle for enlivening those root cellar vegetables that you can still find in the wintertime. Feel free to use any vegetables you have on hand. The beauty of soup is that so many things are interchangeable so long as you have a great stock.

1. In a large pot, combine all of the ingredients, except the cornstarch and water, bring to a boil, and then reduce heat and simmer for 2 hours, until reduced by at least one-third.

2. In a small bowl, combine the cornstarch and water and mix well. Add this slurry to the soup and stir until thickened.

3. Remove the thyme and rosemary stems, season to taste, and serve.

INGREDIENTS:

- 1 LB. COOKED CHICKEN, GRATED
- 2 QUARTS CHICKEN STOCK (SEE PAGE 478)
- 1 LB. CELERIAC, CUBED
- 1 LB. POTATOES, CUBED
- 1 LB. RUTABAGA, CUBED
- 3 SPRIGS FRESH THYME
- 4 BAY LEAVES
- 3 SPRIGS FRESH ROSEMARY
- 2 ONIONS, SLICED
- SALT AND PEPPER, TO TASTE
- ¼ CUP CORNSTARCH
- ¼ CUP WATER

CHICKEN & RICE SOUP

YIELD: 12 SERVINGS / **ACTIVE TIME:** 20 MINUTES / **TOTAL TIME:** 3 HOURS

This simple chicken and rice soup is the perfect cold-weather comfort food. I keep it in the freezer so I can have soup whenever I need it.

1. In a large pot over medium-high heat, combine all of the ingredients except the rice and salt and pepper. Bring to a boil and then lower heat and simmer for at least 2 hours, until reduced by one-third.

2. Add rice 10 minutes before serving.

3. Remove the bay leaves and the thyme and rosemary stems, season to taste, and serve.

INGREDIENTS:

- 1 LB. COOKED CHICKEN, SHREDDED
- 2 QUARTS CHICKEN STOCK (SEE PAGE 478)
- 1 LB. POTATOES, DICED
- ½ LB. CARROTS, DICED
- 1 CUP DICED CELERY
- 2 LARGE ONIONS, SLICED
- 3 SPRIGS FRESH THYME
- 3 SPRIGS FRESH ROSEMARY
- 4 BAY LEAVES
- 4 CUPS WATER
- 1 CUP COOKED RICE
- SALT AND PEPPER, TO TASTE

MULLIGATAWNY CHICKEN CURRY SOUP WITH CURRIED CASHEWS

YIELD: 4 SERVINGS / **ACTIVE TIME:** 20 MINUTES / **TOTAL TIME:** 45 MINUTES

When anglicized, the Tamil words for "pepper water" become the name of this soup. This soup was so popular with the English living in India during the colonial era that it was one of the few Indian dishes mentioned in the literature of the period.

1. In a large stockpot, add the onion, carrots, celery, and butter and cook over medium heat for 5 minutes, or until soft.

2. Stir in the flour, curry, poppy seeds, and cumin and cook for 3 minutes. Pour in the chicken stock and bring to a boil.

3. Add the rice, reduce heat so that the soup simmers, and cook for 15 minutes.

4. Add in the apple, chicken, and thyme and simmer for 10 more minutes.

5. Add the cream, return to a simmer, season with salt and pepper, garnish with cilantro, and serve with Curried Cashews.

CURRIED CASHEWS

1. In a small sauté pan, add the butter and cook over medium heat until melted.

2. Add the cashews and cook for 4 minutes, while stirring constantly.

3. Add the curry powder and cook until the cashews are golden brown.

4. Remove, set on a paper towel, season with salt, and reserve until ready to serve.

INGREDIENTS:

- 1 ONION, CHOPPED
- 2 CARROTS, PEELED AND CHOPPED
- 2 CELERY STALKS, CHOPPED
- ¼ CUP UNSALTED BUTTER
- 2 TABLESPOONS ALL-PURPOSE FLOUR
- 1 TABLESPOON CURRY POWDER
- 1 TABLESPOON POPPY SEEDS
- 1 TEASPOON CUMIN
- 4 CUPS CHICKEN STOCK (SEE PAGE 478)
- ⅓ CUP LONG-GRAIN RICE
- 1 APPLE, PEELED, CORED, AND CHOPPED
- 1 CUP COOKED CHICKEN LEG MEAT, CHOPPED
- ¼ TEASPOON DRIED THYME
- ½ CUP HEAVY CREAM
- SALT AND PEPPER, TO TASTE
- CILANTRO, FOR GARNISH

CURRIED CASHEWS

- 2 TABLESPOONS UNSALTED BUTTER
- ½ CUP UNSALTED RAW CASHEWS
- 1 TEASPOON CURRY POWDER
- SALT, TO TASTE

HOMEY TURKEY & WILD RICE SOUP

YIELD: 6 SERVINGS / **ACTIVE TIME:** 30 MINUTES / **TOTAL TIME:** 2 HOURS

Few things are as comforting as homemade soup, and this recipe is consolation in every bite. Using leftover turkey and homemade stock give this soup a richly delicious depth of flavor.

1. In a large pot, combine all of the ingredients, except the rice, bring to a boil, and then reduce heat and simmer until the soup reduces by one-third.

2. Add the rice and cook until it is tender, about 20 minutes.

3. Season to taste and serve.

INGREDIENTS:

- 2 QUARTS TURKEY STOCK (SEE PAGE 319)
- 1 LB. COOKED TURKEY, CHOPPED
- 3 CARROTS, CHOPPED
- 3 CELERY STALKS, CHOPPED
- SALT AND PEPPER, TO TASTE
- 4 SPRIGS FRESH THYME
- 5 GARLIC CLOVES, MINCED
- 1 ONION, DICED
- 1 CUP WILD RICE

CREAMED VEGETABLE SOUP WITH TURKEY DUMPLINGS

YIELD: 4 SERVINGS / **ACTIVE TIME:** 45 MINUTES / **TOTAL TIME:** 4 HOURS AND 15 MINUTES

Got lots of leftover turkey from your Thanksgiving dinner? Here's a great way to use it up. Feel free to adjust what vegetables are used: this soup is very accommodating.

CREAMED VEGETABLE SOUP

1. In a medium saucepan, warm the oil over medium heat. Add the onion, celery, celeriac, and parsnips and cook for 10 minutes, or until tender.

2. Add the rosemary and cook for 2 minutes. Add the white wine and stock and bring to a boil.

3. Reduce heat so that the soup simmers and cook for 30 minutes, or until the vegetables are very tender.

4. Transfer the soup to a food processor, puree until creamy, and then strain through a fine sieve.

5. Place the soup in a clean pan and bring to a simmer. Add the heavy cream and season with salt and pepper.

6. Serve in warm bowls with the Turkey Dumplings (see next page).

INGREDIENTS:

- 2 TABLESPOONS VEGETABLE OIL
- 1 ONION, CHOPPED
- 2 CELERY STALKS, CHOPPED
- 1 CELERIAC, PEELED AND CHOPPED
- 2 PARSNIPS, PEELED AND CHOPPED
- 1 SPRIG FRESH ROSEMARY, LEAVES REMOVED AND CHOPPED
- ½ CUP WHITE WINE
- 4 CUPS TURKEY STOCK (SEE PAGE 477)
- 1 CUP HEAVY CREAM
- SALT AND PEPPER, TO TASTE

TO SERVE
TURKEY DUMPLINGS

Continued . . .

TURKEY DUMPLINGS

1. Place the bread and parsley in a food processor and pulse until combined. Add the flour and baking powder and blend until combined.

2. Slowly add the milk, egg, and butter to the food processor. Pulse until a paste forms.

3. Fold in the turkey and season with salt and pepper. Refrigerate for 1 hour.

4. Place the stock in a medium saucepan over medium-high heat and bring to a boil.

5. Drop tablespoon-sized balls of the turkey mixture into the stock. Cover and cook at a simmer for 12 minutes. Remove and set aside.

INGREDIENTS (CONTINUED):

DUMPLINGS

- 4 SLICES OF BREAD, CHOPPED
- ½ CUP PARSLEY, LEAVES REMOVED AND CHOPPED
- 1¼ CUP ALL-PURPOSE FLOUR
- 1 TEASPOON BAKING POWDER
- ½ CUP MILK
- 1 EGG
- ¼ CUP UNSALTED BUTTER, MELTED
- 1 CUP COOKED TURKEY LEG MEAT, CHOPPED

    SALT AND PEPPER, TO TASTE

- 4 CUPS TURKEY STOCK (SEE PAGE 477)

DUCK A L'ORANGE BROTH WITH SPICED DUMPLINGS

YIELD: 4 SERVINGS / **ACTIVE TIME:** 45 MINUTES / **TOTAL TIME:** 1 HOUR AND 45 MINUTES

A rich soup inspired by the classic French dish.

1. Place a large saucepan over low heat.

2. Add the duck breasts, fat side down, and cook for 10 minutes, or until the skin is crispy.

3. Flip the breast over and cook for 5 minutes.

4. Remove the breast and set aside. Reserve 2 tablespoons of the rendered fat for the dumplings.

5. Raise heat to medium and add the onion, carrots, celery, and garlic. Cook for 5 minutes, or until soft.

6. Add the wine and cook for 5 minutes, or until it has reduced by half.

7. Add the stock, orange zest, orange juice, honey, thyme, marjoram, coriander, and cumin and bring to a boil.

8. Add the duck legs, reduce heat so that the soup simmers, and cook for 45 minutes, or until the meat is very tender and falling off the bone.

9. Remove the duck legs and let cool for 5 minutes. Remove the meat from the bones and cut into small pieces.

10. Cut the cooked duck breast into ½-inch pieces. Return both the breast and leg meat to the broth.

11. Simmer for 5 minutes, add the Spiced Dumplings, and season with salt and pepper.

12. Serve in warm bowls and garnish with chives.

INGREDIENTS:

- 1 (4 LB.) DUCK, BREASTS AND LEGS REMOVED
- 1 ONION, FINELY CHOPPED
- 2 CARROTS, PEELED AND FINELY CHOPPED
- 2 CELERY STALKS, FINELY CHOPPED
- 1 GARLIC CLOVE, MINCED
- 1 CUP WHITE WINE
- 6 CUPS DUCK STOCK (SEE PAGE 484)
- ZEST AND JUICE OF 2 ORANGES
- 1 TABLESPOON HONEY
- 2 SPRIGS FRESH THYME, LEAVES REMOVED AND CHOPPED
- 2 MARJORAM SPRIGS, LEAVES REMOVED AND CHOPPED
- ½ TEASPOON GROUND CORIANDER
- ½ TEASPOON CUMIN
- SALT AND PEPPER, TO TASTE
- CHIVES, CHOPPED, FOR GARNISH

TO SERVE
SPICED DUMPLINGS

Continued...

SPICED DUMPLINGS

1. Place the bread and the milk in a bowl and soak for 10 minutes.

2. Remove the bread and squeeze out any excess milk.

3. Place the bread and the remaining ingredients in a bowl and stir until well-combined.

4. Spoon out 20 piles of the mixture onto a baking tray. Roll into small balls. (Make sure to taste before you commit to all 20. If the mixture is too dry, add a little milk; if too wet, add a little more flour.)

5. Bring water to boil in a small saucepan. Add the dumplings and cook for 5 minutes.

6. Remove and add to the duck broth. The dumplings can also be prepared in advance and left to cool at room temperature.

INGREDIENTS (CONTINUED):

SPICED DUMPLINGS

- 4 SLICES OF CRUSTY BREAD, CHOPPED
- 4 TABLESPOONS MILK
- 2 STRIPS THICK-CUT BACON, CHOPPED
- 1 SHALLOT, FINELY CHOPPED
- 1 GARLIC CLOVE, MINCED
- 2 SPRIGS FRESH THYME, LEAVES REMOVED AND CHOPPED
- 2 TABLESPOONS DUCK FAT
- 1 EGG YOLK, BEATEN
- ½ TEASPOON PAPRIKA
- ½ TEASPOON CUMIN
- ½ CUP ALL-PURPOSE FLOUR
- SALT AND PEPPER, TO TASTE

PHEASANT, MOREL MUSHROOM & OLIVE BROTH

YIELD: 4 SERVINGS / **ACTIVE TIME:** 20 MINUTES / **TOTAL TIME:** 45 MINUTES

This stew helps keep the pheasant moist, tender, and delicious.

1. Place the flour and the pheasant in a bowl and gently toss until the pheasant is coated.

2. In a large saucepan, warm the vegetable oil over medium-high heat. Add the pheasant breast, skin side down, and cook for 5 minutes, or until evenly browned.

3. Add the onion, mushrooms, and garlic and cook for 5 minutes, or until soft.

4. Add the white wine and cook until reduced by half.

5. Add the stock and bring to a boil. Reduce heat so that the soup simmers and cook for 20 minutes, or until the pheasant is cooked through.

6. Add the olives and simmer for 3 minutes. Season with salt and pepper and serve in warm bowls.

INGREDIENTS:

- ½ CUP FLOUR
- 4 PHEASANT BREASTS
- 2 TABLESPOONS VEGETABLE OIL
- 1 ONION, CHOPPED
- 1½ CUPS MOREL MUSHROOMS, STEMMED AND HALVED
- 3 GARLIC CLOVES, MINCED
- 1 CUP WHITE WINE
- 4 CUPS CHICKEN STOCK (SEE PAGE 478)
- ½ CUP BLACK OLIVES, SLICED
- SALT AND PEPPER, TO TASTE

LEFTOVER TURKEY & CHICKPEA CHILI WITH CORN BREAD & ROSEMARY BUTTER

YIELD: 5 TO 7 SERVING / **ACTIVE TIME:** 35 MINUTES / **TOTAL TIME:** 2 HOURS AND 35 MINUTES

This chili recipe is a great use for all that leftover Thanksgiving turkey. Couple with corn bread and rosemary butter for an amazingly tasty and filling meal.

1. Add olive oil to a Dutch oven over medium-high heat. Once the oil is shimmering, add onion, bell pepper, garlic, oregano, cumin, and chili powder and cook for 5 minutes, stirring often.

2. Add chickpeas, turkey, tomatoes, New Mexican chilies, salt, and stock. Stir the mixture, cover, turn down burner to low, and cook for 2 hours, stirring occasionally.

3. Scoop chili into a bowl and serve with sliced Cornbread and Rosemary Butter. If desired, top the chili with sour cream and grated cheese.

ROSEMARY BUTTER

1. Using a stand mixer or hand mixer, whip the butter, salt, and rosemary until the butter becomes fluffy and lighter.

2. Place the whipped butter in a container and refrigerate until needed.

CORN BREAD

1. Preheat the oven to 450°F and grease an 8-inch baking dish.

2. In a small bowl, combine the milk, eggs, and butter and mix well.

3. In a large bowl, combine the dry ingredients and mix well.

4. Combine the wet ingredients with the dry ingredients and mix until there are no lumps in the batter.

5. Pour the batter into the greased baking dish and bake until the top is brown and the middle is set, about 20 minutes.

INGREDIENTS:

LEFTOVER TURKEY & CHICKPEA CHILI

- 1 TABLESPOON OLIVE OIL
- 1 ONION, DICED
- 1 RED BELL PEPPER, DICED
- 5 GARLIC CLOVES, MINCED
- 1 TABLESPOON FRESH OREGANO
- 1 TABLESPOON CUMIN
- 2 TEASPOONS CHILI POWDER
- 1 (14 OZ.) CAN CHICKPEAS
- 6 OZ. COOKED TURKEY
- ½ LB. TOMATOES, CANNED OR FRESH
- 3 DRIED NEW MEXICAN CHILIES
- SALT, TO TASTE
- 2 CUPS TURKEY STOCK (SEE PAGE 477)
- 2 CUPS GRATED CHEDDAR CHEESE (OPTIONAL)
- 1 CUP SOUR CREAM (OPTIONAL)

ROSEMARY BUTTER

- 4 TABLESPOONS UNSALTED BUTTER, AT ROOM TEMPERATURE
- SALT, TO TASTE
- 2 TABLESPOONS FINELY CHOPPED FRESH ROSEMARY

CORN BREAD

- 1½ CUPS MILK
- 2 EGGS
- 5 TABLESPOONS UNSALTED BUTTER, MELTED
- 1 CUP CORNMEAL
- ¾ CUP ALL-PURPOSE FLOUR
- 1 TABLESPOON SUGAR
- 1 TEASPOON BAKING POWDER
- ½ TEASPOON BAKING SODA
- SALT, TO TASTE

BRAISED CHICKEN CHILI

YIELD: 6 TO 8 SERVINGS / **ACTIVE TIME:** 15 MINUTES / **TOTAL TIME:** 2 HOURS AND 30 MINUTES

This chili uses braised chicken rather than beef and is punched up with the salsa verde.

1. Add olive oil to a Dutch oven over medium-high heat. Once the oil is shimmering, add onion, garlic, oregano, bell pepper, cumin, and chili powder and cook for 5 minutes, stirring often.

2. Add tomatoes, salsa, stock, all chilies, chicken, and salt. Stir, cover, lower the heat, and simmer for 2 hours, stirring occasionally.

3. To serve, scoop chili into a bowl and, if desired, top with cheese and sour cream.

INGREDIENTS:

- 1 TABLESPOON OLIVE OIL
- 1 ONION, DICED
- 4 GARLIC CLOVES, MINCED
- 1 TABLESPOON FRESH OREGANO
- 1 RED BELL PEPPER, DICED
- 1 TABLESPOON CUMIN
- 2 TEASPOONS CHILI POWDER
- ½ LB. TOMATOES (CANNED OR FRESH)
- 1 CUP TOMATILLO SALSA (SEE PAGE 523)
- 2 CUPS CHICKEN STOCK (SEE PAGE 478)
- 3 DRIED NEW MEXICAN CHILIES
- 2 LBS. COOKED CHICKEN, SHREDDED
- SALT, TO TASTE
- 2 CUPS GRATED CHEDDAR CHEESE (OPTIONAL)
- 1 CUP SOUR CREAM (OPTIONAL)

SPICY PAPPARDELLE & DUCK CONFIT WITH BROCCOLI RABE, SHALLOTS & PARMESAN

YIELD: 2 SERVINGS / **ACTIVE TIME:** 30 MINUTES / **TOTAL TIME:** 4 HOURS

This simple pasta with duck is delicious and sure to please even the pickiest of eaters. The pasta will finish cooking in the pan with the duck, so it should be removed from the water when al dente; this can be achieved by draining the pasta 2 minutes before the cooking instructions recommend.

1. In a large pot, bring water to a boil. Once the water is boiling, season with 2 tablespoons of salt, add the pasta, and cook until al dente. Reserve ½ cup of pasta water before draining the pasta.

2. In a sauté pan over medium heat, add the olive oil, shallots, garlic, and red pepper flakes. Season with salt and black pepper, and cook until the garlic starts to slightly brown on the edges. Add the duck and broccoli rabe and cook for 2 minutes, or until the broccoli rabe has wilted.

3. Add the cooked pasta and reserved pasta water and bring to a simmer. When the liquid has reduced by half, add three-quarters of the Parmesan cheese and the butter, remove from the heat, and toss the pasta until a nice creamy sauce coats all the pasta. Season to taste.

4. Divide the pasta into two bowls and garnish with basil leaves and the remaining Parmesan cheese.

INGREDIENTS:

- 1 GALLON WATER
- 2 TABLESPOONS SALT, PLUS MORE TO TASTE
- 7 OZ. FRESH PAPPARDELLE PASTA
- 2 TABLESPOONS OLIVE OIL
- 2 SHALLOTS, MINCED
- 2 GARLIC CLOVES, THINLY SLICED
- 2 TEASPOONS RED PEPPER FLAKES
- BLACK PEPPER, TO TASTE
- 6 OZ. DUCK CONFIT (PULLED OFF THE LEG; SEE PAGE 303)
- 1 BUNCH BROCCOLI RABE, CHOPPED
- 4 OZ. PARMESAN CHEESE, GRATED
- 4 TABLESPOONS UNSALTED BUTTER, ROOM TEMPERATURE
- 6 FRESH BASIL LEAVES, FOR GARNISH

DEVILED EGGS WITH CRISPY CHICKEN SKIN

YIELD: 6 SERVINGS / **ACTIVE TIME:** 15 MINUTES / **TOTAL TIME:** 2 HOURS AND 40 MINUTES

What came first, the chicken or the egg? This recipe answers the age-old quandary by crisping up chicken skins first to elevate this classic snack with delicious, crunchy texture. If you want to serve these at a party, you'll want to quadruple this recipe.

INGREDIENTS:

- 6 CHICKEN SKINS
- 6 EGGS
- SALT AND PEPPER, TO TASTE
- 1 TEASPOON SMOKED PAPRIKA
- ¼ CUP PICKLE JUICE
- 3 TABLESPOONS MAYONNAISE
- 1 TEASPOON HOT SAUCE

1. Preheat the oven to 375°F.

2. Set two sheet pans of equal size on the counter. Flatten out the chicken skins on one of the sheet pans, salt them, and place the second sheet pan on top of the chicken skins to press them down. Bake for 10 to 15 minutes, until crisp. Set aside.

3. Add the eggs to a pot of water over medium-high heat and boil for 10 minutes. Remove the pot from the heat and let the eggs sit for 10 minutes before placing them in an ice water bath to cool. Once the eggs have fully cooled, peel them.

4. Cut the peeled eggs down the middle and scoop out the yolks, placing them in a bowl. Add the remainder of the ingredients to the yolks and mix until smooth. Refrigerate for 2 hours.

5. Place the yolk mixture into a pastry bag or a plastic kitchen bag with a cut in the corner to pipe this mixture onto each egg half. Break the cooled chicken skins into small pieces and top the egg with the chicken skins.

ROASTED LEFTOVER CHICKEN SALAD

YIELD: 4 SERVINGS / **ACTIVE TIME:** 10 MINUTES / **TOTAL TIME:** 10 MINUTES

From our years of dinners on farms we would find creative and exciting ways to use leftover items from prep. This chicken salad recipe is simple but packed with flavor and helps reduce food waste.

1. In a medium mixing bowl combine mayonnaise, onion, celery, mustard and lemon juice. Mix well.

2. Add the chicken and nuts, and salt and pepper to taste.

INGREDIENTS:

- 2–3 CUPS LEFTOVER ROASTED CHICKEN MEAT
- ½ CUP MAYONNAISE
- 1 SPRING ONION, CHOPPED
- ½ CUP CHOPPED CELERY
- 1 TEASPOON MUSTARD
- ½ TEASPOON LEMON JUICE
- ¼ CUP WALNUTS OR PECANS (OPTIONAL)
- SALT AND PEPPER, TO TASTE

CHICKEN MEATBALLS

YIELD: 10 SERVINGS / **ACTIVE TIME:** 15 MINUTES / **TOTAL TIME:** 40 MINUTES

These meatballs provide a good amount of protein and are a great alternative to the traditional beef meatballs.

1. Preheat the oven to 400°F.

2. In a bowl, combine all the ingredients and use your hands to form 1-inch meatballs.

3. Place the meatballs on a sheet pan and bake for 15 to 20 minutes.

INGREDIENTS:

- 1½ LBS. GROUND CHICKEN
- ⅓ CUP ITALIAN BREAD CRUMBS
- ¼ CUP DICED ONION
- 1 TEASPOON ITALIAN SEASONING
- 1 EGG WHITE
- ¼ CUP CHOPPED FRESH PARSLEY
- SALT AND PEPPER, TO TASTE

TURKEY MEATBALLS

YIELD: 10 SERVINGS / **ACTIVE TIME:** 15 MINUTES / **TOTAL TIME:** 40 MINUTES

Turkey is as versatile as chicken, and when ground up it makes for healthy, and delicious, meatballs.

1. Preheat the oven to 400°F.

2. In a bowl, combine all the ingredients and use your hands to form 1-inch meatballs.

3. Place the meatballs on a sheet pan and bake for 15 to 20 minutes.

INGREDIENTS:

- 1½ LBS. GROUND TURKEY
- ⅓ CUP ITALIAN BREAD CRUMBS
- ¼ CUP DICED ONION
- 1 TEASPOON ITALIAN SEASONING
- 1 EGG WHITE
- ¼ CUP CHOPPED FRESH PARSLEY, CHOPPED
- SALT AND PEPPER, TO TASTE

CHICKEN BURGERS

YIELD: 4 SERVINGS / ACTIVE TIME: 15 MINUTES / TOTAL TIME: 30 MINUTES

A chicken burger is another healthier alternative to beef, pork, and lamb burgers, but that doesn't mean it can't be flavorful. Play around with the ratio of white to dark meat in your ground chicken to suit your palate.

1. Preheat your outdoor grill or cast-iron pan. If you don't have either, a frying pan will do. (And if using a pan, lightly grease it with neutral oil.)

2. In a bowl, combine all the ingredients until the mixture is well incorporated.

3. Divide the mixture into four equal parts. Form patties by rolling each piece into a ball in your hands and pressing it flat to form the patty. The patty should be about 1-inch thick.

4. Turn the grill to medium-high heat; if you're using a pan, set your burner on medium-high. Place patties on the grill or pan. Cover the patties and cook for 5 minutes.

5. Flip the patties and cover again, cooking for an additional 5 to 6 minutes, until you have the desired doneness.

INGREDIENTS:

- 1 LB. GROUND CHICKEN
- 1 TEASPOON SALT
- ½ TEASPOON WHITE PEPPER
- 1 EGG WHITE

TURKEY BURGERS

YIELD: 4 SERVINGS / ACTIVE TIME: 15 MINUTES / TOTAL TIME: 30 MINUTES

Don't hate on this healthier burger. The turkey delivers a ton of flavor.

1. Preheat your outdoor grill or cast-iron pan. If you don't have either, a frying pan will do. (And if using a pan, lightly grease it with neutral oil.)

2. In a bowl, combine all the ingredients until the mixture is well incorporated.

3. Divide the mixture into four equal parts. Form patties by rolling each piece into a ball in your hands and pressing it flat to form the patty. The patty should be about 1-inch thick.

4. Turn the grill to medium-high heat; if you're using a pan, set your burner on medium-high. Place patties on the grill or pan. Cover the patties and cook for 5 minutes.

5. Flip the patties and cover again, cooking for an additional 5 to 6 minutes, until you have the desired doneness.

INGREDIENTS:

- 1 LB. GROUND TURKEY
- 1 TEASPOON SALT
- ½ TEASPOON WHITE PEPPER
- 1 EGG WHITE

CHICKEN TERIYAKI BURGERS

YIELD: 4 SERVINGS / **ACTIVE TIME:** 15 MINUTES / **TOTAL TIME:** 30 MINUTES

Teriyaki sauce and pineapple layer in the sweet and savory to this yummy burger.

1. Preheat your outdoor grill or cast-iron pan. If you don't have either, a frying pan will do. (And if using a pan, lightly grease it with neutral oil.)

2. In a bowl, combine all the ingredients, except the pineapple, until each piece is well incorporated.

3. Divide the mixture into four equal parts. Form patties by rolling the mixture into a ball in your hands and pressing it flat to form the patty. The patty should be about 1-inch thick.

4. Turn the grill to medium-high heat; if you're using a pan, set your burner on medium-high. Place patties on the grill or pan. Cover the patties and cook for 5 minutes.

5. While the burgers are cooking, grill or sear the pineapple until it starts to caramelize, about 4 minutes. Flip the pineapple and continue to cook until the other side is caramelized.

6. Flip the patties and cover again, cooking for an additional 5 to 6 minutes, until you have the desired doneness. Top with the pineapple.

INGREDIENTS:

- 1 LB. GROUND CHICKEN
- ½ TEASPOON WHITE PEPPER
- 1 EGG WHITE
- 1 TABLESPOON TERIYAKI SAUCE
- 4 PINEAPPLE RINGS, ½-INCH THICK

TURKEY & KIMCHI BURGERS

YIELD: 4 SERVINGS / **ACTIVE TIME:** 15 MINUTES / **TOTAL TIME:** 30 MINUTES

Delicious as it is, turkey can be a bit dry. Adding the goodness of lacto-fermented kimchi solves that problem with a pungent punch.

1. Preheat your outdoor grill or cast-iron pan. If you don't have either, a frying pan will do. (And if using a pan, lightly grease it with neutral oil.)

2. In a bowl, combine all the ingredients, except the cheese and kimchi, until each piece is well incorporated.

3. Divide the mixture into four equal parts. Form patties by rolling each piece into a ball in your hands and pressing it flat to form the patty. The patty should be about 1-inch thick.

4. Turn the grill to medium-high heat; if you're using a pan, set your burner on medium-high. Place patties on the grill or pan. Cover the patties and cook for 5 minutes.

5. Flip the patties and cover again, cooking for an additional 5 to 6 minutes. About 30 seconds before your burgers are done, top each one with a slice of cheese and cover until the cheese is melted and you have the desired doneness. Top with kimchi.

INGREDIENTS:

- 1 LB. GROUND TURKEY
- 1 TEASPOON SALT
- ½ TEASPOON WHITE PEPPER
- 1 EGG WHITE
- 4 SLICES AMERICAN CHEESE
- ½ CUP KIMCHI

CHICKEN TIKKA BURGER WITH TIKKA MASALA MAYONNAISE

YIELD: 4 SERVINGS / **ACTIVE TIME:** 15 MINUTES / **TOTAL TIME:** 30 MINUTES

As this recipe suggests, I am passionate about Indian food. Feel free to double the Tikka Masala Mayonnaise, as it is an amazing go-to condiment for sandwiches, fries, and raw veggies.

1. Preheat your outdoor grill or cast-iron pan. If you don't have either, a frying pan will do. (And if using a pan, lightly grease it with neutral oil.)

2. In a bowl, combine all the ingredients until the mixture is well incorporated.

3. Divide the mixture into four equal parts. Form patties by rolling each piece into a ball in your hands and pressing it flat to form the patty. The patty should be about 1-inch thick.

4. Turn the grill to medium-high heat; if you're using a pan, set your burner on medium high. Place patties on the grill or pan. Cover the patties and cook for 5 minutes.

5. Flip the patties and cover again, cooking for an additional 5 to 6 minutes, until you have the desired doneness. Top with Tikka Masala Mayonnaise.

INGREDIENTS:

- 1 LB. GROUND CHICKEN
- 1 TEASPOON SALT
- ½ TEASPOON WHITE PEPPER
- 1 EGG WHITE
- ½ TEASPOON CUMIN
- ½ TEASPOON CORIANDER

TIKKA MASALA MAYO

- ¼ CUP TIKKA MASALA SAUCE
- ¼ CUP MAYONNAISE
- JUICE OF ½ LIME
- ½ TEASPOON SALT

TIKKA MASALA MAYONNAISE

1. In a bowl, combine all ingredients until well incorporated. Place in the refrigerator until needed.

THE ULTIMATE LEFTOVER THANKSGIVING SANDWICH

YIELD: 2 SERVINGS / **ACTIVE TIME:** 15 MINUTES / **TOTAL TIME:** 30 MINUTES

The Thanksgiving sandwich. A creation so tasty it is craved beyond the holiday season. In my pursuit of the ultimate leftover Thanksgiving sandwich, I have settled on this recipe. Ideally, you'll have some artisan sourdough available, but any sandwich bread will do. Trust me, you are going to want to make several of these.

1. In a small bowl, combine the mayonnaise and gravy and season with salt and pepper. Set aside.

2. Add oil to a frying pan over medium heat. When the pan warms up, butter one side of each bread slice.

3. Place 2 slices of bread, buttered sides down, in the hot pan and top each one with equal parts gravy mixture, stuffing, and turkey. Spread the cranberry sauce on the unbuttered side of the two remaining bread slices and top the sandwich.

4. When the buttered sides of the bread are golden brown, flip the sandwich and cook until that side is also golden brown.

INGREDIENTS:

- 1 TABLESPOON MAYONNAISE
- 1 TABLESPOON LEFTOVER GRAVY
- SALT AND PEPPER, TO TASTE
- 1 TABLESPOON CANOLA OIL
- 4 SLICES BREAD
- 1 TABLESPOON UNSALTED BUTTER, AT ROOM TEMPERATURE
- ¼ CUP STUFFING
- 1 CUP LEFTOVER TURKEY
- 1 TABLESPOON CRANBERRY SAUCE

CHICKEN PARM SANDWICH

YIELD: 4 SERVINGS / **ACTIVE TIME:** 20 MINUTES / **TOTAL TIME:** 25 HOURS

This sandwich is big, sloppy, and delicious. I finish the chicken in the oven after frying to melt the cheese and make sure it is cooked through. I use brioche buns, but a good sub roll works well, too.

1. In a bowl, combine the chicken and buttermilk and refrigerate, covered, overnight.

2. When ready to cook the chicken, combine panko, salt, Italian seasoning, and red pepper flakes in a shallow dish. In a separate shallow dish, whisk the eggs.

3. Dredge the chicken in the dry mix, then in the eggs, and back into the dry mix. Repeat until all of the chicken is coated.

4. Add oil to a Dutch oven and heat to 350°F. Place the chicken in the pan, making sure to not overcrowd. Work in batches if needed. Fry for 6 to 8 minutes, or until a meat thermometer inserted in the chicken registers 155°F.

5. While the chicken is cooking, warm up the sauce in a small saucepan and preheat the oven to 350°F.

6. Drain the fried chicken on a paper towel–lined plate and salt to taste.

7. Place the drained chicken on a sheet pan, top with cheese, and bake until the cheese is melted, about 5 minutes.

8. To serve, place the chicken on the buns and top with marinara.

INGREDIENTS:

- 2 CHICKEN BREASTS, BUTTERFLIED
- 1 CUP BUTTERMILK
- 1 CUP PANKO
- SALT, TO TASTE
- 2 TEASPOONS ITALIAN SEASONING
- 1 PINCH RED PEPPER FLAKES
- 2 EGGS
- 4 CUPS CANOLA OIL, FOR FRYING
- FRESH MOZZARELLA CHEESE, SLICED
- 4 BUNS
- MARINARA SAUCE, AS NEEDED

PICKLE-FRIED CHICKEN SANDWICH

YIELD: 4 SERVINGS / **ACTIVE TIME:** 15 MINUTES / **TOTAL TIME:** 25 HOURS

Brining this chicken in pickle juice yields an incredibly juicy and delicious sandwich.

1. In a bowl, combine chicken with pickle juice, salt, and pepper. Cover and refrigerate overnight.

2. The next day, combine milk, eggs, and hot sauce and mix well in a bowl.

3. In a separate bowl, combine the flour with the remainder of the ingredients, except the canola oil.

4. Dip the chicken in the milk mixture, then dredge through the spice mixture, return it to the milk mixture, and then back to the spice mixture. Repeat until all of the chicken is coated.

5. Add oil to a Dutch oven over medium-high heat and bring to 325°F. Cook until a meat thermometer inserted in the meat registers 165°F, about 20 minutes.

6. Drain the cooked chicken on a paper towel–lined plate, season to taste, and build sandwiches to your liking.

INGREDIENTS:

- 2 CHICKEN BREASTS, BUTTERFLIED
- 1 CUP PICKLE JUICE
- 1½ TABLESPOONS SALT
- ½ TEASPOON BLACK PEPPER
- 1 CUP WHOLE MILK
- 2 EGGS
- ¼ CUP LOUISIANA HOT SAUCE, OR HOT SAUCE OF YOUR CHOICE
- 2 CUPS ALL-PURPOSE FLOUR
- 2 TABLESPOONS PAPRIKA
- 1 TEASPOON ONION SALT
- 1½ TEASPOONS SALT
- 1½ TEASPOON BLACK PEPPER
- 1 TEASPOON GARLIC POWDER
- 1 TEASPOON CHILI POWDER
- 1 TEASPOON GROUND THYME
- 1 TEASPOON RUBBED SAGE
- 1 TEASPOON GROUND OREGANO
- 1 TEASPOON GROUND BASIL
- 4 CUPS CANOLA OIL, FOR FRYING

CHICKEN KABOB

YIELD: 12 SERVINGS / **ACTIVE TIME:** 15 MINUTES / **TOTAL TIME:** 1 HOUR

This chicken kabob is tasty, juicy, and wonderful with fresh pita bread.

1. Preheat the grill to high and soak skewers in water for 15 to 25 minutes.

2. In a large bowl, combine all of the ingredients and mix with your hands. Form meat around the skewers, making an oblong sausage shape. Refrigerate the kabobs for 30 minutes.

3. Remove the skewers from the refrigerator and spray the grill with nonstick spray or oil. Grill the kabobs until they char on the outside, about 2 to 3 minutes, and rotate to cook evenly. The kabobs will be done in about 6 to 8 minutes.

4. Allow to cool for 5 minutes before removing the meat from the skewers and serving.

INGREDIENTS:

12	(6-INCH) WOODEN SKEWERS
1½	LBS. GROUND CHICKEN
½	CUP MINCED ONION
1	TEASPOON CHOPPED FRESH ROSEMARY
1	TEASPOON CHOPPED FRESH THYME
3	GARLIC CLOVES, MINCED
1	TEASPOON CUMIN
½	TEASPOON DRIED THYME
	SALT AND PEPPER, TO TASTE

POPCORN CHICKEN WITH KIMCHI MAYONNAISE

YIELD: 4 SERVINGS / **ACTIVE TIME:** 25 MINUTES / **TOTAL TIME:** 1 HOUR

This Korean inspired dish is perfect for a snack or anytime!

1. In a bowl, combine the garlic, egg white, soy sauce, sesame oil, white pepper, cornstarch, and salt and mix well. Add the chicken, toss to coat, and refrigerate, covered, for 1 hour.

2. Dust a sheet pan with tapioca starch, add the chicken, and coat well. Add more tapioca starch if necessary.

3. Add canola oil to Dutch oven over medium heat. When the oil registers 350°F, shake any excess starch from the chicken and fry the pieces until golden brown. Do not overcrowd the pot. Drain chicken on a paper towel–lined plate and season to taste.

4. Serve with the Kimchi Mayonnaise.

KIMCHI MAYONNAISE

1. Add oil and kimchi to a small frying pan over medium-high heat and cook until kimchi caramelizes. Add sugar, pepper flakes, and kimchi juice, mix well, and let cool.

2. In a bowl, combine kimchi mixture with lime juice and mayonnaise. Refrigerate to store.

INGREDIENTS:

- 3 GARLIC CLOVES, SMASHED
- 1 EGG WHITE
- 1 TABLESPOON SOY SAUCE
- 1½ TABLESPOONS SESAME OIL
- ½ TEASPOON WHITE PEPPER
- 1 TABLESPOON CORNSTARCH
- SALT, TO TASTE
- 1 LB. BONELESS, SKIN-ON CHICKEN BREAST, CUT INTO BITE-SIZED PIECES
- 7 TABLESPOONS TAPIOCA STARCH, PLUS MORE AS NEEDED
- 2 CUPS CANOLA OIL
- KIMCHI MAYONNAISE (SEE RECIPE), FOR DIPPING

KIMCHI MAYONNAISE

- 1 TABLESPOON CANOLA OIL
- 2 TABLESPOONS FINELY CHOPPED KIMCHI
- 1 TABLESPOON SUGAR
- 1 TABLESPOON RED PEPPER FLAKES
- 2 TABLESPOONS KIMCHI JUICE
- JUICE OF 1 LIME
- ¼ CUP MAYONNAISE

CHICKEN KATSU CURRY

YIELD: 4 SERVINGS / **ACTIVE TIME:** 40 MINUTES / **TOTAL TIME:** 1 HOUR

Tampopo is a small restaurant in Cambridge's Porter Square that I used to visit every weekend. This place introduced me to chicken katsu curry. I became addicted to the sweet curry and the crunchy chicken. I worked this recipe until I recreated that same experience.

1. In a nonstick pan, heat sunflower oil over medium-high heat. Once the oil is shimmering, add the onions and garlic and cook until softened. Stir in carrots and cook over low heat for 10 to 12 minutes.

2. Add flour and curry powder and cook for 1 minute. Gradually stir in stock until well incorporated.

3. Add honey, soy sauce, and bay leaf. Bring to the boil and then turn down heat and simmer for 20 minutes, or until sauce thickens but is still of pouring consistency. Stir in garam masala.

4. Pour the sauce through a sieve, return to the saucepan, and keep on low heat until ready to serve.

5. In a bowl, whisk the eggs. In another bowl, combine panko with a pinch of kosher salt.

6. Dredge the chicken in the eggs, then press the chicken into the panko. Then press the other side of the chicken into the panko so it is well coated. Repeat until all the chicken has been breaded.

7. Add canola oil to the saucepan over medium-high heat. Once the oil is hot, add the chicken, working in batches in order not to overcrowd the pan. Cook for 6 to 8 minutes on each side. Drain the chicken on a paper towel–lined plate. Repeat until all the chicken has been cooked.

8. Serve with rice and curry.

INGREDIENTS:

- 2 TABLESPOONS SUNFLOWER OIL
- 2 ONIONS, SLICED
- 5 GARLIC CLOVES, CHOPPED
- 2 CARROTS, PEELED AND SLICED
- 2 TABLESPOONS ALL-PURPOSE FLOUR
- 4 TEASPOONS JAPANESE CURRY POWDER
- 2½ CUPS CHICKEN STOCK (SEE PAGE 478)
- 2 TEASPOONS HONEY
- 4 TEASPOONS SOY SAUCE
- 1 BAY LEAF
- 1 TEASPOON GARAM MASALA
- 2 EGGS
- 1½ CUPS PANKO
- PINCH OF SALT
- 1 LB. CHICKEN BREAST
- ¼ CUP CANOLA OIL

TO SERVE
COOKED RICE

SOUS VIDE TANDOORI CHICKEN

YIELD: 4 SERVINGS / **ACTIVE TIME:** 30 MINUTES / **TOTAL TIME:** 2 HOURS AND 30 MINUTES

"Tandoori" is a style of Indian cooking that utilizes a "tandoor," a cylindrical clay oven. This recipe isn't about cooking the chicken in a special oven, but it is about sealing in all the flavors that make tandoori so amazing!

1. Set the sous vide cooker to 155°F.

2. Remove the skin from the chicken thighs.

3. In a bowl, mix together the chicken, lemon juice, garam masala, Kashmiri chili powder, and salt. Use your hands to really rub in the seasoning all over the chicken. Wear gloves so your fingers don't turn red.

4. Place the chicken in a vacuum bag and sous vide for 1½ to 2 hours.

5. Place finished chicken in an ice water bath to cool down. Once the meat has cooled down, coat it with oil and Tandoori Paste, and finish the chicken on the grill or in the oven, cooking the chicken until a nice char forms and it achieves an internal temperature of 165°F.

TANDOORI PASTE

1. In a bowl, combine all of the ingredients and mix well.

INGREDIENTS:

- 1 LB. BONE-IN CHICKEN THIGHS
- 1 TEASPOON FRESH LEMON JUICE
- 1 TEASPOON GARAM MASALA
- 1 TEASPOON KASHMIRI CHILI POWDER
- SALT, TO TASTE
- CANOLA OIL, TO COAT CHICKEN
- TANDOORI PASTE (SEE RECIPE)

TANDOORI PASTE

- 2 TABLESPOONS PLAIN YOGURT
- 1 TEASPOON MINCED GARLIC
- 1 TEASPOON TURMERIC
- 1 TEASPOON GARAM MASALA
- 1 TEASPOON CUMIN
- 2 TEASPOON KASHMIRI CHILI POWDER
- 1 TEASPOON FRESH LEMON JUICE

NASHVILLE HOT CHICKEN

YIELD: 6 SERVINGS / **ACTIVE TIME:** 15 MINUTES / **TOTAL TIME:** 45 MINUTES

This isn't just fried chicken; this is fried chicken kicked into rock star mode. It pairs perfectly with dill pickles, so make sure you have some on hand.

1. In a large bowl, coat the chicken with salt and pepper, cover, and refrigerate overnight.

2. The next day, combine milk, eggs, and hot sauce in a bowl.

3. In a bowl, combine the flour with the remainder of the ingredients.

4. Add oil to a Dutch oven and heat it to 325°F for frying.

5. Soak your chicken in the wet mixture and then dredge it in the dry mixture. Return the chicken to the wet mixture and then dredge it in the dry mixture. Repeat until all the chicken is coated.

6. Deep-fry the chicken for 20 minutes. A meat thermometer inserted into the cooked chicken should register 165°F.

7. To serve, ladle fryer oil to taste over the chicken and sprinkle with Nashville Heat Spices.

INGREDIENTS:

- 2 LBS. CHICKEN THIGHS, DRUMSTICKS, BREASTS
- 1½ TABLESPOONS KOSHER SALT
- ½ TEASPOON BLACK PEPPER
- 1 CUP WHOLE MILK
- 2 EGGS
- ¼ CUP LOUISIANA HOT SAUCE, OR HOT SAUCE OF YOUR CHOICE
- 2 CUPS ALL-PURPOSE FLOUR
- 2 TABLESPOONS PAPRIKA
- 1 TEASPOON ONION SALT
- 1½ TEASPOONS SEA SALT
- 1½ TEASPOONS BLACK PEPPER
- 1 TEASPOON GARLIC POWDER
- 1 TEASPOON CHILI POWDER
- 1 TEASPOON GROUND THYME
- 1 TEASPOON RUBBED SAGE
- 1 TEASPOON GROUND OREGANO
- 1 TEASPOON GROUND BASIL
- CANOLA OIL, FOR FRYING

NASHVILLE HEAT SPICES

- 3 TABLESPOONS CAYENNE PEPPER
- 1 TEASPOON PAPRIKA
- 1 TEASPOON GARLIC POWDER
- 1 TABLESPOON BROWN SUGAR

CRISPY CHICKEN WITH CAJUN SAUCE

YIELD: 4 SERVINGS / **ACTIVE TIME:** 20 MINUTES / **TOTAL TIME:** 40 MINUTES

This crispy chicken dish is perfect for a family meal. Even picky eaters will love it!

1. Add butter to a large skillet over medium heat. When the butter foams, add the peppers and onion and sauté for 4 minutes, then add the garlic and red pepper flakes and cook for an additional 3 minutes.

2. Add cream and stock and simmer until the sauce heats through and thickens slightly, about 5 minutes. Add 1 cup Parmesan cheese, stirring to incorporate. Season the sauce to taste with salt and pepper. Reduce heat to low and let reduce at a simmer.

3. Using a meat tenderizer, pound out the chicken thighs, the thinner the better.

4. In a shallow dish, combine the panko, flour, and the remainder of the Parmesan cheese, Old Bay, paprika, cayenne, and salt and pepper and mix well.

5. Pour milk into a separate shallow dish.

6. Dip chicken in bread crumb mixture and then in milk and then back in bread crumbs. Repeat until all of the chicken is breaded.

7. Add canola oil to a cast-iron pan over medium-high heat. When the oil is shimmering, add breaded chicken to the pan and fry for 7 minutes, until crispy and golden brown, and flip and cook for another 5 minutes. Let cooked chicken drain on a paper towel–lined plate. Work in batches and add more oil as needed.

8. Serve with the sauce.

INGREDIENTS:

- 1 TABLESPOON UNSALTED BUTTER
- 1 SMALL YELLOW BELL PEPPER, DICED
- 1 SMALL RED BELL PEPPER, DICED
- 1 SMALL RED ONION, DICED
- 3 GARLIC CLOVES, MINCED
- 2–3 TEASPOONS RED PEPPER FLAKES, PLUS MORE FOR GARNISH
- 2½ CUPS HEAVY CREAM
- 1 CUP CHICKEN STOCK (SEE PAGE 478)
- 1¼ CUPS GRATED PARMESAN CHEESE, DIVIDED (PLUS MORE TO TASTE)
- SALT AND PEPPER, TO TASTE
- 4 BONELESS, SKINLESS CHICKEN THIGHS
- ¾ CUP PANKO
- 2 TABLESPOONS ALL-PURPOSE FLOUR
- OLD BAY SEASONING, TO TASTE
- PAPRIKA, TO TASTE
- CAYENNE PEPPER, TO TASTE
- 1 CUP MILK
- ¼ CUP CANOLA OIL

TUSCAN CHICKEN

YIELD: 4 SERVINGS / **ACTIVE TIME:** 5 MINUTES / **TOTAL TIME:** 50 MINUTES

You'll want bread on hand to mop up the creamy sauce that accompanies these chicken breasts. Or serve with your favorite pasta.

1. Pat dry the chicken with a paper towel and season it with salt, pepper, and paprika.

2. Add a tablespoon of oil to a frying pan over medium-high heat. When the oil begins to shimmer, sear the chicken until golden and cooked through, about 7 minutes per side. Set aside.

3. In the same pan, heat up the rest of the oil and fry the garlic for 1 to 2 minutes, then add the tomatoes and cook for 3 to 4 minutes. Add the mustard and mix it in.

4. Reduce heat, add the cream, bring to a gentle simmer, and season to taste.

5. Add the broccoli and cook for 2 minutes. Then add the Parmesan and cook for another 2 minutes. Add the chicken, garnish with parsley, and serve.

INGREDIENTS:

- 4 CHICKEN BREASTS
- SALT AND PEPPER, TO TASTE
- 1½ TEASPOONS PAPRIKA
- 3 TABLESPOONS OIL FROM JAR OF SUN-DRIED TOMATOES, DIVIDED
- 6 GARLIC CLOVES, MINCED
- 1 JAR SUN-DRIED TOMATOES, DRAINED
- 1 TEASPOON DIJON MUSTARD
- 1 CUP HEAVY CREAM
- 3 CUPS BROCCOLI
- ½ CUP GRATED PARMESAN CHEESE
- FRESH PARSLEY, FINELY CHOPPED, FOR GARNISH

CHICKEN TIKKA MASALA

YIELD: 4 SERVINGS / **ACTIVE TIME:** 15 MINUTES / **TOTAL TIME:** 2 HOURS AND 45 MINUTES

Chicken Tikka Masala is one of my favorite dishes in the world. I have spent the last five years passionately studying Indian food, and this recipe has been an obsession.

1. In a bowl, combine chicken, yogurt, and 2 tablespoons garam masala and mix well. Cover and refrigerate for at least 2 hours; overnight is ideal.

2. Add oil to a large sauté pan over medium-high heat. When the oil begins to shimmer, add onion and cook for 2 minutes before adding the remainder of the garam masala, chilies, ginger garlic paste, and salt. Cook for 2 minutes, stirring constantly.

3. Add brown sugar, turmeric, and chili powder, cooking for 2 minutes, and then stir in tomato puree.

4. Add diced tomatoes, cook for 4 minutes, and remove from heat. Pour this mixture into a blender and puree for 2 minutes.

5. Using the same pan that you used to make the sauce, cook the chicken for 10 minutes, stirring occasionally, and then add the blended sauce, curry leaves, and half the cilantro. Simmer for 20 minutes. If the mixture is too thin, let it reduce, if it is too thick add a little bit of water.

6. Garnish with the remaining cilantro and serve with steamed basmati rice and Naan.

INGREDIENTS:

- 2 LBS. CHICKEN BREAST OR THIGHS, CHOPPED
- 3 TABLESPOONS YOGURT
- 3 TABLESPOONS GARAM MASALA, DIVIDED
- 2 TABLESPOONS CANOLA OIL
- 1 ONION, DICED
- 2 CHILI PEPPERS (KASHMIRI CHILIES, IF YOU CAN FIND THEM)
- 2 TABLESPOON GINGER GARLIC PASTE
- 2 TEASPOONS SALT
- 1 TABLESPOON BROWN SUGAR
- 2 TEASPOONS TURMERIC
- 1 TABLESPOON CHILI POWDER
- 2 TABLESPOONS TOMATO PUREE
- 2 (14 OZ.) CANS DICED TOMATOES
- 10 CURRY LEAVES
- CHOPPED CILANTRO, TO TASTE

TO SERVE

BASMATI RICE

NAAN (SEE PAGE 721)

CHILI CHICKEN

YIELD: 4 SERVINGS / **ACTIVE TIME:** 15 MINUTES / **TOTAL TIME:** 2 HOURS AND 50 MINUTES

This is an Indo-Chinese dish that I made with my friend Aparna when she came to visit from Mumbai. This dish always reminds me of the amazing flavors that fusion cuisine delivers.

1. In a bowl, combine ½ tablespoon chili sauce, ½ tablespoon soy sauce, pepper, and vinegar and mix well. Add the chicken and marinate for at least 45 minutes, though 2 hours is ideal.

2. Sprinkle cornstarch on the chicken along with salt and mix well. Add ¼ teaspoon of the chili powder. If you wish to use egg, you can add it now. Mix well.

3. In a bowl, combine ¾ tablespoon soy sauce, ¼ tablespoon chili sauce, ½ teaspoon chili powder, and ½ teaspoon sugar and mix well.

4. Add oil to the cast-iron skillet over high heat. When the oil reaches 350°F, fry the chicken until it turns golden brown. Drain the fried chicken on a paper towel–lined plate. If you prefer, the chicken can be baked in a 450°F oven for 20 to 25 minutes.

5. Remove excess oil, leaving 1 tablespoon. Add the garlic, fry for about a minute, add onion, scallion (if using), bell pepper, and green chilies, and fry until they turn slightly soft, about 1 minute. Add the sauce that was prepared in Step 3 and bring to a simmer.

6. Add the cooked chicken and heat the mixture for 2 to 3 minutes.

7. Serve with fried rice or noodles.

INGREDIENTS:

- ¾ TABLESPOON CHILI SAUCE
- 1¼ TABLESPOONS SOY SAUCE
- ¼ TEASPOON BLACK PEPPER
- ¾ TEASPOON DISTILLED VINEGAR
- 1 LB. BONELESS CHICKEN
- 2 TABLESPOONS CORNSTARCH
- SALT, TO TASTE
- 2 TABLESPOONS ALL-PURPOSE FLOUR
- ¾ TEASPOON RED CHILI POWDER, DIVIDED
- 1 SMALL EGG (OPTIONAL)
- ½ TEASPOON SUGAR
- 2 CUPS CANOLA OIL
- 1 SCALLION, CHOPPED (OPTIONAL)
- ¾ TABLESPOON MINCED GARLIC
- 1 ONION, THINLY SLICED
- ¼ CUP CHOPPED BELL PEPPER
- 1–2 GREEN CHILIES, HALVED AND DE-SEEDED

TO SERVE
FRIED RICE OR NOODLES

CHICKEN 65

YIELD: 4 SERVINGS / **ACTIVE TIME:** 15 MINUTES / **TOTAL TIME:** 2 HOURS

Chicken 65 is an amazing Indochinese fusion appetizer. Its name originated when a restaurant put it on the menu as #65, and after all these years we still know it as Chicken 65.

1. Wash chicken under running water and pat dry.

2. In a bowl, combine ginger garlic paste, chili powder, lemon juice, black pepper, turmeric, and salt and mix well. Add the chicken and coat. Marinate for at least 1 hour (though refrigerating overnight yields the best end result).

3. When you are ready to cook the chicken, combine the cornstarch and flour in a bowl and then coat the marinated chicken in the mixture.

4. Add chili powder, sugar, garlic, salt, and yogurt to a bowl, mix well, and set aside.

5. Add 1 cup canola oil to a large pan over medium-high heat and fry the chicken in batches until golden, being sure to turn the chicken to cook evenly. Drain the cooked chicken on a paper towel–lined plate.

6. In a clean pan, heat 1 tablespoon canola oil over medium-high heat and add curry leaves, chilies, and cumin and cook until fragrant. Add the prepared seasoning mixture and cook until it begins to simmer.

7. Add the fried chicken to the sauce and cook until the chicken has absorbed most of the liquid.

INGREDIENTS:

- 1 LB. BONELESS CHICKEN THIGHS
- ½ TABLESPOON GINGER GARLIC PASTE
- ½ TEASPOON RED CHILI POWDER
- 1 TEASPOON LEMON JUICE
- ½ TEASPOON BLACK PEPPER
- ⅛ TEASPOON TURMERIC
- SALT, TO TASTE
- 2 TABLESPOONS CORNSTARCH
- 1 TABLESPOON RICE FLOUR
- ¾ TEASPOON RED CHILI POWDER
- ½ TEASPOON SUGAR
- 2 TEASPOONS CHOPPED GARLIC
- 2 TABLESPOONS PLAIN YOGURT
- 1 CUP CANOLA OIL, PLUS 1 TABLESPOON
- 3-5 CURRY LEAVES
- 2 GREEN CHILIES
- ½ TEASPOON CUMIN
- ½ TEASPOON CHOPPED GARLIC
- ½ TEASPOON BLACK PEPPER

CHINESE POTTED CHICKEN

YIELD: 4 TO 6 SERVINGS / **ACTIVE TIME:** 20 MINUTES / **TOTAL TIME:** 24 HOURS

This recipe was handed down by the Chinese father of some dear friends. I am honored to share it now.

1. In a large bowl, whisk together all of the marinade ingredients. Add the chicken pieces, cover the bowl, and refrigerate for at least 2 hours, but ideally for 24 hours.

2. When the chicken is done marinating, remove it from the bowl and pat dry. Reserve the marinade.

3. Grill the chicken on medium-high heat to char, about 4 minutes per side. Do not cook completely. Remove from the grill.

4. In a Dutch oven, place 3 pieces of bacon, add the chicken, and then place the rest of the bacon on top of the chicken. Pour the marinade over the chicken, and add the ginger and garlic.

5. Cover the pot, bring to a boil, and then lower the heat and simmer for 1½ hours.

6. Test chicken; if it pulls apart with a fork, remove it from the pot, along with the bacon. Continue to simmer the sauce. If chicken isn't fork-tender, cook for another 20 minutes before testing again. Tent the chicken and bacon with aluminum foil to keep warm.

7. In a small bowl, combine the cornstarch and water, stir into the sauce, and cook until the sauce thickens.

8. Serve chicken over rice and garnish with sesame seeds.

INGREDIENTS:

MARINADE

- 2 TABLESPOONS SOY SAUCE
- 2 TABLESPOONS WHITE WINE
- ½ TABLESPOON SESAME OIL
- 1 TABLESPOON HOISIN SAUCE
- 1½ TEASPOONS HOT SAUCE
- 1½ TEASPOONS SUGAR
- 4 GARLIC CLOVES, MINCED
- 2 TABLESPOONS MINCED GINGER
- 1 TEASPOON BLACK PEPPER

CHICKEN

- 4–6 LBS. SKIN-ON, BONE-IN CHICKEN PARTS (THIGHS WORK PARTICULARLY WELL); IF USING BREASTS, CUT IN HALF OR THIRDS
- 6 STRIPS OF BACON
- 1 CUP CHICKEN STOCK (SEE PAGE 478)
- ¼ CUP CHOPPED GINGER
- ¼ CUP CHOPPED GARLIC
- 2 TABLESPOONS CORNSTARCH, PLUS 1 TABLESPOON WATER

TO SERVE

COOKED RICE

SESAME SEEDS

OVEN CHICKEN TANDOORI

YIELD: 4 SERVINGS / **ACTIVE TIME:** 30 MINUTES / **TOTAL TIME:** 2 HOURS AND 30 MINUTES

This method uses your oven to create an amazing chicken tandoori that will rival what you get at restaurants.

1. In a bowl, combine all of the ingredients and mix well, making sure to coat the chicken. Wear gloves so your fingers don't turn red.

2. Refrigerate the marinating chicken for at least 2 hours; overnight is ideal.

3. When ready to cook the chicken, let it come to room temperature and preheat the oven to 350°F.

4. Place the chicken on a sheet pan and bake for 15 to 20 minutes, or until a meat thermometer inserted in the thickest part of a thigh registers 160°F.

5. Finish the chicken over high heat on a grill or under the broiler in the oven. The chicken doesn't need long, about 4 minutes or until it starts to char.

TANDOORI PASTE

1. In a bowl, combine all of the ingredients and mix well.

INGREDIENTS:

- TANDOORI PASTE (SEE RECIPE)
- 1 TEASPOON FRESH LEMON JUICE
- 1 TEASPOON GARAM MASALA
- 1 TEASPOON KASHMIRI CHILI POWDER
- SALT, TO TASTE
- 1 LB. SKINLESS CHICKEN THIGHS

TANDOORI PASTE

- 2 TABLESPOONS PLAIN YOGURT
- 1 TEASPOON MINCED GARLIC
- 1 TEASPOON MINCED GINGER
- 1 TEASPOON TURMERIC
- 1 TEASPOON GARAM MASALA
- 1 TEASPOON CUMIN
- 2 TEASPOONS KASHMIRI CHILI POWDER
- 1 TEASPOON FRESH LEMON JUICE

GENERAL TSO CHICKEN

YIELD: 6 SERVINGS / **ACTIVE TIME:** 25 MINUTES / **TOTAL TIME:** 50 MINUTES

I love General Tso Chicken. So much that I have worked on this recipe for a long time and now it is yours. Say goodbye to take out.

1. Prepare the chicken. In a large bowl, beat the egg white until broken down and lightly foamy. Add soy sauce, wine, and vodka and whisk to combine. Set aside half of the marinade in a small bowl. Add baking soda and cornstarch to the large bowl and whisk to combine. Add chicken to a large bowl and turn until coated thoroughly. Cover with plastic wrap and set aside.

2. Prepare the coating. In a large bowl, combine flour, cornstarch, baking powder, and salt and mix well. Add reserved marinade and whisk until mixture has coarse, mealy clumps. Set aside.

3. Prepare the sauce. In a small bowl, combine soy sauce, wine, vinegar, stock, sugar, sesame seed oil, and cornstarch and stir with a fork until cornstarch is dissolved and no lumps remain. Set aside.

4. Add oil to a Dutch oven over medium-high heat and bring to 350°F. As the oil heats up, dredge the marinated chicken in the dry coating. When the oil is ready, fry the chicken, being sure not to overcrowd the pan, and cook until golden all over, about 10 minutes. Set aside.

5. Add the sauce and peanut oil to a wok or large frying pan over medium-high. Add garlic, ginger, scallions, and chilies and cook, stirring constantly for about 3 minutes. Add the chicken to the pan and toss to coat.

6. Serve with rice.

INGREDIENTS:

CHICKEN
- 1 EGG WHITE
- 2 TABLESPOONS DARK SOY SAUCE
- 3 TABLESPOONS SHAOXING WINE
- 2 TABLESPOONS VODKA
- ¼ TEASPOON BAKING SODA
- 3 TABLESPOONS CORNSTARCH
- 1 LB. BONELESS, SKINLESS CHICKEN THIGHS, CUT INTO CHUNKS
- 4 CUPS CANOLA OIL, FOR FRYING

DRY COATING
- ½ CUP ALL-PURPOSE FLOUR
- ½ CUP CORNSTARCH
- ½ TEASPOON BAKING POWDER
- ½ TEASPOON SALT

SAUCE
- 3 TABLESPOONS DARK SOY SAUCE
- 2½ TABLESPOONS SHAOXING WINE
- 2 TABLESPOONS CHINESE RICE VINEGAR OR DISTILLED WHITE VINEGAR
- 3 TABLESPOONS CHICKEN STOCK (SEE PAGE 478)
- 4 TABLESPOONS SUGAR
- 1 TEASPOON ROASTED SESAME SEED OIL
- 1 TABLESPOON CORNSTARCH
- 2 TEASPOONS PEANUT OIL
- 2 TEASPOONS MINCED GARLIC
- 2 TEASPOONS MINCED FRESH GINGER
- 2 TEASPOONS MINCED SCALLION WHITES
- 8 SMALL DRIED RED ARBOL CHILIES

TO SERVE
RICE

THAI DUCK STIR-FRY

YIELD: 5 SERVING / **ACTIVE TIME:** 15 MINUTES / **TOTAL TIME:** 40 MINUTES

Start this Thai-inspired dish by searing duck breast and then add a bunch of fresh vegetables and sauce to make a great stir-fry.

1. Pat the duck dry with paper towels.

2. With a sharp knife, score the skin of the duck breast diagonally, making sure not to cut into the meat. Aim for a ⅛-inch-deep cut. Season with salt.

3. Place the breasts skin side down on a cold cast-iron pan and turn on the heat to medium-low. After 5 minutes, the breasts should begin to bubble a bit as the fat renders. Keep rendering the fat until the skin is golden and a meat thermometer inserted in the breast registers about 125°F.

4. Increase the heat to medium and brown the skin for another 2 minutes, then flip until the duck registers a temperature of 130°F.

5. Transfer the duck to a cutting board and tent with aluminum foil until ready to serve.

6. In a bowl, combine the fish sauce, oyster sauce, sugar, and white pepper and mix well.

7. Place the pan you used to sear the duck over medium-high heat. Pour off all but 1 tablespoon of duck fat from the pan (and reserve the rest of the fat). Add the garlic and ginger, stirring constantly until softened, about a minute. Add the chillies, bell peppers, onion, and peppercorns and cook for about 3 minutes.

8. Add the fish sauce mixture, stir to combine, and add half the Thai basil. Simmer for 2 minutes and then remove from heat. Scatter remaining basil leaves on top of sauce.

9. Serve with slices of duck over rice and generously topped with the sauce.

INGREDIENTS:

- 2 SKIN-ON DUCK BREASTS
- SALT, TO TASTE
- 2 TABLESPOONS FISH SAUCE
- 2 TABLESPOONS OYSTER SAUCE
- 1 TABLESPOON WHITE SUGAR
- 1 PINCH WHITE PEPPER
- 2 GARLIC CLOVES, MINCED
- 1 TABLESPOON MINCED GINGER
- 2 LARGE RED CHILIES, THINLY SLICED
- 1 RED BELL PEPPER, CHOPPED
- 1 GREEN BELL PEPPER, CHOPPED
- 1 ONION, THINLY SLICED
- 2 STRANDS GREEN PEPPERCORNS IN BRINE, DRAINED
- ½ BUNCH FRESH THAI BASIL LEAVES, DIVIDED

TO SERVE
STEAMED JASMINE RICE

THAI RED DUCK CURRY

YIELD: 4 SERVINGS / **ACTIVE TIME:** 15 MINUTES / **TOTAL TIME:** 30 MINUTES

Your local store will likely have precooked duck breasts available for purchase, but it's worth cooking your own just to have access to the rich rendered fat that results from searing in cast iron.

INGREDIENTS:

- 4 BONELESS, SKIN-ON DUCK BREASTS
- ¼ CUP THAI RED CURRY PASTE
- 2½ CUPS COCONUT MILK
- 10 MAKRUT LIME LEAVES (OPTIONAL)
- 1 CUP DICED PINEAPPLE
- 1 TABLESPOON FISH SAUCE, PLUS MORE TO TASTE
- 1 TABLESPOON BROWN SUGAR
- 6 BIRD'S EYE CHILIES, STEMMED
- 20 CHERRY TOMATOES
- 1 CUP BASIL (THAI BASIL STRONGLY PREFERRED)

TO SERVE

- 1½ CUPS COOKED JASMINE RICE

1. Use a very sharp knife to slash the skin on the duck breasts, while taking care not to cut all the way through to the meat.

2. Place a large cast-iron Dutch oven over medium-high heat. Place the duck breasts, skin side down, in the pot and sear until browned, about 4 minutes. This will render a lot of the fat.

3. Turn the duck breasts over and cook until browned on the other side, about 4 minutes. Remove the duck from the pot, let cool, and drain the rendered duck fat. Reserve the duck fat for another use.

4. When the duck breasts are cool enough to handle, remove the skin and discard. Cut each breast into 2-inch pieces.

5. Reduce the heat to medium, add the curry paste, and fry for 2 minutes. Add the coconut milk, bring to a boil, and cook for 5 minutes.

6. Reduce the heat, return the duck to the pot, and simmer for 8 minutes. Add the lime leaves, if using, the pineapple, fish sauce, brown sugar, and chilies, stir to incorporate, and simmer for 5 minutes. Skim to remove any fat from the top as the curry simmers.

7. Taste and add more fish sauce if needed. Stir in the cherry tomatoes and basil and serve alongside the rice.

If you are looking for a spot to utilize the duck fat you reserved in this preparation, use it in place of oil the next time you roast potatoes. It will add a crisp exterior that gives way to a fluffy, flavorful inside.

LAMB, GOAT, VENISON & RABBIT

*T*here is more to red meat than beef. This chapter introduces you to a whole new world of recipes and flavors that you might not have been exposed to before. In this section, you are going to learn the secrets to producing some outstanding dishes using lamb, goat, venison, and rabbit.

ROASTED LAMB CHOPS

YIELD: 2 SERVINGS / **ACTIVE TIME:** 10 MINUTES / **TOTAL TIME:** 25 HOURS

Rosemary makes for a wonderful herbal complement to these roasted lamb chops. Pair with mint jelly or fruit chutney.

INGREDIENTS:

- 2 TABLESPOONS FRESH ROSEMARY
- 2 TEASPOONS SALT
- 1 TEASPOON WHITE PEPPER
- 2 GARLIC CLOVES, MINCED
- ¼ CUP OLIVE OIL, DIVIDED
- 1 LB. LAMB RIB CHOPS

1. In a bowl, combine the rosemary, salt, pepper, garlic, and 2 tablespoons of the olive oil and mix well. Coat the lamb with the mixture and refrigerate, covered, for 24 hours.

2. The next day, let the lamb come to room temperature.

3. Add the remaining 2 tablespoons of olive oil to an ovenproof sauté pan over high heat. When the oil begins to simmer, sear the chops for 2 to 3 minutes per side. This will cook them to medium-rare. For lamb that is more done, place the pan in an oven at 400°F for 3 to 4 minutes, or keep them on the stove at a low heat.

BRAISED LAMB BELLY

YIELD: 6 SERVINGS / **ACTIVE TIME:** 30 MINUTES / **TOTAL TIME:** 4 HOURS

This recipe uses lamb belly to make an amazing braised meal, ranging from tacos to pasta.

1. Preheat the oven to 300°F.

2. Salt the lamb on all sides.

3. Add the canola oil to a Dutch oven over medium-high heat. When the oil begins to shimmer, brown the lamb on all sides.

4. When the lamb is browned, add the remainder of the ingredients, cover, and cook in the oven for 3½ hours, or until the meat is fork-tender.

INGREDIENTS:

- SALT, TO TASTE
- 3 LBS. LAMB BELLY
- 2 TABLESPOONS CANOLA OIL
- 1 ONION, DICED
- 2 CARROTS, DICED
- 3 BAY LEAVES
- 2 TABLESPOONS BLACK PEPPERCORNS
- 2 SPRIGS FRESH ROSEMARY
- 2 CUPS WATER

BRAISED LAMB SHOULDER WITH MINTED PEAS

YIELD: 4 TO 6 SERVINGS / **ACTIVE TIME:** 30 MINUTES / **TOTAL TIME:** 4 HOURS

This recipe was inspired by one of my farmers, Jim Czack of Loundonshire Farms, who raises Black Welsh Lamb. This breed is known for exceptional flavor and quality, and a good fat to protein ratio.

1. Preheat oven to 300°F

2. Season the lamb by rubbing salt all over the meat.

3. Add canola oil to a large saucepan over medium-high heat. When the oil shimmers, brown the lamb on all sides (about 4 minutes per side).

4. Place the remaining ingredients in a Dutch oven. Once the lamb is browned on all sides, add it to the Dutch oven, cover the pot, and place it in the oven. Cook for 3½ hours, or until the meat is fork-tender.

5. Let the lamb rest for at least 25 minutes before serving with the vegetables and Minted Peas.

MINTED PEAS

1. Place the peas and mint in a saucepan, cover with water, and simmer over medium heat until the peas are tender.

2. Drain the water, discard the mint leaves, season to taste, and serve.

INGREDIENTS:

- 5 LBS. BONE-IN LAMB SHOULDER
- ¼ CUP SALT
- 2 TABLESPOONS CANOLA OIL
- 1 SMALL ONION, QUARTERED
- 2 CARROTS, DICED
- 3 BAY LEAVES
- 2 TABLESPOONS BLACK PEPPERCORNS
- 2 SPRIGS FRESH ROSEMARY
- 2 CUPS WATER

MINTED PEAS

- 3 CUPS PEAS
- 3 SPRIGS FRESH MINT
- SALT, TO TASTE

SMOKED LAMB SHOULDER

YIELD: 8 SERVINGS / **ACTIVE TIME:** 30 MINUTES / **TOTAL TIME:** 5 HOURS

Time to break out the smoker, because this lamb shoulder is worth the wait. It is packed with flavor and makes amazing sandwiches if you have anything leftover.

INGREDIENTS:

- 5 LBS. BONELESS LAMB SHOULDER
- 2 CUPS APPLE CIDER VINEGAR, DIVIDED
- 1 CUP APPLE JUICE
- ¼ CUP OLIVE OIL
- 2 TABLESPOONS SALT
- 2 TABLESPOONS BLACK PEPPER
- 1 TABLESPOON DRIED ROSEMARY

1. Bring the smoker to 225°F.

2. Trim excess fat from lamb and inject it with about 1 cup of apple cider vinegar. Make a spritz by combining 1 cup apple cider vinegar with apple juice in a spray bottle and shaking well. Set aside.

3. Pat dry the meat with paper towels and then rub all over with olive oil, salt, pepper, and rosemary.

4. Make sure there is enough water in the drip tray and place the lamb, uncovered, on the smoker for an hour. After the first hour, spray with the spritz every 15 minutes, until a meat thermometer registers an internal temperature of 165°F, about 3½ hours.

5. When the internal temperature reaches 165°F, remove the lamb from the smoker, tightly wrap it in foil, and return it to the smoker until the internal temperature reaches 195°F.

6. Remove from the smoker and let rest for at least 1 hour.

ROSEMARY & LEMON-OILED LEG OF LAMB

YIELD: 4 SERVINGS / **ACTIVE TIME:** 45 MINUTES / **TOTAL TIME:** 14 HOURS

As the rosemary and lemon flavors are relatively soft, they go perfectly when used in a marinade for a leg of lamb.

INGREDIENTS:

- ¾ CUP OLIVE OIL
- ¼ CUP ROSEMARY LEAVES, COARSELY CHOPPED
- 3 LEMONS, JUICED
- 4 GARLIC CLOVES, FINELY CHOPPED
- 6 LBS. BONELESS LEG OF LAMB, BUTTERFLIED
- COARSELY GROUND BLACK PEPPER, TO TASTE
- SEA SALT, TO TASTE

1. The day before you plan to grill, combine the olive oil, rosemary leaves, lemon juice, and garlic in a roasting pan and mix thoroughly.

2. Place the leg of lamb on a large carving board. Season generously with coarsely ground black pepper and sea salt, kneading the lamb so that the pepper and salt are pressed in. Place the seasoned leg of lamb in the roasting pan.

3. Transfer the pan to the refrigerator and let the meat marinate overnight. Note that the olive oil may not cover the meat entirely; in that case, flip the meat once halfway through the marinating process.

4. An hour before grilling, remove the leg of lamb from the refrigerator and let stand at room temperature for at least 1 hour. Reserve the remaining marinade as it will be used for brushing the meat while it is grilled.

5. A half hour before grilling, prepare your gas or charcoal grill to medium-high heat.

6. When the coals are ready, at about 400°F with the coals lightly covered with ash, place the marinated leg of lamb on the grill and cook for about 16 minutes per side for medium-rare, and 17 minutes for medium. While grilling, brush the remaining marinade on top of the lamb. When finished, transfer the lamb to a large carving board and let rest for 15 minutes, allowing for the meat to properly store its juices.

7. Before serving, slice the lamb into ½-inch-thick diagonal strips. Serve warm.

GRILL-ROASTED RACK OF LAMB WITH GARLIC-HERB CRUST

YIELD: 5 TO 6 SERVINGS / **ACTIVE TIME:** 20 MINUTES / **TOTAL TIME:** 14 HOURS

Because the rack of lamb is a very delicate meat, be sure to give it the time to marinate overnight. I suggest pairing this with a glass of red wine.

1. The night before grilling, combine the olive oil, garlic, and lemon zest in a large sealable plastic bag. Pat dry the racks of lamb, and then season them with coarsely ground black pepper and fresh sea salt, kneading the pepper and salt deeply into the meaty sections of the lamb. Add the racks of lamb to a plastic bag and place in the refrigerator. Let marinate overnight.

2. An hour and a half before grilling, remove the racks of lamb from the refrigerator and let rest, uncovered and at room temperature.

3. A half hour before grilling, prepare your gas or charcoal grill to medium heat.

4. While the grill heats, combine all of the ingredients for the garlic-herb crust in a small bowl. Next, take the racks of lamb and generously apply the crust ingredients to it, being sure to apply the majority of the crust on the meaty side of the rack.

5. When the grill is ready, at about 400°F with the coals are lightly covered with ash, place the meat-side of the racks of lamb on the grill and cook for about 3 to 4 minutes. When the crusts are browned, flip the racks of lamb and grill for another 5 minutes for medium-rare.

6. Transfer the racks of lamb from the grill to a large carving board and let rest for about 10 minutes before slicing between the ribs. Serve warm.

INGREDIENTS:

- 2 TABLESPOONS OLIVE OIL
- 2 GARLIC CLOVES, FINELY CHOPPED
- 1 TEASPOON LEMON ZEST
- 2 (8-RIB) RACKS LAMB, ABOUT 1 LB. EACH
- COARSELY GROUND BLACK PEPPER, TO TASTE
- SEA SALT, TO TASTE

GARLIC-HERB CRUST

- 4 GARLIC CLOVES, FINELY CHOPPED
- ½ SMALL SHALLOT, FINELY CHOPPED
- ¼ CUP FLAT-LEAF PARSLEY, COARSELY CHOPPED
- 2 TABLESPOONS ROSEMARY, FINELY CHOPPED
- 1 TABLESPOON THYME, FINELY CHOPPED
- 1 TABLESPOON OLIVE OIL
- COARSELY GROUND BLACK PEPPER, TO TASTE
- SEA SALT, TO TASTE

GO-TO RECIPES

LAMB STEW WITH CANDIED CARROTS AND GARLIC-ROASTED LITTLE CREAMER POTATOES

YIELD: 4 TO 6 SERVINGS / **ACTIVE TIME:** 30 MINUTES / **TOTAL TIME:** 2 HOURS

An ideal, flavorful family meal.

1. Preheat oven to 250°F.

2. In a mixing bowl, add flour and then the lamb pieces. Toss until each piece is evenly coated.

3. In a Dutch oven, add the butter and cook over medium heat until melted.

4. Add the lamb and any additional flour. Cook for 5 minutes, or until the lamb is nicely browned.

5. Add the onion, carrot, leek, and garlic and cook for 5 minutes, or until soft.

6. Add the wine and cook for 5 minutes.

7. Add the stock, bay leaves, and thyme. Raise the heat to medium-high and bring to a boil.

8. Cover the Dutch oven and place it in the oven for 1 hour; remove every 15 minutes to stir the contents.

9. While the stew is cooking, warm the oil in a small sauté pan and then add the sausage. Cook for 5 minutes, or until nicely browned.

10. Increase the oven temperature to 325°F. Remove cover, and add sausage, olives, and sherry vinegar. Cook for an additional 20 minutes, or until lamb is tender.

11. Skim any fat off the top and season with salt and pepper.

12. Remove the thyme and bay leaves and serve in bowls with Candied Carrots and Garlic Roasted Little Creamer Potatoes.

INGREDIENTS:

- ⅓ CUP ALL-PURPOSE FLOUR
- 2 LBS. BUTTERFLIED LEG OF LAMB, FAT AND SILVER SKIN REMOVED, CUT INTO 1-INCH PIECES
- 4 TABLESPOONS BUTTER
- 1 ONION, PEELED AND CHOPPED
- 1 CARROT, PEELED AND CHOPPED
- ½ LEEK, WHITE PART ONLY, CHOPPED
- 5 GARLIC CLOVES, MINCED
- 1 CUP RED WINE
- 2 CUPS BEEF STOCK (SEE PAGE 475)
- 2 BAY LEAVES
- 6 SPRIGS FRESH THYME
- 1 TABLESPOON VEGETABLE OIL
- ½ LB. LAMB OR PORK SAUSAGE, THICKLY SLICED
- ½ CUP KALAMATA OLIVES, PITTED AND SLICED
- 1 TABLESPOON SHERRY VINEGAR
- SALT AND PEPPER, TO TASTE

CANDIED CARROTS

1. In a medium saucepan, bring 6 cups of water to boil. Add the carrots and cook for 8 minutes, or until tender.

2. Drain the carrots. Reduce the heat to the lowest possible setting and return the carrots to the pan. Stir in butter, brown sugar, salt, and pepper. Cook for 3 to 5 minutes, while stirring, until the carrots are evenly coated.

GARLIC-ROASTED LITTLE CREAMER POTATOES

1. Preheat oven to 350°F.

2. Place the potatoes on a baking tray, drizzle with the olive oil, and season with salt and pepper.

3. Place in oven and cook for 10 minutes.

4. Remove the tray from oven and sprinkle the garlic over the potatoes. Return to the oven and cook for an additional 20 minutes, or until the potatoes are cooked through.

5. Remove from pan, toss with the chopped parsley, and serve.

INGREDIENTS:

CANDIED CARROTS

- 5 CARROTS, PEELED AND CUT INTO 1-INCH PIECES
- 2 TABLESPOONS BUTTER
- ¼ CUP LIGHT BROWN SUGAR
- SALT AND PEPPER, TO TASTE

GARLIC-ROASTED LITTLE CREAMER POTATOES

- 1 LB. BABY CREAMER POTATOES
- 2 TABLESPOONS EXTRA VIRGIN OLIVE OIL
- SALT AND PEPPER, TO TASTE
- 6 GARLIC CLOVES, MINCED
- PARSLEY, CHOPPED, FOR GARNISH

VENISON STEW

YIELD: 5 SERVINGS / **ACTIVE TIME:** 10 MINUTES / **TOTAL TIME:** 2 HOURS

This venison stew is perfect to have simmering away on the stove on a cold day.

1. In a bowl, combine flour, garlic powder, salt, and pepper and mix well. Add the venison and toss to coat.

2. Add oil to a large pot over medium-high heat. When the oil begins to shimmer, add the meat and onion and cook until browned, about 10 minutes.

3. Add wine and stock and, using a wooden spatula, scrape the sides of the pot to loosen all the brown bits.

4. Add all of the remaining ingredients, except for the cornstarch, water, and peas. Bring to a boil and then lower the heat and simmer for at least 1 hour, until the meat is fork-tender.

5. Mix equal parts cornstarch and water to create a slurry. Increase the heat and return the stew to a boil and add the slurry, continuing to boil to reach desired consistency. When that is achieved, lower the heat, add the peas, and simmer for 5 minutes before serving.

INGREDIENTS:

- ½ CUP ALL-PURPOSE FLOUR
- 1 TEASPOON GARLIC POWDER
- SALT AND PEPPER, TO TASTE
- 2 LBS. VENISON TIPS
- 4 TABLESPOONS OLIVE OIL
- 1 CUP CHOPPED ONION
- 2 CUPS RED WINE
- 2 CUPS BEEF STOCK (SEE PAGE 475)
- 1 CUP CARROTS, CHOPPED
- 1 CUP CELERY, CHOPPED
- 4 BAY LEAVES
- 1 SPRIG FRESH ROSEMARY, CHOPPED
- 1 CUP CUBED POTATOES
- 3 TABLESPOONS CORNSTARCH
- 3 TABLESPOONS WATER
- 1 CUP PEAS

RABBIT & ROOT VEGETABLE STEW

YIELD: 12 SERVINGS / **ACTIVE TIME:** 20 MINUTES / **TOTAL TIME:** 3 HOURS

Rabbit is a fantastic protein that works well in place of many braised poultry dishes.

1. Add oil and sliced onions to a Dutch oven over medium-high heat and sauté the onions for 3 to 4 minutes. Add the rabbit breasts, making sure to sear on both sides.

2. Add 1 cup stock, cover, and bring to a boil. Once a boil is achieved, lower heat and simmer for 45 minutes, or until the meat is fork-tender.

3. Add the remaining ingredients, except the cornstarch, to the pot, bring to a boil, and then lower the heat. Simmer for 2 hours, until reduced by one-third.

4. In a small bowl, combine the cornstarch and ¼ cup of water and mix well. Add this slurry into the soup and stir until thickened.

5. Remove the herb stems and bay leaves, season to taste, and serve.

INGREDIENTS:

- 1 TABLESPOON CANOLA OIL
- 2 ONIONS, CHOPPED, PLUS 1 CUP SLICED ONIONS FOR BRAISING
- 1 LB. RABBIT BREASTS
- 2 QUARTS, PLUS 1 CUP CHICKEN STOCK (SEE PAGE 478)
- 1 LB. CELERIAC, PEELED AND CHOPPED
- 1 LB. POTATOES, PEELED AND CHOPPED
- 1 LB. RUTABAGA, PEELED AND CHOPPED
- 3 SPRIGS FRESH THYME
- 3 SPRIGS FRESH ROSEMARY
- 4 BAY LEAVES
- 4 CUPS WATER
- SALT AND PEPPER, TO TASTE
- ¼ CUP CORNSTARCH

RABBIT RAGÙ

YIELD: 4 SERVINGS / **ACTIVE TIME:** 30 MINUTES / **TOTAL TIME:** 2 HOURS

Rabbit is the ultimate sustainable protein. It's lean, it tastes great, and it's eco-friendly. Cooking the rabbit on the bone keeps it tender so that it can add even more flavor to an already complex and delicious sauce.

1. Place the olive oil in a large cast-iron Dutch oven and warm over medium heat. Season the pieces of rabbit with salt and pepper, dredge them in the flour, and shake to remove any excess. When the oil starts to smoke, add the rabbit and cook until the pieces are browned on both sides, about 6 minutes per side.

2. Use a slotted spoon to remove the rabbit and set it aside. Add the onion, celery, carrots, and garlic and sauté until browned. Add the tomato paste, stir to coat the vegetables, and cook for approximately 4 minutes.

3. Add the wine and let the mixture come to a boil. Lower the heat and cook until the wine has reduced by half, about 10 minutes.

4. Add the tomatoes, stock, bay leaves, rosemary, and thyme and simmer for 2 minutes before adding the rabbit back to the pot. Simmer for 1½ hours.

5. Use a slotted spoon to remove the rabbit and set it aside. Remove the bay leaves, rosemary, and thyme. When the rabbit is cool enough to handle, pick the meat off the bones and shred it. Discard the bones and return the meat to the sauce.

6. To serve, spoon the ragù over the pappardelle or polenta and garnish with the Parmesan.

INGREDIENTS:

- ¼ CUP OLIVE OIL
- 2½- TO 3½-LB. RABBIT, CUT INTO 8 BONE-IN PIECES
- SALT AND GROUND BLACK PEPPER, TO TASTE
- 1 CUP ALL-PURPOSE FLOUR
- 1 YELLOW ONION, DICED
- 2 CELERY STALKS, DICED
- 2 CARROTS, PEELED AND DICED
- 1 GARLIC CLOVE, MINCED
- 2 TABLESPOONS TOMATO PASTE
- ½ CUP FULL-BODIED RED WINE
- 1 (28 OZ.) CAN CRUSHED TOMATOES
- 1 CUP CHICKEN STOCK (SEE PAGE 478)
- 2 BAY LEAVES
- 2 SPRIGS OF ROSEMARY
- 4 SPRIGS OF THYME
- 2 CUPS WATER
- 2 TABLESPOONS UNSALTED BUTTER, CUT INTO SMALL PIECES
- 1 LB. COOKED PAPPARDELLE OR 4 CUPS COOKED POLENTA
- PARMESAN CHEESE, GRATED, FOR GARNISH

COTTAGE PIE

YIELD: 4 TO 6 SERVINGS / **ACTIVE TIME:** 20 MINUTES / **TOTAL TIME:** 1 HOUR AND 5 MINUTES

Most people are familiar with shepherd's pie, but cottage pie uses lamb and peas to give this minced meat pie some exceptional flavor. This recipe is perfect for those cool spring days when you need a hardy dish.

1. Preheat oven to 350°F.

2. Add the oil and butter to a saucepan over medium-high heat. When the butter begins to foam, add the onion and sauté until slightly golden in color.

3. Add the lamb and continue to cook until brown, then pour in stock and reduce heat to medium-low and simmer for 20 minutes. Add the peas and remove the pan from heat.

4. Place the meat mixture in a casserole dish, cover with mashed potatoes, and bake for 40 minutes, or until potatoes are golden brown.

INGREDIENTS:

- 1 TABLESPOON CANOLA OIL
- 1 TABLESPOON UNSALTED BUTTER
- 1 SPRING ONION, SLICED
- 1 LB. GROUND LAMB
- 1 CUP BEEF STOCK (SEE PAGE 475)
- 1 CUP PEAS
- 5 POTATOES, PEELED, COOKED, AND MASHED
- 1 TABLESPOON CHOPPED FRESH THYME

MUTTON BROTH

YIELD: 6 SERVINGS / **ACTIVE TIME:** 20 MINUTES / **TOTAL TIME:** 2 HOURS AND 15 MINUTES

This spectacular, simple soup is prepared using traditional methods employed in both Ireland and Australia. It pairs well with roasted or boiled potatoes.

1. In a large saucepan, add all of the ingredients and bring to a boil.

2. Reduce heat so that the soup simmers, cover, and cook for 1 to 2 hours or until the mutton is tender.

3. Season with salt and pepper and serve in bowls sprinkled with the parsley.

INGREDIENTS:

- 2 LBS. MUTTON LEG, CUT INTO ½-INCH PIECES
- 2 CARROTS, PEELED AND CHOPPED
- 1 ONION, PEELED AND CHOPPED
- 2 LEEKS, CHOPPED
- 1 TABLESPOON PEARL BARLEY
- 8 CUPS WATER
- SALT AND PEPPER, TO TASTE
- PARSLEY, CHOPPED, FOR GARNISH

BULGARIAN SOUR LAMB SOUP WITH PAPRIKA BUTTER

YIELD: 4 SERVINGS / **ACTIVE TIME:** 25 MINUTES / **TOTAL TIME:** 1 HOUR AND 35 MINUTES

This preparation utilizes lamb, but chicken and pork can also be used. I prefer to place the Paprika Butter in the soup cold, as it adds a nice temperature balance, and the butter's fat pairs nicely with the acidity in the soup.

1. In a medium saucepan, add the oil and cook over medium heat until warm. Add the lamb and cook for 5 minutes, or until it is brown on all sides.

2. Add the onion and cook 5 minutes, or until soft.

3. Sprinkle in the flour and the paprika and cook for 2 minutes. Add the stock, while stirring vigorously, and let soup simmer for 10 minutes.

4. Tie the parsley, scallions, and dill together and add to the soup.

5. Add the rice, bring to a boil, and reduce the heat so that the soup simmers. Cook for 30 to 40 minutes, or until the lamb is tender.

6. Remove the pan from heat and stir in the eggs.

7. Add the vinegar, discard the bunch of herbs, and season to taste.

8. Serve in a warmed bowl. Place a coin of Paprika Butter on top, allow it to melt, and then garnish with the dill.

PAPRIKA BUTTER

1. In a bowl, add the butter and paprika and whisk together. Quenelle the butter and roll it into logs in plastic wrap.

2. Chill until ready to serve.

INGREDIENTS:

- 2 TABLESPOONS VEGETABLE OIL
- 1 LB. LAMB SHOULDER, TRIMMED AND CUBED
- 1 ONION, PEELED AND DICED
- 2 TABLESPOONS ALL-PURPOSE FLOUR
- 1 TABLESPOON PAPRIKA
- 4 CUPS VEAL STOCK (SEE PAGE 476)
- 3 SPRIGS FRESH PARSLEY
- 4 SCALLIONS
- 4 SPRIGS FRESH DILL, PLUS MORE FOR GARNISH
- ¼ CUP LONG-GRAIN RICE
- 2 EGGS, BEATEN
- 2-3 TABLESPOONS DISTILLED VINEGAR
- SALT AND PEPPER, TO TASTE
- PAPRIKA BUTTER (SEE RECIPE)

PAPRIKA BUTTER

- 4 TABLESPOONS BUTTER, SOFTENED
- 2 TEASPOONS PAPRIKA

SMOKED LAMB PÂTÉ

YIELD: 6 SERVINGS / **ACTIVE TIME:** 15 MINUTES / **TOTAL TIME:** 24 HOURS

After you've made that lovely Smoked Lamb Shoulder (see page 385), use the drippings and extra meat to make this incredible pâté, which goes wonderfully on crusty sourdough or crackers.

1. Add the oil and onion to a frying pan over medium-high heat and cook onion until golden.

2. In a blender, combine all of the ingredients and puree. Add extra fat as needed to get the mixture smooth.

3. Place the pâté in a glass jar and pour an extra layer of fat over the top. Cover and refrigerate for 24 hours before serving.

INGREDIENTS:

- 3 TABLESPOONS OLIVE OIL
- 1 ONION, DICED
- ½ CUP SMOKED LAMB SHOULDER (SEE PAGE 385)
- 1½ CUPS LAMB FAT (USE THE DRIPPINGS FROM THE SHOULDER)
- ½ TEASPOON CLOVES
- 1 TEASPOON BLACK PEPPER
- 2 TEASPOONS SALT

LAMB BACON

YIELD: 6 SERVINGS / **ACTIVE TIME:** 10 MINUTES / **TOTAL TIME:** 5 DAYS

Lamb belly makes amazing bacon. This recipe takes some time, but it is really easy and the results are wonderful.

INGREDIENTS:

- 5 TABLESPOONS SALT
- 2 TEASPOONS CURING SALT
- 4 TABLESPOONS DARK BROWN SUGAR
- 4 DRIED BAY LEAVES, CRUSHED
- 1 TABLESPOON BLACK PEPPER
- 1½–2 LBS. BONELESS LAMB BELLY
- APPLEWOOD CHIPS

1. In a bowl, combine the salts, brown sugar, bay leaves, and pepper. Dredge the lamb belly in the rub and massage it into the meat. Place the lamb belly on a sheet pan, uncovered, and refrigerate for 5 days, turning the lamb over once a day.

2. After 5 days, rinse the lamb belly thoroughly, pat it dry with paper towels, and refrigerate, uncovered, overnight.

3. Soak the wood chips in water for 30 minutes, then drain and pat dry. Bring the smoker to 200°F.

4. Add the lamb and smoke at 200°F until a meat thermometer inserted into the center of the meat reads 160°F. Cook for 3 hours. If it registers 160°F before 3 hours have passed, let it cook. This will help develop more flavor.

5. Allow the bacon to cool completely in the refrigerator before using, or wrapping and storing; it will stay good for up to 1 month.

GOAT MEATBALLS

YIELD: 5 SERVINGS / ACTIVE TIME: 15 MINUTES / TOTAL TIME: 35 MINUTES

The red pepper and herbs work really well with the goat in these meatballs.

1. Preheat the oven to 425°F.

2. In bowl, combine all of the ingredients and mix until just incorporated. Form into 1-inch meatballs.

3. Grease a sheet pan, place the meatballs on the pan, and bake for 20 minutes.

INGREDIENTS:

- 1 LB. GROUND GOAT
- ½ ONION, DICED
- 2 EGGS
- 1 TEASPOON DRIED ROSEMARY
- 1 TEASPOON DRIED THYME
- ½ CUP MILK
- 1 TEASPOON CRUSHED RED PEPPER
- 1 TEASPOON PEPPER
- 2 TABLESPOONS GRATED PARMESAN CHEESE

VENISON & PORK MEATBALLS

YIELD: 4 SERVINGS / ACTIVE TIME: 15 MINUTES / TOTAL TIME: 35 MINUTES

Venison is very lean. The addition of pork brings much needed fat and flavor to these amazing meatballs.

1. Preheat the oven to 425°F.

2. In a large bowl, combine the bread crumbs and milk and let sit for 15 minutes.

3. When the soaked bread crumbs are ready, add the remainder of the ingredients and mix until just combined.

4. Using your hands, form 1-inch meatballs.

5. Grease a sheet pan, place the meatballs on the pan, and bake for 20 minutes.

INGREDIENTS:

- ½ CUP BREAD CRUMBS
- ½ CUP MILK
- 1 LB. LEAN GROUND PORK
- 1 LB. GROUND VENISON
- ½ ONION, DICED
- 2 EGGS
- 1 TEASPOON DRIED ROSEMARY
- 1 TEASPOON DRIED THYME
- 1 TEASPOON CRUSHED RED PEPPER
- 1 TEASPOON PEPPER
- 2 TABLESPOONS GRATED PARMESAN CHEESE

GOAT MOMOS

YIELD: 6 SERVINGS / **ACTIVE TIME:** 30 MINUTES / **TOTAL TIME:** 2 HOURS AND 30 MINUTES

A momo is a Nepali-style dumpling that is typically steamed or fried. These goat momos are absolutely delicious. Pair this with a mint or chili chutney for an amazing spin on a dipping sauce.

1. In a large bowl, combine all of the ingredients, except the dough, and mix well. Cover the bowl with plastic wrap and refrigerate for an hour.

2. Knead the dough for 1 minute and then prepare 1-inch balls by rolling pieces of dough into circles between your palms.

3. Dust your work surface with flour and gently flatten the ball with your palm to make a 2-inch circle. Make a few semi-flattened circles and cover with a bowl.

4. Use a rolling pin to roll out each flattened circle into a wrapper. When rolling out the dough, make sure that the middle of the sphere is a bit thicker than the sides; this will help with the structure.

5. Hold the edges of the semi-flattened dough with one hand and with the other hand begin rolling the edges of the dough out, swirling a bit at a time. Continue until the wrapper has a circular shape. Repeat with the remaining semi-flattened dough circles. Cover with a bowl or damp cloth to prevent drying.

6. Hold the wrapper in one palm, put one tablespoon of filling mixture in the wrapper, and with the other hand bring all edges together to the center, making the pleats. Pinch and twist the pleats to close the stuffed dumpling.

7. Heat up a steamer and oil the steamer rack well. Arrange the uncooked dumplings in the steamer. Close the lid and steam until the dumplings are cooked through, about 10 to 13 minutes.

INGREDIENTS:

- 2 LBS. GROUND GOAT
- 1 CUP DICED RED ONION
- 3 TABLESPOONS CHOPPED FRESH CILANTRO
- 1 TABLESPOON MINCED GARLIC
- 1 TABLESPOON MINCED GINGER
- ¼ TEASPOON FRESHLY GRATED NUTMEG
- ½ TEASPOON TURMERIC
- 1 TABLESPOON CURRY POWDER
- 3 FRESH RED CHILIES, MINCED (OR TO TASTE)
- 3 TABLESPOONS CANOLA OIL
- SALT AND PEPPER TO TASTE
- MOMO DOUGH (SEE RECIPE)

MOMO DOUGH

- 4 CUPS ALL-PURPOSE FLOUR
- 1 TABLESPOON OIL
- 1¾ CUPS WATER, PLUS MORE AS NEEDED
- 1 PINCH SALT

MOMO DOUGH

1. In a large bowl, combine flour, oil, water, and salt and mix well.

2. Knead until the dough is smooth, about 8 to 10 minutes. If the dough seems dry, add water a tablespoon at a time.

3. Cover and let it sit out at room temperature for at least 30 minutes. Knead well again for about 5 minutes. Set aside.

LAMB KABOBS

YIELD: 12 SERVINGS / **ACTIVE TIME:** 15 MINUTES / **TOTAL TIME:** 1 HOUR

These simple lamb kabobs are perfect with pita and freshly pickled vegetables.

1. Preheat the grill to high and soak skewers in water for 15 to 25 minutes.

2. In a large bowl, combine all of the ingredients and mix with your hands. Form meat around the skewers, making an oblong sausage shape. Refrigerate the kabobs for 30 minutes.

3. Remove the skewers from the refrigerator and spray the grill with nonstick spray or oil. Grill the kabobs until they char on the outside, about 2 to 3 minutes, and rotate to cook evenly. The kabobs will be done in about 6 to 8 minutes.

4. Allow to cool for 5 minutes before removing the meat from the skewers and serving.

INGREDIENTS:

- 12 (6-INCH) WOODEN SKEWERS
- 1½ LBS. GROUND LAMB
- ½ CUP MINCED ONION
- 1 TEASPOON CHOPPED FRESH ROSEMARY
- 3 GARLIC CLOVES, MINCED
- 1 TEASPOON GROUND CUMIN
- ½ TEASPOON DRIED THYME
- SALT AND FRESHLY GROUND BLACK PEPPER, TO TASTE

MASALA LAMB KABOBS

YIELD: 12 SERVINGS / ACTIVE TIME: 15 MINUTES / TOTAL TIME: 1 HOUR

This Indian-inspired kabob is delicious with naan or roti.

1. Preheat the grill to high and soak skewers in water for 15 to 25 minutes.

2. In a large bowl, combine all of the ingredients and mix with your hands. Form meat around the skewers, making an oblong sausage shape. Refrigerate the kabobs for 30 minutes.

3. Remove the skewers from the refrigerator and spray the grill with nonstick spray or oil. Grill the kabobs until they char on the outside, about 2 to 3 minutes, and rotate to cook evenly. The kabobs will be done in about 6 to 8 minutes.

4. Allow to cool for 5 minutes before removing the meat from the skewer and serving.

INGREDIENTS:

- 12 (6-INCH) WOODEN SKEWERS
- 1½ LBS. GROUND LAMB
- ½ CUP MINCED ONION
- 1 TEASPOON CURRY POWDER
- ½ TEASPOON RED CHILI POWDER
- ½ TABLESPOON MINCED GINGER
- 1 TEASPOON GARAM MASALA
- 3 GARLIC CLOVES, MINCED
- 1 TEASPOON GROUND CUMIN
- ¼ CUP CHOPPED CILANTRO
- 1 TEASPOON GROUND CORIANDER
- SALT AND FRESHLY GROUND BLACK PEPPER, TO TASTE

LAMB BURGERS

YIELD: 4 SERVINGS / **ACTIVE TIME:** 15 MINUTES / **TOTAL TIME:** 30 MINUTES

Keep this burger as simple as possible to really let the lamb be the star.

1. Preheat your outdoor grill or cast-iron pan. If you don't have either, a frying pan will do. (And if using a pan, lightly grease it with neutral oil.)

2. In a bowl, combine all the ingredients until the mixture is well incorporated.

3. Divide the mixture into four equal parts. Form patties by rolling the mixture into a ball in your hands and pressing it flat to form the patty. The patty should be about 1-inch thick.

4. Turn the grill to medium-high heat; if you're using a pan, set your burner on medium-high. Place patties on the grill or pan. Cover the patties and cook for 5 minutes.

5. Flip the patties and cover again, cooking for an additional 5 to 6 minutes, until you have the desired doneness.

INGREDIENTS:

- 1 LB. GROUND LAMB
- 1 TEASPOON SALT
- ½ TEASPOON WHITE PEPPER
- 1 EGG WHITE

LAMB & BEEF BURGERS

YIELD: 4 SERVINGS / **ACTIVE TIME:** 15 MINUTES / **TOTAL TIME:** 30 MINUTES

Lamb gives this burger a richness and delightful gamey flavor that will keep you coming back for more.

1. Preheat your outdoor grill or cast-iron pan. If you don't have either, a frying pan will do. (And if using a pan, lightly grease it with neutral oil.)

2. In a bowl, combine all the ingredients until the mixture is well incorporated.

3. Divide the mixture into four equal parts. Form patties by rolling the mixture into a ball in your hands and pressing it flat to form the patty. The patty should be about 1-inch thick.

4. Turn the grill to medium-high heat; if you're using a pan, set your burner on medium-high. Place patties on the grill or pan. Cover the patties and cook for 5 minutes.

5. Flip the patties and cover again, cooking for an additional 5 to 6 minutes, until you have the desired doneness.

INGREDIENTS:

- ½ LB. GROUND BEEF (80-20)
- ½ LB. GROUND LAMB
- 1 TEASPOON SALT
- ½ TEASPOON WHITE PEPPER
- 1 EGG WHITE
- ⅓ CUP BREAD CRUMBS

GREEK BURGERS WITH FETA CHEESE

YIELD: 4 SERVINGS / **ACTIVE TIME:** 15 MINUTES / **TOTAL TIME:** 30 MINUTES

The salty and savory flavor of feta cheese is the perfect topping for this lamb and beef burger.

1. Preheat your outdoor grill or cast-iron pan. If you don't have either, a frying pan will do. (And if using a pan, lightly grease it with neutral oil.)

2. In a bowl, combine all the ingredients, except the feta cheese, until the mixture is well incorporated.

3. Divide the mixture into four equal parts. Form patties by rolling the mixture into a ball in your hands and pressing it flat to form the patty. The patty should be about 1-inch thick.

4. Turn the grill to medium-high heat; if you're using a pan, set your burner on medium-high. Place patties on the grill or pan. Cover the patties and cook for 5 minutes.

5. Flip the patties and cover again, cooking for an additional 5 to 6 minutes, until you have the desired doneness. Top with feta.

INGREDIENTS:

- ½ LB. GROUND BEEF (80-20)
- ½ LB. GROUND LAMB
- 1 TEASPOON SALT
- ½ TEASPOON WHITE PEPPER
- 1 EGG WHITE
- ⅓ CUP BREAD CRUMBS
- ⅓ CUP FETA CHEESE

GOAT BURGERS

YIELD: 4 SERVINGS / **ACTIVE TIME:** 15 MINUTES / **TOTAL TIME:** 30 MINUTES

A goat burger might sound different depending on where you live in the world, but I think this might change your mind. The deep flavor of goat mixed with the sweet onions works perfectly.

1. Add canola oil to a frying pan over medium heat. When the oil begins to shimmer, add the onion and a pinch of salt and cook until the onions turn golden brown, about 10 minutes. Set aside.

2. Preheat an outdoor grill or cast-iron pan.

3. In a bowl, combine the goat, salt, and white pepper and mix with your hands until it is well incorporated.

4. Divide the mixture into four equal parts. Form patties by rolling the mixture into a ball in your hands and pressing it flat to form the patty. You should aim to have the patty be about 1-inch thick.

5. Cook the patties, covered, for 5 minutes. Flip the burger and cover again, cooking for an additional 5 to 6 minutes, until you have the desired doneness.

6. Top the burger with the browned onions and condiments of your choice.

INGREDIENTS:

2	TABLESPOONS CANOLA OIL
1	ONION, THINLY SLICED
1	TEASPOON SALT, PLUS A PINCH
1	LB. GROUND GOAT
½	TEASPOON WHITE PEPPER

INDIAN-STYLE GOAT CURRY

YIELD: 4 SERVINGS / **ACTIVE TIME:** 15 MINUTES / **TOTAL TIME:** 26 HOURS

This simple goat curry is packed with flavor. Slow simmering ensures super-tender goat that has soaked up all the flavors.

1. In a bowl, combine goat, yogurt, and 2 tablespoons garam masala and mix well. Cover with plastic wrap and refrigerate for at least 2 hours, but ideally overnight.

2. Add oil to a cast-iron skillet over high heat. When the oil begins to shimmer, add the cumin and mustard seeds. When the seeds start to pop, add the onion and curry leaves.

3. After 4 minutes add ginger-garlic paste, salt, and chilies and stir constantly for 2 minutes before adding turmeric, chili powder, cumin, and remaining garam masala.

4. Add tomato puree and cook for 2 minutes, stirring constantly. Add diced tomatoes, cook for 4 minutes, and then remove from heat. Blend the sauce.

5. Sear the goat on all sides in the pan that was used to make the sauce.

6. Once the goat is browned, return the blended sauce to the pan along with 1 cup water. Cover and simmer for 2 hours, or until goat is fork-tender. If the curry is too thin, let it reduce, if it is too thick add a little bit of water. Garnish with the cilantro and serve.

INGREDIENTS:

- 2 LBS. GOAT MEAT, TYPICALLY SHANK, NECK, OR CUBED SHOULDER
- 3 TABLESPOONS YOGURT
- 3 TABLESPOONS GARAM MASALA POWDER, DIVIDED
- 1 TABLESPOON CANOLA OIL
- 2 TEASPOONS CUMIN SEED
- 2 TEASPOONS MUSTARD SEED
- 1 ONION, SLICED
- 10 CURRY LEAVES
- 2 TABLESPOONS GINGER GARLIC PASTE
- 2 TEASPOONS SALT
- 3 CHILIES (KASHMIRI CHILIES IF YOU CAN FIND THEM)
- 2 TEASPOONS TURMERIC
- 1 TABLESPOON CHILI POWDER
- 2 TEASPOONS CUMIN POWDER
- 2 TABLESPOONS TOMATO PUREE
- 2 (16 OZ.) CANS DICED TOMATOES
- 1 CUP WATER
- ½ BUNCH CILANTRO, CHOPPED

BRAISED LAMB NECK WRAP

YIELD: 6 SERVINGS / ACTIVE TIME: 30 MINUTES / TOTAL TIME: 4 HOURS

Lamb neck might seem like something out of your comfort zone, but this recipe will prove you wrong. To really make this cut sing, it needs to be braised down so it is fall-off-the-bone tender. From there it is wrapped in paratha, which is an Indian flaky flatbread.

1. Preheat the oven to 300°F.

2. Salt the lamb on all sides.

3. Add the oil to a Dutch oven over medium-high heat. When the oil begins to shimmer, brown the lamb on all sides.

4. When the lamb is browned, add the remainder of the ingredients, cover, and cook in the oven for 3½ to 4 hours or until the meat falls off the bone.

5. Remove the meat from the pot and let rest until it is cool to the touch. Pull it apart with a fork and serve in a paratha.

INGREDIENTS:

- 3 LBS. LAMB NECK
- SALT, TO TASTE
- 2 TABLESPOONS CANOLA OIL
- 1 ONION, DICED
- 2 CARROTS, DICED
- 3 BAY LEAVES
- 2 TABLESPOONS BLACK PEPPERCORNS
- 2 CUPS WATER
- 1 TABLESPOON GARAM MASALA
- 1 TABLESPOON CUMIN
- 3 TEASPOONS CORIANDER
- PARATHA, TO SERVE (SEE PAGE 717)

VENISON TENDERLOIN WITH BLUEBERRY REDUCTION

YIELD: 4 SERVINGS / **ACTIVE TIME:** 15 MINUTES / **TOTAL TIME:** 40 MINUTES

The earthy flavor so often associated with venison is due to a deer's diet of leaves and other foliage, which is why the meat pairs so well with the sweetness of blueberries.

1. Preheat the oven to 350°F.

2. In a bowl, combine salt, pepper, and nutmeg and mix well. Pat dry the venison and season with the salt mixture.

3. Put an ovenproof pan over high heat. When it is hot, sear the tenderloin, turning every minute or so to make sure all sides are seared.

4. Put the meat in the oven and cook for 8 minutes.

5. Remove the meat from the oven, transfer it to a cutting board, and tent it with aluminum foil. Let rest for 10 minutes.

6. As the meat rests, make the sauce. In a saucepan over medium-high heat, combine berries, sugar, the vinegars, juice, zest, and cloves and bring to boil. Lower the heat and simmer for 5 to 10 minutes to thicken.

7. To serve, slice the tenderloin in medallions, plate, and top with generous portions of sauce.

INGREDIENTS:

- 1 TABLESPOON SALT
- 1 TEASPOON FRESH GROUND PEPPER
- 1 TEASPOON GROUND NUTMEG
- 3 LBS. VENISON TENDERLOIN
- 6 OZ. FRESH OR FROZEN BLACKBERRIES
- ¾ CUP WHITE SUGAR
- ½ CUP BALSAMIC VINEGAR
- ¼ CUP APPLE CIDER VINEGAR
- JUICE AND ZEST OF ONE LEMON
- 4 CLOVES

JAMAICAN-STYLE CURRY GOAT

YIELD: 4 SERVINGS / **ACTIVE TIME:** 15 MINUTES / **TOTAL TIME:** 40 MINUTES

Curry goat is a revered savory dish. Be sure to serve it with rice or roti to soak up all the wonderful gravy.

1. In a bowl, combine the goat with the onions, scotch bonnet peppers, garlic, salt, ground allspice, and curry powder. Cover and refrigerate overnight.

2. The next day, let the goat come to room temperature. Pull some onion and scotch bonnet pepper from the marinade and slice them.

3. Add the oil, sliced onion, and scotch bonnet to a Dutch oven over medium heat and cook for 10 minutes.

4. Turn up the heat to high, add the goat, and brown the meat. After the meat has cooked for 10 minutes, turn the heat down to low. Add coconut milk and potatoes and pour in enough water to cover the meat about halfway. Cover the pot, bring to a boil, and then lower the heat and simmer for 30 minutes.

5. After 30 minutes, add the sugar, stir well, and continue to simmer, covered, for another 40 minutes, or until the meat is fork-tender.

INGREDIENTS:

- 3 LBS. BONE-IN GOAT MEAT, CHOPPED
- 2 RED ONIONS, CHOPPED
- 2 SCOTCH BONNET PEPPERS, CHOPPED
- 2 GARLIC CLOVES, DICED
- 2 TEASPOONS SALT
- 1 TEASPOON GROUND ALLSPICE
- 3 TABLESPOONS CURRY POWDER
- ½ CUP CANOLA OIL
- 1 (13 OZ.) CAN COCONUT MILK
- 2 RUSSET POTATOES
- 2 TEASPOONS BROWN SUGAR

LAMB & SWEET POTATO HASH

YIELD: 4 TO 6 SERVINGS / **ACTIVE TIME:** 20 MINUTES / **TOTAL TIME:** 13 TO 17 HOURS

This meal is the perfect way to use up that leftover leg of lamb from Easter dinner. The sweet, savory, and spicy combination of the lamb and sweet potato puts traditional breakfast hash to shame, and there's nothing wrong with drizzling a little maple sugar over the top to satisfy your midmorning sweet tooth.

1. To prepare the marinade, combine all of the ingredients in a small bowl and transfer to a 1-gallon resealable bag. Place the lamb in the bag, squeeze all of the air out of the bag, and place in the refrigerator for 12 to 16 hours.

2. To prepare the lamb, preheat the oven to 350°F. Place a cast-iron Dutch oven over medium-high heat and add the beef tallow or clarified butter. Remove the lamb from the bag, place in the pot, and sear for 5 minutes on each side.

3. Add the water to the pot, place it in the oven, and cook for 20 minutes, or until the center of the lamb reaches 140°F on an instant-read thermometer. Remove the pot from the oven, set the lamb aside, and drain the liquid from the pot. Let the lamb sit for 15 minutes, then mince.

4. To prepare the sweet potato hash, fill the pot with water and bring to a boil. Add the sweet potatoes and cook until they are just tender, about 5 minutes. Be careful not to overcook them, as you don't want to end up with mashed potatoes. Drain potatoes and set aside.

5. Add the beef tallow or clarified butter, the poblano peppers, onions, garlic, and cumin to the pot and cook over medium heat until all of the vegetables are soft, about 10 minutes.

6. Return the potatoes and the lamb to the pot. Add the salt and cook for another 15 minutes. Add the oregano, season with salt and black pepper, and serve.

INGREDIENTS:

MARINADE

- 4 GARLIC CLOVES, PUREED
- LEAVES FROM 3 SPRIGS OF OREGANO, MINCED
- ¼ CUP DIJON MUSTARD
- ¼ CUP CABERNET SAUVIGNON
- 1 TABLESPOON SALT
- 1 TABLESPOON GROUND BLACK PEPPER

LAMB

- 1½ LBS. LEG OF LAMB, BUTTERFLIED
- 2 TABLESPOONS BEEF TALLOW OR CLARIFIED BUTTER
- 2 CUPS WATER

SWEET POTATO HASH

- 2 SWEET POTATOES, PEELED AND MINCED
- 2 TABLESPOONS BEEF TALLOW OR CLARIFIED BUTTER
- 2 POBLANO PEPPERS, DICED (FOR MORE HEAT, SUBSTITUTE 1 LARGE JALAPEÑO PEPPER FOR ONE OF THE POBLANOS)
- 2 YELLOW ONIONS, MINCED
- 2–3 GARLIC CLOVES, MINCED
- 1 TABLESPOON CUMIN
- 1 TABLESPOON SALT, PLUS MORE TO TASTE
- 1 TABLESPOON CHOPPED FRESH OREGANO
- FRESHLY GROUND BLACK PEPPER, TO TASTE

LAMB VINDALOO

YIELD: 5 SERVINGS / **ACTIVE TIME:** 15 MINUTES / **TOTAL TIME:** 25 HOURS

Vindaloo is a spicy curry that originated in Goa, India. Heavily influenced by the Portuguese, this dish was originally done with pork, but my version uses lamb that cooks slowly in the gravy to become tender and absolutely delicious.

1. In a bowl, combine the lamb, yogurt, and garam masala and mix well. Cover with plastic wrap and refrigerate for 24 hours.

2. Add the canola oil and mustard and cumin seeds to a skillet over medium-high heat. When the seeds begin to pop, add the onion and fry until the onion turns translucent and starts to brown slightly. Add the ginger and garlic and stir, cooking for 1 minute. Add the tomatoes and cook for 1 minute. Add the chilies, cumin, coriander, cinnamon, curry powder, and turmeric and stir, cooking for 2 to 3 minutes.

3. Remove the pan from the heat and place the mixture in a blender with vinegar and ½ cup of water. Blend on high until well blended, about a minute.

4. Place the same skillet over medium-high heat and add 2 tablespoons of canola oil. When it begins to shimmer, add the lamb and cook for 5 to 8 minutes, until browned all over.

5. Add the blended sauce to the skillet. If the mixture is too thick, add a bit of water. Add the chili powder and potatoes and partially cover and simmer for 45 minutes, adding small amounts of water as needed so it doesn't burn.

6. After the 45 minutes, check the lamb. It should be fork-tender; if not, cook longer. Simmer the curry for up to 2 hours.

7. When done, add the teaspoon of garam masala. Serve with basmati rice and Naan (see page 721).

INGREDIENTS:

- 1 LB. BONELESS LAMB SHOULDER, CUT INTO 1-INCH CHUNKS
- 1 CUP YOGURT
- 1 TABLESPOON GARAM MASALA, PLUS 1 TEASPOON
- 3 TABLESPOONS CANOLA OIL
- 1 TEASPOON MUSTARD SEED
- 1 TEASPOON CUMIN SEED
- 1 RED ONION, DICED
- 1-INCH PIECE GINGER, PEELED AND MINCED
- 4 GARLIC CLOVES, MINCED
- 2 WHOLE TOMATOES, DICED
- 2 GREEN CHILIES, DICED
- 4 KASHMIRI CHILIES
- 1 TABLESPOON CUMIN
- 1 TABLESPOON CORIANDER
- ½ TEASPOON CINNAMON
- 1 TABLESPOON CURRY POWDER
- 2 TEASPOONS TURMERIC
- 3 OZ. WHITE DISTILLED VINEGAR
- 1 TEASPOON RED CHILI POWDER
- ½ LB. PEELED AND PARTIALLY COOKED POTATOES, CUT INTO 1-INCH PIECES

TO SERVE

BASMATI RICE

NAAN (SEE PAGE 721)

CHILE-VINEGAR LAMB RIBS

YIELD: 4 SERVINGS / **ACTIVE TIME**: 1 HOUR / **TOTAL TIME**: 7 HOURS

Lamb is used widely in parts of China and Mongolia, making the ingredients in this recipe a natural fit.

INGREDIENTS:

- 2 TABLESPOONS FENNEL SEEDS, LIGHTLY TOASTED
- 2 TABLESPOONS SESAME SEEDS
- 1½ TEASPOONS RED PEPPER FLAKES
- 1 TABLESPOON FLAKED SEA SALT
- 2 TEASPOONS BLACK PEPPERCORNS, LIGHTLY CRUSHED
- 1 (4-LB.) RACK LAMB RIBS, TRIMMED
- ¾ CUPS DARK SOY SAUCE
- 1 CUP RICE VINEGAR
- 2 TEASPOONS GRANULATED SUGAR
- 3 RED THAI CHILE PEPPERS, FINELY CHOPPED
- 2 SCALLIONS, GREENS ONLY, SLICED, FOR GARNISH

1. In a mixing bowl, stir together the fennel seeds, sesame seeds, red pepper flakes, sea salt, and peppercorns.

2. Rub about half the spice mixture onto the meat side of the ribs. Set the ribs on a roasting trivet set inside a large roasting pan lined with aluminum foil. Cover and refrigerate for at least 4 hours, preferably longer.

3. When ready to cook, preheat a gas or charcoal grill to 325°F. Make sure to bank coals to one side if using a charcoal grill. Place the ribs on the grill and cook, turning once, until browned, about 10 minutes.

4. In the meantime, stir together the soy sauce, vinegar, sugar, and chiles in a small bowl until the sugar dissolves. Divide the chile-vinegar between two bowls.

5. Reduce the grill temperature to 250°F, or move the ribs away from the coals on a charcoal grill, and continue grilling until the ribs are tender, turning and basting with chile-vinegar from one bowl every 10 to 15 minutes. The ribs should be ready in about 1 hour and 45 minutes.

6. When ready, places the ribs on a platter and let rest under aluminum foil for 10 minutes. Cut between the bones to separate them.

7. Sprinkle the remaining spice mixture and the scallions over the ribs and serve with the reserved bowl of chile-vinegar.

SPICY LAMB RIBS WITH TZATZIKI & LEMON

YIELD: 4 SERVINGS / **ACTIVE TIME**: 30 MINUTES / **TOTAL TIME**: 3 HOURS

Transport yourself to a Greek island village with these assertively spiced ribs.

1. Stir together the salt, sugar, cumin, coriander, and black pepper in a small bowl. Rub half of the mixture into the ribs and let them sit at room temperature for 1 hour.

2. Whisk the vinegar into the remaining spice mix. Cover and set aside.

3. Preheat a gas or charcoal grill to 325°F. Make sure to bank coals to one side if using a charcoal grill. Place the ribs on the grill and cook, turning once, until browned, about 10 minutes.

4. Reduce the temperature to 250°F, or move the ribs away from the coals on a charcoal grill, and grill until tender, about 1½ hours, turning and basting with the prepared sauce every 10 to 15 minutes.

5. Remove the ribs and let rest under aluminum foil for 10 minutes. Cut into individual ribs. Place the tzatziki in a bowl, drizzle with olive oil, and sprinkle the parsley on top. Serve alongside the ribs and lemon wedges.

INGREDIENTS:

- 2 TABLESPOONS SALT
- 2 TABLESPOONS BROWN SUGAR
- 1 TABLESPOON FRESHLY GROUND CUMIN
- 1 TABLESPOON FRESHLY GROUND CORIANDER
- ½ TEASPOON FRESHLY GROUND BLACK PEPPER
- 1 (4-LB.) RACK LAMB RIBS, CLEANED AND TRIMMED
- ⅔ CUP APPLE CIDER VINEGAR

TO SERVE

TZATZIKI

OLIVE OIL

HANDFUL FRESH PARSLEY

LEMON WEDGES

YUZU SESAME LAMB RIBS

YIELD: 4 SERVINGS / **ACTIVE TIME:** 1 HOUR / **TOTAL TIME:** 4 HOURS

Yuzu is a fragrant and tart citrus fruit that is the base of this Japanese-style *yakiniku* sauce.

1. To begin preparations for the ribs, stir together the salt, sugar, black pepper, and five-spice powder in a small bowl. Rub mixture into the ribs and let them sit at room temperature for 1 hour.

2. To begin preparations for the sauce, in a saucepan, combine everything for the sauce, except the cornstarch slurry and ground sesame seeds. Bring to a boil over medium heat and then reduce to a simmer until slightly thickened, about 5 minutes.

3. Strain the sauce into a small saucepan and return to a simmer. Whisk in the cornstarch slurry and ground sesame seeds, returning to a simmer until the sauce has thickened. Remove from heat and cover until ready to use.

4. Preheat a gas or charcoal grill to 325°F. Make sure to bank coals to one side if using a charcoal grill. Place the ribs on the grill and cook, turning once, until browned, about 10 minutes total.

5. Reduce temperature to 250°F, or move the ribs away from the coals on a charcoal grill, and cook until tender, about 1 hour and 45 minutes, turning and basting with prepared sauce every 10 to 15 minutes.

6. Remove the ribs and let rest under aluminum foil for 10 minutes. Cut the rack into double ribs and serve.

INGREDIENTS:

FOR THE RIBS

- 1 TABLESPOON SALT
- 1 TABLESPOON GRANULATED SUGAR
- 1 TEASPOON FRESHLY GROUND BLACK PEPPER
- 1 TEASPOON CHINESE FIVE-SPICE POWDER
- 1 (4-LB.) RACK LAMB RIBS, TRIMMED

FOR THE YAKINIKU SAUCE

- 1 GARLIC CLOVE, CHOPPED
- ½ CUP YUZU JUICE
- ⅓ CUP SOY SAUCE
- 3 TABLESPOONS MIRIN
- 3 TABLESPOONS GRANULATED SUGAR
- 2 TEASPOONS HONEY
- 2 TEASPOONS SESAME OIL
- 1 TABLESPOON CORNSTARCH, MIXED TO A SLURRY WITH 1 TABLESPOON WATER
- 2 TEASPOONS SESAME SEEDS, TOASTED AND GROUND

BRINES, RUBS, STOCKS & SAUCES

This book would be incomplete without a section on brines, rubs, stocks, and sauces. Sauces add a richness and a depth of flavor to so many meat dishes, while rubs and brines help tenderize meat for the cooking process. Unlock a world of flavor and technique that will help you create recipes on your own, and kick them into overdrive.

MEAT BRINE

YIELD: 1 GALLON / **ACTIVE TIME:** 30 MINUTES / **TOTAL TIME:** 5 TO 7 DAYS

This is the brine I use before making pastrami and Headcheese (see page 236). It works well with beef and pork.

1. In a large stockpot, combine all of the ingredients, except the ice water, bring to a boil, and then remove from heat.

2. Once the brine is cool, add the ice water and submerge meat into the brine. Place a plate on top of the meat to keep it covered with the brine and prevent it from floating. Refrigerate for 5 to 7 days.

3. Remove meat from the brine, rinse it off, and pat dry. Discard the brine.

INGREDIENTS:

1½	CUPS SALT
½	CUP SUGAR
1	CUP BROWN SUGAR
8	TEASPOONS PINK SALT
1	CUP PICKLING SPICE
¼	CUP HONEY
5	GARLIC CLOVES, CRUSHED
2	QUARTS WATER
1	GALLON ICE WATER

MAPLE CHICKEN BRINE

YIELD: 6 SERVINGS / **ACTIVE TIME:** 10 MINUTES / **TOTAL TIME:** 24 HOURS

Maple and chicken are two flavors that when balanced correctly add a sweetness to chicken that is remarkable. For this dry brine the maple sugar helps preserve the tenderness of the chicken while also adding some sweetness to the meat.

1. Remove the innards from the chicken, reserving for gravy if you choose. Rinse the chicken under cold water and pat dry with paper towels.

2. In a bowl, combine all of the other ingredients and mix thoroughly.

3. Rub the mixture under the chicken skin, in the cavity, and all over the outside of the bird. Cover with foil and refrigerate for 24 hours.

4. Follow steps for Roasted Whole Chicken (see page 244).

INGREDIENTS:

1	(3–4 LB.) WHOLE CHICKEN
¾	CUP MAPLE SUGAR
2	TABLESPOONS SALT
2	TEASPOONS PEPPER
2	TEASPOONS LEMON ZEST
2	TEASPOONS PAPRIKA
½	TEASPOON CINNAMON
½	TEASPOON NUTMEG

CITRUS BRINE

YIELD: 6 SERVINGS / **ACTIVE TIME:** 10 MINUTES / **TOTAL TIME:** 24 HOURS

Tangy, sweet, and delicious is how to describe this beautiful citrus brine that makes for really juicy, tender chicken.

1. Remove the innards from the chicken, reserving for gravy if you choose. Rinse the chicken under cold water and pat dry with paper towels.

2. In a bowl, combine all of the other ingredients and mix thoroughly.

3. Rub the mixture under the chicken skin, in the cavity, and all over the outside of the bird. Cover with foil and refrigerate for 24 hours.

4. Follow steps for Roasted Whole Chicken (see page 244).

INGREDIENTS:

1	(3–4 LB.) WHOLE CHICKEN
1	CUP SUGAR
¾	CUP SALT
2	LEMONS
2	ORANGES
2	LIMES
5	BAY LEAVES
2	CINNAMON STICKS
4	GARLIC CLOVES

CAJUN DRY BRINE

YIELD: 6 SERVINGS / **ACTIVE TIME:** 10 MINUTES / **TOTAL TIME:** 24 HOURS

When you want to spice up your chicken and pack a ton of flavor, this is the brine to use: sweet, smoky, and complex.

1. Remove the innards from the chicken, reserving for gravy if you choose. Rinse the chicken under cold water and pat dry with paper towels.

2. In a bowl, combine all of the other ingredients and mix thoroughly.

3. Rub the mixture under the chicken skin, in the cavity, and all over the outside of the bird. Cover with foil and refrigerate for 24 hours.

4. Follow steps for Roasted Whole Chicken (see page 244).

INGREDIENTS:

- 1 (3–4 LB.) WHOLE CHICKEN
- ⅓ CUP SALT
- ¼ CUP BROWN SUGAR
- ½ CUP SMOKED PAPRIKA
- ¼ CUP CHILI POWDER
- 2 TEASPOONS GARLIC POWDER
- 2 TEASPOONS ONION POWDER
- ¼ TEASPOON CAYENNE PEPPER
- 2 TEASPOONS MUSTARD POWDER
- 1 TABLESPOON BLACK PEPPER

HOLIDAY SPICED WET TURKEY BRINE

YIELD: 10 SERVINGS / **ACTIVE TIME:** 45 MINUTES / **TOTAL TIME:** 24 HOURS

This perfect blend of spices adds a warm flavor to holiday turkey.

1. In a pot over medium-high heat, combine all of the ingredients, except the turkey, and bring to a boil. Reduce heat and simmer for 20 minutes. Remove from heat and allow the mixture to cool completely.

2. Remove the innards from the turkey, reserving for gravy if you choose. Rinse the turkey under cold water and pat dry with paper towels.

3. When the brine has cooled down completely, place turkey in the brine, adding water if needed so that the turkey is fully submerged. Cover and refrigerate for 24 hours.

4. Follow steps for Traditional Oven-Roasted Turkey (see page 250).

INGREDIENTS:

¾	CUP SUGAR
½	CUP SALT
5	BAY LEAVES
3	CINNAMON STICKS
4	GARLIC CLOVES
3	SPRIGS FRESH THYME
3	SPRIGS FRESH ROSEMARY
2	TEASPOONS CLOVES
2	TEASPOONS NUTMEG
1	(10–20 LB.) WHOLE TURKEY

HERB BRINED CHICKEN

YIELD: 6 SERVINGS / **ACTIVE TIME:** 10 MINUTES / **TOTAL TIME:** 24 HOURS

F resh herbs and chicken, you can't go wrong.

1. Remove the innards from the chicken, reserving for gravy if you choose. Rinse the chicken under cold water and pat dry with paper towels.

2. In a bowl, combine all of the other ingredients and mix thoroughly.

3. Rub the mixture under the chicken skin, in the cavity, and all over the outside of the bird. Cover with foil and refrigerate for 24 hours.

4. Follow steps for Roasted Whole Chicken (see page 244).

INGREDIENTS:

- 1   (3–4 LB.) WHOLE CHICKEN
- 1   CUP SUGAR
- 2   TABLESPOONS SALT
- 2   TEASPOONS PEPPER
- 2   TEASPOONS LEMON ZEST
- 1   TABLESPOON FRESH ROSEMARY
- 1   TABLESPOON FRESH THYME
- 1   TABLESPOON FRESH PARSLEY

ORANGE & THYME DRY BRINE

YIELD: 10 SERVINGS / **ACTIVE TIME:** 45 MINUTES / **TOTAL TIME:** 24 HOURS

Orange and thyme are two of my favorite flavor combinations. Whether it is savory or sweet, these two flavors yield wonderful results, especially with turkey.

INGREDIENTS:

- 1 (10–20 LB.) WHOLE TURKEY
- 1 TEASPOON BLACK PEPPERCORNS
- ⅓ CUP SALT
- ¼ CUP BROWN SUGAR
- 5 TABLESPOONS ORANGE ZEST
- 1 TABLESPOON DRIED THYME

1. Remove the innards from the turkey, reserving for gravy if you choose. Rinse the turkey under cold water and pat dry with paper towels.

2. Crack the peppercorns using a spice grinder or mortar and pestle, then combine all of the remaining ingredients in a bowl and mix thoroughly.

3. Rub the mixture under the turkey skin, in the cavity, and all over the outside of the bird. Cover with foil and refrigerate for 24 hours.

4. Follow steps for Traditional Oven-Roasted Turkey (see page 250).

BLOOD ORANGE & BOURBON BRINE

YIELD: 10 SERVINGS / **ACTIVE TIME:** 45 MINUTES / **TOTAL TIME:** 24 HOURS

The sweetness of blood orange and the smoky flavor of bourbon make this brine a unique and tasty way of injecting tons of flavor into your next turkey.

1. In a stockpot over medium-high heat, combine all of the ingredients, except the bourbon, cold water, and turkey, being sure to squeeze the oranges into the pot before adding the peels and pulp. Bring to a boil and stir to ensure that the salt and sugar are dissolved.

2. Remove from the heat, add the bourbon and ice water, and allow to come down to room temperature.

3. Remove the innards from the turkey, reserving for gravy if you choose. Rinse the turkey under cold water and pat dry with paper towels.

4. Once the brine is cooled down, place the turkey in the pot, adding water if needed so that the turkey is fully submerged. Cover and refrigerate for 24 hours.

5. Follow steps for Traditional Oven-Roasted Turkey (see page 250).

INGREDIENTS:

- 3 QUARTS HOT WATER
- ¾ CUP SALT
- ½ CUP BROWN SUGAR
- 5 BLOOD ORANGES, HALVED
- 2 CINNAMON STICKS
- 2 WHOLE STAR ANISE
- 3 WHOLE CLOVES
- 2 CUPS BOURBON
- 3 QUARTS ICE-COLD WATER
- 1 (10-20 LBS.) WHOLE TURKEY

SAGE & GARLIC DRY BRINE

YIELD: 10 SERVINGS / **ACTIVE TIME:** 45 MINUTES / **TOTAL TIME:** 24 HOURS

For a more traditional turkey brine I love the simplicity of sage and garlic. A lot of times people tend to oversaturate the turkey with lots of ingredients, but you will be amazed how these two featured ingredients lend themselves to a flavor-packed, mouth watering turkey.

1. Remove the innards from the turkey, reserving for gravy if you choose. Rinse the turkey under cold water and pat dry with paper towels.

2. Crack the peppercorns using a spice grinder or mortar and pestle, then combine all of the remaining ingredients in a bowl and mix thoroughly.

3. Rub the mixture under the turkey skin, in the cavity, and all over the outside of the bird. Cover with foil and refrigerate for 24 hours.

4. Follow steps for Traditional Oven-Roasted Turkey (see page 250).

INGREDIENTS:

- 1 (10–20 LB.) WHOLE TURKEY
- 1 TEASPOON BLACK PEPPERCORNS
- ⅓ CUP SALT
- 2 TABLESPOONS GRANULATED SUGAR
- ¼ CUP CHOPPED FRESH SAGE
- 6 GARLIC CLOVES, CHOPPED
- 2 TEASPOONS DRIED PARSLEY

THREE-PEPPER RUB

YIELD: 10 SERVINGS / **ACTIVE TIME:** 45 MINUTES / **TOTAL TIME:** 24 HOURS

Smoky, spicy, and so good is how you can describe this rub. This is one of my favorite ways to inject a wonderful flavor into turkey skin.

1. Remove the innards from the turkey, reserving for gravy if you choose. Rinse the turkey under cold water and pat dry with paper towels.

2. Crack the peppercorns using a spice grinder or mortar and pestle, then combine all of the remaining ingredients in a bowl and mix thoroughly.

3. Rub the mixture under the turkey skin, in the cavity, and all over the outside of the bird. Cover with foil and refrigerate for 24 hours.

4. Follow steps for Traditional Oven-Roasted Turkey (see page 250).

INGREDIENTS:

- 1 (10–20 LB.) WHOLE TURKEY
- 1 TABLESPOON BLACK PEPPERCORNS
- ⅓ CUP SALT
- 2 TEASPOONS MUSTARD POWDER
- 1 TABLESPOON SMOKED PAPRIKA
- 2 TEASPOONS CHILI FLAKES
- 2 TEASPOONS DRIED PARSLEY
- 2 TEASPOONS DRIED THYME
- 2 TEASPOONS GARLIC POWDER

SMOKED LEMON ZEST RUB

YIELD: 10 SERVINGS / **ACTIVE TIME:** 45 MINUTES / **TOTAL TIME:** 24 HOURS

Black peppercorns and lemon zest give this rub a smoky, citrus flavor that helps bring out the turkey's natural sweetness.

1. Remove the innards from the turkey, reserving for gravy if you choose. Rinse the turkey under cold water and pat dry with paper towels.

2. Fill a smoking gun with your preferred wood chips. Place lemon zest in a bowl and cover the top with plastic wrap, leaving only a small opening for the hose of the smoke gun. Insert the tip of the hose, light the wood chips, and turn on the machine. Allow smoke to sit with the lemon zest for 10 minutes and then repeat the process until the desired smoke flavor is achieved.

3. Grind the coriander and peppercorns using a spice grinder or mortar and pestle, then combine all of the remaining ingredients in a bowl and mix well.

4. Rub the mixture under the turkey skin, in the cavity, and all over the outside of the bird. Cover with foil and refrigerate for 24 hours.

5. Follow steps for Traditional Oven-Roasted Turkey (see page 250).

INGREDIENTS:

- 1 (10–20 LB.) WHOLE TURKEY
- ¼ CUP LEMON ZEST
- 1 TABLESPOON CORIANDER
- 2 TABLESPOONS BLACK PEPPERCORNS
- ⅓ CUP SALT
- 2 TEASPOONS GRANULATED SUGAR
- ¼ CUP CHOPPED PARSLEY
- 2 TABLESPOONS CHOPPED GARLIC
- 1 TABLESPOON CHOPPED OREGANO

JERK SPICE CHICKEN RUB

YIELD: 6 SERVINGS / **ACTIVE TIME:** 15 MINUTES / **TOTAL TIME:** 24 HOURS

Try this rub on your next smoked chicken if you want to make an unforgettable chicken. The deep flavor of these quintessential Jamaican ingredients imparts a smoky spicy flavor that is second to none.

1. Remove the innards from the chicken, reserving for gravy if you choose. Rinse the chicken under cold water and pat dry with paper towels.

2. In a bowl, combine all of the other ingredients and mix thoroughly.

3. Rub the mixture under the chicken skin, in the cavity, and all over the outside of the bird. Cover with foil and refrigerate for 24 hours.

4. Follow the recipe of your choosing to cook the bird.

INGREDIENTS:

- 1 (3–4 LB.) WHOLE CHICKEN
- ⅓ CUP SALT
- ¼ CUP BROWN SUGAR
- 2 TABLESPOONS BLACK PEPPER
- 2 TEASPOONS CINNAMON
- ½ TEASPOON NUTMEG
- ½ TEASPOON CLOVE
- 1 TABLESPOON CHOPPED THYME
- 2 TABLESPOONS SLICED SCALLIONS
- 2 TABLESPOONS CHOPPED GARLIC
- 1 TABLESPOON CHOPPED GINGER
- 2 HABANEROS, STEMMED AND CHOPPED (REMOVE SEEDS TO TAME THE SPICE)
- ¼ CUP FRESH LIME JUICE
- ½ CUP OLIVE OIL

HERB ROASTED POULTRY RUB

YIELD: 10 SERVINGS / **ACTIVE TIME:** 10 MINUTES / **TOTAL TIME:** 24 HOURS

This traditional poultry rub is a simple classic that brings a lot of flavor to the final product of all your hard work. Before chopping the fresh herbs, it's best to remove the leaves from the stems first.

1. Remove the innards from the bird, reserving for gravy if you choose. Rinse the bird under cold water and pat dry with paper towels.

2. Grind the fennel seeds, coriander, bay leaves, and peppercorns using a spice grinder or mortar and pestle.

3. In a bowl, combine all of the ingredients, except the poultry, and mix thoroughly.

4. Rub the mixture under the bird's skin, in the cavity, and all over the outside. Cover with foil and refrigerate for 24 hours.

5. Follow the recipe of your choosing to cook the bird.

INGREDIENTS:

- 1 WHOLE BIRD
- 1 TABLESPOON FENNEL SEEDS
- 1 TABLESPOON CORIANDER
- 2 BAY LEAVES
- 1 TABLESPOON WHITE PEPPERCORNS
- ¼ CUP CHOPPED FRESH PARSLEY
- 2 TABLESPOONS CHOPPED FRESH ROSEMARY
- 2 TABLESPOONS CHOPPED FRESH THYME
- 2 TABLESPOONS CHOPPED FRESH BASIL
- 1 TABLESPOON CHOPPED FRESH OREGANO
- 1 TABLESPOON CHOPPED FRESH SAGE
- 2 TABLESPOONS DRIED LAVENDER
- ⅓ CUP SALT

SPICY POULTRY RUB

YIELD: 10 SERVINGS / **ACTIVE TIME:** 10 MINUTES / **TOTAL TIME:** 24 HOURS

Want to kick up your poultry game with some heat? This spicy rub brings in some big flavors with the addition of cayenne, paprika, and dijon. This is one of my favorite rubs for smoking poultry.

1. Remove the innards from the bird, reserving for gravy if you choose. Rinse the bird under cold water and pat dry with paper towels.

2. In a bowl, combine all of the dry spices and mix well. Set aside.

3. In a bowl, combine the garlic, ginger, hot sauce, and mustard and mix well.

4. Add the dry ingredients to the wet ingredients and mix well.

5. Rub the mixture under the bird's skin, in the cavity, and all over the outside. Cover with foil and refrigerate for 24 hours.

6. Follow the recipe of your choosing to cook the bird.

INGREDIENTS:

- 1 WHOLE BIRD
- ⅓ CUP SALT
- 2 TABLESPOONS BLACK PEPPER
- 1 TABLESPOON SWEET PAPRIKA
- 1 TABLESPOON CAYENNE
- 1 TABLESPOON CHILI FLAKES
- 2 TABLESPOONS CHOPPED GARLIC
- 1 TABLESPOON CHOPPED GINGER
- 1 TABLESPOON HOT SAUCE
- 1 TABLESPOON DIJON MUSTARD

BBQ POULTRY RUB

YIELD: DEPENDS ON APPLICATION / **ACTIVE TIME:** 15 MINUTES / **TOTAL TIME:** 24 HOURS

Never underestimate the power of smoke when it comes to poultry.

1. Remove the innards from the bird, reserving for gravy if you choose. Rinse the bird under cold water and pat dry with paper towels.

2. In a bowl, combine all of the ingredients and mix well.

3. Rub the mixture under the bird's skin, in the cavity, and all over the outside. Cover with foil and refrigerate for 24 hours.

4. Follow the recipe of your choosing to cook the bird.

INGREDIENTS:

- 1 WHOLE BIRD
- ⅓ CUP SALT
- ½ CUP BROWN SUGAR
- ¼ CUP SMOKED PAPRIKA
- 1 TABLESPOON CAYENNE
- 1 TABLESPOON CHILI POWDER
- 2 TEASPOONS CUMIN
- 1 TABLESPOON ONION POWDER
- 2 TABLESPOONS GARLIC POWDER
- 1 TABLESPOON BLACK PEPPER
- 1 TABLESPOON FENNEL SEED
- 1 TABLESPOON CORIANDER
- 1 TABLESPOON MUSTARD POWDER

BROWN SUGAR & ANCHO CHILE POULTRY RUB

YIELD: DEPENDS ON APPLICATION / **ACTIVE TIME:** 10 MINUTES / **TOTAL TIME:** 24 HOURS

Sweet and spicy are two flavor combinations that will always leave people wanting more. This brown sugar and ancho rub delivers the smoky and spicy flavor that we all crave. This rub works well in many different cooking methods.

1. Remove the innards from the bird, reserving for gravy if you choose. Rinse the bird under cold water and pat dry with paper towels.

2. In a bowl, combine all of the ingredients and mix well.

3. Rub the mixture under the bird's skin, in the cavity, and all over the outside. Cover with foil and refrigerate for 24 hours.

4. Follow the recipe of your choosing to cook the bird.

INGREDIENTS:

- 1 WHOLE BIRD
- ⅓ CUP SALT
- ½ CUP BROWN SUGAR
- ½ CUP ANCHO CHILE POWDER
- 1 TABLESPOON BLACK PEPPER
- 2 TABLESPOONS GARLIC POWDER
- 2 TABLESPOONS ONION POWDER
- 1 TABLESPOON MUSTARD POWDER
- 2 TABLESPOONS CORIANDER
- 2 TABLESPOONS CINNAMON

LEMON PEPPER POULTRY RUB

YIELD: 1 WHOLE CHICKEN OR TURKEY / **ACTIVE TIME:** 15 MINUTES / **TOTAL TIME:** 24 HOURS

This simple rub works amazingly on all poultry.

1. Remove the innards from the bird, reserving for gravy if you choose. Rinse the bird under cold water and pat dry with paper towels.

2. In a bowl, combine all of the ingredients and mix well.

3. Rub the mixture under the bird's skin, in the cavity, and all over the outside. Cover with foil and refrigerate for 24 hours.

4. Follow the recipe of your choosing to cook the bird.

INGREDIENTS:

- 1 WHOLE BIRD
- ⅓ CUP SALT
- 3 TABLESPOONS LEMON PEPPER SEASONING
- 1 TABLESPOON ONION POWDER
- 2 TABLESPOONS GARLIC POWDER

TRADITIONAL TURKEY GRAVY

YIELD: 6 CUPS / **ACTIVE TIME:** 15 MINUTES / **TOTAL TIME:** 30 MINUTES

The white wine and shallots give this classic gravy a sweet and savory punch.

1. After the turkey is roasted, transfer to a resting rack and pour the pan drippings into a heat-resistant measuring cup. Allow time for the fat to rise to the surface—placing the drippings in the refrigerator or freezer will speed up the process. Skim off the fat and reserve.

2. Place your roasting pan on the cooktop over medium heat. Add the butter and turkey fat to the pan. When the pan is hot, add the shallots and garlic and cook until translucent, stirring constantly to prevent from sticking, then add the herbs and toast in the pan for 20 seconds until all aromatics are coated.

3. Deglaze the pan with white wine and allow the liquid to evaporate before adding the flour to the pan. Cook for a few minutes while stirring to prevent burning. The consistency should be that of wet sand with a nutty aroma.

4. Slowly add the pan drippings and whisk thoroughly to avoid having lumps. This will look like a paste.

5. Slowly add 1 cup of stock, whisking as you pour. This will form a paste. Slowly pour in the remainder of the stock. If you want thinner gravy, add more stock; if you want thicker gravy, let the gravy reduce at a simmer.

6. Season to taste before serving.

INGREDIENTS:

- ¼ CUP UNSALTED BUTTER
- ½ CUP TURKEY FAT SKIMMED FROM DRIPPINGS
- ¼ CUP CHOPPED SHALLOTS
- 2 TABLESPOONS CHOPPED GARLIC
- 2 TABLESPOONS CHOPPED FRESH THYME
- 1 TABLESPOON CHOPPED SAGE
- 2 TEASPOONS CHOPPED FRESH ROSEMARY
- ¼ CUP WHITE WINE
- ½ CUP ALL-PURPOSE FLOUR, SIFTED
- 2 CUPS PAN DRIPPINGS, FAT SKIMMED OFF AND RESERVED
- 4 CUPS TURKEY STOCK (SEE PAGE 477)
- SALT AND PEPPER, TO TASTE

TRADITIONAL CHICKEN GRAVY

YIELD: 6 CUPS / **ACTIVE TIME:** 15 MINUTES / **TOTAL TIME:** 30 MINUTES

More people should make chicken gravy. They should. Try this recipe the next time you roast a chicken and be one of those people.

1. After chicken is roasted, transfer to a resting rack and pour the pan drippings into a heat-resistant measuring cup. Allow time for the fat to rise to the surface—placing the drippings in the refrigerator or freezer will speed up the process. Skim off the fat and reserve.

2. Place your roasting pan on the cooktop over medium heat. Add the butter and chicken fat to the pan. When the pan is hot, add the shallots and garlic and cook until translucent, stirring constantly to prevent from sticking, then add the herbs and toast in the pan for 20 seconds until all aromatics are coated.

3. Deglaze the pan with white wine and allow the liquid to evaporate before adding the flour to the pan. Cook for a few minutes while stirring to prevent burning. The consistency should be that of wet sand with a nutty aroma.

4. Slowly add the pan drippings and whisk thoroughly to avoid having lumps. This will look like a paste.

5. Slowly add 1 cup of stock, whisking as you pour. This will form a paste. Slowly pour in the remainder of the stock. If you want thinner gravy, add more stock; if you want thicker gravy, let the gravy reduce at a simmer.

6. Season to taste before serving.

INGREDIENTS:

- ¼ CUP UNSALTED BUTTER
- ½ CUP CHICKEN FAT SKIMMED FROM DRIPPINGS
- ¼ CUP CHOPPED SHALLOTS
- 2 TABLESPOONS CHOPPED GARLIC
- 2 TABLESPOONS CHOPPED FRESH THYME
- 1 TABLESPOON CHOPPED FRESH SAGE
- 2 TEASPOONS CHOPPED FRESH ROSEMARY
- ¼ CUP WHITE WINE
- ½ CUP ALL-PURPOSE FLOUR, SIFTED
- 2 CUPS PAN DRIPPINGS, FAT SKIMMED OFF AND RESERVED
- 4 CUPS CHICKEN STOCK, DIVIDED (SEE PAGE 478)
- SALT AND PEPPER, TO TASTE

TURKEY GIBLET GRAVY

YIELD: 5 CUPS / **ACTIVE TIME:** 15 MINUTES / **TOTAL TIME:** 30 MINUTES

Cooking with the innards of the turkey may seem daunting, but it yields the best gravy—plus you're not letting any of the bird go to waste.

INGREDIENTS:

- 2 TABLESPOONS VEGETABLE OIL
- TURKEY GIBLETS: NECK, HEART, LIVER, GIZZARD
- 1 CUP SLICED SHALLOTS
- ½ CUP CELERY
- ½ CUP CARROT
- 4 GARLIC CLOVES, SLICED
- 3 SPRIGS FRESH THYME
- 1 SPRIG FRESH ROSEMARY
- 5 CUPS WATER
- ½ CUP UNSALTED BUTTER
- ½ CUP ALL-PURPOSE FLOUR, SIFTED
- SALT AND PEPPER, TO TASTE

1. Add oil to a large pot over medium-high heat. When the oil begins to shimmer, add the giblets, searing until browned on all sides.

2. Add the shallots, celery, carrot, and garlic to the giblets, stirring every couple of minutes until fully browned.

3. Add the thyme, rosemary, and water; cover; and bring to a boil. After the stock reaches a boil, lower the heat and simmer for about an hour for the flavors to develop; there should be about 4 cups of stock.

4. Strain the stock and set aside. Discard all but the turkey neck from the strainer. Transfer the neck to a cutting board and remove the meat from the neck and set aside. Discard the neck bones.

5. Return the pot to the stove and melt the butter over medium heat. When the butter begins to foam, add the flour, cooking for a few minutes while stirring to prevent burning. The consistency should be that of wet sand with a nutty aroma.

6. Slowly add 1 cup of stock, whisking as you pour. This will form a paste. Slowly pour in the remainder of the stock. If you want thinner gravy, add more stock; if you want thicker gravy, let the gravy reduce at a simmer.

7. Season to taste before serving.

BEEF STOCK

YIELD: 3 QUARTS / **ACTIVE TIME:** 15 MINUTES / **TOTAL TIME:** 6 HOURS

With this recipe you can reduce food waste and make some wonderful stock for soups, stews, and any time you need a burst of flavor. Feel free to use the scraps of vegetables for this stock.

1. Preheat the oven to 400°F. Place the bones on a sheet pan and roast them for 45 minutes.

2. In a large pot over medium-high heat, combine the roasted bones and all the released cooking juices with the remainder of the ingredients and enough cold water to cover the bones. Bring to a boil and then lower heat and simmer.

3. Skim the surface of the stock to get rid of any impurities and continue to simmer until the stock reduces by at least ⅓.

4. Strain the stock and use or store. It can be kept in the refrigerator for up to 2 weeks or in the freezer for 3 months.

INGREDIENTS:

	BEEF BONES (IDEALLY KNUCKLES, NECK, AND/OR SHIN)
2	TABLESPOONS PEPPERCORNS
1	ONION, QUARTERED
2	CARROTS, CUT INTO 1-INCH PIECES
2	CELERY STALKS, CUT INTO 1-INCH PIECES
5	SPRIGS FRESH PARSLEY
3	BAY LEAVES

VEAL STOCK

YIELD: 6 QUARTS / **ACTIVE TIME:** 30 MINUTES / **TOTAL TIME:** 5 HOURS AND 20 MINUTES

When I am making a dish that utilizes a brown stock, I try to use veal bones. Veal are cows aged from six months to a year, and it has a smoother taste than beef. It's also more tender, lighter, and finer in texture. This superiority extends to the bones as well—to me, the gelatin in stock made with veal bones has a better sheen.

1. Preheat oven to 350°F.

2. Lay the bones on a flat baking tray, place in oven, and cook for 30 to 45 minutes, until they are golden brown. Remove and set aside.

3. Meanwhile, in a large stockpot, add the vegetable oil and warm over low heat. Add the vegetables and cook until any additional moisture has evaporated. This allows the flavor of the vegetables to become concentrated.

4. Add the water to the stock pot. Add the bones, aromatics, and tomato paste to the stockpot, raise heat to high, and bring to a boil.

5. Reduce heat so that the stock simmers and cook for a minimum of 2 hours. Skim fat and impurities from the top as the stock cooks. As for when to stop cooking the stock, let the flavor be the judge. I typically like to cook for 4 to 5 hours total.

6. When the stock is finished cooking, strain through a fine strainer or cheesecloth. Place stock in refrigerator to chill.

7. Once cool, skim the fat layer from the top and discard. Use immediately, refrigerate, or freeze.

INGREDIENTS:

- 10 LBS. VEAL BONES
- ½ CUP VEGETABLE OIL
- 1 LEEK, TRIMMED, AND CAREFULLY WASHED, CUT INTO 1-INCH PIECES
- 1 LARGE YELLOW ONION, UNPEELED, CLEANED ROOT, CUT INTO 1-INCH PIECES
- 2 LARGE CARROTS, PEELED AND CUT INTO 1-INCH PIECES
- 1 CELERY STALK WITH LEAVES, CUT INTO 1-INCH PIECES
- 8 OZ. TOMATO PASTE
- 8 SPRIGS FRESH PARSLEY
- 5 SPRIGS FRESH THYME
- 2 BAY LEAVES
- 10 QUARTS WATER
- 1 TEASPOON PEPPERCORNS
- 1 TEASPOON SALT

TURKEY STOCK

YIELD: 2 QUARTS / **ACTIVE TIME:** 15 MINUTES / **TOTAL TIME:** 6 HOURS

Turkey bones make exceptional stock, and you can complement that homey poultry flavor and aroma with vegetable scraps.

1. In a large pot, combine all ingredients along with enough cold water to cover the turkey carcass. Bring to a boil, reduce heat, skim the surface of the stock to remove any impurities, and simmer until stock reduces by one-third.

2. Strain the stock through a cheesecloth-lined mesh strainer and discard any solids. This stock can be kept in the refrigerator for up to 2 weeks or in the freezer for 3 months.

INGREDIENTS:

- 1 TURKEY CARCASS
- 2 TABLESPOON PEPPERCORNS
- 1 ONION, QUARTERED
- 2 CARROTS, CHOPPED
- 2 CELERY STALKS, CHOPPED
- 5 SPRIGS FRESH PARSLEY
- 3 BAY LEAVES

CHICKEN STOCK

YIELD: 2 QUARTS / **ACTIVE TIME:** 15 MINUTES / **TOTAL TIME:** 6 HOURS

Chicken stock is for much more than soup. There is no reason not to always have some on hand in the freezer. Large amounts are necessary for risotto, and smaller portions add flavor to pan sauces and gravies.

1. In a large pot over medium-high heat, combine all of the ingredients with enough cold water to cover the chicken. Bring to a boil and then lower heat and simmer.

2. Skim the surface of the stock to get rid of any impurities and continue to simmer until the stock reduces by at least one-third.

3. Strain the stock and use or store. It can be kept in the refrigerator for up to 2 weeks or in the freezer for 3 months.

INGREDIENTS:

- 1 COOKED CHICKEN CARCASS
- 2 TABLESPOONS PEPPERCORNS
- 1 ONION, QUARTERED
- 2 CARROTS, CUT INTO 1-INCH PIECES
- 2 CELERY STALKS, CUT INTO 1-INCH PIECES
- 5 SPRIGS FRESH PARSLEY
- 3 BAY LEAVES

PORK STOCK

YIELD: 3 QUARTS / **ACTIVE TIME:** 15 MINUTES / **TOTAL TIME:** 6 HOURS

You can get away with making other stocks without roasting the bones, but for pork stock it is a necessary, flavor-producing step.

1. Preheat oven to 400°F.
2. Roast the bones for 25 minutes.
3. Add the bones and all the juices from roasting to a large stock pot.
4. Combine all ingredients along with enough cold water to cover the bones (about 4–5 quarts).
5. Bring to a boil and reduce your heat to a simmer.
6. Skim the surface of the stock to get rid of any impurities.
7. Continue to simmer until the stock reduces by at least one-third.
8. Strain the stock in a mesh strainer and discard any solids. This stock can be kept in the refrigerator for up to 2 weeks or kept in the freezer for 3 months.

INGREDIENTS:

- 3 LBS. PORK BONES (KNUCKLES, NECK, OR SHIN)
- 2 TABLESPOONS PEPPERCORNS
- 1 MEDIUM ONION, QUARTERED
- 2 CARROTS, CUT INTO 1-INCH PIECES
- 2 CELERY STALKS, CUT INTO 1-INCH PIECES
- 5 SPRIGS FRESH PARSLEY
- 3 BAY LEAVES

PORK PHO BROTH

YIELD: 2 QUARTS / **ACTIVE TIME:** 15 MINUTES / **TOTAL TIME:** 2 HOURS

There is so much to be said for low and slow cooking and how it allows incredible flavors to develop. For this broth, I like to let it simmer overnight.

1. Add coriander seeds, cloves, and star anise to a frying pan over medium heat and toast until fragrant, about 4 minutes. Immediately spoon out the spices to avoid burning them.

2. In a large pot, combine all of the ingredients and bring to a boil. Reduce the heat and simmer for at least 2 hours, skimming the surface frequently.

3. Using tongs, remove the pork bones. Season to taste and then strain.

INGREDIENTS:

- 2 TABLESPOONS CORIANDER SEEDS
- 4 CLOVES
- 2 WHOLE STAR ANISE
- 4 LBS. PORK BONES (KNUCKLES AND/OR NECK)
- 2 QUARTS PORK STOCK (SEE PAGE 481)
- ½ ONION
- 1 3-INCH PIECE OF GINGER, DICED
- 1–2 TABLESPOONS SUGAR
- 1–2 TABLESPOONS FISH SAUCE

DUCK STOCK

YIELD: 3 QUARTS / **ACTIVE TIME:** 15 MINUTES / **TOTAL TIME:** 4 HOURS

Duck stock is great for soups, of course, but it can be further reduced for a delicious glaze.

1. Preheat the oven to 400°F.

2. Place the duck carcass on a sheet pan and roast for 20 minutes.

3. In a large pot, combine all of the ingredients with the roasted carcass and the juices it released while being roasted. Cover with enough water to cover the duck, about 5 quarts, and bring to a boil. Reduce the heat and simmer until the stock reduces by at least one-third.

4. Strain the stock and store. It can be refrigerated for up to 2 weeks or kept in the freezer for 3 months.

INGREDIENTS:

- 1 WHOLE COOKED DUCK CARCASS
- 2 TABLESPOONS PEPPERCORNS
- 1 ONION, QUARTERED
- 2 CARROTS, CUT INTO 1-INCH PIECES
- 2 CELERY STALKS, CUT INTO 1-INCH PIECES
- 5 SPRIGS PFRESH PARSLEY
- 3 BAY LEAVES

SWEET & SPICY MARINADE

YIELD: 6 SERVINGS / **ACTIVE TIME:** 5 MINUTES / **TOTAL TIME:** 24 HOURS

For beef or chicken, this marinade works well for so many different applications. Keep this recipe handy because you will use it a lot.

1. In a bowl, combine all of the ingredients and mix well.

2. Add the meat to the bowl, cover with plastic wrap, and refrigerate for 24 hours.

3. Follow the recipe of your choosing to cook.

INGREDIENTS:

¾	CUP PINEAPPLE JUICE
3	JALAPEÑOS, DICED
1½	TABLESPOONS GARLIC POWDER
2	TEASPOONS CHILI POWDER
1	TEASPOON WHITE PEPPER
1	TEASPOON SALT

GO-TO STEAK MARINADE

YIELD: 2 CUPS / **ACTIVE TIME:** 5 MINUTES / **TOTAL TIME:** 24 HOURS

This has been my go-to steak marinade for years, since my days as a sous chef, and I still love it just as much as the day I started using it. This recipe makes enough marinade for about 3 pounds of meat.

1. Blend ingredients together and refrigerate overnight.

INGREDIENTS:

- ⅓ CUP SOY SAUCE
- ½ CUP OLIVE OIL
- ⅓ CUP LEMON JUICE
- ¼ CUP WORCESTERSHIRE SAUCE
- 1½ TABLESPOONS GARLIC POWDER
- 3 TABLESPOONS DRIED BASIL
- 1½ TABLESPOONS FRESH PARSLEY
- 1 TEASPOON WHITE PEPPER
- 1 TEASPOON MINCED GARLIC

BBQ CHICKEN MARINADE

YIELD: 1 CUP / **ACTIVE TIME:** 5 MINUTES / **TOTAL TIME:** 24 HOURS

Who doesn't love grilled chicken? This recipe works perfectly for chicken but is also great for steak. This will make enough marinade for about 2 pounds of meat. You can also do a quick 1-hour brine before marinating the meat, using ½ cup sugar, ½ cup salt, and 4 cups water.

1. Preheat the oven to 400°F.

2. Cut the top off the garlic bulb, exposing the tops of the cloves. Drizzle with olive oil, wrap in aluminum foil, and cook in the oven for about 40 minutes, or until garlic is soft enough to be squeezed out.

3. Blend 3 tablespoons of the roasted garlic with the rest of the ingredients together and refrigerate overnight.

INGREDIENTS:

- 1   HEAD GARLIC
- ¾   CUP OLIVE OIL, PLUS MORE AS NEEDED
- 2   TABLESPOONS FRESH PARSLEY
- 2   TABLESPOONS FRESH OREGANO
- ½   CUP WHITE WINE VINEGAR

LEMON & ROSEMARY CHICKEN MARINADE

YIELD: ¾ CUP / **ACTIVE TIME:** 5 MINUTES / **TOTAL TIME:** 24 HOURS

I use this marinade on chicken breasts before grilling them. This recipe makes enough for 2 pounds of chicken.

INGREDIENTS:

1	GARLIC BULB
½	CUP OLIVE OIL
1	TEASPOON SALT
½	TEASPOON WHITE PEPPER
3	LEMONS, JUICED
¼	CUP FRESH ROSEMARY

1. Preheat the oven to 400°F.

2. Cut the top off a garlic bulb, exposing the tops of the cloves. Drizzle with olive oil, wrap in aluminum foil, and cook in the oven for about 40 minutes, or until garlic is soft enough to be squeezed out.

3. Blend 3 tablespoons of the roasted garlic with the remainder of the ingredients and refrigerate overnight.

COFFEE MARINADE FOR BEEF

YIELD: 6 SERVINGS / **ACTIVE TIME:** 5 MINUTES / **TOTAL TIME:** 24 HOURS

This beef marinade is amazing for grilling. The heat brings out the deep richness of the coffee. This recipe is for about 3 pounds of meat.

INGREDIENTS:

- ¾ CUP STRONG COFFEE
- ½ CUP OLIVE OIL
- 1½ TABLESPOONS GARLIC POWDER
- 3 TABLESPOONS BALSAMIC VINEGAR
- 1½ TEASPOONS DIJON MUSTARD
- 1 TEASPOON WHITE PEPPER
- 1 TEASPOON SALT

1. In a bowl, combine all of the ingredients and mix well.

2. Add the meat to the bowl, cover with plastic wrap, and refrigerate for 24 hours.

3. Follow the recipe of your choosing to cook.

BEEF FAJITA MARINADE

YIELD: 6 SERVINGS / **ACTIVE TIME:** 5 MINUTES / **TOTAL TIME:** 24 HOURS

This sweet and tangy marinade works great for beef, but it also plays well with chicken.

1. In a bowl, combine all of the ingredients and mix well.

2. Add the meat to the bowl, cover with plastic wrap, and refrigerate for 24 hours.

3. Follow the recipe of your choosing to cook.

INGREDIENTS:

- ¾ CUP PINEAPPLE JUICE
- ½ CUP FRESH LIME JUICE
- 1½ TABLESPOONS GARLIC POWDER
- 2 TEASPOONS CUMIN
- 1½ TEASPOONS CHILI POWDER
- 1 TEASPOON WHITE PEPPER
- 1 TEASPOON SALT

ASIAN MARINADE

YIELD: 6 SERVINGS / **ACTIVE TIME:** 5 MINUTES / **TOTAL TIME:** 24 HOURS

Recommended for steak or chicken—especially wings—this marinade packs in wonderful flavor.

1. In a bowl, combine all of the ingredients and mix well.

2. Add the meat to the bowl, cover with plastic wrap, and refrigerate for 24 hours.

3. Follow the recipe of your choosing to cook.

INGREDIENTS:

- ¾ CUP PINEAPPLE JUICE OR PEAR JUICE
- ¼ CUP SOY SAUCE
- 2 TABLESPOONS SHAOXING WINE
- 1½ TABLESPOONS GARLIC POWDER
- 2 TABLESPOONS BROWN SUGAR
- 1½ TEASPOONS SESAME OIL
- 1 TEASPOON WHITE PEPPER
- 1 TEASPOON GROUND GINGER
- 1 TEASPOON SALT

ADOBO MARINADE FOR BEEF

YIELD: 6 SERVINGS / **ACTIVE TIME:** 5 MINUTES / **TOTAL TIME:** 24 HOURS

Want to spice up your beef? Then use this recipe. Add more chilies for a bigger kick.

1. In a blender, combine all of the ingredients and puree.

2. In a bowl, combine the meat and marinade, cover with plastic wrap, and refrigerate for 24 hours.

3. Follow the recipe of your choosing to cook.

INGREDIENTS:

- ½ CUP OLIVE OIL
- 1 (6 OZ.) CAN ADOBO, WITH SAUCE
- 1½ TABLESPOONS GARLIC POWDER
- 2 TEASPOONS CUMIN
- 1 TABLESPOON BROWN SUGAR
- 1½ TEASPOONS CHILI POWDER
- 1 TEASPOON WHITE PEPPER
- 1 TEASPOON SALT

TRADITIONAL BBQ SAUCE

YIELD: 1 QUART / **ACTIVE TIME:** 2 MINUTES / **TOTAL TIME:** 20 MINUTES

This simple sauce is the base for many additional barbecue sauces you will find in this book. It's tangy and sweet and really perfect on all meats.

1. In a pot over medium heat, combine all of the ingredients, whisk together, and cook for 10 minutes, stirring to prevent burning.

2. Reduce heat and simmer on low for an additional 10 minutes.

3. Store in a sealed jar in the fridge.

INGREDIENTS:

- 2 CUPS KETCHUP
- 2 TABLESPOONS YELLOW MUSTARD
- ½ CUP APPLE CIDER VINEGAR
- ¼ CUP BROWN SUGAR
- 2 TABLESPOONS HONEY
- 1 TABLESPOON WORCESTERSHIRE SAUCE
- 1 TABLESPOON FRESH LEMON JUICE

SPICY BBQ SAUCE

YIELD: 1 QUART / **ACTIVE TIME:** 2 MINUTES / **TOTAL TIME:** 20 MINUTES

This is where traditional meets spicy. Two kinds of hot peppers kick this sauce into the atmosphere.

1. In a pot over medium heat, combine all of the ingredients, whisk together, and cook for 10 minutes, stirring to prevent burning.

2. Reduce heat and simmer on low for an additional 10 minutes.

3. Strain the sauce and store in a sealed jar in the fridge.

INGREDIENTS:

- 2 CUPS KETCHUP
- 2 TABLESPOONS YELLOW MUSTARD
- ½ CUP APPLE CIDER VINEGAR
- ¼ CUP BROWN SUGAR
- 2 TABLESPOONS HONEY
- 1 ROASTED JALAPEÑO PEPPER, FINELY CHOPPED
- ½ HABANERO PEPPER, FINELY CHOPPED
- 1 TABLESPOON WORCESTERSHIRE SAUCE
- 1 TABLESPOON FRESH LEMON JUICE

PINEAPPLE BBQ SAUCE

YIELD: 1 QUART / **ACTIVE TIME:** 2 MINUTES / **TOTAL TIME:** 20 MINUTES

I love using this sauce on chicken and pork. It gives a nice sweet and smoky flavor to the meat.

1. In a pot over medium heat, combine all of the ingredients, whisk together, and cook for 10 minutes, stirring to prevent burning.

2. Reduce heat and simmer on low for an additional 10 minutes.

3. Store in a sealed jar in the fridge.

INGREDIENTS:

- ½ CUP PINEAPPLE JUICE
- 2 CUPS KETCHUP
- 2 TABLESPOONS YELLOW MUSTARD
- ½ CUP APPLE CIDER VINEGAR
- ¼ CUP BROWN SUGAR
- 2 TABLESPOONS HONEY
- 1 TABLESPOON WORCESTERSHIRE SAUCE
- 1 TABLESPOON FRESH LEMON JUICE

KC STYLE BBQ SAUCE

YIELD: 1 QUART / **ACTIVE TIME:** 2 MINUTES / **TOTAL TIME:** 20 MINUTES

Kansas City barbecue is known for being sweet and tangy. This sauce is a nod to that tradition.

1. In a pot over medium heat, combine all of the ingredients, whisk together, and cook for 10 minutes, stirring to prevent burning.

2. Reduce heat and simmer on low for an additional 10 minutes.

3. Store in a sealed jar in the fridge.

INGREDIENTS:

- 2 CUPS KETCHUP
- 2 TABLESPOONS YELLOW MUSTARD
- ½ TEASPOON GARLIC POWDER
- ½ TEASPOON ONION POWDER
- ½ TEASPOON PAPRIKA
- 1 TABLESPOON MOLASSES
- ½ CUP WATER
- ½ CUP APPLE CIDER VINEGAR
- 1 CUP BROWN SUGAR
- 1 TABLESPOON WORCESTERSHIRE SAUCE
- 1 TABLESPOON FRESH LEMON JUICE

CHIPOTLE BBQ SAUCE

YIELD: 1 QUART / **ACTIVE TIME:** 2 MINUTES / **TOTAL TIME:** 20 MINUTES

Chipotle peppers define this smoky and spicy sauce that works on everything.

1. In a pot over medium heat, combine all of the ingredients, whisk together, and cook for 10 minutes, stirring to prevent burning.

2. Reduce heat and simmer on low for an additional 10 minutes.

3. Strain the sauce and store in a sealed jar in the fridge.

INGREDIENTS:

- 2 CUPS KETCHUP
- 2 TABLESPOON YELLOW MUSTARD
- ½ TEASPOON GARLIC POWDER
- 1 (4 OZ.) CAN OF CHIPOTLE PEPPERS IN ADOBO
- ½ TEASPOON ONION POWDER
- ½ TEASPOON PAPRIKA
- 1 TABLESPOON MOLASSES
- ½ CUP WATER
- ½ CUP APPLE CIDER VINEGAR
- 1 CUP PACKED BROWN SUGAR
- 1 TABLESPOON WORCESTERSHIRE SAUCE OR COCONUT AMINOS
- 1 TABLESPOON LEMON JUICE

FENNEL LEMON GREMOLATA

YIELD: 1½ CUPS / **ACTIVE TIME:** 10 MINUTES / **TOTAL TIME:** 10 MINUTES

Once you try this on your next grilled steak, you won't ever want steak without it.

1. Mix all of the ingredients in a mortar and pestle or food processor. Blend well and store in an airtight container for up to 3 weeks.

INGREDIENTS:

- ¼ CUP CHOPPED FENNEL FRONDS
- ¼ CUP FINELY CHOPPED FLAT-LEAF PARSLEY LEAVES
- 1 TEASPOON FINELY GRATED LEMON ZEST
- 1 TABLESPOON EXTRA-VIRGIN OLIVE OIL
- SALT, TO TASTE

FRESH CHIMICHURRI SAUCE

YIELD: ½ CUP / **ACTIVE TIME:** 10 MINUTES / **TOTAL TIME:** 10 MINUTES

You can't have a book about meat without including my favorite sauce, fresh chimichurri. This sauce is packed with flavor that cuts through fatty cuts of meat to deliver a flavor bomb!

1. Mix all of the ingredients in a mortar and pestle or food processor. Blend well and store in an airtight container for up to 3 weeks.

INGREDIENTS:

- 1 CUP FRESH PARSLEY LEAVES
- 2 LARGE GARLIC CLOVES, SMASHED
- 1 TEASPOON DRIED THYME
- ¼ TEASPOON DRIED RED-PEPPER FLAKES
- ½ CUP WATER
- ¼ CUP WHITE WINE VINEGAR
- ¼ CUP OLIVE OIL
- 1 TEASPOON SALT
- ⅛ TEASPOON FRESHLY GROUND BLACK PEPPER

STEAK SAUCE

YIELD: 1½ CUPS / **ACTIVE TIME:** 5 MINUTES / **TOTAL TIME:** 5 MINUTES

This classic steak sauce brings the flavor with vinegar and Worcestershire sauce. Try it on your next steak.

1. In a medium bowl, combine all of the ingredients and mix well.
2. Transfer to a jar and refrigerate to store.

INGREDIENTS:

- 1¼ CUPS KETCHUP
- 2 TABLESPOONS PREPARED YELLOW MUSTARD
- 2 TABLESPOONS WORCESTERSHIRE SAUCE
- 1½ TABLESPOONS APPLE CIDER VINEGAR
- 4 DROPS HOT SAUCE
- ½ TEASPOON SALT
- ½ TEASPOON GROUND BLACK PEPPER

BACON JAM

YIELD: 1½ CUPS / **ACTIVE TIME:** 10 MINUTES / **TOTAL TIME:** 30 MINUTES

Bacon jam can be used in a variety of different ways. From topping scallops to a burger, you will love this sweet and savory condiment.

INGREDIENTS:

- 1½ LBS. BACON
- 2 TEASPOONS UNSALTED BUTTER
- 4 YELLOW ONIONS, DICED
- 1 TEASPOON SALT
- ¼ CUP BROWN SUGAR
- ¼ CUP SHERRY VINEGAR
- 1½ TEASPOONS FRESH THYME LEAVES, DIVIDED
- 1 TEASPOON BLACK PEPPER
- 1 PINCH CAYENNE PEPPER
- ½ CUP WATER
- 2 TEASPOONS BALSAMIC VINEGAR
- 2 TEASPOONS EXTRA-VIRGIN OLIVE OIL

1. Add bacon to a large pot over medium heat and cook until crispy and the rendered fat is foaming, about 10 minutes. Pour bacon and rendered fat into a strainer placed over a bowl to drain fat. When fat is drained and bacon is cool enough to handle, remove bacon to a cutting board and finely chop.

2. Return the pot to medium heat and add 2 teaspoons reserved bacon fat and butter. When the butter begins to foam, add onions and salt and sauté until soft and translucent, 7 to 10 minutes.

3. Stir in brown sugar, sherry vinegar, 1 teaspoon thyme leaves, black pepper, and cayenne, then add bacon. Stir water into bacon mixture and cook until a brick-brown color and a jam consistency are achieved, 10 to 15 minutes.

4. Remove jam from the heat and stir in balsamic vinegar, olive oil, and remaining thyme. Stir until shiny and warmed through.

CHIPOTLE CILANTRO CHIMICHURRI FOR MEAT

YIELD: ½ CUP / **ACTIVE TIME:** 5 MINUTES / **TOTAL TIME:** 15 MINUTES

This sauce is perfect for beef, especially steak.

1. If using dried pepper, remove the stems and seeds and place in a bowl. Pour hot water over the pepper and let rehydrate for 5 to 10 minutes.

2. Combine all of the ingredients in the food processor and pulse until well combined. Season to taste.

INGREDIENTS:

- 1 DRIED CHIPOTLE PEPPER OR CHIPOTLE IN ADOBO
- 1 CUP CILANTRO, STEMS INCLUDED
- 2 GARLIC CLOVES, PEELED
- 2 TABLESPOONS OLIVE OIL
- JUICE OF 1 LIME
- ZEST OF ½ LIME
- SALT AND PEPPER, TO TASTE

CILANTRO CHUTNEY

YIELD: ½ CUP / **ACTIVE TIME:** 5 MINUTES / **TOTAL TIME:** 10 MINUTES

Coconut is the secret ingredient in this vibrant condiment that works on pretty much anything.

1. Place all of the ingredients, except for water and salt, in a food processor and puree until smooth, adding water as needed to achieve the desired consistency. Season with salt and refrigerate until ready to serve.

INGREDIENTS:

- 1 BUNCH FRESH CILANTRO
- ½ CUP GRATED FRESH COCONUT
- 15 MINT LEAVES
- 1 TABLESPOON SEEDED AND MINCED JALAPENO PEPPER
- 1 GARLIC CLOVE
- 1 TEASPOON MINCED GINGER
- 1 PLUM TOMATO
- 1 TABLESPOON FRESH LEMON JUICE
- WATER, AS NEEDED
- SALT, TO TASTE

FERMENTED HOT SAUCE

YIELD: 2 CUPS / **ACTIVE TIME:** 10 MINUTES / **TOTAL TIME:** 30 DAYS TO 6 MONTHS

If you're addicted to hot sauce, fermented hot sauce will take that addiction to the next level.

1. Remove the tops of the peppers and split them down the middle.

2. Place the split peppers and the garlic, onion, and salt in a mason jar and cover with the water. Cover the jar and shake well.

3. Place the jar away from direct sunlight and let stand for at least 30 days and up to 6 months. Occasionally unscrew the lid to release some of the gases that build up. Based on my own experience, a longer fermenting time is very much worth it.

4. Once you are ready to make the sauce, reserve most of the brine, transfer the mixture to a blender, and puree to desired thickness. If you want your sauce to be on the thin side, keep adding brine until you have the consistency you want. Season with salt, transfer to a container, cover, and store in the refrigerator for up to 3 months.

INGREDIENTS:

- 2 LBS. CAYENNE PEPPERS
- 1 LB. JALAPEÑO PEPPERS
- 5 GARLIC CLOVES
- 1 RED ONION, QUARTERED
- 3 TABLESPOONS SALT, PLUS MORE TO TASTE
- FILTERED WATER, AS NEEDED

HOMEMADE TOMATO SALSA

YIELD: ABOUT 6 SERVINGS / **ACTIVE TIME:** 35 MINUTES: **TOTAL TIME:** 2 HOURS AND 35 MINUTES

While you can easily buy your own salsa at the store, nothing beats the fresh zing of homemade salsa. This recipe calls for cherry tomatoes. They are slightly tarter in flavor than other tomatoes, depending on the type you purchase, but you can use whatever tomatoes you have on hand.

1. Place a cast-iron skillet on a grill at medium-low heat. Combine the oil, tomatoes, onion, garlic, and jalapeño in a bowl. Next, transfer the mixture to the heated cast-iron skillet and grill for about 20 minutes, stirring every now and then until the tomatoes are charred.

2. Once charred, transfer the contents of the skillet to a food processor along with the remaining ingredients. Pulse until smooth and then cover for at least 2 hours before serving.

INGREDIENTS:

- 2 TABLESPOONS OLIVE OIL
- 2 CUPS CHERRY TOMATOES
- 1 YELLOW ONION, CHOPPED
- 6 GARLIC CLOVES, CRUSHED
- 2 JALAPEÑO PEPPERS, DICED
- ½ CUP FRESH CILANTRO LEAVES
- JUICE FROM 1 LIME
- 1 TEASPOON GROUND CUMIN
- BLACK PEPPER, TO TASTE
- SALT, TO TASTE

TOMATILLO SALSA

YIELD: ABOUT 6 SERVINGS / **ACTIVE TIME:** 15 MINUTES / **TOTAL TIME:** 15 MINUTES

The tomatillos provide a bright, acidic, fruity flavor to this salsa that is unmatched.

1. Place the tomatillos in a saucepan and cover with water. Place over medium-high heat on the stovetop, bring to a boil, and cook for 10 minutes.

2. After 10 minutes, remove the tomatillos from the saucepan and place in a food processor along with the remaining ingredients. Pulse until smooth.

INGREDIENTS:

- 10 TOMATILLOS, HUSKED AND RINSED
- 1 SMALL ONION, CHOPPED
- 2 GARLIC CLOVES, MINCED
- ¼ CUP CHOPPED FRESH CILANTRO
- 1 JALAPEÑO PEPPER, CHOPPED
- BLACK PEPPER, TO TASTE
- SEA SALT, TO TASTE

FUNDAMENTAL PRIME-RIB GRAVY

YIELD: 6 TO 8 SERVINGS / **ACTIVE TIME:** 20 MINUTES / **TOTAL TIME:** 30 MINUTES

The fundamental part to any gravy comes from the juices leftover in the pan after a rib roast has been cooked. Likewise, your gravy goes hand-in-hand with the roasting technique—it's essential that you use a technique that involves a roasting pan so that the juices from the prime rib slowly accumulate at the bottom of the pan.

1. When the roast is finished, remove it and the roasting rack from the pan and transfer to a large carving board. I recommend covering the roast with a sheet or two of aluminum foil so it stays warm. Pour the juices from the roasting pan into a fat separator; discard the fat and return the juices to the original roasting pan. Note that there is still flavor blanketing the roasting pan, so be sure to use the same pan.

2. Next, set the roasting pan over high heat—it may be necessary to use two burners if the roasting rack is too large for one. Add the red wine and stock or broth to the pan and bring to a light simmer, using a wooden spatula to scrape off any bits that still remain on the bottom of the pan so they are incorporated into the sauce.

3. Whisk the butter into the sauce, followed by the flour. Depending on how thick you want the gravy, add more butter and flour in incremental, equal proportions.

4. When thickened, about 1 to 2 minutes later, turn the heat off and stir in the thyme. Taste the gravy and season with the coarsely ground black pepper and fresh sea salt; then transfer to a small gravy boat. Serve hot.

TIP: If you don't have a have fat separator, bring the roasting pan to the sink and tilt to one side. Next, with a large spoon or ladle, slowly spoon away the fat that remains at the top of the dish.

INGREDIENTS:

- 1 CUP DRY RED WINE
- 1½ CUPS BEEF OR VEAL STOCK (SEE PAGE 475 OR 476, RESPECTIVELY)
- 1 TABLESPOON UNSALTED BUTTER
- 1 TABLESPOON ALL-PURPOSE FLOUR
- 1 TABLESPOON FRESH THYME, FINELY CHOPPED
-   COARSELY GROUND BLACK PEPPER, TO TASTE
-   SEA SALT, TO TASTE

MADEIRA SAUCE

YIELD: 6 SERVINGS / **ACTIVE TIME:** 20 MINUTES / **TOTAL TIME:** 25 MINUTES

A very condensed sauce, the most fundamental element to a Madeira is the beef stock, meaning homemade is best.

1. Add the butter to a medium cast-iron skillet and warm over medium heat. Then add the chopped shallot and sauté until translucent, about 4 minutes.

2. Add the flour to the pan and cook for 1 minute. Once incorporated, turn the heat to medium-low and then add the dry red wine, Madeira, beef stock or broth, thyme, and rosemary.

3. Cook until the sauce has been significantly reduced, to your desired consistency, about 15 to 20 minutes.

4. When the sauce is reduced, remove the skillet from the stovetop and season with the coarsely ground black pepper and fresh sea salt. Spoon the Madeira sauce over the cuts of rib roast.

TIP: If you want a stronger sauce, add 1 tablespoon of beef demi-glace to the sauce along with the beef stock. You can find beef demi-glace at the grocery store near the beef broths and stocks.

INGREDIENTS:

- 2 TABLESPOONS UNSALTED BUTTER
- 1 SMALL SHALLOT, FINELY CHOPPED
- 1 TABLESPOON ALL-PURPOSE FLOUR
- ¼ CUP DRY RED WINE
- ¾ CUP MADEIRA
- 1 CUP BEEF STOCK (SEE PAGE 475)
- 2 SPRIGS FRESH THYME, LEAVES REMOVED
- 2 SPRIGS FRESH ROSEMARY, LEAVES REMOVED
- COARSELY GROUND BLACK PEPPER, TO TASTE
- SEA SALT, TO TASTE

HOMEMADE KETCHUP

YIELD: 6 TO 8 SERVINGS / **ACTIVE TIME:** 10 MINUTES / **TOTAL TIME:** 15 MINUTES

You'll be surprised at just how well homemade ketchup pairs with a grilled prime rib, especially when used as a condiment for dipping. Most ketchup made by large-scale manufacturers contains a substantial amount of sucrose. Because of this, when we have natural, homemade ketchup, we tend to think of it as something entirely different, which is a good thing.

1. In a medium bowl, combine the pureed tomatoes, lemon juice, extra-virgin olive oil, onion, garlic, and dark brown sugar. Let rest for 15 minutes.

2. Next, gradually whisk in the apple cider vinegar and water. Season with the coarsely ground black pepper and fresh sea salt.

3. You can serve your ketchup right away or let the flavors meld overnight in the refrigerator.

INGREDIENTS:

- 3 CUPS PUREED TOMATOES
- ¼ MEDIUM LEMON, JUICED
- 2 TABLESPOONS EXTRA-VIRGIN OLIVE OIL
- ½ MEDIUM WHITE ONION, FINELY CHOPPED
- 2 GARLIC CLOVES, MINCED
- ¼ CUP DARK BROWN SUGAR
- ½ CUP APPLE CIDER VINEGAR
- ½ CUP WATER
- COARSELY GROUND BLACK PEPPER, TO TASTE
- SEA SALT, TO TASTE

PIZZAIOLA

YIELD: 6 TO 8 SERVINGS / **ACTIVE TIME:** 35 MINUTES / **TOTAL TIME:** 45 MINUTES

Hailing from Naples, this tangy tomato sauce goes great with steak.

1. Add the extra-virgin olive oil to a cast-iron skillet and place over medium heat. When the oil is hot, add the garlic and cook until golden, about 1 to 2 minutes.

2. Next, add the plum tomatoes, sun-dried tomatoes, oregano, thyme, and red pepper flakes, if using. Simmer for 15 minutes, and then add the wine and basil and season with pepper and salt.

3. Simmer for 20 more minutes, then remove the skillet from the stovetop and serve warm.

INGREDIENTS:

- ¼ CUP EXTRA-VIRGIN OLIVE OIL
- 4 GARLIC CLOVES, FINELY CHOPPED
- 2 LBS. PLUM TOMATOES, CRUSHED BY HAND
- ¼ CUP SUN-DRIED TOMATOES
- 1 SPRIG FRESH OREGANO
- 1 SPRIG FRESH THYME
- 1 TEASPOON RED PEPPER FLAKES (OPTIONAL)
- ¼ CUP DRY WHITE WINE
- ½ CUP FRESH BASIL LEAVES, FINELY CHOPPED
- COARSELY GROUND BLACK PEPPER, TO TASTE
- SEA SALT, TO TASTE

SUN-DRIED TOMATO PESTO

YIELD: 6 TO 8 SERVINGS / **ACTIVE TIME:** 10 MINUTES / **TOTAL TIME:** 10 MINUTES

This is such an easy recipe to make—all you need is a food processor, and the result is a deeply flavorful condiment.

1. In a small food processor, combine all the ingredients except the extra-virgin olive oil and pulse into a thick mixture.

2. Slowly add the extra-virgin olive oil and process until reaching your desired consistency. Serve at room temperature or lightly chilled.

INGREDIENTS:

- 12 SUN-DRIED TOMATOES
- ½ CUP FRESH BASIL LEAVES
- ¼ SMALL SHALLOT
- ¼ CUP PINE NUTS
- 1 GARLIC CLOVE
- 1 TABLESPOON COARSELY GROUND BLACK PEPPER
- 1 TEASPOON SEA SALT
- ½ CUP EXTRA-VIRGIN OLIVE OIL

GARLIC & CHIVE STEAK SAUCE

YIELD: 6 SERVINGS / **ACTIVE TIME:** 10 MINUTES / **TOTAL TIME:** 15 MINUTES

The flavor of chives is very similar to that of a mild onion. Sauces rooted in simple flavors such as garlic and chives always go well with red meat and potato side dishes.

1. In a small bowl, whisk together the sour cream, chives, garlic, and lemon juice, and then season with the coarsely ground black pepper.

TIP: Store in the refrigerator for up to 3 days, and serve at room temperature.

INGREDIENTS:

- ½ CUP SOUR CREAM
- ¼ CUP CHIVES, FINELY CHOPPED
- 3 GARLIC CLOVES, MINCED
- ¼ SMALL LEMON, JUICED
- COARSELY GROUND BLACK PEPPER, TO TASTE
- SEA SALT, TO TASTE

MUSHROOM CREAM SAUCE

YIELD: 6 SERVINGS / **ACTIVE TIME:** 25 MINUTES / **TOTAL TIME:** 30 MINUTES

Mushrooms have a very heavy, hearty flavor that's perfect when sautéed and braised with a cream base.

1. Add the unsalted butter to a medium cast-iron skillet and set over medium heat. When melted, add the shallot and cook until translucent, about 3 to 4 minutes. Add the garlic and sauté until browned, about 2 more minutes.

2. Next, add the mushrooms to the skillet and cook until lightly browned, about 4 to 6 minutes. Then, stir in the heavy whipping cream and thyme. Bring the cream to a boil and then reduce the heat to low and cook for about 10 minutes, until the cream has reduced and the mushrooms are tender.

3. Remove the pan from the heat and season with the coarsely ground black pepper and fresh sea salt. Serve immediately.

INGREDIENTS:

- 1 TABLESPOON UNSALTED BUTTER
- 1 SMALL SHALLOT, MINCED
- 2 GARLIC CLOVES, MINCED
- 2–3 CUPS SLICED WHITE BUTTON MUSHROOMS
- ¾ CUP CREAM
- 1 TABLESPOON FINELY CHOPPED FRESH THYME
- COARSELY GROUND BLACK PEPPER, TO TASTE
- SEA SALT, TO TASTE

HERBED BUTTER

YIELD: 6 SERVINGS / **ACTIVE TIME:** 10 MINUTES / **TOTAL TIME:** 15 MINUTES

You can make this dish up to four days in advance, just be sure to cover and refrigerate it immediately after pulsing in the food processor.

1. In a food processor, pulse the rosemary, thyme, chives, parsley, and garlic, followed by the butter. Season with the coarsely ground black pepper and fresh sea salt, and then cover and transfer to the refrigerator for up to 4 days.

TIP: Before serving, remove the herbed butter from the refrigerator and bring to room temperature by letting it stand for 1 hour.

INGREDIENTS:

- 1 TABLESPOON FRESH ROSEMARY LEAVES
- 1 TABLESPOON FRESH THYME LEAVES
- 1 TABLESPOON FRESH CHIVES
- 1 TABLESPOON FLAT-LEAF PARSLEY LEAVES
- ½ MEDIUM GARLIC CLOVE
- 1 STICK OF UNSALTED BUTTER, AT ROOM TEMPERATURE
- 1 TEASPOON COARSELY GROUND BLACK PEPPER
- 1 TEASPOON SEA SALT

INDUSTRY INSIDERS

*T*he chefs, butchers, and farmers featured in this chapter are wonderful examples of how the meat supply chain should work. They all treat the meat they work with, whether raising it, butchering it, or cooking it, with the utmost respect, which is best for the animals, the environment, and the consumer.

CHEF JUSTIN DAIN

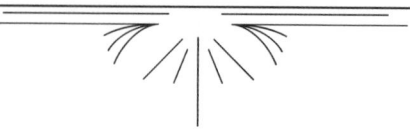

Chef Justin Dain received his training at the prestigious Culinary Institute of America (CIA) in Hyde Park, New York, where he earned an A.O.S. in Culinary Arts in 2000 and then continued his studies at the New England Culinary Institute in Montpelier, Vermont, earning a B.A. in Food and Beverage Management in 2002. His first job in a kitchen was as a line cook at the Boston Harbor Hotel in 2002. From there he worked in several prestigious Boston restaurants.

Since 2010, he has been the Executive Chef at the Hanover Inn in Hanover, New Hampshire, where guests have savored his culinary creations. In 2013, Chef Dain opened PINE, the signature restaurant at the Hanover Inn. PINE embraces a farm-to-table philosophy, utilizing the bounty of ingredients available in and around New England. The restaurant has received numerous accolades in the years it has been open and continues to push the envelope with new inspiring dishes.

Chef Dain has twice been invited to cook at the renowned James Beard House in New York City: a sold-out 2011 dinner, and a 2017 meal that featured local farms from New England.

WALPOLE VALLEY FARM'S CHICKEN

with Fennel, Dunk's Mushrooms, Roasted Shallots Corn, Butternut Squash, Parsnips, & Broccoli Rabe Pesto

YIELD: 4 SERVINGS / **ACTIVE TIME:** 15 MINUTES / **TOTAL TIME:** 1 HOUR AND 30 MINUTES

You might not be able to get your hands on Walpole Valley chicken and Dunk's mushrooms, but so long as you buy products from your local farmers, this dish is sure to wow friends and family.

1. Preheat oven to 400°F degrees.

2. Season the chicken's cavity and entire outside. Rain the salt and pepper down on the chicken to create an even coating over the whole bird.

3. Truss the bird (see page 244).

4. Place the chicken in the oven and cook for 50 to 60 minutes, or until a meat thermometer stuck into the chicken's breast registers 160°F. Remove the chicken from the oven and let rest for 15 minutes before carving.

5. Remove the breast, thighs, drumsticks, and wings. Slice the breast and thighs (remove thigh bones if desired).

6. Serve chicken on a bed of the squash puree with a side of the roasted vegetables and drizzle it all with the pesto.

BUTTERNUT SQUASH PUREE

1. In medium saucepan, combine the heavy cream, butter, cumin, maple syrup, and salt and bring to a simmer. Once hot, remove the pan from the heat.

2. Preheat oven to 325°F. Coat the flesh of the squash with the canola oil and salt. Place the squash flesh-side down on a parchment paper–lined sheet pan.

3. Roast the squash for 40 to 50 minutes, or until very soft.

4. Scoop the squash flesh into a food processor or blender. Puree, slowly adding the cream mixture, until very smooth. Season to taste.

INGREDIENTS:

- 1 (4–5 LB.) WHOLE CHICKEN
- SALT AND PEPPER, TO TASTE

BUTTERNUT SQUASH PUREE

- 1 CUP HEAVY CREAM
- 4 OZ. UNSALTED BUTTER
- 1 TEASPOON CUMIN
- 1 TABLESPOON MAPLE SYRUP
- SALT, TO TASTE
- 2 LBS. BUTTERNUT SQUASH, HALVED AND SEEDED
- 1 TABLESPOON CANOLA OIL

ROASTED VEGETABLES

- 2 CORN EARS
- 4 SHALLOTS, QUARTERED
- 2 FENNEL HEADS, TOPS REMOVED AND QUARTERED
- 1 CUP MUSHROOMS
- 2 LEEKS, HALVED AND CUT INTO 2-INCH PLANKS
- ½ LB. PARSNIPS, PEELED AND DICED
- ½ LB. BUTTERNUT SQUASH, PEELED AND DICED
- 3 TABLESPOONS CANOLA OIL
- SALT, TO TASTE

Continued . . .

ROASTED VEGETABLES

1. Preheat oven to 350°F.

2. Roast the corn whole in the oven for 12 minutes. Once cooked, remove the husk and cut the corn off the cob.

3. In a bow, toss the fennel, mushrooms, leeks, parsnip, and squash with oil and salt. Spread out on a sheet pan and roast for 25 minutes, or until the vegetables are tender.

BROCCOLI RABE PESTO

1. In a pot, bring salted water to a boil. Add the broccoli rabe, cook for 1 minute, then remove and shock in ice water. Chill until fully cooled. Remove from the water and drain on a paper towel–lined plate until dry.

2. Preheat oven to 325°F.

3. On a sheet pan, spread out the pine nuts and toast in the oven for 7 to 8 minutes, or until slightly browned on all edges.

4. In a food processor, combine all the ingredients and puree until very smooth. Add more olive oil to thin out pesto if necessary.

BROCCOLI RABE PESTO

6	CUPS WATER
	SALT AND PEPPER, TO TASTE
1	BUNCH BROCCOLI RABE, LARGE STEMS REMOVED
1	CUP PINE NUTS
4	GARLIC CLOVES
1	CUP BASIL LEAVES
1	CUP GRATED PARMESAN CHEESE
	ZEST OF 1 LEMON
1	CUP EXTRA-VIRGIN OLIVE OIL

SPICED HERITAGE FARM'S PORK TENDERLOIN
with Warm Potato Salad Crossroad's Farm Vegetable Hash, & Harissa Vinaigrette

YIELD: 4 SERVINGS / **ACTIVE TIME:** 15 MINUTES / **TOTAL TIME:** 1 HOUR

Chef Justin's spin on a beautiful pork tenderloin is out of this world.

INGREDIENTS:

- 1 TEASPOON RAS EL HANOUT
- 1 TABLESPOON SALT
- 1 TABLESPOON BROWN SUGAR
- 1 TEASPOON BLACK PEPPER
- 1 TEASPOON TOGARASHI
- 2 PORK TENDERLOINS, SILVER SKIN REMOVED
- 1 TABLESPOON CANOLA OIL

WARM POTATO SALAD
- 3 CUPS FINGERLING POTATOES
- 4 SLICES BACON, COOKED AND CHOPPED
- 1 BUNCH SCALLIONS, THINLY SLICED
- 1 TABLESPOON WHOLE GRAIN MUSTARD
- ¼ CUP RICE WINE VINEGAR
- 3 TABLESPOONS HONEY
- 2 TEASPOONS SALT
- ½ TEASPOON BLACK PEPPER

CROSSROAD FARM'S VEGETABLE HASH
- 2 TABLESPOONS CANOLA OIL
- 2 GARLIC CLOVES, MINCED
- ½ TEASPOON RED PEPPER FLAKES (OPTIONAL)
- 1 ZUCCHINI, DICED
- 1 YELLOW SQUASH, DICED
- 1 CORN EAR, KERNELS CUT OFF COB
- 8 ASPARAGUS STALKS, DICED
- 6 SHISHITO PEPPERS, CUT INTO RINGS
- 1 RED BELL PEPPER, DICED
- 1 PINT CHERRY TOMATOES, HALVED
- 8 BASIL LEAVES, TORN
- SALT, TO TASTE

HARISSA VINAIGRETTE
- 3 TABLESPOONS HARISSA PASTE
- 1 TEASPOON RAS EL HANOUT
- ⅓ CUP RICE WINE VINEGAR
- 1 TABLESPOON FRESH LEMON JUICE
- 3 TABLESPOONS HONEY
- 2 GARLIC CLOVES
- 2 TEASPOONS PEELED AND CHOPPED GINGER
- SALT, TO TASTE
- 1 CUP CANOLA OIL

1. In a small bowl, combine the ras el hanout, salt, brown sugar, black pepper, and togarashi. Coat the pork tenderloins with the canola oil and rub the spice mixture all over the pork. Let the pork marinate for 30 minutes.

2. Set grill to medium-high heat and grill the pork on each side for 5 to 8 minutes, or until a meat thermometer registers an internal temperature of 135°F.

3. Place the pork on a sheet pan with a roasting rack and rest for 10 minutes before slicing.

WARM POTATO SALAD

1. Place the fingerling potatoes in a medium sized pot and fill with water to cover. Bring the potatoes to a simmer and cook for 10 to 12 minutes, or until tender.

2. Drain the potatoes and cut them into a small dice.

3. In a bowl, combine the potatoes with the remainder of the ingredients and mix well. Set aside.

CROSSROAD FARM'S VEGETABLE HASH

1. Add oil to a large sauté pan over medium heat. Once oil is shimmering, add the garlic and red pepper flakes, if using. Cook until garlic begins to have slight browning on the edges.

2. Add the zucchini, squash, corn, asparagus, shishitos, and bell pepper and cook for 4 to 5 minutes. Once the vegetables begin to soften, add the tomatoes and basil, stir, and remove from heat. Season to taste.

HARISSA VINAIGRETTE

1. In a food processor or blender, combine all of the ingredients, except the oil, and puree until smooth.

2. Continue to puree and slowly add the oil until the vinaigrette is emulsified. Season to taste.

THE PINE BURGER

YIELD: 1 SERVING / **ACTIVE TIME:** 15 MINUTES / **TOTAL TIME:** 40 MINUTES

The first time I tried Chef Justin's classic burger I was blown away. Now you have the recipe that I have come to love, and it's very easy to double, triple, or quadruple. He uses beef from Boyden Farm in Cambridge, Vermont. Be sure to use beef from a local farmer close to you.

1. Form the beef into a patty that is slightly larger than the surface of the burger bun.

2. Set the grill to medium-high heat and preheat oven to 325°F.

3. Grill the burger for 45 seconds, then turn the burger a quarter turn and grill for an additional 45 seconds. Flip the burger and repeat. Once the burger is completely marked off on both sides, place the burger in the oven for 7 to 8 minutes, or until medium rare (130°F).

4. Remove the burger from the oven, top with cheese, and place it back in the oven for 30 seconds to melt the cheese.

5. Quickly griddle the bun for added texture and flavor. Place the burger on the bun, add mayonnaise, bacon, and Onion Strings.

ONION STRINGS

1. In a 2-quart saucepan, heat the canola oil to 350°F.

2. In a bowl, combine cornstarch and flour.

3. Using a mandolin, slice the onion very thinly. Break up the rings and toss them in cornstarch-and-flour mixture.

4. Fry the onion rings in the oil for 1 minute or until crispy and golden brown. Remove the onions from the oil and place them on a paper towel–lined plate. Season immediately.

CHIPOTLE MAYONNAISE

1. In a bowl, combine all of the ingredients and mix well. Set aside.

INGREDIENTS:

- 9 OZ. GROUND BEEF (80-20)
- 2 SLICES CHEDDAR CHEESE
- SALT AND PEPPER, TO TASTE
- 1 BRIOCHE BUN, FOR SERVING
- 2 THICK SLICES COOKED APPLEWOOD-SMOKED BACON

ONION STRINGS

- 1 QUART CANOLA OIL, FOR FRYING
- ¼ CUP CORNSTARCH
- ¼ CUP ALL-PURPOSE FLOUR
- 1 SMALL ONION
- SALT AND PEPPER, TO TASTE

CHIPOTLE MAYONNAISE

- 1 CUP MAYONNAISE
- 1 TABLESPOON RICE WINE VINEGAR
- 2 TABLESPOONS CHOPPED CHIPOTLE PEPPERS IN ADOBO SAUCE
- 1 TEASPOON SALT

GNOCCHI BOLOGNESE
with Whipped Ricotta & Parmesan Frico

YIELD: 4 SERVINGS / **ACTIVE TIME:** 15 MINUTES / **TOTAL TIME:** 1 HOUR AND 30 MINUTES

This style of gnocchi skips the potato and uses flour and eggs. Parmesan frico is like a crispy parmesan cracker and adds wonderful texture to this dish.

PARMESAN GNOCCHI

1. In a pot, bring water, butter, and salt to a boil. Add flour and stir with a wooden spoon until fully incorporated. Remove from heat.

2. Place mixture in a stand mixer. Using the paddle attachment, add Parmesan and mix for 30 seconds. Slowly add the eggs, one at a time.

3. In a pot, bring salted water to a boil.

4. Place mixture in a piping bag with a large plain tip. Pipe small amounts of the mixture into the salted boiling water. Cook each batch for 1 minute. Set aside on a sheet pan.

WHIPPED RICOTTA

1. In a small bowl, combine all of the ingredients and whisk until smooth. Set aside.

PARMESAN FRICO

1. Preheat oven to 300°F.

2. Spread out the cheese evenly on a silpat baking sheet. Bake the cheese for 10 to 12 minutes, or until it turns slightly brown and crispy. Remove from the oven and allow to cool to room temperature. Once the cheese has cooled, break it into small pieces.

Continued...

INGREDIENTS:

PARMESAN GNOCCHI
- 2 CUPS WATER
- 2 STICKS UNSALTED BUTTER
- SALT, TO TASTE
- 10 OZ. ALL-PURPOSE FLOUR
- 4 OZ. GRATED PARMESAN CHEESE
- 4 EGGS

WHIPPED RICOTTA
- 1 CUP RICOTTA CHEESE
- 1 TABLESPOON HEAVY CREAM
- 2 TABLESPOONS GRATED PARMESAN CHEESE
- 1 TEASPOON SALT
- ½ TEASPOON FRESHLY CRACKED BLACK PEPPER

PARMESAN FRICO
- 1 CUP GRATED PARMESAN CHEESE

BOLOGNESE

1. In a large pot over medium heat, combine the stocks and pureed tomatoes and reduce by half. Reserve for the sauce.

2. Add olive oil and garlic to a Dutch oven over medium heat and cook for 1 to 2 minutes, until the garlic is fragrant. Remove the garlic.

3. Add the carrot, celery, and onion to the Dutch oven and cook for about 10 minutes, until the vegetables start to caramelize.

4. Add the bacon and herbs and cook for 5 more minutes.

5. Add the ground meats to the vegetable mixture and break up the meat into small pieces. Season with salt and black pepper.

6. Add the half-and-half and cook until it reduces by three-quarters. Once it has reduced, add the white wine and cook until the liquid reduces by half.

7. Add the tomato and stock mixture. Continue to cook the Bolognese until it has thickened, about 2 hours.

8. Add 1 tablespoon of olive oil to a small sauté pan over medium-high heat.

9. Season the chicken livers with salt and black pepper and add to the sauté pan. Cook for 1 to 2 minutes per side, until they are medium rare.

10. Let the chicken livers cool, finely chop, and add the Bolognese sauce. Simmer the sauce for 20 more minutes and turn off the heat. Season to taste.

BOLOGNESE

4	CUPS CHICKEN STOCK (SEE PAGE 478)
4	CUPS VEAL STOCK (SEE PAGE 476)
56	OZ. SAN MARZANO TOMATOES, PUREED
½	CUP EXTRA-VIRGIN OLIVE OIL, PLUS 1 TABLESPOON
3	GARLIC CLOVES, THINLY SLICED
1	CUP DICED CARROT
1	CUP DICED CELERY
2	CUPS DICED SPANISH ONION
8	SLICES BACON, DICED
2	TABLESPOONS CHOPPED FRESH ROSEMARY
2	TABLESPOONS CHOPPED FRESH SAGE
1	LB. GROUND VEAL
1	LB. GROUND BEEF
1	LB. GROUND PORK
	SALT AND PEPPER, TO TASTE
3	CUPS HALF-AND-HALF
2	CUPS WHITE WINE
½	LB. CHICKEN LIVERS

PAN ROASTED DUCK BREAST
Over Warm Farro & Beet Salad with Strawberry & Yuzu Emulsion

YIELD: 2 SERVINGS / **ACTIVE TIME:** 30 MINUTES / **TOTAL TIME:** 1 HOUR

This duck over warm farro and beet salad is a perfect dish to serve at any dinner party.

1. Preheat the oven to 325°F.

2. Using a sharp knife, remove excess fat around the sides of the duck breast. Score the fat of the duck breast.

3. Add oil to a medium-size sauté pan over medium heat. Season the duck breast with salt and black pepper and put the fat-side down in the pan. Render the fat until it is crispy and golden brown.

4. Once the fat is rendered, remove the duck and place, fat-side up, on a sheet pan with a roasting rack. Place the duck in the oven and cook to 130°F. Once the duck has cooked, remove from the oven and let rest for 5 minutes before slicing. Season the duck breast with Maldon sea salt.

FARRO

1. In a pot, bring water to a boil. Add the farro and cook until tender, about 10 minutes. Strain the farro.

2. Add the butter, swiss chard, and thyme to a sauté pan over medium-high heat and cook until the greens wilt. Add the farro and cook until the farro is hot. Season to taste.

BEET & GOAT CHEESE PUREE

1. Combine the water and beet in a pot over medium heat, bring to a boil, and then simmer for 1 hour, until the beat is tender.

2. Remove the beet from the water and reserve 1 cup of the cooking liquid.

3. Once it has cooled to room temperature, peel the beet and place it in a food processor along with the remainder of the ingredients. Puree the mixture until it becomes very smooth. Add more cooking liquid if necessary to achieve a smooth puree.

Continued . . .

INGREDIENTS:
2	(7 OZ.) DUCK BREASTS
1	TABLESPOON CANOLA OIL
	SALT AND PEPPER, TO TASTE
	MALDON SEA SALT, TO TASTE

FARRO
4	CUPS WATER
1	CUP FARRO
1	TABLESPOON UNSALTED BUTTER
2	CUPS CHOPPED RAINBOW SWISS CHARD
1	TABLESPOON CHOPPED FRESH THYME
	SALT AND PEPPER, TO TASTE

BEET & GOAT CHEESE PUREE
4	CUPS WATER
1	LARGE RED BEET, CLEANED
3	OZ. GOAT CHEESE
½	TEASPOON CUMIN
1	CUP BEET COOKING LIQUID
1	TABLESPOON SHERRY VINEGAR
1	TABLESPOON HONEY
	SALT, TO TASTE

BABY BEETS

1. Trim the tops off the baby beets. Place the beets in a small pot and cover with water. Bring the water to a simmer and cook for 20 minutes, or until the beets are tender.

2. Drain the beets, reserving some of the cooking liquid, then peel and quarter them.

3. Combine butter and 1 tablespoon of the reserved cooking liquid in a small sauté pan over low heat. Add the beets and warm them slowly. Season to taste.

STRAWBERRY & YUZU EMULSION

1. Combine the strawberries, yuzu, ginger, honey, vinegar, and salt in a small pot over low heat and cook for 10 minutes, until the strawberries have softened.

2. Cool the mixture to room temperature and place in a food processor or blender. Puree the mixture on medium speed and slowly add the blended olive oil. Add the xanthan gum and turn to high speed. Turn off once the sauce has slightly thickened.

BABY BEETS

4	BABY YELLOW BEETS
4	BABY RED BEETS
1	TABLESPOON UNSALTED BUTTER
	SALT AND PEPPER, TO TASTE

STRAWBERRY & YUZU EMULSION

1½	CUPS QUARTERED STRAWBERRIES
3	OZ. YUZU JUICE
1	TABLESPOON PEELED AND CHOPPED GINGER
3	TABLESPOONS HONEY
½	CUP RICE WINE VINEGAR
	SALT, TO TASTE
1	CUP BLENDED OLIVE OIL
1	TEASPOON XANTHAN GUM

GRILLED TENSEN FARM'S STRIP LOIN OF BEEF,
Dunk's Mushrooms, Charred Scallions, Confit Potatoes & Pickled Herbs

YIELD: 2 SERVINGS / **ACTIVE TIME:** 30 MINUTES / **TOTAL TIME:** 1 HOUR AND 30 MINUTES

Build a full dinner around Chef Justin's beautiful strip loin recipe.

1. Heat grill to high heat and season the steaks and scallions with the blended oil and salt and pepper.

2. Place the steaks and scallions on the grill.

3. Cook the scallions until they begin to char on both sides. Cook steaks to your preference; 128°F degrees is recommended for medium-rare.

4. When the steaks are done, let rest for at least 5 minutes before slicing.

DUNK'S MUSHROOMS

1. Heat a sauté pan over medium-high heat, then add the butter and mushrooms, cooking until the mushrooms begin to slightly caramelize, about 3 minutes. Add the garlic and thyme and sauté for 1 minute. Season to taste and set aside.

CONFIT POTATOES

1. Preheat the oven to 400°F. In a medium-sized bowl, combine the potatoes and salt.

2. Place the potatoes, thyme, rosemary, and bay leaf in a baking dish and cover with the duck fat. Make sure the fat is completely covering the potatoes. Wrap the dish in plastic wrap and then aluminum foil and bake for 1 hour and 15 minutes. Remove the potatoes from the duck fat, strain out the herbs, and reserve the duck fat.

PICKLED HERBS

1. In a food processor, combine all of the ingredients, except the oil, salt, and pepper, and puree until a paste forms.

2. Slowly drizzle in the olive oil and blend until the mixture is smooth. Season with salt and pepper to taste.

INGREDIENTS:

2	(10 OZ.) STRIP LOINS
4	SCALLIONS
1	TABLESPOON BLENDED OLIVE OIL
	SALT AND COARSE BLACK PEPPER, TO TASTE

DUNK'S MUSHROOMS

2	TABLESPOONS UNSALTED BUTTER
½	LB. MUSHROOMS
1	GARLIC, CLOVE MINCED
2	SPRIGS FRESH THYME, LEAVES PULLED FROM STALKS
	SALT, TO TASTE

CONFIT POTATOES

10	OZ. FINGERLING POTATOES, HALVED
	SALT, TO TASTE
5	SPRIGS FRESH THYME
1	SPRIG FRESH ROSEMARY
1	BAY LEAF
2	CUPS DUCK FAT

PICKLED HERBS

1	BUNCH CILANTRO
1	BUNCH PARSLEY
6	GARLIC CLOVES
1	LIME, ZESTED AND JUICED
4	TABLESPOONS RICE WINE VINEGAR
1	TABLESPOON HONEY
1	TEASPOON RED PEPPER FLAKES (OPTIONAL)
1	SHALLOT, QUARTERED
1	CUP BLENDED OLIVE OIL
	SALT AND PEPPER, TO TASTE

GENERAL TSO'S PORK BELLY STEAM BUNS
with Sesame Bok Choy, Radish & Yuzu Mayonnaise

YIELD: 6 SERVINGS / **ACTIVE TIME:** 30 MINUTES / **TOTAL TIME:** 12 HOURS

I remember the first time Chef Justin made these for me. I knew I had to get him to give me the recipe for this book. Get ready to take your appetizers to the next level.

1. In a small bowl, combine the Chinese five spice, garlic, soy sauce, hoisin, and rice wine vinegar and mix until it becomes smooth. Rub the mixture all over the pork belly and let marinate for 12 to 24 hours in the refrigerator.

2. After marinating the pork belly, remove it from the refrigerator. Preheat the oven to 325°F.

3. Put the carrot, onion, celery, salt, and stock in a baking dish and place the pork belly on top. Cover the baking dish with plastic wrap and then aluminum foil. Bake for 2 hours.

4. Take the baking dish out of the oven, remove the plastic and foil, and return the dish to the oven for 45 more minutes, or until the pork belly is slightly caramelized.

5. As the pork cooks, sauté the bok choy with the sesame oil in a frying pan over low heat until tender, about 3 minutes. Season with salt and pepper and set aside.

6. When the pork belly is done, let rest on a sheet pan. Once the pork is cooled, slice it into 2-inch pieces that each weigh approximately 2 oz.

7. Preheat oven to 350°F. To serve, place the sliced pork belly on a sheet pan and warm in the oven for 5 minutes, or until hot throughout. Once hot, rub the pork belly slices with the General Tso's Sauce. Place steam buns in a steamer basket with a lid and steam for 4 minutes. Remove the buns from the steamer and place ½ tablespoon yuzu mayo in the middle of the buns, then add the warmed glazed pork. Top the pork with the sesame bok choy and garnish with thinly sliced radishes, cilantro, and jalapeño, if using.

INGREDIENTS:

- 2 TEASPOONS CHINESE FIVE SPICE
- 2 GARLIC CLOVES, MINCED
- ¼ CUP SOY SAUCE
- ¼ CUP HOISIN SAUCE
- 2 TABLESPOONS RICE WINE VINEGAR
- 2 LBS. PORK BELLY
- 1 CARROT, CHOPPED
- 1 ONION, CHOPPED
- 2 CELERY STALKS, CHOPPED
- 1 TABLESPOON SALT
- 1½ CUP CHICKEN STOCK (SEE PAGE 478)
- 1 CUP BOK CHOY, THINLY SLICED
- 2 TEASPOONS SESAME OIL
- 1 CUP GENERAL TSO'S SAUCE (SEE RECIPE)
- 6 STEAM BUNS (SEE RECIPE)
- ½ CUP YUZU MAYO (CARRIED AT SPECIALTY MARKETS)
- THINLY SLICED RADISHES, JALAPEÑO, AND CILANTRO, FOR GARNISH (OPTIONAL)

Continued...

GENERAL TSO'S SAUCE

1. In a pot over medium heat, combine canola oil, red pepper flakes, minced garlic, ginger, scallions, and jalapeño and sauté for 3 to 5 minutes, until fragrant.

2. Add the soy sauce, hoisin, rice wine vinegar, honey, and stock and simmer for 15 to 20 minutes.

3. In a bowl combine the cornstarch with 4 tablespoons water to make a slurry. Add the slurry to the simmering sauce and cook for 1 minute, or until slightly thickened. Remove from heat. The sauce can be blended to be very smooth or can be used as is.

STEAM BUNS

1. Place the yeast and water in the mixing bowl of a stand mixer fitted with a dough hook and let sit until the mixture becomes foamy, about 10 minutes.

2. Add the remaining ingredients and mix on low for 10 minutes. Cover the bowl with a kitchen towel and place it in a naturally warm spot until it doubles in size, about 45 minutes.

3. Place the risen dough on a flour-dusted work surface and cut into 12 pieces. Roll them into balls, cover with plastic wrap, and let them rise for 30 minutes.

4. Cut a dozen 4" squares of parchment paper. Roll each ball into a 4" oval. Grease a chopstick with shortening, gently press the chopstick down in the middle of each oval, and fold the dough over to create a bun. Place each bun on a square of parchment and let stand for 30 minutes.

5. To cook the buns, place 1" of water in a saucepan and bring to a boil. Place a steaming tray in the pan and, working in batches if necessary, place the buns in the steamer tray, making sure to leave them on the parchment. Steam for 10 minutes.

GENERAL TSO SAUCE

2	TABLESPOONS CANOLA OIL
1	TEASPOON RED PEPPER FLAKES
3	GARLIC CLOVES, MINCED
2	TABLESPOONS FRESH GINGER, MINCED
4	TABLESPOONS SLICED SCALLIONS
1	JALAPEÑO PEPPER, SEEDED AND MINCED
½	CUP SOY SAUCE
½	CUP HOISIN SAUCE
¼	CUP RICE WINE VINEGAR
3	TABLESPOONS HONEY
1	CUP CHICKEN STOCK (SEE PAGE 478)
2	TABLESPOONS CORNSTARCH

STEAM BUNS

1	TEASPOON ACTIVE DRY YEAST
3	TABLESPOONS WATER, AT ROOM TEMPERATURE
1	CUP BREAD FLOUR, PLUS MORE FOR DUSTING
4	TEASPOONS SUGAR
2	TEASPOONS NONFAT DRY MILK POWDER
1	TEASPOON KOSHER SALT
⅛	TEASPOON BAKING POWDER
⅛	TEASPOON BAKING SODA
4	TEASPOONS VEGETABLE SHORTENING, PLUS MORE AS NEEDED

SLOW BRAISED BOYDEN BEEF SHORT RIB,

Celery Root & Parsnip Puree, Confit Shallot, Heirloom Carrots, Tiny Onion Rings, & Gremolata

YIELD: 4 SERVINGS / **ACTIVE TIME:** 1 HOUR / **TOTAL TIME:** 4 HOURS

This short rib recipe is to die for.

1. Cut the short ribs so each portion has 2 bones. Season the short ribs with salt and pepper.

2. Heat canola oil in a Dutch oven over medium-high heat. When the oil starts to smoke, add the short ribs and sear on all sides, about 2 minutes per side. Work in batches and set aside seared short ribs.

3. Add the carrots, onion, and celery to the Dutch oven and cook the vegetables until they caramelize, 10 to 15 minutes. Add the garlic and cook an additional 3 minutes.

4. Add the short ribs back to the pot and deglaze the pan with the wine and vinegar. Reduce the wine by half.

5. Add the tomato puree, stock, thyme, rosemary, and Chinese five spice. Bring to a slow simmer and cook for 3 hours, covered, or put in a 300°F oven, covered, for 3 hours.

6. After 3 hours, check the short ribs. If they are not fork-tender, cook for another 45 minutes. If they are done, remove them from the pot and strain the sauce into a smaller saucepan. Skim off fat from sauce and season to taste.

CELERY ROOT AND PARSNIP PUREE

1. In a small pot, combine the celery root, parsnips, heavy cream, and butter and bring to a simmer. Cover and cook for 30 minutes.

2. When the vegetables are fork-tender put the mixture in a food processor or blender and puree until smooth. Season to taste.

INGREDIENTS:

- 12 BONE-IN SHORT RIBS
  SALT AND PEPPER, TO TASTE
- 2 TABLESPOONS CANOLA OIL
- 2 CARROTS, CHOPPED
- 1 LARGE ONION, CHOPPED
- 3 CELERY STALKS, CHOPPED
- 4 GARLIC CLOVES, THINLY SLICED
- 1 BOTTLE CABERNET SAUVIGNON
- 3 OZ. SHERRY VINEGAR
- 1 CUP PUREED SAN MARZANO TOMATOES
- 4 CUPS VEAL STOCK OR BEEF STOCK (SEE PAGES 475-476)
- 8 SPRIGS FRESH THYME
- 2 SPRIGS FRESH ROSEMARY
- 1 TEASPOON CHINESE FIVE SPICE

CELERY ROOT AND PARSNIP PUREE

- 1 LB. CELERY ROOT, CHOPPED
- 1 LB. PARSNIPS, CHOPPED
- 3 CUPS HEAVY CREAM
- 1 STICK UNSALTED BUTTER
  SALT, TO TASTE

LOCAL CARROTS

- 12 HEIRLOOM BABY CARROTS
- 1 TABLESPOON CANOLA OIL
  SALT AND PEPPER, TO TASTE

Continued . . .

LOCAL CARROTS

1. Preheat oven to 350°F.

2. In a bowl, toss all the ingredients so the carrots are completely coated in seasoning.

3. Place the carrots on a sheet pan and bake for 15 minutes. Set aside.

SHALLOT CONFIT

1. Preheat oven to 325°F.

2. Season the shallots with salt and place them in a small pot with the remainder of the ingredients. Be sure that the oil covers the shallots.

3. Place the covered pot in the oven and cook for 1 hour. Set aside.

TINY ONION RINGS

1. Using a mandolin, thinly slice the onion into rings. Peel and thinly slice the onion into rings.

2. Put the flour, egg, and panko into individual small bowls.

3. Dredge the onion rings in the flour, then the egg, and then the panko. Repeat until all of the onion slices are coated.

4. Add the oil to a medium-sized sauté pan over medium heat. When the oil reaches 310°F carefully add the onion rings, one at a time. Once the onion rings are slightly browned, remove them and drain on a paper towel–lined plate. Season immediately with salt.

GREMOLATA

1. In a small mixing bowl, combine all of the ingredients and mix well.

SHALLOT CONFIT

- 6 SHALLOTS, LEFT WHOLE
- SALT, TO TASTE
- 2 CUPS BLENDED OLIVE OIL.
- 2 SPRIGS FRESH THYME
- 1 SPRIG FRESH ROSEMARY
- 1 BAY LEAF

TINY ONION RINGS

- 1 ONION
- ½ CUP FLOUR
- 1 EGG WHITE, BEATEN
- ½ CUP PANKO
- 2 CUPS CANOLA OIL
- SALT, TO TASTE

GREMOLATA

- 1 BUNCH FLAT-LEAF PARSLEY, FINELY CHOPPED
- 1 LEMON, ZESTED AND JUICED
- 3 GARLIC CLOVES, MINCED
- ½ CUP EXTRA-VIRGIN OLIVE OIL
- 1 TABLESPOON FRESHLY GRATED HORSERADISH
- SALT, TO TASTE

CHEF JEFFREY SCHLISSEL

Chef Jeffrey Schlissel, known as Cheffrey to some, is best known for being the Kingpin to the Bacon Cartel. Like me, he focuses on locally sourced and sustainable products. He works with local farmers to feed their animals better in order to receive a better product and make less of an impact on our natural resources. In his over 30 years of professional experience he has seen too much wasted food, and has tried to get to the point where maximizing profits and limiting waste will be as easy as cracking an egg! I met him when we were on a panel for the American Culinary Federation (ACF). His passion and zeal for this industry struck a chord with me. Since then we have become friends. When I took on this project, I knew I wanted to feature this great chef's work.

Growing up in South Florida exposed him to a Latin pantry and introduced his palate to a broad range of taste experiences. Bold, fiery flavors mesh with sweet, refreshing tropical notes to yield a Zen-like experience for your taste buds. Schlissel thrives on creating dishes that combine Pan-Asian Floribbean flair, or as he calls it "Floribbeanasian."

He is the President of the Palm Beach Chefs Association, winner of *Boca* magazine's best Asian Fusion Restaurant in 2012, named Cutting Edge Chef of 2016 by the National Convention of the ACF, and in 2017 named Chef of the Year for the Palm Beach Chefs Association.

CURED BACON

YIELD: 5 SERVINGS / **ACTIVE TIME:** 1 TO 4 HOURS / **TOTAL TIME:** 10 DAYS

Time to cure some bacon, and Chef Jeffrey has you covered. This is his tried-and-true recipe for curing bacon. "For me," says Schlissel, "smoking our food is the most primal way of cooking. It brings us around the fire to talk, to tell stories, and reconnect with the way things use to be. It is the ultimate challenge to find that sweet spot in the smoker and get the final product to be as close to perfect as possible."

1. Place the pork belly in a shallow pan. Pour the bourbon over the pork belly. Refrigerate, uncovered, for 4 days. Each day flip the pork belly over. If the belly looks like it got thirsty and there is no more bourbon left, add more.

2. On the fourth day, remove the pan from the refrigerator and pat dry the pork belly.

3. In a small bowl, combine the sugar and salt and mix well. Rub it all over the pork belly. Refrigerate the pork belly for another 5 days, uncovered. This helps to "dry" out the pork belly, this is the curing process.

4. At the end of 5 days, freeze the pork belly for 12 to 24 hours.

5. When ready to smoke the pork belly, heat smoker to between 200 and 250°F. Use an ample amount of wood to really smoke the pork belly. Smoke the bacon—it is bacon now!—for at least 1 hour.

6. Cool and slice to use.

INGREDIENTS:

- 3 LBS. PORK BELLY
- 4 OZ. BOURBON, PLUS MORE IF NEEDED
- ½ CUP PALM SUGAR, COCONUT SUGAR, OR BROWN SUGAR
- ¼ CUP SALT

ACAPULCO GOLD RUB

YIELD: 4 CUPS / **ACTIVE TIME:** 5 MINUTES / **TOTAL TIME:** 5 MINUTES

This rub is perfect for steak and ribs.

1. In a bowl, combine all ingredients, mix well, and store in an air tight container.

INGREDIENTS:

- 1 CUP BROWN SUGAR
- ½ CUP SALT
- ½ CUP CHILI POWDER
- ¼ CUP HUNGARIAN PAPRIKA
- ¼ CUP CORIANDER
- 1 TABLESPOON GROUND GINGER
- ¼ CUP CUMIN
- ⅓ CUP GARLIC POWDER
- ⅓ CUP ONION POWDER
- 1 TABLESPOON LEMON ZEST
- ¼ CUP DUTCH COCOA OR FRESH DRIED CHOCOLATE

AGAVE SRIRACHA SWEET GLAZE

YIELD: 1½ CUPS / **ACTIVE TIME:** 5 MINUTES / **TOTAL TIME:** 5 MINUTES

This glaze goes on all things pork, but works especially well with ham.

1. In a bowl, combine all ingredients, mix well, and store in an air tight container.

INGREDIENTS:

- 1 CUP AGAVE
- ¼ CUP GARLIC POWDER
- ¼ CUP SRIRACHA

BRISKET

YIELD: 15 SERVINGS / **ACTIVE TIME:** 20 MINUTES / **TOTAL TIME:** 10 TO 14 HOURS

This is Chef Jeffrey's sure fire brisket recipe.

1. Rub the mustard all over the brisket.

2. In a bowl, combine the salt and pepper and coat the meat with the seasoning mixture.

3. Get the smoker up to 300°F and then smoke the brisket until you reach the stall. The stall can happen 2 to 3 hours in. The stall can last 7 hours, until the temperature starts to rise again.

4. About 4 to 5 hours in, wrap the brisket in butcher paper and continue cooking the brisket.

TIP: Make sure to frequently spray apple cider vinegar over the areas of the brisket that look like they're drying out or starting to "burn."

INGREDIENTS:

- ¼ CUP DIJON MUSTARD
- 12 LBS. PRIME BRISKET, TRIMMED OF SOME, BUT NOT ALL, FAT
- ¼ CUP SALT
- ¼ CUP FRESHLY GROUND BLACK PEPPER
- APPLE CIDER VINEGAR, IN A SPRAY BOTTLE

COLOMBIAN GOLD

YIELD: 4⅓ CUPS / **ACTIVE TIME:** 5 MINUTES / **TOTAL TIME:** 5 MINUTES

This rub is great on any meat.

1. In a bowl, combine all of the ingredients and mix well. Store in an airtight container; the rub will last for 2 months.

INGREDIENTS:

- 1½ CUPS PACKED DARK BROWN SUGAR
- 1 CUP SALT
- 1 CUP GROUND ESPRESSO BEANS
- ¼ CUP FRESHLY GROUND BLACK PEPPER
- ¼ CUP GARLIC POWDER
- 2 TABLESPOONS GROUND CINNAMON
- 2 TABLESPOONS CUMIN
- 2 TABLESPOONS CAYENNE PEPPER

INDIAN KUSH

YIELD: 4 CUPS / **ACTIVE TIME:** 5 MINUTES / **TOTAL TIME:** 5 MINUTES

This Chef Jeffrey rub works particularly well for pulled pork.

1. In a bowl, combine all of the ingredients and mix well. Store in an airtight container; the rub will last for 2 months.

INGREDIENTS:

- 1 CUP BROWN SUGAR
- ½ CUP SALT
- ½ CUP CHILI POWDER
- ¼ CUP HUNGARIAN PAPRIKA
- ¼ CUP CORIANDER
- 1 TABLESPOON GROUND GINGER
- ¼ CUP CUMIN
- ⅓ CUP GARLIC
- ⅓ CUP ONION POWDER
- 1 TABLESPOON LEMON ZEST
- 4 TABLESPOONS GROUND MINT TEA

HARISSA GLAZE

YIELD: 4 SERVINGS / **ACTIVE TIME:** 5 MINUTES / **TOTAL TIME:** 5 MINUTES

Use this glaze to kick up the flavor of meat with the addition of some spicy notes. Try this on ham, and you will never go back.

1. In a bowl, combine all of the ingredients and mix well. Refrigerate in an airtight container to store.

INGREDIENTS:

- 1 CUP HARISSA
- ½ CUP AGAVE SYRUP
- 2 TEASPOONS FRESH LEMON JUICE

SMOKED HARISSA AGAVE CHICKEN WINGS

YIELD: 2 SERVINGS / **ACTIVE TIME:** 1 HOUR AND 20 MINUTES / **TOTAL TIME:** 2 HOURS

This sweet and spicy, slightly smoky glaze takes chicken wings to another level.

1. In a bowl, combine the harissa and agave, adjusting the ratio of heat to sweet based on your preference.

2. Toss the chicken in the sauce and let marinate for at least 1 hour.

3. Get your smoker to 300°F and start cooking the wings. Start "mopping" the wings 20 minutes into cooking and keep mopping them every 10 minutes. This will form a crispy layer.

INGREDIENTS:

- 4 TABLESPOONS HARISSA PASTE
- ¼–½ CUP AGAVE
- 12 JUMBO CHICKEN WINGS

INDIAN KUSH LEG OF LAMB

YIELD: 4 SERVINGS / **ACTIVE TIME:** 1 HOUR / **TOTAL TIME:** 3 HOURS

Chef Jeffrey brings major flavor with his leg of lamb inspired by Indian spices.

1. Let the lamb come to room temperature.

2. Get the smoker to 300°F.

3. Coat the lamb with the mustard and then coat it with the rub.

4. Smoke the lamb for 2 to 3 hours, or until a meat thermometer registers an internal temperature of 145°F.

INGREDIENTS:

- 1 (5-7 LBS.) BONE-IN LEG OF LAMB
- ¼ CUP DIJON MUSTARD
- ¼ CUP INDIAN KUSH RUB (SEE PAGE 570)

MAPLE-CHIPOTLE GLAZE

YIELD: 4 SERVINGS / **ACTIVE TIME:** 5 MINUTES / **TOTAL TIME:** 5 MINUTES

Smoky and sweet, this glaze is a perfect addition to pork or ham.

1. In a bowl, combine all the ingredients and mix well. Refrigerate in an airtight container to store.

INGREDIENTS:

- ¾ CUP REAL MAPLE SYRUP
- 1 CHIPOTLE CHILE, CHOPPED
- 3 TABLESPOONS ADOBO SAUCE (FROM A CAN OF CHIPOTLES IN ADOBO)
- 2 TABLESPOONS KETCHUP
- 1½ TABLESPOONS DIJON MUSTARD
- 1 TABLESPOON APPLE CIDER VINEGAR

MARLEY'S COLLIE

YIELD: 8 SERVINGS / **ACTIVE TIME:** 30 MINUTES / **TOTAL TIME:** 9 HOURS

If you like Jamaican jerk flavors, you'll love how this marinade takes them for a spin.

1. In a food processor, combine the onion, scallions, chiles, garlic, five-spice powder, allspice, pepper, thyme, nutmeg, and salt and process to a coarse paste.

2. With the machine on, add the the soy sauce and oil in a steady stream.

3. Pour the marinade into a large, shallow dish, add meat, and turn to coat. Cover and refrigerate overnight.

INGREDIENTS:

- 1 MEDIUM ONION, CHOPPED
- 3 MEDIUM SCALLIONS, CHOPPED
- 2 SCOTCH BONNET CHILES, CHOPPED
- 2 GARLIC CLOVES, CHOPPED
- FIVE-SPICE POWDER, TO TASTE
- 1 TABLESPOON COARSELY GROUND ALLSPICE BERRIES
- 1 TABLESPOON COARSELY GROUND PEPPER
- 1 TEASPOON CRUMBLED DRIED THYME
- 1 TEASPOON FRESHLY GRATED NUTMEG
- 1 TEASPOON SALT
- ½ CUP SOY SAUCE
- 1 TABLESPOON VEGETABLE OIL

WHOLE JERK CHICKEN

YIELD: 4 SERVINGS / **ACTIVE TIME:** 20 MINUTES / **TOTAL TIME:** 3 TO 4 HOURS

What's better than pieces of jerk chicken? A whole jerked chicken!

1. Let the chicken come to room temperature.

2. Coat the bird, inside and outside, with about ¾ of the marinade.

3. Get the smoker up to 300°F and start to smoke the chicken.

4. Add the agave to the remaining marinade to create a thick "mop." Do not start to mop the chicken too early or the agave will end up caramelizing the chicken, which will result in an unwanted char.

5. Start mopping the bird after about 2½ hours, or when a meat thermometer inserted in the breast registers 145°F.

6. The chicken is done when a meat thermometer inserted in the breast registers 155°F.

TIP: The carry-over cooking as it cools will get the bird to 165°F; if you take the bird to 165°F, it will go to 175°F after letting it cool and you run the risk of tough, dry meat.

INGREDIENTS:

- 1 (3-4 LBS.) WHOLE CHICKEN, WITH BACK SPLIT OFF
- 1 CUP MARLEY'S COLLIE (SEE OPPOSITE PAGE)
- 2-4 TABLESPOONS AGAVE

THE OG RIBS

YIELD: 4 SERVINGS / **ACTIVE TIME:** 20 MINUTES / **TOTAL TIME:** 5 HOURS

Chef Jeffrey's ribs are outstanding.

1. Bring the ribs up to room temperature.

2. Coat the ribs with mustard, followed by the rub.

3. Get the smoker to 300°F.

4. Smoke the ribs for 3 hours.

5. After 3 hours take the ribs at either end with your hands. Pull the ribs upward to test how tender they are. At this point, you can convert to 3-2-1.

TIP: 3-2-1 is a foolproof method for achieving juicy, tender ribs: smoke for 3 hours; wrap the ribs in foil and smoke for another 2 hours; unwrap the ribs and smoke for an additional hour as you mop the ribs with a combination of OG Rub and apple cider vinegar.

INGREDIENTS:

- 1 RACK OF RIBS, CLEANED
- 2 TABLESPOONS MUSTARD
- ¼ CUP OG RUB (SEE BELOW)

THE OG RUB

YIELD: 4 CUPS / **ACTIVE TIME:** 10 MINUTES / **TOTAL TIME:** 10 MINUTES

Use this rub for these ribs and then be sure to experiment with it on other types of meat.

1. Add all of the ingredients to a bowl and mix well.

INGREDIENTS:

- 1 CUP BROWN SUGAR
- ½ CUP SALT
- ½ CUP CHILI POWDER
- ¼ CUP HUNGARIAN PAPRIKA
- ¼ CUP CORIANDER
- ¼ CUP CUMIN
- ⅓ CUP GARLIC POWDER
- ⅓ CUP ONION POWDER
- 1 TABLESPOON GROUND GINGER
- 1 TABLESPOON LEMON ZEST

CHEF GEORGE BEZANSON

George Bezanson hails from the South Shore of Massachusetts and eventually inserted himself into the Boston food scene through his love of Asian-inspired cuisine and tireless quest for great ingredients and techniques. At the age of 23, he held his first executive chef position at the now-closed Rattlesnake Bar & Grill on Boylston Street.

His passion led him to Southwest Florida, where he apprenticed with a sushi master at Bacchus & Co. and expanded his knowledge of Japanese cuisine. He then moved on to be part of the team that opened Cru in Fort Myers, Florida. This James Beard–nominated bistro focused largely on extremely high-end ingredients and wine pairings, obtaining produce from farmers like Lee Jones of the Chef's Garden.

After his time in the Sunshine State, George returned to New England and opened up two successful restaurants, Mint Bistro and the Bridge Café, both in Manchester, New Hampshire. Following many good years providing both fine dining and casual cafe food, he then opened Pressed Cafe in Nashua, where his concept of bringing "feel good everyday food" really began to take shape. An unexpected move to the seacoast brought him to Dover, where he began his adventure at Earth's Harvest Kitchen. The idea was to make feel-good food available every day using techniques he learned in the fine dining world. In doing so, he also took on the daunting task of baking bread. After almost five years, Earth's Harvest Kitchen is a local favorite.

I met George about four years ago when I saw his food pictures pop up on my Instagram feed. The subtle nuances of his cuisine showcased how his talent was far beyond ordinary. I reached out to him and ended up eating at his restaurant. Not only has George become one of my best friends, but he is the person I consult when I have culinary questions. In fact, in 2017 George came on board with my company, The Farmers Dinner, bringing pop-up tasting menus to farms and restaurants all over New England.

Community has always been an incredibly important part of George's culinary journey and always will be. Bringing together farmers, chefs, restaurateurs, diners, and everyone else with a passion for food is really what matters most.

EARTH'S HARVEST KITCHEN PORK BELLY

YIELD: 6 TO 8 SERVINGS / **ACTIVE TIME:** 15 MINUTES / **TOTAL TIME:** 15 HOURS

This is the recipe we use for our pork belly at Earth's Harvest Kitchen. We tested several different recipes, but this simple one beat them all. The end result is so perfect that it can be used in any dish that requires pork belly. In terms of using skin-on or skinless pork belly, it is a matter of preference; both are equally delicious. As with most recipes, ingredients are essential for the best final product. If you can find local pork belly, that is ideal, but any pork belly will do the trick.

1. If using skin-on pork belly, score the skin with a few slices across the surface, trying not pierce the flesh beneath.

2. In a small bowl, mix together the remainder of the ingredients.

3. Coat the pork belly with the seasoning mixture. Really massage the pork belly with the mixture. Place prepared belly in a large bowl, cover with plastic wrap, and refrigerate for 12 hours.

4. Once the pork belly has cured, preheat oven to 425°F and prepare a pan that has higher sides than the height of the pork belly. Rub some oil inside the bottom and sides of the pan until evenly coated.

5. Remove pork belly from fridge and rinse off the seasoning mixture with cold water and pat dry with paper towels. Place belly skin side up in the prepared pan and put it in the oven, uncovered, and cook for 30 to 45 minutes, or until skin becomes golden-brown but not burnt.

6. Remove pork belly from the oven and lower the oven temperature to 300°F. Pour about 1 cup of water into the hot pan, or just enough to cover the bottom of the pan. Make sure not to pour water directly onto the belly—keep that skin crispy! Cover the pan with foil and return it to the oven. Cook for 2 hours, or until the belly is fork-tender but not falling apart.

7. Using a flat metal spatula, lift the pork belly onto a cooling rack or sheet pan and let rest until it is cool enough to be safely sliced, eaten, or stored in the refrigerator or freezer.

TIP: When the cooked pork belly is cool enough to handle, place another sheet pan on top of the belly and use something heavy, like a few cans of beans, to weigh it down and refrigerate it overnight. Doing this will result in uniform slices that crisp up perfectly in the pan when you are ready to cook.

INGREDIENTS:

- 3–4 LBS. SKIN-ON OR SKINLESS PORK BELLY
- ½ CUP SALT
- ¼ CUP SUGAR
- 2 TABLESPOONS FRESHLY GROUND BLACK PEPPER
- CANOLA OIL, AS NEEDED
- 1 CUP WATER

ULTIMATE XO SAUCE

YIELD: 1 QUART / **ACTIVE TIME:** 15 MINUTES / **TOTAL TIME:** 30 MINUTES

There are those moments in life when you eat something and you never forget it. For me, one of those magical experiences happened at Ivan Ramen in New York City. I went there for his world-famous ramen and left gushing over his XO sauce with pickled daikon radish. So simple, yet it was executed so perfectly it absolutely blew my mind. This is my take on the umami-rich condiment that can be put on just about anything, like pork belly.

INGREDIENTS:

- ½ CUP DRIED SHRIMP
- 6 DRIED SHIITAKE MUSHROOM CAPS
- ½ CUP COOKED BACON
- ½ CUP CHINESE SAUSAGE OR SOPRESSATA
- 2 CUPS CANOLA OIL, DIVIDED
- 2 SHALLOTS, MINCED
- ¼ CUP THINLY SLICED GARLIC
- ¼ CUP MINCED GINGER
- ¼ CUP CHILI FLAKES
- 3-4 STAR ANISE
- 1 CINNAMON STICK
- ½ CUP DRIED PORK
- ½ CUP CHINESE SHAOXING WINE OR DRY SHERRY
- ½ CUP SOY SAUCE
- 1 TABLESPOON FISH SAUCE
- ½ CUP SUGAR
- ½ CUP FRIED GARLIC
- ½ CUP FRIED SHALLOT

1. In separate bowls, soak dried shrimp and mushrooms in very hot water until soft but still slightly firm, about 15 to 20 minutes. Drain and set aside.

2. Using a food processor, keeping ingredients separate, chop shrimp, mushrooms, bacon, and Chinese sausage. Set aside.

3. Add 1½ cups of oil to a wok or large sauté pan over medium-high heat. Once the oil is shimmering, add shallots and cook until soft, about 1 minute, continuously moving the shallots around with a wooden spoon to prevent scorching around the edges. Next, add garlic and ginger, cook for another minute and then add chili flakes, star anise, and cinnamon stick and fry for another minute, stirring constantly.

4. Add the remaining oil and mushrooms, cook for 1 minute, stirring and keeping the oil simmering the whole time. Repeat this process with the sausage, shrimp, bacon, and dried pork.

5. Carefully add the wine all at once; please be careful as there may be some flames when adding the alcohol. Simmer for about 30 seconds, then add the soy sauce, fish sauce, and sugar and simmer for another 30 seconds. Now add the fried garlic and fried shallots, mix everything together, and turn off the heat.

6. Don't let cool for more than 5 minutes, in order to prevent further cooking. Pour the sauce into a nonreactive container and store in the refrigerator for up to 1 month.

CHEF RYAN MANNING

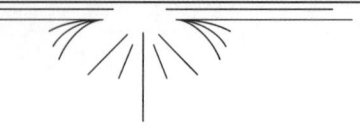

Ryan Manning and I met through the American Culinary Federation and quickly became friends. His passion for Mexican cuisine and culture was one of the many things we bonded over. I had the honor of eating at one of his restaurants in Orlando, Florida, MX Taco, and I was blown away.

Shortly after graduating from the American Culinary Federation Apprenticeship Program at Colonial Williamsburg, Chef Ryan joined the culinary team at Walt Disney World Resort; he then worked his way through banquets at the Grand Floridian to the upscale dining experiences of the California Grill, Flying Fish, and Yachtsman Steakhouse. Eager to highlight his Virginia training, Chef Ryan jumped at the opportunity to open a new Southern-inspired concept called Highball & Harvest at The Ritz-Carlton Orlando. The hotel chain opened doors for him to see the country and the world through varying roles, including opening The Edition Hotel in Miami, working at Navio in The Ritz-Carlton Half Moon Bay, and, ultimately, finding a home as Chef de Cuisine at The Ritz-Carlton Cancun.

It was in Mexico that Chef Ryan fell in love with the cuisine, culture, languages, the people, and, most of all, the genuine hospitality the country offers. With roots that span the pre-Hispanic empires and embrace the influence of European, Asian, African, and Middle Eastern migration, Chef Ryan has set a life goal to become an expert on the cuisine of Mexico.

When he returned to the United States he began to plan a taqueria concept while cooking in Washington, D.C. as the Banquet Chef of The Ritz-Carlton. The first appearance of MX Taco was at the Embassy of Mexico where Chef Ryan was selected to showcase his knowledge of Mexican cuisine as the culinary representative of the embassy for an event of over 4,000 people. The embassy loved the authentic food.

In early 2019 he opened MX Taco in Orlando, where the traditional tacos have found their way into the hearts of the local community.

SOUTHERN BEEF JERKY

YIELD: 6 SERVINGS / **ACTIVE TIME:** 15 MINUTES / **TOTAL TIME:** 40 HOURS

This recipe was developed as we opened a Southern-style restaurant in Florida. We wanted to showcase a touch of Florida citrus in the recipe. Feel free to alter the citrus to give a variety of flavor profiles to this tasty snack.

INGREDIENTS:

- 8 OZ. KETCHUP
- 8 OZ. SOY SAUCE
- 8 OZ. WORCESTERSHIRE SAUCE
- 8 OZ. BROWN SUGAR
- 4 OZ. GARLIC CLOVES
- 2 SHALLOTS
- ½ OZ. GARLIC POWDER
- ½ OZ. ONION POWDER
- ½ OZ. BLACK PEPPER
- ½ OZ. SRIRACHA, OR YOUR FAVORITE HOT SAUCE
- 3 LBS. FLANK STEAK, TRIMMED AND SLICED AGAINST THE GRAIN INTO LONG 1/8-INCH-THICK STRIPS
- ZEST OF 2 ORANGES

1. In a blender, combine all of the ingredients, except the flank steak and orange zest, and blend until smooth.

2. In a large bowl, coat the flank steak with the marinade; reserve the remaining marinade. Cover the steak and refrigerate for 24 to 36 hours, coating thoroughly.

3. Place a single layer of beef on racks on top of sheet pans so the beef is elevated. Do not overlap the beef.

4. Preheat the oven to 165°F. Cook the beef for about 5 hours; the cook time will vary based on the thickness of meat. After each hour, brush the beef jerky with the marinade and flip the pieces and brush the other side.

5. The jerky is ready when it is dry but not crispy or burnt. Let cook before adding zest. Store under refrigeration or eat right away!

CARNE DE CHINAMECA

YIELD: 4 SERVINGS / **ACTIVE TIME:** 10 MINUTES / **TOTAL TIME:** 25 HOURS

Not quite beef jerky, Carne de Chinameca honors the town of Chinameca, Veracruz, and its people. This smoky, bright red, slightly dried, slow-cooked meat is usually cooked over oak wood and enjoyed as a meal or a snack across the state. It's one of those dishes that most families in Chinameca have their own secret recipe for. Traditionally, the meat is hung over an open flame to slowly dry and smoke for hours, but putting it in the smoker and finishing on the grill does the trick.

1. Rehydrate the chilies in hot water for at least 20 minutes.

2. In a blender, combine all of the ingredients, except the steak, and blend until smooth, adding water if necessary. Strain the marinade.

3. In a bowl, combine the beef and the marinade, being sure to coat the meat thoroughly. Cover and refrigerate for at least 24 hours.

4. Bring smoker to 225°F and cook for 30 minutes.

5. Grill over medium-high heat to your desired doneness. Serve with rice and refried beans.

INGREDIENTS:

- 12 GUAJILLO CHILIES, DRIED
- 4 OZ. ACHIOTE PASTE
- 4 GARLIC CLOVES
- 1 OZ. MEXICAN OREGANO
- ½ OZ. CINNAMON
- 4 OZ. WHITE VINEGAR
- 2 OZ. SALT
- 2 LBS. SKIRT STEAK, TRIMMED AND SLICED AGAINST THE GRAIN INTO LONG ¼-INCH-THICK STRIPS

AL PASTOR NEGRO

YIELD: 1½ CUPS / **ACTIVE TIME:** 10 MINUTES / **TOTAL TIME:** 30 MINUTES

Mexico has a strong relationship with indigenous groups that are still spread throughout the country today. In the Yucatan region, formally the heart of the Mayan Empire, the people maintain a strong passion for their heritage through continuing to speak the Mayan language, as well as through their cuisine. This region has three *pastas*, or pastes, that serve as the foundations for many dishes. They are called *recados*—*rojo*, *blanco*, and *negro*. The recipe below is for recado negro, which imparts a smoky quality, as well as a hint of heat, to dishes like relleno negro.

1. Rehydrate the chiles in hot water for at least 20 minutes.

2. Add the garlic to a sauté pan over high heat and char for about 5 minutes, creating a black exterior layer on the cloves. Remove the garlic and set aside.

3. Reduce the heat to medium and toast the achiote with the oregano for about 30 seconds, until lightly toasted.

4. In a blender, combine all of the ingredients and blend until a black paste forms.

5. Reserve in the refrigerator or use right away. The marinade will stay fresh for up to 1 month, or can be frozen for future use.

TIP: Using the al pastor recipe above, steadily fold the paste into the marinade until the marinade takes on a black color. This slightly spicy marinade can then be used to marinate the sliced pork.

INGREDIENTS:

12	DRIED ANCHO CHILIES
4	OZ. GARLIC
1	OZ. ACHIOTE SEEDS
1	OZ. MEXICAN OREGANO
5	ALLSPICE BERRIES
¼	OZ. CUMIN SEEDS
½	OZ. SALT
½	OZ. BLACK PEPPER
4	OZ. WHITE VINEGAR

AL PASTOR PORK MINI TROMPO

YIELD: 5 SERVINGS / **ACTIVE TIME:** 20 MINUTES / **TOTAL TIME:** 28 HOURS

If you travel to Mexico, you will find tacos al pastor everywhere, but especially in Mexico City. You have not eaten tacos al pastor before if it wasn't cut off the *trompo*, a vertical rotisserie that Middle Easterners who began to immigrate to Mexico over 200 years ago brought with them. Over time the traditional Middle Eastern shawarma evolved into something very Mexican!

1. Rehydrate all of the dried chilies in hot water for at least 20 minutes.

2. Break down the pineapple: peel and chop 4 oz. for the marinade; reserve a ½-inch-thick piece of the unpeeled base of the pineapple for the bottom of the trompo; cut a 4x4-inch unpeeled cube of the pineapple for the top of the trompo.

3. In a blender, combine the rehydrated chilies with the remainder of the ingredients, except the pork and unpeeled pineapple pieces, and blend until smooth. Strain the marinade.

4. Place the pork in a bowl and cover it with the marinade. Cover and refrigerate for at least 24 hours.

5. Preheat the oven to 275°F.

6. Place the base of the pineapple on an aluminum foil–lined sheet pan so the skin is face-up. Insert a chopstick or wooden skewer into the middle of the pineapple and begin to skewer the pork slices on the spike one piece at a time, creating layers of pork on top of the base. Rotate the slices so that the trompo does not become one-sided. Once all the pork is on the spit, place the pineapple cube on top. When done, none of the wooden skewer should be exposed.

7. Cover the pork with foil and bake for 4 hours.

8. Remove from the pork from the oven and uncover it. Use a brulee torch to char the outside of the pork, or place the trompo on its side and broil in the oven for 3 to 5 minutes, rotating the pork until it is charred on all sides.

9. Thinly slice and enjoy. Traditionally it is served with corn tortillas, small diced white onion, chopped cilantro, and pineapple slices, but you can also enjoy this meat on its own or with a rice bowl or even inside a quesadilla.

INGREDIENTS:

8	DRIED GUAJILLO CHILIES
2	DRIED CHIPOTLE CHILIES
3	DRIED ANCHO CHILIES
1	PINEAPPLE
5	GARLIC CLOVES
2	WHOLE CLOVES
1	OZ. MEXICAN OREGANO
¼	OZ. CUMIN
	JUICE OF 3 ORANGES
	JUICE OF 3 LIMES
4	OZ. PINEAPPLE JUICE
2	OZ. YELLOW ONION, CHOPPED
2	OZ. SALT
2	LBS. PORK SHOULDER, CUT INTO 1/8-INCH-THICK SLICES EACH ABOUT 4 INCHES LONG

MOJO JERKED PORK BUTT

YIELD: 16 SERVINGS / **ACTIVE TIME:** 2 HOURS / **TOTAL TIME:** 10 HOURS

This recipe makes a lot of food, but once you realize that the next day you're equipped to build the best sandwiches on the planet, it will be gone quick!

1. Place the pork on a large sheet pan and coat it with the Mojo Marinade, followed by the jerk marinade. Let it marinate in the refrigerator for at least 2 hours, or, even better, overnight. If you are using store-bought mojo or jerk, only marinate for 2 hours, as those products are high in salt and will start to "cook" the pork.

2. Let the pork come up to room temperature.

3. Reserve ½ to 1 cup of the marinade and add 2 tablespoons to ¼ cup of agave, depending on how much you reserve—this is your "mop."

4. Bring your smoker up to 300°F and then add your flavoring wood. I like peach/oak/cherry/pecan.

5. Smoke 1 hour per pound, which at 8 lbs. would be close to 8 hours. If you go between 225 to 250°F degrees, you will have to go about 16 hours on the smoker. For the first 4 to 6 hours, smoke the pork butt uncovered, not wrapped. Finish the cooking covered to keep moisture inside the product and to not dry it out. Once the bone slides out clean, you are done.

MOJO MARINADE

1. Pulse the garlic and onion in a blender until very finely chopped. Add orange juice, lime juice, cumin, oregano, lemon-pepper, black pepper, salt, cilantro, and hot pepper sauce. Blend until thoroughly incorporated. Pour in the olive oil and blend until smooth. This preparation will yield about 1 quart of marinade.

INGREDIENTS:

- 8 LBS. BONE-IN PORK BUTT
- 2 CUPS MOJO MARINADE
- 1–4 CUPS JERK MARINADE, DEPENDING ON HOW SPICY YOU WANT IT, OR MARLEY'S COLLIE (SEE PAGE 576)

MOJO MARINADE

- 18 GARLIC CLOVES, COARSELY CHOPPED
- 1½ CUPS MINCED YELLOW ONION
- 3 CUPS FRESHLY SQUEEZED ORANGE JUICE
- 1½ CUPS FRESH LIME JUICE
- 1½ TEASPOONS GROUND CUMIN
- 1 TABLESPOON DRIED OREGANO FLAKES
- 1½ TEASPOONS LEMON PEPPER SEASONING
- 1½ TEASPOONS FRESHLY GROUND BLACK PEPPER
- 1 TABLESPOON SALT
- ¾ CUP CHOPPED CILANTRO
- 1 TABLESPOON HOT PEPPER SAUCE (OPTIONAL)
- 3 CUPS OLIVE OIL

AVEDANO'S HOLLY PARK MARKET

SAN FRANCISCO, CALIFORNIA
ANGELA WILSON, OWNER

Avedano's opened in San Francisco's Bernal Heights in 2007, after Angela Wilson and some of her friends decided the neighborhood needed a market that served their immediate community. Known for its high-quality meat, sandwiches, specialty goods, and butchery classes, the Cortland Avenue space fulfills a demand, and happens to be the former home of a long-gone butcher shop.

How did you come to be a butcher?

I had a chai company. In the early 2000s my downstairs neighbor, who was a waitress, suggested that I open a restaurant with the head chef of her restaurant. We set up a meeting to chat and brainstorm ideas, but there were no sparks. Someone suggested opening a butcher shop and we went for it.

How did you train to be a butcher?

I learned to become a butcher on the job. I was the first employee at Avedano's. I hired a butcher with minimal experience, and we both learned to become butchers on the job. It wasn't an easy thing to learn because learning how to butcher a whole animal can be very expensive if too many mistakes are made. We took our time and learned as much as we could from doing and watching other butchers do demonstrations at the shop.

Your shop practices sustainable, zero-waste butchery. Were these ideas that you were drawn to before becoming a butcher or did you learn to appreciate them by virtue of the work?

I learned to appreciate them by virtue of the work. It's dollars and cents. Anything that goes in the garbage is my money and my profit. A butcher shop has very thin margins, especially if you don't sell alcohol. It's also incredibly costly to sell responsibly raised meat.

How do you cultivate relationships with farmers?

Most of our farmers, who are all local, drop their products off at the shop. I've built relationships with them over time through business and exchanges of knowledge. Another advantage I've also had, which has helped me build relationships with the farmers, is that small farmers need to sell whole animals and there are not a lot of places that will buy the whole animal.

What are you looking for when sourcing meat from farmers?

Pasture raised, small farm, organic feed, as local as we can get, and farmers who care about their animals and sustainable practices.

What cut of meat, from any animal, do you think deserves more attention in home kitchens? Why?

Lamb neck because it has a lot of collagen, and just one cut of lamb neck provides enough meat to feed four people. Some cuts of meat shrink and others expand when you cook them. Lamb neck not only expands, but it's also very versatile.

FARM FRESH WHOLE GUINEA FOWL RUBBED WITH LOCAL SHIO KOJIT

YIELD: 4 SERVINGS / **ACTIVE TIME:** 1 HOUR / **TOTAL TIME:** 3 HOURS

For this recipe "whole" means "whole"—you'll want a guinea fowl with the head and feet attached, and all the giblets saved as well (though removed). Ask your butcher to spatchcock the bird so it can lie flat in the pan.

1. Rinse the bird in cold running water and pat dry. Sprinkle both sides with salt and pepper, then rub with shio koji. Place the bird in a roasting pan, cover, and refrigerate for at least 2 hours, but ideally overnight.

2. Preheat the oven to 400°F. Heat a grill to a high temperature.

3. Remove the bird from the marinade and place it on the grill for about 10 minutes, flipping after 5 minutes. Look for grill marks as an indicator for doneness.

4. While the bird is grilling, add a couple tablespoons of oil to a cast-iron skillet or ovenproof pan over medium-high heat. Add the giblets to the pan and cook, stirring occasionally, until the hen is finished grilling. Once the hen is done, add the bird to the pan with giblets, skin-side down and place in the oven for 30 to 40 minutes, or until the skin is crispy and well browned.

5. Remove the pan from the oven and transfer the bird to a plate, letting it rest for 10 minutes before carving.

6. Put the pan on the stovetop over medium-high heat. Add the garlic and shallots and sauté for no more than 1 minute. Deglaze the pan with the brandy, stirring with a wooden spoon to loosen any browned bits stuck to the bottom. Add the stock and tomato paste and season with salt and pepper. Cook until sauce has reduced by half.

7. Turn off heat and stir in butter until melted. Remove the bird from the pan and portion out the legs, thighs, wings, and breasts to serve. Top the dish with the pan sauce and serve with a wedge of lemon. Serve with polenta, dressing, or rice.

INGREDIENTS:

- 1 WHOLE GUINEA FOWL, SPATCHCOCKED
- 1 TABLESPOON SALT, PLUS MORE TO TASTE
- 1 TABLESPOON FRESHLY GROUND BLACK PEPPER, PLUS MORE TO TASTE
- ½ CUP SHIO KOJI
- EXTRA-VIRGIN OLIVE OIL, AS NEEDED
- 2 GARLIC CLOVES, FINELY CHOPPED
- 1 LARGE SHALLOT, FINELY CHOPPED
- ¼ CUP BRANDY
- ¼ CUP CHICKEN STOCK (SEE PAGE 478)
- 1 TABLESPOON TOMATO PASTE
- 1 TABLESPOON UNSALTED BUTTER
- 1 LEMON, HALVED

PERSIAN BRAISED LAMB NECK WITH SPICED BROTH

YIELD: 4 SERVINGS / **ACTIVE TIME:** 1 HOUR / **TOTAL TIME:** 29 HOURS

Ab goosht is a classic Persian dish made by braising tough cuts of lamb alongside onions, chickpeas, and potatoes with spices and Persian dried limes. After the meat and other ingredients are fully cooked and tender, they're strained from the broth and mashed. This is referred to as *goosh koobideh*, which is served with a side of *ab goosht*, which translates to "meat water." This recipe was introduced to me by my close friend Negar. Without her, I wouldn't fully appreciate Persian cuisine for its beauty and simplicity.

1. Rinse and soak the beans overnight.

2. Add ¼ cup olive oil to a Dutch oven over medium-high heat. When the oil starts to shimmer, sear the lamb on all sides until brown. Once browned, remove and set aside.

3. Add the onions to the pot. There will be brown residue left from searing the meat—scrape it up while browning the onions. This process will take at least 25 minutes. Be patient. The more evenly browned your onions become, the better.

4. Once you have evenly dark-brown onions, add the turmeric and let it bloom for a minute or so, then add the tomato paste and keep cooking for an additional 2 to 3 minutes.

5. Return the lamb to the pan and add about 6 cups water or stock. Only add enough so the meat is three-quarters covered; the top one-quarter should be above the liquid.

6. Preheat the oven to 300°F.

7. Bring the liquid to a boil and add the beans. Poke three or four holes in each dried lime and add those as well. Transfer pot to the oven and cook for 3 hours.

8. After 3 hours, add the potatoes and cook for an additional 45 minutes, or until the beans are soft.

9. Strain the broth from the solids and set aside. Pick all the meat off the bones, then mix with the other solids. With a potato masher, mash the solids until they are almost homogenous.

10. Serve the mashed mixture in a bowl along with a cup or bowl of the broth.

INGREDIENTS:

- ½ CUP DRIED CHICKPEAS
- ½ CUP DRIED WHITE BEANS
- ½ CUP OLIVE OIL
- 1 LAMB NECK
- 2 MEDIUM YELLOW ONIONS, ROUGHLY CHOPPED
- 1 TABLESPOON TURMERIC
- ¼ CUP TOMATO PASTE
- 6 CUPS WATER OR CHICKEN STOCK (SEE PAGE 478)
- 5 PERSIAN DRIED LIMES
- 3 MEDIUM RUSSET POTATOES, CHOPPED

FLEISHERS CRAFT BUTCHERY

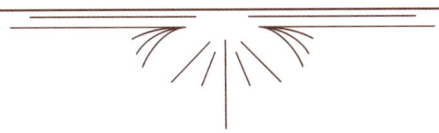

MULTIPLE LOCATIONS IN NEW YORK CITY AND CONNECTICUT
KYLE MCCARTHY, MANAGER

When Fleishers Craft Butchery first opened in 2004, it was founded on the premise that there needed to be a stronger connection between butchers and the meat that they butchered. Fleishers started as a single shop that sourced its meat from trusted partners, but soon found that in order to really champion "Feed to Fork" butchery—in the name of providing top-quality meat that is better for the consumer and the planet—they needed to raise their own livestock. That practice has created a supply chain that has grown into a business with two New York City locations and two Connecticut locations.

How did you come to be a butcher?

My path to becoming a butcher started in restaurants as a chef. After working in and running kitchens for a little over a decade, I was feeling pretty burnt out. I wanted to find a career that was still in the culinary realm and had always been drawn to restaurants that had a whole-animal program for their menu. Naturally, I started looking at local butcher shops in the area. I dropped off a couple resumes and was hired at Fleishers in Brooklyn in 2015.

How did you train to be a butcher?

When I was first hired I started by helping out as counter staff. It was an excellent way to learn the cuts and how to talk about meat more fluently. Even after working in the restaurant industry for over ten years, I had only just scratched the surface of everything there is to learn about meat and butchery. I started volunteering to do any of the less desirable butchery tasks in the shop. Whenever there was a need to debone beef necks or lamb shanks, I would jump at the chance, really anything to get on the butcher block. After a while, I started working up to more complex tasks like tying roasts, using the bandsaw, setting the case until I was able to start fully breaking down whole animals.

How do you cultivate relationships with farmers?

Cultivating a relationship with your farmers is such an important part of this industry. From my experience, it begins with just starting a conversation, asking questions, and learning. Most, if not all, of the farmers I've met are just as interested in what I do as I am about their farming practices. It really becomes a symbiotic relationship where you are both growing together, exchanging knowledge, and learning from each other.

What are you looking for when sourcing meat from farmers?

I really enjoy working with smaller, local farms that share the same commitments to animal welfare and sustainability. Allowing animals to roam free on pasture and maintain a diet that encourages healthy growth go hand-in-hand with producing meat that is both beautiful and delicious. These farmers and animals deserve a great deal of respect.

What cut of meat, from any animal, do you think deserves more attention in home kitchens? Why?

I believe that the sirloin in any species, across the board, is one of the most overlooked cuts. It's so very versatile. It can be tied into a roast, cut into steaks, skin on, skin off, bone-in, bone-out, smoked, braised, grilled, stewed; I really could go on and on. I think one of the most interesting features of the sirloin is the different muscles that make up the sirloin. Each compartment has a slightly different flavor and texture, which makes it extremely exciting to eat.

ROASTED DUCK BREAST & FIVE-SPICE CONFIT DUCK LEGS

YIELD: 4 SERVINGS / **ACTIVE TIME:** 1 HOUR / **TOTAL TIME:** 16 HOURS

This dish really speaks to the versatility of cooking whole ducks. The breast is brined to allow the hoisin, soy, and spices to really penetrate throughout the breast and glaze the skin. The breast is then roasted on the bone while the duck legs are cured overnight and braised in duck fat until they are meltingly tender.

1. Add all of the ingredients, except the duck and duck fat, to a saucepan over medium-high heat and bring to a boil. Cook until the salt and syrup have dissolved. Cool the brine in the refrigerator until it's at or below 40°F.

2. Place the breast in a container and pour the brine over it. Cover the container and refrigerate overnight.

3. Remove the breast from the brine, pat dry, and place the breast skin-side up on a rack over a sheet pan. Salt the skin with a light dusting of salt, roughly 1 teaspoon. This will draw out any excess moisture.

4. Refrigerate, uncovered, for 12 to 24 hours to dry out the skin.

5. Preheat the oven to 380°F.

6. Roast the breast for about 30 minutes, or until the internal temperature of the breast is 115°F. If the skin has not become golden brown, toast under your broiler for 2 to 3 minutes. Keep a close eye on it; the skin can go from golden brown to burnt very quickly.

7. Rest for 10 to 15 minutes. To serve, carefully carve the breast off the bone. Slice along the middle ridge of the breast-plate, allowing the bones to guide your knife to remove the meat. Slice thin at a slight bias.

FIVE-SPICE CURE FOR DUCK LEGS

1. Place the star anise, cinnamon stick, and bay leaves in a small plastic bag and crush them, using the bottom of a heavy pan.

2. Add crushed aromatics, clove, fennel seed, and peppercorns to a dry sauté pan over medium-high heat and toast for 1 to 2 minutes.

INGREDIENTS:

4	CUPS WATER
¼	CUP MAPLE SYRUP
¼	CUP SALT
2	TABLESPOONS HOISIN SAUCE
2	TABLESPOONS SOY SAUCE
1	CINNAMON STICK
3	STAR ANISE PODS
1	TABLESPOON FENNEL SEED
2	BAY LEAVES
1	TEASPOON SZECHUAN PEPPERCORNS
1	TEASPOON MINCED GINGER
1	TEASPOON WHOLE CLOVE
1	(5-6 LBS.) WHOLE DUCK, BROKEN DOWN (SEE PAGE 599)
1	CUP DUCK FAT

FIVE-SPICE CURE

2	STAR ANISE PODS
1	CINNAMON STICK
2	BAY LEAVES
½	TEASPOON CLOVE
1	TEASPOON FENNEL SEED
½	TEASPOON SZECHUAN PEPPERCORNS
¼	CUP SALT
¼	CUP BROWN SUGAR

Continued . . .

3. Once the spices are cooled, thoroughly mix with the salt and sugar in a bowl.

4. Coat the legs generously with the mixture and refrigerate in a covered container overnight.

5. Remove the legs from the cure, rinse, and pat dry. Preheat the oven to 300°F.

6. Place the duck legs skin-side down in a high-sided baking dish or ovenproof saucepan with the duck fat. Cover tightly with a lid or foil and place in the oven for 2½ hours, or until a fork can slide in and out of the leg with little resistance.*

7. To finish, add 1 tablespoon duck fat to a medium sauté pan over medium-high heat and sear the confit legs until crispy on both sides, about 1 to 2 minutes per side.

* This can be done a day in advance and sit submerged in the fat in the refrigerator until ready to use.

BREAKING DOWN A DUCK

1. Remove the backbone from the duck by standing it with the cavity and legs sitting on the cutting board in a stable position. Trace along each side of the backbone with a sharp knife, cutting through the back ribs until you reach the joints of the legs.

2. Use the tip of your knife to sever the connection between the leg joints; you should feel and hear a slight pop. Repeat with the opposite leg.

3. Cut through the rest of the backbone, keeping your knife tight to the spine, releasing each side.

4. To remove the legs, set the duck down, breast-side up. Pinch the fat between the leg and the breast. Pinch closer to the leg to make sure you leave ample skin coverage for the breast. Carefully cut through the skin just above your pinched fingers and follow the contours of the leg to remove.

5. Find the joint between the drumette and flat of the wing and cut down to the cutting board, using the heel of your knife. The joint should have a slight ball shape and your knife should be able to gently cleave through the cartilage.

ROASTED GARLIC PORK BELLY PORCHETTA

YIELD: 6 SERVINGS / **ACTIVE TIME:** 2 HOURS / **TOTAL TIME:** 27 HOURS

Porchetta is a classic Italian boneless pork roast, typically made by rolling the pork belly around the loin. This preparation uses only the belly, creating a rich and decadent centerpiece surrounded with a beautifully crispy crackling. Using the belly for porchetta is slightly more forgiving than the traditional method because you do not have to worry about the loin drying out. Instead, you determine the doneness of the roast by how crispy you like the skin.

INGREDIENTS:

- ½ CUP GARLIC CLOVES, PEELED
- ¼ CUP CANOLA OIL, PLUS MORE AS NEEDED
- ¼ CUP APPLE CIDER VINEGAR
- 1 TEASPOON SALT, PLUS MORE TO TASTE
- 1 TEASPOON CHILI FLAKES
- 2 TABLESPOONS FRESH THYME
- BLACK PEPPER, TO TASTE
- BONELESS PORK BELLY, SPARE RIB SECTION (5-6 LBS.)

1. Place the garlic cloves in a small saucepan and cover with canola oil. Bring to a simmer over medium heat and cook until the garlic becomes golden brown, roughly 10 to 15 minutes.

2. Add all of the remaining ingredients ingredients, except the canola oil and pork belly, to a blender and pulse until smooth. Slowly pour in the canola oil while blending to emulsify the marinade.

3. Trim off any mammary glands from the pork belly, being sure to keep it square. (You can ask a butcher to square off the ventral part of the belly to remove these glands.)

4. Lightly score the skin of the pork belly about half-an-inch apart in one direction. Do not worry about crosshatching; the end result will be better without it.

5. Begin to butterfly the belly open, starting through the side where you removed the mammary glands. (If the butcher removed the glands for you, ask which side they were on for reference.)

6. Slice horizontally through the center of the belly, trying to keep the amount of meat on either side of your blade even. Do not cut all the way through; leave about a half-inch of meat intact. You should be able to open up the pork belly like a book.

7. Season the meat generously with salt and pepper. If you are salt sensitive, the amount of salt in the marinade should suffice.

8. Spread a thin layer of the marinade on all surfaces of the meat, but avoid getting any marinade on the skin.

9. Completely open up the pork belly. Starting with the skinless end, tightly roll the meat until it forms a log shape with the skin on the outside.

Continued . . .

When scoring, lightly slice through the skin with a sharp knife. Do not cut through to the meat. You really just want shallow, barely visible lines in the skin to slightly reveal the fat. This allows for the fat to render and crisp up the skin as it roasts.

10. Using kitchen twine, tightly tie the roast crosswise, starting in the center and working outward at 1-inch intervals. Trim any excess twine.

11. Place the rolled belly on a wire rack over a rimmed sheet pan and season the skin generously with salt. This will draw out any excess moisture, allowing the skin to become very crispy during roasting.

12. Refrigerate, uncovered, for 24 hours.

13. Preheat the oven to 400°F.

14. Roast for 1 hour, then lower the temperature to 325°F. Continue to cook until the skin is nice and crispy and a meat thermometer inserted into the meat registers 190°F, about 2 hours and 30 minutes.

15. Rest for 20 minutes. To serve, remove the twine and slice along that line left behind. This will be the easiest entry point to get a nice, neat slab.

MILES SMITH FARM

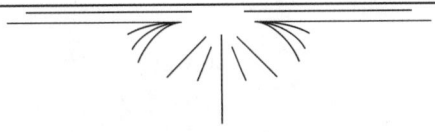

LOUDON, NEW HAMPSHIRE
CAROLE SOULE

When I think of trailblazers in the agricultural world, I think of Carole. I met Carole six years ago when I started hosting a lot of pop-up dinners with my company, The Farmers Dinner. Carole raises Scottish Highlander cattle. These cattle have huge horns, much like the steer you see in Texas, except these cattle are covered in long hair.

To know Carole is to know a woman who is strong, intelligent, and passionate about animals. She names every cow she owns because every cow has an identity and life that she honors. In the countless times I have visited Miles Smith Farm, I am always honored to spend time with Carole and learn from her.

Raising cattle is not an easy job. I have watched Carole regularly deal with life-and-death decisions. She works 90-hour weeks and still she smiles and greets every person as if they are family. Carole is a true testament of what it means to be a steward of the land. She has also started a nonprofit, The Learning Network Foundation, where animals teach humans about farm life.

What type of livestock do you manage?

We manage cattle, Hereford Angus and Scottish Highlander Cross.

Why did you choose Scottish Highlanders?

Honestly, they are cool-looking. In the beginning we needed cattle that would eat a lot of brush because it grows quickly. They would eat all the brush. Over time, we trained them to eat weeds too. Cows learn from each other and their parents so if their parents know to eat weeds, they will teach their offspring. Weeds often have higher proteins than grass so it worked out well for us. We found that highlanders are friendly and have a great temperament so we could train them to be riding cows too!

Cows have two modes when it comes to survival: run or charge, and when you have horns everyone is afraid of you. Highlanders often take a stance and don't run. They fight. This is why we want them to have lots of human interactions so they know humans are okay.

The highlanders are smaller than other breeds. They are a heritage breed. Highlanders are genetically untouched. They carry those survival genes through each generation.

How do you feed them?

We feed them hay and they eat grass. They are free range. Sometimes we substitute some vegetable scraps from gardens as well. In winter, they get more hay to offset the lack of grass compared to what's available in the spring and summer.

Do you use hormones or antibiotics?

We never use hormones and we never use systemic antibiotics. If a cow is sick we use antibiotics because we don't want them to suffer. This is very rare on our farm. If a cow is sick and needs antibiotics, they go through a 30-day detox after the last dose of antibiotics was administered.

What is a systemic antibiotic?

When you put a cow in a feedlot and you give it more grain than it can eat, it gorges itself and gets acidosis, so you give the cow antibiotics to get rid of the symptoms so it eats more. Systemic antibiotics are out of their system when it goes to process, but the problems is that when a cow gets systemic antibiotics, the pathogens have built up resistance to the antibiotics over time.

How do you process your animals?

We use a USDA-certified facility.

What are some of the biggest challenges managing livestock?

Deciding which ones get shipped. You develop a relationship with each cow and it comes down to economics. We have to make decisions and it's hard, but to keep a farm you have to make those decisions. Our cows are healthy and free range so they don't get sick often. Being a farmer is being an observational steward. We must learn to deal with the issues before it becomes a problem. For instance, in winter, water freezes. We can sequester grass, but water freezes and cows drink a lot of water. We have a system of underground lines that go to different pastures to prevent the water from freezing. We installed insulated waterers and the flow of water keeps the water from freezing. Still, even with that system, when it's 20 below for three days you need to get the heater out and make sure they have running water.

How do you choose a good cut of meat?

By flavor. You can get a tough cut of meat and cook it till it is tender and has flavor. You cannot add flavor to the actual cow. The parts of the animal that move the most are the toughest. The front half of the cow holds 80% of the animal's weight. The rear part of the animal doesn't move as much, so it has more flavor and is much more tender. The middle of the animal moves the least, and that is why you find the most tender cuts there. Cuts like the tenderloin don't really move at all. Go for tougher cuts that have more flavor. Expensive cuts don't always mean better.

ARKHIVE FARM

CHESTER, NEW HAMPSHIRE
NOAH BICCHIERI

I met Noah Bicchieri about three years ago when one of my chef friends posted this incredibly marbled beef on his Instagram account. Every other day I would see these different cuts scroll across my screen with the hashtag: #americanwagyu.

Being pretty well connected to the agricultural scene in New England, I wanted to know more about this. I texted my friend, and he told me about Noah. Next thing I knew I was messaging Noah and then on my way to visit the farm.

My first visit to Arkhive was mind-blowing. Noah's knowledge and love for raising cattle is unparalleled. I learned all about how Wagyu is graded in other parts of the world, why it is so revered, and how it is raised. I walked away with a deep appreciation for the breed but also for the legacy and legend that makes this breed so sought after.

When I took on this project, I knew I wanted to share Noah's story with readers. He is an inspiring farmer and steward of the land, and also a friend.

What kind of livestock do you manage?

Wagyu beef cattle, raised American but full-blood Japanese.

Why Wagyu?

I had the experience of eating Wagyu beef in 2017 and it blew me away. So I started doing research and it led me down a path where I discovered that these cattle aren't just raised in Japan. I wanted to find out who was raising them locally. I found a woman out of Vermont named Sheila Patinkin of Vermont Wagyu. I called her and she invited me up to the ranch. I drove up and toured the farm. It's a beautiful cattle, but I quickly realized how expensive this breed is. I decided to take another year and save and make a plan on how I was going to raise them. Then a year later I bought my first full-blood Wagyu at an auction Shelia hosted in 2018.

I'd never been to an auction before. They have a print out of the pedigree and all the info. You meet other ranchers and people fly in from all over. So when I went up to pick up the cow I bought at the auction, I bought another cow at the same time and started crossbreeding them with Wagyu and that is what created American Wagyu. By law, to be called Wagyu, the cow has to be 50% Wagyu by standard. To get the American Wagyu, I crossbred pure Wagyu with Hereford and Charolais.

Why crossbreed with Hereford and Charolais?

Quality of the beef. Hereford are great cattle and it helps the bloodline.

What makes Wagyu beef different?

Wagyu is known for intramuscular marbling. The marbling contributes to tenderness and flavor that isn't found in other breeds to that level.

605

How is it graded?

In the US we have a sliding-grade scale; at the top is prime. On the Japanese and Austrian scale, they measure way beyond what we consider prime. Full-blooded Wagyu beef is just a whole other level beyond prime. The absolute top is A5-BMS 12. Only 2% of the beef in the US grades prime.

What do you feed the cattle?

Feeding depends on where the cow is in its life. Most of their lives they are on pasture, until about 15 to 18 months. After that they are on a more nutritional feed, more specific than just grass. It's all a result of the history of the breed. They weren't raised on pasture in Japan. There is not a lot of grazing land in Japan, so they eat a lot of grain, oats, and barley. This is why the breed has the quality it is known for. After 28 months is when we will start to look at processing, and that is determined by body composition and fat content. The marbling occurs in months 24 to 30, and beyond. I tend to send a cow to process later than the usual 30 months. We wait because we want the cow to develop the marbling and fat content. Low-and-slow growth because they are not pumped full of grain.

Do you use any antibiotics or hormones?

No hormones ever. If an animal is sick and cannot get over it, I will use antibiotics. No systemic antibiotics ever. If you have good animal husbandry practice you won't use antibiotics often, if ever. When they are left to be a cow, they will very rarely get sick.

How do you process your animals?

I process in USDA facilities throughout the Northeast. I choose processors based on quality of the facility and the handling of the animals from moving it through the slaughter process. If an animal is stressed, it can change the meat. We want to take the best care of the animal throughout the slaughter process. We use a lot of facilities that have been designed by Temple Grandin. I handle the trucking of the animals and bring them to the facility. I want to avoid putting stress on the animals at all times.

What makes a good cut of meat?

The quality of the meat, the marbling, the color. If I know the breed, I am going to look for the butchering style. Has the meat been cut right? Is it too thick or too think? Too much fat cap on the rib eye? Butchery skills matter when it comes to good cuts and quality.

What are the biggest challenges you face?

In the Northeast, we don't have thousands of open acres like the Midwest and South have. We are raising these animals on smaller plots of land. We don't always have the ability to raise our own feed. It is hard to do it on a smaller scale. It is very asset heavy. Equipment and feed costs are high. Financially, it is hard in the Northeast. We still aren't looked at as a pinnacle of agriculture as compared to other parts of the country.

My vision is, I want to work in a co-op fashion with a network of other Wagyu ranchers to make sure I can source high-quality cows and feed but also share practices to duplicate the things that work. I would love to put New England on the map as a high-level beef producer, helping other farmers work together to give each other resources and help make this place better for all.

SIDES, SALADS & ACCOMPANIMENTS

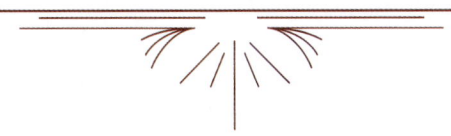

*T*he following recipes will help you make well-rounded meals to balance out all of these meat recipes, from salads and sides to breads and pastas.

ROASTED CAULIFLOWER AU GRATIN

YIELD: 2 SERVINGS / **ACTIVE TIME:** 20 MINUTES / **TOTAL TIME:** 1 HOUR AND 15 MINUTES

One surefire way to get people excited about cauliflower is poaching it in a flavorful stock and then caramelizing mild, nutty cheeses like Emmental and Parmesan on top.

1. Place all of the ingredients, except for the cauliflower and cheeses, in a large saucepan and bring to a boil. Reduce heat so that the mixture simmers gently, add the head of cauliflower, and poach until tender, about 30 minutes.

2. While the cauliflower is poaching, preheat the oven to 450°F. Transfer the tender cauliflower to a baking sheet, place it in the oven, and bake until the top is a deep, golden brown, about 10 minutes.

3. Remove from the oven and spread the cheeses evenly over the top. Return to the oven and bake until the cheeses have browned. Remove from the oven and let cool slightly before cutting it in half and serving.

INGREDIENTS:

- 2 CUPS WHITE WINE
- 2½ CUPS WATER
- ⅓ CUP KOSHER SALT
- 2 STICKS UNSALTED BUTTER
- 6 GARLIC CLOVES, CRUSHED
- 2 SHALLOTS, HALVED
- 1 CINNAMON STICK
- 3 WHOLE CLOVES
- 1 TEASPOON BLACK PEPPERCORNS
- 1 SPRIG FRESH SAGE
- 2 SPRIGS FRESH THYME
- 1 HEAD CAULIFLOWER, LEAVES AND STALK REMOVED
- 1 CUP GRATED EMMENTAL CHEESE
- ¼ CUP GRATED PARMESAN CHEESE

CELERIAC PUREE

YIELD: 4 SERVINGS / **ACTIVE TIME:** 10 MINUTES / **TOTAL TIME:** 45 MINUTES

The key to working with celeriac is to keep it simple and allow the unique flavor to shine.

1. Trim the ends from the celeriac, remove the skin with a vegetable peeler, and use a knife to cut out any recessed or pocked areas. Cut the remainder into thin slices.

2. Place the celeriac, cream, milk, salt, and pepper in a saucepan and bring to a simmer over medium heat, stirring occasionally. Cook until the celery root is fork-tender, about 30 minutes.

3. Transfer the mixture to a blender and puree. Add the butter, season with salt and pepper, and serve.

INGREDIENTS:

- 1½ LBS. CELERIAC
- ½ CUP HEAVY CREAM
- ½ CUP WHOLE MILK
- SALT AND PEPPER, TO TASTE
- 1 STICK UNSALTED BUTTER

RICED CAULIFLOWER

YIELD: 2 SERVINGS / **ACTIVE TIME:** 10 MINUTES / **TOTAL TIME:** 10 MINUTES

This simple dish provides the texture of rice with the nutritional benefits of cauliflower. It can be used in stir-fries or to accompany curries and is excellent on its own with a little butter.

1. Place the cauliflower in a food processor and pulse until it becomes granular.

2. Place the oil in a large skillet and warm over medium heat. When the oil starts to shimmer, add the cauliflower, cover the pan, and cook until tender, 3 to 5 minutes.

3. Season with salt and pepper and serve.

INGREDIENTS:

- 1 LARGE HEAD CAULIFLOWER, TRIMMED AND CHOPPED
- ¼ CUP OLIVE OIL
- SALT AND PEPPER, TO TASTE

CAULIFLOWER MASH

YIELD: 4 SERVINGS / **ACTIVE TIME:** 5 MINUTES / **TOTAL TIME:** 20 MINUTES

Cauliflower is so versatile; it can take on forms that you would never expect. From pizza crusts and hummus to rice and this version, which puts mashed potatoes to shame, cauliflower deserves a gold medal for its contributions to the dinner table.

1. Bring salted water to a boil in a large saucepan. Add the cauliflower and cook until tender, about 10 minutes. Drain and let cool slightly.

2. Place the cauliflower, butter, cream, salt, and pepper in a food processor and blitz until the mixture is rich and smooth. Serve immediately.

INGREDIENTS:

- SALT AND PEPPER, TO TASTE
- 1 HEAD CAULIFLOWER, TRIMMED AND CUT INTO CROWNS
- 3 TABLESPOONS UNSALTED BUTTER, PLUS 1 TEASPOON
- ¼ CUP HEAVY CREAM

HOME STYLE BAKED BEANS

YIELD: 6 TO 8 SERVINGS / **ACTIVE TIME:** 30 MINUTES / **TOTAL TIME:** 1½ TO 2 HOURS

Images of cowboys and campfires will be dancing in your head thanks to this cast-iron skillet version of baked beans.

1. Preheat the oven to 325°F.

2. Warm a 12-inch cast-iron skillet over medium heat and add half of the bacon pieces. Cook until the bacon is just starting to crisp up, about 6 minutes. Transfer to a paper towel–lined plate.

3. Place the remaining bacon in the skillet, raise heat to medium-high, and cook, turning often, until the pieces are browned and crispy, about 10 minutes. Reduce heat to medium. Add the onion and bell pepper and cook, stirring occasionally, until the vegetables start to soften, about 6 minutes.

4. Add the salt, beans, barbecue sauce, mustard, and brown sugar. Stir, season with salt and pepper, and bring to a simmer.

5. Lay the partially cooked pieces of bacon on top and transfer the skillet to the oven. Bake for 1 hour, until the bacon on top is crispy and browned and the sauce is thick. If the consistency seems too thin, cook for an additional 15 to 30 minutes, checking frequently so as not to overcook the beans.

6. Remove from the oven and allow to cool slightly before serving.

INGREDIENTS:

- 6 STRIPS THICK-CUT BACON
- ½ ONION, DICED
- ½ CUP SEEDED AND DICED BELL PEPPER
- 1 TEASPOON SALT, PLUS MORE TO TASTE
- 2 (14 OZ.) CANS PINTO BEANS, RINSED AND DRAINED
- 1 CUP BARBECUE SAUCE (SEE PAGE XXX)
- 1 TEASPOON DIJON MUSTARD
- 2 TABLESPOONS DARK BROWN SUGAR
- BLACK PEPPER, TO TASTE

POTATO & CELERIAC GRATIN WITH GRUYÈRE & FIGS

YIELD: 4 SERVINGS / **ACTIVE TIME:** 25 MINUTES / **TOTAL TIME:** 1 HOUR

When you cook a gratin, I always parboil the vegetables before putting them in the dish with the rest of the ingredients to make sure everything will be cooked through. Plus, the flavor provided by the bay leaf, milk, and crushed garlic is worth the added effort.

1. Preheat the oven to 375°F.

2. Place the potatoes and celeriac in a medium saucepan and cover with water. Add the bay leaf, 1 teaspoon of the salt, the garlic, and milk, bring to a boil, and then reduce heat. Simmer for 1 minute and then drain. Remove the bay leaf and discard.

3. Butter a 10-inch oval gratin dish or casserole dish and add half of the potatoes and celeriac, making sure they are evenly distributed. Sprinkle half of the figs, some salt, half of the Gruyère, and the pinch of nutmeg on top. Repeat with the remaining potatoes and celeriac, seasonings, and Gruyère.

4. Pour the cream over the top and cover the dish with aluminum foil. Place in the oven and bake for 20 minutes. Remove the foil and bake for another 15 minutes, until the top is browned and most of the liquid has cooked off.

5. Remove from the oven and let stand for 15 minutes before serving.

INGREDIENTS:

- 1½ LBS. RUSSET POTATOES, PEELED AND SLICED THIN
- ½ LB. CELERIAC, PEELED AND SLICED THIN
- 1 BAY LEAF
- 2 TEASPOONS KOSHER SALT, DIVIDED
- 2 GARLIC CLOVES, CRUSHED
- 2 TABLESPOONS MILK
- 3 DRIED OR FRESH FIGS, DICED
- 4 OZ. GRUYÈRE CHEESE, GRATED
- 1 PINCH NUTMEG
- ½ CUP HEAVY CREAM

SOUTHERN COLLARD GREENS

YIELD: 4 TO 6 SERVINGS / **ACTIVE TIME:** 30 MINUTES / **TOTAL TIME:** 2 HOURS AND 30 MINUTES

When you think these are done, just keep cooking them.

1. Place the oil in a large saucepan and warm over medium-high heat. When the oil starts to shimmer, add the onion and sauté until translucent, about 3 minutes. Add the ham, reduce heat to medium, and cook until the ham starts to brown, about 5 minutes.

2. Add the remaining ingredients, stir to combine, and cover the pan. Braise the collard greens until they are very tender, about 2 hours. Check on the collards every so often and add water if all of the liquid has evaporated.

INGREDIENTS:

- 2 TABLESPOONS OLIVE OIL
- 1 ONION, DICED
- ½ LB. SMOKED HAM, DICED
- 4 GARLIC CLOVES, DICED
- 3 LBS. COLLARD GREENS, STEMS REMOVED, CHOPPED
- 2 CUPS CHICKEN STOCK (SEE PAGE 478)
- ¼ CUP APPLE CIDER VINEGAR
- 1 TABLESPOON BROWN SUGAR
- 1 TEASPOON RED PEPPER FLAKES

KOHLRABI SLAW WITH MISO DRESSING

YIELD: 4 SERVINGS / **ACTIVE TIME:** 10 MINUTES / **TOTAL TIME:** 10 MINUTES

The Asian flavors of this coleslaw are just perfect alongside grilled meat, especially pork and chicken. If you have a mandoline, it will make quick work of the vegetables. A hand grater will also work.

1. Place the white miso paste, vinegar, sesame oil, ginger, soy sauce, peanut oil, sesame seeds, and maple syrup in a mixing bowl and stir to combine. Set aside.

2. Place the kohlrabies, carrots, and cilantro in a separate bowl and stir to combine.

3. Drizzle a few spoonfuls of the dressing into the coleslaw and stir until evenly coated. Taste, add more dressing if desired, top with the pistachios, and serve.

INGREDIENTS:

- 1 TABLESPOON WHITE MISO PASTE
- 1 TABLESPOON RICE VINEGAR
- 1 TEASPOON SESAME OIL
- 1 TEASPOON MINCED GINGER
- 1 TEASPOON SOY SAUCE
- 3 TABLESPOONS PEANUT OIL
- 1 TABLESPOON SESAME SEEDS
- 1 TEASPOON REAL MAPLE SYRUP
- 3 KOHLRABIES, PEELED AND JULIENNED OR GRATED
- 2 CARROTS, PEELED AND JULIENNED OR GRATED
- ¼ CUP CHOPPED FRESH CILANTRO
- ¼ CUP SHELLED PISTACHIOS, CRUSHED

GERMAN-STYLE KOHLRABI IN BÉCHAMEL

YIELD: 4 SERVINGS / **ACTIVE TIME:** 20 MINUTES / **TOTAL TIME:** 35 MINUTES

This recipe is very simple but creates a rich and comforting dish. The only twist on this traditional version is to include the kohlrabi greens as well, partly to reduce food waste and partly because they add great flavor.

1. Bring water to a boil in a medium saucepan. Add the kohlrabies and a dash of salt and cook until just tender, about 5 minutes. Remove with a slotted spoon and set aside.

2. Add the reserved greens to the pan and cook until tender, about 8 minutes. Remove with a slotted spoon and add to the kohlrabies.

3. Drain the water and dry the pan. Return it to the stove and add the butter. Melt the butter over low heat, add the flour, and whisk to combine. Cook for 1 minute, then slowly add the milk, whisking constantly to prevent lumps from forming. Continue whisking until the sauce has thickened, about 5 minutes. Season with salt and pepper and the nutmeg.

4. Return the kohlrabies and greens to the pan, cook until heated through, and serve.

INGREDIENTS:

- 1 LB. KOHLRABIES, PEELED AND SLICED, GREENS RESERVED AND CHIFFONADE
- SALT AND PEPPER, TO TASTE
- 2 TABLESPOONS UNSALTED BUTTER
- 2 TABLESPOONS ALL-PURPOSE FLOUR
- 1 CUP WHOLE MILK
- 1 PINCH NUTMEG

CHARRED SUMMER LEEKS WITH ROMESCO SAUCE

YIELD: 4 SERVINGS / **ACTIVE TIME:** 15 MINUTES / **TOTAL TIME:** 25 MINUTES

When summer leeks are cooked on a grill, they develop a sweet and smoky flavor. The first step is to steam them quickly on the stove to make sure they get cooked through on the grill. Pairing that sweetness and smoke with the garlicky Romesco Sauce is a slice of heaven.

1. Preheat your grill to medium heat. Cut the dark green sections off of the leeks and remove the roots, keeping the base that holds the layers together. Cut the leeks lengthwise and rinse between each layer to remove dirt, taking care to keep the layers together.

2. Place 1 inch of water in a saucepan, place a steaming tray in the pan, and bring the water to a boil. When boiling, place the leeks in the steaming tray and steam until tender, about 5 minutes.

3. Drain, drizzle oil over the leeks, and season with salt and pepper. When the grill is about 400°F, place the leeks on the grill and cook until browned all over, about 8 minutes per side. When the leeks are nearly charred, transfer to a platter and serve with the Romesco Sauce.

ROMESCO SAUCE

1. Place all of the ingredients, except for the oil, in a blender or food processor and pulse until smooth.

2. Add the oil in a steady stream and puree until emulsified. Season with salt and pepper and serve.

INGREDIENTS:

- 8 SUMMER LEEKS
- OLIVE OIL, TO TASTE
- SALT AND PEPPER, TO TASTE
- ROMESCO SAUCE (SEE RECIPE), FOR SERVING

ROMESCO SAUCE

- 2 LARGE ROASTED RED BELL PEPPERS
- 1 GARLIC CLOVE, SMASHED
- ½ CUP SLIVERED ALMONDS, TOASTED
- ¼ CUP TOMATO PUREE
- 2 TABLESPOONS CHOPPED FRESH FLAT-LEAF PARSLEY
- 2 TABLESPOONS SHERRY VINEGAR
- 1 TEASPOON SMOKED PAPRIKA
- ½ CUP OLIVE OIL
- SALT AND PEPPER, TO TASTE

FRIZZLED LEEKS

YIELD: 2 TO 4 SERVINGS / **ACTIVE TIME:** 15 MINUTES / **TOTAL TIME:** 20 MINUTES

These delectable fried bits are light and airy. While they can be eaten on their own, they are best as a topping for anything from steak to fried rice.

INGREDIENTS:

- 1–2 LARGE LEEKS
- ¼ CUP ALL-PURPOSE FLOUR
- VEGETABLE OIL, FOR FRYING
- SALT, TO TASTE

1. Trim the leeks and cut the white part only into 3-inch sections. Slice the sections in half lengthwise and rinse to remove any dirt. Pat dry and then slice into very thin strips.

2. Place the flour in a mixing bowl, add the leek strips, and toss until evenly coated.

3. Add oil to a Dutch oven until it is about 3 inches deep. Heat the oil to 350°F, or until a pinch of flour dropped in sizzles on contact.

4. Working in batches to ensure that the leeks are fully submerged in the oil, shake the leeks to remove any excess flour and add them to the oil. Fry until browned, transfer to a paper towel–lined plate, and repeat with the remaining leeks.

5. Sprinkle the fried leeks with salt and serve.

POTATO & TOMATO GRATIN

YIELD: 4 TO 6 SERVINGS / **ACTIVE TIME:** 15 MINUTES / **TOTAL TIME:** 45 MINUTES

A testament to the brilliance of French cuisine, this layered dish has all the flavor in the world and is as simple as can be to make. Try serving it with grilled chicken and sautéed kohlrabi.

1. Preheat your oven to 350°F.

2. Place the garlic, parsley, and thyme in a small bowl, stir to combine, and set it aside while you prepare the tomatoes.

3. Lightly oil a 12-inch cast-iron skillet or enameled cast-iron gratin dish and then add a layer of the tomato slices. Season with salt and pepper and add a layer of potatoes and a sprinkle of the garlic-and-parsley mixture. Drizzle with olive oil and continue the layering process until all of the tomatoes, potatoes, and garlic-and-parsley mixture have been used.

4. Cover with foil, place in the oven, and bake for 20 minutes. Remove from the oven and remove the foil. If tomatoes haven't released enough liquid to soften the potatoes, add a bit of the stock. Replace the foil and continue baking for 15 minutes.

5. Remove the foil, cook for an additional 5 minutes, and serve warm.

INGREDIENTS:

- 4 GARLIC CLOVES, MINCED
- LEAVES FROM 1 SMALL BUNCH FRESH PARSLEY, MINCED
- 2 TABLESPOONS MINCED FRESH THYME LEAVES
- OLIVE OIL, TO TASTE
- 2 LBS. TOMATOES, SLICED ¼-INCH THICK
- SALT AND PEPPER, TO TASTE
- 4 WAXY POTATOES, SLICED ¼-INCH THICK
- CHICKEN STOCK (SEE PAGE 478)

PATATAS BRAVAS

YIELD: 4 SERVINGS / **ACTIVE TIME:** 45 MINUTES / **TOTAL TIME:** 1 HOUR

Native to Spain, this smoky potato dish can be found in tapas bars all across that country.

1. Place the potatoes, onion, and 1 tablespoon of the olive oil in a mixing bowl and toss to coat.

2. Line a large cast-iron wok with foil, making sure that the foil extends over the side. Add the soaked wood chips and place the wok over medium heat.

3. When the wood chips are smoking heavily, place a wire rack above the wood chips and add the potatoes, onion, and garlic. Cover the wok with a lid, fold the foil over the lid to seal the wok as best you can, and smoke for 20 minutes. After 20 minutes, remove from heat and keep the wok covered for another 20 minutes.

4. Meanwhile, to make *salsa brava*, combine the tomatoes, paprika, vinegar, and remaining olive oil in a blender and puree. Set the mixture aside.

5. Remove the garlic and onion from the smoker. Peel and roughly chop. Add the garlic and onion to the mixture in the blender and puree until smooth. Season the *salsa brava* with salt. Serve the potatoes with sour cream and the *salsa brava*.

INGREDIENTS:

- 4 MEDIUM POTATOES, CUT INTO THICK PIECES AND PARBOILED
- 1 ONION, WITH SKIN AND ROOT, HALVED
- 3 TABLESPOONS OLIVE OIL, DIVIDED
- 2 CUPS WOOD CHIPS, SOAKED IN COLD WATER FOR 30 MINUTES
- 1 HEAD OF GARLIC, TOP ½ INCH REMOVED
- 1 (14 OZ.) CAN DICED TOMATOES, DRAINED
- 1 TABLESPOON SWEET PAPRIKA
- 1 TABLESPOON SHERRY VINEGAR

  SALT, TO TASTE

  SOUR CREAM, FOR SERVING

POTATO & PARSNIP LATKES

YIELD: 4 SERVINGS / **ACTIVE TIME:** 40 MINUTES / **TOTAL TIME:** 1 HOUR

The secret to making good latkes is to make sure you've removed as much water as possible from the potatoes before cooking. You don't need any special equipment for this, just take up the shredded potato in your hands and squeeze. Parsnips are actually very dry as vegetables go, so they do not need the squeeze put on them. They brown beautifully, though, and will add not only sweet flavor but also a nice, crispy texture to the final product. There are those who like sour cream on their latkes and those who prefer applesauce. Be an accommodating cook and offer both.

1. Preheat the oven to 350°F. Place the grated potatoes in a colander and squeeze one handful at a time until no more liquid can be removed from them. Transfer to a bowl.

2. Add the parsnips, flour, and egg to the potatoes, stir to combine, and season with salt and pepper.

3. Place the oil in a wide sauté pan and warm over medium-high heat. Once the oil is shimmering, add spoonfuls of the latke mixture to the pan and press down to form 3-inch patties, flattening gently with a spatula. Reduce heat to medium-low and cook until browned on both sides, about 8 to 10 minutes per side.

4. When both sides are perfectly browned, test the latkes to see if the interior is fully cooked. If not, place them on a baking sheet and bake in the oven for an additional 10 minutes.

5. Serve hot with the sour cream and applesauce.

INGREDIENTS:

- 2 RUSSET POTATOES, PEELED AND GRATED
- 3 PARSNIPS, PEELED, TRIMMED, CORED, AND GRATED
- 1 TABLESPOON ALL-PURPOSE FLOUR
- 1 EGG
- SALT AND PEPPER, TO TASTE
- 1 TABLESPOON OLIVE OIL
- SOUR CREAM, FOR SERVING
- APPLESAUCE, FOR SERVING

ROASTED PARSNIPS & CARROTS WITH RAS EL HANOUT & HONEY

YIELD: 4 SERVINGS / **ACTIVE TIME:** 20 MINUTES / **TOTAL TIME:** 40 MINUTES

Roasting brings out the best in parsnips and carrots, and adding honey and spice at the end only enhances the deep flavor already there. Ras el hanout is a North African spice blend. Much like Indian curry, there is no official recipe, but it often contains cardamom, cumin, nutmeg, mace, cinnamon, ginger, chilies, allspice, and salt. It's best to adjust your seasonings to taste as you go, to avoid oversalting.

1. Preheat the oven to 400°F. Place the parsnips and carrots in a roasting pan in one layer, add the oil and salt, and toss to coat. Place in the oven and roast for 20 minutes, or until browned.

2. Remove the pan from the oven and pile the vegetables in the center. Drizzle the honey over the top and toss to coat. Sprinkle the ras el hanout over the top and toss to coat.

3. Return the pan to the oven and roast for another 5 to 10 minutes, making sure the vegetables do not burn. Remove from the oven and serve immediately.

INGREDIENTS:

- 4 LARGE PARSNIPS, PEELED, TRIMMED, AND CORED
- 4 LARGE CARROTS, PEELED AND SLICED LENGTHWISE
- 2 TABLESPOONS OLIVE OIL
- SALT AND PEPPER, TO TASTE
- 2 TABLESPOONS HONEY
- 1 TABLESPOON RAS EL HANOUT

CONFIT NEW POTATOES

YIELD: 4 TO 6 SERVINGS / **ACTIVE TIME:** 5 MINUTES / **TOTAL TIME:** 1 HOUR AND 10 MINUTES

New potatoes are young potatoes that are pulled in early spring. They are sweeter than their mature counterparts, since the sugars haven't had time to develop into starches, and are so soft and tender that they don't need to be peeled.

1. Place the oil in a Dutch oven and bring it to 200°F over medium heat.

2. While the oil is warming, wash the potatoes and pat them dry. Carefully place the potatoes in the oil and cook until fork-tender, about 1 hour.

3. Drain the potatoes, season generously with salt and pepper, and stir to ensure that the potatoes are evenly coated. If desired, garnish with rosemary and serve immediately.

TIP: These potatoes should have plenty of flavor, but if you're looking to take them to another level, replace the canola oil with chicken or duck fat.

INGREDIENTS:

- 4 CUPS CANOLA OIL
- 5 LBS. NEW POTATOES
- SALT AND PEPPER, TO TASTE
- FRESH ROSEMARY LEAVES, FOR GARNISH (OPTIONAL)

LOW 'N' SLOW POTATOES

YIELD: 4 SERVINGS / **ACTIVE TIME:** 20 MINUTES / **TOTAL TIME:** 40 MINUTES

This simple dish is on regular rotation in our house because it is delicious and goes with everything. I wanted a sautéed potato dish that didn't involve the extra step of parboiling but still produced cubes that were tender on the inside and crispy on the outside. The trade-off for not fussing with parboiling is that they take a while, but they require nothing beyond an occasional stir. The key to doing it well is to give yourself time—around half an hour—and remain patient while everything browns over low heat. Different starch levels will give you different results, but any type of potato will work.

1. Place the oil in a wide sauté pan and warm over medium heat. When the oil is shimmering, add the potatoes so they sit in one layer. Once you hear them start crackling, reduce heat to low and cook, leaving the potatoes undisturbed, until they have a brown crust on the bottom. This can take up to 10 minutes.

2. Once browned, flip and repeat on another side. Continue until the cubes are brown and crispy all over. Season with salt and serve.

INGREDIENTS:

- 1 TABLESPOON OLIVE OIL
- 2 LBS. POTATOES, DICED
  SALT, TO TASTE

HERBED POTATO SALAD

YIELD: 4 TO 6 SERVINGS / **ACTIVE TIME:** 10 MINUTES / **TOTAL TIME:** 40 MINUTES

The two most common potato salads have either a mayonnaise dressing or, in the German version, a sweet vinegar dressing. The French have a different approach with shallots and herbs and a tangy vinaigrette that lets the natural sweetness of the potatoes come through. The dressing is poured on the potatoes when they are still warm, letting them soak up the flavor.

1. Add the potatoes to a pot of water large enough to hold them all, bring to a boil, reduce heat, and simmer until tender, about 15 minutes.

2. While the potatoes are simmering, whisk together the oil, vinegar, wine, mustard, and teaspoon of salt.

3. When the potatoes are done, drain them and place them in a bowl. Add the vinaigrette and shallot immediately and gently toss, making sure to coat all of the potatoes. Let cool completely.

4. Taste and adjust seasoning as needed. Add the black pepper and fresh herbs, stir to incorporate, and serve.

INGREDIENTS:

- 1½ LBS. LOW-STARCH, NEW, OR RED POTATOES, CUBED
- ½ CUP OLIVE OIL
- 3 TABLESPOONS WHITE WINE VINEGAR
- 2 TABLESPOONS DRY WHITE WINE
- 1 TEASPOON WHOLE-GRAIN DIJON MUSTARD
- 1 TEASPOON KOSHER SALT, PLUS MORE TO TASTE
- 1 SHALLOT, MINCED
-   BLACK PEPPER, TO TASTE
- 2 TABLESPOONS CHOPPED FRESH PARSLEY
- 2 TABLESPOONS CHOPPED FRESH CHIVES
- 2 TABLESPOONS CHOPPED FRESH DILL

BLUE CHEESE GRATIN

YIELD: 4 SERVINGS / **ACTIVE TIME:** 20 MINUTES / **TOTAL TIME:** 45 MINUTES

Thin slices of potato are baked in cream and blue cheese here, resulting in a heavenly, tangy gratin. No need to splurge on Stilton or other high-end blue cheeses for this one; a decent-quality blue from the supermarket will work best. Also, nutmeg is very strong, so best to be too conservative than too liberal with it—otherwise, the potatoes will taste bitter.

1. Preheat the oven to 375°F. While the oven is heating up, grease an 8×5.5-inch gratin dish with butter.

2. Place the potato slices in a saucepan with the garlic, bay leaf, milk, and salt. Bring to a boil over high heat, drain, and discard the bay leaf.

3. Place half of the potatoes in the gratin dish, then sprinkle on half of the blue cheese, a pinch of nutmeg, salt, and pepper. Add the cream and 2 tablespoons of the butter. Place the remaining potatoes on top and repeat with the blue cheese, nutmeg, salt, pepper, and butter.

4. Place the dish in the oven and bake for about 30 minutes, or until the top is brown and crispy. The cream should come halfway up the potatoes at the start and cook down to a rich sauce by the end. If the dish looks too dry, add a little more cream before serving.

INGREDIENTS:

- 4 TABLESPOONS UNSALTED BUTTER, CUBED, PLUS MORE AS NEEDED
- 2 LBS. RUSSET POTATOES, PEELED AND SLICED THIN
- 2 GARLIC CLOVES, SMASHED
- 1 BAY LEAF
- 2 TABLESPOONS WHOLE MILK
- 1 TEASPOON KOSHER SALT, PLUS MORE TO TASTE
- 4 OZ. BLUE CHEESE
- NUTMEG, TO TASTE
- BLACK PEPPER, TO TASTE
- ¾ CUP HEAVY CREAM, PLUS MORE AS NEEDED

ONION RINGS

YIELD: 4 SERVINGS / **ACTIVE TIME:** 15 MINUTES / **TOTAL TIME:** 20 MINUTES

What's better than a burger with fries? A burger with onion rings! For this recipe, skip the kosher salt and use fine-grained, so it will stick to the onion.

1. Place the flour in a shallow bowl, the beaten egg, milk, and paprika in another, and the bread crumbs and Parmesan in another.

2. Place a Dutch oven on the stove and add oil until it is 2 to 3 inches deep. Heat the oil until a few bread crumbs sizzle immediately when dropped in.

3. Dip the onion rings in the flour, then in the egg mixture, and lastly in the bread crumb mixture. Make sure the rings are fully covered by the bread crumb mixture. Carefully drop into the hot oil and fry for several minutes, until brown.

4. Using tongs, turn over to brown the other side (if necessary) and then transfer to a paper towel–lined plate.

5. Sprinkle with fine-grained salt, let cool briefly, and serve with the Creamy Adobo Dip.

CREAMY ADOBO DIP

1. Place all of the ingredients in a bowl, stir to combine, and serve.

INGREDIENTS:

- ½ CUP ALL-PURPOSE FLOUR
- 1 EGG, BEATEN
- ⅓ CUP WHOLE MILK
- ½ TEASPOON PAPRIKA
- ½ CUP PLAIN BREAD CRUMBS
- ½ CUP PANKO BREAD CRUMBS
- 1 TABLESPOON GRATED PARMESAN CHEESE
- VEGETABLE OIL, FOR FRYING
- 2 LARGE YELLOW ONIONS, SLICED INTO THICK RINGS
- FINE-GRAINED SALT, TO TASTE
- CREAMY ADOBO DIP (SEE RECIPE), FOR SERVING

CREAMY ADOBO DIP

- 2 TABLESPOONS MAYONNAISE
- 2 TABLESPOONS SOUR CREAM
- 1 TEASPOON ADOBO SAUCE (FROM A CAN OF CHIPOTLES IN ADOBO SAUCE)

ONION BHAJI

YIELD: 4 SERVINGS (12 FRITTERS) / **ACTIVE TIME:** 20 MINUTES / **TOTAL TIME:** 20 MINUTES

These fritters, which are a popular snack in India, can be varied in dozens of ways. Try adding shredded carrot, chilies, a bit of coconut, or even threads of parsnip to get an idea of what else you might like to incorporate.

1. Place the eggs in a bowl and beat them. Add the onions, flour, coriander, cumin, serrano pepper, and salt and stir to combine.

2. Place the vegetable oil in an 10-inch cast-iron skillet and warm over medium heat. When it starts to shimmer, place a large spoonful of the onion batter and fry until golden brown, about 30 to 45 seconds.

3. Turn the fritter over and fry until it is crisp and golden brown all over, about 30 seconds. Transfer to a paper towel–lined plate to drain.

4. Repeat with the remaining batter, adding and heating more oil if it starts to run low. When all of the fritters have been cooked, serve immediately.

INGREDIENTS:

- 2 EGGS
- 3 LARGE RED ONIONS, SLICED INTO THIN HALF-MOONS
- 5 OZ. ALL-PURPOSE FLOUR
- 1 TEASPOON GROUND CORIANDER
- 1 TEASPOON CUMIN
- 1 SERRANO PEPPER, SEEDED AND MINCED
- ½ TEASPOON SALT
- 1 CUP VEGETABLE OIL, PLUS MORE AS NEEDED

DAL

YIELD: 4 SERVINGS / **ACTIVE TIME:** 20 MINUTES / **TOTAL TIME:** 1 HOUR AND 40 MINUTES

This is an everyday staple in most parts of India, appearing in myriad guises. It's a simple stew of yellow peas, orange lentils, or mung beans that should be slightly soupy so that its deliciousness can seep down into the basmati rice.

1. Place the vegetable oil in a large cast-iron Dutch oven and warm over medium-high heat.

2. Add the onion, garlic, red pepper flakes, curry leaves (if using), and salt and sauté until the onion is slightly translucent, about 2 minutes.

3. Add the yellow split peas, water, and turmeric to a large cast-iron Dutch oven and bring to a simmer. Cover and gently simmer for 1 hour, removing the lid to stir the dal 2 or 3 times.

4. Remove the lid and simmer, while stirring occasionally, until the dal has thickened, about 30 minutes. When the dal has the consistency of porridge, stir in the peas, and cook until they are warmed through.

5. To serve, ladle the dal over the rice.

INGREDIENTS:

- 2 TABLESPOONS VEGETABLE OIL
- 1 YELLOW ONION, DICED
- 2 GARLIC CLOVES, MINCED
- 2 TEASPOONS RED PEPPER FLAKES, OR TO TASTE
- 2 CURRY LEAVES (OPTIONAL)
- 1 TEASPOON SALT
- 1½ CUPS YELLOW SPLIT PEAS, SORTED AND RINSED
- 4 CUPS WATER
- 1 TEASPOON TURMERIC
- 1 CUP FRESH PEAS
- 2 CUPS COOKED BASMATI RICE

BLACK-EYED PEAS WITH COCONUT

YIELD: 4 SERVINGS / **ACTIVE TIME:** 10 MINUTES / **TOTAL TIME:** 9 HOURS

Black-eyed peas have a wonderful starchiness and nutty taste that is utilized far too infrequently. You can use canned peas in this preparation, but dried ones will be better.

1. Drain the black-eyed peas, place them in a large enameled cast-iron Dutch oven, and cover with water. Bring the water to a simmer and cook until the black-eyed peas are tender, about 45 minutes. Drain and set them aside.

2. Place the coconut oil in the Dutch oven and warm over medium heat. When the oil starts to shimmer, add the yellow onion, green onions, tomatoes, habanero, and Berbere Spice and sauté for 2 minutes.

3. Add the coconut milk and stock and bring to a simmer. Reduce the heat to low and gently simmer until the liquid has slightly reduced, about 10 minutes.

4. Return the black-eyed peas to the pot and continue to simmer for 15 minutes.

5. Stir in the cilantro and serve immediately.

TIP: This can be served on its own or over quinoa, millet, or rice.

BERBERE SPICE

Use a mortar and pestle or spice grinder to combine all of the ingredients.

INGREDIENTS:

- 1 CUP DRIED BLACK-EYED PEAS, SOAKED IN COLD WATER FOR 8 HOURS
- ¼ CUP COCONUT OIL
- 1 YELLOW ONION, PEELED AND SLICED
- ½ CUP CHOPPED GREEN ONIONS
- 2 TOMATOES, CHOPPED
- 1 HABANERO PEPPER, STEMMED, SEEDED, AND CHOPPED
- 2 TEASPOONS BERBERE SPICE (SEE RECIPE)
- 1 CUP COCONUT MILK
- 1 CUP CHICKEN STOCK (SEE PAGE 478)
- 1 CUP CILANTRO LEAVES, CHOPPED

BERBERE SPICE

- 1 TEASPOON FENUGREEK SEEDS
- 1 TEASPOON RED PEPPER FLAKES
- 2 TABLESPOONS SWEET PAPRIKA
- ½ TEASPOON GROUND CARDAMOM
- 1 TEASPOON NUTMEG
- ⅛ TEASPOON GARLIC POWDER
- ⅛ TEASPOON GROUND CLOVES
- ⅛ TEASPOON CINNAMON
- ⅛ TEASPOON ALLSPICE

MAC & CHEESE WITH BROWN BUTTER BREAD CRUMBS

YIELD: 6 SERVINGS / **ACTIVE TIME:** 15 MINUTES / **TOTAL TIME:** 1 HOUR

The cheese in this dish will stick to your ribs. Reserve it for those nights when you're especially hungry and can afford to relax after the meal.

1. Preheat oven to 400°F.

2. Fill an enameled cast-iron Dutch oven with water and bring to a boil. Add some salt and then add the macaroni. Cook until slightly under al dente, about 6 to 7 minutes. Drain and set aside.

3. Place the pot over medium heat and add 3 tablespoons of the butter. Cook until the butter starts to give off a nutty smell and browns. Add the bread crumbs, stir, and cook until the bread crumbs start to look like wet sand, 4 to 5 minutes. Remove and set aside.

4. Wipe the Dutch oven out with a paper towel, place over medium-high heat, and add the onion and the remaining butter. Cook, while stirring, until the onion is translucent and soft, about 7 to 10 minutes. Add the flour and whisk until there are no lumps. Add the mustard, turmeric, granulated garlic, and white pepper and whisk until combined. Add the half-and-half or light cream and the milk and whisk until incorporated.

5. Reduce heat to medium and bring the mixture to a simmer. Once you start to see small bubbles forming around the outside of the mixture, add the cheeses one at a time, whisking to combine before adding the next one. When all the cheese has been added and the mixture is smooth, cook until the flour taste is gone, 10 to 15 minutes. Return the pasta to the pot, stir, and top with the bread crumbs.

6. Place in the oven and bake for 10 to 15 minutes. Remove the pot from the oven and serve.

TIP: If you can't find Boursin, whisk some cream cheese and a little softened butter together.

INGREDIENTS:

- SALT AND PEPPER, TO TASTE
- ½ LB. ELBOW MACARONI
- 7 TABLESPOONS UNSALTED BUTTER
- 2 CUPS BREAD CRUMBS (USE PANKO FOR AN EXTRA CRUNCHY TOP)
- ½ YELLOW ONION, MINCED
- 3 TABLESPOONS ALL-PURPOSE FLOUR
- 1 TABLESPOON YELLOW MUSTARD
- 1 TEASPOON TURMERIC
- 1 TEASPOON GRANULATED GARLIC
- 1 TEASPOON WHITE PEPPER
- 2 CUPS HALF-AND-HALF OR LIGHT CREAM
- 2 CUPS WHOLE MILK
- 1 LB. AMERICAN CHEESE, SLICED
- 10 OZ. BOURSIN CHEESE
- 1 LB. EXTRA SHARP CHEDDAR CHEESE, SLICED

CREAMED CORN

YIELD: 8 TO 10 SERVINGS / **ACTIVE TIME:** 30 MINUTES / **TOTAL TIME:** 1 HOUR

When you're tired of corn on the cob and in need of a little comfort, try this fresh spin on a canned classic.

1. Standing each ear of corn up in the middle of a large baking dish, use a sharp knife to cut down the sides and remove all the kernels. With the kernels off, take the blade of a dull knife and press it along each side of the ears to "milk" the cob of its liquid. Discard the milked cobs.

2. Place a 12-inch cast-iron skillet over medium heat and, when hot, lower the heat and add the butter so it melts slowly. When melted, add the corn kernels and milk from the cobs and stir to coat. Increase the heat to medium-high and add the water and half-and-half. Bring to a boil, while stirring constantly, and then reduce to low heat. Add the salt and pepper.

3. In a measuring cup, add the flour and warm water and mix until thoroughly combined. Drizzle the flour mixture into the corn, continuing to stir until the sauce thickens. If it gets too thick, add some more half-and-half. Serve hot.

INGREDIENTS:

- 12 EARS OF FRESH CORN, SHUCKED AND RINSED
- 3 TABLESPOONS UNSALTED BUTTER
- 1 CUP WATER, AT ROOM TEMPERATURE
- 1 CUP HALF-AND-HALF, PLUS MORE AS NEEDED
- ½ TEASPOON SALT
- FRESHLY GROUND BLACK PEPPER, TO TASTE
- 3 TABLESPOONS ALL-PURPOSE FLOUR
- ½ CUP WARM WATER (110°F)

SHERRIED MUSHROOMS WITH PINE NUTS

YIELD: 4 SERVINGS / **ACTIVE TIME:** 10 MINUTES / **TOTAL TIME:** 25 MINUTES

A simple roasted mushroom is a wonder to behold, but it is also a terrific foundation for a ton of great dishes. This dynamic mélange is one, featuring myriad tastes and textures.

INGREDIENTS:

- 2 CUPS PINE NUTS
- 5 TABLESPOONS OLIVE OIL, DIVIDED
- 1 PINCH SALT
- 1 PINCH SUGAR
- 1 LARGE SHALLOT, MINCED
- 1 CUP SHERRY
- ¼ CUP SHERRY VINEGAR
- ZEST OF 1 ORANGE
- 6 CUPS ASSORTED WILD MUSHROOMS (OYSTER, MAITAKE, SHIITAKE, ETC.)
- 4 TABLESPOONS UNSALTED BUTTER
- PARSLEY LEAVES, CHOPPED, FOR GARNISH

1. Place a 12-inch cast-iron skillet over medium heat. When it is hot, add the pine nuts and toast until lightly browned. Remove from the pan and let cool. When they are cool enough to handle, mince and set aside.

2. Add 1 tablespoon of the olive oil, the salt, sugar, and shallot to the skillet and cook until the shallot is translucent. Add the Sherry and vinegar and cook for 5 minutes, until the liquid has reduced. Add the orange zest and then transfer the mixture to a bowl.

3. Add the remaining olive oil to the skillet and warm until it starts to smoke. Add the mushrooms and cook, without stirring, for 3 minutes. Add the butter, stir, and cook until the mushrooms are browned all over and wilted. Remove from the pan and add to the shallot-and-Sherry mixture. Toss to coat.

4. Garnish with the parsley and toasted pine nuts and serve.

MARVELOUS MUSHROOMS

YIELD: 4 SERVINGS / **ACTIVE TIME:** 20 MINUTES / **TOTAL TIME:** 30 MINUTES

There are many kinds of mushrooms available, and you can mix and match them as you desire. Sautéing mushrooms in the skillet with lots of butter yields a rich, earthy stew that is delicious with steak and potatoes. Or simply serve these mushrooms as a topping for burgers or baked polenta with cheese.

1. Place a 12-inch cast-iron skillet over medium-high heat. Add the butter. When melted, add the mushrooms. Cook, while stirring, until the mushrooms begin to soften, about 5 minutes. Reduce the heat to low and let the mushrooms simmer, stirring occasionally, until they cook down, about 15 to 20 minutes.

2. Add the vermouth and stir, then season with salt and pepper. Simmer until the mushrooms are tender. Serve hot.

INGREDIENTS:

- 6 TABLESPOONS UNSALTED BUTTER, CUT INTO SMALL PIECES
- 1 LB. MUSHROOMS, SLICED
- 1 TEASPOON DRY VERMOUTH

SALT AND PEPPER, TO TASTE

SPINACH & SHALLOTS

YIELD: 6 TO 8 SERVINGS / **ACTIVE TIME:** 10 MINUTES / **TOTAL TIME:** 10 MINUTES

Using mellow-flavored shallots instead of the usual garlic and onions keeps the spinach flavor bright in this quick-cooking dish. A splash of balsamic vinegar takes it over the top.

1. Place a 12-inch cast-iron skillet over medium-high heat. Add the olive oil and shallots and cook, while stirring, until shallots are translucent, about 2 minutes.

2. Add the spinach and cook, while stirring, until the leaves are covered by the oil and shallots, about 2 or 3 minutes. The spinach will start to wilt quickly. Reduce the heat and keep stirring so none of it burns. If desired, you can turn the heat to low and cover the skillet so the spinach steams.

3. When the spinach leaves are wilted and still bright green, splash them with the balsamic vinegar, shaking the pan to distribute. Season with salt and pepper and serve.

TIP: This dish works best with more mature spinach. Reserve baby spinach greens for salads and use the larger leaves for this dish. If you prefer a less onion-y dish, use two shallots instead of four.

INGREDIENTS:

- 3 TABLESPOONS OLIVE OIL
- 4 LARGE SHALLOTS, SLICED THIN
- 2 LBS. FRESH SPINACH, STEMMED, RINSED, AND THOROUGHLY DRIED
- 1 TABLESPOON BALSAMIC VINEGAR
- SALT AND PEPPER, TO TASTE

CREAMY SUCCOTASH

YIELD: 8 TO 10 SERVINGS / **ACTIVE TIME:** 30 MINUTES / **TOTAL TIME:** 1 HOUR

Take advantage of the season when fresh corn is plentiful to create this cookout classic. This uses a lot of corn and is a nice alternative to corn on the cob.

1. Bring a medium pot of salted water to a boil over high heat. Add the lima beans and reduce the heat. Cook until they are al dente, about 5 minutes. Drain and set aside.

2. Standing each ear up in the middle of a large baking dish, use a sharp knife to cut down the sides and remove all the kernels. With the kernels off, take the blade of a dull knife and press it along each side of the ears to "milk" the cob of its liquid. Discard the milked cobs.

3. Place the bacon in a nonstick skillet and cook over medium heat until it is crispy, about 8 minutes. Transfer to a paper towel–lined plate and let it drain. When cool enough to handle, chop the bacon into bite-sized pieces.

4. Place a 12-inch cast-iron skillet over medium heat. When hot, lower the heat and add the butter so it melts slowly. When melted, add the corn kernels and "milk" from the cobs and stir to coat the kernels with the butter. Increase the heat to medium-high and add the water and whole milk. Bring to a boil, while stirring constantly, and then reduce heat to low. Add the lima beans, salt, and pepper.

5. Add the flour, cherry tomatoes, and bacon pieces to the skillet, stir to incorporate, and cook over low heat until the sauce thickens. If it gets too thick, add some more whole milk. Serve hot.

TIP: You'll want to freeze some of this to enjoy in the dead of winter. It's easy. Allow the succotash to cool, put it in airtight containers, being sure to push all the air out, seal the container, and place in the freezer. Put the date it was cooked on the container so you remember.

INGREDIENTS:

- 4 CUPS FRESH OR FROZEN LIMA BEANS
- SALT AND FRESHLY GROUND BLACK PEPPER, TO TASTE
- 12 EARS OF CORN, SHUCKED AND RINSED
- ½ LB. THICK-CUT BACON
- 3 TABLESPOONS UNSALTED BUTTER
- 1 CUP WATER, AT ROOM TEMPERATURE
- 1 CUP WHOLE MILK, PLUS MORE AS NEEDED
- 3 TABLESPOONS ALL-PURPOSE FLOUR
- 1 CUP CHERRY TOMATOES, HALVED

ASPARAGUS THREE WAYS, PLUS TWO SAUCES

YIELD: 4 SERVINGS / **ACTIVE TIME:** 10 MINUTES / **TOTAL TIME:** 10 MINUTES

Steamed or grilled asparagus is amenable to many different sauces and dressings. The most traditional would be hollandaise, which adds a rich, lemony accompaniment to the sweet spears. Another option is the nutty tang of tahini dressing. As there are so many options, I've given the basics for cooking asparagus three ways, each as delectable as the next.

1. To prepare any asparagus, rinse well under cold water.

2. Take a spear and bend it close to the end that is opposite the pointy tip; it will snap off at the point where it starts to be too fibrous and tough to eat. Discard any fibrous ends, or reserve them for another preparation.

3. To blanch asparagus, put the spears in salted, boiling water for about 3 minutes, or just tender. Transfer immediately to an ice-water bath to retain the green color.

4. To steam asparagus, arrange the spears in a steaming tray, place the tray above 1 inch of boiling water, and steam for roughly 5 minutes. Transfer immediately to an ice-water bath to retain the green color.

5. To grill asparagus, preheat your grill to medium-high heat. Put the asparagus in a bowl and drizzle on some oil and salt. Toss to coat. Put the asparagus on the grill and cook until it just starts to char, about 4 minutes. Turn, cook the other side for another 4 minutes, then transfer to a plate.

6. Serve any of these preparations with your sauce of choice and season to taste.

INGREDIENTS:

- 1½ LBS. ASPARAGUS
- SALT, TO TASTE
- OLIVE OIL, TO TASTE
- BLENDER HOLLANDAISE (SEE RECIPE), FOR SERVING (OPTIONAL)
- TAHINI DRESSING (SEE RECIPE), FOR SERVING (OPTIONAL)

BLENDER HOLLANDAISE

- 3 LARGE EGG YOLKS
- ¼ TEASPOON KOSHER SALT
- 2 TABLESPOONS FRESH LEMON JUICE
- 1 STICK UNSALTED BUTTER

BLENDER HOLLANDAISE

1. Place the egg yolks, salt, and lemon juice in a blender and turn on high for a few seconds.

2. Melt the butter in a small saucepan over medium-low heat, being careful not to let it brown.

3. While the butter is hot, turn on the blender and, with the top off, slowly drizzle the hot butter into the eggs until fully emulsified.

4. Taste and adjust seasoning if necessary.

TAHINI DRESSING

1. Place the tahini, garlic, oil, and honey in a blender and puree until smooth.

2. Whisk in the lemon juice. If the sauce is too thick, add some water a spoonful at a time.

3. Place the sesame seeds in a small pan and toast over medium heat. Cook for about 3 minutes.

4. Season the dressing with salt and pepper, top with the toasted sesame seeds, and serve.

INGREDIENTS:

TAHINI DRESSING

- ¼ CUP TAHINI
- 1 GARLIC CLOVE, CRUSHED
- 2 TABLESPOONS OLIVE OIL
- 1 TEASPOON HONEY
- JUICE OF 1 LEMON
- WATER, AS NEEDED
- 1 TABLESPOON SESAME SEEDS, FOR GARNISH
- SALT AND BLACK PEPPER, TO TASTE

SIDES, SALADS & ACCOMPANIMENTS | 669

SIMPLE STIR-FRIED BOK CHOY

YIELD: 2 SERVINGS / **ACTIVE TIME:** 10 MINUTES / **TOTAL TIME:** 10 MINUTES

This is sweet and delicate; the perfect accompaniment to any main dish. Adding a splash of mirin at the end of cooking brings out the sweetness of the vegetable without overwhelming it.

1. Place the oil in a small pan and warm over medium-high heat. When it starts to shimmer, add the bok choy and sauté until the green part of the cabbage has wilted, about 5 minutes.

2. Add garlic and cook for 2 minutes, then add the mirin and soy sauce, stir to combine, and cook for 1 more minute.

3. Season with salt and serve.

INGREDIENTS:

- 1 TABLESPOON OLIVE OIL
- ½ LB. BOK CHOY, SLICED
- 2 GARLIC CLOVES, MINCED
- 1 TABLESPOON MIRIN
- 1 TEASPOON SOY SAUCE
- SALT, TO TASTE

KIMCHI

YIELD: 4 CUPS / **ACTIVE TIME:** 30 MINUTES / **TOTAL TIME:** 3 TO 7 DAYS

Simple and versatile, kimchi is the perfect introduction to all that fermentation has to offer.

1. Place the cabbage and salt in a large bowl and stir to combine. Work the mixture with your hands, squeezing to remove as much liquid as possible. Let the mixture rest for 2 hours.

2. Add the remaining ingredients, except for the water. Stir the mixture until well combined and squeeze to remove as much liquid as possible.

3. Transfer the mixture to a container and press down so it is tightly packed. The liquid should be covering the mixture. If it is not, add water until the mixture is covered.

4. Cover the jar and let the mixture sit at room temperature for 3 to 7 days, removing the lid daily to release the gas that has built up. When the taste is to your liking, store in an airtight container in the refrigerator.

INGREDIENTS:

- 1 HEAD NAPA CABBAGE, CUT INTO STRIPS
- ½ CUP KOSHER SALT
- 2 TABLESPOONS MINCED GINGER
- 2 TABLESPOONS MINCED GARLIC
- 1 TEASPOON SUGAR
- 5 TABLESPOONS RED PEPPER FLAKES
- 3 BUNCHES SCALLIONS, TRIMMED AND SLICED

  FILTERED WATER, AS NEEDED

ROASTED BRUSSELS SPROUTS WITH BACON, BLUE CHEESE & PICKLED RED ONION

YIELD: 4 TO 6 SERVINGS / **ACTIVE TIME:** 15 MINUTES / **TOTAL TIME:** 40 MINUTES

Brussels sprouts have a bad reputation with a lot of folks, but when seared and seasoned well, their savory, nutty flavor is a revelation, able to go toe-to-toe with rich ingredients like bacon and blue cheese.

1. Place the vinegar, water, sugar, and salt in a saucepan and bring to a boil. Place the onion in a bowl and pour the boiling liquid over the slices. Cover and allow to cool completely.

2. Place the bacon in a large sauté pan over medium heat and cook, stirring occasionally, until crisp, about 8 minutes. Transfer to a paper towel–lined plate and leave the rendered fat in the pan.

3. Place the Brussels sprouts in the pan cut-side down, season with salt and pepper, and cook over medium heat until they are a deep golden brown, about 7 minutes.

4. Transfer the Brussels sprouts to a platter, top with the pickled onions, bacon, and blue cheese, and serve.

INGREDIENTS:

- 1 CUP CHAMPAGNE VINEGAR
- 1 CUP WATER
- ½ CUP SUGAR
- 2 TEASPOONS SALT, PLUS MORE TO TASTE
- 1 SMALL RED ONION, SLICED
- ½ LB. BACON, CUT INTO 1-INCH PIECES
- 1½ LBS. BRUSSELS SPROUTS, TRIMMED AND HALVED
- BLACK PEPPER, TO TASTE
- 4 OZ. BLUE CHEESE, CRUMBLED

FRIED BRUSSELS SPROUTS WITH MAPLE-CIDER GLAZE

YIELD: 2 TO 4 SERVINGS / **ACTIVE TIME:** 10 MINUTES / **TOTAL TIME:** 15 MINUTES

If you balk at the idea of brussels sprouts, give these deep-fried, maple-glazed ones a shot.

1. Place the maple syrup, vinegar, apple cider, and a pinch of salt in a medium saucepan and cook over medium heat, stirring constantly, until reduced it has by ¼. Remove from heat and set aside.

2. Add oil to a Dutch oven until it is about 3 inches deep. Warm over medium-high heat until it is 350°F or a scrap of Brussels sprout sizzles upon contact. Place the Brussels sprouts in the oil and fry until they are browned, 1 to 2 minutes. Transfer to a paper towel–lined plate to drain.

3. Place the Brussels sprouts in a bowl, season with salt, and add 1 tablespoon of the glaze for every cup of Brussels sprouts. Toss until evenly coated and serve.

TIP: If you prefer not to deep-fry the Brussels sprouts, toss them with oil and salt and roast at 375°F for 20 minutes, until the Brussels sprouts are tender but still have a crunch to them.

INGREDIENTS:

- ¾ CUP REAL MAPLE SYRUP
- ½ CUP APPLE CIDER VINEGAR
- ½ CUP APPLE CIDER
- SALT, TO TASTE
- VEGETABLE OIL, FOR FRYING
- 1 LB. BRUSSELS SPROUTS, TRIMMED AND HALVED

STOVE-TOP BRUSSELS SPROUTS

YIELD: 4 SERVINGS / **ACTIVE TIME:** 10 MINUTES / **TOTAL TIME:** 15 MINUTES

Here is a quick and delicious method for Brussels sprouts.

INGREDIENTS:

- 1 LB. BRUSSELS SPROUTS, TRIMMED AND HALVED
- WATER, AS NEEDED
- OLIVE OIL, AS NEEDED
- SALT AND PEPPER, TO TASTE

1. Warm a wide sauté pan over high heat. When it begins to smoke, add all of the Brussels sprouts and a few tablespoons of water. Place the lid on and steam for 2 minutes.

2. Remove the lid and add enough oil to coat the bottom of the pan. Reduce heat to medium and let the Brussels sprouts brown, turning them every so often to brown on all sides.

3. Continue cooking until the desired tenderness is achieved, adding more oil if the pan starts to look dry. Season with salt and pepper and serve.

GRILLED CABBAGE

YIELD: 4 SERVINGS / **ACTIVE TIME:** 15 MINUTES / **TOTAL TIME:** 45 MINUTES

This deceptively simple preparation of grilled cabbage results in a mellow, tasty side. You could even brush the wedges with oil and place them directly on the grill for a few minutes before placing them in aluminum foil.

1. Preheat your grill to medium heat and cut the head of cabbage into 8 wedges.

2. Remove the core and place the wedges on a large piece of aluminum foil. Season with the garlic powder, salt, and pepper. Create a packet by folding the foil over and crimping the edges.

3. When the grill is about 400°F, place the packet on the grill, cover the grill, and cook until tender, 30 to 40 minutes. Remove from the packet and serve immediately.

INGREDIENTS:

- 1 LARGE HEAD CABBAGE
- 1½ TEASPOONS GARLIC POWDER
- SALT AND PEPPER, TO TASTE

BASIC RED CABBAGE SLAW

YIELD: 2 TO 4 SERVINGS / **ACTIVE TIME:** 10 MINUTES / **TOTAL TIME:** 2 TO 3 HOURS

This is a topper that should be made a few hours ahead of time to give the cabbage time to soften. Once it is ready, it works on top of tacos and is a nice complement to grilled chicken and steak.

1. Place the cabbage in a large bowl, sprinkle the salt on top, and toss to distribute. Use your hands to work the salt into the cabbage, then let it sit for 2 to 3 hours.

2. Once it has rested, taste to gauge the saltiness: if too salty, rinse under cold water and let drain; if just right, add the lime juice and cilantro, stir to combine, and serve.

INGREDIENTS:

- 1 SMALL RED CABBAGE, CORED AND SLICED AS THINLY AS POSSIBLE
- 1 TEASPOON KOSHER SALT, PLUS MORE TO TASTE
-   JUICE OF 1 LIME
- 1 BUNCH FRESH CILANTRO, CHOPPED

MELON, CUCUMBER & PROSCIUTTO SALAD WITH MINT VINAIGRETTE

YIELD: 4 TO 6 SERVINGS / **ACTIVE TIME:** 15 MINUTES / **TOTAL TIME:** 40 MINUTES

The versatile melon can comfortably straddle the sweet-savory divide. Here it pairs up with crispy, cured prosciutto and creamy feta to carry this dynamic salad.

1. Preheat the oven to 350°F.

2. Place the prosciutto on a parchment-lined baking sheet. Cover with another sheet of parchment and place another baking sheet that is the same size on top. Place in the oven and bake until the prosciutto is crisp, about 12 minutes. Remove from the oven and let cool. When the prosciutto is cool enough to handle, chop it into bite-sized pieces.

3. Place the cantaloupe, honeydew melon, and cucumber in a salad bowl, season with salt and pepper, and toss to combine. Add the jalapeño and vinaigrette and toss until evenly coated. Plate the salad, top with the chopped prosciutto and feta, and garnish with the mint leaves.

MINT VINAIGRETTE

1. Place all of the ingredients in a mixing bowl and whisk until thoroughly combined.

INGREDIENTS:

8	SLICES PROSCIUTTO
3	CUPS DICED CANTALOUPE
3	CUPS DICED HONEYDEW MELON
1	CUCUMBER, SLICED
	SALT AND PEPPER, TO TASTE
1	JALAPEÑO PEPPER, STEMMED, SEEDED TO TASTE, AND SLICED
	MINT VINAIGRETTE (SEE RECIPE), TO TASTE
⅔	CUP CRUMBLED FETA CHEESE
	FRESH MINT LEAVES, CHOPPED, FOR GARNISH

MINT VINAIGRETTE

3	TABLESPOONS CHOPPED FRESH MINT
¼	CUP OLIVE OIL
3	TABLESPOONS APPLE CIDER VINEGAR
1	TABLESPOON HONEY
2	TEASPOONS DICED SHALLOT
1	TEASPOON KOSHER SALT
¼	TEASPOON BLACK PEPPER

THAI BEEF & CUCUMBER SALAD

YIELD: 2 SERVINGS / **ACTIVE TIME:** 15 MINUTES / **TOTAL TIME:** 1 HOUR 30 MINUTES

When the weather is hot, even the grill seems too much to bear. This is a light but filling salad, and it is a perfect use for leftover roast beef or steak, or deli-sliced roast beef. The whole recipe requires no cooking save for boiling water for the noodles.

1. Bring 6 cups of water to a boil in a medium saucepan and place the noodles in a baking pan. Pour the water over the noodles and let sit until tender, about 20 minutes. Drain well and place them in a bowl.

2. Add the remaining ingredients, except for the sesame seeds, and toss to combine. Chill in the refrigerator for 1 hour, garnish with the sesame seeds, and serve.

INGREDIENTS:

- 2 OZ. PACKAGE MUNG BEAN OR THIN RICE NOODLES
- 1 CARROT, PEELED AND GRATED
- 1 SMALL CUCUMBER, SEEDED AND DICED
- ZEST AND JUICE OF 1 LIME
- 10 FRESH MINT LEAVES, CHOPPED
- 1–2 TABLESPOONS SOY SAUCE
- 1 TEASPOON PALM SUGAR OR MAPLE SYRUP
- ½ TEASPOON KOSHER SALT
- 1 TABLESPOON THAI FISH SAUCE
- 2–4 OZ. THINLY SLICED ROAST BEEF OR LEFTOVER STEAK, TORN INTO BITE-SIZED PIECES
- RICE VINEGAR, TO TASTE
- HOT SAUCE, TO TASTE
- SESAME SEEDS, FOR GARNISH

ROASTED BABY BEET, RADISH & APPLE SALAD WITH BLUE CHEESE MOUSSE

YIELD: 4 TO 6 SERVINGS / **ACTIVE TIME:** 30 MINUTES / **TOTAL TIME:** 1 HOUR AND 20 MINUTES

Baby beets are picked in the spring to thin the field and leave room for other beets to grow, an early harvest that results in a rich and delicate flavor. When you're working with something so uniquely delicious, it's important to keep it simple, and roasting these beets with a few aromatics is all they require.

1. Preheat the oven to 400°F.

2. Form three sheets of aluminum foil into pouches. Group the beets according to color and place each group into a pouch. Drizzle each with the oil and sprinkle with salt. Divide the whole sprigs of thyme, garlic, and water between the pouches and seal them. Place the pouches on a baking sheet, place in the oven, and cook until fork-tender, 45 minutes to 1 hour depending on the size of the beets. Remove the pouches from the oven and let cool. When cool enough to handle, peel the beets, cut into bite-sized pieces, and set aside.

3. Bring a pot of salted water to a boil and prepare an ice-water bath in a mixing bowl. Remove the greens from the radishes, wash them thoroughly, and set aside. Quarter the radishes.

4. Place the radishes in the boiling water, cook for 1 minute, and then transfer to the water bath until completely cool. Drain and set aside.

5. Place the blue cheese, heavy cream, ricotta, and thyme leaves in a food processor and puree until smooth. Set the mousse aside.

6. Place the beets, except for the red variety, in a salad bowl. Add the radishes, radish greens, and apples and toss to combine. Add half of the vinaigrette, season with salt and pepper, and toss to coat.

7. Spread the mousse on the serving dishes. Place the salad on top, sprinkle the red beets over the salad, drizzle with the remaining vinaigrette, and garnish with the honeycomb.

HONEY MUSTARD VINAIGRETTE

1. Place all of the ingredients, except for the oil, in a small mixing bowl and whisk to combine. Add the oil in a slow stream and whisk until incorporated.

INGREDIENTS:

- 9 BABY BEETS (3 EACH OF RED, GOLDEN, AND PINK)
- 3 TABLESPOONS OLIVE OIL
- 1 TABLESPOON KOSHER SALT, PLUS MORE TO TASTE
- 9 SPRIGS FRESH THYME, 6 LEFT WHOLE, LEAVES REMOVED FROM 3
- 6 GARLIC CLOVES
- 6 TABLESPOONS WATER
- 8 RADISHES WITH TOPS
- ¾ CUP BLUE CHEESE, AT ROOM TEMPERATURE
- ½ CUP HEAVY CREAM
- ½ CUP RICOTTA CHEESE, AT ROOM TEMPERATURE
- 2 APPLES, PEELED, CORED, AND DICED
- ¼ CUP HONEY MUSTARD VINAIGRETTE (SEE RECIPE)
- BLACK PEPPER, TO TASTE
- 2 OZ. HONEYCOMB, FOR GARNISH

HONEY MUSTARD VINAIGRETTE

- ¼ CUP HONEY
- 2 TABLESPOONS WHOLE GRAIN MUSTARD
- 3 TABLESPOONS APPLE CIDER VINEGAR
- 1 TEASPOON KOSHER SALT
- ½ TEASPOON BLACK PEPPER
- ⅓ CUP OLIVE OIL

ROASTED BRASSICA SALAD WITH PICKLED RAMPS & BUTTERMILK CAESAR DRESSING

YIELD: 4 TO 6 SERVINGS / **ACTIVE TIME:** 20 MINUTES / **TOTAL TIME:** 35 MINUTES

Broccoli, Brussels sprouts, and cauliflower are only a few of the fine members of the brassica family. Charring them brings out their sweet side, which pairs wonderfully with the creamy and slightly acidic buttermilk dressing.

1. Bring a large pot of salted water to a boil. Add the cauliflower, cook for 1 minute, remove with a slotted spoon, and transfer to a paper towel–lined plate. Wait for the water to return to a boil, add the broccoli, and cook for 30 seconds. Use a slotted spoon to remove the broccoli and let the water drip off before transferring it to the paper towel–lined plate.

2. Place the oil and Brussels sprouts, cut-side down, in a large cast-iron skillet. Add the broccoli and cauliflower, season with salt and pepper, and cook over high heat without moving the vegetables. Cook until charred, turn over, and cook until charred on that side. Remove and transfer to a bowl.

3. Add the Pickled Ramps and Buttermilk Caesar Dressing to the bowl and toss to evenly coat. Garnish with Parmesan cheese and red pepper flakes and serve.

PICKLED RAMPS

1. Place all of the ingredients, except for the ramps, in a small saucepan and bring to a boil over medium heat.

2. Add the ramps, reduce heat, and simmer for 1 minute. Transfer to a mason jar, cover with aluminum foil, and let cool completely. Once cool, cover with a lid and store in the refrigerator for up to 1 week.

BUTTERMILK CAESAR DRESSING

1. Place all of the ingredients in a food processor and puree until combined. Season to taste and serve.

INGREDIENTS:

- 1 SMALL HEAD CAULIFLOWER, TRIMMED AND CUT INTO BITE-SIZED PIECES
- 1 HEAD BROCCOLI, CUT INTO FLORETS
- ¼ CUP OLIVE OIL
- ¼ LB. BRUSSELS SPROUTS, TRIMMED AND HALVED
- SALT AND PEPPER, TO TASTE
- 10 PICKLED RAMPS (SEE RECIPE)
- BUTTERMILK CAESAR DRESSING (SEE RECIPE)
- PARMESAN CHEESE, GRATED, FOR GARNISH
- RED PEPPER FLAKES, FOR GARNISH

PICKLED RAMPS

- ½ CUP CHAMPAGNE VINEGAR
- ½ CUP WATER
- ¼ CUP SUGAR
- 1½ TEASPOONS KOSHER SALT
- ¼ TEASPOON FENNEL SEEDS
- ¼ TEASPOON CORIANDER SEEDS
- ⅛ TEASPOON RED PEPPER FLAKES
- 10 SMALL RAMP BULBS

BUTTERMILK CAESAR DRESSING

- 1 LARGE GARLIC CLOVE, MINCED
- 2 ANCHOVY FILLETS
- ⅔ CUP MAYONNAISE
- ¼ CUP BUTTERMILK
- ¼ CUP GRATED PARMESAN CHEESE
- ZEST OF 1 LEMON
- 1 TEASPOON WORCESTERSHIRE SAUCE
- 1 TEASPOON KOSHER SALT, PLUS MORE TO TASTE
- ½ TEASPOON BLACK PEPPER, PLUS MORE TO TASTE

SHAVED BRUSSELS SPROUTS & KALE SALAD WITH BLOOD ORANGE VINAIGRETTE

YIELD: 4 TO 6 SERVINGS / **ACTIVE TIME:** 10 MINUTES / **TOTAL TIME:** 25 MINUTES

Brussels sprouts are as delicious raw as they are cooked, and pairing their robust, savory flavor with bright citrus is the perfect way to play up that attribute.

1. Place the bacon in a sauté pan and cook over medium heat until crisp, about 8 minutes. Transfer to a paper towel–lined plate to drain. When cool enough to handle, chop into bite-sized pieces.

2. Remove the skin from the segments of blood orange and cut each segment in half. Place in a mixing bowl, add the Brussels sprouts and kale, season with salt and pepper, and toss to combine. Add the Blood Orange Vinaigrette, toss to evenly coat, and season to taste.

3. Plate the salad, top with the bacon, garnish with the toasted pecans and Parmesan cheese, and serve with the remaining vinaigrette on the side.

BLOOD ORANGE VINAIGRETTE

1. Place all of the ingredients, except for the oil, in a blender. Puree on high and add the oil in a slow stream. Puree until the mixture has emulsified and season to taste.

INGREDIENTS:

- ½ LB. BACON, SLICED
- 3 BLOOD ORANGES, PEELED
- 1 LB. BRUSSELS SPROUTS, TRIMMED AND SLICED VERY THIN WITH A MANDOLINE
- 2 CUPS PACKED BABY KALE
- SALT AND PEPPER, TO TASTE
- ⅔ CUP BLOOD ORANGE VINAIGRETTE (SEE RECIPE)
- ½ CUP TOASTED PECANS, FOR GARNISH
- PARMESAN CHEESE, SHAVED, FOR GARNISH

BLOOD ORANGE VINAIGRETTE

- ½ CUP BLOOD ORANGE JUICE (ABOUT 2 BLOOD ORANGES)
- ½ TEASPOON KOSHER SALT
- ¼ TEASPOON BLACK PEPPER
- 1½ TABLESPOONS APPLE CIDER VINEGAR
- 1 TABLESPOON HONEY
- 1 ICE CUBE
- 1 CUP OLIVE OIL

MIDSUMMER CORN & BEAN SALAD

YIELD: 4 TO 6 SERVINGS / **ACTIVE TIME:** 15 MINUTES / **TOTAL TIME:** 12 HOURS

This is a great make-ahead recipe when the local corn is ripe. In fact, if it is really fresh with great flavor, you can skip the cooking part all together and use raw kernels. I had some lovely dried beans on hand, but if you are pressed for time, use canned white or black beans. The maple syrup is meant to accentuate the sweetness of the corn, so add according to your personal preference.

1. Place the oil in a wide sauté pan, add the corn, and cook over medium-high heat until slightly brown, about 5 minutes. Remove from heat and let cool.

2. Drain the beans and place in a saucepan. Cover with water. Bring to a boil, reduce heat to a simmer, and cook until the beans are tender, about 45 minutes. Drain and cool.

3. Place all of the ingredients in a salad bowl, toss to combine, and chill in the refrigerator for 2 hours.

4. Taste, adjust seasoning as needed, and serve.

INGREDIENTS:

- 1 TABLESPOON OLIVE OIL
- 4 CUPS CORN KERNELS (PREFERABLY FRESH)
- ½ CUP DRIED BEANS, SOAKED OVERNIGHT
- 1 SMALL RED BELL PEPPER, DICED
- 1 SMALL GREEN BELL PEPPER, DICED
- ½ RED ONION, DICED
- JUICE OF ½ LIME
- 1 TEASPOON CUMIN
- TABASCO, TO TASTE
- 3 TABLESPOONS CHOPPED FRESH CILANTRO
- 1 TABLESPOON MAPLE SYRUP, PLUS MORE TO TASTE
- SALT AND PEPPER, TO TASTE

YU CHOY WITH GARLIC & SOY

YIELD: 4 SERVINGS / **ACTIVE TIME:** 10 MINUTES / **TOTAL TIME:** 15 MINUTES

Steaming yu choy keeps it tender and light. If the stalks are large, leave them to cook a little longer.

1. Place the yu choy in a sauté pan large enough to fit all of the stalks, cover with the water, cover the pan, and cook over high heat.

2. After about 5 minutes, check the thickest stalk to see if it is tender. If not, cook until it is. Once tender, add the oil and the garlic. Sauté until the garlic is fully cooked but not browned, about 2 minutes.

3. Add the vinegar and soy sauce, toss to combine, and serve.

INGREDIENTS:

- 1½ LBS. YU CHOY (IF ESPECIALLY LONG, CUT THEM IN HALF)
- ¼ CUP WATER
- 1 TABLESPOON OLIVE OIL
- 2 GARLIC CLOVES, CHOPPED
- ½ TABLESPOON RICE VINEGAR
- 1 TABLESPOON SOY SAUCE

YU CHOY WITH BLACK BEAN GARLIC SAUCE & EXTRA GARLIC

YIELD: 4 SERVINGS / **ACTIVE TIME:** 15 MINUTES / **TOTAL TIME:** 20 MINUTES

Black bean garlic sauce is made from fermented black beans and soy sauce, and you can find it in Asian markets or in the Asian section of most grocery stores. It is perfect with steamed yu choy because a spoonful makes for an intense, instant sauce.

1. Place the yu choy in a sauté pan large enough to fit all the greens, add the water, cover the pan, and cook over high heat.

2. After about 5 minutes, remove the lid and cook until most of the water cooks off.

3. Add the oil and garlic and stir-fry until the garlic is fragrant, about 2 minutes.

4. Add the black bean garlic sauce, stir to coat, and cook until heated through. Serve immediately.

INGREDIENTS:

- 1½ LBS. YU CHOY, CHOPPED INTO 3-INCH PIECES
- ¼ CUP WATER
- ½ TABLESPOON OLIVE OIL
- 1 GARLIC CLOVE, MINCED
- 1 TABLESPOON BLACK BEAN GARLIC SAUCE

ZUCCHINI WITH TOMATOES, FETA, GARLIC & LEMON

YIELD: 4 SERVINGS / **ACTIVE TIME:** 25 MINUTES / **TOTAL TIME:** 45 MINUTES

The key to this dish is to not crowd the squash in the pan; otherwise, they will steam instead of brown. You can substitute yellow summer squash or pattypan squash, just cut them into pieces of a similar size so they cook at the same rate.

1. Place the oil in a large sauté pan and warm over medium-high heat. When it starts to shimmer, add the zucchini, making sure not to overcrowd the pan. Let the squash brown, flip, then brown on the other sides. Season with salt and pepper. If it is necessary to cook in batches, set the browned zucchini aside and repeat, adding oil if the pan starts to look dry.

2. Return the zucchini to the pan and add the garlic. Cook until the garlic starts to soften, about 3 minutes. Add the tomato, cook for 1 more minute to heat everything through, and transfer the mixture to a platter.

3. Sprinkle the feta and parsley on top, season with lemon juice, and serve.

INGREDIENTS:

- 1 TABLESPOON OLIVE OIL, PLUS MORE AS NEEDED
- 3 ZUCCHINI, CHOPPED
- SALT AND PEPPER, TO TASTE
- 2 GARLIC CLOVES, CHOPPED
- 1 LARGE TOMATO, DICED
- 2 OZ. FETA CHEESE, CRUMBLED
- 2 TABLESPOONS CHOPPED FRESH PARSLEY
- FRESH LEMON JUICE, TO TASTE

RUSTIC WHOLE WHEAT BREAD

YIELD: 1 LOAF / **ACTIVE TIME:** 30 MINUTES / **TOTAL TIME:** 21 HOURS

Bread making is a delicate art, as the wrong measurements can lead to a flat loaf and a disappointed baker. That being said, this whole wheat masterpiece will leave no one disappointed, especially when it's served while still warm with plenty of farm-fresh butter.

1. Place the flours and water in a large mixing bowl and use your hands to combine the mixture into a dough. Cover the bowl with a kitchen towel and let the mixture set for 45 minutes to 1 hour.

2. Sprinkle the yeast and salt over the dough and fold until they have been incorporated. Cover the bowl with the kitchen towel and let stand for 30 minutes. Remove the towel, fold a corner of the dough into the center, and cover. Repeat every 30 minutes until all of the corners have been folded in.

3. After the last fold, cover the dough with the kitchen towel and let it sit for 12 to 14 hours.

4. Dust a work surface lightly with flour and place the dough on it. Fold each corner of the dough into the center, flip the dough over, and roll it into a smooth ball. Dust your hands with flour as needed. Be careful not to roll or press the dough too hard, as this will prevent the dough from expanding properly. Dust a bowl with flour and place the dough, seam side down, in the bowl. Let stand until it has roughly doubled in size, about 1 hour and 15 minutes.

5. Cut a round piece of parchment paper that is 1 inch larger than the circumference of your cast-iron Dutch oven. When the dough has approximately 1 hour left in its rise (this is also known as "proofing"), preheat the oven to 475°F and place the covered Dutch oven in the oven as it warms.

6. When the dough has roughly doubled in size, invert it onto a lightly floured work surface. Use a very sharp knife to score one side of the loaf. Using oven mitts, remove the Dutch oven from the oven. Use a bench scraper to transfer the dough onto the piece of parchment, scored side up. Hold the sides of the parchment and carefully lower the dough into the Dutch oven. Cover the Dutch oven and place it in the oven for 20 minutes.

7. Remove the lid and bake the loaf for an additional 20 minutes. Remove from the oven and let cool on a wire rack for at least 2 hours before slicing.

INGREDIENTS:

- 3¼ CUPS ALL-PURPOSE FLOUR, PLUS MORE FOR DUSTING
- 1¼ CUPS WHOLE WHEAT FLOUR
- 1½ CUPS WATER (90°F)
- JUST UNDER ¼ TEASPOON ACTIVE DRY YEAST
- 2¼ TEASPOONS SALT

RUSTIC WHITE BREAD

YIELD: 1 LOAF / **ACTIVE TIME:** 30 MINUTES / **TOTAL TIME:** 21 HOURS

Don't be thrown by the "white" in the name. Letting the dough rest overnight allows an incredible amount of flavor to develop, resulting in a loaf that is anything but bland.

INGREDIENTS:

- 4½ CUPS ALL-PURPOSE FLOUR, PLUS MORE FOR DUSTING
- 1½ CUPS WATER (90°F)
- JUST UNDER ¼ TEASPOON ACTIVE DRY YEAST
- 2¼ TEASPOONS SALT

1. Place the flour and water in a large mixing bowl and use your hands to combine the mixture into a dough. Cover the bowl with a kitchen towel and let the mixture set for 45 minutes to 1 hour.

2. Sprinkle the yeast and salt over the dough and fold until they have been incorporated. Cover the bowl with the kitchen towel and let stand for 30 minutes. Remove the towel, fold a corner of the dough into the center, and cover. Repeat every 30 minutes until all of the corners have been folded in.

3. After the last fold, cover the dough with the kitchen towel and let it sit for 12 to 14 hours.

4. Dust a work surface lightly with flour and place the dough on it. Fold each corner of the dough to the center, flip the dough over, and roll it into a smooth ball. Dust your hands with flour as needed. Be careful not to roll or press the dough too hard, as this will prevent the dough from expanding properly. Dust a bowl with flour and place the dough, seam side down, in the bowl. Let stand until it has roughly doubled in size, about 1 hour and 15 minutes.

5. Cut a round piece of parchment paper that is 1 inch larger than the circumference of your cast-iron Dutch oven. When the dough has approximately 1 hour left in its rise (this is also known as "proofing"), preheat the oven to 475°F and place the covered Dutch oven in the oven as it warms.

6. When the dough has roughly doubled in size, invert it onto a lightly floured work surface. Use a very sharp knife to score one side of the loaf. Using oven mitts, remove the Dutch oven from the oven. Use a bench scraper to transfer the dough onto the piece of parchment, scored side up. Hold the sides of the parchment and carefully lower the dough into the Dutch oven. Cover the Dutch oven and place it in the oven for 20 minutes.

7. Remove the lid and bake the loaf for an additional 20 minutes. Remove from the oven and let cool on a wire rack for at least 2 hours before slicing.

SOURDOUGH BREAD

YIELD: 1 LARGE LOAF / **ACTIVE TIME:** 20 MINUTES / **TOTAL TIME:** 30 HOURS

Sourdough is just four ingredients—water, starter, salt, and flour—but it has a complexity that is unmatched. If you have not made real artisan bread before, here is your chance to learn. Be warned: once you try a slice of this bread, you'll never go back to the pre sliced stuff again.

INGREDIENTS:

- 1⅔ CUPS FILTERED WATER (78°F), PLUS 1 TEASPOON
- 5 CUPS BREAD FLOUR, PLUS MORE FOR DUSTING
- ¾ CUP SOURDOUGH STARTER (SEE PAGE 709)
- 1½ TEASPOONS SALT

1. Combine the 1⅔ cups water and the flour in a bowl and stir until no dry clumps remain and the dough has come together slightly. Cover with plastic wrap and let rest for 30 minutes.

2. Add the Sourdough Starter, salt, and the additional teaspoon of water to the dough. Knead for 10 minutes, until the dough is smooth and elastic. Place the dough in a bowl, cover with plastic wrap, and store in a naturally warm place for 4 hours.

3. Place the dough on a flour-dusted work surface and fold the left side of the dough to the right, fold the right side of the dough to the left, and fold the bottom toward the top. Form into a rough ball, return to the bowl, cover with plastic wrap, and let rest for 30 minutes.

4. After 30 minutes, place the ball of dough on a floured surface and repeat the folds made in Step 3. Form the dough into a ball, dust it with flour, and place it in a bowl with the seam facing up. Dust a clean kitchen towel with flour, cover the bowl with it, and place the bowl in the refrigerator overnight.

5. Approximately 2 hours before you are ready to bake the bread, remove it from the refrigerator and allow it to come to room temperature.

6. Preheat oven to 500°F. Place a covered cast-iron Dutch oven in the oven as it warms.

7. When the dough is at room temperature and the oven is ready, remove the Dutch oven from the oven and carefully place the ball of dough into the Dutch oven. Score the top of the dough with a very sharp knife or razor blade, making a long cut across the middle. Cover the Dutch oven, place it in the oven, and bake for 25 minutes.

8. Remove the Dutch oven, lower the oven temperature to 480°F, remove the lid, and bake the bread for another 25 minutes. Remove from the oven and let cool on a wire rack for 2 hours before slicing.

COUNTRY SOURDOUGH BREAD

YIELD: 1 LARGE LOAF / **ACTIVE TIME:** 20 MINUTES / **TOTAL TIME:** 2 DAYS

This sourdough uses a few different flours to give the bread a rustic, wheaty taste. Keep in mind that the hydration level in your levain needs to increase when you bake with whole wheat flour to balance out the additional density.

1. Place 2¾ cups of the water and the flours in a bowl and mix until incorporated. Cover with plastic wrap and let stand for 1 hour.

2. Add the salt, the levain, and the remaining teaspoon of water to the dough. Transfer the dough to a flour-dusted work surface and knead until the dough is smooth and elastic, about 10 minutes.

3. Place the kneaded dough in a bowl and cover with plastic wrap. Place the bowl in a naturally warm spot and let stand for 4 hours.

4. Transfer the dough to a flour-dusted work surface. Fold the left side of the dough to the right, fold the right side of the dough to the left, and fold the bottom toward the top. Form into a rough ball, return to the bowl, cover with plastic wrap, and let rest for 30 minutes.

5. After 30 minutes, place the ball of dough on a floured surface and repeat the folds made in Step 4. Form the dough into a ball, dust it with flour, and place it in a bowl with the seam facing up. Dust a clean kitchen towel with flour, cover the bowl with it, and place the bowl in the refrigerator overnight.

6. Remove the dough from the refrigerator 2 hours before baking and allow it to come to room temperature.

7. Preheat the oven to 500°F. Place a covered cast-iron Dutch oven in the oven as it warms.

8. When the oven is ready, remove the Dutch oven and carefully place the dough into it. Score the top with a very sharp knife or razor blade, making one long cut across the middle. Cover the Dutch oven and place the bread in the oven.

9. Cook for 25 minutes, remove the Dutch oven, and lower the oven's temperature to 480°F. Remove the Dutch oven's cover, return the bread to the oven, and bake for another 25 minutes, until it sounds hollow when tapped.

10. Remove the bread from the oven, transfer to a wire rack, and allow to cool for 2 hours before slicing.

COUNTRY SOURDOUGH LEVAIN

1. Place the ingredients in a mixing bowl and stir until well combined. Cover the bowl and let stand at room temperature for 8 hours.

INGREDIENTS:

- 2¾ CUPS FILTERED WATER (78°F), PLUS 1 TEASPOON
- 4 CUPS BREAD FLOUR, PLUS MORE FOR DUSTING
- 1 CUP WHOLE WHEAT FLOUR
- ½ CUP ORGANIC RYE FLOUR
- 2 TEASPOONS SALT
- 1 CUP COUNTRY SOURDOUGH LEVAIN (SEE RECIPE)

COUNTRY SOURDOUGH LEVAIN

- ½ CUP BREAD FLOUR
- ¼ CUP WHOLE WHEAT FLOUR
- 3⅓ TABLESPOONS ORGANIC RYE FLOUR
- 3 TABLESPOONS SOURDOUGH STARTER (SEE PAGE 709)
- ⅓ CUP FILTERED WATER (85 TO 90°F)

SOURDOUGH STARTER

YIELD: APPROXIMATELY ½ CUP / **ACTIVE TIME:** 1 HOUR / **TOTAL TIME:** 1 WEEK

The success of any sourdough loaf lies in the quality of the starter. While that's a lot of pressure, all you need to handle it is time—the longer the starter has to develop, the more flavor it will impart to the bread.

INGREDIENTS:

- 1 CUP ORGANIC RYE FLOUR
- 7 CUPS WATER
- 6 CUPS BREAD FLOUR

1. Combine the rye flour and 1 cup of the water in a large mason jar or bowl and stir until thoroughly combined. Put in a dark, naturally warm place and let stand for 24 hours.

2. The next day, discard three-quarters of the mixture. Add 1 cup of the water and 1 cup of the bread flour and stir until thoroughly combined. Let stand for 24 hours.

3. Repeat Step 2 for another 5 days. On the fourth or fifth day, the mixture should start to bubble.

4. After 1 week, you will have a viable sourdough starter. Store the starter in the refrigerator. If you will not be baking bread during a particular week, discard three-quarters of the starter and feed what remains with 1 cup water and 1 cup bread flour. When you are ready to make bread, start bulking up the starter 24 hours ahead of time, adding equal parts bread flour and water every 8 hours.

NO-KNEAD BREAD

YIELD: 1 SMALL LOAF / **ACTIVE TIME:** 20 MINUTES / **TOTAL TIME:** 24 HOURS

Use a 7-quart cast-iron Dutch oven for this recipe. This delicious bread is a great way to upgrade a pimento cheese sandwich—there is really nothing easier. Just remember that it takes up to two days to make, so plan ahead!

1. In a large bowl, add the yeast and sugar and top with the warm water. Stir to dissolve the yeast. Cover the measuring cup with plastic wrap and set it aside for about 15 minutes. If the yeast doesn't foam, it is not alive and you'll need to start over.

2. When the yeast is proofed, add the salt and flour. Stir until just blended with the yeast, sugar, and water. The dough will be sticky.

3. Cover the bowl with plastic wrap and set aside for at least 15 hours and up to 18 hours, preferably in a place that's 65 to 70°F.

4. The dough will be bubbly when you go to work with it. Lightly dust a work surface with flour and scoop the dough out onto it. Dust your fingers with flour so they don't stick to the dough. Fold it gently once or twice.

5. Transfer the dough to a clean, room-temperature bowl and cover with a kitchen towel. Let stand until doubled in size, another 1 to 2 hours.

6. While the dough is on its final rise, preheat the oven to 450°F, placing a cast-iron Dutch oven inside with the lid on so it gets hot. When the oven is ready and the dough has risen, carefully remove the lid and gently scoop the dough from the bowl into the Dutch oven. Cover and bake for 20 minutes. Remove the lid and continue to bake for another 25 minutes, until the top is golden and it sounds hollow when tapped.

7. Remove the Dutch oven from the oven and use kitchen towels to carefully transfer bread to a rack or cutting board. Allow to cool at least 20 minutes before serving.

INGREDIENTS:

- ½ TABLESPOON ACTIVE DRY YEAST
- ¼ TEASPOON SUGAR
- 1½ CUPS WATER (110 TO 115°F)
- 1½ TEASPOONS KOSHER SALT
- 3 CUPS ALL-PURPOSE FLOUR, PLUS MORE FOR DUSTING

CLASSIC CORN BREAD

YIELD: 4 TO 6 SERVINGS / **ACTIVE TIME:** 1 HOUR / **TOTAL TIME:** 3 TO 4 HOURS

If you're going to make bread in a cast-iron skillet, you have to make corn bread. In fact, many restaurants now serve corn bread right in a cast-iron pan.

1. In a large bowl, combine the cornmeal, sugar, salt, and boiling water. Stir to combine and let sit for several hours in a cool, dark place or overnight in the refrigerator. Stir occasionally while the batter is resting.

2. When ready to make, preheat oven to 450°F.

3. Add flour, the 1 tablespoon of melted butter, eggs, baking powder, baking soda, and milk to the batter. Stir to thoroughly combine.

4. Heat the skillet over medium-high heat and melt the teaspoon of butter in it. Add the batter.

5. Transfer the skillet to the oven and cook for 15 minutes.

6. Reduce the heat to 250°F and cook another 40 minutes, or until the bread is golden brown on top and set in the center.

INGREDIENTS:

- 4 CUPS FINELY GROUND YELLOW CORNMEAL
- ¾ CUP SUGAR
- 1 TABLESPOON SALT
- 4 CUPS BOILING WATER
- 1 CUP ALL-PURPOSE FLOUR
- 1 TABLESPOON UNSALTED BUTTER, MELTED, PLUS 1 TEASPOON
- 2 EGGS, LIGHTLY BEATEN
- 2 TEASPOONS BAKING POWDER
- 1 TEASPOON BAKING SODA
- 1 CUP WHOLE MILK

CORN TORTILLAS

YIELD: 20 TORTILLAS / **ACTIVE TIME:** 50 MINUTES / **TOTAL TIME:** 50 MINUTES

You really should be making your own corn tortillas, as a warm tortilla lifted straight from a cast-iron griddle or skillet is a thing of beauty. The main ingredient, masa harina, is a corn flour that is available in most grocery stores.

INGREDIENTS:

- 2 CUPS MASA HARINA, PLUS MORE AS NEEDED
- ½ TEASPOON SALT
- 1 CUP WARM WATER (110°F), PLUS MORE AS NEEDED
- 2 TABLESPOONS VEGETABLE OIL OR MELTED LARD

1. Place the masa harina and salt in a bowl and stir to combine. Slowly add the warm water and oil (or lard) and stir until they are incorporated and a soft dough forms. The dough should be quite soft and not at all sticky. If it is too dry, add more water. If the dough is too wet, add more masa harina.

2. Wrap the dough in plastic (or place it in a resealable bag) and let it rest at room temperature for 30 minutes. It can be stored in the refrigerator for up to 24 hours; just be careful not to let it dry out.

3. Cut a 16-inch piece of plastic wrap and lay half of it across the bottom plate of a tortilla press.

4. Place a large cast-iron griddle across two burners and warm over high heat.

5. Pinch off a small piece of the dough and roll it into a ball. Place in the center of the lined tortilla press, fold the plastic over the top of the dough, and press down the top plate to flatten the dough. Do not use too much force. If the tortilla is too thin, you will have a hard time getting it off of the plastic. Open the press and carefully peel off the disk of dough. Reset the plastic.

6. Place the disk on the hot, dry griddle and toast for 30 to 45 seconds. Flip over and cook for another minute. Remove from the griddle and set aside. Repeat the process with the remaining dough.

PARATHA

YIELD: 8 SERVINGS / **ACTIVE TIME:** 25 MINUTES / **TOTAL TIME:** 30 MINUTES

There is something joyful in making flatbreads, don't you think? It harkens back to childhood. The soft dough. Gently rolling it into shape. It's like playing, but the results are edible, which is where the real joy comes in. We should all revel in a chance to play with our food.

1. Place the flours and salt in the bowl of a stand mixer. Turn on low and slowly add the warm water. Mix until incorporated and then slowly add the vegetable oil. When the oil has been incorporated, place the dough on a lightly floured work surface and knead until it is quite smooth, about 8 minutes.

2. Divide the dough into 8 small balls and dust them with flour.

3. Use your hands to roll out each ball into a long rope. Spiral each rope into a large disk.

4. Use a rolling pin to flatten the spiraled disks until they are no more than ¼-inch thick. Lightly brush each disk with a small amount of vegetable oil.

5. Place a cast-iron *tava*, cast-iron skillet, or griddle over very high heat for about 4 minutes. Brush the surface with some of the ghee or melted butter and place a disk of the dough on the surface. Cook until it is blistered and brown, about 1 minute. Turn over and cook the other side. Transfer the cooked paratha to a plate and repeat with the remaining disks. Serve warm or at room temperature.

NOTE: If you want to freeze any extras, make sure to place parchment paper between them to prevent them from melding together.

INGREDIENTS:

- 2 CUPS PASTRY FLOUR, PLUS MORE FOR DUSTING
- 1 CUP WHOLE WHEAT FLOUR
- ¼ TEASPOON SALT
- 1 CUP WARM WATER (110°F)
- 5 TABLESPOONS VEGETABLE OIL, PLUS MORE AS NEEDED
- 5 TABLESPOONS GHEE OR MELTED UNSALTED BUTTER

PITA BREAD

YIELD: 16 PITAS / **ACTIVE TIME:** 1 HOUR / **TOTAL TIME:** 2 HOURS

Pitas are delicious, somewhat chewy bread pockets that originated in the Mediterranean region. They can be filled with just about anything and are popular around the world, but are especially prevalent in Middle Eastern cuisine.

INGREDIENTS:

- 1 PACKET OF ACTIVE DRY YEAST (2¼ TEASPOONS)
- 2½ CUPS WARM WATER (110 TO 115°F)
- 3 CUPS ALL-PURPOSE FLOUR, PLUS MORE FOR DUSTING
- 1 TABLESPOON OLIVE OIL, PLUS MORE FOR THE SKILLET
- 1 TABLESPOON SALT
- 3 CUPS WHOLE WHEAT FLOUR
- UNSALTED BUTTER, FOR GREASING THE BOWL

1. Proof the yeast by mixing with the warm water. Let sit for about 10 minutes until foamy.

2. In a large bowl, add the yeast mix into the all-purpose flour and stir until it forms a stiff dough. Cover and let the dough rise for about 1 hour.

3. Add the oil and salt to the dough and stir in the whole wheat flour in ½-cup increments. When finished, the dough should be soft. Turn onto a lightly floured surface and knead it until it is smooth and elastic, about 10 minutes.

4. Coat the bottom and sides of a large mixing bowl (ceramic is best) with butter. Place the ball of dough in the bowl, cover loosely with plastic wrap, put it in a naturally warm, draft-free location, and let it rise until doubled in size, about 45 minutes to 1 hour.

5. On a lightly floured surface, punch down the dough and cut into 16 pieces. Put the pieces on a baking sheet and cover with a kitchen towel while working with individual pieces.

6. Roll out the pieces with a rolling pin until they are approximately 7 inches across. Stack them between sheets of plastic wrap.

7. Warm a 10-inch cast-iron skillet over high heat and lightly oil the bottom. Cook the individual pitas for about 20 seconds on one side, then flip and cook for about a minute on the other side, until bubbles form. Turn again and continue to cook until the pita puffs up, another minute or so. Keep the skillet lightly oiled while cooking, and store the pitas on a plate under a clean kitchen towel until ready to serve.

NAAN

YIELD: 8 PIECES / **ACTIVE TIME:** 1 HOUR / **TOTAL TIME:** 3 TO 4 HOURS

This is the bread that is traditionally served with Indian cuisine. It's usually cooked in a tandoor (clay oven) in India, but the cast-iron skillet works just fine.

1. Proof the yeast by mixing it with the sugar and ½ cup of the warm water. Let sit for 10 minutes until foamy.

2. In a bowl, add the remaining water, flour, salt, baking powder, and yeast mix. Stir to combine. Add the yogurt and 2 tablespoons of the butter and stir to form a soft dough.

3. Transfer to a lightly floured surface and knead the dough until it is springy and elastic, about 10 minutes.

4. Coat the bottom and sides of a large mixing bowl (ceramic is best) with butter. Place the ball of dough in the bowl, cover loosely with plastic wrap, put it in a naturally warm, draft-free location, and let it rise until doubled in size, about 1 to 2 hours.

5. Punch down the dough. Lightly flour a work surface again, take out the dough and, using a rolling pin, make a circle of it. Cut it into 8 slices (like a pie).

6. Heat the skillet over high heat until it is very hot, about 5 minutes. Working with individual pieces of dough, roll them out to soften the sharp edges and make the pieces look more like teardrops. Brush both sides with olive oil and, working one at a time, place the pieces in the skillet.

7. Cook for 1 minute, turn the dough with tongs, cover the skillet, and cook the other side for about a minute (no longer). Transfer cooked naan to a plate and cover with foil to keep warm while making the additional pieces. Serve warm.

TIP: You can add herbs or spices to the dough or the pan to make naan with different flavors, like adding ¼ cup chopped fresh parsley to the dough, or sprinkling the skillet lightly with cumin, coriander, or turmeric (or a combination) before cooking the pieces of naan. You can also use a seasoned olive oil to brush the pieces before cooking—one that has been infused with hot pepper flakes or roasted garlic.

INGREDIENTS:

- 1½ TEASPOONS ACTIVE DRY YEAST
- ½ TABLESPOON SUGAR
- 1 CUP WARM WATER (110 TO 115°F)
- 3 CUPS ALL-PURPOSE FLOUR OR 1½ CUPS ALL-PURPOSE AND 1½ CUPS WHOLE WHEAT PASTRY FLOUR, PLUS MORE FOR DUSTING
- ¼ TEASPOON SALT
- 1 TEASPOON BAKING POWDER
- ½ CUP PLAIN YOGURT
- 4 TABLESPOONS UNSALTED BUTTER, MELTED, PLUS MORE FOR GREASING THE BOWL
- ¼ CUP OLIVE OIL

GLUTEN-FREE BREAD

YIELD: 1 SMALL LOAF / **ACTIVE TIME:** 25 MINUTES / **TOTAL TIME:** 3 HOURS

We are fortunate to live in a time when gluten-free options are numerous. If you love bread and can't or don't want to eat gluten, make this recipe and dig in! You'll be amazed at the result—an equally crusty and fluffy loaf that tastes great.

INGREDIENTS:

- ½ TEASPOON INSTANT YEAST
- ¼ TEASPOON SUGAR
- 1½ CUPS WATER (110 TO 115° F), PLUS MORE AS NEEDED
- 1 TEASPOON KOSHER SALT
- 1½ TEASPOONS XANTHAN GUM
- 3 CUPS BOB'S RED MILL GLUTEN-FREE FLOUR, PLUS MORE FOR DUSTING
- ⅓ CUP BOB'S RED MILL SWEET WHITE RICE FLOUR

1. Put the yeast and sugar in a measuring cup and drizzle in about ½ cup warm water. Hot water will kill the yeast, so it's important that the water be warm without being hot. Cover the measuring cup with plastic wrap and set it aside for about 15 minutes. If the yeast doesn't foam, it is not alive and you'll need to start over.

2. When the yeast is proofed, pour it into a large bowl and add remaining warm water. Stir gently to combine. Combine the salt and xanthan gum with the flours, and add the dry mixture to the yeast mixture. Stir with a wooden spoon until combined. Add up to an additional cup of warm water to accommodate the rice flour, which is tackier than regular flour. The dough should be wet and sticky.

3. Put a dusting of flour on a flat surface and lift out the dough. With flour on your hands and more at the ready, begin kneading the dough so that it loses its stickiness. Don't overdo it, and don't use too much flour, just enough that the dough starts to become more cohesive.

4. Place the dough in a large bowl, cover the bowl with plastic wrap, and allow to rise untouched for at least 1 hour and up to several hours. Gently punch it down, re-cover with the plastic, and allow to rise again for another 30 minutes or so.

5. While the dough is on its final rise, preheat the oven to 450°F. Put a piece of parchment paper on the bottom of a cast-iron Dutch oven and put it in the oven with the lid on so it gets hot. When the oven is ready and dough has risen, carefully remove the lid and gently scoop the dough from the bowl into the pot. Cover and bake for 15 minutes. Remove the lid and continue to bake for another 15 to 20 minutes, until the top is golden and it sounds hollow when tapped.

6. Remove the pot from the oven and use kitchen towels to carefully remove the bread. Allow to cool before slicing.

WHOLE WHEAT PASTA DOUGH

YIELD: 1¾ LBS. / **ACTIVE TIME:** 1 HOUR / **TOTAL TIME:** 2 TO 3 HOURS

Whether you're looking for healthier pasta alternatives or are a fan of chewier pastas like linguine, whole wheat pasta is an excellent conduit for thick, creamy sauces.

INGREDIENTS:

- 4 CUPS FINELY GROUND WHOLE WHEAT FLOUR, PLUS MORE AS NEEDED
- 1½ TEASPOONS SALT
- 4 LARGE EGG YOLKS
- 1 TABLESPOON OLIVE OIL
- 2 TABLESPOONS WATER, PLUS MORE AS NEEDED

1. On a flat work surface, combine the flour and salt and form it into a mound. Create a well in the center, then add the egg yolks, oil, and the 2 tablespoons of water. Using a fork or your fingertips, gradually start pulling the flour into the pool of egg, beginning with the flour at the inner rim of the well. Continue to gradually add flour until the dough starts holding together in a single floury mass, adding more water—1 tablespoon at a time—if the mixture is too dry to stick together. Once the dough feels firm and dry and can form a craggy-looking ball, it is time to start kneading.

2. Begin by working the remaining flour on the work surface into the ball of dough. Using the heel of your hand, push the ball of dough away from you in a downward motion. Turn the dough 45 degrees each time you repeat this motion, as doing so incorporates the flour more evenly. The dough should have a smooth, elastic texture. If the dough still feels wet, tacky, or sticky, dust it with flour and continue kneading. If it feels too dry and is not completely sticking together, wet your hands with water and continue kneading. Wet your hands as many times as you need in order to help the dough shape into a ball. Knead for 8 to 10 minutes to create a dough that is smooth and springy and to eliminate any air bubbles and bits of unincorporated flour in the dough. The dough has been sufficiently kneaded when it is very smooth and gently pulls back into place when stretched.

3. Wrap the ball of dough tightly in plastic wrap and let rest for at least 1 hour and up to 2 hours. If using within a few hours, leave it out on the kitchen counter; otherwise, refrigerate it (it will keep for up to 3 days). If refrigerated, the dough may experience some discoloration, but it won't affect the flavor.

4. Cut the dough into four even pieces. Set one piece on a smooth work surface and wrap up the rest in plastic wrap to prevent drying. Shape the dough into a ball, place it on the work surface, and, with the palm of your hand, push down on it so that it looks like a thick pita. Using a rolling pin, roll the dough to ½ inch thick. Try as much as possible to keep the thickness and width of the dough "patty" even, as it will help the dough fit through the pasta maker more easily.

ALL-YOLK PASTA DOUGH

YIELD: ¾ LB. / **ACTIVE TIME:** 1 HOUR / **TOTAL TIME:** 2 TO 3 HOURS

The use of only egg yolks produces a rich, golden dough that creates tender pasta. It is perfect for thin, fragile, or small, filled pastas.

INGREDIENTS:

- 1½ CUPS ALL-PURPOSE FLOUR
- ⅓ CUP FINELY MILLED "00" FLOUR, PLUS MORE AS NEEDED
- 8 LARGE EGG YOLKS
- 2 TABLESPOONS LUKEWARM WATER (90°F), PLUS MORE AS NEEDED

1. On a flat work surface, form the flours into a mound. Create a well in the center, then add the egg yolks and the 2 tablespoons of water. Using a fork or your fingertips, gradually start pulling the flour into the pool of egg, beginning with the flour at the inner rim of the well. Continue to gradually add flour until the dough starts holding together in a single floury mass, adding more water—1 tablespoon at a time—if the mixture is too dry to stick together. Once the dough feels firm and dry and can form a craggy-looking ball, it is time to start kneading.

2. Begin by working the remaining flour on the work surface into the ball of dough. Using the heel of your hand, push the ball of dough away from you in a downward motion. Turn the dough 45 degrees each time you repeat this motion, as doing so incorporates the flour more evenly. The dough should have a smooth, elastic texture. If the dough still feels wet, tacky, or sticky, dust it with flour and continue kneading. If it feels too dry and is not completely sticking together, wet your hands with water and continue kneading. Wet your hands as many times as you need in order to help the dough shape into a ball. Knead for 8 to 10 minutes to create a dough that is smooth and springy and to eliminate any air bubbles and bits of unincorporated flour in the dough. The dough has been sufficiently kneaded when it is very smooth and gently pulls back into place when stretched.

3. Wrap the ball of dough tightly in plastic wrap and let rest for at least 1 hour and up to 2 hours. If using within a few hours, leave it out on the kitchen counter; otherwise, refrigerate it (it will keep for up to 3 days). If refrigerated, the dough may experience some discoloration, but it won't affect the flavor.

4. Cut the dough into four even pieces. Set one piece on a smooth work surface and wrap up the rest in plastic wrap to prevent drying. Shape the dough into a ball, place it on the work surface, and, with the palm of your hand, push down on it so that it looks like a thick pita. Using a rolling pin, roll the dough to ½ inch thick. Try as much as possible to keep the thickness and width of the dough "patty" even, as it will help the dough fit through the pasta maker more easily.

TIP: This dough is suitable for popular noodles such as linguine and spaghetti.

THREE-EGG BASIC PASTA DOUGH

YIELD: ABOUT 1 LB. / **ACTIVE TIME:** 1 HOUR / **TOTAL TIME:** 2 TO 3 HOURS

This simple, delicious pasta recipe is sure to become your go-to.

1. On a flat work surface, form the flour into a mound. Create a well in the center, then add the eggs, egg yolk, and the 2 tablespoons of water. Using a fork or your fingertips, gradually start pulling the flour into the pool of egg, beginning with the flour at the inner rim of the well. Continue to gradually add flour until the dough starts holding together in a single floury mass, adding more water—1 tablespoon at a time—if the mixture is too dry to stick together. Once the dough feels firm and dry and can form a craggy-looking ball, it is time to start kneading.

2. Begin by working the remaining flour on the work surface into the ball of dough. Using the heel of your hand, push the ball of dough away from you in a downward motion. Turn the dough 45 degrees each time you repeat this motion, as doing so incorporates the flour more evenly. The dough should have a smooth, elastic texture. If the dough still feels wet, tacky, or sticky, dust it with flour and continue kneading. If it feels too dry and is not completely sticking together, wet your hands with water and continue kneading. Wet your hands as many times as you need in order to help the dough shape into a ball. Knead for 8 to 10 minutes to create a dough that is smooth and springy, and to eliminate any air bubbles and bits of unincorporated flour in the dough. The dough has been sufficiently kneaded when it is very smooth and gently pulls back into place when stretched.

3. Wrap the ball of dough tightly in plastic wrap and let rest for at least 1 hour and up to 2 hours. If using within a few hours, leave it out on the kitchen counter; otherwise, refrigerate it (it will keep for up to 3 days). If refrigerated, the dough may experience some discoloration (but it won't affect the flavor at all).

4. Cut the dough into four even pieces. Set one piece on a smooth work surface and wrap up the rest in plastic wrap to prevent drying. Shape the dough into a ball, place it on the work surface, and, with the palm of your hand, push down on it so that it looks like a thick pita. Using a rolling pin, roll the dough to ½ inch thick. Try as much as possible to keep the thickness and width of the dough "patty" even, as it will help the dough fit through the pasta maker more easily.

TIP: This dough is suitable for popular noodles such as fettuccine, pappardelle, and tagliatelle.

INGREDIENTS:

- 2¾ CUPS ALL-PURPOSE FLOUR, PLUS MORE FOR DUSTING
- 3 LARGE EGGS
- 1 EGG YOLK
- 2 TABLESPOONS LUKEWARM WATER (90°F), PLUS MORE AS NEEDED

FARFALLE

YIELD: ¾ LB. PASTA / **ACTIVE TIME:** 45 MINUTES / **TOTAL TIME:** 1 TO 3 HOURS

Farfalle means "butterfly" in Italian, and these lighter-than-air pasta shapes are well worth their elegant name. They work well with tomato- or cream-based sauces, but feel free to experiment with your favorites.

INGREDIENTS:

- ALL-YOLK PASTA DOUGH (SEE PAGE 726)
- SEMOLINA FLOUR, FOR DUSTING
- SALT, TO TASTE

1. Prepare the dough as directed, rolling the dough to the second-thinnest setting (generally notch 4) for pasta sheets that are about ⅛ inch thick. Lay the pasta sheets on lightly floured parchment-lined baking sheets and cover loosely with plastic wrap. Work quickly to keep the pasta sheets from drying out, which makes it harder for the pasta to stick together.

2. Working with one pasta sheet at a time, place it on a lightly floured work surface and trim both ends to create a rectangle. Using a pastry cutter, cut the pasta sheet lengthwise into 1- to 1¼-inch-wide ribbons. Carefully separate the ribbons from each other, then, using a ridged pastry cutter, cut the ribbons into 2-inch pieces. To form the butterfly shape, place the index finger of your nondominant hand on the center of the piece of pasta. Then place the thumb and index finger of your dominant hand on the sides of the rectangle—right in the middle—and pinch the dough together to create a butterfly shape. Firmly pinch the center again to help it hold its shape. Leave the ruffled ends of the farfalle untouched. Repeat with all the pasta sheets.

3. Set the farfalle on lightly floured parchment-lined baking sheets so they are not touching. Allow them to air-dry for at least 30 minutes and up to 3 hours, and then cook. Alternatively, you can place them, once air-dried, in a bowl, cover with a kitchen towel, and refrigerate for up to 3 days. Or freeze on the baking sheets, transfer to freezer bags, and store in the freezer for up to 2 months. Do not thaw them prior to cooking (they will become mushy) and add an extra minute or so to their cooking time.

4. To cook the farfalle, bring a large pot of salted water to a boil. Add the farfalle and cook until the pasta is tender but still chewy, 2 to 3 minutes. Drain and serve with the sauce of your choice.

FAZZOLETTI

YIELD: ABOUT 1 LB. / **ACTIVE TIME:** 30 MINUTES / **TOTAL TIME:** 1 HOUR AND 30 MINUTES

These thin, square, or rectangular pasta shapes resemble handkerchiefs, which is where they get their name. Don't worry about making perfect squares; they taste just as good when uneven.

INGREDIENTS:

- THREE-EGG BASIC PASTA DOUGH (SEE PAGE 727)
- SEMOLINA FLOUR, FOR DUSTING
- SALT, TO TASTE

1. Prepare the dough as directed, rolling the dough to the thinnest setting (generally notch 5) for pasta sheets that are about $\frac{1}{16}$ inch thick. Lay the pasta sheets on lightly floured parchment-lined baking sheets and let them air-dry for 15 minutes.

2. Cut each pasta sheet into as many 2½-inch squares or 1½×2½-inch rectangles as possible. Set them on lightly floured parchment-covered baking sheets so they are not touching. Gather any scraps together into a ball, put it through the pasta maker to create additional pasta sheets, and cut those as well. Allow them to air-dry for 1 hour, turning them over once halfway through, and then cook. Alternatively, you can place them, once air-dried, in a bowl, cover with a kitchen towel, and refrigerate for up to 3 days.

3. To cook the fazzoletti, cook for about 1 minute in a pot of boiling, salted water, until they are tender but still chewy.

GARGANELLI

YIELD: 1½ LBS. / **ACTIVE TIME:** 1 HOUR AND 30 MINUTES / **TOTAL TIME:** 4 HOURS AND 30 MINUTES

This Bolognese pasta requires a delicate touch and a bit of patience, but once you get the hang of rolling the squares around the chopstick, you're sure to want it for every meal.

INGREDIENTS:

- 2¼ CUPS SEMOLINA FLOUR, PLUS MORE FOR DUSTING
- 1½ TEASPOONS SALT, PLUS MORE FOR THE PASTA WATER
- 3 LARGE EGGS
- 2 TABLESPOONS OLIVE OIL
- 2 TABLESPOONS WATER

1. Combine all of the ingredients and prepare the dough as directed on page 726. Then use a pasta maker to roll the dough to the second-thinnest setting (generally notch 4) for pasta sheets that are about ⅛ inch thick. Lay the pasta sheets on flour-dusted, parchment-lined baking sheets and cover them loosely with plastic wrap.

2. Working with one pasta sheet at a time, lightly dust it with flour. Cut it into 1½-inch-wide strips and then cut the strips into 1½-inch squares. Repeat with the remaining pasta sheets. Cover the squares loosely with plastic wrap. Gather any scraps together into a ball, put it through the pasta maker to create additional pasta sheets, and cut those as well.

3. To make each garganello, place one square of pasta dough on a lightly floured work surface with one of the corners pointing toward you. Using a chopstick, gently roll the square of pasta around the chopstick, starting from the corner closest to you, until a tube forms. Once completely rolled, press down slightly as you seal the ends together, then carefully slide the pasta tube off the chopstick and lightly dust with flour. Set them on flour-dusted, parchment-lined baking sheets and allow them to air-dry for 1 hour, turning them over halfway through.

4. To cook the garganelli, cook for 2 to 3 minutes in a pot of boiling, salted water, until they are tender but still chewy.

NODI

YIELD: ABOUT 1 LB. / **ACTIVE TIME:** 1 HOUR AND 30 MINUTES / **TOTAL TIME:** 3 TO 4 HOURS

This gondola-inspired pasta is formed by creating a knot in the center of a thin noodle.

INGREDIENTS:

- 1¾ CUPS SEMOLINA FLOUR, PLUS MORE FOR DUSTING
- 1 TEASPOON SALT, PLUS MORE FOR THE PASTA WATER
- ½ TEASPOON FENNEL SEEDS, FINELY GROUND
- ⅔ CUP WARM WATER

1. Put the flour, salt, and fennel seeds in a large bowl and add the water. Begin mixing with a fork until the mixture starts to roughly stick together and look coarse. Gather it together with your hands and transfer it to a lightly floured work surface.

2. Using the heel of your hand, push the ball of dough away from you in a downward motion. Turn the dough 45 degrees each time you repeat this motion, as doing so incorporates the flour more evenly. If the dough feels too dry, wet your hands as many times as you need in order to help shape the dough into a ball. Knead for 10 minutes.

3. Cover the dough tightly with plastic wrap to keep it from drying out and let rest for at least 1 hour, but 2 hours is even better. If using within a few hours, leave out on the kitchen counter. Otherwise, put it in the refrigerator, where it will keep for up to 3 days.

4. Between the palms of your hands or on a lightly floured work surface, roll the dough into a 2-inch-thick log and cut it across into 18 rounds of even thickness (the easiest way to do this is to cut the roll in half and continue cutting each piece in half until you have 18 pieces). Cover all the pieces but the one you are working with to keep them from drying out.

5. With the palms of your hands, roll the piece of dough left out into a long rope ⅛ inch thick. Now make the knots. Starting on one end of the rope, tie a simple knot, gently pull on both ends to slightly tighten the knot, then cut the knot off the rope, leaving a tail on each side of about ⅜-inch long. Keep making and cutting off knots in this manner until you use up all of the rope. Repeat with the remaining pieces of dough. Set the finished knots on lightly floured, parchment-lined baking sheets so they are not touching. Allow them to air-dry for 2 hours, turning them over once halfway through, and then cook. Alternatively, you can place them, once air-dried, in a bowl, cover with a kitchen towel, and refrigerate for up to 3 days.

6. To cook the nodi, place in a large pot of boiling, salted water for 2 to 3 minutes, until they are tender but still firm.

ORECCHIETTE

YIELD: ABOUT 1 LB. / **ACTIVE TIME:** 1 HOUR AND 30 MINUTES / **TOTAL TIME:** 4 HOURS

*O*recchiette means "little ears" and refers to this pasta's flattened shape. Orecchiette provide an interesting texture contrast when cooked, as the center is soft while the outer edge is just the slightest bit chewy.

INGREDIENTS:

- 2 CUPS SEMOLINA FLOUR, PLUS MORE FOR DUSTING
- 1 TEASPOON SALT, PLUS MORE FOR THE PASTA WATER
- ¾ CUP WATER, PLUS MORE AS NEEDED

1. Combine the flour and salt in a large bowl. Add the water a little at a time while mixing with a fork. Continue mixing the dough until it starts holding together in a single floury mass. If it is still too dry to stick together, add more water, 1 teaspoon at a time, until it does. Work the dough with your hands until it feels firm and dry and can be formed into a craggy-looking ball.

2. Transfer the dough to a lightly floured work surface and knead it for 10 minutes. Because it is made with semolina flour, the dough can be quite stiff and hard. You can also mix and knead this in a stand mixer; don't try it with a handheld mixer—the dough is too stiff and could burn the motor out. Using the heel of your hand, push the ball of dough away from you in a downward motion. Turn the dough 45 degrees each time you repeat this motion, as doing so incorporates the flour more evenly. Wet your hands as needed if the dough is too sticky. After 10 minutes of kneading, the dough will only be slightly softer (most of the softening is going to occur when the dough rests, which is when the gluten network within the dough will relax). Shape into a ball, cover tightly with plastic wrap, and let rest in the refrigerator for at least 2 hours and up to 2 days.

3. Cut the dough into four equal sections. Take one dough section and shape it into an oval with your hands. Cover the remaining sections with plastic wrap to prevent it from drying out. Place on a lightly floured work surface and, with the palms of your hands, roll it against the surface until it becomes a long ½-inch-thick rope. Using a sharp paring knife, cut the rope into ¼-inch discs, lightly dusting with semolina flour so they don't stick together.

4. To form the orecchiette, place a disc on the work surface. Stick your thumb in flour, place it on top of the disc, and, applying a little pressure, drag your thumb, and the accompanying dough, across to create an ear-like shape. Flour your thumb before making each orecchiette for best results. Lightly dust the orecchiette with flour and set them on lightly floured, parchment-covered baking sheets so they are not touching. Allow them to air-dry for 1 hour, turning them over once halfway through, and then cook. Alternatively, you can place them, once air-dried, in a bowl, cover with a kitchen towel, and refrigerate for up to 3 days.

5. To cook the orecchiette, place in a large pot of boiling, salted water until they are tender but still chewy, 3 to 4 minutes.

RIBBON PASTA

YIELD: ABOUT 1 LB. / **ACTIVE TIME:** 20 MINUTES / **TOTAL TIME:** 1 HOUR

Ribbon pasta includes things like pappardelle, tagliatelle, and the classic fettuccine, and it is one of the easiest pasta types to make on your own. With a little practice, you can have these churned out in as little as a half hour—so long as you don't mind using a little elbow grease.

INGREDIENTS:

- 1 RECIPE THREE-EGG BASIC PASTA DOUGH (SEE PAGE 727)
- SEMOLINA FLOUR, FOR DUSTING
- SALT, PLUS MORE FOR THE PASTA WATER

1. Prepare the dough as directed, rolling the dough to the second thinnest setting (generally notch 4) or thinnest setting (generally notch 5 or 6) to form pasta sheets that are, respectively, ⅛ or 1/16 inch thick. Lay the pasta sheets on lightly floured, parchment-lined baking sheets. Let the sheets air-dry for 15 minutes, turning them over halfway through (doing this will make them easier to cut).

2. Lightly flour the surface of a pasta sheet and gently roll it up, starting from a short end, to create a pasta roll. Use a very sharp knife to gently slice the roll across to your preferred width. Cut into 1- to 1½-inch-wide strips for pappardelle; ¾-inch-wide strips for tagliatelle; and ½- to ¼-inch-wide strips for fettuccine.

3. Lightly dust the cut roll with flour, then begin to gently unfold the strips, one by one, as you shake off any excess flour. Arrange them straight and spread out or lay them down by shaping them in a coil (referred to as a bird's nest). Repeat with all the pasta sheets. Let air-dry for 30 minutes and then cook. Alternatively, cover them with a kitchen towel and refrigerate for up to 3 days.

4. To cook the pasta ribbons, bring a large pot of water to a boil. Once it is boiling, add salt (1 tablespoon for every 4 cups water) and stir. Add the pasta and stir for the first minute to prevent any sticking and to untangle the strands of pasta. Tagliolini will not require additional cooking, so drain them as soon as they hit the water and you stir to disentangle them. Cook the remaining pasta ribbons until tender but still chewy, anywhere from 1 to 3 minutes. Drain and add them to the sauce of your choice.

TAJARIN

YIELD: ½ LB. / **ACTIVE TIME:** 20 MINUTES / **TOTAL TIME:** 1 HOUR

These delicate, flat noodles are best broken up in a soup or served alongside roasted proteins and vegetables.

INGREDIENTS:

- ALL-YOLK PASTA DOUGH (SEE PAGE 726)
- SEMOLINA FLOUR, FOR DUSTING
- SALT, TO TASTE

1. Prepare the dough as directed, rolling the dough to the thinnest setting (generally notch 5) for pasta sheets that are about 1/16 inch thick. Cut into 8-inch-long sheets. Lay the pasta sheets on lightly floured parchment-lined baking sheets. Air-dry for 15 minutes.

2. Working with one pasta sheet at a time, lightly dust it with semolina flour, then gently roll it up, starting from a short end. Using a very sharp knife, gently slice the roll across into 1/12-inch-wide strips. Lightly dust the cut roll with flour, then gently begin unfolding the strips, one by one, as you shake off any excess flour. Arrange them either straight and spread out or curled in a coil. Repeat with all the pasta sheets. Allow them to air-dry for 30 minutes and then cook. Alternatively, you can place them, once air-dried, on a baking sheet, cover with a kitchen towel, and refrigerate for up to 3 days.

3. To cook the tajarin, bring a large pot of salted water to a boil. Cook until the pasta is tender but still chewy, typically no more than 2 minutes. Drain and serve with the sauce of your choice.

TROFIE

YIELD: ABOUT 1 LB. / **ACTIVE TIME:** 40 MINUTES / **TOTAL TIME:** 4 TO 5 HOURS

Trofie are thin, spiraled noodles that work best with strong sauces, including pesto and ragùs.

INGREDIENTS:

- 2¾ CUPS ALL-PURPOSE FLOUR
- 1 TEASPOON SALT, PLUS MORE TO TASTE
- 1 CUP WATER
- SEMOLINA FLOUR, FOR DUSTING

1. Put the flour and salt in a large bowl, mix well with a fork, and add the water. Mix with the fork until all the water has been absorbed, then start working the dough with your hands. In a few minutes the crumbly mixture will begin to come together as a grainy dough.

2. Transfer the dough, along with any bits stuck to the bowl, to a lightly floured work surface. Begin to knead the dough. Using the heel of your hand, push the ball of dough away from you in a downward motion. Turn the dough 45 degrees each time you repeat this motion, as doing so incorporates the flour more evenly. Knead the dough for about 10 minutes. Cover the dough with plastic wrap to keep it from drying out and let rest at room temperature for 1 hour, but 2 hours is even better.

3. Between the palms of your hands or on a lightly floured work surface, roll the dough into a 2-inch-thick log and cut it across into eight pieces (the easiest way to do this is to cut the roll in half and continue cutting each piece in half until you have eight pieces). Cover all the dough pieces but the one you are working with to keep them from drying out. Shape each piece of dough into a ball, and then roll it until it is a long, ½-inch-thick rope. Cut into ½-inch pieces and dust them with flour.

4. Working with one piece at a time, press down on the dough with your fingertips and roll the dough down the palm of your other hand. This action will cause the piece of dough to turn into a narrow spiral with tapered ends. Repeat with the remaining pieces of dough. Dust the spirals with flour, set them on flour-dusted, parchment-lined baking sheets, and allow them to air-dry for 2 hours, turning them over halfway through.

5. To cook the trofie, cook for 3 to 4 minutes in a pot of boiling, salted water, until they are tender but still chewy.

CHINESE EGG NOODLES

YIELD: ABOUT 1 LB. / **ACTIVE TIME:** 45 MINUTES / **TOTAL TIME:** 2 HOURS

This noodle is a staple of recipes like lo mein and chow mein, as well as many Asian soups. It is incredibly easy to make and, of course, delicious.

INGREDIENTS:

- 2 CUPS ALL-PURPOSE FLOUR, PLUS MORE FOR DUSTING
- 1 TEASPOON SALT, PLUS MORE FOR THE PASTA WATER
- 2 LARGE EGGS, LIGHTLY BEATEN
- 3 TABLESPOONS WATER, PLUS MORE AS NEEDED

1. Mix the flour and salt together in a large bowl. Add the eggs and mix until a floury dough forms. Add the 3 tablespoons of water and continue to mix until you almost cannot see any remaining traces of flour. If you find, even after adding the water, that your dough is still very floury, add more water, 1 tablespoon at a time, and continue mixing it with your hands until the dough starts coming together more easily. Start kneading the dough in the bowl with your dominant hand. Continue kneading in the bowl until a smooth ball forms; this may take about 10 minutes. Wrap the dough tightly in plastic wrap and let rest at room temperature for 40 to 50 minutes to allow the gluten in the dough to relax.

2. Unwrap the dough and place it on a lightly floured work surface. Using a rolling pin, begin "beating" the dough, turning it over after every 10 whacks or so. Continue doing this for 6 minutes. Then, shape the dough into a ball, cover with plastic wrap, and let rest at room temperature for another 30 minutes.

3. Return the dough to the work surface (no need to flour again). Cut it in half and wrap one half in plastic wrap to prevent drying. Roll the other half into a large, thin sheet about twice the length and breadth of the length of your rolling pin (you should be able to almost see your hand through it). Lightly flour both sides of the sheet of dough and then fold the sheet of dough twice over itself to create a three-layered fold (like a letter).

4. Using a very sharp knife, slice across the roll into evenly spaced strands. You can make them as thin or thick as you'd like. As you cut the dough, be sure to hold the knife perpendicular to the surface and lightly push the newly cut strip away from the roll with the knife to completely separate it. Continue until you have cut the entire roll, then lightly dust the slivered noodles with flour to prevent any sticking. Transfer the noodles to a parchment-lined baking sheet, shaking off any excess flour if necessary. You can leave them nested or unspool them according to your preference, as they will unravel and straighten once boiled. Repeat with the remaining dough. Cook in a pot of boiling, salted water for 3 to 4 minutes, or cover and refrigerate for up to 1 day.

UDON NOODLES

YIELD: 1 LB. / **ACTIVE TIME:** 1 HOUR / **TOTAL TIME:** 2 TO 3 HOURS

Japanese udon noodles are best when homemade, as the packaged versions lack some of the chewiness and bulk of the fresh variety.

INGREDIENTS:

- ¼ CUP WARM WATER, PLUS MORE AS NEEDED
- 1 TEASPOON FINE SEA SALT
- 2¼ CUPS CAKE FLOUR OR FINELY MILLED "00" FLOUR
- POTATO STARCH OR CORNSTARCH, FOR DUSTING

1. Stir the water and salt together in a small bowl until the salt dissolves. Put the flour in a large bowl and make a well in the center. Add the salted water in a stream while stirring the flour. Once all the water has been added, begin working the dough with your hands to incorporate all the flour. If the dough is too dry, add water in 1-teaspoon increments until the dough sticks together.

2. Transfer the dough to a work surface that you have dusted very lightly with the potato starch or cornstarch. Knead the dough with the palm of your dominant hand, turning it 45 degrees with each pressing, until the dough becomes uniformly smooth and slowly springs back when pressed by a finger, about 10 minutes. Cover the dough tightly with plastic wrap and let rest for 1 to 2 hours to relax the gluten.

3. Cut the dough into two pieces. Set one on a lightly dusted work surface and wrap the other in plastic wrap to prevent drying. Pat the piece of dough into a rectangular shape and, using a lightly dusted rolling pin, roll the dough into a ⅛-inch-thick rectangle. Lightly dust the dough and then fold it twice over itself to create a three-layered fold, as you would a letter.

4. Using a very sharp knife, slice the roll into ⅛-inch-wide strands. As you cut the dough, be sure to hold the knife perpendicular to the surface and lightly push the newly cut strip away from the roll with the knife to completely separate it. Continue until you have cut the entire roll, then lightly dust the slivered pasta to prevent any sticking. Transfer the noodles to a parchment-lined baking sheet, shaking off any excess starch if necessary. Repeat with the remaining dough. Udon noodles quickly turn brittle and break when handled, so cook as soon as you finish making them.

5. To cook the udon, cook for about 1 minute in a pot of boiling, salted water, until they are tender but still chewy.

APPENDIX

LITH. & PUBL. BY KIMMEL & VOIGT 254 & 256 CANAL ST. N.Y.

"THE BEST IN

"THE MARKET."

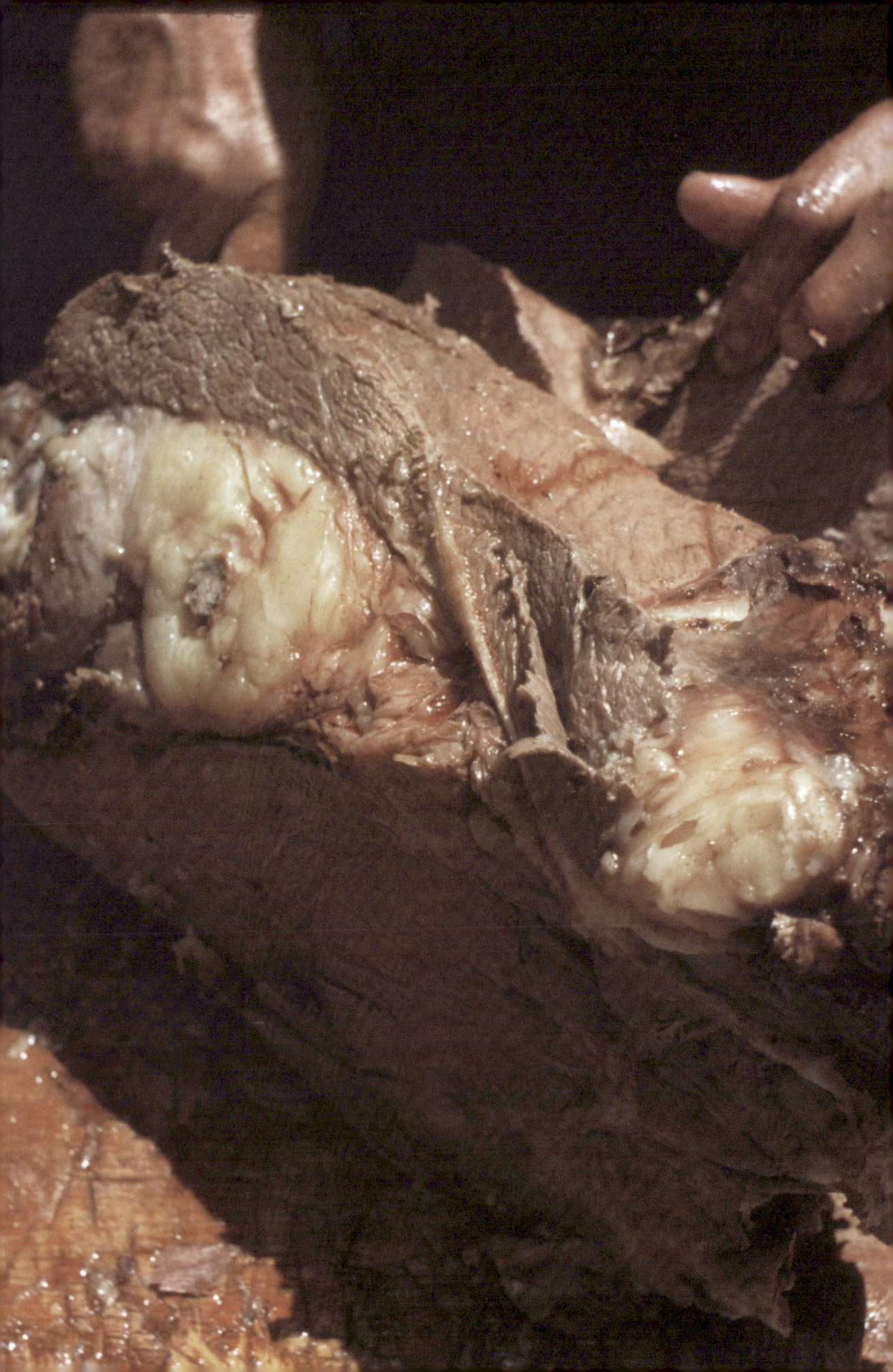

CARVING

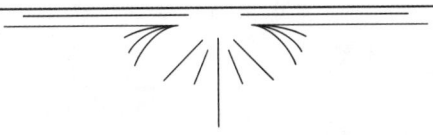

The following is taken from *The Boston Cook Book*, by D. A. Lincoln, an acclaimed instructor in all things culinary and domestic who wrote this in 1886. As the *Independent* wrote in reviewing the book: "It is so perfectly and generously up to everything culinary, that it cannot help spilling over a little into sciences and philosophy. It is the trimmest, best arranged, best illustrated, most intelligible, manual of cookery as a high art, and as an economic art, that has appeared." The buttoned-up tone of the prose makes for an amusing read, especially in Lincoln's introductory text, but there is no denying the expertise evident in her detailed instructions for carving and serving all of these cuts of meat.

CARVING AND SERVING

BY MRS. D.A. LINCOLN
AUTHOR OF "THE BOSTON COOK BOOK"

GENERAL DIRECTIONS

"Do you teach your pupils how to carve?"

"Please give us a lecture on carving; my husband says he will come if you will."

I have been so frequently addressed in this way that I have decided to publish a manual on the Art of Carving. Instruction in this art cannot be given at a lecture with any profit to my pupils or satisfaction to myself. One cannot learn by simply seeing a person carve a few times. As much as any other art, it requires study; and success is not attainable without much practice. There are certain rules which should be thoroughly understood; if followed faithfully in daily practice, they will help more than mere observation.

This manual is not offered as a guide for special occasions, company dinners, etc., nor for those whose experience renders it unnecessary, or whose means allow them to employ one skilled in the art. But it is earnestly hoped that the suggestions here offered will aid those who desire, at their own table in everyday home life, to acquire that ease and perfection of manner which, however suddenly it may be confronted with obstacles, will be equal to every occasion.

Printed rules for carving are usually accompanied with cuts showing the position of the joint or fowl on the platter, and having lines indicating the method of cutting. But this will not be attempted in this manual, as such illustrations seldom prove helpful; for the actual thing before us bears faint resemblance to the pictures, which give us only the surface, with no hint of what may be inside.

It is comparatively a slight matter to carve a solid mass of lean meat. It is the bones, tough gristle, and tendons, that interfere with the easy progress of the knife. To expect any one to carve well without any conception of the internal structure of what may be placed before him is

as absurd as to expect one to amputate a limb successfully who has no knowledge of human anatomy.

Some notion of the relative position of bones, joints, fat, tough and tender muscles, is the first requisite to good carving. All agree that skill in carving may be acquired by practice; and so it may. Any one can divide a joint if he cut and hack at it long enough, and so learn after a time just where to make the right cut. But a more satisfactory way is to make a careful study before the material is cooked, and thus learn the exact position of every joint, bone, and muscle. Become familiar with a shoulder or a leg of mutton; locate the joints by moving the bones in the joints, or by cutting it into sections, some time when it is to be used for a stew. Or remove the bone in the leg by scraping the meat away at either end. Learn to distinguish the different cuts of meat. The best way to learn about carving poultry and game is to cut them up for a stew or fricassee, provided care be taken not to chop them, but to disjoint them skillfully.

Then, when you attempt to carve, do the best you can every time. Never allow yourself to be careless about it, even should the only spectators be your wife and children. But do not make your first effort in the art at a company dinner. Every lady should learn the art. There is no reason why she may not excel in it, as she has every opportunity to study the joint or fowl before cooking. Strength is not required, so much as neatness and care. A firm, steady hand, a cool, collected manner, and confidence in one's ability will help greatly. Children also should be taught this accomplishment, and should be taught it as soon as they can handle a knife safely. If parents would allow the children to share their duties at the daily family table, and occasionally when company is present, a graceful manner would soon be acquired. When called upon to preside over their own homes there would less frequently be heard the apology, "Father always carved at home, and I have had no practice." The only recollection that I now have of a dinner at a friend's some years ago is the easy and skillful way a young son of my hostess presided at the head of the table, while the father occupied the place of guest at the mother's right hand.

One must learn first of all to carve neatly, without scattering crumbs or splashing gravy over the cloth or platter; also to cut straight, uniform slices. This may seem an easy matter; but do we often see pressed beef, tongue, or even bread cut as it should be? Be careful to divide the material in such a manner that each person may be served equally well. Have you never received all flank, or a hard dry wing, while another guest had all tenderloin, or the second joint? After a little experience you can easily distinguish between the choice portions and the inferior. Lay each portion on the plate with the browned or best side up. Keep it compact, not mussy; and serve a good portion of meat, not a bone with hardly any meat on it.

After all are served, the portion on the platter should not be left jagged, rough, and sprawling, but should look inviting enough to tempt one to desire a second portion.

Care should be taken to carve in such a way as to get the best effect. A nice joint is often made less inviting from having been cut with the grain, while meat of rather poor quality is made more tender and palatable if divided across the grain. Where the whole of the joint is not required, learn to carve economically, that it may be left in good shape for another dinner.

After you have learned to do the simplest work neatly and gracefully, much painstaking will be necessary in acquiring the power to accomplish with elegance the more difficult

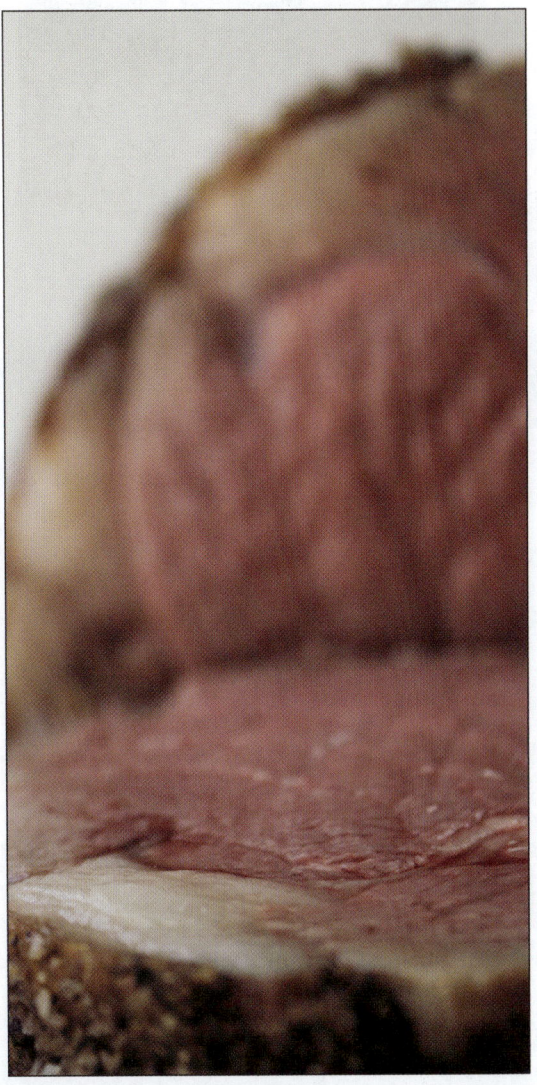

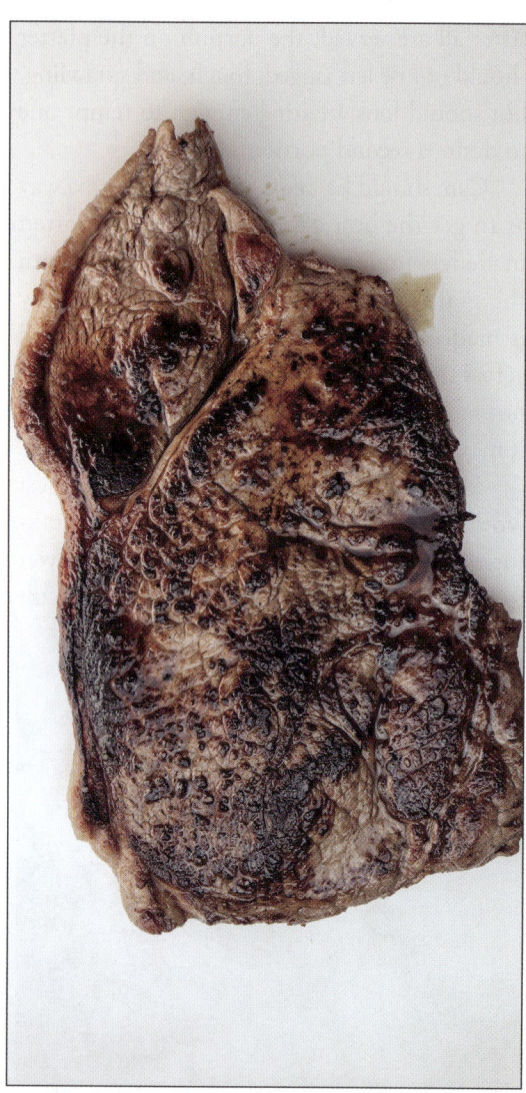

tasks. For to reach the highest degree of excellence in the art, one must be able to carve the most difficult joint with perfect skill and ease.

But after all this study and a great amount of practice failure often happens, and blame is laid upon the carver which really belongs to some other person,—the butcher, the cook, the table-girl, or the guest. Not all men who sell meat know or practice the best way of cutting up meat. Much may be done by the butcher and by the cook to facilitate the work of the carver. These helps will be noticed more particularly under the head of special dishes.

An essential aid to easy carving, and one often overlooked, is that the platter be large enough to hold not merely the joint or fowl while whole, but also the several portions as they are detached.

The joint should be placed in the middle of the platter, in the position indicated under special directions. There should be sufficient space on either side for the portions of meat as they are carved; that is, space on the bottom, none of the slices being allowed to hang over the edge of the dish. If necessary, provide an extra dish. The persistency with which some housekeepers cling to a small dish for fear the meat will look lost on a larger one often makes successful carving impossible.

The platter should be placed near the carver, that he may easily reach any part of the joint. The cook should see that all skewers, strings, etc., be removed before sending the meat or fish to the table. It is extremely awkward to find one's knife impeded by a bit of twine.

The carver may stand or sit, as suits his convenience. Anything that is done easily is generally done gracefully, but when one works at a disadvantage awkwardness is always the result.

A very important matter is the condition of the knife. It should have a handle easy to grasp, a long, thin, sharp, pointed blade, and be of a size adapted to the article to be carved and to the person using it. A lady or a child will prefer a small knife. Be as particular to have the knife sharp as to have it bright and clean; and always sharpen it before announcing the dinner. It is very annoying for a person to be obliged to wait and sharpen the knife, or to turn the meat round to get it into the right position. Never allow a carving-knife to be used to cut bread, or for any other than its legitimate purpose.

The fork should be strong, with long tines, and should have a guard. Place the fork deep enough in the meat so that you can hold it firmly in position. Hold the knife and fork in an easy, natural way. Many persons grasp the fork as if it were a dagger, and stab it into the meat; but such a display of force is unnecessary and clownish. The hand should be over the handle of the fork, the palm down, and the forefinger extended.

1. Loin: best end.
2. Loin: chump end.
3. Fillet.
4. Hind Knuckle.
5. Fore Knuckle.
6. Neck: best end.
7. Neck: scrag end.
8. Blade bone.
9. Breast: best end.
10. Breast: brisket end.

1. The Spare rib. 2. Hand. 3. Belly or Spring. 4. Fore Loin. 5. Hind Loin. 6. Leg.

1. Leg.
2. Loin: best end.
3. Loin: chump end.
4. Neck: best end.
5. Neck: scrag end.
6. Shoulder.
7. Breast.
A Chine is two Loins.
A Saddle is two Necks.

Do not appear to make hard work of carving. Avoid all scowling or contortion of the mouth if a difficult spot be touched. Don't let your countenance betray the toughness of the joint or your own lack of skill. Work slowly but skillfully, and thus avoid the danger of landing the joint in your neighbor's lap.

Do not be guilty of the discourtesy of asking each guest, before you begin to carve, to choose between roast lamb and warmed-over beef, or between pie and pudding, or whatever you may have, and thus cause a guest who may have chosen the lamb or the pie the discomfort of knowing that it has been cut solely for her. Such economy may be excusable in the privacy of one's own family, but not in the presence of invited guests. First divide or carve what you have to serve, and then offer the choice to your guests.

"To carve and serve decently and in good order" is indeed mainly the duty of the host; but there is sometimes an unfortunate lack of skill on the part of the hostess in her share of the serving. A certain pride is permitted to her, and is expected of her, in serving neatly her tea, coffee, and soup, in dividing appropriately her pies and puddings, and even in cutting and arranging deftly the bread upon her board.

A word to the guest, and then we will proceed to explicit directions. Never stare at the carver. Remember you are invited to dine, not to take a lesson in carving. Appear perfectly unconscious of his efforts; a glance now and then will give you sufficient insight into his method. There often seems to be an irresistible fascination about carving which silences all tongues and draws all eyes to the head of the table. The most skillful carver will sometimes fail if conscious of being watched. With a little tact the hostess can easily engage the attention of her guests, that the carver may not be annoyed.

Should your preference be asked, and you have any, name it at once, provided there is also enough for others who may prefer the same kind. Remember there are only two fillets, or side-bones, or second joints; if you are the first to be served, do not test the skill of the carver by preferring a portion difficult to obtain.

Many of these cautions may seem uncalled for, but they have been suggested by personal observation of their necessity. People of good breeding would never err in any of these ways; but alas, not all people are well bred, and innate selfishness often crops out in small matters.

The following explicit directions have not been taken from books. They were given to the writer a few years ago by one who was an adept in the art, who had received her instruction from a skillful surgeon, and who at her own table gave a practical demonstration of the fact that a lady can not only "carve decently and in good order," but with ease and elegance.

SPECIAL DIRECTIONS

TIP OF THE SIRLOIN, OR RIB ROAST
It is easier to carve this joint by cutting across the ribs, parallel with the backbone, but that is cutting with the grain; and meat, especially beef, seems more tender if cut across the grain.

Place it on the platter with the backbone at the right. If the backbones be not removed before cooking, place the fork in the middle and cut close to the backbone down to the ribs. Shave off the thick, gristly cord near the backbone, as this, if left on, interferes with cutting thin slices. Then cut, from the side nearest you, thin uniform slices parallel with the ribs. Run the knife under and separate them from the bone. Many prefer to remove the bone and skewer the meat into a roll before cooking. It may then be laid, flesh down, on the dish, and carved across the top horizontally in thin slices; or if you find it easier, place it with the skin surface up, and carve down from the flesh side nearest you.

This style of serving is generally preferred, but there are advantages in retaining the bone; for the thin end when rolled under is not cooked to such a nice degree of crispness, and the slices are usually larger than desired. Again, the ribs, by keeping the meat in position, secure for it a clean cut, and not one broken and jagged, and the thin end may be served or not, as you please.

SIRLOIN ROAST

The backbone or thickest end should be at the right end of the dish. Carve a sirloin roast by cutting several thin slices parallel with the ribs. Then cut down across the ribs near the backbone, and also at the flank end, and separate the slices.

The slices should be as thin as possible and yet remain slices, not shavings. Turn the meat over and cut out the tenderloin and slice it in the same manner across the grain; or turn the meat over and remove the tenderloin first. Many prefer to leave the tenderloin to be served cold. Cut slices of the crisp fat on the flank in the same way, and serve to those who wish it. This is a part which many dislike, but some persons consider it very choice. Always offer it unless you know the tastes of those whom you are serving.

THE BACK OF THE RUMP

A roast from the back of the rump, if cooked without removing the bone, should be placed on the platter with the backbone on the farther side. Cut first underneath to loosen the meat from the bone. Then, if the family be large and all the meat is to be used, the slices may be cut lengthwise; but should only a small quantity be needed, cut crosswise and only from the small end. It is then in better shape for the second day.

It is more economical to serve the poorer parts the first day, as they are never better than when hot and freshly cooked. Reserve the more tender meat to be served cold.

FILLET OF BEEF OR TENDERLOIN

Before cooking, remove all the fat, and every fibre of the tough white membrane. Press it into shape again and lard it, or cover it with its own fat. If this fibre be not removed, the sharpest knife will fail to cut through it. Place it on the platter with the larger end at the right; or if two short fillets be used, place the thickest ends in the middle. Carve from the thickest part, in thin, uniform slices.

ROUND OF BEEF, FILLET OF VEAL, OR FRICANDEAU OF VEAL

These are placed on the platter, flesh side up, and carved in horizontal slices, care being taken to carve evenly, so that the portion remaining may be in good shape. As the whole of the browned outside comes off with the first slices, divide this into small pieces, to be served if desired with the rare, juicy, inside slices.

BEEF STEAK

It may seem needless to direct one how to carve a sirloin steak, but it sometimes appears to require more skill than to carve poultry, as those who have been so unfortunate as to receive only the flank can testify.

I believe most strongly, as a matter of economy, in removing the bone, and any tough membrane or gristle that will not be eaten, before cooking the steak. If there be a large portion of the flank, cook that in some other way. With a small, sharp knife cut close to the rib on each side, round the backbone, and remove the tough white membrane on the edge of the tenderloin. Leave the fat on the upper edge, and the kidney fat also, or a part of it, if it be very thick. There need be no waste or escape of juices if the cutting be done quickly, neatly, and just before cooking. Press the tenderloin—that is, the small portion on the under side of the bone—close to the upper part, that the shape may not be changed.

In serving place it on the dish with the tenderloin next to the carver. Cut in long narrow strips from the fat edge down through the tenderloin. Give each person a bit of tenderloin, upper part, and fat. If the bone be not removed before cooking, remove the tenderloin first by cutting close to the bone, and divide it into narrow pieces; then remove the meat from the upper side of the bone and cut in the same manner. A long, narrow strip about as wide as the steak is thick is much more easily managed on one's plate than a square piece. Serve small portions, and then, if more be desired, help again.

In carving large rump steaks or round steaks, cut always across the grain, in narrow strips. Carving-knives are always sharper than table-knives, and should do the work of cutting the fibers of the meat; then the short fibers may easily be separated by one's own knife. There is a choice in the several muscles of a large rump steak, and it is quite an art to serve it equally.

LEG OF MUTTON OR LAMB, OR KNUCKLE OF VEAL

Before cooking, remove the rump-bones at the larger end. For a small family it is more economical to remove all the bones and fill the cavity with stuffing. Tie or skewer it into compact shape; there is then less waste, as the meat that is not used at the first dinner does not become dry and hard by keeping.

In serving, the thickest part of the leg should be toward the back of the platter. Put the fork in at the top, turn the leg toward you to bring the thickest part up, and cut through to the bone. Cut several slices of medium thickness, toward the thickest part, then slip the knife under and cut them away from the bone. A choice bit of crisp fat may be found on the larger end, and there is a sweet morsel near the knuckle or lower joint. If more be required, slice from the under side of the bone in the same manner.

LEG OF VENISON

This is carved in the same way as a leg of mutton,—through the thickest part down to the bone.

SADDLE OF MUTTON

Remove the ends of the ribs and roll the flank under before cooking. Place it on the platter with the tail end at the left. Put the fork in firmly near the centre, and carve down to the ribs in long slices, parallel with the backbone, and the whole length. Slip the knife under and separate the slices from the ribs; do the same on the other side of the back. Divide the slices if very long. Cut the crisp fat from the sides in slanting slices. Turn partly over and remove the choice bit of tenderloin and kidney fat under the ribs.

Carving a saddle of mutton in this way is really cutting with the grain of the meat, but it is the method adopted by the best authorities. It is only the choicest quality of mutton, and that which has been kept long enough to be very tender, that is prepared for cooking in this way. The fibers are not so tough as those of beef; there is no perceptible difference in the tenderness of the meat when cut in this manner, and there is an advantage in obtaining slices which are longer, and yet as thin as those from cutting across the grain.

SADDLE OF VENISON

Carve the same as a saddle of mutton. Serve some of the dish gravy with each portion. Venison and mutton soon become chilled, the fat particularly, thus losing much of their delicacy. Send them to the table very hot, on hot platters; carve quickly, and serve at once on warm plates.

HAUNCH OF VENISON OR MUTTON

This is the leg and loin undivided, or, as more commonly called, the hind quarter. The butcher should split the whirl-bone, disjoint the backbone, and split the ribs in the flank. The rump-bone and aitch-bone may be removed before cooking. Place it on the platter with the loin or backbone nearest the carver. Separate the leg from the loin; this is a difficult joint to divide when the bones have not been removed, but it can be done with practice. When the leg has been taken off, carve

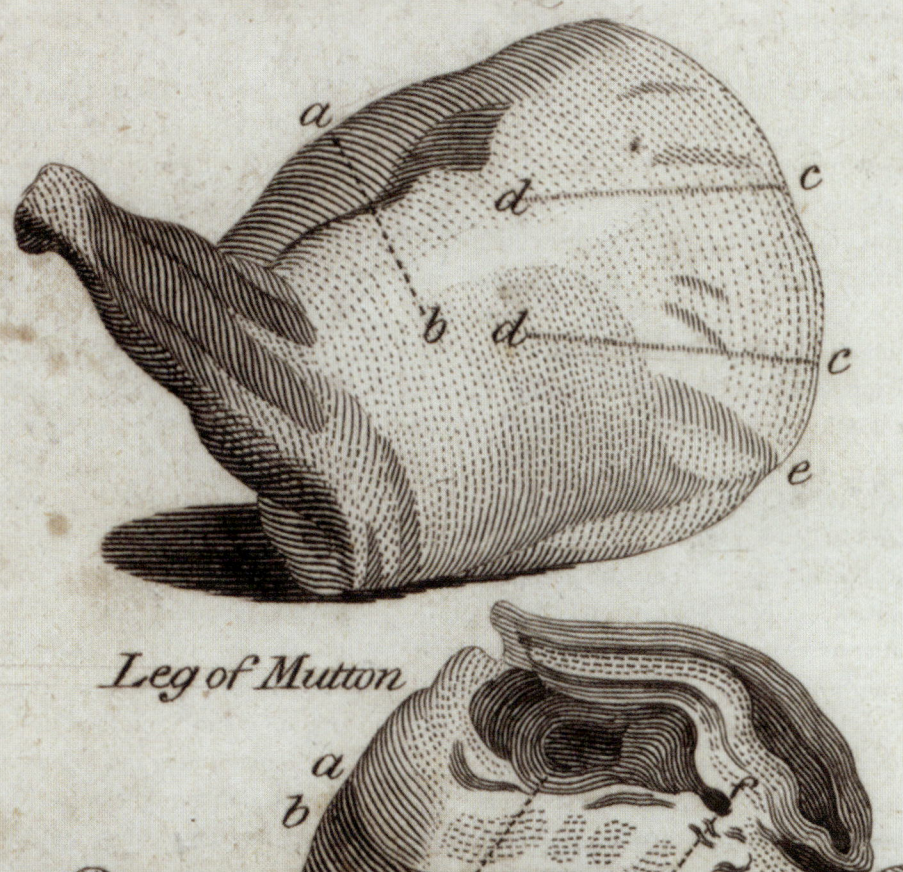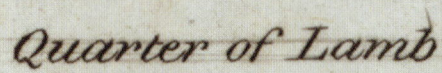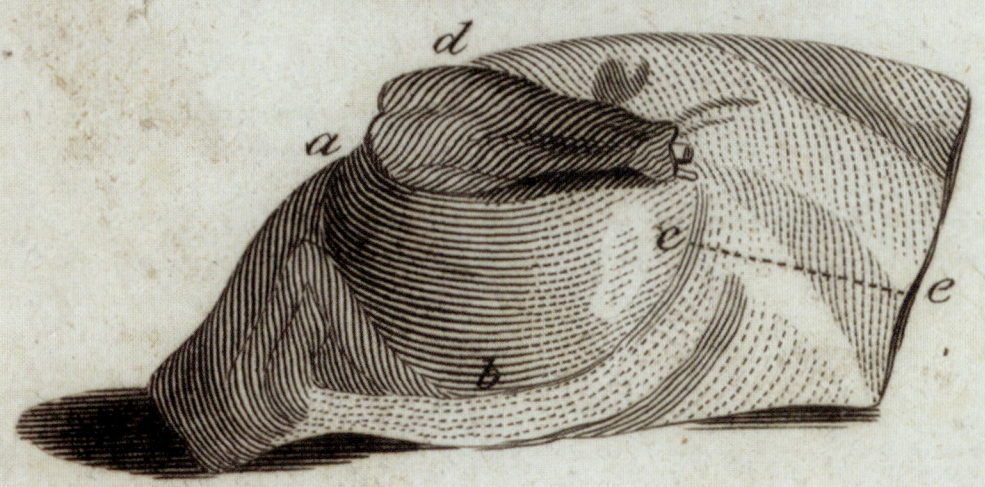

that as directed on page 762. Carve the loin by first cutting off the flank and dividing it, then divide between each rib in the loin, or cut long slices parallel with the backbone, in the same way as directed for a saddle of mutton. Some English authorities recommend cutting perpendicularly through the thickest part of the leg near the knuckle, and then cutting across at right angles with this first cut, in long thin slices, the entire length of the joint; the slices are then separated from the bone and divided as desired. When carved in this way the loin and leg are not divided. This is not so economical as the first method.

LOIN OF MUTTON, LAMB, VEAL, PORK, OR VENISON

These should always be divided at the joints in the backbone by the butcher; then it is an easy matter to separate the ribs, serving one to each person, with a portion of the kidney and fat if desired. But if the butcher neglect to do this, and you have no cleaver with which to do it, it is better to cut slices down to the ribs parallel with the backbone, as directed in the saddle of mutton, than to suffer the annoyance of hacking at the joints.

Before cooking a loin of pork, gash through the fat between the ribs; this will give more of the crisp fat, and will aid in separating the ribs.

SHOULDER OF MUTTON OR VEAL

Place it on the platter with the thickest part up. From the thickest part cut thin slices, slanting down to the knuckle; then make several cuts across to the larger end, and remove these slices from the shoulder-blade. Separate the blade at the shoulder-joint, and remove it. Cut the meat under the blade in perpendicular slices.

Any part of the forequarter of mutton is more tender and palatable, and more easily carved, if before cooking it be boned and stuffed. Or it may be boned, rolled, and corned.

FOREQUARTER OF LAMB OR VEAL

This is a difficult joint for a beginner, but after a little study and practice one may manipulate it with dexterity. Some time when a lamb stew or fricassee is to be prepared, study the joint carefully and practice cutting it up, and thus become familiar with the position of the shoulder-blade joint,—the only one difficult to reach. The backbone should always be disjointed. The ribs should be divided across the breast and at the junction of the breast-bone, and the butcher should also remove the shoulder-blade and the bone in the leg. Unless the joint be very young and tender, it is better to use the breast portion for a stew or fricassee; but when nice and tender the breast may be roasted with the other portions, as the choice gelatinous morsels near the breast-bones are preferred by many. This joint consists of three portions,—the shoulder or knuckle, the breast or brisket, and the ribs. Put it on the platter with the backbone up. Put the fork in near the knuckle. Cut through the flesh clear round the leg and well up on the shoulder, but not too far on the breast. With the fork lift the leg away from the shoulder, cutting in till you come to the joint, after separating which, remove the leg to a separate dish, to be afterward cut into thin slices through the thickest part. Cut across from left to right where the ribs have been broken, separating the gristly breast from the upper portion. Then remove the blade if it has not been done before cooking. Divide each of these portions between the ribs, and serve a piece of the rib, the breast, or a slice from the leg, as preferred.

NECK OF VEAL

The vertebra should be disjointed, and the ribs cut on the inside through the bone only, on the thin end. Place it on the platter with the back up and cut across from left to right, where the ribs were divided, separating the small ends of the ribs from the thicker upper portion; then cut between each short rib. Carve from the back down in slanting slices, then slip the knife under close to the ribs and remove the slices.

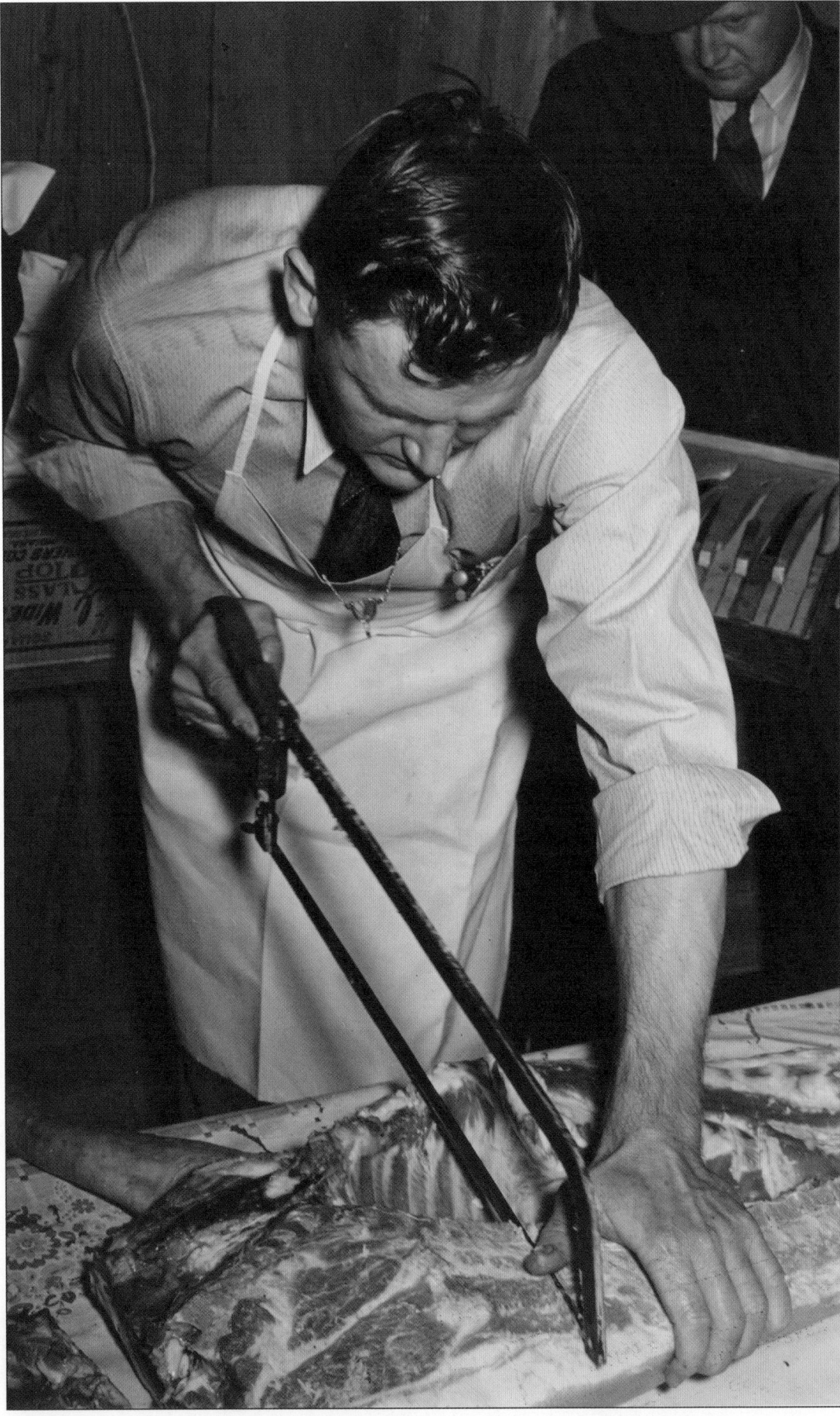

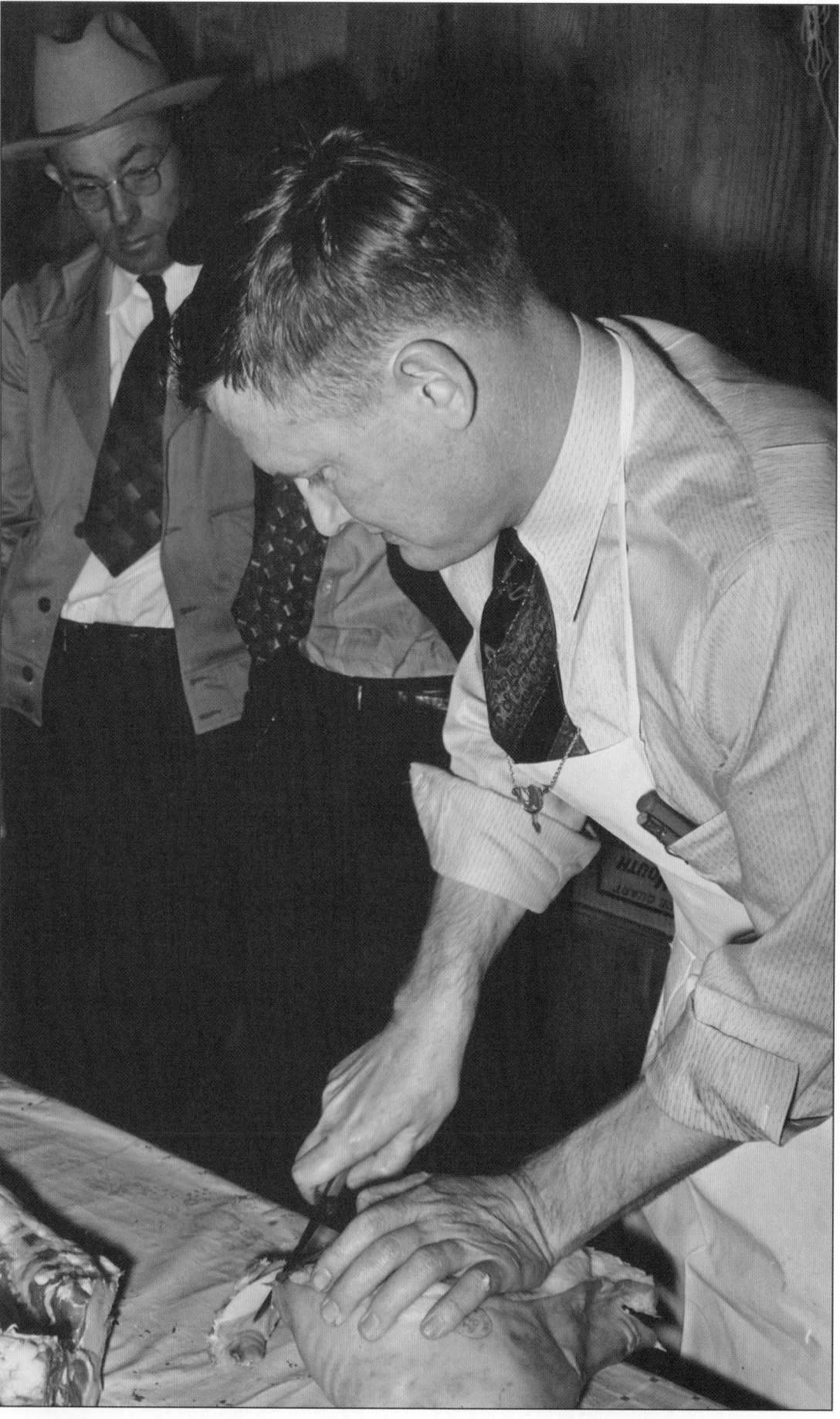

This gives a larger portion than the cutting of the slices straight would give, and yet not so large as if each were helped to a whole rib. Serve a short rib with each slice.

BREAST OF VEAL

Place it on the dish with the breast-bone or brisket nearest you. Cut off the gristly brisket, then separate it into sections. Cut the upper part parallel with the ribs, or between each rib if very small. Slice the sweetbread, and serve a portion of brisket, rib, and sweetbread to each person.

CALF'S HEAD

Calf's head served whole is a favorite dish in England, but seldom seen on American tables. For those who have this preference a few hints about carving may be desirable. Place it on the platter with the face toward the right. Cut from left to right, through the middle of the cheek down to the bone, in several parallel slices of medium thickness; then separate them from the bone. Cut down at the back of the throat and slice the throat sweetbread. With the point of the knife cut out the gelatinous portion near the eye, and serve to those who desire it. There is a small portion of delicate lean meat to be found after removing the jawbone. Some are fond of the palate, which lies under the head. The tongue should be sliced, and a portion of this and of the brains offered to each person.

ROAST PIG

This is sometimes partly divided before serving. Cut off the head and divide it through the middle; then divide through the backbone. Place it on the platter back to back, with half the head on each end of the dish.

If the pig be very young, it is in better style to serve it whole. Before cooking, truss the forelegs forward and the hind legs backward. Place the pig on the platter with the head at the left. Cut off the head, separating the neck-joint with the point of the knife, then cut through the flesh on either side. Take off the shoulders by cutting in a circle from under the foreleg round nearly to the backbone and down again. Bend it forward and cut through the joint. Cut off the hams in the same way. Then split the backbone the entire length and divide between each rib. Cut slices from the thickest part of the hams and the shoulders. The ribs are the choice portion, but those who like it at all consider any part of it a delicacy.

HAM

If the ham is not to be served whole, the simplest and most economical way is to begin near the smaller end and cut in very thin slices, on each side of the bone. Divide the slices and arrange them neatly on the dish, one lapping over another, with the fat edge outside.

Where the whole ham is to appear on the table it should be trimmed neatly, and the end of the bone covered with a paper ruffle. The thickest part should be on the further side of the platter. Make an incision through the thickest part, a little way from the smaller end. Shave off in very thin slices, cutting toward the larger end and down to the bone at every slice. The knife should be very sharp to make a clean cut, and each slice should have a portion of the fat with the crisp crust. To serve it hot a second day, fill the cavity with a bread stuffing, cover it with buttered crumbs, and brown it in the oven. If it is to be served cold, brown the crumbs first and then sprinkle them over the stuffing. If this be done the edges will not dry and the symmetry of the ham is preserved. Carve as before, toward the larger end, and if more be needed, cut also from the other side of the bone.

By filling the cavity again with stuffing, a ham may be served as a whole one the third time and look as inviting as when first served. Should there be two or three inches of the thickest end left for another serving, saw off the bone, lay the meat flesh side up, with the fat on the further side of the platter, and carve horizontally in thin slices.

TONGUE

The center of the tongue is the choicest portion. Cut across in slices as thin as a wafer. The tip of the tongue is more delicate when cut lengthwise in thin slices, though this is not the usual practice.

CORNED BEEF

Corned beef should be put while hot into a pan or mould, in layers of fat and lean, with the fibers running the long way of the pan. After pressing it, place it on the platter and slice thinly from one end. This gives uniform slices, cut across the grain, each one having a fair proportion of fat and lean.

CHARTREUSE, OR PRESSED MEAT

Any moulds of meat, either plain or in jelly or rice, should be cut from one end, or in the middle and toward either end, in uniform slices, the thickness varying with the kind of meat. Be careful not to break them in serving. If only a part of a slice be desired, divide it neatly. Help also to the rice or jelly.

TO CUT UP A CHICKEN FOR A STEW OR FRICASSEE

Nothing is more unsightly and unappetizing than a portion of chicken with the bones chopped at all sorts of angles, and with splinters of bone in the meat. All bones will separate easily at the joint when the cord or tendon and gristly portion connecting them have been cut.

After the chicken has been singed and wiped, and the crop removed from the end of the neck, place it in front of you with the breast up and the neck at the left. With a small sharp knife make an incision in the thin skin between the inside of the legs and the body. Cut through the skin only, down toward the right side of the leg, and then on the left. Bend the leg over toward you, and you will see where the flesh joins the body and also where the joint is, for the bone will move in the joint. Cut through the flesh close to the body, first on the right of the joint and then on the left, and as you bend the leg over, cut the cord and gristle in the joint, and this will free the leg from the body. Find the joint in the leg and divide it neatly. Work the wing until you see where the joint is, then cut through the flesh on the shoulder, bend the wing up and cut down through the gristle and cord. Make a straight clean cut, leaving no jagged edges. Divide the wing in the joint, and then remove the leg and wing from the opposite side, and divide in the same way. Make an incision in the skin near the vent, cut through the membrane lying between the breast and the tail down to the backbone on each side, remove the entrails, and break off the backbone just below the ribs. Separate the side-bones from the back by cutting close to the backbone from one end to the other on each side. This is a little difficult to do; and in your first experiment it would be better not to divide it until after boiling it, as it separates more easily after the connecting gristle has been softened by cooking. Take off the neck close to the back by cutting through the flesh and twisting or wringing it until the bone is disjointed.

Cut off the wish-bone in a slanting direction from the front of the breast-bone down to the shoulder on each side. Cut through the cartilage between the end of the collar-bone and the breast. Cut between the end of the shoulder-blade and the back down toward the wing-joint, turn the blade over toward the neck, and cut through the joint.

This joint in the wing, collar-bone, and shoulder-blade is the hardest to separate. Remove the breast from the back by cutting through the cartilage connecting the ribs; this can be seen from the inside. The breast should be left whole and the bone removed after stewing; but if the chicken is to be fried you may remove the bone first.

It is not necessary in boiling a chicken to divide it so minutely, for the wings and legs can be disjointed, and the side-bones and breast separated from the back more easily after cooking; but it is valuable practice, and if one learns to do it neatly it will help in carving a boiled fowl or roast turkey.

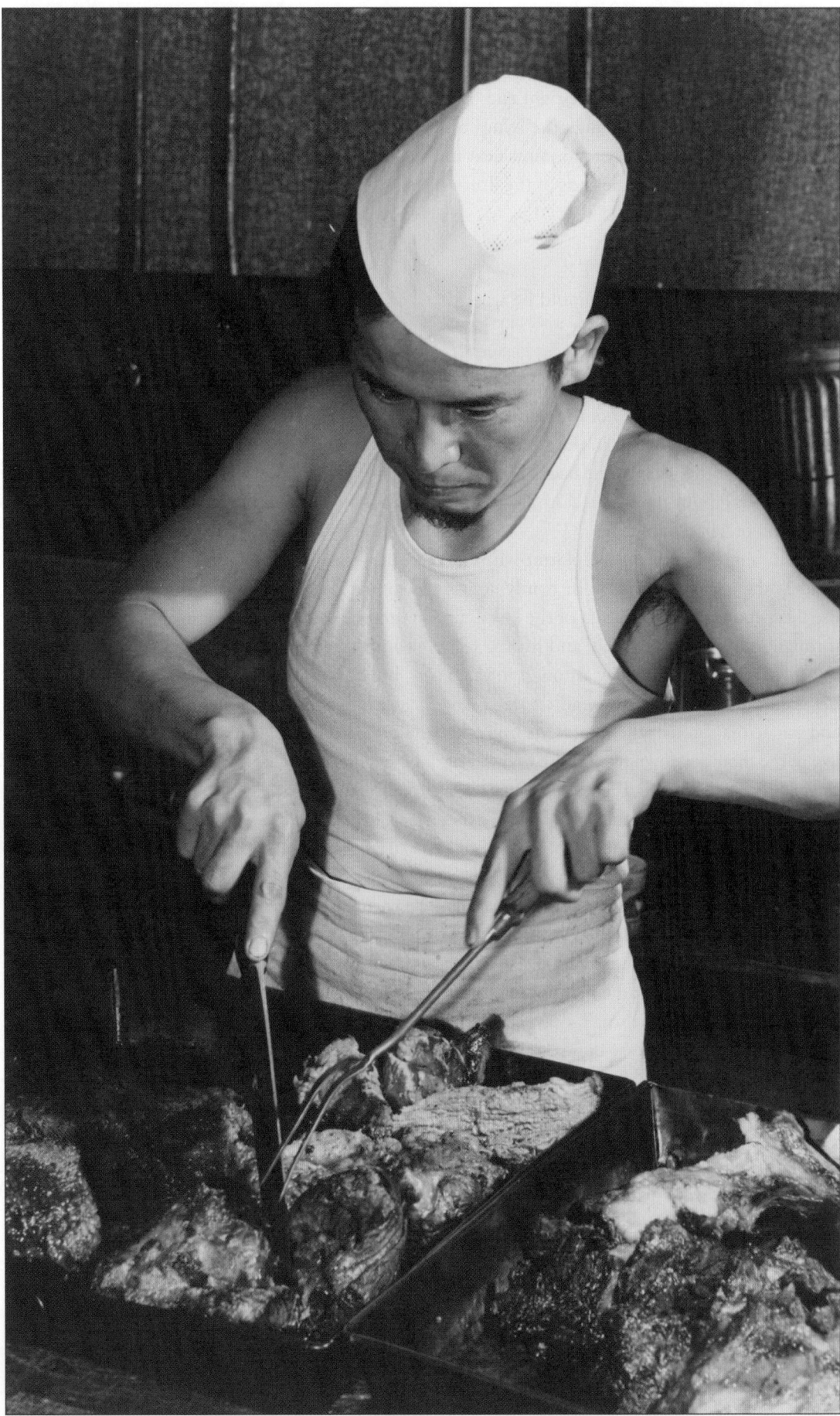

In arranging a fricasseed chicken on the platter, put the neck and ribs at the left end of the dish and the backbone at the right end. Put the breast over the ribs, arrange the wings on each side of the breast, the second joints next to the side-bones, and cross the ends of the drumsticks over the tail.

BOILED FOWL OR TURKEY

Fowls or turkeys for boiling should be trussed with the ends of the legs drawn into the body through a slit in the skin, and kept in place with a small skewer. Turn the tip of the wing over on the back. Cut off the neck, not the skin, close to the body, and after putting in the stuffing, fasten the skin of the neck to the back. Put strips of cloth round it, or pin it in a cloth, to keep it white and preserve the shape.

In carving, place it on the platter with the head at the left. Put the fork in firmly across the breast-bone. With the point of the knife cut through the skin near the tail, and lift the legs out from the inside. Then cut through the skin between the legs and body, bend the leg over, and cut across through the joint. Cut from the top of the shoulder down toward the body until the wing-joint is exposed, then cut through this, separating the wing from the body. Remove the leg and wing from the other side. Shave off a thin slice on the end of the breast toward each wing-joint, slip the knife under at the top of the breast-bone, and turn back the wish-bone.

Capons and large fowls may be sliced thinly across the breast in the same manner as a roast turkey. But if the fowl be small, draw the knife along the edge of the breast-bone on each side, and lay the meat away from the bone; the fillets will separate easily. Then divide the meat across the grain. Separate the collar-bone from the breast. Slip the knife under the shoulder-blade, turn it over, and separate at the joint. Cut through the cartilage connecting the ribs; this will separate the breast from the back. Now remove the fork from the breast, turn the back over, place the knife midway, and with the fork lift up the tail end, separating the back from the body. Place the fork in the middle of the backbone, cut close to the backbone from one end to the other on each side, freeing the side-bones.

The wing and breast of a boiled fowl are the favorite portions. It is important that the fowl be cooked just right. If underdone, the joints will not separate readily; and if overdone they will fall apart so quickly that carving is impossible. Unless the knife be very sharp, and the work done carefully, the skin of the breast will come off with the leg or wing.

BROILED CHICKEN

Split the chicken down the back and remove the backbone. If the chicken be very young and tender—and only such are suitable for broiling—remove the breast-bone before cooking, or cut the bone through the middle, lengthwise and crosswise from the inside, without cutting into the meat. In serving, divide through the breast from the neck down, and serve half to each person; or if a smaller portion be desired, divide each half crosswise through the breast, leaving the wing on one part and the leg on the other.

If the chicken be large, break the joints of the legs, thighs, and wings, without breaking through the skin; cut the tendons on the thighs from the inside, cut the membrane on the inside of the collar-bone and wing-joint, and remove the breast-bone. This may all be done before cooking, and will not injure the appearance of the outside.

In serving, separate the legs and wings at the joints, then separate the breast from the lower part, and divide the breast lengthwise and crosswise.

Carving-scissors are convenient for cutting any kind of broiled game or poultry.

ROAST TURKEY

Turkeys should be carefully trussed. The wings and thighs should be brought close to the body and kept in position by skewers. The ends of

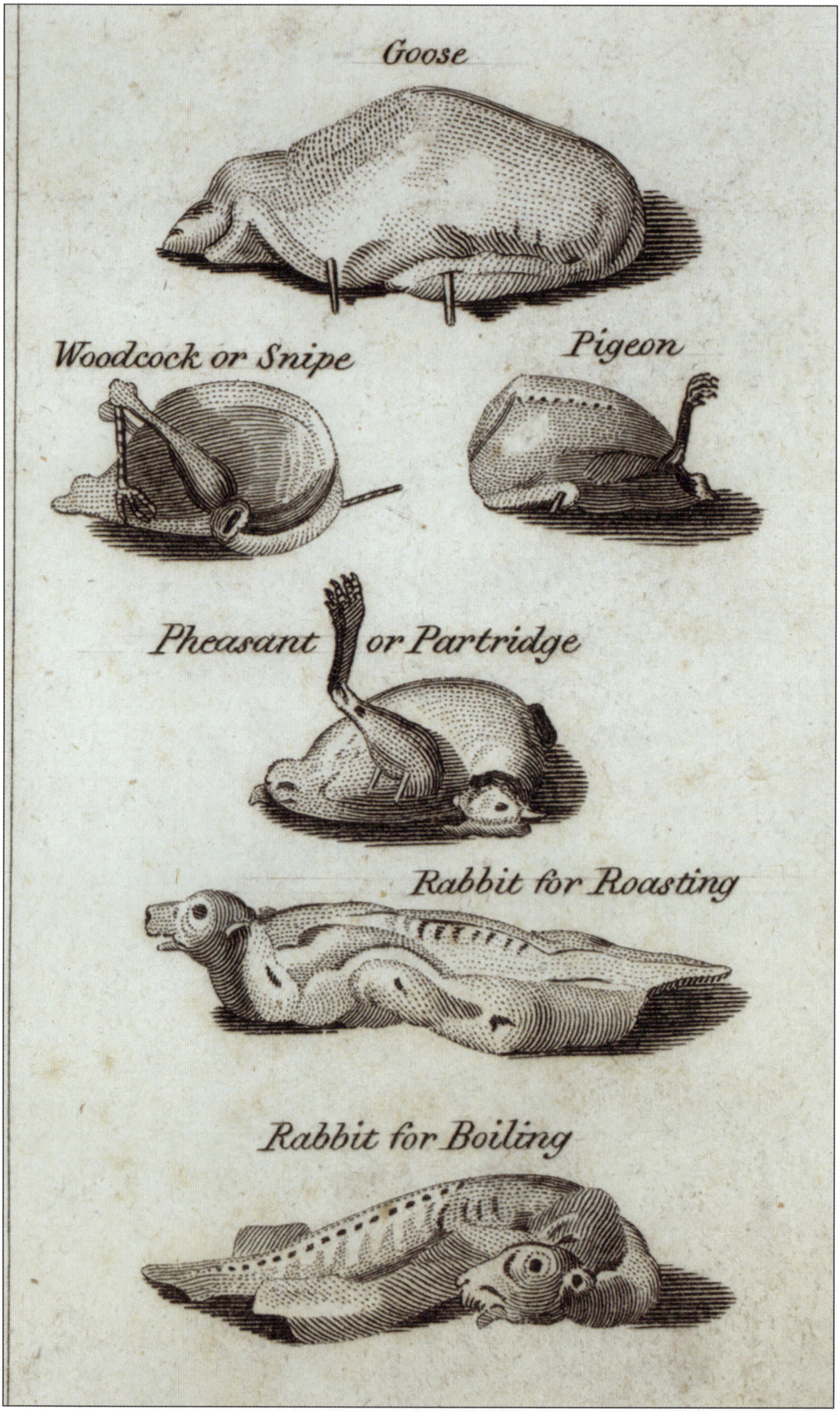

the drumsticks may be drawn into the body or crossed over the tail and tied firmly.

After cooking, free the ends of the drumsticks from the body and trim them with a paper ruffle. This will enable the carver to touch them if necessary without soiling his hands. Place the turkey on the platter with the head at the left. Unless the platter be very large, provide an extra dish, also a fork for serving.

Insert the carving-fork across the middle of the breast-bone. Cut through the skin between the breast and the thigh. Bend the leg over, and cut off close to the body and through the joint. Cut through the top of the shoulder down through the wing-joint. Shave off the breast in thin slices, slanting from the front of the breast-bone down toward the wing-joint.

If the family be small and the turkey is to be served for a second dinner, carve only from the side nearest you. Tip the bird over slightly, and with the point of the knife remove the oyster and the small dark portion found on the side-bone. Then remove the fork from the breast and divide the leg and wing. Cut through the skin between the body and breast, and with a spoon remove a portion of the stuffing. Serve light or dark meat and stuffing, as preferred. If carved in this way, the turkey will be left with one half entire, and if placed on a clean platter with the cut side nearest the carver, and garnished with parsley, will present nearly as fine an appearance, to all but the carver, as when first served.

When there are many to be served, take off the leg and wing from each side and slice the whole of the breast before removing the fork; then divide as required.

It is not often necessary to cut up the whole body of the turkey; but where every scrap of the meat will be needed, or you wish to exercise your skill, proceed to carve in this manner.

Put the fork in firmly across the middle of the breast-bone. Cut through the skin between the leg and body. Bend the leg over and cut off at the joint. If the turkey be very tender or overcooked, the side-bone will separate from the back and come away with the second joint, making it more difficult to separate the thigh from the side-bone. Cut through the top of the shoulder and separate the wing at the joint. Cut off the leg and wing from the other side. Carve the breast on each side, in thin slices, slanting slightly toward the wing. Be careful to take a portion of crisp outside with each slice. Shave off the crisp skin near the neck, in order to reach the stuffing. Insert the point of the knife at the front of the breast-bone, turn back the wish-bone and separate it. Cut through the cartilage on each side, separating the collar-bones from the breast. Tip the body slightly over and slip the knife under the end of the shoulder-blade; turn it over toward the wing. Repeat this process on the opposite side. Cut through the cartilage which divides the ribs, separating the breast-bone from the back. Lay the breast on one side and remove the fork from it. Take the stuffing from the back. Turn the back over, place the knife midway just below the ribs, and with the fork lift up the tail end, separating the back from the body. Place the fork in the middle of the backbone, and cut close to the backbone from one end to the other, on each side, freeing the side-bone. Then divide the legs and wings at the joints. The joint in the leg is not quite in the middle of the bend, but a trifle nearer the thigh. It requires some practice to strike these joints in the right spot. Cut off the meat from each side of the bone in the second joint and leg, as these when large are more than one person requires, and it is inconvenient to have so large bones on one's plate.

It is easier to finish the carving before beginning to serve. An expert carver will have the whole bird disjointed and literally in pieces with a very few strokes of the knife.

ROAST GOOSE

A green goose neatly trussed and "done to a turn" looks very tempting on the platter; but there is so little meat in proportion to the size of the bird that unless it be skillfully carved only a small number can be served. The breast of a goose is broader and flatter then that of a turkey. It should be carved in a different manner, although many writers give the same directions for carving both.

Place it on the platter with the head at the left. Insert the fork firmly across the ridge of the breast-bone. Begin at the wing and cut down through the meat to the bone, the whole length of the breast. Cut down in the same way in parallel slices, as thin as can be cut, until you come to the ridge of the breast-bone. Slip the knife under the meat at the end of the breast, and remove the slices from the bone. Cut in the same manner on the other side of the breast. Cut through the skin below the breast, insert a spoon and help to the stuffing. If more be required, cut the wing off at the joint. Then tip the body over slightly and cut off the leg. This thigh-joint is tougher, and requires more skill in separating, than the second joint of a turkey. It lies nearer the backbone. But practice and familiarity with its location will enable one to strike it accurately. The wish-bone, shoulder-blade, and collar-bone may be removed according to the directions given for carving roast turkey. Some prefer to remove the wing and leg before slicing the breast.

ROAST DUCK
Place it in the same position and carve in the same way as a goose. Begin at the wing, and cut down to the bone in long thin slices, parallel with the breast-bone; then remove them from the bone. The breast is the favorite portion; but the "wing of a flyer and the leg of a swimmer" are esteemed by epicures.

The stuffing is not often desired, but if so it may be found by cutting across below the end of the breast.

Geese and ducks are seldom entirely cut up at the table, as there is very little meat on the back. But often from a seemingly bare carcass enough may be obtained to make a savory entrée.

PIGEONS
These, if small, are served whole. If large, cut through the middle from the neck to the end of the breast and down through the backbone. The bones are thin, and may easily be divided with a sharp knife. When smaller portions are required, cut from the shoulder down below the leg, separating the wing and leg from the body.

PARTRIDGES
Cut through above the joint of the wing, down below the leg, and remove the wing and leg in one portion. Cut under the breast from the lower end through the ribs to the neck and remove the breast entire. Then divide it through the middle, and, if very plump, divide again. When very small they may be divided through the breast and back into two equal parts.

LARDED GROUSE
Turn the legs over and free them from the body. Cut slices down to the bone the entire length of the breast; then slip the knife under and remove the slices. Cut off the wing and leg, and separate the backbone from the body. There are some morsels on the back which are considered choice by those who like the peculiar flavor of this game. As this is a dry meat, help generously to the bread sauce which should always accompany it.

Where this is the principal dish, or where a larger portion is required, divide it through the breast, as directed for small pigeons.

Woodcock, Snipe, and other small birds are usually served whole. But if only a portion be desired, divide them through the breast.

RABBIT
A rabbit should be trussed, with the forelegs turned toward the back, and the hind legs forward. Place it on the platter with the back up and head at the left. Remove the shoulders by cutting round between them and the body, carrying the knife up nearly to the backbone. Turn them back and cut through the joint. Remove

the hind legs in the same manner. Then place the fork in the middle of the back and cut several slices from each side of the loin parallel with the backbone. The loin is the choicest part.

SWEETBREADS, CHOPS, AND CUTLETS
These are not divided, one being served to each person.

CONVERSION TABLE

WEIGHTS
1 oz. = 28 grams
2 oz. = 57 grams
4 oz. (¼ lb.) = 113 grams
8 oz. (½ lb.) = 227 grams
16 oz. (1 lb.) = 454 grams

VOLUME MEASURES
⅛ teaspoon = 0.6 ml
¼ teaspoon = 1.23 ml
½ teaspoon = 2.5 ml
1 teaspoon = 5 ml
1 tablespoon (3 teaspoons) = ½ fluid oz. = 15 ml
2 tablespoons = 1 fluid oz. = 29.5 ml
¼ cup (4 tablespoons) = 2 fluid oz. = 59 ml
⅓ cup (5 ⅓ tablespoons) = 2.7 fluid oz. = 80 ml
½ cup (8 tablespoons) = 4 fluid oz. = 120 ml
⅔ cup (10 ⅔ tablespoons) = 5.4 fluid oz. = 160 ml
¾ cup (12 tablespoons) = 6 fluid oz. = 180 ml
1 cup (16 tablespoons) = 8 fluid oz. = 240 ml

TEMPERATURE EQUIVALENTS

°F	°C	Gas Mark
225	110	¼
250	130	½
275	140	1
300	150	2
325	170	3
350	180	4
375	190	5
400	200	6
425	220	7
450	230	8
475	240	9
500	250	10

LENGTH MEASURES
1/16 inch = 1.6 mm
⅛ inch = 3 mm
¼ inch = 1.35 mm
½ inch = 1.25 cm
¾ inch = 2 cm
1 inch = 2.5 cm

IMAGE CREDITS

Pages 8, 10-11, 33 (bottom), 759, 776-777 by Doug Levy; pages 14-19, 750-752, 754, 757, 763, 765-766, 768, 771-772 courtesy of Library of Congress; pages 21-31, 119-120, 145, 228, 273, 296, 300, 314, 317-318, 323-324, 394-395, 403-404, 430 courtesy of Cider Mill Press; pages 33, 39-40, 59, 71, 99, 113, 124, 128, 133-134, 170, 172, 185, 191-192, 195-196, 199, 203, 207, 211, 218, 221, 225, 233, 251, 254, 257, 261-262, 265, 278, 281, 285, 289, 293, 299, 302, 305, 308, 330, 340, 343, 347, 351, 355, 359-360, 363-364, 367, 370, 380, 383, 397, 407, 411, 416, 424, 426, 429, 433, 451, 465, 470, 477, 479, 496, 499-500, 515, 604, 606, 612, 639-640, 674, 685, 689-690, 693, 704-705, 707-708, 755-756 by Keith Sarasin; pages 546, 549-550, 553, 557-558, 561 courtesy of Justin Dain; pages 579-580, 583 courtesy of George Bezanson; page 589 courtesy of Ryan Manning; pages 591-592 courtesy of Angela Wilson; Pages 595, 598, 601 courtesy of Fleishers Craft Butchery.

All other images used under official license from Shutterstock.com.

ACKNOWLEDGMENTS

This book is dedicated to the memory of Steven P. Mahoney Jr.
  May the fire of your passions never die in those you've left behind.
  I carry your light with me every day.
  I miss you . . . I love you.
  May this book honor your memory and inspire those who read it as you continue to inspire us.

SPECIAL THANKS
Marie Collins—Your support means the world to me. I am so proud of you.
Carole Soule
Noah Bicchieri
The Farms & Farmers
Nancy M. Sarasin
Matt Jackson—The irony isn't misplaced here.
Dan Duquette
David Withham
Jimmy McDonald
Patrick Soucy
Jim Challenger—The Challenger Bread Pan. You are amazing, thank you!
Doug Levy—Doug Levy Photography, thank you for everything!

ABOUT THE AUTHOR

Chef Keith Sarasin is an author, chef, speaker, and restaurateur. His love for food was developed at a young age when he would cook for his mother using old cookbooks that were given to him by his grandmother. He began his culinary career at the age of 15, washing dishes and making subs. As the years went on he worked with a variety of restaurants, working his way up from Sous Chef to Executive Chef. Sarasin worked as a private chef before founding The Farmers Dinner in 2012. The Farmers Dinner hosts upscale dinners on New England farms. Since 2012, The Farmers Dinner has hosted over 87 farm-to-table events across New England and fed more than 17,000 customers, raising over $125,000 for local farms.

Sarasin is the author of two other books, *The Perfect Turkey* and *The Farmers Dinner Cookbook*. In 2019, he opened his first restaurant built around locally sourced, farm-to-table style cuisine. Sarasin has been the keynote speaker for a variety of organizations, including The American Culinary Federation.

INDEX

Acapulco Gold Rub, 567
achiote paste, Carne de Chinameca, 586
achiote seeds, Al Pastor Negro, 587
adobo chilies/sauce
   Adobo Marinade for Beef, 495
   Adobo Pulled Pork, 217
   Creamy Adobo Dip, 649
   Maple-Chipotle Glaze, 575
agave/agave syrup
   Agave Sriracha Sweet Glaze, 568
   Harissa Glaze, 573
   Smoked Harissa Agave Chicken Wings, 573
Al Pastor Negro, 587
Al Pastor Pork Mini Trompo, 588
All-Yolk Pasta Dough, 698
almonds, Charred Summer Leeks with Romesco Sauce, 631
American cheese
   Bacon Cheeseburgers, 105
   Turkey & Kimchi Burgers, 339
American Chop Suey, 140
ancho chilies
   Al Pastor Negro, 587
   Al Pastor Pork Mini Trompo, 588

anchovy fillets, Buttermilk Caesar Dressing, 672
Andouille Sausage, Jambalaya with Bacon &, 209. *See also* sausage
apple juice/apple cider
   Applewood Smoked Ribs with Molasses BBQ Sauce, 166
   Fried Brussels Sprouts with Maple-Cider Glaze, 659
   Rotisserie Pig Roast, 168
   Smoked Lamb Shoulder, 385
apples
   Beef and Braised Cabbage Soup with Horseradish Cream, 118
   Mulligatawny Chicken Curry Soup with Curried Cashews, 315
   Pork & Apple Casserole, 223
   Roasted Baby Beet, Radish & Apple Salad with Blue Cheese Mousse, 671
   Sausage, Apple & Pecan Stuffing, 197
   Applewood Smoked Ribs with Molasses BBQ Sauce, 166
Arkhive Farm, 9, 605–607

Arugula, Pork Burgers with Pepper Jack &, 227
Asian Marinade, 494
Asian Meatballs, 232
asparagus
   Asparagus Three Ways, Plus Two Sauces, 650–651
   Crossroad Farm's Vegetable Hash, 548–549
Avedano's Holly Park Market, 591–594

baby back pork ribs, Killer BBQ Spare Ribs, 182–183
bacon
   Bacon, Pear & Rosemary Stuffing, 194
   Bacon & Pea Soup, 205
   Bacon Cheeseburgers, 105
   Bacon Jam, 513
   Candied Bacon, 190
   Chinese Potted Chicken, 368
   Cured Bacon, 566
   Gnocchi Bolognese, 554
   Home-Style Baked Beans, 620
   Jambalaya with Bacon & Andouille Sausage, 209
   Jowl Bacon, 190
   Lamb Bacon, 409

Pork Belly & Sweet Corn Chowder, 201
Roasted Brussels Sprouts with Bacon, Blue Cheese & Pickled Red Onion, 656
Roasted Corn & Red Pepper Bisque with Bacon, 202
Shaved Brussels Sprouts & Kale Salad with Blood Orange Vinaigrette, 675
Spiced Dumplings, 322
Texas BBQ Burgers, 112
PINE Burger, 551
Ultimate XO Sauce, 582
Warm Potato Salad, 548–549
Balsamic Glaze, Crispy Pork Belly with, 171
Barbacoa, Slow-Cooker, 98
Barley, and Portobello Mushroom Soup, Beef, 121
Basic Burgers, 102
Basic Flank Steak, 49
Basic Grilled Double-Cut Rib Chops, 189
Basic Grilled Roast Pork Loin, 176
Basic Red Cabbage Slaw, 664
basil
　Broccoli Rabe Pesto, 547
　Pizzaiola, 531
　Sausage Ragù, 231
　Sun-Dried Tomato Pesto, 532
BBQ Chicken Marinade, 489
BBQ Glazed Meat Loaf, 155
BBQ Poultry Rub, 464
BBQ Sauce

BBQ Glazed Meat Loaf, 155
BBQ Turkey Wings, 291
Chipotle BBQ Sauce, 505
KC-Style BBQ Sauce, 502
Molasses BBQ Sauce, 166
Pineapple BBQ Sauce, 501
Rotisserie Pig Roast, 168
Spicy BBQ Sauce, 498
Texas BBQ Burgers, 112
Traditional BBQ Sauce, 497
BBQ Turkey Wings, 291
beans
　Home-Style Baked Beans, 620
　Hot Dog with Curry Spiced Lentils, 130
　Midsummer Corn & Bean Salad, 676
　Persian Braised Lamb Neck with Spiced Broth, 594
　Twice-Cooked Pork Belly & Bean Soup with Brown Bread Croutons, 210
beef
　Adobo Marinade for Beef, 495
　American Chop Suey, 140
　Asian Meatballs, 232
　Bacon Cheeseburgers, 105
　Basic Burgers, 102
　Basic Flank Steak, 49
　BBQ Glazed Meat Loaf, 155
　Beef, Barley, and Portobello Mushroom Soup, 121
　Beef & Pork Burgers with Caramelized Onion Mayo, 107

Beef & Vegetable Soup, 116
Beef and Braised Cabbage Soup with Horseradish Cream, 118
Beef Fajita Marinade, 492
Beef Kabob, 126
Beef Stew, 115
Bolognese, 148
Braised Beef Cheek, 97
Brisket, 91
Burnt Ends, 92
Carne Asada, 88
Carne de Chinameca, 586
carving, 730–733
Cheese-Stuffed Meatballs, 137
Chickpea & Beef Chili, 117
Chile-Rubbed London Broil, 50
Chipotle T-Bone Steak, 65
Coffee-Rubbed Sirloin, 58
Cola Meatballs, 135
Corndog, 132
Crowned Roast of Beef, 78–79
cuts of, 21–23
Filet Mignon, 46
Five-Alarm Beef Chili, 122
Frenched Prime Rib, 76–77
Garlic & Thyme-Rubbed Sirloin, 62
Gnocchi Bolognese, 554
Greek Burgers with Feta Cheese, 421
Grilled Filet, 66
Grilled New York Strip Steak, 54
Grilled Rib Eye Steak, 57
Grilled Tensen Farm's Strip Loin of Beef, 559

ground, 23
Hot Dog with Curry Spiced Lentils, 131
Italian Meatballs, 141
Jalapeño Burgers with Red Pepper Jam, 111
Kefta with Warm Chickpeas and Salad, 144
Lamb & Beef Burgers, 418
Masala Braised Short Rib Sandwich with Cilantro Chutney, 129
Meatballs with Arrabiata, 136
Mexican Burgers with Cilantro Mayonnaise, 106
Mushroom & Swiss Burgers, 108
Old Fashioned Meat Loaf, 154
Oven-Roasted New York Strip Steak with Whipped Potatoes and Swiss Chard, 87
Perfect Prime Rib, 73
Porterhouse, 45
Roast Beef au Jus with Vegetables, 101
Seared Filet Mignon, 69
Seared Filet Mignon with Peppercorn Cream Sauce, 70
Seared New York Strip Steak, 42
Seared Rib Eye, 38
Shepherd's Pie, 147
Short Rib Ramen with Beef Broth, 125
Slow Braised Boyden Beef Short Rib, 563–564
Slow-Cooker Barbacoa, 98
Slow-Cooker Corned Beef, 151
Smoked Prime Rib, 83
Sous Vide New York Strip Steak, 53
Sous Vide Rib Eye, 41
Southern Beef Jerky, 585
steak temperature guide, 57
Sweet & Sour Meatballs, 138
Texas BBQ Burgers, 112
Thai Beef & Cucumber Salad, 668
PINE Burger, 551
Wagyu Hot Dog, 131
Wood-Fired Prime Rib, 80
Yankee Short Ribs with Roasted Potatoes & Carrots, 61
beef cheeks
Braised Beef Cheek, 97
Slow-Cooker Barbacoa, 98
Beef Stock
Beef, Barley, and Portobello Mushroom Soup, 121
Beef & Vegetable Soup, 116
Beef Stew, 115
Bolognese, 148
Braised Beef Cheek, 97
Brisket, 91
Cottage Pie, 401
Fundamental Prime-Rib Gravy, 524
Lamb Stew with Candied Carrots and Garlic Roasted Little Creamer Potatoes, 392–393
Madeira Sauce, 528
Masala Braised Short Rib Sandwich with Cilantro Chutney, 129
recipe, 475
Roast Beef au Jus with Vegetables, 101
Shepherd's Pie, 147
Short Rib Ramen with Beef Broth, 125
Swedish Meatballs, 143
Venison Stew, 396
Yankee Short Ribs with Roasted Potatoes & Carrots, 61
beef tips
Beef & Vegetable Soup, 116
Beef Stew, 115
beer
Beer-Brined Turkey, 266
Five-Alarm Beef Chili, 122
Jambalaya with Bacon & Andouille Sausage, 209
Slow-Cooker Barbacoa, 98
beets
Baby Beets, 555–556
Beet & Goat Cheese Puree, 555–556
Roast Beef au Jus with Vegetables, 101
Roasted Baby Beet, Radish & Apple Salad with Blue Cheese Mousse, 671
Bergeron, Gary, 9–10
Bezanson, George, 579–583
Bicchieri, Noah, 9, 605–607
Black Bean Garlic Sauce & Extra Garlic, Yu Choy with, 680
Blender Hollandaise, 650
blood oranges
Blood Orange & Bourbon Brine, 454

Blood Orange Vinaigrette, 675

See also oranges/orange juice

blue cheese
- Blue Cheese Gratin, 646
- Habanero, Calabrian Chile & Pineapple Chicken Wings with Chipotle Blue Cheese, 279
- Pulled Pork with Blue Cheese Polenta & Roasted Peach Hot Sauce, 220
- Roasted Baby Beet, Radish & Apple Salad with Blue Cheese Mousse, 671
- Roasted Brussels Sprouts with Bacon, Blue Cheese & Pickled Red Onion, 656

Blueberry Reduction, Venison Tenderloin, 427

bok choy
- General Tso's Pork Belly Steam Buns, 560, 562
- Simple Stir-Fried Bok Choy, 652

Bolognese, 148

Bone-In Pork Roast, 172

bourbon
- Blood Orange & Bourbon Brine, 454
- Cured Bacon, 566

Braised and Grilled Pork with Rosemary, 180

Braised Beef Cheek, 97

Braised Chicken Chili, 328

Braised Jowl Pulled Pork, 216

Braised Lamb Belly, 381

Braised Lamb Neck Wrap, 424

Braised Lamb Shoulder with Minted Peas, 382

breads
- Country Sourdough Bread, 689
- Gluten-Free Bread, 694
- No-Knead Bread, 693
- Rustic White Bread, 685
- Rustic Whole Wheat Bread, 682
- Sourdough Bread, 686
- Sourdough Starter, 690

breast, goat, 31

brines
- Blood Orange & Bourbon Brine, 454
- Cajun Dry Brine, 449
- Citrus Brine, 448
- Herb Brined Chicken, 452
- Holiday Spiced Wet Turkey Brine, 450
- Maple Chicken Brine, 447
- Meat Brine, 444
- Orange & Thyme Dry Brine, 453
- Sage & Garlic Dry Brine, 455

brisket
- Brisket (basic recipe), 91
- Burnt Ends, 92
- Schlissel's recipe for, 569
- Slow-Cooker Corned Beef, 151

broccoli, Roasted Brassica Salad with Pickled Ramps & Buttermilk Caesar Dressing, 672

broccoli rabe
- Broccoli Rabe Pesto, 547
- Spicy Pappardelle and Duck Confit with

Broccoli Rabe, Shallots & Parmesan, 329

Brown Bread Croutons, Twice-Cooked Pork Belly & Bean Soup with, 210

Brown Sugar & Ancho Chile Poultry Rub, 466

Brussels sprouts
- Cornish Game Hens with Baby Brussels Sprouts & Caramelized Onions, 272
- Fried Brussels Sprouts with Maple-Cider Glaze, 659
- Roasted Brassica Salad with Pickled Ramps & Buttermilk Caesar Dressing, 672
- Roasted Brussels Sprouts with Bacon, Blue Cheese & Pickled Red Onion, 656
- Shaved Brussels Sprouts & Kale Salad with Blood Orange Vinaigrette, 675
- Stove-Top Brussels Sprouts, 660

Bulgarian Sour Lamb Soup with Paprika Butter, 405

burgers
- Bacon Cheeseburgers, 105
- Basic Burgers, 102
- Beef & Pork Burgers with Caramelized Onion Mayo, 107
- Chicken Burgers, 336
- Chicken Teriyaki Burgers, 338
- Chicken Tikka Burger with Tikka Masala Mayonnaise, 341
- Goat Burgers, 422

Greek Burgers with Feta Cheese, 421
Jalapeño Burgers with Red Pepper Jam, 111
Lamb & Beef Burgers, 418
Lamb Burgers, 418
Mexican Burgers with Cilantro Mayonnaise, 106
Mushroom & Swiss Burgers, 108
Pork Burgers with Pepper Jack & Arugula, 227
Texas BBQ Burgers, 112
PINE Burger, 551
Turkey & Kimchi Burgers, 339
Turkey Burgers, 336
See also sandwiches
Burnt Ends, 92
butchery, history of, 15–19
Butter, Herbed, 539
buttermilk, Chicken Parm Sandwich, 345
butternut squash
　Butternut Squash and Chorizo Bisque, 206
　Butternut Squash Puree, 545, 547
cabbage
　Basic Red Cabbage Slaw, 664
　Beef and Braised Cabbage Soup with Horseradish Cream, 118
　Boiled Dinner, 224
　Grilled Cabbage, 663
　Kimchi, 655
　Korean Chicken Thighs with Sweet Potato Vermicelli, 301
Cajun Dry Brine, 449
Candied Bacon, 190

Caramelized Onion Mayo, Beef & Pork Burgers with, 107
Carne Asada, 88
Carne de Chinameca, 586
carrots
　Beef, Barley, and Portobello Mushroom Soup, 121
　Beef & Vegetable Soup, 116
　Beef Stew, 115
　Beef Stock, 475
　Boiled Dinner, 224
　Bolognese, 148
　Braised Beef Cheek, 97
　Braised Lamb Belly, 381
　Braised Lamb Neck Wrap, 424
　Braised Lamb Shoulder with Minted Peas, 382
　Candied Carrots, 393
　Chicken & Rice Soup, 312
　Chicken Katsu Curry, 353
　Chicken Stock, 478
　Duck a l'Orange Broth with Spiced Dumplings, 321–322
　Duck Stock, 484
　General Tso's Pork Belly Steam Buns, 560, 562
　Gnocchi Bolognese, 554
　Habanero, Calabrian Chile & Pineapple Chicken Wings with Chipotle Blue Cheese, 279
　Head Cheese, 236
　Homey Turkey & Wild Rice Soup, 316
　Indian Chicken and Rice Soup, 309

Jambalaya with Bacon & Andouille Sausage, 209
Kohlrabi Slaw with Miso Dressing, 627
Lamb Stew with Candied Carrots and Garlic Roasted Little Creamer Potatoes, 392–393
Mulligatawny Chicken Curry Soup with Curried Cashews, 315
Mutton Broth, 402
Pork Stock, 481
Rabbit Ragù, 400
Roast Beef au Jus with Vegetables, 101
Roasted Parsnips & Carrots with Ras el Hanout & Honey, 638
Slow Braised Boyden Beef Short Rib, 563–564
Smoked Chicken Noodle Soup, 310
Thai Beef & Cucumber Salad, 668
Turkey Giblet Gravy, 474
Turkey Stock, 319, 477
Twice-Cooked Pork Belly & Bean Soup with Brown Bread Croutons, 210
Venison Stew, 396
Yankee Short Ribs with Roasted Potatoes & Carrots, 61
carry-over cooking, 32
carving
　beef, 758, 762
　chartreuse, 768
　chicken, 768, 770
　corned beef, 768
　duck, 774

general directions, 753-758
ham, 767
lamb, 764, 767
mutton, 762, 764
poultry, 768, 770, 773-774
rabbit, 774-775
roast pig, 767
tongue, 768
tools for, 32
turkey, 253, 770, 773
veal, 762, 764, 767
venison, 762, 764
Cashews, Mulligatawny Chicken Curry Soup with Curried, 315
cauliflower
   Cauliflower Mash, 619
   Marinated Chicken with Chipotle Cauliflower, 247
   Riced Cauliflower, 616
   Roasted Brassica Salad with Pickled Ramps & Buttermilk Caesar Dressing, 672
   Roasted Cauliflower au Gratin, 612
celeriac/celery root
   Celeriac Puree, 615
   Celery Root and Parsnip Puree, 563–564
   Chicken & Root Vegetable Stew, 311
   Creamed Vegetable Soup with Turkey Dumplings, 319–320
   Potato & Celeriac Gratin with Gruyère Figs, 623
   Rabbit & Root Vegetable Stew, 399

celery
   Bacon, Pear & Rosemary Stuffing, 194
   Beef, Barley, and Portobello Mushroom Soup, 121
   Beef & Vegetable Soup, 116
   Beef Stew, 115
   Beef Stock, 475
   Bolognese, 148
   Chicken & Rice Soup, 312
   Chicken Stock, 478
   Creamed Vegetable Soup with Turkey Dumplings, 319–320
   Crowned Roast of Beef, 78–79
   Duck a l'Orange Broth with Spiced Dumplings, 321–322
   Duck Stock, 484
   General Tso's Pork Belly Steam Buns, 560, 562
   Gnocchi Bolognese, 554
   Head Cheese, 236
   Homey Turkey & Wild Rice Soup, 316
   Indian Chicken and Rice Soup, 309
   Jambalaya with Bacon & Andouille Sausage, 209
   Mulligatawny Chicken Curry Soup with Curried Cashews, 315
   Pork Belly & Sweet Corn Chowder, 201
   Pork Stock, 481
   Roast Beef au Jus with Vegetables, 101
   Roasted Leftover Chicken Salad, 332

Sausage, Apple & Pecan Stuffing, 197
Slow Braised Boyden Beef Short Rib, 563–564
Smoked Chicken Noodle Soup, 310
Turkey Giblet Gravy, 474
Turkey Stock, 319, 477
Twice-Cooked Pork Belly & Bean Soup with Brown Bread Croutons, 210
Venison Stew, 396
Char Siu, 184
chard
   Oven-Roasted New York Strip Steak with Whipped Potatoes and Swiss Chard, 87
   Pan Roasted Duck Breast, 555–556
   Pork Roulade with Oranges, Rainbow Chard, and Wild Rice, 163
Charred Summer Leeks with Romesco Sauce, 631
chartreuse, carving, 740
cheddar cheese
   Braised Chicken Chili, 328
   Cheese-Stuffed Meatballs, 137
   Chickpea & Beef Chili, 117
   Five-Alarm Beef Chili, 122
   Leftover Turkey & Chickpea Chili with Corn Bread & Rosemary Butter, 326
   PINE Burger, 551
cheese. *See individual cheese types*

Cheese-Stuffed Meatballs, 137
chicken
Braised Chicken Chili, 328
Cajun Dry Brine, 449
carving, 740, 745
Chicken & Rice Soup, 312
Chicken & Root Vegetable Stew, 311
Chicken 65, 366
Chicken Burgers, 336
Chicken Katsu Curry, 353
Chicken Kebob, 349
Chicken Legs with Potatoes & Fennel, 297
Chicken Meatballs, 335
Chicken Parm Sandwich, 345
Chicken Stock, 478
Chicken Teriyaki Burgers, 338
Chicken Tikka Burger with Tikka Masala Mayonnaise, 341
Chicken Tikka Masala, 362
Chili Chicken, 365
Chinese Potted Chicken, 368
Citrus Brine, 448
Crispy Chicken with Cajun Sauce, 358
cuts from, 26–27
Deviled Eggs with Crispy Chicken Skin, 331
General Tso Chicken, 371
Habanero, Calabrian Chile & Pineapple Chicken Wings with Chipotle Blue Cheese, 279
Herb Brined Chicken, 452
Hot & Spicy Grilled Wings, 283
Indian Chicken and Rice Soup, 309
Jerk Spice Chicken Rub, 460
Korean Chicken Thighs with Sweet Potato Vermicelli, 301
Maple Chicken Brine, 447
Marinated Chicken with Chipotle Cauliflower, 247
Marley's Collie, 576
Masala Wings, 280
Mulligatawny Chicken Curry Soup with Curried Cashews, 315
Nashville Hot Chicken, 357
Oven Chicken Tandoori, 369
Pickle Fried Chicken Sandwich, 346
Popcorn Chicken with Kimchi Mayonnaise, 350
Roasted Chicken Thighs with Tabbouleh, 298
Roasted Leftover Chicken Salad, 332
Roasted Whole Chicken, 244
Seared Chicken Breast, 294
Smoked Chicken Noodle Soup, 310
Smoked Chicken Thighs, 295
Smoked Harissa Agave Chicken Wings, 573
Sous Vide Cajun Chicken Wings, 284
Sous Vide Chicken Breast, 292
Sous Vide Tandoori Chicken, 354
Sweet & Smoky Roasted Chicken, 246
Sweet & Sticky Chicken Wings, 276
Tuscan Chicken, 361
Walpole Valley Farm's Chicken, 545, 547
Whole Jerk Chicken, 577
chicken livers, Gnocchi Bolognese, 554
Chicken Stock
Adobo Pulled Pork, 217
Bacon & Pea Soup, 205
Sausage-Stuffed Peppers, 213
Bolognese, 148
Braised Chicken Chili, 328
Braised Jowl Pulled Pork, 216
Chicken & Rice Soup, 312
Chicken & Root Vegetable Stew, 311
Chicken Katsu Curry, 353
Chinese Eggplant with Sausage, 239
Chinese Potted Chicken, 368
Crispy Chicken with Cajun Sauce, 358
Farm Fresh Whole Guinea Fowl Rubbed with Local Shio Kojit, 593
General Tso Chicken, 371
General Tso's Pork Belly Steam Buns, 560, 562
General Tso's Sauce, 562

INDEX | 793

Gnocchi Bolognese, 554
Indian Chicken & Rice
 Soup, 309
Jambalaya with Bacon &
 Andouille Sausage, 209
Kefta with Warm
 Chickpeas & Salad,
 144
Mulligatawny Chicken
 Curry Soup with
 Curried Cashews, 315
Persian Braised Lamb
 Neck with Spiced Broth,
 594
Pheasant, Morel
 Mushroom, & Olive
 Broth, 325
Pork Roulade with Oranges,
 Rainbow Chard, & Wild
 Rice, 163
Potato & Tomato Grain,
 633
Pulled Pork, 214
Pulled Pork with Blue
 Cheese Polenta &
 Roasted Peach Hot
 Sauce, 220
Rabbit & Root Vegetable
 Stew, 399
Rabbit Ragù, 400
recipe, 478
Smoked Chicken Noodle
 Soup, 310
Thick-Cut Pork Chops
 with Stone Fruit &
 Bulgur Wheat, 228
Traditional Chicken
 Gravy, 472
Twice-Cooked Pork Belly
 & Bean Soup with
 Brown Bread Croutons,
 210
chickpeas

Chickpea & Beef Chili,
 117
Kefta with Warm
 Chickpeas and Salad,
 144
Leftover Turkey &
 Chickpea Chili
 with Corn Bread &
 Rosemary Butter, 326
Persian Braised Lamb
 Neck with Spiced Broth,
 594
Chile-Rubbed London Broil,
 50
Chile-Vinegar Lamb Ribs,
 435
chili
 Braised Chicken Chili,
  328
 Chickpea & Beef Chili,
  117
 Chili Chicken, 365
 Five-Alarm Beef Chili, 122
 Leftover Turkey &
  Chickpea Chili
  with Corn Bread &
  Rosemary Butter, 326
chilies
 Adobo Pulled Pork, 217
 Al Pastor Negro, 587
 Al Pastor Pork Mini
  Trompo, 588
 Braised Chicken Chili,
  328
 Carne de Chinameca, 586
 Chicken 65, 366
 Chicken Tikka Masala,
  362
 Chickpea & Beef Chili,
  117
 Chili Chicken, 365
 Five-Alarm Beef Chili, 122
 General Tso Chicken, 371

 Goat Momos, 413
 Habanero, Calabrian Chile
  & Pineapple Chicken
  Wings with Chipotle
  Blue Cheese, 279
 Indian-Style Goat Curry,
  423
 Lamb Vindaloo, 432
 Leftover Turkey &
  Chickpea Chili
  with Corn Bread &
  Rosemary Butter, 326
 Thai Duck Stir Fry, 372
chimichurri sauce
 Chipotle Cilantro
  Chimichurri for Meat,
  514
 Fresh Chimichurri Sauce,
  509
Chinese Egg Noodles, 716
Chinese Eggplant with
 Sausage, 239
Chinese Potted Chicken,
 368
chipotle chilies/peppers
 Chipotle BBQ Sauce, 505
 Chipotle Cilantro
  Chimichurri for Meat,
  514
 Chipotle Mayonnaise, 551
 Habanero, Calabrian Chile
  & Pineapple Chicken
  Wings with Chipotle
  Blue Cheese, 279
 Maple-Chipotle Glaze,
  575
 Slow-Cooker Barbacoa, 98
 PINE Burger, 551
Chipotle T-Bone Steak, 65
chives, Garlic & Chive Steak
 Sauce, 535
chop/rack, lamb, 28
Chorizo Bisque, Butternut

Chorizo Bisque, Butternut Squash and, 206. *See also* sausage
chuck roast, Slow-Cooker Barbacoa, 98
cilantro
   Basic Red Cabbage Slaw, 664
   Cilantro Chutney, 514
   Chipotle Cilantro Chimichurri for Meat, 514
   Corn Salad, 132
   Goat Momos, 413
   Grilled Tensen Farm's Strip Loin of Beef, 559
   Homemade Tomato Salsa, 520
   Indian-Style Goat Curry, 423
   Kefta with Warm Chickpeas and Salad, 144
   Kohlrabi Slaw with Miso Dressing, 627
   Masala Braised Short Rib Sandwich with Cilantro Chutney, 129
   Masala Lamb Kabobs, 417
   Mexican Burgers with Cilantro Mayonnaise, 106
   Midsummer Corn & Bean Salad, 676
   Tomatillo Salsa, 523
Citrus Brine, 448
Classic Smoked Ribs, 164
Classic Corn Bread, 713
coconut, Cilantro Chutney, 514
coconut milk
   Jamaican-Style Curry Goat, 428

Thai Red Duck Curry, 373
Coffee Marinade for Beef, 491
Coffee-Rubbed Sirloin, 58
Cola Meatballs, 135
Cola-Braised Pulled Pork, 219
Collard Greens, Southern, 624
Colombian Gold, 570
Confit Duck Legs, 303
Confit New Potatoes, 641
Confit Potatoes, 559
Confit Turkey Legs, 306
corn
   Beef & Vegetable Soup, 116
   Corn Salad, 132
   Crossroad Farm's Vegetable Hash, 548–549
   Midsummer Corn & Bean Salad, 676
   Pork Belly & Sweet Corn Chowder, 201
   Roasted Corn & Red Pepper Bisque with Bacon, 202
   Shepherd's Pie, 147
   Walpole Valley Farm's Chicken, 545, 547
Corn bread, 713
Corndog, 132
corned beef
   carving, 740
   Slow-Cooker Corned Beef, 151
Cornish Game Hens with Baby Brussels Sprouts & Caramelized Onions, 272
cotija cheese, Corn Salad, 132
Cottage Pie, 401

Country Sourdough Bread, 689
Country Sourdough Levain, 689
Creamed Vegetable Soup with Turkey Dumplings, 319
Creamy Adobo Dip, 649
Crispy Chicken with Cajun Sauce, 358
Crispy Pork Belly with Balsamic Glaze, 171
Crossroad Farm's Vegetable Hash, 548–549
Crowned Roast of Beef, 78–79
cucumbers
   Kefta with Warm Chickpeas and Salad, 144
   Melon, Cucumber & Prosciutto Salad with Mint Vinaigrette, 667
   Roasted Chicken Thighs with Tabbouleh, 298
   Spicy Lamb Ribs with Tzatziki & Lemon, 436
   Thai Beef & Cucumber Salad, 668
Curried Cashews, Mulligatawny Chicken Curry Soup with, 315

Dain, Justin, 544–564
Deep-Fried Turkey, 259
double-cut rib chops, Basic Grilled Double-Cut Rib Chops, 189
dressings
   Blood Orange Vinaigrette, 675
   Buttermilk Caesar Dressing, 672

Harissa Vinaigrette, 548–549
Honey Mustard Vinaigrette, 671
Mint Vinaigrette, 667
duck
　breaking down, 599
　carving, 746–747
　Confit Duck Legs, 303
　Duck a l'Orange Broth with Spiced Dumplings, 321–322
　Duck Stock, 484
　Pan Roasted Duck Breast, 555–556
　Roasted Duck Breast and Five-Spice Confit Duck Legs, 597, 599
　Roasted Whole Duck, 268
　Seared Duck Breast, 304
　Spicy Pappardelle and Duck Confit with Broccoli Rabe, Shallots & Parmesan, 329
　Thai Duck Stir Fry, 372
　Thai Red Duck Curry, 373
Duck Stock
　Duck a l'Orange Broth with Spiced Dumplings, 321–322
　recipe, 484
dumplings
　Creamed Vegetable Soup with Turkey Dumplings, 319–320
　Duck a l'Orange Broth with Spiced Dumplings, 321–322
　Goat Momos, 413
　Pork Gyoza, 198
Dunk's Mushrooms, 559

Earth's Harvest Kitchen Pork Belly, 581
eggplants, Chinese Eggplant with Sausage, 239
eggs, Deviled Eggs with Crispy Chicken Skin, 331
Eggs with Crispy Chicken Skin, Deviled, 331
Emmental cheese, Roasted Cauliflower au Gratin, 612
eye of round beef roast, Roast Beef au Jus with Vegetables, 101

Fajita Marinade, Beef, 492
Farfalle, 700
Farm Fresh Whole Guinea Fowl Rubbed with Local Shio Kojit, 593
farro, Pan Roasted Duck Breast, 555–556
Fazzoletti, 703
fennel
　Chicken Legs with Potatoes and Fennel, 297
　Fennel Lemon Gremolata, 506
　Pork Roulade with Oranges, Rainbow Chard, and Wild Rice, 163
　Walpole Valley Farm's Chicken, 545, 547
Fermented Hot Sauce, 519
fermented red bean curd, Char Siu, 184
feta cheese
　Corn Salad, 132
　Greek Burgers with Feta Cheese, 421

Melon, Cucumber & Prosciutto Salad with Mint Vinaigrette, 667
　Zucchini with Tomatoes, Feta, Garlic & Lemon, 681
Figs, Potato & Celeriac Gratin with Gruyère, 623
filet mignon
　Filet Mignon (basic recipe), 46
　Grilled Filet, 66
　Seared Filet Mignon, 69
　Seared Filet Mignon with Peppercorn Cream Sauce, 70
fish sauce
　Korean Chicken Thighs with Sweet Potato Vermicelli, 301
　Pork Pho Broth, 482
　Thai Beef & Cucumber Salad, 668
　Thai Duck Stir Fry, 372
　Ultimate XO Sauce, 582
Five-Alarm Beef Chili, 122
Five-Spice Cure for Duck Legs, 597, 599
flank steak
　Basic Flank Steak, 49
　Southern Beef Jerky, 585
Fleishers Craft Butchery, 595–602
forequarter beef chuck, 21
forequarter beef plate, 21
forequarter brisket, 21
forequarter ribs, 21
forequarter shank, 21
foreshank, goat, 31
Frenched Prime Rib, 76–77
Fresh Chimichurri Sauce, 509

Fried Brussels Sprouts with Maple-Cider Glaze, 659
Frizzled Leeks, 632
Fundamental Prime-Rib Gravy, 524

Garganelli, 704
garlic
   Al Pastor Negro, 587
   Al Pastor Pork Mini Trompo, 588
   Asian Meatballs, 232
   Sausage-Stuffed Peppers, 213
   Braised Beef Cheek, 97
   Braised Chicken Chili, 328
   Braised Jowl Pulled Pork, 216
   Broccoli Rabe Pesto, 547
   Carne de Chinameca, 586
   Chicken Katsu Curry, 353
   Chicken Kebob, 349
   Chickpea & Beef Chili, 117
   Chinese Potted Chicken, 368
   Citrus Brine, 448
   Cornish Game Hens with Baby Brussels Sprouts & Caramelized Onions, 272
   Crispy Chicken with Cajun Sauce, 358
   Crowned Roast of Beef, 78–79
   Fermented Hot Sauce, 519
   Five-Alarm Beef Chili, 122
   Garlic & Chive Steak Sauce, 535
   Garlic & Thyme-Rubbed Sirloin, 62
   General Tso's Sauce, 562
   Gnocchi Bolognese, 554
   Gremolata, 564
   Grilled Tensen Farm's Strip Loin of Beef, 559
   Grill-Roasted Rack of Lamb with Garlic-Herb Crust, 389
   Homemade Tomato Salsa, 520
   Homey Turkey & Wild Rice Soup, 316
   Jambalaya with Bacon & Andouille Sausage, 209
   Killer BBQ Spare Ribs, 182–183
   Lamb & Sweet Potato Hash, 431
   Lamb Kebobs, 414
   Lamb Stew with Candied Carrots and Garlic Roasted Little Creamer Potatoes, 392–393
   Lamb Vindaloo, 432
   Leftover Turkey & Chickpea Chili with Corn Bread & Rosemary Butter, 326
   Marinated Chicken with Chipotle Cauliflower, 247
   Masala Braised Short Rib Sandwich with Cilantro Chutney, 129
   Masala Lamb Kabobs, 417
   Meat Brine, 444
   Oven-Roasted New York Strip Steak with Whipped Potatoes and Swiss Chard, 87
   Patatas Bravas, 634
   Pheasant, Morel Mushroom, and Olive Broth, 325
   Pizzaiola, 531
   Popcorn Chicken with Kimchi Mayonnaise, 350
   Potato & Tomato Grain, 633
   Pulled Pork with Blue Cheese Polenta & Roasted Peach Hot Sauce, 220
   Roast Beef au Jus with Vegetables, 101
   Roasted Baby Beet, Radish & Apple Salad with Blue Cheese Mousse, 671
   Roasted Cauliflower au Gratin, 612
   Roasted Garlic Pork Belly Porchetta, 600, 602
   Roasted Whole Duck, 268
   Rosemary and Lemon Oiled Leg of Lamb, 386
   Sage & Garlic Dry Brine, 455
   Short Rib Ramen with Beef Broth, 125
   Slow Braised Boyden Beef Short Rib, 563–564
   Slow-Cooker Barbacoa, 98
   Slow-Cooker Corned Beef, 151
   Smoked Prime Rib, 83
   Southern Beef Jerky, 585
   Southern Collard Greens, 624
   Turkey Giblet Gravy, 474
   Ultimate XO Sauce, 582
   Yu Choy with Black Bean Garlic Sauce & Extra Garlic, 680
   Yu Choy with Garlic & Soy, 679

Zucchini with Tomatoes, Feta, Garlic & Lemon, 681
General Tso Chicken, 371
General Tso's Pork Belly Steam Buns, 560, 562
General Tso's Sauce, 560, 562
ginger
   Chicken Tikka Masala, 362
   Chinese Potted Chicken, 368
   General Tso's Sauce, 562
   Goat Momos, 413
   Habanero, Calabrian Chile & Pineapple Chicken Wings with Chipotle Blue Cheese, 279
   Jerk Spice Chicken Rub, 460
   Kimchi, 655
   Korean Chicken Thighs with Sweet Potato Vermicelli, 301
   Lamb Vindaloo, 432
   Masala Lamb Kabobs, 417
   Pan Roasted Duck Breast, 555–556
   Pork Pho Broth, 482
   Spicy Poultry Rub, 463
   Sweet & Sticky Chicken Wings, 276
   Thai Duck Stir Fry, 372
   Ultimate XO Sauce, 582
Gluten-Free Bread, 694
Gnocchi Bolognese, 552, 554
goat
   cuts from, 30–31
   Goat Burgers, 422
   Goat Meatballs, 410
   Goat Momos, 413
   Indian-Style Goat Curry, 423
   Jamaican-Style Curry Goat, 428
Goat Cheese Puree, Beet &, 555–556
Go-To Steak Marinade, 486
gravies
   Fundamental Prime-Rib Gravy, 524
   Traditional Chicken Gravy, 472
   Traditional Turkey Gravy, 471
   Turkey Giblet Gravy, 474
   See also sauces
Greek Burgers with Feta Cheese, 421
green beans, Beef & Vegetable Soup, 116
Gremolata, 564
Grilled Cabbage, 663
Grilled Filet, 66
Grilled New York Strip Steak, 54
Grilled Rib Eye Steak, 57
Grilled Roast Pineapple Pork Loin, 177
Grilled Roast Pork with Orange Rind, 179
Grilled Tensen Farm's Strip Loin of Beef, 559
Grilled Whole Turkey, 260
Grill-Roasted Rack of Lamb with Garlic-Herb Crust, 389
ground beef
   American Chop Suey, 140
   Asian Meatballs, 232
   Bacon Cheeseburgers, 105
   Basic Burgers, 102
   BBQ Glazed Meat Loaf, 155
   Beef & Pork Burgers with Caramelized Onion Mayo, 107
   Beef Kabob, 126
   Bolognese, 148
   Cheese-Stuffed Meatballs, 137
   Chickpea & Beef Chili, 117
   Cola Meatballs, 135
   fat ratios for, 23
   Five-Alarm Beef Chili, 122
   Gnocchi Bolognese, 554
   Greek Burgers with Feta Cheese, 421
   Italian Meatballs, 141
   Jalapeño Burgers with Red Pepper Jam, 111
   Kefta with Warm Chickpeas and Salad, 144
   Lamb & Beef Burgers, 418
   Meatballs with Arrabiata, 136
   Mexican Burgers with Cilantro Mayonnaise, 106
   Mushroom & Swiss Burgers, 108
   Old Fashioned Meat Loaf, 154
   Shepherd's Pie, 147
   Swedish Meatballs, 143
   Sweet & Sour Meatballs, 138
   Texas BBQ Burgers, 112
   PINE Burger, 551
ground chicken
   Chicken Burgers, 336
   Chicken Kebob, 349
   Chicken Meatballs, 335
   Chicken Teriyaki Burgers, 338

Chicken Tikka Burger with Tikka Masala Mayonnaise, 341
ground lamb
 Greek Burgers with Feta Cheese, 421
 Lamb & Beef Burgers, 418
 Lamb Burgers, 418
 Lamb Kebobs, 414
 Masala Lamb Kabobs, 417
ground pork
 Asian Meatballs, 232
 Beef & Pork Burgers with Caramelized Onion Mayo, 107
 Bolognese, 148
 Chinese Eggplant with Sausage, 239
 Gnocchi Bolognese, 554
 Italian Meatballs, 141
 Pork Burgers with Pepper Jack & Arugula, 227
 Pork Gyoza, 198
 Swedish Meatballs, 143
 Venison & Pork Meatballs, 410
ground turkey
 Turkey & Kimchi Burgers, 339
 Turkey Burgers, 336
 Turkey Meatballs, 335
ground veal, Gnocchi Bolognese, 554
Gruyère Figs, Potato & Celeriac Gratin with, 623
Guinea Fowl Rubbed with Local Shio Kojit, Farm Fresh Whole, 593

Habanero, Calabrian Chile & Pineapple Chicken Wings with Chipotle Blue Cheese, 279
ham
 Boiled Dinner, 224
 carving, 736, 739
 cut, 24
 Head Cheese, 236
 Southern Collard Greens, 624
harissa
 Harissa Glaze, 573
 Harissa Vinaigrette, 548–549
 Smoked Harissa Agave Chicken Wings, 573
hash
 Crossroad Farm's Vegetable Hash, 548–549
 Lamb & Sweet Potato Hash, 431
Head Cheese, 236
Herb Brined Chicken, 452
Herb Roasted Poultry Rub, 461
Herbed Butter, 539
Herbed Potato Salad, 645
hindquarter beef flank, 23
hindquarter beef round, 23
hindquarter short loin, 21
hindquarter sirloin, 23
hindquarter tenderloin, 23
hindshank, goat, 31
Hoisin Sauce
 Asian Meatballs, 232
 Chinese Potted Chicken, 368
 General Tso's Pork Belly Steam Buns, 560, 562
 General Tso's Sauce, 562 recipe, 232
 Roasted Duck Breast and Five-Spice Confit Duck Legs, 597, 599

Holiday Spiced Wet Turkey Brine, 450
Hollandaise, Blender, 650
Homemade Ketchup, 529
Homemade Tomato Salsa, 520
Home-Style Baked Beans, 620
Homey Turkey & Wild Rice Soup, 316
honey
 Grilled Roast Pineapple Pork Loin, 177
 Honey Mustard Vinaigrette, 671
 Killer BBQ Spare Ribs, 182–183
 Meat Brine, 444
 Roasted Parsnips & Carrots with Ras el Hanout & Honey, 638
 Roasted Whole Duck, 2 68
 Horseradish Cream, Beef and Braised Cabbage Soup with, 118
Hot & Spicy Grilled Wings, 283
hot dogs
 Corndog, 132
 Hot Dog with Curry Spiced Lentils, 130
 Wagyu Hot Dog, 131

Indian Chicken and Rice Soup, 309
Indian Kush, 570
Indian Kush Leg of Lamb, 574
Indian-Style Goat Curry, 423
Italian Meatballs, 141

Jalapeño Burgers with Red Pepper Jam, 111
Jamaican-Style Curry Goat, 428
Jambalaya with Bacon & Andouille Sausage, 209
Jerk Spice Chicken Rub, 460
jowl, pork
  about, 24
  Braised Jowl Pulled Pork, 216
  Jowl Bacon, 190

Kale Salad with Blood Orange Vinaigrette, Shaved Brussels Sprouts &, 675
KC-Style BBQ Sauce, 502
Kefta with Warm Chickpeas and Salad, 144
Ketchup
  Homemade Ketchup, 529
Killer BBQ Spare Ribs, 182–183
kimchi
  Kimchi (basic recipe), 655
  Popcorn Chicken with Kimchi Mayonnaise, 350
  Turkey & Kimchi Burgers, 339
kohlrabies
  German-Style Kohlrabi in Béchamel, 628
  Kohlrabi Slaw with Miso Dressing, 627
Korean Chicken Thighs with Sweet Potato Vermicelli, 301

lamb
  Braised Lamb Belly, 381
  Braised Lamb Neck Wrap, 424
  Braised Lamb Shoulder with Minted Peas, 382
  Bulgarian Sour Lamb Soup with Paprika Butter, 405
  carving, 733–736
  Chile-Vinegar Lamb Ribs, 435
  Cottage Pie, 401
  cuts from, 28–29
  Greek Burgers with Feta Cheese, 421
  Grill-Roasted Rack of Lamb with Garlic-Herb Crust, 389
  Indian Kush Leg of Lamb, 574
  Kefta with Warm Chickpeas and Salad, 144
  Lamb & Sweet Potato Hash, 431
  Lamb Bacon, 409
  Lamb Burgers, 418
  Lamb Kebobs, 414
  Lamb Stew with Candied Carrots and Garlic Roasted Little Creamer Potatoes, 392–393
  Lamb Vindaloo, 432
  Masala Lamb Kabobs, 417
  Persian Braised Lamb Neck with Spiced Broth, 594
  Roasted Lamb Chops, 378
  Rosemary and Lemon Oiled Leg of Lamb, 386
  Smoked Lamb Pâté, 406
  Smoked Lamb Shoulder, 385
  Spicy Lamb Ribs with Tzatziki & Lemon, 436
  Yuzu Sesame Lamb Ribs, 439
  See also mutton
Lamb Stock
  Bulgarian Sour Lamb Soup with Paprika Butter, 405
  Cottage Pie, 401
  Lamb Stew with Candied Carrots and Garlic Roasted Little Creamer Potatoes, 392–393
leeks
  Charred Summer Leeks with Romesco Sauce, 631
  Frizzled Leeks, 632
  Head Cheese, 236
  Lamb Stew with Candied Carrots and Garlic Roasted Little Creamer Potatoes, 392–393
  Mutton Broth, 402
  Walpole Valley Farm's Chicken, 545, 547
Leftover Giblet Gravy, 342
Leftover Turkey & Chickpea Chili with Corn Bread & Rosemary Butter, 326
leg, goat, 31
leg, lamb
  about, 28
  Indian Kush Leg of Lamb, 574
  Lamb & Sweet Potato Hash, 431
  Lamb Stew with Candied Carrots and Garlic Roasted Little Creamer Potatoes, 392
  Rosemary and Lemon Oiled Leg of Lamb, 386
Lemon Pepper Poultry Rub, 467

lemons
　Citrus Brine, 448
　Fennel Lemon Gremolata, 506
　Gremolata, 564
　Lemon & Rosemary Chicken Marinade, 490
　Roasted Pheasant, 271
　Roasted Whole Duck, 268
　Rosemary and Lemon Oiled Leg of Lamb, 386
　Smoked Lemon Zest Rub, 459
　Sous Vide Lemon-Pepper Turkey Wings, 287
　Spicy Lamb Ribs with Tzatziki & Lemon, 436
　Traditional Oven-Roasted Turkey, 250
　Zucchini with Tomatoes, Feta, Garlic & Lemon, 681
Lentils, Hot Dog with Curry Spiced, 130
limes/lime juice
　Basic Red Cabbage Slaw, 664
　Beef Fajita Marinade, 492
　Citrus Brine, 448
　Grilled Tensen Farm's Strip Loin of Beef, 559
　Jerk Spice Chicken Rub, 460
　Marinated Chicken with Chipotle Cauliflower, 247
　Persian Braised Lamb Neck with Spiced Broth, 594
　Roasted Chicken Thighs with Tabbouleh, 298
　Slow-Cooker Barbacoa, 98
loin, goat, 31

loin, pork
　about, 24
　Basic Grilled Roast Pork Loin, 176
　Braised and Grilled Pork with Rosemary, 180
　Grilled Roast Pineapple Pork Loin, 177
　Grilled Roast Pork with Orange Rind, 179
　Porchetta, 160
　Pork Roulade with Oranges, Rainbow Chard, and Wild Rice, 163
loin chops, lamb, 28
London Broil, Chile-Rubbed, 50
Low 'n' Slow Potatoes, 642

Madeira Sauce, 528
Manning, Ryan, 584–590
Maple Chicken Brine, 447
maple syrup
　Fried Brussels Sprouts with Maple-Cider Glaze, 659
　Habanero, Calabrian Chile & Pineapple Chicken Wings with Chipotle Blue Cheese, 279
　Killer BBQ Spare Ribs, 182–183
　Maple-Chipotle Glaze, 575
　Roasted Duck Breast and Five-Spice Confit Duck Legs, 597, 599
marinades
　Adobo Marinade for Beef, 495
　Asian Marinade, 494
　BBQ Chicken Marinade, 489

　Beef Fajita Marinade, 492
　Coffee Marinade for Beef, 491
　Go-To Steak Marinade, 486
　Lemon & Rosemary Chicken Marinade, 490
　Sweet & Spicy Marinade, 485
Marinated Chicken with Chipotle Cauliflower, 247
Marinara Sauce, Chicken Parm Sandwich, 345
Marley's Collie, 576
Masala Braised Short Rib Sandwich with Cilantro Chutney, 129
Masala Lamb Kabobs, 417
Masala Wings, 280
McCarthy, Kyle, 595–596
meat
　carving, 32, 725–747
　cuts of, 20–31
　cutting, 32
　grain of, 32
　identifying quality, 19–20
　letting rest, 32
　serving, 32
　See also individual meat varieties
Meat Brine, 444
meatballs
　Asian Meatballs, 232
　Cheese-Stuffed Meatballs, 137
　Chicken Meatballs, 335
　Cola Meatballs, 135
　Goat Meatballs, 410
　Italian Meatballs, 141
　Meatballs with Arrabiata, 136

INDEX | 801

Swedish Meatballs, 143
Sweet & Sour Meatballs, 138
Turkey Meatballs, 335
Venison & Pork Meatballs, 410
Melon, Cucumber & Prosciutto Salad with Mint Vinaigrette, 667
Mexican Burgers with Cilantro Mayonnaise, 106
mezcal, Carne Asada, 88
Midsummer Corn & Bean Salad, 676
Miles Smith Farm, 603–604
mint
   Braised Lamb Shoulder with Minted Peas, 382
   Melon, Cucumber & Prosciutto Salad with Mint Vinaigrette, 667
   Thai Beef & Cucumber Salad, 668
mirin
   Short Rib Ramen with Beef Broth, 125
   Simple Stir-Fried Bok Choy, 652
   Sweet & Sticky Chicken Wings, 276
   Yuzu Sesame Lamb Ribs, 439
Mojo Jerked Pork Butt, 590
Molasses BBQ Sauce, 166
Momos, Goat, 413
mozzarella cheese, Chicken Parm Sandwich, 345
Mulligatawny Chicken Curry Soup with Curried Cashews, 315
mushrooms
   Beef, Barley, and Portobello Mushroom Soup, 121
   Crowned Roast of Beef, 78–79
   Dunk's Mushrooms, 559
   Grilled Tensen Farm's Strip Loin of Beef, 559
   Korean Chicken Thighs with Sweet Potato Vermicelli, 301
   Mushroom & Swiss Burgers, 108
   Mushroom Cream Sauce, 536
   Pheasant, Morel Mushroom, and Olive Broth, 325
   Ultimate XO Sauce, 582
   Walpole Valley Farm's Chicken, 545, 547
mutton
   carving, 733–735
   Mutton Broth, 402

napa cabbage
   Kimchi, 655
   Korean Chicken Thighs with Sweet Potato Vermicelli, 301
   See also cabbage
Nashville Hot Chicken, 357
neck, goat, 31
neck, lamb
   about, 28
   Braised Lamb Neck Wrap, 424
   Persian Braised Lamb Neck with Spiced Broth, 594
New Mexican chilies
   Braised Chicken Chili, 328
   Chickpea & Beef Chili, 117
   Five-Alarm Beef Chili, 122
   Leftover Turkey & Chickpea Chili with Corn Bread & Rosemary Butter, 326
New York strip
   Grilled New York Strip Steak, 54
   Oven-Roasted New York Strip Steak with Whipped Potatoes and Swiss Chard, 87
   Seared New York Strip Steak, 42
   Sous Vide New York Strip Steak, 53
Nodi, 707
No-Knead Bread, 693
nuts. *See individual nut types*

OG Ribs, The, 578
Old Fashioned Meat Loaf, 154
olives
   Lamb Stew with Candied Carrots and Garlic Roasted Little Creamer Potatoes, 392–393
   Pheasant, Morel Mushroom, and Olive Broth, 325
onions
   Bacon Jam, 513
   Beef & Pork Burgers with Caramelized Onion Mayo, 107
   Beef and Braised Cabbage Soup with Horseradish Cream, 118
   Sausage-Stuffed Peppers, 213
   Braised Beef Cheek, 97

Braised Jowl Pulled Pork, 216
Chicken & Root Vegetable Stew, 311
Chicken Katsu Curry, 353
Cornish Game Hens with Baby Brussels Sprouts & Caramelized Onions, 272
Gnocchi Bolognese, 554
Head Cheese, 236
Jamaican-Style Curry Goat, 428
Lamb & Sweet Potato Hash, 431
Onion Rings, 649
Onion Strings, 551
Persian Braised Lamb Neck with Spiced Broth, 594
Pork Pâté aka Gorton, 193
Pork Terrine, 235
Rabbit & Root Vegetable Stew, 399
Roasted Brussels Sprouts with Bacon, Blue Cheese & Pickled Red Onion, 656
Slow Braised Boyden Beef Short Rib, 563–564
Smoked Chicken Noodle Soup, 310
Tiny Onion Rings, 564
oranges/orange juice
  Blood Orange & Bourbon Brine, 454
  Citrus Brine, 448
  Duck a l'Orange Broth with Spiced Dumplings, 321–322
  Grilled Roast Pork with Orange Rind, 179
  Orange & Thyme Dry Brine, 453
  Pork Roulade with Oranges, Rainbow Chard, and Wild Rice, 163
  Shaved Brussels Sprouts & Kale Salad with Blood Orange Vinaigrette, 675
  Slow-Cooker Barbacoa, 98
Orecchiette, 708
Oven Chicken Tandoori, 369
Oven-Roasted New York Strip Steak with Whipped Potatoes and Swiss Chard, 87
oyster sauce, Thai Duck Stir Fry, 372

Pan Roasted Duck Breast, 555–556
Paprika Butter, Bulgarian Sour Lamb Soup with, 405
Paratha, Braised Lamb Neck Wrap, 424
Parmesan cheese
  Sausage-Stuffed Peppers, 213
  Bolognese, 148
  Broccoli Rabe Pesto, 547
  Buttermilk Caesar Dressing, 672
  Crispy Chicken with Cajun Sauce, 358
  Goat Meatballs, 410
  Italian Meatballs, 141
  Onion Rings, 649
  Parmesan Frico, 552
  Parmesan Gnocchi, 552
  Roasted Brassica Salad with Pickled Ramps & Buttermilk Caesar Dressing, 672
  Roasted Cauliflower au Gratin, 612
  Shaved Brussels Sprouts & Kale Salad with Blood Orange Vinaigrette, 675
  Spicy Pappardelle and Duck Confit with Broccoli Rabe, Shallots & Parmesan, 329
  Tuscan Chicken, 361
  Venison & Pork Meatballs, 410
  Whipped Ricotta, 552
parsley
  Cheese-Stuffed Meatballs, 137
  Cola Meatballs, 135
  Fennel Lemon Gremolata, 506
  Fresh Chimichurri Sauce, 509
  Gremolata, 564
  Grilled Tensen Farm's Strip Loin of Beef, 559
  Grill-Roasted Rack of Lamb with Garlic-Herb Crust, 389
  Herb Roasted Poultry Rub, 461
  Kefta with Warm Chickpeas and Salad, 144
  Meatballs with Arrabiata, 136
  Smoked Lemon Zest Rub, 459
  Spicy Lamb Ribs with Tzatziki & Lemon, 436
  Turkey Dumplings, 320
  Turkey Meatballs, 335

parsnips
    Celery Root and Parsnip Puree, 563–564
    Creamed Vegetable Soup with Turkey Dumplings, 319–320
    Potato & Parsnip Latkes, 637
    Roast Beef au Jus with Vegetables, 101
    Roasted Parsnips & Carrots with Ras el Hanout & Honey, 638
    Slow Braised Boyden Beef Short Rib, 563–564
    Walpole Valley Farm's Chicken, 545, 547

pasta
    All-Yolk Pasta Dough, 698
    American Chop Suey, 140
    Chinese Egg Noodles, 716
    Crispy Chicken with Cajun Sauce, 358
    Farfalle, 700
    Fazzoletti, 703
    Garganelli, 704
    Nodi, 707
    Orecchiette, 708
    Ribbon Pasta, 711
    Sausage Ragù, 231
    Smoked Chicken Noodle Soup, 310
    Spicy Pappardelle and Duck Confit with Broccoli Rabe, Shallots & Parmesan, 329
    Tajarin, 712
    Three-Egg Basic Pasta Dough, 699
    Trofie, 715
    Udon Noodles, 719
    Whole Wheat Pasta Dough, 697

Pasta Sauce, Meatballs with Arrabiata, 136
Peach Hot Sauce, Pulled Pork with Blue Cheese Polenta & Roasted, 220
pears/pear juice
    Asian Marinade, 494
    Bacon, Pear & Rosemary Stuffing, 194
peas
    Bacon & Pea Soup, 205
    Beef Stew, 115
    Braised Lamb Shoulder with Minted Peas, 382
    Cottage Pie, 401
    Venison Stew, 396
pecans
    Roasted Leftover Chicken Salad, 332
    Sausage, Apple & Pecan Stuffing, 197
    Shaved Brussels Sprouts & Kale Salad with Blood Orange Vinaigrette, 675
Pepper Jack & Arugula, Pork Burgers with, 227
Peppercorn Cream Sauce, Seared Filet Mignon with, 70
peppers, bell
    Sausage-Stuffed Peppers, 213
    Braised Chicken Chili, 328
    Charred Summer Leeks with Romesco Sauce, 631
    Chickpea & Beef Chili, 117
    Chili Chicken, 365
    Crispy Chicken with Cajun Sauce, 358

    Crossroad Farm's Vegetable Hash, 548–549
    Five-Alarm Beef Chili, 122
    Home-Style Baked Beans, 620
    Jambalaya with Bacon & Andouille Sausage, 209
    Leftover Turkey & Chickpea Chili with Corn Bread & Rosemary Butter, 326
    Midsummer Corn & Bean Salad, 676
    Roasted Corn & Red Pepper Bisque with Bacon, 202
    Sausage Ragù, 231
    Thai Duck Stir Fry, 372
peppers, hot
    Chicken Tikka Masala, 362
    Corn Salad, 132
    Crossroad Farm's Vegetable Hash, 548–549
    Fermented Hot Sauce, 519
    Five-Alarm Beef Chili, 122
    General Tso's Sauce, 562
    Habanero, Calabrian Chile & Pineapple Chicken Wings with Chipotle Blue Cheese, 279
    Homemade Tomato Salsa, 520
    Jalapeño Burgers with Red Pepper Jam, 111
    Jamaican-Style Curry Goat, 428
    Jerk Spice Chicken Rub, 460
    Lamb & Sweet Potato Hash, 431
    Marley's Collie, 576
    Melon, Cucumber &

Prosciutto Salad with
   Mint Vinaigrette, 667
Mexican Burgers with
   Cilantro Mayonnaise, 106
Pulled Pork with Blue
   Cheese Polenta &
   Roasted Peach Hot
   Sauce, 220
Roast Beef au Jus with
   Vegetables, 101
Slow-Cooker Barbacoa, 98
Spicy BBQ Sauce, 498
Sweet & Spicy Marinade,
   485
Texas BBQ Burgers, 112
Thai Red Duck Curry, 373
Tomatillo Salsa, 523
Perfect Prime Rib, 73
Persian Braised Lamb Neck
   with Spiced Broth, 594
pesto
   Broccoli Rabe Pesto, 547
   Sun-Dried Tomato Pesto,
   532
pheasant
   Pheasant, Morel
      Mushroom, and Olive
      Broth, 325
   Roasted Pheasant, 271
Pickle Fried Chicken
   Sandwich, 346
Pickled Herbs, 559
Pickled Ramps, 672
pig, cuts from, 24–25. See
   also bacon; ham; pork
PINE Burger, The, 551
pine nuts
   Broccoli Rabe Pesto, 547
   Sun-Dried Tomato Pesto,
   532
pineapple/pineapple juice
   Al Pastor Pork Mini
      Trompo, 588

Asian Marinade, 494
Beef Fajita Marinade, 492
Chicken Teriyaki Burgers,
   338
Grilled Roast Pineapple
   Pork Loin, 177
Habanero, Calabrian Chile
   & Pineapple Chicken
   Wings with Chipotle
   Blue Cheese, 279
Killer BBQ Spare Ribs,
   182–183
Pineapple BBQ Sauce, 501
Sweet & Spicy Marinade,
   485
Thai Red Duck Curry, 373
pistachios, Kohlrabi Slaw
   with Miso Dressing, 627
Pizzaiola, 531
Polenta & Roasted Peach
   Hot Sauce, Pulled Pork
   with Blue Cheese, 220
Popcorn Chicken with
   Kimchi Mayonnaise,
   350
porchetta
   Porchetta (basic recipe),
   160
   Roasted Garlic Pork Belly
      Porchetta, 600, 602
pork
   Adobo Pulled Pork, 217
   Al Pastor Pork Mini
      Trompo, 588
   Asian Meatballs, 232
   Bacon, Pear & Rosemary
      Stuffing, 194
   Bacon & Pea Soup,
      205
   Basic Grilled Double-Cut
      Rib Chops, 189
   Basic Grilled Roast Pork
      Loin, 176

Beef & Pork Burgers with
   Caramelized Onion
   Mayo, 107
Boiled Dinner, 224
Bolognese, 148
Bone-In Pork Roast, 172
Braised and Grilled Pork
   with Rosemary, 180
Braised Jowl Pulled Pork,
   216
Butternut Squash and
   Chorizo Bisque, 206
Candied Bacon, 190
Char Siu, 184
Chinese Eggplant with
   Sausage, 239
Classic Smoked Ribs, 164
Cola-Braised Pulled Pork,
   219
Crispy Pork Belly with
   Balsamic Glaze, 171
Cured Bacon, 566
cuts of, 24–25
Earth's Harvest Kitchen
   Pork Belly, 581
General Tso's Pork Belly
   Steam Buns, 560, 562
Gnocchi Bolognese, 554
Grilled Roast Pineapple
   Pork Loin, 177
Grilled Roast Pork with
   Orange Rind, 179
Head Cheese, 236
Italian Meatballs, 141
Jambalaya with Bacon &
   Andouille Sausage, 209
Jowl Bacon, 190
Killer BBQ Spare Ribs,
   182–183
Lamb Stew with Candied
   Carrots and Garlic
   Roasted Little Creamer
   Potatoes, 392–393

Mojo Jerked Pork Butt, 590
Porchetta, 160
Pork & Apple Casserole, 223
Pork Belly & Sweet Corn Chowder, 201
Pork Burgers with Pepper Jack & Arugula, 227
Pork Gyoza, 198
Pork Pâté aka Gorton, 193
Pork Pho Broth, 482
Pork Roulade with Oranges, Rainbow Chard, and Wild Rice, 163
Pork Stock, 481
Pork Terrine, 235
Pulled Pork, 214
Pulled Pork with Blue Cheese Polenta & Roasted Peach Hot Sauce, 220
Roasted Corn & Red Pepper Bisque with Bacon, 202
Roasted Garlic Pork Belly Porchetta, 600, 602
Rotisserie Pig Roast, 168
Sausage, Apple & Pecan Stuffing, 197
Spiced Heritage Farm's Pork Tenderloin, 548–549
Swedish Meatballs, 143
Thick-Cut Pork Chops with Stone Fruit & Bulgur Wheat, 228
Twice-Cooked Pork Belly & Bean Soup with Brown Bread Croutons, 210
Venison & Pork Meatballs, 410

pork belly
Char Siu, 184
Crispy Pork Belly with Balsamic Glaze, 171
Cured Bacon, 566
Earth's Harvest Kitchen Pork Belly, 581
General Tso's Pork Belly Steam Buns, 560, 562
Porchetta, 160
Pork Belly & Sweet Corn Chowder, 201
Roasted Garlic Pork Belly Porchetta, 600, 602
Twice-Cooked Pork Belly & Bean Soup with Brown Bread Croutons, 210
pork butt (Boston butt)
about, 24
Mojo Jerked Pork Butt, 590
Pork Pâté aka Gorton, 193
Pork Terrine, 235
pork foot, 24
pork jowl, 24
pork loin
Basic Grilled Roast Pork Loin, 176
Braised and Grilled Pork with Rosemary, 180
Grilled Roast Pineapple Pork Loin, 177
Grilled Roast Pork with Orange Rind, 179
Porchetta, 160
Pork Roulade with Oranges, Rainbow Chard, and Wild Rice, 163
pork shoulder
Adobo Pulled Pork, 217
Al Pastor Pork Mini Trompo, 588

Cola-Braised Pulled Pork, 219
Pulled Pork, 214
Pulled Pork with Blue Cheese Polenta & Roasted Peach Hot Sauce, 220
pork side, 24
Pork Stock
Pork Pho Broth, 482
recipe, 481
Twice-Cooked Pork Belly & Bean Soup with Brown Bread Croutons, 210
pork tenderloin
Pork & Apple Casserole, 223
Spiced Heritage Farm's Pork Tenderloin, 548–549
Porterhouse, 45
potatoes
Beef & Vegetable Soup, 116
Beef Stew, 115
Blue Cheese Gratin, 646
Boiled Dinner, 224
Chicken & Rice Soup, 312
Chicken & Root Vegetable Stew, 311
Chicken Legs with Potatoes & Fennel, 297
Confit New Potatoes, 641
Confit Potatoes, 559
Cottage Pie, 401
Garlic Roasted Little Creamer Potatoes, 393
Grilled Tensen Farm's Strip Loin of Beef, 559
Herbed Potato Salad, 645

Indian Chicken & Rice
  Soup, 309
Jamaican-Style Curry
  Goat, 428
Lamb Stew with Candied
  Carrots and Garlic
  Roasted Little Creamer
  Potatoes, 392–393
Lamb Vindaloo, 432
Low 'n' Slow Potatoes, 642
Oven-Roasted New
  York Strip Steak with
  Whipped Potatoes and
  Swiss Chard, 87
Patatas Bravas, 634
Persian Braised Lamb
  Neck with Spiced Broth,
  594
Pork Belly & Sweet Corn
  Chowder, 201
Potato & Celeriac Gratin
  with Gruyère Figs, 623
Potato & Parsnip Latkes,
  637
Potato & Tomato Grain,
  633
Rabbit & Root Vegetable
  Stew, 399
Shepherd's Pie, 147
Smoked Chicken Noodle
  Soup, 310
Venison Stew, 396
Warm Potato Salad,
  548–549
Yankee Short Ribs with
  Roasted Potatoes &
  Carrots, 61
poultry
  BBQ Poultry Rub, 464
  Brown Sugar & Ancho
    Chile Poultry Rub, 466
  carving, 740, 742, 745–
    747

Herb Roasted Poultry
  Rub, 461
Lemon Pepper Poultry
  Rub, 467
Spicy Poultry Rub, 463
*See also individual poultry
  types*
prime rib
  Frenched Prime Rib,
    76–77
  Perfect Prime Rib, 73
  Smoked Prime Rib, 83
  Wood-Fired Prime Rib, 80
Prosciutto Salad with Mint
  Vinaigrette, Melon,
  Cucumber &, 667
pulled pork
  Adobo Pulled Pork, 217
  Braised Jowl Pulled Pork,
    216
  Cola-Braised Pulled Pork,
    219
  Pulled Pork (basic recipe),
    214
  Pulled Pork with Blue
    Cheese Polenta &
    Roasted Peach Hot
    Sauce, 220

quality meat, identifying,
  19–20

rabbit
  carving, 747
  Rabbit & Root Vegetable
    Stew, 399
  Rabbit Ragù, 400
rack, goat, 31
Radish & Apple Salad with
  Blue Cheese Mousse,
  Roasted Baby Beet,, 671
Ramen with Beef Broth,
  Short Rib, 125

Red Pepper Jam, Jalapeño
  Burgers with, 111
rib eye
  Grilled Rib Eye Steak, 57
  Seared Rib Eye, 38
  Sous Vide Rib Eye, 41
rib roast
  Crowned Roast of Beef,
    78–79
  Frenched Prime Rib,
    76–77
  Slow Braised Boyden Beef
    Short Rib, 563–564
  Smoked Prime Rib, 83
  OG Ribs, 578
  Wood-Fired Prime Rib, 80
Ribbon Pasta, 711
ribs
  Applewood Smoked Ribs
    with Molasses BBQ
    Sauce, 166
  Chile-Vinegar Lamb Ribs,
    435
  Classic Smoked Ribs, 164
  Killer BBQ Spare Ribs,
    182–183
  Masala Braised Short Rib
    Sandwich with Cilantro
    Chutney, 129
  Spicy Lamb Ribs with
    Tzatziki & Lemon, 436
  OG Ribs, 578
  Yankee Short Ribs with
    Roasted Potatoes &
    Carrots, 61
  Yuzu Sesame Lamb Ribs,
    439
rice
  Sausage-Stuffed Peppers,
    213
  Bulgarian Sour Lamb Soup
    with Paprika Butter,
    405

Chicken & Rice Soup, 312
Chicken Katsu Curry, 353
Chinese Potted Chicken, 368
General Tso Chicken, 371
Homey Turkey & Wild Rice Soup, 316
Indian Chicken and Rice Soup, 309
Jambalaya with Bacon & Andouille Sausage, 209
Mulligatawny Chicken Curry Soup with Curried Cashews, 315
Pork Roulade with Oranges, Rainbow Chard, and Wild Rice, 163
Thai Duck Stir Fry, 372
Riced Cauliflower, 616
Rickety Ranch, 9–10
ricotta cheese
  Roasted Baby Beet, Radish & Apple Salad with Blue Cheese Mousse, 671
  Whipped Ricotta, 552
Roast Beef au Jus with Vegetables, 101
roast pig, carving, 736
Roasted Baby Beet, Radish & Apple Salad with Blue Cheese Mousse, 671
Roasted Brassica Salad with Pickled Ramps & Buttermilk Caesar Dressing, 672
Roasted Brussels Sprouts with Bacon, Blue Cheese & Pickled Red Onion, 656

Roasted Cauliflower au Gratin, 612
Roasted Chicken Thighs with Tabbouleh, 298
Roasted Corn & Red Pepper Bisque with Bacon, 202
Roasted Duck Breast & Five-Spice Confit Duck Legs, 597, 599
Roasted Garlic Pork Belly Porchetta, 600
Roasted Lamb Chops, 378
Roasted Leftover Chicken Salad, 332
Roasted Parsnips & Carrots with Ras el Hanout & Honey, 638
Roasted Pheasant, 271
Roasted Whole Chicken, 244
Roasted Whole Duck, 268
Romesco Sauce, Charred Summer Leeks with, 631
rosemary
  Bacon, Pear & Rosemary Stuffing, 194
  Bolognese, 148
  Braised and Grilled Pork with Rosemary, 180
  Cornish Game Hens with Baby Brussels Sprouts & Caramelized Onions, 272
  Crowned Roast of Beef, 78–79
  Gnocchi Bolognese, 552, 554
  Grill-Roasted Rack of Lamb with Garlic-Herb Crust, 389
  Herb Brined Chicken, 452
  Herb Roasted Poultry Rub, 461

  Herbed Butter, 539
  Leftover Turkey & Chickpea Chili with Corn Bread & Rosemary Butter, 326
  Lemon & Rosemary Chicken Marinade, 490
  Porchetta, 160
  Pork & Apple Casserole, 223
  Roasted Lamb Chops, 378
  Rosemary & Lemon Oiled Leg of Lamb, 386
  Smoked Lamb Shoulder, 385
  Smoked Prime Rib, 83
  Wood-Fired Prime Rib, 80
Rotisserie Pig Roast, 168
rubs
  Acapulco Gold Rub, 567
  BBQ Poultry Rub, 464
  Brown Sugar & Ancho Chile Poultry Rub, 466
  Colombian Gold, 570
  Herb Roasted Poultry Rub, 461
  Indian Kush, 570
  Jerk Spice Chicken Rub, 460
  Lemon Pepper Poultry Rub, 467
  Spicy Poultry Rub, 463
  Three Pepper Rub, 458
rump, lamb, 28
Rustic White Bread, 685
Rustic Whole Wheat Bread, 682
rutabaga
  Chicken & Root Vegetable Stew, 311
  Rabbit & Root Vegetable Stew, 399

Smoked Lemon Zest Rub, 459

Sage & Garlic Dry Brine, 455
salads
   Basic Red Cabbage Slaw, 664
   Corn Salad, 132
   Herbed Potato Salad, 645
   Kefta with Warm Chickpeas and Salad, 144
   Melon, Cucumber & Prosciutto Salad with Mint Vinaigrette, 667
   Midsummer Corn & Bean Salad, 676
   Roasted Brassica Salad with Pickled Ramps & Buttermilk Caesar Dressing, 672
   Shaved Brussels Sprouts & Kale Salad with Blood Orange Vinaigrette, 675
   Thai Beef & Cucumber Salad, 668
   Warm Potato Salad, 548–549
salsa
   Homemade Tomato Salsa, 520
   Tomatillo Salsa, 523
Salt & Pepper Grilled Turkey Wings, 288
sandwiches
   Braised Lamb Neck Wrap, 424
   Chicken Parm Sandwich, 345
   General Tso's Pork Belly Steam Buns, 560, 562
   Pickle Fried Chicken Sandwich, 346
   Ultimate Leftover Thanksgiving Sandwich, 342
   See also burgers
sauces
   Blender Hollandaise, 650
   Chipotle BBQ Sauce, 505
   Chipotle Cilantro Chimichurri for Meat, 514
   Fennel Lemon Gremolata, 506
   Fermented Hot Sauce, 519
   Fresh Chimichurri Sauce, 509
   Garlic & Chive Steak Sauce, 535
   Hoisin Sauce, 232
   Homemade Ketchup, 529
   Homemade Tomato Salsa, 520
   KC-Style BBQ Sauce, 502
   Leftover Giblet Gravy, 342
   Madeira Sauce, 528
   Molasses BBQ Sauce, 166
   Mushroom Cream Sauce, 536
   Pizzaiola, 531
   Spicy BBQ Sauce, 498
   Steak Sauce, 510
   Sun-Dried Tomato Pesto, 532
   Tahini Dressing, 651
   Tomatillo Salsa, 523
   Traditional BBQ Sauce, 497
   Ultimate XO Sauce, 582
   See also gravies
sausage
   Sausage-Stuffed Peppers, 213
   Butternut Squash and Chorizo Bisque, 206
   Chinese Eggplant with Sausage, 239
   Chipotle Blue Cheese Sauce, 279
   Crowned Roast of Beef, 78–79
   Habanero, Calabrian Chile & Pineapple Hot Sauce, 279
   Jambalaya with Bacon & Andouille Sausage, 209
   Lamb Stew with Candied Carrots and Garlic Roasted Little Creamer Potatoes, 392–393
   Pineapple BBQ Sauce, 501
   Roasted Peach Hot Sauce, 220
   Sausage, Apple & Pecan Stuffing, 197
   Sausage Ragù, 231
   Traditional Chicken Gravy, 472
   Traditional Turkey Gravy, 471
   Turkey Giblet Gravy, 474
   Ultimate XO Sauce, 582
scallions
   Asian Meatballs, 232
   Bulgarian Sour Lamb Soup with Paprika Butter, 405
   Chile-Vinegar Lamb Ribs, 435
   General Tso's Sauce, 562
   Grilled Tensen Farm's Strip Loin of Beef, 559
   Jerk Spice Chicken Rub, 460
   Kimchi, 655
   Korean Chicken Thighs with Sweet Potato Vermicelli, 301
   Marley's Collie, 576

Warm Potato Salad, 548–549
Schlissel, Jeffrey, 565–578
Seared Chicken Breast, 294
Seared Duck Breast, 304
Seared Filet Mignon, 69
Seared Filet Mignon with Peppercorn Cream Sauce, 70
Seared New York Strip Steak, 42
Seared Rib Eye, 38
Seared Turkey Breast, 307
serving meat, 32
shallots
  Chicken Legs with Potatoes and Fennel, 297
  Farm Fresh Whole Guinea Fowl Rubbed with Local Shio Kojit, 593
  Grilled Tensen Farm's Strip Loin of Beef, 559
  Grill-Roasted Rack of Lamb with Garlic-Herb Crust, 389
  Herbed Potato Salad, 645
  Korean Chicken Thighs with Sweet Potato Vermicelli, 301
  Madeira Sauce, 528
  Mushroom Cream Sauce, 536
  Roasted Cauliflower au Gratin, 612
  Roasted Chicken Thighs with Tabbouleh, 298
  Shallot Confit, 564
  Slow Braised Boyden Beef Short Rib, 563–564
  Southern Beef Jerky, 585
  Spiced Dumplings, 322
  Spicy Pappardelle and

Duck Confit with Broccoli Rabe, Shallots & Parmesan, 329
Sun-Dried Tomato Pesto, 532
Traditional Chicken Gravy, 472
Traditional Turkey Gravy, 471
Turkey Giblet Gravy, 474
Ultimate XO Sauce, 582
Walpole Valley Farm's Chicken, 545, 547
shank, lamb, 28
Shaved Brussels Sprouts & Kale Salad with Blood Orange Vinaigrette, 675
Shepherd's Pie, 147
short ribs
  Masala Braised Short Rib Sandwich with Cilantro Chutney, 129
  Short Rib Ramen with Beef Broth, 125
  Slow Braised Boyden Beef Short Rib, 563–564
  Yankee Short Ribs with Roasted Potatoes & Carrots, 61
shoulder, goat, 31
shoulder, lamb, 28
shoulder picnic, 24
shrimp, Ultimate XO Sauce, 582
sides
  Asparagus Three Ways, Plus Two Sauces, 650–651
  Basic Red Cabbage Slaw, 664
  Blue Cheese Gratin, 646
  Cauliflower Mash, 619
  Celeriac Puree, 615

  Charred Summer Leeks with Romesco Sauce, 631
  Confit New Potatoes, 641
  Fried Brussels Sprouts with Maple-Cider Glaze, 659
  Frizzled Leeks, 632
  German-Style Kohlrabi in Béchamel, 628
  Grilled Cabbage, 663
  Herbed Potato Salad, 645
  Home-Style Baked Beans, 620
  Kimchi, 655
  Kohlrabi Slaw with Miso Dressing, 627
  Low 'n' Slow Potatoes, 642
  Melon, Cucumber & Prosciutto Salad with Mint Vinaigrette, 667
  Midsummer Corn & Bean Salad, 676
  Onion Rings, 649
  Patatas Bravas, 634
  Potato & Celeriac Gratin with Gruyère Figs, 623
  Potato & Parsnip Latkes, 637
  Potato & Tomato Grain, 633
  Riced Cauliflower, 616
  Roasted Baby Beet, Radish & Apple Salad with Blue Cheese Mousse, 671
  Roasted Brassica Salad with Pickled Ramps & Buttermilk Caesar Dressing, 672
  Roasted Brussels Sprouts with Bacon, Blue Cheese & Pickled Red Onion, 656

Roasted Cauliflower au Gratin, 612
Roasted Parsnips & Carrots with Ras el Hanout & Honey, 638
Shaved Brussels Sprouts & Kale Salad with Blood Orange Vinaigrette, 675
Simple Stir-Fried Bok Choy, 652
Southern Collard Greens, 624
Stove-Top Brussels Sprouts, 660
Yu Choy with Black Bean Garlic Sauce & Extra Garlic, 680
Yu Choy with Garlic & Soy, 679
Zucchini with Tomatoes, Feta, Garlic & Lemon, 681
Simple Stir-Fried Bok Choy, 652
sirloin steak
  Beef and Braised Cabbage Soup with Horseradish Cream, 118
  Coffee-Rubbed Sirloin, 58
  Garlic & Thyme-Rubbed Sirloin, 62
skirt steak
  Carne Asada, 88
  Carne de Chinameca, 586
Slow Braised Boyden Beef Short Rib, 563–564
Slow-Cooker Barbacoa, 98
Slow-Cooker Corned Beef, 151
Smoked Chicken Noodle Soup, 310
Smoked Chicken Thighs, 295

Smoked Harissa Agave Chicken Wings, 573
Smoked Lamb Pâté, 406
Smoked Lamb Shoulder, 385
Smoked Lemon Zest Rub, 459
Smoked Prime Rib, 83
Smoked Whole Turkey, 256
Soule, Carole, 603–604
soups and stews
  Bacon & Pea Soup, 205
  Beef, Barley, and Portobello Mushroom Soup, 121
  Beef & Vegetable Soup, 116
  Beef and Braised Cabbage Soup with Horseradish Cream, 118
  Beef Stew, 115
  Beef Stock, 475
  Braised Chicken Chili, 328
  Bulgarian Sour Lamb Soup with Paprika Butter, 405
  Butternut Squash and Chorizo Bisque, 206
  Chicken & Rice Soup, 312
  Chicken & Root Vegetable Stew, 311
  Chicken Stock, 478
  Chickpea & Beef Chili, 117
  Chili Chicken, 365
  Creamed Vegetable Soup with Turkey Dumplings, 319–320
  Duck a l'Orange Broth with Spiced Dumplings, 321–322
  Duck Stock, 484

  Five-Alarm Beef Chili, 122
  Homey Turkey & Wild Rice Soup, 316
  Indian Chicken and Rice Soup, 309
  Jambalaya with Bacon & Andouille Sausage, 209
  Lamb Stew with Candied Carrots and Garlic Roasted Little Creamer Potatoes, 392–393
  Leftover Turkey & Chickpea Chili with Corn Bread & Rosemary Butter, 326
  Mulligatawny Chicken Curry Soup with Curried Cashews, 315
  Mutton Broth, 402
  Pheasant, Morel Mushroom, and Olive Broth, 325
  Pork Belly & Sweet Corn Chowder, 201
  Pork Pho Broth, 482
  Pork Stock, 481
  Rabbit & Root Vegetable Stew, 399
  Roasted Corn & Red Pepper Bisque with Bacon, 202
  Smoked Chicken Noodle Soup, 310
  Turkey Stock, 477
  Twice-Cooked Pork Belly & Bean Soup with Brown Bread Croutons, 210
  Venison Stew, 396
Sourdough Bread, 686
Sourdough Starter, 690
Sous Vide Cajun Chicken Wings, 284

Sous Vide Chicken Breast, 292
Sous Vide Lemon-Pepper Turkey Wings, 287
Sous Vide New York Strip Steak, 53
Sous Vide Rib Eye, 41
Sous Vide Tandoori Chicken, 354
Sous Vide Turkey, 264
Sous Vide Turkey Wings, 287
Southern Beef Jerky, 585
Southern Collard Greens, 624
spare ribs, 24
Spatchcock Grilled Turkey, 263
Spatchcock Turkey, 255
Spiced Dumplings, 322
Spiced Heritage Farm's Pork Tenderloin, 548–549
Spicy BBQ Sauce, 498
Spicy Grilled Turkey Wings, 288
Spicy Lamb Ribs with Tzatziki & Lemon, 436
Spicy Pappardelle and Duck Confit with Broccoli Rabe, Shallots & Parmesan, 329
Spicy Poultry Rub, 463
spinach
  Crowned Roast of Beef, 78–79
  Tuscan Chicken, 361
squash, summer, Crossroad Farm's Vegetable Hash, 548–549
squash, winter
  Butternut Squash and Chorizo Bisque, 206
  Walpole Valley Farm's Chicken, 545, 547
Sriracha Sweet Glaze, Agave, 568
St. Louis cut ribs
  Applewood Smoked Ribs with Molasses BBQ Sauce, 166
  Classic Smoked Ribs, 164
steak sauce
  Garlic & Chive Steak Sauce, 535
  Steak Sauce (basic recipe), 510
steak temperature guide, 57
Stove-Top Brussels Sprouts, 660
Strawberry & Yuzu Emulsion, 556
strip loins, Grilled Tensen Farm's Strip Loin of Beef, 559
stuffing
  Bacon, Pear & Rosemary Stuffing, 194
  Sausage, Apple & Pecan Stuffing, 197
  Ultimate Leftover Thanksgiving Sandwich, 342
sun-dried tomatoes
  Chicken Legs with Potatoes and Fennel, 297
  Pizzaiola, 531
  Sun-Dried Tomato Pesto, 532
  Tuscan Chicken, 361
Swedish Meatballs, 143
Sweet & Smoky Roasted Chicken, 246
Sweet & Sour Meatballs, 138
Sweet & Spicy Marinade, 485
Sweet & Sticky Chicken Wings, 276
sweet potatoes
  Korean Chicken Thighs with Sweet Potato Vermicelli, 301
  Lamb & Sweet Potato Hash, 431
Swiss Burgers, Mushroom &, 108
Swiss Chard, Oven-Roasted New York Strip Steak with Whipped Potatoes and, 87

Tabbouleh, Roasted Chicken Thighs with, 298
Tahini Dressing, 651
Tajarin, 712
Tandoori Paste
  Oven Chicken Tandoori, 369
  recipe, 354, 369
  Sous Vide Tandoori Chicken, 354
T-Bone Steak, Chipotle, 65
Texas BBQ Burgers, 112
Thai Beef & Cucumber Salad, 668
Thai Duck Stir Fry, 372
Thai Red Duck Curry, 373
Thick-Cut Pork Chops with Stone Fruit & Bulgur Wheat, 228
Three Pepper Rub, 458
Three-Egg Basic Pasta Dough, 699
Thyme-Rubbed Sirloin, Garlic and, 62
Tikka Masala Mayonnaise, Chicken Tikka Burger with, 341

Tomatillo Salsa, 523
tomatoes
   American Chop Suey, 140
   Beef & Vegetable Soup, 116
   Sausage-Stuffed Peppers, 213
   Bolognese, 148
   Braised Chicken Chili, 328
   Chicken Tikka Masala, 362
   Chickpea & Beef Chili, 117
   Crossroad Farm's Vegetable Hash, 548–549
   Five-Alarm Beef Chili, 122
   Gnocchi Bolognese, 554
   Homemade Ketchup, 529
   Homemade Tomato Salsa, 520
   Indian Chicken and Rice Soup, 309
   Indian-Style Goat Curry, 423
   Jambalaya with Bacon & Andouille Sausage, 209
   Kefta with Warm Chickpeas and Salad, 144
   Killer BBQ Spare Ribs, 182–183
   Lamb Vindaloo, 432
   Leftover Turkey & Chickpea Chili with Corn Bread & Rosemary Butter, 326
   Masala Braised Short Rib Sandwich with Cilantro Chutney, 129
   Meatballs with Arrabiata, 136
   Patatas Bravas, 634
   Pizzaiola, 531
   Pork Roulade with Oranges, Rainbow Chard, and Wild Rice, 163
   Potato & Tomato Grain, 633
   Roasted Chicken Thighs with Tabbouleh, 298
   Sausage Ragù, 231
   Slow Braised Boyden Beef Short Rib, 563–564
   Sun-Dried Tomato Pesto, 532
   Zucchini with Tomatoes, Feta, Garlic & Lemon, 681
   See also sun-dried tomatoes
tongue, carving, 739
Traditional BBQ Sauce, 497
Traditional Chicken Gravy, 472
Traditional Oven-Roasted Turkey, 250
Traditional Turkey Gravy, 471
Trofie, 715
turkey
   BBQ Turkey Wings, 291
   Beer-Brined Turkey, 266
   carving, 253, 742, 745–746
   Confit Turkey Legs, 306
   Creamed Vegetable Soup with Turkey Dumplings, 319–320
   Deep-Fried Turkey, 259
   Grilled Whole Turkey, 260
   Holiday Spiced Wet Turkey Brine, 450
   Homey Turkey & Wild Rice Soup, 316
   Leftover Turkey & Chickpea Chili with Corn Bread & Rosemary Butter, 326
   Orange & Thyme Dry Brine, 453
   Sage & Garlic Dry Brine, 455
   Salt & Pepper Grilled Turkey Wings, 288
   Seared Turkey Breast, 307
   Smoked Lemon Zest Rub, 459
   Smoked Whole Turkey, 256
   Sous Vide Lemon-Pepper Turkey Wings, 287
   Sous Vide Turkey, 264
   Sous Vide Turkey Wings, 287
   Spatchcock Grilled Turkey, 263
   Spatchcock Turkey, 255
   Spicy Grilled Turkey Wings, 288
   thawing, 252
   Ultimate Leftover Thanksgiving Sandwich, 342
   Three Pepper Rub, 458
   Traditional Oven-Roasted Turkey, 250
   trussing, 252
   Turkey & Kimchi Burgers, 339
   Turkey Burgers, 336
   Turkey Dumplings, 320
   Turkey Giblet Gravy, 474
   Turkey Meatballs, 335
   Turkey Stock, 319
Turkey Stock
   Bacon, Pear & Rosemary Stuffing, 194

Chickpea & Beef Chili, 117
Creamed Vegetable Soup with Turkey Dumplings, 319–320
Homey Turkey & Wild Rice Soup, 316
Leftover Turkey & Chickpea Chili with Corn Bread & Rosemary Butter, 326
recipe, 319, 477
Sausage, Apple & Pecan Stuffing, 197
Traditional Turkey Gravy, 471
Turkey Dumplings, 320
Tuscan Chicken, 361
Twice-Cooked Pork Belly & Bean Soup with Brown Bread Croutons, 210
Tzatziki & Lemon, Spicy Lamb Ribs with, 436

Udon Noodles, 719
Ultimate Leftover Thanksgiving Sandwich, The, 342
Ultimate XO Sauce, 582
veal
  Bolognese, 148
  carving, 733–736
  Gnocchi Bolognese, 552, 554
  See also beef
Veal Stock
  Beef, Barley, & Portobello Mushroom Soup, 121
  Beef and Braised Cabbage Soup with Horseradish Cream, 118
  Bolognese, 148
  Bulgarian Sour Lamb Soup with Paprika Butter, 405
  Fundamental Prime-Rib Gravy, 524
  Gnocchi Bolognese, 554
  recipe, 476
  Slow Braised Boyden Beef Short Rib, 563–564

Vegetable Stock
  Butternut Squash and Chorizo Bisque, 206
  Southern Collard Greens, 624
venison
  carving, 734–735
  Venison & Pork Meatballs, 410
  Venison Stew, 396
  Venison Tenderloin with Blueberry Reduction, 427

Wagyu Hot Dog, 131
walnuts, Roasted Leftover Chicken Salad, 332
Walpole Valley Farm's Chicken, 545, 547
Whipped Ricotta, 552
Whole Jerk Chicken, 577
Whole Wheat Pasta Dough, 697
Wilson, Angela, 591–592
wine, red
  Beef, Barley, and Portobello Mushroom Soup, 121
  Beef Stew, 115
  Braised Beef Cheek, 97
  Fundamental Prime-Rib Gravy, 524
  Lamb & Sweet Potato Hash, 431
  Lamb Stew with Candied Carrots and Garlic Roasted Little Creamer Potatoes, 392–393
  Madeira Sauce, 528
  Masala Wings, 280
  Rabbit Ragù, 400
  Roast Beef au Jus with Vegetables, 101
  Slow Braised Boyden Beef Short Rib, 563–564
  Venison Stew, 396
wine, shaoxing
  Asian Marinade, 494
  Char Siu, 184
  Chinese Eggplant with Sausage, 239
  General Tso Chicken, 371
  Ultimate XO Sauce, 582
wine, white
  Braised and Grilled Pork with Rosemary, 180
  Chicken Legs with Potatoes and Fennel, 297
  Chinese Potted Chicken, 368
  Creamed Vegetable Soup with Turkey Dumplings, 319–320
  Duck a l'Orange Broth with Spiced Dumplings, 321–322
  Gnocchi Bolognese, 554
  Pheasant, Morel Mushroom, and Olive Broth, 325
  Pizzaiola, 531
  Roasted Cauliflower au Gratin, 612
  Roasted Chicken Thighs with Tabbouleh, 298
  Traditional Chicken Gravy, 472

Traditional Turkey Gravy, 471

wings
- BBQ Turkey Wings, 291
- Habanero, Calabrian Chile & Pineapple Chicken Wings with Chipotle Blue Cheese, 279
- Hot & Spicy Grilled Wings, 283
- Salt & Pepper Grilled Turkey Wings, 288
- Smoked Harissa Agave Chicken Wings, 573
- Sous Vide Cajun Chicken Wings, 284
- Sous Vide Lemon-Pepper Turkey Wings, 287
- Sous Vide Turkey Wings, 287
- Spicy Grilled Turkey Wings, 288
- Sweet & Sticky Chicken Wings, 276

Wood-Fired Prime Rib, 80

Yankee Short Ribs with Roasted Potatoes & Carrots, 61

yogurt
- Chicken 65, 366
- Chicken Tikka Masala, 362
- Indian-Style Goat Curry, 423
- Lamb Vindaloo, 432
- Oven Chicken Tandoori, 369
- Sous Vide Tandoori Chicken, 354

yu choy
- Yu Choy with Black Bean Garlic Sauce & Extra Garlic, 680
- Yu Choy with Garlic & Soy, 679

yuzu
- Strawberry & Yuzu Emulsion, 556
- Yuzu Sesame Lamb Ribs, 439

zucchini
- Crossroad Farm's Vegetable Hash, 548–549
- Zucchini with Tomatoes, Feta, Garlic & Lemon, 681

ABOUT CIDER MILL PRESS BOOK PUBLISHERS

Good ideas ripen with time. From seed to harvest, Cider Mill Press brings fine reading, information, and entertainment together between the covers of its creatively crafted books. Our Cider Mill bears fruit twice a year, publishing a new crop of titles each spring and fall.

"Where Good Books Are Ready for Press"

501 Nelson Place
Nashville, Tennessee 37214

cidermillpress.com